AF696999

# FUNCTIONS MODELING CHANGE:
# A Preparation for Calculus

**Fifth Edition**

Produced by the Calculus Consortium and initially funded by a National Science Foundation Grant.

Eric Connally
*Harvard University Extension*

Deborah Hughes-Hallett
*The University of Arizona*

Andrew M. Gleason
*Harvard University*

Philip Cheifetz
*Nassau Community College*

Ann Davidian
*Gen. Douglas MacArthur HS*

Daniel E. Flath
*Macalester College*

Selin Kalaycıoğlu
*New York University*

Brigitte Lahme
*Sonoma State University*

Patti Frazer Lock
*St. Lawrence University*

Guadalupe I. Lozano
*The University of Arizona*

William G. McCallum
*The University of Arizona*

Jerry Morris
*Sonoma State University*

Karen Rhea
*University of Michigan*

Ellen Schmierer
*Nassau Community College*

Pat Shure
*University of Michigan*

Adam H. Spiegler
*Loyola University Chicago*

Carl Swenson
*Seattle University*

Aaron D. Wootton
*University of Portland*

Elliot J. Marks

with the assistance of

Frank Avenoso
*Nassau Community College*

Douglas Quinney
*University of Keele*

Katherine Yoshiwara
*Los Angeles Pierce College*

WILEY

**Dedicated to Ben, Jonah, and Isabel**

| | |
|---|---|
| PUBLISHER | Laurie Rosatone |
| ACQUISITIONS EDITOR | Joanna Dingle |
| ASSOCIATE EDITOR | Jacqueline Sinacori |
| MARKETING MANAGER | Kimberly Kanakes |
| SENIOR PRODUCT DESIGNER | Tom Kulesa |
| SENIOR CONTENT MANAGER | Karoline Luciano |
| SENIOR PRODUCTION EDITOR | Ken Santor |
| FREELANCE PROJECT EDITOR | Anne Scanlan-Rohrer |
| EDITORIAL PROGRAM ASSISTANT | Courtney Welsh |
| MARKETING ASSISTANT | Jesse Adler |
| COVER DESIGNER | Madelyn Lesure |
| COVER AND CHAPTER OPENING PHOTO | ©Patrick Zephyr/Patrick Zephyr Nature Photography |

This book was set in Stix by the Consortium using TeX, Mathematica, and the package AsTeX, which was written by Alex Kasman. The process was managed by Elliot Marks.

Copyright ©2015, 2012, 2011, 2006, 2003 John Wiley & Sons, Inc. All rights reserved.

Founded in 1807, John Wiley & Sons, Inc. has been a valued source of knowledge and understanding for more than 200 years, helping people around the world meet their needs and fulfill their aspirations. Our company is built on a foundation of principles that include responsibility to the communities we serve and where we live and work. In 2008, we launched a Corporate Citizenship Initiative, a global effort to address the environmental, social, economic, and ethical challenges we face in our business. Among the issues we are addressing are carbon impact, paper specifications and procurement, ethical conduct within our business and among our vendors, and community and charitable support. For more information, please visit our website: www.wiley.com/go/citizenship.

No part of this publication may be reproduced, stored in a retrieval system or transmitted in any form or by any means, electronic, mechanical, photocopying, recording, scanning or otherwise, except as permitted under Sections 107 or 108 of the 1976 United States Copyright Act, without either the prior written permission of the Publisher, or authorization through payment of the appropriate per-copy fee to the Copyright Clearance Center, 222 Rosewood Drive, Danvers, MA 01923, (508) 750-8400, fax (508) 750-4470. Requests to the Publisher for permission should be addressed to the Permissions Department, John Wiley & Sons, Inc., 111 River Street, Hoboken, NJ 07030, (201) 748-6011, fax (201) 748-6008,wbsite: http://www.wiley.com/go/permissions.

This material is based upon work supported by the National Science Foundation under Grant No. DUE-9352905. Opinions expressed are those of the authors and not necessarily those of the Foundation.

ISBN-13 978-1-118-58319-7 (Paper)
ISBN-13 978-1-118-98630-1 (Cloth)
ISBN-13 978-1-118-94258-1 (BRV)

# PREFACE

In the 21st century, students need the ability to solve problems; innovation depends on citizens who can think critically. The fifth edition of *Functions Modeling Change: A Preparation for Calculus* reaffirms our effort to refocus the teaching of mathematics in a balanced blend of concepts and procedures. Our approach encourages students to develop their problem-solving skills while acquiring the mathematical background needed to learn calculus or pursue their careers.

## Fifth Edition: Focus

This edition of Precalculus continues to stress conceptual understanding and the connections among mathematical ideas. Understanding exponential change, for example, means being able to relate the size of a percent growth to the shape of a function's graph, as well as assessing the reasonability of a model.

Functions are the foundation of much of mathematics. Our approach encourages a robust understanding of functions in traditional and novel contexts. Skills are introduced and reinforced throughout the book. This balance of skills and understanding promotes students' critical thinking.

## Fifth Edition: Flexibility

Precalculus courses are taken by a wide range of students and are taught in a wide variety of styles. As instructors ourselves, we know that the balance we choose depends on the students we have: sometimes a focus on conceptual understanding is best; sometimes more skill-building is needed.

To enable instructors to select the balance appropriate for their students, the fifth edition expands the options available for customizing material. For example, we have integrated examples involving linear inequalities in our treatment of linear equations, and offered more examples on rates of change in several chapters. The sections involving limit notation and phase shift have been restructured so that these topics can be easily skipped. Transformations of functions are now introduced in Chapter 2, while the full treatment remains in Chapter 6, providing instructors with more freedom about when to introduce them and how deeply to cover this topic.

## Origin of Text: The Calculus Consortium for Higher Education

This book is the work of faculty at a diverse consortium of institutions, and was originally generously supported by the National Science Foundation. It represents the first consensus among such a diverse group of faculty to have shaped a mainstream precalculus text. Bringing together the results of research and experience with the views of many users, this text is designed to be used in a wide range of institutions.

## Guiding Principles: Varied Problems and the Rule of Four

Conceptual understanding is enhanced when students engage in non-procedural problem-solving. Problems classified into clear-cut types tend to develop proficiency at finding answers, but not necessarily a robust conceptual understanding. Strong problem-solvers feel capable of making progress on problems they have not seen before, not just those of known type. Consequently, we are guided by the following principles:

- Problems are varied and often challenging. Many of our problems cannot be answered by following a template in the text.
- Text examples are diverse and represent the natural integration of skills and concepts.
- The Rule of Four: each concept and function is represented symbolically, numerically, graphically, and verbally. This principle, originally introduced by the Consortium, promotes multiple representations.

- We provide students and instructors quick means of assessing comprehension before moving on by using the true-false Strengthen Your Understanding problems at the end of each chapter.
- Our central theme, functions as models of change, links the components of our precalculus curriculum. Algebra is integrated where appropriate.
- Topics are fewer in number than it is customary so they can be treated in greater depth. The topics are those essential to the study of calculus.
- Problems involving real data give students practice with modeling.
- We include both classic and current applications to prepare students to model with mathematics in a variety of contexts.
- Our problems allow students to become proficient in the use of technology, including symbolic manipulators, computers, tablets, and online software, as appropriate.
- Our precalculus materials allow for a broad range of teaching styles. They are flexible enough for use in large lecture halls, small classes, or in group or lab settings.

## Changes in the Fifth Edition

The fifth edition reflects the many helpful suggestions from users while preserving the focus and guiding principles of previous editions. We have made the following changes:

- **Many examples and problems** are new or have been rewritten. Data has been updated and new data introduced.
- **The three chapters on trigonometry** have been reorganized and rewritten:
  - **Trigonometric functions on the circle** are introduced both in radians and degrees in Chapter 7, highlighting the natural relationship between radians and the unit circle as soon as periodic functions are introduced.
  - **Sinusoidal functions** modeling periodic phenomena and **simple trigonometric equations** are also introduced in Chapter 7, while more involved trigonometric models and equations requiring relationships such as **double-angle identities** are treated in Chapter 9.
  - **Phase shift** has been made an optional topic in Chapter 7.
  - **Triangle trigonometry** is discussed separately in Chapter 8, after trigonometric and sinusoidal functions are treated in Chapter 7.
- A brief exploration of **linear inequalities** has been integrated with the material on solving linear equations in Chapter 1.
- **Vertical and horizontal shifts** are introduced in Chapter 2, and referenced in Chapter 3 when introducing the vertex form of a quadratic equation. Shifts are reviewed at the beginning of Chapter 6 and considered in combination with other transformations.
- **Odd and even functions** are introduced by looking at invariance of certain functions under reflections, and thus better integrated with the transformation focus of Chapter 6.
- The effect of **changing the order of transformations**, the last section in Chapter 6 in the fourth edition, has been shortened and included in the previous section.
- The section on **power functions** in Chapter 11 has been rewritten to increase the focus on graphical behavior and proportionality.
- **Examples on average rate of change** have been added throughout the book, such as for exponential functions (in Chapter 5) and periodic functions (in Chapters 7 and 9).
- **WileyPLUS**, the primary online resource suite paired with the textbook, has been updated with improved problems and hints, including many new problems from the fifth edition.

## What Student Background is Expected?

Students using this book should have successfully completed a course in intermediate algebra or high school algebra II. The book is thought-provoking for well-prepared students while still accessible to students with weaker backgrounds. Providing numerical, graphical, and algebraic approaches builds on different student strengths and provides students with a variety of ways to master the material. Multiple representations give students tools to persist, lowering failure rates.

## Our Experiences

The first four editions of this book were used at hundreds of schools around the country in a wide variety of settings. It has been used successfully in both semester and quarter systems, in large lectures and small classes as well as in full-year courses in secondary schools. It has also been used in computer labs and small groups, often with the integration of a number of different technologies.

## Content

The central theme of this book is functions as models of change. We emphasize that functions can be grouped into families and that functions can be used as models. We explore how function characteristics connect to difference quotients and rates of change, naturally previewing key calculus ideas.

Because linear, quadratic, exponential, power, and periodic functions are most frequently used to model physical phenomena, they are introduced before polynomial and rational functions. Once introduced, a family of functions is compared and contrasted with other families of functions.

A large number of the examples that students see in this precalculus course are real-world problems. By the end of the course, we hope that students will use functions to help them understand the world in which they live. We include non-routine problems to emphasize that such problems are not only part of mathematics, but in some sense are the reason for doing mathematics.

### Technology

The book does not require any specific software or technology. Instructors have used the material with graphing calculators, graphing software, or scientific calculators.

### Chapter 1: Linear Functions and Change

This chapter introduces the concept of a function as well as graphical, tabular, symbolic, and verbal representations of functions, discussing the advantages and disadvantages of each representation. It introduces rates of change and uses them to characterize linear functions. Examples on modeling with linear functions, including interpreting linear inequalities, are discussed. A section on fitting a linear function to data is included.

The **Skills Refresher** section for Chapter 1 reviews linear equations, linear inequalities, and the coordinate plane.

### Chapter 2: Functions

This chapter studies functions in more detail. It introduces domain, range, previews function shifts and the concepts of composite and inverse functions, and investigates the idea of concavity using rates of change. A section on piecewise functions is included.

### Chapter 3: Quadratic Functions

This chapter introduces the standard, factored, and vertex forms of a quadratic function and explores their relationship to graphs, including shifts. The family of quadratic functions provides an opportunity to see the effect of parameters on functional behavior.

The **Skills Refresher** section for Chapter 3 reviews factoring, completing the square, and quadratic equations.

## Chapter 4: Exponential Functions

This chapter introduces the family of exponential functions and the number $e$. It compares exponential and linear functions, solves exponential equations graphically, and gives applications to compound interest.

The **Skills Refresher** section for Chapter 4 reviews the properties of exponents.

## Chapter 5: Logarithmic Functions

This chapter introduces logarithmic functions with base 10 and base $e$, both in order to solve exponential equations and as inverses of exponential functions. After discussing manipulations with logarithms, the chapter focuses on modeling with exponential functions and logarithms. Logarithmic scales and a section on linearizing data conclude the chapter.

The **Skills Refresher** section for Chapter 5 reviews the properties of logarithms.

## Chapter 6: Transformations of Functions and Their Graphs

This chapter investigates transformations. It revisits shifts, introduces reflections and stretches, and explores even and odd symmetry using transformations. The effect of ordering when combining transformations is explored.

## Chapter 7: Trigonometry and Periodic Functions

This chapter introduces trigonometric functions in both radians and degrees as models for periodic motion. After exploring graphs and formulas of sine, cosine, and tangent, general sinusoidal functions are introduced. This chapter introduces the inverse trigonometric functions to solve trigonometric equations.

The **Skills Refresher** section for Chapter 7 reviews sine and cosine values of special angles measured in both radians and degrees.

## Chapter 8: Triangle Trigonometry and Polar Coordinates

This chapter develops right-triangle trigonometry, and introduces the Law of Sines and the Law of Cosines. It also defines polar coordinates, explores their relationship to Cartesian coordinates, and considers the graphs of polar inequalities.

Note: Chapter 8 can be skipped by instructors who prefer not to emphasize triangle trigonometry.

## Chapter 9: Trigonometric Identities, Models and Complex Numbers

This chapter opens with a discussion of trigonometric equations, and then explores trigonometric identities and their role in trigonometric models, such as damped oscillations and acoustic beats. The chapter closes with complex numbers, including Euler's Formula and de Moivre's theorem.

## Chapter 10: Compositions, Inverses, and Combinations of Functions

This chapter discusses combinations of functions. It investigates composite and inverse functions, which were introduced in Chapter 2, in more detail.

## Chapter 11: Polynomial and Rational Functions

This chapter discusses power functions, polynomials, and rational functions. The chapter explores dominance and long-run behavior and concludes by comparing polynomial and exponential functions, and by fitting functions to data.

The **Skills Refresher** section for Chapter 11 reviews algebraic fractions.

## Chapter 12: Vectors

This chapter contains material on vectors and operations involving vectors. An introduction to matrices is included in the last section.

### Chapter 13: Sequences and Series

This chapter introduces arithmetic and geometric sequences and series and their applications.

### Chapter 14: Parametric Equations and Conic Sections

The concluding chapter looks at parametric equations, implicit functions, hyperbolic functions, and conic sections: circles, ellipses, and hyperbolas. The chapter includes a section on the geometrical properties of the conic sections and their applications to orbits.

Note: Chapter 14 is available online only.

## Supplementary Materials

The following supplementary materials are available for the fifth edition:

- **The Instructor's Manual** contains teaching tips, lesson plans, syllabi, and worksheets. It has been expanded and revised to include worksheets, identification of technology-oriented problems, and new syllabi. (ISBN 978-1-119-01383-9)
- **The Printed Test Bank** contains test questions arranged by section.
- **The Instructor's Solution Manual** has complete solutions to all problems. (ISBN 978-1-118-94162-1)
- **The Student Solution Manual** has complete solutions to half of the odd-numbered problems. (ISBN 978-1-118-94163-8)
- **The Computerized Test Bank**, available in both PC and Macintosh formats, allows instructors to create, customize, and print a test containing any combinations of questions from a large bank of questions. Instructors can also customize the questions or create their own.
- **Classroom Activities** are posted at the book companion website. These activities were developed to facilitate in-class group work as well as to introduce new concepts and to practice skills. In addition to the blank copies for each activity that can be handed out to the students, a copy of the activity with fully worked out solutions is also available.
- **The Book Companion Site** at www.wiley.com/college/connally contains all instructor supplements.
- **WileyPLUS** is a powerful online suite of teaching and learning resources tightly integrated with the text. WileyPLUS enables instructors to assign, deliver and grade individually customized homework assignments using exercises and problems from the text. Students receive immediate feedback on their homework and access to full solutions to assigned problems electronically. Students may also access hints to the problems. In addition to online homework, WileyPLUS provides student tutorials, an instructor gradebook, integrated links to the electronic version of the text, and all of the text's supplemental materials. For more information, visit www.wiley.com/college/wileyplus or contact your local Wiley representative for more details.
- **Mini-lecture Videos** linked with examples in the WileyPLUS student version of the text provide greater detail to the solution of examples in each section of the text. These may assist students in reading the text prior to class or in reviewing material after class.
- **The Faculty Network** is a peer-to-peer network of academic faculty dedicated to the effective use of technology in the classroom. This group can help you apply innovative classroom techniques, implement specific software packages, and tailor the technology experience to the specific needs of each individual class. Visit www.wherefacultyconnect.com or ask your Wiley representative for details.

### ConcepTests

ConcepTests, modeled on the pioneering work of Harvard physicist Eric Mazur, are questions designed to promote active learning during class, particularly (but not exclusively) in large lectures. Our evaluation data show students taught with ConcepTests outperformed students taught by traditional lecture methods 73% versus 17% on conceptual questions, and 63% versus 54% on computational problems. ConcepTests arranged by section are available in PowerPoint and Classroom Response System-ready formats from your Wiley representative. (ISBN 978-1-118-94161-4)

## Acknowledgments

We would like to thank the many people who made this book possible. First, we would like to thank the National Science Foundation for their trust and their support; we are particularly grateful to Jim Lightbourne and Spud Bradley.

We are also grateful to our Advisory Board for their guidance: Benita Albert, Lida Barrett, Simon Bernau, Robert Davis, Lovenia Deconge-Watson, John Dossey, Ronald Douglas, Eli Fromm, Bill Haver, Don Lewis, Seymour Parter, John Prados, and Stephen Rodi.

Working with Laurie Rosatone, Anne Scanlan-Rohrer, Ken Santor, Shannon Corliss, Joanna Dingle, Kimberly Kanakes, Jacqueline Sinacori, Courtney Walsh, and Madelyn Lesure at John Wiley is a pleasure. We appreciate their patience and imagination.

Many people have contributed significantly to this text. They include: Lauren Akers, Fahd Alshammari, David Arias, Tim Bean, Charlotte Bonner, Bill Bossert, Brian Bradie, Noah S. Brannen, Mike Brilleslyper, Donna Brouillette, Matthew M. Campbell, Jo Cannon, Ray Cannon, Kenny Ching, Anna Chung, Pierre Cressant, Marcia Crump, Laurie Delitsky, Bob Dobrow, Helen M. Doerr, Ian Dowker, Carolyn Edmond, Maryann Faller, Aidan Flanagan, Brendan Fry, Brad Garner, Carrie Garner, John Gerke, Christie Gilliland, Wynne Guy, Donnie Hallstone, David Halstead, Larry Henly, Dean Hickerson, Jo Ellen Hillyer, Bob Hoburg, Phil Hotchkiss, Mike Huffman, Mac Hyman, Rajini Jesudason, Loren Johnson, David Jones, Scott Kaplowitch, Nazanin Karimi, Thomas Kershaw, Mary Kilbride, Steve Kinholt, Kandace Kling, Angela Kubena, Rob LaQuaglia, Barbara Leasher, Kenneth Lemp, Richard Little, David Lovelock, Guadalupe Lozano Terán, Nicholas Lyktey, Chaimaa Makoudi, Len Malinowski, Nancy Marcus, Kate McGivney, Gowri Meda, Bob Megginson, Deborah Moore, Eric Motylinski, Bill Mueller, Michael Nakamaye, Kyle Niedzwiecki, Karina Nielsen, Kathryn Oswald, Igor Padure, Bridget Neale Paris, Janet Ray, Ritam Ray, Ken Richardson, Halip Saifi, Sharon Sanders, Mary Schumacher, Mike Seery, Mike Sherman, Donna Sherrill, Max Shuchman, Fred Shure, Kanwal Singh, Myra Snell, Regina Souza, Natasha Speer, Sonya Stanley, Michael Steuer, Jim Stone, Peggy Tibbs, Jeff Taft, Elias Toubassi, Jerry Uhl, Pat Wagener, Benjamin West, Dale Winter, Timothy Wisecarver, and Xianbao Xu.

Reports from the following reviewers were most helpful in shaping the second, third, fourth, and fifth editions: Victor Akatsa, Jeffrey Anderson, Beth Borel, Linda Braddy, Michael Brassington, Ingrid Brown-Scott, Linda Casper, Kim Chudnick, Ted Coe, Ray Collings, Joe Coyle, Pam Crawford, Marcia Crump, Monica Davis, Laurie Delitsky, Phyllis Desormeaux, Helen Doerr, Diane Downie, Peter Dragnev, Patricia Dueck, Julie Fisher, Jennifer Fowler, Alyne Fulte, David Gillette, Jack Green, Zdenka Guadarrama, Donnie Hallstone, Jeff Hoherz, Majid Hosseini, Rick Hough, Ann Humes, Bryan Ingham, Mary Lou Jumonville, Pallavi Ketkar, William Kiele, Mile Krajcevski, John LaMaster, Richard Lane, Debbie Lee, Phyllis Leonard, Daphne MacLean, Diane Mathios, Krista Maxson, Michael McAfee, Vince McGarry, Maria Miles, Laura Moore-Mueller, Ellen Musen, Dave Nolan, Linda O'Brien, Chris Parks, Scott Perry, Jeffrey S. Powell, Anne Praderas, Michael Price, Mary Rack, Gregory Rhoads, Joe Rody, Emily Roth, John Sawka, Amy N. Schachle, Barbara Shabell Deirdre Smith, Ernie Solheid, Regina P. Souza, Sandy Spears, Diana Staats, Jonathan Stadler, Mary Jane Sterling, Dennis Stramiello, Debbie Strance, Allison Sutton, John Thomason, Diane Van Nostrand, Jim Vicich, Linda Wagner, Nicole Williams, Jim Winston, Vauhn Wittman-Grahler, Bruce Yoshiwara, Fang Zhao, and Xinyun Zhu. Special thanks are owed to Faye Riddle for administering the project and to Alex Kasman for his software support.

| | | | |
|---|---|---|---|
| Eric Connally | Deborah Hughes-Hallett | Andrew M. Gleason | Philip Cheifetz |
| Ann Davidian | Dan Flath | Selin Kalaycıoğlu | Brigitte Lahme |
| Patti Frazer Lock | Guadalupe I. Lozano | Elliot J. Marks | William G. McCallum |
| Jerry Morris | Karen Rhea | Ellen Schmierer | Pat Shure |
| Adam H. Spiegler | Carl Swenson | Aaron D. Wootton | |

## To Students: How to Learn from this Book

- This book may be different from other math textbooks that you have used, so it may be helpful to know about some of the differences in advance. At every stage, this book emphasizes the *meaning* (in practical, graphical or numerical terms) of the symbols you are using. There is much less emphasis on "plug-and-chug" and using formulas, and much more emphasis on the interpretation of these formulas than you may expect. You will often be asked to explain your ideas in words or to explain an answer using graphs.
- The book contains the main ideas of precalculus in plain English. Success in using this book will depend on reading, questioning, and thinking hard about the ideas presented. It will be helpful to read the text in detail, not just the worked examples.
- There are few examples in the text that are exactly like the homework problems, so homework problems can't be done by searching for similar-looking "worked out" examples. Success with the homework will come by grappling with the ideas of precalculus.
- Many of the problems in the book are open-ended. This means that there is more than one correct approach and more than one correct solution. Sometimes, solving a problem relies on common-sense ideas that are not stated in the problem explicitly but which you know from everyday life.
- This book assumes that you have access to a calculator or computer that can graph functions and find (approximate) roots of equations. There are many situations where you may not be able to find an exact solution to a problem, but can use a calculator or computer to get a reasonable approximation. An answer obtained this way can be as useful as an exact one. However, the problem does not always state that a calculator is required, so use your own judgment.
- This book attempts to give equal weight to four methods for describing functions: graphical (a picture), numerical (a table of values), algebraic (a formula) and verbal (words). Sometimes it's easier to translate a problem given in one form into another. For example, you might replace the graph of a parabola with its equation, or plot a table of values to see its behavior. It is important to be flexible about your approach: if one way of looking at a problem doesn't work, try another.
- Students using this book have found discussing these problems in small groups helpful. There are a great many problems that are not cut-and-dried; it can help to attack them with the other perspectives your colleagues can provide. If group work is not feasible, see if your instructor can organize a discussion session in which additional problems can be worked on.
- You are probably wondering what you'll get from the book. The answer is, if you put in a solid effort, you will get a real understanding of functions as well as a real sense of how mathematics is used in the age of technology.

# Table of Contents

## 7 TRIGONOMETRY AND PERIODIC FUNCTIONS 265



## 8 TRIANGLE TRIGONOMETRY AND POLAR COORDINATES 327



## 9 TRIGONOMETRIC IDENTITIES, MODELS, AND COMPLEX NUMBERS 353

Chapter One

# LINEAR FUNCTIONS AND CHANGE

© Patrick Zephyr/Patrick Zephyr Nature Photography

## Contents

# 1.1 FUNCTIONS AND FUNCTION NOTATION

In everyday language, the word *function* expresses the notion of dependence. For example, a person might say that election results are a function of the economy, meaning that the winner of an election is determined by how the economy is doing. Someone else might claim that car sales are a function of the weather, meaning that the number of cars sold on a given day is affected by the weather.

In mathematics, the meaning of the word *function* is more precise, but the basic idea is the same. A function is a relationship between two quantities. If the value of the first quantity determines exactly one value of the second quantity, we say the second quantity is a function of the first. We make the following definition:

> A **function** is a rule that takes certain numbers as inputs and assigns to each input number exactly one output number. The output is a function of the input.

The inputs and outputs are also called *variables*.

## Representing Functions: Words, Tables, Graphs, and Formulas

A function can be described using words, data in a table, points on a graph, or a formula.

**Example 1** It is a surprising biological fact that most crickets chirp at a rate that increases as the temperature increases. For the snowy tree cricket (*Oecanthus fultoni*), the relationship between temperature and chirp rate is so reliable that this type of cricket is called the thermometer cricket. We can estimate the temperature (in degrees Fahrenheit) by counting the number of times a snowy tree cricket chirps in 15 seconds and adding 40. For instance, if we count 20 chirps in 15 seconds, then a good estimate of the temperature is $20 + 40 = 60°F$.[1]

The rule used to find the temperature $T$ (in °F) from the chirp rate $R$ (in chirps per minute) is an example of a function. The input is chirp rate and the output is temperature. Describe this function using words, a table, a graph, and a formula.

**Solution**

- **Words**: To estimate the temperature, we count the number of chirps in fifteen seconds and add forty. Alternatively, we can count $R$ chirps per minute, divide $R$ by four and add forty. This is because there are one-fourth as many chirps in fifteen seconds as there are in sixty seconds. For instance, 80 chirps per minute works out to $\frac{1}{4} \cdot 80 = 20$ chirps every 15 seconds, giving an estimated temperature of $20 + 40 = 60°F$.
- **Table**: Table 1.1 gives the estimated temperature, $T$, as a function of $R$, the number of chirps per minute. Notice the pattern in Table 1.1: each time the chirp rate, $R$, goes up by 20 chirps per minute, the temperature, $T$, goes up by 5°F.

**Table 1.1** *Chirp rate and temperature*

| $R$, chirp rate (chirps/minute) | $T$, predicted temperature (°F) |
|---|---|
| 20 | 45 |
| 40 | 50 |
| 60 | 55 |
| 80 | 60 |
| 100 | 65 |
| 120 | 70 |
| 140 | 75 |
| 160 | 80 |

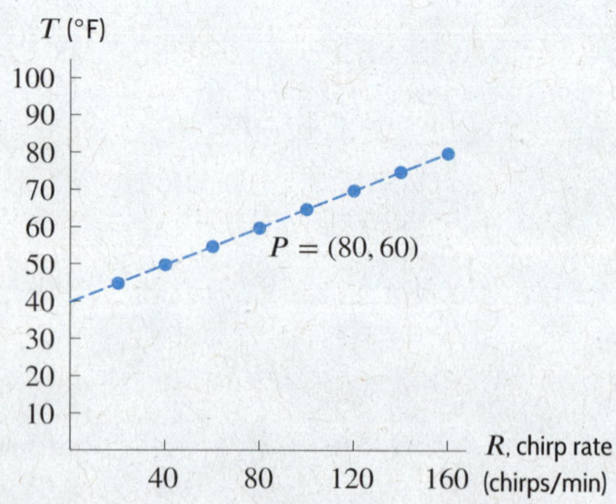


**Figure 1.1**: Chirp rate and temperature

[1]This relationship is often called Dolbear's Law, as it was first proposed in Amos Dolbear, "The Cricket as a Thermometer," in *The American Naturalist*, 31(1897), pp. 970–971.

- **Graph**: The data from Table 1.1 are plotted in Figure 1.1. For instance, the pair of values $R = 80$, $T = 60$ is plotted as the point $P$, which is 80 units along the horizontal axis and 60 units up the vertical axis. Data represented in this way are said to be plotted on the *Cartesian plane*. The precise position of $P$ is shown by its coordinates, written $P = (80, 60)$.
- **Formula**: A formula is an equation giving $T$ in terms of $R$. Dividing the chirp rate by four and adding forty gives the estimated temperature, so:

$$\underbrace{\text{Estimated temperature (in °F)}}_{T} = \frac{1}{4} \cdot \underbrace{\text{Chirp rate (in chirps/min)}}_{R} + 40.$$

Rewriting this using the variables $T$ and $R$ gives the formula:

$$T = \frac{1}{4}R + 40.$$

Let's check the formula. Substituting $R = 80$, we have

$$T = \frac{1}{4} \cdot 80 + 40 = 60,$$

which agrees with point $P = (80, 60)$ in Figure 1.1. The formula $T = \frac{1}{4}R + 40$ also tells us that if $R = 0$, then $T = 40$. Thus, the dashed line in Figure 1.1 crosses (or intersects) the $T$-axis at $T = 40$; we say the $T$-*intercept* is 40.

All the descriptions given in Example 1 provide the same information, but each description has a different emphasis. A relationship between variables is often given in words, as at the beginning of Example 1. Table 1.1 is useful because it shows the predicted temperature for various chirp rates. Figure 1.1 is more suggestive of a trend than the table, although it is harder to read exact values of the function. For example, you might have noticed that every point in Figure 1.1 falls on a straight line that slopes up from left to right. In general, a graph can reveal a pattern that might otherwise go unnoticed. Finally, the formula has the advantage of being both compact and precise. However, this compactness can also be a disadvantage since it may be harder to gain as much insight from a formula as from a table or a graph.

## Mathematical Models

When we use a function to describe an actual situation, the function is referred to as a **mathematical model**. The formula $T = \frac{1}{4}R + 40$ is a mathematical model of the relationship between the temperature and the cricket's chirp rate. Such models can be powerful tools for understanding phenomena and making predictions. For example, this model predicts that when the chirp rate is 80 chirps per minute, the temperature is 60°F. In addition, since $T = 40$ when $R = 0$, the model predicts that the chirp rate is 0 at 40°F. Whether the model's predictions are accurate for chirp rates down to 0 and temperatures as low as 40°F is a question that mathematics alone cannot answer; an understanding of the biology of crickets is needed. However, we can safely say that the model does not apply for temperatures below 40°F, because the chirp rate would then be negative. For the range of chirp rates and temperatures in Table 1.1, the model is remarkably accurate.

In everyday language, saying that $T$ is a function of $R$ suggests that making the cricket chirp faster would somehow make the temperature change. Clearly, the cricket's chirping does not cause the temperature to be what it is. In mathematics, saying that the temperature "depends" on the chirp rate means only that knowing the chirp rate is sufficient to tell us the temperature.

## Function Notation

To indicate that a quantity $Q$ is a function of a quantity $t$, we abbreviate

$$Q \text{ is a function of } t \quad \text{to} \quad Q \text{ equals "} f \text{ of } t\text{"}$$

and, using function notation, to

$$Q = f(t).$$

Thus, applying the rule $f$ to the input value, $t$, gives the output value, $f(t)$, which is a value of $Q$. Here $Q$ is called the *dependent variable* and $t$ is called the *independent variable*. In other words,

$$\text{Output} = f(\text{Input})$$

or

$$\text{Dependent} = f(\text{Independent}).$$

We could have used any letter, not just $f$, to represent the rule.

The expressions "$Q$ depends on $t$" or "$Q$ is a function of $t$" do *not* imply a cause-and-effect relationship, as the snowy tree cricket example illustrates.

**Example 2** Example 1 gives the following formula for estimating air temperature, $T$, based on the chirp rate, $R$, of the snowy tree cricket:

$$T = \frac{1}{4}R + 40.$$

In this formula, $T$ depends on $R$. Writing $T = f(R)$ indicates that the relationship is a function.

**Example 3** The number of gallons of paint needed to paint a house depends on the size of the house. A gallon of paint typically covers 250 square feet. Thus, the number of gallons of paint, $n$, is a function of the area to be painted, $A$ ft$^2$. We write $n = f(A)$.

(a) Find a formula for $f$.
(b) Explain in words what the statement $f(10{,}000) = 40$ tells us about painting houses.

**Solution** (a) If $A = 250$, the house requires one gallon of paint. If $A = 500$, it requires $500/250 = 2$ gallons of paint, if $A = 750$ it requires $750/250 = 3$ gallons of paint, and so on. We see that a house of area $A$ requires $A/250$ gallons of paint, so $n$ and $A$ are related by the formula

$$n = f(A) = \frac{A}{250}.$$

(b) The input of the function $n = f(A)$ is an area and the output is an amount of paint. The statement $f(10{,}000) = 40$ tells us that an area of $A = 10{,}000$ ft$^2$ requires $n = 40$ gallons of paint.

## Functions Don't Have to Be Defined by Formulas

People sometimes think that functions are always represented by formulas. However, other representations, such as tables or graphs, can be useful.

**Example 4** The average monthly rainfall, $R$, at Chicago's O'Hare airport is given in Table 1.2, where time, $t$, is in months and $t = 1$ is January, $t = 2$ is February, and so on. The rainfall is a function of the month, so we write $R = f(t)$. However, there is no formula that gives $R$ when $t$ is known. Evaluate $f(1)$ and $f(11)$. Explain what your answers mean.

**Table 1.2** *Average monthly rainfall at Chicago's O'Hare airport*

| Month, $t$ | 1 | 2 | 3 | 4 | 5 | 6 | 7 | 8 | 9 | 10 | 11 | 12 |
|---|---|---|---|---|---|---|---|---|---|---|---|---|
| Rainfall, $R$ (inches) | 1.8 | 1.8 | 2.7 | 3.1 | 3.5 | 3.7 | 3.5 | 3.4 | 3.2 | 2.5 | 2.4 | 2.1 |

**Solution** The value of $f(1)$ is the average rainfall in inches at Chicago's O'Hare airport in a typical January. From the table, $f(1) = 1.8$ inches. Similarly, $f(11) = 2.4$ means that in a typical November, there are 2.4 inches of rain at O'Hare.

## When Is a Relationship Not a Function?

It is possible for two quantities to be related and yet for neither quantity to be a function of the other.

**Example 5** A national park contains foxes that prey on rabbits. Table 1.3 gives the two populations, $F$ and $R$, over a 12-month period, where $t = 0$ means January 1, $t = 1$ means February 1, and so on.

**Table 1.3** *Number of foxes and rabbits in a national park, by month*

| $t$, month | 0 | 1 | 2 | 3 | 4 | 5 | 6 | 7 | 8 | 9 | 10 | 11 |
|---|---|---|---|---|---|---|---|---|---|---|---|---|
| $R$, rabbits | 1000 | 750 | 567 | 500 | 567 | 750 | 1000 | 1250 | 1433 | 1500 | 1433 | 1250 |
| $F$, foxes | 150 | 143 | 125 | 100 | 75 | 57 | 50 | 57 | 75 | 100 | 125 | 143 |

(a) Is $F$ a function of $t$? Is $R$ a function of $t$?
(b) Is $F$ a function of $R$? Is $R$ a function of $F$?

**Solution** (a) Remember that for a relationship to be a function, an input can only give a single output. Both $F$ and $R$ are functions of $t$. For each value of $t$, there is exactly one value of $F$ and exactly one value of $R$. For example, Table 1.3 shows that if $t = 5$, then $R = 750$ and $F = 57$. This means that on June 1 there are 750 rabbits and 57 foxes in the park. If we write $R = f(t)$ and $F = g(t)$, then $f(5) = 750$ and $g(5) = 57$.

(b) No, $F$ is not a function of $R$. For example, suppose $R = 750$, meaning there are 750 rabbits. This happens both at $t = 1$ (February 1) and at $t = 5$ (June 1). In the first instance, there are 143 foxes; in the second instance, there are 57 foxes. Since there are $R$-values which correspond to more than one $F$-value, $F$ is not a function of $R$.

Similarly, $R$ is not a function of $F$. At time $t = 5$, we have $R = 750$ when $F = 57$, while at time $t = 7$, we have $R = 1250$ when $F = 57$ again. Thus, the value of $F$ does not uniquely determine the value of $R$.

### How to Tell if a Graph Represents a Function: Vertical Line Test

What does it mean graphically for $y$ to be a function of $x$? Look at a graph of $y$ against $x$, with $y$ on the vertical axis and $x$ on the horizontal axis. For a function, each $x$-value corresponds to exactly one $y$-value. This means that the graph intersects any vertical line at most once (either once or not at all). If a vertical line cuts the graph twice, the graph contains two points with different $y$-values but the same $x$-value; this violates the definition of a function. Thus, we have the following criterion:

> **Vertical Line Test**: If there is a vertical line that intersects a graph in more than one point, then the graph does not represent a function.

**Example 6** In which of the graphs in Figures 1.2 and 1.3 could $y$ be a function of $x$?

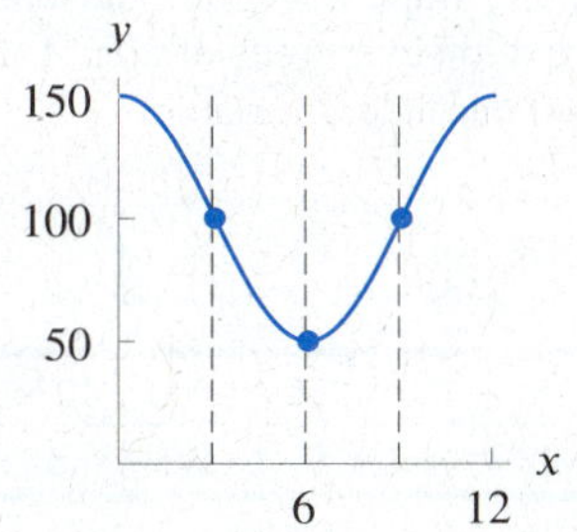


**Figure 1.2**: Since no vertical line intersects this curve at more than one point, $y$ could be a function of $x$

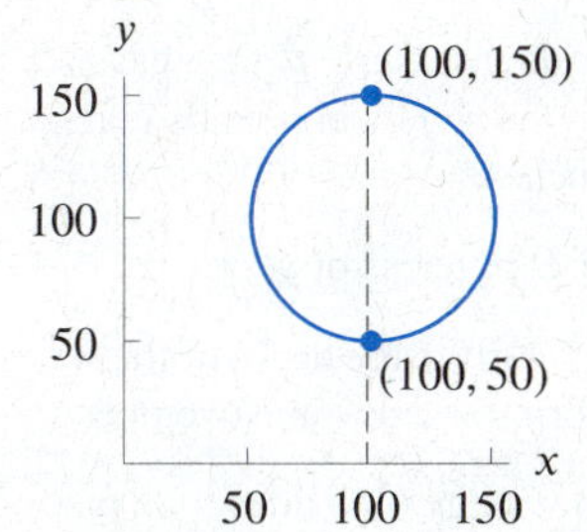


**Figure 1.3**: Since one vertical line intersects this curve at more than one point, $y$ is not a function of $x$

**Solution** The graph in Figure 1.2 could represent $y$ as a function of $x$ because no vertical line intersects this curve in more than one point. The graph in Figure 1.3 does not represent a function because the vertical line shown intersects the curve at two points.

A graph fails the vertical line test if at least one vertical line cuts the graph more than once, as in Figure 1.3. However, if a graph represents a function, then *every* vertical line must intersect the graph at no more than one point.

## Exercises and Problems for Section 1.1

### Skill Refresher

In Exercises S1–S4, simplify each expression.

**S1.** $c + \frac{1}{2}c$

**S2.** $P + 0.07P + 0.02P$

**S3.** $2\pi r^2 + 2\pi r \cdot 2r$

**S4.** $\dfrac{12\pi - 2\pi}{6\pi}$

In Exercises S5–S8, find the value of the expressions for the given value of $x$ and $y$.

**S5.** $x - 5y$ for $x = \frac{1}{2}$, $y = -5$.

**S6.** $1 - 12x + x^2$ for $x = 3$.

**S7.** $\dfrac{3}{2 - x^3}$ for $x = -1$.

**S8.** $\dfrac{4}{1 + 1/x}$ for $x = -\frac{3}{4}$.

The figures in Exercises S9–S10 are parallelograms. Find the coordinates of the point $A$.

**S9.**

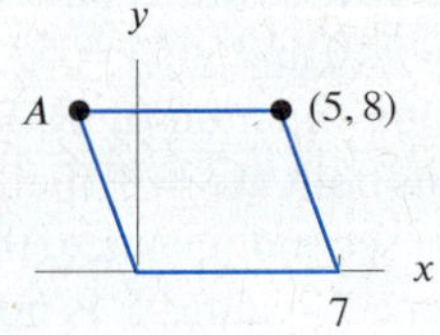


**S10.**

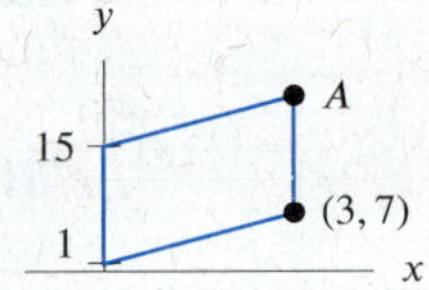


### Exercises

In Exercises 1–2, write the relationship using function notation (that is, $y$ is a function of $x$ is written $y = f(x)$).

**1.** Number of molecules, $m$, in a gas, is a function of the volume of the gas, $v$.

**2.** Weight, $w$, is a function of caloric intake, $c$.

In Exercises 3–6, label the axes for a sketch to illustrate the given statement.

**3.** "Over the past century we have seen changes in the population, $P$ (in millions), of the city. . ."

**4.** "Sketch a graph of the cost of manufacturing $q$ items. . ."

**5.** "Graph the pressure, $p$, of a gas as a function of its volume, $v$, where $p$ is in pounds per square inch and $v$ is in cubic inches."

**6.** "Graph $D$ in terms of $y$. . ."

**7.** Figure 1.4 gives the depth of the water at Montauk Point, New York, for a day in November.

(a) How many high tides took place on this day?
(b) How many low tides took place on this day?
(c) How much time elapsed in between high tides?

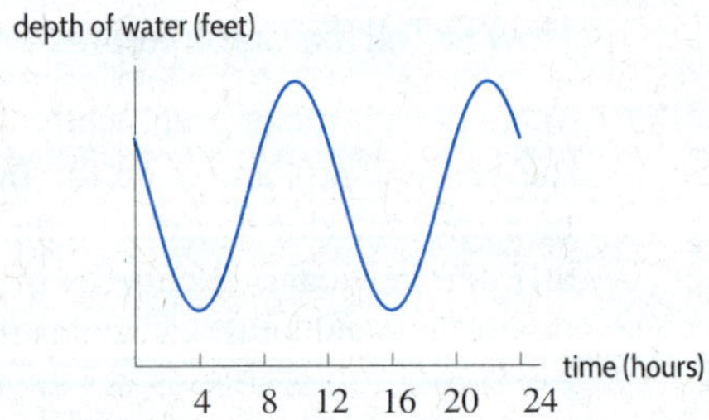


Figure 1.4

**8.** Using Table 1.4, graph $n = f(A)$, the number of gallons of paint needed to cover walls of area $A$. Identify the independent and dependent variables.

Table 1.4

| $A$ | 0 | 250 | 500 | 750 | 1000 | 1250 | 1500 |
|---|---|---|---|---|---|---|---|
| $n$ | 0 | 1 | 2 | 3 | 4 | 5 | 6 |

**9.** Use Figure 1.5 to fill in the missing values:

**(a)** $f(0) = ?$ **(b)** $f(?) = 0$

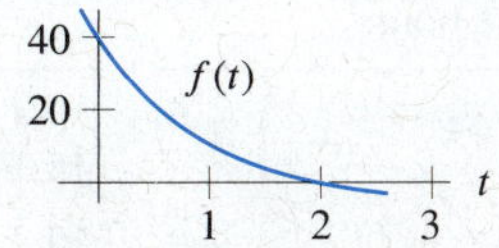


Figure 1.5

**10.** Use Table 1.5 to fill in the missing values. (There may be more than one answer.)

**(a)** $f(0) = ?$ **(b)** $f(?) = 0$
**(c)** $f(1) = ?$ **(d)** $f(?) = 1$

Table 1.5

| $x$ | 0 | 1 | 2 | 3 | 4 |
|---|---|---|---|---|---|
| $f(x)$ | 4 | 2 | 1 | 0 | 1 |

Exercises 11–14 use Figure 1.6.

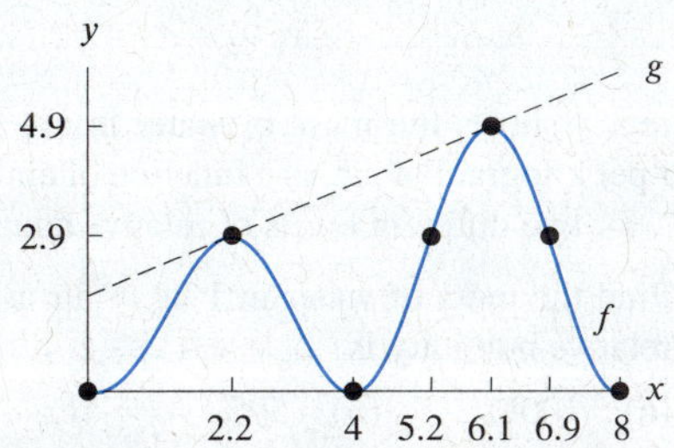


Figure 1.6

**11.** Find $f(6.9)$.

**12.** Give the coordinates of two points on the graph of $g$.

**13.** Solve $f(x) = 0$ for $x$.

**14.** Solve $f(x) = g(x)$ for $x$.

**15.** **(a)** You are going to graph $p = f(w)$. Which variable goes on the horizontal axis?
**(b)** If $10 = f(-4)$, give the coordinates of a point on the graph of $f$.
**(c)** If 6 is a solution of the equation $f(w) = 1$, give a point on the graph of $f$.

In Exercises 16–19 a relationship is given between two quantities. Are both quantities functions of the other one, or is one or neither a function of the other? Explain.

**16.** $7w^2 + 5 = z^2$ **17.** $y = x^4 - 1$ **18.** $m = \sqrt{t}$

**19.** The number of gallons of gas, $g$, at \$3 per gallon and the number of pounds of coffee, $c$, at \$10 per pound that can be bought for a total of \$100.

In Exercises 20–25, could the graph represent $y$ as a function of $x$?

**20.**

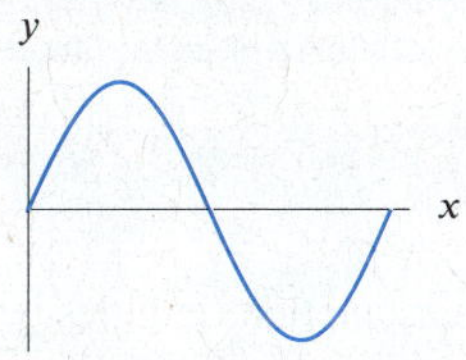


**21.**

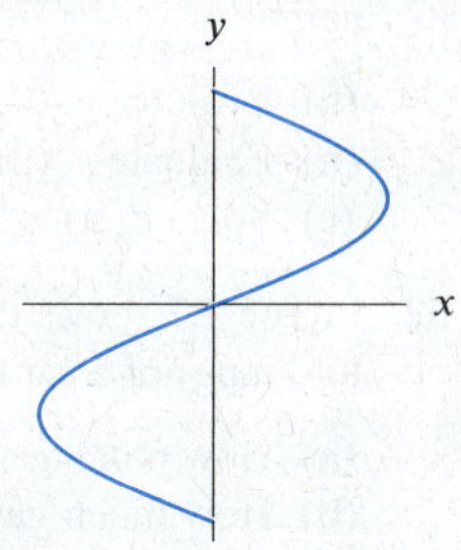


**22.**

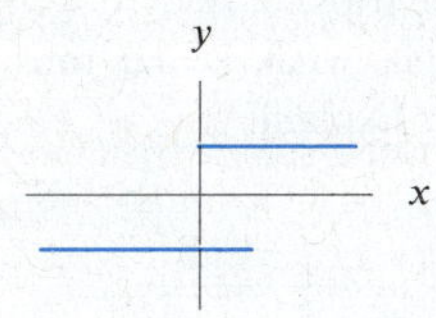


**23.**

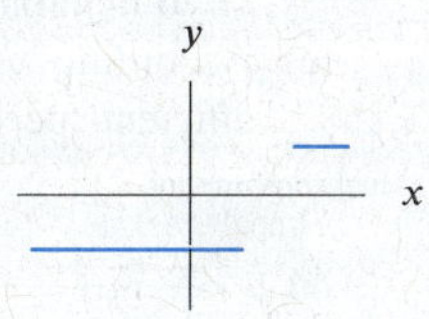


**24.**

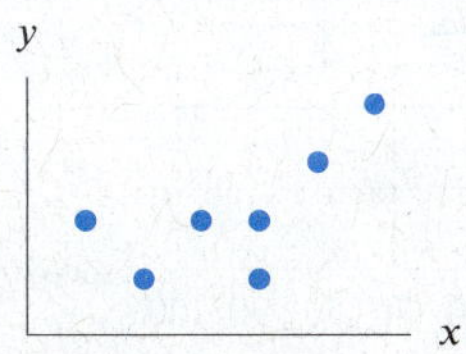


**25.**

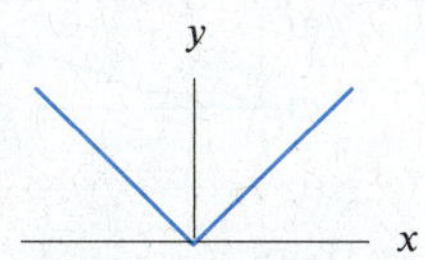


## Problems

**26.** At the end of a semester, students' math grades are listed in a table which gives each student's ID number in the left column and the student's grade in the right column. Let $N$ represent the ID number and $G$ represent the grade. Which quantity, $N$ or $G$, must necessarily be a function of the other?

**27.** A person's blood sugar level at a particular time of the day is partially determined by the time of the most recent meal. After a meal, blood sugar level increases rapidly, then slowly comes back down to a normal level. Sketch a person's blood sugar level as a function of time over the course of a day. Label the axes to indicate normal blood sugar level and the time of each meal.

**28.** When a parachutist jumps out of a plane, the speed of her fall increases until she opens her parachute, at which time her falling speed suddenly decreases and stays constant until she reaches the ground. Sketch a possible graph of the height $H$ of the parachutist as a function of time $t$, from the time when she jumps from the plane to the time when she reaches the ground.

**29.** A buzzard is circling high overhead when it spies some road kill. It swoops down, lands, and eats. Later it takes off sluggishly, and resumes circling overhead, but at a lower altitude. Sketch a possible graph of the height of the buzzard as a function of time.

**30.** Table 1.6 gives the ranking $r$ for three different names—Hannah, Alexis, and Madison. Of the three names, which was most popular and which was least popular in

**(a)** 1995? **(b)** 2004?

**Table 1.6** *Ranking of names—Hannah ($r_h$), Alexis ($r_a$), and Madison ($r_m$)—for girls born between 1995 ($t = 0$) and 2004 ($t = 9$)*[2]

| $t$ | 0 | 1 | 2 | 3 | 4 | 5 | 6 | 7 | 8 | 9 |
|---|---|---|---|---|---|---|---|---|---|---|
| $r_h$ | 7 | 7 | 5 | 2 | 2 | 2 | 3 | 3 | 4 | 5 |
| $r_a$ | 14 | 8 | 8 | 6 | 3 | 6 | 5 | 5 | 7 | 11 |
| $r_m$ | 29 | 15 | 10 | 9 | 7 | 3 | 2 | 2 | 3 | 3 |

**31.** Table 1.6 gives information about the popularity of the names Hannah, Madison, and Alexis. Describe in words what your answers to parts (a)–(c) tell you about these names.

**(a)** Evaluate $r_m(0) - r_h(0)$.
**(b)** Evaluate $r_m(9) - r_h(9)$.
**(c)** Solve $r_m(t) < r_a(t)$.

**32.** Figure 1.7 shows the fuel consumption (in miles per gallon, mpg) of a car traveling at various speeds (in mph).

**(a)** How much gas is used on a 300-mile trip at 40 mph?
**(b)** How much gas is saved by traveling 60 mph instead of 70 mph on a 200-mile trip?
**(c)** According to this graph, what is the most fuel-efficient speed to travel? Explain.

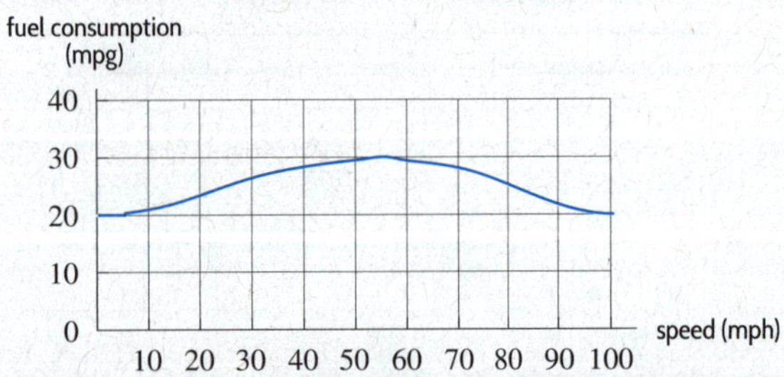


**Figure 1.7**

**33.** Match each story about a bike ride to one of the graphs (i)–(v), where $d$ represents distance from home and $t$ is time in hours since the start of the ride. (A graph may be used more than once.)

**(a)** Starts 5 miles from home and rides 5 miles per hour away from home.
**(b)** Starts 5 miles from home and rides 10 miles per hour away from home.
**(c)** Starts 10 miles from home and arrives home one hour later.
**(d)** Starts 10 miles from home and is halfway home after one hour.
**(e)** Starts 5 miles from home and is 10 miles from home after one hour.

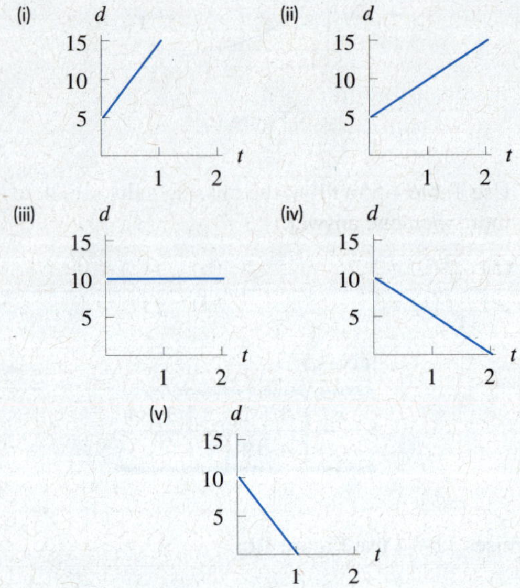


**34.** Figure 1.8 shows the mass of water in air, in grams of water per kilogram of air, as a function of air temperature in °C, for two different levels of relative humidity.

**(a)** Find the mass of water in 1 kg of air at 30°C if the relative humidity is

**(a)** 100% **(b)** 50% **(c)** 75%

**(b)** How much water is in a room containing 300 kg of air if the relative humidity is 50% and the temperature is 20°C?
**(c)** The density of air is approximately 1.2 kg/m$^3$. If the relative humidity in your classroom is 50% and the temperature is 20°C, estimate the amount of water in the air.

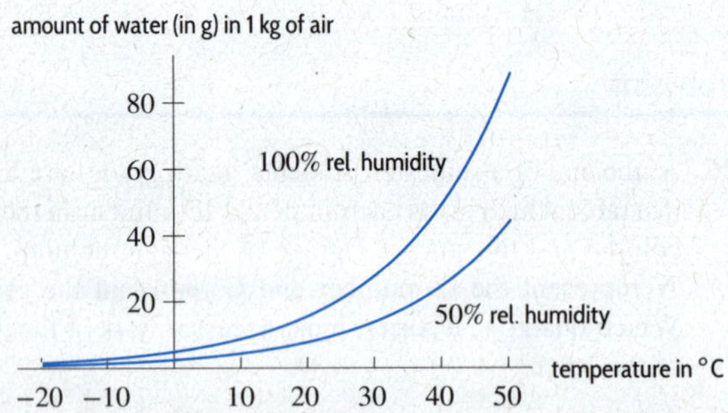


**Figure 1.8**

[2]Data from the SSA website at www.ssa.gov, accessed January 12, 2006.

**35.** Let $f(t)$ be the number of people, in millions, who own cell phones $t$ years after 1990. Explain the meaning of the following statements.

**(a)** $f(10) = 100.3$ **(b)** $f(a) = 20$
**(c)** $f(20) = b$ **(d)** $n = f(t)$

**36.** **(a)** Ten inches of snow is equivalent to about one inch of rain.[3] Write an equation for the amount of precipitation, measured in inches of rain, $r = f(s)$, as a function of the number of inches of snow, $s$.
**(b)** Evaluate and interpret $f(5)$.
**(c)** Find $s$ such that $f(s) = 5$ and interpret your result.

**37.** An 8-foot-tall cylindrical water tank has a base of diameter 6 feet.

**(a)** How much water can the tank hold?
**(b)** How much water is in the tank if the water is 5 feet deep?
**(c)** Write a formula for the volume of water as a function of its depth in the tank.

**38.** Table 1.7 gives $A = f(d)$, the amount of money in bills of denomination $d$ circulating in US currency in 2013.[4] For example, there were \$74.5 billion worth of \$50 bills in circulation.

**(a)** Find $f(100)$. What does this tell you about money?
**(b)** Are there more \$1 bills or \$5 bills in circulation?

**Table 1.7**

| Denomination (\$) | 1 | 2 | 5 | 10 | 20 | 50 | 100 |
|---|---|---|---|---|---|---|---|
| Circulation (\$bn) | 10.6 | 2.1 | 12.7 | 18.5 | 155.0 | 74.5 | 924.7 |

**39.** Table 1.8 shows the daily low temperature for a one-week period in New York City during July.

**(a)** What was the low temperature on July 19?
**(b)** When was the low temperature 73°F?
**(c)** Is the daily low temperature a function of the date?
**(d)** Is the date a function of the daily low temperature?

**Table 1.8**

| Date | 17 | 18 | 19 | 20 | 21 | 22 | 23 |
|---|---|---|---|---|---|---|---|
| Low temp (°F) | 73 | 77 | 69 | 73 | 75 | 75 | 70 |

**40.** Use the data from Table 1.3 on page 5.

**(a)** Plot $R$ on the vertical axis and $t$ on the horizontal axis. Use this graph to explain why you believe that $R$ is a function of $t$.
**(b)** Plot $F$ on the vertical axis and $t$ on the horizontal axis. Use this graph to explain why you believe that $F$ is a function of $t$.
**(c)** Plot $F$ on the vertical axis and $R$ on the horizontal axis. From this graph show that $F$ is not a function of $R$.
**(d)** Plot $R$ on the vertical axis and $F$ on the horizontal axis. From this graph show that $R$ is not a function of $F$.

**41.** Since Roger Bannister broke the 4-minute mile on May 6, 1954, the record has been lowered by over sixteen seconds. Table 1.9 shows the year and times (as min:sec) of new world records for the one-mile run.[5] (Official records for the mile ended in 1999.)

**(a)** Is the time a function of the year? Explain.
**(b)** Is the year a function of the time? Explain.
**(c)** Let $y(r)$ be the year in which the world record, $r$, was set. Explain what is meant by the statement $y(3:47.33) = 1981$.
**(d)** Evaluate and interpret $y(3:51.1)$.

**Table 1.9**

| Year | Time | Year | Time | Year | Time |
|---|---|---|---|---|---|
| 1954 | 3:59.4 | 1966 | 3:51.3 | 1981 | 3:48.53 |
| 1954 | 3:58.0 | 1967 | 3:51.1 | 1981 | 3:48.40 |
| 1957 | 3:57.2 | 1975 | 3:51.0 | 1981 | 3:47.33 |
| 1958 | 3:54.5 | 1975 | 3:49.4 | 1985 | 3:46.32 |
| 1962 | 3:54.4 | 1979 | 3:49.0 | 1993 | 3:44.39 |
| 1964 | 3:54.1 | 1980 | 3:48.8 | 1999 | 3:43.13 |
| 1965 | 3:53.6 | | | | |

**42.** The sales tax on an item is 6%. Express the total cost, $C$, in terms of the price of the item, $P$.

**43.** A price increases 5% due to inflation and is then reduced 10% for a sale. Express the final price as a function of the original price, $P$.

**44.** Write a formula for the area of a circle as a function of its radius and determine the percent increase in the area if the radius is increased by 10%.

**45.** There are $x$ male job-applicants at a certain company and $y$ female applicants. Suppose that 15% of the men are accepted and 18% of the women are accepted. Write an expression in terms of $x$ and $y$ representing each of the following quantities:

**(a)** The total number of applicants to the company.
**(b)** The total number of applicants accepted.
**(c)** The percentage of all applicants accepted.

[3] http://mo.water.usgs.gov/outreach/rain, accessed May 7, 2006.
[4] http://www.federalreserve.gov/paymentsystems/coin_currcircvalue.htm, accessed February 16, 2014.
[5] www.infoplease.com/ipsa/A0112924.html, accessed January 15, 2006.

**46.** A chemical company spends $2 million to buy machinery before it starts producing chemicals. Then it spends $0.5 million on raw materials for each million liters of chemical produced.

**(a)** The number of liters produced ranges from 0 to 5 million. Make a table showing the relationship between the number of million liters produced, $l$, and the total cost, $C$, in millions of dollars, to produce that number of million liters.

**(b)** Find a formula that expresses $C$ as a function of $l$.

**47.** A person leaves home and walks due west for a time and then walks due north.

**(a)** The person walks 10 miles in total. If $w$ represents the (variable) distance west she walks, and $D$ represents her (variable) distance from home at the end of her walk, is $D$ a function of $w$? Why or why not?

**(b)** Suppose now that $x$ is the distance that she walks in total. Is $D$ a function of $x$? Why or why not?

## 1.2 RATE OF CHANGE

The worldwide annual sales of smartphones have increased each year from when they were first introduced. To measure how fast sales increase, we calculate a *rate of change* in number of units sold

$$\frac{\text{Change in units sold}}{\text{Change in time}}.$$

Due to the rising popularity of smartphones, annual sales of regular cell phones, or *feature phones*, have been declining. See Table 1.10.

Let us calculate the rate of change of smartphone and feature phone sales between 2010 and 2013. Table 1.10 gives

$$\begin{array}{c}\text{Average rate of change of}\\ \text{smartphone sales from 2010 to 2013}\end{array} = \frac{\text{Change in sales}}{\text{Change in time}} = \frac{968-301}{2013-2010} \approx 222.3 \ \frac{\text{mn}}{\text{year}}.$$

Thus, the number of smartphones sold has increased on average by 222.3 million units per year between 2010 and 2013. See Figure 1.9. Similarly, Table 1.10 gives

$$\begin{array}{c}\text{Average rate of change of feature phone}\\ \text{sales from 2010 to 2013}\end{array} = \frac{\text{Change in sales}}{\text{Change in time}} = \frac{838-1079}{2013-2010} \approx -80.3 \ \frac{\text{mn}}{\text{year}}.$$

Thus, the number of feature phones sold has decreased on average by 80.3 million units per year between 2010 and 2013. See Figure 1.10.

**Table 1.10** *Worldwide annual sales of smartphones and feature phones*[6]

| Year | 2010 | 2011 | 2012 | 2013 |
|---|---|---|---|---|
| Smartphone sales (millions of units) | 301 | 480 | 661 | 968 |
| Feature phone sales (millions of units) | 1079 | 1075 | 914 | 838 |

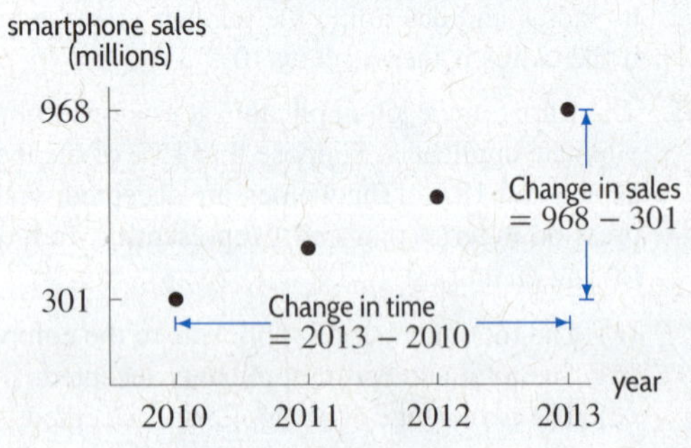


**Figure 1.9**: Smartphone sales

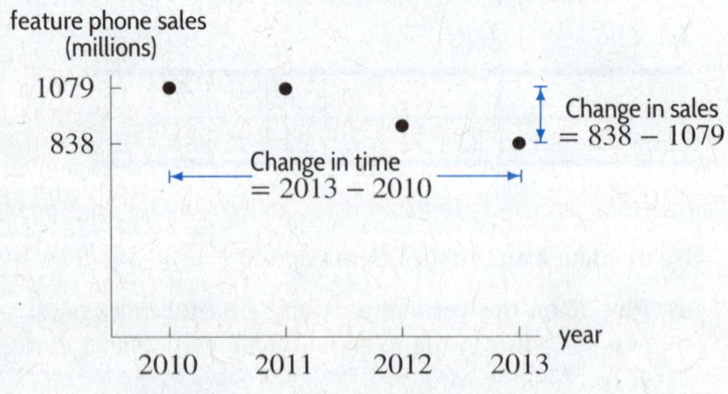


**Figure 1.10**: Feature phone sales

[6] Accessed Feb. 14 2014, www.fiercewireless.com/europe/story/gartner-smartphones-outsell-feature-phones-2013/2014-02-13 and www.fiercewireless.com/europe/special-reports/analyzing-worlds-14-biggest-handset-makers-q2-2013

## Rate of Change of a Function

The rate of change of sales is an example of the rate of change of a function. In general, if $Q = f(t)$, we write $\Delta Q$ for a change in $Q$ and $\Delta t$ for a change in $t$. We define:[7]

> The **average rate of change**, or **rate of change**, of $Q$ with respect to $t$ over an interval is
>
> $$\text{Average rate of change over an interval} = \frac{\text{Change in } Q}{\text{Change in } t} = \frac{\Delta Q}{\Delta t}.$$

The average rate of change of the function $Q = f(t)$ over an interval tells us how much $Q$ changes, on average, for each unit change in $t$ within that interval. On some parts of the interval, $Q$ may be changing rapidly, while on other parts $Q$ may be changing slowly. The average rate of change evens out these variations.

## Increasing and Decreasing Functions

In the previous example, the average rate of change of smartphone sales is positive on the interval from 2010 to 2013 since the number of smartphones sold increased over this interval. Similarly, the average rate of change of feature phone sales is negative on the same interval since the number of feature phones sold decreased over this interval. The annual sales of smartphones is an example of an *increasing function* and the annual sales of feature phones is an example of a *decreasing function*. In general we say the following:

> If $Q = f(t)$ for $t$ in the interval $a \le t \le b$,
> - $f$ is an **increasing function** if the values of $f$ increase as $t$ increases in this interval.
> - $f$ is a **decreasing function** if the values of $f$ decrease as $t$ increases in this interval.

Looking at smartphone sales, we see that an increasing function has a positive rate of change. From the feature phone sales, we see that a decreasing function has a negative rate of change. In general:

> If $Q = f(t)$,
> - If $f$ is an increasing function, then the average rate of change of $Q$ with respect to $t$ is positive on every interval.
> - If $f$ is a decreasing function, then the average rate of change of $Q$ with respect to $t$ is negative on every interval.

**Example 1** The function $A = q(r) = \pi r^2$ gives the area, $A$, of a circle as a function of its radius, $r$. Graph $q$. Explain how the fact that $q$ is an increasing function can be seen on the graph.

**Solution** The area increases as the radius increases, so $A = q(r)$ is an increasing function. We can see this in Figure 1.11 because the graph climbs as we move from left to right and the average rate of change, $\Delta A/\Delta r$, is positive on every interval.

[7]The Greek letter $\Delta$, delta, is often used in mathematics to represent change. In this book, we use rate of change to mean average rate of change across an interval. In calculus, rate of change means something called instantaneous rate of change.

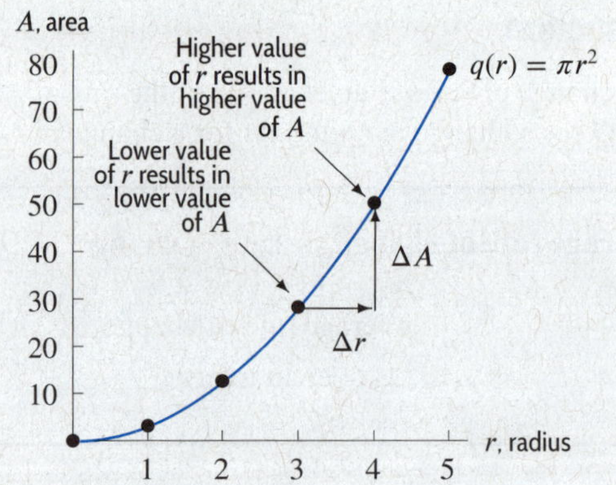


**Figure 1.11:** The graph of an increasing function, $A = q(r)$, rises when read from left to right

**Example 2** Carbon-14 is a radioactive element that exists naturally in the atmosphere and is absorbed by living organisms. When an organism dies, the carbon-14 present at death begins to decay. Let $L = g(t)$ represent the quantity of carbon-14 (in micrograms, $\mu$g) in a tree $t$ years after its death. See Table 1.11. Explain why we expect $g$ to be a decreasing function of $t$ and how the graph displays this.

**Table 1.11** *Quantity of carbon-14 as a function of time*

| $t$, time (years) | 0 | 1000 | 2000 | 3000 | 4000 | 5000 |
|---|---|---|---|---|---|---|
| $L$, quantity of carbon-14 ($\mu$g) | 200 | 177 | 157 | 139 | 123 | 109 |

**Solution** Since the amount of carbon-14 is decaying over time, $g$ is a decreasing function. In Figure 1.12, the graph falls as we move from left to right and the average rate of change in the level of carbon-14 with respect to time, $\Delta L/\Delta t$, is negative on every interval.

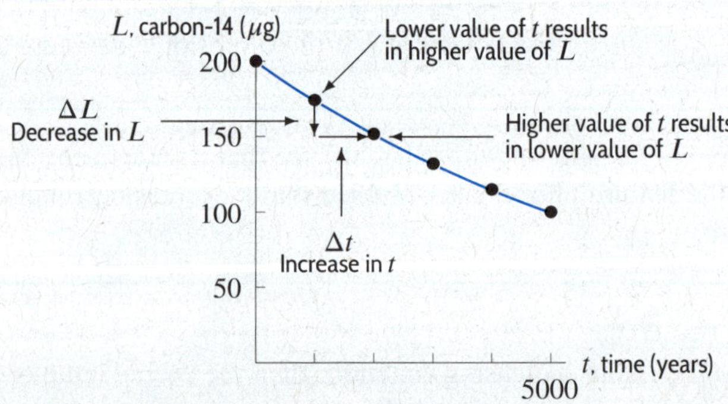


**Figure 1.12:** The graph of a decreasing function, $L = g(t)$, falls when read from left to right

In general, we can identify an increasing or decreasing function from its graph as follows:

- The graph of an increasing function rises when read from left to right.
- The graph of a decreasing function falls when read from left to right.

Many functions have some intervals on which they are increasing and other intervals on which they are decreasing. These intervals can often be identified from the graph.

**Example 3** On what intervals is the function graphed in Figure 1.13 increasing? Decreasing?

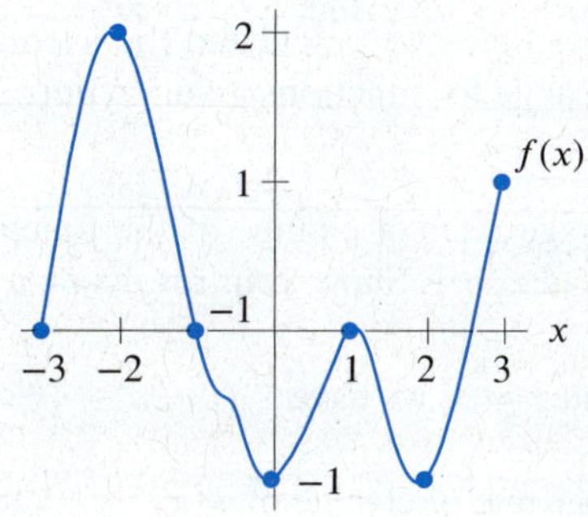


Figure 1.13: Graph of a function that is increasing on some intervals and decreasing on others

**Solution** The function appears to be increasing for values of $x$ between −3 and −2, for $x$ between 0 and 1, and for $x$ between 2 and 3. The function appears to be decreasing for $x$ between −2 and 0 and for $x$ between 1 and 2. Using inequalities, we say that $f$ is increasing for $-3 < x < -2$, for $0 < x < 1$, and for $2 < x < 3$. Similarly, $f$ is decreasing for $-2 < x < 0$ and $1 < x < 2$.

## Function Notation for the Average Rate of Change

Suppose we want to find the average rate of change of a function $Q = f(t)$ over the interval $a \le t \le b$. On this interval, the change in $t$ is given by

$$\Delta t = b - a.$$

At $t = a$, the value of $Q$ is $f(a)$, and at $t = b$, the value of $Q$ is $f(b)$. Therefore, the change in $Q$ is given by

$$\Delta Q = f(b) - f(a).$$

Using function notation, we express the average rate of change as follows:

$$\begin{array}{c}\text{Average rate of change of } Q = f(t)\\ \text{over the interval } a \le t \le b\end{array} = \frac{\text{Change in } Q}{\text{Change in } t} = \frac{\Delta Q}{\Delta t} = \frac{f(b) - f(a)}{b - a}.$$

In Figure 1.14, notice that the average rate of change is given by the ratio of the rise, $f(b) - f(a)$, to the run, $b - a$. This ratio is also called the *slope* of the dashed line segment.[8]

In the future, we may drop the word "average" and talk about the rate of change over an interval.

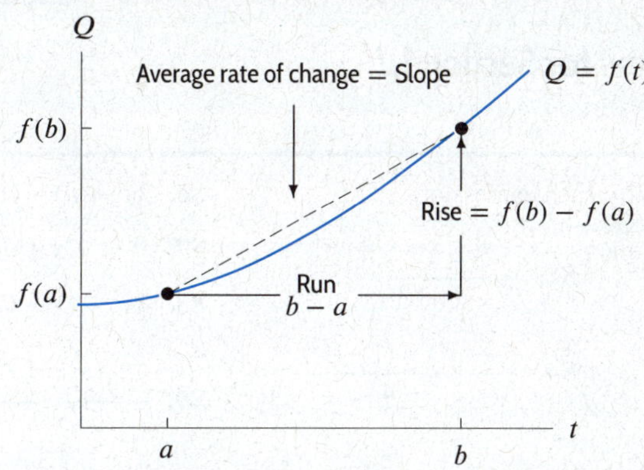


Figure 1.14: The average rate of change is the ratio Rise/Run

[8]See Section 1.3 for further discussion of slope.

In previous examples we calculated the average rate of change from data. We now calculate average rates of change for functions given by formulas.

**Example 4** Calculate the average rates of change of the function $f(x) = x^2$ between $x = 1$ and $x = 3$ and between $x = -2$ and $x = 1$. Show your results on a graph.

**Solution** Between $x = 1$ and $x = 3$, we have

$$\begin{aligned}\text{Average rate of change of } f(x) \text{ over the interval } 1 \le x \le 3 &= \frac{\text{Change in } f(x)}{\text{Change in } x} = \frac{f(3) - f(1)}{3 - 1}\\ &= \frac{3^2 - 1^2}{3 - 1} = \frac{9 - 1}{2} = 4.\end{aligned}$$

Between $x = -2$ and $x = 1$, we have

$$\begin{aligned}\text{Average rate of change of } f(x) \text{ over the interval } -2 \le x \le 1 &= \frac{\text{Change in } f(x)}{\text{Change in } x} = \frac{f(1) - f(-2)}{1 - (-2)}\\ &= \frac{1^2 - (-2)^2}{1 - (-2)} = \frac{1 - 4}{3} = -1.\end{aligned}$$

The average rate of change between $x = 1$ and $x = 3$ is positive because $f(x)$ is increasing on this interval. See Figure 1.15. However, on the interval from $x = -2$ and $x = 1$, the function is partly decreasing and partly increasing. The average rate of change on this interval is negative because the decrease on the interval is larger than the increase.

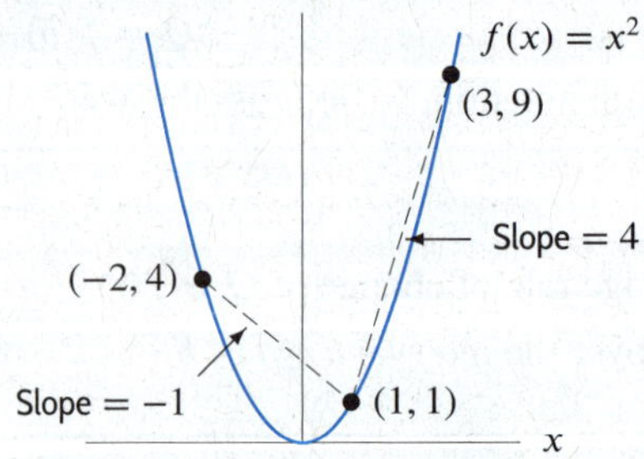


**Figure 1.15**: Average rate of change of $f(x)$ on an interval is the slope of the dashed line on that interval

## Exercises and Problems for Section 1.2

### Skill Refresher

In Exercises S1–S10, simplify each expression.

**S1.** $\dfrac{4-6}{3-2}$

**S2.** $\dfrac{1-3}{2^2-(-3)^2}$

**S3.** $\dfrac{-3-(-9)}{-1-2}$

**S4.** $\dfrac{(1-3^2)-(1-4^2)}{3-4}$

**S5.** $\dfrac{\frac{1}{2}-(-4)^2-\left(\frac{1}{2}-(5^2)\right)}{-4-5}$

**S6.** $2(x+a)-3(x-b)$

**S7.** $x^2-(2x+a)^2$

**S8.** $4x^2-(x-b)^2$

**S9.** $\dfrac{x^2-\frac{3}{4}-\left(y^2-\frac{3}{4}\right)}{x-y}$

**S10.** $\dfrac{2(x+h)^2-2x^2}{(x+h)-x}$

## Exercises

In Exercises 1–4, on what intervals is the function increasing? Decreasing?

**1.**

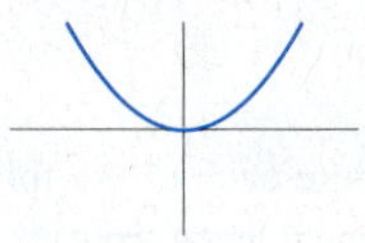


**2.**

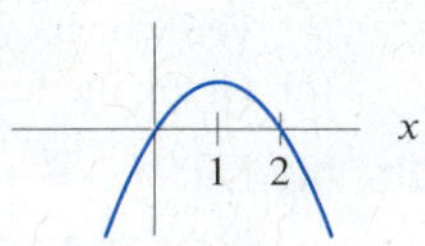


**3.**

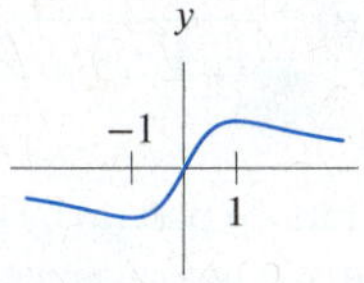


**4.**

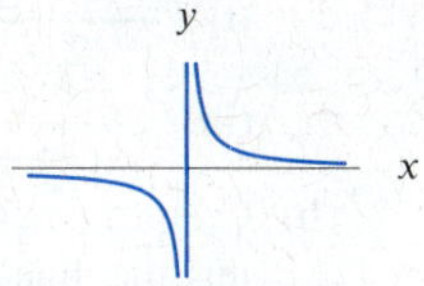


**5.** Table 1.10 on page 10 gives the annual sales (in millions) of smartphones and feature phones. What was the average rate of change of annual sales of each of them between

**(a)** 2010 and 2012? **(b)** 2012 and 2013?

**(c)** Interpret these results in terms of sales.

**6.** Table 1.10 on page 10 shows that feature phone sales are a function of smartphone sales. Is this function increasing or decreasing?

**7.** In 2007, you have 40 songs in your favorite iTunes playlist. In 2010, you have 120 songs. In 2014, you have 40. What is the average rate of change per year in the number of songs in your favorite iTunes playlist between

**(a)** 2007 and 2010? **(b)** 2010 and 2014?

**(c)** 2007 and 2014?

**8.** Table 1.12 gives the populations of two cities (in thousands) over a 17-year period.

**(a)** Find the average rate of change of each population on the following intervals:

(i) 1996 to 2006 (ii) 1996 to 2013

(iii) 2001 to 2013

**(b)** What do you notice about the average rate of change of each population? Explain what the average rate of change tells you about each population.

**Table 1.12**

| Year | 1996 | 1998 | 2001 | 2006 | 2013 |
|---|---|---|---|---|---|
| $P_1$ | 42 | 46 | 52 | 62 | 76 |
| $P_2$ | 82 | 80 | 77 | 72 | 65 |

**9.** Figure 1.16 shows distance traveled as a function of time.

**(a)** Find $\Delta D$ and $\Delta t$ between:

(i) $t = 2$ and $t = 5$ (ii) $t = 0.5$ and $t = 2.5$

(iii) $t = 1.5$ and $t = 3$

**(b)** Compute the rate of change, $\Delta D/\Delta t$, over each of the intervals in part (a), and interpret its meaning.

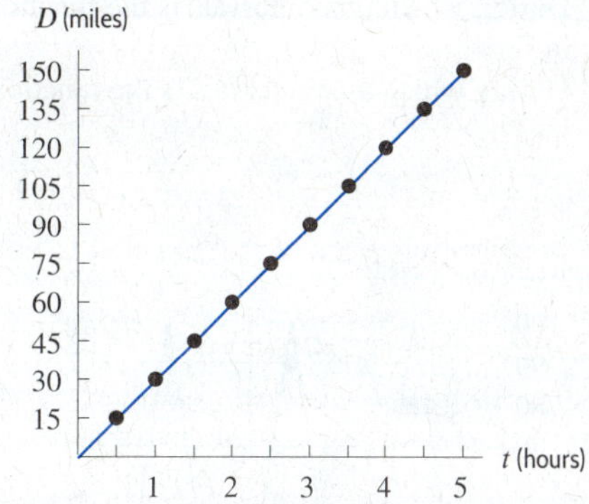


**Figure 1.16**

Exercises 10–14 use Figure 1.17.

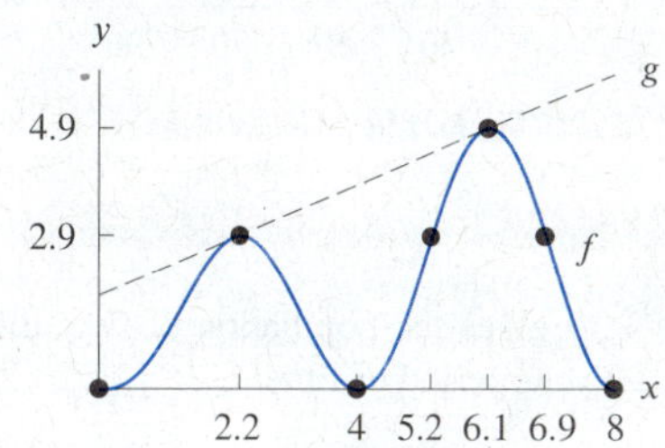


**Figure 1.17**

**10.** Find the average rate of change of $f$ for $2.2 \leq x \leq 6.1$.

**11.** Give two different intervals on which $\Delta f(x)/\Delta x = 0$.

**12.** What is the average rate of change of $g$ between $x = 2.2$ and $x = 6.1$?

**13.** What is the relation between the average rate of change of $f$ and the average rate of change of $g$ between $x = 2.2$ and $x = 6.1$?

**14.** Is the rate of change of $f$ positive or negative on the following intervals?

**(a)** $2.2 \leq x \leq 4$ **(b)** $5 \leq x \leq 6$

**15.** If $F$ is a decreasing function, what can you say about $F(-2)$ compared to $F(2)$?

**16.** If $G$ is an increasing function, what can you say about $G(3) - G(-1)$?

## Problems

**17.** Figure 1.18 shows the percent of the side of the moon toward the earth illuminated by the sun at different times during the year 2008. Use the figure to answer the following questions.

**(a)** Give the coordinates of the points $A, B, C, D, E$.
**(b)** Plot the point $F = (15, 60)$ and $G = (60, 15)$. Which point is on the graph?
**(c)** During which time intervals is the function increasing?
**(d)** During which time intervals is the function decreasing?

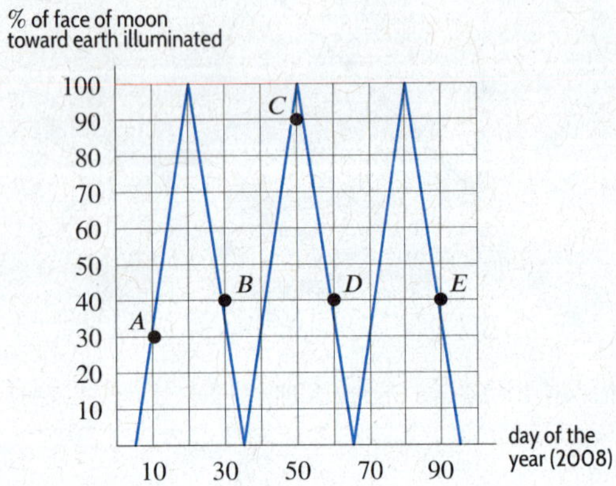


**Figure 1.18:** Moon phases

**18.** Figure 1.19 gives the population of two different towns over a 50-year period of time.

**(a)** Which town starts (in year $t = 0$) with the most people?
**(b)** Which town is growing faster over these 50 years?

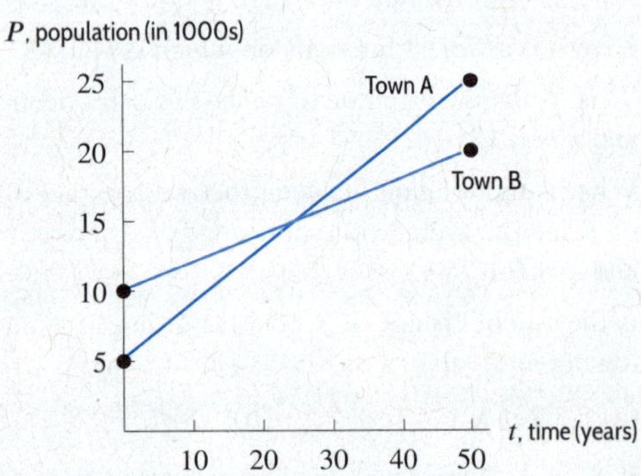


**Figure 1.19**

**19.** **(a)** What is the average rate of change of $g(x) = 2x - 3$ between the points $(-2, -7)$ and $(3, 3)$?
**(b)** The function $g$ is either increasing or decreasing everywhere. Explain how your answer to part (a) tells you which.
**(c)** Graph the function.

**20.** **(a)** Let $f(x) = 16 - x^2$. Compute each of the following expressions, and interpret each as an average rate of change.

(i) $\dfrac{f(2) - f(0)}{2 - 0}$ (ii) $\dfrac{f(4) - f(2)}{4 - 2}$

(iii) $\dfrac{f(4) - f(0)}{4 - 0}$

**(b)** Graph $f(x)$. Illustrate each ratio in part (a) by sketching the line segment with the given slope. Over which interval is the average rate of decrease the greatest?

**21.** Imagine you constructed a list of the world record times for a particular event—such as the mile footrace, or the 100-meter freestyle swimming race—in terms of when they were established. Is the world record time a function of the date when it was established? If so, is this function increasing or decreasing? Explain. Could a world record be established twice in the same year? Is the world record time a function of the year it was established?

**22.** The function $P = f(t)$ gives the population of a town, in thousands, after $t$ years. A graph of $f$ is given in Figure 1.20.

**(a)** Find the average rate of change of the population of the town during the first 10 years.
**(b)** Does the population of the town grow more between $t = 5$ and $t = 10$ years, or between $t = 15$ and $t = 30$ years? Explain.
**(c)** Does the population of the town grow faster between $t = 5$ and $t = 10$ years, or between $t = 15$ and $t = 30$ years? Explain.

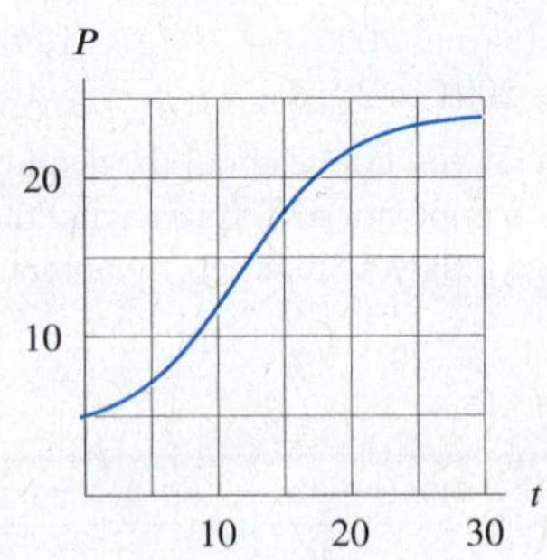


**Figure 1.20**

**23.** The most freakish change in temperature ever recorded was from −4°F to 45°F between 7:30 am and 7:32 am on January 22, 1943 at Spearfish, South Dakota.[9] What was the average rate of change of the temperature for this time period?

**24.** You have zero dollars now and the average rate of change in your net worth is \$5000 per year. How much money will you have in forty years?

**25.** The surface of the sun has dark areas known as sunspots, that are cooler than the rest of the sun's surface. The number of sunspots[10] fluctuates with time, as shown in Figure 1.21.

**(a)** Explain how you know the number of sunspots, $s$, in year $t$ is a function of $t$.
**(b)** Approximate the time intervals on which $s$ is an increasing function of $t$.

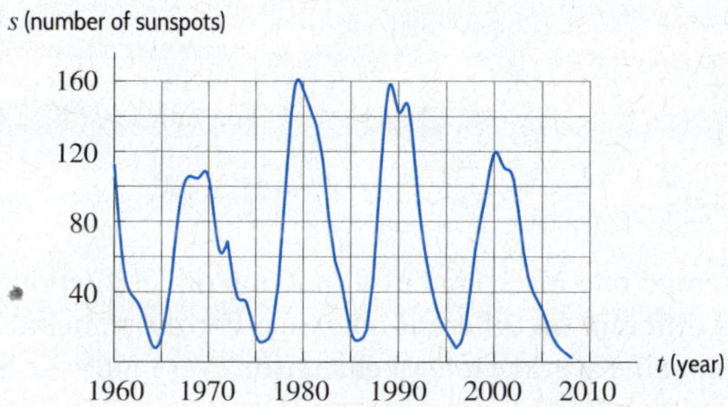


**Figure 1.21**

**26.** Table 1.13 shows the number of calories used per minute as a function of body weight for three sports.[11]

**(a)** Determine the number of calories that a 200-lb person uses in one half-hour of walking.
**(b)** Who uses more calories, a 120-lb person swimming for one hour or a 220-lb person bicycling for a half-hour?
**(c)** Does the number of calories used by a person walking increase or decrease as weight increases?

**Table 1.13**

| Activity | 100 lb | 120 lb | 150 lb | 170 lb | 200 lb | 220 lb |
|---|---|---|---|---|---|---|
| Walking | 2.7 | 3.2 | 4.0 | 4.6 | 5.4 | 5.9 |
| Bicycling | 5.4 | 6.5 | 8.1 | 9.2 | 10.8 | 11.9 |
| Swimming | 5.8 | 6.9 | 8.7 | 9.8 | 11.6 | 12.7 |

**27.** Because scientists know how much carbon-14 a living organism should have in its tissues, they can measure the amount of carbon-14 present in the tissue of a fossil and then calculate how long it took for the original amount to decay to the current level, thus determining the time of the organism's death. A tree fossil is found to contain 130 $\mu$g of carbon-14, and scientists determine from the size of the tree that it would have contained 200 $\mu$g of carbon-14 at the time of its death. Using Table 1.11 on page 12, approximately how long ago did the tree die?

**28.** Figure 1.22 shows the graph of the function $g(x)$.

**(a)** Estimate $\dfrac{g(4) - g(0)}{4 - 0}$.
**(b)** The ratio in part (a) is the slope of a line segment joining two points on the graph. Sketch this line segment on the graph.
**(c)** Estimate $\dfrac{g(b) - g(a)}{b - a}$ for $a = -9$ and $b = -1$.
**(d)** On the graph, sketch the line segment whose slope is given by the ratio in part (c).

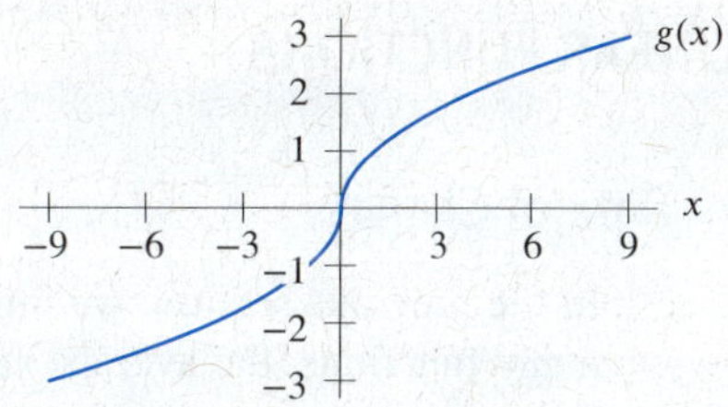


**Figure 1.22**

**29.** Find the average rate of change of $f(x) = 3x^2 + 1$ between the points

**(a)** $(1, 4)$ and $(2, 13)$ **(b)** $(j, k)$ and $(m, n)$
**(c)** $(x, f(x))$ and $(x+h, f(x+h))$

**30.** A water company employee measured the water level (in centimeters) in a reservoir during March and computed the average rates of change in Table 1.14.

**(a)** What are the units of the average rates of change in Table 1.14?
**(b)** Interpret the average rates of change in context.
**(c)** For each time interval, what was the total change in water level?
**(d)** Draw a possible graph of the water level as a function of time.

**Table 1.14**

| Interval (days) | $1 \le t < 11$ | $11 \le t < 16$ | $16 \le t < 31$ |
|---|---|---|---|
| Average rate of change | 7 | 5 | 3 |

[9] *The Guinness Book of Records.* 1995.
[10] ftp://ftp.ngdc.noaa.gov/STP/SOLAR DATA/SUNSPOT NUMBERS/YEARLY.PLT, accessed November 30, 2009.
[11] From *1993 World Almanac.*

**31.** Table 1.15 gives the amount of garbage, $G$, in millions of tons, produced[12] in the US in year $t$.

**(a)** What is the value of $\Delta t$ for consecutive entries in this table?
**(b)** Calculate the value of $\Delta G$ for each pair of consecutive entries in this table.
**(c)** Are all the values of $\Delta G$ you found in part (b) the same? What does this tell you?
**(d)** The function $G$ changed from increasing to decreasing between 2005 and 2010. To what might this be attributed?

Table 1.15

| $t$ | 1960 | 1970 | 1980 | 1990 | 2000 | 2005 | 2010 |
|---|---|---|---|---|---|---|---|
| $G$ | 88.1 | 121.1 | 151.6 | 208.3 | 242.5 | 252.7 | 249.9 |

**32.** Table 1.16 shows the times, $t$, in sec, achieved every 10 meters by Carl Lewis in the 100-meter final of the World Championship in Rome in 1987.[13] Distance, $d$, is in meters.

**(a)** For each successive time interval, calculate the average rate of change of distance. What is a common name for the average rate of change of distance?
**(b)** Where did Carl Lewis attain his maximum speed during this race? Some runners are running their fastest as they cross the finish line. Does that seem to be true in this case?

Table 1.16

| $t$ | 0.00 | 1.94 | 2.96 | 3.91 | 4.78 | 5.64 |
|---|---|---|---|---|---|---|
| $d$ | 0 | 10 | 20 | 30 | 40 | 50 |
| $t$ | 6.50 | 7.36 | 8.22 | 9.07 | 9.93 | |
| $d$ | 60 | 70 | 80 | 90 | 100 | |

# 1.3 LINEAR FUNCTIONS

## Constant Rate of Change

In the previous section, we introduced the average rate of change of a function on an interval. For many functions, the average rate of change is different on different intervals. For the remainder of this chapter, we consider functions that have the same average rate of change on every interval. Such a function has a graph that is a line and is called *linear*.

### Population Growth

Mathematical models of population growth are used by city planners to project the growth of towns and states. Biologists model the growth of animal populations and physicians model the spread of an infection in the bloodstream. One possible model, a linear model, assumes that the population changes at the same average rate on every time interval.

**Example 1** A town of 30,000 people grows by 2000 people every year. Since the population, $P$, is growing at the constant rate of 2000 people per year, $P$ is a linear function of time, $t$, in years.

(a) What is the average rate of change of $P$ over every time interval?
(b) Make a table that gives the town's population every five years over a 20-year period. Graph the population.
(c) Find a formula for $P$ as a function of $t$.

**Solution** (a) The average rate of change of population with respect to time is 2000 people per year.
(b) The initial population in year $t = 0$ is $P = 30{,}000$ people. Since the town grows by 2000 people every year, after five years it has grown by

$$\frac{2000 \text{ people}}{\text{year}} \cdot 5 \text{ years} = 10{,}000 \text{ people}.$$

Thus, in year $t = 5$ the population is given by

$$P = \text{Initial population} + \text{New population} = 30{,}000 + 10{,}000 = 40{,}000.$$

[12] http://http://www.epa.gov/osw/nonhaz/municipal/pubs/msw_2010_rev_factsheet.pdf, accessed January, 2013.
[13] W. G. Pritchard, "Mathematical Models of Running", *SIAM Review*. 35, 1993, pp. 359–379.

In year $t = 10$ the population is given by

$$P = 30{,}000 + \underbrace{2000 \text{ people/year} \cdot 10 \text{ years}}_{20{,}000 \text{ new people}} = 50{,}000.$$

Similar calculations for year $t = 15$ and year $t = 20$ give the values in Table 1.17. See Figure 1.23; the dashed line shows the trend in the data.

**Table 1.17** *Population over 20 years*

| $t$, years | $P$, population |
|---|---|
| 0 | 30,000 |
| 5 | 40,000 |
| 10 | 50,000 |
| 15 | 60,000 |
| 20 | 70,000 |

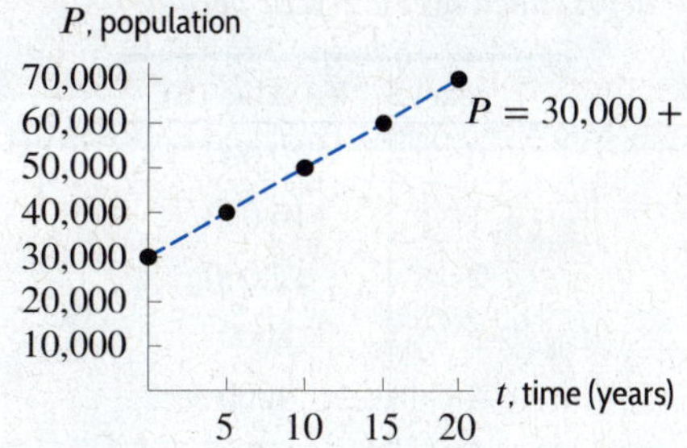


**Figure 1.23:** Town's population over 20 years

(c) From part (b), we see that the size of the population is given by

$$\begin{aligned} P &= \text{Initial population} + \text{Number of new people} \\ &= 30{,}000 + 2000 \text{ people/year} \cdot \text{Number of years}, \end{aligned}$$

so a formula for $P$ in terms of $t$ is

$$P = 30{,}000 + 2000t.$$

The graph of the population data in Figure 1.23 is a straight line. The average rate of change of the population over every interval is the same, namely 2000 people per year. Any linear function has the same average rate of change over every interval. Thus, we talk about *the* rate of change of a linear function. In general:

- A **linear function** has a constant rate of change.
- The graph of any linear function is a straight line.

## Financial Models

Economists and accountants use linear functions for *straight-line depreciation*. For tax purposes, the value of certain equipment is considered to decrease, or depreciate, over time. For example, computer equipment may be state-of-the-art today, but after several years it is outdated. Straight-line depreciation assumes that the rate of change of value with respect to time is constant.

**Example 2** A small business spends \$20,000 on new computer equipment and, for tax purposes, chooses to depreciate it to \$0 at a constant rate over a five-year period.

(a) Make a table and a graph showing the value of the equipment over the five-year period.
(b) Give a formula for value as a function of time.

**Solution** (a) After five years, the equipment is valued at \$0. If $V$ is the value in dollars and $t$ is the number of years, we see that

$$\begin{array}{c}\text{Rate of change of value} \\ \text{from } t = 0 \text{ to } t = 5\end{array} = \frac{\text{Change in value}}{\text{Change in time}} = \frac{\Delta V}{\Delta t} = \frac{-\$20{,}000}{5 \text{ years}} = -\$4000 \text{ per year.}$$

Thus, the value drops at the constant rate of \$4000 per year. (Notice that $\Delta V$ is negative because the value of the equipment decreases.) See Table 1.18 and Figure 1.24. Since $V$ changes at a constant rate, $V = f(t)$ is a linear function and its graph is a straight line. The rate of change, −\$4000 per year, is negative because the function is decreasing and the graph slopes down.

**Table 1.18** *Value of equipment depreciated over a 5-year period*

| $t$, year | $V$, value (\$) |
|---|---|
| 0 | 20,000 |
| 1 | 16,000 |
| 2 | 12,000 |
| 3 | 8000 |
| 4 | 4000 |
| 5 | 0 |

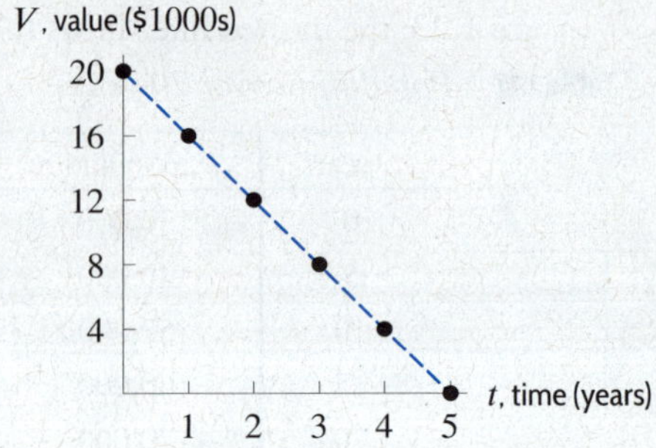


**Figure 1.24:** Value of equipment depreciated over a 5-year period

(b) After $t$ years have elapsed,

$$\text{Decrease in value of equipment} = \$4000 \cdot \text{Number of years} = \$4000t.$$

The initial value of the equipment is \$20,000, so at time $t$,

$$V = 20{,}000 - 4000t.$$

## A General Formula for the Family of Linear Functions

Example 1 involved a town whose population is growing at a constant rate with formula

$$\text{Current population} = \underbrace{\text{Initial population}}_{\text{30,000 people}} + \underbrace{\text{Growth rate}}_{\text{2000 people per year}} \times \underbrace{\text{Number of years}}_{t}$$

so

$$P = 30{,}000 + 2000t.$$

In Example 2, the value, $V$, as a function of $t$ is given by

$$\text{Total cost} = \underbrace{\text{Initial value}}_{\$20{,}000} + \underbrace{\text{Change per year}}_{-\$4000 \text{ per year}} \times \underbrace{\text{Number of years}}_{t}$$

so

$$V = 20{,}000 + (-4000)t.$$

Using the symbols $x$, $y$, $b$, $m$, we see formulas for both of these linear functions follow the same pattern:

$$\underbrace{\text{Output}}_{y} = \underbrace{\text{Initial value}}_{b} + \underbrace{\text{Rate of change}}_{m} \times \underbrace{\text{Input}}_{x}.$$

Summarizing, we get the following results:

> If $y = f(x)$ is a linear function, then for some constants $b$ and $m$:
>
> $$y = b + mx.$$
>
> - $m$ is called the **slope**, and gives the rate of change of $y$ with respect to $x$. Thus,
>
> $$m = \frac{\Delta y}{\Delta x}.$$
>
> If $(x_0, y_0)$ and $(x_1, y_1)$ are any two distinct points on the graph of $f$, then
>
> $$m = \frac{\Delta y}{\Delta x} = \frac{y_1 - y_0}{x_1 - x_0}.$$
>
> - $b$ is called the **vertical intercept**, or ***y*-intercept**, and gives the value of $y$ for $x = 0$. In mathematical models, $b$ typically represents an initial, or starting, value of the output.

Every linear function can be written in the form $y = b + mx$. Different linear functions have different values for $m$ and $b$. These constants are known as *parameters*; the set of all linear functions is called a *family*.

**Example 3** In Example 1, the population function, $P = 30{,}000 + 2000t$, has slope $m = 2000$ and vertical intercept $b = 30{,}000$. In Example 2, the value of the computer equipment, $V = 20{,}000 - 4000t$, has slope $m = -4000$ and vertical intercept $b = 20{,}000$.

## Tables for Linear Functions

A table of values could represent a linear function if the rate of change is constant, for all pairs of points in the table; that is,

$$\text{Rate of change of linear function} = \frac{\text{Change in output}}{\text{Change in input}} = \text{Constant}.$$

Thus, if the value of $x$ goes up by equal steps in a table for a linear function, then the value of $y$ goes up (or down) by equal steps as well. We say that changes in the value of $y$ are *proportional* to changes in the value of $x$.

**Example 4** Table 1.19 gives values of two functions, $p$ and $q$. Could either of these functions be linear?

**Table 1.19** *Values of two functions p and q*

| $x$ | 50 | 55 | 60 | 65 | 70 |
|---|---|---|---|---|---|
| $p(x)$ | 0.10 | 0.11 | 0.12 | 0.13 | 0.14 |
| $q(x)$ | 0.01 | 0.03 | 0.06 | 0.14 | 0.15 |

**Solution** The value of $x$ goes up by equal steps of $\Delta x = 5$. The value of $p(x)$ also goes up by equal steps of $\Delta p = 0.01$, so $\Delta p/\Delta x$ is a constant. See Table 1.20. Thus, $p$ could be a linear function.

**Table 1.20** *Values of $\Delta p/\Delta x$*

| $x$ | $p(x)$ | $\Delta p$ | $\Delta p/\Delta x$ |
|---|---|---|---|
| 50 | 0.10 | | |
| | | 0.01 | 0.002 |
| 55 | 0.11 | | |
| | | 0.01 | 0.002 |
| 60 | 0.12 | | |
| | | 0.01 | 0.002 |
| 65 | 0.13 | | |
| | | 0.01 | 0.002 |
| 70 | 0.14 | | |

**Table 1.21** *Values of $\Delta q/\Delta x$*

| $x$ | $q(x)$ | $\Delta q$ | $\Delta q/\Delta x$ |
|---|---|---|---|
| 50 | 0.01 | | |
| | | 0.02 | 0.004 |
| 55 | 0.03 | | |
| | | 0.03 | 0.006 |
| 60 | 0.06 | | |
| | | 0.08 | 0.016 |
| 65 | 0.14 | | |
| | | 0.01 | 0.002 |
| 70 | 0.15 | | |

In contrast, the value of $q(x)$ does not go up by equal steps. The value climbs by 0.02, then by 0.03, and so on. See Table 1.21. This means that $\Delta q/\Delta x$ is not constant. Thus, $q$ could not be a linear function.

It is possible to have data from a linear function in which neither the $x$-values nor the $y$-values go up by equal steps. However the rate of change must be constant, as in the following example.

**Example 5** The former Republic of Yugoslavia exported cars called Yugos to the US between 1985 and 1989. The car is now a collector's item.[14] Table 1.22 gives the quantity of Yugos sold, $Q$, and the price, $p$, for each year from 1985 to 1988.

(a) Using Table 1.22, explain why $Q$ could be a linear function of $p$.

(b) What does the rate of change of this function tell you about Yugos?

**Table 1.22** *Price and sales of Yugos in the US*

| Year | Price in \$, $p$ | Number sold, $Q$ |
|---|---|---|
| 1985 | 3990 | 49,000 |
| 1986 | 4110 | 43,000 |
| 1987 | 4200 | 38,500 |
| 1988 | 4330 | 32,000 |

**Solution** (a) We are interested in $Q$ as a function of $p$, so we plot $Q$ on the vertical axis and $p$ on the horizontal axis. The data points in Figure 1.25 appear to lie on a straight line, suggesting a linear function.

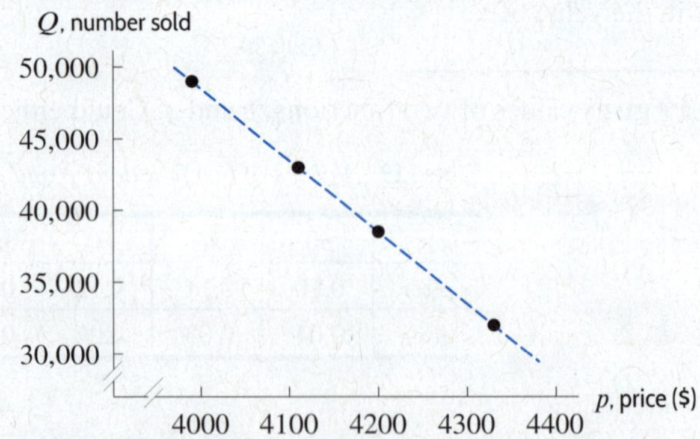


**Figure 1.25:** Since the data from Table 1.22 falls on a straight line, the table could represent a linear function

[14]www.inet.hr/~pauric/epov.htm, accessed January 16, 2006.

To provide further evidence that $Q$ is a linear function, we check that the rate of change of $Q$ with respect to $p$ is constant for the points given. When the price of a Yugo rose from \$3990 to \$4110, sales fell from 49,000 to 43,000. Thus,

$$\Delta p = 4110 - 3990 = 120,$$

$$\Delta Q = 43{,}000 - 49{,}000 = -6000.$$

Since the number of Yugos sold decreased, $\Delta Q$ is negative. Thus, as the price increased from \$3990 to \$4110,

$$\text{Rate of change of quantity as price increases} = \frac{\Delta Q}{\Delta p} = \frac{-6000}{120} = -50 \text{ cars per dollar.}$$

Next, we calculate the rate of change as the price increased from \$4110 to \$4200 to see if the rate remains constant:

$$\text{Rate of change} = \frac{\Delta Q}{\Delta p} = \frac{38{,}500 - 43{,}000}{4200 - 4110} = \frac{-4500}{90} = -50 \text{ cars per dollar,}$$

and as the price increased from \$4200 to \$4330:

$$\text{Rate of change} = \frac{\Delta Q}{\Delta p} = \frac{32{,}000 - 38{,}500}{4330 - 4200} = \frac{-6500}{130} = -50 \text{ cars per dollar.}$$

Since the rate of change, $-50$, is constant, $Q$ could be a linear function of $p$. Given additional data, $\Delta Q/\Delta p$ might not remain constant. However, based on the table, it appears that the function is linear.

(b) Since $\Delta Q$ is the change in the number of cars sold and $\Delta p$ is the change in price, the rate of change is $-50$ cars per dollar. Thus the number of Yugos sold decreased by 50 each time the price increased by \$1.

## Warning: Not All Graphs That Look Like Lines Represent Linear Functions

The graph of any linear function is a line. However, a function's graph can look like a line without actually being one. Consider the following example.

**Example 6** The function $P = 100(1.02)^t$ approximates the population of Mexico in the early 2000s. Here $P$ is the population (in millions) and $t$ is the number of years since 2000. Table 1.23 and Figure 1.26 show values of $P$ over a 5-year period. Is $P$ a linear function of $t$?

**Table 1.23** *Population of Mexico t years after 2000*

| $t$ (years) | $P$ (millions) |
|---|---|
| 0 | 100 |
| 1 | 102 |
| 2 | 104.04 |
| 3 | 106.12 |
| 4 | 108.24 |
| 5 | 110.41 |

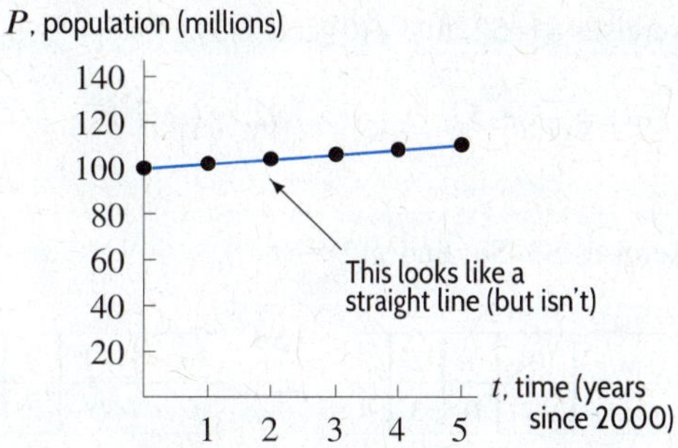


**Figure 1.26**: Graph of $P = 100(1.02)^t$ over 5-year period: Looks linear (but is not)

**Solution** The formula $P = 100(1.02)^t$ cannot be written in the form $P = b+mt$, so $P$ is not a linear function of $t$. However, the graph of $P$ in Figure 1.26 appears to be a straight line. We check $P$'s rate of change in Table 1.23. When $t = 0$, $P = 100$ and when $t = 1$, $P = 102$. Thus, between 2000 and 2001,

$$\text{Rate of change of population} = \frac{\Delta P}{\Delta t} = \frac{102 - 100}{1 - 0} = 2.$$

For the interval from 2001 to 2002, we have

$$\text{Rate of change} = \frac{\Delta P}{\Delta t} = \frac{104.04 - 102}{2 - 1} = 2.04,$$

and for the interval from 2004 to 2005, we have

$$\text{Rate of change} = \frac{\Delta P}{\Delta t} = \frac{110.41 - 108.24}{5 - 4} = 2.17.$$

Thus, $P$'s rate of change is not constant. In fact, $P$ appears to be increasing at a faster and faster rate. Table 1.24 and Figure 1.27 show values of $P$ over a longer (60-year) period. On this scale, these points do not appear to fall on a straight line. However, the graph of $P$ curves upward so gradually at first that over the short interval shown in Figure 1.26, it barely curves at all. The graphs of many nonlinear functions, when viewed on a small scale, appear to be linear.

**Table 1.24** *Population over 60 years*

| $t$ (years since 2000) | $P$ (millions) |
|---|---|
| 0 | 100 |
| 10 | 121.90 |
| 20 | 148.59 |
| 30 | 181.14 |
| 40 | 220.80 |
| 50 | 269.16 |
| 60 | 328.10 |

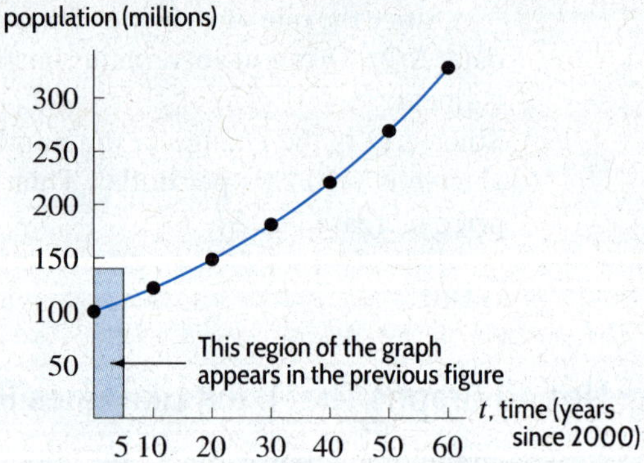


**Figure 1.27**: Graph of $P = 100(1.02)^t$ over 60 years: Not linear

## Exercises and Problems for Section 1.3

### Skill Refresher

In Exercises S1–S2, find $f(0)$ and $f(3)$.

**S1.** $f(x) = \frac{2}{3}x + 5$

**S2.** $f(t) = 17 - 4t$

In Exercises S3–S4, find $f(2) - f(0)$.

**S3.**

| $x$ | 0 | 1 | 2 | 3 |
|---|---|---|---|---|
| $f(x)$ | −2 | 0 | 3 | 4 |

**S4.**

| $t$ | −1 | 0 | 1 | 2 |
|---|---|---|---|---|
| $f(t)$ | 0 | 2 | 7 | −1 |

In Exercises S5–S6, find the coordinates of the $x$ and $y$ intercepts.

**S5.** $y = -4x + 3$

**S6.** $5x - 2y = 4$

For each of the linear expressions in $x$ in Exercises S7–S10, give the constant term and the coefficient of $x$.

**S7.** $3 - 2x + \frac{1}{2}$

**S8.** $4 - 3(x + 2) + 6(2x - 1)$

**S9.** $ax - ab - 3x + a + 3$

**S10.** $5(x - 1) + 3$

## Exercises

In Exercises 1–3, which line has the greater

**(a)** Slope? **(b)** $y$-intercept?

**1.** $y = -1 + 2x;\quad y = -2 + 3x$

**2.** $y = 3 + 4x;\quad y = 5 - 2x$

**3.** $y = \frac{1}{4}x;\quad y = 1 - 6x$

In Exercises 4–9, could the data represent a linear function? If so, give the rate of change.

**4.**

| $x$ | 0 | 5 | 10 | 15 |
|---|---|---|---|---|
| $f(x)$ | 10 | 20 | 30 | 40 |

**5.**

| $x$ | 0 | 10 | 20 | 30 |
|---|---|---|---|---|
| $h(x)$ | 20 | 40 | 50 | 55 |

**6.**

| $x$ | 0 | 100 | 300 | 600 |
|---|---|---|---|---|
| $g(x)$ | 50 | 100 | 150 | 200 |

**7.**

| $t$ | 1 | 2 | 3 | 4 | 5 |
|---|---|---|---|---|---|
| $g(t)$ | 5 | 4 | 5 | 4 | 5 |

**8.**

| $x$ | −3 | −1 | 0 | 3 |
|---|---|---|---|---|
| $j(x)$ | 5 | 1 | −1 | −7 |

**9.**

| $\gamma$ | 9 | 8 | 7 | 6 | 5 |
|---|---|---|---|---|---|
| $p(\gamma)$ | 42 | 52 | 62 | 72 | 82 |

In Exercises 10–13, identify the vertical intercept and the slope, and explain their meanings in practical terms.

**10.** The population of a town can be represented by the formula $P(t) = 54.25 - \frac{2}{7}t$, where $P(t)$ represents the population, in thousands, and $t$ represents the time, in years, since 1970.

**11.** A stalactite grows according to the formula $L(t) = 17.75 + \frac{1}{250}t$, where $L(t)$ represents the length of the stalactite, in inches, and $t$ represents the time, in years, since the stalactite was first measured.

**12.** The profit, in dollars, of selling $n$ items is given by $P(n) = 0.98n - 3000$.

**13.** A phone company charges according to the formula $C(n) = 29.99 + 0.05n$, where $n$ is the number of minutes, and $C(n)$ is the monthly phone charge, in dollars.

## Problems

**14.** Table 1.25 shows the cost $C$, in dollars, of selling $x$ cups of coffee per day from a cart.

**(a)** Using the table, show that the relationship appears to be linear.
**(b)** Plot the data in the table.
**(c)** Find the slope of the line. Explain what this means in the context of the given situation.
**(d)** Why should it cost \$50 to serve zero cups of coffee?

**Table 1.25**

| $x$ | 0 | 5 | 10 | 50 | 100 | 200 |
|---|---|---|---|---|---|---|
| $C$ | 50.00 | 51.25 | 52.50 | 62.50 | 75.00 | 100.00 |

**15.** Table 1.26 gives the proposed fine $r = f(v)$ to be imposed on a motorist for speeding, where $v$ is the motorist's speed and 55 mph is the speed limit.

**(a)** Decide whether $f$ appears to be linear.
**(b)** What would the rate of change represent in practical terms for the motorist?
**(c)** Plot the data points.

**Table 1.26**

| $v$ (mph) | 60 | 65 | 70 | 75 | 80 | 85 |
|---|---|---|---|---|---|---|
| $r$ (dollars) | 75 | 100 | 125 | 150 | 175 | 200 |

**16.** In 2003, the number, $N$, of cases of SARS (Severe Acute Respiratory Syndrome) reported in Hong Kong[15] was initially approximated by $N = 78.9 + 30.1t$, where $t$ is the number of days since March 17. Interpret the constants 78.9 and 30.1.

**17.** A new Toyota RAV4 costs \$23,500. The car's value depreciates linearly to \$18,823 in three years time.[16] Write a formula which expresses its value, $V$, in terms of its age, $t$, in years.

**18.** In 2012, the population of a town was 21,510 and growing by 63 people per year. Find a formula for $P$, the town's population, in terms of $t$, the number of years since 2012.

**19.** A flight costs \$10,000 to operate, regardless of the number of passengers. Each ticket costs \$127. Express profit, $\pi$, as a linear function of the number of passengers, $n$, on the flight.

[15] World Health Organization, www.who.int/csr/sars/country/en, accessed September, 2005
[16] http://www.motortrend.com/used_cars/11/toyota/rav4/pricing/, accessed January 2013

**20.** Owners of an inactive quarry in Australia have decided to resume production. They estimate that it will cost them \$10,000 per month to maintain and insure their equipment and that monthly salaries will be \$30,000. It costs \$800 to mine a ton of rocks. Write a formula that expresses the total cost each month, $c$, as a function of $r$, the number of tons of rock mined per month.

**21.** In each case, graph a linear function with the given rate of change. Label and put scales on the axes.

**(a)** Increasing at 2.1 inches/day
**(b)** Decreasing at 1.3 gallons/mile

**22.** A small café sells coffee for \$3.50 per cup. On average, it costs the café \$0.50 to make a cup of coffee (for grounds, hot water, filters). The café also has a fixed daily cost of \$450 (for rent, wages, utilities).

**(a)** Let $R$, $C$, and $P$ be the café's daily revenue, costs, and profit, respectively, for selling $x$ cups of coffee in a day. Find formulas for $R$, $C$, and $P$ as functions of $x$. [Hint: The revenue, $R$, is the total amount of money that the café brings in. The cost, $C$, includes the fixed daily cost as well as the cost for all $x$ cups of coffee sold. $P$ is the café's profit after costs have been accounted for.]
**(b)** Plot $P$ against $x$. For what $x$-values is the graph of $P$ below the $x$-axis? Above the $x$-axis? Interpret your results.
**(c)** Interpret the slope and both intercepts of your graph in practical terms.

**23.** Table 1.27 gives the area and perimeter of a square as a function of the length of its side.

**(a)** From the table, decide if either area or perimeter could be a linear function of side length.
**(b)** From the data make two graphs, one showing area as a function of side length, the other showing perimeter as a function of side length. Connect the points.
**(c)** If you find a linear relationship, give its corresponding rate of change and interpret its significance.

**Table 1.27**

| Length of side | 0 | 1 | 2 | 3 | 4 | 5 | 6 |
|---|---|---|---|---|---|---|---|
| Area of square | 0 | 1 | 4 | 9 | 16 | 25 | 36 |
| Perimeter of square | 0 | 4 | 8 | 12 | 16 | 20 | 24 |

**24.** Make two tables, one comparing the radius of a circle to its area, the other comparing the radius of a circle to its circumference. Repeat parts (a), (b), and (c) from Problem 23, this time comparing radius with circumference, and radius with area.

**25.** Sri Lanka is an island that experienced approximately linear population growth from 1950 to 2000. On the other hand, Afghanistan was torn by warfare in the 1980s and did not experience linear or near-linear growth.[17]

**(a)** Table 1.28 gives the population of these two countries, in millions. Which of these two countries is A and which is B? Explain.
**(b)** What is the approximate rate of change of the linear function? What does the rate of change represent in practical terms?
**(c)** Estimate the population of Sri Lanka in 1988.

**Table 1.28**

| Year | 1950 | 1960 | 1970 | 1980 | 1990 | 2000 |
|---|---|---|---|---|---|---|
| Population of country A | 8.2 | 9.8 | 12.4 | 15.1 | 14.7 | 23.9 |
| Population of country B | 7.5 | 9.9 | 12.5 | 14.9 | 17.2 | 19.2 |

**26.** Table 1.29 gives the average temperature, $T$, at a depth $d$, in a borehole in Belleterre, Quebec.[18] Evaluate $\Delta T/\Delta d$ on the following intervals, and explain what your answers tell you about borehole temperature.

**(a)** $25 \le d \le 150$
**(b)** $25 \le d \le 75$
**(c)** $100 \le d \le 200$

**Table 1.29**

| $d$, depth (m) | 25 | 50 | 75 | 100 |
|---|---|---|---|---|
| $T$, temp (°C) | 5.50 | 5.20 | 5.10 | 5.10 |
| $d$, depth (m) | 125 | 150 | 175 | 200 |
| $T$, temp (°C) | 5.30 | 5.50 | 5.75 | 6.00 |
| $d$, depth (m) | 225 | 250 | 275 | 300 |
| $T$, temp (°C) | 6.25 | 6.50 | 6.75 | 7.00 |

**27.** Table 1.29 gives the temperature-depth profile, $T = f(d)$, in a borehole in Belleterre, Quebec, where $T$ is the average temperature at a depth $d$.

**(a)** Could $f$ be linear?
**(b)** Graph $f$. What do you notice about the graph for $d \ge 150$?
**(c)** What can you say about the average rate of change of $f$ for $d \ge 150$?

[17] www.census.gov/ipc/www/idbsusum.html, accessed January 12, 2006.

[18] Hugo Beltrami of St. Francis Xavier University and David Chapman of the University of Utah posted this data at http://geophysics.stfx.ca/public/borehole/borehole.html, accessed November 10, 2005.

**28.** A company finds that there is a linear relationship between the amount of money that it spends on advertising and the number of units it sells. If it spends no money on advertising, it sells 300 units. For each additional \$5000 spent, an additional 20 units are sold.

**(a)** If $x$ is the amount of money that the company spends on advertising, find a formula for $y$, the number of units sold as a function of $x$.
**(b)** How many units does the firm sell if it spends \$25,000 on advertising? \$50,000?
**(c)** How much advertising money must be spent to sell 700 units?
**(d)** What is the slope of the line you found in part (a)? Give an interpretation of the slope that relates units sold and advertising costs.

**29.** Tuition cost $T$ (in dollars) for part-time students at Stonewall College is given by $T = 300 + 200C$, where $C$ represents the number of credits taken.

**(a)** Find the tuition cost for eight credits.
**(b)** How many credits were taken if the tuition was \$1700?
**(c)** Make a table showing costs for taking from one to twelve credits. For each value of $C$, give both the tuition cost, $T$, and the cost per credit, $T/C$. Round to the nearest dollar.
**(d)** Which of these values of $C$ has the smallest cost per credit?
**(e)** What does the 300 represent in the formula for $T$?
**(f)** What does the 200 represent in the formula for $T$?

**30.** The summit of Africa's largest peak, Mt. Kilimanjaro, consists of the northern and southern ice fields and the Furtwangler glacier. An article in the Proceedings of the National Academy of Sciences[19] indicates that in 2000 ($t = 0$) the area of the ice cover at the peak of Mt. Kilimanjaro was approximately 1951 m$^2$. By 2007, the area had shrunk to approximately 1555 m$^2$.

**(a)** If this decline is modeled by a linear function, find $A = f(t)$, the equation of the ice-cover area as a function of time. Explain what the slope and $A$-intercept mean in terms of the ice cover.
**(b)** Evaluate $f(11)$.
**(c)** If this model is correct, when would you expect the ice cover to disappear?

**31.** When each of the following equations are written in the form $y = b + mx$, the result is $y = 5 + 4x$. Find the constants $r, s, k, j$ in these equations.

**(a)** $y = 2r + x\sqrt{s}$
**(b)** $y = \dfrac{1}{k} - (j - 1)x$.

**32.** Graph the following function in the window $-10 \le x \le 10, -10 \le y \le 10$. Is this graph a line? Explain.

$$y = -x\left(\frac{x - 1000}{900}\right)$$

**33.** Graph $y = 2x + 400$ using the window $-10 \le x \le 10, -10 \le y \le 10$. Describe what happens, and how you can fix it by using a better window.

**34.** Graph $y = 200x + 4$ using the window $-10 \le x \le 10, -10 \le y \le 10$. Describe what happens and how you can fix it by using a better window.

**35.** Figure 1.28 shows the graph of $y = x^2/1000 + 5$ in the window $-10 \le x \le 10, -10 \le y \le 10$. Discuss whether this is a linear function.

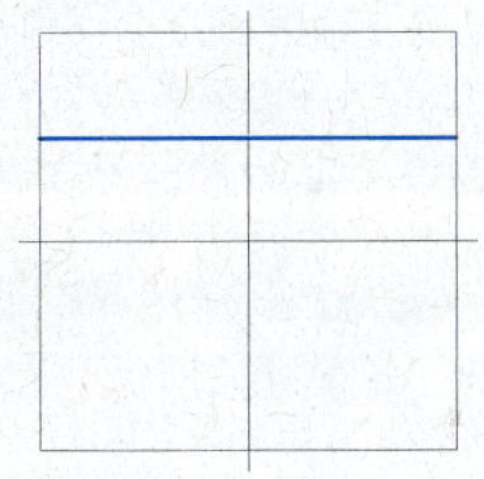

Figure 1.28

**36.** The cost of a cab ride is given by the function $C = 2.50 + 2d$, where $d$ is the number of miles traveled and $C$ is in dollars. Choose an appropriate window and graph the cost of a ride for a cab that travels no farther than a 10-mile radius from the center of the city.

## 1.4 FORMULAS FOR LINEAR FUNCTIONS

To find a formula for a linear function we find values for the slope, $m$, and the vertical intercept, $b$, in the formula $y = b + mx$.

### Finding a Formula for a Linear Function from a Table of Data

If a table of data represents a linear function, we first calculate $m$ and then determine $b$.

[19] http://www.pnas.org/content/early/2009/10/30/0906029106.full.pdf+html, accessed November 27, 2009.

**Example 1** A grapefruit is thrown into the air. Its velocity, $v$, is a linear function of $t$, the time since it was thrown. A positive velocity indicates the grapefruit is rising and a negative velocity indicates it is falling. Check that the data in Table 1.30 corresponds to a linear function. Find a formula for $v$ in terms of $t$.

**Table 1.30** *Velocity of a grapefruit t seconds after being thrown into the air*

| $t$, time (sec) | 1 | 2 | 3 | 4 |
|---|---|---|---|---|
| $v$, velocity (ft/sec) | 48 | 16 | −16 | −48 |

**Solution** Figure 1.29 shows the data in Table 1.30. The points appear to fall on a line. To check that the velocity function is linear, calculate the rates of change of $v$ and see that they are constant. From time $t = 1$ to $t = 2$, we have

$$\text{Average rate of change of velocity with time} = \frac{\Delta v}{\Delta t} = \frac{16 - 48}{2 - 1} = -32.$$

For the next second, from $t = 2$ to $t = 3$, we have

$$\text{Average rate of change} = \frac{\Delta v}{\Delta t} = \frac{-16 - 16}{3 - 2} = -32.$$

You can check that the rate of change from $t = 3$ to $t = 4$ is also −32.

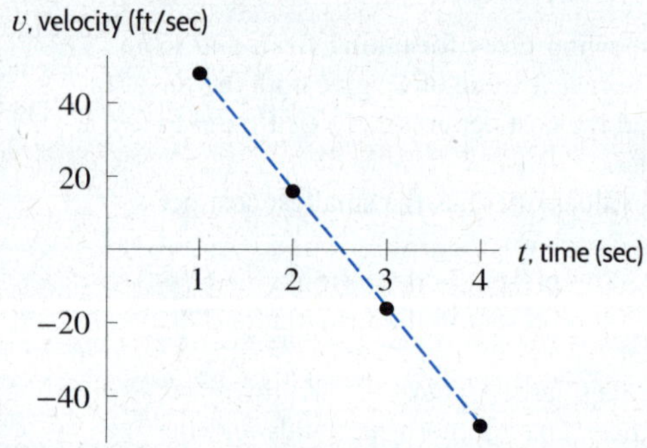


**Figure 1.29:** Velocity of a grapefruit is a linear function of time

A formula for $v$ is of the form $v = b + mt$. Since $m$ is the rate of change, we have $m = -32$ so $v = b - 32t$. The initial velocity (at $t = 0$) is represented by $b$. We are not given the value of $v$ when $t = 0$, but we can use any data point to calculate $b$. For example, $v = 48$ when $t = 1$, so

$$48 = b - 32 \cdot 1,$$

which gives

$$b = 80.$$

Thus, a formula for the velocity is $v = 80 - 32t$.

What does the rate of change, $m$, in Example 1 tell us about the grapefruit? Think about the units:

$$m = \frac{\Delta v}{\Delta t} = \frac{\text{Change in velocity}}{\text{Change in time}} = \frac{-32 \text{ ft/sec}}{1 \text{ sec}} = -32 \text{ ft/sec per second.}$$

The value of $m$, −32 ft/sec per second, tells us that the grapefruit's velocity is decreasing by 32 ft/sec for every second that goes by. We say the grapefruit is accelerating at −32 ft/sec per second. (The units ft/sec per second are often written ft/sec$^2$. Negative acceleration is also called deceleration.[20])

[20]The notation ft/sec$^2$ is shorthand for ft/sec per second; it does not mean a "square second" in the same way that areas are measured in square feet or square meters.

## Finding a Formula for a Linear Function from a Graph

We can calculate the slope, $m$, of a linear function using two points on its graph. Having found $m$, we can use either of the points to calculate $b$, the vertical intercept.

**Example 2** Figure 1.30 shows oxygen consumption as a function of heart rate (that is, pulse) for two people.

(a) Assuming linearity, find formulas for these two functions.

(b) Interpret the slope of each graph in terms of oxygen consumption.

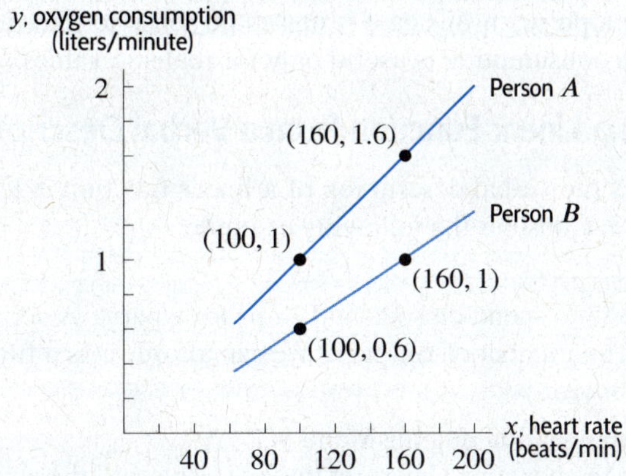


Figure 1.30: Oxygen consumption of two people running on treadmills

**Solution** (a) Let $x$ be heart rate and let $y$ be oxygen consumption. Since we are assuming linearity, $y = b+mx$. The two points on person $A$'s line, (100, 1) and (160, 1.6), give

$$\text{Slope of A's line } = m = \frac{\Delta y}{\Delta x} = \frac{1.6-1}{160-100} = 0.01.$$

Thus $y = b + 0.01x$. To find $b$, use the fact that $y = 1$ when $x = 100$:

$$\begin{aligned}1 &= b + 0.01(100)\\ 1 &= b+1\\ b &= 0,\end{aligned}$$

so $y = 0.01x$. Alternatively, $b$ can be found using the fact that $x = 160$ if $y = 1.6$.

For person $B$, we again begin with the formula $y = b + mx$. In Figure 1.30, two points on B's line are (100, 0.6) and (160, 1), so

$$\text{Slope of B's line } = m = \frac{\Delta y}{\Delta x} = \frac{1-0.6}{160-100} = \frac{0.4}{60} \approx 0.0067.$$

To find $b$, use the fact that $y = 1$ when $x = 160$:

$$\begin{aligned}1 &= b + (0.4/60)\cdot 160\\ 1 &= b + 1.067\\ b &= -0.067.\end{aligned}$$

Thus, for person $B$, we have $y = -0.067 + 0.0067x$.

(b) The slope for person $A$ is $m = 0.01$, so

$$m = \frac{\text{Change in oxygen consumption}}{\text{Change in heart rate}} = \frac{\text{Change in liters/min}}{\text{Change in beats/min}} = 0.01\frac{\text{liters}}{\text{heartbeat}}.$$

Every additional heartbeat (per minute) for person $A$ translates to an additional 0.01 liters (per minute) of oxygen consumed.

The slope for person $B$ is $m = 0.0067$. Thus, for every additional beat (per minute), person $B$ consumes an additional 0.0067 liter of oxygen (per minute). Since the slope for person $B$ is smaller than for person $A$, person $B$ consumes less additional oxygen than person $A$ for the same increase in pulse.

What do the $y$-intercepts of the functions in Example 2 say about oxygen consumption? Often the $y$-intercept of a function is a starting value. In this case, the $y$-intercept would be the oxygen consumption of a person whose pulse is zero (i.e. $x = 0$). Since a person running on a treadmill must have a pulse, in this case it makes no sense to interpret the $y$-intercept this way. The formula for oxygen consumption is useful only for realistic values of the pulse.

## Finding a Formula for a Linear Function from a Verbal Description

Sometimes the verbal description of a linear function is less straightforward than those we saw in Section 1.3. Consider the following example.

**Example 3** We have \$48 to spend on soda and chips for a party. A six-pack of soda costs \$6 and a bag of chips costs \$4. The number of six-packs we can afford, $y$, is a function of the number of bags of chips we decide to buy, $x$.

(a) Find an equation relating $x$ and $y$.
(b) Graph the equation. Interpret the intercepts and the slope in the context of the party.

**Solution** (a) If we spend all \$48 on soda and chips, then we have the following equation:

$$\text{Amount spent on chips} + \text{Amount spent on soda} = \$48.$$

If we buy $x$ bags of chips at \$4 per bag, then the amount spent on chips is $\$4x$. Similarly, if we buy $y$ six-packs of soda at \$6 per six-pack, then the amount spent on soda is $\$6y$. Thus,

$$4x + 6y = 48.$$

We can solve for $y$, giving

$$6y = 48 - 4x$$
$$y = 8 - \frac{2}{3}x.$$

This is a linear function with slope $m = -2/3$ and $y$-intercept $b = 8$.

(b) The graph of this function is a discrete set of points, since the number of bags of chips and the number of six-packs of soda must be (nonnegative) integers.

To find the $y$-intercept, we set $x = 0$, giving

$$4 \cdot 0 + 6y = 48.$$

So $6y = 48$, giving $y = 8$.

Substituting $y = 0$ gives the $x$-intercept,

$$4x + 6 \cdot 0 = 48.$$

So $4x = 48$, giving $x = 12$. Thus the points $(0, 8)$ and $(12, 0)$ are on the graph.

The point $(0, 8)$ indicates that we can buy 8 six-packs of soda if we buy no chips. The point $(12, 0)$ indicates that we can buy 12 bags of chips if we buy no soda. The other points on the line describe affordable options between these two extremes. For example, the point $(6, 4)$ is on the line, because

$$4 \cdot 6 + 6 \cdot 4 = 48.$$

This means that if we buy 6 bags of chips, we can afford 4 six-packs of soda.

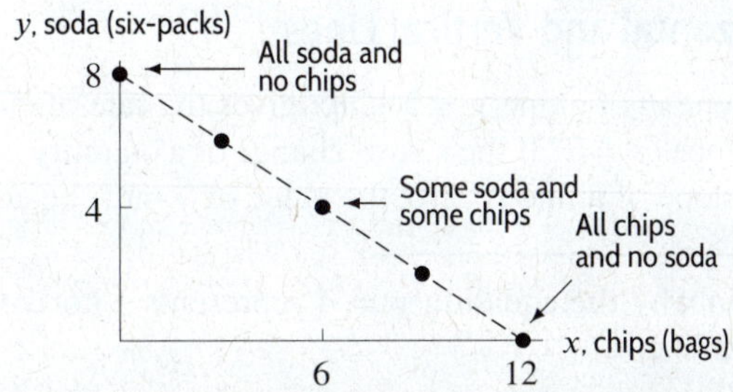


**Figure 1.31:** Relation between the number of six-packs, $y$, and the number of bags of chips, $x$

The points marked in Figure 1.31 represent affordable options. All affordable options lie on or below the line $4x + 6y = 48$. Not all points on the line are affordable options. For example, suppose we purchase one six-pack of soda for \$6. That leaves \$42 to spend on chips, meaning we would have to buy 10.5 bags of chips, which is not possible. Therefore, the point (10.5, 1) is not an option, although it is a point on the line $4x + 6y = 48$.

To interpret the slope, notice that

$$m = \frac{\Delta y}{\Delta x} = \frac{\text{Change in number of six-packs}}{\text{Change in number of bags of chips}},$$

so the units of $m$ are six-packs of soda per bags of chips. The fact that $m = -2/3$ means that for each additional 3 bags of chips purchased, we can purchase 2 fewer six-packs of soda. This occurs because 2 six-packs cost \$12, the same as 3 bags of chips. Thus, $m = -2/3$ is the rate at which the amount of soda we can buy decreases as we buy more chips.

## Alternative Forms for the Equation of a Line

In Example 3, the equation $2x + 3y = 24$ represents a linear relationship between $x$ and $y$ even though the equation is not in the form $y = b + mx$. The following equations represent lines.

- The *slope-intercept form* is
  $y = b + mx$ where $m$ is the slope and $b$ is the $y$-intercept.
- The *point-slope form* is
  $y - y_0 = m(x - x_0)$ where $m$ is the slope and $(x_0, y_0)$ is a point on the line.
- The *standard form* is
  $Ax + By + C = 0$ where $A$, $B$, and $C$ are constants.

If we know the slope of a line and the coordinates of a point on the line, it is often convenient to use the point-slope form of the equation.

**Example 4** Use the point-slope form to find the equation of the line for the oxygen consumption of person $A$ in Example 2.

**Solution** In Example 2, we found the slope of person $A$'s line to be $m = 0.01$. Since the point (100, 1) lies on the line, the point-slope form gives the equation

$$y - 1 = 0.01(x - 100).$$

To check that this gives the same equation we got in Example 2, we multiply out and simplify:

$$y - 1 = 0.01x - 1$$
$$y = 0.01x.$$

Alternatively, we could have used the point (160, 1.6) instead of (100, 1), giving

$$y - 1.6 = 0.01(x - 160).$$

Multiplying out again gives $y = 0.01x$.

## Equations of Horizontal and Vertical Lines

The slope $m$ of a line $y = b + mx$ gives the rate of change of $y$ with respect to $x$. What about a line with slope $m = 0$? If the rate of change of a quantity is zero, then the quantity does not change. Thus, if the slope of a line is zero, the value of $y$ must be constant. Such a line is horizontal.

**Example 5** Explain why the equation $y = 4$ represents a horizontal line and the equation $x = 4$ represents a vertical line.

**Solution** The equation $y = 4$ represents a linear function with slope $m = 0$. To see this, notice that this equation can be rewritten as $y = 4 + 0 \cdot x$. Thus, the value of $y$ is 4 no matter what the value of $x$ is. See Figure 1.32. Similarly, the equation $x = 4$ means that $x$ is 4 no matter what the value of $y$ is. Every point on the line in Figure 1.33 has $x$ equal to 4, so this line is the graph of $x = 4$.

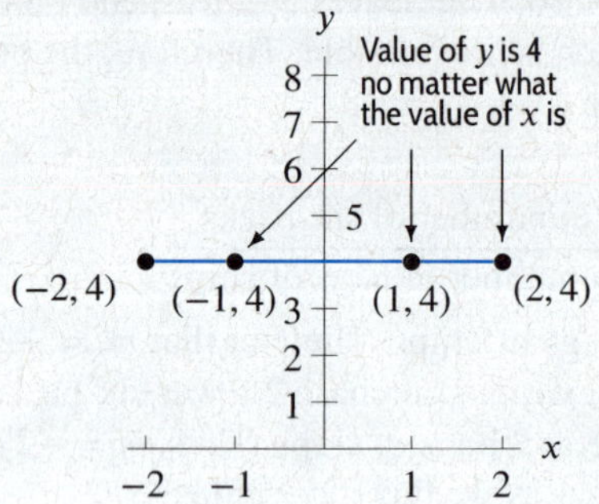


Figure 1.32: The horizontal line $y = 4$ has slope 0

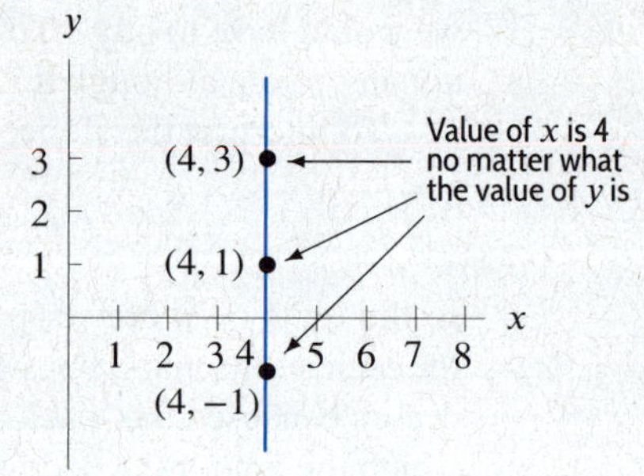


Figure 1.33: The vertical line $x = 4$ has an undefined slope

What is the slope of a vertical line? Figure 1.33 shows three points, $(4, -1)$, $(4, 1)$, and $(4, 3)$ on a vertical line. Calculating the slope, gives

$$m = \frac{\Delta y}{\Delta x} = \frac{3-1}{4-4} = \frac{2}{0}.$$

The slope is undefined because the denominator, $\Delta x$, is 0. The slope of every vertical line is undefined for the same reason. All the $x$-values on such a line are equal, so $\Delta x$ is 0, and the denominator of the expression for the slope is 0. A vertical line is not the graph of a function, since it fails the vertical line test. It does not have an equation of the form $y = b + mx$.

In summary,

> For any constant $k$:
> - The graph of the equation $y = k$ is a horizontal line and its slope is zero.
> - The graph of the equation $x = k$ is a vertical line and its slope is undefined.

## Slopes of Parallel and Perpendicular Lines

Figure 1.34 shows two parallel lines. These lines are parallel because they have equal slopes.

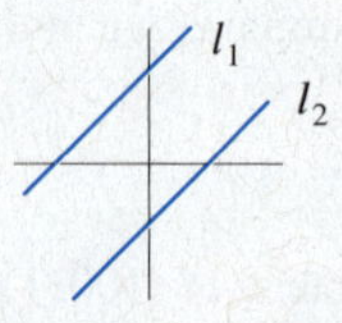


Figure 1.34: Parallel lines: $l_1$ and $l_2$ have equal slopes

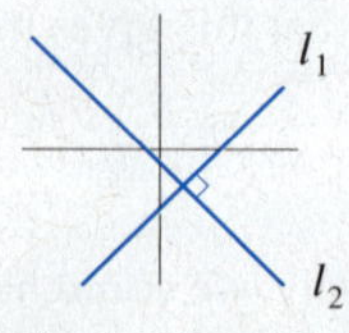


Figure 1.35: Perpendicular lines: Slopes of $l_1$ and $l_2$ are negative reciprocals of each other

What about perpendicular lines? Two perpendicular lines are graphed in Figure 1.35. We can see that if one line has a positive slope, then any perpendicular line must have a negative slope. In fact, we show that if $l_1$ and $l_2$ are two perpendicular lines with slopes, $m_1$ and $m_2$, then $m_1$ is the negative reciprocal of $m_2$. If $m_1$ and $m_2$ are not zero, we have the following result:

> Let $l_1$ and $l_2$ be two lines having slopes $m_1$ and $m_2$, respectively. Then:
> - These lines are parallel if and only if $m_1 = m_2$.
> - These lines are perpendicular if and only if $m_1 = -\dfrac{1}{m_2}$.

In addition, any two horizontal lines are parallel and $m_1 = m_2 = 0$. Any two vertical lines are parallel and $m_1$ and $m_2$ are undefined. A horizontal line is perpendicular to a vertical line. See Figures 1.36–1.38.

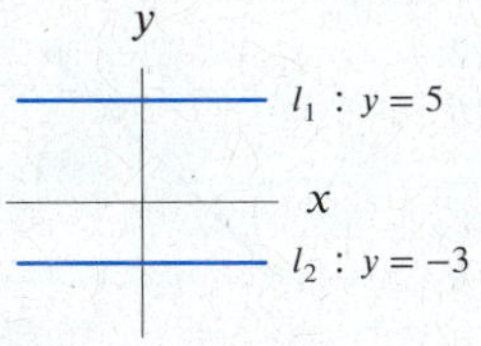


**Figure 1.36:** Any two horizontal lines are parallel

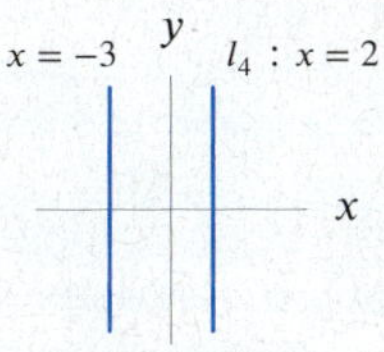


**Figure 1.37:** Any two vertical lines are parallel

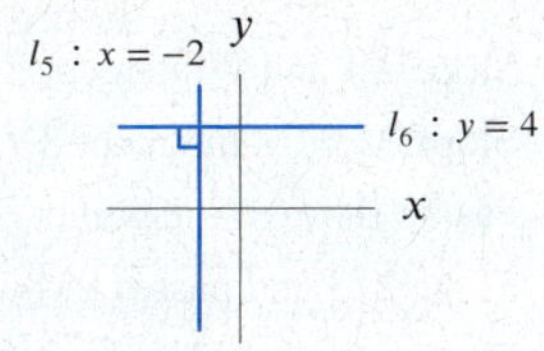


**Figure 1.38:** A horizontal line and a vertical line are perpendicular

## Justification of Formula for Slopes of Perpendicular Lines

Figure 1.39 shows $l_1$ and $l_2$, two perpendicular lines with slopes $m_1$ and $m_2$. Neither line is horizontal or vertical, so $m_1$ and $m_2$ are both defined and nonzero. We will show that

$$m_2 = -\frac{1}{m_1}.$$

Using right triangle $\Delta PQR$ with side lengths $a$ and $b$, we see that

$$m_1 = \frac{b}{a}.$$

Rotating $\Delta PQR$ by 90° about the point $P$ produces triangle $\Delta PST$. Using $\Delta PST$, we see that

$$m_2 = -\frac{a}{b} = -\frac{1}{b/a} = -\frac{1}{m_1}.$$

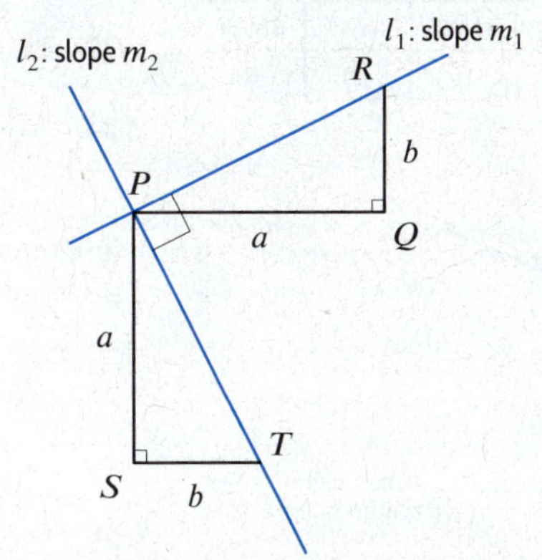


**Figure 1.39:** Perpendicular lines

## Exercises and Problems for Section 1.4

### Skill Refresher

Solve the equations in Exercises S1–S5.

**S1.** $y - 5 = 21$

**S2.** $2x - 5 = 13$

**S3.** $2x - 5 = 4x - 9$

**S4.** $17 - 28y = 13y + 24$

**S5.** $\frac{5}{3}(y+2) = \frac{1}{2} - y$

In Exercises S6–S10, solve for the indicated variable.

**S6.** $I = Prt$, for $P$.

**S7.** $C = \frac{5}{9}(F - 32)$, for $F$.

**S8.** $C = 2\pi r$, for $r$.

**S9.** $ab + ax = c - ax$, for $x$.

**S10.** $by - d = ay + c$, for $y$.

### Exercises

Find formulas for the linear functions in Exercises 1–8.

1. Slope $-4$ and $x$-intercept 7
2. Slope 3 and $y$-intercept 8
3. Passes through the points $(-1, 5)$ and $(2, -1)$
4. Slope $2/3$ and passes through the point $(5, 7)$
5. Has $x$-intercept 3 and $y$-intercept $-5$
6. Slope 0.1, passes through $(-0.1, 0.02)$
7. Function $f$ has $f(0.3) = 0.8$ and $f(0.8) = -0.4$
8. Function $f$ has $f(-2) = 7$ and $f(3) = -3$

Exercises 9–15 give data from a linear function. Find a formula for the function.

**9.**

| Year, $t$ | 0 | 1 | 2 |
|---|---|---|---|
| Value of computer, $\$V = f(t)$ | 2000 | 1500 | 1000 |

**10.**

| Price per bottle, $p$ (\$) | 0.50 | 0.75 | 1.00 |
|---|---|---|---|
| Number of bottles sold, $q = f(p)$ | 1500 | 1000 | 500 |

**11.**

| Temperature, $y = f(x)$ (°C) | 0 | 5 | 20 |
|---|---|---|---|
| Temperature, $x$ (°F) | 32 | 41 | 68 |

**12.**

| Temperature, $y = f(x)$, (°R) | 459.7 | 469.7 | 489.7 |
|---|---|---|---|
| Temperature, $x$ (°F) | 0 | 10 | 30 |

**13.**

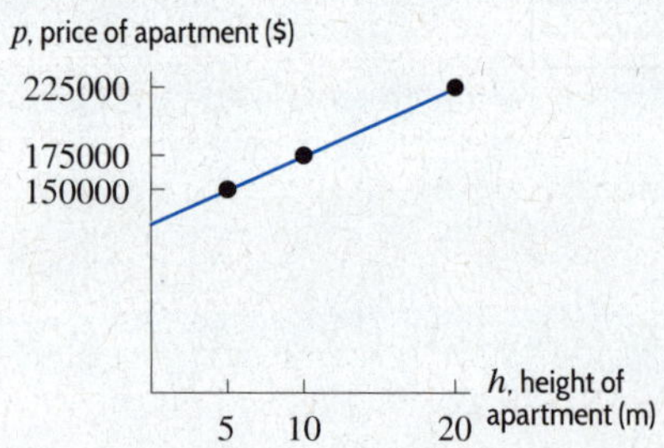


**14.**

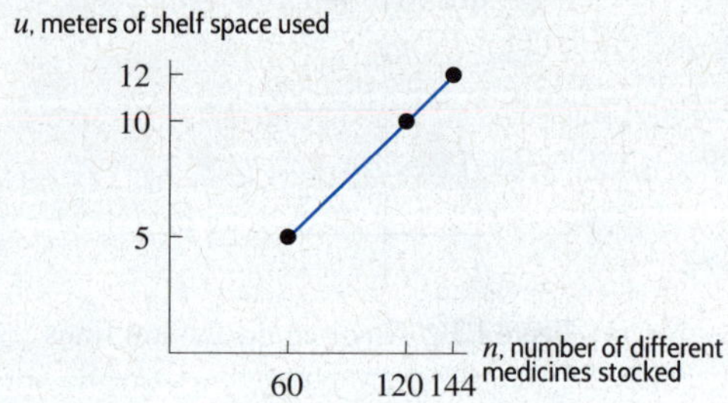


**15.**

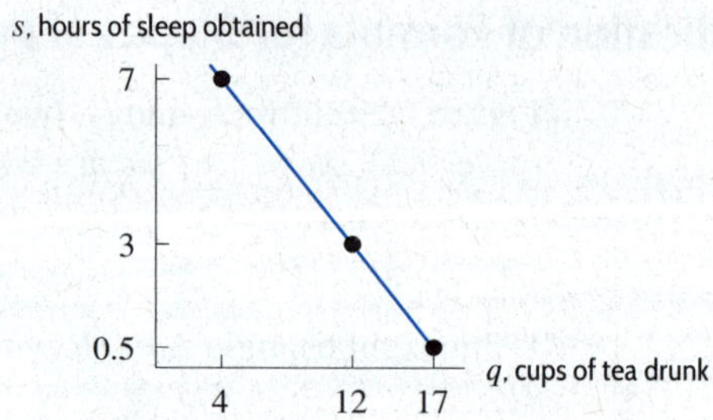


In Exercises 16–24, if possible rewrite the equation in slope-intercept form, $y = b + mx$.

**16.** $5(x + y) = 4$

**17.** $3x + 5y = 20$

**18.** $0.1y + x = 18$

**19.** $5x - 3y + 2 = 0$

**20.** $y - 0.7 = 5(x - 0.2)$

**21.** $y = 5$

**22.** $3x + 2y + 40 = x - y$

**23.** $x = 4$

**24.** $\frac{x + y}{7} = 3$

In Exercises 25–30, is the function linear? If so, rewrite it in slope-intercept form.

**25.** $g(w) = -\frac{1 - 12w}{3}$

**26.** $F(P) = 13 - \frac{2^{-1}}{4}P$

**27.** $j(s) = 3s^{-1} + 7$

**28.** $C(r) = 2\pi r$

**29.** $h(x) = 3^x + 12$

**30.** $f(x) = m^2x + n^2$

In Exercises 31–36, are the lines perpendicular? Parallel? Neither?

**31.** $y = 5x - 7; \quad y = 5x + 8$

**32.** $y = 4x + 3; \quad y = 13 - \frac{1}{4}x$

**33.** $y = 2x + 3; \quad y = 2x - 7$

**34.** $y = 4x + 7; \quad y = \frac{1}{4}x - 2$

**35.** $f(q) = 12q + 7; \quad g(q) = \frac{1}{12}q + 96$

**36.** $2y = 16 - x; \quad 4y = -8 - 2x$

## Problems

Find formulas for the linear functions in Problems 37–40.

**37.** The graph of $f$ contains $(-3, -8)$ and $(5, -20)$.

**38.** $g(100) = 2000$ and $g(400) = 3800$

**39.** $P = h(t)$ gives the size of a population that begins with 12,000 members and grows by 225 members each year.

**40.** The graph of $h$ intersects the graph of $y = x^2$ at $x = -2$ and $x = 3$.

**41.** **(a)** By hand, graph $y = 3$ and $x = 3$.
**(b)** Can the equations in part (a) be written in slope-intercept form?

**42.** Find the equation of the line parallel to $3x + 5y = 6$ and passing through the point $(0, 6)$.

**43.** Find the equation of the line passing through the point $(2, 1)$ and perpendicular to the line $y = 5x - 3$.

**44.** Find the equations of the lines parallel to and perpendicular to the line $y + 4x = 7$, and through the point $(1, 5)$.

**45.** In Table 1.31, data for two functions $f$ and $g$ are given. One of the functions is linear, and the other is not.

**(a)** Which of the two functions is linear? Explain how you know.
**(b)** Find the equation of the linear function. Write your answer in slope-intercept form.

**Table 1.31**

| $x$ | 1 | 1.5 | 2 | 2.5 | 3 |
|---|---|---|---|---|---|
| $f(x)$ | −1.22 | −0.64 | −0.06 | 0.52 | 1.10 |
| $g(x)$ | 9.71 | 4.86 | 2.43 | 1.22 | 0.61 |

**46.** Line $l$ is given by $y = 3 - \frac{2}{3}x$ and point $P$ has coordinates $(6, 5)$.

**(a)** Find the equation of the line containing $P$ and parallel to $l$.
**(b)** Find the equation of the line containing $P$ and perpendicular to $l$.
**(c)** Graph the equations in parts (a) and (b).

**47.** An empty champagne bottle is tossed from a hot-air balloon. Its upward velocity is measured every second and recorded in Table 1.32.

**(a)** Describe the motion of the bottle in words. What do negative values of $v$ represent?
**(b)** Find a formula for $v$ in terms of $t$.
**(c)** Explain the physical significance of the slope of your formula.
**(d)** Explain the physical significance of the $t$-axis and $v$-axis intercepts.

**Table 1.32**

| $t$ (sec) | 0 | 1 | 2 | 3 | 4 | 5 |
|---|---|---|---|---|---|---|
| $v$ (ft/sec) | 40 | 8 | −24 | −56 | −88 | −120 |

Table 1.33 gives the cost, $C(n)$, of producing a certain good as a linear function of $n$, the number of units produced. Use the table to answer Problems 48–50.

**Table 1.33**

| $n$ (units) | 100 | 125 | 150 | 175 |
|---|---|---|---|---|
| $C(n)$ (dollars) | 11000 | 11125 | 11250 | 11375 |

**48.** Evaluate the following expressions. Give economic interpretations for each.

**(a)** $C(175)$ **(b)** $C(175) - C(150)$
**(c)** $\frac{C(175) - C(150)}{175 - 150}$

**49.** Estimate $C(0)$. What is the economic significance of this value?

**50.** The *fixed cost* of production is the cost incurred before any goods are produced. The *unit cost* is the cost of producing an additional unit. Find a formula for $C(n)$ in terms of $n$, given that

$$\text{Total cost} = \text{Fixed cost} + \text{Unit cost} \cdot \text{Number of units}$$

**51.** John wants to buy a dozen rolls. The local bakery sells sesame and poppy-seed rolls for the same price.

**(a)** Make a table of all the possible combinations of rolls if he buys a dozen, where $s$ is the number of sesame seed rolls and $p$ is the number of poppy-seed rolls.
**(b)** Find a formula for $p$ as a function of $s$.
**(c)** Graph this function.

**52.** In a college meal plan you pay a membership fee; then all your meals are at a fixed price per meal.

(a) If 90 meals cost $1005 and 140 meals cost $1205, write a linear function that describes the cost of a meal plan, $C$, in terms of the number of meals, $n$.
(b) What is the cost per meal and what is the membership fee?
(c) Find the cost for 120 meals.
(d) Find $n$ in terms of $C$.
(e) Use part (d) to determine the maximum number of meals you can buy on a budget of $1285.

**53.** The solid waste generated each year in the cities of the US is increasing.[21] The solid waste generated, in millions of tons, was 88.1 in 1960 and 249.9 in 2010. The trend appears linear during this time.

(a) Construct a formula for the amount of municipal solid waste generated in the US by finding the equation of the line through these two points.
(b) Use this formula to predict the amount of municipal solid waste generated in the US, in millions of tons, in the year 2020.

**54.** The demand for gasoline can be modeled as a linear function of price. If the price of gasoline is $p = \$3.10$ per gallon, the quantity demanded in a fixed period is $q = 65$ gallons. If the price rises to $3.50 per gallon, the quantity demanded falls to 45 gallons in that period.

(a) Find a formula for $q$ in terms of $p$.
(b) Explain the economic significance of the slope of your formula.
(c) Explain the economic significance of the $q$-axis and $p$-axis intercepts.

**55.** Find a formula for the linear function $h(t)$ whose graph intersects the graph of $j(t) = 30(0.2)^t$ at $t = -2$ and $t = 1$.

**56.** Find the equation of line $l$ in Figure 1.40.

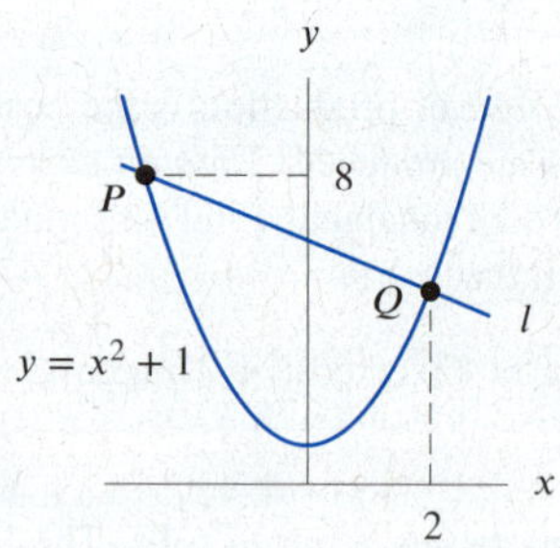


Figure 1.40

**57.** Find the equation of the line $l$, shown in Figure 1.41, if its slope is $m = 4$.

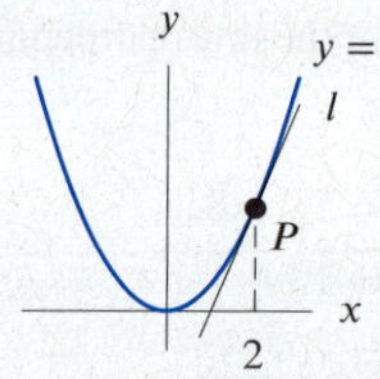


Figure 1.41

**58.** Find an equation for the line $l$ in Figure 1.42 in terms of the constant $A$ and values of the function $f$.

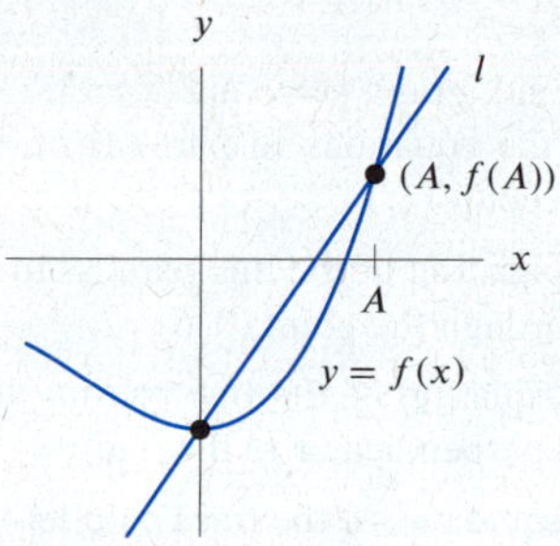


Figure 1.42

**59.** A dose-response function can be used to describe the increase in risk associated with the increase in exposure to various hazards. For example, the risk of contracting lung cancer depends, among other things, on the number of cigarettes a person smokes per day. This risk can be described by a linear dose-response function. For example, it is known that smoking 10 cigarettes per day increases a person's probability of contracting lung cancer by a factor of 25, while smoking 20 cigarettes a day increases the probability by a factor of 50.

(a) Find a formula for $i(x)$, the increase in the probability of contracting lung cancer for a person who smokes $x$ cigarettes per day as compared to a non-smoker.
(b) Evaluate $i(0)$.
(c) Interpret the slope of the function $i$.

**60.** Wire is sold by gauge size, where the diameter of the wire is a decreasing linear function of gauge. Gauge 2 wire has a diameter of 0.2656 inches and gauge 8 wire has a diameter of 0.1719 inches. Find the diameter for wires of gauge 12.5 and gauge 0. What values of the gauge do not make sense in this model?

[21] http://http://www.epa.gov/osw/nonhaz/municipal/pubs/msw_2010_rev_factsheet.pdf, accessed January, 2013.

**61.** You can type four pages in 50 minutes and nine pages in an hour and forty minutes.

**(a)** Find a linear function for the number of pages typed, $p$, as a function of time, $t$. If time is measured in minutes, what values of $t$ make sense in this example?
**(b)** How many pages can be typed in two hours?
**(c)** Interpret the slope of the function in practical terms.
**(d)** Use the result in part (a) to solve for time as a function of the number of pages typed.
**(e)** How long does it take to type a 15-page paper?
**(f)** Write a short paragraph explaining why it is useful to know both of the formulas obtained in part (a) and part (d).

In Problems 62–63, write the functions in slope-intercept form. Identify the values of $b$ and $m$.

**62.** $v(s) = \pi x^2 - 3xr - 4rs - s\sqrt{x}$

**63.** $w(r) = \pi x^2 - 3xr - 4rs - s\sqrt{x}$

**64.** The development time, $t$, of an organism is the number of days required for the organism to mature, and the development rate is defined as $r = 1/t$. In cold-blooded organisms such as insects, the development rate depends on temperature: the colder it is, the longer the organism takes to develop. For such organisms, the degree-day model[22] assumes that the development rate $r$ is a linear function of temperature $H$ (in °C):

$$r = b + kH.$$

**(a)** According to the degree-day model, there is a minimum temperature $H_{\text{min}}$ below which an organism never matures. Find a formula for $H_{\text{min}}$ in terms of the constants $b$ and $k$.
**(b)** Define $S$ as $S = (H - H_{\text{min}})t$, where $S$ is the number of degree-days. That is, $S$ is the number of days $t$ times the number of degrees between $H$ and $H_{\text{min}}$. Use the formula for $r$ to show that $S$ is a constant. In other words, find a formula for $S$ that does not involve $H$. Your formula will involve $k$.
**(c)** A certain organism requires $t = 25$ days to develop at a constant temperature of $H = 20$°C and has $H_{\text{min}} = 15$°C. Using the fact that $S$ is a constant, how many days does it take for this organism to develop at a temperature of 25°C?
**(d)** In part (c) we assumed that the temperature $H$ is constant throughout development. If the temperature varies from day to day, the number of degree-days can be accumulated until they total $S$, at which point the organism completes development. For instance, suppose on the first day the temperature is $H = 20$°C and that on the next day it is $H = 22$°C. Then for these first two days

$$\text{Total number of degree days} = (20 - 15) \cdot 1 + (22 - 15) \cdot 1 = 12.$$

Based on Table 1.34, on what day does the organism reach maturity?

**Table 1.34**

| Day | 1 | 2 | 3 | 4 | 5 | 6 | 7 | 8 | 9 | 10 | 11 | 12 |
|---|---|---|---|---|---|---|---|---|---|---|---|---|
| $H$ (°C) | 20 | 22 | 27 | 28 | 27 | 31 | 29 | 30 | 28 | 25 | 24 | 26 |

**65.** (Continuation of Problem 64.) Table 1.35 gives the development time $t$ (in days) for an insect as a function of temperature $H$ (in °C).

**(a)** Find a linear formula for $r$, the development rate, in terms of $H$.
**(b)** Find the value of $S$, the number of degree-days required for the organism to mature.

**Table 1.35**

| $H$, °C | 20 | 22 | 24 | 26 | 28 | 30 |
|---|---|---|---|---|---|---|
| $t$, days | 14.3 | 12.5 | 11.1 | 10.0 | 9.1 | 8.3 |

# 1.5 MODELING WITH LINEAR FUNCTIONS

## Interpreting the Parameters of a Linear Function

The slope-intercept form for a linear function is $y = b + mx$, where $b$ is the $y$-intercept and $m$ is the slope. The parameters $b$ and $m$ can be used to compare linear functions.

[22] Information drawn from a web site created by Dr. Alexei A. Sharov at the Virginia Polytechnic Institute, http://www.ento.vt.edu/ sharov/PopEcol/popecol.html, accessed September 2002.

**Example 1** With time, $t$, in years, the populations of four towns, $P_A$, $P_B$, $P_C$ and $P_D$, are given by the following formulas:

$$P_A = 20{,}000 + 1600t, \quad P_B = 50{,}000 - 300t, \quad P_C = 650t + 45{,}000, \quad P_D = 15{,}000(1.07)^t.$$

(a) Which populations are represented by linear functions?
(b) Describe in words what each linear model tells you about that town's population. Of the towns that grow linearly, which town starts out with the most people? Which is growing fastest?

**Solution** (a) The populations of towns $A$, $B$, and $C$ are represented by linear functions because they are written in the form $P = b + mt$. Town $D$'s population does not grow linearly since its formula, $P_D = 15{,}000(1.07)^t$, cannot be expressed in the form $P_D = b + mt$.
(b) For town $A$, we have

$$P_A = \underbrace{20{,}000}_{b} + \underbrace{1600}_{m} \cdot t,$$

so $b = 20{,}000$ and $m = 1600$. This means that in year $t = 0$, town $A$ has 20,000 people. It grows by 1600 people per year.

For town $B$, we have

$$P_B = \underbrace{50{,}000}_{b} + \underbrace{(-300)}_{m} \cdot t,$$

so $b = 50{,}000$ and $m = -300$. This means that town $B$ starts with 50,000 people. The negative slope indicates that the population is decreasing at the rate of 300 people per year.

For town $C$, we have

$$P_C = \underbrace{45{,}000}_{b} + \underbrace{650}_{m} \cdot t,$$

so $b = 45{,}000$ and $m = 650$. This means that town $C$ begins with 45,000 people and grows by 650 people per year.

Town $B$ starts out with the most people, 50,000, but town $A$, with a rate of change of 1600 people per year, grows the fastest of the three towns that grow linearly.

## The Effect of the Parameters on the Graph of a Linear Function

The graph of a linear function is a line. Changing the values of $b$ and $m$ gives different members of the family of linear functions. In summary:

Let $y = b + mx$. Then the graph of $y$ against $x$ is a line.

- The $y$-intercept, $b$, tells us where the line crosses the $y$-axis.
- If the slope, $m$, is positive, the line climbs from left to right. If the slope, $m$, is negative, the line falls from left to right.
- The slope, $m$, tells us how fast the line is climbing or falling.
- The larger the magnitude of $m$ (either positive or negative), the steeper the graph of $f$.

**Example 2** (a) Graph the three linear functions $P_A$, $P_B$, $P_C$ from Example 1 and show how to identify the values of $b$ and $m$ from the graph.

(b) Graph $P_D$ from Example 1 and explain how the graph shows $P_D$ is not a linear function.

**Solution** (a) Figure 1.43 gives graphs of the three functions:

$$P_A = 20{,}000 + 1600t, \qquad P_B = 50{,}000 - 300t, \quad \text{and} \quad P_C = 45{,}000 + 650t.$$

The values of $b$ identified in Example 1 tell us the vertical intercepts. Figure 1.43 shows that the graph of $P_A$ crosses the $P$-axis at $P = 20{,}000$, the graph of $P_B$ crosses at $P = 50{,}000$, and the graph of $P_C$ crosses at $P = 45{,}000$.

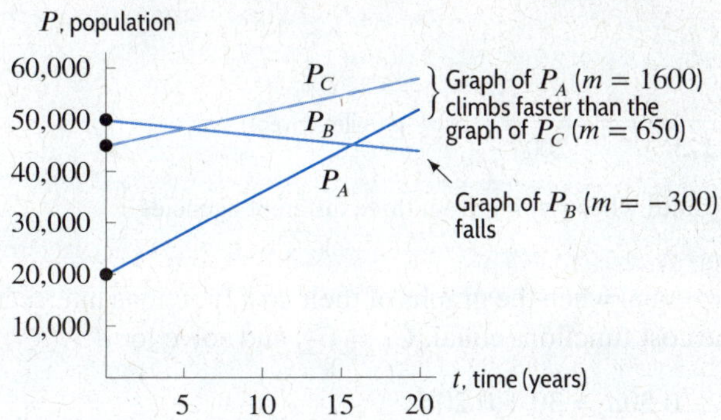


**Figure 1.43:** Graphs of three linear functions, $P_A$, $P_B$, and $P_C$, showing starting values and rates of climb

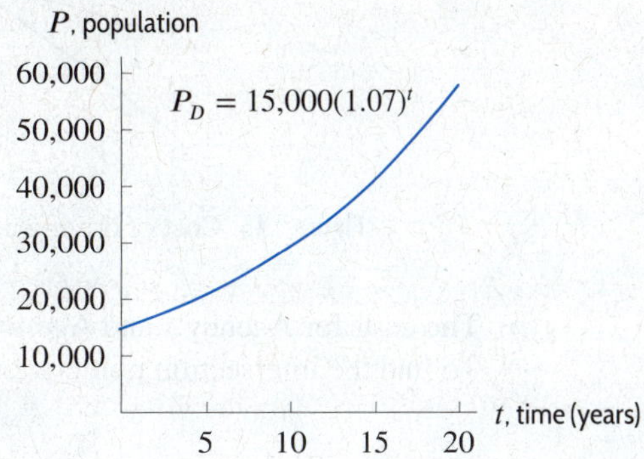


**Figure 1.44:** Graph of $P_D = 15{,}000(1.07)^t$ is not a line

Notice that the graphs of $P_A$ and $P_C$ are both climbing and that $P_A$ climbs faster than $P_C$. This corresponds to the fact that the slopes of these two functions are positive ($m = 1600$ for $P_A$ and $m = 650$ for $P_C$) and the slope of $P_A$ is larger than the slope of $P_C$.

The graph of $P_B$ falls when read from left to right, indicating that population decreases over time. This corresponds to the fact that the slope of $P_C$ is negative ($m = -300$).

(b) Figure 1.44 gives a graph of $P_D$. Since it is not a line, $P_D$ is not a linear function.

## Intersection of Two Lines

The point $(x, y)$ at which two lines intersect satisfies the equations for both lines. Thus, to find this point, we solve the equations simultaneously.[23] When modeling real phenomena, the point of intersection often has a practical meaning.

**Example 3** The cost, in dollars, of renting a car for a day from three different rental agencies and driving it $d$ miles is given by the following functions:

$$C_1 = 0.50d, \qquad C_2 = 30 + 0.20d, \qquad C_3 = 50 + 0.10d.$$

(a) For each agency, describe the rental agreement in words and graph the cost function. Use one set of axes for all three functions.

(b) For what driving distance does Agency 1 cost the same as Agency 2? If you drive 50 miles or less, which agency is cheapest?

(c) How do you determine which agency provides the cheapest car rental for any given driving distance?

[23]The algebra in this section is reviewed in the Skills Refresher on page 62.

**Solution**

(a) See Figure 1.45. Agency 1 charges \$50 plus \$0.10 per mile driven. Agency 2 charges \$30 plus \$0.20 per mile. Agency 3 charges \$0.50 per mile driven.

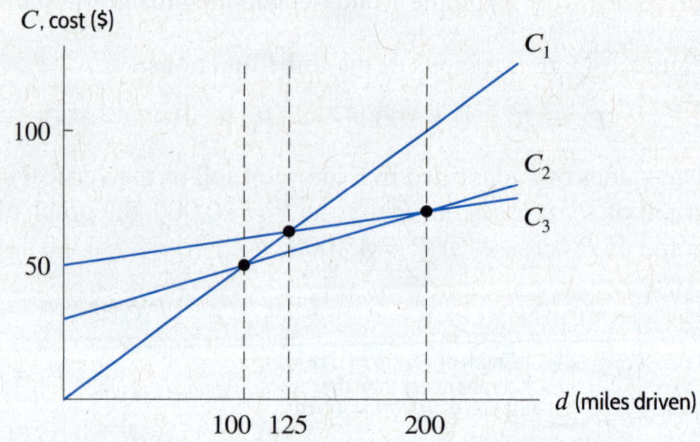


**Figure 1.45:** Cost of driving a car $d$ miles when renting from three different agencies

(b) The costs for Agency 1 and Agency 2 are equal when the graphs of their cost functions intersect. To find the intersection point, we set the cost functions equal, $C_1 = C_2$, and solve for $d$:

$$\begin{aligned} 0.50d &= 30 + 0.20d \\ 0.50d - 0.20d &= 30 \\ 0.30d &= 30 \\ d &= 100. \end{aligned}$$

So, if we drive 100 miles, the rental cost is the same for Agency 1 and Agency 2, as Figure 1.45 shows.

Figure 1.45 also shows that the graph of $C_1$ is below the graph of $C_2$ for $d < 100$. This means that it is cheaper to rent from Agency 1 than from Agency 2 if we drive less than 100 miles. For $d < 100$, the graph of $C_1$ is also below the graph of $C_3$, so renting from Agency 1 is the cheapest option if we drive 50 miles or less.

(c) Which agency is cheapest depends on how far we drive. If we are not driving far, Agency 1 may be cheapest because it only charges for miles driven and has no other fees. If we want to drive a long way, Agency 3 may be cheapest (even though it charges \$50 up front) because it has the lowest per-mile rate.

Figure 1.45 and part (b) show that for $d$ up to 100 miles, Agency 1 costs the least.

For $d$ between 100 and about 200 miles, Figure 1.45 suggests the values of $C_2$ are less than the values of $C_1$ and $C_3$. To find the exact value of $d$ where $C_2$ and $C_3$ intersect, we solve the equation $C_2 = C_3$:

$$\begin{aligned} 30 + 0.20d &= 50 + 0.10d \\ 0.10d &= 20 \\ d &= 200. \end{aligned}$$

Thus Agency 2 is cheapest when we drive between 100 and 200 miles.

For $d$ more than 200 miles, the values of $C_3$ are less than the values of $C_1$ and $C_2$, meaning Agency 3 is cheapest for large distances—as expected since $C_3$ has the lowest per-mile rate.

Notice that the point where $C_1$ and $C_3$ intersect is not important for deciding which agency is the cheapest: $C_2$ still has the lowest cost value where $C_1$ and $C_3$ intersect.

## Linear Inequalities in Two Variables

We now see how to solve problems using linear inequalities rather than linear equations.[24]

**Example 4** The temperature at the base of a mountain is 90°F and drops by 5°F for every thousand-foot increase in elevation. On what part of the mountain is the temperature cooler than 40°F?

**Solution** Since the temperature, $T$, drops at a constant rate as the elevation, $h$, increases, if $T$ is in degrees Fahrenheit and $h$ is in thousands of feet from the base of the mountain, we have

$$T = 90 - 5h.$$

We want to know for what values of $h$ we have $90 - 5h < 40$.

One way to solve the inequality is to subtract 40 from each side of the inequality to obtain

$$50 - 5h < 0,$$

and then add $5h$ to each side to get

$$50 < 5h.$$

Dividing each side by 5 gives

$$10 < h.$$

This means that the temperature is below 40° at any point more than 10,000 feet above the base of the mountain.

A second approach is to subtract 90 from each side of $90 - 5h < 40$ to get

$$-5h < -50.$$

At this point, we divide each side of the inequality by $-5$. Since $-5$ is negative, this division switches the direction of the inequality, giving

$$h > 10.$$

Note that if we forget to switch the direction of the inequality, we get the incorrect solution $h < 10$, suggesting that the temperature is cooler than 40°F below the 10,000-foot line. However, the temperature is cooler near the top of the mountain, not near the bottom.

**Example 5** When people consume more calories than required, they gain weight; when they consume fewer calories than required, they lose weight. For an active 20-year-old, six-foot-tall man weighing $w$ pounds, the Harris-Benedict equation[25] states that the number of calories, $C$, needed daily to maintain weight is $C = 1321 + 9.44w$.

(a) The points satisfying $C = 1321 + 9.44w$ lie on a line. If the $w$-axis is horizontal, the points satisfying

$$C > 1321 + 9.44w \qquad \text{or} \qquad C < 1321 + 9.44w$$

lie above or below the line. Which of these two inequalities does the point $w = 170$, $C = 3500$ satisfy? Is the point above or below the line? How about the point $w = 200$, $C = 3000$?

(b) Explain what your answers to part (a) mean in the context of weight and caloric intake.

(c) Graph the set of all points that satisfy the inequality $C < 1321 + 9.44w$. What does this region represent in terms of calories and weight?

(d) Sketch the region where $C \leq 3000$ and $C > 1321 + 9.44w$. What does this region represent in terms of calories and weight?

[24]Methods for solving linear inequalities are included in the Skills Refresher on page 62.

[25]*The Harris Benedict equation reevaluated.* A.M. Roza and H.M. Shizgal. American Journal of Clinical Nutrition. Vol. 40, No. 1 (July 1984): 168-182.

**Solution** (a) To determine which inequality is satisfied, we substitute the value of $w$ into the expression $1321 + 9.44w$ and compare the result with the value of $C$.

If $w = 170$, $C = 3500$, substituting $w$ gives $1321 + 9.44 \cdot 170 = 2925.8$. Since $C > 2925.8$, we have

$$C > 1321 + 9.44w. \qquad \text{Thus, the point (170, 3500) is above the line.}$$

If $w = 200$, $C = 3000$, substituting $w$ gives $1321 + 9.44 \cdot 200 = 3209$. Since $C < 3209$, we have

$$C < 1321 + 9.44w. \qquad \text{Thus, the point (200, 3000) is below the line.}$$

(b) A man weighing 170 pounds who eats 3500 calories per day is eating more calories than needed, so he will gain weight. On the other hand, a man weighing 200 pounds who eats 3000 calories daily will lose weight, because he is eating fewer calories than needed.

(c) Figure 1.46 shows the line $C = 1321 + 9.44w$, the point (170, 3500) above the line and the point (200, 3000) below the line.

The points satisfying $C < 1321 + 9.44w$ are those whose $C$-coordinates are lower than $1321 + 9.44w$; that is, all the points below the line. The region represents all of the combinations of weights and caloric intakes at which an active, 6-foot-tall, 20 year-old man loses weight.

Note that the line itself is not included in the region, since on this line $C$ is equal to, not less than, $1321 + 9.44w$. We use a dashed line to indicate this.

(d) The region, shaded in Figure 1.47, shows the combinations of weights and calories—up to 3000 calories—for which an active 6-foot, 20 year-old man gains weight.

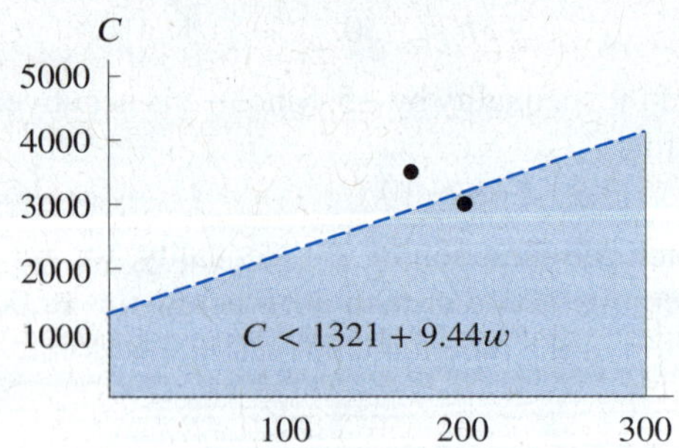


**Figure 1.46**: Shaded region is solution to $C < 1321 + 9.44w$

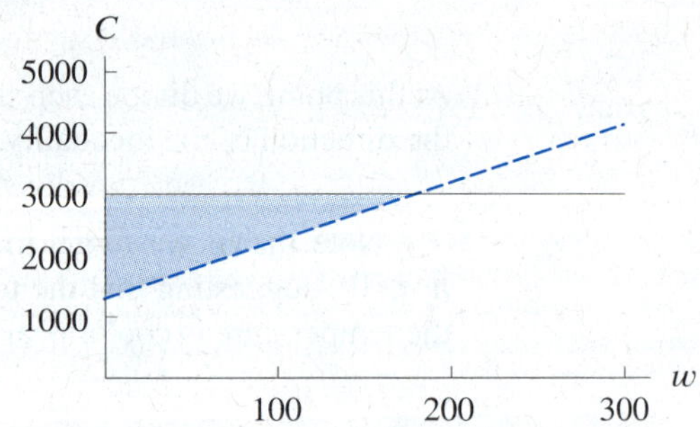


**Figure 1.47**: Shaded region is $C \leq 3000$ and $C < 1321 + 9.44w$

## Exercises and Problems for Section 1.5

### Skill Refresher

In Exercises S1–S6, solve the equations simultaneously if possible.

**S1.** $\begin{cases} x + y = 3 \\ y = 5 \end{cases}$

**S2.** $\begin{cases} x + y = 3 \\ x - y = 5 \end{cases}$

**S3.** $\begin{cases} x + y = 2 \\ 2x + 2y = 7 \end{cases}$

**S4.** $\begin{cases} y = x - 3 \\ 2y - 2x = -6 \end{cases}$

**S5.** $\begin{cases} 2x - y = 10 \\ x + 2y = 15 \end{cases}$

**S6.** $\begin{cases} 2(x + y) = 3 \\ x = y + 3(x - 5) \end{cases}$

Determine the points of intersection for Exercises S7–S8.

**S7.**

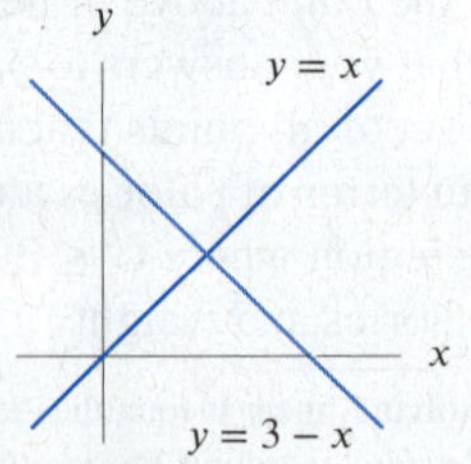


**S8.**

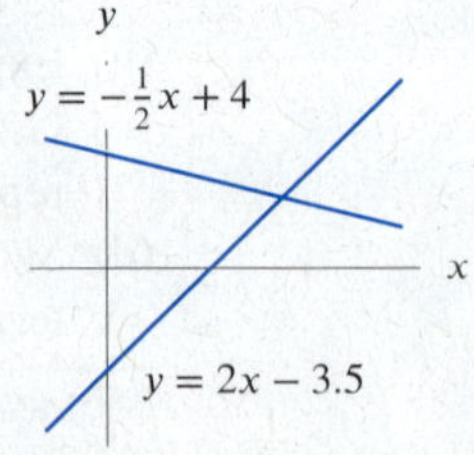

## Exercises

**1.** Without using a calculator, match the equations (a) – (f) to the graphs (I) – (VI).

**(a)** $y = -2.72x$ **(b)** $y = 0.01 + 0.001x$
**(c)** $y = 27.9 - 0.1x$ **(d)** $y = 0.1x - 27.9$
**(e)** $y = -5.7 - 200x$ **(f)** $y = x/3.14$

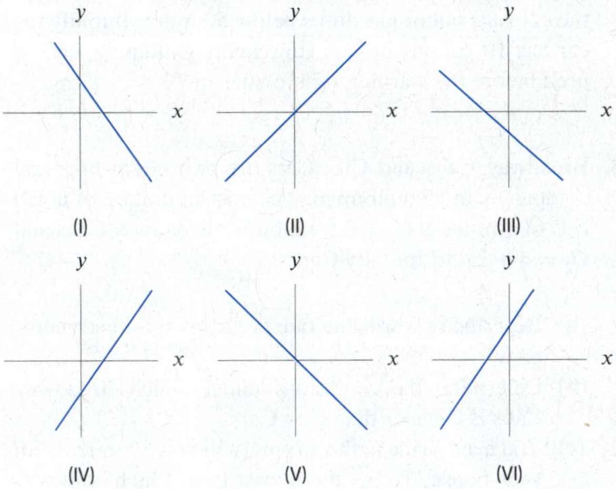


**2.** Without a calculator, match the equations (a)–(g) to the graphs (I)–(VII).

**(a)** $y = x - 5$ **(b)** $-3x + 4 = y$
**(c)** $5 = y$ **(d)** $y = -4x - 5$
**(e)** $y = x + 6$ **(f)** $y = x/2$
**(g)** $5 = x$

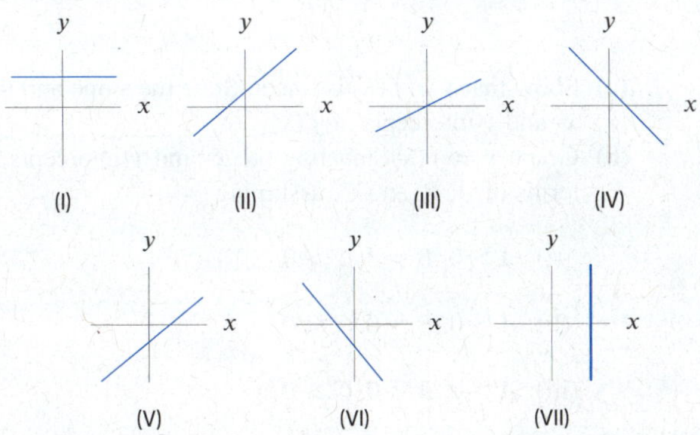


**3.** Figure 1.48 gives lines $A$, $B$, $C$, $D$, and $E$. Without a calculator, match each line to $f, g, h, u$ or $v$:

$$f(x) = 20 + 2x$$
$$g(x) = 20 + 4x$$
$$h(x) = 2x - 30$$
$$u(x) = 60 - x$$
$$v(x) = 60 - 2x$$

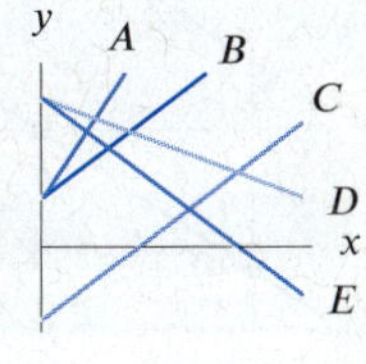


Figure 1.48

**4.** Without a calculator, match the following functions to the lines in Figure 1.49:

$$f(x) = 5 + 2x$$
$$g(x) = -5 + 2x$$
$$h(x) = 5 + 3x$$
$$j(x) = 5 - 2x$$
$$k(x) = 5 - 3x$$

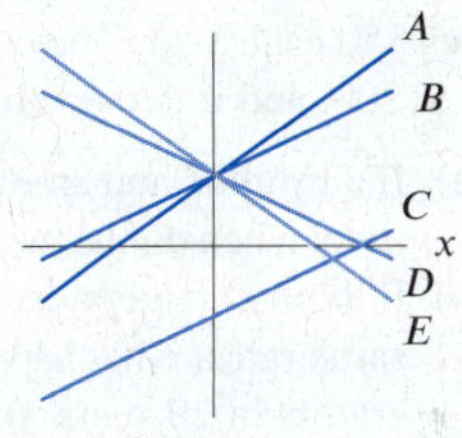


Figure 1.49

**5.** Without a calculator, match the systems of inequalities (a)–(d) to the graphs (I)–(IV). Give reasons.

**(a)** $y \geq x$ and $y \leq 2x$ **(b)** $y \geq 1+2x$ and $y \leq 10-x$
**(c)** $y \geq 2$ and $y \leq 12 - 2x$ **(d)** $y \geq 3x + 1$ and $y \leq 10$

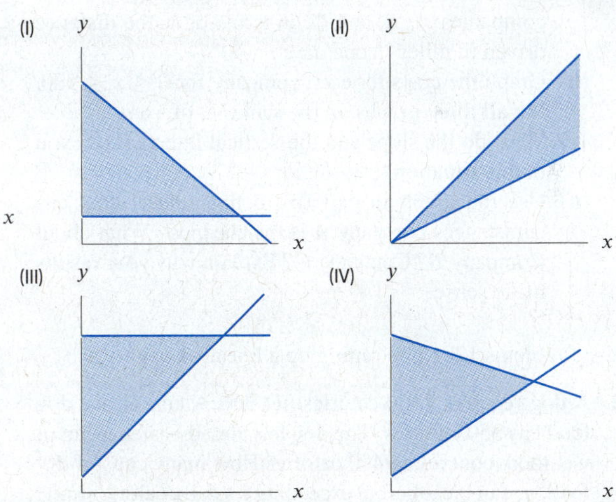


## Problems

**6.** Sketch five different functions in the family $y = -2 - ax$ for $a < 0$.

**7.** Estimate the slope of the line in Figure 1.50 and find an approximate equation for the line.

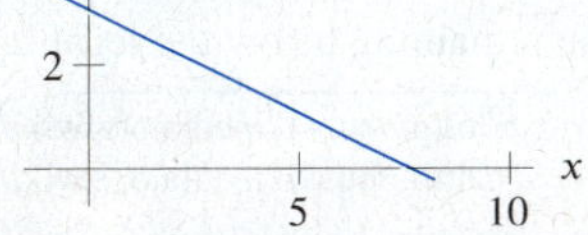


Figure 1.50

**8.** Find the equation of the line $l_2$ in Figure 1.52.

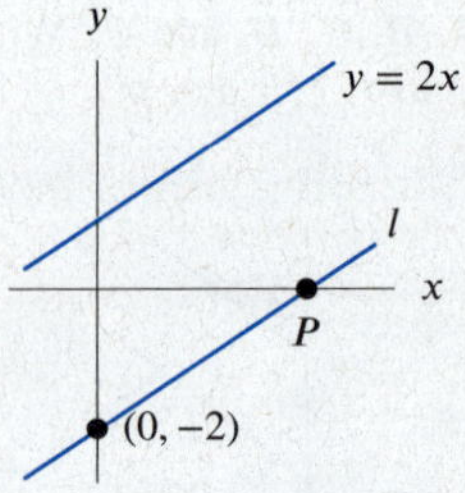


Figure 1.51

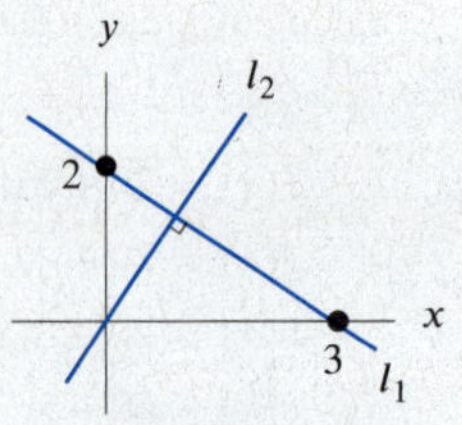


Figure 1.52

**9.** Line $l$ in Figure 1.51 is parallel to the line $y = 2x + 1$. Find the coordinates of the point $P$.

**10.** The cost of a Frigbox refrigerator is \$950, and it depreciates \$50 each year. The cost of an Arctic Air refrigerator is \$1200, and it depreciates \$100 per year.

**(a)** If a Frigbox and an Arctic Air are bought at the same time, when do the two refrigerators have equal value?

**(b)** If both refrigerators continue to depreciate at the same rates, what happens to the values of the refrigerators in 20 years' time? What does this mean?

**11.** You need to rent a car and compare the charges of three different companies. Company $A$ charges 20 cents per mile plus \$20 per day. Company $B$ charges 10 cents per mile plus \$35 per day. Company $C$ charges \$70 per day with no mileage charge.

**(a)** Find formulas for the cost of driving cars rented from companies $A$, $B$, and $C$, in terms of $x$, the distance driven in miles in one day.

**(b)** Graph the costs for each company for $0 \le x \le 500$. Put all three graphs on the same set of axes.

**(c)** What do the slope and the vertical intercept tell you in this situation?

**(d)** Use the graph in part (b) to find under what circumstances company $A$ is the cheapest. What about Company $B$? Company $C$? Explain why your results make sense.

Solve Problems 12–15 by setting up a linear inequality.

**12.** A dog requires 1300 calories per day. A cup of dry dog food has 350 calories. The dog has already eaten a can of wet food, containing 400 calories How many cups of dry food, $n$, can it eat before exceeding its daily caloric limit?

**13.** Blood alcohol concentration (BAC) is often used as a measure of intoxication.[26] For a 180-pound man, the BAC increases linearly by 0.02% for every drink taken in a one-hour period. If at the start of the hour he has a BAC of $B = 0.01\%$, how many drinks, $n$, can he have during the hour while remaining below the legal limit of 0.08%?

**14.** For a 180-pound man, blood alcohol concentration (BAC) decreases linearly by 0.015% every hour if he does not consume any further drinks.[27] If at the start of the hour he has a BAC of $B = 0.13\%$, how many hours, $t$, must he wait before his BAC falls below the legal limit of 0.08%?

**15.** A car's fuel economy is 30 mpg. A warning light is displayed if the remaining distance that can be driven before the car runs out of gas drops below 50 miles. Initially the car has 10 gallons of gas. How many gallons, $g$, can be used before the warning light comes on?

**16.** Broadway Cabs and City Cars are two competing taxi companies in a small town. The cost in dollars of a cab ride of $d$ miles is $C_1 = 2.5+2.6d$ for Broadway Cabs and $C_2 = 4 + 2.45d$ for City Cars.

**(a)** Describe in words the fare breakdown for each company.

**(b)** Determine the conditions under which Broadway Cabs is cheaper than City Cars.

**(c)** You need a ride to the airport, which is 12 miles from your house. To get the lowest fare, which company should you call and how much will it cost?

**17.** Assume $A$, $B$, $C$ are constants with $A \neq 0$, $B \neq 0$. Consider the equation

$$Ax + By = C.$$

**(a)** Show that $y = f(x)$ is linear. State the slope and the $x$- and $y$-intercepts of $f(x)$.

**(b)** Graph $y = f(x)$, labeling the $x$- and $y$-intercepts in terms of $A$, $B$, and $C$, assuming

(i) $A > 0, B > 0, C > 0$

(ii) $A > 0, B > 0, C < 0$

(iii) $A > 0, B < 0, C > 0$

**18.** Fill in the missing coordinates for the points in the following figures.

**(a)** The triangle in Figure 1.53.

**(b)** The parallelogram in Figure 1.54.

[26]http://en.wikipedia.org/wiki/Blood_alcohol_content#cite_ref-4, accessed February 8, 2013.
[27]http://en.wikipedia.org/wiki/Blood_alcohol_content#cite_ref-4, accessed February 8, 2013.

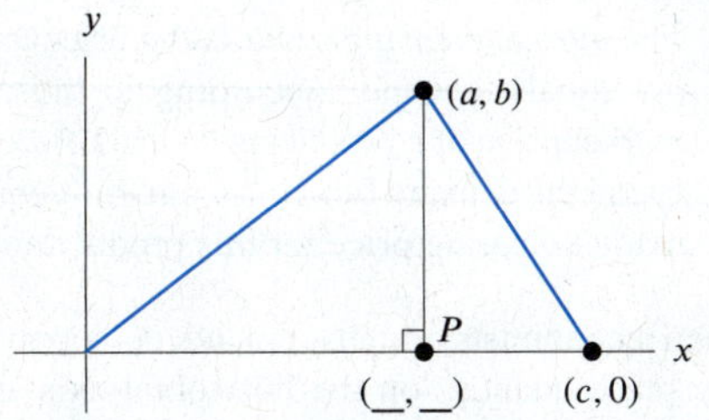


Figure 1.53

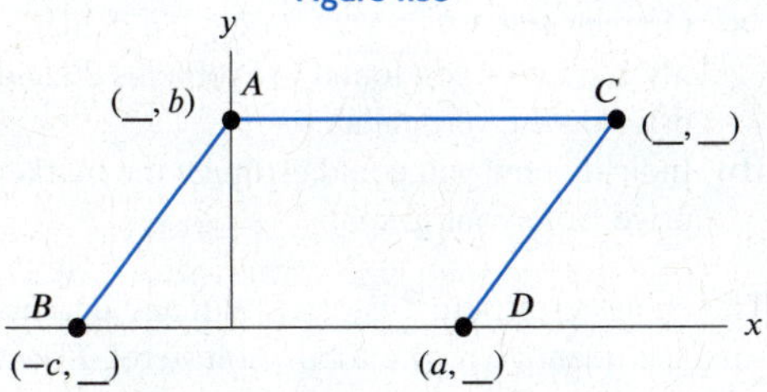


Figure 1.54

**19.** Fill in the missing coordinates in Figure 1.55. Write an equation for the line connecting the two points.

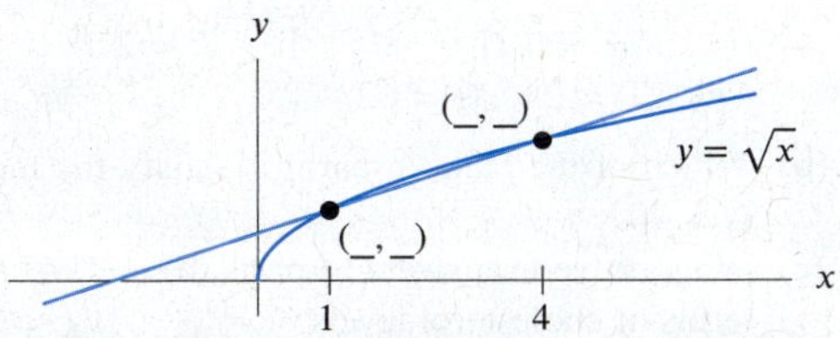


Figure 1.55

**20.** Using the window $-10 \le x \le 10, -10 \le y \le 10$, graph $y = x$, $y = 10x$, $y = 100x$, and $y = 1000x$.

**(a)** Explain what happens to the graphs of the lines as the slopes become large.
**(b)** Write an equation of a line that passes through the origin and is horizontal.

**21.** Graph $y = x + 1$, $y = x + 10$, and $y = x + 100$ in the window $-10 \le x \le 10, -10 \le y \le 10$.

**(a)** Explain what happens to the graph of a line, $y = b + mx$, as $b$ becomes large.
**(b)** Write a linear equation whose graph cannot be seen in the window $-10 \le x \le 10, -10 \le y \le 10$ because all its $y$-values are less than the $y$-values shown.

**22.** The graphical interpretation of the slope is that it shows steepness. Using a calculator or a computer, graph the function $y = 2x - 3$ in the following windows:

**(a)** $-10 \le x \le 10$ by $-10 \le y \le 10$
**(b)** $-10 \le x \le 10$ by $-100 \le y \le 100$
**(c)** $-10 \le x \le 10$ by $-1000 \le y \le 1000$
**(d)** Write a sentence about how steepness is related to the window being used.

**23.** A circle of radius 2 is centered at the origin and goes through the point $(-1, \sqrt{3})$.

**(a)** Find an equation for the line through the origin and the point $(-1, \sqrt{3})$.
**(b)** Find an equation for the tangent line to the circle at $(-1, \sqrt{3})$. [Hint: A tangent line is perpendicular to the radius at the point of tangency.]

**24.** Find an equation for the altitude through point $A$ of the triangle $ABC$, where $A$ is $(-4, 5)$, $B$ is $(-3, 2)$, and $C$ is $(9, 8)$. [Hint: The altitude of a triangle is perpendicular to the base.]

**25.** Two lines are given by $y = b_1 + m_1 x$ and $y = b_2 + m_2 x$, where $b_1$, $b_2$, $m_1$, and $m_2$ are constants.

**(a)** What conditions are imposed on $b_1$, $b_2$, $m_1$, and $m_2$ if the two lines have no points in common?
**(b)** What conditions are imposed on $b_1$, $b_2$, $m_1$, and $m_2$ if the two lines have all points in common?
**(c)** What conditions are imposed on $b_1$, $b_2$, $m_1$, and $m_2$ if the two lines have exactly one point in common?
**(d)** What conditions are imposed on $b_1$, $b_2$, $m_1$, and $m_2$ if the two lines have exactly two points in common?

In Problems 26–27, what is true about the constant $\beta$ in the following linear equation if its graph has the given property?

$$y = \frac{x}{\beta - 3} + \frac{1}{6 - \beta}.$$

**26.** Positive slope, positive $y$-intercept.

**27.** Perpendicular to the line $y = (\beta - 7)x - 3$.

**28.** A theater manager graphed weekly profits as a function of the number of patrons and found that the relationship was linear. One week the profit was \$11,328 when 1324 patrons attended. Another week 1529 patrons produced a profit of \$13,275.50.

**(a)** Find a formula for weekly profit, $y$, as a function of the number of patrons, $x$.
**(b)** Interpret the slope and the $y$-intercept.
**(c)** What is the break-even point (the number of patrons for which there is zero profit)?
**(d)** Find a formula for the number of patrons as a function of profit.
**(e)** If the weekly profit was \$17,759.50, how many patrons attended the theater?

**29.** In economics, the *demand* for a product is the amount of that product that consumers are willing to buy at a given price. The quantity demanded of a product usually decreases if the price of that product increases. Suppose that a company believes there is a linear relationship between the demand for its product and its price. The company knows that when the price of its product was \$3 per unit, the quantity demanded weekly was 500 units, and that when the unit price was raised to \$4, the quantity demanded weekly dropped to 300 units. Let $D$ represent the quantity demanded weekly at a unit price of $p$ dollars.

**(a)** Calculate $D$ when $p = 5$. Interpret your result.
**(b)** Find a formula for $D$ in terms of $p$.
**(c)** The company raises the price of the good and the new quantity demanded weekly is 50 units. What is the new price?
**(d)** Give an economic interpretation of the slope of the function you found in part (b).
**(e)** Find $D$ when $p = 0$. Find $p$ when $D = 0$. Give economic interpretations of both these results.

**30.** In economics, the *supply* of a product is the quantity of that product suppliers are willing to provide at a given price. In theory, the quantity supplied of a product increases if the price of that product increases. Suppose that there is a linear relationship between the quantity supplied, $S$, of the product described in Problem 29 and its price, $p$. The quantity supplied weekly is 100 when the price is \$2 and the quantity supplied rises by 50 units when the price rises by \$0.50.

**(a)** Find a formula for $S$ in terms of $p$.
**(b)** Interpret the slope of your formula in economic terms.
**(c)** Is there a price below which suppliers will not provide this product?
**(d)** The *market clearing price* is the price at which supply equals demand. According to theory, the free-market price of a product is its market clearing price. Using the demand function from Problem 29, find the market clearing price for this product.

**31.** When economists graph demand or supply equations, they place quantity on the horizontal axis and price on the vertical axis.

**(a)** On the same set of axes, graph the demand and supply equations you found in Problems 29 and 30, with price on the vertical axis.
**(b)** Indicate how you could estimate the market clearing price from your graph.

**32.** The cholesterol ratio[28] is used by many doctors to measure the quantity of the total cholesterol $T$ in the body, measured in milligrams per deciliter of blood (mg/dl), relative to the quantity of good cholesterol $G$, also measured in mg/dl. In order to minimize risk of cholesterol-related illnesses, the good cholesterol level should always exceed $T/5$ but never be less than $G = 40$.

**(a)** Which of the following pairs of $T$ and $G$ values satisfy the inequality $G > T/5$?

(i) $T = 100, G = 38$ (ii) $T = 220, G = 42$
(iii) $T = 240, G = 60$

**(b)** Which of the pairs in part (a) satisfy the inequality $G > 40$?
**(c)** What do your answers to parts (a) and (b) mean in terms of cholesterol levels?
**(d)** With $T$ on the horizontal axis, graph all the points that satisfy both the inequalities $G > T/5$ and $G > 40$. What do the points in this region represent?

# 1.6 FITTING LINEAR FUNCTIONS TO DATA

When real data are collected in the laboratory or the field, they are often subject to experimental error. Even if there is an underlying linear relationship between two quantities, real data may not fit this relationship perfectly. However, even if a data set does not perfectly conform to a linear function, we may still be able to use a linear function to help us analyze the data.

## Laboratory Data: The Viscosity of Motor Oil

The viscosity of a liquid, or its resistance to flow, depends on the liquid's temperature. Pancake syrup is a familiar example: straight from the refrigerator, it pours very slowly. When warmed on the stove, its viscosity decreases and it becomes quite runny.

The viscosity of motor oil is a measure of its effectiveness as a lubricant in the engine of a car. Thus, the effect of engine temperature is an important determinant of motor-oil performance. Table 1.36 gives the viscosity, $v$, of motor oil as measured in the lab at different temperatures, $T$.

[28] http://www.heart.org/HEARTORG/Conditions/Cholesterol/AboutCholesterol/What-Your-Cholesterol-Levels-Mean_UCM_305562_Article.jsp, accessed February 20, 2014

**Table 1.36** *The measured viscosity, $v$, of motor oil as a function of the temperature, $T$*

| $T$, temperature (°F) | $v$, viscosity (lbs·sec/in$^2$) |
|---|---|
| 160 | 28 |
| 170 | 26 |
| 180 | 24 |
| 190 | 21 |
| 200 | 16 |
| 210 | 13 |
| 220 | 11 |
| 230 | 9 |

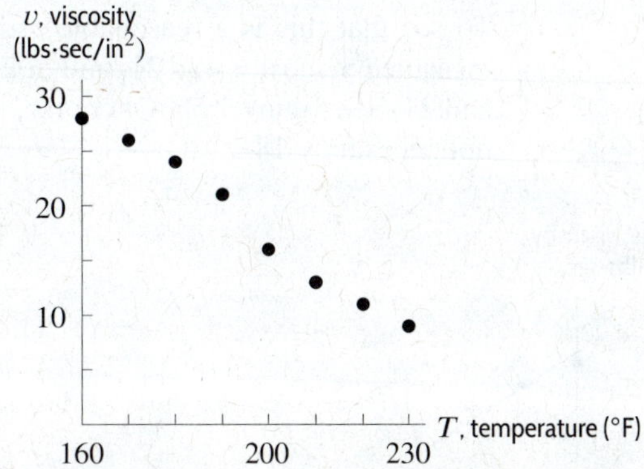


**Figure 1.56:** The viscosity data from Table 1.36

The *scatter plot* of the data in Figure 1.56 shows that the viscosity of motor oil decreases, approximately linearly, as its temperature rises. To find a formula relating viscosity and temperature, we fit a line to these data points.

Fitting the best line to a set of data is called *linear regression*. One way to fit a line is to draw a line "by eye." Alternatively, many computer programs and calculators compute regression lines. Figure 1.57 shows the data from Table 1.36 together with the computed regression line,

$$v = 75.6 - 0.293T.$$

Notice that none of the data points lies exactly on the regression line, although it fits the data well.

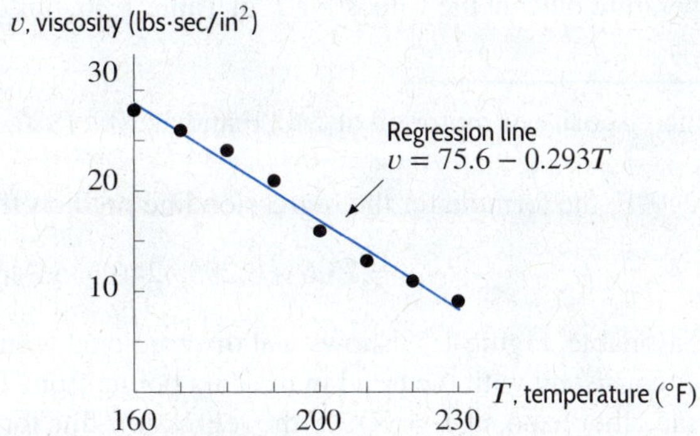


**Figure 1.57:** A graph of the viscosity data from Table 1.36, together with a regression line (provided by a calculator)

### The Assumptions Involved In Finding a Regression Line

When we find a regression line for the data in Table 1.36, we are assuming that the value of $v$ is related to the value of $T$. However, there may be experimental errors in our measurements. For example, if we measure viscosity twice at the same temperature, we may get two slightly different values. Alternatively, something besides engine temperature could be affecting the oil's viscosity (the oil pressure, for example). Thus, even if we assume that the temperature readings are exact, the viscosity readings include some degree of uncertainty.

## Interpolation and Extrapolation

The formula for viscosity can be used to make predictions. Suppose we want to know the viscosity of motor oil at $T = 196$°F. The formula gives

$$v = 75.6 - 0.293 \cdot 196 \approx 18.2 \text{ lb} \cdot \text{sec/in}^2.$$

To see that this is a reasonable estimate, compare it to the entries in Table 1.36. At 190°F, the measured viscosity was 21, and at 200°F, it was 16; the predicted viscosity of 18.2 is between 16 and 21. See Figure 1.58. Of course, if we measured the viscosity at $T = 196$°F in the lab, we might not get exactly 18.2.

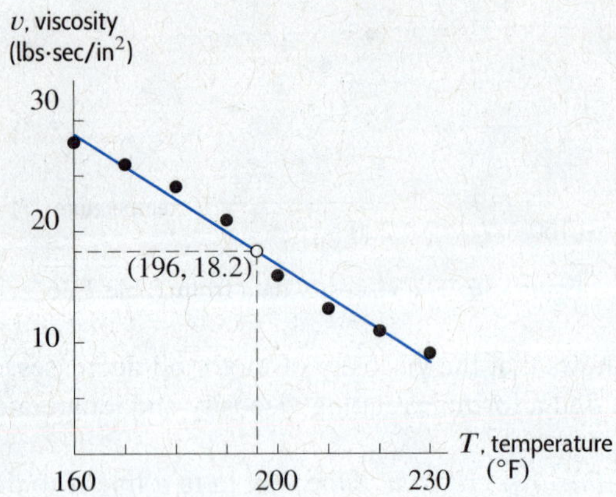


**Figure 1.58:** Regression line used to predict the viscosity at 196°F

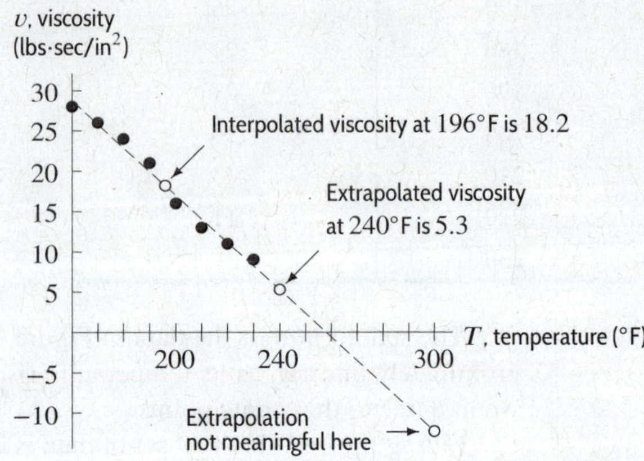


**Figure 1.59:** The data from Table 1.36 together with the predicted viscosity at $T = 196$°F, $T = 240$°F, and $T = 300$°F

Since the temperature $T = 196$°F is between two temperatures for which $v$ is known (190°F and 200°F), the estimate of 18.2 is said to be an *interpolation*. If instead we estimate the value of $v$ at a temperature outside the values for $T$ in Table 1.36, our estimate is called an *extrapolation*.

**Example 1** Predict the viscosity of motor oil at 240°F and at 300°F.

**Solution** At $T = 240$°F, the formula for the regression line predicts that the viscosity of motor oil is

$$v = 75.6 - 0.293 \cdot 240 = 5.3 \text{ lb} \cdot \text{sec/in}^2.$$

This is reasonable. Figure 1.59 shows that the predicted point—represented by an open circle on the graph—is consistent with the trend in the data points from Table 1.36.

On the other hand, at $T = 300$°F the regression-line formula gives

$$v = 75.6 - 0.293 \cdot 300 = -12.3 \text{ lb} \cdot \text{sec/in}^2.$$

This is unreasonable because viscosity cannot be negative. To understand what went wrong, notice that in Figure 1.59, the open circle representing the point (300, −12.3) is far from the plotted data points. By making a prediction at 300°F, we have assumed—incorrectly—that the trend observed in laboratory data extended as far as 300°F.

In general, interpolation tends to be more reliable than extrapolation because we are making a prediction on an interval we already know something about instead of making a prediction beyond the limits of our knowledge.

## How Regression Works

How does a calculator or computer decide which line fits the data best? We assume that the value of $y$ is related to the value of $x$, although other factors could influence $y$ as well. Thus, we assume that we can pick the value of $x$ exactly but that the value of $y$ may be only partially determined by this $x$-value.

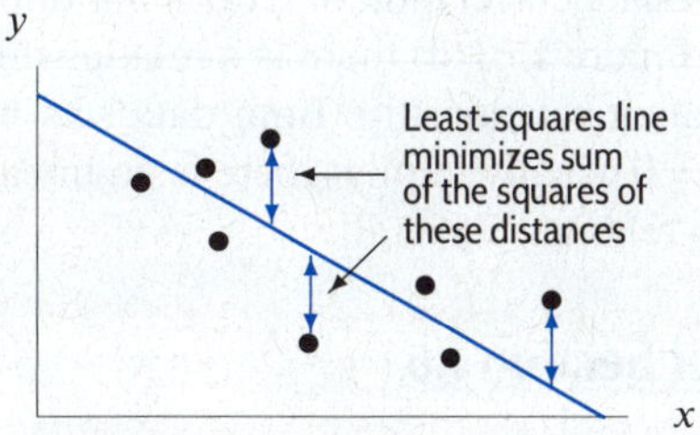


**Figure 1.60:** A given set of data and the corresponding least-squares regression line

One way to fit a line to the data is shown in Figure 1.60. The line shown was chosen to minimize the sum of the squares of the vertical distances between the data points and the line. Such a line is called a *least-squares line*. There are formulas which a calculator or computer uses to calculate the slope, $m$, and the $y$-intercept, $b$, of the least-squares line.

## Correlation

When a computer or calculator calculates a regression line, it also gives a *correlation coefficient*, $r$. This number lies between $-1$ and $+1$ and measures how well a particular regression line fits the data. If $r = 1$, the data lie exactly on a line of positive slope. If $r = -1$, the data lie exactly on a line of negative slope. If $r$ is close to 0, the data may be completely scattered, or there may be a nonlinear relationship between the variables. (See Figure 1.61.)

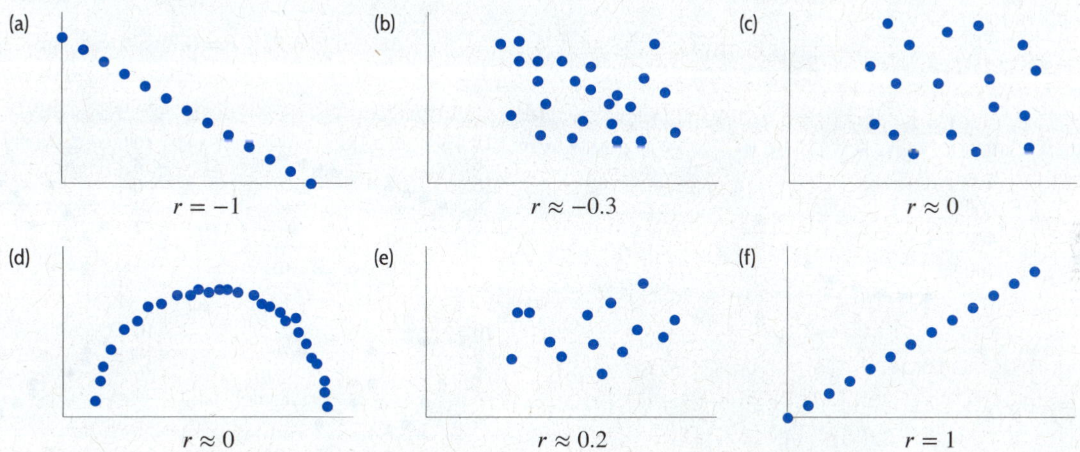


**Figure 1.61:** Various data sets and correlation coefficients

**Example 2** The correlation coefficient for the viscosity data in Table 1.36 on page 47 is $r \approx -0.99$. The fact that $r$ is negative tells us that the regression line has negative slope. The fact that $r$ is close to $-1$ tells us that the regression line fits the data well.

### The Differences Between Relation, Correlation, and Causation

It is important to understand that a high correlation (either positive or negative) between two quantities does *not* imply causation. For example, there is a high correlation between children's reading level and shoe size.[29] However, large feet do not cause a child to read better (or vice versa). Larger feet and improved reading ability are both a consequence of growing older.

Notice also that a correlation of 0 does not imply that there is no relationship between $x$ and $y$. For example, in Figure 1.61(d) there is a relationship between $x$ and $y$-values, while Figure 1.61(c) exhibits no apparent relationship. Both data sets have a correlation coefficient of $r \approx 0$. Thus a correlation of $r = 0$ usually implies there is no linear relationship between $x$ and $y$, but this does not mean there is no relationship at all.

## Exercises and Problems for Section 1.6

In Problems 1–6, is the given value of $r$ reasonable? Give a reason for your answer.

**1.** $r = 1$

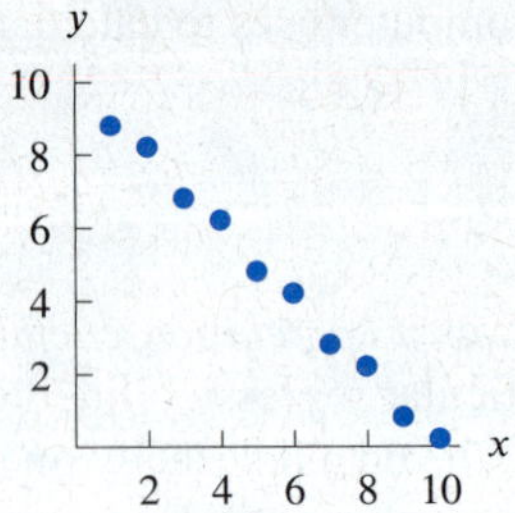


**2.** $r = 0.7$

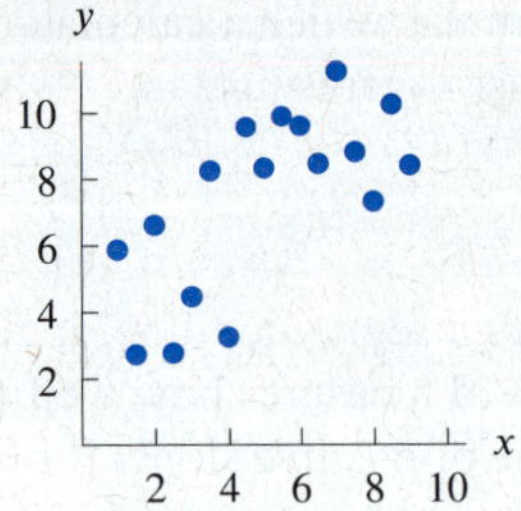


**3.** $r = 0.9$

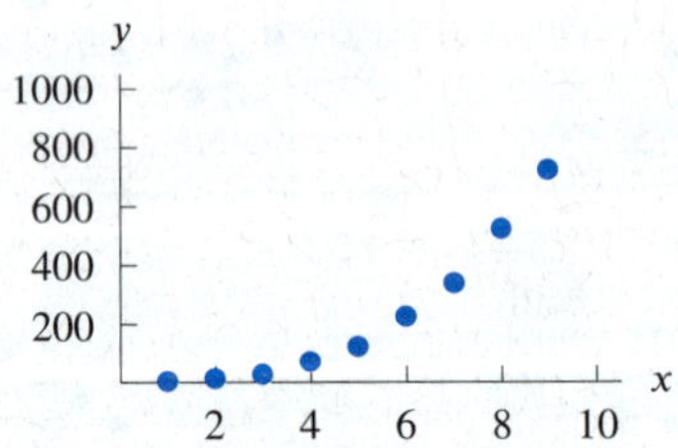


**4.** $r = -0.9$

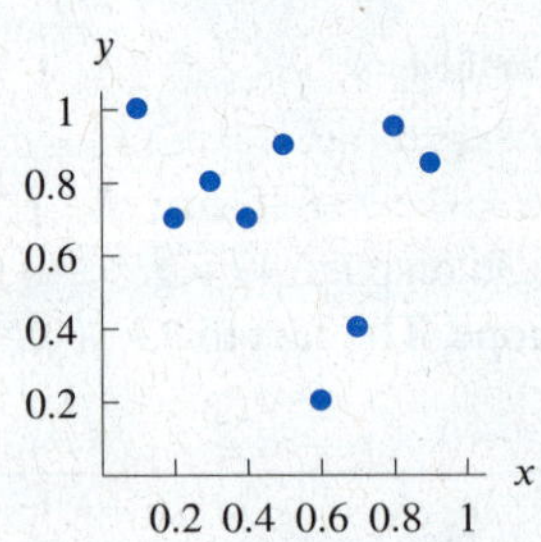


**5.** $r = 0.9$

| $x$ | 1 | 2 | 3 | 4 | 5 |
|---|---|---|---|---|---|
| $y$ | 3.8 | 3.2 | 1.8 | 1.2 | −0.2 |

**6.** $r = 0.3$

| $x$ | 1 | 2 | 3 | 4 | 5 |
|---|---|---|---|---|---|
| $y$ | 3.477 | 5.531 | 14.88 | 5.924 | 8.049 |

**7.** Match the $r$ values with scatter plots in Figure 1.62.

$$r = -0.98, \quad r = -0.5, \quad r = -0.25,$$

$$r = 0, \quad r = 0.7, \quad r = 1.$$

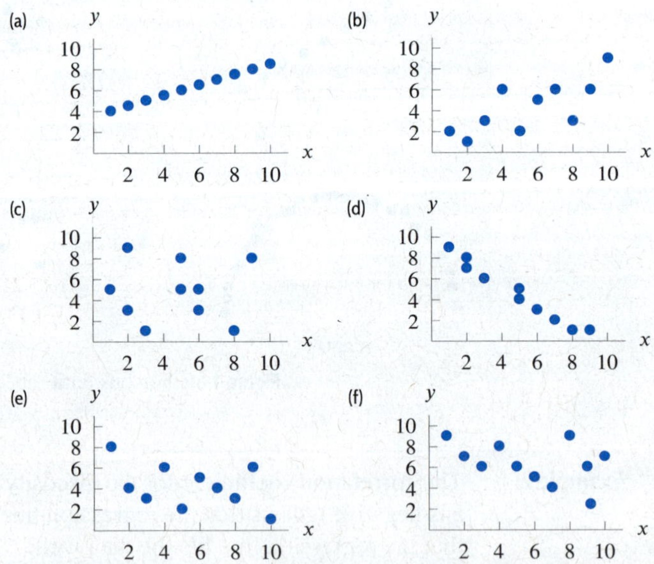


**Figure 1.62**

[29]From *Statistics*, 2nd edition, by David Freedman. Robert Pisani, Roger Purves, Ani Adhikari, p. 142 (New York: W.W. Norton, 1991).

**8.** Table 1.37 shows the number of calories burned per minute by a person walking at 3 mph.

**(a)** Make a scatter plot of this data.
**(b)** Draw a regression line by eye.
**(c)** Roughly estimate the correlation coefficient by eye.

**Table 1.37**

| Body weight (lb) | 100 | 120 | 150 | 170 | 200 | 220 |
|---|---|---|---|---|---|---|
| Calories | 2.7 | 3.2 | 4.0 | 4.6 | 5.4 | 5.9 |

**9.** The manager of a small nature park found the regression line for the weekly profits, $p$ in dollars, as a function of the number of visitors, $n$, was $p = 4n - 2200$.

**(a)** How much profit is predicted if 700 people visit the park in a week?
**(b)** Explain the meaning of the slope of the regression line in the context of this problem. Include units.
**(c)** Explain the meaning of the vertical intercept of the regression line in the context of this problem. Include units.
**(d)** How many weekly visitors does the park need if it is not to lose money?

**10.** An ecologist tracked 290 topi that were born in 2004. The number of topi, $T$, living each subsequent year is recorded in Table 1.38.

**(a)** Make a scatter plot of this data. Let $t = 0$ represent 2004.
**(b)** Draw by eye a line of good fit and estimate its equation. (Round the coefficients to integers.)
**(c)** Use a calculator or computer to find the equation of the least squares line. (Round the coefficients to integers.)
**(d)** Interpret the slope and each intercept of the line.
**(e)** Interpret the correlation between the year and the number of topi born in 2004 that are still alive.

**Table 1.38**

| Year | 2004 | 2005 | 2006 | 2007 | 2008 | 2009 | 2010 | 2011 | 2012 |
|---|---|---|---|---|---|---|---|---|---|
| Topi | 290 | 288 | 268 | 206 | 140 | 90 | 64 | 44 | 8 |

**11.** Students in a chemistry lab are measuring the volume and mass of samples of distilled water and salt water. The results are in Tables 1.39 and 1.40.

**(a)** Make scatter plots of the data, with volume on the horizontal axis.
**(b)** Find the lines of best fit for each data set and graph them together with the scatter plot.
**(c)** What are the units of the slopes of the lines? Interpret the slopes in the context of the problem.
**(d)** What do the slopes tell you about the difference between distilled water and salt water?

**Table 1.39** *Distilled water*

| Volume (ml) | 20 | 40 | 60 | 80 | 100 |
|---|---|---|---|---|---|
| Mass (grams) | 19 | 41 | 58 | 82 | 97 |

**Table 1.40** *Salt water*

| Volume (ml) | 20 | 40 | 60 | 80 | 100 |
|---|---|---|---|---|---|
| Mass (grams) | 22 | 44.5 | 63 | 89 | 109 |

**12.** The rate of oxygen consumption, in microliters per gram per hour, for Colorado potato beetles increases with temperature.[30] See Table 1.41.

**(a)** Make a scatter plot of this data.
**(b)** Draw an estimated regression line by eye.
**(c)** Use a calculator or computer to find the equation of the regression line. Round constants in the equation to the nearest integer.
**(d)** Interpret the slope and each intercept of the regression equation.
**(e)** Interpret the correlation between temperature and oxygen rate.

**Table 1.41**

| °C | 10 | 15 | 20 | 25 | 30 |
|---|---|---|---|---|---|
| Oxygen consumption rate | 90 | 125 | 200 | 300 | 375 |

**13.** Table 1.42 gives the data on hand strength collected from college freshman using a grip meter.

**(a)** Make a scatter plot of these data treating the strength of the preferred hand as the independent variable.
**(b)** Draw a line on your scatter plot that is a good fit for these data and use it to find an approximate equation for the regression line.
**(c)** Using a graphing calculator or computer, find the equation of the least squares line.
**(d)** What would the predicted grip strength in the nonpreferred hand be for a student with a preferred hand strength of 37?
**(e)** Discuss interpolation and extrapolation using specific examples in relation to this regression line.

[30] Adapted from R. E. Rickefs and G. L. Miller, *Ecology,* 4th ed. W. H. Freeman, New York, 1999.

**(f)** Discuss why $r$, the correlation coefficient, is both positive and close to 1.

**(g)** Why do the points tend to cluster into two groups on your scatter plot?

**Table 1.42** *Hand strength for 20 students in kilograms*

| Preferred | 28 | 27 | 45 | 20 | 40 | 47 | 28 | 54 | 52 | 21 |
|---|---|---|---|---|---|---|---|---|---|---|
| Nonpreferred | 24 | 26 | 43 | 22 | 40 | 45 | 26 | 46 | 46 | 22 |
| Preferred | 53 | 52 | 49 | 45 | 39 | 26 | 25 | 32 | 30 | 32 |
| Nonpreferred | 47 | 47 | 41 | 44 | 33 | 20 | 27 | 30 | 29 | 29 |

**14.** Table 1.43 shows men's and women's world records for swimming distances from 50 meters to 1500 meters.[31]

**(a)** What values would you add to Table 1.43 to represent the time taken by both men and women to swim 0 meters?

**(b)** Plot men's time against distance, with time $t$ in seconds on the vertical axis and distance $d$ in meters on the horizontal axis. It is claimed that a straight line models this behavior well. What is the equation for that line? What does its slope represent? On the same graph, plot women's time against distance and find the equation of the straight line that models this behavior well. Is this line steeper or flatter than the men's line? What does that mean in terms of swimming? What are the values of the vertical intercepts? Do these values have a practical interpretation?

**(c)** On another graph plot the women's times against the men's times, with women's times, $w$, on the vertical axis and men's times, $m$, on the horizontal axis. It should look linear. How could you have predicted this linearity from the equations you found in part (b)? What is the slope of this line and how can it be interpreted? A newspaper reporter claims that the women's records are about 8% slower than the men's. Do the facts support this statement? What is the value of the vertical intercept? Does this value have a practical interpretation?

**Table 1.43** *Men's and women's world swimming records*

| Distance (m) | 50 | 100 | 200 | 400 | 800 | 1500 |
|---|---|---|---|---|---|---|
| Men (sec) | 21.64 | 47.84 | 104.06 | 220.08 | 458.65 | 874.56 |
| Women (sec) | 24.13 | 53.62 | 116.64 | 243.85 | 496.22 | 952.10 |

## CHAPTER SUMMARY

- **Functions**
  Definition: a rule which takes certain numbers as inputs and assigns to each input exactly one output number.
  Function notation, $y = f(x)$.
  Use of vertical line test.
- **Average Rate of Change**
  Average rate of change of $Q = f(t)$ on $a \le t \le b$ is
  $$\frac{\Delta Q}{\Delta t} = \frac{f(b) - f(a)}{b - a}.$$
  Increasing, decreasing functions; identifying from average rate of change.
- **Linear Functions**
  Value of $y$ changes at constant rate.
  Tables for linear functions.
- **Formulas for Linear Functions**
  Slope-intercept form: $y = b + mx$.
  Point-slope form: $y - y_0 = m(x - x_0)$.
  Standard form: $Ax + By + C = 0$.
- **Properties of Linear Functions**
  Interpretation of slope, vertical and horizontal intercepts.
  Intersection of lines: Solution of equations.
  Horizontal and vertical lines.
  Parallel lines: $m_1 = m_2$.
  Perpendicular lines: $m_1 = -\frac{1}{m_2}$.
- **Fitting Lines to Data**
  Linear regression; correlation. Interpolation, extrapolation; dangers of extrapolation.

## REVIEW EXERCISES AND PROBLEMS FOR CHAPTER ONE

### Exercises

**1. (a)** Suppose $x$ and $y$ are the coordinates of a point on the circle $x^2 + y^2 = 1$. Is $y$ a function of $x$? Why or why not?

**(b)** Suppose $x$ and $y$ are the coordinates of a point on the part of the circle $x^2 + y^2 = 1$ that is above the $x$-axis. Is $y$ a function of $x$? Why or why not?

[31] Data from *The World Almanac and Book of Facts: 2006*, World Almanac Education Group, Inc., New York, 2006.

**2.** **(a)** Is the area, $A$, of a square a function of the length of one of its sides, $s$?

**(b)** Is the area, $A$, of a rectangle a function of the length of one of its sides, $s$?

**3.** **(a)** Which of the graphs in Figure 1.63 represent $y$ as a function of $x$? (Note that an open circle indicates a point that is not included in the graph; a solid dot indicates a point that is included in the graph.)

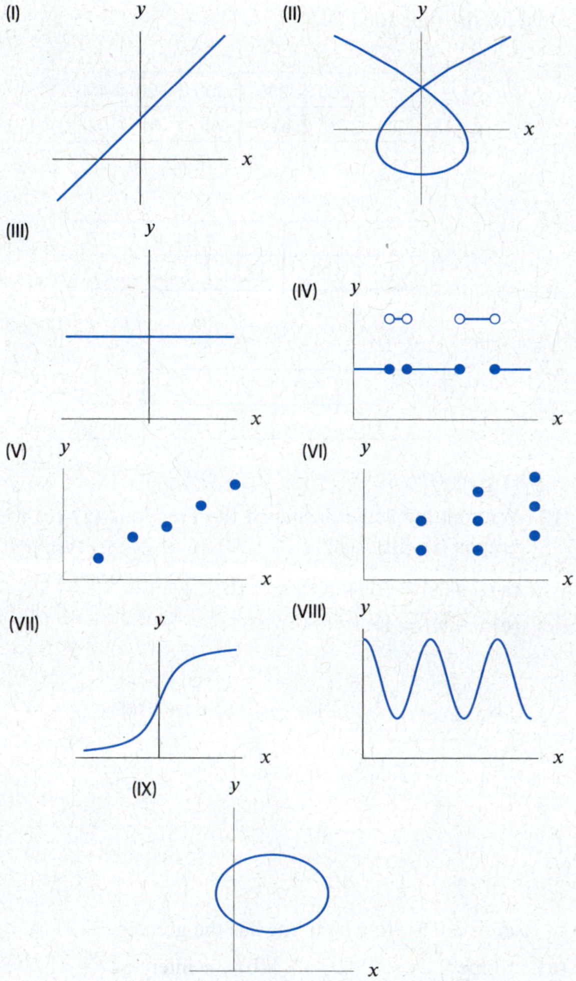


Figure 1.63

**(b)** Which of the graphs in Figure 1.63 could represent the following situations? Give reasons.

(i) SAT Math score versus SAT Verbal score for a small number of students.

(ii) Total number of daylight hours as a function of the day of the year, shown over a period of several years.

**(c)** Among graphs (I)–(IX) in Figure 1.63, find two that could give the cost of train fare as a function of the time of day. Explain the relationship between cost and time for both choices.

**4.** **(a)** Make a table of values for $f(x) = 10/(1 + x^2)$ for $x = 0, 1, 2, 3$.

**(b)** What $x$-value gives the largest $f(x)$ value in your table? How could you have predicted this before doing any calculations?

**5.** Table 1.44 shows data for two populations (in hundreds) for five different years. Find the average rate of change per year of each population over the following intervals.

**(a)** 1995 to 2005 **(b)** 2000 to 2012
**(c)** 1995 to 2012

**Table 1.44**

| Year | 1995 | 1997 | 2000 | 2005 | 2012 |
|---|---|---|---|---|---|
| $P_1$ | 53 | 64 | 75 | 86 | 97 |
| $P_2$ | 85 | 81 | 77 | 73 | 69 |

**6.** The following tables represent the relationship between the button number, $N$, that you push, and the snack, $S$, delivered by three different vending machines.[32]

**(a)** One of these vending machines is not a good one to use, because $S$ is not a function of $N$. Which one?

**(b)** For which vending machine(s) is $S$ a function of $N$?

**(c)** For which of the vending machines is $N$ not a function of $S$?

*Vending Machine #1*

| $N$ | $S$ |
|---|---|
| 1 | M&Ms |
| 2 | pretzels |
| 3 | dried fruit |
| 4 | Hershey's |
| 5 | fat-free cookies |
| 6 | Snickers |

*Vending Machine #2*

| $N$ | $S$ |
|---|---|
| 1 | M&Ms or dried fruit |
| 2 | pretzels or Hershey's |
| 3 | Snickers or fat-free cookies |

*Vending Machine #3*

| $N$ | $S$ |
|---|---|
| 1 | M&Ms |
| 2 | M&Ms |
| 3 | pretzels |
| 4 | dried fruit |
| 5 | Hershey's |
| 6 | Hershey's |
| 7 | fat-free cookies |
| 8 | Snickers |
| 9 | Snickers |

[32]For each $N$, vending machine #2 dispenses one or the other product at random.

**7.** Figure 1.64 shows the average monthly temperature in Albany, New York, over a twelve-month period. (January is month 1.)

**(a)** Make a table showing average temperature as a function of the month of the year.

**(b)** What is the warmest month in Albany?

**(c)** Over what interval of months is the temperature increasing? Decreasing?

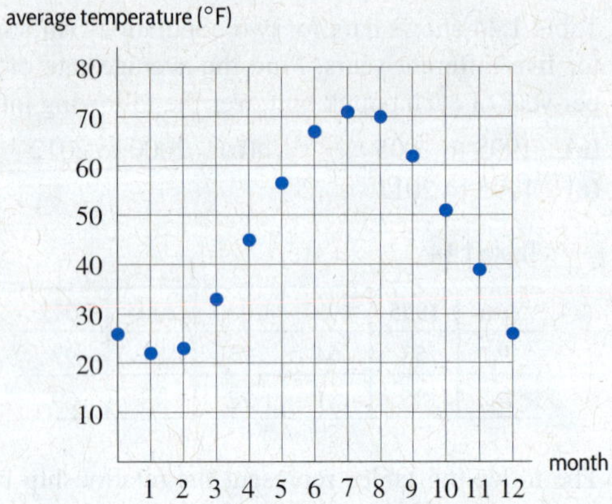


Figure 1.64

**8.** In 1947, Jesse Owens, the US gold medal track star of the 1930s and 1940s, ran a 100-yard race against a horse. The race, "staged" in Havana, Cuba, is filled with controversy; some say Owens received a head start, others claim the horse was drugged. Owens himself revealed some years later that the starting gun was placed next to the horse's ear, causing the animal to rear and remain at the gate for a few seconds. Figure 1.65 depicts speeds measured against time for the race.

**(a)** How fast were Owens and the horse going at the end of the race?

**(b)** When were the participants both traveling at the same speed?

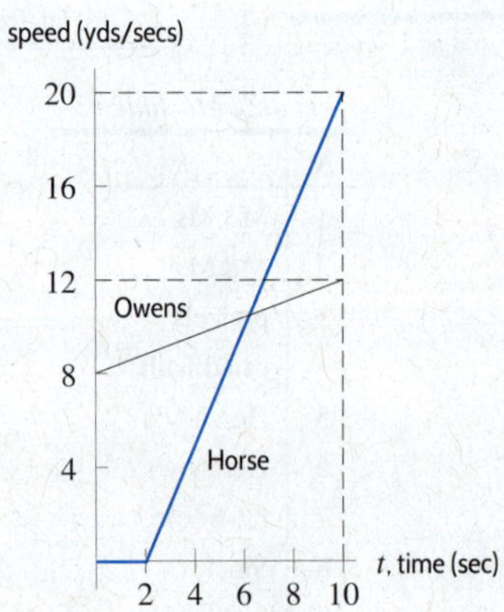


Figure 1.65

In Exercises 9–10, could the table represent a linear function?

**9.**

| $\lambda$ | 1 | 2 | 3 | 4 | 5 |
|---|---|---|---|---|---|
| $q(\lambda)$ | 2 | 4 | 8 | 16 | 32 |

**10.**

| $t$ | 3 | 6 | 9 | 12 | 15 |
|---|---|---|---|---|---|
| $a(t)$ | 2 | 4 | 6 | 8 | 10 |

Exercises 11–13 give data from a linear function. Find a formula for the function.

**11.**

| $x$ | 200 | 230 | 300 | 320 | 400 |
|---|---|---|---|---|---|
| $g(x)$ | 70 | 68.5 | 65 | 64 | 60 |

**12.**

| $t$, | 1.2 | 1.3 | 1.4 | 1.5 |
|---|---|---|---|---|
| $f(t)$ | 0.736 | 0.614 | 0.492 | 0.37 |

**13.**

| $t$ | 5.2 | 5.3 | 5.4 | 5.5 |
|---|---|---|---|---|
| $f(t)$ | 73.6 | 61.4 | 49.2 | 37 |

**14.** Without a calculator, match the functions (a)–(c) to the graphs (i)–(iii).

**(a)** $f(x) = 3x + 1$ **(b)** $g(x) = -2x + 1$

**(c)** $h(x) = 1$

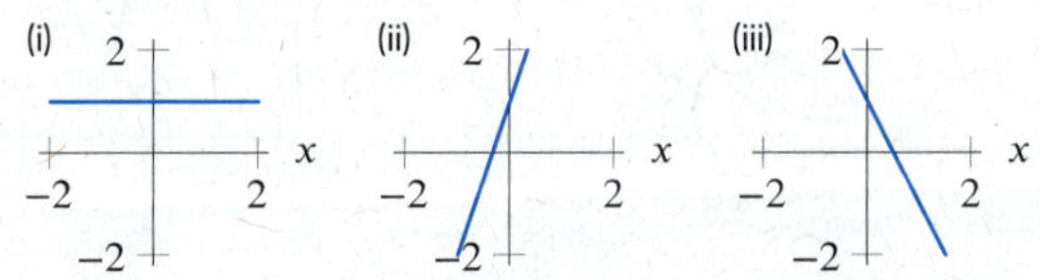


In Exercises 15–16, which line has the greater

**(a)** Slope? **(b)** $y$-intercept?

**15.** $y = 7 + 2x, \quad y = 8 - 15x$

**16.** $y = 5 - 2x, \quad y = 7 - 3x$

Are the lines in Exercises 17–20 perpendicular? Parallel? Neither?

**17.** $y = 5x + 2; \quad y = 2x + 5$

**18.** $y = 14x - 2; \quad y = -\frac{1}{14}x + 2$

**19.** $y = 3x + 3; \quad y = -\frac{1}{3}x + 3$

**20.** $7y = 8 + 21x; 9y = 77 - 3x$

## Problems

In Problems 21–23, use Table 1.45, which gives values of $v = r(s)$, the eyewall wind profile of a typical hurricane.[33] The eyewall of a hurricane is the band of clouds that surrounds the eye of the storm. The eyewall wind speed $v$ (in mph) is a function of the height above the ground $s$ (in meters).

**Table 1.45**

| $s$ | 0 | 100 | 200 | 300 | 400 | 500 |
|---|---|---|---|---|---|---|
| $v$ | 90 | 110 | 116 | 120 | 121 | 122 |
| $s$ | 600 | 700 | 800 | 900 | 1000 | 1100 |
| $v$ | 121 | 119 | 118 | 117 | 116 | 115 |

**21.** Evaluate and interpret $r(300)$.

**22.** At what altitudes does the eyewall wind speed appear to equal or exceed 116 mph?

**23.** At what height is the eyewall wind speed greatest?

**24.** You are looking at the graph of $y$, a function of $x$.

**(a)** What is the maximum number of times that the graph can intersect the $y$-axis? Explain.

**(b)** Can the graph intersect the $x$-axis an infinite number of times? Explain.

**25.** A bug starts out ten feet from a light, flies closer to the light, then farther away, then closer than before, then farther away. Finally the bug hits the bulb and flies off. Sketch the distance of the bug from the light as a function of time.

**26.** Although there were 20 women in the Senate in 2013, the first woman ever elected to the Senate was Hattie Wyatt Caraway of Arkansas. She was appointed to fill the vacancy caused by the death of her husband, then won election in 1932, was reelected in 1938, and served until 1945. Table 1.46 shows the number of female senators at the beginning of the first session of each Congress.[34]

**(a)** Is the number of female senators a function of the Congress's number, $c$? Explain.

**(b)** Is the Congress's number a function of the number of female senators? Explain.

**(c)** Let $S(c)$ represent the number of female senators serving in the $c^{\text{th}}$ Congress. What does the statement $S(110) = 16$ mean?

**(d)** Evaluate and interpret $S(112)$.

**Table 1.46** *Female senators, S, in Congress c*

| $c$ | 98 | 100 | 102 | 104 | 106 | 108 | 110 | 112 | 113 |
|---|---|---|---|---|---|---|---|---|---|
| $S$ | 2 | 2 | 2 | 8 | 9 | 14 | 16 | 17 | 20 |

**27.** A light is turned off for several hours. It is then turned on. After a few hours it is turned off again. Sketch the light bulb's temperature as a function of time.

**28.** According to Charles Osgood, CBS news commentator, it takes about one minute to read 15 double-spaced typewritten lines on the air.[35]

**(a)** Construct a table showing the time Charles Osgood is reading on the air in seconds as a function of the number of double-spaced lines read for $0, 1, 2, \ldots, 10$ lines. From your table, how long does it take Charles Osgood to read 9 lines?

**(b)** Plot this data on a graph with the number of lines on the horizontal axis.

**(c)** From your graph, estimate how long it takes Charles Osgood to read 9 lines. Estimate how many lines Charles Osgood can read in 30 seconds.

**(d)** Construct a formula which relates the time $T$ to $n$, the number of lines read.

**29.** The distance between Cambridge and Wellesley is 10 miles. A person walks part of the way at 5 miles per hour, then jogs the rest of the way at 8 mph. Find a formula that expresses the total amount of time for the trip, $T(d)$, as a function of $d$, the distance walked.

**30.** A cylindrical can is closed at both ends and its height is twice its radius. Express its surface area, $S$, as a function of its radius, $r$. [Hint: The surface of a can consists of a rectangle plus two circular disks.]

**31.** A lawyer does nothing but sleep and work during a day. There are 1440 minutes in a day. Write a linear function relating minutes of sleep, $s$, to minutes of work, $w$.

For the functions in Problems 32–34:

**(a)** Find the average rate of change between the points

(i) $(-1, f(-1))$ and $(3, f(3))$

(ii) $(a, f(a))$ and $(b, f(b))$

(iii) $(x, f(x))$ and $(x + h, f(x + h))$

**(b)** What pattern do you see in the average rate of change between the three pairs of points?

**32.** $f(x) = 5x - 4$

**33.** $f(x) = \frac{1}{2}x + \frac{5}{2}$

**34.** $f(x) = x^2 + 1$

[33]Data from the National Hurricane Center, www.nhc.noaa.gov/aboutwindprofile.shtml, accessed October 7, 2004.
[34]http://www.cawp.rutgers.edu/fast_facts/levels_of_office/Congress-Current.php, accessed February 15, 2014
[35]T. Parker, *Rules of Thumb* (Boston: Houghton Mifflin, 1983).

**35.** The population, $P(t)$, in millions, of a country in year $t$, is given by the formula $P(t) = 22 + 0.3t$.

**(a)** Construct a table of values for $t = 0, 10, 20, \ldots, 50$.
**(b)** Plot the points you found in part (a).
**(c)** What is the country's initial population?
**(d)** What is the average rate of change of the population, in millions of people/year?

**36.** A woodworker sells rocking horses. His start-up costs, including tools, plans, and advertising, total \$5000. Labor and materials for each horse cost \$350.

**(a)** Calculate the woodworker's total cost, $C$, to make 1, 2, 5, 10, and 20 rocking horses. Graph $C$ against $n$, the number of rocking horses that he carves.
**(b)** Find a formula for $C$ in terms of $n$.
**(c)** What is the rate of change of the function $C$? What does the rate of change tell us about the woodworker's expenses?

**37.** Outside the US, temperature readings are usually given in degrees Celsius; inside the US, they are often given in degrees Fahrenheit. The exact conversion from Celsius, $C$, to Fahrenheit, $F$, uses the formula

$$F = \frac{9}{5}C + 32.$$

An approximate conversion is obtained by doubling the temperature in Celsius and adding 30° to get the equivalent Fahrenheit temperature.

**(a)** Write a formula using $C$ and $F$ to express the approximate conversion.
**(b)** How far off is the approximation if the Celsius temperature is −5°, 0°, 15°, 30°?
**(c)** For what temperature (in Celsius) does the approximation agree with the actual formula?

**38.** Find a formula for the line intersecting the graph of $f(x)$ at $x = 1$ and $x = 3$, where

$$f(x) = \frac{10}{x^2 + 1}.$$

**39.** Find the equation of the line $l$ in Figure 1.66. The shapes under the line are squares.

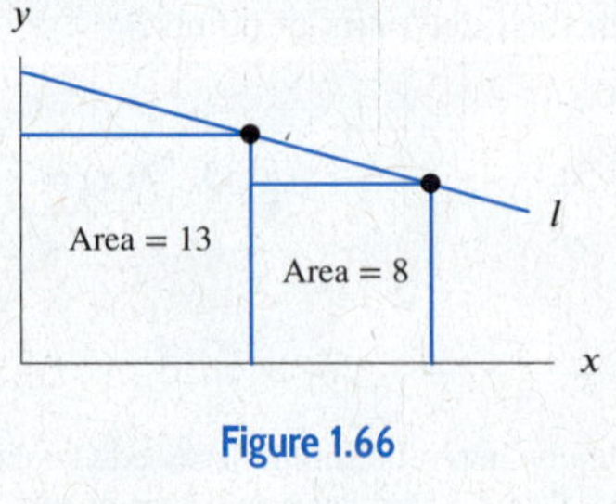


Figure 1.66

**40.** A bullet is shot straight up into the air from ground level. After $t$ seconds, the velocity of the bullet, in meters per second, is approximated by the formula

$$v = f(t) = 1000 - 9.8t.$$

**(a)** Evaluate the following: $f(0), f(1), f(2), f(3), f(4)$. Compile your results in a table.
**(b)** Describe in words what is happening to the speed of the bullet. Discuss why you think this is happening.
**(c)** Evaluate and interpret the slope and both intercepts of $f(t)$.
**(d)** The gravitational field near the surface of Jupiter is stronger than that near the surface of the earth, which, in turn, is stronger than the field near the surface of the moon. How is the formula for $f(t)$ different for a bullet shot from Jupiter's surface? From the moon?

**41.** Describe a linear (or nearly linear) relationship that you have encountered outside the classroom. Determine the rate of change and interpret it in practical terms.

**42.** The figure gives graphs of $g$, a linear function, and of $f(x) = 12 - 0.5(x + 4)^2$. Find a possible formula for $g$.

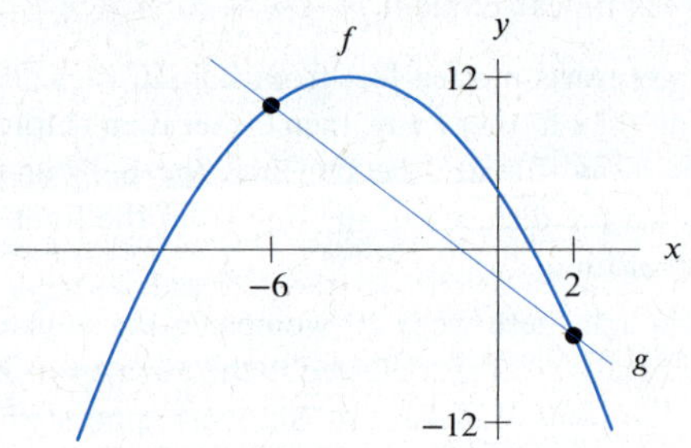


**43.** Write in slope-intercept form and identify the values of $b$ and $m$:

$$f(r) = rx^3 + 3rx^2 + 2r + 4sx + 7s + 3.$$

**44.** Find an equation for the line intersecting the graph of $f$ at $x = -2$ and $x = 5$ given that $f(x) = 2 + \dfrac{3}{x + 5}$.

**45.** A business consultant works 10 hours a day, 6 days a week. She divides her time between meetings with clients and meetings with co-workers. A client meeting requires 3 hours while a co-worker meeting requires 2 hours. Let $x$ be the number of co-worker meetings the consultant holds during a given week. If $y$ is the number of client meetings for which she has time remaining, then $y$ is a function of $x$. Assume this relationship is linear and that meetings can be split up and continued on different days.

**(a)** Graph the relationship between $y$ and $x$. [Hint: Consider the maximum number of client and co-worker meetings that can be held.]

**(b)** Find a formula for $y$ as a function of $x$.

**(c)** Explain what the slope and the $x$- and $y$-intercepts represent in the context of the consultant's meeting schedule.

**(d)** A change is made so that co-worker meetings take 90 minutes instead of 2 hours. Graph this situation. Describe those features of this graph that have changed from the one sketched in part (a) and those that have remained the same.

**46.** You start 60 miles east of Pittsburgh and drive east at a constant speed of 50 miles per hour. (Assume that the road is straight and permits you to do this.) Find a formula for $d$, your distance from Pittsburgh, as a function of $t$, the number of hours of travel.

**47.** Find a formula for the line parallel to the line $y = 20 - 4x$ and containing the point (3, 12).

**48.** Find the equation of the linear function $g$ whose graph is perpendicular to the line $5x - 3y = 6$; the two lines intersect at $x = 15$.

**49.** Find the coordinates of point $P$ in Figure 1.67.

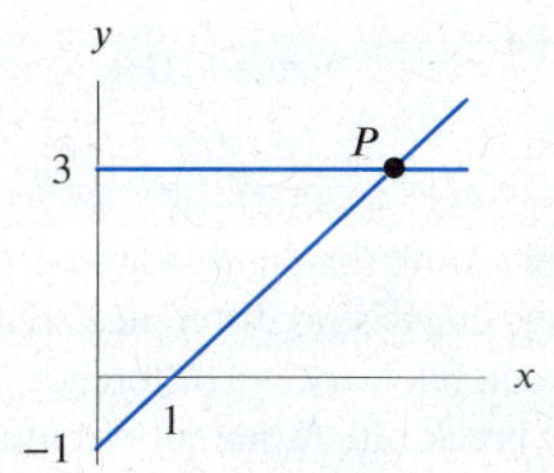


Figure 1.67

**50.** A cell phone company offers three different "pay as you go" daily cell phone plans.

- Plan A charges \$0.25 per minute.
- Plan B charges \$.99 per day plus \$0.10 per minute.
- Plan C charges a fixed rate of \$1.99 per day.

Let $Y_A$, $Y_B$, $Y_C$ represent the daily charges using plans A, B, and C, respectively. Let $x$ be the number of minutes per day spent on the phone.

**(a)** Find formulas for $Y_A$, $Y_B$, $Y_C$ as functions of $x$.

**(b)** Figure 1.68 shows the functions in part (a). Which function corresponds to which graph?[36]

**(c)** Find the $x$-values for which plan B is cheapest.

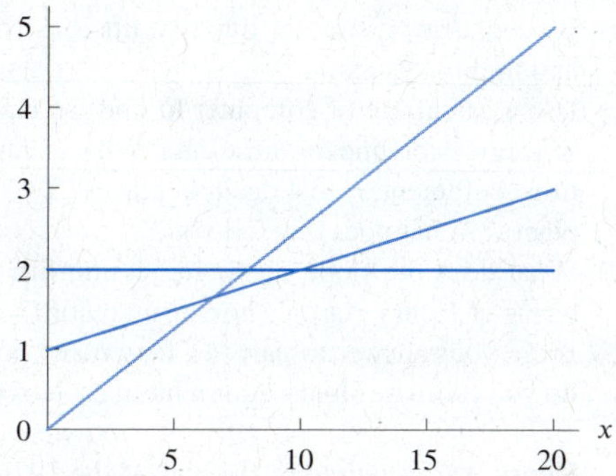


Figure 1.68

**51.** A commission is a payment made to an employee based on a percentage of sales made. For example, car salespeople earn commission on the selling price of a car. In parts (a)–(d), explain how to choose between the options for different levels of sales.

**(a)** A weekly salary of \$100 or a weekly salary of \$50 plus 10% commission.

**(b)** A weekly salary of \$175 plus 7% commission or a weekly salary of \$175 plus 8% commission.

**(c)** A weekly salary of \$145 plus 7% commission or a weekly salary of \$165 plus 7% commission.

**(d)** A weekly salary of \$225 plus 3% commission or a weekly salary of \$180 plus 6% commission.

**52.** Table 1.47 shows the IQ of ten students and the number of hours of TV each watches per week.

**(a)** Make a scatter plot of the data.

**(b)** By eye, make a rough estimate of the correlation coefficient.

**(c)** Use a calculator or computer to find the least squares regression line and the correlation coefficient. Your values should be correct to four decimal places.

**Table 1.47**

| IQ | 110 | 105 | 120 | 140 | 100 | 125 | 130 | 105 | 115 | 110 |
|---|---|---|---|---|---|---|---|---|---|---|
| TV | 10 | 12 | 8 | 2 | 12 | 10 | 5 | 6 | 13 | 3 |

**53.** For 35 years, major league baseball Hall of Fame member Henry "Hank" Aaron held the record for the greatest number of career home runs. His record was broken by Barry Bonds in 2007. Table 1.48 shows Aaron's cumulative yearly record[37] from the start of his career, 1954, until 1973.

**(a)** Plot Aaron's cumulative number of home runs $H$ on the vertical axis, and the time $t$ in years along the horizontal axis, where $t = 1$ corresponds to 1954.

[36] Actual charges are typically rounded up to the last minute.

[37] Adapted from "Graphing Henry Aaron's home-run output" by H. Ringel, *The Physics Teacher*, January 1974, page 43.

**(b)** By eye, draw a straight line that fits these data well and find its equation.

**(c)** Use a calculator or computer to find the equation of the regression line for these data. What is the correlation coefficient, $r$, to 4 decimal places? To 3 decimal places? What does this tell you?

**(d)** What does the slope of the regression line mean in terms of Henry Aaron's home-run record?

**(e)** From your answer to part (d), how many home runs do you estimate Henry Aaron hit in each of the years 1974, 1975, 1976, and 1977? If you were told that Henry Aaron retired at the end of the 1976 season, would this affect your answers?

**Table 1.48** *Henry Aaron's cumulative home-run record, H, from* 1954 *to* 1973, *with t in years since 1953*

| $t$ | 1 | 2 | 3 | 4 | 5 | 6 | 7 | 8 | 9 | 10 |
|---|---|---|---|---|---|---|---|---|---|---|
| $H$ | 13 | 40 | 66 | 110 | 140 | 179 | 219 | 253 | 298 | 342 |
| $t$ | 11 | 12 | 13 | 14 | 15 | 16 | 17 | 18 | 19 | 20 |
| $H$ | 366 | 398 | 442 | 481 | 510 | 554 | 592 | 639 | 673 | 713 |

**54.** The graph of a linear function $y = f(x)$ passes through the two points $(a, f(a))$ and $(b, f(b))$, where $a < b$ and $f(a) < f(b)$.

**(a)** Graph the function labeling the two points.

**(b)** Find the slope of the line in terms of $f$, $a$, and $b$.

**55.** Let $f(x) = 0.003 - (1.246x + 0.37)$.

**(a)** Calculate the following average rates of change:

(i) $\dfrac{f(2) - f(1)}{2 - 1}$ (ii) $\dfrac{f(1) - f(2)}{1 - 2}$

(iii) $\dfrac{f(3) - f(4)}{3 - 4}$

**(b)** Rewrite $f(x)$ in the form $f(x) = b + mx$.

Write the linear function $y = -3 - x/2$ in the forms given in Problems 56–57, assuming all constants are positive.

**56.** $y = \dfrac{p}{p-1} - r^2x$ **57.** $y = \dfrac{x+k}{z}$

**58.** You spend $c$ dollars on $x$ apples and $y$ bananas. In Figure 1.69, line $l$ gives $y$ as a function of $x$.

**(a)** If apples cost $p$ dollars each and bananas cost $q$ dollars each, label the $x$- and $y$-intercepts of $l$. [Note: Your labels will involve the constants $p, q$ or $c$.]

**(b)** What is the slope of $l$?

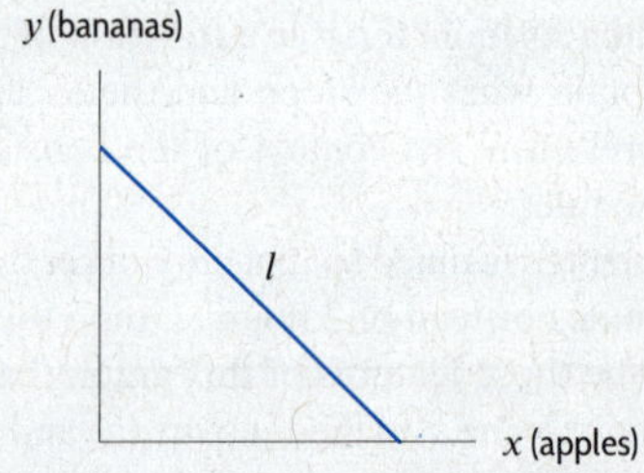


**Figure 1.69**: Axes not necessarily to scale

**59.** The apples in Problem 58 cost more than bananas, so $p > q$. Which of the two lines, $l_1$ or $l_2$, in Figure 1.70 could represent $y = f(x)$?

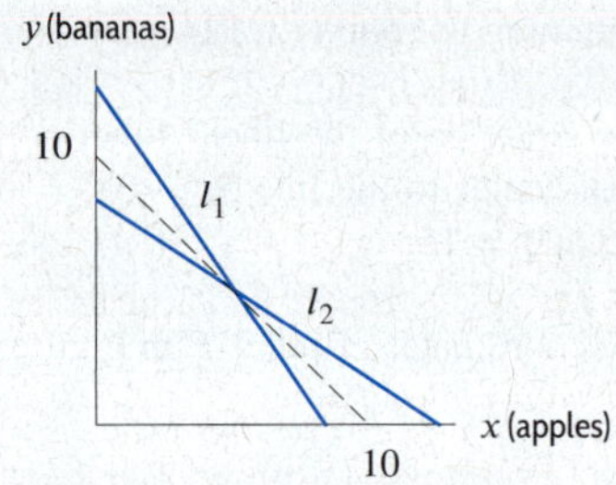


**Figure 1.70**

**60.** Many people think that hair growth is stimulated by haircuts. In fact, there is no difference in the rate hair grows after a haircut, but there *is* a difference in the rate at which hair's ends break off. A haircut eliminates dead and split ends, thereby slowing the rate at which hair breaks. However, even with regular haircuts, hair will not grow to an indefinite length. The average life cycle of human scalp hair is 3-5 years, after which the hair is shed.[38]

Judy trims her hair once a year, when its growth is slowed by split ends. She cuts off just enough to eliminate dead and split ends, and then lets it grow another year. After 5 years, she realizes her hair won't grow any longer. Graph the length of her hair as a function of time. Indicate when she receives her haircuts.

**61.** Academics have suggested that loss of worker productivity can result from sleep deprivation. An article in the September 26, 1993, *New York Times* quotes David Poltrack, the senior vice president for planning and research at CBS, as saying that seven million Americans are staying up an hour later than usual to watch talk show host David Letterman. The article goes on to quote Timothy Monk, a professor at the University of Pittsburgh School of Medicine, as saying, "... my hunch is that the effect [on productivity due to sleep deprivation among

[38] *Britannica Micropaedia,* vol. 5 (Chicago: Encyclopaedia Britannica, Inc., 1989).

this group] would be in the area of a 10 percent decrement." The article next quotes Robert Solow, a Nobel prize-winning professor of economics at MIT, who suggests the following procedure to estimate the impact that this loss in productivity will have on the US economy—an impact he dubbed "the Letterman loss." First, Solow says, we find the percentage of the work force who watch the program. Next, we determine this group's contribution to the gross domestic product (GDP). Then we reduce the group's contribution by 10% to account for the loss in productivity due to sleep deprivation. The amount of this reduction is "the Letterman loss."

**(a)** The article estimated that the GDP is \$6.325 trillion, and that 7 million Americans watch the show. Assume that the nation's work force is 118 million people and that 75% of David Letterman's audience belongs to this group. What percentage of the work force is in Dave's audience?

**(b)** What percent of the GDP would be expected to come from David Letterman's audience? How much money would they have contributed if they had not watched the show?

**(c)** How big is "the Letterman loss"?

**62.** Judging from the graph of $y = f(x)$ in the figure, find a possible formula for the line intersecting it at $x = -2\pi$ and $x = 3\pi$.

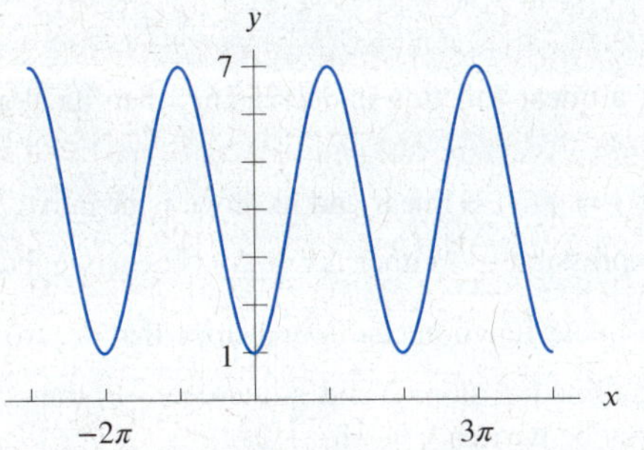


Problems 63–65 refer to Table 1.49, which describes a *boustrophedonic pairing function*.[39] We can use this table to list all the positive roots of the positive integers, such as $\sqrt{2}, \sqrt{3}, \sqrt[3]{2}, \sqrt[3]{3}, \sqrt[4]{2}, \sqrt[4]{3}, \ldots$, without omitting any. If $n$ is an entry in the table, we let $c$ stand for the column number and $r$ the row number of the entry and define the function $g$ by

$$g(n) = c^{1/r}.$$

For instance, if $n = 6$, then $c = 2$ and $r = 3$ so we have $g(6) = 2^{1/3}$, or $\sqrt[3]{2}$.

**Table 1.49**

| | 1 | 2 | 3 | 4 | 5 | 6 | ⋯ |
|---|---|---|---|---|---|---|---|
| 1 | 1 | 2 | 9 | 10 | 25 | 26 | ⋯ |
| 2 | 4 | 3 | 8 | 11 | 24 | 27 | ⋯ |
| 3 | 5 | 6 | 7 | 12 | 23 | 28 | ⋯ |
| 4 | 16 | 15 | 14 | 13 | 22 | 29 | ⋯ |
| 5 | 17 | 18 | 19 | 20 | 21 | 30 | ⋯ |
| 6 | 36 | 35 | 34 | 33 | 32 | 31 | ⋯ |
| ⋮ | ⋮ | ⋮ | ⋮ | ⋮ | ⋮ | ⋮ | ⋱ |

**63.** Evaluate $g(22)$.

**64.** Evaluate $g(54)$.

**65.** Find a solution to $g(n) = \sqrt{3}$. Is this solution unique? Explain your reasoning.

Problems 66–67 are about the A-series of paper. Many countries use A4 paper, which is different from the 8.5 by 11 inch paper used in the US.[40] Two sheets of A4 paper, if laid side by side (not end to end), are the same size as one sheet of A3 paper. Likewise, two sheets of A3 are the same as one sheet of A2, and so on. See Figure 1.71. Each sheet in the series has the same proportions, length to width, and the largest sheet, A0, has an area of exactly 1 $m^2$.

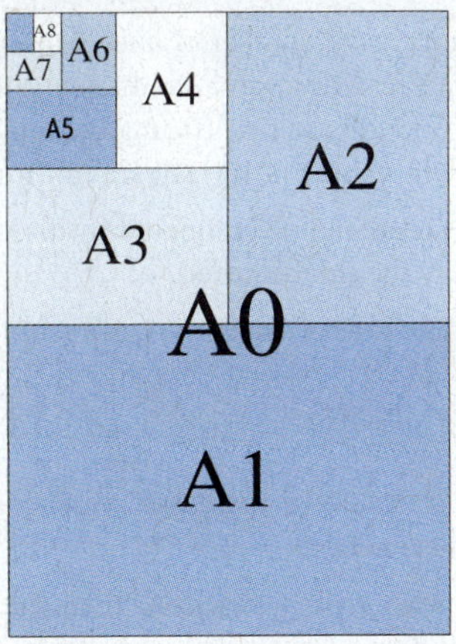


**Figure 1.71**

[39]The term *boustrophedonic* means "as the ox turns," as oxen do when plowing: Notice how the entries in the table, 1, 2, 3, 4 …, turn back and forth, like an ox plowing a field.

[40]http://en.wikipedia.org/wiki/ISO_216, accessed January 30, 2008. Note that the actual sizes of A-series paper are rounded to the nearest millimeter.

**66.** Let $f(n)$ be the area in $cm^2$ of the size $n$ sheet in the A-series, so that $f(1)$ is the area of A1 paper, $f(2)$ of A2 paper, and so on. Since 1 m equals 100 cm, we know that $f(0) = (100\text{ cm})^2 = 10{,}000\text{ cm}^2$. Complete the table of values of $f$.

**Table 1.50**

| $n$ | 0 | 1 | 2 | 3 | 4 | 5 |
|---|---|---|---|---|---|---|
| $f(n)$ | 10,000 | | | | | |

**67.** We know that A1 paper has the same proportions (length over width) as A0. Further, the length of A1 paper is the same as the width $W$ of A0, and the width of A1 is half the length $L$ of A0. (See Figure 1.71.)

**(a)** Find a formula for $L$ in terms of $W$. What does your formula tell you about the shape of A-series paper? How does the shape compare to US letter paper?

**(b)** Given that A4 paper has area 625 $cm^2$, find its width and height.

## STRENGTHEN YOUR UNDERSTANDING

Are the statements in Problems 1–54 true or false? Give an explanation for your answer.

**1.** $Q = f(t)$ means $Q$ is equal to $f$ times $t$.

**2.** A function must be defined by a formula.

**3.** If $P = f(x)$ then $P$ is called the dependent variable.

**4.** Independent variables are always denoted by the letter $x$ or $t$.

**5.** It is possible for two quantities to be related and yet neither be a function of the other.

**6.** A function is a rule that takes certain values as inputs and assigns to each input value exactly one output value.

**7.** It is possible for a table of values to represent a function.

**8.** If $Q$ is a function of $P$, then $P$ is a function of $Q$.

**9.** The graph of a circle is not the graph of a function.

**10.** If $n = f(A)$ is the number of angels that can dance on the head of a pin whose area is $A$ square millimeters, then $f(10) = 100$ tells us that 10 angels can dance on the head of a pin whose area is 100 square millimeters.

**11.** Average speed can be computed by dividing the distance traveled by the time elapsed.

**12.** The average rate of change of a function $Q$ with respect to $t$ over an interval can be symbolically represented as $\frac{\Delta t}{\Delta Q}$.

**13.** If $y = f(x)$ and as $x$ increases, $y$ increases, then $f$ is an increasing function.

**14.** If $f$ is a decreasing function, then the average rate of change of $f$ on any interval is negative.

**15.** The average rate of change of a function over an interval is the slope of a line connecting two points of the graph of the function.

**16.** The average rate of change of $y = 3x - 4$ between $x = 2$ and $x = 6$ is 7.

**17.** The average rate of change of $f(x) = 10 - x^2$ between $x = 1$ and $x = 2$ is the ratio $\frac{10 - 2^2 - 10 - 1^2}{2 - 1}$.

**18.** If $y = x^2$ then the slope of the line connecting the point $(2, 4)$ to the point $(3, 9)$ is the same as the slope of the line connecting the point $(-2, 4)$ to the point $(-3, 9)$.

**19.** A linear function can have different rates of change over different intervals.

**20.** The graph of a linear function is a straight line.

**21.** If a line has the equation $3x + 2y = 7$, then the slope of the line is 3.

**22.** A table of values represents a linear function if

$$\frac{\text{Change in output}}{\text{Change in input}} = \text{constant}.$$

**23.** If a linear function is decreasing, then its slope is negative.

**24.** If $y = f(x)$ is linear and its slope is negative, then in the expression $\frac{\Delta y}{\Delta x}$ either $\Delta x$ or $\Delta y$ is negative, but not both.

**25.** A linear function can have a slope that is zero.

**26.** If a line has slope 2 and $y$-intercept $-3$, then its equation may be written $y = -3x + 2$.

**27.** The line $3x + 5y = 7$ has slope $3/5$.

**28.** A line that goes through the point $(-2, 3)$ and whose slope is 4 has the equation $y = 4x + 5$.

**29.** The line $4x + 3y = 52$ intersects the $x$-axis at $x = 13$.

**30.** If $f(x) = -2x + 7$ then $f(2) = 3$.

**31.** The line that passes through the points $(1, 2)$ and $(4, -10)$ has slope 4.

**32.** The linear equation $y - 5 = 4(x + 1)$ is equivalent to the equation $y = 4x + 6$.

**33.** The line $y - 4 = -2(x + 3)$ goes through the point $(4, -3)$.

**34.** The line whose equation is $y = 3 - 7x$ has slope $-7$.

**35.** The line $y = -5x + 8$ intersects the $y$-axis at $y = 8$.

**36.** The equation $y = -2 - \frac{2}{3}x$ represents a linear function.

37. The lines $y = 8 - 3x$ and $-2x + 16y = 8$ both cross the $y$-axis at $y = 8$.
38. The graph of $f(x) = 6$ is a line whose slope is six.
39. The lines $y = -\frac{4}{5}x + 7$ and $4x - 5y = 8$ are parallel.
40. The lines $y = 7 + 9x$ and $y - 4 = -\frac{1}{9}(x + 5)$ are perpendicular.
41. The lines $y = -2x + 5$ and $y = 6x - 3$ intersect at the point $(1, 3)$.
42. If two lines never intersect then their slopes are equal.
43. The equation of a line parallel to the $y$-axis could be $y = -\frac{3}{4}$.
44. A line parallel to the $x$-axis has slope zero.
45. The slope of a vertical line is undefined.
46. Fitting the best line to a set of data is called linear regression.
47. The process of estimating a value within the range for which we have data is called interpolation.ă
48. Extrapolation tends to be more reliable than interpolation.
49. If two quantities have a high correlation then one quantity causes the other.
50. If the correlation coefficient is zero, there is not a relationship between the two quantities.
51. A correlation coefficient can have a value of $-\frac{3}{7}$.
52. A value of a correlation coefficient is always between negative and positive 1.
53. A correlation coefficient of one indicates that all the data points lie on a straight line.
54. A regression line is also referred to as a least squares line.

# SKILLS REFRESHER FOR CHAPTER ONE: LINEAR EQUATIONS AND THE COORDINATE PLANE

## Solving Linear Equations

To solve a linear equation, we isolate the variable.

**Example 1** Solve $22 + 1.3t = 31.1$ for $t$.

**Solution** We subtract 22 from both sides. Since $31.1 - 22 = 9.1$, we have

$$1.3t = 9.1.$$

We divide both sides by 1.3, so

$$t = \frac{9.1}{1.3} = 7.$$

**Example 2** Solve $3 - [5.4 + 2(4.3 - x)] = 2 - (0.3x - 0.8)$ for $x$.

**Solution** We begin by clearing the innermost parentheses on each side. Using the distributive law, this gives

$$3 - [5.4 + 8.6 - 2x] = 2 - 0.3x + 0.8.$$

Then

$$\begin{aligned} 3 - 14 + 2x &= 2 - 0.3x + 0.8 \\ 2.3x &= 13.8, \\ x &= 6. \end{aligned}$$

**Example 3** Solve $ax = c + bx$ for $x$. Assume $a \neq b$.

**Solution** To solve for $x$, we first get all the terms involving $x$ on the left side by subtracting $bx$ from both sides

$$ax - bx = c.$$

Factoring on the left, $ax - bx = (a - b)x$, enables us to solve for $x$ by dividing both sides by $(a - b)$:

$$\begin{aligned} x(a - b) &= c \\ x &= \frac{c}{(a - b)}. \end{aligned}$$

Since $a \neq b$, division by $(a - b)$ is possible.

**Example 4** Solve for $q$ if $p^2q + r(-q - 1) = 4(p + r)$.

**Solution** We first collect all the terms containing $q$ on the left side of the equation.

$$p^2q - rq - r = 4p + 4r$$
$$p^2q - rq = 4p + 5r.$$

To solve for $q$, we factor and then divide by the coefficient of $q$.

$$q(p^2 - r) = 4p + 5r$$
$$q = \frac{4p + 5r}{p^2 - r}.$$

## Solving Exactly Versus Solving Approximately

Some equations can be solved exactly, often by using algebra. For example, the equation $7x - 1 = 0$ has the exact solution $x = 1/7$. Other equations can be hard or even impossible to solve exactly. However, it is often possible, and sometimes easier, to find an approximate solution to an equation by using a graph or a numerical method on a calculator. The equation $7x - 1 = 0$ has the approximate solution $x \approx 0.14$ (since $1/7 = 0.142857\ldots$). We use the sign $\approx$, meaning approximately equal, when we want to emphasize that we are making an approximation.

## Solving Linear Inequalities

We can solve a linear inequality like we solve a linear equation, performing operations on both sides of the inequality to isolate the variable. Recall that, when we multiply or divide both sides of an inequality by a positive number, the direction of the inequality does not change. However, multiplying or dividing both sides by a negative number switches the direction of the inequality.

**Example 5** Solve the inequalities

(a) $10 + 0.2d \leq 30$

(b) $10 + 0.3y < 0.1y$.

**Solution** (a) We first subtract 10 from each side of the inequality to get

$$0.2d \leq 20.$$

Dividing both sides of the inequality by 0.2 does not switch the direction, so we have

$$d \leq \frac{20}{0.2}.$$

Thus the solution is $d \leq 100$, all values of $d$ smaller than or equal to 100.

(b) We first subtract $0.3y$ from each side of the inequality to get

$$10 < -0.2y$$

Dividing both sides of the inequality by $-0.2$ switches the direction of the inequality, so we get

$$\frac{10}{-0.2} > y$$

Thus the solution is $-50 > y$, all values of $y$ smaller than $-50$.

**Example 6** Solve the inequality $ax + k \geq bx$ for $x$ if $a < b$.

**Solution** We begin by subtracting $ax$ from each side to obtain $k \geq bx - ax$. Since $x$ is common factor of the terms on the right side, we can rewrite this expression as a product,

$$k \geq (b - a)x.$$

We now divide each side of the inequality by $b - a$. Since we are given that $a < b$, we know that $b - a > 0$, so dividing both sides by $b - a$ does not switch the direction of the inequality. Thus we have

$$\frac{k}{b - a} \geq x.$$

Therefore, the solution of the inequality is $x \leq k/(b - a)$.

## Systems of Linear Equations

To solve for two unknowns, we must have two equations—that is, two relationships between the unknowns. Similarly, three unknowns require three equations, and $n$ unknowns ($n$ an integer) require $n$ equations. The group of equations is known as a *system* of equations. To solve the system, we find the *simultaneous* solutions to all equations in the system.

We can solve these equations either by *substitution* (see Example 7) or by *elimination* (see Example 8).

**Example 7** Solve for $x$ and $y$ in the following system of equations using substitution.

$$\begin{cases} y + \dfrac{x}{2} = 3 \\ 2(x + y) = 1 - y \end{cases}$$

**Solution** Solving the first equation for $y$, we write $y = 3 - x/2$. Substituting for $y$ in the second equation gives

$$2\left(x + \left(3 - \frac{x}{2}\right)\right) = 1 - \left(3 - \frac{x}{2}\right).$$

Then

$$\begin{aligned} 2x + 6 - x &= -2 + \frac{x}{2} \\ x + 6 &= -2 + \frac{x}{2} \\ 2x + 12 &= -4 + x \\ x &= -16. \end{aligned}$$

Using $x = -16$ in the first equation to find the corresponding $y$, we have

$$\begin{aligned} y - \frac{16}{2} &= 3 \\ y &= 3 + 8 = 11. \end{aligned}$$

Thus, the solution that simultaneously solves both equations is $x = -16$, $y = 11$.

**Example 8** Solve for $x$ and $y$ in the following system of equations using elimination.

$$\begin{cases} 8x - 5y = 11 \\ -2x + 10y = -1. \end{cases}$$

**Solution** To eliminate $y$, we observe that if we multiply the first equation by 2, the coefficients of $y$ are $-10$ and 10:

$$\begin{cases} 16x - 10y = 22 \\ -2x + 10y = -1. \end{cases}$$

Adding these two equations gives

$$\begin{aligned} 14x &= 21 \\ x &= 3/2. \end{aligned}$$

We can substitute this value for $x$ in either of the original equations to find $y$. For example,

$$\begin{aligned} 8\left(\frac{3}{2}\right) - 5y &= 11 \\ 12 - 5y &= 11 \\ -5y &= -1 \\ y &= 1/5 \end{aligned}$$

Thus, the solution is $x = 3/2$, $y = 1/5$.

## Intersection of Two Lines

The coordinates of the point of intersection of two lines satisfy the equations of both lines. Thus, the point can be found by solving the equations simultaneously.

**Example 9** Find the point of intersection of the lines $y = 3 - \frac{2}{3}x$ and $y = -4 + \frac{3}{2}x$.

**Solution** Since the $y$-values of the two lines are equal at the point of intersection, we have

$$-4 + \frac{3}{2}x = 3 - \frac{2}{3}x.$$

Notice that we have converted a pair of equations into a single equation by eliminating one of the two variables. This equation can be simplified by multiplying both sides by 6:

$$\begin{aligned} 6\left(-4 + \frac{3}{2}x\right) &= 6\left(3 - \frac{2}{3}x\right) \\ -24 + 9x &= 18 - 4x \\ 13x &= 42 \\ x &= \frac{42}{13}. \end{aligned}$$

We can find the corresponding $y$-value using either equation:

$$\text{First equation:} \quad y = 3 - \frac{2}{3} \cdot \frac{42}{13} = \frac{11}{13}$$
$$\text{Second equation:} \quad y = -4 + \frac{3}{2} \cdot \frac{42}{13} = \frac{11}{13}.$$

We see that the point of intersection, $\left(\frac{42}{13}, \frac{11}{13}\right) = (3.2308, 0.8462)$, satisfies both equations. The lines and their point of intersection are shown in Figure 1.72.

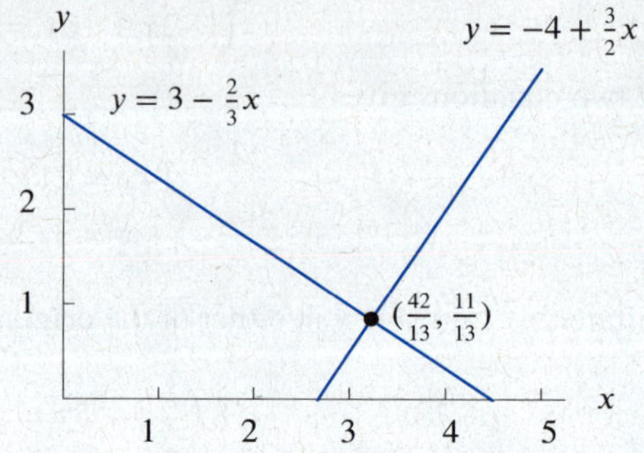


**Figure 1.72**: Intersection of lines is solution to simultaneous equations

## Exercises on Skills for Chapter 1

Solve the equations in Exercises 1–12.

**1.** $3x = 15$

**2.** $-2y = 12$

**3.** $4z = 22$

**4.** $x + 3 = 10$

**5.** $w - 23 = -34$

**6.** $7 - 3y = -14$

**7.** $13t + 2 = 47$

**8.** $0.5x - 3 = 7$

**9.** $3t - \dfrac{2(t-1)}{3} = 4$

**10.** $2(r + 5) - 3 = 3(r - 8) + 21$

**11.** $B - 4[B - 3(1 - B)] = 42$

**12.** $1.06s - 0.01(248.4 - s) = 22.67s$

In Exercises 13–22, solve for the indicated variable.

**13.** $A = l \cdot w$, for $l$.

**14.** $l = l_0 + \dfrac{k}{2}w$, for $w$.

**15.** $h = v_0 t + \dfrac{1}{2}at^2$, for $a$.

**16.** $3xy + 1 = 2y - 5x$, for $y$.

**17.** $u(v + 2) + w(v - 3) = z(v - 1)$, for $v$.

**18.** $S = \dfrac{rL - a}{r - 1}$, for $r$.

**19.** $\dfrac{a - cx}{b + dx} + a = 0$, for $x$.

**20.** $\dfrac{At - B}{C - B(1 - 2t)} = 3$, for $t$.

**21.** $y'y^2 + 2xyy' = 4y$, for $y'$.

**22.** $2x - (xy' + yy') + 2yy' = 0$, for $y'$.

In Exercises 23–32, solve the inequalities.

**23.** $800 + 3n < 500 + 5n$

**24.** $7z - 2 > 3z + 5$

**25.** $3 - 2.1x < 5 - 0.1x$

**26.** $4 - 0.2t \leq 11 + 0.3t$

**27.** $12 - 9k \geq 7 - 14k$

**28.** $2y \geq 6 + 5.32y$

**29.** $1 + 0.02y \leq 0.001y$

**30.** $3(5 - t) > 4(t - 3)$

**31.** $0.5(t - 4) < 0.3(t - 7)$

**32.** $5(x + 0.2) \geq x - 3$

**33.** Solve for $x$ in the inequality

$$ax - b > a + 3x.$$

Assume $a$ and $b$ are constants where

**(a)** $a$ is larger than 10.
**(b)** $a$ is negative.

In Problems 34–36 solve the inequality for $x$. Assume $a$ and $b$ are constants, and $a < 0$.

**34.** $5ax + a < 6x$

**35.** $3x + 2b < b - 5x + a$

**36.** $ax + 2b > x + 5$

In Exercises 37–41, solve the systems of equations for $x$ and $y$.

**37.** $\begin{cases} 3x - 2y = 6 \\ y = 2x - 5 \end{cases}$

**38.** $\begin{cases} x = 7y - 9 \\ 4x - 15y = 26 \end{cases}$

**39.** $\begin{cases} 2x + 3y = 7 \\ y = -\frac{3}{5}x + 6 \end{cases}$

**40.** $\begin{cases} 3x - y = 17 \\ -2x - 3y = -4 \end{cases}$

**41.** $\begin{cases} ax + y = 2a \\ x + ay = 1 + a^2 \end{cases}$

Determine the points of intersection for Exercises 42–43.

**42.**

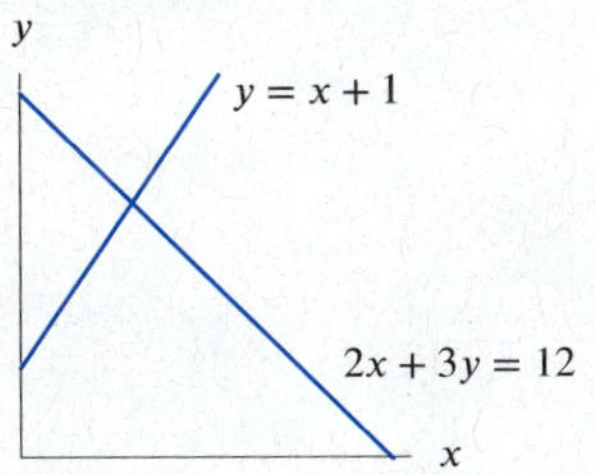


**43.**

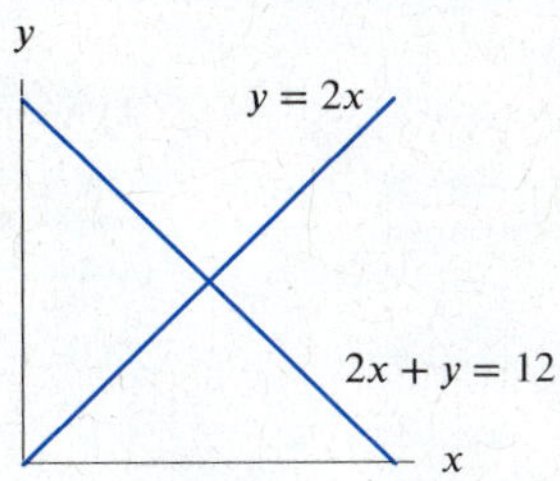


In Exercises 44–49, find the point of intersection of the given lines. Sketch and label a graph that illustrates your solution.

**44.** $y = 20 - 3x$ and $y = 50 - 5x$

**45.** $y = 2 + 0.3x$ and $y = -5 - 0.5x$

**46.** $y = -40 - 3x$ and $30 - 5x$

**47.** $y = 0.1x - 0.397$ and $y = 0.2x - 0.394$

**48.** $y = 3x - 20$ and $y = 0.2x - 4$

**49.** $y = 3000 + 18x$ and $y = 4000 + 37x$

The figures in Exercises 50–51 are parallelograms. Find the coordinates of the labeled point(s).

**50.**

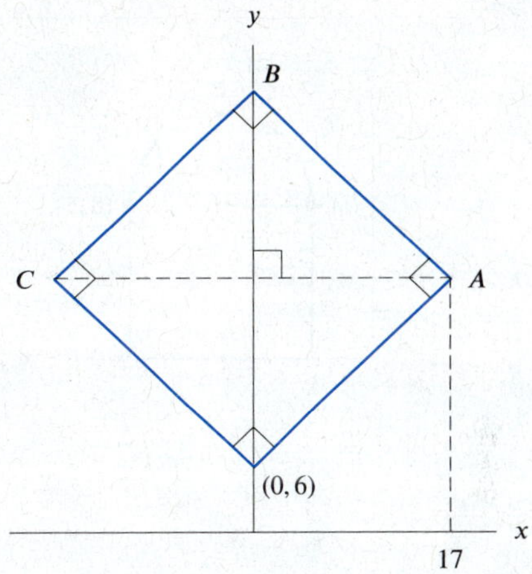


**51.**

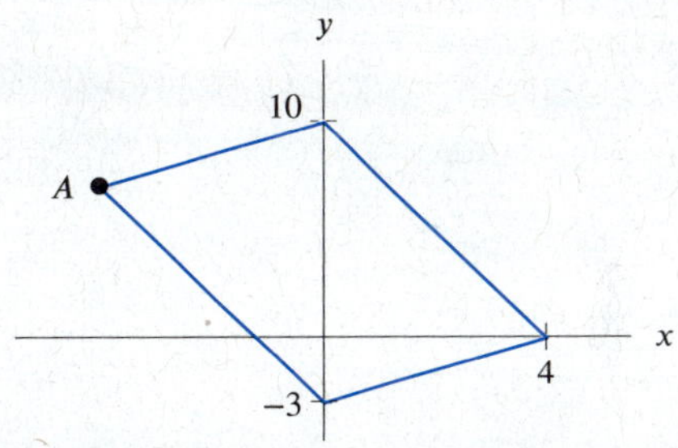


The figures in Exercises 52–55 contain a semicircle with the center marked. Find the coordinates of A, a point on the diameter, and B, an extreme point (highest, lowest, or farthest to the right).

**52.**

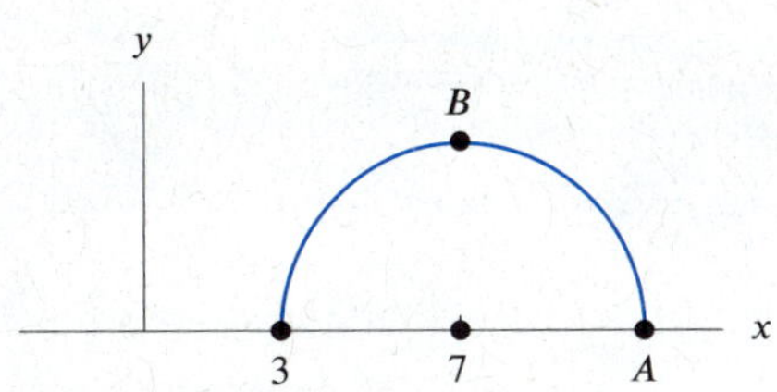

**53.**

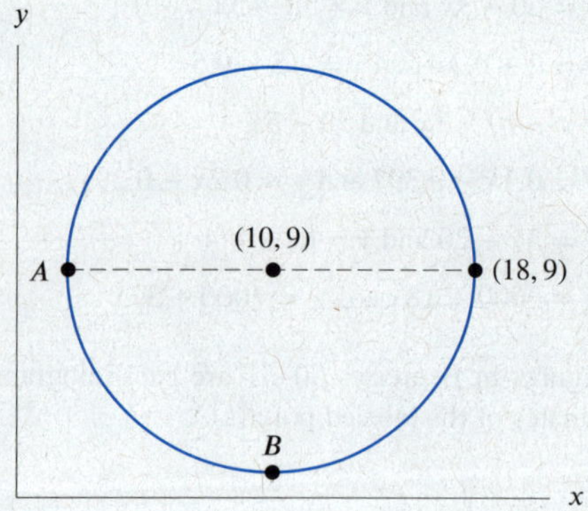


**54.**

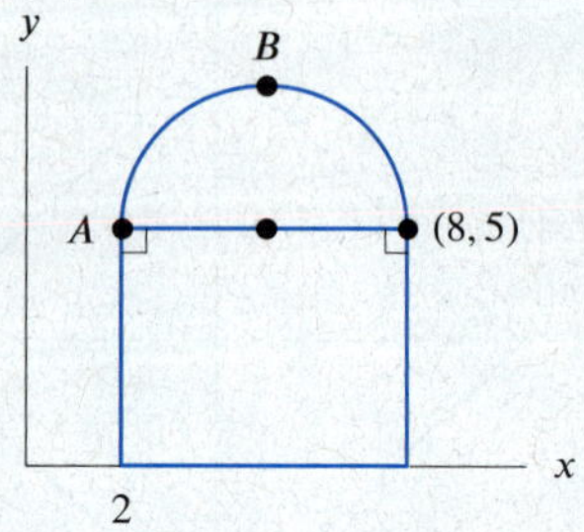


**55.**

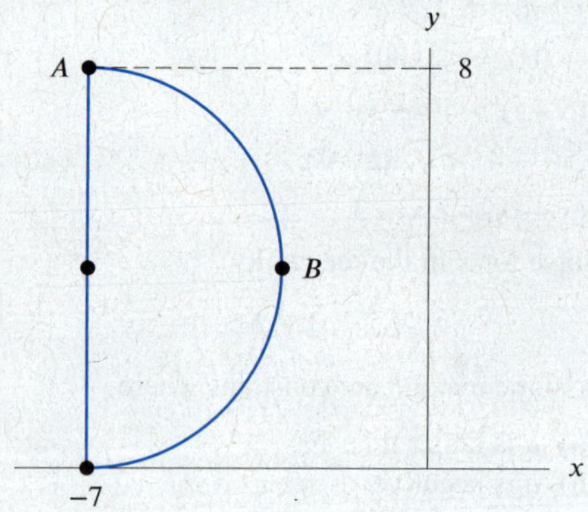

Chapter Two

# FUNCTIONS

© Patrick Zephyr/Patrick Zephyr Nature Photography

## Contents

# 2.1 INPUT AND OUTPUT

## Finding Output Values: Evaluating a Function

Evaluating a function means calculating the value of a function's output from a particular value of the input.

In the housepainting example on page 4, the notation $n = f(A)$ indicates that the number of gallons of paint $n$ is a function of area $A$. The expression $f(A)$ represents the output of the function—specifically, the amount of paint required to cover an area of $A$ ft$^2$. For example, $f(20{,}000)$ represents the number of gallons of paint required to cover an area of 20,000 ft$^2$.

**Example 1** Using the fact that 1 gallon of paint covers 250 ft$^2$, evaluate the expression $f(20{,}000)$.

**Solution** To evaluate $f(20{,}000)$, calculate the number of gallons required to cover 20,000 ft$^2$:

$$f(20{,}000) = \frac{20{,}000 \text{ ft}^2}{250 \text{ ft}^2/\text{gallon}} = 80 \text{ gallons of paint.}$$

### Evaluating a Function Using a Formula

If we have a formula for a function, we evaluate it by substituting the input value into the formula.

**Example 2** The formula for the area of a circle of radius $r$ is $A = q(r) = \pi r^2$. Use the formula to evaluate $q(10)$ and $q(20)$. What do your results tell you about circles?

**Solution** In the expression $q(10)$, the value of $r$ is 10, so

$$q(10) = \pi \cdot 10^2 = 100\pi \approx 314.$$

Similarly, substituting $r = 20$, we have

$$q(20) = \pi \cdot 20^2 = 400\pi \approx 1257.$$

The statements $q(10) \approx 314$ and $q(20) \approx 1257$ tell us that a circle of radius 10 cm has an area of approximately 314 cm$^2$ and a circle of radius 20 cm has an area of approximately 1257 cm$^2$.

**Example 3** Let $g(x) = \dfrac{x^2+1}{5+x}$. Evaluate the following expressions.

(a) $g(3)$ (b) $g(-1)$ (c) $g(a)$

**Solution** (a) To evaluate $g(3)$, replace every $x$ in the formula with 3:

$$g(3) = \frac{3^2+1}{5+3} = \frac{10}{8} = 1.25.$$

(b) To evaluate $g(-1)$, replace every $x$ in the formula with $(-1)$:

$$g(-1) = \frac{(-1)^2+1}{5+(-1)} = \frac{2}{4} = 0.5.$$

(c) To evaluate $g(a)$, replace every $x$ in the formula with $a$:

$$g(a) = \frac{a^2+1}{5+a}.$$

Evaluating a function may involve algebraic simplification, as the following example shows.

**Example 4** Let $h(x) = x^2 - 3x + 5$. Evaluate and simplify the following expressions.

(a) $h(2)$ (b) $h(a-2)$ (c) $h(a) - 2$ (d) $h(a) - h(2)$

**Solution** Notice that $x$ is the input and $h(x)$ is the output. It is helpful to rewrite the formula as

$$\text{Output} = h(\text{Input}) = (\text{Input})^2 - 3 \cdot (\text{Input}) + 5.$$

(a) For $h(2)$, we have Input $= 2$, so

$$h(2) = (2)^2 - 3 \cdot (2) + 5 = 3.$$

(b) In this case, Input $= a - 2$. We substitute and multiply out

$$\begin{aligned} h(a-2) &= (a-2)^2 - 3(a-2) + 5 \\ &= a^2 - 4a + 4 - 3a + 6 + 5 \\ &= a^2 - 7a + 15. \end{aligned}$$

(c) First input $a$, then subtract 2:

$$\begin{aligned} h(a) - 2 &= a^2 - 3a + 5 - 2 \\ &= a^2 - 3a + 3. \end{aligned}$$

(d) Since we found $h(2) = 3$ in part (a), we subtract from $h(a)$:

$$\begin{aligned} h(a) - h(2) &= a^2 - 3a + 5 - 3 \\ &= a^2 - 3a + 2. \end{aligned}$$

## Finding Input Values: Solving Equations

Given an input, we evaluate the function to find the output. Sometimes the situation is reversed: we know the output and we want to find a corresponding input. If the function is given by a formula, the input values are solutions to an equation.

**Example 5** Use the cricket function $T = \frac{1}{4}R + 40$, introduced on page 3, to find the rate, $R$, at which the snowy tree cricket chirps when the temperature, $T$, is 76°F.

**Solution** We want to find $R$ when $T = 76$. Substitute $T = 76$ into the formula and solve the equation

$$\begin{aligned} 76 &= \frac{1}{4}R + 40 \\ 36 &= \frac{1}{4}R && \text{subtract 40 from both sides} \\ 144 &= R. && \text{multiply both sides by 4} \end{aligned}$$

The cricket chirps at a rate of 144 chirps per minute when the temperature is 76°F.

**Example 6** Suppose $f(x) = \dfrac{1}{\sqrt{x-4}}$.

(a) Find an $x$-value that results in $f(x) = 2$.
(b) Is there an $x$-value that results in $f(x) = -2$?

**Solution** (a) To find an $x$-value that results in $f(x) = 2$, solve the equation

$$2 = \frac{1}{\sqrt{x-4}}.$$

Square both sides:

$$4 = \frac{1}{x-4}.$$

Now multiply by $(x-4)$:

$$4(x-4) = 1$$
$$4x - 16 = 1$$
$$x = \frac{17}{4} = 4.25.$$

The $x$-value is 4.25. (Note that the simplification $(x-4)/(x-4) = 1$ in the second step was valid because $x - 4 \neq 0$.)

(b) Since $\sqrt{x-4}$ is nonnegative if it is defined, its reciprocal, $f(x) = \dfrac{1}{\sqrt{x-4}}$ is also nonnegative if it is defined. Thus, $f(x)$ is not negative for any $x$ input, so there is no $x$-value that results in $f(x) = -2$.

In the next example, we solve an equation for a quantity that is being used to model a physical quantity; we must choose the solutions that make sense in the context of the model.

**Example 7** Let $A = q(r)$ be the area of a circle of radius $r$, where $r$ is in cm. What is the radius of a circle whose area is 100 $\text{cm}^2$?

**Solution** The output $q(r)$ is an area. Solving the equation $q(r) = 100$ for $r$ gives the radius of a circle whose area is 100 $\text{cm}^2$. Since the formula for the area of a circle is $q(r) = \pi r^2$, we solve

$$q(r) = \pi r^2 = 100$$
$$r^2 = \frac{100}{\pi}$$
$$r = \pm\sqrt{\frac{100}{\pi}} = \pm 5.642.$$

We have two solutions for $r$, one positive and one negative. Since a circle cannot have a negative radius, we take $r = 5.642$ cm. A circle of area 100 $\text{cm}^2$ has a radius of 5.642 cm.

## Finding Output and Input Values From Tables and Graphs

The following two examples use function notation with a table and a graph respectively.

**Example 8** Table 2.1 shows the revenue, $R = f(t)$, received by the National Football League,[1] from network TV as a function of the year, $t$, since 1975. Contracts are signed with all networks at once for periods ranging from 2 to 8 years. The table shows the annual revenue under the contract holding that year.

(a) Evaluate and interpret $f(20)$. (b) Solve and interpret $f(t) = 2600$.

**Table 2.1**

| $t$ (years) | 0 | 5 | 10 | 15 | 20 | 25 | 30 | 35 |
|---|---|---|---|---|---|---|---|---|
| $R$ (m. \$) | 55 | 162 | 420 | 900 | 1097 | 2600 | 2600 | 3735 |

**Solution** (a) Table 2.1 shows $f(20) = 1097$. Since $t = 20$ in the year 1995, we know that NFL's revenue from TV was \$1097 million in the year 1995.

(b) Solving $f(t) = 2600$ means finding a year in which TV revenues were \$2600 million. There are two values, $t = 25$ and $t = 30$. In other words, in 2000 and 2005, NFL's TV revenues were \$2600 million. (The same contract was in effect in 2000 and 2005.)

[1]John Vrooman, "The economic structure of the NFL," in *The Economics of the National Football League*, K.G. Quinn (Ed.), Springer, 2011.

**Example 9** A man drives from his home to a store and back. The entire trip takes 30 minutes. Figure 2.1 gives his velocity $v(t)$ (in mph) as a function of the time $t$ (in minutes) since he left home. A negative velocity indicates that he is traveling away from the store back to his home.

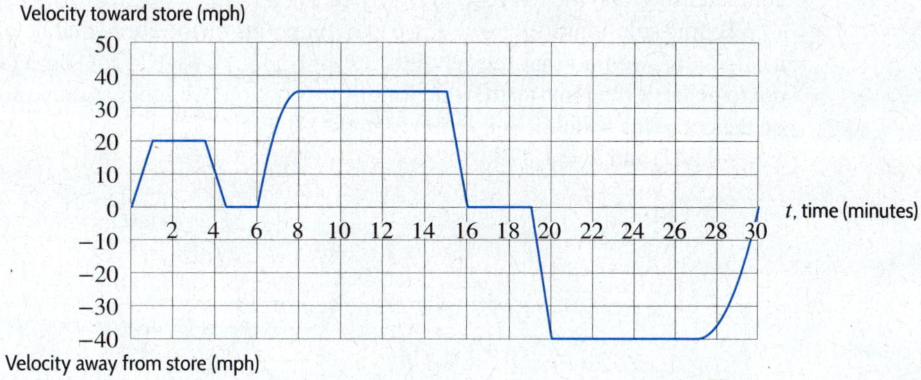


**Figure 2.1:** Velocity of a man on a trip to the store and back

Evaluate and interpret:

(a) $v(5)$ (b) $v(24)$ (c) $v(8) - v(6)$ (d) $v(-3)$

Solve for $t$ and interpret:

(e) $v(t) = 15$ (f) $v(t) = -20$ (g) $v(t) = v(7)$ (h) $v(t) > 30$

**Solution**

(a) To evaluate $v(5)$, look on the graph where $t = 5$ minutes. Five minutes after he left home, his velocity is 0 mph. Thus, $v(5) = 0$. Perhaps he had to stop at a light.

(b) The graph shows that $v(24) = -40$ mph. After 24 minutes, he is traveling at 40 mph away from the store, back to his home.

(c) From the graph, $v(8) = 35$ mph and $v(6) = 0$ mph. Thus, $v(8) - v(6) = 35 - 0 = 35$. This shows that the man's speed increased by 35 mph in the interval between $t = 6$ minutes and $t = 8$ minutes.

(d) The quantity $v(-3)$ is not defined since the graph only gives velocities for nonnegative times.

(e) To solve for $t$ when $v(t) = 15$, look on the graph where the velocity is 15 mph. This occurs at $t \approx 0.75$ minute, 3.75 minutes, 6.5 minutes, and 15.5 minutes. At each of these four times the man's velocity was 15 mph.

(f) To solve $v(t) = -20$ for $t$, we see that the velocity is $-20$ mph (that is, 20 mph toward home) at $t \approx 19.5$ and $t \approx 29$ minutes.

(g) First we evaluate $v(7) \approx 30$. To solve $v(t) = 30$, we look for the values of $t$ making the velocity 30 mph. One such $t$ is of course $t = 7$; the other $t$ is $t \approx 15$ minutes. These are the two times when the velocity is the same as it is at 7 minutes.

(h) To solve for $t$ when $v(t) > 30$, look on the graph where the velocity toward the store is greater than 30 mph. This occurs for $t$ such that $7 < t < 15$. Therefore, between 7 and 15 minutes after he left home, the man was driving faster than 30 mph toward the store. Note that there are also times when the man is driving faster than 30 mph away from the store; however, these times are not solutions to the inequality $v(t) > 30$, since $v(t)$ is positive only when the man is driving toward the store.

If we are given a formula for a function, we can use the graph of the function to find input values that satisfy a certain equation or inequality.

**Example 10** Let $f(x) = x^3 - 3x$. Find approximate values of the input $x$ such that

(a) $f(x) = 1$ (b) $f(x) > 1$.

**Solution** (a) We can determine for what values of $x$ the output $f(x)$ is equal to 1 by using a calculator or computer to graph the curve $y = f(x)$. See Figure 2.2.

To find solutions to $f(x) = 1$, we look for points on the curve that lie on the line $y = 1$. Using a calculator, we find that these points are $(-1.532, 1)$, $(-0.347, 1)$, and $(1.879, 1)$. By zooming out to observe the long-term behavior of the function, we see that these are the only three points on the curve for which $y = 1$. Therefore, the approximate solutions of $f(x) = 1$ are $x = -1.532$, $x = -0.347$, and $x = 1.879$.

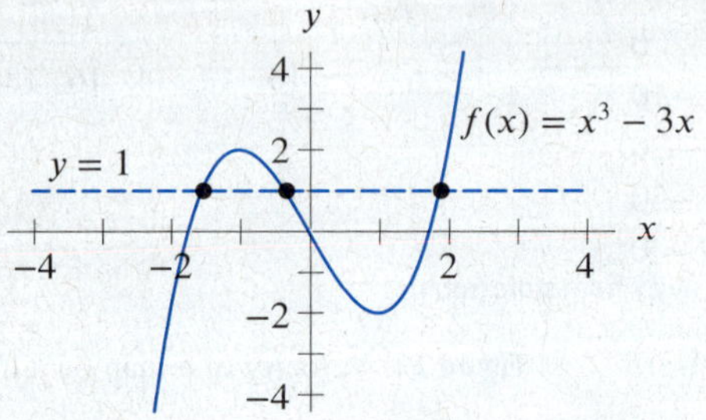


**Figure 2.2:** The function $f(x) = x^3 - 3x$

(b) We can solve $f(x) > 1$ by referring to Figure 2.2. We look for points on the graph where the $y$-coordinate is greater than 1; these are points on the graph that lie above the line $y = 1$.

We know that the graph crosses the line $y = 1$ at $x = -1.532$, $x = -0.347$, and $x = 1.879$. We see from the figure that the graph of $f(x)$ lies above the line $y = 1$ for all $x$ such that $-1.532 < x < -0.347$ or $x > 1.879$. Therefore, the approximate solutions of $f(x) > 1$ are

$$-1.532 < x < -0.347 \quad \text{or} \quad x > 1.879.$$

Occasionally, looking at a graph helps us locate exact solutions to an equation or inequality.

**Example 11** Let $f(x) = 1 + \frac{1}{2}x$ and $g(x) = 3/(1 + x^2)$. Find all values of the input $x$ such that $f(x) \leq g(x)$.

**Solution** We begin by using a calculator or computer to graph the curves $y = f(x)$ and $y = g(x)$, as in Figure 2.3.

The two graphs appear to intersect at the point $(1, 3/2)$. We can confirm that this is an intersection point by computing $f(1)$ and $g(1)$:

$$f(1) = 1 + \tfrac{1}{2}(1) = \tfrac{3}{2} \quad \text{and} \quad g(1) = \frac{3}{1 + 1^2} = \frac{3}{2}.$$

Thus the point $(1, 3/2)$ is the intersection point of the two graphs. Since the graph of $f(x)$ lies below the graph of $g(x)$ only to the left of $x = 1$, the solution to the inequality $f(x) \leq g(x)$ is $x \leq 1$.

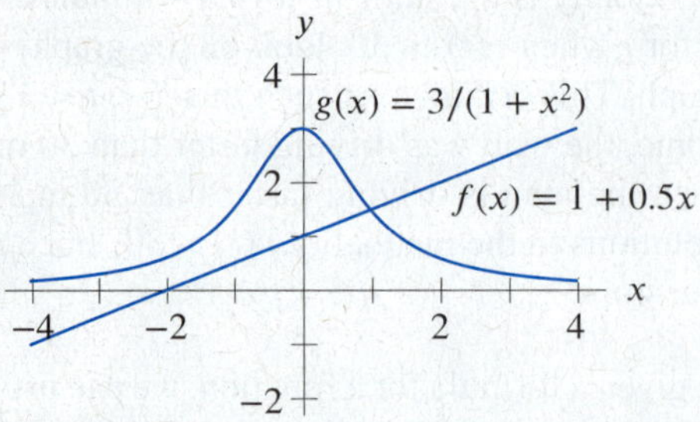


**Figure 2.3:** The functions $f(x) = 1 + \frac{1}{2}x$ and $g(x) = 3/(1 + x^2)$

## Exercises and Problems for Section 2.1

### Skill Refresher

For Exercises S1–S6, expand and simplify.

**S1.** $5(x-3)$

**S2.** $a(2a+5)$

**S3.** $(m-5)(4(m-5)+2)$

**S4.** $(x+2)(3x-8)$

**S5.** $3\left(1+\frac{1}{x}\right)$

**S6.** $3+2\left(\frac{1}{x}\right)^2-x$

Solve the equations in Exercises S7–S10.

**S7.** $x^2-9=0$

**S8.** $\sqrt{2x-1}+3=9$

**S9.** $\frac{21}{z-5}-\frac{13}{z^2-5z}=3$

**S10.** $2x^{3/2}-1=7$

### Exercises

**1.** If $g(x)=-\frac{1}{2}x^{1/3}$, find $g(-27)$.

**2.** Let $f(x)=\frac{2x+1}{x+1}$. For what value of $x$ is $f(x)=0.3$?

**3.** If $f(t)=t^2-4$, **(a)** Find $f(0)$ **(b)** Solve $f(t)=0$.

**4.** If $g(x)=x^2-5x+6$, **(a)** Find $g(0)$ **(b)** Solve $g(x)=0$.

**5.** If $g(t)=\frac{1}{t+2}-1$, **(a)** Find $g(0)$ **(b)** Solve $g(t)=0$.

**6.** If $h(x)=ax^2+bx+c$, find $h(0)$.

If $p(r)=r^2+5$, evaluate the expressions in Exercises 7–8.

**7.** $p(7)$

**8.** $p(x)+p(8)$

In Figure 2.4, mark the point(s) representing the statements in Exercises 9–12 and label their coordinates.

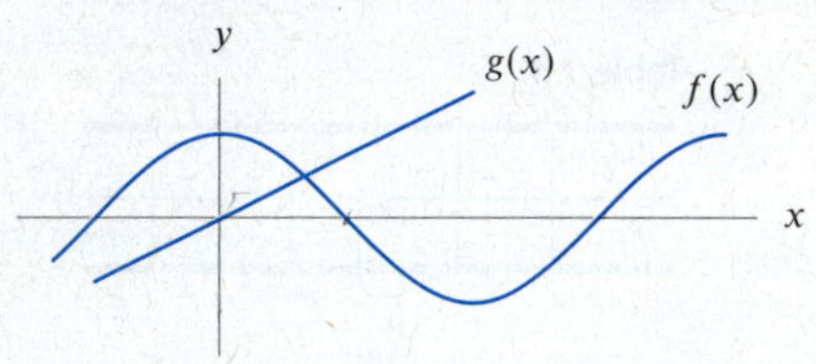


**Figure 2.4**

**9.** $f(0)=2$

**10.** $f(-3)=f(3)=f(9)=0$

**11.** $f(2)=g(2)$

**12.** $g(x)>f(x)$ for $x>2$

In Exercises 13–16, let $f(x)=x^2+2x-2$ and $g(x)=2x+2$. Find all $x$ values satisfying the following.

**13.** $g(x)=2$

**14.** $g(x)<2$

**15.** $f(x)<x^2$

**16.** $f(x)>g(x)$

### Problems

**17.** Let $F=g(t)$ be the number of foxes in a park as a function of $t$, the number of months since January 1. Evaluate $g(9)$ using Table 1.3 on page 5. What does this tell us about the fox population?

**18.** Let $F=g(t)$ be the number of foxes in month $t$ in the national park described in Example 5 on page 5. Solve the equation $g(t)=75$. What does your solution tell you about the fox population?

**19.** Table 2.2 shows the base salary of members of Congress[2], $S=f(t)$, in thousands of dollars, as a function of the year, $t$, since 1992.

**(a)** Evaluate and interpret $f(12)$.
**(b)** Solve and interpret $f(t)=169.3$.

**Table 2.2**

| Year, $t$ | 0 | 4 | 8 | 12 | 16 | 20 |
|---|---|---|---|---|---|---|
| Salary, $S$ | 129.5 | 133.6 | 141.3 | 158.1 | 169.3 | 174.0 |

**20.** Let $f(x)=3+2x^2$. Find $f\left(\frac{1}{3}\right)$ and $\frac{f(1)}{f(3)}$. Are they equal?

**21.** If $g(x)=x\sqrt{x}+100x$, evaluate without a calculator

**(a)** $g(100)$ **(b)** $g(4/25)$ **(c)** $g(1.21\cdot 10^4)$

**22.** Let $g(x)=x^2+x$. Find formulas for the following functions. Simplify your answers.

**(a)** $g(-3x)$ **(b)** $g(1-x)$ **(c)** $g(x+\pi)$
**(d)** $g(\sqrt{x})$ **(e)** $g(1/(x+1))$ **(f)** $g(x^2)$

**23.** Let $f(x)=\frac{x}{x-1}$.

**(a)** Find and simplify

(i) $f\left(\frac{1}{t}\right)$ (ii) $f\left(\frac{1}{t+1}\right)$

**(b)** Solve $f(x)=3$.

[2] https://hobnobblog.com/congress-by-the-numbers/pay-and-perquisites-of-members-of-congress-including-a-history-of-house-and-senate-salaries/, accessed October 24, 2013.

**24.** Let $f(x) = x^2$. Find and simplify the following.

**(a)** $f(2)$ **(b)** $f(2+h)$

**(c)** $f(2+h) - f(2)$ **(d)** $\dfrac{f(2+h) - f(2)}{h}$

**25.** Let $g(t) = 2t - 1$. Find and simplify

$$\frac{g(t+h) - g(t)}{h}.$$

**26.** **(a)** Using Figure 2.5, fill in Table 2.3.

**Table 2.3**

| $x$ | −2 | −1 | 0 | 1 | 2 | 3 |
|---|---|---|---|---|---|---|
| $h(x)$ | | | | | | |

(b) Evaluate $h(3) - h(1)$ (c) Evaluate $h(2) - h(0)$

(d) Evaluate $2h(0)$ (e) Evaluate $h(1) + 3$

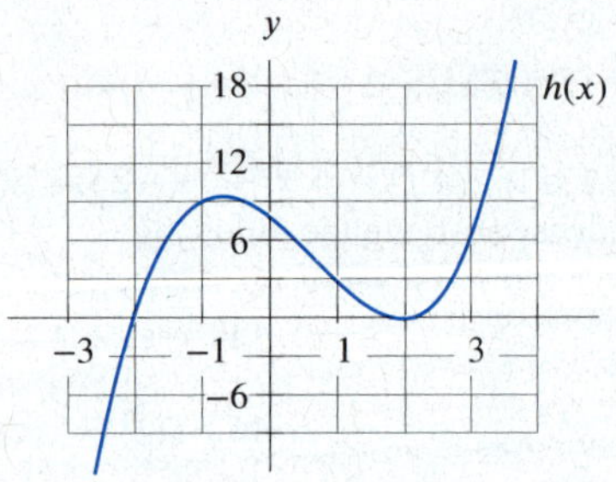


**Figure 2.5**

**27.** A ball is thrown up from the ground with initial velocity 64 ft/sec. Its height at time $t$ is

$$h(t) = -16t^2 + 64t.$$

**(a)** Evaluate $h(1)$ and $h(3)$. What does this tell us about the height of the ball?

**(b)** Sketch this function. Using a graph, determine when the ball hits the ground and the maximum height of the ball.

**28.** The profit, in dollars, made by a theater when $n$ tickets are sold is $P(n) = 20n - 500$.

**(a)** Calculate $P(0)$, and explain what this number means for the theater.

**(b)** Under what circumstances will the profit equal 0?

**(c)** What is the meaning of the quantity $P(100)$? What are its units?

**29.** Let $v(t) = t^2 - 2t$ be the velocity, in ft/sec, of an object at time $t$, in seconds.

**(a)** What is the initial velocity, $v(0)$?

**(b)** When does the object have a velocity of zero?

**(c)** What is the meaning of the quantity $v(3)$? What are its units?

**30.** Let $s(t) = 11t^2 + t + 100$ be the position, in miles, of a car driving on a straight road at time $t$, in hours. The car's velocity at any time $t$ is given by $v(t) = 22t + 1$.

**(a)** Use function notation to express the car's position after 2 hours. Where is the car then?

**(b)** Use function notation to express the question, "When is the car going 65 mph?"

**(c)** Where is the car when it is going 67 mph?

**31.** Let $v = f(t)$ be the speed of a braking car, in feet per second, $t$ seconds after the brakes are first applied. A graph of $f$ is given in Figure 2.6. Complete the following, and explain the meaning of your answers in terms of the car.

**(a)** If $t = 1$, estimate $f(t+5)$.

**(b)** If $t = 1$, estimate $f(t) + 5$.

**(c)** Solve $f(t+2) = 40$ for $t$.

**(d)** Solve $f(t) + 10 = 40$ for $t$.

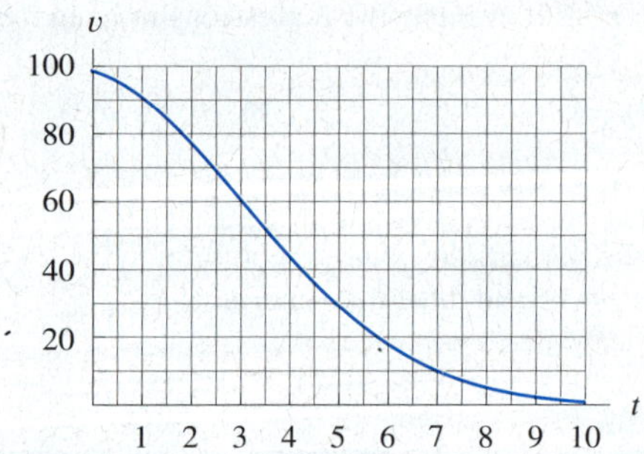


**Figure 2.6**

**32.** Use the letters $a$, $b$, $c$, $d$, $e$, $h$ in Figure 2.7 to answer the following questions.

**(a)** What are the coordinates of the points $P$ and $Q$?

**(b)** Evaluate $f(b)$.

**(c)** Solve $f(x) = e$ for $x$.

**(d)** Suppose $c = f(z)$ and $z = f(x)$. What is $x$?

**(e)** Suppose $f(b) = -f(d)$. What additional information does this give you?

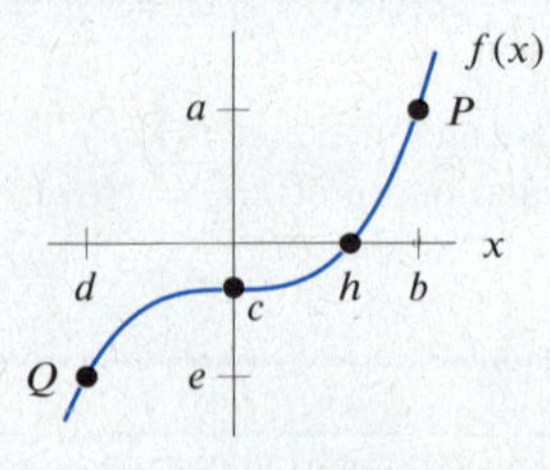


**Figure 2.7**

**33.** New York state income tax is based on taxable income, which is part of a person's total income. The tax owed to the state is calculated using the taxable income (not total income). In 2013, for a single person with a taxable income between \$77,150 and \$205,850, the tax owed was \$4650.63 plus 6.65% of the taxable income over \$77,150.[3]

**(a)** Compute the tax owed by a lawyer whose taxable income is \$88,000.
**(b)** Consider a lawyer whose taxable income is 80% of her total income, $x$, where $x$ is between \$100,000 and \$150,000. Write a formula for $T(x)$, the taxable income.
**(c)** Write a formula for $L(x)$, the amount of tax owed by the lawyer in part (b).
**(d)** Use $L(x)$ to evaluate the tax owed for $x = 110{,}000$ and compare your results to part (a).

**34.** Table 2.4 shows $N(s)$, the number of sections of Economics 101, as a function of $s$, the number of students in the course. If $s$ is between two numbers listed in the table, then $N(s)$ is the higher number of sections.

**(a)** Evaluate and interpret:
(i) $N(150)$ (ii) $N(80)$ (iii) $N(55.5)$
**(b)** Solve for $s$ and interpret:
(i) $N(s) = 4$ (ii) $N(s) = N(125)$

Table 2.4

| $s$ | 50 | 75 | 100 | 125 | 150 | 175 | 200 |
|---|---|---|---|---|---|---|---|
| $N(s)$ | 4 | 4 | 5 | 5 | 6 | 6 | 7 |

**35.** **(a)** Complete Table 2.5 using

$$f(x) = 2x(x-3) - x(x-5) \quad \text{and} \quad g(x) = x^2 - x.$$

What do you notice? Graph these two functions. Are the two functions the same? Explain.

**(b)** Complete Table 2.6 using

$$h(x) = x^5 - 5x^3 + 6x + 1 \quad \text{and} \quad j(x) = 2x + 1.$$

What do you notice? Graph these two functions. Are the two functions the same? Explain.

Table 2.5

| $x$ | −2 | −1 | 0 | 1 | 2 |
|---|---|---|---|---|---|
| $f(x)$ | | | | | |
| $g(x)$ | | | | | |

Table 2.6

| $x$ | −2 | −1 | 0 | 1 | 2 |
|---|---|---|---|---|---|
| $h(x)$ | | | | | |
| $j(x)$ | | | | | |

**36.** Values of $f$ and $g$ are given in Table 2.7.

**(a)** Evaluate $f(1)$ and $g(3)$.
**(b)** Describe in full sentences the patterns you see in the values for each function.
**(c)** Assuming that the patterns you observed in part (b) hold true for all values of $x$, calculate $f(5)$, $f(-2)$, $g(5)$, and $g(-2)$.
**(d)** Find possible formulas for $f(x)$ and $g(x)$.

Table 2.7

| $x$ | −1 | 0 | 1 | 2 | 3 | 4 |
|---|---|---|---|---|---|---|
| $f(x)$ | −4 | −1 | 2 | 5 | 8 | 11 |
| $g(x)$ | 4 | 1 | 0 | 1 | 4 | 9 |

Problems 37–38 concern $v = r(s)$, the eyewall wind profile of a hurricane at landfall, where $v$ is the eyewall wind speed (in mph) as a function of $s$, the height (in meters) above the ground. (The eyewall is the band of clouds that surrounds the eye of the storm.) Let $s_0$ be the height at which the wind speed is greatest, and let $v_0 = r(s_0)$. Interpret the following in terms of the hurricanes.

**37.** $r(0.5s_0)$ **38.** $r(s) = 0.75v_0$

**39.** Let $h(x) = x^2 + bx + c$. Evaluate and simplify:
**(a)** $h(1)$ **(b)** $h(b+1)$

In Problems 40–42, if $f(x) = \dfrac{ax}{a+x}$, find and simplify

**40.** $f(a)$ **41.** $f(1-a)$ **42.** $f\left(\dfrac{1}{1-a}\right)$

**43.** Figure 2.8 shows $y = f(x)$. Label the coordinates of any points on the graph where

**(a)** $f(c) = 0$.
**(b)** $f(0) = d$.

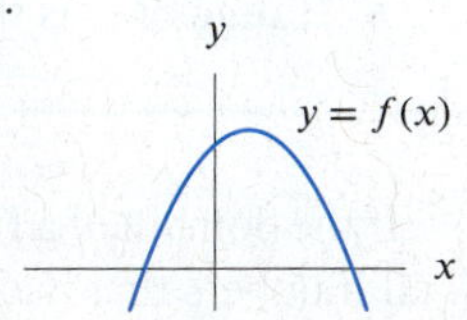


Figure 2.8

**44.** An epidemic of influenza spreads through a city. Figure 2.9 is the graph of $I = f(w)$, where $I$ is the number of individuals (in thousands) infected $w$ weeks after the epidemic begins.

**(a)** Evaluate $f(2)$ and explain its meaning in terms of the epidemic.
**(b)** Approximately how many people were infected at the height of the epidemic? When did that occur? Write your answer in the form $f(a) = b$.
**(c)** Solve $f(w) = 4.5$ and explain what the solutions mean in terms of the epidemic.

[3]www.nystax.gov, accessed February 21, 2014.

**(d)** The graph used $f(w) = 6w(1.3)^{-w}$. Use the graph to estimate the solution of the inequality $6w(1.3)^{-w} \geq 6$. Explain what the solution means in terms of the epidemic.

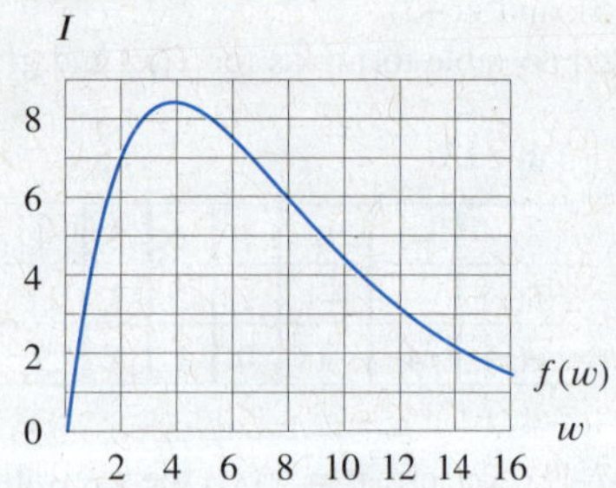


Figure 2.9

Problems 45–46 concern studies suggesting that as carbon dioxide ($CO_2$) levels rise, hurricanes will become more intense.[4] Hurricane intensity is measured in terms of the minimum central pressure $P$ (in mb): the lower the pressure, the more powerful the storm. Since warm ocean waters fuel hurricanes, $P$ is a decreasing function of $H$, sea surface temperature in °C. Let $P = n(H)$ be the hurricane-intensity function for present-day $CO_2$ levels, and let $P = N(H)$ be the hurricane-intensity function for future projected $CO_2$ levels. If $H_0$ is the average temperature in the Caribbean Sea, what do the following quantities tell you about hurricane intensity?

**45.** $N(H_0) - n(H_0)$

**46.** $n(H_0 + 1) - n(H_0)$

## 2.2 DOMAIN AND RANGE

In Example 4 on page 4, we defined $R$ to be the average monthly rainfall at Chicago's O'Hare airport in month $t$. Although $R$ is a function of $t$, the value of $R$ is not defined for every possible value of $t$. For instance, it makes no sense to consider the value of $R$ for $t = -3$, or $t = 8.21$, or $t = 13$ (since a year has 12 months). Thus, although $R$ is a function of $t$, this function is defined only for $t = 1, t = 2, t = 3, \ldots, t = 12$. Notice also that $R$, the output value of this function, takes only the values $\{1.8, 2.1, 2.4, 2.5, 2.7, 3.1, 3.2, 3.4, 3.5, 3.7\}$.

A function is often defined only for certain values of the independent variable. Also, the dependent variable often takes on only certain values. This leads to the following definitions:

If $Q = f(t)$, then the
- **Domain** of $f$ is the set of input values, $t$, which yield an output value.
- **Range** of $f$ is the corresponding set of output values, $Q$.

If the domain of a function is not specified, we usually assume that it is as large as possible—that is, all numbers that make sense as inputs for the function. For example, if there are no restrictions, the domain of the function $f(x) = x^2$ is the set of all real numbers, because we can substitute any real number into the formula $f(x) = x^2$. Sometimes, however, we may restrict the domain to suit a particular application. If the function $f(x) = x^2$ is used to represent the area of a square of side $x$, we restrict the domain to positive numbers.

**Example 1** The house-painting function $n = f(A)$ in Example 3 on page 4 has domain $A > 0$ because all houses have some positive paintable area. There is a practical upper limit to $A$ because houses cannot be infinitely large, but in principle, $A$ can be as large or as small as we like, as long as it is positive. Therefore we take the domain of $f$ to be $A > 0$.

The range of this function is $n \geq 0$, because we cannot use a negative amount of paint.

[4] *Journal of Climate*, September 14, 2004, pages 3477–3495 or www.theguardian.com/environment/2013/aug/23/climate-change-carbon-emissions-ipcc-extreme, accessed May 21, 2014.

## Choosing Realistic Domains and Ranges

When a function is used to model a real situation, we may need to modify the domain and range.

**Example 2** Algebraically speaking, the formula

$$T = \frac{1}{4}R + 40$$

can be used for all values of $R$. If we know nothing more about this function than its formula, its domain is all real numbers. The formula for $T = \frac{1}{4}R + 40$ can return any value of $T$ when we choose an appropriate $R$-value. (See Figure 2.10.) Thus, the range of the function is also all real numbers. However, if we use this formula to represent the temperature, $T$, as a function of a cricket's chirp rate, $R$, as we did in Example 1 on page 2, some values of $R$ cannot be used. For example, it does not make sense to talk about a negative chirp rate. Also, there is some maximum chirp rate $R_{max}$ that no cricket can physically exceed. Thus, to use this formula to express $T$ as a function of $R$, we must restrict $R$ to the interval $0 \leq R \leq R_{max}$ shown in Figure 2.11.

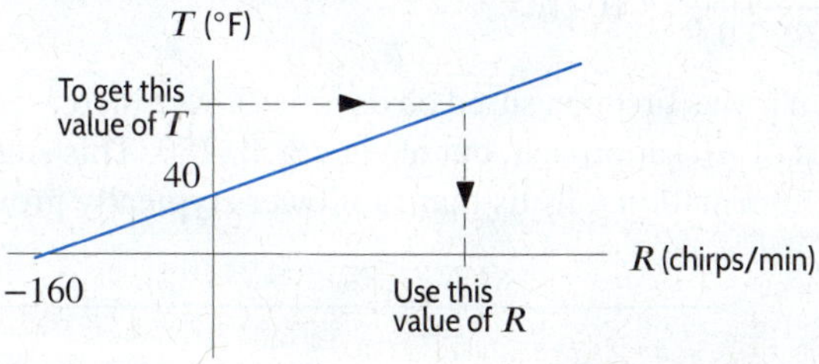

**Figure 2.10**: Graph showing that any $T$ value can be obtained from some $R$ value

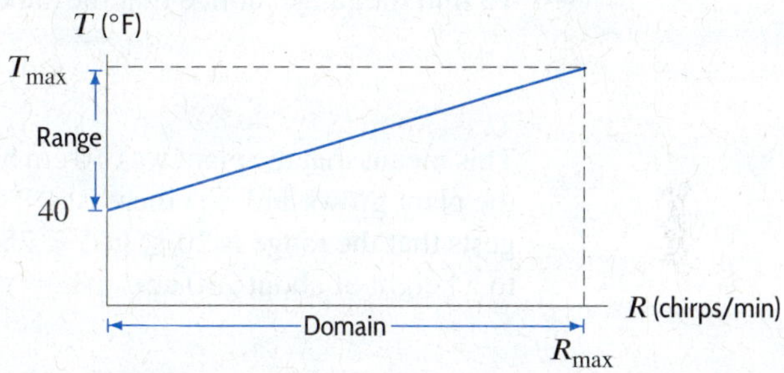

**Figure 2.11**: Graph showing that if $0 \leq R \leq R_{max}$, then $40 \leq T \leq T_{max}$

The range of the cricket function is also restricted. Since the chirp rate is nonnegative, the smallest value of $T$ occurs when $R = 0$. This happens at $T = 40$. On the other hand, if the temperature gets too hot, the cricket will not be able to keep chirping faster. If the temperature $T_{max}$ corresponds to the chirp rate $R_{max}$, then the values of $T$ are restricted to the interval $40 \leq T \leq T_{max}$.

## Using a Graph to Find the Domain and Range of a Function

A good way to estimate the domain and range of a function is to examine its graph. The domain is the set of input values on the horizontal axis that give rise to a point on the graph; the range is the corresponding set of output values on the vertical axis.

**Example 3** A sunflower plant is measured every day $t$, for $t \geq 0$. The height, $h(t)$ centimeters, of the plant[5] can be modeled by the *logistic function*, which is graphed in Figure 2.12:

$$h(t) = \frac{250}{1 + 24(0.9)^t}.$$

(a) What is the domain of the function in this model?

[5] Adapted from H.S. Reed and R.H. Holland, "Growth of an Annual Plant Helianthus," *Proc. Nat. Acad. Sci.*, 5, 1919.

(b) What is the range of the function? What does this tell you about the height of the sunflower?

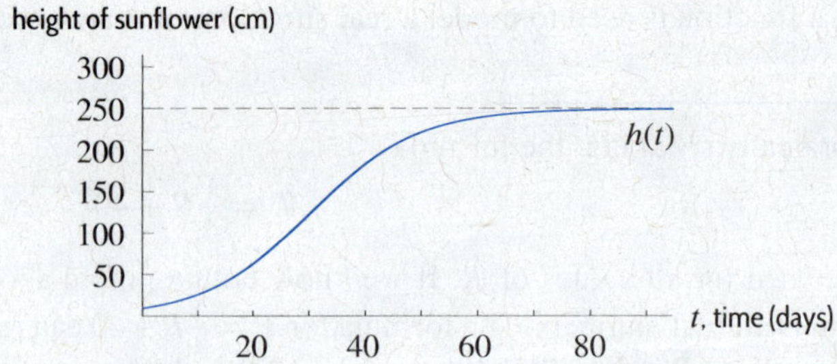


**Figure 2.12:** Height of sunflower as a function of time

**Solution**

(a) Measurements can be made at any time after $t = 0$. The graph starts on the vertical axis and extends to the right, so the domain of the function is $t \geq 0$. If we consider the fact that the sunflower dies on some day $T$, then the domain is $0 \leq t \leq T$.

(b) To find the range, notice that the smallest value of $h$ occurs at $t = 0$. Evaluating gives

$$h(0) = \frac{250}{1 + 24(0.9)^0} = 10 \text{ cm}.$$

This means that the plant was 10 cm high when it was first measured on day $t = 0$. As $t$ increases, the plant grows and $h(t)$ increases. The values of $h(t)$ approach, but never reach, 250. This suggests that the range is $10 \leq h(t) < 250$. This information tells us that sunflowers typically grow to a height of about 250 cm.

## Using Formulas and Graphs to Find Domains and Ranges

When a function is defined by a formula, its domain and range can often be determined by examining the formula algebraically.

**Example 4** State the domain and range of $g$, where

$$g(x) = \frac{1}{x - 2}.$$

**Solution** The domain is all real numbers except those which do not yield an output value. The expression $1/(x - 2)$ is defined for any real number $x$ except $x = 2$ (division by 0 is undefined). Therefore,

$$\text{Domain: all real } x, \quad x \neq 2.$$

The range is all real numbers that the formula can return as output values. It is not possible for $g(x)$ to equal zero, since 1 divided by a real number is never zero. All real numbers except 0 are possible output values, since all nonzero real numbers have reciprocals. To see this algebraically, suppose we ask if a particular value of $y$ is in the range. Then we solve for $y$ in

$$y = \frac{1}{x - 2} \quad \text{giving} \quad x = \frac{1}{y} + 2,$$

so any value of $y$ is possible except $y = 0$. Thus

$$\text{Range: all real values,} \quad g(x) \neq 0.$$

The graph in Figure 2.13 reflects this the domain and this range.

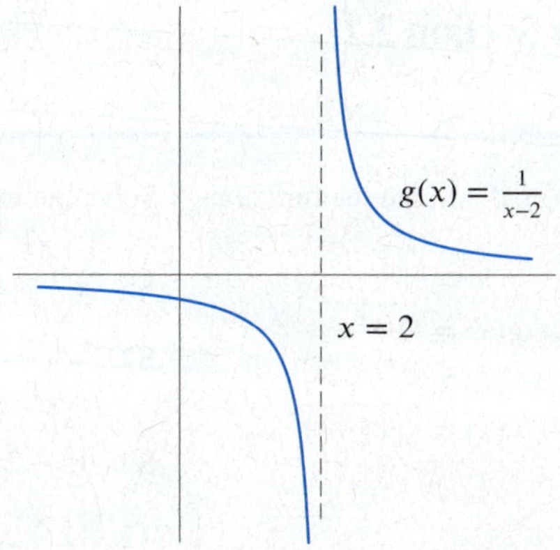


**Figure 2.13**: Domain and range of $g(x) = 1/(x - 2)$

---

**Example 5** Find the domain of the function $f(x) = \dfrac{1}{\sqrt{x-4}}$ by examining its formula and its graph.

**Solution** The domain is all real numbers except those for which the function is undefined. The square root of a negative number is undefined (if we restrict ourselves to real numbers), and so is division by zero. Therefore we need

$$x - 4 > 0.$$

Thus, the domain is all real numbers greater than 4.

$$\text{Domain:} \quad x > 4.$$

From the formula $f(x) = 1/\sqrt{x-4}$ and the graph in Figure 2.14, we see that

$$\text{Range:} \quad f(x) > 0.$$

Notice that in Example 6 on page 71 we saw that the output of $f(x) = 1/\sqrt{x-4}$ cannot be negative. Exercise 21 asks for an algebraic argument.

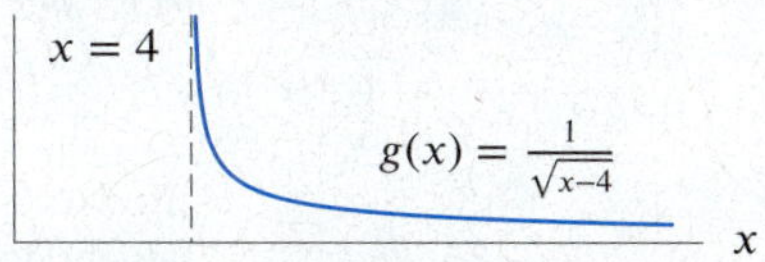


**Figure 2.14**: Domain and range of $f(x) = 1/\sqrt{x-4}$

---

**Example 6** Find the domain of the function $h(x) = \dfrac{1}{\sqrt{4-x^2}}$ by examining its formula.

**Solution** As in Example 5, the domain consists of all $x$-values making the denominator positive, that is:

$$4 - x^2 > 0.$$

Thus, to find the domain, we solve for the $x$-values satisfying $4 > x^2$. That is, we want all the $x$-values whose squares are smaller then 4. These are the numbers between $-2$ and 2, so we have

$$\text{Domain:} \quad -2 < x < 2.$$

---

Rational functions can lead to more complicated domains and ranges, such as those resulting from "holes" in the graph. See Section 11.5, page 469.

## Exercises and Problems for Section 2.2

### Skill Refresher

In Exercises S1–S4, for what value(s), if any, are the functions undefined?

**S1.** $f(x) = \dfrac{x-2}{x-3}$

**S2.** $g(x) = \dfrac{1}{x(x-3)}$

**S3.** $h(x) = \sqrt{x-15}$

**S4.** $k(x) = \sqrt{15-x}$

Solve the inequalities in Exercises S5–S12.

**S5.** $x - 8 > 0$

**S6.** $-x + 5 > 0$

**S7.** $-3(n-4) > 12$

**S8.** $12 \leq 24 - 4a$

**S9.** $x^2 - 25 > 0$

**S10.** $36 - x^2 \geq 0$

**S11.** $12 - 2a^2 \leq a^2$

**S12.** $y^2 - 3 > 15 - y^2$

### Exercises

In Exercises 1–4, estimate the domain and range of the function. Assume the entire graph is shown.

**1.**

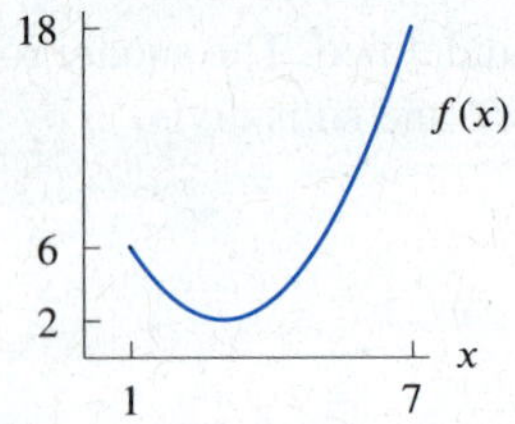


**2.**

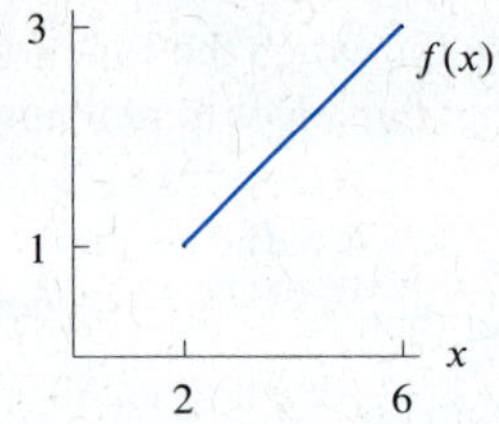


**3.**

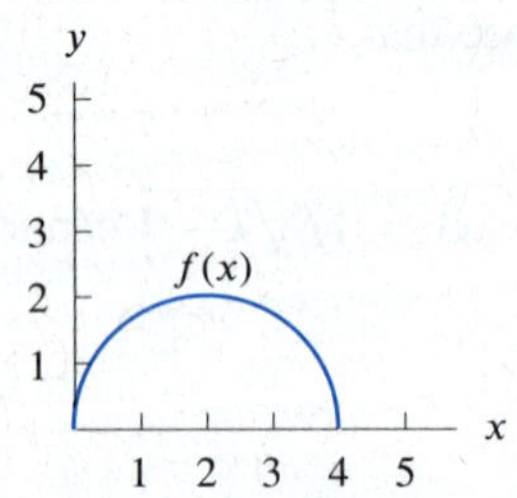


**4.**

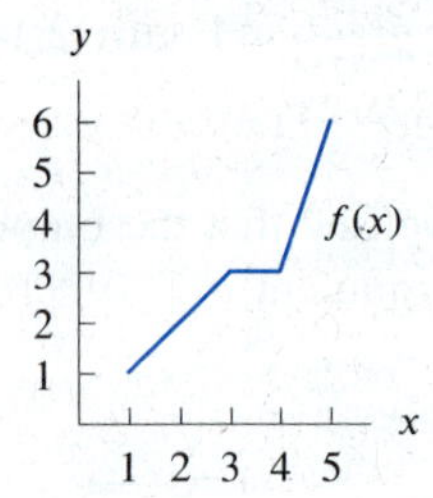


In Exercises 5–8, use a graph to find the range of the function on the given domain.

**5.** $f(x) = \dfrac{1}{x}, \quad -2 \leq x \leq 2$

**6.** $f(x) = \dfrac{1}{x^2}, \quad -1 \leq x \leq 1$

**7.** $f(x) = x^2 - 4, \quad -2 \leq x \leq 3$

**8.** $f(x) = \sqrt{9 - x^2}, \quad -3 \leq x \leq 1$

In Exercises 9–18, find the domain of the function algebraically.

**9.** $f(x) = \dfrac{1}{x+3}$

**10.** $p(t) = \dfrac{1}{t^2 - 4}$

**11.** $f(t) = \dfrac{t-3}{3t+9}$

**12.** $n(q) = \dfrac{1}{q^4 + 2}$

**13.** $f(x) = \dfrac{1}{\sqrt{9+x}}$

**14.** $y(t) = \dfrac{1}{t^4}$

**15.** $f(x) = \sqrt{x^2 - 4}$

**16.** $q(r) = \sqrt[3]{r^2 - 16}$

**17.** $m(x) = x^2 - 9$

**18.** $t(a) = \sqrt[4]{a-2}$

In Exercises 19–22, find the domain and range of the function algebraically.

**19.** $m(q) = \dfrac{1}{5}q - 4$

**20.** $f(x) = \sqrt{15 - 4x}$

**21.** $f(x) = \dfrac{1}{\sqrt{x-4}}$

**22.** $f(x) = \dfrac{1}{\sqrt{9-x}}$

### Problems

In Problems 23–26, you are given the domain $D$ of the function. Where applicable, find possible values for the constants $a$ and $b$.

**23.** $f(x) = \dfrac{1}{x-a}$, D: all real numbers $\neq 3$

**24.** $p(t) = \dfrac{1}{(2t-a)(t+b)}$, D: all real numbers except 4 and 5.

**25.** $n(q) = \sqrt{q^2 + a}$, D: all real numbers

**26.** $m(r) = \sqrt{r-a}$, D: all real numbers $\geq -3$

**27.** Give a formula for a function that is undefined for $x = -2$ and for $x < -4$, but is defined everywhere else.

**28.** Give a formula for a function whose domain is all negative values of $x$ except $x = -5$.

**29.** A restaurant is open from 2 pm to 2 am each day, and a maximum of 200 clients can fit inside. If $f(t)$ is the number of clients in the restaurant $t$ hours after 2 pm each day, what are a reasonable domain and range for $f(t)$?

**30.** A car gets the best mileage at intermediate speeds. Graph the gas mileage as a function of speed. Determine a reasonable domain and range for the function and justify your reasoning.

**31.** A movie theater seats 200 people. For any particular show, the amount of money the theater makes is a function of the number of people, $n$, in attendance. If a ticket costs \$4.00, find the domain and range of this function. Sketch its graph.

**32.** **(a)** Table 2.8 shows the relationship between the number of calories used per minute of walking and a person's weight in pounds.[6] Use the table to

(i) Determine the number of calories that a person weighing 200 lb uses in a half-hour of walking.

(ii) Describe in words the relationship between weight and the number of calories used. Identify the dependent and independent variables and specify whether the function is increasing or decreasing.

**(b)** The function described in part (a)(ii) is approximately linear.

(i) Estimate the equation for this function and graph it.

(ii) Give a reasonable domain and range for your function based on the data in the table.

(iii) Use your function to estimate how many calories per minute a person who weighs 135 lb uses per minute of walking.

**Table 2.8** *Calories per minute as a function of weight*

| Weight | 100 lb | 120 lb | 150 lb | 170 lb | 200 lb | 220 lb |
|---|---|---|---|---|---|---|
| Walking | 2.7 | 3.2 | 4.0 | 4.6 | 5.4 | 5.9 |

In Exercises 33–34, find the domain and range of the function.

**33.** $g(x) = a + 1/x$, where $a$ is a constant

**34.** $q(x) = (x - b)^{1/2} + 6$, where $b$ is a constant

**35.** The last digit, $d$, of a phone number is a function of $n$, its position in the phone book. Table 2.9 gives $d$ for the first 10 listings in the 2009 New York State telephone directory.[7] The table shows that the last digit of the first listing is 1, the last digit of the second listing is 5, and so on. In principle we could use a phone book to figure out other values of $d$. For instance, if $n = 300$, we could count down to the 300th listing in order to determine $d$. So we write $d = f(n)$.

**(a)** What is the value of $f(6)$?

**(b)** Explain how you could use the phone book to find the domain of $f$.

**(c)** What is the range of $f$?

**Table 2.9**

| $n$ | 1 | 2 | 3 | 4 | 5 | 6 | 7 | 8 | 9 | 10 |
|---|---|---|---|---|---|---|---|---|---|---|
| $d$ | 1 | 2 | 1 | 5 | 9 | 9 | 0 | 1 | 1 | 7 |

**36.** Let $t$ be time in seconds and let $r(t)$ be the rate, in gallons/second, that water enters a reservoir:

$$r(t) = 800 - 40t.$$

**(a)** Evaluate the expressions $r(0)$, $r(15)$, $r(25)$, and explain their physical significance.

**(b)** Graph $y = r(t)$ for $0 \le t \le 30$, labeling the intercepts. What is the physical significance of the slope and the intercepts?

**(c)** For $0 \le t \le 30$, when does the reservoir have the most water? When does it have the least water?

**(d)** What are the domain and range of $r(t)$?

**37.** In month $t = 0$, a small group of rabbits escapes from a ship onto an island where there are no rabbits. The island rabbit population, $p(t)$, in month $t$ is given by

$$p(t) = \frac{1000}{1 + 19(0.9)^t}, \quad t \ge 0.$$

**(a)** Evaluate $p(0)$, $p(10)$, $p(50)$, and explain their meaning in terms of rabbits.

**(b)** Graph $p(t)$ for $0 \le t \le 100$. Describe the graph in words. Does it suggest the growth in population you would expect among rabbits on an island?

**(c)** Estimate the range of $p(t)$. What does this tell you about the rabbit population?

**(d)** Explain how you can find the range of $p(t)$ from its formula.

**38.** Bronze is an alloy or mixture of the metals copper and tin. The properties of bronze depend on the percentage of copper in the mix. A chemist decides to study the properties of a given alloy of bronze as the proportion of copper is varied. She starts with 9 kg of bronze that contain 3 kg of copper and 6 kg of tin and either adds or removes copper. Let $f(x)$ be the percentage of copper in the mix if $x$ kg of copper are added ($x > 0$) or removed ($x < 0$).

**(a)** State the domain and range of $f$. What does your answer mean in the context of bronze?

**(b)** Find a formula in terms of $x$ for $f(x)$.

**(c)** If the formula you found in part (b) was not intended to represent the percentage of copper in an alloy of bronze, but instead simply defined an abstract mathematical function, what would be the domain and range of this function?

[6] Source: 1993 World Almanac. Speeds assumed are 3 mph for walking, 10 mph for bicycling, and 2 mph for swimming.

[7] www6.oft.state.ny.us/telecom/phones/. Accessed February 10, 2010.

**39.** The surface area of a cylindrical aluminum can is a measure of how much aluminum the can requires. If the can has radius $r$ and height $h$, its surface area $A$ and its volume $V$ are given by the equations:

$$A = 2\pi r^2 + 2\pi rh \quad \text{and} \quad V = \pi r^2 h.$$

**(a)** The volume, $V$, of a 12-oz cola can is 355 cm$^3$. A cola can is approximately cylindrical. Express its surface area $A$ as a function of its radius $r$, where $r$ is measured in centimeters. [Hint: First solve for $h$ in terms of $r$.]

**(b)** Graph $A = s(r)$, the surface area of a cola can whose volume is 355 cm$^3$, for $0 \le r \le 10$.

**(c)** What is the domain of $s(r)$? Based on your graph, what, approximately, is the range of $s(r)$?

**(d)** The manufacturers wish to use the smallest amount of aluminum (in cm$^2$) necessary to make a 12-oz cola can. Use your answer in (c) to find the minimum amount of aluminum needed. State the values of $r$ and $h$ that minimize the amount of aluminum used.

**(e)** The radius of a real 12-oz cola can is about 3.25 cm. Show that real cola cans use more aluminum than necessary to hold 12 oz of cola. Why do you think real cola cans are made in this way?

## 2.3 PIECEWISE-DEFINED FUNCTIONS

A function may employ different formulas on different parts of its domain. Such a function is said to be *piecewise defined*. For example, the function graphed in Figure 2.15 has the following formulas:

$$y = x^2 \quad \text{for } x \le 2 \qquad\qquad \text{or more compactly} \quad y = \begin{cases} x^2 & \text{for } x \le 2 \\ 6 - x & \text{for } x > 2. \end{cases}$$
$$y = 6 - x \quad \text{for } x > 2$$

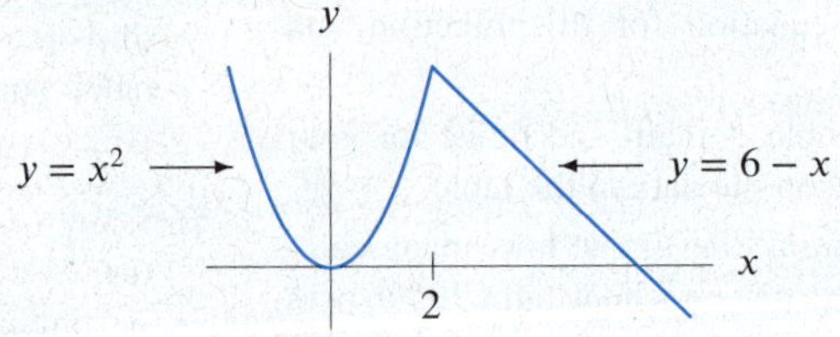


Figure 2.15: Piecewise defined function

**Example 1** Graph the function $y = g(x)$ given by the following formulas:

$$g(x) = x + 1 \quad \text{for} \quad x \le 2 \qquad \text{and} \qquad g(x) = 1 \quad \text{for} \quad x > 2.$$

Using bracket notation, this function is written:

$$g(x) = \begin{cases} x + 1 & \text{for } x \le 2 \\ 1 & \text{for } x > 2. \end{cases}$$

**Solution** For $x \le 2$, graph the line $y = x + 1$. The solid dot at the point $(2, 3)$ shows that it is included in the graph. For $x > 2$, graph the horizontal line $y = 1$. See Figure 2.16. The open circle at the point $(2, 1)$ shows that it is not included in the graph. (Note that $g(2) = 3$, and $g(2)$ cannot have more than one value.)

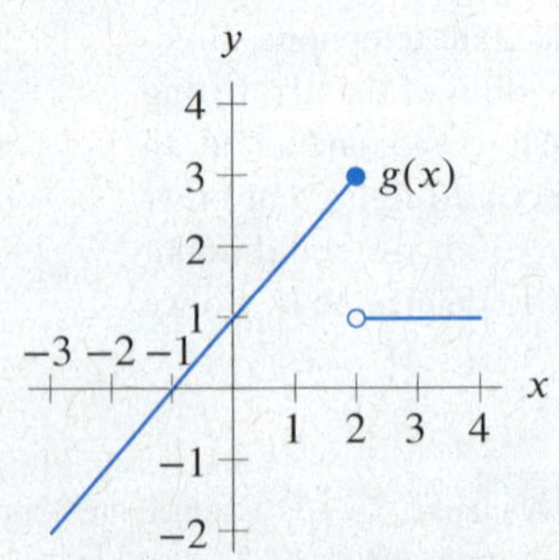


Figure 2.16: Graph of the piecewise defined function $g$

**Example 2** A data plan for a smart phone charges 99 cents per day for data usage up to 20 megabytes and 7 cents for each additional megabyte or part of a megabyte.

(a) Use bracket notation to write a formula for the daily cost, $C$, of the data plan as a function of the number of megabytes $m$ used.
(b) Graph the function.
(c) State the domain and range of the function.

**Solution** (a) For $0 \leq m \leq 20$, the value of $C$ is 99 cents. If $m > 20$, we subtract 20 to find the additional number of megabytes used and multiply by the rate, 7 cents per megabyte.[8] The cost function in cents is thus

$$C = f(m) = \begin{cases} 99 & \text{for } 0 \leq m \leq 20 \\ 99 + 7(m-20) & \text{for } m > 20, \end{cases}$$

or, after simplifying,

$$C = f(t) = \begin{cases} 99 & \text{for } 0 \leq m \leq 20 \\ 7m - 41 & \text{for } m > 20. \end{cases}$$

(b) See Figure 2.17.
(c) Because negative data usage does not make sense, the domain is $m \geq 0$. From the graph, we see that the range is $C \geq 99$.

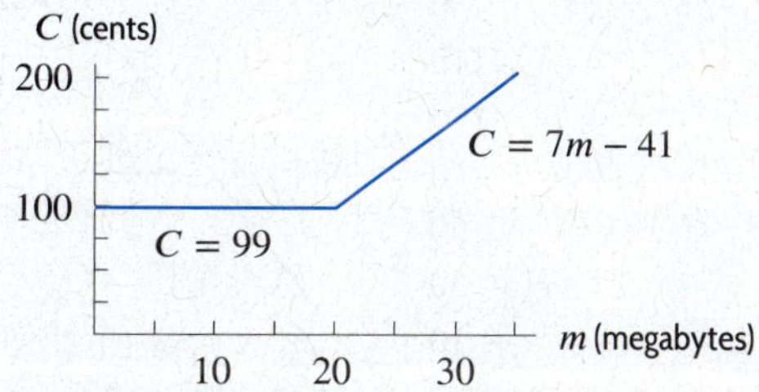


**Figure 2.17:** Daily cost of a data plan

**Example 3** The Ironman Triathlon is a race that consists of three parts: a 2.4-mile swim followed by a 112-mile bike race and then a 26.2-mile marathon. A participant swims steadily at 2 mph, cycles steadily at 20 mph, and then runs steadily at 9 mph.[9] Assuming that no time is lost during the transition from one stage to the next, find a formula for the distance covered, $d$, in miles, as a function of the elapsed time $t$ in hours, from the beginning of the race. Graph the function.

**Solution** For each leg of the race, we use the formula Distance = Rate · Time. First, we calculate how long it took for the participant to cover each of the three parts of the race. The first leg took $2.4/2 = 1.2$ hours, the second leg took $112/20 = 5.6$ hours, and the final leg took $26.2/9 \approx 2.91$ hours. Thus, the participant finished the race in $1.2 + 5.6 + 2.91 = 9.71$ hours.

During the first leg, $t \leq 1.2$ and the speed is 2 mph, so

$$d = 2t \qquad \text{for} \qquad 0 \leq t \leq 1.2.$$

During the second leg, $1.2 < t \leq 1.2 + 5.6 = 6.8$ and the speed is 20 mph. The length of time spent in the second leg is $(t - 1.2)$ hours. Thus, by time $t$,

$$\text{Distance covered in the second leg} = 20(t - 1.2) \quad \text{for } 1.2 < t \leq 6.8.$$

When the participant is in the second leg, the total distance covered is the sum of the distance covered in the first leg (2.4 miles) plus the part of the second leg that has been covered by time $t$:

$$\begin{aligned} d &= 2.4 + 20(t - 1.2) \\ &= 20t - 21.6 \qquad \text{for } 1.2 < t \leq 6.8. \end{aligned}$$

[8] In actuality, most plans round data usage to whole megabytes or blocks of megabytes.
[9] Personal communication Susan Reid, Athletics Department, University of Arizona.

In the third leg, $6.8 < t \le 9.71$ and the speed is 9 mph. Since 6.8 hours were spent on the first two parts of the race, the length of time spent on the third leg is $(t - 6.8)$ hours. Thus, by time $t$,

$$\text{Distance covered in the third leg} = 9(t - 6.8) \quad \text{for } 6.8 < t \le 9.71.$$

When the participant is in the third leg, the total distance covered is the sum of the distances covered in the first leg (2.4 miles) and the second leg (112 miles), plus the part of the third leg that has been covered by time $t$:

$$\begin{aligned} d &= 2.4 + 112 + 9(t - 6.8) \\ &= 9t + 53.2 \qquad \text{for } 6.8 < t \le 9.71. \end{aligned}$$

The formula for $d$ is different on different intervals of $t$:

$$d = \begin{cases} 2t & \text{for} \quad 0 \le t \le 1.2 \\ 20t - 21.6 & \text{for} \quad 1.2 < t \le 6.8 \\ 9t + 53.2 & \text{for} \quad 6.8 < t \le 9.71. \end{cases}$$

Figure 2.18 gives a graph of the distance covered, $d$, as a function of time, $t$. Notice the three pieces.

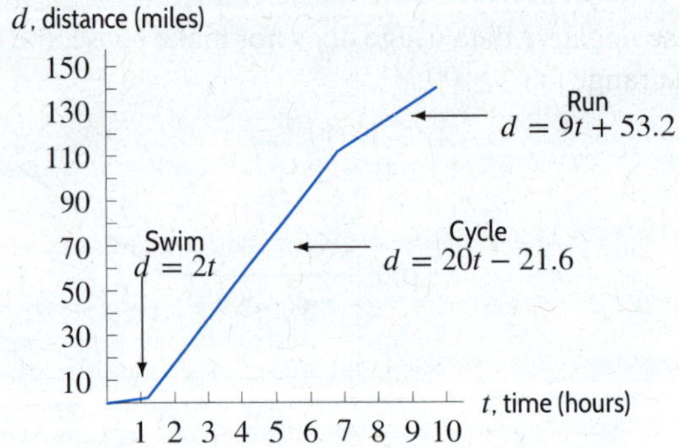


**Figure 2.18**: Ironman Triathlon: $d$ as a function of $t$

## The Absolute Value Function

The absolute value of $x$, written $|x|$, is defined piecewise:

$$\text{For nonnegative } x, \quad |x| = x.$$

$$\text{For negative } x, \quad |x| = -x.$$

(Remember that $-x$ is a positive number if $x$ is a negative number.) For example, if $x = -3$, then

$$|-3| = -(-3) = 3.$$

For $x = 0$, we have $|0| = 0$. This leads to the following two-part definition:

> The **absolute value function** is defined by
>
> $$f(x) = |x| = \begin{cases} x & \text{for} \quad x \ge 0 \\ -x & \text{for} \quad x < 0 \end{cases}.$$

Table 2.10 gives values of $f(x) = |x|$ and Figure 2.19 shows a graph of $f(x)$.

**Table 2.10** *Absolute value function*

| $x$ | $\lvert x\rvert$ |
|---|---|
| −3 | 3 |
| −2 | 2 |
| −1 | 1 |
| 0 | 0 |
| 1 | 1 |
| 2 | 2 |
| 3 | 3 |

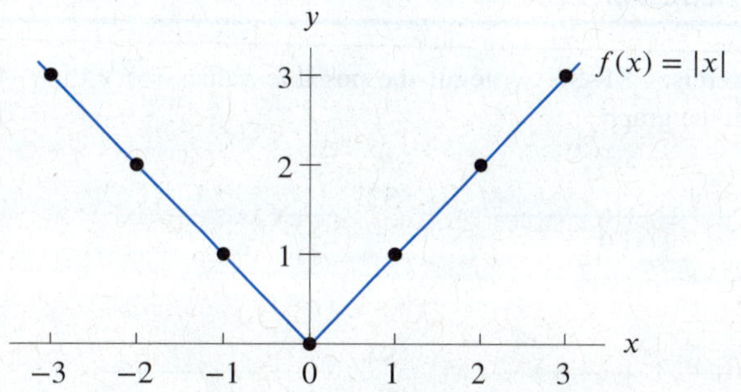


**Figure 2.19**: Graph of absolute value function

## Visualizing Absolute Value on the Number Line

Notice that the absolute value of a number is always the distance between that number and zero on the number line. For examples, see Figure 2.20.

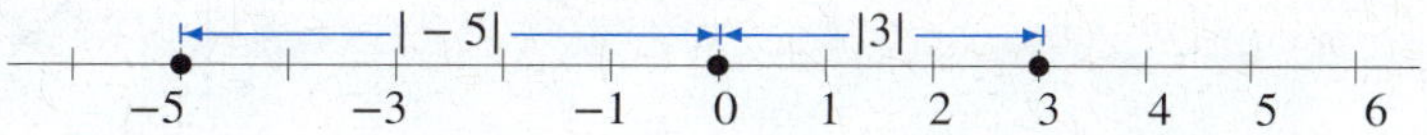


**Figure 2.20**: Distance from 3 to 0 is $|3|$. Distance from −5 to 0 is $|-5|$

Similarly, we can think of $|x-4|$ as the distance between $x$ and 4, as in Figure 2.21.

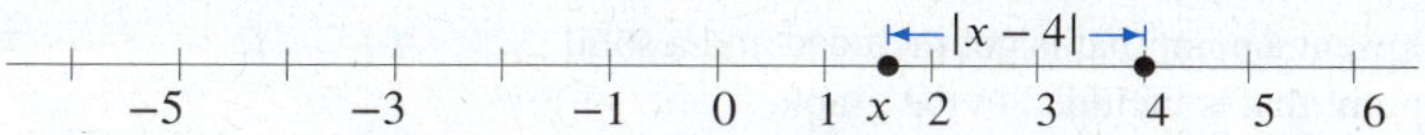


**Figure 2.21**: Distance from $x$ to 4 is $|x-4|$

In general:

> If $a$ and $b$ are real numbers, then $|a-b|$ is the distance between $a$ and $b$ on the number line.

**Example 4** Let $f(x) = |2x-7|$.

(a) What are the domain and range of $f(x)$? (b) Find all values of $x$ such that $f(x) = 3$.

**Solution**

(a) The function $f(x)$ gives the distance between $2x$ and 7 on the number line. This distance can be calculated for any real number $x$, so the domain is all real numbers. The distance can be any nonnegative real number, so the range of is all real numbers greater than or equal to 0.

(b) We want to find all numbers $x$ such that $|2x-7| = 3$. That is, we want the distance between $2x$ and 7 to be 3. Thus, $2x$ must be three units to the left or three units to the right of 7; that is,

$$2x = 4 \qquad \text{or} \qquad 2x = 10.$$

In the first case, $x = 2$; in the second, $x = 5$. So the values of $x$ such that $f(x) = 3$ are $x = 2$ and $x = 5$.

# Exercises and Problems for Section 2.3

## Skill Refresher

In Exercises S1–S6, write all the possible values for $x$ that match the graph.

**S1.** 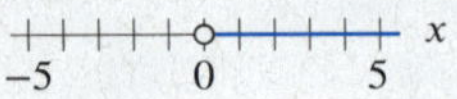

**S2.** 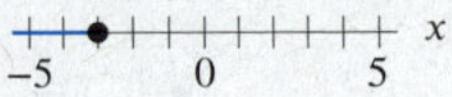

**S3.** 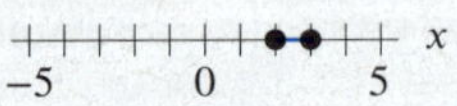

**S4.** 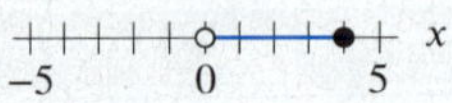

**S5.** 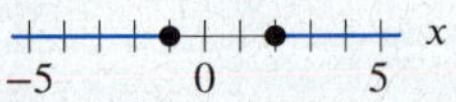

**S6.** 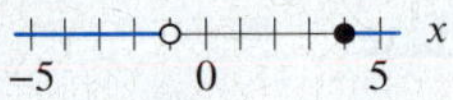

In Exercises S7–S10, determine the domain and range of the function.

**S7.** 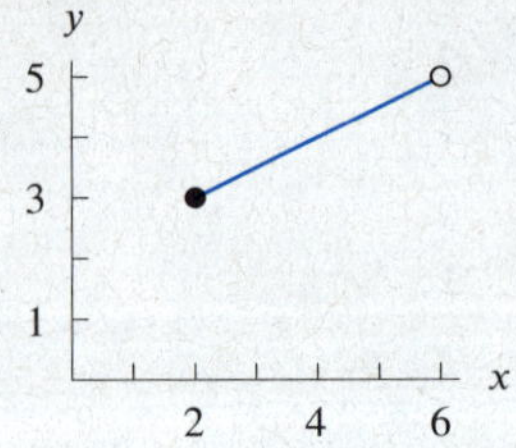

**S8.** 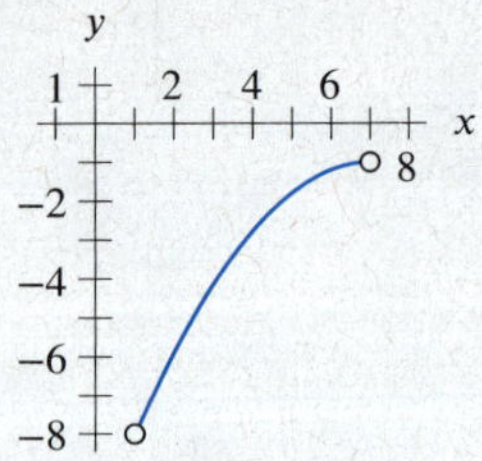

**S9.** 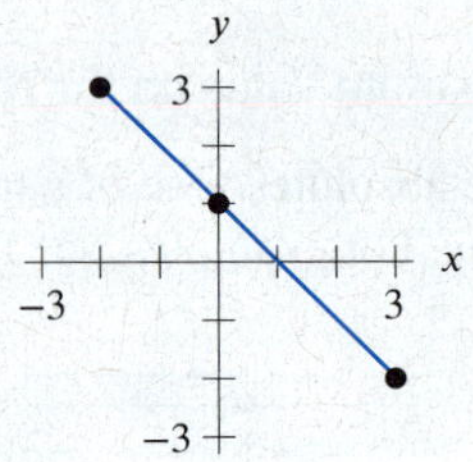

**S10.** 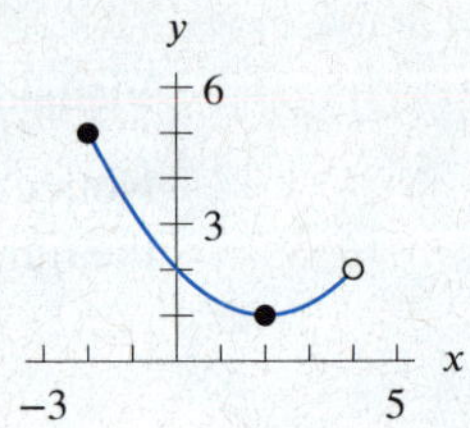

## Exercises

In Exercises 1–4, graph the piecewise defined function. Use an open circle to represent a point that is not included and a solid dot to indicate a point that is included in the graph.

**1.** $f(x) = \begin{cases} -1, & -1 \leq x < 0 \\ 0, & 0 \leq x < 1 \\ 1, & 1 \leq x < 2 \end{cases}$

**2.** $f(x) = \begin{cases} x+1, & -2 \leq x < 0 \\ x-1, & 0 \leq x < 2 \\ x-3, & 2 \leq x < 4 \end{cases}$

**3.** $f(x) = \begin{cases} x+4, & x \leq -2 \\ 2, & -2 < x < 2 \\ 4-x, & x \geq 2 \end{cases}$

**4.** $f(x) = \begin{cases} x^2, & x \leq 0 \\ \sqrt{x}, & 0 < x < 4 \\ x/2, & x \geq 4 \end{cases}$

In Exercises 5–8, write formulas for the functions.

**5.** 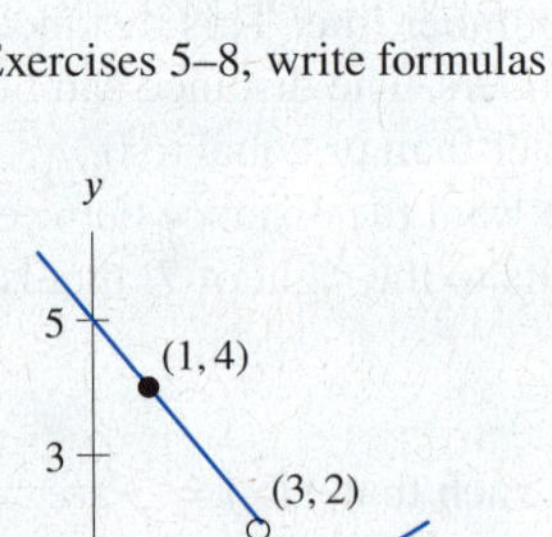


**6.** 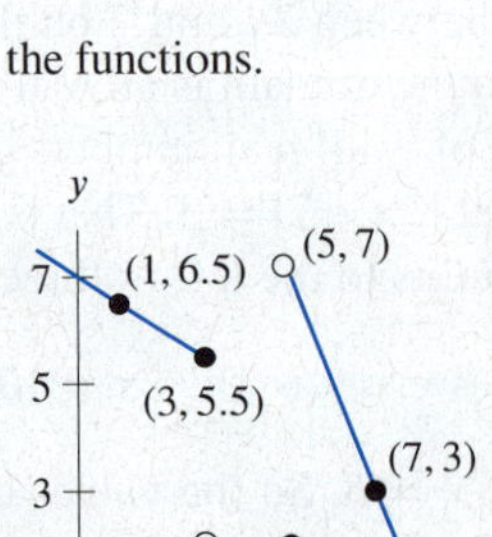


**7.** 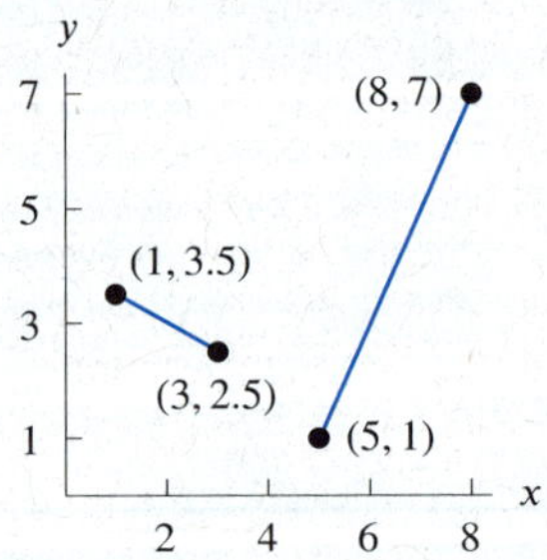


**8.** 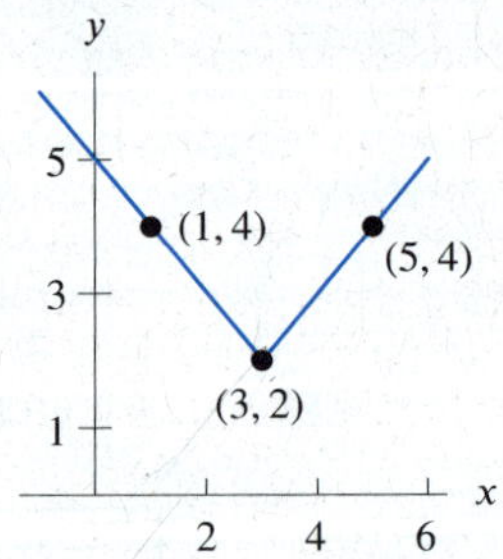


For Exercises 9–10, find the domain and range.

**9.** $G(x) = \begin{cases} x+1 & \text{for } x < -1 \\ x^2+3 & \text{for } x \geq -1 \end{cases}$

**10.** $F(x) = \begin{cases} x^3 & \text{for } x \leq 1 \\ 1/x & \text{for } x > 1 \end{cases}$

Find all $x$ values satisfying the equations in Exercises 11–16.

**11.** $|x| = 5$

**12.** $2|y| - 20 = 0$

**13.** $|x-3| = 7$

**14.** $20 = 10|x-5|$

**15.** $|y-3| = 8 - |y-3|$

**16.** $|2y-6| = 8$

## Problems

**17.** Figure 2.22 shows an entire graph. An open circle represents a point that is not included.

**(a)** Is $y$ a function of $x$? Explain.
**(b)** Is $x$ a function of $y$? Explain.
**(c)** If you identified a function in part (a) or part (b), what is the range of that function?

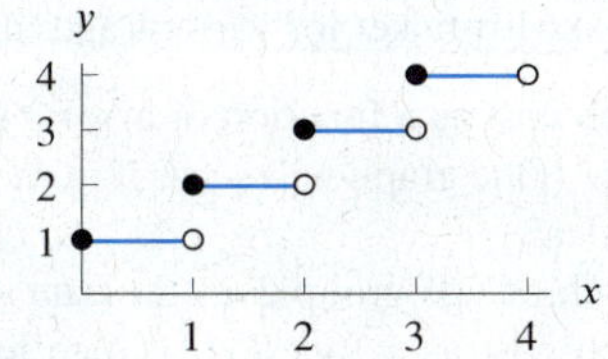


Figure 2.22

**18.** Many people believe that $\sqrt{x^2} = x$. We will investigate this claim graphically and numerically.

**(a)** Graph the two functions $x$ and $\sqrt{x^2}$ in the window $-5 \le x \le 5$, $-5 \le y \le 5$. Based on what you see, do you believe that $\sqrt{x^2} = x$? What function does the graph of $\sqrt{x^2}$ remind you of?
**(b)** Complete Table 2.11. Based on this table, do you believe that $\sqrt{x^2} = x$? What function does the table for $\sqrt{x^2}$ remind you of? Is this the same function you found in part (a)?

Table 2.11

| $x$ | $-5$ | $-4$ | $-3$ | $-2$ | $-1$ | 0 | 1 | 2 | 3 | 4 | 5 |
|---|---|---|---|---|---|---|---|---|---|---|---|
| $\sqrt{x^2}$ | | | | | | | | | | | |

**(c)** Explain how you know that $\sqrt{x^2}$ is the same as the function $|x|$.
**(d)** Graph the function $\sqrt{x^2} - |x|$ in the window $-5 \le x \le 5$, $-5 \le y \le 5$. Explain what you see.

**19.** **(a)** Graph $u(x) = |x|/x$ in the window $-5 \le x \le 5$, $-5 \le y \le 5$. Explain what you see.
**(b)** Complete Table 2.12. Does this table agree with what you found in part (a)?

Table 2.12

| $x$ | $-5$ | $-4$ | $-3$ | $-2$ | $-1$ | 0 | 1 | 2 | 3 | 4 | 5 |
|---|---|---|---|---|---|---|---|---|---|---|---|
| $\|x\|/x$ | | | | | | | | | | | |

**(c)** Identify the domain and range of $u(x)$.
**(d)** Comment on the claim that $u(x)$ can be written as

$$u(x) = \begin{cases} -1 & \text{if } x < 0, \\ 0 & \text{if } x = 0, \\ 1 & \text{if } x > 0. \end{cases}$$

**20.** Let $g(x) = \begin{cases} -1 & \text{for } x < 0 \\ x^3 & \text{for } x \ge 0 \end{cases}$

**(a)** Find $g(-2)$, $g(2)$, and $g(0)$.
**(b)** Find the domain and range of $g(x)$.

**21.** Let $f(x) = \begin{cases} 3x & \text{for } -1 \le x \le 1 \\ -x+4 & \text{for } 1 < x \le 5 \end{cases}$

**(a)** Find $f(0)$ and $f(3)$.
**(b)** Find the domain and range of $f(x)$.

**22.** Let $f(x) = \begin{cases} 1/x & \text{for } x < -1 \\ x^2 & \text{for } -1 \le x \le 1 \\ \sqrt{x} & \text{for } x > 1. \end{cases}$

**(a)** Evaluate $f(-2)$ and $f(2)$.
**(b)** What is the range of $f$?

**23.** Use bracket notation to write a formula for the piecewise function that is defined by $y = x^2$ for negative values of $x$ and by $y = x - 1$ for $x$ values that are greater than or equal to zero.

**24.** A floor-refinishing company charges \$1.83 per square foot to strip and refinish a tile floor for up to 1000 square feet. There is an additional charge of \$350 for toxic waste disposal for any job that includes more than 150 square feet of tile.

**(a)** Express the cost, $y$, of refinishing a floor as a function of the number of square feet, $x$, to be refinished.
**(b)** Graph the function. Give the domain and range.

**25.** The charge for a taxi ride in New York City is \$2.50 upon entry and \$0.40 for each $1/5$ of a mile traveled (rounded up to the nearest $1/5$ mile), when the taxicab is traveling at 6 mph or more. In addition, a New York State Tax Surcharge of \$0.50 is added to the fare.[10]

**(a)** Make a table showing the cost of a trip as a function of its length. Your table should start at zero and go up to two miles in $1/5$-mile intervals.
**(b)** What is the cost for a 1.2-mile trip?
**(c)** How far can you go for \$5.80?
**(d)** Graph the cost function in part (a).

[10]www.nyc.gov/html/tlc/html/passenger/taxicab_rate.shtml, accessed February 11, 2010.

**26.** A museum charges \$40 for a group of 10 or fewer people. A group of more than 10 people must, in addition to the \$40, pay \$2 per person for the number of people above 10. For example, a group of 12 pays \$44 and a group of 15 pays \$50. The maximum group size is 50.

**(a)** Draw a graph that represents this situation.
**(b)** What are the domain and range of the cost function?

**27.** At a supermarket checkout, a scanner records the prices of the foods you buy. In order to protect consumers, the state of Michigan passed a "scanning law" that says something similar to the following:

> If there is a discrepancy between the price marked on the item and the price recorded by the scanner, the consumer is entitled to receive 10 times the difference between those prices; this amount given must be at least \$1 and at most \$5. Also, the consumer will be given the difference between the prices, in addition to the amount calculated above.

For example: If the difference is 5¢, you should receive \$1 (since 10 times the difference is only 50¢ and you are to receive at least \$1), plus the difference of 5¢. Thus, the total you should receive is $\$1.00 + \$0.05 = \$1.05$.
If the difference is 25¢, you should receive 10 times the difference in addition to the difference, giving $(10)(0.25) + 0.25 = \$2.75$.
If the difference is 95¢, you should receive \$5 (because $10(.95) = \$9.50$ is more than \$5, the maximum penalty), plus 95¢, giving $5 + 0.95 = \$5.95$.

**(a)** What is the lowest possible refund?
**(b)** Suppose $x$ is the difference between the price scanned and the price marked on the item, and $y$ is the amount refunded to the customer. Write a formula for $y$ in terms of $x$. [Hint: Look at the sample calculations.]
**(c)** What would the difference between the price scanned and the price marked have to be in order to obtain a \$9.00 refund?
**(d)** Graph $y$ as a function of $x$.

**28.** Seattle City Light[11] charges residents for electricity on a daily basis. There is a basic daily charge of 15.70 cents. In addition, for each day the first 10 kWh cost 4.66 cents per kWh and any additional kWhs are 10.71 cents per kWh. (A kWh is a unit of energy.)

**(a)** Make a table showing the cost in dollars of usage from 0 to 40 kWh in increments of 5 kWh.
**(b)** Write a piecewise defined function to describe the usage rate.
**(c)** What is the cost for 33 kWh?
**(d)** How many kWh can you burn on a day for \$3?

**29.** Gore Mountain is a ski resort in the Adirondack mountains in upstate New York. Table 2.13 shows the cost of a weekday ski-lift ticket for various ages and dates.

**(a)** Graph cost as a function of age for each time period given. (One graph will serve for times when rates are identical.)
**(b)** For which age group does the date affect cost?
**(c)** Graph cost as a function of date for the age group mentioned in part (b).
**(d)** Why does the cost fluctuate as a function of date?

**Table 2.13** *Ski-lift ticket prices at Gore Mountain, 1998–1999*[12]

| Age | Opening-Dec 12 | Dec 13-Dec 24 | Dec 25-Jan 3 | Jan 4-Jan 15 | Jan 16-Jan 18 |
|---|---|---|---|---|---|
| Up to 6 | Free | Free | Free | Free | Free |
| 7–12 | \$19 | \$19 | \$19 | \$19 | \$19 |
| 13–69 | \$29 | \$34 | \$39 | \$34 | \$39 |
| 70+ | Free | Free | Free | Free | Free |

| Age | Jan 19-Feb 12 | Feb 13-Feb 21 | Feb 22-Mar 28 | Mar 29-Closing |
|---|---|---|---|---|
| Up to 6 | Free | Free | Free | Free |
| 7–12 | \$19 | \$19 | \$19 | \$19 |
| 13–69 | \$34 | \$39 | \$34 | \$29 |
| 70+ | Free | Free | Free | Free |

In Problems 30–31:

**(a)** Use the definition of absolute value to write a piecewise formula for $f$.
**(b)** Graph $f$.

**30.** $f(x) = |x^2 - 4|$

**31.** $f(x) = |2x - 6|$

## 2.4 PREVIEW OF TRANSFORMATIONS: SHIFTS

Suppose we shift the graph of a function vertically or horizontally, giving the graph of a new function. In this section we see the relationship between the formulas for the original function and the new function.

[11] http://www.seattle.gov, accessed April 2013
[12] The Olympic Regional Development Authority.

## Vertical Shift: The Heating Schedule for an Office Building

We start with an example of a vertical shift in the context of the heating schedule for a building.

**Example 1** To save money, an office building is kept warm only during business hours. Figure 2.23 shows the temperature, $H$, in °F, as a function of time, $t$, in hours after midnight. At midnight ($t = 0$), the building's temperature is 50°F. This temperature is maintained until 4 am. Then the building begins to warm up so that by 8 am the temperature is 70°F. At 4 pm the building begins to cool. By 8 pm, the temperature is again 50°F.

Suppose that the building's superintendent decides to keep the building 5°F warmer than before. Sketch a graph of the new function.

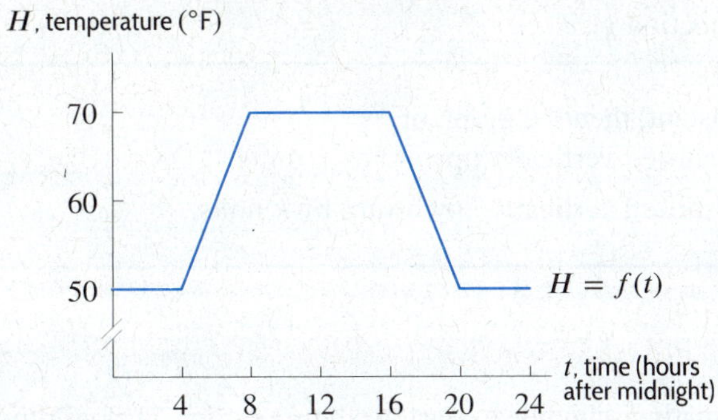


Figure 2.23: The heating schedule at an office building, $H = f(t)$

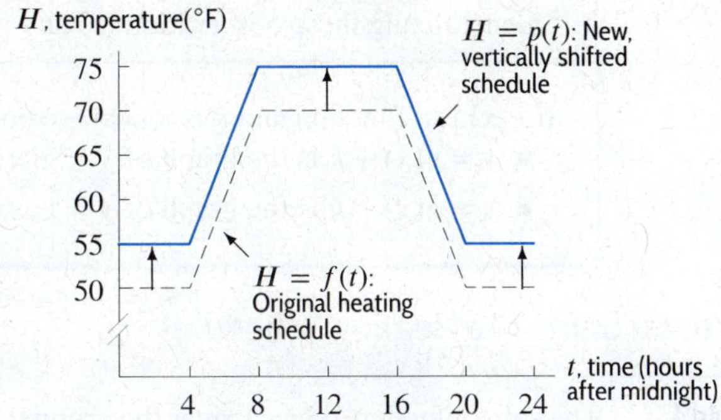


Figure 2.24: Graph of new heating schedule, $H = p(t)$, obtained by shifting original graph, $H = f(t)$, upward by 5 units

**Solution** The graph of $f$, the heating schedule function of Figure 2.23, is shifted upward by 5 units. The new heating schedule, $H = p(t)$, is graphed in Figure 2.24. The building's overnight temperature is now 55°F instead of 50°F and its daytime temperature is 75°F instead of 70°F. The 5°F increase in temperature corresponds to the 5-unit vertical shift in the graph.

**Example 2** What is the relationship between the formula for $f(t)$, the original heating schedule and $p(t)$, the new heating schedule?

**Solution** The temperature under the new schedule, $p(t)$, is always 5°F warmer than the temperature under the old schedule, $f(t)$. Thus, at any time $t$,

$$\underbrace{\text{New temperature}}_{p(t)} = \underbrace{\text{Old temperature}}_{f(t)} + 5.$$

So the relationship between $p$ and $f$ is given by the equation

$$p(t) = f(t) + 5.$$

We can get information from the relationship $p(t) = f(t) + 5$ even though we do not have an explicit formula for $f$ or $p$.

Suppose we want to know the temperature at 6 am under the schedule $p(t)$. The graph of $f(t)$ shows that under the old schedule $f(6) = 60$. Substituting $t = 6$ into the equation relating $f$ and $p$ gives

$$p(6) = f(6) + 5 = 60 + 5 = 65.$$

Thus, at 6 am the temperature under the new schedule is 65°F.

**Example 3** Find $q(t)$, the formula for the heating schedule if at each time the temperature is 2°F lower than the original temperature.

**Solution** At any time $t$

$$\underbrace{\text{New temperature}}_{q(t)} = \underbrace{\text{Old temperature}}_{f(t)} - 2.$$

So we have

$$q(t) = f(t) - 2.$$

The graph of $q(t)$ is the graph of $f$ shifted down by 2 units.

Generalizing these observations to any function $g$:

> If $g(x)$ is a function and $k$ is a positive constant, then the graph of
> - $y = g(x) + k$ is the graph of $y = g(x)$ shifted vertically upward by $k$ units.
> - $y = g(x) - k$ is the graph of $y = g(x)$ shifted vertically downward by $k$ units.

## Horizontal Shift: The Heating Schedule

**Example 4** The superintendent then changes the original heating schedule to start two hours earlier. The building now begins to warm at 2 am instead of 4 am, reaches 70°F at 6 am instead of 8 am, begins cooling off at 2 pm instead of 4 pm, and returns to 50°F at 6 pm instead of 8 pm. How are these changes reflected in the graph of the heating schedule?

**Solution** Figure 2.25 gives a graph of $H = r(t)$, the new heating schedule, which is obtained by shifting the graph of the original heating schedule, $H = f(t)$, two units to the left.

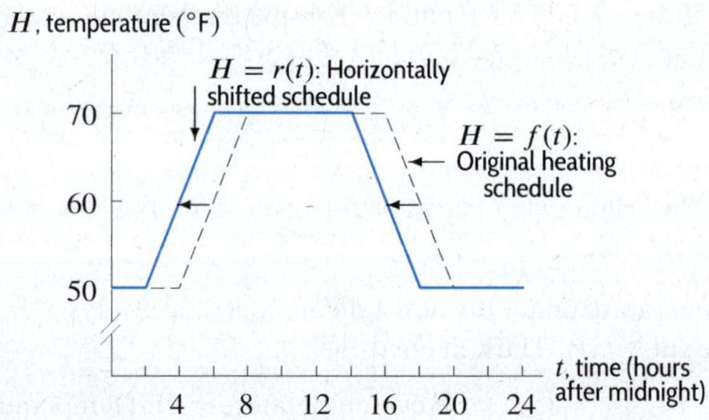


**Figure 2.25:** Graph of new heating schedule, $H = r(t)$, found by shifting $f$, the original graph, 2 units to the left

Notice that the upward shift in Example 1 results in a warmer temperature, whereas the leftward shift in Example 4 results in an earlier schedule.

**Example 5** In Example 4 the heating schedule was changed to 2 hours earlier, shifting the graph horizontally 2 units to the left. Find a formula for $r$, this new schedule, in terms of $f$, the original schedule.

**Solution** The old schedule always reaches a given temperature 2 hours after the new schedule. For example, at 4 am the temperature under the new schedule reaches 60°. The temperature under the old schedule reaches 60° at 6 am, 2 hours later. In general, we see that

$$\begin{array}{c} \text{Temperature under new schedule} \\ \text{at time } t \end{array} = \begin{array}{c} \text{Temperature under old schedule} \\ \text{at time } (t+2)\text{, two hours later.} \end{array}$$

Algebraically, we have

$$r(t) = f(t+2).$$

This is a formula for $r$ in terms of $f$.

Let's check the formula from Example 5 by using it to calculate $r(14)$, the temperature under the new schedule at 2 pm. The formula gives

$$r(14) = f(14+2) = f(16).$$

Figure 2.23 shows that $f(16) = 70$. Thus, $r(14) = 70$. This agrees with Figure 2.25.

**Example 6** Suppose now the heating schedule is made 1 hour later than it was originally. Find the function $s(t)$ that describes this schedule.

**Solution** The new schedule reaches a particular temperature 1 hour later than the original. For example, under the old schedule, the temperature reaches 60°F at 6 am, while it reaches 60°F at 7 am under the new schedule. Thus

$$\begin{array}{c}\text{Temperature under new schedule}\\ \text{at time } t\end{array} = \begin{array}{c}\text{Temperature under old schedule}\\ \text{at time } (t-1)\text{, one hour earlier.}\end{array}$$

Thus, we have

$$s(t) = f(t-1).$$

The graph of $s(t)$ is the graph of $f$ shifted to the right by 1 unit.

Generalizing these observations to any function $g$:

> If $g(x)$ is a function and $h$ is a positive constant, then the graph of
> - $y = g(x+h)$ is the graph of $y = g(x)$ shifted horizontally to the left by $h$ units.
> - $y = g(x-h)$ is the graph of $y = g(x)$ shifted horizontally to the right by $h$ units.

A vertical or horizontal shift of the graph of a function is called a *translation* because it does not change the shape of the graph, but simply translates it to another position in the plane. Shifts and translations are examples of *transformations* of a function. We will see others in Chapter 6.

## Inside Versus Outside Changes

Since the horizontal shift in the heating schedule, $q(t) = f(t+2)$, involves a change to the input value, it is called an *inside change* to $f$. Similarly the vertical shift, $p(t) = f(t) + 5$, is called an *outside change* because it involves changes to the output value.

**Example 7** If $n = f(A)$ gives the number of gallons of paint needed to cover a house of area $A$ ft$^2$, explain the meaning of the expressions $f(A+10)$ and $f(A)+10$ in the context of painting.

**Solution** These two expressions are similar in that they both involve adding 10. However, for $f(A+10)$, the 10 is added on the inside, so 10 is added to the area, $A$. Thus,

$$n = f(\underbrace{A+10}_{\text{Area}}) = \begin{array}{c}\text{Amount of paint needed}\\ \text{to cover an area of } (A+10)\text{ ft}^2\end{array} = \begin{array}{c}\text{Amount of paint needed to cover}\\ \text{an area 10 ft}^2\text{ larger than } A.\end{array}$$

The expression $f(A) + 10$ represents an outside change. We are adding 10 to $f(A)$, which represents an amount of paint, not an area. We have

$$n = \underbrace{f(A)}_{\text{Amount of paint}} + 10 = \begin{matrix}\text{Amount of paint needed}\\ \text{to cover region of area } A\end{matrix} + 10 \text{ gals} = \begin{matrix}\text{10 gallons more paint than}\\ \text{amount needed to cover area } A.\end{matrix}$$

In $f(A+10)$, we added 10 square feet on the inside of the function, which means that the area to be painted is now 10 ft$^2$ larger. In $f(A)+10$, we added 10 gallons to the outside, which means that we have 10 more gallons of paint than we need.

## Combining Horizontal and Vertical Shifts

We have seen how a function's formula changes when we shift its graph horizontally or vertically. What happens when we shift it both horizontally and vertically?

**Example 8** Let $g$ be the transformation of the heating schedule function, $H = f(t)$, given by

$$g(t) = f(t-2) - 5.$$

(a) Sketch the graph of $H = g(t)$.
(b) Describe in words the heating schedule determined by $g$.

**Solution** (a) To graph $g$, we break the transformation into two steps. First, we sketch $H = f(t-2)$. This is an inside change to the function $f$, which shifts the graph of $f$ to the right 2 units. Next, we sketch $H = f(t-2) - 5$ by shifting our sketch of $H = f(t-2)$ down 5 units. The result is shown in Figure 2.26.

(b) The function $g$ represents a schedule that is both 2 hours later and 5 degrees cooler than the original schedule.

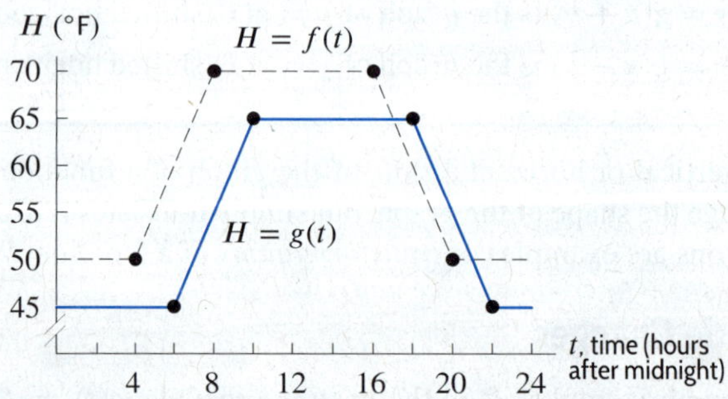


**Figure 2.26**: Graph of $g(t) = f(t-2) - 5$ is graph of $H = f(t)$ shifted right by 2 and down by 5

## Exercises and Problems for Section 2.4

### Skill Refresher

In Exercises S1–S4, evaluate each function at $x = 4$.

**S1.** $f(x) = \sqrt{x}$

**S2.** $g(x) = \sqrt{x} + 6$

**S3.** $h(x) = \sqrt{x} - 3$

**S4.** $k(x) = \sqrt{x+5}$

In Exercises S5–S8, solve for $x$.

**S5.** $x^2 = 4$

**S6.** $x^2 - 5 = 4$

**S7.** $(x-1)^2 = 4$

**S8.** $(x+1)^2 = 4$

## Exercises

**1.** Using Table 2.14, complete the tables for $g$, $h$, $k$, $m$, where:

**(a)** $g(x) = f(x-1)$ **(b)** $h(x) = f(x+1)$
**(c)** $k(x) = f(x)+3$ **(d)** $m(x) = f(x-1)+3$

Explain how the graph of each function relates to the graph of $f(x)$.

**Table 2.14**

| $x$ | −2 | −1 | 0 | 1 | 2 |
|---|---|---|---|---|---|
| $f(x)$ | −3 | 0 | 2 | 1 | −1 |

| $x$ | −1 | 0 | 1 | 2 | 3 |
|---|---|---|---|---|---|
| $g(x)$ | | | | | |

| $x$ | −3 | −2 | −1 | 0 | 1 |
|---|---|---|---|---|---|
| $h(x)$ | | | | | |

| $x$ | −2 | −1 | 0 | 1 | 2 |
|---|---|---|---|---|---|
| $k(x)$ | | | | | |

| $x$ | −1 | 0 | 1 | 2 | 3 |
|---|---|---|---|---|---|
| $m(x)$ | | | | | |

**2.** Figure 2.27 shows $f(x)$. Graph $y = f(x+3)+3$. Label all important features.

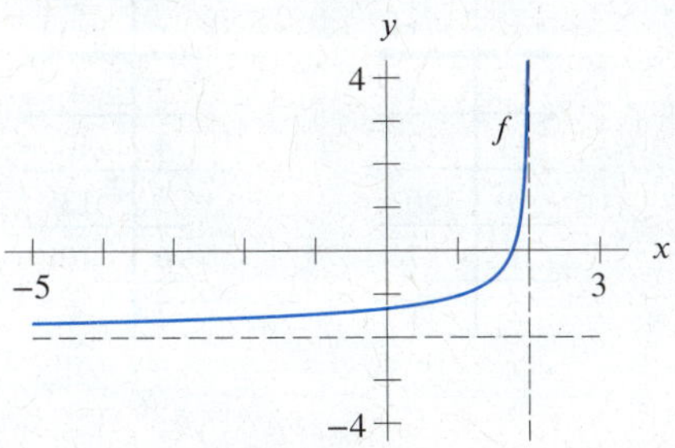


**Figure 2.27**

In Exercises 3–6, use Figure 2.28 to graph the transformation of $f$.

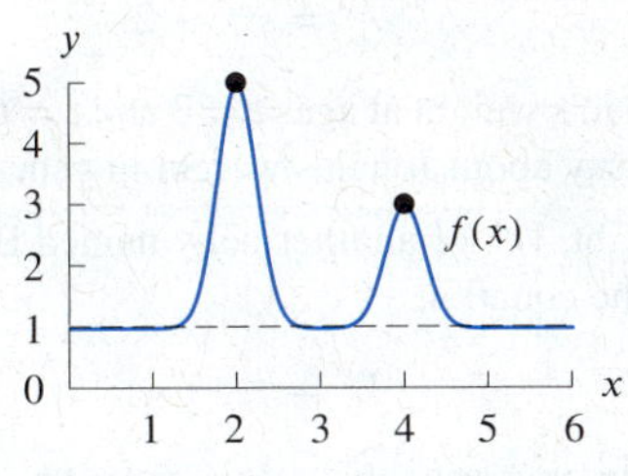


**Figure 2.28**

**3.** $y = f(x+2)$

**4.** $y = f(x)+2$

**5.** $y = f(x-1)-5$

**6.** $y = f(x+6)-4$

**7.** The graph of $f(x)$ contains the point $(3, -4)$. What point must be on the graph of

**(a)** $f(x)+5$? **(b)** $f(x+5)$?
**(c)** $f(x-3)-2$?

**8.** The domain of the function $g(x)$ is $-2 < x < 7$. What is the domain of $g(x-2)$?

**9.** The range of the function $R(s)$ is $100 \le R(s) \le 200$. What is the range of $R(s) - 150$?

**10.** **(a)** Using Table 2.15, evaluate

(i) $f(x)$ for $x = 6$
(ii) $f(5) - 3$
(iii) $f(5-3)$
(iv) $g(x)+6$ for $x = 2$
(v) $g(x+6)$ for $x = 2$
(vi) $3g(x)$ for $x = 0$
(vii) $f(3x)$ for $x = 2$
(viii) $f(x) - f(2)$ for $x = 8$
(ix) $g(x+1) - g(x)$ for $x = 1$

**(b)** Using the values in the table, solve

(i) $g(x) = 6$ (ii) $f(x) = 574$
(iii) $g(x) = 281$

**(c)** The values in the table were obtained using the formulas $f(x) = x^3+x^2+x-10$ and $g(x) = 7x^2-8x-6$. Use the table to find two solutions to the equation $x^3 + x^2 + x - 10 = 7x^2 - 8x - 6$.

**Table 2.15**

| $x$ | 0 | 1 | 2 | 3 | 4 | 5 | 6 | 7 | 8 | 9 |
|---|---|---|---|---|---|---|---|---|---|---|
| $f(x)$ | −10 | −7 | 4 | 29 | 74 | 145 | 248 | 389 | 574 | 809 |
| $g(x)$ | −6 | −7 | 6 | 33 | 74 | 129 | 198 | 281 | 378 | 489 |

## Problems

**11.** Figure 2.29 shows $y = f(x)$. Give a formula in terms of $f$ for the function in Figure 2.30. Your formula should be of the form $y = f(x-h)+k$ for appropriate constants $h$ and $k$.

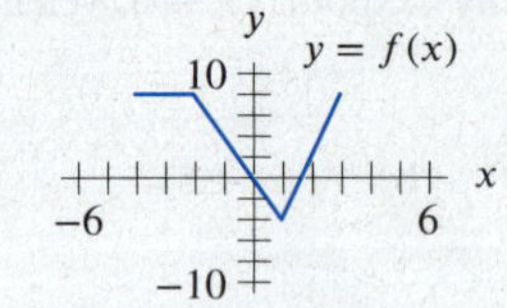


Figure 2.29

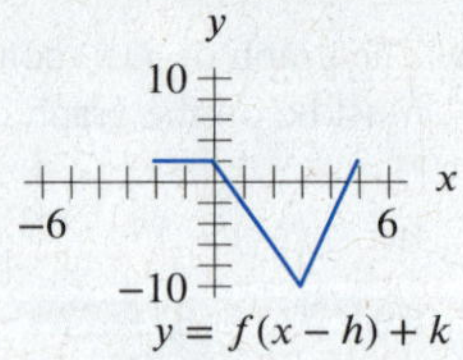


Figure 2.30

**12.** Figure 2.31 shows $y = m(r)$. Each graph in parts (a)–(d) is a translation of the graph of $y = m(r)$. Give a formula for each of these functions in terms of $m$.

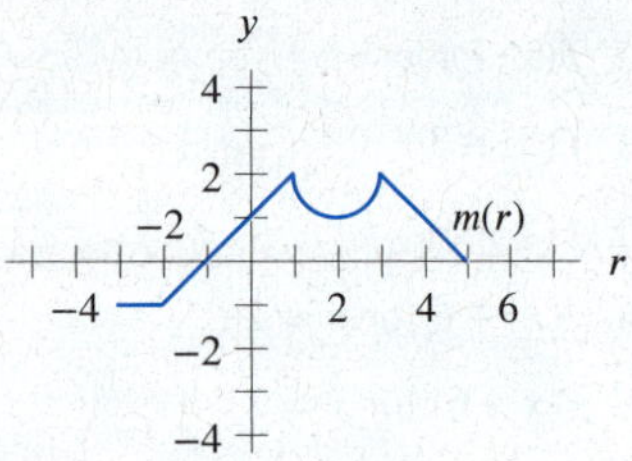


Figure 2.31

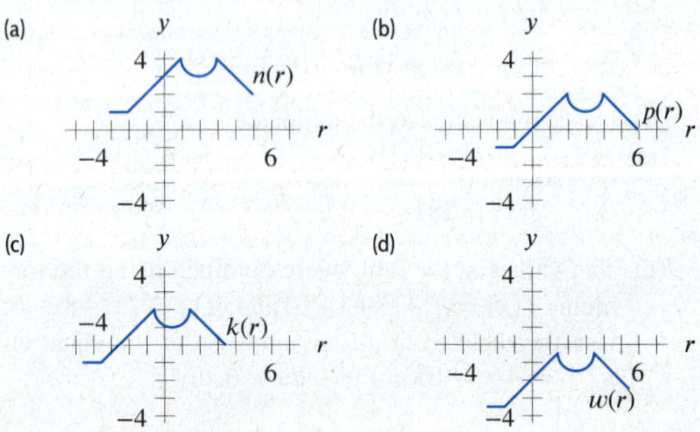


**13.** The graph of $g(x)$ contains the point $(-2, 5)$. Write a formula for a translation of $g$ whose graph contains the point

**(a)** $(-2, 8)$ **(b)** $(0, 5)$

**14.** Table 2.16 contains values of $f(x)$. Each function in parts (a)–(c) is a translation of $f(x)$. Find a possible formula for each of these functions in terms of $f$. For example, given the data in Table 2.17, you could say that $k(x) = f(x)+1$.

**Table 2.16**

| $x$ | 0 | 1 | 2 | 3 | 4 | 5 | 6 | 7 |
|---|---|---|---|---|---|---|---|---|
| $f(x)$ | 0 | 0.5 | 2 | 4.5 | 8 | 12.5 | 18 | 24.5 |

**Table 2.17**

| $x$ | 0 | 1 | 2 | 3 | 4 | 5 | 6 | 7 |
|---|---|---|---|---|---|---|---|---|
| $k(x)$ | 1 | 1.5 | 3 | 5.5 | 9 | 13.5 | 19 | 25.5 |

**(a)**

| $x$ | 0 | 1 | 2 | 3 | 4 | 5 | 6 | 7 |
|---|---|---|---|---|---|---|---|---|
| $h(x)$ | −2 | −1.5 | 0 | 2.5 | 6 | 10.5 | 16 | 22.5 |

**(b)**

| $x$ | 0 | 1 | 2 | 3 | 4 | 5 | 6 | 7 |
|---|---|---|---|---|---|---|---|---|
| $g(x)$ | 0.5 | 2 | 4.5 | 8 | 12.5 | 18 | 24.5 | 32 |

**(c)**

| $x$ | 0 | 1 | 2 | 3 | 4 | 5 | 6 | 7 |
|---|---|---|---|---|---|---|---|---|
| $i(x)$ | −1.5 | 0 | 2.5 | 6 | 10.5 | 16 | 22.5 | 30 |

**15.** Tables 2.18 and 2.19 give values of functions $v$ and $w$. Given that $w(x) = v(x-h)+k$, find the constants $h$ and $k$.

**Table 2.18**

| $x$ | $v(x)$ |
|---|---|
| −2 | 11 |
| −1 | 17 |
| 0 | 20 |
| 1 | 17 |
| 2 | 11 |

**Table 2.19**

| $x$ | $w(x)$ |
|---|---|
| 3 | 4 |
| 4 | 10 |
| 5 | 13 |
| 6 | 10 |
| 7 | 4 |

In Problems 16–17, let $s(t)$ denote the average weight (in pounds) of a baby at age $t$ months.

**16.** The weight, $V$, of a particular baby named Jonah is related to the average weight function $s(t)$ by the equation

$$V = s(t) + 2.$$

Find Jonah's weight at ages $t = 3$ and $t = 6$ months. What can you say about Jonah's weight in general?

**17.** The weight, $W$, of another baby named Ben is related to $s(t)$ by the equation

$$W = s(t+4).$$

What can you say about Ben's weight at age $t = 3$ months? At $t = 6$ months? Assuming that babies increase in weight over the first year of life, decide if Ben is of average weight for his age, above average, or below average.

**18.** The function $P(t)$ gives the number of people in a certain population in year $t$. Interpret in terms of population:

**(a)** $P(t) + 100$ **(b)** $P(t + 100)$

In Problems 19–24, explain in words the effect of the transformation on the graph of $q(z)$ for positive constants $a$, $b$.

**19.** $q(z) + 3$ **20.** $q(z) - a$

**21.** $q(z + 4)$ **22.** $q(z - a)$

**23.** $q(z + b) - a$ **24.** $q(z - 2b) + ab$

**25.** Let $T(d)$ give the average temperature in your hometown on the $d^{\text{th}}$ day of last year (so $d = 1$ is January 1, etc).

**(a)** Graph $T(d)$ for $1 \le d \le 365$.
**(b)** Give a possible value for each of the following: $T(6)$; $T(100)$; $T(215)$; $T(371)$.
**(c)** What is the relationship between $T(d)$ and $T(d + 365)$? Explain.
**(d)** If you graph $w(d) = T(d + 365)$ on the same axes as $T(d)$, how would the two graphs compare?
**(e)** Do you think the function $T(d) + 365$ has any practical significance? Explain.

**26.** Let $S(d)$ give the height of high tide in Seattle on a specific day, $d$, of the year. Use shifts of the function $S(d)$ to find formulas for each of the following functions:

**(a)** $T(d)$, the height of high tide in Tacoma on day $d$, if we assume that high tide in Tacoma is always one foot higher than high tide in Seattle.
**(b)** $A(d)$, the height of high tide in Astoria on day $d$, if we assume that high tide in Astoria is the same height as the previous day's high tide in Seattle.

**27.** Let $H(t)$ be the thermometer reading (Celsius) of a person $t$ hours after onset of an illness. Normal body temperature is 37°C. Find a formula for the fever, the number $f(t)$ of degrees greater than normal, after $t$ hours.

# 2.5 PREVIEW OF COMPOSITE AND INVERSE FUNCTIONS

## Composition of Functions

Two functions may be connected by the fact that the output of one is the input of the other. For example, to find the cost, $C$, in dollars, to paint an area $A$ square feet, we need to know the number, $n$, of gallons of paint required. Since one gallon covers 250 square feet, we have the function $n = f(A) = A/250$. If paint is \$30.50 a gallon, we have the function $C = g(n) = 30.5n$. We substitute $n = f(A)$ into $g(n)$ to find the cost $C$ as a function of $A$.

**Example 1** Find a formula for cost, $C$, as a function of area, $A$, to be painted.

**Solution** Since we have

$$C = 30.5n \quad \text{and} \quad n = \frac{A}{250},$$

substituting for $n$ in the formula for $C$ gives

$$C = 30.5\frac{A}{250} = 0.122A.$$

We say that $C$ is a "function of a function", or *composite function*. If the function giving $C$ in terms of $A$ is called $h$, so $C = h(A)$, then we write

$$C = h(A) = g(f(A)).$$

The function $h$ is said to be the *composition* of the functions $f$ and $g$. We say $f$ is the *inside* function and $g$ is the *outside* function. In this example, the composite function $C = h(A) = g(f(A))$ tells us the cost of painting an area of $A$ square feet.

> For two functions $f(t)$ and $g(t)$, the function $f(g(t))$ is said to be a **composition** of $f$ with $g$. The function $f(g(t))$ is defined by using the output of the function $g$ as the input to $f$.

The composite function $f(g(t))$ is defined only for values in the domain of $g$ whose $g(t)$ values are in the domain of $f$.

**Example 2** The air temperature, $T$, in °F, is given in terms of the chirp rate, $R$, in chirps per minute, of a snowy tree cricket by the function

$$T = f(R) = \frac{1}{4}R + 40.$$

If the chirp rate varies with the number of hours since midnight, $x$, according to the function

$$R = g(x) = 20 + x^2,$$

find how temperature varies with time by obtaining a formula for $h$, where $T = h(x)$.

**Solution** Since $f(R)$ is a function of $R$ and $R = g(x)$, we see that $g$ is the inside function and $f$ is the outside function. Thus we substitute $R = g(x)$ into $f$:

$$T = f(R) = f(g(x)) = \frac{1}{4}g(x) + 40 = \frac{1}{4}(20 + x^2) + 40 = \frac{1}{4}x^2 + 45.$$

Thus, for $0 \le x \le 10$, we have

$$T = h(x) = \frac{1}{4}x^2 + 45.$$

Example 3 shows another example of composition.

**Example 3** Let $f(x) = 2x + 1$ and $g(x) = x^2 - 3$.

(a) Calculate $f(g(3))$ and $g(f(3))$.
(b) Find formulas for $f(g(x))$ and $g(f(x))$.

**Solution** (a) We want

$$f(g(3)).$$

We start by evaluating $g(3)$. The formula for $g$ gives $g(3) = 3^2 - 3 = 6$, so

$$f(g(3)) = f(6).$$

The formula for $f$ gives $f(6) = 2 \cdot 6 + 1 = 13$, so

$$f(g(3)) = 13.$$

To calculate $g(f(3))$, we have

$$\begin{aligned} g(f(3)) &= g(7) && \text{Because } f(3) = 2 \cdot 3 + 1 = 7 \\ &= 46 && \text{Because } g(7) = 7^2 - 3 \end{aligned}$$

Notice that, $f(g(3)) \neq g(f(3))$. The functions $f(g(x))$ and $g(f(x))$ are different.

(b) In the formula for $f(g(x))$,

$$\begin{aligned} f(\underbrace{\ g(x)\ }_{\text{Input for } f}) &= f(x^2 - 3) && \text{Because } g(x) = x^2 - 3 \\ &= 2(x^2 - 3) + 1 && \text{Because } f(\text{Input}) = 2 \cdot \text{Input} + 1 \\ &= 2x^2 - 5. \end{aligned}$$

Check this formula by evaluating $f(g(3))$, which we know to be 13:

$$f(g(3)) = 2 \cdot 3^2 - 5 = 13.$$

In the formula for $g(f(x))$,

$$g(\underbrace{f(x)}_{\text{Input for } g}) = g(2x+1) \qquad \text{Because } f(x) = 2x+1$$

$$= (2x+1)^2 - 3 \qquad \text{Because } g(\text{Input}) = \text{Input}^2 - 3$$

$$= 4x^2 + 4x - 2.$$

Check this formula by evaluating $g(f(3))$, which we know to be 46:

$$g(f(3)) = 4 \cdot 3^2 + 4 \cdot 3 - 2 = 46.$$

## Inverse Functions

The roles of a function's input and output can sometimes be reversed. For example, the population, $P$, of birds on an island is given, in thousands, by $P = f(t)$, where $t$ is the number of years since 2007. In this function, $t$ is the input and $P$ is the output. If the population is increasing, knowing the population enables us to calculate the year. Thus we can define a new function, $t = g(P)$, which tells us the value of $t$ given the value of $P$ instead of the other way round. For this function, $P$ is the input and $t$ is the output.The functions $f$ and $g$ are called *inverses* of each other. A function which has an inverse is said to be *invertible*.

The fact that $f$ and $g$ are inverse functions means that they go in "opposite directions." The function $f$ takes $t$ as input and outputs $P$, while $g$ takes $P$ as input and outputs $t$.

### Inverse Function Notation

In the preceding discussion, there was nothing about the names of the two functions that stressed their special relationship. If we want to emphasize that $g$ is the inverse of $f$, we call it $f^{-1}$ (read "$f$-inverse"). To express the fact that the population of birds, $P$, is a function of time, $t$, we write

$$P = f(t).$$

To express the fact that the time $t$ is also determined by $P$, so that $t$ is a function of $P$, we write

$$t = f^{-1}(P).$$

The symbol $f^{-1}$ is used to represent the function that gives the output $t$ for a given input $P$.

**Warning**: The $-1$ that appears in the symbol $f^{-1}$ for the inverse function is not an exponent. Unfortunately, the notation $f^{-1}(x)$ might lead us to interpret it as $(f(x))^{-1} = \frac{1}{f(x)}$. The two expressions are not the same in general: $f^{-1}(x)$ is the output when $x$ is fed into the inverse of $f$, while $(f(x))^{-1} = \frac{1}{f(x)}$ is the reciprocal of the number we get when $x$ is fed into $f$.

**Example 4** Using $P = f(t)$, where $P$ represents the population, in thousands, of birds on an island and $t$ is the number of years since 2007:

(a) What does $f(4)$ represent? (b) What does $f^{-1}(4)$ represent?

**Solution** (a) The expression $f(4)$ is the bird population (in thousands) in the year 2011.

(b) Since $f^{-1}$ is the inverse function, $f^{-1}$ is a function which takes population as input and returns time as output. Therefore, $f^{-1}(4)$ is the number of years after 2007 at which there were 4,000 birds on the island.

**Example 5** Suppose that $g$ is an invertible function, with $g(10) = -26$ and $g^{-1}(0) = 7$. What other values of $g$ and $g^{-1}$ do you know?

**Solution** Because $g(10) = -26$, we know that $g^{-1}(-26) = 10$; because $g^{-1}(0) = 7$, we know that $g(7) = 0$.

Not all functions are invertible. We revisit this issue in Chapter 10 where we see how to tell whether a function has an inverse.

### Finding a Formula for the Inverse Function

In the next example, we find the formula for an inverse function.

**Example 6** The cricket function, which gives temperature, $T$, in terms of chirp rate, $R$, is

$$T = f(R) = \frac{1}{4} \cdot R + 40.$$

Find a formula for the inverse function, $R = f^{-1}(T)$.

**Solution** The inverse function gives the chirp rate in terms of the temperature, so we solve the following equation for $R$:

$$T = \frac{1}{4} \cdot R + 40,$$

giving

$$T - 40 = \frac{1}{4} \cdot R$$
$$R = 4(T - 40).$$

Thus, $R = f^{-1}(T) = 4(T - 40)$.

### Domain and Range of an Inverse Function

The input values of the inverse function $f^{-1}$ are the output values of the function $f$. Thus, the domain of $f^{-1}$ is the range of $f$. For the cricket function, $T = f(R) = \frac{1}{4}R + 40$, if a realistic domain is $0 \leq R \leq 160$, then the range of $f$ is $40 \leq T \leq 80$. The domain of $f^{-1}$ is then $40 \leq T \leq 80$ and the range is $0 \leq R \leq 160$. See Figure 2.32.

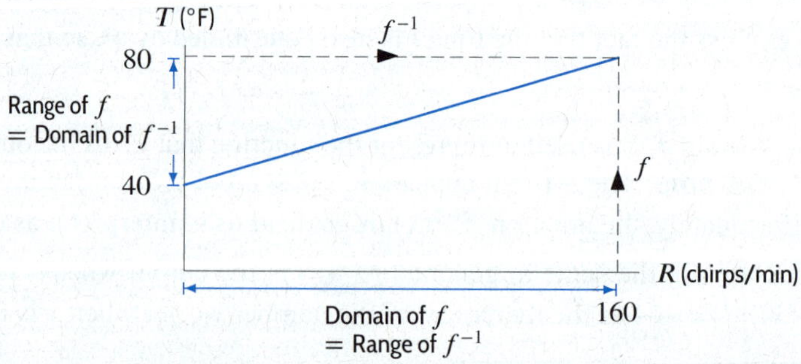


**Figure 2.32:** Domain and range of an inverse function

## Relation Between Composition and Inverses: A Function and Its Inverse Undo Each Other

Composing a function with its inverse gives a striking result, as we see in the next example.

**Example 7** Calculate the composite functions $f^{-1}(f(R))$ and $f(f^{-1}(T))$ for the cricket example. Interpret the results.

**Solution** Since $f(R) = \frac{1}{4}R + 40$. and $f^{-1}(T) = 4(T - 40)$, we have

$$f^{-1}(f(R)) = f^{-1}\left(\frac{1}{4} \cdot R + 40\right) = 4\left(\left(\frac{1}{4} \cdot R + 40\right) - 40\right) = R.$$
$$f(f^{-1}(T)) = f(4(T - 40)) = \frac{1}{4}(4(T - 40)) + 40 = T.$$

To interpret these results, we use the fact that $f(R)$ gives the temperature corresponding to chirp rate $R$, and $f^{-1}(T)$ gives the chirp rate corresponding to temperature $T$. Thus $f^{-1}(f(R))$ gives the chirp rate at temperature $f(R)$, which is $R$. Similarly, $f(f^{-1}(T))$ gives the temperature at chirp rate $f^{-1}(T)$, which is $T$.

In Example 7, we see that $f^{-1}(f(R)) = R$ and $f(f^{-1}(T)) = T$. This illustrates the following result, which we see is true in general in Chapter 10.

> The functions $f$ and $f^{-1}$ are called inverses because they "undo" each other when composed.

**Example 8** Let $y = f(x) = 2x + 8$.

(a) Find a formula for the inverse function $x = f^{-1}(y)$.
(b) Show that $f^{-1}(f(x)) = x$ and $f(f^{-1}(y)) = y$.

**Solution** (a) We have $y = 2x + 8$, so we solve this equation for $x$:

$$\begin{aligned} y &= 2x + 8 \\ y - 8 &= 2x \\ 0.5y - 4 &= x \end{aligned}$$

We see that $x = f^{-1}(y) = 0.5y - 4$.

(b) We have:

$$f^{-1}(f(x)) = f^{-1}(2x + 8) = 0.5(2x + 8) - 4 = (x + 4) - 4 = x.$$

and

$$f(f^{-1}(y)) = f(0.5y - 4) = 2(0.5y - 4) + 8 = (y - 8) + 8 = y.$$

We see that this result is always true in Chapter 10.

## Exercises and Problems for Section 2.5

### Skill Refresher

Solve the equations in Exercises S1–S7 for $y$.

**S1.** $x = 3y - 4$

**S2.** $4x - 3y = 7$

**S3.** $x = \dfrac{2y + 1}{y - 2}$

**S4.** $x = \dfrac{4y}{1 - y}$

**S5.** $x = \sqrt{y} - 2$

**S6.** $x = y^3 - 4$

**S7.** $x = 5 - (2y)^3$

Simplify the expressions in Exercises S8–S11.

**S8.** $5\left(\frac{1}{5}x - 1\right) + 5$

**S9.** $(2x + 1)^2 - 4$

**S10.** $3(y - 2)^2 - 7$

**S11.** $(1 - t)^2 - (1 - t)$

## Exercises

In Exercises 1–4, use the complete graph of the invertible function $f(x)$ shown to estimate the domain of $f^{-1}(y)$.

**1.**

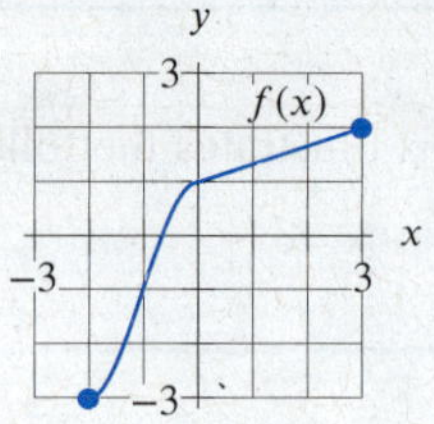

**2.**

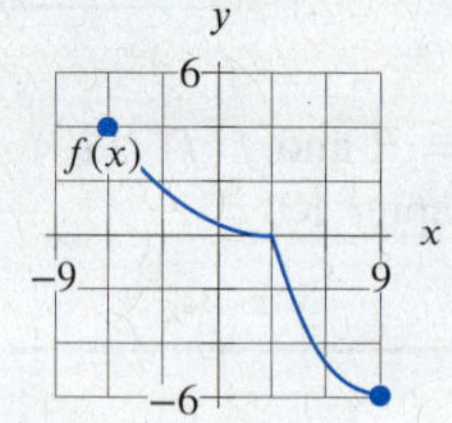

**3.**

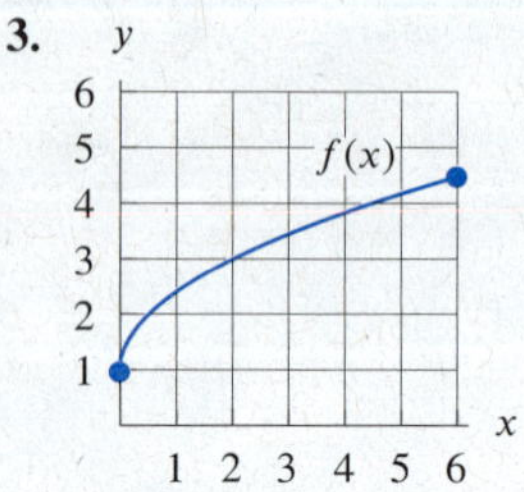

**4.**

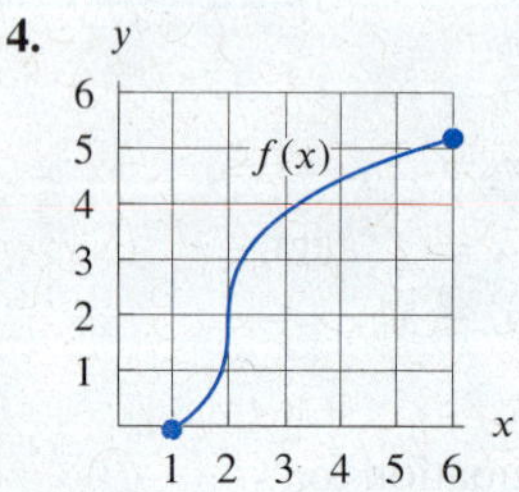

In Exercises 5–12, use $f(x) = 3x - 1$ and $g(x) = 1 - x^2$.

**5.** $f(g(0))$

**6.** $g(f(0))$

**7.** $g(f(2))$

**8.** $f(g(2))$

**9.** $f(g(x))$

**10.** $g(f(x))$

**11.** $f(f(x))$

**12.** $g(g(x))$

In Exercises 13–15, give the meaning and units of the composite function.

**13.** $A(f(t))$, where $r = f(t)$ is the radius, in centimeters, of a circle at time $t$ minutes, and $A(r)$ is the area, in square centimeters, of a circle of radius $r$ centimeters.

**14.** $R(f(p))$, where $Q = f(p)$ is the number of barrels of oil sold by a company when the price is $p$ dollars/barrel and $R(Q)$ is the revenue earned in millions of dollars from a sale of $Q$ barrels.

**15.** $C(A(d))$, where $A(d)$ is the area of a circular pizza with diameter $d$, and $C(x)$ is the price of pizza with area $x$.

In Exercises 16–20, give the meaning and units of the inverse function. (Assume $f$ is invertible.)

**16.** $P = f(t)$ is population in millions in year $t$.

**17.** $T = f(H)$ is time in minutes to bake a cake at $H$°F.

**18.** $N = f(t)$ is number of inches of snow in the first $t$ days of January.

**19.** $x = C(w)$ is calories for $w$ ounces of almonds.

**20.** $c = P(d)$ is the price, in dollars, for a pizza of diameter $d$ inches.

In Exercises 21–27, find the inverse function.

**21.** $y = f(t) = 2t + 3$

**22.** $y = f(x) = 3x - 7$

**23.** $y = g(x) = x^3 + 1$

**24.** $Q = f(x) = x^3 + 3$

**25.** $A = f(r) = \pi r^2, r \geq 0$

**26.** $y = g(s) = 1 + \dfrac{1}{s}$

**27.** $P = f(x) = \dfrac{3x}{2x+1}$

**28.** For $f(x) = 5x$, evaluate $f(4)$ and $f^{-1}(15)$.

**29.** For $k(x) = 4x - 10$, evaluate $k(3)$ and $k^{-1}(14)$.

**30.** For $h(x) = \dfrac{x}{4} + 2$, evaluate $h\left(-\frac{1}{2}\right)$ and $h^{-1}(8)$.

**31.** For $g(x) = 2x^3 - 1$, evaluate $g(\frac{3}{2})$ and $g^{-1}(-17)$.

**32.** Use the graph in Figure 2.33 to fill in the missing values:

**(a)** $f(0) = ?$ **(b)** $f(?) = 0$
**(c)** $f^{-1}(0) = ?$ **(d)** $f^{-1}(?) = 0$

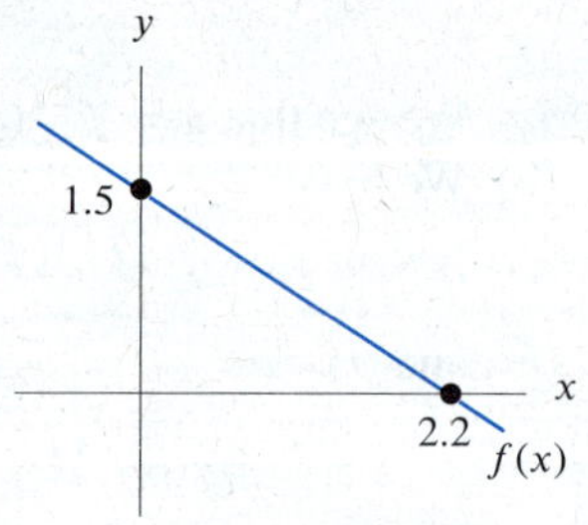

Figure 2.33

**33.** Use the graph in Figure 2.34 to fill in the missing values:

**(a)** $f(0) = ?$ **(b)** $f(?) = 0$
**(c)** $f^{-1}(0) = ?$ **(d)** $f^{-1}(?) = 0$

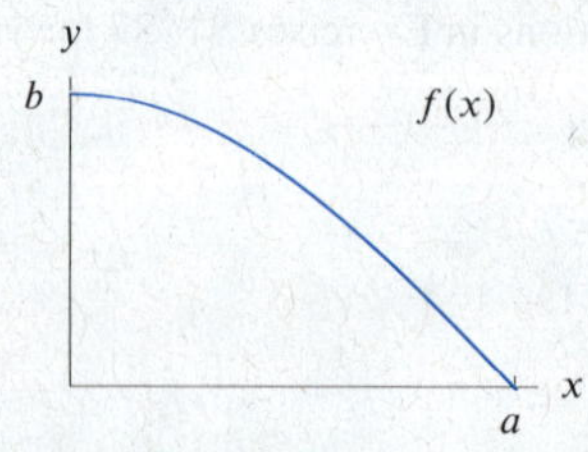

Figure 2.34

## Problems

**34.** The cost (in dollars) of producing $x$ air conditioners is $C = g(x) = 600 + 45x$. Find a formula for the inverse function $g^{-1}(C)$.

In Problems 35–37, let $n = f(A) = A/250$, where $n$ is the number of gallons of paint needed for an area of $A$ square feet.

**35.** Find a formula for the inverse function $A = f^{-1}(n)$.

**36.** Interpret and evaluate $f(100)$ and $f^{-1}(100)$.

**37.** Calculate the composite functions $f^{-1}(f(A))$ and $f(f^{-1}(n))$. Explain the results.

**38.** The daily cost (in dollars) of renting a car and driving up to 500 miles a day is $C = f(m) = 32 + 0.19m$, where $m$ is the number of miles driven.

(a) Find the domain and range of $f^{-1}(C)$.
(b) Find a formula for the inverse function $f^{-1}(C)$.

**39.** Table 2.20 gives values of an invertible function, $f$.

(a) Using the table, fill in the missing values:

(i) $f(0) = ?$ (ii) $f(?) = 0$
(iii) $f^{-1}(0) = ?$ (iv) $f^{-1}(?) = 0$

(b) How do the answers to (i)–(iv) in part (a) relate to one another? In particular, how could you have obtained the answers to (iii) and (iv) from the answers to (i) and (ii)?

**Table 2.20**

| $x$ | −2 | −1 | 0 | 1 | 2 |
|---|---|---|---|---|---|
| $f(x)$ | 5 | 4 | 2 | 0 | −3 |

**40.** Suppose that $j(x) = h^{-1}(x)$ and that both $j$ and $h$ are defined for all values of $x$. Let $h(4) = 2$ and $j(5) = -3$. Evaluate if possible:

(a) $j(h(4))$ (b) $j(4)$ (c) $h(j(4))$
(d) $j(2)$ (e) $h^{-1}(-3)$ (f) $j^{-1}(-3)$
(g) $h(5)$ (h) $(h(-3))^{-1}$ (i) $(h(2))^{-1}$

In Problems 41–44, find the domain and range of the function by first finding the formula for the inverse function.

**41.** $t(a) = \sqrt[3]{a+1}$

**42.** $n(r) = r^3 + 2$

**43.** $m(x) = \dfrac{1}{\sqrt{x-2}}$

**44.** $p(x) = \dfrac{1}{\sqrt{3-x}}$

**45.** The gross domestic product (GDP) of the US is given by $G(t)$ where $t$ is the number of years since 2000 and the units of $G$ are billions of dollars.[13]

(a) What is meant by $G(13) = 16{,}011.2$?
(b) What is meant by $G^{-1}(13{,}776) = 7$?

**46.** The cost of producing $q$ thousand loaves of bread is $C(q)$ dollars. Interpret the following statements in terms of bread; give units.

(a) $C(5) = 653$
(b) $C^{-1}(80) = 0.62$
(c) The solution to $C(q) = 790$ is 6.3
(d) The solution to $C^{-1}(x) = 1.2$ is 150

**47.** The perimeter of a square of side $s$ is given by $P = f(s) = 4s$. Find and interpret

(a) $f(3)$ (b) $f^{-1}(20)$ (c) $f^{-1}(P)$

**48.** The cost, $C$, in thousands of dollars, of producing $q$ kg of a chemical is given by $C = f(q) = 100 + 0.2q$. Find and interpret

(a) $f(10)$ (b) $f^{-1}(200)$ (c) $f^{-1}(C)$

In Problems 49–51, let $H = f(t) = \frac{5}{9}(t - 32)$, where $H$ is temperature in degrees Celsius and $t$ is in degrees Fahrenheit.

**49.** Find and interpret the inverse function, $f^{-1}(H)$.

**50.** Using the results of Problem 49, evaluate and interpret:

(a) $f(0)$ (b) $f^{-1}(0)$
(c) $f(100)$ (d) $f^{-1}(100)$

**51.** The temperature, $t = g(n) = 68 + 10 \cdot 2^{-n}$, in degrees Fahrenheit of a room is a function of the number, $n$, of hours that the air conditioner has been running. Find and interpret $f(g(n))$. Give units.

**52.** The period, $T$, of a pendulum of length $l$ is given by $T = f(l) = 2\pi\sqrt{l/g}$, where $g$ is a constant. Find a formula for $f^{-1}(T)$ and explain its meaning.

**53.** The area, in square centimeters, of a circle whose radius is $r$ cm is given by $A = \pi r^2$.

(a) Write this formula using function notation, where $f$ is the name of the function.
(b) Evaluate $f(0)$.
(c) Evaluate and interpret $f(r + 1)$.
(d) Evaluate and interpret $f(r) + 1$.
(e) What are the units of $f^{-1}(4)$?

[13] http://www.businessinsider.com/us-q1-2013-gdp-first-estimate-2013-4. Accessed April, 2013.

**54.** The radius, $r$, in centimeters, of a melting snowball is given by $r = 50 - 2.5t$, where $t$ is time in hours. The snowball is spherical, with volume $V = \frac{4}{3}\pi r^3$ cm$^3$. Find a formula for $V = f(t)$, the volume of the snowball as a function of time.

**55.** The area, $A = f(d)$ in$^2$, of a circular pizza is a function of the diameter $d$, in inches. A package of pepperoni costs \$2.99 and covers 250 square inches of pizza.

**(a)** Write the formula for $f(d)$.
**(b)** Find a formula for $C = g(A)$, the cost in dollars of adding pepperoni to a pizza of area $A$ in$^2$.
**(c)** Find and interpret $C = g(f(d))$.
**(d)** Evaluate and interpret $f(11)$ and $g(11)$.

**56.** A circular oil slick is expanding with radius, $r$ in yards, at time $t$ in hours given by $r = 2t - 0.1t^2$, for $0 \le t \le 10$. Find a formula for the area in square yards, $A = f(t)$, as a function of time.

**57.** Table 2.21 gives values for $d = f(v)$, where $d$, in meters, is the stopping distance for a car traveling at velocity $v$ kilometers per hour.

**(a)** Evaluate and interpret $f(60)$.
**(b)** Estimate $f(70)$.
**(c)** Evaluate and interpret $f^{-1}(70)$.

**Table 2.21**

| $v$ (km/hr) | 20 | 30 | 40 | 50 | 60 | 80 | 90 | 100 | 120 |
|---|---|---|---|---|---|---|---|---|---|
| $d$ (m) | 6 | 10 | 16 | 23 | 30 | 50 | 60 | 70 | 100 |

**58.** Let $f(a)$ be the cost in dollars of $a$ pounds of organic apples at Fresh Abundance[14] in November 2009. What do the following statements tell you? What are the units of each of the numbers?

**(a)** $f(2) = 2.80$ **(b)** $f(0.5) = 0.70$
**(c)** $f^{-1}(0.35) = 0.25$ **(d)** $f^{-1}(7) = 5$

**59.** Carbon dioxide is one of the greenhouse gases that are believed to affect global warming. Between 2004 and 2008, the concentration of carbon dioxide in the earth's atmosphere increased steadily from 375 parts per million (ppm) to 383 ppm.[15] Let $C(t)$ be the concentration in ppm of carbon dioxide $t$ years after 2004 during these four years.

**(a)** State the domain and range of $C(t)$.
**(b)** What is the practical meaning of $C(4)$? What is its value?
**(c)** What does $C^{-1}(381)$ represent?

# 2.6 CONCAVITY

## Concavity and Rates of Change

The graph of a linear function is a straight line because the average rate of change is constant. However, not all graphs are straight lines; they may bend up or down. The function giving salary $S$, in \$1000s, as a function of $t$, time in years since being hired, is shown in Table 2.22 and Figure 2.35. Since the rate of change increases with time, the slope of the graph increases as $t$ increases, so the graph bends upward. We say such graphs are *concave up*.

**Table 2.22** *Salary: Increasing rate of change*

| $t$ (years) | $S$ (\$1000s) | Rate of change $\Delta S/\Delta t$ |
|---|---|---|
| 0 | 40 | |
| | | 3.2 |
| 10 | 72 | |
| | | 5.6 |
| 20 | 128 | |
| | | 10.2 |
| 30 | 230 | |
| | | 18.1 |
| 40 | 411 | |

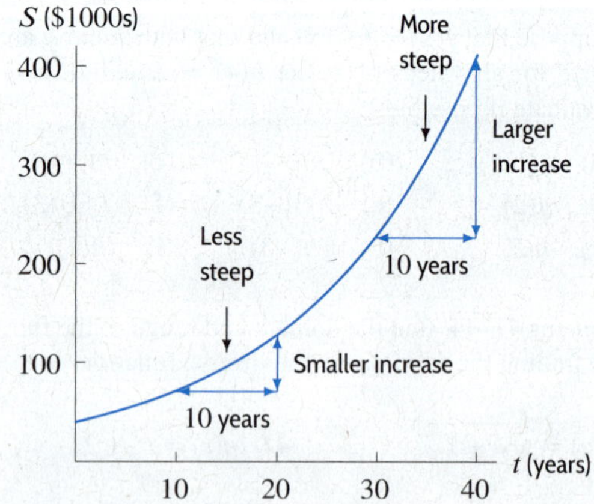


**Figure 2.35**: Graph of salary function is concave up because rate of change increases

14www.freshabundance.com/index.php?main_page=product_info&products_id=254, accessed November 5, 2009.
15*The World Almanac and Book of Facts 2010*, page 297.

The next example shows that a decreasing function can also be concave up.

**Example 1** Table 2.23 shows $Q$, the quantity of carbon-14 (in $\mu$g) in a 200 $\mu$g sample remaining after $t$ thousand years. We see from Figure 2.36 that $Q$ is a decreasing function of $t$, so its rate of change is always negative. What can we say about the concavity of the graph, and what does this mean about the rate of change of the function?

**Table 2.23** *Carbon-14: Increasing rate of change*

| $t$ (thousand years) | $Q$ ($\mu$g) | Rate of change $\Delta Q/\Delta t$ |
|---|---|---|
| 0 | 200 | |
| | | −18.2 |
| 5 | 109 | |
| | | −9.8 |
| 10 | 60 | |
| | | −5.4 |
| 15 | 33 | |

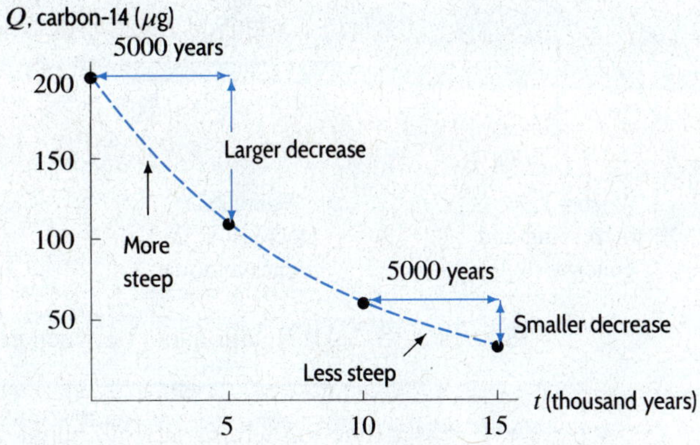


Figure 2.36: Graph of the quantity of carbon-14 is concave up

**Solution** The graph bends upward, so it is concave up. Table 2.24 shows that the rate of change of the function is increasing, because the rate is becoming less negative. Figure 2.36 shows how the increasing rate of change can be visualized on the graph: the slope is negative and increasing.

Graphs can bend downward; we call such graphs *concave down.*

**Example 2** Table 2.24 gives the distance traveled by a cyclist, Karim, as a function of time. What is the concavity of the graph? Was Karim's speed (that is, the rate of change of distance with respect to time) increasing, decreasing, or constant?

**Solution** Table 2.24 shows Karim's speed was decreasing throughout the trip. Figure 2.37 shows how the decreasing speed leads to a decreasing slope and a graph which bends downward; thus the graph is concave down..

**Table 2.24** *Karim's distance as a function of time, with the average speed for each hour*

| $t$, time (hours) | $d$, distance (miles) | Average speed, $\Delta d/\Delta t$ (mph) |
|---|---|---|
| 0 | 0 | |
| | | 20 mph |
| 1 | 20 | |
| | | 15 mph |
| 2 | 35 | |
| | | 10 mph |
| 3 | 45 | |
| | | 7 mph |
| 4 | 52 | |
| | | 5 mph |
| 5 | 57 | |

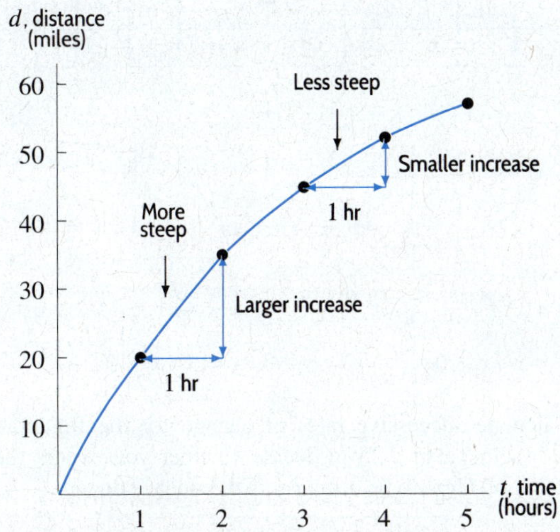


Figure 2.37: Karim's distance as a function of time

## Summary: Increasing and Decreasing Functions; Concavity

Knowing that a function is either increasing or decreasing and either concave up or concave down at a point allows us to determine the basic shape of the graph at that point. The four basic shapes are in Figures 2.38–2.41:

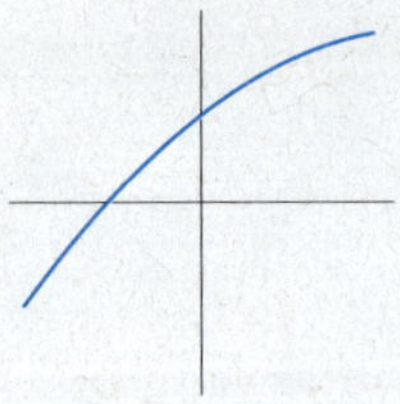

**Figure 2.38:** Increasing and concave down

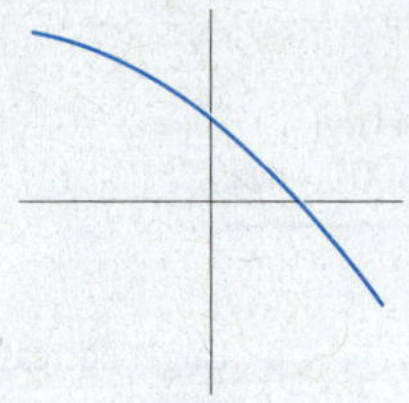

**Figure 2.39:** Decreasing and concave down

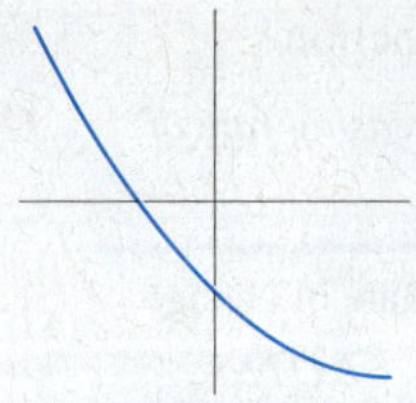

**Figure 2.40:** Decreasing and concave up

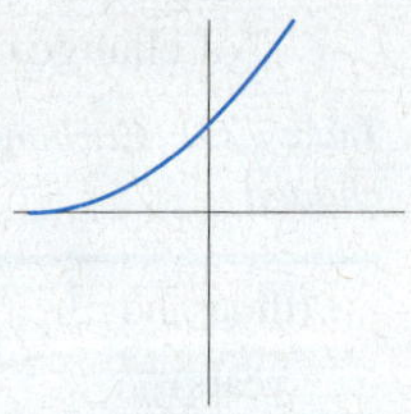

**Figure 2.41:** Increasing and concave up

Figures 2.38–2.41: Relationship between concavity and rate of change:

- If $f$ is a function whose rate of change increases (gets less negative or more positive as we move from left to right[16]), then the graph of $f$ is **concave up**. That is, the graph bends upward.
- If $f$ is a function whose rate of change decreases (gets less positive or more negative as we move from left to right), then the graph of $f$ is **concave down**. That is, the graph bends downward.

If a function has a constant rate of change, its graph is a line and it is neither concave up nor concave down.

## Exercises and Problems for Section 2.6

### Exercises

Do the graphs of the functions in Exercises 1–8 appear to be concave up, concave down, or neither?

**1.**

| $x$ | 0 | 1 | 3 | 6 |
|---|---|---|---|---|
| $f(x)$ | 1.0 | 1.3 | 1.7 | 2.2 |

**2.**

| $t$ | 0 | 1 | 2 | 3 | 4 |
|---|---|---|---|---|---|
| $f(t)$ | 20 | 10 | 6 | 3 | 1 |

**3.**

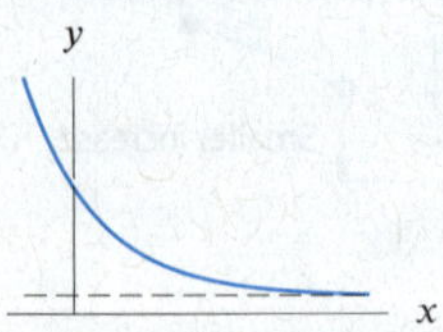


**4.**

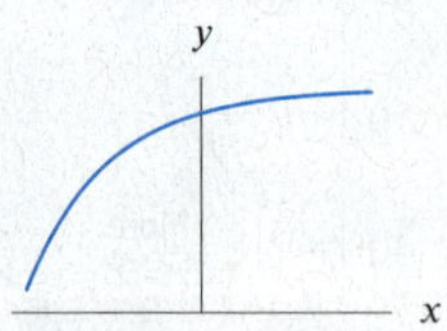


**5.** $y = x^2$

**6.** $y = -x^2$

**7.** $y = x^3, x > 0$

**8.** $y = x^3, x < 0$

**9.** Calculate successive rates of change for the function, $H(x)$, in Table 2.25 to decide whether you expect the graph of $H(x)$ to be concave up or concave down.

**Table 2.25**

| $x$ | 12 | 15 | 18 | 21 |
|---|---|---|---|---|
| $H(x)$ | 21.40 | 21.53 | 21.75 | 22.02 |

**10.** Calculate successive rates of change for the function, $R(t)$, in Table 2.26 to decide whether you expect the graph of $R(t)$ to be concave up or concave down.

**Table 2.26**

| $t$ | 1.5 | 2.4 | 3.6 | 4.8 |
|---|---|---|---|---|
| $R(t)$ | −5.7 | −3.1 | −1.4 | 0 |

**11.** Sketch a graph that is everywhere negative, increasing, and concave down.

**12.** Sketch a graph that is everywhere positive, increasing, and concave up.

[16] In fact, we need to take the average rate of change over an arbitrarily small interval.

## Problems

In Problems 13–16, estimate the intervals where the graph is
**(a)** Increasing and concave down
**(b)** Decreasing and concave down
**(c)** Increasing and concave up
**(d)** Decreasing and concave up

**13.**

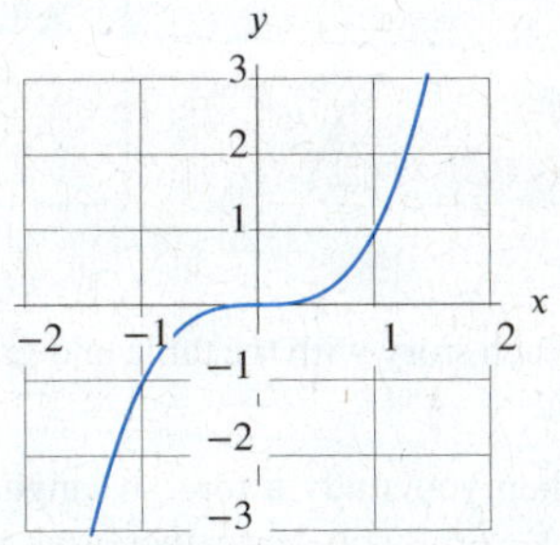


**14.**

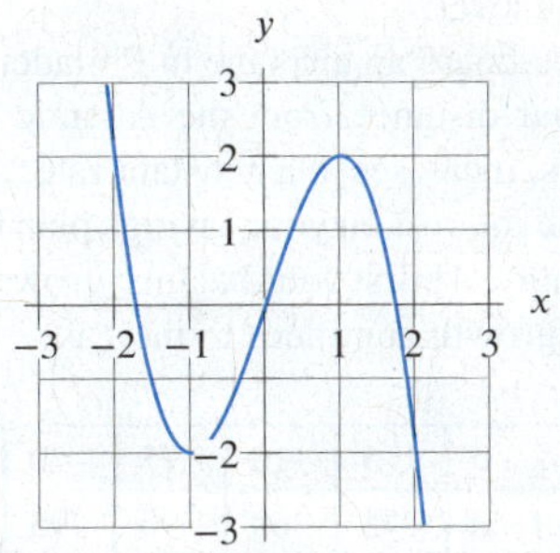


**15.**

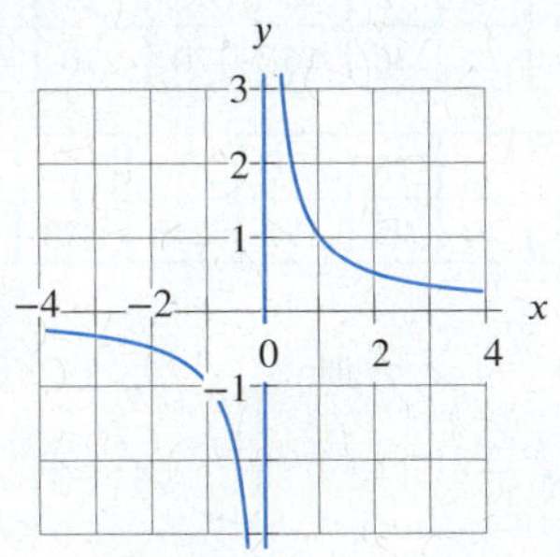


**16.**

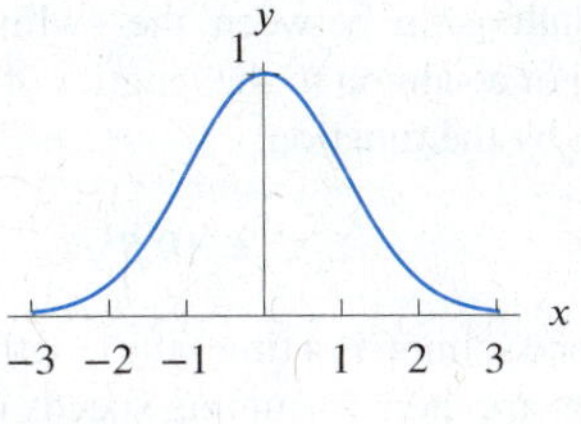


Are the functions in Problems 17–21 increasing or decreasing? What does the scenario tell you about the concavity of the graph modeling it?

**17.** When a drug is injected into a person's bloodstream, the amount of the drug present in the body increases rapidly at first. If the person receives daily injections, the body metabolizes the drug so that the amount of the drug present in the body continues to increase, but at a decreasing rate. Eventually, the quantity levels off at a saturation level.

**18.** After a cup of hot chocolate is poured, the temperature cools off very rapidly at first, and then cools off more slowly, until the temperature of the hot chocolate eventually reaches room temperature.

**19.** When a rumor begins, the number of people who have heard the rumor increases slowly at first. As the rumor spreads, the rate of increase gets greater (as more people continue to tell their friends the rumor), and then slows down again (when almost everyone has heard the rumor).

**20.** When money is deposited in the bank, the amount of money increases slowly at first. As the size of the account increases, the amount of money increases more rapidly, since the account is earning interest on the new interest, as well as on the original amount.

**21.** When a new product is introduced, the number of people who use the product increases slowly at first, and then the rate of increase is faster (as more and more people learn about the product). Eventually, the rate of increase slows down again (when most people who are interested in the product are already using it).

**22.** Graph $f(x)$ with all of these properties:

- $f(0) = 4$
- $f$ is decreasing and concave up for $-\infty < x < 0$
- $f$ is increasing and concave up for $0 < x < 6$
- $f$ is increasing and concave down for $6 < x < 8$
- $f$ is decreasing and concave down for $x > 8$

**23.** An incumbent politician running for reelection declared that the number of violent crimes is no longer rising and is presently under control. Does the graph shown in Figure 2.42 support this claim? Why or why not?

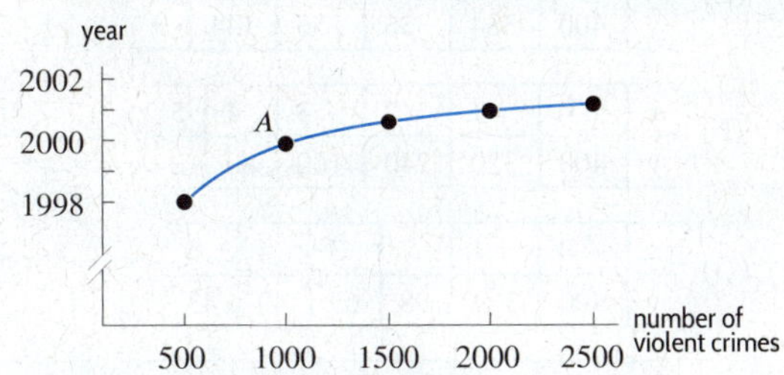


Figure 2.42

**24.** The rate at which water is entering a reservoir is given for time $t > 0$ by the graph in Figure 2.43. A negative rate means that water is leaving the reservoir. In parts (a)–(d), give the largest interval on which:

**(a)** The volume of water is increasing.
**(b)** The volume of water is constant.
**(c)** The volume of water is increasing fastest.
**(d)** The volume of water is decreasing.

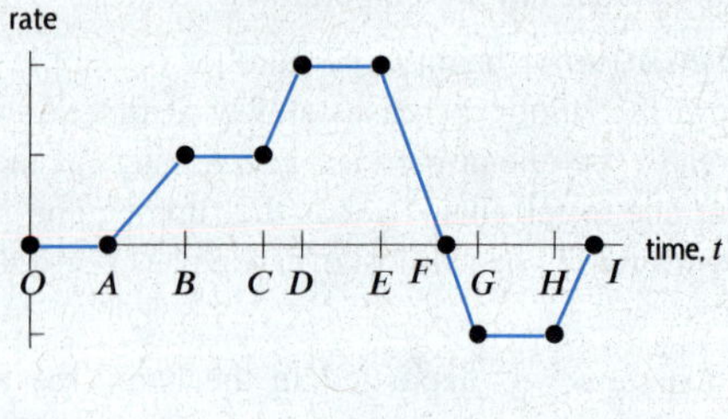


Figure 2.43

**25.** Match each of the following descriptions with an appropriate graph and table of values.

**(a)** The weight of your jumbo box of Fruity Flakes decreases by an equal amount every week.
**(b)** The machinery depreciated rapidly at first, but its value declined more slowly as time went on.
**(c)** In free fall, your distance from the ground decreases faster and faster.
**(d)** For a while it looked as if the decline in profits was slowing down, but then they began declining ever more rapidly.

(E)

| $x$ | 0 | 1 | 2 | 3 | 4 | 5 |
|---|---|---|---|---|---|---|
| $y$ | 400 | 384 | 336 | 256 | 144 | 0 |

(F)

| $x$ | 0 | 1 | 2 | 3 | 4 | 5 |
|---|---|---|---|---|---|---|
| $y$ | 400 | 320 | 240 | 160 | 80 | 0 |

(G)

| $x$ | 0 | 1 | 2 | 3 | 4 | 5 |
|---|---|---|---|---|---|---|
| $y$ | 400 | 184 | 98 | 63 | 49 | 43 |

(H)

| $x$ | 0 | 1 | 2 | 3 | 4 | 5 |
|---|---|---|---|---|---|---|
| $y$ | 412 | 265 | 226 | 224 | 185 | 38 |

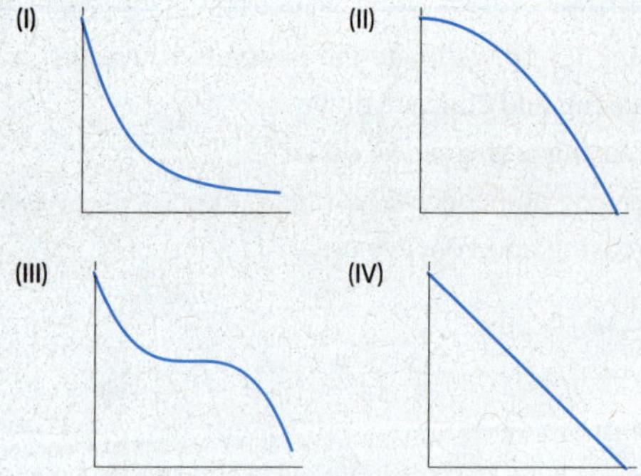


**26.** Match each story with the table and graph which best represent it.

**(a)** When you study a foreign language, the number of new verbs you learn increases rapidly at first, but slows almost to a halt as you approach your saturation level.
**(b)** You board an airplane in Philadelphia heading west. Your distance from the Atlantic Ocean, in kilometers, increases at a constant rate.
**(c)** The interest on your savings plan is compounded annually. At first your balance grows slowly, but its rate of growth continues to increase.

(E)

| $x$ | 0 | 5 | 10 | 15 | 20 | 25 |
|---|---|---|---|---|---|---|
| $y$ | 20 | 275 | 360 | 390 | 395 | 399 |

(F)

| $x$ | 0 | 5 | 10 | 15 | 20 | 25 |
|---|---|---|---|---|---|---|
| $y$ | 20 | 36 | 66 | 120 | 220 | 400 |

(G)

| $x$ | 0 | 5 | 10 | 15 | 20 | 25 |
|---|---|---|---|---|---|---|
| $y$ | 20 | 95 | 170 | 245 | 320 | 395 |

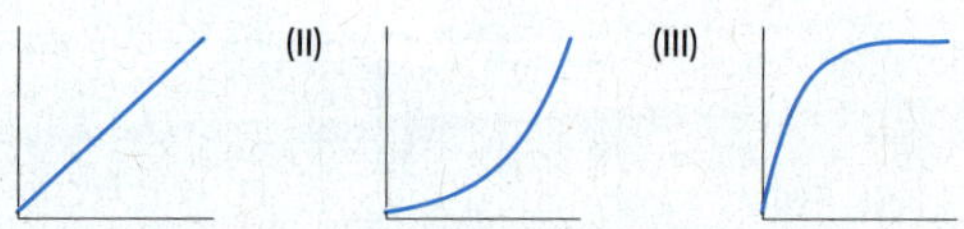


**27.** The relationship between the swimming speed $U$ (in cm/sec) of a salmon to the length $l$ of the salmon (in cm) is given by the function[17]

$$U = 19.5\sqrt{l}.$$

**(a)** If one salmon is 4 times the length of another salmon, how are their swimming speeds related?
**(b)** Graph the function $U = 19.5\sqrt{l}$. Describe the graph using words such as increasing, decreasing, concave up, concave down.

[17] From K. Schmidt-Nielsen, *Scaling, Why is Animal Size so Important?* (Cambridge: CUP, 1984).

**(c)** Using a property that you described in part (b), answer the question "Do larger salmon swim faster than smaller ones?"

**(d)** Using a property that you described in part (b), answer the question "Imagine four salmon—two small and two large. The smaller salmon differ in length by 1 cm, as do the two larger. Is the difference in speed between the two smaller fish, greater than, equal to, or smaller than the difference in speed between the two larger fish?"

**28.** The graph of $f$ is concave down for $0 \le x \le 6$. Which is bigger: $\dfrac{f(3)-f(1)}{3-1}$ or $\dfrac{f(5)-f(3)}{5-3}$? Why?

## CHAPTER SUMMARY

- **Input and Output**
  Evaluating functions: finding $f(a)$ for given $a$.
  Solving equations: finding $x$ if $f(x) = b$ for given $b$.
- **Domain and Range**
  Domain: set of input values.
  Range: set of output values.
  Piecewise defined functions: different formulas on different intervals.
- **Horizontal and Vertical Shifts**
- **Composite Functions**
  Finding $f(g(x))$ given $f(x)$ and $g(x)$.
- **Inverse Functions**
  If $y = f(x)$, then $f^{-1}(y) = x$.
  Evaluating $f^{-1}(b)$. Interpretation of $f^{-1}(b)$.
  Formula for $f^{-1}(y)$ given formula for $f(x)$.
- **Concavity**
  Concave up: increasing rate of change.
  Concave down: decreasing rate of change.

## REVIEW EXERCISES AND PROBLEMS FOR CHAPTER TWO

### Exercises

In Exercises 1–2, evaluate the function for $x = -7$.

**1.** $f(x) = x/2 - 1$ **2.** $f(x) = x^2 - 3$

For Exercises 3–6, calculate exactly the values of $y$ when $y = f(4)$ and of $x$ when $f(x) = 6$.

**3.** $f(x) = \dfrac{6}{2 - x^3}$ **4.** $f(x) = \sqrt{20 + 2x^2}$

**5.** $f(x) = 4x^{3/2}$ **6.** $f(x) = x^{-3/4} - 2$

**7.** If $f(x) = 2x + 1$, **(a)** Find $f(0)$ **(b)** Solve $f(x) = 0$.

**8.** If $f(x) = \dfrac{x}{1 - x^2}$, find $f(-2)$.

**9.** If $P(t) = 170 - 4t$, find $P(4) - P(2)$.

**10.** Let $h(x) = 1/x$. Find

**(a)** $h(x + 3)$ **(b)** $h(x) + h(3)$

**11.** **(a)** Using Table 2.27, evaluate $f(1)$, $f(-1)$, and $-f(1)$.
**(b)** Solve $f(x) = 0$ for $x$.

Table 2.27

| $x$ | −1 | 0 | 1 | 2 |
|---|---|---|---|---|
| $f(x)$ | 0 | −1 | 2 | 1 |

**12.** **(a)** In Figure 2.44, estimate $f(0)$.
**(b)** For what $x$-value(s) is $f(x) = 0$?
**(c)** For what $x$-value(s) is $f(x) > 0$?

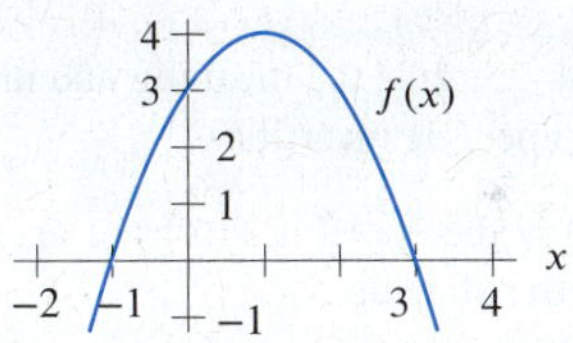


Figure 2.44

Find the domain and range of functions in Exercises 13–18 algebraically.

**13.** $q(x) = \sqrt{x^2 - 9}$ **14.** $m(r) = \dfrac{1}{\sqrt{r^2 - 1}}$

**15.** $m(x) = 9 - x$ **16.** $n(x) = 9 - x^4$

**17.** $m(t) = \dfrac{t}{3} + 2$ **18.** $s(q) = \dfrac{2q + 3}{5 - 4q}$

**19.** **(a)** How can you tell from the graph of a function that an $x$-value is not in the domain? Sketch an example.
**(b)** How can you tell from the formula for a function that an $x$-value is not in the domain? Give an example.

**20.** Let $g(x) = x^2 + x$. Evaluate and simplify the following.

(a) $-3g(x)$ (b) $g(1) - x$
(c) $g(x) + \pi$ (d) $\sqrt{g(x)}$
(e) $g(1)/(x + 1)$ (f) $(g(x))^2$

**21.** Let $f(x) = 1 - x$. Evaluate and simplify the following.

(a) $2f(x)$ (b) $f(x) + 1$ (c) $f(1 - x)$
(d) $(f(x))^2$ (e) $f(1)/x$ (f) $\sqrt{f(x)}$

In Exercises 22–23, let $f(x) = 3x - 7$ and $g(x) = x^3 + 1$ to find a formula for the function.

**22.** $f(g(x))$ **23.** $g(f(x))$

In Exercises 24–25, give the meaning and units of the composite function.

**24.** $a(g(w))$, where $F = g(w)$ is the force, in newtons, on a rocket when the wind speed is $w$ meters/sec and $a(F)$ is the acceleration, in meters/sec$^2$, when the force is $F$ newtons.

**25.** $P(f(t))$, where $l = f(t)$ is the length, in centimeters, of a pendulum at time $t$ minutes, and $P(l)$ is the period, in seconds, of a pendulum of length $l$.

In Exercises 26–33, use $f(x) = x^2 + 1$ and $g(x) = 2x + 3$.

**26.** $f(g(0))$ **27.** $f(g(1))$ **28.** $g(f(0))$ **29.** $g(f(1))$

**30.** $f(g(x))$ **31.** $g(f(x))$ **32.** $f(f(x))$ **33.** $g(g(x))$

In Exercises 34–35, give the meaning and units of the inverse function. (Assume $f$ is invertible.)

**34.** $V = f(t)$ is the speed in km/hr of an accelerating car $t$ seconds after starting.

**35.** $I = f(r)$ is the interest earned, in dollars, on a $10,000 deposit at an interest rate of $r$% per year, compounded annually.

In Exercises 36–37, find the domain and range of the function.

**36.** $h(x) = \dfrac{a}{\sqrt{x}}$, where $a$ is a constant

**37.** $p(x) = |x - b| + 6$, where $b$ is a constant

In Exercises 38–39, find the inverse function.

**38.** $y = g(t) = \sqrt{t} + 1$ **39.** $P = f(q) = 14q - 2$

In Exercises 40–42, let $P = f(t)$ be the population, in millions, of a country at time $t$ in years and let $E = g(P)$ be the daily electricity consumption, in megawatts, when the population is $P$. Give the meaning and units of the function. Assume both $f$ and $g$ are invertible.

**40.** $g(f(t))$ **41.** $f^{-1}(P)$ **42.** $g^{-1}(E)$

**43.** Calculate successive rates of change for the function, $p(t)$, in Table 2.28 to decide whether you expect the graph of $p(t)$ to be concave up or concave down.

**Table 2.28**

| $t$ | 0.2 | 0.4 | 0.6 | 0.8 |
|---|---|---|---|---|
| $p(t)$ | −3.19 | −2.32 | −1.50 | −0.74 |

**44.** If $p(x) = \dfrac{12}{\sqrt{x}}$, evaluate $p(8)$ and $p^{-1}(\sqrt{2})$

**45.** For $f(x) = 12 - \sqrt{x}$, evaluate $f(16)$ and $f^{-1}(3)$.

In Exercises 46–47, graph the function.

**46.** $f(x) = \begin{cases} x^2 & \text{for} \quad x \le 1 \\ 2 - x & \text{for} \quad x > 1 \end{cases}$

**47.** $g(x) = \begin{cases} x + 5 & \text{for} \quad x < 0 \\ x^2 + 1 & \text{for} \quad 0 \le x \le 2 \\ 3 & \text{for} \quad x > 2 \end{cases}$

## Problems

**48.** If $V = \frac{1}{3}\pi r^2 h$ gives the volume of a cone, what is the value of $V$ when $r = 3$ inches and $h = 2$ inches? Give units.

**49.** Let $q(x) = 3 - x^2$. Evaluate and simplify:

(a) $q(5)$ (b) $q(a)$
(c) $q(a - 5)$ (d) $q(a) - 5$
(e) $q(a) - q(5)$

**50.** Let $p(x) = x^2 + x + 1$. Find $p(-1)$ and $-p(1)$. Are they equal?

**51.** Using Figure 2.45, find a formula for $g(x)$ in terms of $f(x)$.

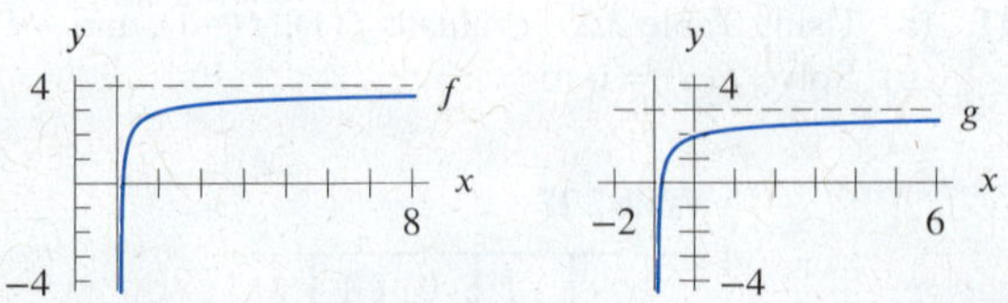


**Figure 2.45**

**52.** Chicago's average monthly rainfall, $R = f(t)$ inches, is given as a function of month, $t$, in Table 2.29. (January is $t = 1$.) Solve and interpret:

**(a)** $f(t) = 3.7$ **(b)** $f(t) = f(2)$

**Table 2.29**

| $t$ | 1 | 2 | 3 | 4 | 5 | 6 | 7 | 8 |
|---|---|---|---|---|---|---|---|---|
| $R$ | 1.8 | 1.8 | 2.7 | 3.1 | 3.5 | 3.7 | 3.5 | 3.4 |

**53.** Let $f(x) = \sqrt{x^2 + 16} - 5$.

**(a)** Find $f(0)$.
**(b)** For what values of $x$ is $f(x)$ zero?
**(c)** Find $f(3)$.
**(d)** What is the vertical intercept of the graph of $f(x)$?
**(e)** Where does the graph cross the $x$-axis?

**54.** Use the graph of $f(x)$ in Figure 2.46 to estimate:

**(a)** $f(0)$ **(b)** $f(1)$ **(c)** $f(b)$ **(d)** $f(c)$ **(e)** $f(d)$

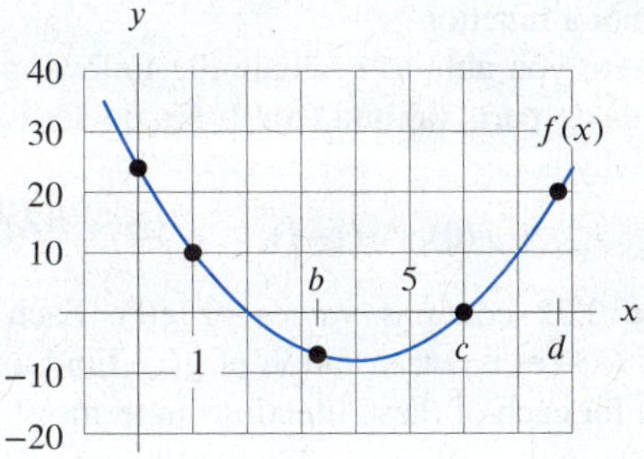


Figure 2.46

**55.** Let $f$ be the function shown in Figure 2.47.

**(a)** Estimate all intervals on which $f$ is increasing.
**(b)** Estimate all intervals on which $f$ is concave up.

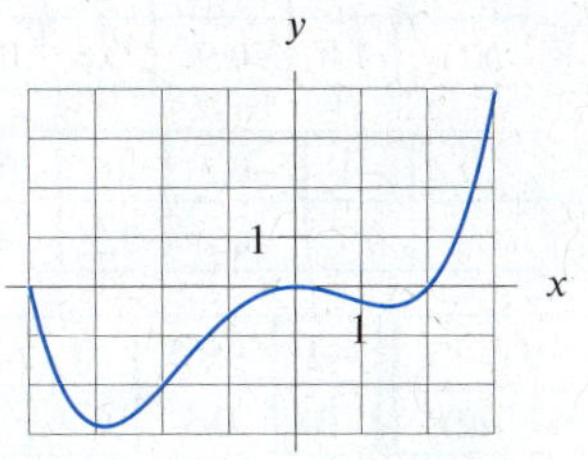


Figure 2.47

**56.** Use the graph in Figure 2.48 to fill in the missing values:

**(a)** $f(0) = ?$ **(b)** $f(?) = 0$
**(c)** $f^{-1}(0) = ?$ **(d)** $f^{-1}(?) = 0$

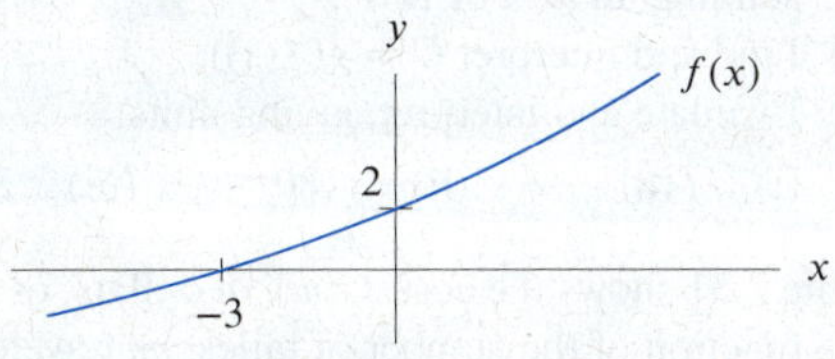


Figure 2.48

**57.** In Figure 2.49, the values $c$ and $d$ are labeled on the $x$-axis. On the $y$-axis, locate the following quantities:

**(a)** $h(c)$ **(b)** $h(d)$
**(c)** $h(c + d)$ **(d)** $h(c) + h(d)$

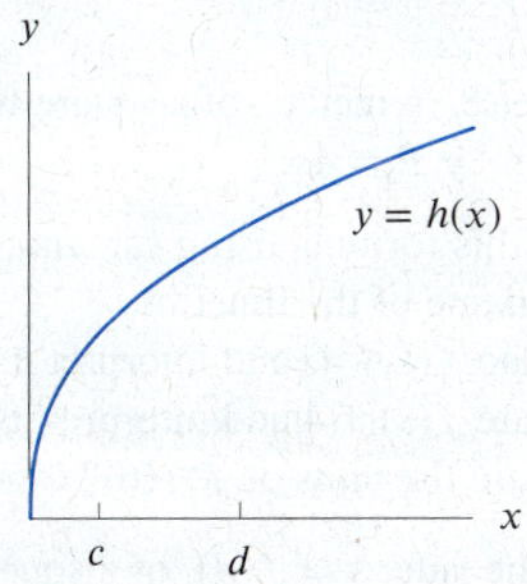


Figure 2.49

**58.** Use the values of the invertible function in Table 2.30 to find as many values of $g^{-1}$ as possible.

**Table 2.30**

| $t$ | 1 | 2 | 3 | 4 | 5 |
|---|---|---|---|---|---|
| $y = g(t)$ | 7 | 12 | 13 | 19 | 22 |

**59.** The formula $V = f(r) = \frac{4}{3}\pi r^3$ gives the volume of a sphere of radius $r$. Find a formula for the inverse function, $f^{-1}(V)$, giving radius as a function of volume.

**60.** The formula for the volume of a cube with side $s$ is $V = s^3$. The formula for the surface area of a cube is $A = 6s^2$.

**(a)** Find and interpret the formula for the function $s = f(A)$.
**(b)** If $V = g(s)$, find and interpret the formula for $g(f(A))$.

**61.** The area, $A = f(s)$ ft$^2$, of a square wooden deck is a function of the side $s$ feet. A can of stain costs \$29.50 and covers 200 square feet of wood.

**(a)** Write the formula for $f(s)$.
**(b)** Find a formula for $C = g(A)$, the cost in dollars of staining an area of $A$ ft$^2$.
**(c)** Find and interpret $C = g(f(s))$.
**(d)** Evaluate and interpret, giving units:

(i) $f(8)$ (ii) $g(80)$ (iii) $g(f(10))$

**62.** Table 2.31 shows the cost, $C(m)$, in dollars, of a taxi ride as a function of the number of miles, $m$, traveled.

**(a)** Estimate and interpret $C(3.5)$ in practical terms.
**(b)** Assume $C$ is invertible. What does $C^{-1}(3.5)$ mean in practical terms? Estimate $C^{-1}(3.5)$.

**Table 2.31**

| $m$ | 0 | 1 | 2 | 3 | 4 | 5 |
|---|---|---|---|---|---|---|
| $C(m)$ | 0 | 2.50 | 4.00 | 5.50 | 7.00 | 8.50 |

**63.** The perimeter, in meters, of a square whose side is $s$ meters is given by $P = 4s$.

**(a)** Write this formula using function notation, where $f$ is the name of the function.
**(b)** Evaluate $f(s + 4)$ and interpret its meaning.
**(c)** Evaluate $f(s) + 4$ and interpret its meaning.
**(d)** What are the units of $f^{-1}(6)$?

**64.** **(a)** Find the side, $s = f(d)$, of a square as a function of its diagonal $d$.
**(b)** Find the area, $A = g(s)$, of a square as a function of its side $s$.
**(c)** Find the area $A = h(d)$ as a function of $d$.
**(d)** What is the relation between $f$, $g$, and $h$?

**65.** Suppose that $f(x)$ is invertible and that both $f$ and $f^{-1}$ are defined for all values of $x$. Let $f(2) = 3$ and $f^{-1}(5) = 4$. Evaluate the following expressions, or, if the given information is insufficient, write unknown.

(a) $f^{-1}(3)$ (b) $f^{-1}(4)$ (c) $f(4)$

**66.** Let $k(x) = 6 - x^2$.

**(a)** Find a point on the graph of $k(x)$ whose $x$-coordinate is $-2$.
**(b)** Find two points on the graph whose $y$-coordinates are $-2$.
**(c)** Graph $k(x)$ and locate the points in parts (a) and (b).
**(d)** Let $p = 2$. Calculate $k(p) - k(p - 1)$.

**67.** **(a)** Find a point on the graph of $h(x) = \sqrt{x + 4}$ whose $x$-coordinate is 5.
**(b)** Find a point on the graph whose $y$-coordinate is 5.
**(c)** Graph $h(x)$ and mark the points in parts (a) and (b).
**(d)** Let $p = 2$. Calculate $h(p + 1) - h(p)$.

**68.** Let $t(x)$ be the time required, in seconds, to melt 1 gram of a compound at $x$°C.

**(a)** Express the following statement as an equation using $t(x)$: It takes 272 seconds to melt 1 gram of the compound at 400°C.
**(b)** Explain the following equations in words:

(i) $t(800) = 136$ (ii) $t^{-1}(68) = 1600$

**(c)** Above a certain temperature, doubling the temperature, $x$, halves the melting time. Express this fact with an equation involving $t(x)$.

**69.** **(a)** The Fibonacci sequence is a sequence of numbers that begins $1, 1, 2, 3, 5, \ldots$. Each term in the sequence is the sum of the two preceding terms. For example,

$$2 = 1 + 1, \quad 3 = 2 + 1, \quad 5 = 2 + 3, \ldots.$$

Based on this observation, complete the following table of values for $f(n)$, the $n^{\text{th}}$ term in the Fibonacci sequence.

| $n$ | 1 | 2 | 3 | 4 | 5 | 6 | 7 | 8 | 9 | 10 | 11 | 12 |
|---|---|---|---|---|---|---|---|---|---|---|---|---|
| $f(n)$ | 1 | 1 | 2 | 3 | 5 | | | | | | | |

**(b)** The table of values in part (a) can be completed even though we don't have a formula for $f(n)$. Does the fact that we don't have a formula mean that $f(n)$ is not a function?
**(c)** Are you able to evaluate the following expressions using parts (a) and (b)? If so, do so; if not, explain why not.

$$f(0), \quad f(-1), \quad f(-2), \quad f(0.5).$$

**70.** Table 2.32 contains values of $g(t)$. Each function in parts (a)–(e) is a translation of $g(t)$. Find a possible formula for each of these functions in terms of $g$.

**Table 2.32**

| $t$ | −1 | −0.5 | 0 | 0.5 | 1 |
|---|---|---|---|---|---|
| $g(t)$ | 0.5 | 0.8 | 1.0 | 0.9 | 0.6 |

**(a)**

| $t$ | −1 | −0.5 | 0 | 0.5 | 1 |
|---|---|---|---|---|---|
| $a(t)$ | 1.0 | 1.3 | 1.50 | 1.4 | 1.1 |

**(b)**

| $t$ | −1 | −0.5 | 0 | 0.5 | 1 |
|---|---|---|---|---|---|
| $b(t)$ | 1.0 | 0.9 | 0.6 | 0.1 | −0.4 |

**(c)**

| $t$ | −1 | −0.5 | 0 | 0.5 | 1 |
|---|---|---|---|---|---|
| $c(t)$ | 0.7 | 0.6 | 0.3 | −0.2 | −0.7 |

**(d)**

| $t$ | −1 | −0.5 | 0 | .5 | 1 |
|---|---|---|---|---|---|
| $d(t)$ | 0 | 0.5 | 0.8 | 1.0 | 0.9 |

**(e)**

| $t$ | −1 | −0.5 | 0 | 0.5 | 1 |
|---|---|---|---|---|---|
| $e(t)$ | 1.2 | 1.7 | 2.0 | 2.2 | 2.1 |

**71.** A psychologist conducts an experiment to determine the effect of sleep loss on job performance. Let $p = f(t)$ be the number of minutes it takes the average person to complete a particular task if he or she has lost $t$ minutes of sleep, where $t = 0$ represents exactly 8 hours of sleep. For instance, $f(60)$ is the amount of time it takes the average person to complete the task after sleeping for only 7 hours. Let $p_0 = f(0)$ and let $t_1$, $t_2$, and $t_3$ be positive constants. Explain what the following statements tell you about sleep loss and job performance.

**(a)** $f(30) = p_0 + 5$
**(b)** $f(t_1) = 2p_0$
**(c)** $f(2t_1) = 1.5f(t_1)$
**(d)** $f(t_2 + 60) = f(t_2 + 30) + 10$

**72.** Give a formula for a function whose domain is all non-negative values of $x$ except $x = 3$.

**73.** Give a formula for a function that is undefined for $x = 8$ and for $x < 4$, but is defined everywhere else.

**74.** Many printing presses are designed with large plates that print a fixed number of pages as a unit. Each unit is called a signature. A particular press prints signatures of 16 pages each. Suppose $C(p)$ is the cost of printing a book of $p$ pages, assuming each signature printed costs \$0.14.

**(a)** What is the cost of printing a book of 128 pages? 129 pages? $p$ pages?
**(b)** What are the domain and range of $C$?
**(c)** Graph $C(p)$ for $0 \le p \le 128$.

**75.** Table 2.33 shows the population, $P$, in millions, of Ireland[18] at various times between 1780 and 1910, with $t$ in years since 1780.

**(a)** When was the population increasing? Decreasing?
**(b)** For each successive time interval, construct a table showing the average rate of change of the population.
**(c)** From the table you constructed in part (b), when is the graph of the population concave up? Concave down?
**(d)** When was the average rate of change of the population the greatest? The least? How is this related to part (c)? What does this mean in human terms?
**(e)** Graph the data in Table 2.33 and join the points by a curve to show the trend in the data. From this graph identify where the curve is increasing, decreasing, concave up and concave down. Compare your answers to those you got in parts (a) and (c). Identify the region you found in part (d).
**(f)** Something catastrophic happened in Ireland between 1780 and 1910. When? What happened in Ireland at that time to cause this catastrophe?

**Table 2.33** *The population of Ireland from 1780 to 1910, where $t = 0$ corresponds to 1780*

| $t$ | 0 | 20 | 40 | 60 | 70 | 90 | 110 | 130 |
|---|---|---|---|---|---|---|---|---|
| $P$ | 4.0 | 5.2 | 6.7 | 8.3 | 6.9 | 5.4 | 4.7 | 4.4 |

## STRENGTHEN YOUR UNDERSTANDING

Are the statements in Problems 1–46 true or false? Give an explanation for your answer.

**1.** If $f(t) = 3t^2 - 4$ then $f(2) = 0$.

**2.** If $f(x) = x^2 - 9x + 10$ then $f(b) = b^2 - 9b + 10$.

**3.** If $f(x) = x^2$ then $f(x + h) = x^2 + h^2$.

**4.** If $q = \dfrac{1}{\sqrt{z^2 + 5}}$ then the values of $z$ that make $q = \frac{1}{3}$ are $z = \pm 2$.

**5.** If $W = \dfrac{t+4}{t-4}$ then when $t = 8$, $W = 1$.

**6.** If $f(t) = t^2 + 64$ then $f(0) = 64$.

**7.** If $f(x) = 0$ then $x = 0$.

**8.** If $f(x) = x^2 + 2x + 7$ then $f(-x) = f(x)$.

**9.** If $g(x) = \dfrac{3}{\sqrt{x^2 + 4}}$ then $g(x)$ can never be zero.

**10.** If $h(p) = -6p + 9$ then $h(3) + h(4) = h(7)$.

**11.** The domain of a function is the set of input values.

**12.** If a function is being used to model a real-world situation, the domain and range are often determined by the constraints of the situation being modeled.

**13.** The domain of $f(x) = \dfrac{4}{x-3}$ consists of all real numbers $x$, $x \neq 0$.

**14.** If $f(x) = \sqrt{2 - x}$, the domain of $f$ consists of all real numbers $x \ge 2$.

**15.** The range of $f(x) = \dfrac{1}{x}$ is all real numbers.

**16.** The range of $y = 4 - \dfrac{1}{x}$ is $0 < y < 4$.

**17.** If $f(x) = \frac{2}{5}x + 6$ and its domain is $15 \le x \le 20$ then the range of $f$ is $12 \le x \le 14$.

**18.** The domain of $f(x) = \dfrac{x}{\sqrt{x^2 + 1}}$ is all real numbers.

[18] Adapted from D. N. Burghes and A. D. Wood, *Mathematical Models in the Social, Management and Life Science*, p. 104 (Ellis Horwood, 1980).

19. The graph of the absolute value function $y = |x|$ has a V shape.

20. The domain of $f(x) = |x|$ is all real numbers.

21. If $f(x) = |x|$ and $g(x) = |-x|$ then for all $x$, $f(x) = g(x)$.

22. If $f(x) = |x|$ and $g(x) = -|x|$ then for all $x$, $f(x) = g(x)$.

23. If $y = \frac{x}{|x|}$ then $y = 1$ for $x \neq 0$.

24. If $f(x) = \begin{cases} 3 & \text{if } x < 0 \\ x^2 & \text{if } 0 \leq x \leq 4 \\ 7 & \text{if } x > 4 \end{cases}$, then $f(3) = 0$.

25. Let $f(x) = \begin{cases} x & \text{if } x < 0 \\ x^2 & \text{if } 0 \leq x \leq 4. \\ -x & \text{if } x > 4 \end{cases}$ If $f(x) = 4$ then $x = 2$.

26. If $g(x) = f(x)+3$ then the graph of $g(x)$ is a vertical shift of the graph of $f$.

27. If $g(t) = f(t-2)$ then the graph of $g(t)$ can be obtained by shifting the graph of $f$ two units to the left.

28. Figure 2.50 suggests that $g(x) = f(x+2)+1$.

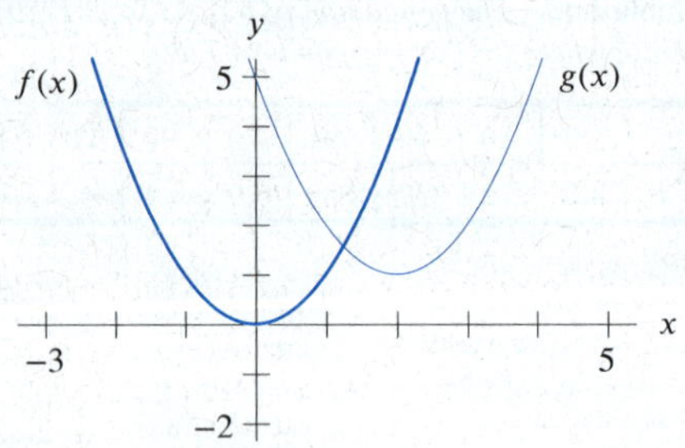


Figure 2.50

29. Figure 2.51 could be the graph of $f(x) = |x-1| - 2$.

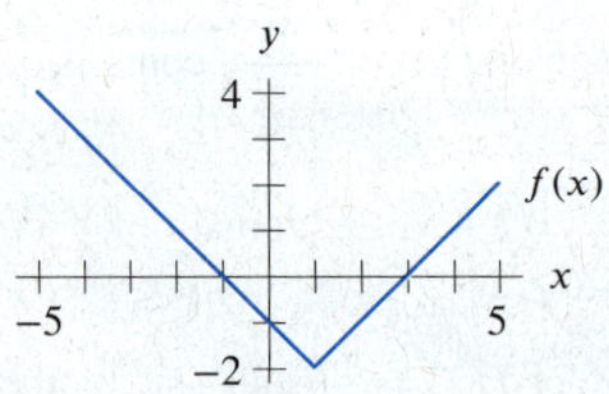


Figure 2.51

30. If $f(3) = 5$ and $f$ is invertible, then $f^{-1}(3) = 1/5$.

31. If $h(7) = 4$ and $h$ is invertible, then $h^{-1}(4) = 7$.

32. If $f(x) = \frac{3}{4}x - 6$ then $f^{-1}(8) = 0$.

33. If $R = f(S) = \frac{2}{3}S + 8$ then $S = f^{-1}(R) = \frac{3}{2}(R-8)$.

34. In general $f^{-1}(x) = (f(x))^{-1}$.

35. If $f(x) = \frac{x}{x+1}$ then $f(t^{-1}) = \frac{1/t}{1/t+1}$.

36. The units of the output of a function are the same as the units of output of its inverse.

37. The functions $f(x) = 2x + 1$ and $g(x) = \frac{1}{2}x - 1$ are inverses.

38. If $q = f(x)$ is the quantity of rice in tons required to feed $x$ million people for a year and $p = g(q)$ is the cost, in dollars, of $q$ tons of rice, then $g(f(x))$ is the dollar cost of feeding $x$ million people for a year.

39. If $f(t) = t + 2$ and $g(t) = 3t$, then $g(f(t)) = 3(t+2) = 3t + 6$.

40. A fireball has radius $r = f(t)$ meters $t$ seconds after an explosion. The volume of the ball is $V = g(r)$ meter$^3$ when it has radius $r$ meters. Then the units of measurement of $g(f(t))$ are meter$^3$/sec.

41. If the graph of a function is concave up, then the average rate of change of a function over an interval of length 1 increases as the interval moves from left to right.

42. The function $f$ in the table could be concave up.

| $x$ | −2 | 0 | 2 | 4 |
|---|---|---|---|---|
| $f(x)$ | 5 | 6 | 8 | 12 |

43. The function $g$ in the table could be concave down.

| $t$ | −1 | 1 | 3 | 5 |
|---|---|---|---|---|
| $g(t)$ | 9 | 8 | 6 | 3 |

44. A straight line is concave up.

45. A function can be both decreasing and concave down.

46. If a function is concave up, it must be increasing.

Chapter Three

# QUADRATIC FUNCTIONS

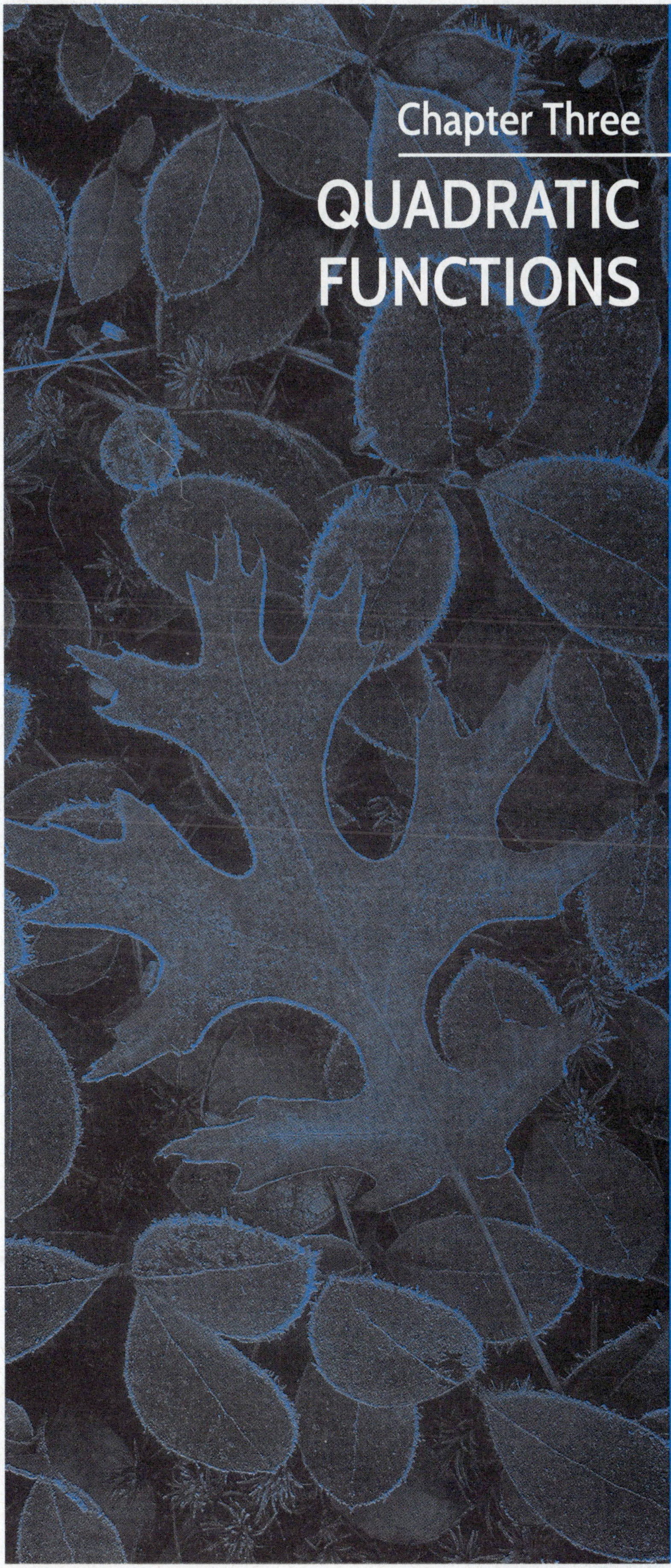

© Patrick Zephyr/Patrick Zephyr Nature Photography

## Contents

# 3.1 INTRODUCTION TO THE FAMILY OF QUADRATIC FUNCTIONS

A baseball is "popped" straight up by a batter. The height of the ball above the ground is given by the function $y = f(t) = -16t^2 + 47t + 3$, where $t$ is time in seconds after the ball leaves the bat and $y$ is in feet. See Figure 3.1. Note that the path of the ball is straight up and down, although the graph of height against time is a curve. The ball goes up fast at first and then more slowly because of gravity; thus the graph of its height as a function of time is concave down.

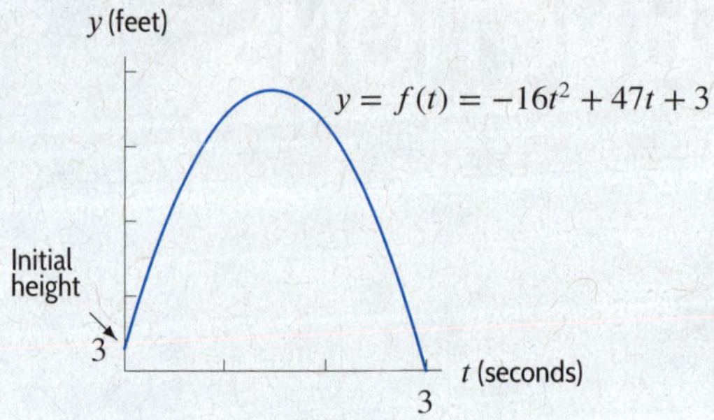


**Figure 3.1:** The height of a ball $t$ seconds after being "popped up". (Note: This graph does not represent the ball's path.)

The baseball height function is an example of a *quadratic function* whose standard form is

$$y = ax^2 + bx + c.$$

The graph of a quadratic function is called a *parabola*. Notice that the function in Figure 3.1 is concave down and has a maximum corresponding to the time at which the ball stops rising and begins to fall back to the earth. The maximum point on the parabola is called the *vertex*. The intersection of the parabola with the horizontal axis gives the times when the height of the ball is zero; these times are the *zeros* of the height function. The horizontal intercepts of a graph occur at the zeros of the function. In this section we examine the zeros and concavity of quadratic functions, and in the next section we see how to find the vertex.

## Finding the Zeros of a Quadratic Function

A natural question to ask is when the ball hits the ground. The graph suggests that $y = 0$ when $t$ is approximately 3. In symbols, the question is: For what value(s) of $t$ does $f(t) = 0$? These are the zeros of the function.

It is easy to find the zeros of a quadratic function if it is expressed in *factored form*,

$$q(x) = a(x - r)(x - s), \qquad \text{where } a,\ r, s \text{ are constants, } a \neq 0.$$

Then $r$ and $s$ are zeros of the function $q$.

The baseball function factors as

$$y = -1(16t + 1)(t - 3),$$

so the zeros of $f$ are $t = 3$ and $t = -1/16$. We are interested in positive values of $t$, so the ball hits the ground 3 seconds after it was hit. (For more on factoring, see the Skills Review on page 131.)

**Example 1** Find the zeros of $f(x) = x^2 - x - 6$.

**Solution** To find the zeros, set $f(x) = 0$ and solve for $x$ by factoring:

$$x^2 - x - 6 = 0$$
$$(x - 3)(x + 2) = 0.$$

Thus the zeros are $x = 3$ and $x = -2$.

We can also find the zeros of a quadratic function by using the quadratic formula. (See the Skills Review on page 131 to review the quadratic formula.)

**Example 2** Find the zeros of $f(x) = x^2 - x - 6$ by using the quadratic formula.

**Solution** We solve the equation $x^2 - x - 6 = 0$. For this equation, $a = 1$, $b = -1$, and $c = -6$. Thus

$$x = \frac{-b \pm \sqrt{b^2 - 4ac}}{2a} = \frac{-(-1) \pm \sqrt{(-1)^2 - 4(1)(-6)}}{2(1)}$$
$$= \frac{1 \pm \sqrt{25}}{2} = 3 \text{ or } -2.$$

The zeros are $x = 3$ and $x = -2$, the same as we found by factoring.

Since the zeros of a function occur at the horizontal intercepts of its graph, quadratic functions without horizontal intercepts (such as in the next example) have no zeros.

**Example 3** Figure 3.2 shows a graph of $h(x) = -\frac{1}{2}x^2 - 2$. What happens if we try to use algebra to find the zeros of $h$?

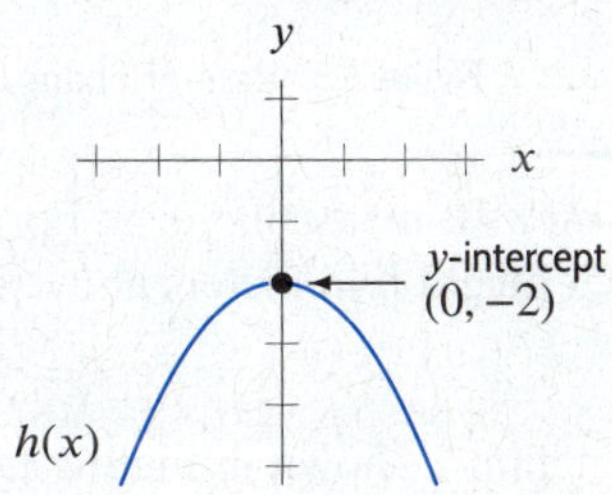


**Figure 3.2**: Zeros of $h(x) = -(x^2/2) - 2$?

**Solution** To find the zeros, we solve the equation

$$-\frac{1}{2}x^2 - 2 = 0$$
$$-\frac{1}{2}x^2 = 2$$
$$x^2 = -4$$
$$x = \pm\sqrt{-4}.$$

Since $\sqrt{-4}$ is not a real number, there are no real solutions, so $h$ has no real zeros. This corresponds to the fact that the graph of $h$ in Figure 3.2 does not cross the $x$-axis.

## Concavity and Rates of Change for Quadratic Functions

A quadratic function has a graph that is either concave up or concave down. Recall that for a graph that is concave up, rates of change increase as we move right. For a graph that is concave down, rates of change decrease as we move right.

**Example 4** Let $f(x) = x^2$. Find the average rate of change of $f$ over the four consecutive intervals of length 2 between $x = -4$ and $x = 4$. What do these rates tell you about the concavity of the graph of $f$?

**Solution** Between $x = 0$ and $x = 2$, we have

$$\text{Average rate of change of } f = \frac{f(2) - f(0)}{2 - 0} = \frac{2^2 - 0^2}{2 - 0} = 2.$$

Between $x = 2$ and $x = 4$, we have

$$\text{Average rate of change of } f = \frac{f(4) - f(2)}{4 - 2} = \frac{4^2 - 2^2}{4 - 2} = 6.$$

Similarly, between $x = -4$ and $x = -2$, we can show the average rate of change is $-6$; between $x = -2$ and $x = 0$, the rate of change is $-2$. See Figure 3.3. Since the rates of change are increasing, the graph is concave up.

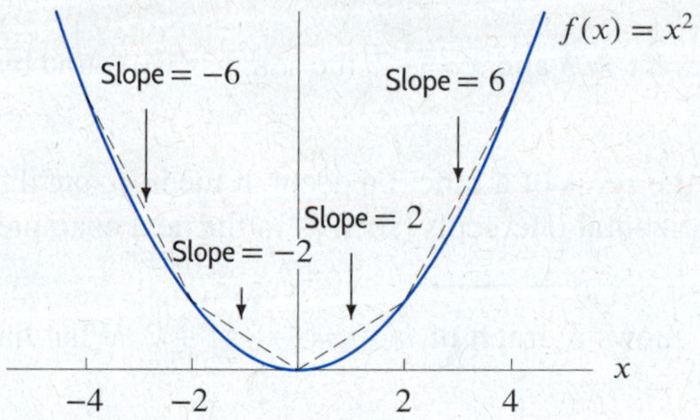


**Figure 3.3**: Rate of change and concavity of $f(x) = x^2$

**Example 5** A high-diver jumps off a 10-meter platform. For $t$ in seconds after the diver leaves the platform until she hits the water, her height $h$ in meters above the water is given by

$$h = f(t) = -4.9t^2 + 8t + 10.$$

The graph of this function is shown in Figure 3.4.

(a) Estimate and interpret the domain and range of the function, and the intercepts of the graph.
(b) Identify the concavity.

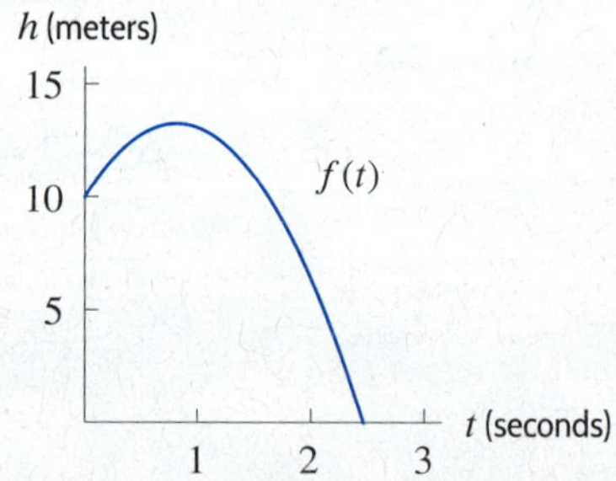


**Figure 3.4**: Height of diver above water as a function of time

**Solution** (a) The diver enters the water when her height above the water is 0. This occurs when

$$h = f(t) = -4.9t^2 + 8t + 10 = 0.$$

Using the quadratic formula to solve this equation, we find the only positive solution is $t = 2.462$ seconds. The domain is the interval of time the diver is in the air, which is approximately $0 \leq t \leq 2.462$. To find the range of $f$, we look for the largest and smallest outputs for $h$. From the graph, the diver's maximum height appears to occur at about $t = 1$, so we estimate the largest output value for $f$ to be about

$$f(1) = -4.9 \cdot 1^2 + 8 \cdot 1 + 10 = 13.1 \text{ meters}.$$

Thus, the range of $f$ is approximately $0 \leq f(t) \leq 13.1$. Methods to find the exact maximum height are in Section 3.2.

The vertical intercept of the graph is

$$f(0) = -4.9 \cdot 0^2 + 8 \cdot 0 + 10 = 10 \text{ meters}.$$

The diver's initial height is 10 meters (the height of the diving platform). The horizontal intercept is the point estimated earlier where $f(t) = 0$. This intercept, $t = 2.462$ seconds, gives the time after leaving the platform that the diver enters the water.

(b) In Figure 3.4, we see that the graph is bending downward over its entire domain, so it is concave down. This is reflected in Table 3.1, where the rate of change, $\Delta h/\Delta t$, is decreasing.

**Table 3.1** *Slope of* $f(t) = -4.9t^2 + 8t + 10$

| $t$ (sec) | $h$ (meters) | Rate of change $\Delta h/\Delta t$ |
|---|---|---|
| 0 | 10 | |
| | | 5.55 |
| 0.5 | 12.775 | |
| | | 0.65 |
| 1.0 | 13.100 | |
| | | −4.25 |
| 1.5 | 10.975 | |
| | | −9.15 |
| 2.0 | 6.400 | |

The previous two examples illustrate that the concavity of the graph of $y = ax^2 + bx + c$ is determined by the sign of the coefficient $a$:

- If $a > 0$, the parabola is concave up.
- If $a < 0$, the parabola is concave down.

**Example 6** Determine the concavity of the graphs of the following quadratic functions:

(a) $y = -3x^2 - 3x + 2$

(b) $y = -3(-2x + 1)(x - 4)$

**Solution** (a) Since the coefficient of $x^2$ is negative, this function has a graph which is concave down.

(b) Expanding we have

$$y = -3(-2x + 1)(x - 4) = -3(-2x^2 + 8x + x - 4) = 6x^2 - 27x + 12.$$

Since the coefficient of $x^2$ is positive, the graph of this function is concave up.

## Finding a Formula From the Zeros and Vertical Intercept

If we know the zeros and the vertical intercept of a quadratic function, we can use the factored form to find a formula for the function.

**Example 7** Find the equation of the parabola in Figure 3.5 using the factored form.

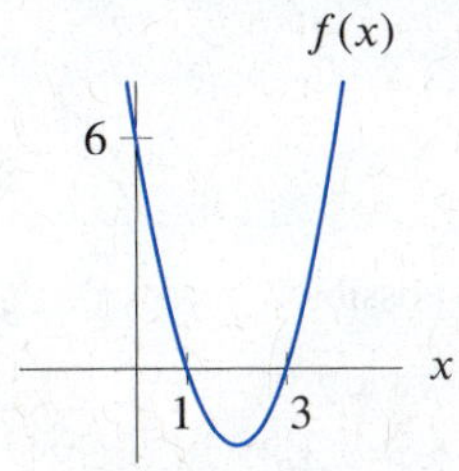


**Figure 3.5**: Finding a formula for a quadratic from the zeros

**Solution** Since the parabola has $x$-intercepts at $x = 1$ and $x = 3$, its formula can be written as

$$f(x) = a(x-1)(x-3).$$

To find $a$, we substitute $x = 0, y = 6$, giving

$$6 = a \cdot 3$$
$$a = 2.$$

Thus, the formula is

$$f(x) = 2(x-1)(x-3).$$

Multiplying out gives $f(x) = 2x^2 - 8x + 6$.

## Formulas for Quadratic Functions

The function $f$ in Example 7 can be written in at least two different ways. The standard form $f(x) = 2x^2-8x+6$ shows that the parabola opens upward, since the coefficient of $x^2$ is positive; the constant 6 gives the vertical intercept. The factored form $f(x) = 2(x-1)(x-3)$ shows that the parabola crosses the $x$-axis at $x = 1$ and $x = 3$. In general, we have the following:

> The graph of a **quadratic function** is a **parabola**.
>
> The **standard form** for a quadratic function is
>
> $$y = ax^2 + bx + c, \quad \text{where } a,\ b,\ c \text{ are constants, } a \neq 0.$$
>
> The parabola is concave up (opens upward) if $a > 0$ or concave down (opens downward) if $a < 0$, and intersects the $y$-axis at $c$.
>
> The **factored form**, when it exists, is
>
> $$y = a(x-r)(x-s), \quad \text{where } a,\ r,\ s \text{ are constants, } a \neq 0.$$
>
> The parabola intersects the $x$-axis at $x = r$ and $x = s$.

In the next section we look at another form for quadratic functions which shows how to find the vertex of the graph.

## Exercises and Problems for Section 3.1

### Skill Refresher

Multiply and write the expressions in Exercises S1–S2 without parentheses. Gather like terms.

**S1.** $(t^2+1)\,50t - (25t^2+125)\,2t$

**S2.** $(A^2 - B^2)^2$

For Exercises S3–S8, factor completely if possible.

**S3.** $u^2 - 2u$

**S4.** $x^2 + 3x + 2$

**S5.** $3x^2 - x - 4$

**S6.** $(s+2t)^2 - 4p^2$

**S7.** $16x^2 - 1$

**S8.** $y^3 - y^2 - 12y$

Solve the equations in Exercises S9–S10.

**S9.** $x^2 + 7x + 6 = 0$

**S10.** $2w^2 + w - 10 = 0$

## Exercises

Are the functions in Exercises 1–8 quadratic? If so, write the function in the form $f(x) = ax^2 + bx + c$.

1. $g(t) = 3(t-2)^2 + 7$
2. $f(x) = (2x-3)(5-x)$
3. $w(n) = n(n-3)(n-2) - n^2(n-8)$
4. $h(t) = -16(t-3)(t+1)$
5. $R(q) = \dfrac{1}{q^2}(q^2+1)^2$
6. $K(x) = 13^2 + 13^x$
7. $T(n) = \sqrt{5} + \sqrt{3n^4} - \sqrt{\dfrac{n^4}{4}}$
8. $r(v) = \dfrac{v^2+\sqrt{2}}{3} + \dfrac{v-3}{5} + \pi v^2$
9. Find the zeros of $Q(r) = 2r^2 - 6r - 36$ by factoring.
10. Find the zeros of $Q(x) = 5x - x^2 + 3$ using the quadratic formula.
11. Solve for $x$ using the quadratic formula and demonstrate your solution graphically:
    (a) $6x - \frac{1}{3} = 3x^2$ (b) $2x^2 + 7.2 = 5.1x$

In Exercises 12–19, find the zeros (if any) of the function algebraically.

12. $y = (2-x)(3-2x)$
13. $y = 9x^2 + 6x + 1$
14. $y = 6x^2 - 17x + 12$
15. $N(t) = t^2 - 7t + 10$
16. $y = 89x^2 + 55x + 34$
17. $Q(r) = 2r^2 - 6r - 36$
18. $y = x^4 + 5x^2 + 6$
19. $y = x - \sqrt{x} - 12$

In Exercises 20–23, is the graph of the quadratic function concave up or concave down?

20. $y = 2x - 3x^2 + 1$
21. $y = 5(x-3)^2$
22. $y = 3(1-2x)(x-4)$
23. $y = -2(5x-1)(4-x)$

## Problems

24. Use the quadratic formula to find the time at which the baseball in Figure 3.1 on page 116 hits the ground.

In Problems 25–26, find a formula for the quadratic function whose graph has the given properties.

25. A $y$-intercept of $y = 7$ and the only zero at $x = -2$.
26. A $y$-intercept of $y = 7$ and $x$-intercepts at $x = 1, 4$.
27. Is there a quadratic function with zeros $x = 1$, $x = 2$ and $x = 3$?
28. Graph a quadratic function which has all the following properties: concave up, $y$-intercept is $-6$, zeros at $x = -2$ and $x = 3$.
29. Can you graph a quadratic function which has all the following properties: concave down, $y$-intercept is $-6$, zeros at $x = -2$ and $x = 3$?
30. Determine the concavity of the graph of $f(x) = 4 - x^2$ between $x = -1$ and $x = 5$ by calculating average rates of change over intervals of length 2.
31. The graph of a quadratic function, $f(x)$, passes through the points $(1,2)$, $(3,4)$ and $(5,2)$. Is the graph of $f(x)$ concave up or concave down? Give a reason for your answer.
32. The quadratic function, $f(x)$, has no zeros and its graph passes through the point $(1,1)$. Is the graph of $f(x)$ concave up or concave down? Give a reason for your answer.
33. Without a calculator, graph the following function by factoring and plotting zeros:
$$y = -4cx + x^2 + 4c^2 \quad \text{for} \quad c > 0.$$
34. Without a calculator, graph $y = 3x^2 - 16x - 12$ by factoring and plotting zeros.

In Problems 35–36, find a formula for the parabola.

35.

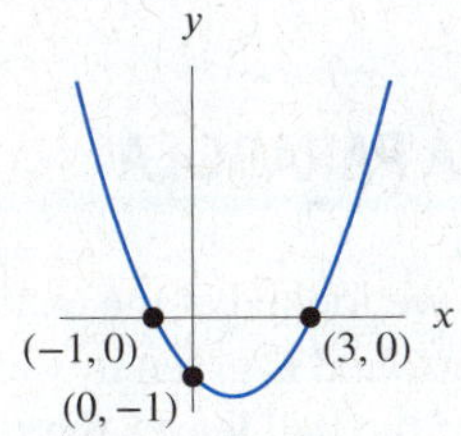


36.

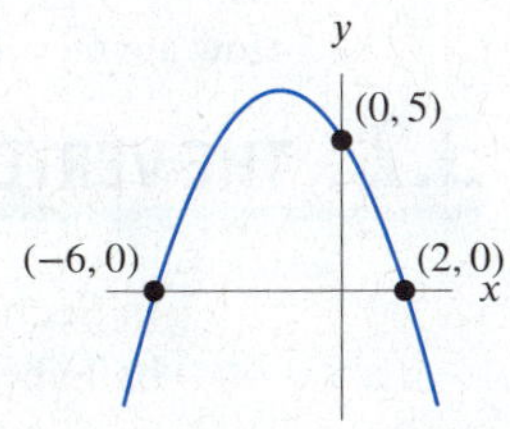


37. Using the factored form, find the formula for the parabola whose zeros are $x = -1$ and $x = 5$, and which passes through the point $(-2, 6)$.
38. Find two different quadratic functions with zeros $x = 1$, $x = 2$.
39. Write $y = 3(0.5x-4)(4-20x)$ in the form $y = k(x-r)(x-s)$ and give the values of $k, r, s$.

**40.** A ball is thrown into the air. Its height (in feet) $t$ seconds later is given by $h(t) = 80t - 16t^2$.

**(a)** Evaluate and interpret $h(2)$.
**(b)** Solve the equation $h(t) = 80$. Interpret your solutions and illustrate them on a graph of $h(t)$.

**41.** A snowboarder slides up from the bottom of a half-pipe and comes down again, sliding with little resistance on the snow. Her height above the top edge of the pipe $t$ seconds after starting up the side is $-4.9t^2 + 14t - 5$ meters.

**(a)** What is her height at $t = 0$?
**(b)** After how many seconds does she reach the top edge? Return to the edge of the pipe?
**(c)** How long is she in the air?

**42.** Let $V(t) = t^2 - 4t + 4$ represent the velocity after $t$ seconds of an object in meters per second.

**(a)** What is the object's initial velocity?
**(b)** When is the object not moving?
**(c)** Identify the concavity of the velocity graph.

**43.** Table 3.2 gives the approximate number of cell phone subscribers, $S$, in millions, in the US.[1] If $x$ is in years since 2005, show that this data can be approximated by the quadratic function $p(x) = -1.9x^2 + 24.4x + 208.8$. What does this model predict for the year 2015? How good is this model for predicting the future?

**Table 3.2**

| Year | 2005 | 2006 | 2007 | 2008 | 2009 |
|---|---|---|---|---|---|
| $S$ (millions) | 207.9 | 233.0 | 250.0 | 262.7 | 276.6 |

**44.** Let $f(x) = x^2$ and $g(x) = x^2 + 2x - 8$.

**(a)** Graph $f$ and $g$ in the window $-10 \le x \le 10$, $-10 \le y \le 10$. How are the two graphs similar? How are they different?
**(b)** Graph $f$ and $g$ in the window $-10 \le x \le 10$, $-10 \le y \le 100$. Why do the two graphs appear more similar on this window than on the window from part (a)?
**(c)** Graph $f$ and $g$ in the window $-20 \le x \le 20$, $-10 \le y \le 400$, the window $-50 \le x \le 50$, $-10 \le y \le 2500$, and the window $-500 \le x \le 500$, $-2500 \le y \le 250{,}000$. Describe the change in appearance of $f$ and $g$ on these three successive windows.

**45.** A relief package is dropped from a moving airplane. The package has an initial forward horizontal velocity and follows a quadratic graph path (instead of dropping straight down). Figure 3.6 shows the height of the package, $h$, in km, as a function of the horizontal distance, $d$, in meters, as it drops.

**(a)** From what height was the package released?
**(b)** How far away from the spot above which it was released does the package hit the ground?
**(c)** Write a formula for $h(d)$. [Hint: The package starts falling at the highest point on the graph.]

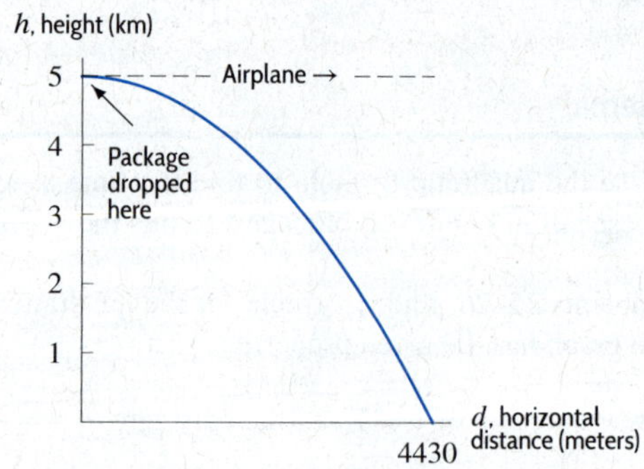


Figure 3.6

## 3.2 THE VERTEX OF A PARABOLA

In Section 3.1, we looked at the example of a baseball popped upward by a batter. The height of the ball above the ground is given by the quadratic function $y = f(t) = -16t^2 + 47t + 3$, where $t$ is time in seconds after the ball leaves the bat, and $y$ is in feet. See Figure 3.7.

The point on the graph with the largest $y$ value appears to be approximately (1.5, 37.5). (See Example 5 on page 126 for the exact values.) This means that the baseball reaches its maximum height of about 37.5 feet about 1.5 seconds after being hit. The maximum point on the parabola is called the *vertex*. For quadratic functions the vertex shows where the function reaches either its maximum value or, in the case of a concave-up parabola, its minimum value.

The vertex of a parabola can be determined exactly if the quadratic function is written in the form

$$y = a(x - h)^2 + k.$$

[1] http://www.infoplease.com/ipa/A0933563.html. Last accessed January 13, 2014.

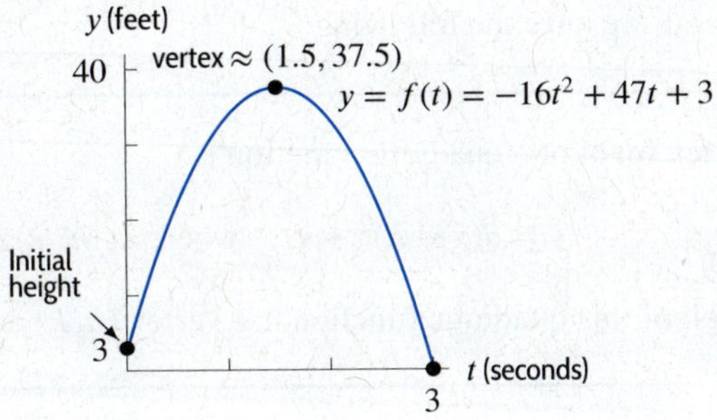


Figure 3.7: Height of baseball at time $t$

This form arises when we shift the parabola $y = x^2$ horizontally by $h$ and vertically by $k$, moving the vertex from $(0, 0)$ to $(h, k)$.

**Example 1**

(a) Sketch $f(x) = (x + 3)^2 - 4$, and indicate the vertex.
(b) Estimate the coordinates of the vertex from the graph.
(c) Explain how the formula for $f$ can be used to obtain the minimum of $f$.
(d) Explain how vertical and horizontal shifts can be used to obtain the coordinates of the vertex.

**Solution**

(a) Figure 3.8 shows a sketch of $f(x)$; the vertex gives the minimum value of the function.
(b) The vertex of $f$ appears to be about at the point $(-3, -4)$.
(c) Note that $(x+3)^2$ is always positive or zero, so $(x+3)^2$ takes on its smallest value when $x+3 = 0$, that is, at $x = -3$. At this point $(x + 3)^2 - 4$ takes on its smallest value,

$$f(-3) = (-3 + 3)^2 - 4 = 0 - 4 = -4.$$

Thus, we see that the vertex is exactly at the point $(-3, -4)$ and the minimum is $-4$.
(d) The function $f(x) = (x + 3)^2 - 4$ is a transformation of the function $f(x) = x^2$, with an inside change of $-3$ and an outside change of $-4$. This indicates that $f(x) = x^2$ has been shifted 3 units to the left and 4 units down. Thus, the vertex is $(-3, -4)$.

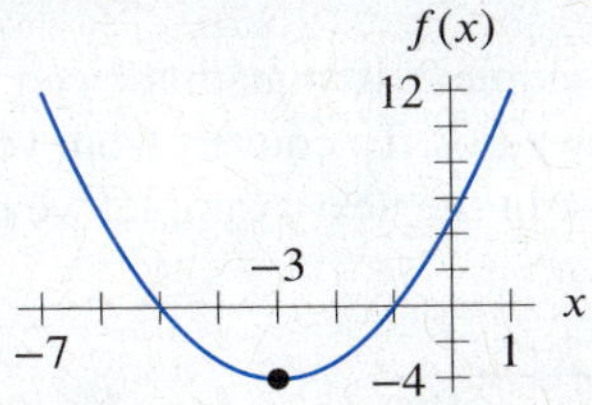


Figure 3.8: A graph of $f(x) = (x + 3)^2 - 4$

Notice that if we select $x$-values that are equally spaced to the left and the right of the vertex, the $y$-values of the function are equal. For example, $f(-2) = f(-4) = -3$. The graph is symmetric about a vertical line that passes through the vertex. This line is called the *axis of symmetry*. The function in Example 1 has axis of symmetry $x = -3$.

## The Vertex Form of a Quadratic Function

Writing the function $f$ in Example 1 in the form

$$f(x) = (x + 3)^2 - 4$$

enabled us to find the vertex of the graph and the location and value for the minimum of the function.

In general, we have the following:

> The **vertex form** of a quadratic function is
>
> $$y = a(x-h)^2 + k, \quad \text{where } a,\ h,\ k \text{ are constants, } a \neq 0.$$
>
> The graph of this quadratic function has vertex $(h, k)$ and axis of symmetry $x = h$.

A quadratic function can always be expressed in both standard form and vertex form.

**Example 2** (a) Convert the quadratic $f(x) = (x+3)^2 - 4$ in Example 1 to standard form.
(b) Explain how the vertical intercept can be found from the standard form.
(c) Find the zeros of $f$.
(d) Explain how the axis of symmetry of the parabola is related to the zeros.

**Solution** (a) Expanding $(x+3)^2$ and gathering like terms converts $f$ to standard form:

$$f(x) = x^2 + 6x + 9 - 4 = x^2 + 6x + 5.$$

Since the coefficient of $x^2$ is $a = 1$, the parabola opens upward.
(b) The vertical intercept is $f(0) = 5$. In the standard form $y = ax^2 + bx + c$, the $y$-intercept is the constant $c$.
(c) To find the zeros, we factor $f$ and write it as

$$f(x) = (x+1)(x+5).$$

This form shows that the zeros are $x = -1$ and $x = -5$.
(d) The axis of symmetry is $x = -3$, which is the vertical line through the midpoint of the zeros: $(-1-5)/2 = -3$.

As we see in Example 2, if a quadratic can be written in factored form, its axis of symmetry is at the midpoint of the zeros. To convert from vertex to standard form, we expand the squared term and gather like terms. In the next example, we convert from from standard form to vertex form by *completing the square*.[2]

**Example 3** Put each quadratic function into vertex form by completing the square and then graph it.
(a) $s(x) = x^2 - 6x + 8$ (b) $t(x) = -4x^2 - 12x - 8$

**Solution** (a) To complete the square, find the square of half of the coefficient of the $x$-term, $(-6/2)^2 = 9$. Add and subtract this number after the $x$-term:

$$s(x) = \underbrace{x^2 - 6x + 9}_{\text{Perfect square}} - 9 + 8,$$

so

$$s(x) = (x-3)^2 - 1.$$

The vertex of $s$ is $(3, -1)$ and the axis of symmetry is the vertical line $x = 3$. The parabola opens upward. See Figure 3.9.

[2] A more detailed explanation of this method is in the Skills Review on page 136.

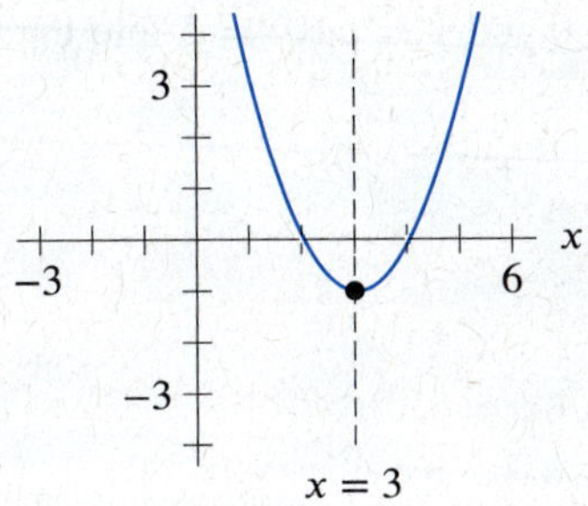


Figure 3.9: $s(x) = x^2 - 6x + 8$

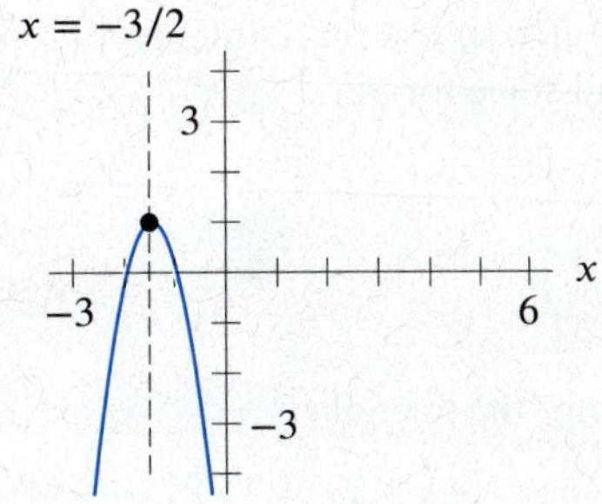


Figure 3.10: $t(x) = -4x^2 - 12x - 8$

(b) To complete the square, first factor out −4, the coefficient of $x^2$, giving

$$t(x) = -4(x^2 + 3x + 2).$$

Now add and subtract the square of half the coefficient of the $x$-term, $(3/2)^2 = 9/4$, inside the parentheses. This gives

$$t(x) = -4\Big(\underbrace{x^2 + 3x + \frac{9}{4}}_{\text{Perfect square}} - \frac{9}{4} + 2\Big)$$

$$t(x) = -4\left(\left(x + \frac{3}{2}\right)^2 - \frac{1}{4}\right)$$

$$t(x) = -4\left(x + \frac{3}{2}\right)^2 + 1.$$

The vertex of $t$ is $(-3/2, 1)$, the axis of symmetry is $x = -3/2$, and the parabola opens downward. See Figure 3.10.

## Finding a Formula Given the Vertex and Another Point

If we know the vertex of a quadratic function and one other point, we can use the vertex form to find its formula.

**Example 4** Find the formula for the quadratic function graphed in Figure 3.11.

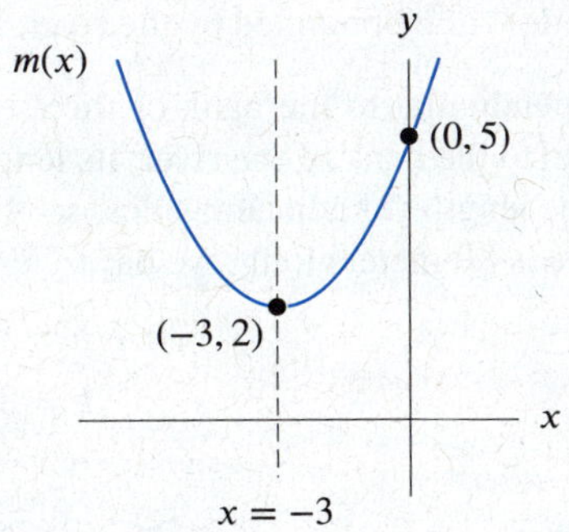


Figure 3.11: Finding a formula for a quadratic from the vertex

**Solution** Since the vertex is given, we use the form $m(x) = a(x - h)^2 + k$ to find $a$, $h$, and $k$. The vertex is $(-3, 2)$, so $h = -3$ and $k = 2$. Thus,

$$m(x) = a(x - (-3))^2 + 2,$$

so

$$m(x) = a(x + 3)^2 + 2.$$

To find $a$, use the $y$-intercept $(0, 5)$. Substitute $x = 0$ and $y = m(0) = 5$ into the formula for $m(x)$ and solve for $a$:

$$5 = a(0+3)^2 + 2$$
$$3 = 9a$$
$$a = \frac{1}{3}.$$

Thus, the formula is

$$m(x) = \frac{1}{3}(x+3)^2 + 2.$$

If we want the formula in standard form, we multiply out:

$$m(x) = \frac{1}{3}x^2 + 2x + 5.$$

## Modeling with Quadratic Functions

In applications, it is often useful to find the maximum or minimum value of a quadratic function. The next example returns to the baseball that started this chapter. Problem 25 on page 128 asks you to derive the vertex form of the baseball height function; here we see what this form can tell us.

**Example 5** For $t$ in seconds, the height of a baseball in feet is given both in standard and in vertex form by

$$y = f(t) = -16t^2 + 47t + 3 = -16\left(t - \frac{47}{32}\right)^2 + \frac{2401}{64}.$$

Find the maximum height reached by the baseball and the time at which that height is reached.

**Solution** From the vertex form, we see the vertex is at the point $\left(\frac{47}{32}, \frac{2401}{64}\right) = (1.469, 37.516)$. This means that the ball reaches it maximum height of 37.516 feet at $t = 1.469$ seconds. Note that these values are close to the graphical estimates $(1.5, 37.5)$ made at the beginning of this section.

**Example 6** A city decides to make a park by fencing off a section of riverfront property. Funds are allotted for 80 meters of fence. The area enclosed will be a rectangle, but only three sides will be enclosed by a fence—the other side will be bounded by the river. What is the maximum area that can be enclosed?

**Solution** Two sides are perpendicular to the bank of the river and have equal length, which we call $h$. The other side is parallel to the bank of the river; its length is $b$. See Figure 3.12. The area, $A$, of the park is the product of the lengths of adjacent sides, so $A = bh$.

Since the fence is 80 meters long, we have

$$2h + b = 80$$
$$b = 80 - 2h.$$

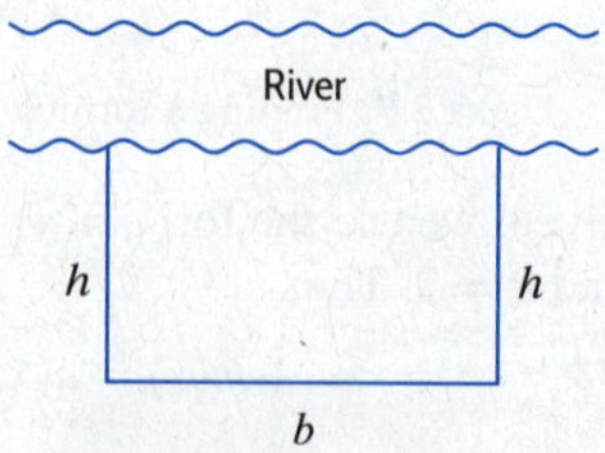


Figure 3.12: A park next to a river

Thus,

$$A = bh = (80 - 2h)h$$
$$A = -2h^2 + 80h.$$

The function $A = -2h^2 + 80h$ is quadratic. Since the coefficient of $h^2$ is negative, the parabola opens downward and we have a maximum at the vertex. The factored form of the quadratic is $A = -2h(h - 40)$, so the zeros are $h = 0$ and $h = 40$. The vertex of the parabola occurs on its axis of symmetry, midway between the zeros, at $h = 20$. Substituting $h = 20$ gives the maximum area:

$$A = -2 \cdot 20(20 - 40) = -40(-20) = 800 \text{ meters}^2.$$

## Exercises and Problems for Section 3.2

### Skill Refresher

For Exercises S1–S4, complete the square for each expression.

**S1.** $y^2 - 12y$

**S2.** $s^2 + 6s - 8$

**S3.** $c^2 + 3c - 7$

**S4.** $4s^2 + s + 2$

In Exercises S5–S7, solve by completing the square.

**S5.** $r^2 - 6r + 8 = 0$

**S6.** $n^2 = 3n + 18$

**S7.** $5q^2 - 8 = 2q$

In Exercises S8–S10, solve using factoring, completing the square, or the quadratic formula.

**S8.** $-3t^2 + 4t + 9 = 0$

**S9.** $n^2 + 4n - 3 = 2$

**S10.** $2q^2 + 4q - 5 = 8$

### Exercises

For the quadratic functions in Exercises 1–2, state the coordinates of the vertex, the axis of symmetry, and whether the parabola opens upward or downward.

1. $f(x) = 3(x - 1)^2 + 2$
2. $g(x) = -(x + 3)^2 - 4$
3. Find the vertex and axis of symmetry of the graph of $v(t) = t^2 + 11t - 4$.
4. Find the vertex and axis of symmetry for the parabola whose equation is $y = 3x^2 - 6x + 5$.
5. Sketch the quadratic functions given in standard form. Identify the values of the parameters $a$, $b$, and $c$. Label the zeros, axis of symmetry, vertex, and $y$-intercept.

   **(a)** $g(x) = x^2 + 3$ **(b)** $f(x) = -2x^2 + 4x + 16$

In Exercises 6–9, find a formula for the parabola.

**6.**

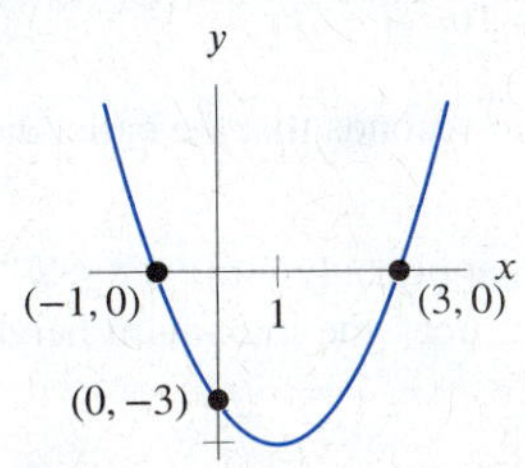


**7.**

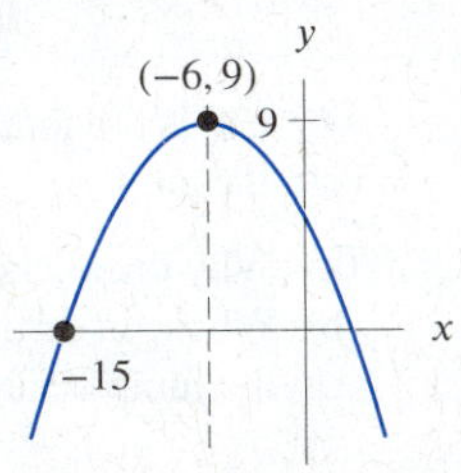


**8.**

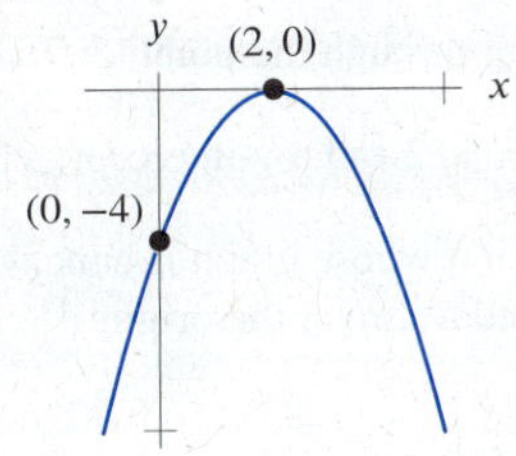


**9.**

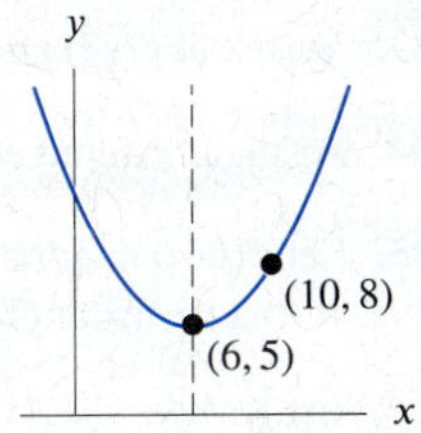


For Exercises 10–13, convert the quadratic functions to vertex form by completing the square. Identify the vertex and the axis of symmetry.

10. $f(x) = x^2 + 8x + 3$
11. $g(x) = -2x^2 + 12x + 4$
12. $p(t) = 2t^2 - 0.12t + 0.1$
13. $w(z) = -3z^2 + 9z - 2$

In Exercises 14–16, write the quadratic function in its standard, vertex, and factored forms.

14. $y = -6 + \dfrac{x^2 - x}{2}$
15. $f(t) = \dfrac{5t^2 - 20}{2}$
16. $g(s) = (s - 5)(2s + 3)$

In Exercises 17–24, write a formula and graph the transformation of $m(n) = \frac{1}{2}n^2$.

**17.** $y = m(n) + 1$

**18.** $y = m(n + 1)$

**19.** $y = m(n) - 3.7$

**20.** $y = m(n - 3.7)$

**21.** $y = m(n) + \sqrt{13}$

**22.** $y = m(n + 2\sqrt{2})$

**23.** $y = m(n + 3) + 7$

**24.** $y = m(n - 17) - 159$

## Problems

**25.** Find the vertex form of $f(t) = -16t^2 + 47t + 3$.

**26.** Show that $y = x^2 - x + 41$ has no real zeros.

**27.** Find the value of $k$ so that the graph of $y = (x - 3)^2 + k$ passes through the point $(6, 13)$.

**28.** The parabola $y = ax^2 + k$ has vertex $(0, -2)$ and passes through the point $(3, 4)$. Find its equation.

**29.** Using the vertex form, find a formula for the parabola with vertex $(2, 5)$ that passes through the point $(1, 2)$.

In Problems 30–34, find a formula for the quadratic function whose graph has the given properties.

**30.** Vertex at $(4, 2)$ and $y$-intercept of $y = 6$.

**31.** Vertex at $(4, 2)$ and $y$-intercept of $y = -4$.

**32.** Vertex at $(4, 2)$ and zeros at $x = -3, 11$.

**33.** Vertex at $(7, 3)$ and passing through the point $(3, 7)$.

**34.** A single $x$-intercept at $x = 1/2$ and a $y$-intercept at 3.

**35.** Let $f$ be a quadratic function whose graph is concave up with a vertex at $(1, -1)$, and a zero at the origin.

- **(a)** Graph $y = f(x)$.
- **(b)** Determine a formula for $f(x)$.
- **(c)** Determine the range of $f$.
- **(d)** Find any other zeros.

**36.** Find the vertex of $y = 26 + 0.4x - 0.01x^2$ exactly. Graph the function, labeling all intercepts.

**37.** Find the vertex of $y = 0.03x^2 + 1.8x + 2$ exactly. Graph the function, labeling all intercepts.

**38.** If we know a quadratic function $f$ has a zero at $x = -1$ and vertex at $(1, 4)$, do we have enough information to find a formula for this function? If your answer is yes, find it; if not, give your reasons.

**39.** Figure 3.13 shows $f(x) = x^2$. Define $g$ by shifting the graph of $f$ to the right 2 units and down 1 unit; see Figure 3.14. Find a formula for $g$ in terms of $f$. Find a formula for $g$ in terms of $x$.

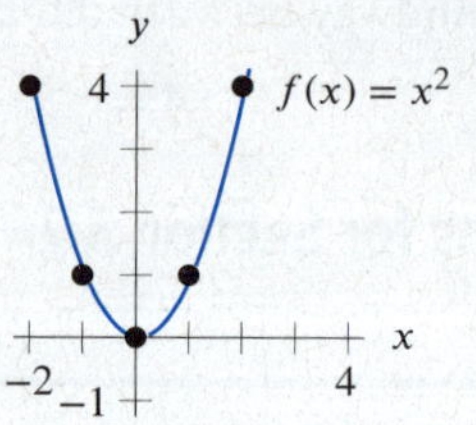


Figure 3.13

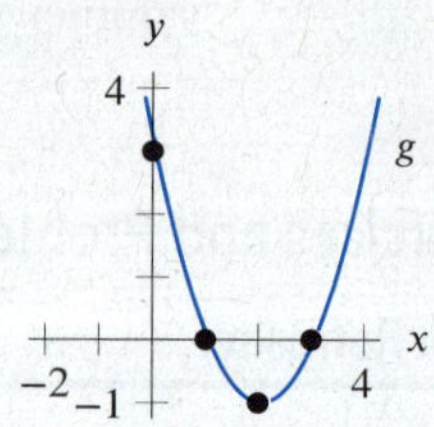


Figure 3.14

**40.** The function $g(x)$ is obtained by shifting the graph of $y = x^2$. If $g(3) = 16$, give a possible formula for $g$ when

- **(a)** $g$ is the result of applying only a horizontal shift to $y = x^2$.
- **(b)** $g$ is the result of applying only a vertical shift to $y = x^2$.
- **(c)** $g$ is the result of applying a horizontal shift right 2 units and an appropriate vertical shift of $y = x^2$.

**41.** Graph $y = x^2 - 10x + 25$ and $y = x^2$. Use a shift transformation to explain the relationship between the two graphs.

**42.** Let $f(x) = x^2$ and let $g(x) = (x - 3)^2 + 2$.

- **(a)** Give the formula for $g$ in terms of $f$, and describe the relationship between $f$ and $g$ in words.
- **(b)** Is $g$ a quadratic function? If so, find its standard form and the parameters $a$, $b$, and $c$.
- **(c)** Graph $g$, labeling all important features.

**43.** If you have a string of length 50 cm, what are the dimensions of the rectangle of maximum area that you can enclose with your string? Explain your reasoning. What about a string of length $k$ cm?

**44.** A ballet dancer jumps in the air. The height, $h(t)$, in feet, of the dancer at time $t$, in seconds since the start of the jump, is given by[3]

$$h(t) = -16t^2 + 16Tt,$$

where $T$ is the total time in seconds that the ballet dancer is in the air.

- **(a)** Why does this model apply only for $0 \le t \le T$?
- **(b)** When, in terms of $T$, does the maximum height of the jump occur?

[3] K. Laws, *The Physics of Dance* (Schirmer, 1984).

**(c)** Show that the time, $T$, that the dancer is in the air is related to $H$, the maximum height of the jump, by the equation

$$H = 4T^2.$$

**45.** A football player kicks a ball at an angle of 37° above the ground with an initial speed of 20 meters/second. The height in meters, $h$, as a function of the horizontal distance traveled, $d$, is given by:

$$h = 0.75d - 0.0192d^2.$$

**(a)** Graph the path the ball follows.
**(b)** When the ball hits the ground, how far is it from the spot where the football player kicked it?
**(c)** What is the maximum height the ball reaches during its flight?
**(d)** What is the horizontal distance the ball has traveled when it reaches its maximum height?[4]

## CHAPTER SUMMARY

- **General Formulas for Quadratic Functions**
  Standard form:
  $$y = ax^2 + bx + c, a \neq 0$$
  Factored form:
  $$y = a(x - r)(x - s)$$
  Vertex form:
  $$y = a(x - h)^2 + k$$
- **Graphs of Quadratic Functions**
  Graphs are parabolas
  Vertex $(h, k)$
  Axis of symmetry, $x = h$
  Effect of parameter $a$
  Opens upward (concave up) if $a > 0$, minimum at $(h, k)$
  Opens downward (concave down) if $a < 0$, maximum at $(h, k)$
  Factored form displays zeros at $x = r$ and $x = s$
- **Solving Quadratic Equations**
  Factoring
  Quadratic formula
  $$x = \frac{-b \pm \sqrt{b^2 - 4ac}}{2a}$$
  Completing the square

## REVIEW EXERCISES AND PROBLEMS FOR CHAPTER THREE

### Exercises

Are the functions in Exercises 1–3 quadratic? If so, write the function in standard form.

**1.** $f(x) = 2(7 - x)^2 + 1$

**2.** $L(P) = (P + 1)(1 - P)$

**3.** $g(m) = m(m^2 - 2m) + 3\left(14 - \frac{m^3}{3}\right) + \sqrt{3}m$

In Exercises 4–9, find the zeros (if any) of the function algebraically.

**4.** $y = 2x^2 + 5x + 2$

**5.** $y = 4x^2 - 4x - 8$

**6.** $y = 7x^2 + 16x + 4$

**7.** $y = 5x^2 + 2x - 1$

**8.** $y = -17x^2 + 23x + 19$

**9.** $y = 3x^2 - 2x + 6$

**10.** Show that the function $y = -x^2 + 7x - 13$ has no real zeros.

**11.** Find the vertex and axis of symmetry of the graph of $w(x) = -3x^2 - 30x + 31$.

In Exercises 12–16, find a possible formula for the parabola with the given conditions.

**12.** The vertex is $(1, -2)$ and the $y$-intercept is $y = -5$.

**13.** The vertex is $(4, -2)$ and the $y$-intercept is $y = -3$.

**14.** The vertex is $(-7, -3)$ and it passes through the point $(-3, -7)$.

**15.** The parabola goes through the origin and its vertex is $(1, -1)$.

**16.** The $x$-intercepts are at $x = -1$ and $x = 2$ and $(-2, 16)$ is on the function's graph.

[4]Adapted from R. Halliday, D. Resnick, and K. Krane, *Physics* (New York: Wiley, 1992), p. 58.

In Exercises 17–20, find a formula for the parabola.

**17.**

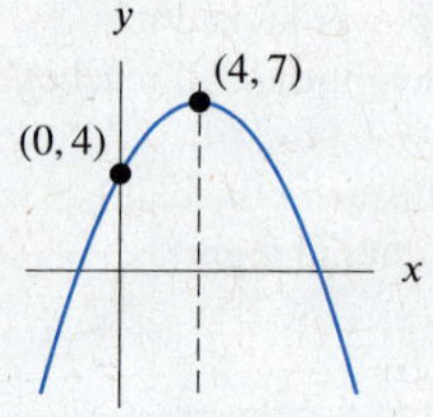


**18.**

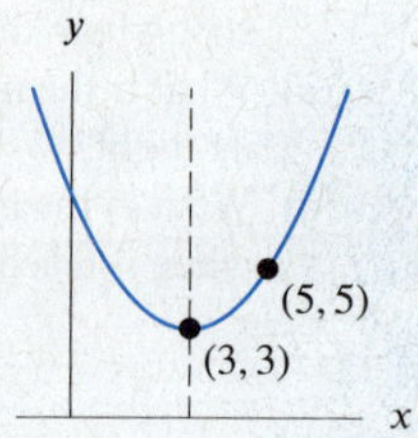


**19.**

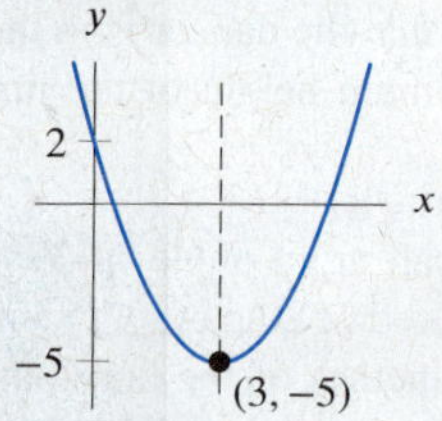


**20.**

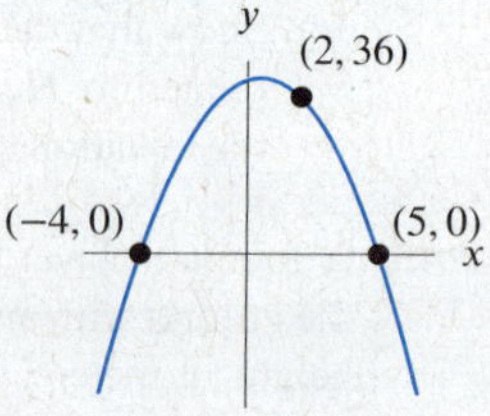


## Problems

For each parabola in Problems 21–24, state the coordinates of the vertex, the axis of symmetry, the $y$-intercept, and whether the curve is concave up or concave down.

**21.** $y = 2(x - 3/4)^2 - 2/3$ **22.** $y = -1/2(x+6)^2 - 4$

**23.** $y = (x - 0.6)^2$ **24.** $y = -0.3x^2 - 7$

In Problems 25–28, write each function in factored form or vertex form and then state its vertex and zeros.

**25.** $y = 0.3x^2 - 0.6x - 7.2$

**26.** $y = 2x^2 - 4x - 2$

**27.** $y = -3x^2 + 24x - 36$

**28.** $y = 2x^2 + (7/3)x + 1$

**29.** For $x$ between $x = -2$ and $x = 4$, determine the concavity of the graph of $f(x) = (x-1)^2 + 2$ by calculating average rates of change over intervals of length 2.

**30.** A tomato is thrown vertically into the air at time $t = 0$. Its height, $d(t)$ (in feet), above the ground at time $t$ (in seconds) is given by

$$d(t) = -16t^2 + 48t.$$

(a) Graph $d(t)$.
(b) Find $t$ when $d(t) = 0$. What is happening to the tomato the first time $d(t) = 0$? The second time?
(c) When does the tomato reach its maximum height?
(d) What is the maximum height that the tomato reaches?

**31.** When slam-dunking, a basketball player seems to hang in the air at the height of his jump. The height $h(t)$, in feet above the ground, of a basketball player at time $t$, in seconds since the start of a jump, is given by

$$h(t) = -16t^2 + 16Tt,$$

where $T$ is the total time in seconds that it takes to complete the jump. For a jump that takes 1 second to complete, how much of this time does the basketball player spend at the top 25% of the trajectory? [Hint: Find the maximum height reached. Then find the times at which the height is 75% of this maximum.]

## STRENGTHEN YOUR UNDERSTANDING

Are the statements in Problems 1–15 true or false? Give an explanation for your answer.

1. The quadratic function $f(x) = x(x+2)$ is in factored form.
2. If $f(x) = (x+1)(x+2)$, then the zeros of $f$ are 1 and 2.
3. A quadratic function whose graph is concave up has a maximum.
4. All quadratic equations have the form $f(x) = ax^2$.
5. If the height above the ground of an object at time $t$ is given by $s(t) = at^2 + bt + c$, then $s(0)$ tells us when the object hits the ground.
6. To find the zeros of $f(x) = ax^2 + bx + c$, solve the equation $ax^2 + bx + c = 0$ for $x$.
7. Every quadratic equation has two real solutions.
8. There is only one quadratic function with zeros at $x = -2$ and $x = 2$.
9. A quadratic function has exactly two zeros.
10. The graph of every quadratic function is a parabola.
11. The maximum or minimum point of a parabola is called its vertex.
12. If a parabola is concave up its vertex is a maximum point.
13. If the equation of a parabola is written as $y = a(x-h)^2 + k$, then the vertex is located at the point $(-h, k)$.
14. If the equation of a parabola is written as $y = a(x-h)^2 + k$, then the axis of symmetry is found at $x = h$.
15. If the equation of a parabola is $y = ax^2 + bx + c$ and $a < 0$, then the parabola opens downward.

# SKILLS REFRESHER FOR CHAPTER 3: QUADRATIC EQUATIONS

## SKILLS FOR FACTORING

### Expanding an Expression

The *distributive property* for real numbers $a$, $b$, and $c$ tells us that

$$a(b+c) = ab + ac,$$

and

$$(b+c)a = ba + ca.$$

We use the distributive property and the rules of exponents to multiply algebraic expressions involving parentheses. This process is sometimes referred to as *expanding* the expression.

**Example 1** Multiply the following expressions and simplify.

(a) $3x^2\left(x + \frac{1}{6}x^{-3}\right)$ (b) $\left((2t)^2 - 5\right)\sqrt{t}$

**Solution** (a) $3x^2\left(x + \frac{1}{6}x^{-3}\right) = \left(3x^2\right)(x) + \left(3x^2\right)\left(\frac{1}{6}x^{-3}\right) = 3x^3 + \frac{1}{2}x^{-1}$.

(b) $\left((2t)^2 - 5\right)\sqrt{t} = (2t)^2(\sqrt{t}) - 5\sqrt{t} = \left(4t^2\right)\left(t^{1/2}\right) - 5t^{1/2} = 4t^{5/2} - 5t^{1/2}$.

If there are two terms in each factor, then there are four terms in the product:

$$(a+b)(c+d) = a(c+d) + b(c+d) = ac + ad + bc + bd.$$

The following special cases of the above product occur frequently. Learning to recognize their forms aids in factoring.

$$(a+b)(a-b) = a^2 - b^2$$
$$(a+b)^2 = a^2 + 2ab + b^2$$
$$(a-b)^2 = a^2 - 2ab + b^2$$

**Example 2** Expand the following and simplify by gathering like terms.

(a) $(5x^2+2)(x-4)$ (b) $(2\sqrt{r}+2)(4\sqrt{r}-3)$ (c) $\left(3 - \frac{1}{2}x\right)^2$

**Solution** (a) $\left(5x^2+2\right)(x-4) = \left(5x^2\right)(x) + \left(5x^2\right)(-4) + (2)(x) + (2)(-4) = 5x^3 - 20x^2 + 2x - 8$.

(b) $(2\sqrt{r}+2)(4\sqrt{r}-3) = (2)(4)(\sqrt{r})^2 + (2)(-3)(\sqrt{r}) + (2)(4)(\sqrt{r}) + (2)(-3) = 8r + 2\sqrt{r} - 6$.

(c) $\left(3 - \frac{1}{2}x\right)^2 = 3^2 - 2(3)\left(\frac{1}{2}x\right) + \left(-\frac{1}{2}x\right)^2 = 9 - 3x + \frac{1}{4}x^2$.

### Factoring

To write an expanded expression in factored form, we "un-multiply" the expression. Some techniques for factoring are given in this section. We can check factoring by multiplying the factors.

#### Removing a Common Factor

It is sometimes useful to factor out the same factor from each of the terms in an expression. This is basically the distributive law in reverse:

$$ab + ac = a(b+c).$$

One special case is removing a factor of $-1$, which gives

$$-a - b = -(a + b)$$

Another special case is

$$(a - b) = -(b - a)$$

**Example 3** Factor the following:

(a) $\frac{2}{3}x^2y + \frac{4}{3}xy$ (b) $(2p + 1)p^3 - 3p(2p + 1)$ (c) $-\frac{s^2t}{8w} - \frac{st^2}{16w}$

**Solution** (a) $\frac{2}{3}x^2y + \frac{4}{3}xy = \frac{2}{3}xy(x + 2)$.

(b) $(2p + 1)p^3 - 3p(2p + 1) = (p^3 - 3p)(2p + 1) = p(p^2 - 3)(2p + 1)$.

(Note that the expression $(2p + 1)$ was one of the factors common to both terms.)

(c) $-\frac{s^2t}{8w} - \frac{st^2}{16w} = -\frac{st}{8w}\left(s + \frac{t}{2}\right)$.

### Grouping Terms

Even though all the terms may not have a common factor, we can sometimes factor by first grouping the terms and then removing a common factor.

**Example 4** Factor $x^2 - hx - x + h$.

**Solution** $x^2 - hx - x + h = \left(x^2 - hx\right) - (x - h) = x(x - h) - (x - h) = (x - h)(x - 1)$.

### Factoring Quadratics

One way to factor quadratics is to mentally multiply out the possibilities.

**Example 5** Factor $t^2 - 4t - 12$.

**Solution** If the quadratic factors, it is of the form

$$t^2 - 4t - 12 = (t + ?)(t + ?).$$

We are looking for two numbers whose product is $-12$ and whose sum is $-4$. By trying combinations, we find

$$t^2 - 4t - 12 = (t - 6)(t + 2).$$

**Example 6** Factor $4 - 2M - 6M^2$.

**Solution** By a similar method to the previous example, we find $4 - 2M - 6M^2 = (2 - 3M)(2 + 2M)$.

### Perfect Squares and the Difference of Squares

Recognition of the special products $(x + y)^2$, $(x - y)^2$ and $(x + y)(x - y)$ in expanded form is useful in factoring. Reversing the results given on page 131, we have

$$a^2 + 2ab + b^2 = (a + b)^2,$$

$$a^2 - 2ab + b^2 = (a - b)^2,$$

$$a^2 - b^2 = (a - b)(a + b).$$

When we see squared terms in an expression to be factored, it is often useful to look for one of these forms. The difference of squares identity (the third one listed previously) is especially useful.

**Example 7** Factor: (a) $16y^2 - 24y + 9$ (b) $25S^2R^4 - T^6$ (c) $x^2(x-2) + 16(2-x)$

**Solution** (a) $16y^2 - 24y + 9 = (4y-3)^2$.
(b) $25S^2R^4 - T^6 = \left(5SR^2\right)^2 - \left(T^3\right)^2 = \left(5SR^2 - T^3\right)\left(5SR^2 + T^3\right)$.
(c) $x^2(x-2) + 16(2-x) = x^2(x-2) - 16(x-2) = (x-2)\left(x^2-16\right) = (x-2)(x-4)(x+4)$.

## Solving Quadratic Equations

**Example 8** Give exact and approximate solutions to $x^2 = 3$.

**Solution** The exact solutions are $x = \pm\sqrt{3}$; approximate ones are $x \approx \pm 1.73$, or $x \approx \pm 1.732$, or $x \approx \pm 1.73205$ (since $\sqrt{3} = 1.732050808\ldots$). Notice that the equation $x^2 = 3$ has only two exact solutions, but many possible approximate solutions, depending on how much accuracy is required.

### Solving by Factoring

Some equations can be put into factored form such that the product of the factors is zero. Then we solve by using the fact that if $a \cdot b = 0$, then either $a$ or $b$ (or both) is zero.

**Example 9** Solve $(x+1)(x+3) = 15$.

**Solution** Although it is true that if $a \cdot b = 0$, then $a = 0$ or $b = 0$, it is *not* true that $a \cdot b = 15$ means that $a = 15$ or $b = 15$, or that $a$ and $b$ are 3 and 5. To solve this equation, we expand the left-hand side and rearrange so that the right-hand side is equal to zero:

$$x^2 + 4x + 3 = 15,$$

$$x^2 + 4x - 12 = 0.$$

Then, factoring gives

$$(x-2)(x+6) = 0.$$

Thus $x = 2$ and $x = -6$ are the solutions.

**Example 10** Solve $2(x+3)^2 = 5(x+3)$.

**Solution** You might be tempted to divide both sides by $(x+3)$. However, if you do this you will overlook one of the solutions. Instead, write

$$\begin{aligned} 2(x+3)^2 - 5(x+3) &= 0 \\ (x+3)(2(x+3) - 5) &= 0 \\ (x+3)(2x+6-5) &= 0 \\ (x+3)(2x+1) &= 0. \end{aligned}$$

Thus, $x = -1/2$ and $x = -3$ are solutions.

Note that if we had divided by $(x+3)$ at the start, we would have lost the solution $x = -3$, which was obtained by setting $x + 3 = 0$.

### Solving with the Quadratic Formula

Instead of factoring, we can solve the equation $ax^2 + bx + c = 0$ by using the quadratic formula:

$$x = \frac{-b \pm \sqrt{b^2 - 4ac}}{2a}.$$

The quadratic formula is derived by completing the square for $y = ax^2 + bx + c$. See page 136.

**Example 11** Solve $11 + 2x = x^2$.

**Solution** The equation is

$$-x^2 + 2x + 11 = 0.$$

The expression on the left does not factor using integers, so we use

$$x = \frac{-2 + \sqrt{4 - 4(-1)(11)}}{2(-1)} = \frac{-2 + \sqrt{48}}{-2} = \frac{-2 + \sqrt{16 \cdot 3}}{-2} = \frac{-2 + 4\sqrt{3}}{-2} = 1 - 2\sqrt{3},$$

$$x = \frac{-2 - \sqrt{4 - 4(-1)(11)}}{2(-1)} = \frac{-2 - \sqrt{48}}{-2} = \frac{-2 - \sqrt{16 \cdot 3}}{-2} = \frac{-2 - 4\sqrt{3}}{-2} = 1 + 2\sqrt{3}.$$

The exact solutions are $x = 1 - 2\sqrt{3}$ and $x = 1 + 2\sqrt{3}$.

The decimal approximations to these numbers $x = 1 - 2\sqrt{3} = -2.464$ and $x = 1 + 2\sqrt{3} = 4.464$ are approximate solutions to this equation. The approximate solutions could also be found directly from a graph or calculator.

## Exercises for Skills for Factoring

For Exercises 1–15, expand and simplify.

1. $2(3x - 7)$
2. $-4(y + 6)$
3. $12(x + y)$
4. $-7(5x - 8y)$
5. $x(2x + 5)$
6. $3z(2x - 9z)$
7. $-10r(5r + 6rs)$
8. $x(3x - 8) + 2(3x - 8)$
9. $5z(x - 2) - 3(x - 2)$
10. $(x + 1)(x + 3)$
11. $(x - 2)(x + 6)$
12. $(5x - 1)(2x - 3)$
13. $(y + 1)(z + 3)$
14. $(12y - 5)(8w + 7)$
15. $(5z - 3)(x - 2)$

Multiply and write the expressions in Problems 16–22 without parentheses. Gather like terms.

16. $-(x - 3) - 2(5 - x)$
17. $(x - 5)6 - 5(1 - (2 - x))$
18. $\left(3x - 2x^2\right)4 + (5 + 4x)(3x - 4)$
19. $P(p - 3q)^2$
20. $4(x - 3)^2 + 7$
21. $-\left(\sqrt{2x} + 1\right)^2$
22. $u\left(u^{-1} + 2^u\right)2^u$

For Exercises 23–67, factor completely if possible.

23. $2x + 6$
24. $3y + 15$
25. $5z - 30$
26. $4t - 6$
27. $10w - 25$
28. $3u^4 - 4u^3$
29. $3u^7 + 12u^2$
30. $12x^3y^2 - 18x$
31. $14r^4s^2 - 21rst$
32. $x^2 + 3x - 2$
33. $x^2 - 3x + 2$
34. $x^2 - 3x - 2$
35. $x^2 + 2x + 3$
36. $x^2 - 2x - 3$
37. $x^2 - 2x + 3$
38. $x^2 + 2x - 3$
39. $2x^2 + 5x + 2$
40. $2x^2 - 10x + 12$
41. $x^2 + 3x - 28$
42. $x^3 - 2x^2 - 3x$
43. $x^3 + 2x^2 - 3x$

44. $ac + ad + bc + bd$
45. $x^2 + 2xy + 3xz + 6yz$
46. $x^2 - 1.4x - 3.92$
47. $a^2x^2 - b^2$
48. $\pi r^2 + 2\pi rh$
49. $B^2 - 10B + 24$
50. $c^2 + x^2 - 2cx$
51. $x^2 + y^2$
52. $a^4 - a^2 - 12$
53. $(t+3)^2 - 16$
54. $x^2 + 4x + 4 - y^2$
55. $a^3 - 2a^2 + 3a - 6$
56. $b^3 - 3b^2 - 9b + 27$
57. $c^2d^2 - 25c^2 - 9d^2 + 225$
58. $hx^2 + 12 - 4hx - 3x$
59. $r(r-s) - 2(s-r)$
60. $y^2 - 3xy + 2x^2$
61. $x^2e^{-3x} + 2xe^{-3x}$
62. $t^2e^{5t} + 3te^{5t} + 2e^{5t}$
63. $P(1+r)^2 + P(1+r)^2r$
64. $x^2 - 6x + 9 - 4z^2$
65. $dk + 2dm - 3ek - 6em$
66. $\pi r^2 - 2\pi r + 3r - 6$
67. $8gs - 12hs + 10gm - 15hm$

Solve the equations in Exercises 68–93.

68. $y^2 - 5y - 6 = 0$
69. $4s^2 + 3s - 15 = 0$
70. $\dfrac{2}{x} + \dfrac{3}{2x} = 8$
71. $\dfrac{3}{x-1} + 1 = 5$
72. $\sqrt{y-1} = 13$
73. $-16t^2 + 96t + 12 = 60$
74. $g^3 - 4g = 3g^2 - 12$
75. $8 + 2x - 3x^2 = 0$
76. $2p^3 + p^2 - 18p - 9 = 0$
77. $N^2 - 2N - 3 = 2N(N-3)$
78. $\dfrac{1}{64}t^3 = t$
79. $x^2 - 1 = 2x$
80. $4x^2 - 13x - 12 = 0$
81. $60 = -16t^2 + 96t + 12$
82. $n^5 + 80 = 5n^4 + 16n$
83. $5a^3 + 50a^2 = 4a + 40$
84. $y^2 + 4y - 2 = 0$
85. $\dfrac{2}{z-3} + \dfrac{7}{z^2 - 3z} = 0$
86. $\dfrac{x^2 + 1 - 2x^2}{(x^2+1)^2} = 0$
87. $4 - \dfrac{1}{L^2} = 0$
88. $2 + \dfrac{1}{q+1} - \dfrac{1}{q-1} = 0$
89. $\sqrt{r^2 + 24} = 7$
90. $\dfrac{1}{\sqrt[3]{x}} = -2$
91. $3\sqrt{x} = \dfrac{1}{2}x$
92. $10 = \sqrt{\dfrac{v}{7\pi}}$
93. $\dfrac{(3x+4)(x-2)}{(x-5)(x-1)} = 0$

In Exercises 94–97, solve for the indicated variable.

94. $T = 2\pi\sqrt{\dfrac{l}{g}}$, for $l$.
95. $Ab^5 = C$, for $b$.
96. $|2x + 1| = 7$, for $x$.
97. $\dfrac{x^2 - 5mx + 4m^2}{x - m} = 0$, for $x$

Solve the systems of equations in Exercises 98–102.

98. $\begin{cases} y = 2x - x^2 \\ y = -3 \end{cases}$

99. $\begin{cases} y = 1/x \\ y = 4x \end{cases}$

100. $\begin{cases} x^2 + y^2 = 36 \\ y = x - 3 \end{cases}$

101. $\begin{cases} y = 4 - x^2 \\ y - 2x = 1 \end{cases}$

102. $\begin{cases} y = x^3 - 1 \\ y = e^x \end{cases}$

103. Let $\ell$ be the line of slope 3 passing through the origin. Find the points of intersection of the line $\ell$ and the parabola whose equation is $y = x^2$. Sketch the line and the parabola, and label the points of intersection.

Determine the points of intersection for Problems 104–105.

104.

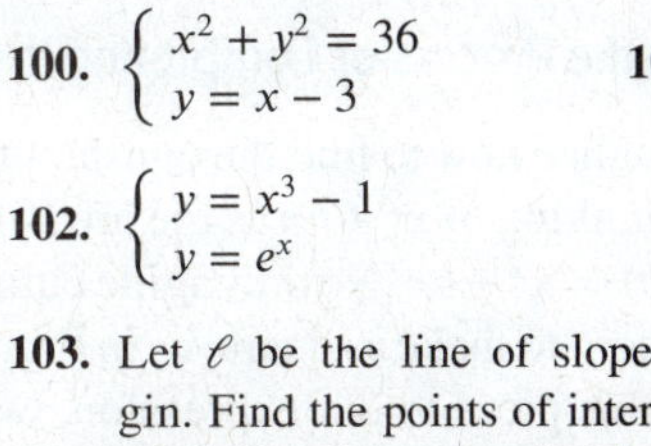


105.

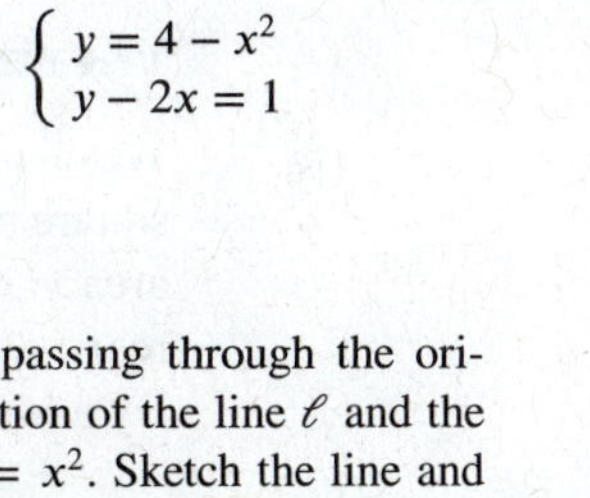

## COMPLETING THE SQUARE

An example of changing the form of an expression is the conversion of $ax^2 + bx + c$ into the form $a(x - h)^2 + k$. We make this conversion by *completing the square*, a method for producing a perfect square within a quadratic expression. A perfect square is an expression of the form:

$$(x + n)^2 = x^2 + 2nx + n^2.$$

In order to complete the square in an expression, we must find that number $n$, which is half the coefficient of $x$. Before giving a general procedure, let's work through an example.

**Example 1** Complete the square to rewrite $x^2 - 10x + 4$ in the form $a(x - h)^2 + k$.

**Solution** Step 1: We divide the coefficient of $x$ by 2, giving $\frac{1}{2}(-10) = -5$.
Step 2: We square the result of step 1, giving $(-5)^2 = 25$.
Step 3: Now add and subtract the 25 after the $x$-term:

$$\begin{aligned} x^2 - 10x + 4 &= x^2 - 10x + 25 - 25 + 4 \\ &= \underbrace{(x^2 - 10x + 25)}_{\text{Perfect square}} -25 + 4. \end{aligned}$$

Step 4: Notice that we have created a perfect square, $x^2 - 10x + 25$. The next step is to factor the perfect square and combine the constant terms, $-25 + 4$, giving the final result:

$$x^2 - 10x + 4 = (x - 5)^2 - 21.$$

Thus, $a = +1$, $h = +5$, and $k = -21$.

### Visualizing the Process of Completing the Square

We can visualize how to find the constant that needs to be added to $x^2 + bx$ in order to obtain a perfect square by thinking of $x^2 + bx$ as the area of a rectangle. For example, the rectangle in Figure 3.15 has area $x(x+b) = x^2 + bx$. Now imagine cutting the rectangle into pieces as in Figure 3.16 and trying to rearrange them to make a square, as in Figure 3.17. The corner piece, whose area is $(b/2)^2$, is missing. By adding this piece to our expression, we "complete" the square: $x^2 + bx + (b/2)^2 = (x + b/2)^2$.

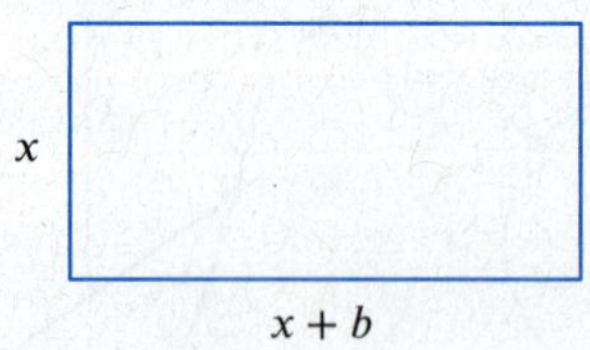


Figure 3.15: Rectangle with sides $x$ and $x + b$

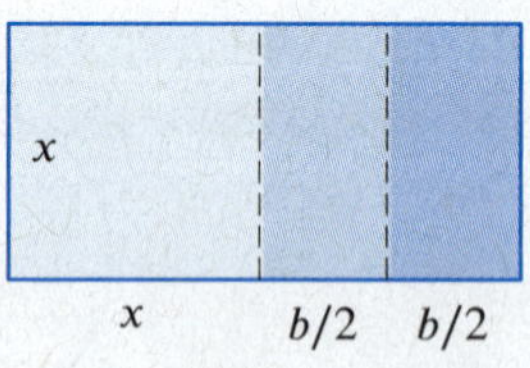


Figure 3.16: Cut off 2 strips of width $b/2$

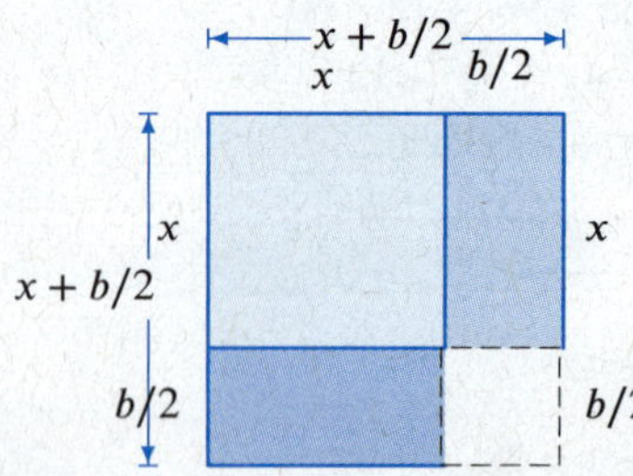


Figure 3.17: Rearrange to see a square with missing corner of area $(b/2)^2$

The procedure we followed can be summarized as follows:

**To complete the square** in the expression $x^2 + bx + c$, divide the coefficient of $x$ by 2, giving $b/2$. Then add and subtract $(b/2)^2 = b^2/4$ and factor the perfect square:

$$x^2 + bx + c = \left(x + \frac{b}{2}\right)^2 - \frac{b^2}{4} + c.$$

To complete the square in the expression $ax^2 + bx + c$, factor out $a$ first.

The next example has a coefficient $a$ with $a \neq 1$. After factoring out the coefficient, we follow the same steps as in Example 1.

**Example 2** Complete the square in the formula $h(x) = 5x^2 + 30x - 10$.

**Solution** We first factor out 5:

$$h(x) = 5(x^2 + 6x - 2).$$

Now we complete the square in the expression $x^2 + 6x - 2$.
Step 1: Divide the coefficient of $x$ by 2, giving 3.
Step 2: Square the result: $3^2 = 9$.
Step 3: Add the result after the $x$ term, then subtract it:

$$h(x) = 5(\underbrace{x^2 + 6x + 9}_{\text{Perfect square}} - 9 - 2).$$

Step 4: Factor the perfect square and simplify the rest:

$$h(x) = 5\left((x+3)^2 - 11\right).$$

Now that we have completed the square, we can multiply by the 5:

$$h(x) = 5(x+3)^2 - 55.$$

## Deriving the Quadratic Formula

We derive a general formula to find the zeros of $q(x) = ax^2 + bx + c$, with $a \neq 0$, by completing the square. To find the zeros, set $q(x) = 0$:

$$ax^2 + bx + c = 0.$$

Before we complete the square, we factor out the coefficient of $x^2$:

$$a\left(x^2 + \frac{b}{a}x + \frac{c}{a}\right) = 0.$$

Since $a \neq 0$, we can divide both sides by $a$:

$$x^2 + \frac{b}{a}x + \frac{c}{a} = 0.$$

To complete the square, we add and then subtract $((b/a)/2)^2 = b^2/(4a^2)$:

$$\underbrace{x^2 + \frac{b}{a}x + \frac{b^2}{4a^2}}_{\text{Perfect square}} - \frac{b^2}{4a^2} + \frac{c}{a} = 0.$$

We factor the perfect square and simplify the constant term, giving:

$$\left(x + \frac{b}{2a}\right)^2 - \left(\frac{b^2 - 4ac}{4a^2}\right) = 0 \qquad \text{since } \frac{-b^2}{4a^2} + \frac{c}{a} = \frac{-b^2}{4a^2} + \frac{4ac}{4a^2} = -\left(\frac{b^2-4ac}{4a^2}\right)$$

$$\left(x + \frac{b}{2a}\right)^2 = \frac{b^2 - 4ac}{4a^2} \qquad \text{adding } \frac{b^2-4ac}{4a^2} \text{ to both sides}$$

$$x + \frac{b}{2a} = \pm\sqrt{\frac{b^2 - 4ac}{4a^2}} = \frac{\pm\sqrt{b^2 - 4ac}}{2a} \qquad \text{taking the square root}$$

$$x = \frac{-b}{2a} \pm \frac{\sqrt{b^2 - 4ac}}{2a} \qquad \text{subtracting } b/2a$$

$$x = \frac{-b \pm \sqrt{b^2 - 4ac}}{2a}.$$

## Exercises for Skills for Completing the Square

For Exercises 1–8, complete the square for each expression.

**1.** $x^2 + 8x$
**2.** $w^2 + 7w$
**3.** $2r^2 + 20r$
**4.** $3t^2 + 24t - 13$
**5.** $a^2 - 2a - 4$
**6.** $n^2 + 4n - 5$
**7.** $3r^2 + 9r - 4$
**8.** $12g^2 + 8g + 5$

In Exercises 9–12, rewrite in the form $a(x - h)^2 + k$.

**9.** $x^2 - 2x - 3$
**10.** $10 - 6x + x^2$
**11.** $-x^2 + 6x - 2$
**12.** $3x^2 - 12x + 13$

In Exercises 13–22, complete the square to find the vertex of the parabola.

**13.** $y = x^2 + 6x + 3$
**14.** $y = x^2 - x + 4$
**15.** $y = -x^2 - 8x + 2$
**16.** $y = x^2 - 3x - 3$
**17.** $y = -x^2 + x - 6$
**18.** $y = 3x^2 + 12x$
**19.** $y = -4x^2 + 8x - 6$
**20.** $y = 5x^2 - 5x + 7$
**21.** $y = 2x^2 - 7x + 3$
**22.** $y = -3x^2 - x - 2$

In Exercises 23–29, solve by completing the square.

**23.** $g^2 = 2g + 24$
**24.** $p^2 - 2p = 6$
**25.** $d^2 - d = 2$
**26.** $2r^2 + 4r - 5 = 0$
**27.** $2s^2 = 1 - 10s$
**28.** $7r^2 - 3r - 6 = 0$
**29.** $5p^2 + 9p = 1$

In Exercises 30–35, solve by using the quadratic formula.

**30.** $n^2 - 4n - 12 = 0$
**31.** $2y^2 + 5y = -2$
**32.** $6k^2 + 11k = -3$
**33.** $w^2 + w = 4$
**34.** $z^2 + 4z = 6$
**35.** $2q^2 + 6q - 3 = 0$

In Exercises 36–46, solve using factoring, completing the square, or the quadratic formula.

**36.** $r^2 - 2r = 8$
**37.** $s^2 + 3s = 1$
**38.** $z^3 + 2z^2 = 3z + 6$
**39.** $25u^2 + 4 = 30u$
**40.** $v^2 - 4v - 9 = 0$
**41.** $3y^2 = 6y + 18$
**42.** $2p^2 + 23 = 14p$
**43.** $2w^3 + 24 = 6w^2 + 8w$
**44.** $4x^2 + 16x - 5 = 0$
**45.** $49m^2 + 70m + 22 = 0$
**46.** $8x^2 - 1 = 2x$

Chapter Four

# EXPONENTIAL FUNCTIONS

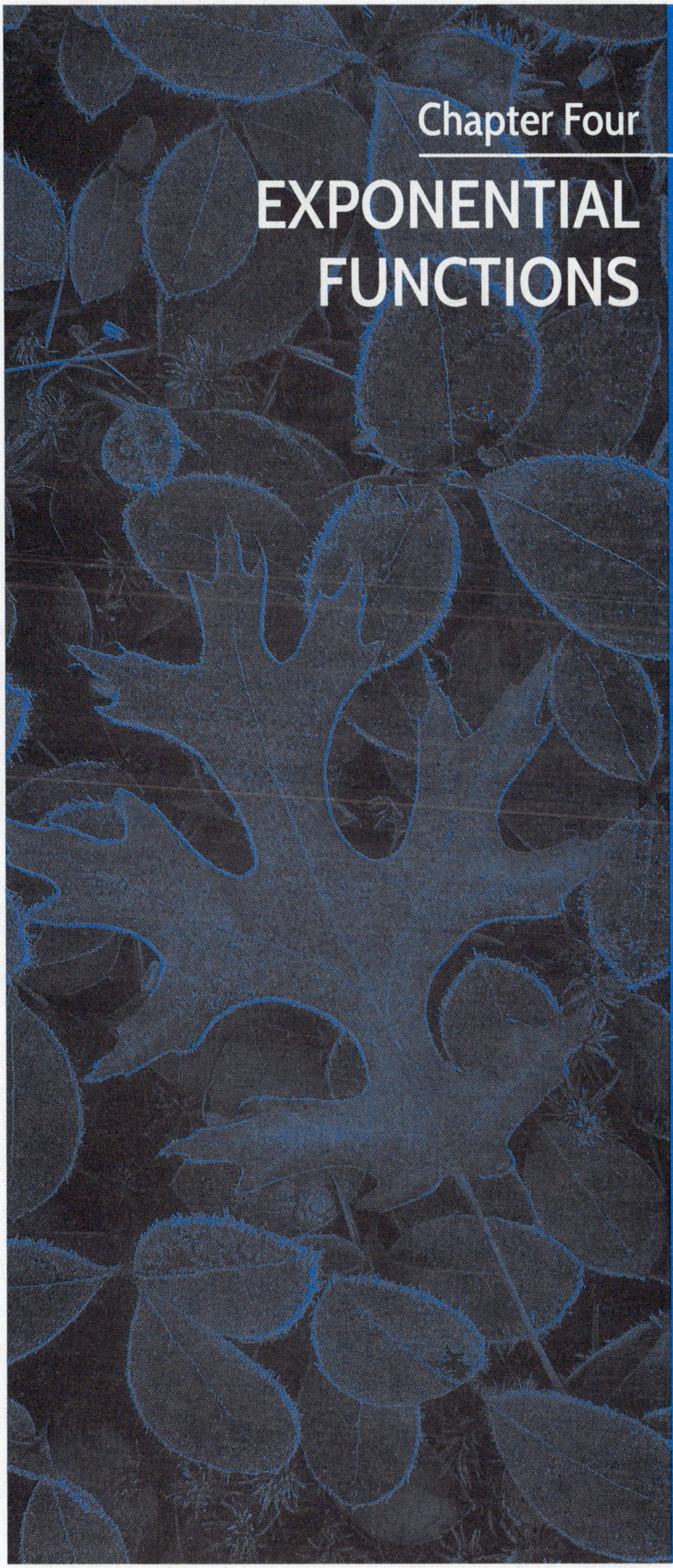

© Patrick Zephyr/Patrick Zephyr Nature Photography

## Contents

# 4.1 INTRODUCTION TO THE FAMILY OF EXPONENTIAL FUNCTIONS

## Increasing at a Constant Percent Rate: Exponential Growth

Linear functions represent quantities that change at a constant rate. In this section we introduce functions that represent change at a constant *percent* rate, the *exponential functions*.

**Example 1** After graduation from college, you will probably be looking for a job. Suppose you are offered a job at a starting salary of \$40,000 per year. To strengthen the offer, the company promises annual raises of 6% per year for at least the first four years after you are hired. Compute and graph your salary for the first four years.

**Solution** If $t$ represents the number of years since the beginning of your contract, then for $t = 0$, your salary is \$40,000. At the end of the first year, when $t = 1$, your salary increases by 6% so

$$\begin{aligned} \text{Salary when } t=1 &= \text{Original salary} + 6\% \text{ of Original salary} \\ &= 40{,}000 + 0.06 \cdot 40{,}000 \\ &= 42{,}400 \text{ dollars.} \end{aligned}$$

After the second year, your salary again increases by 6%, so

$$\begin{aligned} \text{Salary when } t=2 &= \text{Former salary} + 6\% \text{ of Former salary} \\ &= 42{,}400 + 0.06 \cdot 42{,}400 \\ &= 44{,}944 \text{ dollars.} \end{aligned}$$

Notice that your raise is higher in the second year than in the first since the second 6% increase applies both to the original \$40,000 salary and to the \$2400 raise given in the first year.

Salary calculations for four years have been rounded and recorded in Table 4.1. At the end of the third and fourth years your salary again increases by 6%, and your raise is larger each year. Not only are you given the 6% increase on your original salary, but your raises earn raises as well.

**Table 4.1** *Raise amounts and resulting salaries for a person earning 6% annual salary increases*

| Year | Raise amount (\$) | Salary (\$) |
|---|---|---|
| 0 | | 40,000.00 |
| 1 | 2400.00 | 42,400.00 |
| 2 | 2544.00 | 44,944.00 |
| 3 | 2696.64 | 47,640.64 |
| 4 | 2858.44 | 50,499.08 |

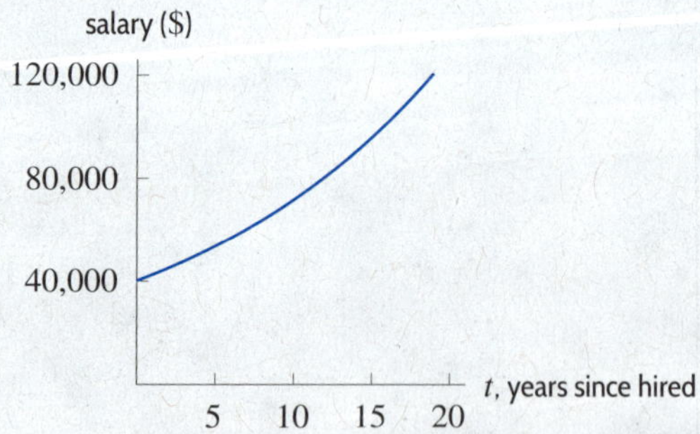


**Figure 4.1:** Salary over a 20-year period

Figure 4.1 shows your salary over a 20-year period assuming the annual increase remains 6%. Since your salary increases by a larger amount each year, the graph is not a line, but bends upward; it is concave up.

### Decreasing at a Constant Percent Rate: Exponential Decay

A quantity which decreases at a constant percent rate is said to be decreasing exponentially.

**Example 2** Carbon-14 is used to estimate the age of organic compounds. Over time, radioactive carbon-14 decays into a stable form. The decay rate is 11.4% every 1000 years. Starting with 200-microgram ($\mu$g) of carbon-14, compute and graph the quantity of carbon-14 remaining every thousand years for the next 3000 years.

**Solution** Since 11.4% of the carbon-14 decays in the first 1000 years, we have

$$\begin{aligned}\text{Quantity remaining after 1000 years} &= \text{Initial quantity} - 11.4\% \text{ of Initial quantity}\\ &= 200 - 0.114 \cdot 200\\ &= 177.2\ \mu\text{g}.\end{aligned}$$

Similarly,

$$\begin{aligned}\text{Quantity remaining after 2000 years} &= \text{Quantity remaining after 1000 years} - 11.4\% \text{ of Quantity remaining after 1000 years}\\ &= 177.2 - 0.114 \cdot 177.2 \approx 156.999\ \mu\text{g},\end{aligned}$$

and

$$\begin{aligned}\text{Quantity remaining after 3000 years} &= \text{Quantity remaining after 2000 years} - 11.4\% \text{ of Quantity remaining after 2000 years}\\ &= 156.999 - 0.114 \cdot 156.999 \approx 139.101\ \mu\text{g}.\end{aligned}$$

See Table 4.2. Notice that during each 1000-year period, the amount of carbon-14 that decays is smaller than in the previous period. This is because we take 11.4% of a smaller quantity each time.

**Table 4.2** *Quantity of carbon-14 remaining*

| Years elapsed | Quantity decayed ($\mu$g) | Quantity remaining ($\mu$g) |
|---|---|---|
| 0 | — | 200.0 |
| 1000 | 22.8 | 177.2 |
| 2000 | 20.201 | 156.999 |
| 3000 | 17.898 | 139.101 |

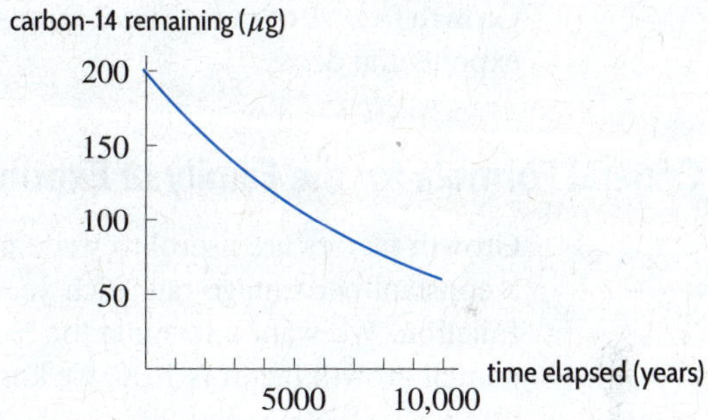


**Figure 4.2:** Quantity of carbon-14 over 10,000 years

Figure 4.2 shows the quantity of carbon-14 left from a 200 $\mu$g sample over 10,000 years. Because the amount decreases by a smaller amount over each successive thousand-year interval, the graph is not linear but bends upward; it is concave up.

## Growth Factors and Percent Growth Rates

Instead of using percent growth or decay rates, there is another way to think about exponentially changing quantities—the growth factor, which we introduce next.

### The Growth Factor of an Increasing Exponential Function

The salary in Example 1 increases by 6% every year. We say that the annual percent growth rate is 6%. We know that each year,

$$\text{New salary} = \text{Old salary} + 6\% \text{ of Old salary}.$$

We can rewrite this as follows:

$$\text{New salary} = 100\% \text{ of Old salary} + 6\% \text{ of Old salary} = 106\% \text{ of Old salary}.$$

Since $106\% = 1.06$, we have

$$\text{New salary} = 1.06 \cdot \text{Old salary}.$$

We call 1.06 the *growth factor*.

### The Growth Factor of a Decreasing Exponential Function

In Example 2, the carbon-14 changes by $-11.4\%$ every 1000 years. The negative growth rate tells us that the quantity of carbon-14 decreases over time. We have

$$\text{New quantity} = \text{Old quantity} - 11.4\% \text{ of Old quantity},$$

which can be rewritten as

$$\text{New quantity} = 100\% \text{ of Old quantity} - 11.4\% \text{ of Old quantity} = 88.6\% \text{ of Old quantity}.$$

Since $88.6\% = 0.886$, we have

$$\text{New quantity} = 0.886 \cdot \text{Old quantity}.$$

We call 0.886 the growth factor even though the amount of carbon-14 is decreasing. Although it may sound strange, we use "growth factor" to describe both increasing and decreasing quantities. Growth factors of more than 1 represent exponential increase; growth factors of less than 1 represent exponential decay.

## A General Formula for the Family of Exponential Functions

Growth factors are useful to write formulas for exponential growth and decay. Because it grows at a constant percentage rate each year, the salary, $S$, in Example 1 is an example of an exponential function. We want a formula for $S$ in terms of $t$, the number of years since being hired. Since the annual growth factor is 1.06, we know that for each year,

$$\text{New salary} = \text{Old salary} \cdot 1.06.$$

Thus, after one year, or when $t = 1$,

$$S = \underbrace{40{,}000}_{\text{Old salary}}(1.06).$$

Similarly, when $t = 2$,

$$S = \underbrace{40{,}000(1.06)}_{\text{Old salary}}(1.06) = 40{,}000(1.06)^2.$$

Here there are *two* factors of 1.06 because the salary has increased by 6% twice. After $t$ years the salary has increased by a factor of 1.06 a total of $t$ times. Thus,

$$S = 40{,}000(1.06)^t.$$

In general, we have:

An **exponential function** $Q = f(t)$ has the standard form

$$f(t) = ab^t, \quad b > 0,$$

where $a$ is the initial value of $Q$ (at $t = 0$) and $b$, the base, is the growth factor. The growth factor is given by

$$b = 1 + r$$

where $r$ is the growth rate. (For example, $b = 1.06$ and $r = 0.06$ for a 6% growth rate.) For $a > 0$:

- If there is exponential growth, then $r > 0$ and $b > 1$.
- If there is exponential decay, then $r < 0$ and $0 < b < 1$.

See Figure 4.3 and Figure 4.4.

The constants $a$ and $b$ are called *parameters*. The base $b$ is restricted to positive values because if $b < 0$ then $b^t$ is undefined for some exponents $t$, for example, $t = 1/2$. The values $a = 0$ and $b = 1$ are usually excluded because they give constant functions which are not generally considered exponential.

## Rate of Change

First suppose $a > 0$. The average rate of change of $f(t) = ab^t$ between two points is positive when $b > 1$, since $f(t)$ is increasing. (See Figure 4.3.) The rate of change is negative when $0 < b < 1$ since $f(t)$ is decreasing. (See Figure 4.4.) The graphs for both $b > 1$ and $0 < b < 1$ bend upward—they are concave up. Thus, for $a > 0$, the rate of change of the exponential function increases as the points move to the right.

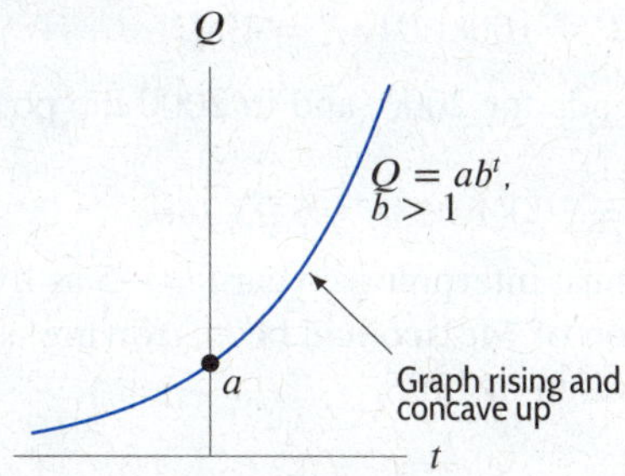


Figure 4.3: Exponential growth: $b > 1$

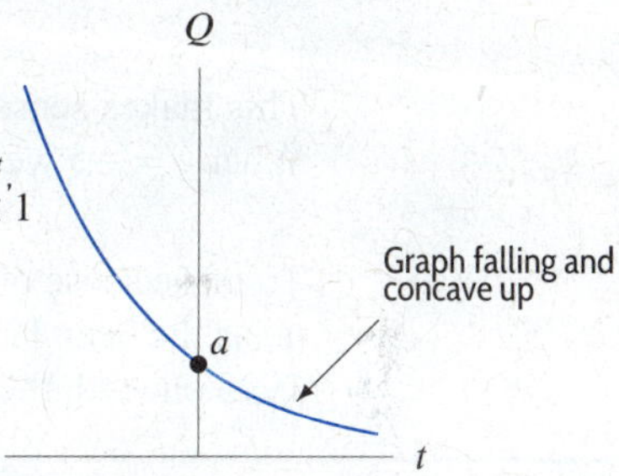


Figure 4.4: Exponential decay: $0 < b < 1$

If $a < 0$, the rate of change of $f(t)$ is positive for $0 < b < 1$ and negative for $b > 1$. In both cases, the graph is concave down; the rate of change decreases as we move right.

**Example 3** Carbon-14 decays at a rate of 11.4% every 1000 years. Write a formula for the quantity, $Q$, of a 200-$\mu$g sample remaining as a function of time, $t$, in thousands of years.

**Solution** The growth factor of carbon-14 over 1000 years is $1 - 0.114 = 0.886$. Originally, there are 200 $\mu$g, so the quantity remaining after $t$ thousand years is given by

$$Q = 200(0.886)^t.$$

## Projections into the Future

We can use a mathematical model, such as an exponential function, to make predictions, such as future salaries or population levels.

**Example 4** Use the formula $S = 40{,}000(1.06)^t$ to calculate your salary after 4 years, 12 years, and 40 years.

**Solution** After 4 years, $t = 4$, and we have

$$S = 40{,}000(1.06)^4 \approx \$50{,}499.08.$$

Notice that this agrees with Table 4.1 on page 140. After 12 years, $t = 12$, and we have

$$S = 40{,}000(1.06)^{12} \approx \$80{,}487.86.$$

After 12 years, the salary has more than doubled from the initial salary of $40,000. When $t = 40$ we have

$$S = 40{,}000(1.06)^{40} \approx \$411{,}428.72.$$

Thus if you work for 40 years and consistently earn 6% annual raises, your salary will be over $400,000 a year.

**Example 5** Mexico's population was 100 million in the year 2000; since then the population has grown an average of 1.3% per year.[1]

(a) Find a formula for $P$, the population of Mexico (in millions) $t$ years since 2000.

(b) What does the formula predict when $t = 0$? When $t = -5$? What do these values tell you about the population of Mexico?

**Solution** (a) In 2000, the population of Mexico was 100 million and growing at a constant 1.3% annual rate. The growth factor is $b = 1 + 0.013 = 1.013$, and $a = 100$, so

$$P = 100(1.013)^t.$$

Because the growth factor will change eventually, this formula may not give accurate results for large values of $t$.

(b) If $t = 0$, then, since $(1.013)^0 = 1$, we have

$$P = 100(1.013)^0 = 100.$$

This makes sense because $t = 0$ stands for 2000, and in 2000 the population was 100 million. When $t = -5$ we have

$$P = 100(1.013)^{-5} \approx 93.746.$$

To make sense of this number, we must interpret the year $t = -5$ as five years before 2000; that is, as the year 1995. If the population of Mexico had been growing at a 1.3% annual rate from 1995 onward, then it was 93.746 million in 1995.

**Example 6** On August 2, 1988, a US District Court judge imposed a fine on the city of Yonkers, New York, for defying a federal court order involving housing desegregation.[2] The fine started at $100 for the first day and was to double daily until the city chose to obey the court order.

(a) What was the daily percent growth rate of the fine?

(b) Find a formula for the fine as a function of $t$, the number of days since August 2, 1988.

(c) If Yonkers waited 30 days before obeying the court order, what would the fine have been?

**Solution** (a) Since the fine increased each day by a factor of 2, the fine grew exponentially with growth factor $b = 2$. To find the percent growth rate, we set $b = 1 + r = 2$, from which we find $r = 1$, or 100%. Thus the daily percent growth rate is 100%. This makes sense because when a quantity increases by 100%, it doubles in size.

(b) If $t$ is the number of days since August 2, the formula for the fine, $P$ in dollars, is

$$P = 100 \cdot 2^t.$$

(c) After 30 days, the fine is $P = 100 \cdot 2^{30} \approx 1.074 \cdot 10^{11}$ dollars, or $107,374,182,400.

[1] http://www.census.gov/population/international/data/idb/region.php, accessed March, 2014.

[2] *The Boston Globe,* August 27, 1988.

## Exercises and Problems for Section 4.1

### Skill Refresher

In Exercises S1–S2, express the percentages in decimal form.

**S1.** 6%

**S2.** 0.6%

In Exercises S3–S4, express the decimals as a percent.

**S3.** 0.0012

**S4.** 1.23

### Exercises

Are the functions in Exercises 1–9 exponential? If so, write the function in the form $f(t) = ab^t$.

**1.** $g(w) = 2(2^{-w})$

**2.** $q(r) = \dfrac{-4}{3^r}$

**3.** $f(x) = \dfrac{3^{2x}}{4}$

**4.** $m(t) = (2 \cdot 3^t)^3$

**5.** $G(t) = 3(t)^t$

**6.** $j(x) = 2^x 3^x$

**7.** $Q(t) = 8^{t/3}$

**8.** $p(r) = 2^r + 3^r$

**9.** $K(x) = \dfrac{2^x}{3 \cdot 3^x}$

In Exercises 10–13, give the starting value $a$, the growth factor $b$, and the growth rate $r$ if $Q = ab^t = a(1+r)^t$.

**10.** $Q = 1750(1.593)^t$

**11.** $Q = 34.3(0.788)^t$

**12.** $Q = 79.2(1.002)^t$

**13.** $Q = 0.0022(2.31)^{-3t}$

What is the growth factor in Exercises 14–17? Assume time is measured in the units given.

**14.** Water usage is increasing by 3% per year.

**15.** A city grows by 28% per decade.

**16.** A diamond mine is depleted by 1% per day.

**17.** A forest shrinks 80% per century.

**18.** Without a calculator, match each of the formulas to one of the graphs in Figure 4.5.

**(a)** $y = 0.8^t$ **(b)** $y = 5(3)^t$

**(c)** $y = -6(1.03)^t$ **(d)** $y = 15(3)^{-t}$

**(e)** $y = -4(0.98)^t$ **(f)** $y = 82(0.8)^{-t}$

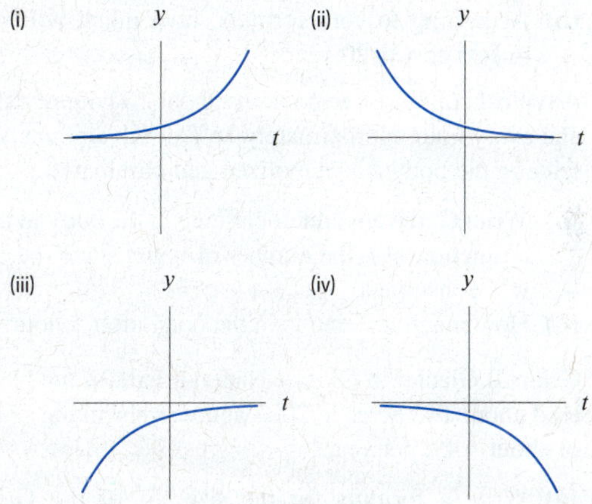


Figure 4.5

### Problems

**19.** The populations, $P$, of six towns with time $t$ in years are given by

(i) $P = 1000(1.08)^t$ (ii) $P = 600(1.12)^t$

(iii) $P = 2500(0.9)^t$ (iv) $P = 1200(1.185)^t$

(v) $P = 800(0.78)^t$ (vi) $P = 2000(0.99)^t$

**(a)** Which towns are growing in size? Which are shrinking?

**(b)** Which town is growing the fastest? What is the annual percent growth rate for that town?

**(c)** Which town is shrinking the fastest? What is the annual percent decay rate for that town?

**(d)** Which town has the largest initial population (at $t = 0$)? Which town has the smallest?

**20.** The value, $V$, of a \$100,000 investment that earns 3% annual interest is given by $V = f(t)$ where $t$ is in years. How much is the investment worth in 3 years?

**21.** An investment decreases by 5% per year for 4 years. By what total percent does it decrease?

**22.** The value \$$V$ of an investment in year $t$ is given by $V = 2500(1.0325)^t$. Describe the investment in words.

In Problems 23–28, the initial value (at year $t = 0$) and the percent change per year of a quantity $Q$ are given.

**(a)** Write a formula for $Q$ as a function of $t$.

**(b)** What is the value of $Q$ when $t = 10$?

**23.** Initial amount 2000; increasing by 5% per year

**24.** Initial amount 35; decreasing by 8% per year

**25.** Initial amount 112.8; decreasing by 23.4% per year

**26.** Initial amount 5.35; increasing by 0.8% per year

**27.** Initial amount 5; increasing by 100% per year

**28.** Initial amount 0.2; decreasing by 0.5% per year

**29.** In 2014, the population of a country was 70 million and growing at a rate of 1.9% per year. Assuming the percentage growth rate remains constant, express the population, $P$, as a function of $t$, the number of years after 2014.

**30.** Every year, a lake becomes more polluted, and 2% fewer organisms can live in it. If in 2015 there are one million organisms, write an equation relating $O$, the number of organisms, to time, $t$, in years since 2015.

**31.** The mass, $Q$, of a sample of tritium (a radioactive isotope of hydrogen) decays at a rate of 5.626% per year. Write a function giving the mass of a 726-gram sample after a time, $t$, in years. Graph this decay function.

**32.** **(a)** The annual inflation rate in the US is 1.5% per year. If a movie ticket currently costs \$10, find a formula for $p$, the price of the ticket $t$ years from today, assuming that movie tickets keep up with inflation.
**(b)** According to your formula, how much will movie tickets cost in 20 years?

**33.** A typical cup of coffee contains about 100 mg of caffeine and every hour approximately 16% of the amount of caffeine in the body is metabolized and eliminated.

**(a)** Write $C$, the amount of caffeine in the body in mg, as a function of $t$, the number of hours since the coffee was consumed.
**(b)** How much caffeine is in the body after 5 hours?

**34.** Grinnell Glacier in Glacier National Park in the US covered about 142 acres in 2007 and was shrinking at a rate of about 4.4% per year.[3]

**(a)** Write a formula for the size, $S$, of the Grinnell Glacier, in acres, as a function of years $t$ since 2007.
**(b)** Use the model to predict the size of the glacier in the year 2020.
**(c)** According to the model, how many acres of ice were lost from the glacier between 2007 and 2010?

**35.** Polluted water is passed through a series of filters. Each filter removes 85% of the remaining impurities. Initially, the untreated water contains impurities at a level of 420 parts per million (ppm). Find a formula for $L$, the remaining level of impurities, after the water has been passed through a series of $n$ filters.

**36.** In 2014 the number of people infected by a virus was $P_0$. Due to a new vaccine, the number of infected people has decreased by 20% each year since 2014. In other words, only 80% as many people are infected each year as were infected the year before. Find a formula for $P = f(n)$, the number of infected people $n$ years after 2014. Graph $f(n)$. Explain, in terms of the virus, why the graph has the shape it does.

**37.** A cold yam is placed in a hot oven. Newton's Law of Heating tells us that the difference between the oven's temperature and the yam's temperature decays exponentially with time. The yam's temperature is initially 0°F, the oven's temperature is 300°F, and the temperature difference decreases by 3% per minute. Find a formula for $Y(t)$, the yam's temperature at time $t$.

**38.** On July 1, 2012, the population of California was 38.04 million[4] and growing at an annual rate of 2.1%. Assume that growth continues at the same rate.

**(a)** By how much will the population increase between 2012 and 2030? Between 2030 and 2048?
**(b)** Explain how you can tell before doing the calculations which of the two answers in part (a) is larger.

**39.** In 2007, world solar photovoltaic (PV) market installations totaled 2826 megawatts and were growing at a rate of 62% per year. In the same year, Japan's PV market installations totaled 230 megawatts and were declining at a rate of 23% per year.[5]

**(a)** Use the information given to predict PV market installations in the world and in Japan in the year 2015.
**(b)** What percent of world PV market installations were in Japan in 2007? In 2015?

**40.** The population of India was about 1.22 billion people in 2013 and was growing at a rate of about 1.28% per year.[6]

**(a)** Write a formula for the population, $P$, of India, in billions, as a function of years $t$ since 2013.
**(b)** If the growth rate stays constant, predict the population of India in the year 2015 and the year 2020.
**(c)** Find the rate of change of India's population, in million people per year, during the year 2013.
**(d)** Find the rate of change of India's population, in people per minute, during the year 2013.

In Problems 41–44, calculate the rate of change of the functions over the interval $1 \leq x \leq 2$.

**41.** $f(x) = 3^x$

**42.** $g(x) = \left(\frac{1}{3}\right)^x$

**43.** $h(x) = 5^x$

**44.** $k(x) = \left(\frac{1}{5}\right)^x$

**45.** Let $f(x) = 2^x$.

**(a)** Calculate the rate of change of $f(x)$ over the following intervals

(i) $0 \leq x \leq 1$ (ii) $1 \leq x \leq 2$
(iii) $2 \leq x \leq 3$ (iv) $3 \leq x \leq 4$

[3]"Warming climate shrinking Glacier Park's glaciers," www.usatoday.com, October 15, 2007.
[4]http://www.census.gov/popest/data/state/totals/2012, accessed April, 2013.
[5]www.solarbuzz.com/marketbuzz2008, accessed February, 2010.
[6]https://www.cia.gov/library/publications/the-world-factbook/geos/in.html, accessed February 20, 2014.

**(b)** What pattern do you notice? Use this pattern to guess the value of the rate of change over $19 \leq x \leq 20$ and then check your answer.

In Problems 46–47, use the following model for the annual revenue for Amazon[7] between 2009 and 2011:

$$R(t) = 24.48(1.4)^t,$$

where $t$ is in years since 2009, and $R(t)$ is in billion dollars.

**46.** **(a)** How much did Amazon's revenue grow between 2009 and 2010? Between 2010 and 2011?
**(b)** Explain what the answers to part (a) tell us about the concavity of the graph of $R(t)$ between 2009 and 2011, and what this means in practical terms.

**47.** **(a)** Assume Amazon's annual revenue continues to rise at the same rate after 2011. Without doing any calculations, arrange in increasing order the average rates of change of $R(t)$ from:

(i) 2011 to 2012 (ii) 2013 to 2014
(iii) 2010 to 2011

**(b)** Suppose Amazon's annual revenue continues to grow after 2011 but the size of the yearly increase shrinks after 2014. What does this say about the concavity and shape of the graph of annual revenue after 2014?

Write the exponential function $y = 5(0.5)^{t/3}$ in the forms given in Problems 48–49. Give the values of all constants.

**48.** $y = ab^t$

**49.** $y = a \cdot 4^{kt}$

**50.** In the year 2012, a total of 17 million passengers took a cruise vacation.[8] The global cruise industry has been growing at a rate of approximately 4% per year for the last five years.

**(a)** Write a formula to approximate the number, $N$, of cruise passengers (in millions) $t$ years after 2012.
**(b)** How many cruise passengers are predicted in the year 2015? Approximately how many passengers went on a cruise in the year 2009?

**51.** Write the following exponential function in standard form, $f(t) = ab^t$, giving the values of $a$ and $b$:

$$f(t) = \frac{60}{5 \cdot 2^{t/11.2}}.$$

**52.** The amount (in milligrams) of a drug in the body $t$ hours after taking a pill is given by $A(t) = 25(0.85)^t$.

**(a)** What is the initial dose given?
**(b)** What percent of the drug leaves the body each hour?
**(c)** What is the amount of drug left after 10 hours?
**(d)** After how many hours is there less than 1 milligram left in the body?

**53.** Every year, teams from 64 colleges qualify to compete in the NCAA women's basketball playoffs. For each round, every team is paired with an opponent. A team is eliminated from the tournament once it loses a round. So, at the end of a round, only half of the teams move on to the next round. Let $N(r)$ be the number of teams remaining in competition after $r$ rounds of the tournament have been played.

**(a)** Find a formula for $N(r)$ and graph $y = N(r)$.
**(b)** In 2013 the Connecticut Huskies defeated the Louisville Cardinals 93–60 in the final round. How many rounds did Connecticut have to go through to win the championship game?

**54.** The UN Food and Agriculture Organization estimates that 2.17% of the world's natural forests existing in 1990 were gone by the end of the decade. In 1990, the world's forest cover stood at 4077 million hectares.[9]

**(a)** How many million hectares of natural forests were lost during the 1990s?
**(b)** How many million hectares of natural forests existed in the year 2000?
**(c)** Write an exponential formula approximating the number of million hectares of natural forest in the world $t$ years after 1990.
**(d)** What was the annual percent decay rate during the 1990s?
**(e)** During the years 2000 through 2005, the world's natural forests decreased by approximately 0.18% per year. Approximately how many million hectares of natural forests existed in the year 2005?

**55.** The population of a small town increases by a growth factor of 1.134 over a two-year period.

**(a)** By what percent does the town increase in size during the two-year period?
**(b)** If the town grows by the same percent each year, what is its annual percent growth rate?

**56.** Forty percent of a radioactive substance decays in five years. By what percent does the substance decay each year?

**57.** The *Home* section of many Sunday newspapers includes a mortgage table similar to Table 4.3.[10] The table gives the

[7] http://www.marketwatch.com/investing/stock/amzn/financials, accessed March, 2014.
[8] http://www.f-cca.com/downloads/2012-Cruise-Industry-Overview-Statistics.pdf, accessed May 10, 2013.
[9] www.fao.org/forestry/fra/fra/2005/en/, accessed January 2, 2010.
[10] http://www.drcalculator.com, accessed January 1, 2010.

monthly payment per $1000 borrowed for loans at various interest rates and time periods.

**(a)** Determine the monthly payment on a
  (i) $60,000 mortgage at 4% for fifteen years.
  (ii) $60,000 mortgage at 4% for thirty years.
  (iii) $60,000 mortgage at 6% for fifteen years.

**(b)** Over the life of the loan, how much money would be saved on a 15-year mortgage of $60,000 if the rate were 4% instead of 6%?

**(c)** Over the life of the loan, how much money would be saved on an 4% mortgage of $60,000 if the term of the loan was fifteen years rather than thirty years?

**Table 4.3**

| Interest rate (%) | 15-year loan | 20-year loan | 25-year loan | 30-year loan |
|---|---|---|---|---|
| 4.00 | 7.40 | 6.06 | 5.28 | 4.77 |
| 4.50 | 7.65 | 6.33 | 5.56 | 5.07 |
| 5.00 | 7.91 | 6.60 | 5.85 | 5.37 |
| 5.50 | 8.17 | 6.88 | 6.14 | 5.68 |
| 6.00 | 8.44 | 7.16 | 6.44 | 6.00 |
| 6.50 | 8.71 | 7.46 | 6.75 | 6.32 |
| 7.00 | 8.99 | 7.75 | 7.07 | 6.65 |
| 7.50 | 9.27 | 8.06 | 7.39 | 6.99 |
| 8.00 | 9.56 | 8.36 | 7.72 | 7.34 |

**58.** You owe $2000 on a credit card. The card charges 1.5% monthly interest on your balance, and requires a minimum monthly payment of 2.5% of your balance. All transactions (payments and interest charges) are recorded at the end of the month. You make only the minimum required payment every month and incur no additional debt.

**(a)** Complete Table 4.4 for a twelve-month period.

**(b)** What is your unpaid balance after one year has passed? At that time, how much of your debt have you paid off? How much money in interest charges have you paid your creditors?

**Table 4.4**

| Month | Balance | Interest | Minimum payment |
|---|---|---|---|
| 0 | $2000.00 | $30.00 | $50.00 |
| 1 | $1980.00 | $29.70 | $49.50 |
| 2 | $1960.20 | | |
| ⋮ | | | |

**59.** A one-page letter is folded into thirds to go into an envelope. If it were possible to repeat this kind of tri-fold 20 times, how many miles thick would the letter be? (A stack of 150 pieces of stationery is one inch thick; 1 mile = 5280 feet.)

Problems 60–61 concern ISO $A$-series paper, commonly used in many countries. $A0$ is the largest sheet, $A1$ next largest, and so on, with $A4$ being somewhat similar to the 8.5 by 11 inch paper standard in the US. The width in millimeters (mm) of a sheet of $A_n$ paper in this series is given by the formula[11]

$$f(n) = 1000 \cdot 2^{-\frac{1}{4}-\frac{n}{2}}.$$

**60.** Show that this is an exponential function by writing it in standard form, $f(n) = ab^n$. What do the values of $a$ and $b$ tell you about $A$-series paper?

**61.** Evaluate and simplify the ratio $f(n+2)/f(n)$. What does your answer tell you about $A$-series paper?

**62.** Figure 4.6 is the graph of $f(x) = 4 \cdot b^x$. Find the slope of the line segment $PQ$ in terms of $b$.

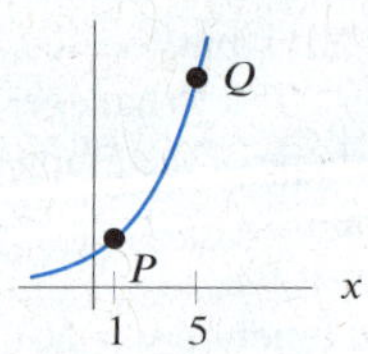


Figure 4.6

**63.** The figure shows graphs of two functions:

$$f(t) = a_0 \cdot 2^{-t/\tau_0}$$
$$g(t) = a_1 \cdot 2^{-t/\tau_1}.$$

Which is larger, $a_0$ or $a_1$? $\tau_0$ or $\tau_1$?

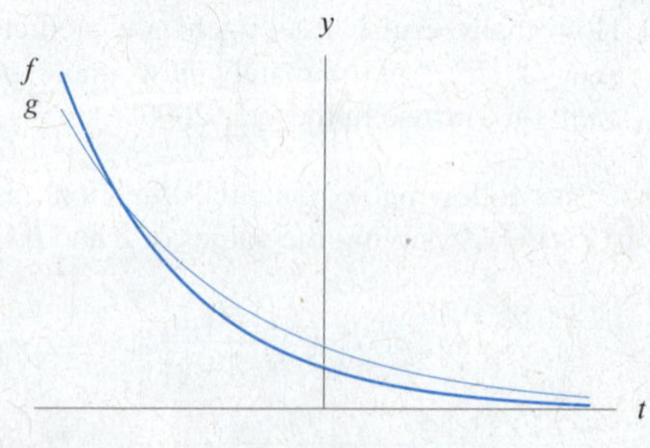


[11] Actual paper sizes are rounded to the nearest mm, so this formula is only approximate. See /www.cl.cam.ac.uk/˜mgk25/iso-paper.html, accessed February 24, 2008.

Use Figure 4.7 in Problems 64–67.

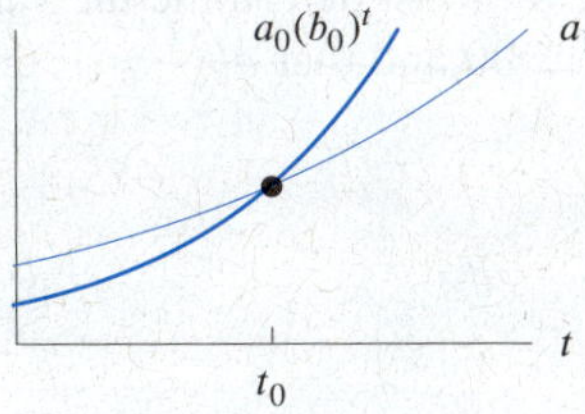


Figure 4.7

**64.** Which is greater, $a_0$ or $a_1$?

**65.** Which is greater, $b_0$ or $b_1$?

**66.** What happens to $t_0$ if $a_0$ is increased while the other quantities remain fixed?

**67.** What happens to $t_0$ if $b_1$ is decreased while the other quantities remain fixed?

**68.** Let $P$ be the number of students in a school district, $N$ be the size of the tax base (in households), and $r$ be the average annual tax rate (in \$/household).

**(a)** Find a formula for $R$, the total tax revenue, in terms of $N$ and $r$.

**(b)** Find a formula for $A$, the average revenue per student.

**(c)** Suppose the tax base goes up by 2% and the tax rate is raised by 3%. Find formulas for the new tax base and tax rate in terms of $N$ and $r$.

**(d)** Using your answer to part (c), find a formula for the new total tax revenue in terms of $R$. By what percent did $R$ increase?

**(e)** Over the time period in part (c), the student population rises by 8%. Find a formula for the new average revenue in terms of $A$. Did the average revenue rise or fall? By how much?

# 4.2 COMPARING EXPONENTIAL AND LINEAR FUNCTIONS

The exponential function $Q = ab^t$ represents a quantity changing at a constant percent rate. In this section we compare exponential and linear models and we fit exponential models to data from tables and graphs.

## Identifying Linear and Exponential Functions From a Table

Table 4.5 gives values of a linear and an exponential function. Notice that the value of $x$ changes by equal steps of $\Delta x = 5$. The function $f$ could be linear because the difference between consecutive values of $f(x)$ is constant: $f(x)$ increases by 15 each time $x$ increases by 5.

**Table 4.5** *Two functions, one linear and one exponential*

| $x$ | 20 | 25 | 30 | 35 | 40 | 45 |
|---|---|---|---|---|---|---|
| $f(x)$ | 30 | 45 | 60 | 75 | 90 | 105 |
| $g(x)$ | 1000 | 1200 | 1440 | 1728 | 2073.6 | 2488.32 |

On the other hand, the difference between consecutive values of $g(x)$ is *not* constant:

$$1200 - 1000 = 200$$
$$1440 - 1200 = 240$$
$$1728 - 1440 = 288.$$

Thus, $g$ is not linear. However, the *ratio* of consecutive values of $g(x)$ is constant:

$$\frac{1200}{1000} = 1.2, \quad \frac{1440}{1200} = 1.2, \quad \frac{1728}{1440} = 1.2,$$

and so on. Note that $1200 = 1.2(1000)$, $1440 = 1.2(1200)$, $1728 = 1.2(1440)$. Thus, each time $x$ increases by 5, the value of $g(x)$ increases by a factor of 1.2. This pattern of constant ratios is indicative of exponential functions. In general:

> For a table of data that gives $y$ as a function of $x$ and in which $\Delta x$ is constant:
> - If the *difference* of consecutive $y$-values is constant, the table could represent a linear function.
> - If the *ratio* of consecutive $y$-values is constant, the table could represent an exponential function.

## Finding a Formula for an Exponential Function

To find a formula for the exponential function in Table 4.5, we must determine the values of $a$ and $b$ in the formula $g(x) = ab^x$. The table tells us that $ab^{20} = 1000$ and that $ab^{25} = 1200$. Taking the ratio gives

$$\frac{ab^{25}}{ab^{20}} = \frac{1200}{1000} = 1.2.$$

Notice that the value of $a$ cancels in this ratio, so

$$\frac{ab^{25}}{ab^{20}} = b^5 = 1.2.$$

We solve for $b$ by raising each side to the $1/5$ power:

$$(b^5)^{1/5} = b = 1.2^{1/5} \approx 1.03714.$$

Now that we have the value of $b$, we can substitute into the formula for $g$ and solve for $a$. Since $g(20) = ab^{20} = 1000$, we have

$$a(1.03714)^{20} = 1000$$
$$a = \frac{1000}{1.03714^{20}} \approx 482.253.$$

Thus, a formula for $g$ is $g(x) = 482.253(1.037)^x$. (Note that we could have used any two points from the table to find $a$ and $b$.)

### Modeling Linear and Exponential Growth Using Two Data Points

If we are given two data points, we can fit either a linear or an exponential function to the points. The following example compares the predictions made by a linear model and an exponential model fitted to the same data.

**Example 1** At time $t = 0$ years, some turtles are released into a wetland. When $t = 4$ years, a biologist estimates there are 300 turtles in the wetland. Three years later, the biologist estimates there are 450 turtles. Let $P$ represent the size of the turtle population in year $t$.

(a) Find a formula for $P = f(t)$ assuming linear growth. Interpret the slope and $P$-intercept of your formula in terms of the turtle population.

(b) Now find a formula for $P = g(t)$ assuming exponential growth. Interpret the parameters of your formula in terms of the turtle population.

(c) In year $t = 12$, the biologist estimates that there are 900 turtles in the wetland. What does this indicate about the two population models?

**Solution** (a) Assuming linear growth, we have $P = f(t) = b + mt$, and

$$m = \frac{\Delta P}{\Delta t} = \frac{450 - 300}{7 - 4} = \frac{150}{3} = 50.$$

Calculating $b$ gives

$$300 = b + 50 \cdot 4$$
$$b = 100,$$

so $P = f(t) = 100 + 50t$. This formula tells us that 100 turtles were originally released into the wetland and that the number of turtles increases at the constant rate of 50 turtles per year.

(b) Assuming exponential growth, we have $P = g(t) = ab^t$. The values of $a$ and $b$ are calculated from the ratio

$$\frac{ab^7}{ab^4} = \frac{450}{300},$$

so

$$b^3 = 1.5.$$

Thus,

$$b = (1.5)^{1/3} \approx 1.145.$$

Using the fact that $g(4) = ab^4 = 300$ to find $a$ gives

$$a(1.145)^4 = 300$$

$$a = \frac{300}{1.145^4} \approx 175, \quad \text{Rounding to the nearest whole turtle}$$

so $P = g(t) = 175(1.145)^t$. This formula tells us that 175 turtles were originally released into the wetland and the number increases at about 14.5% per year.

(c) In year $t = 12$, there are approximately 900 turtles. The linear function from part (a) predicts

$$P = 100 + 50 \cdot 12 = 700 \text{ turtles.}$$

The exponential formula from part (b), however, predicts

$$P = 175(1.145)^{12} \approx 889 \text{ turtles.}$$

The fact that 889 is closer to the observed value of 900 turtles suggests that, during the first 12 years, exponential growth is a better model of the turtle population than linear growth. The two models are graphed in Figure 4.8.

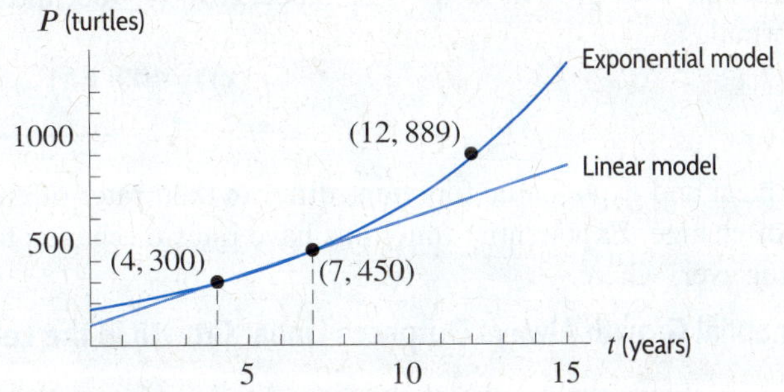


**Figure 4.8:** Comparison of the linear and exponential models of the turtle population

## Formulas and Rates of Change: Linear Versus Exponential

In some ways the formulas for linear and exponential functions are similar. If $y$ is a linear function of $x$ and $x$ is a positive integer, we can write $y = b + mx$ as

$$y = b + \underbrace{m + m + m + \ldots + m}_{x \text{ times}}.$$

Similarly, if $y$ is an exponential function of $x$, so that $y = a \cdot b^x$ and $x$ is a positive integer, we can write

$$y = a \cdot \underbrace{b \cdot b \cdot b \cdot \ \ldots \ \cdot b}_{x \text{ times}}.$$

So linear functions involve repeated sums whereas exponential functions involve repeated products. In both cases, $x$ determines the number of repetitions.

There are other similarities between the formulas for linear and exponential functions. In the formula $y = b + mx$, $b$ tells us the $y$-intercept and $m$ tells us how the function is changing. In the formula $y = a \cdot b^x$, $a$ tells us the $y$-intercept and $b$ tells us how the function is changing.

**Example 2** The tables in parts (a) and (b) contain values from an exponential or linear function. For each table, decide if the function is linear or exponential, and find a possible formula for the function.

(a)

| $x$ | $f(x)$ |
|---|---|
| 0 | 65 |
| 1 | 75 |
| 2 | 85 |
| 3 | 95 |
| 4 | 105 |

(b)

| $x$ | $g(x)$ |
|---|---|
| 0 | 400 |
| 1 | 600 |
| 2 | 900 |
| 3 | 1350 |
| 4 | 2025 |

**Solution**

(a) The function values increase by 10 as $x$ increases by 1, so this is a linear function with slope $m = 10$. Since $f(0) = 65$, the vertical intercept is 65. A possible formula is

$$f(x) = 65 + 10x.$$

(b) The function is not linear, since $g(x)$ increases by different amounts as $x$ increases by 1. To check that $g$ is exponential, we look at ratios of consecutive values:

$$\frac{600}{400} = 1.5, \quad \frac{900}{600} = 1.5, \quad \frac{1350}{900} = 1.5, \quad \frac{2025}{1350} = 1.5.$$

Each time $x$ increases by 1, the value of $g(x)$ increases by a factor of 1.5. This is an exponential function with growth factor 1.5. Since $g(0) = 400$, the vertical intercept is 400. A possible formula is

$$g(x) = 400(1.5)^x.$$

Linear and exponential functions differ in their rates of change. Linear functions have constant rates of change. Exponential functions have rates of change that are increasing everywhere or decreasing everywhere.

## Exponential Growth Always Outpaces Linear Growth in the Long Run

Figure 4.8 shows the graphs of the linear and exponential models for the turtle population from Example 1. The graphs highlight the fact that, although these two graphs remain fairly close for the first ten or so years, the exponential model predicts explosive growth in the long run. This is because the rate of change of this exponential function is increasing whereas the rate of change of a linear function is constant.

It can be shown that an exponentially increasing quantity will, in the long run, always grow faster than a linearly increasing quantity. This fact led the $19^{\text{th}}$-century clergyman and economist Thomas Malthus to make some rather gloomy predictions, which are illustrated in the next example.

**Example 3** The population of a country is initially 2 million people and is increasing at 4% per year. The country's annual food supply is initially adequate for 4 million people and is increasing at a constant rate adequate for an additional 0.5 million people per year.

(a) Based on these assumptions, in approximately what year will this country first experience shortages of food?

(b) If the country doubled its initial food supply, would shortages still occur? If so, when? (Assume the other conditions do not change.)

(c) If the country doubled the rate at which its food supply increases, in addition to doubling its initial food supply, would shortages still occur? If so, when? (Again, assume the other conditions do not change.)

**Solution** Let $P$ represent the country's population (in millions) and $N$ the number of people the country can feed (in millions). The population increases at a constant percent rate, so it can be modeled by an exponential function. The initial population is $a = 2$ million people and the annual growth factor is $b = 1 + 0.04 = 1.04$, so a formula for the population is

$$P = 2(1.04)^t.$$

In contrast, the food supply increases by a constant amount each year and is therefore modeled by a linear function. The initial food supply is adequate for $b = 4$ million people and the growth rate is $m = 0.5$ million per year, so the number of people that can be fed is

$$N = 4 + 0.5t.$$

(a) Figure 4.9(a) gives the graphs of $P$ and $N$ over a 105-year span. For many years, the food supply is far in excess of the country's needs. However, after about 78 years the population has begun to grow so rapidly that it catches up to the food supply and then outstrips it. After that time, the country will suffer from shortages.

(b) If the country can initially feed eight million people rather than four, the formula for $N$ is

$$N = 8 + 0.5t.$$

However, as we see from Figure 4.9(b), this measure only buys the country three or four extra years with an adequate food supply. After 81 years, the population is growing so rapidly that the head start given to the food supply makes little difference.

(c) If the country doubles the rate at which its food supply increases, from 0.5 million per year to 1.0 million per year, the formula for $N$ is

$$N = 8 + 1.0t.$$

Unfortunately the country still runs out of food eventually. Judging from Figure 4.9(c), this happens in about 102 years.

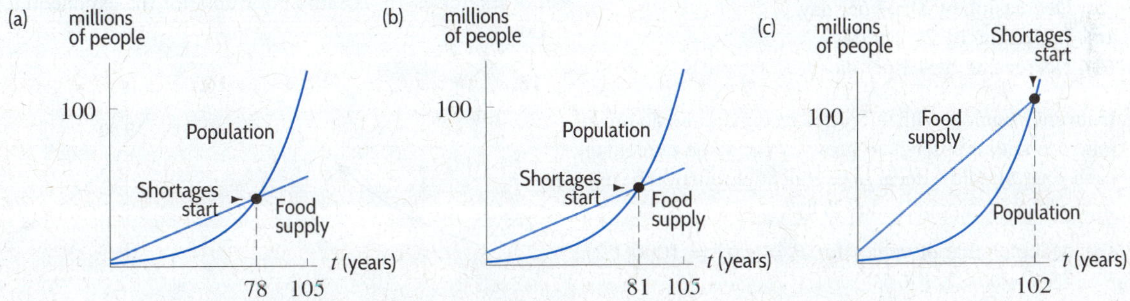


**Figure 4.9:** These graphs illustrate the fact that an exponentially growing population eventually outstrips a linearly growing food supply

Malthus believed that populations increase exponentially while food production increases linearly. The last example explains his gloomy prediction: Any population eventually outstrips its food supply, leading to famine and war.

## Exercises and Problems for Section 4.2

### Skill Refresher

In Exercises S1–S4, write each of the following with single positive exponents.

**S1.** $b^4 \cdot b^6$

**S2.** $8g^3 \cdot (-4g)^2$

**S3.** $\dfrac{18a^{10}b^6}{6a^3b^{-4}}$

**S4.** $\dfrac{(2a^3b^2)^3}{(4ab^{-4})^2}$

In Exercises S5–S6, evaluate the functions for $t = 0$ and $t = 3$.

**S5.** $f(t) = 5.6(1.043)^t$

**S6.** $g(t) = 12{,}837(0.84)^t$

In Exercises S7–S10, solve for $x$.

**S7.** $4x^3 = 20$

**S8.** $\dfrac{5x^3}{x^5} = 125$

**S9.** $\dfrac{4x^8}{3x^3} = 7$

**S10.** $\sqrt{4x^3} = 5$

## Exercises

**1.** The following formulas give the populations (in 1000s) of four different cities, $A$, $B$, $C$, and $D$, where $t$ is in years. Which are changing exponentially? Describe in words how each of these populations is changing over time. Graph those that are exponential.

$$P_A = 200 + 1.3t, \quad P_B = 270(1.021)^t,$$

$$P_C = 150(1.045)^t, \quad P_D = 600(0.978)^t.$$

**2.** A population has size 5000 at time $t = 0$, with $t$ in years.

**(a)** If the population decreases by 100 people per year, find a formula for the population, $P$, at time $t$.

**(b)** If the population decreases by 8% per year, find a formula for the population, $P$, at time $t$.

**3.** Write a formula for the price $p$ of a gallon of gas in $t$ days if the price is \$2.50 on day $t = 0$ and the price is:

**(a)** Increasing by \$0.03 per day.

**(b)** Decreasing by \$0.07 per day.

**(c)** Increasing by 2% per day.

**(d)** Decreasing by 4% per day.

**4.** In an environment with unlimited resources and no predators, a population tends to grow by the same percentage each year. Should a linear or exponential function be used to model such a population? Why?

**5. (a)** Make a table of values for $P = f(t) = 1000(1.2)^t$, for $t = 0, 1, 2, 3, 4, 5$.

**(b)** Calculate each of the ratios of successive terms:

$$\frac{f(1)}{f(0)}, \quad \frac{f(2)}{f(1)}, \quad \frac{f(3)}{f(2)}, \quad \frac{f(4)}{f(3)}, \quad \frac{f(5)}{f(4)}.$$

**(c)** Discuss the results. Why is this outcome to be expected?

The tables in Exercises 6–9 contain values from an exponential or a linear function. In each problem:

**(a)** Decide if the function is linear or exponential.

**(b)** Find a possible formula for each function and graph it.

**6.**

| $x$ | $f(x)$ |
|---|---|
| 0 | 12.5 |
| 1 | 13.75 |
| 2 | 15.125 |
| 3 | 16.638 |
| 4 | 18.301 |

**7.**

| $x$ | $g(x)$ |
|---|---|
| 0 | 0 |
| 1 | 2 |
| 2 | 4 |
| 3 | 6 |
| 4 | 8 |

**8.**

| $x$ | $h(x)$ |
|---|---|
| 0 | 14 |
| 1 | 12.6 |
| 2 | 11.34 |
| 3 | 10.206 |
| 4 | 9.185 |

**9.**

| $x$ | $i(x)$ |
|---|---|
| 0 | 18 |
| 1 | 14 |
| 2 | 10 |
| 3 | 6 |
| 4 | 2 |

**10.** Find $g(t) = ab^t$ if $g(10) = 50$ and $g(30) = 25$.

**11.** Find a formula for $f(x)$, an exponential function such that $f(-8) = 200$ and $f(30) = 580$.

**12.** Suppose that $f(x)$ is exponential and that $f(-3) = 54$ and $f(2) = \frac{2}{9}$. Find a formula for $f(x)$.

**13.** Find a formula for $f(x)$, an exponential function such that $f(2) = 1/27$ and $f(-1) = 27$.

**14.** Find the equation of an exponential curve through the points $(-1, 2)$, $(1, 0.3)$.

For Exercises 15–20, find a formula for the exponential function.

**15.**

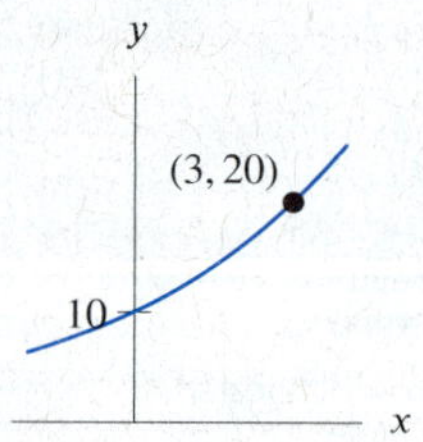


**16.**

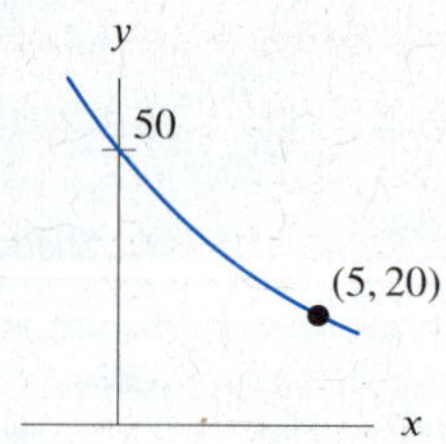


**17.**

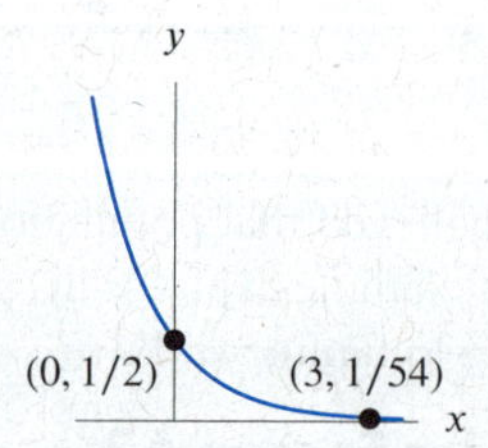


**18.**

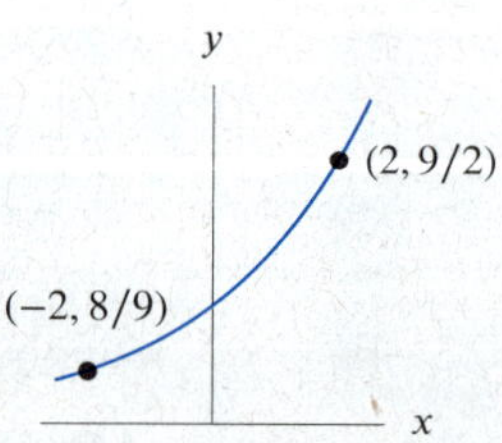


**19.**

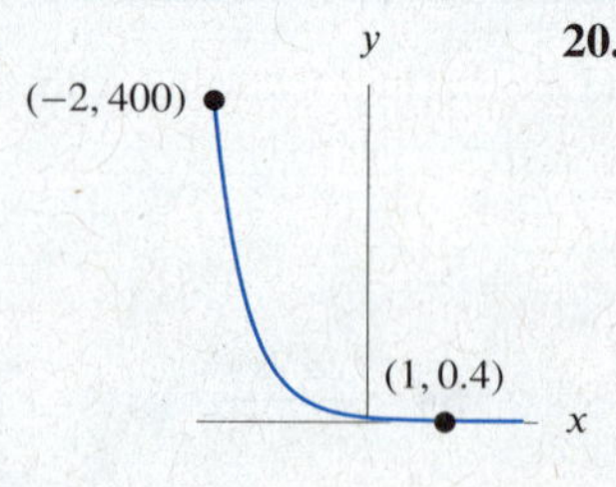


**20.**

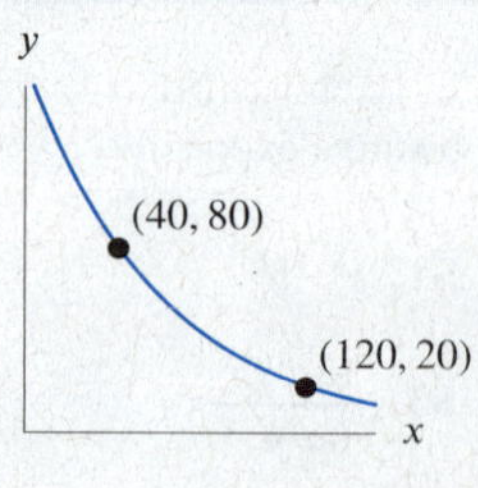

In Exercises 21–22, find an exponential function for the graph, or explain why this is not possible.

**21.**

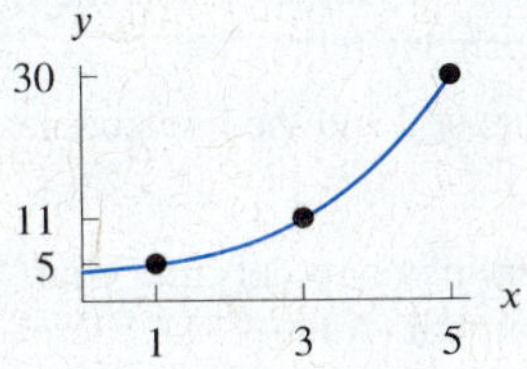


**22.**

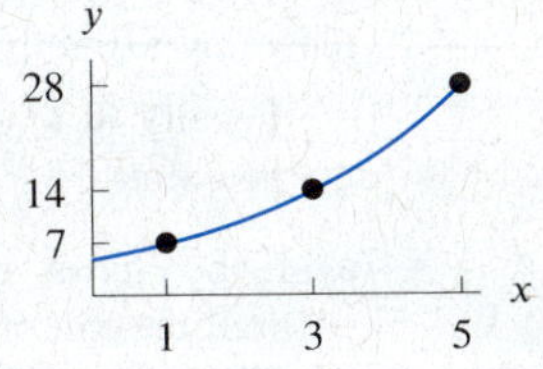


**23.** Find a formula for the exponential function $g$ whose graph contains the points $(2.3, 0.4)$ and $(3.5, 0.1)$.

**24.** Find a formula for the exponential function $f$ given $f(-5) = 22$ and $f(17) = 46$.

**25.** In Table 4.6, data for the functions $f, g$ and $h$ are given, rounded to the nearest 0.01. One of them is linear, one is exponential, and one is neither.

**(a)** Which function is linear, and which is exponential? Give a reason for each answer.

**(b)** Find a formula for the exponential function.

**Table 4.6**

| $x$ | 0.00 | 3.00 | 6.00 | 9.00 | 12.00 |
|---|---|---|---|---|---|
| $f(x)$ | 2.23 | 2.45 | 2.70 | 2.97 | 3.26 |
| $g(x)$ | 2.23 | 3.83 | 4.53 | 4.94 | 5.23 |
| $h(x)$ | 2.23 | 3.70 | 5.17 | 6.64 | 8.11 |

**26.** Table 4.7 gives values for two functions $f$ and $g$, one linear and one exponential.

**(a)** For each function, calculate the rates of change over the intervals $2 \leq x \leq 8$ and $8 \leq x \leq 11$.

**(b)** Determine which is the linear function and which is the exponential function.

**Table 4.7**

| $x$ | 2 | 8 | 11 |
|---|---|---|---|
| $f(x)$ | 1.728 | 4.299 | 7.43 |
| $g(x)$ | 1.464 | 5.088 | 6.9 |

**27.** Which (if any) of the functions in the table could be linear? Which could be exponential? (Note that table values may reflect rounding.)

| $x$ | 0 | 5 | 10 | 15 | 20 |
|---|---|---|---|---|---|
| $f(x)$ | 95.4 | 85.9 | 77.3 | 69.6 | 62.6 |
| $g(x)$ | 44.8 | 40.9 | 36.8 | 32.5 | 28.0 |
| $h(x)$ | 37.3 | 36.6 | 35.9 | 35.2 | 34.5 |

## Problems

In Problems 28–30, could the function be approximately linear or approximately exponential? Is it neither? For those that could be nearly linear or nearly exponential, find a formula.

**28.**

| $t$ | 3 | 10 | 14 |
|---|---|---|---|
| $Q(t)$ | 7.51 | 8.7 | 9.39 |

**29.**

| $t$ | 5 | 9 | 15 |
|---|---|---|---|
| $R(t)$ | 2.32 | 2.61 | 3.12 |

**30.**

| $t$ | 5 | 12 | 16 |
|---|---|---|---|
| $S(t)$ | 4.35 | 6.72 | 10.02 |

**31.** Graphs of a linear and an exponential function are shown in Figure 4.10. Find formulas for each of the functions.

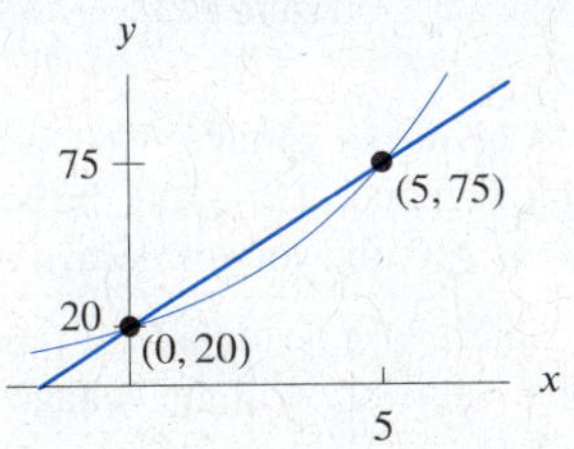


**Figure 4.10**

**32.** Find formulas for the exponential functions in Figure 4.11.

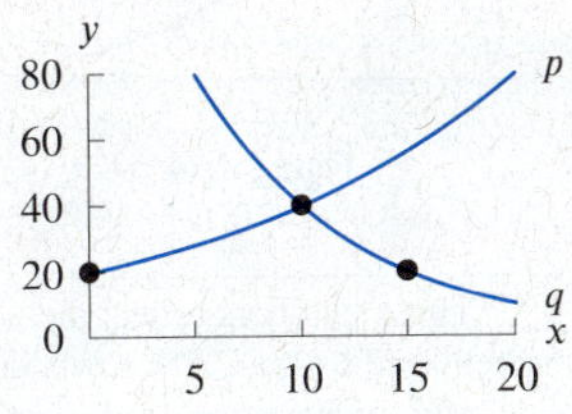


**Figure 4.11**

**33.** Let $p(x) = 2 + x$ and $q(x) = 2^x$. Estimate the values of $x$ such that $p(x) < q(x)$.

**34.** If $f(x) = 12 + 20x$ and $g(x) = \frac{1}{2} \cdot 3^x$, for what values of $x$ is $g(x) < f(x)$?

**35.** Figure 4.12 shows the balance, $P$, in a bank account.

**(a)** Find a possible formula for $P = f(t)$ assuming the balance grows exponentially.
**(b)** What was the initial balance?
**(c)** What annual interest rate does the account pay?

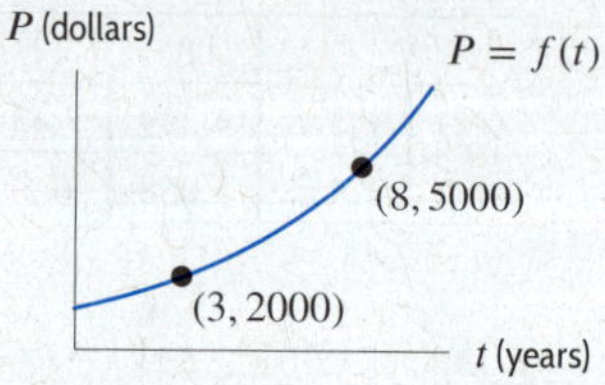


Figure 4.12

**36.** A 1987 treaty to protect the ozone layer produced dramatic declines in global production, $P$, of chlorofluorocarbons (CFCs). See Figure 4.13.[12] Find a formula for $P$ as an exponential function of the number of years, $t$, since 1989. What was the annual percent decay rate?

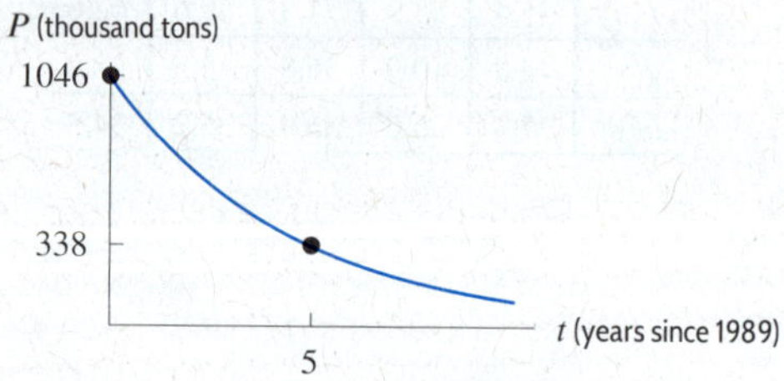


Figure 4.13

**37.** Figure 4.14 gives the voltage, $V(t)$, across a circuit element at time $t$ seconds. For $t < 0$, the voltage is a constant 80 volts; for $t \geq 0$, the voltage decays exponentially.

**(a)** Find a piecewise formula for $V(t)$.
**(b)** At what value of $t$ will the voltage reach 0.1?

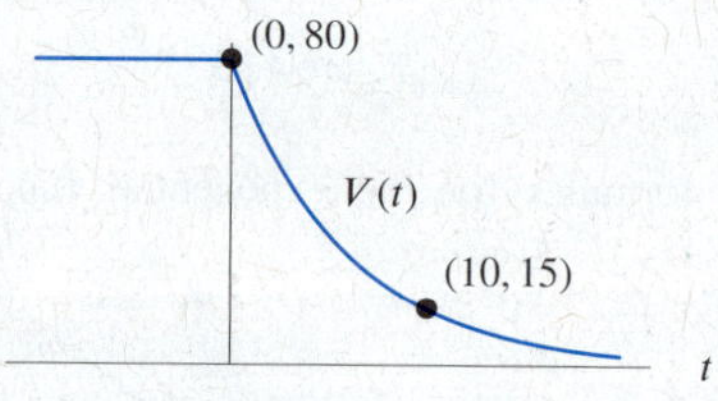


Figure 4.14

**38.** Suppose the city of Yonkers is offered two alternative fines by the judge. (See Example 6 on page 144.)

Penalty A: \$1 million on August 2 and the fine increases by \$10 million each day thereafter.

Penalty B: 1¢ on August 2 and the fine doubles each day thereafter.

**(a)** If the city of Yonkers plans to defy the court order until the end of the month (August 31), compare the fines incurred under Penalty A and Penalty B.
**(b)** If $t$ represents the number of days after August 2, express the fine incurred as a function of $t$ under
(i) Penalty A (ii) Penalty B
**(c)** Assuming your formulas in part (b) hold for $t \geq 0$, is there a time such that the fines incurred under both penalties are equal? If so, estimate that time.

**39.** There were 191 million licensed drivers in the US in 2000 and 210 million in 2010.[13] Find a formula for the number, $N$, of licensed drivers in the US as a function of $t$, the number of years since 2000, assuming growth is

**(a)** Linear **(b)** Exponential

**40.** According to the College Board,[14] the average cost of tuition (including fees, room and board) at private four-year nonprofit colleges has risen from \$24,071 in 2003 to \$30,094 in 2013. What will the average cost of tuition be in 2023 assuming:

**(a)** Linear growth? **(b)** Exponential growth?

**41.** Short-track 500-m speed skating became a Winter Olympic event in 1994, and Chae Ji-Hoon of Korea won the event that year with a time of 43.45 seconds. In 2010, Charles Hamelin from Canada won the event with a time of 40.981 seconds. Find a formula for the predicted winning time in the 500-m speed-skating event as a function of the number of years since 1994, and predict the winning time in 2018, if we assume the decrease in time is

**(a)** Linear **(b)** Exponential

**42.** In year $t = 0$ a lake is estimated to have about 3500 trout in it. Ten years later, at $t = 10$, the population of trout is believed to be about 1700.

**(a)** Write a formula for the size of the population $P$ as a function of year $t$ if we assume the decrease is linear. What is the rate of change, in fish per year, of the function over the ten-year period?

[12] These numbers reflect the volume of the major CFCs multiplied by their respective ozone-depleting potentials (ODPs), as reported by the U. N. Environmental Programme Ozone Secretariat. See hq.uncp.org/ozone, accessed November, 2001.
[13] http://www.fhwa.dot.gov/policyinformation/statistics/2010/dv1c.cfm, accessed May 18, 2013.
[14] http://trends.collegeboard.org/college-pricing/figures-tables/published-prices-national, page accessed March 16, 2014. All amounts are in 2013 dollars.

(b) Write a formula for the size of the population $P$ as a function of year $t$ if we assume the decrease is exponential. What is the percent rate of change, in percent per year, of the function over the ten-year period?

(c) Graph the two functions on the same coordinate system. Indicate the points at $t = 0$ and $t = 10$.

43. Cocoa production[15] is shown in Table 4.8 for the world and the Ivory Coast, in millions of tons, as a function of the number of years since 2000. In each case, determine if production is better modeled with a linear or an exponential function and find a formula for the function.

**Table 4.8**

| Years since 2000 | 0 | 1 | 2 | 3 | 4 |
|---|---|---|---|---|---|
| World cocoa production | 3.1 | 3.875 | 4.844 | 6.055 | 7.568 |
| Ivory Coast cocoa production | 1.3 | 1.34 | 1.38 | 1.42 | 1.46 |

44. The relative humidity of an air mass is the ratio of the quantity of water in the air to the maximum quantity of water that the air could hold at the same temperature, expressed as a percent. If no water is added or removed from the air, the relative humidity is halved for every 10°C increase in temperature.

(a) In one air mass, the relative humidity is 100% at 10°C. Sketch a graph of the relative humidity as a function of temperature in °C for temperatures between 10°C and 50°C.

(b) Explain how your graph shows that warm air can hold more water than cold air.

(c) Find a formula for the relative humidity in % as a function of the temperature $T$ in °C for $T \geq 10$.

45. The average gain in life expectancy at birth in the US has remained almost constant for 150 years, at an increase of 3 months in life expectancy per year.[16] Assume that this rate of increase continues. Life expectancy in the US in 2011 was 78.7 years.

(a) Is life expectancy increasing linearly or exponentially?

(b) Find a formula for life expectancy, $L$, in years, in the US at birth as a function of the number of years $t$ since 2000.

(c) What is the predicted life expectancy for babies born in 2050?

46. The number of asthma sufferers in the world was about 84 million in 1990 and 300 million in 2012.[17] Let $N$ represent the number of asthma sufferers (in millions) worldwide $t$ years after 1990.

(a) Write $N$ as a linear function of $t$. What is the slope? What does it tell you about asthma sufferers?

(b) Write $N$ as an exponential function of $t$. What is the growth factor? What does it tell you about asthma sufferers?

(c) How many asthma sufferers are predicted worldwide in the year 2020 with the linear model? With the exponential model?

47. With an initial population $P_0$, what is the value of a population at the end of 10 years, given each of the following assumptions? Graph each population against time.

(a) The population decreases linearly and the decrease is 10% in the first year.

(b) The population decreases exponentially at the rate of 10% a year.

48. In 2000, the population of a town was 20,000, and it grew by 4.14% that year. By 2010, the town's population had reached 30,000.

(a) Can this population be best described by a linear or an exponential model, or neither? Explain.

(b) If possible, find a formula for $P(t)$, this population $t$ years after 2000.

49. In 2003, the population of a town was 18,500 and it grew by 250 people by the end of the year. By 2013, its population had reached 22,500.

(a) Can this population be best described by a linear or an exponential model, or neither? Explain.

(b) If possible, find a formula for $P(t)$, the population $t$ years after 2003.

## 4.3 GRAPHS OF EXPONENTIAL FUNCTIONS

As with linear functions, an understanding of the significance of the parameters $a$ and $b$ in the formula $Q = ab^t$ helps us analyze and compare exponential functions.

[15] Adapted from www.treecrops.org/crops/cocoaoutlook, accessed November, 2009.
[16] http://www.cdc.gov/nchs/fastats/lifexpec.htm, accessed May 19, 2013.
[17] www.healthylifestyle-solutions.blogspot.com, accessed January 2, 2010; www.moh.gov.sa/en/HealthAwareness/healthDay/1433/Pages/013.aspx, accessed May 19, 2013.

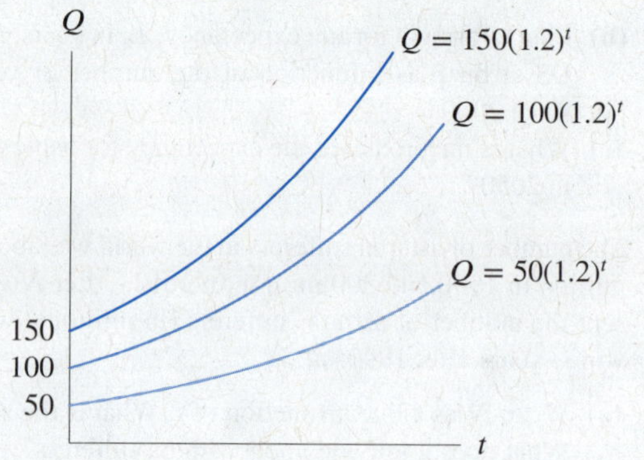


**Figure 4.15:** Graphs of $Q = a(1.2)^t$ for $a = 50$, 100, and 150

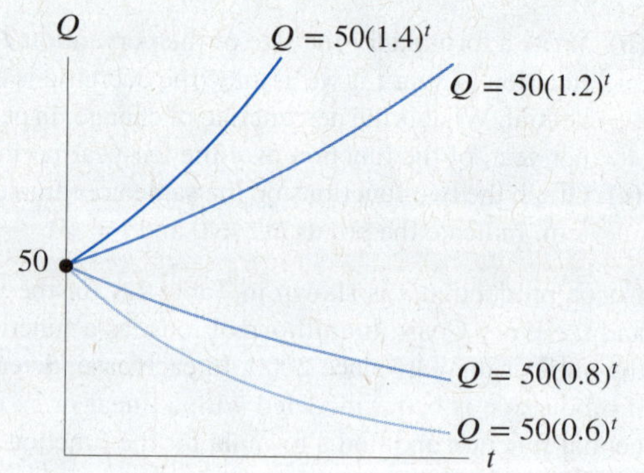


**Figure 4.16:** Graphs of $Q = 50b^t$ for $b = 0.6, 0.8, 1.2$ and 1.4

## Graphs of the Exponential Family: The Effect of the Parameter *a*

In the formula $Q = ab^t$, the value of $a$ tells us where the graph crosses the $Q$-axis, since $a$ is the value of $Q$ when $t = 0$. In Figure 4.15 each graph has the same value of $b$ but different values of $a$ and thus different vertical intercepts.

## Graphs of the Exponential Family: The Effect of the Parameter *b*

Each graph in Figure 4.16 has a different value of the base, $b$, but the same value of $a$ (and the same $Q$-intercept). If $b > 1$, the graph climbs when read from left to right (the function is increasing), and if $0 < b < 1$, the graph falls when read from left to right (the function is decreasing).

Figure 4.16 also shows how the value of $b$ affects the steepness of the graph of $Q = ab^t$. If $b > 1$, the greater the value of $b$, the more rapidly the graph rises. If $0 < b < 1$, the smaller the value of $b$, the more rapidly the graph falls. In every case, however, the graph is concave up.

### Horizontal Asymptotes

The $t$-axis is a *horizontal asymptote* for the graph of $Q = ab^t$, because $Q$ approaches 0 as $t$ gets large, either positively or negatively. For exponential decay, such as $Q = f(t) = a(0.6)^t$ in Figure 4.16, the value of $Q$ approaches 0 as $t$ gets large and positive. We write

$$Q \to 0 \quad \text{as} \quad t \to \infty.$$

This means that $Q$ is as close to 0 as we like for all sufficiently large values of $t$. For exponential growth, the value of $Q$ approaches zero as $t$ grows more negative. (See Figure 4.28 on page 168.) In this case, we write

$$Q \to 0 \quad \text{as} \quad t \to -\infty.$$

This means that $Q$ is as close to 0 as we like for all sufficiently large negative values of $t$. We make the following definition:

The horizontal line $y = k$ is a **horizontal asymptote** of a function, $f$, if the function values get arbitrarily close to $k$ as $x$ gets large (either positively or negatively or both). We describe this behavior using the notation

$$f(x) \to k \quad \text{as} \quad x \to \infty$$

or

$$f(x) \to k \quad \text{as} \quad x \to -\infty.$$

The following example shows this notation used in an application.

**Example 1** A capacitor is the part of an electrical circuit that stores electric charge. The quantity of charge stored decreases exponentially with time. Stereo amplifiers provide an example: When an amplifier is turned off, the display lights fade slowly because it takes time for the capacitors to discharge. (Thus, it can be unsafe to open a stereo or a computer immediately after it is turned off.)

In a particular circuit $t$ seconds after being switched off, the quantity of stored charge (in micro-coulombs) is given by

$$Q = 200(0.9)^t, \quad t \geq 0.$$

(a) Describe in words how the stored charge changes over time.
(b) What quantity of charge remains after 10 seconds? 20 seconds? 30 seconds? 1 minute? 2 minutes? 3 minutes?
(c) Graph the charge over the first minute. What does the horizontal asymptote of the graph tell you about the charge?

**Solution** (a) The charge is initially 200 micro-coulombs. Since $b = 1 + r = 0.9$, we have $r = -0.10$, which means that the charge level decreases by 10% each second.

(b) Table 4.9 gives the value of $Q$ at $t = 0, 10, 20, 30, 60, 120$, and 180. Notice that as $t$ increases, $Q$ gets closer and closer to, but does not quite reach, zero. The charge stored by the capacitor is getting smaller, but never completely vanishes. That is,

$$Q \to 0 \quad \text{as} \quad t \to \infty.$$

(c) Figure 4.17 shows $Q$ over a 60-second interval. The horizontal asymptote at $Q = 0$ corresponds to the fact that the charge gets very small as $t$ increases. After 60 seconds, for all practical purposes, the charge is zero.

**Table 4.9** *Charge (in micro-coulombs) stored by a capacitor over time*

| $t$ (seconds) | $Q$, charge level |
|---|---|
| 0 | 200 |
| 10 | 69.736 |
| 20 | 24.315 |
| 30 | 8.478 |
| 60 | 0.359 |
| 120 | 0.000646 |
| 180 | 0.00000116 |

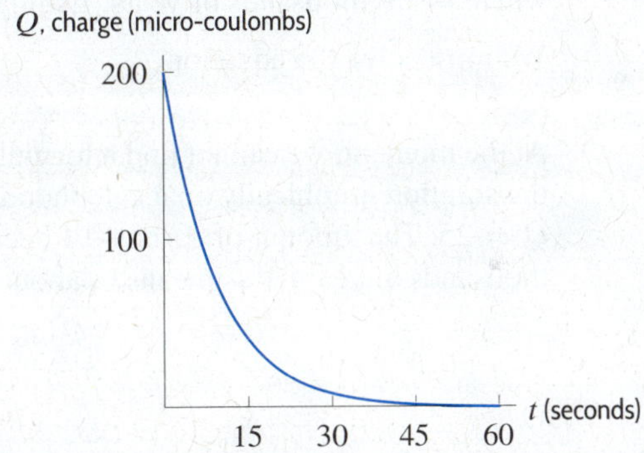


**Figure 4.17**: The charge stored by a capacitor over one minute

## Solving Exponential Equations Graphically

We are often interested in solving equations involving exponential functions. In the following examples, we do this graphically. In Section 5.1, we will see how to solve equations using logarithms.

**Example 2** In Example 6 on page 144, the fine, $P$, imposed on the city of Yonkers is given by $P = 100 \cdot 2^t$ dollars, where $t$ is the number of days after August 2. In 1988, the annual budget of the city was \$337 million. If the city chose to disobey the court order, on what day would the fine have wiped out the entire annual budget?

**Solution** We need to find the day on which the fine reaches \$337 million. That is, we must solve the equation

$$100 \cdot 2^t = 337{,}000{,}000.$$

Using a computer or graphing calculator, we can graph $P = 100 \cdot 2^t$ to find the point at which the fine reaches 337 million. From Figure 4.18, we see that this occurs between $t = 21$ and $t = 22$.

At day $t = 21$, August 23, the fine is:

$$P = 100 \cdot 2^{21} = 209{,}715{,}200$$

or just over \$200 million. On day $t = 22$, the fine is

$$P = 100 \cdot 2^{22} = 419{,}430{,}400$$

or almost \$420 million—quite a bit more than the city's entire annual budget!

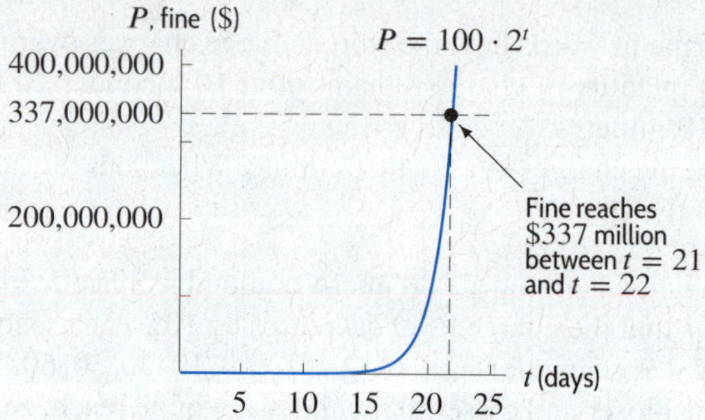


**Figure 4.18:** The fine imposed on Yonkers exceeds \$337 million after 22 days

**Example 3** A 200-$\mu$g sample of carbon-14 decays according to the formula

$$Q = 200(0.886)^t$$

where $t$ is in thousands of years. Estimate when there is 25 $\mu$g of carbon-14 left.

**Solution** We must solve the equation

$$200(0.886)^t = 25.$$

At the moment, we cannot find a formula for the solution to this equation. However, we can estimate the solution graphically on a calculator. Figure 4.19 shows a graph of $Q = 200(0.886)^t$ and the line $Q = 25$. The amount of carbon-14 decays to 25 micrograms at $t \approx 17.180$. Since $t$ is measured in thousands of years, this means in about 17,180 years.

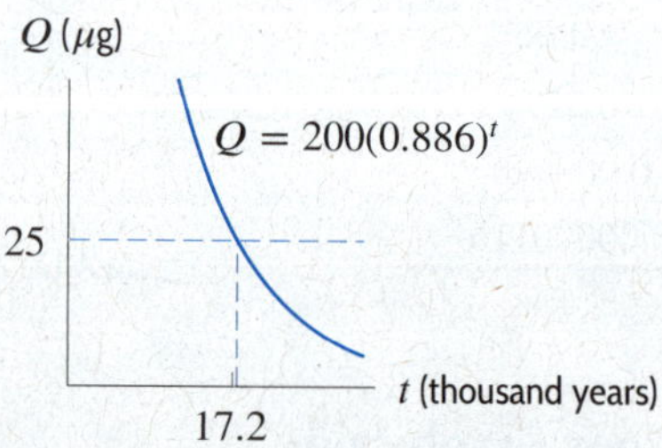


**Figure 4.19:** Solving the equation $200(0.886)^t = 25$

## Finding an Exponential Function for Data

Table 4.10 gives population data for the Houston Metro Area since 1900.[18] In Section 1.6, we saw how to fit a linear function to data, but Figure 4.20 suggests that an exponential function may fit the data better.

Exponential regression on a calculator or computer gives the following exponential function:[19]

$$P = 190(1.034)^t.$$

[18] Based on data from from http://www.census.gov/compendia/statab/2012/tables/12s0020.pdf, accessed January, 2014.

[19] From a graphing calculator or spreadsheet. Depending on the algorithm used, other programs may give different results.

**Table 4.10** *Population (in thousands) of Houston Metro Area, t years after 1900*

| $t$ | $N$ | $t$ | $N$ |
|---|---|---|---|
| 0 | 184 | 60 | 1583 |
| 10 | 236 | 70 | 2183 |
| 20 | 332 | 80 | 3122 |
| 30 | 528 | 90 | 3767 |
| 40 | 737 | 100 | 4715 |
| 50 | 1070 | 110 | 5947 |

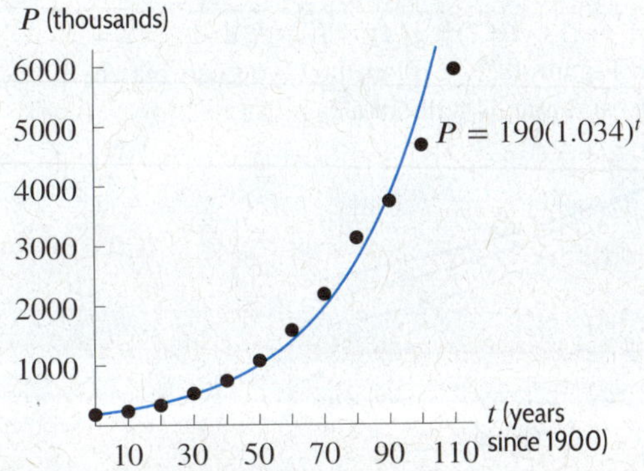


**Figure 4.20**: Houston population data with an exponential model

This equation fits the data well, with $a = 190$ being close to the initial data value of 184.

The graph in Figure 4.20 shows that $b = 1.034$ is a reasonable growth factor, so that, on average, the population was increasing at a rate of about 3.4% per year between 1900 and 2010.

## Optional: Horizontal Asymptotes using Limit Notation

We can also represent the fact that $Q \to 0$ as $t \to \infty$ using the idea of a *limit*. We say that the limit of $f(t)$ is 0 as $t$ goes to infinity, and we use limit notation to write

$$\lim_{t\to\infty} f(t) = 0.$$

For exponential growth, the value of $Q$ approaches zero as $t$ grows more negative. Using limit notation, we write

$$\lim_{t\to-\infty} f(t) = 0.$$

To describe the behavior of a function, $f$, with a horizontal asymptote $y = k$ using limit notation, we write

$$\lim_{x\to\infty} f(x) = k \quad \text{or} \quad \lim_{x\to-\infty} f(x) = k.$$

# Exercises and Problems for Section 4.3

## Exercises

In Exercises 1–4, could the function graphed be exponential? If so, does the growth factor, $b$, satisfy $b > 1$ or $0 < b < 1$?

**1.**

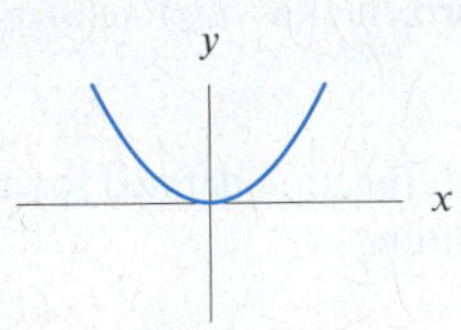


**2.**

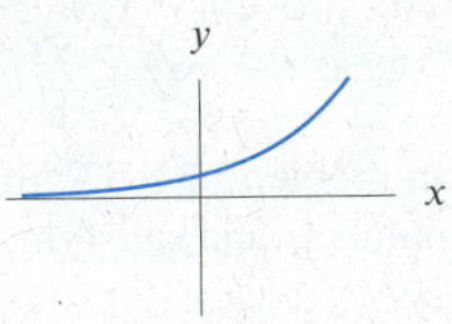


**3.**

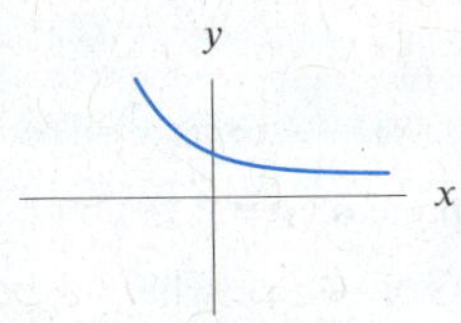


**4.**

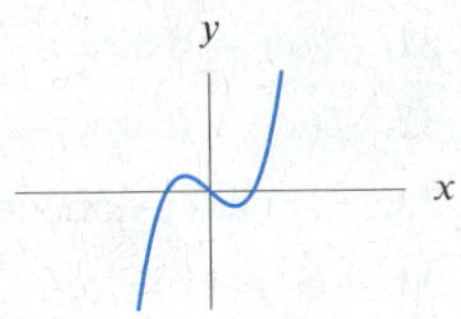


**5.** **(a)** Make a table of values for $f(x) = 2^x$ for $x = -3, -2, -1, 0, 1, 2, 3$.

**(b)** Graph $f(x)$. Describe the graph in words.

**6.** **(a)** Make a table of values for $f(x) = \left(\frac{1}{2}\right)^x$ for $x = -3, -2, -1, 0, 1, 2, 3$.

**(b)** Graph $f(x)$. Describe the graph in words.

**7.** The graphs of $f(x) = (1.1)^x$, $g(x) = (1.2)^x$, and $h(x) = (1.25)^x$ are in Figure 4.21. Explain how you can match these formulas and graphs without a calculator.

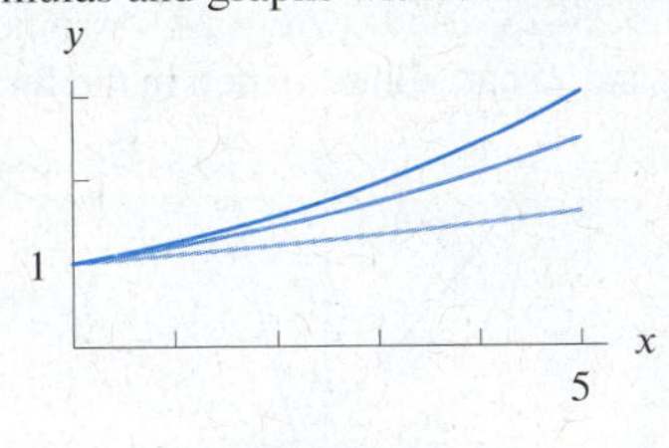


**Figure 4.21**

**8.** The graphs of $f(x) = (0.7)^x$, $g(x) = (0.8)^x$, and $h(x) = (0.85)^x$ are in Figure 4.22. Explain how you can match these formulas and graphs without a calculator.

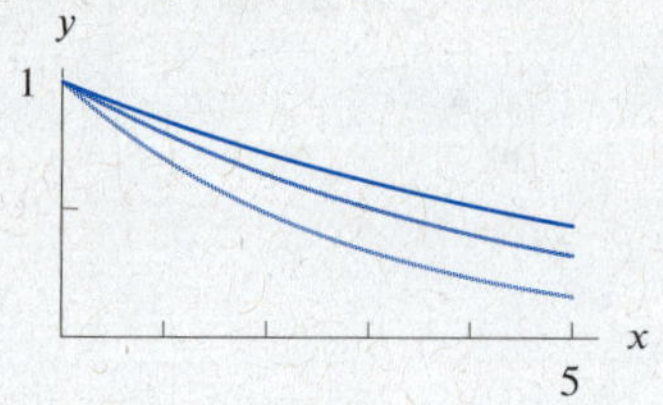


Figure 4.22

In Exercises 9–14, will the graphs of the two functions cross in the first quadrant? Explain your reasoning without using a calculator.

**9.** $f(x) = 10(1.03)^x$; $g(x) = 2(1.05)^x$

**10.** $f(x) = 250(1.2)^x$; $g(x) = 300(1.3)^x$

**11.** $f(x) = 10(1.05)^x$; $g(x) = 9(0.95)^x$

**12.** $f(x) = 5(0.8)^x$; $g(x) = 2(1.05)^x$

**13.** $f(x) = 500(0.8)^x$; $g(x) = 450(0.7)^x$

**14.** $f(x) = 1000(0.9)^x$; $g(x) = 875(0.95)^x$

**15.** The volume of biodegradable material in a compost pile is shown in Figure 4.23, with time $t$ measured in weeks. Use the graph to estimate:

**(a)** The volume after 5 weeks.

**(b)** The number of weeks until the volume is 20 ft$^3$.

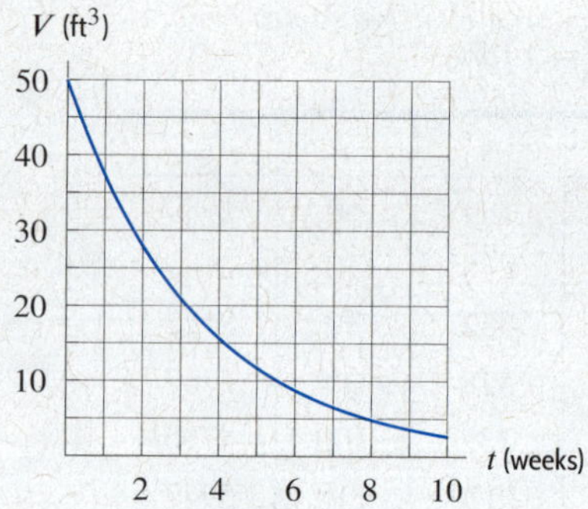


Figure 4.23

**16.** Let $f(x) = 4^x$, $g(x) = 4^x + 2$, and $h(x) = 4^x - 3$. What is the relationship between the graph of $f(x)$ and the graphs of $h(x)$ and $g(x)$?

**17.** Let $f(x) = \left(\frac{1}{3}\right)^x$, $g(x) = \left(\frac{1}{3}\right)^{x+4}$, and $h(x) = \left(\frac{1}{3}\right)^{x-2}$. How do the graphs of $g(x)$ and $h(x)$ compare to the graph of $f(x)$?

**18.** Solve $y = 46(1.1)^x$ graphically for $x$ if $y = 91$.

**19.** Solve $p = 22(0.87)^q$ graphically for $q$ if $p = 10$.

**20.** Solve $4m = 17(2.3)^w$ graphically for $w$ if $m = 12$.

**21.** Solve $P/7 = (0.6)^t$ graphically for $t$ if $P = 2$.

## Problems

**22.** On a graphing calculator, set a window of $-4 \le x \le 4$, $-1 \le y \le 6$ and graph the following functions using several different values of $a$ for each. Restrict to $0 < a < 5$, but include some values of $a$ with $a < 1$. In each case, describe what the graphs have in common.

**(a)** $y = a2^x$ **(b)** $y = 2a^x$

**23.** For which value(s) of $a$ and $b$ is $y = ab^x$ an increasing function? A decreasing function? Concave up?

**24.** What are the domain and range of the exponential function $Q = ab^t$ where $a$ and $b$ are both positive constants?

For Problems 25–28, use Figure 4.24. Assume the equations for $A$, $B$, $C$, and $D$ can all be written in the form $y = ab^t$.

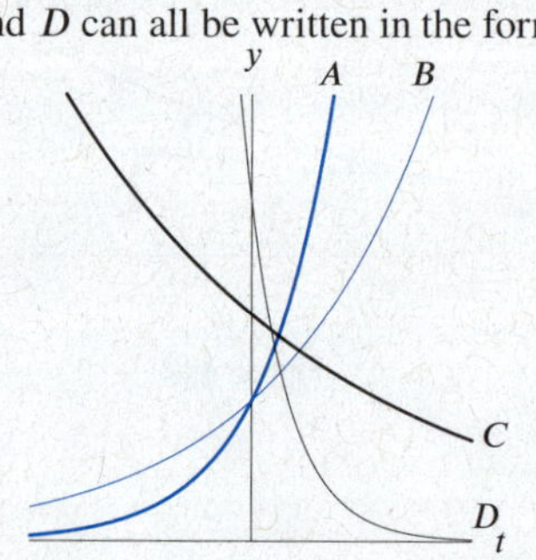


Figure 4.24

**25.** Which function has the largest value for $a$?

**26.** Which two functions have the same value for $a$?

**27.** Which function has the smallest value for $b$?

**28.** Which function has the largest value for $b$?

**29.** If $b > 1$, what is the horizontal asymptote of $y = ab^t$ as $t \to -\infty$?

**30.** If $0 < b < 1$, what is the horizontal asymptote of $y = ab^t$ as $t \to \infty$?

In Problems 31–33, graph $f(x)$, a function defined for all real numbers and satisfying the condition.

**31.** $f(x) \to 3$ as $x \to -\infty$

**32.** $f(x) \to 0$ as $x \to -\infty$ and $f(x) \to -\infty$ as $x \to \infty$

**33.** $f(x)$ has a horizontal asymptote of $y = 5$.

**34.** Let $f(x) = 5 + 3(0.9)^x$. As $x \to \infty$, what happens to $f(x)$? Does this function have a horizontal asymptote, and if so, what is it? Justify your answer both analytically and graphically.

**35.** Consider the exponential functions graphed in Figure 4.25 and the six constants $a, b, c, d, p, q$.

**(a)** Which of these constants are definitely positive?
**(b)** Which of these constants are definitely between 0 and 1?
**(c)** Which of these constants could be between 0 and 1?
**(d)** Which two of these constants are definitely equal?
**(e)** Which one of the following pairs of constants could be equal?

$a$ and $p$ $b$ and $d$ $b$ and $q$ $d$ and $q$

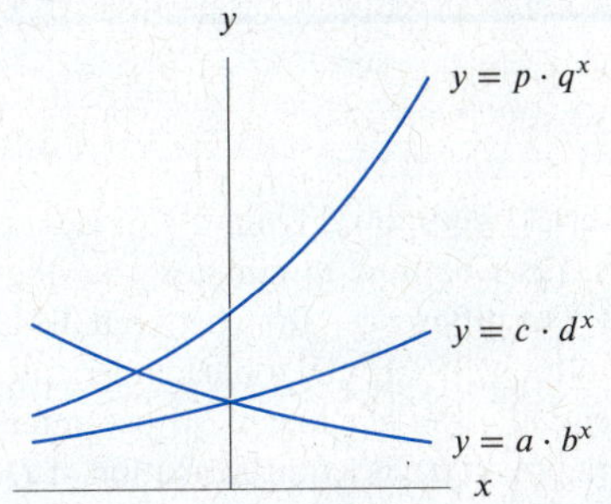


Figure 4.25

**36.** For $t \geq 0$, let $H(t) = 68 + 93(0.91)^t$ give the temperature of a cup of coffee in degrees Fahrenheit $t$ minutes after it is brought to class.

**(a)** Find formulas for $H(t + 15)$ and $H(t) + 15$.
**(b)** Graph $H(t)$, $H(t + 15)$, and $H(t) + 15$.
**(c)** Describe in practical terms a situation modeled by the function $H(t + 15)$. What about $H(t) + 15$?
**(d)** Which function, $H(t+15)$ or $H(t)+15$, approaches the same final temperature as the function $H(t)$? What is that temperature?

Problems 37–38 use Figure 4.26, where $y_0$ is the $y$-coordinate of the point of intersection of the graphs. Describe what happens to $y_0$ if the following changes are made, assuming the other quantities remain the same.

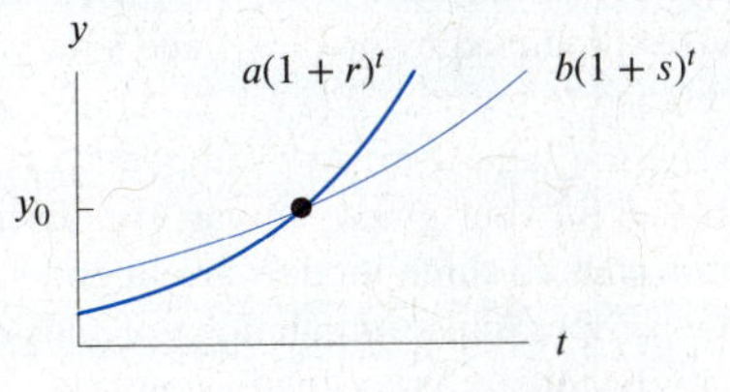


Figure 4.26

**37.** $r$ is increased

**38.** $a$ is increased

**39.** In several European countries, the population is declining. The population of Bulgaria was estimated to be 6.92 million in 2014 and decreasing at 0.83% per year.[20] If this trend continues:

**(a)** Give a formula for the population, $P$, of Bulgaria, in millions, as a function of years, $t$, since 2014.
**(b)** What is the predicted population in 2030?
**(c)** By what percent is the population predicted to drop between 2014 and 2030?
**(d)** To one decimal place, estimate $t$ when the population is predicted to fall below 4 million.

**40.** Let $P = f(t) = 1000(1.04)^t$ be the population of a community in year $t$.

**(a)** Evaluate $f(0)$ and $f(10)$. What do these expressions represent in terms of the population?
**(b)** Using a calculator or a computer, find appropriate viewing windows on which to graph the population for the first 10 years and for the first 50 years. Give the viewing windows you used and sketch the resulting graphs.
**(c)** If the percentage growth rate remains constant, approximately when will the population reach 2500 people?

**41.** You use a calculator to graph $y = 1.04^{5x}$. After correctly entering y = 1.04^(5x), you see the graph in Figure 4.27. A friend graphed the function by entering y = 1.04^5x and said, "The graph is a straight line, so I must have the wrong window." Explain why changing the window will not correct your friend's error.

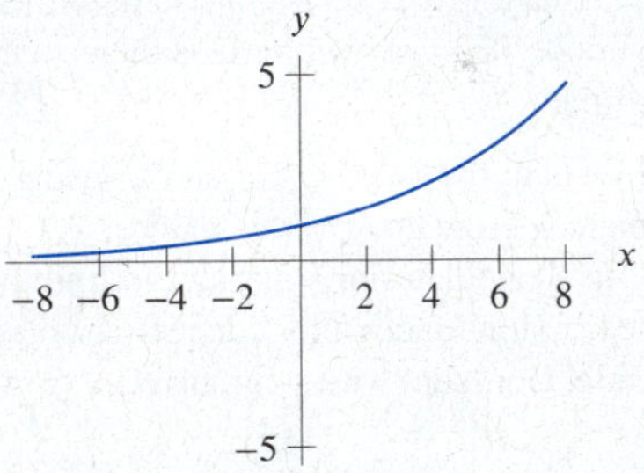


Figure 4.27

**42.** The number of cases of a disease, $y$, is reduced by 10% each year.

**(a)** If there are initially 10,000 cases, express $y$ as a function of $t$, the number of years elapsed.
**(b)** How many cases will there be 5 years from now?
**(c)** How long does it take to reduce the number of cases to 1000?

[20] https://www.cia.gov/library/publications/the-world-factbook/rankorder/2002rank.html, accessed April, 2014.

**43.** The earth's atmospheric pressure, $P$, in terms of height above sea level, is often modeled by an exponential decay function. The pressure at sea level is 1013 millibars and the pressure decreases by 14% for every kilometer above sea level.

**(a)** What is the atmospheric pressure at 50 km?
**(b)** Estimate the altitude $h$ at which the pressure equals 900 millibars.

**44.** The population of a colony of rabbits grows exponentially. The colony begins with 10 rabbits; five years later there are 340 rabbits.

**(a)** Give a formula for the population of the colony of rabbits as a function of the time.
**(b)** Use a graph to estimate how long it takes for the population of the colony to reach 1000 rabbits.

**45.** Let $f$ be a piecewise-defined function given by

$$f(x) = \begin{cases} 2^x, & x < 0 \\ 0, & x = 0 \\ 1 - \frac{1}{2}x, & x > 0. \end{cases}$$

**(a)** Graph $f$ for $-3 \leq x \leq 4$.
**(b)** The domain of $f(x)$ is all real numbers. What is its range?
**(c)** What are the intercepts of $f$?
**(d)** What happens to $f(x)$ as $x \to \infty$ and $x \to -\infty$?
**(e)** Over what intervals is $f$ increasing? Decreasing?

**46.** Table 4.11 shows the concentration of theophylline, a common asthma drug, in the blood stream as a function of time after injection of a 300-mg initial dose.[21] It is claimed that this data set is consistent with an exponential decay model $C = ab^t$ where $C$ is the concentration and $t$ is the time.

**(a)** Estimate the values of $a$ and $b$, using ratios to estimate $b$. How good is this model?
**(b)** Use a calculator or computer to find the exponential regression function for concentration as a function of time. Compare with your answers from part (a).

**Table 4.11**

| Time (hours) | 0 | 1 | 3 | 5 | 7 | 9 |
|---|---|---|---|---|---|---|
| Concentration (mg/l) | 12.0 | 10.0 | 7.0 | 5.0 | 3.5 | 2.5 |

**47.** Table 4.12 shows global wind-energy-generating capacity, $W$ (in megawatts), as a function of the number of years, $t$, since 1995.[22]

**(a)** Plot the data and explain why it is reasonable to approximate these data with an exponential function.
**(b)** Use a calculator or computer to fit an exponential function to these data.
**(c)** What annual percent growth rate does the exponential model show?

**Table 4.12**

| $t$ | 0 | 1 | 2 | 3 | 4 |
|---|---|---|---|---|---|
| $W$ | 4780 | 6070 | 7640 | 10,150 | 13,930 |
| $t$ | 5 | 6 | 7 | 8 | 9 |
| $W$ | 18,450 | 24,930 | 32,037 | 39,664 | 47,760 |

**48.** A hot brick is removed from a kiln and set on the floor to cool. Let $t$ be time in minutes after the brick was removed. The difference, $D(t)$, between the brick's temperature, initially 350°F, and room temperature, 70°F, decays exponentially over at a rate of 3% per minute. The brick's temperature, $H(t)$, is a transformation of $D(t)$. Find a formula for $H(t)$. Compare the graphs of $D(t)$ and $H(t)$, paying attention to the asymptotes.

**49.** Three scientists, working independently of each other, arrive at the following formulas to model the spread of a species of mussel in a system of fresh water lakes:

$$f_1(x) = 3(1.2)^x, \quad f_2(x) = 3(1.21)^x, \quad f_3(x) = 3.01(1.2)^x,$$

where $f_n(x)$, $n = 1, 2, 3$, is the number of individual mussels (in 1000s) predicted by model number $n$ to be living in the lake system after $x$ months have elapsed.

**(a)** Graph these three functions for $0 \leq x \leq 60$, $0 \leq y \leq 40{,}000$.
**(b)** The graphs of these three models do not seem all that different from each other. But do the three functions make significantly different predictions about the future mussel population? To answer this, graph the difference function, $f_2(x) - f_1(x)$, of the population sizes predicted by models 1 and 2, as well as the difference functions, $f_3(x) - f_1(x)$ and $f_3(x) - f_2(x)$. (Use the same window as in part (a).)
**(c)** Based on your graphs in part (b), discuss the assertion that all three models are in good agreement as far as long-range predictions of mussel population are concerned. What conclusions can you draw about exponential functions in general?

**50.** If the exponential function $ab^x$ has the property that $\lim_{x\to\infty} ab^x = 0$, what can you say about the value of $b$?

[21]Based on D. N. Burghes, I. Huntley, and J. McDonald, *Applying Mathematics* (Ellis Horwood, 1982).
[22]The Worldwatch Institute, *Vital Signs* 2005 (New York: W.W. Norton & Company, 2005), p. 35.

In Problems 51–52, graph $f(x)$, a function defined for all real numbers and satisfying the condition.

**51.** $\lim_{x\to\infty} f(x) = 5$

**52.** $\lim_{x\to-\infty} f(x) = 2$ and $\lim_{x\to\infty} f(x) = -1$

In Problems 53–54, assume that all important features are shown in the graph of $y = f(x)$. Estimate

**(a)** $\lim_{x\to-\infty} f(x)$ **(b)** $\lim_{x\to\infty} f(x)$

**53.**

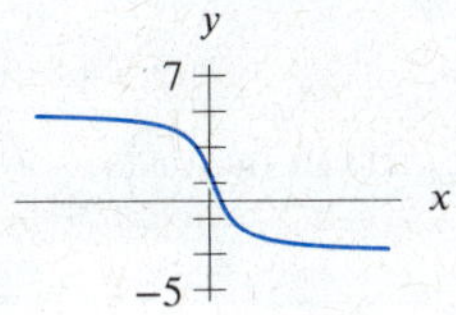


**54.**

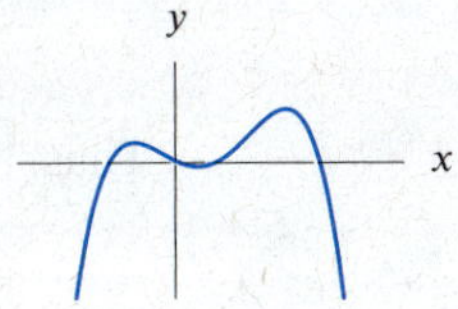


## 4.4 APPLICATIONS TO COMPOUND INTEREST

What is the difference between a bank account that pays 12% interest once per year and one that pays 1% interest every month?[23] Imagine we deposit \$1000 into the first account. Then, after 1 year, we have (assuming no other deposits or withdrawals)

$$\$1000(1.12) = \$1120.$$

But if we deposit \$1000 into the second account, then after 1 year, or 12 months, we have

$$\$1000\underbrace{(1.01)(1.01)\ldots(1.01)}_{\text{12 months of 1\% monthly interest}} = 1000(1.01)^{12} = \$1126.83.$$

Thus, we earn \$6.83 more in the second account than in the first. To see why this happens, notice that the 1% interest we earn in January itself earns interest at a rate of 1% per month. Similarly, the 1% interest we earn in February earns interest, and so does the interest earned in March, April, May, and so on. The extra \$6.83 comes from interest earned on interest. This effect is known as *compounding*. We say that the first account earns 12% interest per year *compounded annually* and the second account earns 12% interest per year *compounded monthly*.

### Nominal Versus Effective Rate

The expression 12% interest per year compounded monthly means that interest is added twelve times a year and that $12\%/12 = 1\%$ of the current balance is added each time. We refer to the 12% per year as the *nominal annual rate*, or *nominal rate* (nominal means "in name only"). When the interest is compounded more frequently than once a year, the account effectively earns more than the nominal rate. Thus, we distinguish between nominal rate and *effective annual rate*, or *effective rate*. The effective annual rate tells you how much interest the investment actually earns in a year. In the US, the effective annual rate is sometimes called the APY (annual percentage yield).

**Example 1** What are the nominal and effective annual rates of an account paying 12% interest per year, compounded annually? Compounded monthly?

**Solution** Since an account paying 12% interest per year, compounded annually, grows by exactly 12% in one year, we see that its nominal rate is the same as its effective rate: both are 12%.

The account paying 12% interest per year, compounded monthly, also has a nominal rate of 12%. On the other hand, since it pays 1% interest every month, after 12 months, its balance increases by a factor of

$$\underbrace{(1.01)(1.01)\ldots(1.01)}_{\text{12 months of 1\% monthly growth}} = 1.01^{12} = 1.1268250.$$

Thus, effectively, the account earns 12.683% interest in a year, so its effective rate is 12.683%.

[23]Interest rates are currently much lower.

**Example 2** What is the effective annual rate of an account that pays interest at the nominal rate of 6% per year, compounded daily? Compounded hourly?

**Solution** Since there are 365 days in a year, daily compounding pays interest at the rate of

$$\frac{6\%}{365} = 0.0164384\% \text{ per day.}$$

Thus, the daily growth factor is

$$1 + \frac{0.06}{365} = 1.000164384.$$

If at the beginning of the year the account balance is $P$, after 365 days the balance is

$$P \cdot \underbrace{\left(1 + \frac{0.06}{365}\right)^{365}}_{\substack{\text{365 days of} \\ \text{0.0164384\% daily interest}}} = P \cdot (1.061831).$$

Thus, this account earns interest at the effective annual rate of 6.1831%.

Notice that daily compounding results in a higher effective rate than yearly compounding (6.1831% versus 6%), because with daily compounding the interest has the opportunity to earn interest.

If interest is compounded hourly, since there are $24 \cdot 365$ hours in a year, the balance at year's end is

$$P \cdot \left(1 + \frac{0.06}{24 \cdot 365}\right)^{24 \cdot 365} = P \cdot (1.061836).$$

The effective rate is now 6.1836% instead of 6.1831%—that is, just slightly better than the effective rate with daily compounding. The effective rate increases with the frequency of compounding.

In the previous examples, we computed the growth factor for one year. We now compute the growth factor over $t$ years.

**Example 3** At the beginning of the year you deposit $P$ dollars in an account paying interest at the nominal rate of 4% per year compounded quarterly. By what factor does $P$ grow in 3 years?

**Solution** Compounding quarterly pays interest at the rate of

$$\frac{4\%}{4} = 1\% \text{ per quarter.}$$

Thus, the quarterly growth factor is $1+0.01 = 1.01$. At the beginning of the year the account balance is $P$. After one year the balance is

$$P \cdot \underbrace{\left(1 + \frac{0.04}{4}\right)^{4}}_{\substack{\text{4 quarters (1 year) of} \\ \text{1\% quarterly interest}}} = P \cdot (1.0406).$$

So after three years the balance is

$$P \cdot \left(1 + \frac{0.04}{4}\right)^{4} \cdot \left(1 + \frac{0.04}{4}\right)^{4} \cdot \left(1 + \frac{0.04}{4}\right)^{4} = P \cdot \underbrace{\left(1 + \frac{0.04}{4}\right)^{4 \cdot 3}}_{\substack{\text{3 years of} \\ \text{1\% quarterly interest}}} = P \cdot (1.126825).$$

After three years the initial account balance $P$ grows by a factor of 1.126825.

To summarize:

If interest at an annual rate of $r$ is compounded $n$ times a year, then $r/n$ times the current balance is added $n$ times a year. Therefore, with an initial deposit of $\$P$, the balance $t$ years later is

$$B = P \cdot \left(1 + \frac{r}{n}\right)^{nt}.$$

Note that $r$ is the nominal rate; for example, $r = 0.05$ if the annual rate is 5%.

## Exercises and Problems for Section 4.4

### Exercises

1. What is the effective annual yield if a deposit of \$1000 grows to \$3500 in 15 years?
2. Suppose \$1000 is invested in an account yielding interest at a nominal rate of 8% per year. Find the balance three years later if the interest is compounded
   (a) Monthly (b) Weekly
   (c) Daily
3. An account pays interest at a nominal rate of 8% per year. Find the effective annual yield if interest is compounded
   (a) Monthly (b) Weekly
   (c) Daily
4. An investment grows by 5% per year for 20 years. By what percent does it increase over the 20-year period?
5. An investment decreases by 50% over an 8-year period. At what effective annual percent rate does it decrease?
6. A bank pays interest at the nominal rate of 1.3% per year. What is the effective annual rate if compounding is:
   (a) Annual (b) Monthly

In Exercises 7–10, what is the balance after 1 year if an account containing \$500 earns the stated yearly nominal interest, compounded

(a) Annually? (b) Weekly (52 weeks per year)?

(c) Every minute (525,600 per year)?

7. 1% 8. 3% 9. 5% 10. 8%

In Exercises 11–14, what are the nominal and effective annual rates for an account paying the stated annual interest, compounded

(a) Annually? (b) Quarterly?
(c) Daily?

11. 1% 12. 100% 13. 3% 14. 6%

### Problems

15. An investment grows by 30% over a 5-year period. What is its effective annual percent growth rate?
16. An investment decreases by 60% over a 12-year period. At what effective annual percent rate does it decrease?
17. An investment grows by 3% per year for 10 years. By what percent does it increase over the 10-year period?
18. A sum of \$850 is invested for 10 years and the interest is compounded quarterly. There is \$1000 in the account at the end of 10 years. What is the nominal annual rate?
19. If you need \$25,000 six years from now, what is the minimum amount of money you need to deposit into a bank account that pays 5% annual interest, compounded:
   (a) Annually (b) Monthly (c) Daily
   (d) Your answers get smaller as the number of times of compounding increases. Why is this?
20. If the balance, $M$, at time $t$ in years, of a bank account that compounds its interest payments monthly is given by
   $$M = M_0(1.07763)^t,$$
   (a) What is the effective annual rate for this account?
   (b) What is the nominal annual rate?
21. Suppose \$3000 was invested into one of five accounts and $t$ is time in years. For each verbal description (i)–(v), state which formulas (a)–(e) could represent it.
   (a) $B = 3000(1.2)^t$ (b) $B = 3000(1.12)^t$
   (c) $B = 3000(1.06)^{2t}$ (d) $B = 3000(1.06)^{t/2}$
   (e) $B = 3000(1.03)^{4t}$

(i) This investment earned 12% per year, compounded annually.
(ii) This investment earned more than 1% each month.
(iii) This investment earned 12% per year, compounded twice annually.
(iv) This investment earned less than 3% each quarter.
(v) This investment earned more than 6% every 6 months.

**22.** Without any calculation, describe in words what the following formulas tell you about the value of the investments they describe. The units are dollars and years.

**(a)** $V = 1500(1.077)^t$
**(b)** $V = 9500(0.945)^t$
**(c)** $V = 1000 \cdot 3^{t/5}$
**(d)** $V = 500\left(1 + \frac{0.04}{12}\right)^{12t}$

## 4.5 THE NUMBER *e*

An irrational number, introduced by Euler[24] in 1727, is so important that it is given a special name, $e$. Its value is approximately $e \approx 2.71828\ldots$. It is often used for the base, $b$, of the exponential function. Base $e$ is called the *natural base*. This may seem mysterious, as what could possibly be natural about using an irrational base such as $e$? The answer is that the formulas of calculus are much simpler if $e$ is used as the base for exponentials. Some of the remarkable properties of the number $e$ are introduced in Problems 51–52 on page 176. Since $2 < e < 3$, the graph of $Q = e^t$ lies between the graphs of $Q = 3^t$ and $Q = 2^t$. See Figure 4.28.

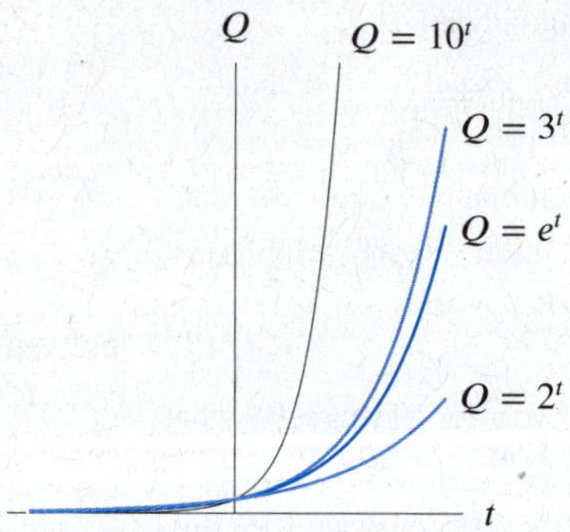


Figure 4.28: Graphs of exponential functions with various bases

### Exponential Functions with Base *e*

Since any positive base $b$ can be written as a power of $e$:

$$b = e^k,$$

any exponential function $Q = ab^t$ can be rewritten in terms of $e$:

$$Q = ab^t = a\left(e^k\right)^t = ae^{kt}.$$

If $b > 1$, then $k$ is positive; if $0 < b < 1$, then $k$ is negative. The constant $k$ is called the *continuous growth rate*. In general:

For the exponential function $Q = ab^t$, the **continuous growth rate**, $k$, is given by solving $e^k = b$. Then

$$Q = ae^{kt}.$$

If $a$ is positive,

- If $k > 0$, then $Q$ is increasing.
- If $k < 0$, then $Q$ is decreasing.

[24]Leonhard Euler (1707–1783), a Swiss mathematician, introduced $e$, $f(x)$ notation, $\pi$, and $i$ (for $\sqrt{-1}$).

The value of the continuous growth rate, $k$, may be given as a decimal or a percent. If $t$ is in years, for example, then the units of $k$ are given per year; if $t$ is in minutes, then $k$ is given per minute.

**Example 1** Give the continuous growth rate of each of the following functions and graph each function:

$$P = 5e^{0.2t}, \qquad Q = 5e^{0.3t}, \qquad \text{and} \quad R = 5e^{-0.2t}.$$

**Solution** The function $P = 5e^{0.2t}$ has a continuous growth rate of 20%, and $Q = 5e^{0.3t}$ has a continuous 30% growth rate. The function $R = 5e^{-0.2t}$ has a continuous growth rate of −20%. The negative sign in the exponent tells us that $R$ is decreasing instead of increasing.

Because $a = 5$ in all three formulas, all three functions cross the vertical axis at 5. Note that the graphs of these functions in Figure 4.29 have the same shape as the exponential functions in Section 4.3. They are concave up and have horizontal asymptotes of $y = 0$. (Note that $P \to 0$ and $Q \to 0$ as $t \to -\infty$, whereas $R \to 0$ as $t \to \infty$.)

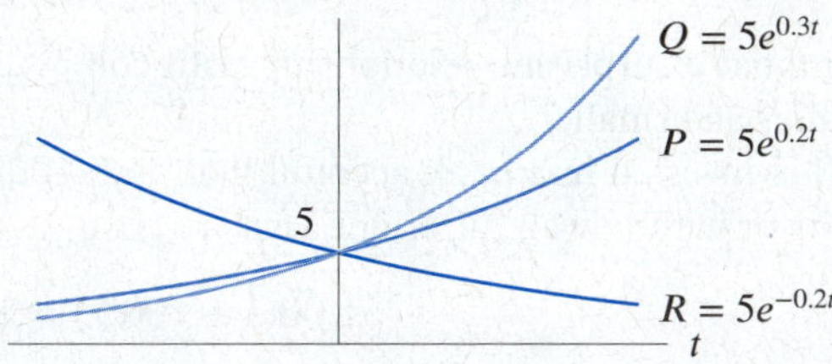


**Figure 4.29:** Exponential functions with different continuous growth rates

**Example 2** A population increases from 7.3 million at a continuous rate of 2.2% per year. Write a formula for the population, and estimate graphically when the population reaches 10 million.

**Solution** We express the formula in base $e$ since the continuous growth rate is given. If $P$ is the population (in millions) in year $t$, then

$$P = 7.3e^{0.022t}.$$

See Figure 4.30. We see that $P = 10$ when $t \approx 14.3$. Thus, it takes about 14.3 years for the population to reach 10 million.

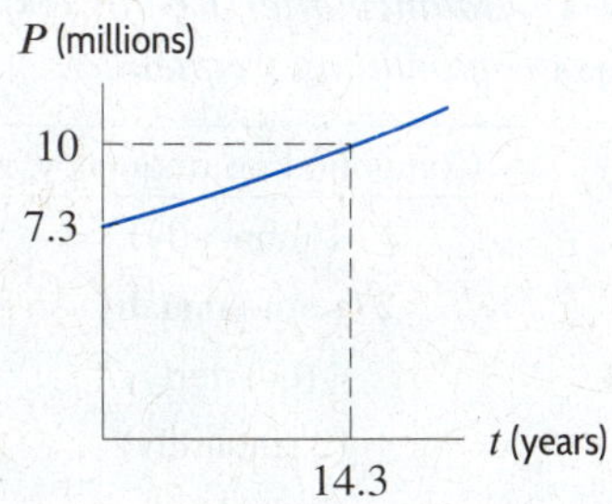


**Figure 4.30:** Estimating when the population reaches 10 million

**Example 3** Caffeine leaves the body at a continuous rate of 17% per hour. How much caffeine is left in the body 8 hours after drinking a cup of coffee containing 100 mg of caffeine?

**Solution** If $A$ is the amount of caffeine in the body $t$ hours after drinking the coffee, then

$$A = 100e^{-0.17t}.$$

Note that the continuous growth rate is −17% since $A$ is decreasing. After 8 hours, we have $A = 100e^{-0.17(8)} = 25.67$ mg.

### The Difference Between Annual and Continuous Growth Rates

If $P = P_0(1.07)^t$, with $t$ in years, we say that $P$ is growing at an *annual* rate of 7%. If $P = P_0e^{0.07t}$, with $t$ in years, we say that $P$ is growing at a *continuous* rate of 7% per year. Since $e^{0.07} = 1.0725\ldots$, we can rewrite $P_0e^{0.07t} = P_0(1.0725)^t$. In other words, a 7% continuous rate and a 7.25% annual rate generate the same increases in $P$. We say the two rates are equivalent.

We can check that $e^{0.0677} = 1.07\ldots$, so a 7% annual growth rate is equivalent to a 6.77% continuous growth rate. The continuous growth rate is always smaller than the equivalent annual rate.

The bank account example at the start of Section 4.4 illustrates why a quantity growing at continuous rate of 7% per year increases faster than a quantity growing at an annual rate of 7%: More frequent compounding earns more interest. We can think of continuous interest as earning interest very frequently.

## Connection: The Number $e$ and Compound Interest

The number $e$ has a surprising relationship with compound interest when the compounding period is made smaller and smaller.

If \$1.00 is invested in a bank account that pays 100% interest once a year, then, assuming no other deposits or withdrawals, after one year we have

$$\$1.00(1 + 100\%) = \$2.00.$$

Now we decrease the compounding period. Suppose \$1.00 is invested in a bank account that pays 100% nominal interest compounded $n$ times a year; then

$$\text{Balance after one year} = \$1.00\left(1 + \frac{100\%}{n}\right)^n.$$

How large can the balance be? Table 4.13 shows the balance after one year as the interest is calculated more and more frequently. Remarkably, as $n$ increases, the balance approaches $e = 2.71828182\ldots$.

Assuming that the initial balance of \$1 is growing at a continuous rate of 100%, the balance after one year would be $1 \cdot e^1 = e$ dollars, so the surprising relationship arises from the fact that by increasing the compounding frequency we get the continuous growth introduced earlier.

**Table 4.13** *Balance after 1 year, 100% nominal interest, various compounding frequencies*

| Compounding frequency, $n$ | Approximate balance |
|---|---|
| 1 (annually) | \$2.00 |
| 2 (semi-annually) | \$2.25 |
| 4 (quarterly) | \$2.441406 |
| 12 (monthly) | \$2.613035 |
| 365 (daily) | \$2.714567 |
| 8760 (hourly) | \$2.718127 |
| 525,600 (each minute) | \$2.718279 |
| 31,536,000 (each second) | \$2.718282 |

In Example 2 of Section 4.4 we calculated the effective interest rates for two accounts with a 6% per year nominal interest rate, but different compounding periods. We see that the account with more frequent compounding earns a higher effective rate, though the increase is small. Compounding more and more frequently—every minute or every second or many times per second—increases the effective rate still further. However, there is a limit to how much an account can earn by increasing the frequency of compounding.

**Table 4.14** *Effect of increasing the frequency of compounding, 6% nominal interest*

| Compounding frequency | Annual growth factor | Effective annual rate |
|---|---|---|
| Annually | 1.0600000 | 6% |
| Monthly | 1.0616778 | 6.16778% |
| Daily | 1.0618313 | 6.18313% |
| Hourly | 1.0618363 | 6.18363% |
| ⋮ | ⋮ | ⋮ |
| Continuously | $e^{0.06} \approx 1.0618365$ | 6.18365% |

Table 4.14 shows several compounding periods with their annual growth factors and effective annual rates. As the compounding periods become shorter, we discover that the growth factor approaches $e^{0.06}$. Using a calculator, we check that the final value for the annual growth factors in Table 4.14 is given by

$$e^{0.06} \approx 1.0618365,$$

If an account with a 6% nominal interest rate per year delivers an annual growth factor of $e^{0.06}$, we say that the interest has been *compounded continuously*. In general:

> If interest on an initial deposit of $\$P$ is *compounded continuously* at a nominal rate of $r$ per year, the balance $t$ years later can be calculated using the formula
>
> $$B = Pe^{rt}.$$
>
> For example, if the nominal rate is 6%, then $r = 0.06$.

It is important to realize that the functions $B = Pe^{0.06t}$ and $B = P(1.0618365)^t$ both give the balance in a bank account growing at a continuous rate of 6% per year. These formulas both represent the *same* exponential function—they just describe it in different ways.[25]

**Example 4** An investment has a yield equivalent to a 2.323% continuous yearly rate. Find the effective annual rate.

**Solution** Since $e^{0.02323} = 1.0235$, the effective annual rate is 2.35%. As expected, the effective annual rate is larger than the continuous yearly rate.

**Example 5** Which is better: An account that pays 8% annual interest compounded quarterly or an account that pays 7.95% annual interest compounded continuously?

**Solution** To decide, we compare effective annual rates. The account that pays 8% interest compounded quarterly pays 2% interest 4 times a year. Thus, in one year the balance is

$$P(1.02)^4 \approx P(1.08243),$$

which means the effective annual rate is 8.243%.

The account that pays 7.95% interest compounded continuously has a year-end balance of

$$Pe^{0.0795} \approx P(1.08275),$$

so the effective annual rate is 8.275%. Thus, 7.95% compounded continuously pays more than 8% compounded quarterly.

[25] Actually, this is not precisely true, because we rounded off when we found $b = 1.0618365$. However, we can find $b$ to as many digits as we want, and to this extent the two formulas are the same.

## Exercises and Problems for Section 4.5

### Skill Refresher

In Exercises S1–S4, using a calculator, approximate to 3 decimal places.

**S1.** $e^{0.07}$ **S2.** $10e^{-0.14}$ **S3.** $\dfrac{2}{\sqrt[3]{e}}$ **S4.** $e^{3e}$

In Exercises S5–S8, evaluate the functions for $t = 0$ and $t = 4$.

**S5.** $f(t) = 2.3e^{0.3t}$

**S6.** $g(t) = 4.2e^{-0.12t}$

**S7.** $h(t) = 153 + 8.6e^{0.43t}$

**S8.** $k(t) = 289 - 4.7e^{-0.0018t}$

In Exercises S9–S15, write the function in the form $f(t) = ae^{kt}$.

**S9.** $f(t) = \left(3e^{0.04t}\right)^3$ **S10.** $g(z) = 5e^{7z} \cdot e^{4z} \cdot 3e^{z}$

**S11.** $Q(t) = e^{7-3t}$ **S12.** $Q(t) = \sqrt{e^{3+6t}}$

**S13.** $m(x) = \dfrac{7e^{0.2x}}{\sqrt{3e^{x}}}$ **S14.** $P(t) = \left(2\sqrt[3]{e^{5t}}\right)^4$

**S15.** $H(r) = \dfrac{\left(e^{0.4r}\right)^2}{6e^{0.15r}}$

### Exercises

**1.** Without a calculator, match the functions $y = 2^x$, $y = 3^x$, and $y = e^x$ with the graphs in Figure 4.31.

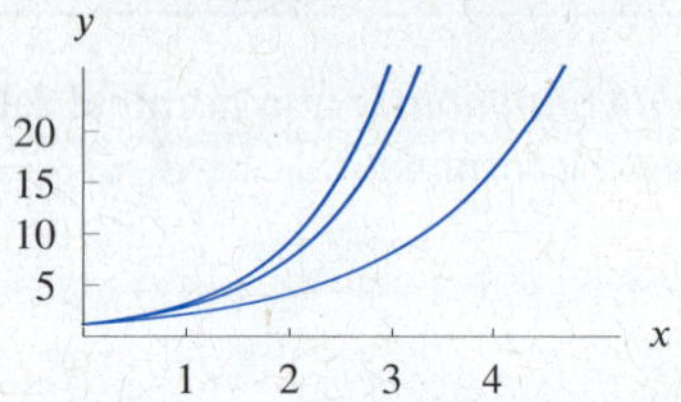


Figure 4.31

**2.** Without a calculator, match the functions $y = e^x$, $y = 2e^x$, and $y = 3e^x$ to the graphs in Figure 4.32.

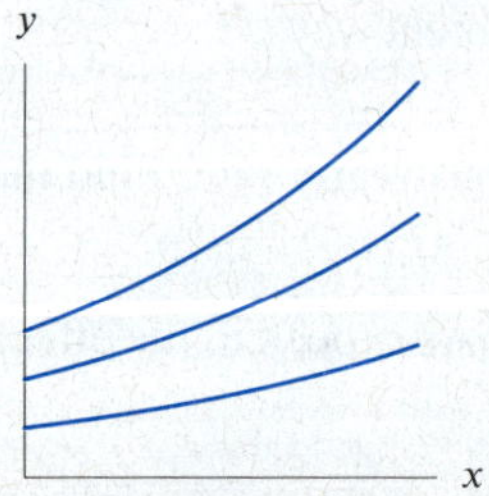


Figure 4.32

**3.** Without a calculator, match each formula to one of the graphs (I)–(IV) in Figure 4.33.

**(a)** $e^{-0.01t}$ **(b)** $e^{0.05t}$ **(c)** $e^{-0.10t}$ **(d)** $e^{0.20t}$

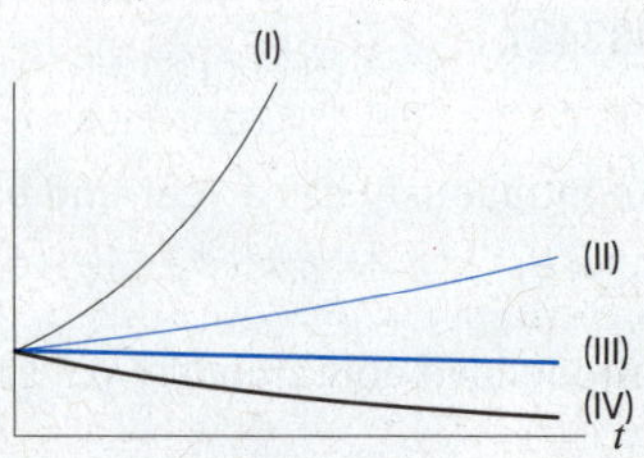


Figure 4.33

**4.** Without a calculator, match the functions (a)–(d) with the graphs (I)–(IV) in Figure 4.34.

**(a)** $e^{0.25t}$ **(b)** $(1.25)^t$ **(c)** $(1.2)^t$ **(d)** $e^{0.3t}$

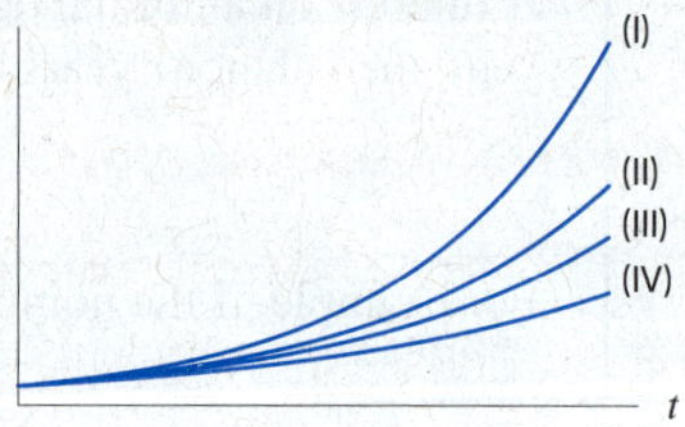


Figure 4.34

**5.** Without a calculator, match the functions (a)–(d) with the graphs (I)–(IV) in Figure 4.35.

**(a)** $1.5^x$ **(b)** $e^{0.45x}$ **(c)** $e^{0.47x}$ **(d)** $e^{0.5x}$

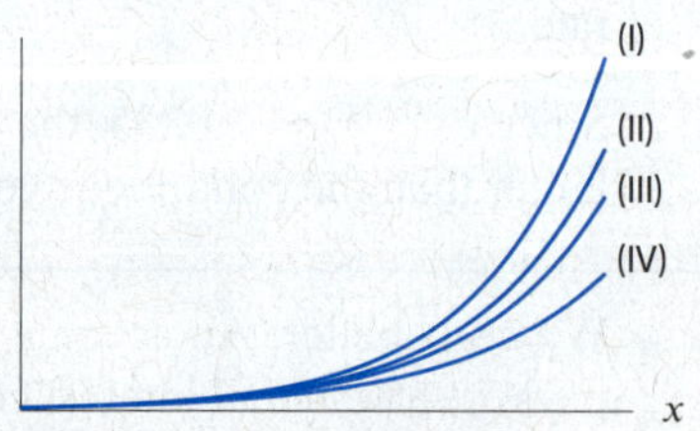


Figure 4.35

**6.** Without a calculator, match the functions $y = e^x$, $y = e^{-x}$, and $y = -e^x$ to the graphs in Figure 4.36.

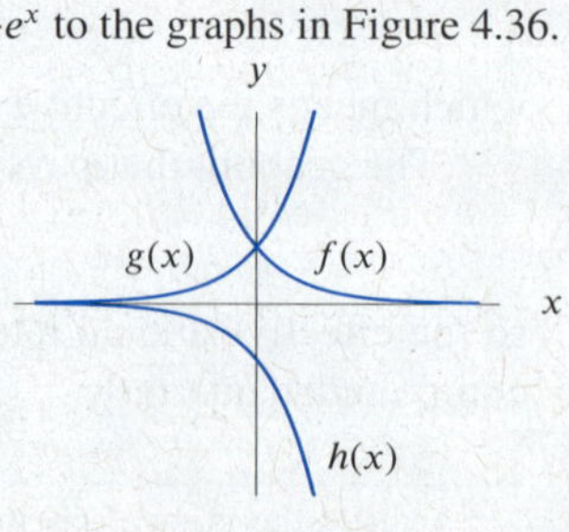


Figure 4.36

**7.** Without a calculator, match the functions (a)–(d) with the graphs (I)–(IV) in Figure 4.37.

**(a)** $y = e^x$ **(b)** $y = e^{-x}$
**(c)** $y = e^{-2x}$ **(d)** $y = e^{-3x}$

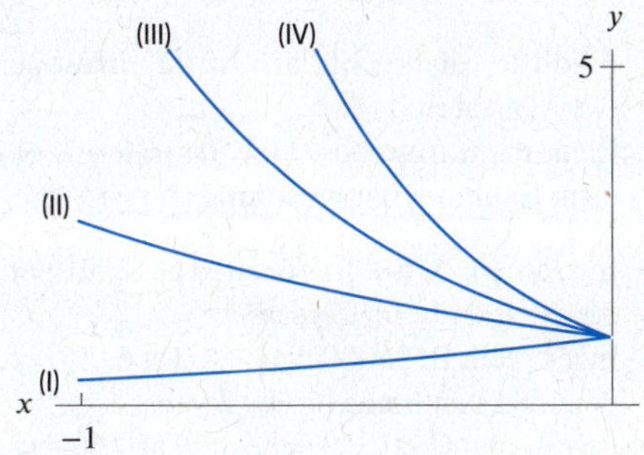


Figure 4.37

In Exercises 8–13, the expression shows how the quantity $Q$ is changing over time $t$.

**(a)** What is the quantity at time $t = 0$?
**(b)** Is the quantity increasing or decreasing over time?
**(c)** What is the percent per unit time growth or decay rate?
**(d)** Is the growth rate continuous?

**8.** $Q = 50e^{1.05t}$ **9.** $Q = 2.7(0.12)^t$

**10.** $Q = 0.01e^{-0.2t}$ **11.** $Q = 158(1.137)^t$

**12.** $Q = 25e^{0.032t}$ **13.** $Q = 2^t$

**14.** Without a calculator, arrange the following quantities in ascending order:

**(a)** $3^{2.2}, e^{2.2}, (\sqrt{2})^{2.2}$ **(b)** $3^{-2.2}, e^{-2.2}$

## Problems

In Problems 15–17, an initial quantity $Q_0$ and a growth rate are given. Give a formula for quantity $Q$ as a function of time $t$, and find the value of the quantity at $t = 10$, if we assume that the growth rate is:

**(a)** Not continuous **(b)** Continuous

**15.** $Q_0 = 8$; growth rate of 12% per unit time

**16.** $Q_0 = 100$; growth rate of 5% per unit time

**17.** $Q_0 = 500$; decay rate of 7% per unit time

**18.** The following formulas each describe the size of an animal population, $P$, in $t$ years since the start of the study. Describe the growth of each population in words.

**(a)** $P = 200(1.028)^t$ **(b)** $P = 50e^{-0.17t}$
**(c)** $P = 1000(0.89)^t$ **(d)** $P = 600e^{0.20t}$
**(e)** $P = 2000 - 300t$ **(f)** $P = 600 + 50t$

**19.** A population of 3.2 million grows at a constant percentage rate.

**(a)** What is the population one century later if there is:
(i) An annual growth rate of 2%
(ii) A continuous growth rate of 2% per year

**(b)** Explain how you can tell which of the answers is larger before doing the calculations.

**20.** A town starts with 10,000 residents and grows at a continuous rate of 4% per year.

**(a)** Write a formula for the population of the town after $t$ years.
**(b)** What is the population of the town after 5 years have passed?
**(c)** Find the annual growth rate of the population of the town.

**21.** A population is 25,000 in year $t = 0$ and grows at a continuous rate of 7.5% per year.

**(a)** Find a formula for $P(t)$, the population in year $t$.
**(b)** By what percent does the population increase each year? Why is this more than 7.5%?

**22.** A town has population 3000 people at year $t = 0$. Write a formula for the population, $P$, in year $t$ if the town

**(a)** Grows by 200 people per year.
**(b)** Grows by 6% per year.
**(c)** Grows at a continuous rate of 6% per year.
**(d)** Shrinks by 50 people per year.
**(e)** Shrinks by 4% per year.
**(f)** Shrinks at a continuous rate of 4% per year.

**23.** A population grows from its initial level of 22,000 at a continuous growth rate of 7.1% per year.

**(a)** Find a formula for $P(t)$, the population in year $t$.
**(b)** By what percent does the population increase each year?

**24.** City A has a continuous growth rate of 15% per year, and City B has an annual growth rate of 16% per year. Which city is growing the fastest? Justify your answer.

**25.** A bacteria colony that starts with 100 bacteria has a continuous growth rate of 25% per hour.

**(a)** How many bacteria are in the colony after 4 hours?
**(b)** By what hourly percentage does the population of the colony grow?

**26.** Without any calculation, describe in words what the following formulas tell you about the value of three different investments. The units are dollars and years.

**(a)** $V = 1000e^{0.115t}$ **(b)** $V = 1000 \cdot 2^{t/6}$
**(c)** $V = 1000(1.122)^t$

**27.** An investment of \$7000 increases in value at a continuous rate of 5.2% per year. What is its value in 7 years?

**28.** From time $t = 0$, with $t$ in years, a \$1200 deposit in a bank account grows according to the formula

$$B = 1200e^{0.03t}.$$

**(a)** What is the balance in the account at the end of 100 years?

**(b)** When does the balance first go over \$50,000?

**29.** Calculate the amount of money in a bank account if \$5000 is deposited for 10 years at an interest rate of

**(a)** 2% annually

**(b)** 2% continuously per year

**30.** Find the effective annual yield and the continuous growth rate if $Q = 5500\, e^{0.19t}$.

**31.** At time $t$ in years, the value, $V$, of an investment of \$1000 is given by $V = 1000e^{0.02t}$. When is the investment worth \$3000?

**32.** How long does it take an investment to double if it grows according to the formula $V = 537e^{0.015t}$? Assume $t$ is in years.

**33.** The same amount of money is deposited into two different bank accounts paying the same nominal rate, one compounded annually and the other compounded continuously. Which curve in Figure 4.38 corresponds to which compounding method? What is the initial deposit?

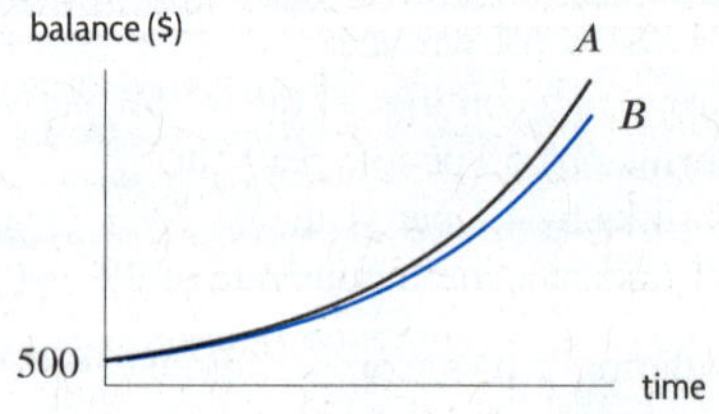


Figure 4.38

**34.** Find the effective annual rate if \$5000 is deposited at 2% annual interest, compounded continuously.

**35.** An investment of \$500 earns interest at a 6.75% continuous annual rate. Give a formula $V = ab^t$ for the investment's value in year $t$, and state the values of $a$, $b$, and the annual growth rate $r$.

**36.** A bank account pays 3% annual interest. As the number of compounding periods increases, the effective interest rate earned also rises.

**(a)** Find the annual interest rate earned by the account if the interest is compounded:

(i) Quarterly (ii) Monthly
(iii) Weekly (iv) Daily

**(b)** Evaluate $e^{0.03}$, where $e = 2.71828\ldots$ Explain what your result tells you about the bank account.

**37.** Which is better, an account paying 5.3% interest compounded continuously or an account paying 5.5% interest compounded annually? Justify your answer.

**38.** Three different investments are given.

**(a)** Find the balance of each of the investments after a two-year period.

**(b)** Rank them from best to worst in terms of rate of return. Explain your reasoning.

- Investment A: \$875 deposited at 13.5% per year compounded daily for 2 years.
- Investment B: \$1000 deposited at 6.7% per year compounded continuously for 2 years.
- Investment C: \$1050 deposited at 4.5% per year compounded monthly for 2 years.

**39.** Rank the following three bank-deposit options from best to worst.

- Bank A: 2% compounded daily
- Bank B: 2.1% compounded monthly
- Bank C: 2.05% compounded continuously

**40.** Which is larger after 5 years: an investment of \$1000 earning 5% per year compounded monthly or an investment of \$1100 earning 4% per year compounded continuously? Which is larger after 10 years? Justify your answers.

**41.** One investment pays 5% interest compounded annually and another investment has a yield equivalent to 5% interest compounded continuously. Given a deposit of \$10,000, what is the difference in the balance between the two banks in 8 years?

**42.** An investment of \$1000 earns 8% interest compounded continuously. An investment of \$1500 earns 6% interest compounded annually. Figure 4.39 shows the balance of the two investments over time.

**(a)** Which graph goes with which investment?

**(b)** Use a graph of the two functions to estimate the time until the balances are equal.

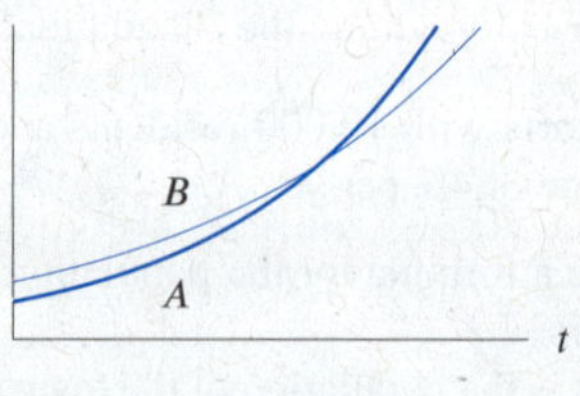


Figure 4.39

**43.** The GDP of Chile was 145.8 billion dollars in 2007 and was growing at a continuous rate of 5.1% per year.[26]

**(a)** Find a formula for $G$, the GDP of Chile in billion dollars, as a function of $t$, in years since 2007.
**(b)** By what percent does the GDP increase each year?
**(c)** Use your answer to part (b) to write a formula for $G$ as a function of $t$ using the form $G = ab^t$
**(d)** Graph your answers to part (a) and part (c) on the same axes. Explain what you see.

**44.** What can you say about the value of the constants $a, k, b, l$ in Figure 4.40?

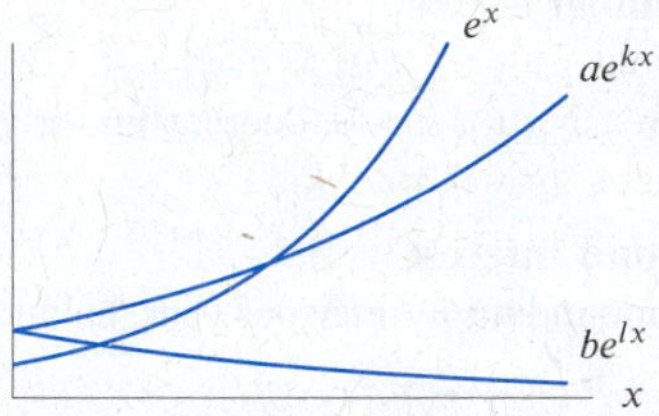


Figure 4.40

**45.** World poultry production was anticipated to rise to 106 million tons in the year 2013 and increase at a continuous rate of 1.81% per year.[27] Assume that this growth rate continues.

**(a)** Write an exponential formula for world poultry production, $P$, in million tons, as a function of the number of years, $t$, since 2013.
**(b)** Use the formula to estimate world poultry production in the year 2018.
**(c)** Use a graph to estimate the year in which world poultry production goes over 120 million tons.

**46.** During the first week of an epidemic, the number of people in a metropolitan area who are infected by the disease is growing at a continuous rate of 60% per day.

**(a)** If 10 people are initially infected, how many people will be infected after a week has passed?
**(b)** By what daily percentage does the number of infected people grow?

**47.** A radioactive substance decays at a continuous rate of 14% per year, and 50 mg of the substance is present in the year 2009.

**(a)** Write a formula for the amount present, $A$ (in mg), $t$ years after 2009.
**(b)** How much will be present in the year 2019?
**(c)** Estimate when the quantity drops below 5 mg.

**48.** The radioactive substance Cesium-134 decays at a continuous rate of 33.6% per year, while Thulium-171 decays at an annual rate of 30.3% per year. Which substance decays faster? Justify your answer.

**49.** The annual inflation rate, $r$, for a five-year period is given in Table 4.15.

**(a)** By what total percent did prices rise between the start of 2000 and the end of 2004?
**(b)** What is the average annual inflation rate for this time period?
**(c)** At the beginning of 2000, a shower curtain costs \$20. Make a prediction for the good's cost at the beginning of 2010, using the average inflation rate found in part (b).

**Table 4.15**

| $t$ | 2000 | 2001 | 2002 | 2003 | 2004 |
|---|---|---|---|---|---|
| $r$ | 3.4% | 2.8% | 1.6% | 2.3% | 2.7% |

**50.** In the 1980s a northeastern bank experienced an unusual robbery. Each month an armored car delivered cash deposits from local branches to the main office, a trip requiring only one hour. One day, however, the delivery was six hours late. This delay turned out to be a scheme devised by an employee to defraud the bank. The armored car drivers had lent the money, a total of approximately \$200,000,000, to arms merchants, who then used it as collateral against the purchase of illegal weapons. The interest charged for this loan was 20% per year compounded continuously. How much was the fee for the six-hour period?

**51.** This problem explores the value of $(1 + 1/n)^n$ for integer values of $n$ as $n$ gets large.

**(a)** Use a calculator or computer to evaluate $(1 + 1/n)^n$, correct to seven decimal places, for $n$ = 1000, 10,000, 100,000, and 1,000,000. Does $(1+1/n)^n$ appear to be an increasing or decreasing function of $n$?
**(b)** The limit of the sequence of values in part (a) is $e = 2.718281828\ldots$. What power of 10 is needed to give a value of $e$ correct to 6 decimal places?
**(c)** What happens if you evaluate $(1+1/n)^n$ using much larger values of $n$? For example, try $n = 10^{16}$ on a calculator.

**52.** With more terms giving a better approximation, it can be shown that

$$e = 1 + \frac{1}{1} + \frac{1}{1\cdot 2} + \frac{1}{1\cdot 2\cdot 3} + \frac{1}{1\cdot 2\cdot 3\cdot 4} + \cdots.$$

**(a)** Use a calculator to sum the five terms shown.

[26] www.nationmaster.com, accessed February 2010.
[27] http://www.thepoultrysite.com/articles/2871/fao-food-outlook-june-2013-poultry-meat, accessed July, 2013.

**(b)** Find the sum of the first seven terms.

**(c)** Compare your sums with the calculator's displayed value for $e$ (which you can find by entering e^ 1) and state the number of correct digits in the five- and seven-term sum.

**(d)** How many terms of the sum are needed in order to give a nine-decimal-digit approximation equal to the calculator's displayed value for $e$?

In Problems 53–56, find the limits.

**53.** $\lim_{x\to\infty} e^{-3x}$

**54.** $\lim_{t\to-\infty} 5e^{0.07t}$

**55.** $\lim_{t\to\infty}(2-3e^{-0.2t})$

**56.** $\lim_{t\to-\infty} 2e^{-0.1t+6}$

**57.** If $\lim_{t\to\infty} ae^{kt}=\infty$, what can you say about the values of $a$ and $k$?

## CHAPTER SUMMARY

- **Exponential Functions**
  Value of $f(t)$ changes at constant percent rate with respect to $t$.
- **General Formula for Exponential Functions**
  Exponential function: $f(t)=ab^t$, $\quad b>0$.
  $f$ increasing for $b>1$, decreasing for $0<b<1$.
  Growth factor: $b=1+r$.
  Growth rate: $r$, percent change as a decimal.
- **Comparing Linear and Exponential Functions**
  An increasing exponential function eventually overtakes any linear function.
- **Graphs of Exponential Functions**
  Concavity; asymptotes; effect of parameters $a$ and $b$.
  Solving exponential equations graphically; finding an exponential function for data.
- **The Number $e$**
  Continuous growth: $f(t)=ae^{kt}$.
  $f$ is increasing for $k>0$, decreasing for $k<0$.
  Continuous growth rate: $k$.
- **Compound Interest**
  For compounding $n$ times per year, balance,
  $$B=P\left(1+\frac{r}{n}\right)^{nt}.$$
  For continuous compounding, $B=Pe^{kt}$.
  Nominal rate, $r$ or $k$, versus effective rate earned over one year.
- **Horizontal Asymptotes and Limits to Infinity**

## REVIEW EXERCISES AND PROBLEMS FOR CHAPTER FOUR

### Exercises

In Exercises 1–6, you start with 500 items. How many do you have after the following change?

**1.** 10% increase

**2.** 100% increase

**3.** 1% decrease

**4.** 42% decrease

**5.** 42% increase followed by 42% decrease

**6.** 42% decrease followed by 42% increase

**7.** Find a formula for $P=f(t)$, the size of a population that begins in year $t=0$ with 2200 members and decreases at a 3.2% annual rate.

**8.** Without a calculator or computer, match each exponential formula to one of the graphs I–VI.

**(a)** $10(1.2)^t$ **(b)** $10(1.5)^t$ **(c)** $20(1.2)^t$

**(d)** $30(0.85)^t$ **(e)** $30(0.95)^t$ **(f)** $30(1.05)^t$

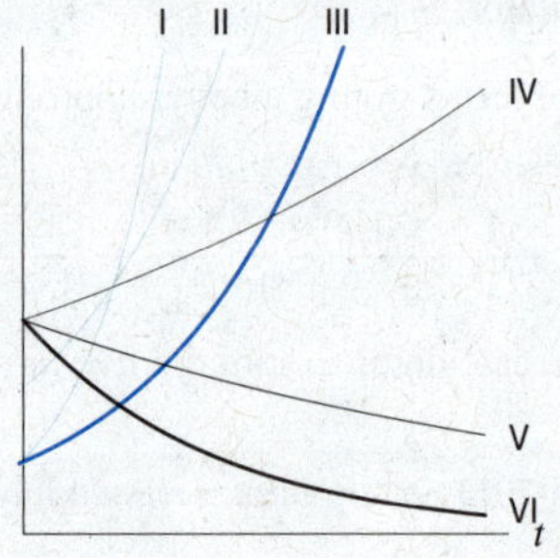


**9.** A quantity increases from 10 to 12. By what percent has it increased? Now suppose that it had increased from 100 to 102. What is the percent increase in this case?

**10.** Determine whether the function with values in Table 4.16 could be exponential.

**Table 4.16**

| $x$ | 1 | 2 | 4 | 5 | 8 | 9 |
|---|---|---|---|---|---|---|
| $f(x)$ | 4096 | 1024 | 64 | 16 | 0.25 | 0.0625 |

In Problems 11–14, could the function be linear or exponential or is it neither? Write possible formulas for the linear or exponential functions.

**11.**

| $r$ | 1 | 3 | 7 | 15 | 31 |
|---|---|---|---|---|---|
| $p(r)$ | 13 | 19 | 31 | 55 | 103 |

**12.**

| $x$ | 6 | 9 | 12 | 18 | 24 |
|---|---|---|---|---|---|
| $q(x)$ | 100 | 110 | 121 | 146.41 | 177.16 |

**13.**

| $x$ | 10 | 12 | 15 | 16 | 18 |
|---|---|---|---|---|---|
| $f(x)$ | 1 | 2 | 4 | 8 | 16 |

**14.**

| $t$ | 1 | 2 | 3 | 4 | 5 |
|---|---|---|---|---|---|
| $g(t)$ | 512 | 256 | 128 | 64 | 32 |

## Problems

**15.** In 2013, the cost of a morning train ticket from Boston to New York was \$126.[28] Assume that the price rises by 5% per year. Make a table showing the price of tickets each year until 2017.

**16.** Radioactive gallium-67 decays by 1.48% every hour; there are 100 milligrams initially.

**(a)** Find a formula for the amount of gallium-67 remaining after $t$ hours.
**(b)** How many milligrams are left after 24 hours? After 1 week?

**17.** A bank pays interest at the nominal rate of 4.2% per year. What is the effective annual rate if compounding is:

**(a)** Annual **(b)** Monthly **(c)** Continuous

**18.** Explain the difference between linear and exponential growth. That is, without writing down any formulas, describe how linear and exponential functions progress differently from one value to the next.

In Problems 19–24, find formulas for the exponential functions satisfying the given conditions.

**19.** $h(0) = 3$ and $h(1) = 15$

**20.** $f(3) = -3/8$ and $f(-2) = -12$

**21.** $g(1/2) = 4$ and $g(1/4) = 2\sqrt{2}$

**22.** $g(0) = 5$ and $g(-2) = 10$

**23.** $g(1.7) = 6$ and $g(2.5) = 4$

**24.** $f(1) = 4$ and $f(3) = d$

**25.** Suppose $f(-3) = 5/8$ and $f(2) = 20$. Find a formula for $f$ assuming it is:

**(a)** Linear **(b)** Exponential

For Problems 26–31, find formulas for the exponential functions.

**26.**

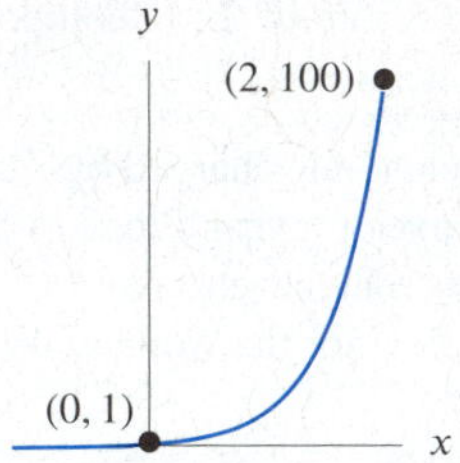


**27.**

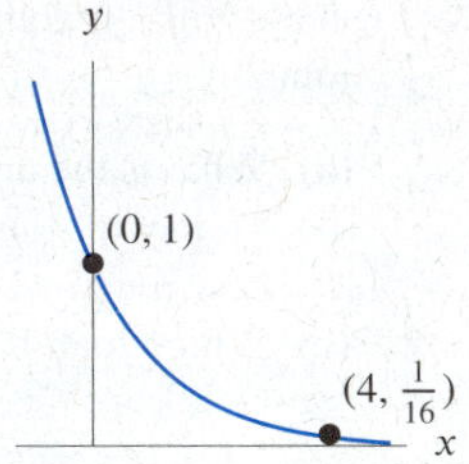


**28.**

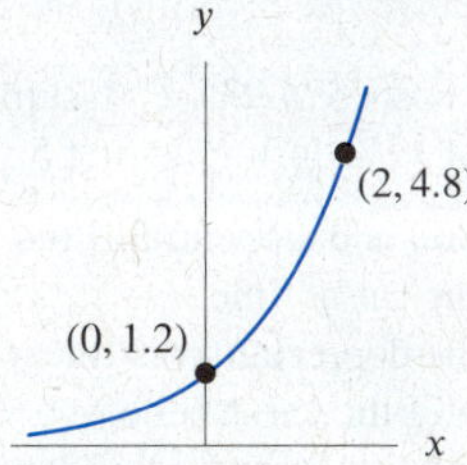


**29.**

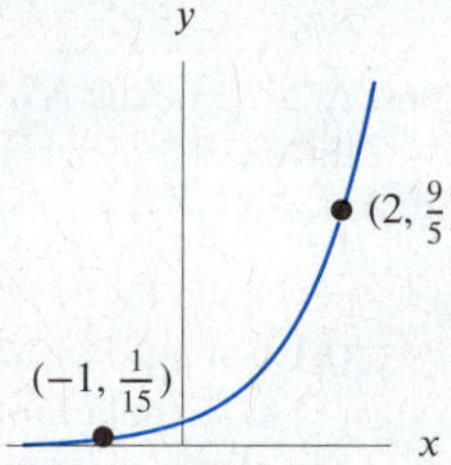


**30.**

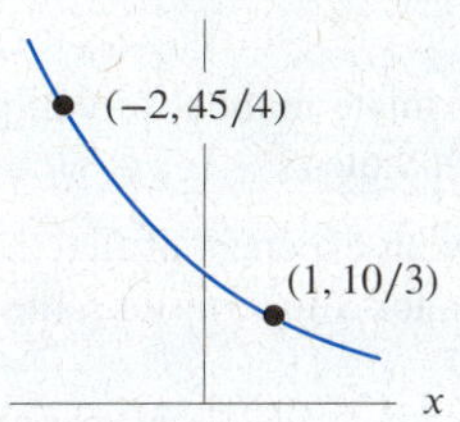


**31.**

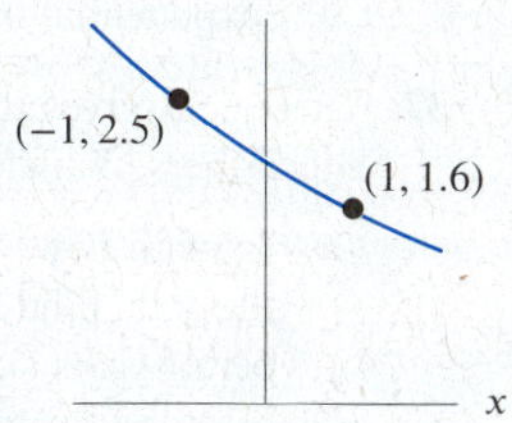


**32.** Find possible formulas for the functions in Figure 4.41.

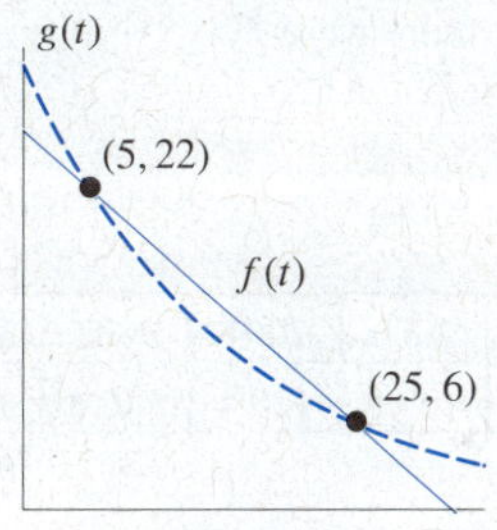


**Figure 4.41**

**33.** Let $P(t)$ be the population of a country, in millions, $t$ years after 1990, with $P(7) = 3.21$ and $P(13) = 3.75$.

**(a)** Find a formula for $P(t)$ assuming it is linear. Describe in words the country's annual population growth given this assumption.
**(b)** Find a formula for $P(t)$ assuming it is exponential. Describe in words the country's annual population growth given this assumption.

**34.** A population has size 100 at time $t = 0$, with $t$ in years.

**(a)** If the population grows by 10 people per year, find a formula for the population, $P$, at time $t$.
**(b)** If the population grows by 10% per year, find a formula for the population, $P$, at time $t$.
**(c)** Graph both functions on the same axes.

[28]http://www.amtrak.com, accessed July, 2013.

**35.** In 1940, there were about 10 brown tree snakes per square mile on the island of Guam, and in 2002, there were about 20,000 per square mile.[29] Find an exponential formula for the number, $N$, of brown tree snakes per square mile on Guam $t$ years after 1940. What was, on average, the annual percent increase in the population during this period?

**36.** A 2013 Lexus AWD sedan costs \$74,935.[30] Assume that the car depreciates a total of 42% during its first 5 years.

**(a)** Suppose the depreciation is exponential. Find a formula for the value of the car at time $t$.
**(b)** Suppose instead that the depreciation is linear. Find a formula for the value of the car at time $t$.
**(c)** If this were your car and you were trading it in after 4 years, which depreciation model would you prefer (exponential or linear)?

**37.** Table 4.17 gives the approximate number of cell phone subscribers, $S$, in the United States.[31]

**(a)** Explain how you know an exponential function fits the data. Find a formula for $S$ in terms of $t$, the number of years since 2000.
**(b)** Interpret the growth rate in terms of cell phone subscribers.
**(c)** In 2006, there were 233 million subscribers. Does this fit the pattern?
**(d)** In 2010, there were 300.5 million subscribers. Does this fit the pattern?

**Table 4.17**

| Year | 2000 | 2001 | 2002 | 2003 | 2004 | 2005 |
|---|---|---|---|---|---|---|
| Subscribers (m.) | 109.5 | 128.4 | 140.8 | 158.7 | 182.1 | 207.9 |

**38.** Find the annual growth rates of a quantity which:

**(a)** Doubles in size every 7 years
**(b)** Triples in size every 11 years
**(c)** Grows by 3% per month
**(d)** Grows by 18% every 5 months

In Problems 39–40, graph $f(x)$, a function defined for all real numbers and satisfying the condition.

**39.** $f(x) \to 5$ as $x \to \infty$

**40.** $f(x) \to 3$ as $x \to \infty$ and $f(x) \to -2$ as $x \to -\infty$

**41.** The functions $f(x) = (\frac{1}{2})^x$ and $g(x) = 1/x$ are similar in that they both tend toward zero as $x$ becomes large. Using a calculator, determine which function, $f$ or $g$, approaches zero faster.

In Problems 42–43, graph the function to find horizontal asymptotes.

**42.** $f(x) = 8 - 2^x$

**43.** $f(x) = 3^{-x^2} + 2$

**44.** Without making any calculations, briefly describe the following investments, where $t$ is in years. Be specific.

**(a)** $V = 2500e^{-0.0434t}$
**(b)** $V = 4000(1.005)^{12t}$
**(c)** $V = 8000 \cdot 2^{-t/14}$
**(d)** $V = 5000 + 250(t - 10)$

**45.** Without making any calculations, briefly describe in words what the following formulas tell you about the size of the animal populations they describe. Be specific. Note that $t$ is in years.

**(a)** $P = 5200(1.118)^t$
**(b)** $P = 4600(1.01)^{12t}$
**(c)** $P = 3800\left(\frac{1}{2}\right)^{t/12}$
**(d)** $P = 8000e^{0.0778t}$
**(e)** $P = 1675 - 25(t - 30)$

**46.** Without a calculator, match each of the following formulas to one of the graphs in Figure 4.5 on page 145.

**(a)** $y = 8.3e^{-t}$ **(b)** $y = 2.5e^t$ **(c)** $y = -4e^{-t}$

**47.** If $t$ is in years, the formulas for dollar balances of two different investment accounts are:

$$f(t) = 1100(1.05)^t \quad \text{and} \quad g(t) = 1500e^{0.05t}.$$

**(a)** Describe in words the investment account modeled by $f$.
**(b)** Describe the account modeled by $g$. State the effective annual yield.

**48.** Accion is a non-profit microlending organization which makes small loans to entrepreneurs who do not qualify for bank loans.[32] A New York woman who sells clothes from a cart has the choice of a \$1000 loan from Accion to be repaid by \$1150 a year later and a \$1000 loan from a loan shark with an annual interest rate of 22%, compounded annually.

**(a)** What is the annual interest rate charged by Accion?
**(b)** To pay off the loan shark for a year's loan of \$1000, how much would the woman have to pay?
**(c)** Which loan is a better deal for the woman? Why?

[29] *Science News*, Vol. 162, August 10, 2002, p. 85.
[30] http://usnews.rankingsandreviews.com/cars-trucks/Lexus_LS/prices/, accessed May, 2013.
[31] http://www.infoplease.com/ipa/A0933563.html. Last accessed January 13, 2014.
[32] http://www.accionusa.org/, accessed November, 2005.

**49.** Write $p(x) = \dfrac{7e^{6x} \cdot \sqrt{e} \cdot (2e^x)^{-1}}{10e^{4x}}$ in the form $p(x) = ae^x$. All constants should be expressed exactly.

In Problems 50–51, write each function in standard form. Note that one of them is linear and one exponential.

**50.** $r(v) = vj^w - 4tj^w + kvj^w$

**51.** $s(w) = vj^w - 4tj^w + kvj^w$

Problems 52–53 concern ISO $A$- and $B$-series paper, commonly used in many countries. The width in millimeters (mm) of a sheet of $A_n$ paper in this series is given by the formula[33]

$$f(n) = 1000 \cdot 2^{-\frac{1}{4}-\frac{n}{2}}.$$

The width $g(n)$ of a sheet of $B_n$ paper is the geometric mean of the widths of $A_n$ paper and the next larger size, $A_{n-1}$ paper. Since the geometric mean of two quantities $p$ and $q$ is the square root of their product, this means

$$g(n) = \sqrt{f(n) \cdot f(n-1)}.$$

**52.** Evaluate $g(1)$. What does your answer tell you about $B$-series paper?

**53.** Show that $g$ is an exponential function by writing it in standard form.

**54.** The figure gives graphs of two functions, $f$ and $g$. Explain why not both of these functions can be exponential.

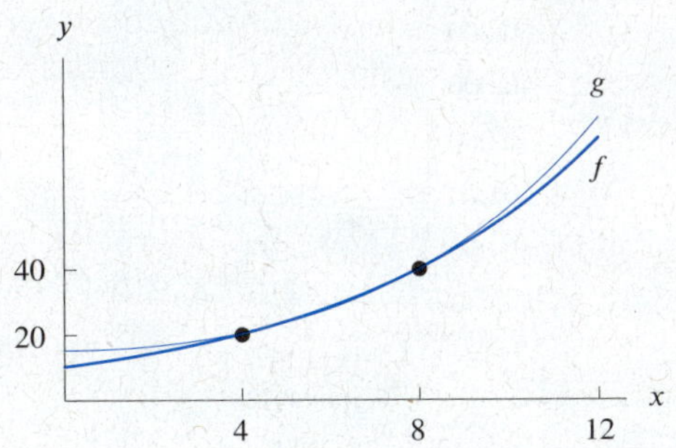


**55.** Write a paragraph that compares the function $f(x) = a^x$, where $a > 1$, and $g(x) = b^x$, where $0 < b < 1$. Include graphs in your answer.

Problems 56–59 use Figure 4.42, which shows $f(x) = ab^x$ and $g(x) = cd^x$ on three different scales. Their point of intersection is marked.

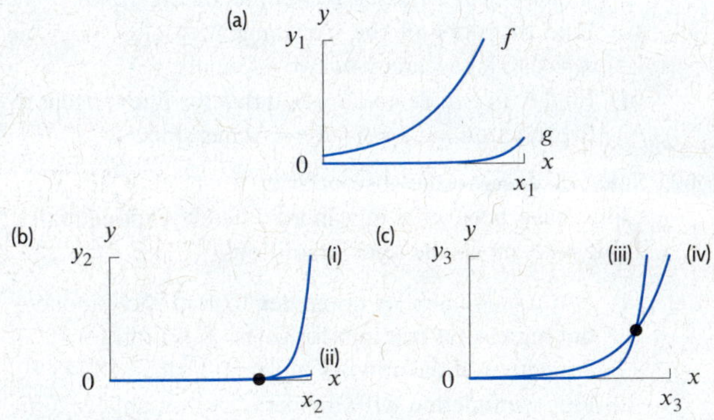


**Figure 4.42**

**56.** Which is larger, $a$ or $c$?

**57.** Which is larger, $b$ or $d$?

**58.** Rank in order from least to greatest: $x_1, x_2, x_3$.

**59.** Match $f$ and $g$ to the graphs labeled (i)–(iv) in (b) and (c).

**60.** On November 27, 1993, the *New York Times* reported that wildlife biologists have found a direct link between the increase in the human population in Florida and the decline of the local black bear population. From 1953 to 2013, the human population increased, on average, at a rate of about 3% per year. However, between 1953 and 2006 the black bear population decreased at a rate of 6% per year. In 1953 the black bear population was 11,000.

**(a)** The 2013 human population of Florida was 19.3 million. What was the human population in 1953?

**(b)** Find the black bear population for 2000.

**(c)** Had this trend continued,[34] when would the black bear population have numbered less than 100?

**(d)** In 2000 conservation measures[35] were introduced to link disconnected Black Bear populations and encourage movement of the bears between habitats. By 2012, the population of Black Bears in Florida[36] had reached 3000. Find the average rate of increase since 2000.

[33] Actual paper sizes are rounded to the nearest mm, so this formula is only approximate. See http://www.cl.cam.ac.uk/~mgk25/iso-paper.html, accessed February 24, 2008.

[34] Since 1993, the black bear population has started to increase again: www.myfwc.com/bear, accessed January, 2014.

[35] http://ww2.valdosta.edu/ aesanfor/black bear corridor.pdf. Last accessed January 13, 2014.

[36] http://myfwc.com/wildlifehabitats/managed/bear/. Last accessed January 13, 2014.

**61.** **(a)** Using a computer or calculator, graph $f(x) = 2^x$.
**(b)** Find the slope of the line tangent to $f$ at $x = 0$ to an accuracy of two decimals. [Hint: Zoom in on the graph until it is indistinguishable from a line and estimate the slope using two points on the graph.]
**(c)** Find the slope of the line tangent to $g(x) = 3^x$ at $x = 0$ to an accuracy of two decimals.
**(d)** Find $b$ (to two decimals) such that the line tangent to the function $h(x) = b^x$ at $x = 0$ has slope 1.

**62.** Sales of energy-efficient compact fluorescent lamps in China have been growing approximately exponentially. Table 4.18 shows the sales in millions.[37]

**(a)** Use a calculator or computer to find the exponential regression function for sales, $S$ (in millions), as a function of the number of years, $t$, since 1994.
**(b)** Plot the function with the data. Does it appear to fit the data well?
**(c)** What annual percent growth rate does the exponential model show?
**(d)** If this growth rate continues, what sales are predicted in the year 2010?

**Table 4.18**

| Year | 1994 | 1996 | 1998 | 2000 | 2002 | 2003 |
|---|---|---|---|---|---|---|
| Sales (millions) | 20 | 30 | 60 | 125 | 295 | 440 |

Find possible formulas for the functions in Problems 63–65.

**63.** $V$ gives the value of an investment that begins in year $t = 0$ with \$12,000 and yields 4.2% annual interest, compounded continuously.

**64.** The exponential function $p(t)$ given that $p(20) = 300$ and $p(50) = 40$.

**65.** The linear function $q(x)$ whose graph intersects the graph of $y = 5000e^{-x/40}$ at $x = 50$ and $x = 150$.

**66.** An investment worth $V = \$2500$ in year $t = 0$ earns 4.2% annual interest, compounded continuously. Find a formula for $V$ in terms of $t$.

**67.** If \$5000 is deposited in an account paying a nominal interest rate of 4% per year, how much is in the account 10 years later if interest is compounded

**(a)** Annually? **(b)** Continuously?

**68.** A population is represented by $P = 12{,}000e^{-0.122t}$. Give the values of $a$, $k$, $b$, and $r$, where $P = ae^{kt} = ab^t$. What do these values tell you about the population?

**69.** This problem uses a calculator or computer to explore graphically the value of $(1 + 1/x)^x$ as $x$ gets large.

**(a)** Graph $y = (1 + 1/x)^x$ for $1 \le x \le 10$.
**(b)** Are the values of $y$ in part (a) increasing or decreasing?
**(c)** Do the values of $y$ in part (a) appear to approach a limiting value?
**(d)** Graph $y = (1 + 1/x)^x$, for $1 \le x \le 100$, and then for $1 \le x \le 1000$. Do the $y$ values appear to approach a limiting value? If so, approximately what is it?
**(e)** Graph $y = (1 + 1/x)^x$ and $y = e$ on the same axes, for $1 \le x \le 10{,}000$. What does the graph suggest?
**(f)** By checking $x = 10{,}000$, $x = 20{,}000$, and so on, decide how large (as a multiple of 10,000) $x$ should be to give a value of $e$ correct to 4 decimal places.

**70.** Hong Kong shifted from British to Chinese rule in 1997. Figure 4.43 shows[38] the number of people who emigrated from Hong Kong during each of the years from 1980 to 1992.

**(a)** Find an exponential function that approximates the data.
**(b)** What does the model predict about the number of emigrants in 1997?
**(c)** Briefly explain why this model is or is not useful to predict emigration in the year 2010.

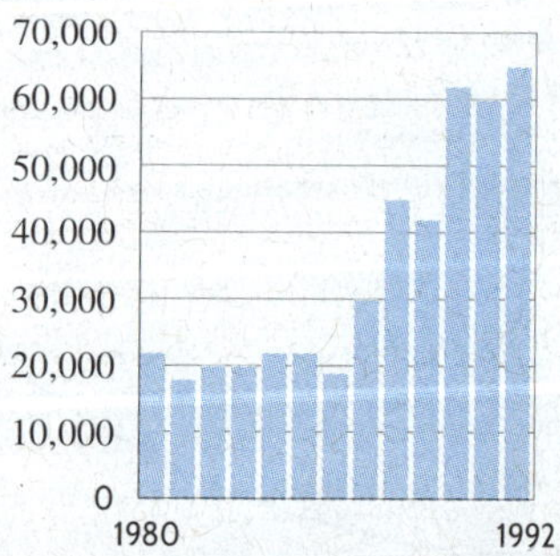


Figure 4.43

**71.** Before the AIDS epidemic, Botswana[39] had a rapidly growing population, as shown in Table 4.19. In 2005, the population started falling.

**(a)** Fit an exponential growth model, $P = ab^t$, to this data set, where $P$ is the population in millions and $t$ measures the years since 1975 in 5-year intervals—so $t = 1$ corresponds to 1980. Estimate $a$ and $b$. Plot the data set and $P = ab^t$ on the same graph.

[37] S. Nadel and Hong, "Market Data on Efficient Lighting," Right Light 6 Conference, Session 8, May, 2005.
[38] Adapted from the *New York Times*, July 5, 1995.
[39] N. Keyfitz, *World Population Growth and Aging* (Chicago: University of Chicago Press), 1990.

**(b)** Starting from 1975, how long does it take for the population of Botswana to double? When is the population of Botswana projected to exceed 214 million, the 1975 population of the US?

**Table 4.19**

| Year | 1975 | 1980 | 1985 | 1990 |
|---|---|---|---|---|
| Population (millions) | 0.755 | 0.901 | 1.078 | 1.285 |

Problems 72–73 use Figure 4.44, where $t_0$ is the $t$-coordinate of the point of intersection of the graphs. Describe what happens to $t_0$ if the following changes are made, assuming the other quantities remain the same.

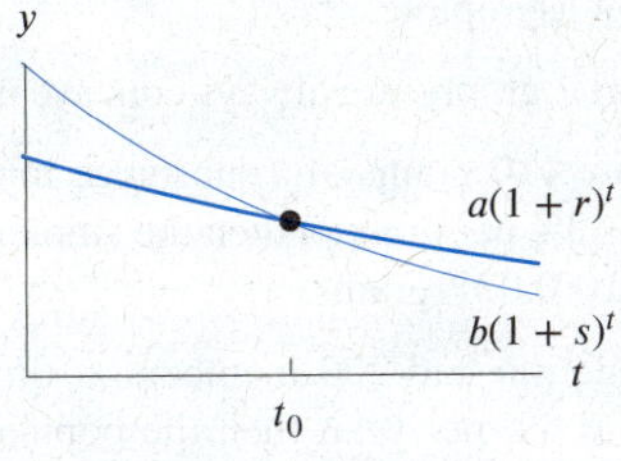


**Figure 4.44**

**72.** $b$ is decreased

**73.** $r$ is increased

**74.** It is a well-documented fact that the earning power of men is higher than that of women.[40] Table 4.20 gives the median income of year-round full-time workers in the US in dollars.

**(a)** Plot the data and connect the points.
**(b)** Let $t$ be the year. Construct two functions of the form $W(t) = ae^{b(t-1960)}$, one each for the men's and women's earning power data.
**(c)** Graph the two functions from 1960 to 2010 and again from 2010 to 2080.
**(d)** Do the graphs in part (c) predict women's salaries will catch up with men's? If so, when?
**(e)** Comment on your predictions in part (d).

**Table 4.20**

| Year | 1960 | 1970 | 1980 | 1990 | 2000 | 2010 |
|---|---|---|---|---|---|---|
| Female | 1261 | 2237 | 4920 | 10,070 | 16,063 | 38,757 |
| Male | 4080 | 6670 | 12,530 | 20,293 | 28,343 | 50,074 |

**75.** According to a letter to the *New York Times* on April 10, 1993, "... the probability of [a driver's] involvement in a single-car accident increases exponentially with increasing levels of blood alcohol." The letter goes on to state that when a driver's blood-alcohol content (BAC) is 0.15, the risk of such an accident is about 25 times greater than for a nondrinker.

**(a)** Let $p_0$ be a nondrinker's probability of being involved in a single-car accident. Let $f(x)$ be the probability of an accident for a driver whose blood alcohol level is $x$. Find a formula for $f(x)$. (This only makes sense for some values of $x$.)
**(b)** At the time of the letter, the legal definition of intoxication was a BAC of 0.1 or higher. According to your formula for $f(x)$, how many times more likely to be involved in a single-car accident was a driver at the legal limit than a nondrinker?
**(c)** Suppose that new legislation is proposed to change the definition of legal intoxication. The new definition states that a person is legally intoxicated when their likelihood of involvement in a single-car accident is three times that of a non-drinker. To what BAC would the new definition of legal intoxication correspond?

Pure water is not perfectly clear—it has a bluish cast—because it absorbs light differently at different wavelengths. The table gives values of the *absorption coefficient* $\mu(\lambda)$, in units of $cm^{-1}$, as a function of the wavelength $\lambda$ in nanometers or nm.[41] If light of wavelength $\lambda$ nanometers (nm) travels through $l$ cm of water, the percent transmitted (without being absorbed) is given by

$$T_\lambda(l) = e^{-\mu(\lambda)\cdot l}.$$

Note that here, $T_\lambda$ is the name of the function. Answer Problems 76–77.

| color | $\lambda$ | $\mu(\lambda)$ | color | $\lambda$ | $\mu(\lambda)$ |
|---|---|---|---|---|---|
| violet | 400 | 0.000066 | yellow | 570 | 0.000695 |
| indigo | 445 | 0.000075 | orange | 590 | 0.001351 |
| blue | 475 | 0.000114 | red | 650 | 0.003400 |
| green | 510 | 0.000325 | — | — | — |

**76.** What percent of red light will be transmitted after passing through 200 cm of water?

**77.** What percent of blue light will be transmitted over the same distance?

[40] http://www.politifact.com/truth-o-meter/statements/2011/oct/05/david-axelrod/david-axelrod-says-typical-white-male-worker-earns/. Accessed January 12, 2014.

[41] Note that aside from absorption, optical scattering also plays a role. See http://omlc.ogi.edu/spectra/water/data/pope97.dat, http://eosweb.larc.nasa.gov/EDDOCS/Wavelengths_for_Colors.html, and http://en.wikipedia.org/wiki/Color_of_water, accessed February 29, 2008.

Find the limits in Problems 78–83.

**78.** $\lim_{x\to\infty} 257(0.93)^x$

**79.** $\lim_{t\to\infty} 5.3e^{-0.12t}$

**80.** $\lim_{x\to-\infty} (15 - 5e^{3x})$

**81.** $\lim_{t\to-\infty} (21(1.2)^t + 5.1)$

**82.** $\lim_{x\to\infty} (7.2 - 2e^{3x})$

**83.** $\lim_{x\to-\infty} (5e^{-7x} + 1.5)$

In Problems 84–85, graph $f(x)$, a function defined for all real numbers and satisfying the condition.

**84.** $\lim_{x\to-\infty} f(x) = -4$

**85.** $\lim_{x\to-\infty} f(x) = 0$ and $\lim_{x\to\infty} f(x) = -\infty$

## STRENGTHEN YOUR UNDERSTANDING

Are the statements in Problems 1–32 true or false? Give an explanation for your answer.

**1.** Exponential functions are functions that increase or decrease at a constant percent rate.

**2.** The independent variable in an exponential function is always found in the exponent.

**3.** If $y = 40(1.05)^t$ then $y$ is an exponential function of $t$.

**4.** The following table shows a function that could be exponential.

| $x$ | 1 | 2 | 4 | 5 | 6 |
|---|---|---|---|---|---|
| $y$ | 1 | 2 | 4 | 7 | 11 |

**5.** If your salary, $S$, grows by 4% each year, then $S = S_0(0.04)^t$ where $t$ is in years.

**6.** If $f(t) = 4(2)^t$ then $f(2) = 64$.

**7.** If $f(t) = 3(\frac{2}{5})^t$ then $f$ is a decreasing function.

**8.** If $Q = f(t) = 1000(0.5)^t$ then when $Q = 125, t = 3$.

**9.** If $Q = f(t) = ab^t$ then $a$ is the initial value of $Q$.

**10.** If we are given two data points, we can find a linear function and an exponential function that go through these points.

**11.** A population that has 1000 members and decreases at 10% per year can be modeled as $P = 1000(0.10)^t$.

**12.** A positive increasing exponential function always becomes larger than any increasing linear function in the long run.

**13.** A possible formula for an exponential function that passes through the point (0, 1) and the point (2, 10) is $y = 4.5t + 1$.

**14.** If a population increases by 50% each year, then in two years it increases by 100%.

**15.** In the formula $Q = ab^t$, the value of $a$ tells us where the graph crosses the $Q$-axis.

**16.** In the formula $Q = ab^t$, ifă $a > 1$, the graph always rises as we read from left to right.

**17.** The symbol $e$ represents a constant whose value is approximately 2.71828.

**18.** If $f(x) \to k$ as $x \to \infty$ we say that the line $y = k$ is a horizontal asymptote.

**19.** Exponential graphs are always concave up.

**20.** If there are 110 grams of a substance initially and its decay rate is 3% per minute, then the amount after $t$ minutes is $Q = 110(0.03)^t$ grams.

**21.** If a population had 200 members at time zero and was growing at 4% per year, then the population size after $t$ years can be expressed as $P = 200(1.04)^t$.

**22.** If $P = 5e^{0.2t}$, we say the continuous growth rate of the function is 2%.

**23.** If $P = 4e^{-0.90t}$, we say the continuous growth rate of the function is 10%.

**24.** If $Q = 3e^{0.2t}$, then when $t = 5$, $Q = 3$.

**25.** If $Q = Q_0e^{kt}$, with $Q_0$ positive and $k$ negative, then $Q$ is decreasing.

**26.** If an investment earns 5% compounded monthly, its effective rate will be more than 5%.

**27.** If a \$500 investment earns 6% per year, compounded quarterly, we can find the balance after three years by evaluating the formula $B = 500(1 + \frac{6}{4})^{3\cdot 4}$.

**28.** If interest on a \$2000 investment is compounded continuously at 3% per year, the balance after five years is found by evaluating the formula $B = 2000e^{(0.03)(5)}$.

**29.** Investing \$10,000 for 20 years at 5% earns more if interest is compounded quarterly than if it is compounded annually.

**30.** Investing \$$P$ for $T$ years always earns more if interest is compounded continuously than if it is compounded annually.

**31.** There is no limit to the amount a twenty-year \$10,000 investment at 5% interest can earn if the number of times the interest is compounded becomes greater and greater.

**32.** If you put \$1000 into an account that earns 5.5% compounded continuously, then it takes about 18 years for the investment to grow to \$2000.

# SKILLS REFRESHER FOR CHAPTER 4: EXPONENTS

We list the definition and properties that are used to manipulate exponents.

**Definition of Zero, Negative, and Fractional Exponents**

If $m$ and $n$ are positive integers:[42]

- $a^0 = 1$
- $a^{-n} = \dfrac{1}{a^n}$
- $a^{1/n} = \sqrt[n]{a}$, the $n^{\text{th}}$ root of $a$
- $a^{m/n} = \sqrt[n]{a^m} = (\sqrt[n]{a})^m$

**Properties of Exponents**

- $a^m \cdot a^n = a^{m+n}$ For example, $2^4 \cdot 2^3 = (2\cdot2\cdot2\cdot2)\cdot(2\cdot2\cdot2) = 2^7$.
- $\dfrac{a^m}{a^n} = a^{m-n}, a \neq 0$ For example, $\dfrac{2^4}{2^3} = \dfrac{2\cdot2\cdot2\cdot2}{2\cdot2\cdot2} = 2^1$.
- $(a^m)^n = a^{mn}$ For example, $(2^3)^2 = 2^3 \cdot 2^3 = 2^6$.
- $(ab)^n = a^n b^n$
- $\left(\dfrac{a}{b}\right)^n = \dfrac{a^n}{b^n}, \quad b \neq 0$

Be aware of the following notational conventions:

$$
\begin{aligned}
ab^n &= a(b^n), &\quad &\text{but } ab^n \neq (ab)^n,\\
-b^n &= -(b^n), &\quad &\text{but } -b^n \neq (-b)^n,\\
-ab^n &= (-a)(b^n).
\end{aligned}
$$

For example, $-2^4 = -(2^4) = -16$, but $(-2)^4 = (-2)(-2)(-2)(-2) = +16$. Also, be sure to realize that for $n \neq 1$,

$$(a+b)^n \neq a^n + b^n \qquad \text{Power of a sum} \neq \text{Sum of powers.}$$

**Example 1** Evaluate without a calculator:

(a) $(27)^{2/3}$ (b) $(4)^{-3/2}$ (c) $8^{1/3} - 1^{1/3}$

**Solution**

(a) We have $(27)^{2/3} = \sqrt[3]{27^2} = \sqrt[3]{729} = 9$, or, equivalently, $(27)^{2/3} = \left(27^{1/3}\right)^2 = \left(\sqrt[3]{27}\right)^2 = 3^2 = 9$.

(b) We have $(4)^{-3/2} = (2)^{-3} = \dfrac{1}{2^3} = \dfrac{1}{8}$.

(c) We have $8^{1/3} - 1^{1/3} = 2 - 1 = 1$.

[42] We assume that the base is restricted to the values for which the power is defined.

**Example 2** Use the rules of exponents to simplify the following:

(a) $\dfrac{100x^2y^4}{5x^3y^2}$ (b) $\dfrac{y^4(x^3y^{-2})^2}{2x^{-1}}$ (c) $\sqrt[3]{-8x^6}$ (d) $\left(\dfrac{M^{1/5}}{3N^{-1/2}}\right)^2$

**Solution** (a) We have

$$\frac{100x^2y^4}{5x^3y^2} = 20(x^{2-3})(y^{4-2}) = 20x^{-1}y^2 = \frac{20y^2}{x}.$$

(b) We have

$$\frac{y^4\left(x^3y^{-2}\right)^2}{2x^{-1}} = \frac{y^4x^6y^{-4}}{2x^{-1}} = \frac{y^{(4-4)}x^{(6-(-1))}}{2} = \frac{y^0x^7}{2} = \frac{x^7}{2}.$$

(c) We have

$$\sqrt[3]{-8x^6} = \sqrt[3]{-8}\cdot\sqrt[3]{x^6} = -2x^2.$$

(d) We have

$$\left(\frac{M^{1/5}}{3N^{-1/2}}\right)^2 = \frac{\left(M^{1/5}\right)^2}{\left(3N^{-1/2}\right)^2} = \frac{M^{2/5}}{3^2N^{-1}} = \frac{M^{2/5}N}{9}.$$

**Example 3** Solve for $x$:

(a) $\dfrac{10x^7}{4x^2} = 37$ (b) $\dfrac{x^2}{3x^5} = 10$ (c) $\sqrt{9x^5} = 10$

**Solution** (a) We have

$$\begin{aligned}\frac{10x^7}{4x^2} &= 37\\ 2.5x^5 &= 37\\ x^5 &= 14.8\\ x &= (14.8)^{1/5} = 1.714.\end{aligned}$$

(b) We have

$$\begin{aligned}\frac{x^2}{3x^5} &= 10\\ \frac{1}{3}x^{-3} &= 10\\ \frac{1}{x^3} &= 30\\ x^3 &= \frac{1}{30}\\ x &= \left(\frac{1}{30}\right)^{1/3} = 0.322.\end{aligned}$$

(c) We have

$$\begin{aligned}\sqrt{9x^5} &= 10\\ 3x^{5/2} &= 10\\ x^{5/2} &= \frac{10}{3}\\ x &= \left(\frac{10}{3}\right)^{2/5} = 1.619.\end{aligned}$$

## Exercises to Skills for Chapter 4

For Exercises 1–33, evaluate without a calculator.

**1.** $(-5)^2$ **2.** $11^2$ **3.** $10^4$

**4.** $(-1)^{13}$ **5.** $\frac{5^3}{5^2}$ **6.** $\frac{10^8}{10^5}$

**7.** $\frac{6^4}{6^4}$ **8.** $\sqrt{4}$ **9.** $\sqrt{4^2}$

**10.** $\sqrt{4^4}$ **11.** $\sqrt{(-4)^2}$ **12.** $\frac{1}{7^{-2}}$

**13.** $\frac{2^7}{2^3}$ **14.** $(-1)^{445}$ **15.** $-11^2$

**16.** $\left(5^0\right)^3$ **17.** $2.1\left(10^3\right)$ **18.** $16^{1/2}$

**19.** $16^{1/4}$ **20.** $16^{3/4}$ **21.** $16^{5/4}$

**22.** $16^{5/2}$ **23.** $100^{5/2}$ **24.** $\sqrt{(-4)^2}$

**25.** $(-1)^3\sqrt{36}$ **26.** $(0.04)^{1/2}$ **27.** $(-8)^{2/3}$

**28.** $3^{-1}$ **29.** $3^{-3/2}$ **30.** $25^{-1}$

**31.** $25^{-2}$ **32.** $(1/27)^{-1/3}$ **33.** $(0.125)^{1/3}$

Simplify the expressions in Exercises 34–55 and leave without radicals if possible. Assume all variables are positive.

**34.** $\sqrt{x^4}$ **35.** $\sqrt{y^8}$

**36.** $\sqrt{w^8z^4}$ **37.** $\sqrt{x^5y^4}$

**38.** $\sqrt{49w^9}$ **39.** $\sqrt{25x^3z^4}$

**40.** $\sqrt{r^2}$ **41.** $\sqrt{r^3}$

**42.** $\sqrt{r^4}$ **43.** $\sqrt{64s^7}$

**44.** $\sqrt{50x^4y^6}$ **45.** $\sqrt{48u^{10}v^{12}y^5}$

**46.** $\sqrt{6s^2t^3v^5}\sqrt{6st^5v^3}$ **47.** $\left(S\sqrt{16xt^2}\right)^2$

**48.** $\sqrt{e^{2x}}$ **49.** $(3AB)^{-1}\left(A^2B^{-1}\right)^2$

**50.** $e^{kt}\cdot e^3\cdot e$ **51.** $\sqrt{M+2}(2+M)^{3/2}$

**52.** $\left(y^{-2}e^y\right)^2$ **53.** $\frac{a^{n+1}3^{n+1}}{a^n3^n}$

**54.** $\left(a^{-1}+b^{-1}\right)^{-1}$

**55.** $\left(\frac{35(2b+1)^9}{7(2b+1)^{-1}}\right)^2$ (Do not expand $(2b+1)^9$.)

If possible, evaluate the quantities in Exercises 56–64. Check your answers with a calculator.

**56.** $(-32)^{3/5}$ **57.** $-32^{3/5}$ **58.** $-625^{3/4}$

**59.** $(-625)^{3/4}$ **60.** $(-1728)^{4/3}$ **61.** $64^{-3/2}$

**62.** $-64^{3/2}$ **63.** $(-64)^{3/2}$ **64.** $81^{5/4}$

In Exercises 65–66, solve for $x$.

**65.** $7x^4 = 20x^2$ **66.** $2(x+2)^3 = 100$

In Exercises 67–68, use algebra to find the point of intersection.

**67.**

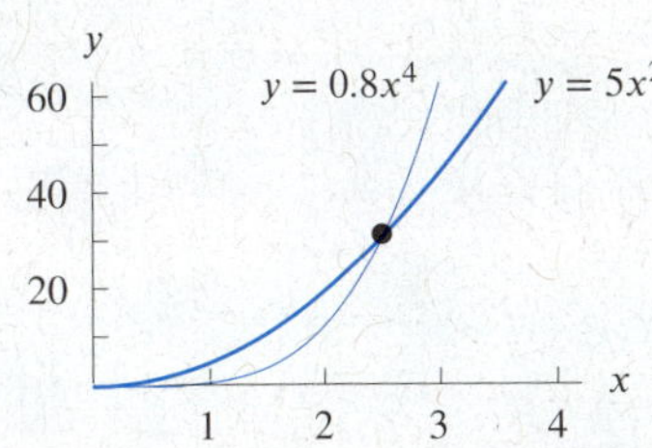

**68.**

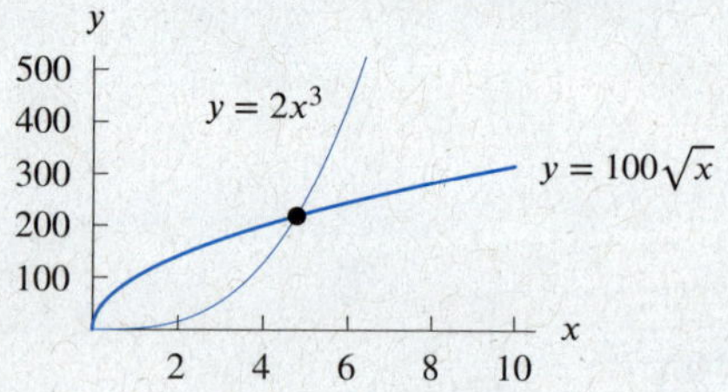


Are the statements in Exercises 69–74 true or false?

**69.** $x^2y^5 = (xy)^{10}$

**70.** $5u^2 + 5u^3 = 10u^5$

**71.** $(3r)^2 9s^2 = 81r^2s^2$

**72.** $\sqrt[3]{-64b^3c^6} = -4bc^2$

**73.** $-4w^2 - 3w^3 = -w^2(4 + 3w)$

**74.** $(u+v)^{-1} = \dfrac{1}{u} + \dfrac{1}{v}$

Solve the equations in Exercises 75–76 in terms of $r$ and $s$, given that

$$2^r = 5 \quad \text{and} \quad 2^s = 7.$$

**75.** $2^x = 35.$

**76.** $2^x = 140.$

Let $2^a = 5$ and $2^b = 7$. Using exponent rules, solve the equations in Exercises 77–82 in terms of $a$ and $b$.

**77.** $5^x = 32$

**78.** $7^x = \dfrac{1}{8}$

**79.** $25^x = 64$

**80.** $14^x = 16$

**81.** $5^x = 7$

**82.** $0.4^x = 49$

Chapter Five

# LOGARITHMIC FUNCTIONS

© Patrick Zephyr/Patrick Zephyr Nature Photography

## Contents

# 5.1 LOGARITHMS AND THEIR PROPERTIES

## What Is a Logarithm?

A population grows according to the formula $P = 10^t$, where $P$ is its size at time $t$, in hours. When will the population be 2500? To find out, we solve the following equation for $t$:

$$10^t = 2500.$$

In Section 4.2, we used a graphical method to approximate $t$. Now we introduce a function that gives precisely the exponent of 10 we need.

Since $10^3 = 1000$ and $10^4 = 10{,}000$, and $1000 < 2500 < 10{,}000$, the exponent we are looking for is between 3 and 4. But how do we find the exponent exactly? To answer this question, we define the *common logarithm function*, or simply the *log function*, written $\log_{10} x$, or $\log x$, as follows.

> If $x$ is a positive number,
>
> $$\log x \text{ is the exponent of 10 that gives } x.$$
>
> In other words,
>
> $$y = \log x \qquad \text{means} \qquad 10^y = x.$$

For example, $\log 100 = 2$, because 2 is the exponent of 10 that gives 100, or $10^2 = 100$.

Logarithms with base 10 and base $e$ are most commonly used, but any positive base not equal to 1 can be used. For instance, $\log_2 x$ (read "log, base 2, of $x$") is often used in computer science.

To solve the equation $10^t = 2500$, we must find the power of 10 that gives 2500. Using the log button on a calculator, we can approximate this exponent. We find

$$\log 2500 \approx 3.398, \qquad \text{which means that} \qquad 10^{3.398} \approx 2500.$$

As predicted, this exponent is between 3 and 4. The precise exponent is $\log 2500$; the approximate value is 3.398. Thus, it takes roughly 3.4 hours for the population to reach 2500.

**Example 1** Rewrite the following statements using exponents instead of logs.

(a) $\log 100 = 2$ (b) $\log 0.01 = -2$ (c) $\log 30 = 1.477$

**Solution** For each statement, we use the fact that if $y = \log x$ then $10^y = x$.

(a) Since $\log 100 = 2$ we have $10^2 = 100$.
(b) Since $\log 0.01 = -2$ we have $10^{-2} = 0.01$.
(c) Since $\log 30 = 1.477$ we have $10^{1.477} = 30$. (Actually, this is only an approximation. Using a calculator, we see that $10^{1.477} = 29.9916\ldots$ and that $\log 30 = 1.47712125\ldots$.)

**Example 2** Rewrite the following statements using logs instead of exponents.

(a) $10^5 = 100{,}000$ (b) $10^{-4} = 0.0001$ (c) $10^{0.8} = 6.3096$.

**Solution** For each statement, we use the fact that if $10^y = x$, then $y = \log x$.

(a) $10^5 = 100{,}000$ means that $\log 100{,}000 = 5$.
(b) $10^{-4} = 0.0001$ means that $\log 0.0001 = -4$.
(c) $10^{0.8} = 6.3096$ means that $\log 6.3096 = 0.8$. (This, too, is only an approximation because $10^{0.8}$ actually equals $6.30957344\ldots$.)

### Logarithms Are Exponents

Note that logarithms are just exponents! Thinking in terms of exponents is often a good way to answer a logarithm problem.

**Example 3** Without a calculator, evaluate the following, if possible:

(a) $\log 1$ (b) $\log 10$ (c) $\log 1{,}000{,}000$

(d) $\log 0.001$ (e) $\log \dfrac{1}{\sqrt{10}}$ (f) $\log(-100)$

**Solution**

(a) We have $\log 1 = 0$, since $10^0 = 1$.
(b) We have $\log 10 = 1$, since $10^1 = 10$.
(c) Since $1{,}000{,}000 = 10^6$, the exponent of 10 that gives 1,000,000 is 6. Thus, $\log 1{,}000{,}000 = 6$.
(d) Since $0.001 = 10^{-3}$, the exponent of 10 that gives 0.001 is $-3$. Thus, $\log 0.001 = -3$.
(e) Since $1/\sqrt{10} = 10^{-1/2}$, the exponent of 10 that gives $1/\sqrt{10}$ is $-\frac{1}{2}$. Thus $\log(1/\sqrt{10}) = -\frac{1}{2}$.
(f) Since 10 to any power is positive, $-100$ cannot be written as a power of 10. Thus, $\log(-100)$ is undefined.

## Logarithmic and Exponential Functions Are Inverses

The operation of taking a logarithm "undoes" the exponential function; the logarithm and the exponential are inverse functions. For example, $\log(10^6) = 6$ and $10^{\log 6} = 6$. In general,

For any $N$,

$$\log(10^N) = N$$

and for $N > 0$,

$$10^{\log N} = N.$$

**Example 4** Evaluate without a calculator: (a) $\log\left(10^{8.5}\right)$ (b) $10^{\log 2.7}$ (c) $10^{\log(x+3)}$

**Solution** Using $\log(10^N) = N$ and $10^{\log N} = N$, we have:

(a) $\log\left(10^{8.5}\right) = 8.5$ (b) $10^{\log 2.7} = 2.7$ (c) $10^{\log(x+3)} = x + 3$

You can check the first two results on a calculator.

## Properties of Logarithms

In Chapter 4, we saw how to solve exponential equations such as $100 \cdot 2^t = 337{,}000{,}000$ graphically. To use logarithms to solve these equations, we use the properties of logarithms, which are justified on page 192.

**Properties of the Common Logarithm**

- By definition, $y = \log x$ means $10^y = x$.
- In particular,

$$\log 1 = 0 \quad \text{and} \quad \log 10 = 1.$$

- The functions $10^x$ and $\log x$ are inverses, so they "undo" each other:

$$\begin{aligned} \log(10^x) &= x && \text{for all } x, \\ 10^{\log x} &= x && \text{for } x > 0. \end{aligned}$$

- For $a$ and $b$ both positive and any value of $t$,

$$\log(ab) = \log a + \log b$$

$$\log\left(\frac{a}{b}\right) = \log a - \log b$$

$$\log(b^t) = t \cdot \log b.$$

We can now use logarithms to solve the equation that we solved graphically in Section 4.2.

**Example 5** Solve $100 \cdot 2^t = 337{,}000{,}000$ for $t$.

**Solution** Dividing both sides of the equation by 100 gives

$$2^t = 3{,}370{,}000.$$

Taking logs of both sides gives

$$\log\left(2^t\right) = \log(3{,}370{,}000).$$

Since $\log(2^t) = t \cdot \log 2$, we have

$$t \log 2 = \log(3{,}370{,}000),$$

so, solving for $t$, we have

$$t = \frac{\log(3{,}370{,}000)}{\log 2} = 21.684.$$

In Example 2 on page 159, we found, by graphical approximation, that the fine faced by the city of Yonkers exceeded the city's annual budget between day 21 and day 22.

## The Natural Logarithm

When $e$ is used as the base for exponential functions, computations are easier with the use of another logarithm function, called log base $e$. The log base $e$ is used so frequently that it has its own notation: $\ln x$, read as the *natural log of* $x$. We make the following definition:

For $x > 0$,

$$\ln x \text{ is the power of } e \text{ that gives } x$$

or, in symbols,

$$y = \ln x \quad \text{means} \quad e^y = x,$$

and $y$ is called the **natural logarithm** of $x$.

Just as the functions $10^x$ and $\log x$ are inverses, so are $e^x$ and $\ln x$. The function $\ln x$ has similar properties to the common log function:

**Properties of the Natural Logarithm**

- By definition, $y = \ln x$ means $x = e^y$.
- In particular,
$$\ln 1 = 0 \quad \text{and} \quad \ln e = 1.$$
- The functions $e^x$ and $\ln x$ are inverses, so they "undo" each other:
$$\begin{aligned} \ln(e^x) &= x \qquad \text{for all } x \\ e^{\ln x} &= x \qquad \text{for } x > 0. \end{aligned}$$
- For $a$ and $b$ both positive and any value of $t$,
$$\begin{aligned} \ln(ab) &= \ln a + \ln b \\ \ln\left(\frac{a}{b}\right) &= \ln a - \ln b \\ \ln(b^t) &= t \cdot \ln b. \end{aligned}$$

**Example 6** Solve for $x$:

(a) $5e^{2x} = 50$ (b) $3^x = 100$.

**Solution** (a) We first divide both sides by 5 to obtain
$$e^{2x} = 10.$$
Taking the natural log of both sides, we have
$$\begin{aligned} \ln(e^{2x}) &= \ln 10 \\ 2x &= \ln 10 \\ x &= \frac{\ln 10}{2} = 1.151. \end{aligned}$$

(b) Taking natural logs of both sides,
$$\begin{aligned} \ln(3^x) &= \ln 100 \\ x \ln 3 &= \ln 100 \\ x &= \frac{\ln 100}{\ln 3} = 4.192. \end{aligned}$$

In part (a), using the natural log, rather than log base 10, simplified the calculation because the base was $e$. In part (b), we used natural logs, but we could equally well have used logs to base 10.

For more practice with logarithms, see the Skills Review on page 229.

## Misconceptions and Calculator Errors Involving Logs

It is important to know how to use the properties of logarithms. It is equally important to recognize statements that are *not* true. Beware of the following:

- $\log(a+b)$ is not the same as $\log a + \log b$
- $\log(a-b)$ is not the same as $\log a - \log b$
- $\log(ab)$ is not the same as $(\log a)(\log b)$
- $\log\left(\frac{a}{b}\right)$ is not the same as $\frac{\log a}{\log b}$
- $\log\left(\frac{1}{a}\right)$ is not the same as $\frac{1}{\log a}$.

There are no formulas to simplify either $\log(a+b)$ or $\log(a-b)$. Also, the expression $\log 5x^2$ is not the same as $2 \cdot \log 5x$, because the exponent, 2, applies only to the $x$ and not to the 5. However, it is correct to write

$$\log 5x^2 = \log 5 + \log x^2 = \log 5 + 2\log x.$$

Using a calculator to evaluate expressions like $\log(\frac{17}{3})$ requires care. On some calculators, entering log 17/3 gives 0.410, which is incorrect. This is because the calculator assumes that you mean $(\log 17)/3$, which is not the same as $\log(17/3)$. Notice also that

$$\frac{\log 17}{\log 3} = \frac{1.230}{0.477} = 2.579,$$

which is not the same as either $(\log 17)/3$ or $\log(17/3)$. Thus, the following expressions are all different:

$$\log\frac{17}{3} = 0.753, \qquad \frac{\log 17}{3} = 0.410, \qquad \text{and} \qquad \frac{\log 17}{\log 3} = 2.579.$$

## Justification of $\log(a \cdot b) = \log a + \log b$ and $\log(a/b) = \log a - \log b$

If $a$ and $b$ are both positive, we can write $a = 10^m$ and $b = 10^n$, so $\log a = m$ and $\log b = n$. Then, the product $a \cdot b$ can be written

$$a \cdot b = 10^m \cdot 10^n = 10^{m+n}.$$

Therefore $m+n$ is the power of 10 needed to give $a \cdot b$, so

$$\log(a \cdot b) = m + n,$$

which gives

$$\boxed{\log(a \cdot b) = \log a + \log b.}$$

Similarly, the quotient $a/b$ can be written as

$$\frac{a}{b} = \frac{10^m}{10^n} = 10^{m-n}.$$

Therefore, $m-n$ is the power of 10 needed to give $a/b$, so

$$\log\left(\frac{a}{b}\right) = m - n,$$

and thus

$$\boxed{\log\left(\frac{a}{b}\right) = \log a - \log b.}$$

## Justification of $\log(b^t) = t \cdot \log b$

Suppose that $b$ is positive, so we can write $b = 10^k$ for some value of $k$. Then

$$b^t = (10^k)^t.$$

We have rewritten the expression $b^t$ so that the base is a power of 10. Using a property of exponents, we can write $(10^k)^t$ as $10^{kt}$, so

$$b^t = (10^k)^t = 10^{kt}.$$

Therefore $kt$ is the power of 10 which gives $b^t$, so

$$\log(b^t) = kt.$$

But since $b = 10^k$, we know $k = \log b$. This means

$$\log(b^t) = (\log b)t = t \cdot \log b.$$

Thus, for $b > 0$ we have

$$\boxed{\log\left(b^t\right) = t \cdot \log b.}$$

## Exercises and Problems for Section 5.1

### Skill Refresher

Without using logs or a calculator, solve the equations in Exercises S1–S10 if possible.

**S1.** $10^x = 1{,}000{,}000$

**S2.** $10^t = 0.01$

**S3.** $e^z = \sqrt{e^3}$

**S4.** $10^x = 1$

**S5.** $e^w = 0$

**S6.** $e^{3x} = \dfrac{1}{e^5}$

**S7.** $\sqrt{e^{9t}} = e^7$

**S8.** $10^{-x} = -1000$

**S9.** $10^{2t} = \sqrt[4]{0.1}$

**S10.** $e^{3x} = \sqrt[3]{e^5}$

### Exercises

Rewrite the statements in Exercises 1–6 using exponents instead of logs.

**1.** $\log 19 = 1.279$

**2.** $\log 4 = 0.602$

**3.** $\ln 26 = 3.258$

**4.** $\ln(0.646) = -0.437$

**5.** $\log P = t$

**6.** $\ln q = z$

Rewrite the statements in Exercises 7–10 using logs.

**7.** $10^8 = 100{,}000{,}000$

**8.** $e^{-4} = 0.0183$

**9.** $10^v = \alpha$

**10.** $e^a = b$

In Exercises 11–28, evaluate without a calculator.

**11.** $\log 1000$

**12.** $\log\sqrt{1000}$

**13.** $\log 1$

**14.** $\log 0.1$

**15.** $\log(10^0)$

**16.** $\log\sqrt{10}$

**17.** $\log(10^5)$

**18.** $\log(10^2)$

**19.** $10^{\log 100}$

**20.** $10^{\log 1}$

**21.** $10^{\log(0.01)}$

**22.** $\ln 1$

**23.** $\ln e^0$

**24.** $\ln e^5$

**25.** $\ln\sqrt{e}$

**26.** $e^{\ln 2}$

**27.** $\log\left(\dfrac{1}{\sqrt{10}}\right)$

**28.** $\ln\left(\dfrac{1}{\sqrt{e}}\right)$

Solve the equations in Exercises 29–34 using logs.

**29.** $2^x = 11$

**30.** $(1.45)^x = 25$

**31.** $e^{0.12x} = 100$

**32.** $10 = 22(0.87)^q$

**33.** $48 = 17(2.3)^w$

**34.** $2/7 = (0.6)^{2t}$

## Problems

**35.** Express the following in terms of $x$ without logs.

**(a)** $\log 100^x$ **(b)** $1000^{\log x}$ **(c)** $\log 0.001^x$

**36.** Express the following in terms of $x$ without natural logs.

**(a)** $\ln e^{2x}$ **(b)** $e^{\ln(3x+2)}$

**(c)** $\ln\left(\frac{1}{e^{5x}}\right)$ **(d)** $\ln\sqrt{e^x}$

**37.** Evaluate the following pairs of expressions without using a calculator. What do you notice?

**(a)** $\log(10\cdot 100)$ and $\log 10 + \log 100$
**(b)** $\log(100\cdot 1000)$ and $\log 100 + \log 1000$
**(c)** $\log\left(\frac{10}{100}\right)$ and $\log 10 - \log 100$
**(d)** $\log\left(\frac{100}{1000}\right)$ and $\log 100 - \log 1000$
**(e)** $\log(10^2)$ and $2\log 10$
**(f)** $\log(10^3)$ and $3\log 10$

**38.** **(a)** Write the general formulas reflected in what you observed in Problem 37.

**(b)** Apply these formulas to rewrite $\log\left(\frac{AB}{C}\right)^p$ at least two different ways.

In Problems 39–44, is the statement true or false?

**39.** $\log AB = \log A + \log B$

**40.** $\frac{\log A}{\log B} = \log A - B$

**41.** $\log A \log B = \log A + \log B$

**42.** $p\cdot\log A = \log A^p$

**43.** $\log\sqrt{x} = \frac{1}{2}\log x$

**44.** $\sqrt{\log x} = \log(x^{1/2})$

In Problems 45–50, use properties of logarithms to find a value for $x$. Assume $a$, $b$, $M$, and $N$ are constants.

**45.** $\log(3\cdot 2^x) = 8$ **46.** $\ln(25(1.05)^x) = 6$

**47.** $\ln(ab^x) = M$ **48.** $\log(MN^x) = a$

**49.** $\ln(3x^2) = 8$ **50.** $\log(5x^3) = 2$

**51.** Let $u = \log 2$ and $v = \log 3$. Evaluate the following expressions in terms of $u$ and/or $v$. For example, $\log 9 = \log(3^2) = 2\log 3 = 2v$.

**(a)** $\log 6$ **(b)** $\log 0.08$ **(c)** $\log\sqrt{\frac{3}{2}}$ **(d)** $\log 5$

**52.** Without using a calculator, write the following quantities in terms of $\log 15$ and/or $\log 5$.

**(a)** $\log 3$ **(b)** $\log 25$ **(c)** $\log 75$

**53.** A graph of $P = 25(1.075)^t$ is given in Figure 5.1.

**(a)** What is the initial value of $P$ (when $t = 0$)? What is the percent growth rate?
**(b)** Use the graph to estimate the value of $t$ when $P = 100$.
**(c)** Use logs to find the value of $t$ when $P = 100$.

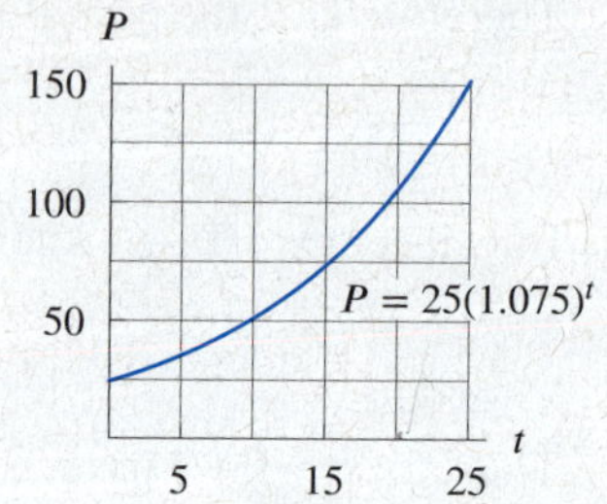


Figure 5.1

**54.** A graph of $Q = 10e^{-0.15t}$ is given in Figure 5.2.

**(a)** What is the initial value of $Q$ (when $t = 0$)? What is the continuous percent decay rate?
**(b)** Use the graph to estimate the value of $t$ when $Q = 2$.
**(c)** Use logs to find the value of $t$ when $Q = 2$.

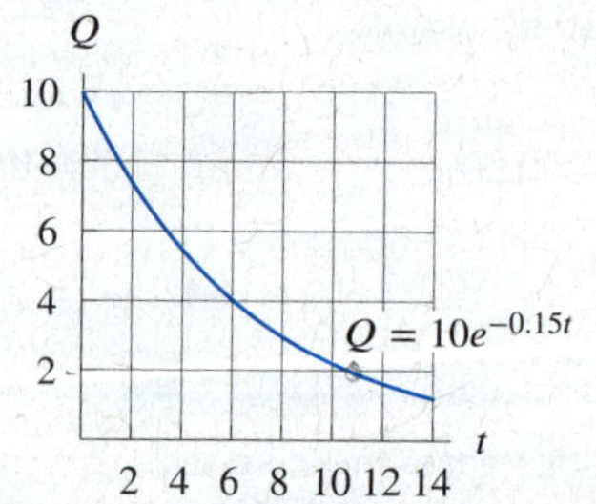


Figure 5.2

**55.** Find a possible formula for the exponential function $S$ in Figure 5.3, if $R(x) = 5.1403(1.1169)^x$.

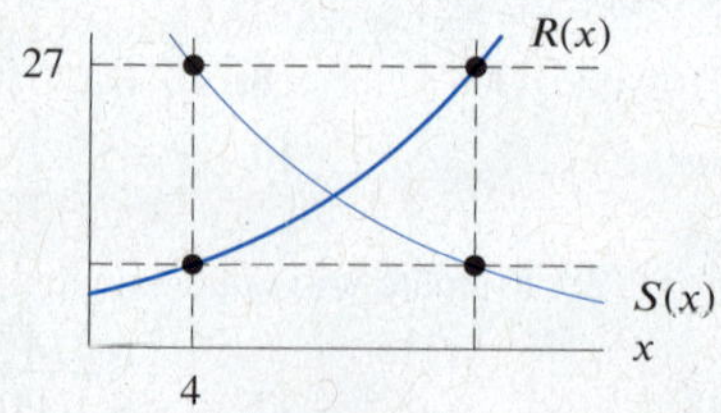


Figure 5.3

**56.** Find a possible formula for the exponential function $P$ in Figure 5.4, if $Q(x) = 117.7181(0.7517)^x$.

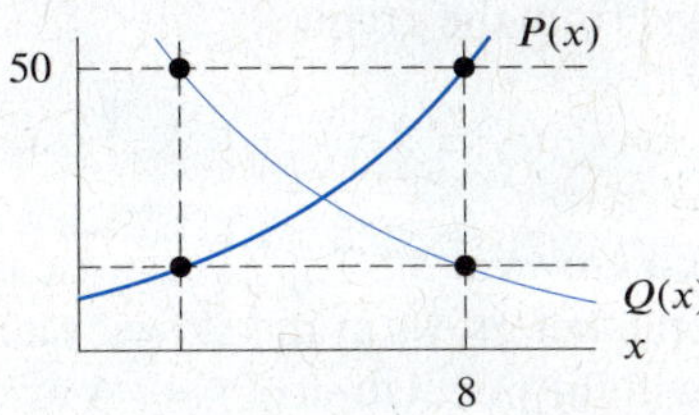


Figure 5.4

In Problems 57–76, solve the equations exactly for $x$ or $t$.

**57.** $3(1.081)^t = 14$

**58.** $84(0.74)^t = 38$

**59.** $40e^{-0.2t} = 12$

**60.** $200 \cdot 2^{t/5} = 355.$

**61.** $6000\left(\frac{1}{2}\right)^{t/15} = 1000$

**62.** $e^{x+4} = 10$

**63.** $e^{x+5} = 7 \cdot 2^x$

**64.** $100^{2x+3} = \sqrt[3]{10{,}000}$

**65.** $3^{x+4} = 10$

**66.** $0.4(\frac{1}{3})^{3x} = 7 \cdot 2^{-x}$

**67.** $\log(10^{5x+1}) = 2$

**68.** $400e^{0.1x} = 500e^{0.08x}$

**69.** $58e^{4t+1} = 30$

**70.** $ab^x = c$

**71.** $Pe^{kx} = Q$

**72.** $P - P_0e^{kx} = 0$

**73.** $\log(2x + 5) \cdot \log(9x^2) = 0$

**74.** $\log(1 - x) - \log(1 + x) = 2$

**75.** $\ln(2x + 3) \cdot \ln x^2 = 0$

**76.** $\dfrac{\log x^2 + \log x^3}{\log(100x)} = 3$

**77.** Solve each of the following equations exactly for $x$.

**(a)** $e^{2x} + e^{2x} = 1$ **(b)** $2e^{3x} + e^{3x} = b$

**78.** If we square a positive quantity $n$, we double its log: $\log\left(n^2\right) = 2\log n$. Briefly describe what happens to the log when we double a positive quantity $n$.

**79.** Because they are so large, it is impossible to compare directly the two numbers[1]

$$A = 2^{2^{2^{83}}} \quad \text{and} \quad B = 3^{3^{3^{52}}}.$$

Instead, first simplify $\ln(\ln A)$ and $\ln(\ln B)$, and then use a calculator to determine which is larger, $A$ or $B$.

**80.** Both of these numbers are slightly larger than 1:

$$A = 5^{3^{-47}} \quad \text{and} \quad B = 7^{5^{-32}}.$$

Which is larger? Explain your reasoning.

**81.** Consider the exponential function $Q = r \cdot s^t$. Letting $q = \ln Q$, show that $q$ is a linear function of $t$ by writing it in the form $q = b + mt$. State the values of $m$ and $b$.

**82.** The *arithmetic mean* of two numbers is half their sum, and the *geometric mean* is the square root of their product. Using log properties, show how we can think of the log of the geometric mean of $v$ and $w$ as the arithmetic mean of two related numbers, $p$ and $q$.

## 5.2 LOGARITHMS AND EXPONENTIAL MODELS

Logarithms are useful for solving some exponential equations symbolically, often giving more precise answers than graphical methods.

**Example 1** In Example 3 on page 160 we solved the equation $200(0.886^t) = 25$ graphically, where $t$ is in thousands of years. We found that a 200-microgram sample of carbon-14 decays to 25 micrograms in approximately 17,200 years. Now solve $200(0.886)^t = 25$ using logarithms.

**Solution** First, isolate the power on one side of the equation:

$$200(0.886^t) = 25$$
$$0.886^t = 0.125.$$

Take the log of both sides, and use the fact that $\log(0.886^t) = t\log 0.886$. Then

$$\log(0.886^t) = \log 0.125$$
$$t\log 0.886 = \log 0.125,$$

[1] Adapted from Robert P. Munafo's website on large numbers. These numbers are given as examples of Class-4 numbers. A Class-3 number is too large to be evaluated exactly on a computer; a Class-4 number is one whose *logarithm* is too large to be evaluated exactly. See http://www.mrob.com/pub/index.html, accessed April 7, 2008.

so

$$t = \frac{\log 0.125}{\log 0.886} = 17.180 \text{ thousand years.}$$

This answer is close to the value of 17,200 that we found from the graph.

**Example 2** The population of City A begins with 50,000 people and grows at 3.5% per year. The population of City B begins with a larger population of 250,000 people but grows at the slower rate of 1.6% per year. Assuming that these growth rates hold constant, will the population of City A ever catch up to the population of City B? If so, when?

**Solution** If $t$ is time measured in years and $P_A$ and $P_B$ are the populations of these two cities, then

$$P_A = 50{,}000(1.035)^t \quad \text{and} \quad P_B = 250{,}000(1.016)^t.$$

We want to solve the equation

$$50{,}000(1.035)^t = 250{,}000(1.016)^t.$$

We first get the exponential terms together by dividing both sides of the equation by $50{,}000(1.016)^t$:

$$\frac{(1.035)^t}{(1.016)^t} = \frac{250{,}000}{50{,}000} = 5.$$

Since $\frac{a^t}{b^t} = \left(\frac{a}{b}\right)^t$, this gives

$$\left(\frac{1.035}{1.016}\right)^t = 5.$$

Taking logs of both sides and using $\log b^t = t \log b$, we have

$$\log\left(\frac{1.035}{1.016}\right)^t = \log 5$$

$$t \log\left(\frac{1.035}{1.016}\right) = \log 5$$

$$t = \frac{\log 5}{\log(1.035/1.016)} \approx 86.865.$$

Thus, the cities' populations will be equal in just under 87 years. To check this, notice that when $t = 86.865$,

$$P_A = 50{,}000(1.035)^{86.865} = 992{,}575$$

and

$$P_B = 250{,}000(1.016)^{86.865} = 992{,}572.$$

The answers are not exactly equal because we rounded off the value of $t$. Rounding can introduce significant errors, especially when logs and exponentials are involved. Using $t = 86.86480867$, the computed values of $P_A$ and $P_B$ agree to three decimal places.

## Doubling Time

Eventually, any exponentially growing quantity doubles, or increases by 100%. Since its percent growth rate is constant, the time it takes for the quantity to grow by 100% is also a constant. This time period is called the *doubling time*.

**Example 3** (a) Find the time needed for the turtle population described by the function $P = 175(1.145)^t$ to double its initial size.

(b) How long does this population take to quadruple its initial size? To increase by a factor of 8?

Solution (a) The initial size is 175 turtles; doubling this gives 350 turtles. We need to solve the following equation for $t$:

$$\begin{aligned} 175(1.145)^t &= 350 \\ 1.145^t &= 2 \\ \log\left(1.145^t\right) &= \log 2 \\ t \cdot \log 1.145 &= \log 2 \\ t &= \frac{\log 2}{\log 1.145} \approx 5.119 \text{ years.} \end{aligned}$$

We check this by noting that

$$175(1.145)^{5.119} = 350,$$

which is double the initial population. In fact, at any time it takes the turtle population about 5.119 years to double in size.

(b) Since the population function is exponential, it increases by 100% every 5.119 years. Thus it doubles its initial size in the first 5.119 years, quadruples its initial size in two 5.119-year periods, or 10.238 years, and increases by a factor of 8 in three 5.119-year periods, or 15.357 years. We check this by noting that

$$175(1.145)^{10.238} = 700,$$

or 4 times the initial size, and that

$$175(1.145)^{15.357} = 1400,$$

or 8 times the initial size.

**Example 4** A population doubles in size every 20 years. What is its continuous growth rate per year?

Solution We are not given the initial size of the population, but we do not need this information. Let $P_0$ represent the initial size of the population and $P = P_0e^{kt}$. After 20 years, $P = 2P_0$, so we solve

$$P_0e^{k\cdot 20} = 2P_0.$$

Canceling $P_0$ on both sides we have

$$\begin{aligned} e^{20k} &= 2 \\ 20k &= \ln 2 \quad \text{Taking ln of both sides} \\ k &= \frac{\ln 2}{20} = 0.03466. \end{aligned}$$

Thus, the population grows at the continuous rate of 3.466% per year.

## Half-Life

Just as an exponentially growing quantity doubles in a fixed amount of time, an exponentially decaying quantity decreases by a factor of 2 in a fixed amount of time, called the *half-life*.

**Example 5** Carbon-14 decays radioactively at a constant annual rate of 0.0121%. Check that the half-life of carbon-14 is about 5728 years and show how this value is obtained from the decay rate.

Solution We are not given an initial amount of carbon-14, but we do not need it. Let the symbol $Q_0$ represent the initial quantity of carbon-14. The growth rate is −0.000121 because carbon-14 is decaying. So the growth factor is $b = 1 - 0.000121 = 0.999879$. Thus, after $t$ years the amount left is

$$Q = Q_0(0.999879)^t.$$

Substituting $t = 5728$ years, we see that the quantity remaining is half the initial quantity:

$$Q = Q_0(0.999879)^{5728} = 0.50Q_0.$$

Thus the half-life of carbon-14 is 5728 years.

If we did not know the half-life, we could find it by determining how long it takes for the quantity to drop to half its initial level. We solve for $t$ in the equation

$$\frac{1}{2}Q_0 = Q_0(0.999879)^t.$$

Dividing each side by $Q_0$, we have

$$\frac{1}{2} = 0.999879^t.$$

Taking logs, we have

$$\begin{aligned} \log\frac{1}{2} &= \log\left(0.999879^t\right) \\ \log 0.5 &= t \cdot \log 0.999879 \\ t &= \frac{\log 0.5}{\log 0.999879} = 5728.143. \end{aligned}$$

Thus, no matter how much carbon-14 there is initially, after about 5728 years, half will remain.

Similarly, we can determine the growth rate given the half-life or doubling time.

**Example 6** The quantity, $Q$, of a substance decays according to the formula $Q = Q_0e^{-kt}$, where $t$ is in minutes. The half-life of the substance is 11 minutes. What is the value of $k$?

**Solution** We know that after 11 minutes, $Q = \frac{1}{2}Q_0$. Thus, solving for $k$, we get

$$\begin{aligned} Q_0e^{-k\cdot 11} &= \frac{1}{2}Q_0 \\ e^{-11k} &= \frac{1}{2} \\ -11k &= \ln\frac{1}{2} \\ k &= \frac{\ln(1/2)}{-11} = 0.06301. \end{aligned}$$

This substance decays at the continuous rate of 6.301% per minute.

## Converting Between $Q = ab^t$ and $Q = ae^{kt}$

Any exponential function can be written in either of the two forms:

$$Q = ab^t \qquad \text{or} \qquad Q = ae^{kt}.$$

Logarithms enable us to convert between the two. If $b = e^k$, so $k = \ln b$, the two formulas represent the same function.

**Example 7** Convert the exponential function $P = 175(1.145)^t$ to the form $P = ae^{kt}$.

**Solution** Since the new formula represents the same function, we want $P = 175$ when $t = 0$. Thus, substituting $t = 0$ gives $175 = ae^{k(0)} = a$, so $a = 175$. The parameter $a$ in both functions represents the initial population. For all $t$,

$$175(1.145)^t = 175(e^k)^t,$$

so we must find $k$ such that

$$e^k = 1.145.$$

Therefore $k$ is the power of $e$ that gives 1.145. Applying the natural log to both sides, we have

$$k = \ln 1.145 = 0.1354.$$

Therefore,

$$P = 175e^{0.1354t}.$$

**Example 8** Convert the formula $Q = 7e^{0.3t}$ to the form $Q = ab^t$.

**Solution** Using the properties of exponents,

$$Q = 7e^{0.3t} = 7(e^{0.3})^t.$$

Using a calculator, we find $e^{0.3} = 1.3499$, so

$$Q = 7(1.3499)^t.$$

Notice that logs are needed to convert $Q = ab^t$ to $Q = ae^{kt}$, but not the other way.

**Example 9** Assuming $t$ is in years, find the continuous and annual percent growth rates in Examples 7 and 8.

**Solution** In Example 7, the annual percent growth rate is 14.5% and the continuous percent growth rate per year is 13.54%. In Example 8, the continuous percent growth rate is 30% and the annual percent growth rate is 34.99%.

**Example 10** Find the continuous percent growth rate of $Q = 200(0.886)^t$, where $t$ is in thousands of years.

**Solution** Since this function describes exponential decay, we expect a negative value for $k$. We want

$$e^k = 0.886.$$

Solving for $k$ gives

$$k = \ln(0.886) = -0.12104.$$

So we have $Q = 200e^{-0.12104t}$ and the continuous growth rate is $-12.104\%$ per thousand years.

## Exponential Growth Problems That Cannot Be Solved by Logarithms

Some equations with the variable in the exponent cannot be solved using logarithms.

**Example 11** With $t$ in years, the population of a country (in millions) is given by $P = 2(1.02)^t$, while the food supply (in millions of people that can be fed) is given by $N = 4 + 0.5t$. Determine the year in which the country first experiences food shortages.

**Solution** The country starts to experience shortages when the population equals the number of people that can be fed—that is, when $P = N$. We attempt to solve the equation $P = N$ by using logs:

$$\begin{aligned} 2(1.02)^t &= 4 + 0.5t \\ 1.02^t &= 2 + 0.25t \quad \text{Dividing by 2} \\ \log 1.02^t &= \log(2 + 0.25t) \\ t \log 1.02 &= \log(2 + 0.25t). \end{aligned}$$

Unfortunately, we cannot isolate $t$, so this equation cannot be solved using logs. However, we can approximate the solution of the original equation numerically or graphically.[2] See Figure 5.5. The values of $P$ and $N$ are equal when $t \approx 199.381$. Thus, it is almost 200 years before shortages occur.

[2]Compare Example 3 in Section 4.2.

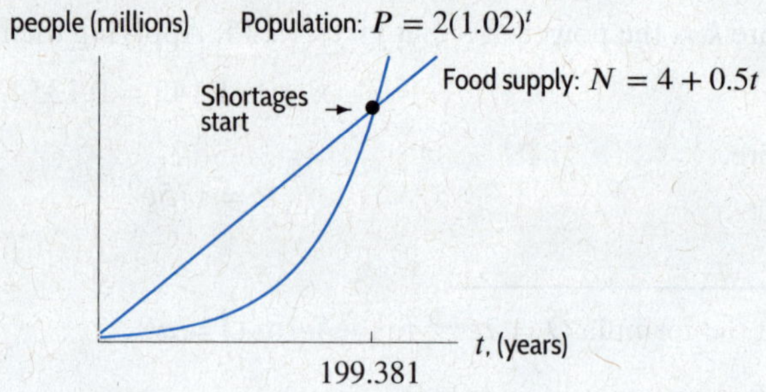


**Figure 5.5:** Finding the intersection of linear and exponential graphs

## Exercises and Problems for Section 5.2

### Skill Refresher

For Exercises S1–S4, simplify the expression if possible.

**S1.** $10^{-\log 5x}$ **S2.** $e^{-3\ln t}$

**S3.** $t \ln e^{t/2}$ **S4.** $10^{2+\log x}$

In Exercises S5–S10, solve for $x$.

**S5.** $4^x = 9$ **S6.** $e^x = 8$

**S7.** $2e^x = 13$ **S8.** $e^{7x} = 5e^{3x}$

**S9.** $\log(2x + 7) = 2$ **S10.** $\log(2x) = \log(x + 10)$

### Exercises

In Exercises 1–4, convert to the form $Q = ab^t$.

**1.** $Q = 4e^{7t}$ **2.** $Q = 0.3e^{0.7t}$

**3.** $Q = \frac{14}{5}e^{0.03t}$ **4.** $Q = e^{-0.02t}$

For Exercises 5–6, write the exponential function in the form $y = ab^t$. Find $b$ accurate to four decimal places. If $t$ is measured in years, give the percent annual growth or decay rate and the continuous percent growth or decay rate per year.

**5.** $y = 25e^{0.053t}$ **6.** $y = 100e^{-0.07t}$

In Exercises 7–10, convert to the form $Q = ae^{kt}$.

**7.** $Q = 12(0.9)^t$ **8.** $Q = 16(0.487)^t$

**9.** $Q = 14(0.862)^{1.4t}$ **10.** $Q = 721(0.98)^{0.7t}$

For Exercises 11–12, write the exponential function in the form $y = ae^{kt}$. Find $k$ accurate to four decimal places. If $t$ is measured in years, give the percent annual growth rate and the continuous percent growth rate per year.

**11.** $y = 6000(0.85)^t$ **12.** $y = 5(1.12)^t$

In Exercises 13–20, give the starting value $a$, the growth rate $r$, and the continuous growth rate $k$.

**13.** $Q = 230(1.182)^t$ **14.** $Q = 0.181\left(e^{0.775}\right)^t$

**15.** $Q = 0.81(2)^t$ **16.** $Q = 5 \cdot 2^{t/8}$

**17.** $Q = 12.1 \cdot 10^{-0.11t}$ **18.** $Q = 40e^{(t-5)/12}$

**19.** $Q = 2e^{(1-3t/4)}$ **20.** $Q = 2^{-(t-5)/3}$

### Problems

Find the doubling time in Exercises 21–24.

**21.** A bank account is growing by 2.7% per year.

**22.** A population is growing according to $P = P_0e^{0.2t}$.

**23.** The population of a city is growing by 26% per year.

**24.** A company's profits are increasing by an annual growth factor of 1.12.

Find the half-lives of the substances in Exercises 25–27.

**25.** Einsteinium-253, which decays at a rate of 3.406% per day.

**26.** Tritium, which decays at a rate of 5.471% per year.

**27.** A radioactive substance that decays at a continuous rate of 11% per minute.

**28.** A population grows from 11000 to 13000 in three years. Assuming the growth is exponential, find the:

**(a)** Annual growth rate **(b)** Continuous growth rate

**(c)** Why are your answers to parts (a) and (b) different?

**29.** A population doubles in size every 15 years. Assuming exponential growth, find the

**(a)** Annual growth rate **(b)** Continuous growth rate

**30.** A population increases from 5.2 million at an annual rate of 3.1%. Find the continuous growth rate.

**31.** You place \$800 in an account that earns 4% annual interest, compounded annually. How long will it be until you have \$2000?

**32.** A \$5000 investment earns 7.2% annual interest, and an \$8000 investment earns 5.4%, both compounded annually. How long will it take for the smaller investment to catch up to the larger one?

**33.** A \$9000 investment earns 5.6% annual interest, and a \$4000 investment earns 8.3%, both compounded continuously. When will the smaller catch up to the larger?

**34.** **(a)** What annual interest rate, compounded continuously, is equivalent to an annual rate of 8%, compounded annually?

**(b)** What annual interest rate, compounded annually, is equivalent to an annual rate of 6%, compounded continuously?

**35.** If 17% of a radioactive substance decays in 5 hours, what is the half-life of the substance?

**36.** The half-life of nicotine in the body is 2 hours. What is the continuous decay rate?

**37.** Sketch the exponential function $y = u(t)$ given that it has a starting value of 0.8 and a doubling time of 12 years. Label the axes and indicate the scale.

**38.** Total power generated by wind worldwide doubles every 3 years.[3] In 2008, world wind-energy generating capacity was about 90 thousand megawatts. Find the continuous growth rate and give a formula for wind-generating capacity $W$ (in thousand megawatts) as a function of $t$, the number of years since 2008.

**39.** A town has 5000 people in year $t = 0$. Calculate how long it takes for the population $P$ to double once, twice, and three times, assuming that the town grows at a constant rate of

**(a)** 500 people per year.

**(b)** 5% per year.

**40.** **(a)** Find the time required for an investment to triple in value if it earns 4% annual interest, compounded continuously.

**(b)** Now find the time required assuming that the interest is compounded annually.

[3] World Wind Energy Report 2008.

**41.** **(a)** Estimate the doubling time of the exponential function shown in Figure 5.6.

**(b)** Use the doubling time to find the continuous percent growth rate and give a formula for the function.

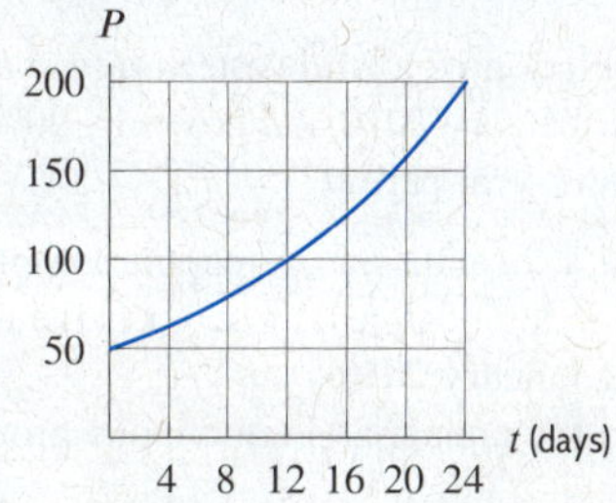


Figure 5.6

**42.** **(a)** The quantity of caffeine in the body after drinking a cup of coffee is shown in Figure 5.7. Estimate the half-life of caffeine.

**(b)** Use the half-life to find the continuous percent decay rate and give a formula for $Q$ as function of $t$.

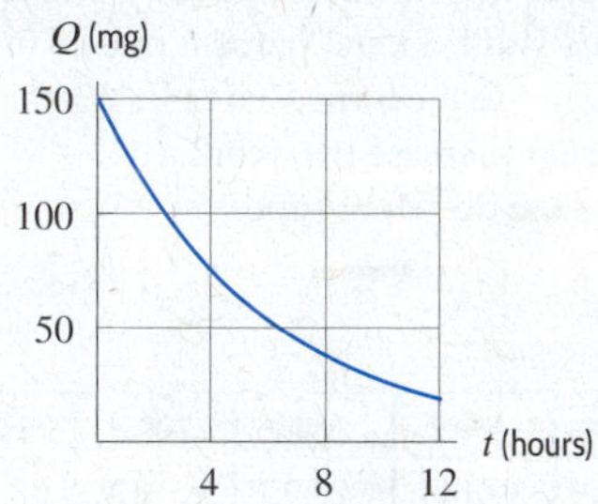


Figure 5.7

**43.** The temperature, $H$, in °F, of a cup of coffee $t$ hours after it is set out to cool is given by the equation:

$$H = 70 + 120(1/4)^t.$$

**(a)** What is the coffee's temperature initially (that is, at time $t = 0$)? After 1 hour? 2 hours?

**(b)** How long does it take the coffee to cool down to 90°F? 75°F?

**44.** Use algebra to show that the time it takes for a quantity growing exponentially to double is independent of the starting quantity and the time. To do this, let $d$ represent the time it takes for $P$ to double. Show that if $P$ becomes $2P$ at time $t+d$, then $d$ depends only on the growth factor $b$, but not on the starting quantity $a$ and time $t$. (Assume $P \neq 0$.)

**45.** Prices climb at a constant 3% annual rate.

**(a)** By what percent will prices have climbed after 5 years?

**(b)** How long will it take for prices to climb 25%?

**46.** The growth of an animal population, $P$, is described by the function $P = 300 \cdot 2^{t/20}$.

**(a)** How large is this population in year $t = 0$? $t = 20$?
**(b)** When does this population reach 1000?

**47.** The population of a Midwestern city starts with 25,000 people in January 2010 and grows by 20% over each successive four-year period.

**(a)** Find a formula for the population of the city, $P$, as a function of time, $t$, where $t$ is the number of years after January 2010.
**(b)** Find the annual and continuous growth rates of the city.
**(c)** In what year does the population of the city reach 40,000?

**48.** The world's population is aging. The approximate world population age 80 or older[4] is given in Table 5.1.

**(a)** Find a formula for $P$, the number of people in the world age 80 or older, in millions, as a function of time, $t$, in years since 2005. Use the form $P = ab^t$. What is the annual percent rate of increase?
**(b)** Convert to the form $P = ae^{kt}$. What is the continuous percent increase per year?
**(c)** Find the doubling time.

**Table 5.1**

| $t$ (year) | 2005 | 2006 | 2007 | 2008 | 2009 |
|---|---|---|---|---|---|
| $P$ (millions) | 89.144 | 92.175 | 95.309 | 98.550 | 101.901 |

**49.** The US census[5] projects the population of the state of Washington using the function $N(t) = 6.72e^{0.026t}$, where $N(t)$ is in millions and $t$ is in years since 2012.

**(a)** What is the population's continuous growth rate?
**(b)** What is the population of Washington in year $t = 0$?
**(c)** How many years is it before the population triples?
**(d)** In what year does this model indicate a population of only one person? Is this reasonable or unreasonable?

**50.** Technetium-99m is a radioactive substance used to diagnose brain diseases. Its half-life is approximately 6 hours. Initially you have 200 mg of technetium-99m.

**(a)** Write an equation that gives the amount of technetium-99m remaining after $t$ hours.
**(b)** Determine the number of hours needed for your sample to decay to 120 mg.
**(c)** Determine the concavity of the graph that models the half-life of technetium-99m using average rates of change over intervals of length 2 between $t = 0$ and $t = 6$.

**51.** A manager at Saks Fifth Avenue wants to estimate the number of customers to expect on the last shopping day before Christmas. She collects data from three previous years, and determines that the crowds follow the same general pattern. When the store opens at 10 am, 500 people enter, and the total number in the store doubles every 40 minutes. When the number of people in the store reaches 10,000, security guards need to be stationed at the entrances to control the crowds. At what time should the guards be commissioned?

**52.** In 1991, the body of a man was found in melting snow in the Alps of Northern Italy. An examination of the tissue sample revealed that 46% of the carbon-14 present in his body at the time of his death had decayed. The half-life of carbon-14 is approximately 5728 years. How long ago did this man die?

**53.** Figure 5.8 shows the graphs of the exponential functions $f$ and $g$, and the linear function, $h$.

**(a)** Find formulas for $f$, $g$, and $h$.
**(b)** Find the exact value(s) of $x$ such that $f(x) = g(x)$.
**(c)** Estimate the value(s) of $x$ such that $f(x) = h(x)$.

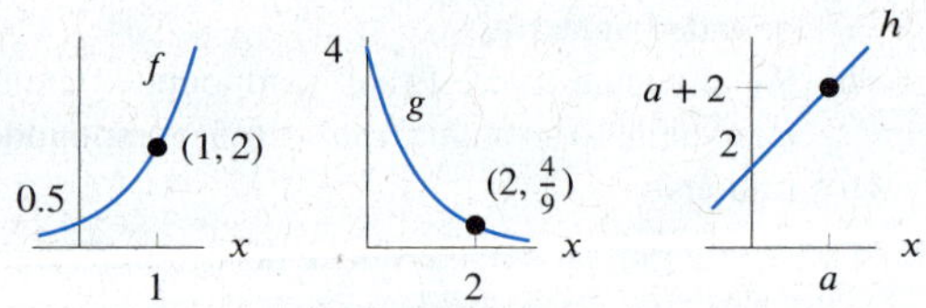


**Figure 5.8**

**54.** A person's blood alcohol content (BAC) is a measure of how much alcohol is in the blood stream. When a person stops drinking, the BAC declines over time as the alcohol is metabolized. The BAC, $Q$, of a person $t$ minutes after the he stops drinking is given by

$$Q = Q_0 e^{-t/\tau},$$

where $Q_0$ is the person's initial BAC and $\tau$ is known as the *elimination time*. How long does it take for a person's BAC to drop from 0.10 to 0.04 if the elimination time is 2.5 hours?

**55.** The size of a population, $P$, of toads $t$ years after it is introduced into a wetland is given by

$$P = \frac{1000}{1 + 49(1/2)^t}.$$

**(a)** How many toads are there in year $t = 0$? $t = 5$? $t = 10$?

[4]UN Department of Economic and Social Affairs, 2009.
[5]http://quickfacts.census.gov/qfd/states/53000.html, accessed July 23, 2013.

**(b)** How long does it take for the toad population to reach 500? 750?

**(c)** What is the maximum number of toads that the wetland can support?

Problems 56–57 involve the Rule of 70, which gives quick estimates of the doubling time of an exponentially growing quantity. If $r\%$ is the annual growth rate of the quantity, then the Rule of 70 says

$$\text{Doubling time in years} \approx \frac{70}{r}.$$

**56.** Use the Rule of 70 to estimate how long it takes a \$1000 investment to double if it grows at the following annual rates: 1%, 2%, 5%, 7%, 10%. Compare with the actual doubling times.

**57.** Using natural logs, solve for the doubling time for $Q = ae^{kt}$. Use your result to explain why the Rule of 70 works.

**58.** A food chain connects groups of living beings that feed on each other. The bottom level, perhaps grass, stores 25,000 kilocalories (kcals) per square meter per year. The next level, rabbits, feeds on the grass. The third level, foxes, feeds on the rabbits. Each level extracts some of the kilocalories from the previous level; the *ecological efficiency* of a food chain is the proportion of energy transferred from one level to the next. A food chain may have more than three levels.

**(a)** How much energy is transferred to the third level if the ecological efficiency is

(i) 10%? (ii) 8%? (iii) 5%?

**(b)** The predator at the top level requires 30 kcal per square meter per year. What is the highest level that the top predator can occupy if the ecological efficiency is

(i) 12%? (ii) 10%? (iii) 8%?

**(c)** What ecological efficiency enables this top predator to occupy the fourth level of the food chain?

**59.** Find values for $a, b, k, s$ where

$$f(t) = ab^t = ae^{kt} = a \cdot 2^{t/s},$$

given that $f(-20) = 5$ and $f(40) = 30$.

**60.** Write the exponential function $y = ab^t$ in the form

$$y = e^{k(t-t_0)}.$$

Give $k$ and $t_0$ in terms of $a$ and $b$.

**61.** *Gompertz functions* can be used to model population growth.[6] Solve $f(t) = 3$ for $t$ for the particular Gompertz function

$$f(t) = 6e^{-0.5e^{-0.1t}}.$$

**62.** **(a)** Rewrite the equation $23(1.36)^t = 85$ in the form $e^{k+rt} = e^s$. State the values of the constants $k, r$, and $s$.

**(b)** Solve the original equation in terms of $k, r$, and $s$, then give a numerical approximation.

**63.** **(a)** Rewrite the equation $1.12^t = 6.3$ in the form $10^{vt} = 10^w$. State the values of the constants $v$ and $w$.

**(b)** Solve the original equation in terms of $v$ and $w$, then give a numerical approximation.

# 5.3 THE LOGARITHMIC FUNCTION AND ITS APPLICATIONS

## The Graph, Domain, and Range of the Common Logarithm

In Section 5.1 we defined $y = \log x$, the log function (to base 10), for all positive numbers. Thus,

$$\text{Domain of } \log x \text{ is all positive numbers.}$$

We next determine the range of $y = \log x$, which is graphed in Figure 5.9, The log graph crosses the $x$-axis at $x = 1$, because $\log 1 = \log(10^0) = 0$. The graph climbs to $y = 1$ at $x = 10$, because $\log 10 = \log(10^1) = 1$. In order for the log graph to climb to $y = 2$, the value of $x$ must reach 100, or $10^2$, and in order for it to climb to $y = 3$, the value of $x$ must be $10^3$, or 1000. To reach the modest height of $y = 20$ requires $x$ to equal $10^{20}$, or 100 billion billion! The log function increases so slowly that it often serves as a benchmark for other slow-growing functions. Nonetheless, the graph of $y = \log x$ eventually climbs to any value we choose.

[6] See http://en.wikipedia.org/wiki/Gompertz_curve, accessed April 13, 2008.

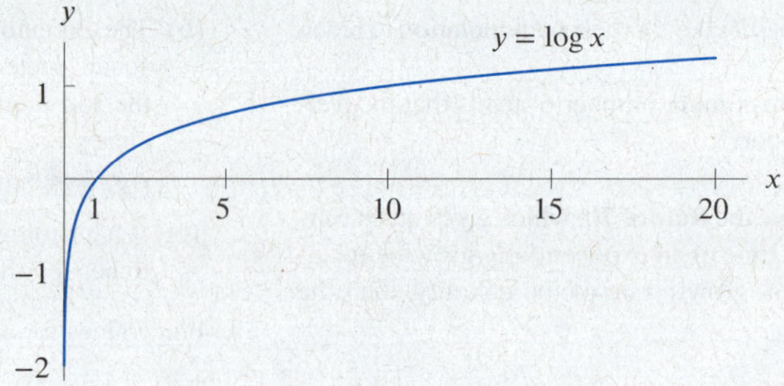


**Figure 5.9**: The log function grows very rapidly for $0 < x < 1$ and very slowly for $x > 1$. It has a vertical asymptote at $x = 0$ and never touches the $y$-axis

Although $x$ cannot equal zero in the log function, we can choose $x > 0$ to be as small as we like. As $x$ decreases toward zero, the values of $\log x$ get large and negative. For example,

$$\begin{aligned} \log 0.1 &= \log 10^{-1} = -1, \\ \log 0.01 &= \log 10^{-2} = -2, \\ &\vdots \\ \log 0.0000001 &= \log 10^{-7} = -7, \qquad \text{and so on.} \end{aligned}$$

So small positive values of $x$ give exceedingly large negative values of $y$. We say that the graph of $\log x$ has a *vertical asymptote* at $x = 0$ and

$$\text{Range of } \log x \text{ is all real numbers.}$$

## Rate of Change and Concavity

The rate of change of the log function between any two points is positive since the function is increasing. Notice also that the rate of change decreases as the two points move to the right, because the curve bends downward. The graph is concave down.

## Graphs of the Inverse Functions $y = \log x$ and $y = 10^x$

The fact that $y = \log x$ and $y = 10^x$ are inverses means that their graphs are related. Looking at Tables 5.2 and 5.3, we see that the point $(0.01, -2)$ is on the graph of $y = \log x$ and the point $(-2, 0.01)$ is on the graph of $y = 10^x$. In general, if the point $(a, b)$ is on the graph of $y = \log x$, the point $(b, a)$ is on the graph of $y = 10^x$. Thus, the graph of $y = \log x$ is the graph of $y = 10^x$ with $x$- and $y$-axes interchanged. If the $x$- and $y$-axes have the same scale, this is equivalent to reflecting the graph of $y = 10^x$ across the diagonal line $y = x$. See Figure 5.10.

**Table 5.2** *Log function*

| $x$ | $y = \log x$ |
|---|---|
| 0.01 | −2 |
| 0.1 | −1 |
| 1 | 0 |
| 10 | 1 |
| 100 | 2 |
| 1000 | 3 |

**Table 5.3** *Exponential function*

| $x$ | $y = 10^x$ |
|---|---|
| −2 | 0.01 |
| −1 | 0.1 |
| 0 | 1 |
| 1 | 10 |
| 2 | 100 |
| 3 | 1000 |

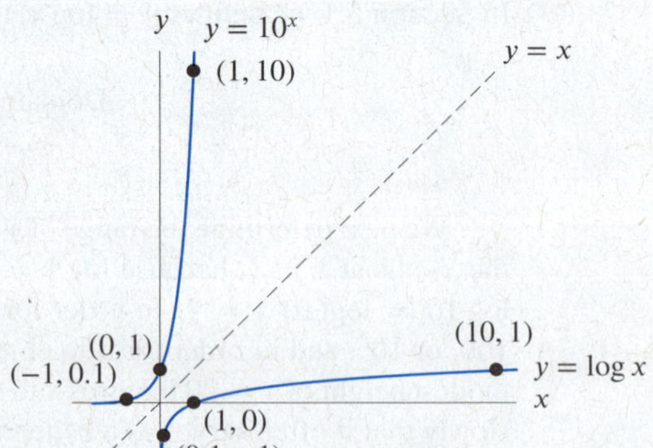


**Figure 5.10**: The functions $y = \log x$ and $y = 10^x$ are inverses of one another

## Graph of Natural Logarithm: Rate of Change and Concavity

In addition to similar algebraic properties, the natural log and the common log have similar graphs. Like the common log, the natural log is an increasing function with a graph that is concave down. Thus, the rates of change of the natural log decrease as we move from left to right.

**Example 1** Graph $y = \ln x$ for $0 < x < 10$.

**Solution** Values of $\ln x$ are in Table 5.4. Like the common log, the natural log is only defined for $x > 0$ and has a vertical asymptote at $x = 0$. The graph is slowly increasing and concave down.

**Table 5.4** *Values of* $\ln x$ *(rounded)*

| $x$ | $\ln x$ |
|---|---|
| 0 | Undefined |
| 1 | 0 |
| 2 | 0.7 |
| $e$ | 1 |
| 3 | 1.1 |
| 4 | 1.4 |
| ⋮ | ⋮ |

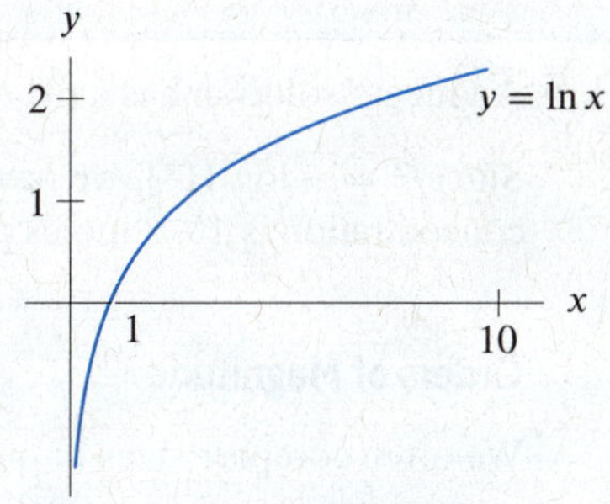


**Figure 5.11**: Graph of the natural logarithm

The functions $y = \ln x$ and $y = e^x$ are inverses. If the scales on the axes are the same, their graphs are reflections of one another across the line $y = x$. See Figure 5.12. For example, the vertical asymptote of the logarithm is the reflection of the horizontal asymptote of the exponential. On page 209, we see how to write asymptotes in limit notation.

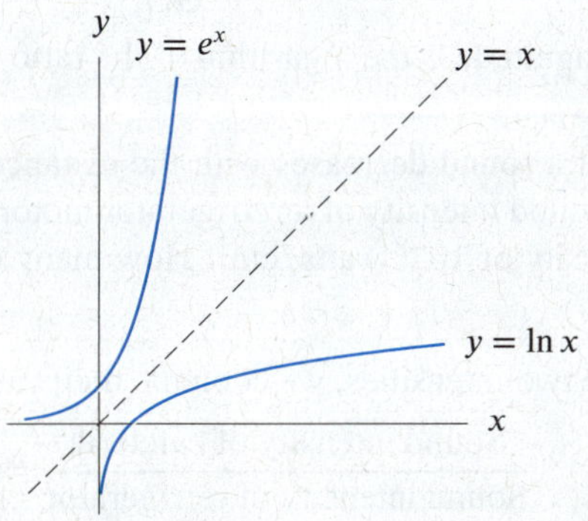


**Figure 5.12**: The functions $y = \ln x$ and $y = e^x$ are inverses of one another

## Applications of Logarithms

Figure 5.9 shows that, due to their slow growth, logs can shrink large ranges of numbers into smaller ranges: The log of every number between 1 and 1 billion is between 0 and 9. The figure also shows that logs grow fast between 0 and 1: Numbers packed between $10^{-9}$ and 1 have logs spread out between $-9$ to 0.

This property of logs makes them useful in measuring quantities whose magnitudes vary widely over large ranges (sound intensity and earthquakes) or over small ranges (chemical acidity).

### Chemical Acidity

In chemistry, the acidity of a liquid is expressed using pH. The acidity depends on the hydrogen ion concentration in the liquid (in moles per liter); this concentration, written $[H^+]$, varies in a small range between 0 and 1. The greater the hydrogen ion concentration, the more acidic the solution. The pH is defined as:

$$\text{pH} = -\log[H^+].$$

**Example 2** The hydrogen ion concentration of seawater is $[H^+] = 1.1 \cdot 10^{-8}$ moles per liter. Estimate the pH of seawater. Then check your answer with a calculator.

**Solution** We want to estimate $\text{pH} = -\log(1.1 \cdot 10^{-8})$. Since $1.1 \cdot 10^{-8} \approx 10^{-8}$ and $\log 10^{-8} = -8$, we know that

$$\text{pH} = -\log(1.1 \cdot 10^{-8}) \approx -(-8) = 8.$$

Using a calculator, we have

$$\text{pH} = -\log(1.1 \cdot 10^{-8}) = 7.959.$$

**Example 3** A vinegar solution has a pH of 3. Determine the hydrogen ion concentration.

**Solution** Since $3 = -\log[H^+]$, we have $-3 = \log[H^+]$. This means that $10^{-3} = [H^+]$. So the hydrogen ion concentration is $10^{-3}$ moles per liter.

## Orders of Magnitude

We often compare sizes or quantities by computing their ratios. If $A$ is twice as tall as $B$, then

$$\frac{\text{Height of } A}{\text{Height of } B} = 2.$$

If one object is 10 times heavier than another, we say it is an *order of magnitude* heavier. If one quantity is two factors of 10 greater than another, we say it is two orders of magnitude greater, and so on. For example, the value of a dollar is two orders of magnitude greater than the value of a penny, because we have

$$\frac{\$1}{\$0.01} = 100 = 10^2.$$

The order of magnitude is the logarithm of the ratio of their magnitudes.

**Example 4** The intensity of a sound decreases with the distance from its source and is measured in watts per unit area. The sound intensity of a refrigerator motor is $10^{-11}$ watts/cm$^2$. A typical school cafeteria has sound intensity of $10^{-8}$ watts/cm$^2$. How many orders of magnitude more intense is the sound of the cafeteria?

**Solution** To compare the two intensities, we compute their ratio:

$$\frac{\text{Sound intensity of cafeteria}}{\text{Sound intensity of refrigerator}} = \frac{10^{-8}}{10^{-11}} = 10^{-8-(-11)} = 10^3.$$

Thus, the sound intensity of the cafeteria is 1000 times greater than the sound intensity of the refrigerator. The log of this ratio is 3. We say that the sound intensity of the cafeteria is three orders of magnitude greater than the sound intensity of the refrigerator.

## Decibels

The intensity of audible sound varies over an enormous range. The range is so enormous that we consider the logarithm of the sound intensity. This is the idea behind the *decibel* (abbreviated dB). To measure a sound in decibels, the sound's intensity, $I$, is compared to the intensity of a standard benchmark sound, $I_0$. The intensity of $I_0$ is defined to be $10^{-16}$ watts/cm$^2$, roughly the lowest intensity audible to humans. The comparison between a sound intensity $I$ and the benchmark sound intensity $I_0$ is made as follows:

$$\text{Noise level in decibels} = 10 \cdot \log\left(\frac{I}{I_0}\right).$$

For instance, let's find the decibel rating of the refrigerator in Example 4. First, we find how many orders of magnitude more intense the refrigerator sound is than the benchmark sound:

$$\frac{I}{I_0} = \frac{\text{Sound intensity of refrigerator}}{\text{Benchmark sound intensity}} = \frac{10^{-11}}{10^{-16}} = 10^5.$$

Thus, the refrigerator's intensity is 5 orders of magnitude more than $I_0$, the benchmark intensity. We have

$$\text{Decibel rating of refrigerator} = 10 \cdot \underbrace{\text{Number of orders of magnitude}}_{5} = 50 \text{ dB}.$$

Note that 5, the number of orders of magnitude, is the log of the ratio $I/I_0$. We use the log function because it "counts" the number of powers of 10. Thus if $N$ is the decibel rating, then

$$N = 10 \log \left( \frac{I}{I_0} \right).$$

---

**Example 5** (a) If a sound doubles in intensity, by how many units does its decibel rating increase?

(b) Loud music can measure 110 dB whereas normal conversation measures 50 dB. How many times more intense is loud music than normal conversation?

**Solution** (a) Let $I$ be the sound's intensity before it doubles. Once doubled, the new intensity is $2I$. The decibel rating of the original sound is $10 \log(I/I_0)$, and the decibel rating of the new sound is $10 \log(2I/I_0)$. The difference in decibel ratings is given by

$$\begin{aligned}
\text{Difference in decibel ratings} &= 10 \log \left( \frac{2I}{I_0} \right) - 10 \log \left( \frac{I}{I_0} \right) \\
&= 10 \left( \log \left( \frac{2I}{I_0} \right) - \log \left( \frac{I}{I_0} \right) \right) && \text{Factoring out 10} \\
&= 10 \cdot \log \left( \frac{2I/I_0}{I/I_0} \right) && \text{Using the property } \log a - \log b = \log(a/b) \\
&= 10 \cdot \log 2 && \text{Canceling } I/I_0 \\
&\approx 3.010 \text{ dB}. && \text{Because } \log 2 \approx 0.3
\end{aligned}$$

Thus, if the sound intensity is doubled, the decibel rating goes up by approximately 3 dB.

(b) If $I_M$ is the sound intensity of loud music, then

$$10 \log \left( \frac{I_M}{I_0} \right) = 110 \text{ dB}.$$

Similarly, if $I_C$ is the sound intensity of conversation, then

$$10 \log \left( \frac{I_C}{I_0} \right) = 50 \text{ dB}.$$

Computing the difference of the decibel ratings gives

$$10 \log \left( \frac{I_M}{I_0} \right) - 10 \log \left( \frac{I_C}{I_0} \right) = 60.$$

Dividing by 10 gives

$$\log\left(\frac{I_M}{I_0}\right) - \log\left(\frac{I_C}{I_0}\right) = 6$$

$$\log\left(\frac{I_M/I_0}{I_C/I_0}\right) = 6 \quad \text{Using the property } \log b - \log a = \log(b/a)$$

$$\log\left(\frac{I_M}{I_C}\right) = 6 \quad \text{Canceling } I_0$$

$$\frac{I_M}{I_C} = 10^6. \quad \log x = 6 \text{ means that } x = 10^6$$

So $I_M = 10^6 I_C$, which means that loud music is $10^6$ times, or one million times, as intense as normal conversation.

## Vertical Asymptotes

In Section 4.3 we saw that the graph of an exponential function has a horizontal asymptote. In Figure 5.10 on page 204, we see that $y = 10^x$ has a horizontal asymptote at $y = 0$, because

$$\text{as } x \to -\infty, \quad 10^x \to 0.$$

In the same figure, we see that $\log x$ has a vertical asymptote at $x = 0$. This is because as $x$ gets closer to zero, $y = \log x$ takes on larger and larger negative values. We write

$$\text{as } x \to 0^+, \quad \log x \to -\infty.$$

The notation $x \to 0^+$ is read "$x$ approaches zero from the right" and means that we are choosing smaller and smaller positive values of $x$—that is, we are sliding toward $x = 0$ through small positive values. In general, to describe vertical asymptotes, we use the notation

$$x \to a^+$$

to mean that $x$ slides toward $a$ from the right (that is, through values larger than $a$) and

$$x \to a^-$$

to mean that $x$ slides toward $a$ from the left (that is, through values smaller than $a$).

We summarize the information about both horizontal and vertical asymptotes:

Let $y = f(x)$ be a function and let $a$ be a finite number.

- The graph of $f$ has a **horizontal asymptote** of $y = a$ if

$$\text{as } x \to \infty, \quad f(x) \to a \quad \text{or} \quad \text{as } x \to -\infty, \quad f(x) \to a \quad \text{or both.}$$

- The graph of $f$ has a **vertical asymptote** of $x = a$ if

$$\text{as } x \to a^+, \quad f(x) =\to \infty \quad \text{or} \quad \text{as } x \to a^+, \quad f(x) =\to -\infty \quad \text{or}$$

$$\text{as } x \to a^-, \quad f(x) =\to \infty \quad \text{or} \quad \text{as } x \to a^-, \quad f(x) \to -\infty.$$

Notice that the method for finding a vertical asymptote is different from the method for finding a horizontal asymptote. Vertical asymptotes occur where the function values grow very large, either positive or negative, as $x$ approaches a finite value (that is, where $f(x) \to \infty$ or $f(x) \to -\infty$ as $x \to a$). Horizontal asymptotes occur where the function values approach a finite number as $x$ takes on large positive or large negative values (that is, as $x \to \infty$ or $x \to -\infty$).

### Optional: Asymptotes and Limit Notation

If $f(x) \to \infty$ as $x \to a^+$, we say the *limit* of $f(x)$ as $x$ approaches $a$ from the right is infinity,[7] and write

$$\lim_{x\to a^+} f(x) = \infty.$$

If $f(x) \to \infty$ as $x \to a^-$, we write

$$\lim_{x\to a^-} f(x) = \infty.$$

If both $\lim_{x\to a^+} f(x) = \infty$ and $\lim_{x\to a^-} f(x) = \infty$, we say the limit of $f(x)$ as $x$ approaches $a$ is infinity, and write

$$\lim_{x\to a} f(x) = \infty.$$

Similarly, we can write

$$\lim_{x\to a^+} f(x) = -\infty \quad \text{or} \quad \lim_{x\to a^-} f(x) = -\infty \quad \text{or} \quad \lim_{x\to a} f(x) = -\infty.$$

Using limit notation, the information about both horizontal and vertical asymptotes for a function $y = f(x)$ and $a$ a finite number is given as:

- The graph of $f$ has a **horizontal asymptote** of $y = a$ if

$$\lim_{x\to \infty} f(x) = a \quad \text{or} \quad \lim_{x\to -\infty} f(x) = a \quad \text{or both.}$$

- The graph of $f$ has a **vertical asymptote** of $x = a$ if we have one or more of the following:

$$\lim_{x\to a^+} f(x) = \infty \quad \text{or} \quad \lim_{x\to a^+} f(x) = -\infty \quad \text{or} \quad \lim_{x\to a^-} f(x) = \infty \quad \text{or} \quad \lim_{x\to a^-} f(x) = -\infty.$$

## Exercises and Problems for Section 5.3

### Skill Refresher

For Exercises S1–S2, evaluate without a calculator.

**S1.** $\log 0.0001$

**S2.** $\dfrac{\log 100^6}{\log 100^2}$

For Exercises S3–S4, rewrite the exponential equation in equivalent logarithmic form.

**S3.** $10^5 = 100{,}000$

**S4.** $e^2 = 7.389$

For Exercises S5–S6, rewrite the logarithmic equation in equivalent exponential form.

**S5.** $-\ln x = 12$

**S6.** $\log(x + 3) = 2$

For Exercises S7–S8, if possible, write the expression using sums and/or differences of logarithmic expressions that do not contain the logarithms of products, quotients or powers.

**S7.** $\ln(x(7 - x)^3)$

**S8.** $\ln\left(\dfrac{xy^2}{z}\right)$

For Exercises S9–S10, rewrite the expression as a single logarithm.

**S9.** $\ln x^3 + \ln x^2$

**S10.** $\frac{1}{3}\log 8 - \frac{1}{2}\log 25$

[7] Some authors say that these limits do not exist.

## Exercises

In Exercises 1–2, find the domain and range of the function.

**1.** $y = \ln(x-3)$ **2.** $y = \ln(x+1)$

In Exercises 3–6, is the rate of change between two points on the graph positive or negative? Does the rate of change increase or decrease as you move right? Is the graph concave up or concave down?

**3.**

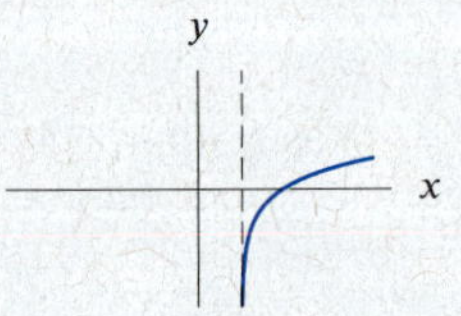


**4.**

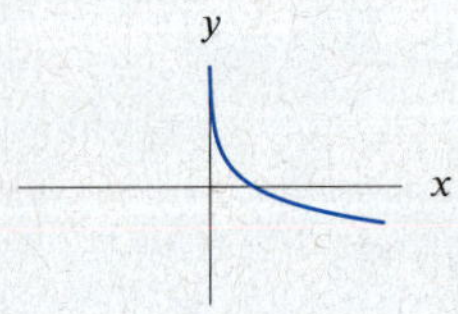


**5.**

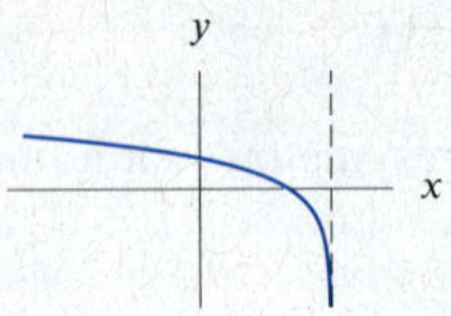


**6.**

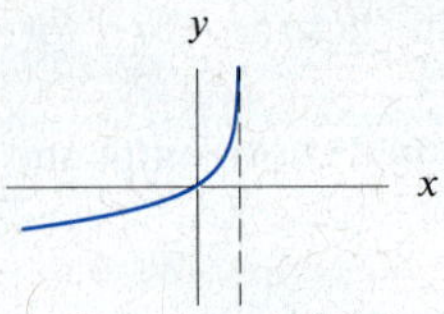


**7.** Without a calculator, match the functions $y = 10^x$, $y = e^x$, $y = \log x$, $y = \ln x$ with the graphs in Figure 5.13.

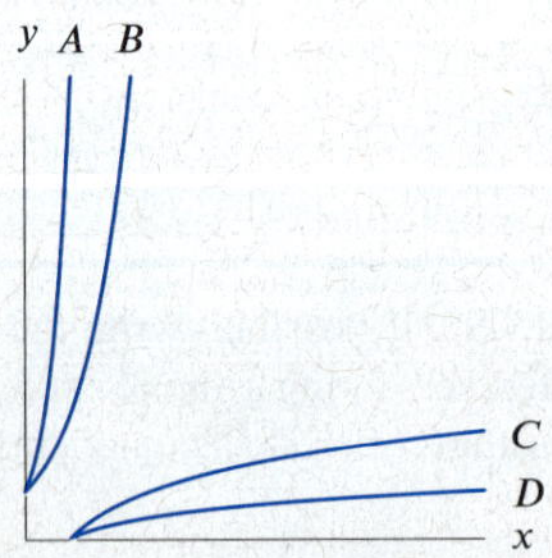


Figure 5.13

**8.** Without a calculator, match the functions $y = 2^x$, $y = e^{-x}$, $y = 3^x$, $y = \ln x$, $y = \log x$ with the graphs in Figure 5.14.

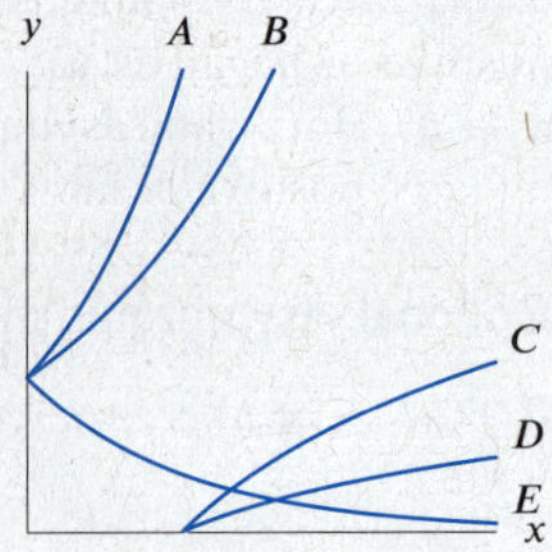


Figure 5.14

**9.** What is the equation of the asymptote of the graph of $y = 10^x$? Of the graph of $y = 2^x$? Of the graph of $y = \log x$?

**10.** What is the equation for the asymptote of the graph of $y = e^x$? Of the graph of $y = e^{-x}$? Of the graph of $y = \ln x$?

Graph the functions in Exercises 11–14. Label all asymptotes and intercepts.

**11.** $y = 2 \cdot 3^x + 1$ **12.** $y = -e^{-x}$

**13.** $y = \log(x-4)$ **14.** $y = \ln(x+1)$

In Exercises 15–16, graph the function. Identify any vertical asymptotes. State the domain of the function.

**15.** $y = 2\ln(x-3)$ **16.** $y = 1 - \ln(2-x)$

In Exercises 17–21, find the hydrogen ion concentration, $[H^+]$, for the substances.[8] [Hint: $\text{pH} = -\log[H^+]$.]

**17.** Hydrochloric acid, with a pH of 0.

**18.** Battery acid, with a pH of 1.

**19.** Lye, with a pH of 13.

**20.** Baking soda, with a pH of 8.3.

**21.** Tomatoes, with a pH of 4.5.

**22.** What is the value (if any) of the following?

(a) $10^{-x}$ as $x \to \infty$ (b) $\log x$ as $x \to 0^+$

**23.** What is the value (if any) of the following?

(a) $e^x$ as $x \to -\infty$ (b) $\ln x$ as $x \to 0^+$

[8]Data from www.miamisci.org/ph/hhoh.html, accessed November, 2001.

## Problems

**24.** Immediately following the gold medal performance of the US women's gymnastic team in the 1996 Olympic Games, an NBC commentator, John Tesh, said of one team member: "Her confidence and performance have grown logarithmically." He clearly thought this was an enormous compliment. Is it a compliment? Is it realistic?

**25.** Match the graphs (a)–(c) to one of the functions $r(x)$, $s(x)$, $t(x)$ whose values are in the tables.

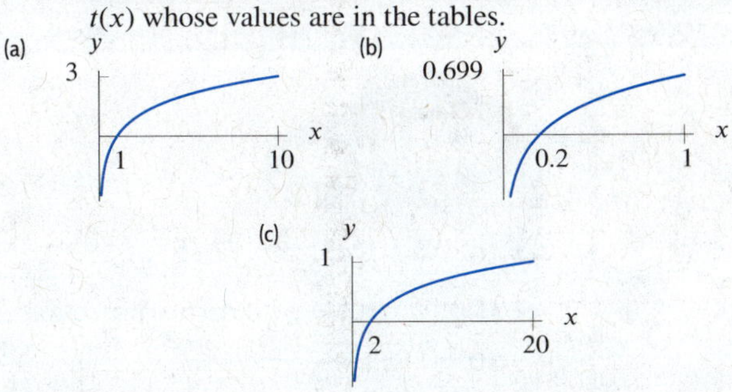


| $x$ | 2 | 4 | 10 |
|---|---|---|---|
| $r(x)$ | 1 | 1.301 | 1.699 |

| $x$ | 0.5 | 5 | 10 |
|---|---|---|---|
| $s(x)$ | −0.060 | 0.379 | 0.699 |

| $x$ | 0.1 | 2 | 100 |
|---|---|---|---|
| $t(x)$ | −3 | 0.903 | 6 |

In Problems 26–31, find possible formulas for the functions using logs or exponentials.

**26.**

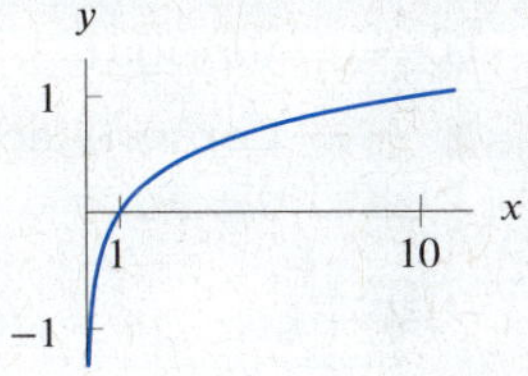


**27.**

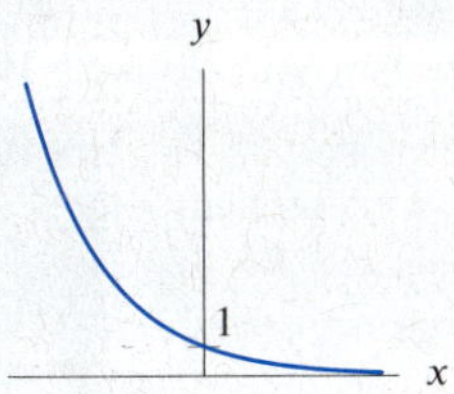


**28.**

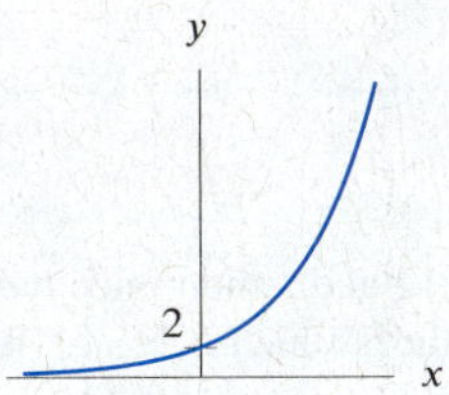


**29.**

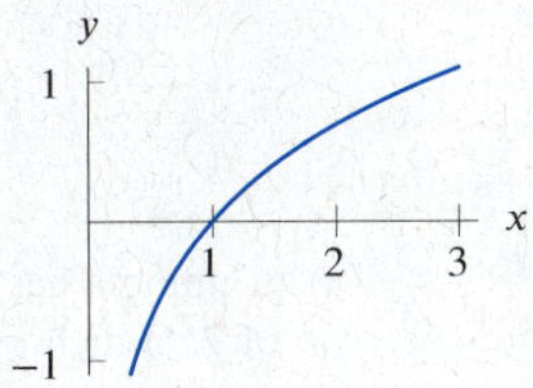


**30.**

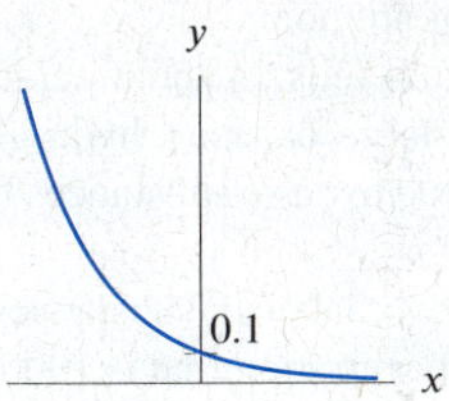


**31.**

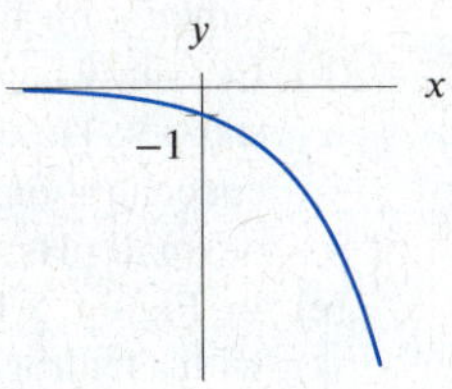


**32.** Match the statements (a)–(d) with one or more of the functions (I)–(IV).

**(a)** The graph has a vertical asymptote at $x = 0$.
**(b)** The graph has a nonzero horizontal asymptote.
**(c)** The graph has no asymptotes
**(d)** The graph tends toward the $x$-axis as $x$ gets larger and larger.

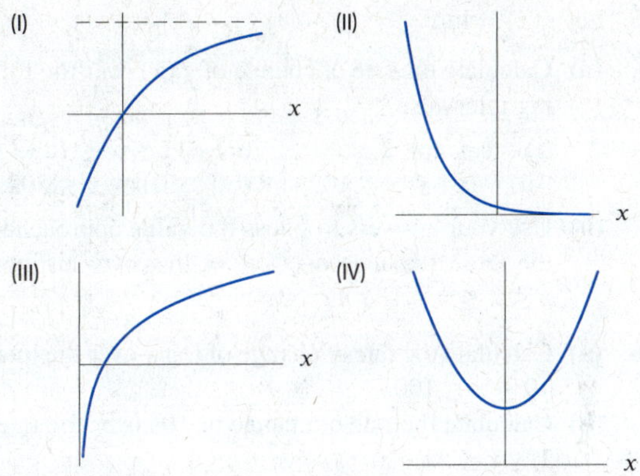


**33.** Match the statements (a)–(d) with one or more of the functions (I)–(IV).

**(a)** The graph has two horizontal asymptotes.
**(b)** The graph has both a vertical asymptote and a horizontal asymptote.
**(c)** The graph tends towards infinity both as $x$ tends to 0 and as $x$ gets larger and larger.
**(d)** The graph has a vertical asymptote at $x = 0$.

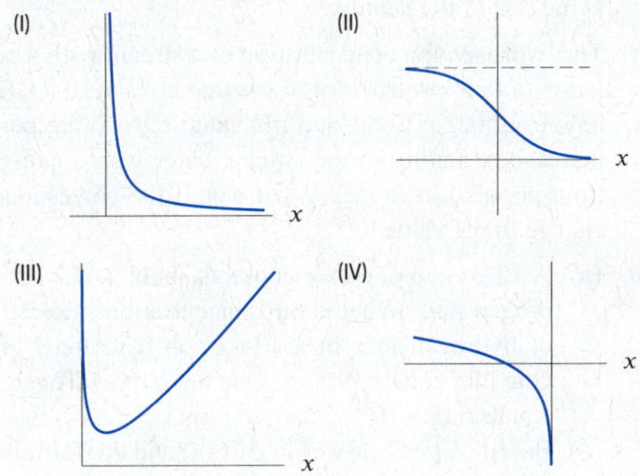

**34.** Table 5.5 gives values for a function $f(x)$.

**(a)** Calculate the rate of change of $f(x)$ over the intervals $4 \le x \le 6$ and $6 \le x \le 8$.

**(b)** A friend tells you that $f(x) = \ln x$. Use your answer to part (a) to explain why your friend is incorrect.

**Table 5.5**

| $x$ | 4 | 6 | 8 |
|---|---|---|---|
| $f(x)$ | 1.386 | 1.792 | 3.078 |

**35.** Let $f(x) = \ln x$.

**(a)** Calculate the rate of change of $f(x)$ over the following intervals:

(i) $1 \le x \le 2$  (ii) $11 \le x \le 12$
(iii) $101 \le x \le 102$  (iv) $501 \le x \le 502$

**(b)** Use your answers to guess the value approached by the rate of change of $f(x) = \ln x$ over an interval $n \le x \le n+1$ as $n$ grows large.

**36.** **(a)** Calculate the rate of change of $\log x$ over the interval $10 \le x \le 100$.

**(b)** Calculate the rate of change of $10^x$ over the interval $1 \le x \le 2$.

**(c)** What do you notice about your answers to parts (a) and (b)? Explain why this occurs.

**37.** **(a)** Using the definition of pH on page 205, find the concentrations of hydrogen ions in solutions with

(i) pH = 2  (ii) pH = 4  (iii) pH = 7

**(b)** A high concentration of hydrogen ions corresponds to an acidic solution. From your answer to part (a), decide if solutions with high pHs are more or less acidic than solutions with low pHs.

**38.** In an interview an oceanographer states that the seawater off the coast of Washington is projected to increase 150% in acidity by the end of the century. If we assume that the current average pH level is 8.1, find the projected pH level at the end of the century.

**39.** The hydrogen ion concentration of a stream with a population of rainbow trout was measured at $7.94(10^{-7})$. Rainbow trout begin to die at a pH value of 6. What percent increase in acidity in the stream water would cause the trout population to die? What would the corresponding change in pH value be?

**40.** **(a)** A 12-oz cup of coffee contains about $2.41 \cdot 10^{18}$ hydrogen ions. What is the concentration (moles/liter) of hydrogen ions in a 12-oz cup of coffee? [Hint: One liter equals 30.3 oz. One mole of hydrogen ions equals $6.02 \cdot 10^{23}$ hydrogen ions.]

**(b)** Based on your answer to part (a) and the formula for pH, what is the pH of a 12-oz cup of coffee?

**41.** **(a)** The pH of lime juice is about 2.3. What is the concentration of hydrogen ions in lime juice?

**(b)** A person squeezes 2 oz of lime juice into a cup. Based on your answer to part (a), how many hydrogen ions does this juice contain?

**42.** A biology book[9] has the diagram in Figure 5.15 showing the relationship between the hydrogen ion concentration, $[H^+]$, in moles per liter, and the pH scale:

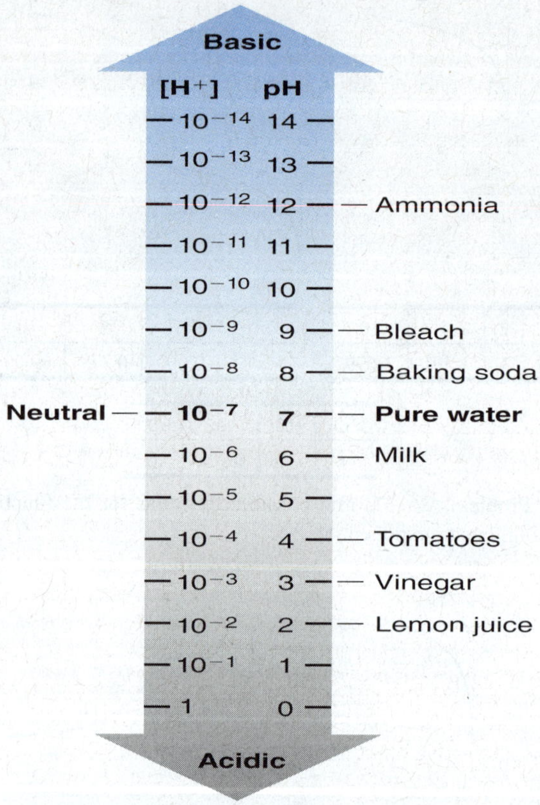


**Figure 5.15**

**(a)** Ammonia has a pH of 12 and lemon juice has a pH of 2. Which one is acidic? Which is basic? What is the hydrogen ion concentration of each? How many times greater is the hydrogen ion concentration of lemon juice than that of ammonia?

**(b)** Explain what the diagram tells us about acidic substances: Do they have large or small hydrogen ion concentrations, compared to basic substances? Large or small pHs?

**(c)** In Figure 5.15, find two labeled substances, one with a hydrogen ion concentration that is 1000 times greater than the other.

[9] *Principles of Biology*, www.nature.com/principles, accessed February 9, 2014.

**(d)** An article states that seawater has increased in hydrogen ion concentration by 30%. The initial sea water concentration was pH $= 7.96$ and $[H^+] = 1.1(10^{-8})$ moles/liter. What is the current pH?

**(e)** In two consecutive years, scientists measured the pH in a pond as 6.8 and 6.4, respectively. Find the percent change of the hydrogen ion concentration in the pond between the two years.

In Problems 43–47, sound in decibels is measured by comparing the sound intensity, $I$, to a benchmark sound $I_0$ with intensity $10^{-16}$ watts/cm$^2$. Then,

$$\text{Noise level in decibels} = 10\log(I/I_0).$$

**43.** Death of hearing tissue begins to occur at a noise level of 180 dB. Compute the sound's intensity at this noise level.

**44.** The noise level of a whisper is 30 dB. Compute the sound intensity of a whisper.

**45.** The loudest crowd roar at a sports stadium was 131.76 dB and was achieved by Galatasaray Spor Kulubu (Turkey) in Istanbul, Turkey, on 18 March 2011.[10] The previous world record of 128.7 dB was set in 2000 by Denver Broncos fans.[11] How many times more intense was the roar of the crowd of Turkish soccer fans than the roar of the Broncos fans?

**46.** Sound $A$ measures 30 decibels and sound $B$ is 5 times as loud as sound $A$. What is the decibel rating of sound B to the nearest integer?

**47.** **(a)** Let $D_1$ and $D_2$ represent the decibel ratings of sounds of intensity $I_1$ and $I_2$, respectively. Using log properties, find a simplified formula for the difference between the two ratings, $D_2 - D_1$, in terms of the two intensities, $I_1$ and $I_2$.

**(b)** If a sound's intensity doubles, how many decibels louder does the sound become?

Problems 48–51 use the *Richter scale* for the strength of an earthquake. The strength, $W$, of the seismic waves of an earthquake are compared to the strength, $W_0$, of the seismic waves of a standard earthquake. The Richter scale rating, $M$, is

$$M = \log\left(\frac{W}{W_0}\right).$$

**48.** In 2013 the Khash earthquake in Iran had a Richter-scale rating of 7.8.[12] How many times more powerful were the seismic waves of the Khash earthquake than standard seismic waves?

**49.** In 1986 the worst nuclear power plant accident in history occurred in Chernobyl, Ukraine. The explosion resulted in seismic waves with a Richter scale rating of 3.5. How many times stronger were the seismic waves of the Chernobyl disaster than standard seismic waves?

**50.** Let $M_1$ and $M_2$ be the magnitude of two earthquakes whose seismic waves are of sizes $W_1$ and $W_2$, respectively. Using log properties, find a simplified formula for the difference $M_2 - M_1$ in terms of $W_1$ and $W_2$.

**51.** The 2010 earthquake off California had a rating of 7.2 on the Richter scale.[13] How many times larger than the California earthquake were the seismic waves in the March 2005 earthquake off the coast of Sumatra, which measured 8.7 on the Richter scale? Give your answer to the nearest integer.

**52.** Give the domain of $y = \dfrac{1}{2 - \sqrt{7 - e^{2t}}}$.

**53.** Find **(a)** $\lim\limits_{x\to 0^+} \log x$ **(b)** $\lim\limits_{x\to 0^-} \ln(-x)$

**54.** Match the statements (a)–(d) with the functions (I)–(IV).

**(a)** $\lim\limits_{x\to 0^+} f(x) = -\infty$ **(b)** $\lim\limits_{x\to 0^-} f(x) = 0$

**(c)** $\lim\limits_{x\to \infty} f(x) = \infty$ **(d)** $\lim\limits_{x\to -\infty} f(x) = 0$

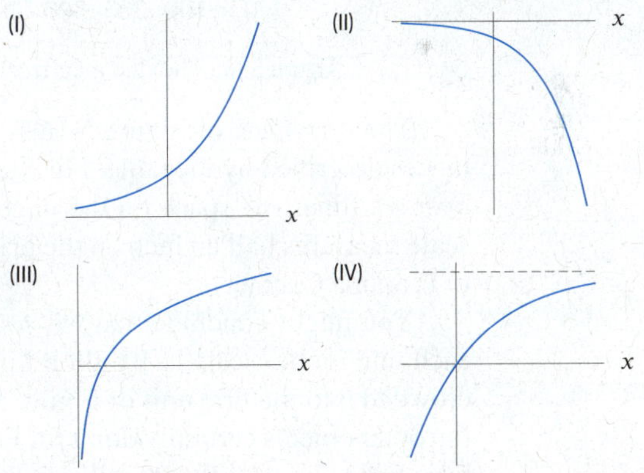


[10] See http://www.encyclopedia.com/doc/1G1-65629077.html, accessed July 24, 2013.
[11] See http://www.encyclopedia.com/doc/1G1-65629077.html, accessed January 8, 2010.
[12] http://comcat.cr.usgs.gov/earthquakes/eventpage/usb000g7x7, accessed July 24, 2013
[13] http://en.wikipedia.org/wiki/List_of_earthquakes_in_California, accessed July 24, 2013.

# 5.4 LOGARITHMIC SCALES

## The Solar System and Beyond

Table 5.6 gives the distance from the sun to a number of different astronomical objects. The planet Mercury is 58,000,000 km from the sun, the earth is 149,000,000 km from the sun, and Pluto is 5,900,000,000 km, or almost 6 billion kilometers from the sun. The table also gives the distance to Proxima Centauri, the star closest to the sun, and to the Andromeda Galaxy, the spiral galaxy closest to our own galaxy, the Milky Way.

**Table 5.6** *Distance from the sun to various astronomical objects*

| Object | Distance (million km) | Object | Distance (million km) |
|---|---|---|---|
| Mercury | 58 | Saturn | 1426 |
| Venus | 108 | Uranus | 2869 |
| Earth | 149 | Neptune | 4495 |
| Mars | 228 | Pluto | 5900 |
| Jupiter | 778 | Proxima Centauri | $4.1 \cdot 10^7$ |
| | | Andromeda Galaxy | $2.4 \cdot 10^{13}$ |

### Linear Scales

We can represent the information in Table 5.6 graphically in order to get a better feel for the distances involved. Figure 5.16 shows the distance from the sun to the first five planets on a *linear scale*, which means that the evenly spaced units shown in the figure represent equal distances. In this case, each unit represents 100 million kilometers.

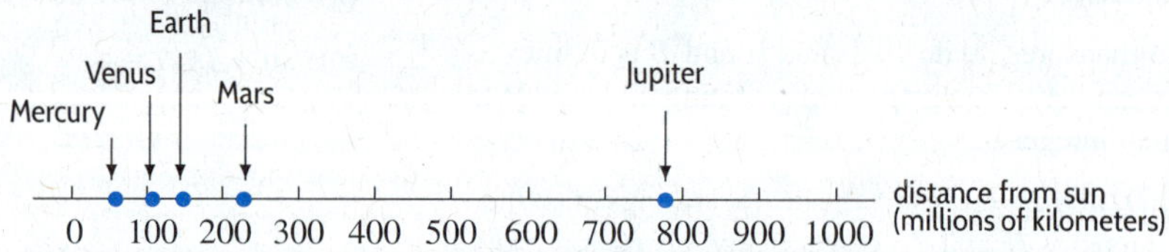


**Figure 5.16**: The distance from the sun of the first five planets (in millions of kilometers)

The drawback of Figure 5.16 is that the scale is too small to show all of the astronomical distances described by the table. For example, to show the distance to Pluto on this scale would require over six times the space on the page. Even worse, assuming that each 100 million km unit on the scale measures half an inch on the printed page, we would need 3 miles of paper to show the distance to Proxima Centauri!

You might conclude that we could fix this problem by choosing a larger scale. In Figure 5.17 each unit on the scale is 1 billion kilometers. Notice that all five planets shown by Figure 5.16 are crowded into the first unit of Figure 5.17; even so, the distance to Pluto barely fits. The distances to the other objects certainly don't fit. For instance, to show the Andromeda Galaxy, Figure 5.17 would have to be almost 200,000 miles long. Choosing an even larger scale will not improve the situation.

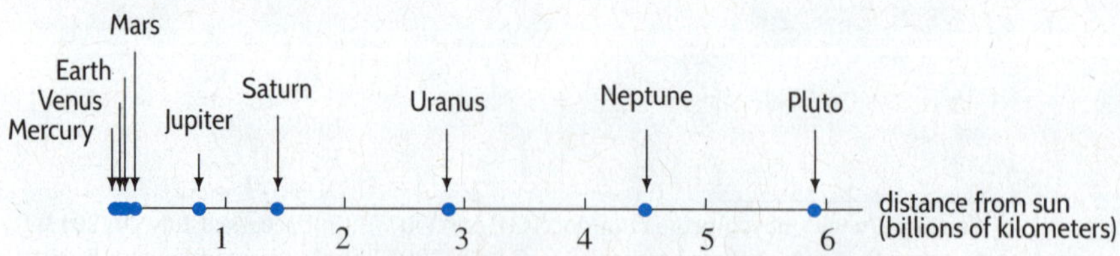


**Figure 5.17**: The distance to all nine planets (in billions of kilometers)

## Logarithmic Scales

We conclude that the data in Table 5.6 cannot easily be represented on a linear scale. If the scale is too small, the more distant objects do not fit; if the scale is too large, the less distant objects are indistinguishable. The problem is not that the numbers are too big or too small; the problem is that the numbers vary too greatly in size.

We consider a different type of scale on which equal distances are not evenly spaced. All the objects from Table 5.6 are represented in Figure 5.18. The nine planets are still cramped, but it is possible to tell them apart. Each tick mark on the scale in Figure 5.18 represents a distance ten times larger than the one before it. This kind of scale is called *logarithmic*.

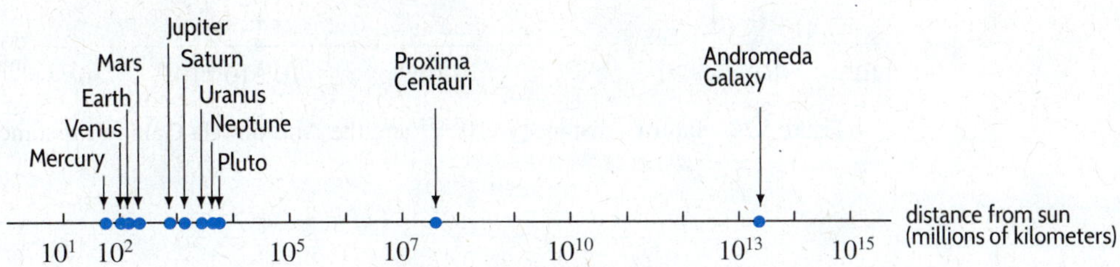


**Figure 5.18**: The distance from the sun (in millions of kilometers)

## How Do We Plot Data on a Logarithmic Scale?

A logarithmic scale is marked with increasing powers of 10: $10^1$, $10^2$, $10^3$, and so on. Notice that even though the distances in Figure 5.18 are not evenly spaced, the exponents are evenly spaced. Therefore the distances in Figure 5.18 are spaced according to their logarithms.

In order to plot Mercury's distance from the sun, 58 million kilometers, we use the fact that

$$10 < 58 < 100,$$

so Mercury's distance is between $10^1$ and $10^2$, as shown in Figure 5.18. To plot Mercury's distance more precisely, calculate $\log 58 = 1.763$, so $10^{1.763} = 58$, and use the exponent 1.763 to represent Mercury's position. See Figure 5.19.

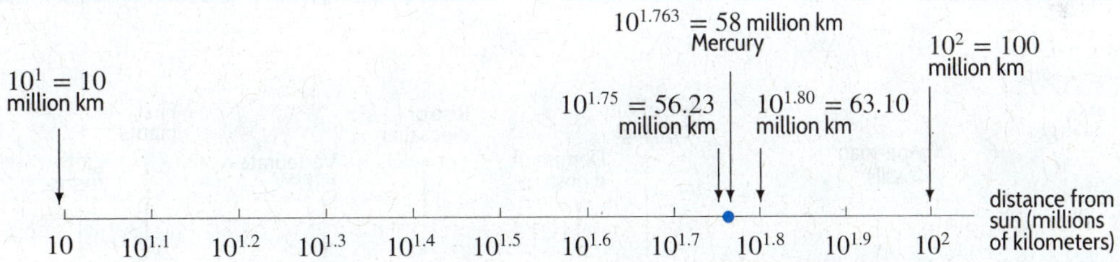


**Figure 5.19**: Mercury's distance, 58 million kilometers, falls between $10^{1.75} = 56.23$ and $10^{1.80} = 63.10$ million kilometers

**Example 1** Where should Saturn be on the logarithmic scale? What about the Andromeda Galaxy?

**Solution** Saturn's distance is 1426 million kilometers, so we want the exponent of 10 that gives 1426, which is

$$\log 1426 \approx 3.154119526.$$

Thus $10^{3.154} \approx 1426$, so we use the exponent 3.154 to indicate Saturn's distance.

Similarly, the distance to the Andromeda Galaxy is $2.4 \cdot 10^{13}$ million kilometers, and since

$$\log(2.4 \cdot 10^{13}) \approx 13.38,$$

we use the exponent 13.38 to represent the galaxy's distance. See Figure 5.20.

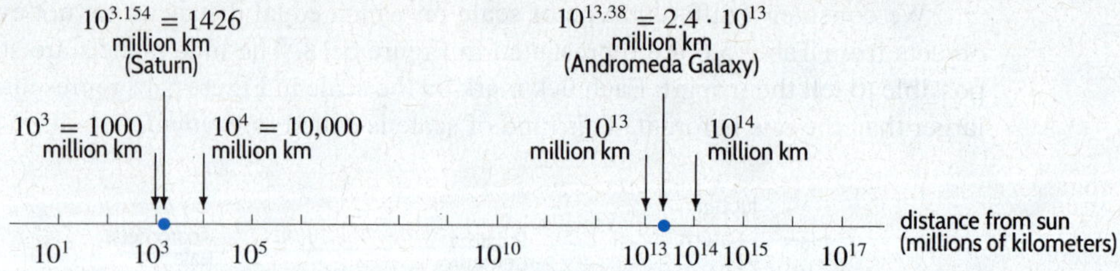


**Figure 5.20**: Saturn's distance is $10^{3.154}$ and the Andromeda Galaxy's distance is $10^{13.38}$

## Logs of Small Numbers

The history of the world, like the distance to the stars and planets, involves numbers of vastly different sizes. Table 5.7 gives the ages of certain events[14] and the logarithms of their ages. The logarithms have been used to plot the events in Figure 5.21.

**Table 5.7** *Ages of various events in earth's history and logarithms of the ages*

| Event | Age (millions of years) | log (age) | Event | Age (millions of years) | log (age) |
|---|---|---|---|---|---|
| Man emerges | 1 | 0 | Rise of dinosaurs | 245 | 2.39 |
| Ape-man fossils | 5 | 0.70 | Vertebrates appear | 570 | 2.76 |
| Rise of cats, dogs, pigs | 37 | 1.57 | First plants | 2500 | 3.40 |
| Demise of dinosaurs | 67 | 1.83 | Earth forms | 4450 | 3.65 |

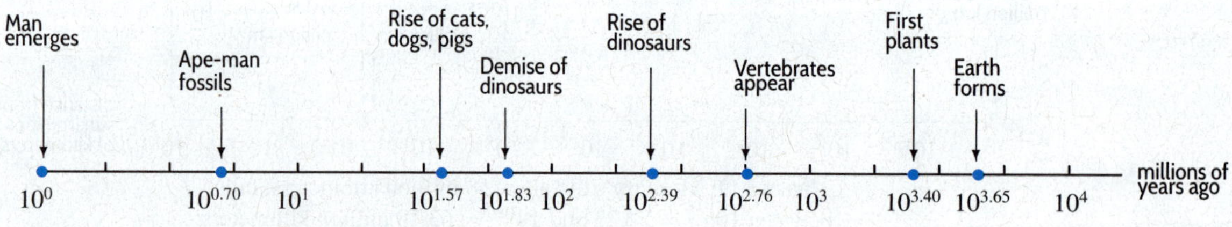


**Figure 5.21**: Logarithmic scale showing the ages of various events (in millions of years ago)

The events described by Table 5.7 all happened at least 1 million years ago. How do we indicate events which occurred less than 1 million years ago on the log scale?

**Example 2** Where should the building of the pyramids be indicated on the log scale?

[14] *CRC Handbook,* 75th ed., sec. 14-8.

**Solution** The pyramids were built about 5000 years ago, or

$$\frac{5000}{1{,}000{,}000} = 0.005 \text{ million years ago.}$$

Notice that 0.005 is between 0.001 and 0.01, that is,

$$10^{-3} < 0.005 < 10^{-2}.$$

Since

$$\log 0.005 \approx -2.30,$$

we use the exponent −2.30 for the pyramids. See Figure 5.22.

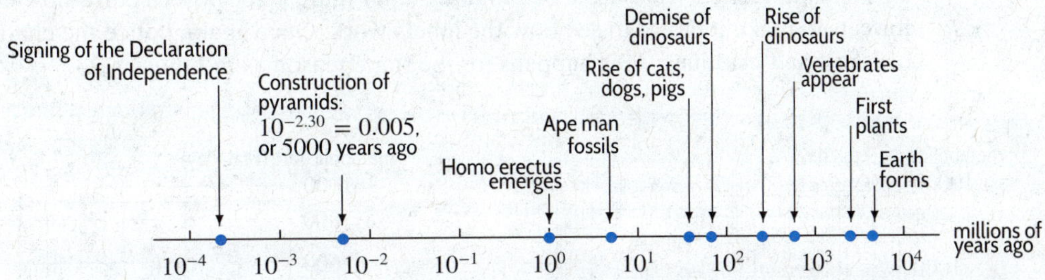


**Figure 5.22**: Logarithmic scale showing the ages of various events. Note that events that are less than 1 million years old are indicated by negative exponents

## Another Way to Label a Log Scale

In Figures 5.21 and 5.22, the log scale has been labeled so that exponents are evenly spaced. Another way to label a log scale is with the values themselves instead of the exponents. See Figure 5.23.

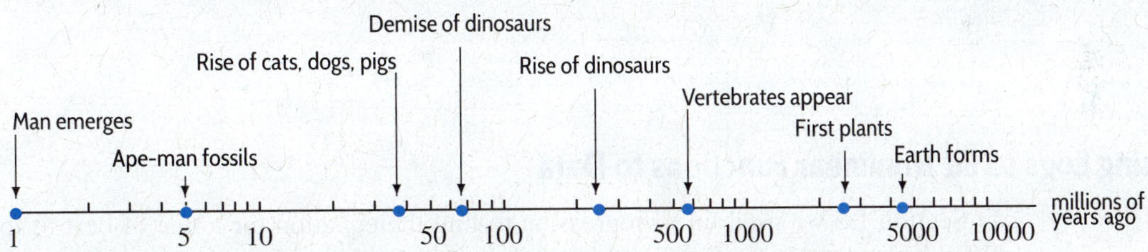


**Figure 5.23**: Some data from Figure 5.22 with axis labeled using actual values, not logs

In Figure 5.22, consecutive powers of 10 are evenly spaced because exponents are evenly spaced. However, in Figure 5.23, where the scale is labeled with the actual values ($10^1 = 10$, $10^2 = 100$, $10^3 = 1000$, etc.), the numbers appear to "pile up" in a way characteristic of log scales.

# Log-Log Scales

Table 5.8 shows the average metabolic rate in kilocalories per day (kcal/day) for animals of different weights.[15] (A kilocalorie is the same as a standard nutritional Calorie.) For instance, a 1-lb rat consumes about 35 kcal/day, whereas a 1750-lb horse consumes almost 9500 kcal/day.

[15] *The New York Times*, January 11, 1999.

**Table 5.8** *The metabolic rate (in kcal/day) for animals of different weights*

| Animal | Weight (lbs) | Rate (kcal/day) |
|---|---|---|
| Rat | 1 | 35 |
| Cat | 8 | 166 |
| Human | 150 | 2000 |
| Horse | 1750 | 9470 |

It is not practical to plot these data on an ordinary set of axes. Both input and output values span too broad a range. However, we can plot the data using log scales for both the horizontal (weight) axis and the vertical (rate) axis. See Figure 5.24. Figure 5.25 shows a close-up view of the data point for cats to make it easier to see how the labels work. Once again, notice the characteristic piling up of labels and gridlines. This happens for the same reason as in Figure 5.23.

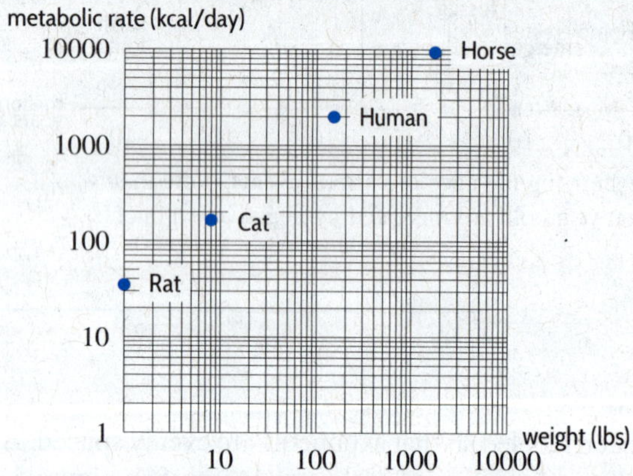


**Figure 5.24:** Metabolic rate (in kcal/hr) plotted against body weight

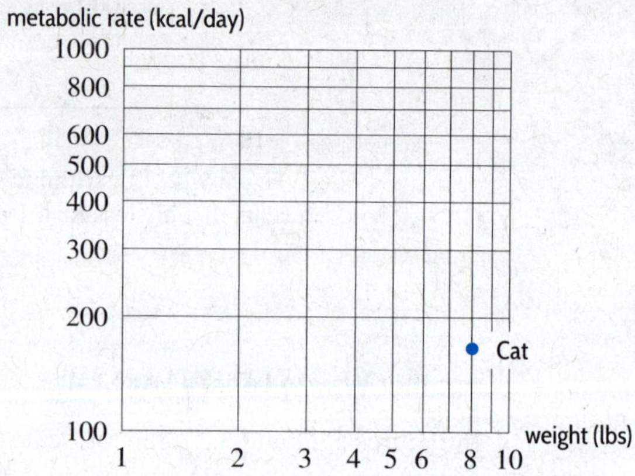


**Figure 5.25:** A close-up view of the Cat data point

## Using Logs to Fit Nonlinear Functions to Data

In Section 1.6 we used linear regression to find the equation for a line of best fit for a set of data. What if the data do not lie close to a straight line, but instead approximate the graph of some other function? In this section we see how logarithms help us fit data to nonlinear functions.

## The iTunes Store

On February 24, 2010, a 71-year-old man from Georgia named Louie Sulcer downloaded the ten-billionth song from the iTunes store (a recording by Johnny Cash). Mr. Sulcer received a call from Apple co-founder Steve Jobs and a $10,000 gift card.

Table 5.9 shows the total number of iTunes sales (in millions of songs) where $t$ is the number of days since the iTunes Store opened on April 28, 2003. From Figure 5.26, we see that sales have grown at an ever-increasing rate. The trend is not linear.[16]

[16]Sales until Feb. 6, 2013, from http://en.wikipedia.org/wiki/ITunes_Store#Music_2, accessed May 18, 2014.

**Table 5.9** *Number of songs sold*

| $t$, Days since April 28, 2003 | $N$, number of songs sold (millions) |
|---|---|
| 440 | 100 |
| 598 | 200 |
| 637 | 250 |
| 812 | 500 |
| 1032 | 1000 |
| 1353 | 2000 |
| 1442 | 2500 |
| 1879 | 5000 |
| 2494 | 10000 |
| 3425 | 20000 |
| 3572 | 25000 |

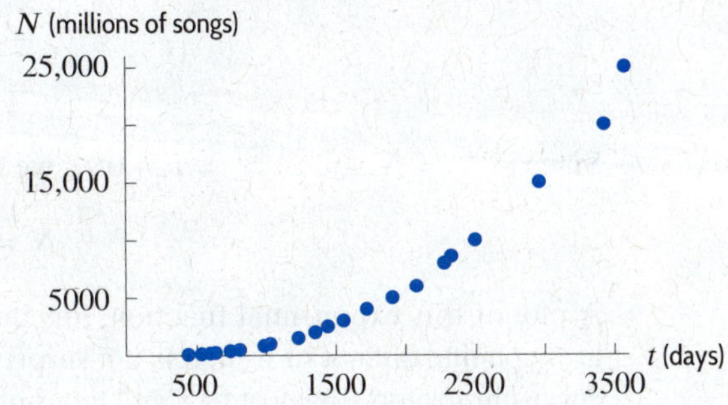


**Figure 5.26:** The number of songs sold, $N$, as a function of days, $t$, since iTunes store opened

## Using a Log Scale to Linearize Data

How do we decide what functions fit the data in Figure 5.26? A common method is to take logs of one or both variables and see if the resulting data is close to linear. First we take logs of $N$, in millions of songs sold. Table 5.10 contains values of $\ln N$; Figure 5.27 shows these plotted against $t$, time in days since the store opened. The fact that the plot of the data still shows a curve, and is not scattered evenly around the line, indicates that we have not yet found a good fit. To confirm this, we use the regression line, which relates $\ln N$ and $t$, to find an equation relating $N$ and $t$.

**Table 5.10** *Number of songs sold*

| $t$, days | $y = \ln N$ |
|---|---|
| 440 | 4.61 |
| 598 | 5.30 |
| 637 | 5.52 |
| 812 | 6.21 |
| 1032 | 6.91 |
| 1353 | 7.60 |
| 1442 | 7.82 |
| 1879 | 8.52 |
| 2494 | 9.21 |
| 3425 | 9.90 |
| 3572 | 10.13 |

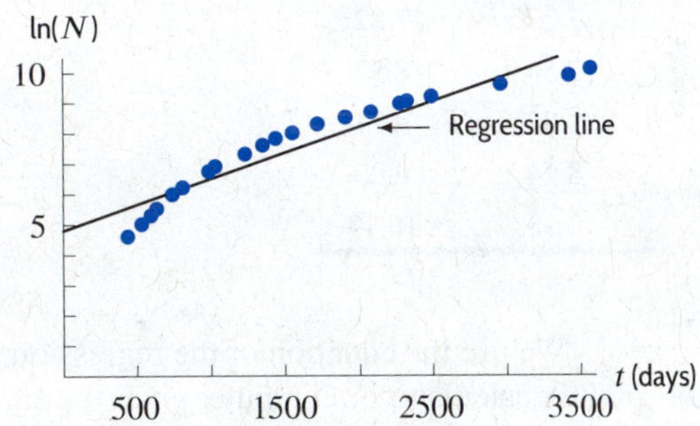


**Figure 5.27:** The vertical axis is the natural log of the number of songs sold

## Finding a Formula for the Curve

We say the data in the second column of Table 5.10 have been *transformed*. A calculator or computer gives a regression line for the transformed data in Figure 5.27 as:[17]

$$y = 4.821 + 0.0016t.$$

[17]The values obtained by a computer or another calculator may vary slightly from the ones given.

To transform the equation back to our original variables, $N$ and $t$, we substitute $y = \ln N$, giving

$$\ln N = 4.821 + 0.0016t.$$

We solve for $N$ by raising $e$ to both sides:

$$\begin{aligned} e^{\ln N} &= e^{4.821+0.0016t} \\ &= \left(e^{4.821}\right)\left(e^{0.0016t}\right). \end{aligned}$$

Since $e^{\ln N} = N$ and $e^{4.821} \approx 124.089$, we have

$$N = 124.089e^{0.0016t}.$$

A plot of this exponential function, together with the data, is shown in Figure 5.29. Since the regression line did not fit well, it is not surprising that the exponential fit is not particularly good—the curve climbs too slowly at first and too rapidly later.

## Finding Another Possible Formula for the Curve

Now we try taking logs of time, $t$, as well as of $N$. Table 5.11 contains values of $\ln t$ and $\ln N$; Figure 5.28 shows their plot. A regression line fits this data well; we say the data has been *linearized*.

**Table 5.11** *Songs sold*

| $x = \ln t$, days | $y = \ln N$ |
|---|---|
| 6.09 | 4.61 |
| 6.39 | 5.30 |
| 6.46 | 5.52 |
| 6.70 | 6.21 |
| 6.94 | 6.91 |
| 7.21 | 7.60 |
| 7.27 | 7.82 |
| 7.54 | 8.52 |
| 7.82 | 9.21 |
| 8.14 | 9.90 |
| 8.18 | 10.13 |

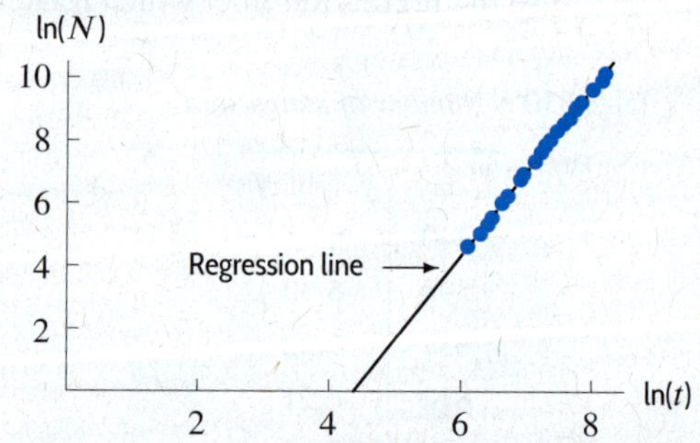


**Figure 5.28:** Vertical axis is natural log of songs sold; horizontal axis is natural log of time

We use the equation of the regression line in Figure 5.28 to find another equation relating $N$ and $t$. A calculator or computer gives the equation of the line as

$$y = -11.5496 + 2.6515x.$$

To transform back to our original variables, $N$ and $t$, we substitute $y = \ln N$ and $x = \ln t$, giving

$$\ln N = -11.5496 + 2.6515 \ln t.$$

Raising $e$ to both sides gives:

$$\begin{aligned} e^{\ln N} &= e^{-11.5496+2.6515 \ln t} \\ &= \left(e^{-11.5496}\right)\left(e^{2.6515 \ln t}\right) \end{aligned}$$

$$= 0.00000964 \underbrace{\left(e^{\ln t}\right)}_{t}{}^{2.6515}$$

$$= 0.00000964t^{2.6515}.$$

A plot of this power function is shown in Figure 5.30. The fit appears to be excellent.

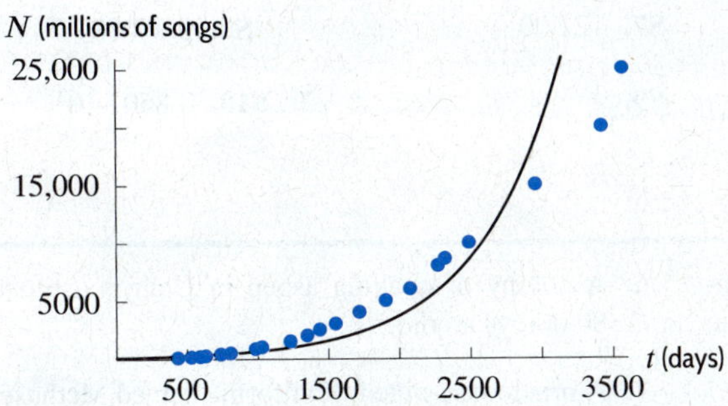


**Figure 5.29**: An exponential function fit to the data

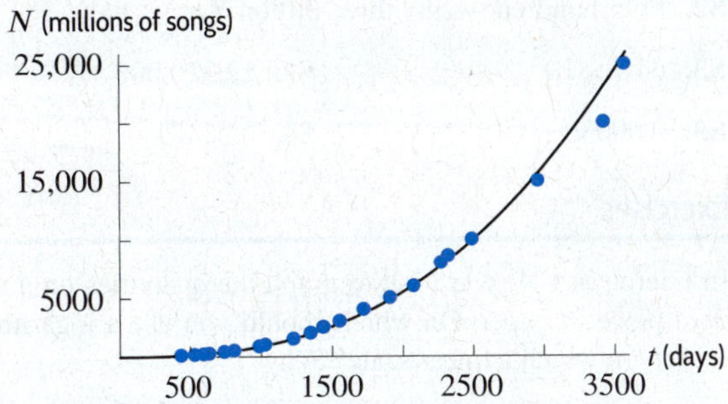


**Figure 5.30**: A power function fit to the data

## Summary: Fitting Exponential and Power Functions to Data

Taking logs is a powerful tool because it allows us to fit two families of functions—exponentials and powers—to data.

To fit an exponential function, $N = ae^{kt}$, to a set of data of the form $(t, N)$, we use three steps. First, we transform the relationship by taking the natural log of both sides and making the substitution $y = \ln N$. This leads to the equation

$$\begin{aligned} y = \ln N &= \ln\left(ae^{kt}\right) \\ &= \ln a + \ln e^{kt} \\ &= \ln a + kt. \end{aligned}$$

Setting $b = \ln a$ gives a linear equation with $k$ as the slope and $b$ as the $y$-intercept:

$$y = b + kt.$$

Secondly, we can now use linear regression on the variables $t$ and $y$. (Remember that $y = \ln N$.) Last, as step three, we transform the linear regression equation back into our original variables by substituting $\ln N$ for $y$ and solving for $N$.

To fit a power function, $N = kt^p$, to a set of data of the form $(t, N)$, we take similar steps. We first transform the relationship by the substitutions $x = \ln t$ and $y = \ln N$. This leads to the equation

$$\begin{aligned} y = \ln N &= \ln\left(kt^p\right) \\ &= \ln k + \ln t^p \\ &= \ln k + p \ln t. \end{aligned}$$

Setting $b = \ln k$ and and $m = p$ gives a linear equation with $p$ as the slope and $b$ as the $y$-intercept:

$$y = b + m \underbrace{\ln t}_{x} = b + mx.$$

Now we can use linear regression on the variables $x$ and $y$. (Remember that $x = \ln t$ and $y = \ln N$.) Last, we transform the linear regression equation back into our original variables by substituting $\ln N$ for $y$ and $\ln t$ for $x$. Then we solve for $N$.

## Exercises and Problems for Section 5.4

### Skill Refresher

In Exercises S1–S6, write the numbers in scientific notation.

**S1.** One million four hundred fifty-five thousand

**S2.** Four hundred twenty three billion

**S3.** $64.7 \times 10^3$

**S4.** 12,310,000

**S5.** 0.00036

**S6.** 0.00471

In Exercises S7–S10, without a calculator, determine between which two powers of ten the following numbers lie.

**S7.** 12,500

**S8.** 0.000881

**S9.** $\frac{1}{3}$

**S10.** $3{,}850 \cdot 10^8$

### Exercises

In Exercises 1–4, you wish to graph the quantities on a standard piece of paper. On which should you use a logarithmic scale? On which a linear scale? Why?

**1.** The wealth of 20 different people, one of whom is a multi-billionaire.

**2.** The number of diamonds owned by 20 people, one of whom is a multi-billionaire.

**3.** The number of meals per week eaten in restaurants for a random sample of 20 people worldwide.

**4.** The number of tuberculosis bacteria in 20 different people, some never exposed to the disease, some slightly exposed, some with mild cases, and some dying of it.

**5.** **(a)** Use a calculator to fill in the following tables (round to 4 decimal digits).

| $n$ | 1 | 2 | 3 | 4 | 5 | 6 | 7 | 8 | 9 |
|---|---|---|---|---|---|---|---|---|---|
| $\log n$ | | | | | | | | | |

| $n$ | 10 | 20 | 30 | 40 | 50 | 60 | 70 | 80 | 90 |
|---|---|---|---|---|---|---|---|---|---|
| $\log n$ | | | | | | | | | |

**(b)** Using the results of part (a), plot the integer points 2 through 9 and the multiples of 10 from 20 to 90 on the log-scaled axis shown in Figure 5.31.

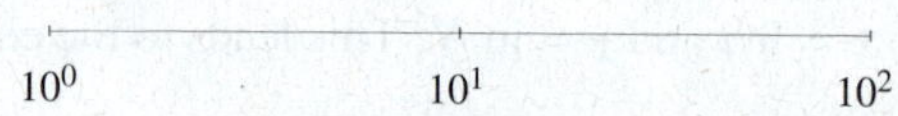


Figure 5.31

In Exercises 6–11, say where you would mark the given animal lifespan on an inch scale, where 0 inches represents $10^0 = 1$ year and 5 inches represents $10^5 = 100{,}000$ years. Give your answer to the nearest tenth of an inch.[18]

**6.** A colony of quaking aspen in Utah is estimated to be 80,000 years old.

**7.** A bristlecone pine in California named Methuselah is estimated by ring count to be 4838 years old.

**8.** A specimen of antarctic sponge is estimated to be 1550 years old.

**9.** The Puget Sound saltwater clam called a geoduck can live for 160 years.

**10.** The oldest recorded dog lived 29 years.

**11.** The common house mouse can live 4 years in captivity.

For the tables in Exercises 12–14,

**(a)** Use linear regression to find a linear function $y = b + mx$ that fits the data. Record the correlation coefficient.

**(b)** Use linear regression on the values $x$ and $\ln y$ to fit a function of the form $\ln y = b + mx$. Record the correlation coefficient. Convert to an exponential function $y = ae^{kx}$.

**(c)** Compare the correlation coefficients. Graph the data and the two functions to assess which function fits best.

**12.**

| $x$ | $y$ |
|---|---|
| 30 | 70 |
| 85 | 120 |
| 122 | 145 |
| 157 | 175 |
| 255 | 250 |
| 312 | 300 |

**13.**

| $x$ | $y$ |
|---|---|
| 8 | 23 |
| 17 | 150 |
| 23 | 496 |
| 26 | 860 |
| 32 | 2720 |
| 37 | 8051 |

**14.**

| $x$ | $y$ |
|---|---|
| 3.2 | 35 |
| 4.7 | 100 |
| 5.1 | 100 |
| 5.5 | 150 |
| 6.8 | 200 |
| 7.6 | 300 |

[18] From http://en.wikipedia.org/wiki/Maximum_life_span and http://en.wikipedia.org/wiki/List_of_long-living_organisms, accessed June, 2010.

## Problems

**15.** The signing of the Declaration of Independence is marked on the log scale in Figure 5.22 on page 217. To two decimal places, what is its position?

**16.** Figure 5.32 shows the prices of seven different items, with the scale markings representing the logarithm of the price in dollars. Give the approximate price of each item, and explain why a log scale was necessary.

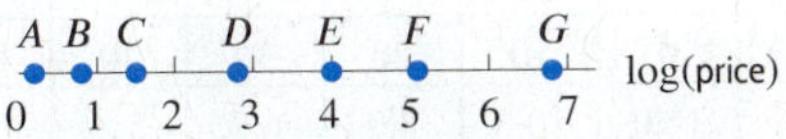


Figure 5.32

**17.** **(a)** Draw a line segment about 5 inches long. On it, choose an appropriate linear scale and mark points that represent the integral powers of two from zero to the sixth power. What is true about the location of the points as the exponents get larger?
**(b)** Draw a second line segment. Repeat the process in (a) but this time use a logarithmic scale so that the units are now powers of ten. What do you notice about the location of these points?

**18.** The usual distances for track (running) events are 100 meters, 200 meters, 400 meters, 800 meters, 1500 meters, 3000 meters, 5000 meters, and 10,000 meters.

**(a)** Plot the length of each track event on a linear scale.
**(b)** Plot the length of each track event on a logarithmic scale.
**(c)** Which scale, (a) or (b), is more useful to the runner?
**(d)** On each figure identify the point corresponding to 50 meters.

**19.** Microfinance refers to financial services, such as loans, offered to people with very low incomes. Table 5.12 shows the number of microborrowers in 2006.[19]

**(a)** Plot the data (in millions of borrowers) on a linear scale.
**(b)** Plot the data on a logarithmic scale.
**(c)** Which scale is more appropriate? Why?

**Table 5.12**

| Region | Borrowers (millions) |
|---|---|
| A: Africa | 8.4 |
| B: Asia | 112.7 |
| C: Eastern Europe and Central Asia | 3.4 |
| D: Latin America and the Caribbean | 6.8 |
| E: Middle East and North Africa | 1.7 |
| F: North America and Western Europe | 0.05 |

**20.** Table 5.13 shows the numbers of deaths in 2010 due to various causes in the US.[20]

**(a)** Explain why a log scale is necessary to plot the data from Table 5.13.
**(b)** Find the log of each value given.
**(c)** Plot the data using a log scale. Label each point with the related cause.

**Table 5.13**

| Cause | Deaths |
|---|---|
| Scarlet fever | 3 |
| Whooping cough | 25 |
| Asthma | 3404 |
| HIV | 8,369 |
| Kidney diseases | 50,476 |
| Accidents | 120,839 |
| Malignant neoplasms | 574,743 |
| Cardiovascular disease | 780,213 |
| All causes | 2,468,435 |

**21.** Table 5.14 shows the dollar value of some items in 2012. Plot and label these values on a log scale.

**Table 5.14**

| Item | Dollar value | Item | Dollar value |
|---|---|---|---|
| Pack of gum | 0.50 | New house | 152,000 |
| Movie ticket | 8 | Lottery ticket sales[21] | 60 bn |
| New laptop computer | 513 | Bill Gates's[22] worth | 72.7 bn |
| Year at college | 22,261 | US GDP | 15.87 tn |
| Luxury car | 106,000 | National debt | 16 tn |

**22.** Table 5.15 shows the sizes of various organisms. Plot and label these values on a log scale.

**Table 5.15**

| Animal | Size (cm) | Animal | Size (cm) |
|---|---|---|---|
| Virus | 0.0000005 | Domestic cat | 60 |
| Bacterium | 0.0002 | Wolf (with tail) | 200 |
| Human cell | 0.002 | Thresher shark | 600 |
| Ant | 0.8 | Giant squid | 2200 |
| Hummingbird | 12 | Sequoia | 7500 |

[19] Based on data in Daley-Harris, *State of the Microsummit Campaign Report 2007*, p. 22.
[20] http://www.cdc.gov/nchs/data/dvs/deaths_2010_release.pdf, accessed July 24, 2013.
[21] http://www.gamblingdata.com/node/50847, accessed July 26, 2013.
[22] http://www.celebritynetworth.com/dl/bill-gates-net-worth-2012, accessed July 26, 2013.

**23.** **(a)** Complete Table 5.16 with values of $y = 3^x$.
**(b)** Complete Table 5.17 with values for $y = \log(3^x)$. What kind of function is this?
**(c)** Complete tables for $f(x) = 2 \cdot 5^x$ and $g(x) = \log(2 \cdot 5^x)$. What kinds of functions are these?
**(d)** What seems to be true about a function which is the logarithm of an exponential function? Is this true in general?

**Table 5.16**

| $x$ | 0 | 1 | 2 | 3 | 4 | 5 |
|---|---|---|---|---|---|---|
| $y = 3^x$ | | | | | | |

**Table 5.17**

| $x$ | 0 | 1 | 2 | 3 | 4 | 5 |
|---|---|---|---|---|---|---|
| $y = \log(3^x)$ | | | | | | |

**24.** Repeat part (b) and (c) of Problem 23 using the natural log function. Is your answer to part (d) the same?

**25.** **(a)** Plot the data in Table 5.18.
**(b)** What kind of function might the data from part (a) represent?
**(c)** Now plot $\log y$ versus $x$ instead of $y$ versus $x$. What do you notice?

**Table 5.18**

| $x$ | 0.2 | 1.3 | 2.1 | 2.8 | 3.4 | 4.5 |
|---|---|---|---|---|---|---|
| $y$ | 5.7 | 12.3 | 21.4 | 34.8 | 52.8 | 113.1 |

**26.** Table 5.19 shows the value, $y$, of US imports from China with $x$ in years since 2006.[23]

**(a)** Find a formula for a linear function $y = b + mx$ that fits the data.
**(b)** Find $\ln y$ for each $y$ value, and use the $x$ and $\ln y$ values to find a formula for a linear function $\ln y = b + mx$ that fits the data.
**(c)** Use the equation in part (b) to find an exponential function of the form $y = ae^{kx}$ that fits the data.

**Table 5.19** *Value of US imports from China in billions of dollars*

| | 2006 | 2007 | 2009 | 2009 | 2010 | 2011 | 2012 |
|---|---|---|---|---|---|---|---|
| $x$ | 0 | 1 | 2 | 3 | 4 | 5 | 6 |
| $y$ | 288 | 321 | 338 | 296 | 365 | 399 | 426 |

**27.** Table 5.20 shows newspapers' share of the expenditure of national advertisers. Using the method of Problem 26, fit an exponential function of the form $y = ae^{kx}$ to the data, where $y$ is percent share and $x$ is the number of years since 1950.

**Table 5.20** *Newspapers' share of advertising*

| | 1950 | 1960 | 1970 | 1980 | 1990 | 1992 |
|---|---|---|---|---|---|---|
| $x$ | 0 | 10 | 20 | 30 | 40 | 42 |
| $y$ | 16.0 | 10.8 | 8.0 | 6.7 | 5.8 | 5.0 |

**28.** To study how recognition memory decreases with time, the following experiment was conducted. The subject read a list of 20 words slowly aloud, and later, at different time intervals, was shown a list of 40 words containing the 20 words that he or she had read. The percentage, $P$, of words recognized was recorded as a function of $t$, the time elapsed in minutes. Table 5.21 shows the averages for 5 different subjects.[24] This is modeled by $P = a \ln t + b$.

**(a)** Find $\ln t$ for each value of $t$, and then use regression on a calculator or computer to estimate $a$ and $b$.
**(b)** Graph the data points and regression line on a coordinate system of $P$ against $\ln t$.
**(c)** When does this model predict that the subjects will recognize no words? All words?
**(d)** Graph the data points and curve $P = a \ln t + b$ on a coordinate system with $P$ against $t$, with $0 \le t \le 10{,}500$.

**Table 5.21** *Percentage of words recognized*

| $t$, min | 5 | 15 | 30 | 60 | 120 | 240 |
|---|---|---|---|---|---|---|
| $P$% | 73.0 | 61.7 | 58.3 | 55.7 | 50.3 | 46.7 |
| $t$, min | 480 | 720 | 1440 | 2880 | 5760 | 10,080 |
| $P$% | 40.3 | 38.3 | 29.0 | 24.0 | 18.7 | 10.3 |

**29.** Table 5.22 gives the length $\ell$ (in cm) and weight $w$ (in gm) of 16 different fish known as threadfin bream (*Nemipterus marginatus*) found in the South China Sea.[25]

**(a)** Let $W = \ln w$ and $L = \ln \ell$. For these sixteen data points, plot $W$ on the vertical axis and $L$ on the horizontal axis. Describe the resulting scatterplot.

[23] http://www.census.gov/foreign-trade/balance/c5700.html, accessed July 27, 2013.
[24] Adapted from D. Lewis, *Quantitative Methods in Psychology* (New York: McGraw-Hill, 1960).
[25] Data taken from *Introduction to Tropical Fish Stock Assessment* by Per Sparre, Danish Institute for Fisheries Research, and Siebren C. Venema, FAO Fisheries Department, available at http://www.fao.org/docrep/W5449E/w5449e00.htm. This source cites the following original reference: Pauly, D., 1983. Some simple methods for the assessment of tropical fish stocks.

**(b)** Fitting a line to the scatterplot you drew in part (a), find a possible formula for $W$ in terms of $L$.

**(c)** Based on your formula for part (b), find a possible formula for $w$ in terms of $\ell$.

**(d)** Comment on your formula, keeping in mind what you know about units as well the typical relationship between weight, volume, and length.

**Table 5.22** *Length and weight of fish*

| Type | 1 | 2 | 3 | 4 | 5 | 6 | 7 | 8 |
|---|---|---|---|---|---|---|---|---|
| $\ell$ | 8.1 | 9.1 | 10.2 | 11.9 | 12.2 | 13.8 | 14.8 | 15.7 |
| $w$ | 6.3 | 9.6 | 11.6 | 18.5 | 26.2 | 36.1 | 40.1 | 47.3 |
| Type | 9 | 10 | 11 | 12 | 13 | 14 | 15 | 16 |
| $\ell$ | 16.6 | 17.7 | 18.7 | 19.0 | 20.6 | 21.9 | 22.9 | 23.5 |
| $w$ | 65.6 | 69.4 | 76.4 | 82.5 | 106.6 | 119.8 | 169.2 | 173.3 |

**30.** A light, flashing regularly, consists of cycles, each cycle having a dark phase and a light phase. The frequency of this light is measured in cycles per second. As the frequency is increased, the eye initially perceives a series of flashes of light, then a coarse flicker, a fine flicker, and ultimately a steady light. The frequency at which the flickering disappears is called the fusion frequency.[26] Table 5.23 shows the results of an experiment[27] in which the fusion frequency $F$ was measured as a function of the light intensity $I$. It is modeled by $F = a \ln I + b$.

**(a)** Find $\ln I$ for each value of $I$, and then use linear regression on a calculator or computer to estimate $a$ and $b$ in the equation $F = a \ln I + b$.

**(b)** Plot $F$ against $\ln I$, showing the data points and the line.

**(c)** Plot $F$ against $I$, showing the data points and the curve and give its equation.

**(d)** The units of $I$ are arbitrary, that is, not given. If the units of $I$ were changed, which of the constants $a$ and $b$ would be affected, and in what way?

**Table 5.23** *Fusion frequency, F, as a function of the light intensity, I*

| $I$ | 0.8 | 1.9 | 4.4 | 10.0 | 21.4 | 48.4 | 92.5 | 218.7 | 437.3 | 980.0 |
|---|---|---|---|---|---|---|---|---|---|---|
| $F$ | 8.0 | 12.1 | 15.2 | 18.5 | 21.7 | 25.3 | 28.3 | 31.9 | 35.2 | 38.2 |

## CHAPTER SUMMARY

- **Logarithms**
  Common log: $y = \log x$ means $10^y = x$.
  $\log 10 = 1, \log 1 = 0$.
  Natural log: $y = \ln x$ means $e^y = x$.
  $\ln e = 1, \ln 1 = 0$.
- **Properties of Logs**

$$\log(ab) = \log a + \log b \qquad \ln(ab) = \ln a + \ln b.$$
$$\log(a/b) = \log a - \log b \qquad \ln(a/b) = \ln a - \ln b.$$
$$\log(b^t) = t \log b \qquad \ln(b^t) = t \ln b.$$
$$\log(10^x) = 10^{\log x} = x \qquad \ln(e^x) = e^{\ln x} = x.$$

- **Converting Between Base $b$ and Base $e$**
  If $Q = ab^t$ and $Q = ae^{kt}$, then $k = \ln b$.
- **Solving Equations Using Logs**
  Solve equations such as $ab^t = c$ and $ae^{kt} = c$ using logs. Not all exponential equations can be solved with logs, e.g. $2^t = 3 + t$.
- **Logarithmic Functions**
  Graph; domain; range; concavity; asymptotes.
- **Applications of Logarithms**
  Doubling time; half life;
  Chemical acidity; orders of magnitude; decibels.
- **Logarithmic Scales**
  Plotting data; log-log scales. Linearizing data and fitting curves to data using logs.
- **Limits and Limits from the Right and from the Left**

## REVIEW EXERCISES AND PROBLEMS FOR CHAPTER FIVE

### Exercises

In Exercises 1–2, convert to the form $Q = ab^t$.

**1.** $Q = 7e^{-10t}$

**2.** $Q = 5e^t$

In Exercises 3–6, convert to the form $Q = ae^{kt}$.

**3.** $Q = 4 \cdot 7^t$

**4.** $Q = 2 \cdot 3^t$

**5.** $Q = 4 \cdot 8^{1.3t}$

**6.** $Q = 973 \cdot 6^{2.1t}$

[26] R. S. Woodworth, *Experimental Psychology* (New York: Holt and Company, 1948).
[27] D. Lewis, *Quantitative Methods in Psychology* (New York: McGraw-Hill, 1960).

Solve the equations in Exercises 7–20 exactly if possible.

**7.** $1.04^t = 3$

**8.** $e^{0.15t} = 25$

**9.** $91 = 46(1.1)^x$

**10.** $e^{0.044t} = 6$

**11.** $5(1.014)^{3t} = 12$

**12.** $5(1.15)^t = 8(1.07)^t$

**13.** $5(1.031)^x = 8$

**14.** $4(1.171)^x = 7(1.088)^x$

**15.** $3\log(2x+6) = 6$

**16.** $1.7(2.1)^{3x} = 2(4.5)^x$

**17.** $3^{4\log x} = 5$

**18.** $13e^{0.081t} = 25e^{0.032t}$

**19.** $87e^{0.066t} = 3t + 7$

**20.** $\log x + \log(x-1) = \log 2$

In Exercises 21–23, simplify fully.

**21.** $\log\left(100^{x+1}\right)$

**22.** $\ln\left(e \cdot e^{2+M}\right)$

**23.** $\ln(A+B) - \ln(A^{-1} + B^{-1})$

In Exercises 24–29, state the domain of the function and identify any vertical asymptote of its graph. You need not graph the function.

**24.** $y = \ln(x+8)$

**25.** $y = \log(x-20)$

**26.** $y = \log(12-x)$

**27.** $y = \ln(300-x)$

**28.** $y = \ln\left(x - e^2\right)$

**29.** $y = \log(x+15)$

## Problems

**30.** Suppose that $x = \log A$ and that $y = \log B$. Write the following expressions in terms of $x$ and $y$.

(a) $\log(AB)$ (b) $\log(A^3 \cdot \sqrt{B})$

(c) $\log(A-B)$ (d) $\dfrac{\log A}{\log B}$

(e) $\log \dfrac{A}{B}$ (f) $AB$

**31.** Let $p = \ln m$ and $q = \ln n$. Write the following expressions in terms of $p$ and/or $q$ without using logs.

(a) $\ln(nm^4)$ (b) $\ln\left(\dfrac{1}{n}\right)$

(c) $\dfrac{\ln m}{\ln n}$ (d) $\ln(n^3)$

**32.** Let $x = 10^U$ and $y = 10^V$. Write the following expressions in terms of $U$ and/or $V$ without using logs.

(a) $\log xy$ (b) $\log\left(\dfrac{x}{y}\right)$

(c) $\log x^3$ (d) $\log\left(\dfrac{1}{y}\right)$

**33.** Solve the following equations. Give approximate solutions if exact ones can't be found.

(a) $e^{x+3} = 8$ (b) $4(1.12^x) = 5$

(c) $e^{-0.13x} = 4$ (d) $\log(x-5) = 2$

(e) $2\ln(3x) + 5 = 8$ (f) $\ln x - \ln(x-1) = 1/2$

(g) $e^x = 3x + 5$ (h) $3^x = x^3$

(i) $\ln x = -x^2$

**34.** Solve for $x$ exactly.

(a) $\dfrac{3^x}{5^{x-1}} = 2^{x-1}$

(b) $-3 + e^{x+1} = 2 + e^{x-2}$

(c) $\ln(2x-2) - \ln(x-1) = \ln x$

(d) $9^x - 7 \cdot 3^x = -6$

(e) $\ln\left(\dfrac{e^{4x}+3}{e}\right) = 1$

(f) $\dfrac{\ln(8x) - 2\ln(2x)}{\ln x} = 1$

In Problems 35–38, the Richter scale ratings for two earthquakes are $M_1$ and $M_2$, with $M_2 > M_1$. If the earthquakes have seismic waves of sizes $W_1$ and $W_2$, respectively, then

$$M_2 - M_1 = \log\left(\frac{W_2}{W_1}\right).$$

How many times greater than the smaller one are the seismic waves for the larger one?

**35.** $M_1 = 4.2$ and $M_2 = 6.4$

**36.** $M_1 = 5.3$ and $M_2 = 5.8$

**37.** $M_1 = 4.4$ and $M_2 = 5.6$

**38.** $M_1 = 5.7$ and $M_2 = 8.1$

**39.** With $t$ in years, the formulas for dollar balances of two bank accounts are:

$$f(t) = 1100(1.05)^t \quad \text{and} \quad g(t) = 1500e^{0.05t}.$$

**(a)** Describe in words the bank account modeled by $f$.

**(b)** Describe the account modeled by $g$. State the effective annual rate.

**(c)** What continuous interest rate has the same effective growth rate as $f$?

**40.** **(a)** Let $B = 5000(1.06)^t$ give the balance of a bank account after $t$ years. If the formula for $B$ is written $B = 5000e^{kt}$, estimate the value of $k$ correct to four decimal places. What is the financial meaning of $k$?

**(b)** The balance of a bank account after $t$ years is given by the formula $B = 7500e^{0.072t}$. If the formula for $B$ is written $B = 7500b^t$, find $b$ exactly, and give the value of $b$ correct to four decimal places. What is the financial meaning of $b$?

**41.** The number of bacteria present in a culture after $t$ hours is given by the formula $N = 1000e^{0.69t}$.

**(a)** How many bacteria will there be after 1/2 hour?
**(b)** How long before there are 1,000,000 bacteria?
**(c)** What is the doubling time?

**42.** In 2010, the population of the country Erehwon was 50 million people and increasing by 2.9% every year. The population of the country Ecalpon, on other hand, was 45 million people and increasing by 3.2% every year.

**(a)** For each country, write a formula expressing the population as a function of time $t$, where $t$ is the number of years since 2010.
**(b)** Find the value(s) of $t$, if any, when the two countries have the same population.
**(c)** When is the population of Ecalpon double that of Erehwon?

**43.** The price $P(t) = 5(2)^{t/7}$ of a good is rising due to inflation, where $t$ is time in years.

**(a)** What is the doubling time?
**(b)** What is the annual inflation rate?

**44.** Let $P = 15(1.04)^t$ give the population (in thousands) of a town, with $t$ in years.

**(a)** Describe the population growth in words.
**(b)** If the formula for $P$ is written $P = 15(b)^{12t}$, find $b$ exactly. What is the meaning of $b$ in the context of the population?
**(c)** If the formula for $P$ is written $P = 15(2)^{t/c}$, find the value of $c$ correct to 2 decimals. What is the meaning of $c$ in this context?

In Problems 45–47, use $v(t) = 20e^{0.2t}$ and $w(t) = 12e^{0.22t}$.

**45.** Solve $v(t) = 30$ exactly.

**46.** Solve $3v(2t) = 2w(3t)$ exactly.

**47.** Find the doubling time of $w$.

**48.** A calculator confirms that $5 \approx 10^{0.7}$. Show how to use this fact to approximate the value of log 25.

**49.** **(a)** What are the domain and range of $f(x) = 10^x$? What is the asymptote of $f(x) = 10^x$?
**(b)** What does your answer to part (a) tell you about the domain, range, and asymptotes of $g(x) = \log x$?

**50.** What is the domain of $y = \ln(x^2 - x - 6)$?

**51.** **(a)** Plot the data given by Table 5.24. What kind of function might fit this data well?
**(b)** Using the substitution $z = \ln x$, transform the data in Table 5.24, and compile your results into a new table. Plot the transformed data as $y$ versus $\ln x$.
**(c)** What kind of function gives a good fit to the plot you made in part (b)? Find a formula for $y$ in terms of $z$ that fits the data well.
**(d)** Using the formula from part (c), find a formula for $y$ in terms of $x$ that gives a good fit to the data in Table 5.24.
**(e)** What does your formula from part (d) tell you about $x$ as a function of $y$ (as opposed to $y$ as a function of $x$)?

**Table 5.24**

| $x$ | 0.21 | 0.55 | 1.31 | 3.22 | 5.15 | 12.48 |
|---|---|---|---|---|---|---|
| $y$ | −11 | −2 | 6.5 | 16 | 20.5 | 29 |

**52.** Radioactive carbon-14 decays according to the function $Q(t) = Q_0e^{-0.000121t}$ where $t$ is time in years, $Q(t)$ is the quantity remaining at time $t$, and $Q_0$ is the amount of present at time $t = 0$. Estimate the age of a skull if 23% of the original quantity of carbon-14 remains.

**53.** Suppose 2 mg of a drug is injected into a person's bloodstream. As the drug is metabolized, the quantity diminishes at the continuous rate of 4% per hour.

**(a)** Find a formula for $Q(t)$, the quantity of the drug remaining in the body after $t$ hours.
**(b)** By what percent does the drug level decrease during any given hour?
**(c)** The person must receive an additional 2 mg of the drug whenever its level has diminished to 0.25 mg. When must the person receive the second injection?
**(d)** When must the person receive the third injection?

**54.** A rubber ball is dropped onto a hard surface from a height of 6 feet, and it bounces up and down. At each bounce it rises to 90% of the height from which it fell.

**(a)** Find a formula for $h(n)$, the height reached by the ball on bounce $n$.
**(b)** How high will the ball bounce on the $12^{\text{th}}$ bounce?
**(c)** How many bounces before the ball rises no higher than an inch?

**55.** Oil leaks from a tank. At hour $t = 0$ there are 250 gallons of oil in the tank. Each hour after that, 4% of the oil leaks out.

**(a)** What percent of the original 250 gallons has leaked out after 10 hours? Why is it less than $10 \cdot 4\% = 40\%$?
**(b)** If $Q(t) = Q_0e^{kt}$ is the quantity of oil remaining after $t$ hours, find the value of $k$. What does $k$ tell you about the leaking oil?

**56.** Before the advent of computers, logarithms were calculated by hand. Various tricks were used to evaluate different logs. One such trick exploits the fact that $2^{10} \approx 1000$. (It actually equals 1024.) Using this fact and the log properties, show that

**(a)** $\log 2 \approx 0.3$ **(b)** $\log 7 \approx 0.85$

**57.** A googol is the number 1 followed by 100 zeros, or $10^{100}$. A googolplex is the number 1 followed by a googol zeros, or $10^{\text{googol}}$. Evaluate:

**(a)** $\sqrt{\log(\text{googol})}$ **(b)** $\log\sqrt{\text{googol}}$ **(c)** $\sqrt{\log(\text{googolplex})}$

**58.** Since $e = 2.718\ldots$ we know that $2 < e < 3$, which means that $2^2 < e^2 < 3^2$. Without using a calculator, explain why

**(a)** $1 < \ln 3 < 2$ **(b)** $1 < \ln 4 < 2$

**59.** Simplify the expression $\sqrt{1000^{\frac{1}{12}\cdot\log k}}$. Your answer should be exact and should not involve exponents or logs, though it may involve radicals.

## STRENGTHEN YOUR UNDERSTANDING

Are the statements in Problems 1–37 true or false? Give an explanation for your answer.

**1.** The log of 2000 is less than 3.

**2.** The inverse of $y = \ln x$ is $y = e^x$.

**3.** If $2^x = 1024$ then $x = 10$.

**4.** If a quantity grew to 4 times its original amount in 8 hours then its doubling time is one-half hour.

**5.** If the function $y = ab^t$ is converted to the form $y = ae^{kt}$, $k$ is always equal to $\ln b$.

**6.** If $x$ is a positive number, $\log x$ is the exponent of 10 that gives $x$.

**7.** If $10^y = x$ then $\log x = y$.

**8.** The quantity $10^{-k}$ is a negative number when $k$ is positive.

**9.** For any $n$, we have $\log(10^n) = n$.

**10.** If $n > 0$, then $10^{\log n} = n$.

**11.** If $a$ and $b$ are positive, $\log\left(\frac{a}{b}\right) = \frac{\log a}{\log b}$.

**12.** If $a$ and $b$ are positive, $\ln(a + b) = \ln a + \ln b$.

**13.** For any value $a$, $\log a = \ln a$.

**14.** For any value $x$, $\ln(e^{2x}) = 2x$.

**15.** The function $y = \log x$ has an asymptote at $y = 0$.

**16.** The graph of the function $y = \log x$ is concave down.

**17.** The reflected graph of $y = \log x$ across the line $y = x$ is the graph of $y = 10^x$.

**18.** If $y = \log\sqrt{x}$ then $y = \frac{1}{2}\log x$.

**19.** The function $y = \log(b^t)$ is always equal to $y = (\log b)^t$.

**20.** The values of $\ln e$ and $\log 10$ are both 1.

**21.** If $7.32 = e^t$ then $t = \frac{7.32}{e}$.

**22.** If $50(0.345)^t = 4$, then $t = \frac{\log(4/50)}{\log 0.345}$.

**23.** If $ab^t = n$, then $t = \frac{\log(n/a)}{\log b}$.

**24.** The doubling time of a quantity $Q = Q_0e^{kt}$ is the time it takes for any $t$-value to double.

**25.** The half-life of a quantity is the time it takes for the quantity to be reduced by half.

**26.** If the half-life of a substance is 5 hours then there will be $\frac{1}{4}$ of the substance in 25 hours.

**27.** If $y = 6(3)^t$, then $y = 6e^{(\ln 3)t}$.

**28.** If a population doubles in size every 20 years, its annual continuous growth rate is 20%.

**29.** If $Q = Q_0e^{kt}$, then $t = \frac{\ln(Q/Q_0)}{k}$.

**30.** Log scales provide a way to graph quantities that have vastly different magnitudes.

**31.** An elephant weighs about 8000 pounds. Its weight, plotted on a log scale, would be a little before 4.

**32.** A virus has a cell size of about 0.0000005 cm. On a log scale, its size would be plotted at about 7.

**33.** The closest star outside our solar system is about 26,395,630,000,000 miles away from Earth. This distance, plotted on a log scale, would occur a little after 26.

**34.** In a graph made using a log-log scale, consecutive powers of 10 are equally spaced on the horizontal axis and on the vertical axis.

**35.** One million and one billion differ by one order of magnitude.

**36.** After fitting a data set with both an exponential function, $y = Ae^{kx}$, and a power function, $y = Bx^n$, we must have $B = A$.

**37.** Given the points on a cubic curve, (1, 1), (2, 8), (3, 27) and (4, 64) it is not possible to fit an exponential function to this data.

# SKILLS REFRESHER FOR CHAPTER 5: LOGARITHMS

We list the definitions and properties of the common and natural logarithms.

**Properties of Logarithms** If $M, N > 0$:

- Logarithm of a product: $\log MN = \log M + \log N$ $\quad$ $\ln MN = \ln M + \ln N$
- Logarithm of a quotient: $\log M/N = \log M - \log N$ $\quad$ $\ln M/N = \ln M - \ln N$
- Logarithm of a power: $\log M^P = P\log M$ $\quad$ $\ln M^P = P\ln M$
- Logarithm of 1: $\log 1 = 0$ $\quad$ $\ln 1 = 0$
- Logarithm of the base: $\log 10 = 1$ $\quad$ $\ln e = 1$

Be aware of the following two common errors,

$$\log(M+N) \neq (\log M)(\log N)$$

and

$$\log(M-N) \neq \frac{\log M}{\log N}.$$

**Relationships Between Logarithms and Exponents**

- $\log N = x$ if and only if $10^x = N$ $\quad$ $\ln N = x$ if and only if $e^x = N$
- $\log 10^x = x$ $\quad$ $\ln e^x = x$
- $10^{\log x} = x$, for $x > 0$ $\quad$ $e^{\ln x} = x$, for $x > 0$

**Example 1** Evaluate without a calculator:

(a) $\log 10{,}000$ (b) $\ln 1$

**Solution**

(a) Common logarithms are powers of 10. The power of 10 needed to get 10,000 is 4, so $\log 10{,}000 = 4$.

(b) Natural logarithms are powers of $e$. The power of $e$ needed to get 1 is zero, so $\ln 1 = 0$.

**Example 2** Write the equation in exponential form:

(a) $\log x = -3$ (b) $\ln x = \sqrt{2}$

**Solution**

(a) By definition $\log x = -3$ means $10^{-3} = x$.

(b) By definition $\ln x = \sqrt{2}$ means $e^{\sqrt{2}} = x$.

**Example 3** Write the equation in logarithmic form:

(a) $10^x = 1000$ (b) $10^{-2} = 0.01$ (c) $e^{-1} = 0.368$

**Solution**

(a) By definition $10^x = 1000$ means $\log 1000 = x$.

(b) By definition $10^{-2} = 0.01$ means $\log 0.01 = -2$.

(c) By definition $e^{-1} = 0.368$ means $\ln 0.368 = -1$.

**Example 4** Write the expression using sums and/or differences of logarithmic expressions that do not contain the logarithms of products, quotients or powers.

(a) $\log(10x)$ (b) $\ln\left(\frac{e^2}{\sqrt{x}}\right)$

**Solution** (a)

$$\begin{aligned}\log(10x) &= \log 10 + \log x && \text{Logarithm of a product}\\ &= 1 + \log x. && \text{Logarithm of the base}\end{aligned}$$

(b)

$$\begin{aligned}\ln\left(\frac{e^2}{\sqrt{x}}\right) &= \ln e^2 - \ln\sqrt{x} && \text{Logarithm of a quotient}\\ &= \ln e^2 - \ln x^{1/2}\\ &= 2\ln e - \frac{1}{2}\ln x && \text{Logarithm of a power}\\ &= 2\cdot 1 - \frac{1}{2}\ln x && \text{Logarithm of the base}\\ &= 2 - \frac{1}{2}\ln x.\end{aligned}$$

**Example 5** Write the expression as a single logarithm.

(a) $\ln x - 2\ln y$ (b) $3(\log x + \frac{4}{3}\log y)$

**Solution** (a)

$$\begin{aligned}\ln x - 2\ln y &= \ln x - \ln y^2 && \text{Logarithm of a power}\\ &= \ln\left(\frac{x}{y^2}\right). && \text{Logarithm of a quotient}\end{aligned}$$

(b)

$$\begin{aligned}3\left(\log x + \frac{4}{3}\log y\right) &= 3\log x + 4\log y\\ &= \log x^3 + \log y^4 && \text{Logarithm of a power}\\ &= \log(x^3y^4). && \text{Logarithm of a product}\end{aligned}$$

**Example 6** Express in terms of $x$ without logarithms.

(a) $e^{3\ln x}$ (b) $\log 10^{2x}$

**Solution** (a)

$$\begin{aligned}e^{3\ln x} &= e^{\ln x^3} && \text{Logarithm of a power}\\ &= x^3. && \text{Logarithm of the base}\end{aligned}$$

(b) $\log 10^{2x} = 2x.$ Logarithm of the base

**Example 7** Solve the equation for $x$.

(a) $12e^x = 5$ (b) $2^{-3x} = 17$

**Solution** (a)

$$\begin{aligned} 12e^x &= 5 \\ e^x &= \frac{5}{12} && \text{Dividing by 12} \\ \ln e^x &= \ln\left(\frac{5}{12}\right) && \text{Taking ln of both sides} \\ x &= \ln\left(\frac{5}{12}\right). && \text{Logarithm of the base} \end{aligned}$$

(b)

$$\begin{aligned} 2^{-3x} &= 17 \\ \log 2^{-3x} &= \log 17 && \text{Taking logs of both sides} \\ -3x \log 2 &= \log 17 && \text{Logarithm of a power} \\ -3x &= \frac{\log 17}{\log 2} && \text{Dividing by log 2} \\ x &= -\frac{\log 17}{3\log 2}. && \text{Dividing by -3} \end{aligned}$$

**Example 8** Solve the equation for $x$.

(a) $2(\log(2x+50)) - 4 = 0$ (b) $\ln(x+2) = 3$

**Solution** (a)

$$\begin{aligned} 2(\log(2x+50)) - 4 &= 0 \\ 2(\log(2x+50)) &= 4 \\ \log(2x+50) &= 2 \\ 10^{\log(2x+50)} &= 10^2 && \text{Converting to exponential form} \\ 2x+50 &= 10^2 \\ 2x+50 &= 100 \\ 2x &= 50 \\ x &= 25. \end{aligned}$$

(b)

$$\begin{aligned} \ln(x+2) &= 3 \\ e^{\ln(x+2)} &= e^3 && \text{Raising } e \text{ to each side} \\ x+2 &= e^3 \\ x &= e^3 - 2. \end{aligned}$$

## Exercises to Skills for Chapter 5

For Exercises 1–8, evaluate without a calculator.

**1.** $\log(\log 10)$

**2.** $\ln(\ln e)$

**3.** $2\ln e^4$

**4.** $\ln\left(\frac{1}{e^5}\right)$

**5.** $\dfrac{\log 1}{\log 10^5}$

**6.** $e^{\ln 3} - \ln e$

**7.** $\sqrt{\log 10{,}000}$

**8.** $10^{\log 7}$

For Exercises 9–12, rewrite the exponential equation in equivalent logarithmic form.

**9.** $10^{-4} = 0.0001$

**10.** $10^{0.477} = 3$

**11.** $e^{-2} = 0.135$

**12.** $e^{2x} = 7$

For Exercises 13–15, rewrite the logarithmic equation in equivalent exponential form.

**13.** $\log 0.01 = -2$

**14.** $\ln x = -1$

**15.** $\ln 4 = x^2$

For Exercises 16–24, if possible, write the expression using sums and/or differences of logarithmic expressions that do not contain the logarithms of products, quotients, or powers.

**16.** $\log 2x$

**17.** $\dfrac{\ln x}{2}$

**18.** $\log\left(\frac{x}{5}\right)$

**19.** $\log\left(\dfrac{x^2+1}{x^3}\right)$

**20.** $\ln\sqrt{\dfrac{x-1}{x+1}}$

**21.** $\log(x^2 + y^2)$

**22.** $\log(x^2 - y^2)$

**23.** $(\log x)(\log y)$

**24.** $\dfrac{\ln x^2}{\ln(x+2)}$

For Exercises 25–31, rewrite the expression as a single logarithm.

**25.** $\log 12 + \log x$

**26.** $\ln x^2 - \ln(x+10)$

**27.** $\frac{1}{2}\log x + 4\log y$

**28.** $\log 3 + 2\log\sqrt{x}$

**29.** $3\left(\log(x+1) + \frac{2}{3}\log(x+4)\right)$

**30.** $\ln x + \ln\left(\frac{y}{2}(x+4)\right) + \ln z^{-1}$

**31.** $2\log(9 - x^2) - (\log(3+x) + \log(3-x))$

For Exercises 32–39, simplify the expression if possible.

**32.** $2\ln e^{\sqrt{x}}$

**33.** $\log(A^2 + B^2)$

**34.** $\log 10x - \log x$

**35.** $2\ln x^{-2} + \ln x^4$

**36.** $\ln\sqrt{x^2+16}$

**37.** $\log 100^{2z}$

**38.** $\dfrac{\ln e}{\ln e^2}$

**39.** $\ln\dfrac{1}{e^x+1}$

For Exercises 40–45, solve for $x$ using logarithms.

**40.** $12^x = 7$

**41.** $3 \cdot 5^x = 9$

**42.** $4 \cdot 13^{3x} = 17$

**43.** $e^{-5x} = 9$

**44.** $12^{5x} = 3 \cdot 15^{2x}$

**45.** $19^{6x} = 77 \cdot 7^{4x}$

In Exercises 46–49, solve for $x$.

**46.** $3\log(4x+9) - 6 = 2$

**47.** $4\log(9x+17) - 5 = 1$

**48.** $\ln(3x+4) = 5$

**49.** $2\ln(6x-1) + 5 = 7$

Chapter Six

# TRANSFORMATIONS OF FUNCTIONS AND THEIR GRAPHS

© Patrick Zephyr/Patrick Zephyr Nature Photography

**Contents**

# 6.1 SHIFTS, REFLECTIONS, AND SYMMETRY

In Section 2.4 we saw how a function and its graph change under horizontal and vertical shifts. Shifts translate the graph of a function without changing its shape or orientation in the $xy$-plane.

**Example 1**

(a) Figure 6.1 shows graphs of $f(x) = e^x$, $g(x)$, and $h(x)$, where $g(x)$ is $f(x)$ shifted 2 units upward and $h(x)$ is $f(x)$ shifted 1 unit downward. Find formulas for $g(x)$ and $h(x)$.

(b) Figure 6.2 shows graphs of $p(x) = \ln x$, $q(x)$, and $r(x)$, where $q(x)$ is $p(x)$ shifted 1 to the right and $r(x)$ is $p(x)$ shifted 2 to the left. Find formulas for $p(x)$ and $r(x)$.

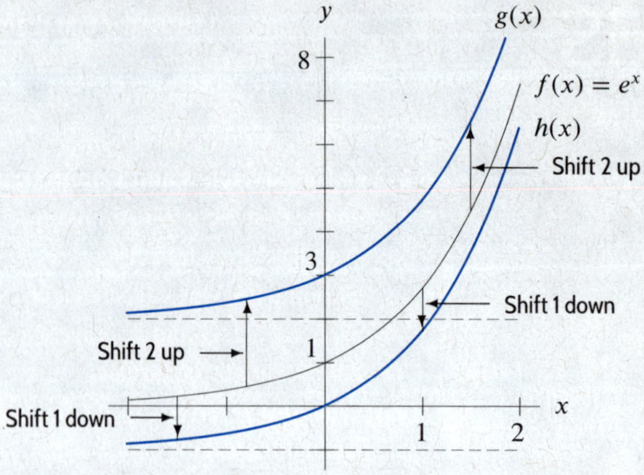


**Figure 6.1**: $g(x)$ and $h(x)$ are vertical shifts of $f(x)$

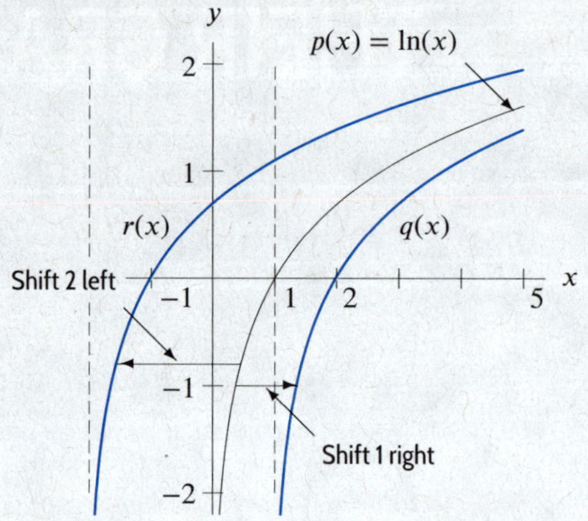


**Figure 6.2**: $q(x)$ and $r(x)$ are horizontal shifts of $p(x)$

**Solution**

(a) In Figure 6.1, the graph of $g(x)$ is the graph of $f(x) = e^x$ shifted 2 units vertically upward. The graph of $h(x)$ is the graph of $f(x) = e^x$ shifted 1 unit vertically downward. Thus, the formulas are

$$g(x) = f(x) + 2 \qquad \text{and} \qquad h(x) = f(x) - 1.$$

(b) In Figure 6.2, the graph of $q(x)$ is the graph of $p(x) = \ln x$ shifted horizontally 1 unit to the right. The graph of $r(x)$ is the graph of $p(x) = \ln x$ shifted horizontally 2 units to the left. Thus, the formulas are

$$q(x) = p(x - 1) \qquad \text{and} \qquad r(x) = p(x + 2).$$

In Example 1, the graphs of $g(x)$ and $h(x)$ are vertical shifts of the graph of $f(x) = e^x$. Vertical shifts involve adding or subtracting a constant to or from a function's output; they are called *outside changes*. The graphs of $q(x)$ and $r(x)$ are horizontal shifts of the graph of $p(x) = \ln x$. Horizontal shifts involve adding or subtracting a constant to a function's input; they are called *inside changes*.

A translation of a function may be a combination of vertical and horizontal shifts.

**Example 2**

Figure 6.3 shows a graph of $f(x)$ and a translation $g(x)$. The point $(0, 0)$ on the graph of $y = f(x)$ moves to the point $(5, 1)$ on the graph of the $g(x)$. Give a formula for $g(x)$ in terms of $f(x)$.

**Solution**

The point $(0, 0)$ on the graph of $y = f(x)$ moves 5 units to the right and 1 unit up to the point $(5, 1)$. This means that all points on the graph of $f$ must move 5 units right and 1 unit up to form the graph of $g$. We have

$$g(x) = f(x - 5) + 1.$$

Notice that $g(5) = f(0) + 1 = 1$, as expected.

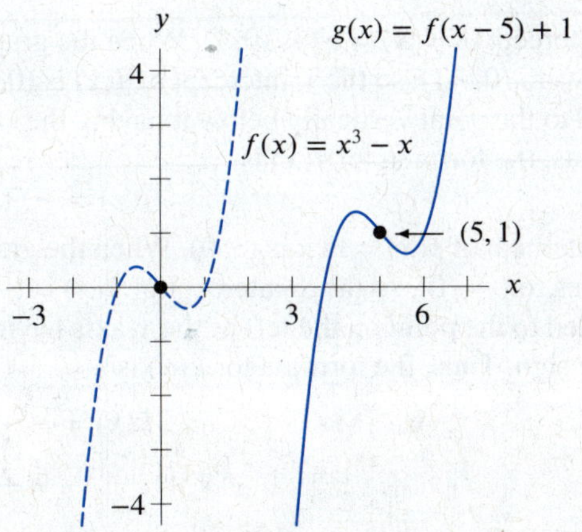


**Figure 6.3**: $g(x)$ is a translation of $f(x)$

To summarize, we combine the two formulas introduced in Section 2.4 with the following result:

> Translations of a function $f(x)$ can be expressed in the form
>
> $$y = f(x - h) + k.$$
>
> - Outside change by a constant $k$ shifts the graph up by $k$ units if $k > 0$ (down if $k < 0$)
> - Inside change by a constant $h$ shifts the graph right by $h$ units if $h > 0$ (left if $h < 0$)

## Reflections

Now we look at transformations which are not shifts. Reflections are transformations that flip a graph in the $xy$-plane without changing its shape.

**Example 3**

(a) Figure 6.4 shows the graph of $j(x)$, the reflection of the graph of $f(x) = e^x$ about the $x$-axis. By looking at how the $y$-values change in going from $f(x)$ to $j(x)$, find the formula for $j(x)$.

(b) Figure 6.5 shows the graph of $k(x)$, the reflection of the graph of $p(x) = \ln x$ about the $y$-axis. By looking at how the $x$-values change in going from $p(x)$ to $k(x)$, find the formula for $k(x)$.

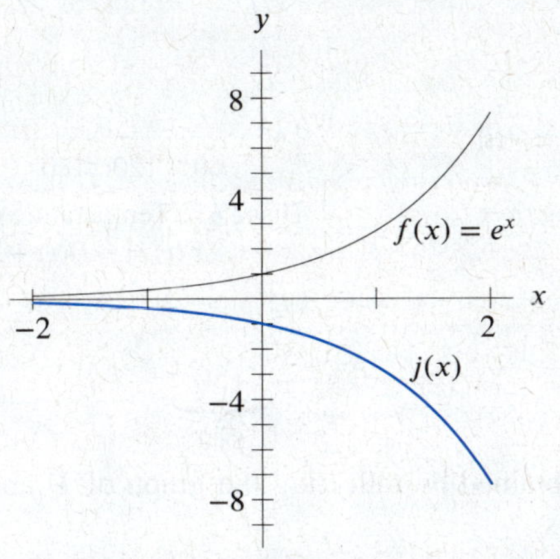


**Figure 6.4**: $j(x)$ is a reflection of $f(x)$ about the $x$-axis

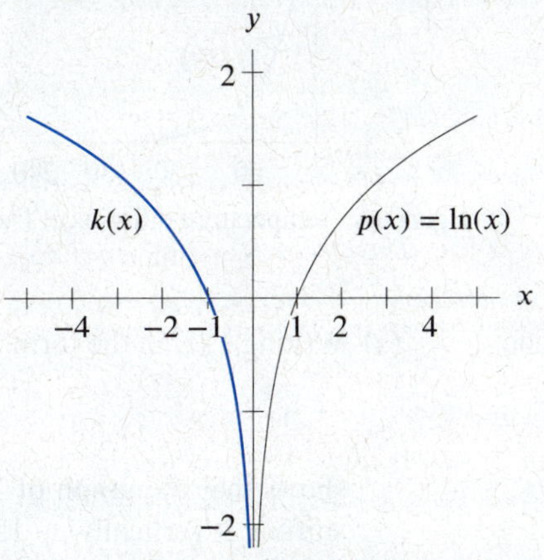


**Figure 6.5**: $k(x)$ is a reflection of $p(x)$ about the $y$-axis

**Solution**

(a) The $y$-intercept of $f(x) = e^x$ is $(0, 1)$. When the graph of $f(x)$ is reflected about the $x$-axis, this point goes to $(0, -1)$, so the $y$-intercept of $j(x)$ is $(0, -1)$. In general, each point on $f(x) = e^x$ is reflected to the point vertically below it having the same $x$-value but a $y$-value with the opposite sign. Thus, the formula for $j(x)$ is

$$j(x) = -f(x).$$

(b) The $x$-intercept of $p(x) = \ln x$ is $(1, 0)$. When the graph of $p(x)$ is reflected about the $y$-axis, this point goes to $(-1, 0)$, so the $x$-intercept of $k(x)$ is $(-1, 0)$. In general, each point on $p(x) = \ln x$ is reflected to the point on the left of the $y$-axis having the same $y$-value but an $x$-value with the opposite sign. Thus, the formula for $k(x)$ is

$$k(x) = p(-x).$$

As we see in Example 3, a reflection about the $x$-axis corresponds to an outside change; a reflection about the $y$-axis corresponds to an inside change.

## Combining Shifts and Reflections

We can combine horizontal and vertical shifts with the horizontal and vertical reflections to make more complex transformations of functions. When a function models real-life phenomena, its intercepts and asymptotes often have practical meaning—both before and after transformations.

**Example 4** When a yam is taken from a refrigerator at 0°C and put into an oven at 150°C, the yam's temperature rises toward that of the oven.[1] Let $Y(t)$ be the temperature in °C of the yam $t$ minutes after it is put in the oven. Let $D(t) = 150 - Y(t)$ be the temperature difference between the oven and the yam at time $t$. Figure 6.6 gives the graph of $D(t)$.

(a) From the formula relating $Y(t)$ and $D(t)$, describe the transformations we apply to the graph of $D(t)$ to obtain the graph of $Y(t)$.
(b) Use the graph of $D(t)$ to sketch a graph of $Y(t)$.
(c) Explain the significance of the vertical intercept of $D$, then of $Y$.
(d) Explain the significance of the horizontal asymptote of $D$, then of $Y$.

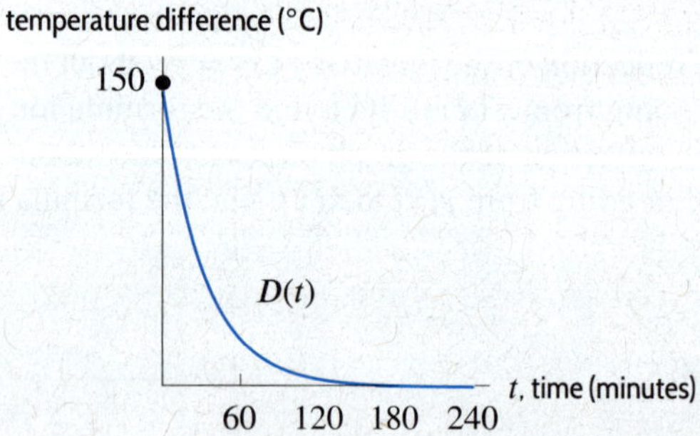


**Figure 6.6**: Temperature difference: Decreasing over time

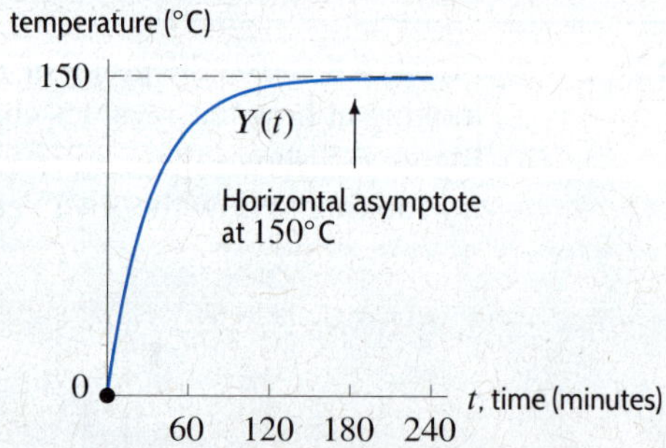


**Figure 6.7**: Temperature of the yam, $Y(t) = -D(t) + 150$

**Solution**

(a) Writing $Y(t)$ in the form

$$Y(t) = \underbrace{-D(t)}_{\text{Reflect}} + \underbrace{150}_{\text{Shift}}$$

shows that the graph of $Y$ is obtained by reflecting the graph of $D$ about the $t$-axis and then shifting it vertically up 150 units.

(b) See Figure 6.7.

[1] In practice, refrigerators keep food slightly above 0°C—at about 2–3°C—to prevent freezing.

(c) The vertical intercept $D = 150$ tells us that the initial temperature difference between the yam and the oven is 150°C. The vertical intercept $Y = 0$ says that the temperature of the yam is 0°C when it is first put into the oven.

(d) The horizontal asymptote, $D = 0$, tell us that eventually there will be no difference in temperature between the yam and the oven. The horizontal asymptote, $Y = 150$, represents the temperature of the oven. As the yam sits in the oven, its temperature approaches the temperature of the oven—eventually there will be virtually no temperature difference.

## Symmetry Under Reflection: Even and Odd Functions

The graphs of some functions remain unchanged under reflection; these graphs have *symmetry*. This property of graphs is important because it often makes calculations with the function easier. We now see what functions give rise to graphs with symmetry.

### Even Functions

The graph of $p(x) = x^2$ in Figure 6.8 is *symmetric* about the $y$-axis. In other words, the part of the graph to the left of the $y$-axis is the mirror image of the part to the right of the $y$-axis. Reflecting the graph of $p(x)$ about the $y$-axis gives the graph of $p(x)$ again. Functions whose graphs are symmetric about the $y$-axis are called *even functions*, because power functions with even exponents, such as $y = x^2, y = x^4, y = x^6, \ldots$ have this property.

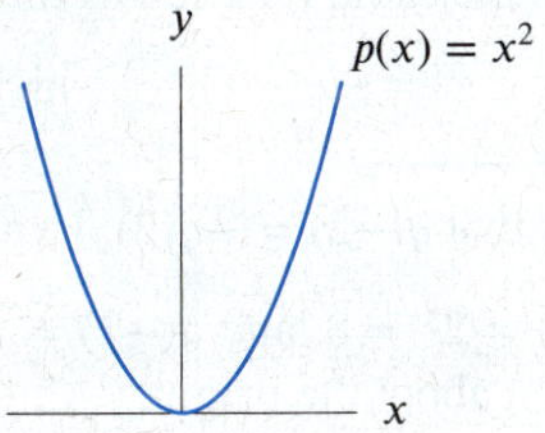


Figure 6.8: Reflecting the graph of $p(x) = x^2$ about the $y$-axis does not change its appearance

Since $y = p(-x)$ is a reflection of the graph of $p$ about the $y$-axis and $p(x)$ is even, we have

$$p(-x) = p(x).$$

**Example 5** For $p(x) = x^2$, check that $p(-2) = p(2)$. Explain what this tells us about the graph of $p$.

**Solution** Let $x = 2$. Then $p(2) = 2^2 = 4$, and $p(-2) = (-2)^2 = 4$, so $p(-2) = p(2)$. This means that the point $(2, 4)$ and its reflection about the $y$-axis, $(-2, 4)$, are both on the graph of $p(x)$.

**Example 6** For the function $p(x) = x^2$, check algebraically that $p(-x) = p(x)$ for all $x$.

**Solution** Substitute $-x$ into the formula for $p(x)$, giving

$$\begin{aligned} p(-x) &= (-x)^2 = (-x) \cdot (-x) \\ &= x^2 \\ &= p(x). \end{aligned}$$

Thus, $p(-x) = p(x)$.

## Odd Functions

Figure 6.9 shows the graph of $q(x) = x^3$. Figure 6.10 shows the graph of $q$ reflected first about the $y$-axis and then about the $x$-axis (or vice versa)—giving the graph of $q$ again. This kind of symmetry is called *symmetry about the origin*. Functions whose graphs are symmetric about the origin are called *odd functions*, because power functions with odd exponents, such as $y = x^3$, $y = x^5$, $y = x^7$, … have this property.

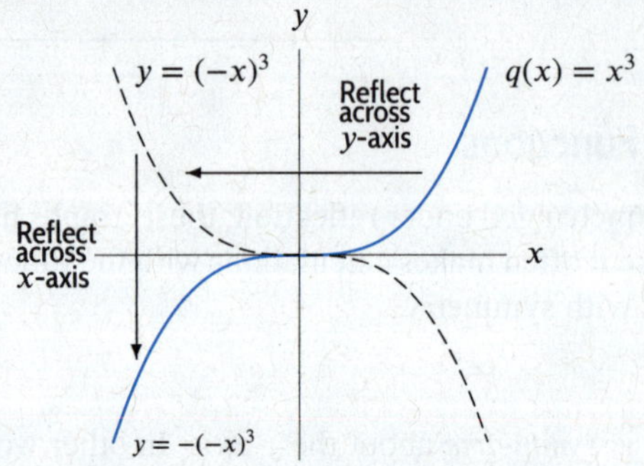


Figure 6.9: If the graph is reflected about the $y$-axis and then about the $x$-axis, it does not change

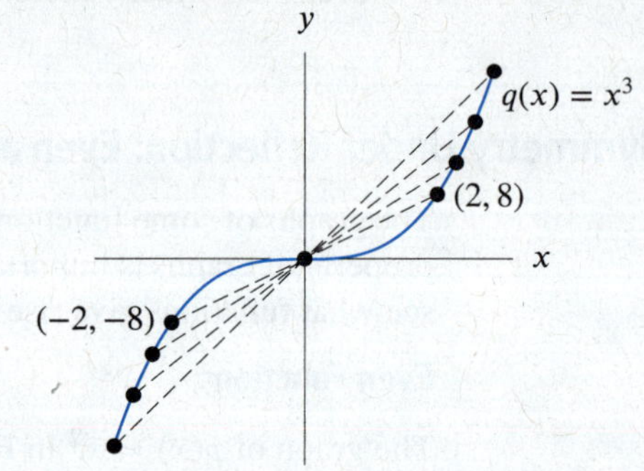


Figure 6.10: If every point on this graph is reflected about the origin, the graph is unchanged

Since $y = -q(-x)$ is a reflection about the $y$-axis followed by a reflection about the $x$-axis, and $q(x) = x^3$ is odd, $q$ is the same function as this double reflection. That is,

$$q(x) = -q(-x), \qquad \text{which means that} \qquad q(-x) = -q(x).$$

**Example 7** For $q(x) = x^3$, check that $q(-2) = -q(2)$. Explain what this tells us about the graph of $q$.

**Solution** Let $x = 2$. Then $q(2) = 2^3 = 8$, and $q(-2) = (-2)^3 = -8$, so $q(-2) = -q(2)$. This means the point $(2, 8)$ and its reflection about the origin, $(-2, -8)$, are both on the graph of $q$. See Figure 6.10.

**Example 8** For the function $q(x) = x^3$, check algebraically that $q(-x) = -q(x)$ for all $x$.

**Solution** We evaluate $q(-x)$, giving

$$\begin{aligned} q(-x) &= (-x)^3 = (-x)\cdot(-x)\cdot(-x) \\ &= -x^3 \\ &= -q(x). \end{aligned}$$

Thus, $q(-x) = -q(x)$.

In summary, we have the following general results:

> If $f$ is a function and $x$ is in its domain,
> - $f$ is called an **even function** if $f(-x) = f(x)$ for all values of $x$
> - $f$ is called an **odd function** if $f(-x) = -f(x)$ for all values of $x$
>
> The graph of $f$ is
> - Symmetric about the $y$-axis if $f$ is an even function
> - Symmetric about the origin if $f$ is an odd function

In Example 9, we see how knowing a function is even or odd can be useful.

**Example 9** The equation $f(x) = x^4 - 24x^2 - 25 = 0$ has solution $x = 5$. Write down another solution to this equation.

**Solution** We could factor $f(x)$ and find all the solutions. However, the function $f(x) = x^4 - 24x^2 - 25$ is even as it contains only even powers of $x$. Thus, since $f(5) = 0$, we know $f(-5) = 0$ also, so $x = -5$ is another solution.

**Example 10** Determine whether the following functions are even, odd, or neither.
(a) $f(x) = |x|$ (b) $g(x) = 1/x$ (c) $h(x) = -x^3 - 3x^2 + 2$

**Solution** The graphs of the functions in Figures 6.11, 6.12, and 6.13 are helpful in identifying symmetry.

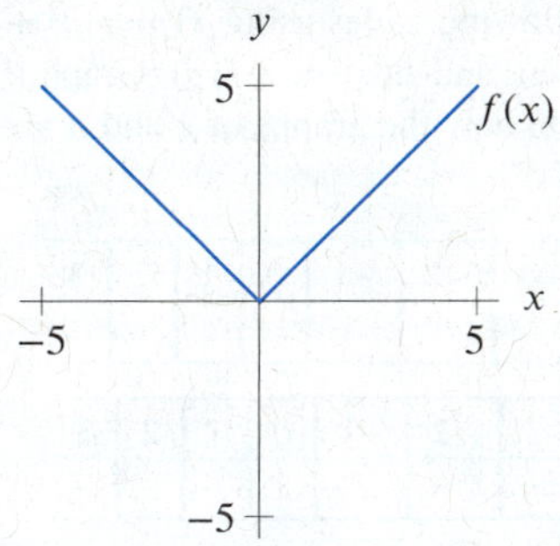


**Figure 6.11**: The graph of $f(x) = |x|$ appears to be symmetric about the $y$-axis

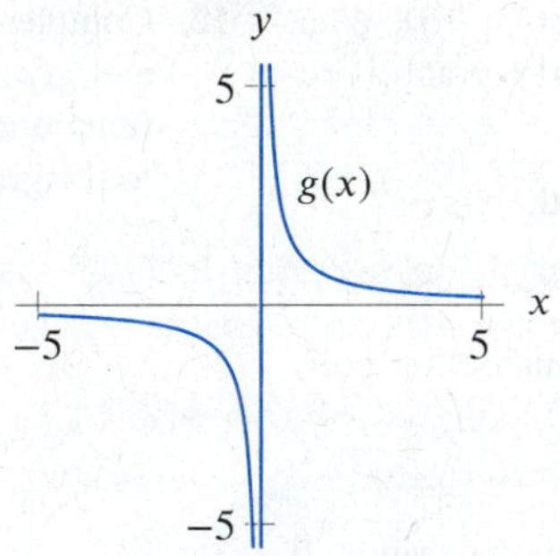


**Figure 6.12**: The graph of $g(x) = 1/x$ appears to be symmetric about the origin

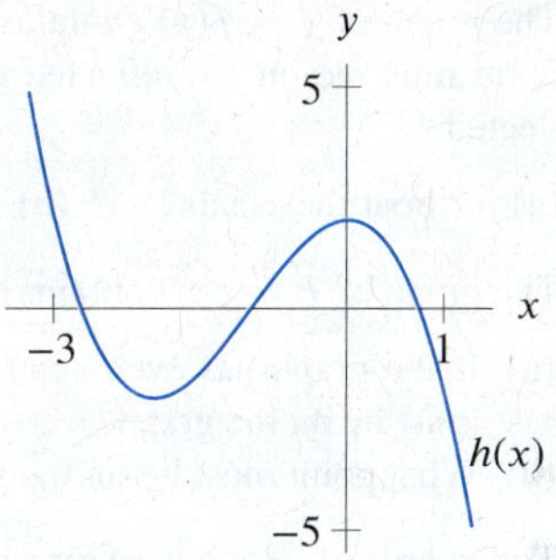


**Figure 6.13**: The graph of $h(x) = -x^3 - 3x^2 + 2$ is symmetric neither about the $y$-axis nor about the origin

It appears that the graph of $f$ is symmetric about the $y$-axis, the graph of $g$ is symmetric about the origin, and the graph of $h$ has neither type of symmetry. This means that $f$ is an even function, $g$ is odd function, and $h$ is neither even nor odd. However, to be sure, check algebraically.

If $f(-x) = f(x)$, then $f$ is an even function. We check by substituting $-x$ in for $x$:

$$\begin{aligned} f(-x) &= |-x| \\ &= |x| \\ &= f(x). \end{aligned}$$

Thus, $f$ is an even function.

If $g(-x) = -g(x)$, then $g$ is an odd function. We check by substituting $-x$ for $x$:

$$\begin{aligned} g(-x) &= \frac{1}{-x} \\ &= -\frac{1}{x} \\ &= -g(x). \end{aligned}$$

Thus, $g$ is an odd function.

The graph of $h$ does not appear to be symmetric. To confirm, look at an example, say $x = 1$:

$$h(1) = -1^3 - 3 \cdot 1^2 + 2 = -2.$$

Now substitute $x = -1$, giving

$$h(-1) = -(-1)^3 - 3 \cdot (-1)^2 + 2 = 0.$$

Thus $h(1) \neq h(-1)$, so $h$ is not an even function. Also, $h(-1) \neq -h(1)$, so $h$ is not an odd function.

# Exercises and Problems for Section 6.1

## Skill Refresher

In Exercises S1–S4, evaluate each function at $x = 3$. Round your answer to 3 decimal places.

**S1.** $f(x) = e^x$

**S2.** $g(x) = -e^x$

**S3.** $h(x) = e^{-x}$

**S4.** $k(x) = -e^{-x}$

In Exercises S5–S10, give a formula for the functions:

**(a)** $f(-x)$ **(b)** $-f(x)$

**S5.** $f(x) = 2x^2$

**S6.** $f(x) = \dfrac{1}{x}$

**S7.** $f(x) = 2x^3 - 3$

**S8.** $f(x) = 4x^3 + 9x$

**S9.** $f(x) = 3x^4 - 2x$

**S10.** $f(x) = \dfrac{3x^3}{x^2 - 1}$

## Exercises

**1.** The graph of $y = f(x)$ contains the point $(2, -3)$. What point must lie on the reflected graph if the graph is reflected

**(a)** About the $y$-axis? **(b)** About the $x$-axis?

**2.** The graph of $P = g(t)$ contains the point $(-1, -5)$.

**(a)** If the graph has even symmetry, which other point must lie on the graph?

**(b)** What point must lie on the graph of $-g(t)$?

**3.** The graph of $H(x)$ is symmetric about the origin. If $H(-3) = 7$, what is $H(3)$?

**4.** The range of $Q(x)$ is $-2 \le Q(x) \le 12$. What is the range of $-Q(x)$?

In Exercises 5–8, the function $Q(t)$ has domain $t > 0$ and range $-4 \le Q(t) \le 7$. Give the domain and range for the transformation of $Q(t)$.

**5.** $y = Q(-t)$

**6.** $y = -Q(t)$

**7.** $y = -Q(-t)$

**8.** $y = -Q(t - 4)$

**9.** If the graph of $y = e^x$ is reflected about the $y$-axis, what is the formula for the resulting graph? Check by graphing both functions together.

**10.** Graph $y = f(x) = 4^x$ and $y = f(-x)$ on the same set of axes. How are these graphs related? Give an explicit formula for $y = f(-x)$.

**11.** Graph $y = g(x) = \left(\frac{1}{3}\right)^x$ and $y = -g(x)$ on the same set of axes. How are these graphs related? Give an explicit formula for $y = -g(x)$.

**12.** Complete the following tables using $f(p) = p^2 + 2p - 3$, and $g(p) = f(-p)$, and $h(p) = -f(p)$. Graph the three functions. Explain how the graphs of $g$ and $h$ are related to the graph of $f$.

| $p$ | −3 | −2 | −1 | 0 | 1 | 2 | 3 |
|---|---|---|---|---|---|---|---|
| $f(p)$ | | | | | | | |

| $p$ | −3 | −2 | −1 | 0 | 1 | 2 | 3 |
|---|---|---|---|---|---|---|---|
| $g(p)$ | | | | | | | |

| $p$ | −3 | −2 | −1 | 0 | 1 | 2 | 3 |
|---|---|---|---|---|---|---|---|
| $h(p)$ | | | | | | | |

Give a formula and graph for each of the transformations of $m(n) = n^2 - 4n + 5$ in Exercises 13–16.

**13.** $y = m(-n)$

**14.** $y = -m(n)$

**15.** $y = -m(-n)$

**16.** $y = m(-n) + 3$

Give a formula and graph for each of the transformations of $k(w) = 3^w$ in Exercises 17–20.

**17.** $y = k(-w)$

**18.** $y = -k(w)$

**19.** $y = -k(-w)$

**20.** $y = -k(w - 2)$

In Exercises 21–24, show that the function is even, odd, or neither.

**21.** $f(x) = 7x^2 - 2x + 1$

**22.** $f(x) = 4x^7 - 3x^5$

**23.** $f(x) = 8x^6 + 12x^2$

**24.** $f(x) = x^5 + 3x^3 - 2$

## Problems

**25.** **(a)** Graph the function obtained from $f(x) = x^3$ by first reflecting about the $x$-axis, then translating up two units. Write a formula for the resulting function.

**(b)** Graph the function obtained from $f$ by first translating up two units, then reflecting about the $x$-axis. Write a formula for the resulting function.

**(c)** Are the functions in parts (a) and (b) the same?

**26.** **(a)** Graph the function obtained from $g(x) = 2^x$ by first reflecting about the $y$-axis, then translating down three units. Write a formula for the resulting function.

**(b)** Graph the function obtained from $g$ by first translating down three units, then reflecting about the $y$-axis. Write a formula for the resulting function.

**(c)** Are the functions in parts (a) and (b) the same?

**27.** Using Figure 6.14, evaluate

**(a)** $f(-x)$ for $x = -4$
**(b)** $-f(x)$ for $x = -6$
**(c)** $-f(-x)$ for $x = -4$
**(d)** $-f(x+2)$ for $x = 0$
**(e)** $f(-x)+4$ for $x = -6$

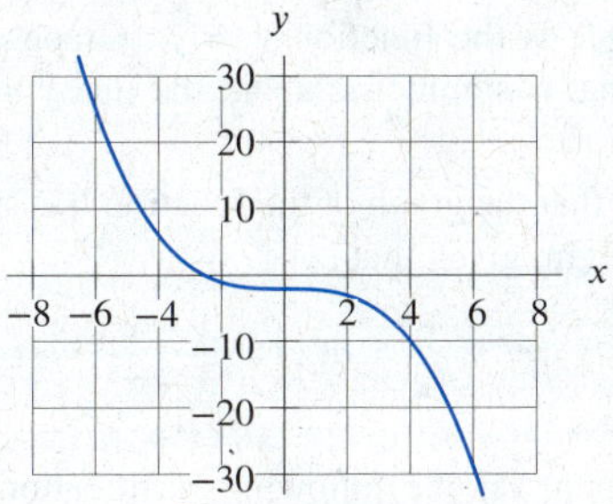


Figure 6.14

**28.** **(a)** If $g(x) = \sqrt[3]{x}$, find a formula for $g(-x)$.
**(b)** Graph $y = g(x) = \sqrt[3]{x}$, $y = g(-x)$, and $y = -g(x)$ on the same axes.
**(c)** Is $g(x) = \sqrt[3]{x}$ even, odd, or neither?

**29.** If the graph of a line $y = b + mx$ is reflected about the $y$-axis, what are the slope and intercepts of the resulting line?

**30.** Graph $y = \log(1/x)$ and $y = \log x$ on the same axes. How are the two graphs related? Use the properties of logarithms to explain the relationship algebraically.

**31.** Using Figure 6.15, graph the following transformations of $f$ on separate axes.

**(a)** $y = f(x) - 2$ **(b)** $y = f(x-2)$
**(c)** $y = -f(x)$ **(d)** $y = f(-x)$

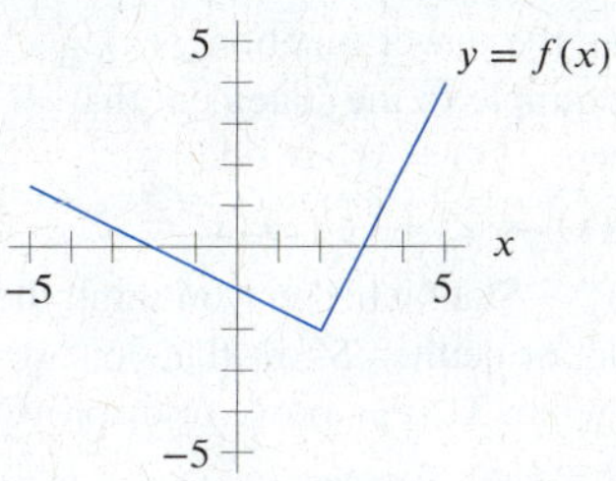


Figure 6.15

**32.** Using Figure 6.16, match the functions (i)-(v) with a graph (a)-(e).

(i) $y = f(-x)$ (ii) $y = -f(x)$
(iii) $y = f(-x)+3$ (iv) $y = -f(x-1)$
(v) $y = -f(-x)$

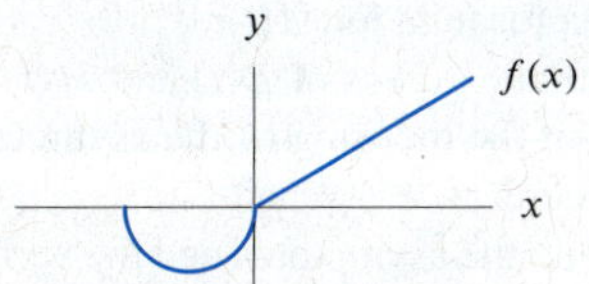


Figure 6.16

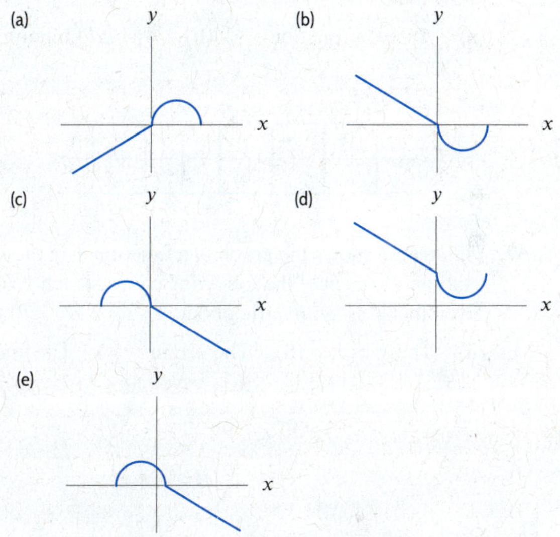


**33.** For $t \geq 0$, let $H(t) = 20 + 54(0.91)^t$ give the temperature of a bowl of soup in degrees Celsius $t$ minutes after it is taken out of the microwave and let $k$ be a positive constant.

**(a)** Find a formula for $H(t) + k$.
**(b)** Graph $H(t) + k$ for $k = 0$, 10 and 20.
**(c)** What final temperature does the function $H(t) + 10$ approach? What is the practical meaning of this temperature? What about $H(t) + k$?

**34.** For $t \geq 0$, let $H(t) = 20 + 40(0.91)^t$ give the temperature of a bowl of soup in degrees Celsius $t$ minutes after it is taken out of the microwave and let $h$ be a positive constant.

**(a)** Find a formula for $H(t-h)$.
**(b)** Graph $H(t-h)$ for $h = 0$, 10 and 20.
**(c)** Describe in practical terms a situation modeled by the function $H(t-10)$. What about $H(t-h)$?

**35.** A chicken is removed from a freezer and placed in a refrigerator to thaw. Let $t$ be time in minutes after the chicken was removed. The difference, $D(t)$, between the temperature of the refrigerator, 3°C, and the chicken's temperature, initially −18°C, decreases exponentially over time at a rate of 3% per minute. The chicken's temperature, $H(t)$, is a transformation of $D(t)$.

**(a)** Find a formula for $H(t)$.
**(b)** Sketch the graphs of $D(t)$ and $H(t)$. For each graph, explain the meaning of the asymptote and the vertical intercept.
**(c)** How do the asymptotes and the vertical intercepts of the graphs of $D(t)$ and $H(t)$ differ if the refrigerator is instead set to 1°C?

**36.** In Table 6.1, fill in as many $y$-values as you can if you know that $f$ is

**(a)** An even function **(b)** An odd function.

**Table 6.1**

| $x$ | −3 | −2 | −1 | 0 | 1 | 2 | 3 |
|---|---|---|---|---|---|---|---|
| $y$ | 5 | | −4 | | | −8 | |

**37.** Figure 6.17 shows the graph of a function $f$ in the second quadrant. In each of the following cases, sketch $y = f(x)$, given that $f$ is symmetric about

**(a)** The $y$-axis. **(b)** The origin. **(c)** The line $y = x$.

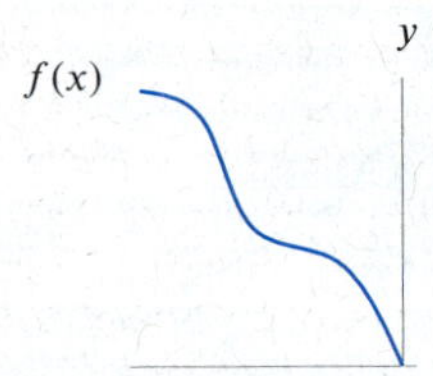


**Figure 6.17**

**38.** For each table, decide whether the function could be symmetric about the $y$-axis, about the origin, or neither.

**(a)**

| $x$ | −3 | −2 | −1 | 0 | 1 | 2 | 3 |
|---|---|---|---|---|---|---|---|
| $f(x)$ | 6 | 1 | −2 | −3 | −2 | 1 | 6 |

**(b)**

| $x$ | −3 | −2 | −1 | 0 | 1 | 2 | 3 |
|---|---|---|---|---|---|---|---|
| $g(x)$ | −8.1 | −2.4 | −0.3 | 0 | 0.3 | 2.4 | 8.1 |

**(c)**

| $x$ | −3 | −2 | −1 | 0 | 1 | 2 | 3 |
|---|---|---|---|---|---|---|---|
| $f(x)+g(x)$ | −2.1 | −1.4 | −2 | −3 | −1 | 3.4 | 14.1 |

**(d)**

| $x$ | −3 | −2 | −1 | 0 | 1 | 2 | 3 |
|---|---|---|---|---|---|---|---|
| $f(x+1)$ | 1 | −2 | −3 | −2 | 1 | 6 | 13 |

**39.** Let $f(x)$ be an even function, and let $g(x)$ be an odd function. If possible, determine whether each function must be even or odd.

**(a)** $h(x) = f(x)g(x)$ **(b)** $k(x) = f(x) + g(x)$
**(c)** $m(x) = g(f(x))$

**40.** Let both $f(x)$ and $g(x)$ be odd functions. If possible, determine whether each function must be even or odd.

**(a)** $h(x) = f(x)g(x)$ **(b)** $k(x) = f(x) - g(x)$
**(c)** $m(x) = f(g(x))$

**41.** A function is called symmetric about the line $y = x$ if interchanging $x$ and $y$ gives the same graph. The simplest example is the function $y = x$. Graph another straight line that is symmetric about the line $y = x$ and give its equation.

**42.** Show that the graph of the function $h$ is symmetric about the origin, given that

$$h(x) = \frac{1 + x^2}{x - x^3}.$$

**43.** Comment on the following justification that the function $f(x) = x^3 - x^2 + 1$ is an even function: Because $f(0) = 1 \neq -f(0)$, we know that $f(x)$ is not odd. If a function is not odd, it must be even.

**44.** Let $f(x)$ be a function that is always increasing and concave down. Determine whether each function is always increasing or always decreasing, and whether each function is always concave up or always concave down.

**(a)** $f(-x)$ **(b)** $-f(x)$ **(c)** $-f(-x)$

**45.** Is it possible for an odd function whose domain is all real numbers to be strictly concave up?

**46.** Let $f(x) = b + mx$.

**(a)** Can $f(x)$ be even? How?
**(b)** Can $f(x)$ be odd? How?
**(c)** Can $f(x)$ be both odd and even? How?

**47.** If $f$ is an odd function and defined at $x = 0$, what is the value of $f(0)$? Explain how you can use this result to show that $c(x) = x + 1$ and $d(x) = 2^x$ are not odd.

**48.** In the first quadrant an even function is increasing and concave down. What can you say about the function's behavior in the second quadrant?

**49.** Show that the power function $f(x) = x^{1/3}$ is odd. Give a counterexample to the statement that all power functions of the form $f(x) = x^p$ are odd.

**50.** Graph $s(x) = 2^x + (\frac{1}{2})^x$, $c(x) = 2^x - (\frac{1}{2})^x$, and $n(x) = 2^x - (\frac{1}{2})^{x-1}$. State whether you think these functions are even, odd, or neither. Show that your statements are true using algebra. That is, prove or disprove statements such as $s(-x) = s(x)$.

**51.** There are functions that are *neither* even nor odd. Is there a function that is *both* even and odd?

**52.** Some functions are symmetric about the $y$-axis. Is it possible for a function to be symmetric about the $x$-axis?

# 6.2 VERTICAL STRETCHES AND COMPRESSIONS

We have studied translations and reflections of graphs. In this section, we consider vertical stretches and compressions of graphs. As with a vertical translation, a vertical stretch or compression of a function is represented by an outside change to its formula.

## Vertical Stretch: An Amplifier

An amplifier takes a weak signal from a recording and transforms it into a stronger signal to power a set of speakers.

Figure 6.18 shows a graph of an audio signal (in volts) as a function of time, $t$, both before and after amplification. Notice that the wave crests of the amplified signal are 3 times as high as those of the original signal; similarly, the amplified wave troughs are 3 times deeper than the original wave troughs. The amplifier has boosted the strength of the signal by a factor of 3. If $f$ is the original signal function and $V$ is the amplified signal function, then

$$\underbrace{\text{Amplified signal strength at time } t}_{V(t)} = 3 \cdot \underbrace{\text{Original signal strength at time } t}_{f(t)},$$

so we have

$$V(t) = 3 \cdot f(t).$$

This formula tells us that the values of the amplified signal are 3 times the values of the original signal. The graph of $V$ is the graph of $f$ stretched vertically by a factor of 3. As expected, a vertical stretch of the graph of $f(t)$ corresponds to an outside change in the formula.

Notice that the $t$-intercepts remain fixed under a vertical stretch, because the $f$-value of these points is 0, which is unchanged when multiplied by 3.

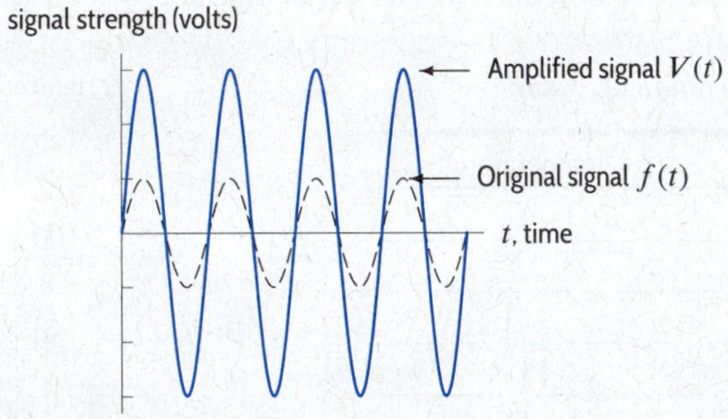


**Figure 6.18:** An amplifier transforms a weak signal into a signal 3 times as strong

Notice also that if we regard $V(t)$ as the initial function, then $f(t)$ is a *compression* of $V(t)$ by a factor of $1/3$:

$$f(t) = \frac{1}{3}V(t).$$

## Negative Stretch Factor

What happens if we multiply a function by a negative stretch factor? Figure 6.19 gives a graph of a function $y = f(x)$, together with a graph of $y = -2 \cdot f(x)$. The stretch factor is $k = -2$. We think of $y = -2f(x)$ as a combination of two separate transformations of $y = f(x)$. First, the graph is stretched by a factor of 2, then it is reflected across the $x$-axis.

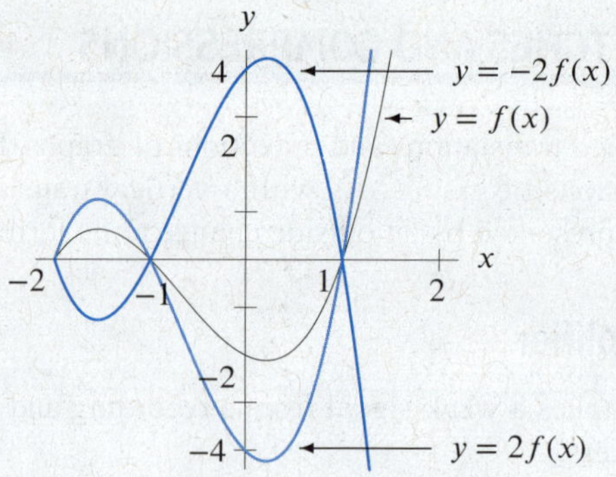


**Figure 6.19**: The graph of $y = -2f(x)$ is a vertically stretched version of the graph of $y = f(x)$ that has been reflected across the $x$-axis

## Formula for Vertical Stretch or Compression

Generalizing the examples gives the following result:

> If $f$ is a function and $k$ is a constant, then the graph of $y = k \cdot f(x)$ is the graph of $y = f(x)$
> - Vertically stretched by a factor of $k$, if $k > 1$.
> - Vertically compressed by a factor of $k$, if $0 < k < 1$.
> - Vertically stretched or compressed by a factor $|k|$ and reflected across $x$-axis, if $k < 0$.

**Example 1** A yam is placed in a 150°C oven. Table 6.2 gives values of $H = r(t)$, the yam's temperature $t$ minutes after being placed in the oven. Figure 6.20 shows these data points joined by a curve.

(a) Describe the function $r$ in words. What do the data tell you about the yam's temperature?

(b) Make a table of values for $q(t) = 1.5r(t)$. Graph the function $q$.

(c) How are the functions $q$ and $r$ related? Under what condition might $q$ describe a yam's temperature?

**Table 6.2** *Temperature of a yam*

| $t$, time (min) | $r(t)$, temperature (°C) |
|---|---|
| 0 | 0 |
| 10 | 75 |
| 20 | 113 |
| 30 | 131 |
| 40 | 141 |
| 50 | 145 |
| 60 | 148 |

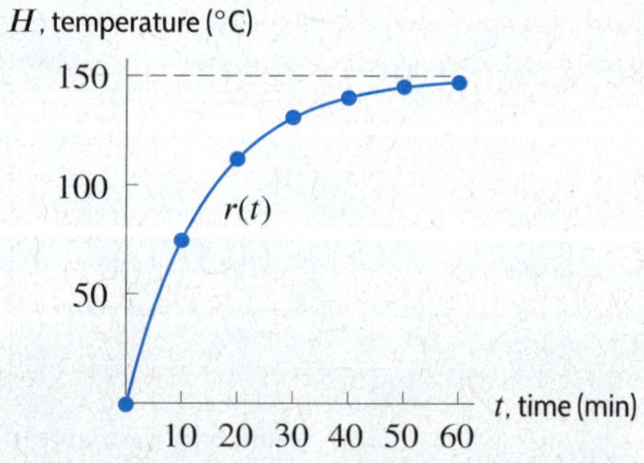


**Figure 6.20**: The temperature of a yam at time $t$

**Solution**

(a) The function $r$ is increasing and concave down. The yam starts out at 0°C and warms up quickly at first. Later, the yam heats up more slowly. The temperature levels off at 150°C, the oven's temperature, represented by the horizontal asymptote.

(b) We calculate values of $q(t)$ from values of $r$. For example, Table 6.2 gives $r(0) = 0$ and $r(10) = 75$. Thus,

$$q(0) = 1.5r(0) = 1.5 \cdot 0 = 0,$$

$$q(10) = 1.5r(10) = 1.5(75) = 112.5, \text{ and so on.}$$

The values for $q(t)$ are 1.5 times as large as the corresponding values for $r(t)$. See Table 6.3. The data are plotted in Figure 6.21.

(c) The graph of $q$ is the graph of $r$ stretched vertically by a factor of $k = 1.5$. The horizontal asymptote of $r$ was $H = 150$, so the horizontal asymptote of $q$ is $H = 1.5 \cdot 150 = 225$. This suggests that the yam has been placed in a 225°C oven instead of a 150°C oven.

**Table 6.3** *Values of $q(t) = 1.5r(t)$*

| $t$, time (min) | $q(t)$, temperature (°C) |
|---|---|
| 0 | 0 |
| 10 | 112.5 |
| 20 | 169.5 |
| 30 | 196.5 |
| 40 | 211.5 |
| 50 | 217.5 |
| 60 | 222 |

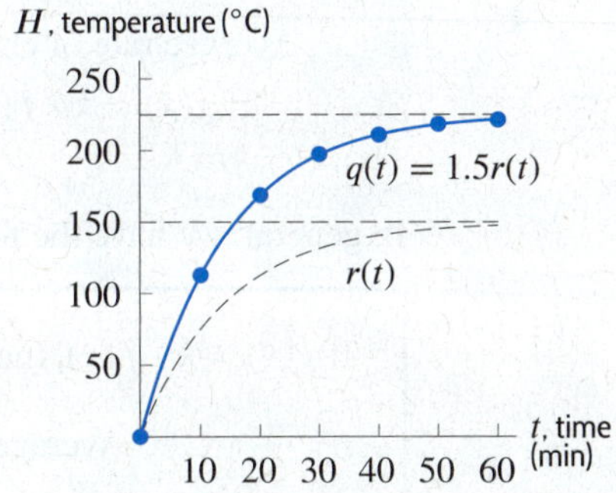


**Figure 6.21**: Graph of $q(t) = 1.5r(t)$

## Stretch Factors and Average Rates of Change

Consider again the graph of the audio signal and its amplification in Figure 6.18. Notice that the amplified signal, $V$, is increasing on the same intervals as the original signal, $f$. Similarly, both functions decrease on the same intervals.

Stretching or compressing a function vertically does not change the intervals on which the function increases or decreases. However, the average rate of change of a function, visible in the steepness of the graph, is altered by a vertical stretch or compression.

**Example 2** In Example 1, the function $H = r(t)$ gives the temperature (in °C) of a yam placed in a 150°C oven. The function $q(t) = 1.5r(t)$ gives the temperature of the yam placed in a 225°C oven. In both cases, the temperature starts at 0°C. After 10 minutes, $r(10) = 75$°C and $q(10) = 112.5$°C.

(a) For each of the two yams, find the rate of change of temperature over ten-minute intervals from $t = 0$ to $t = 60$.

(b) For each ten-minute interval, what is the relationship between the two values you found in part (a) for the yams?

**Solution** (a) For the yam with temperature $r(t)$, from $t = 0$ to $t = 10$, we have

$$\text{Average rate of change of temperature, } r = \frac{\Delta H}{\Delta t} = \frac{r(10) - r(0)}{10 - 0} = \frac{75 - 0}{10} = 7.5\text{°C/min.}$$

Thus, in the 150°C oven, the yam's temperature increased at an average rate of 7.5°C per minute. For the yam with temperature $q(t)$, from $t = 0$ to $t = 10$,

$$\text{Average rate of change of temperature, } q = \frac{\Delta H}{\Delta t} = \frac{112.5 - 0}{10} = 11.25\text{°C/min.}$$

Thus, in the 225°C oven, the yam's temperature increased at an average rate of 11.25°C per minute. Similar calculations for other time intervals give the values in Table 6.4.

(b) Notice that $11.25 = 1.5(7.5)$, and each rate of change in the 225°C oven is 1.5 times the corresponding rate of change in the 150°C oven. This reflects the fact that $q(t) = 1.5r(t)$.

**Table 6.4** *The average rate of change of yams' temperatures, $r(t)$ and $q(t)$*

| Time interval (min) | 0 – 10 | 10 – 20 | 20 – 30 | 30 – 40 | 40 – 50 | 50 – 60 |
|---|---|---|---|---|---|---|
| Average rate of change of $r$ (°C/min) | 7.5 | 3.8 | 1.8 | 1 | 0.4 | 0.3 |
| Average rate of change of $q$ (°C/min) | 11.25 | 5.7 | 2.7 | 1.5 | 0.6 | 0.45 |

In the last example, multiplying a function by a factor of 1.5 had the effect of multiplying the function's average rate of change on each interval by the same factor. More generally, if $g(x) = k \cdot f(x)$, between $x = a$ and $x = b$,

$$\begin{aligned}\text{Average rate of change of } g(x) = k \cdot f(x) &= \frac{\Delta y}{\Delta x} = \frac{g(b) - g(a)}{b - a} = \frac{k \cdot f(b) - k \cdot f(a)}{b - a} = k \cdot \frac{f(b) - f(a)}{b - a} \\ &= k \cdot (\text{Average rate of change of } f).\end{aligned}$$

In general, we have the following result:

> If $g(x) = k \cdot f(x)$, then on any interval,
>
> $$\text{Average rate of change of } g = k \cdot (\text{Average rate of change of } f).$$

## Combining Stretches and Shifts

Stretches and shifts can be combined to give more general transformations of functions.

**Example 3** The function $y = f(x)$ is graphed in Figure 6.22. Graph the function $g(x) = -\frac{1}{2}f(x+3) - 1$.

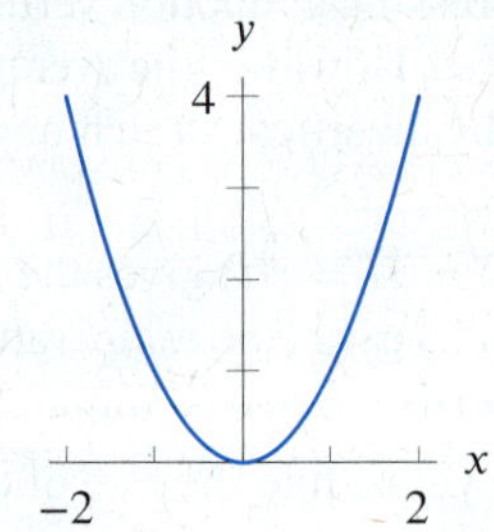


Figure 6.22: Graph of $y = f(x)$

**Solution** To combine the transformations, we work from inside outward in Figure 6.23. The graphs corresponding to each step are shown in Figure 6.24. Note that we did not need a formula for $f$ to graph $g$.

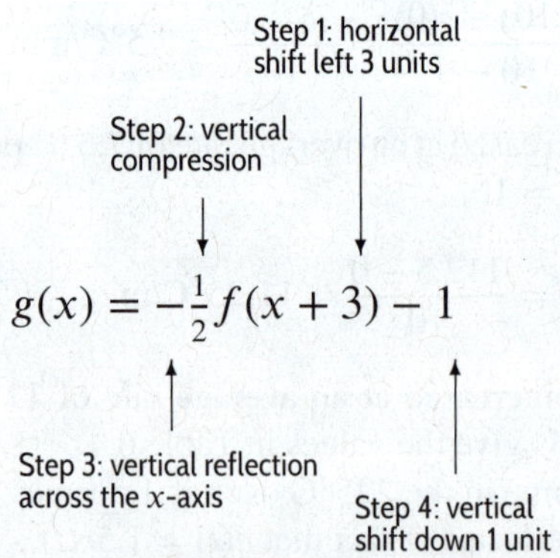


Figure 6.23: Steps in transforming $f(x)$ to $g(x) = -(1/2)f(x+3) - 1$

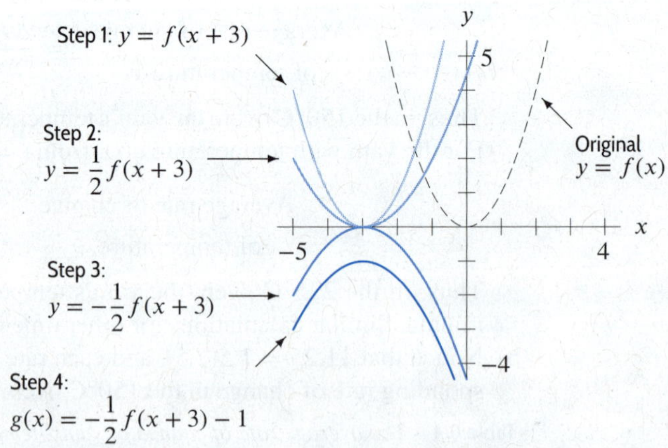


Figure 6.24: The graph of $y = f(x)$ transformed in four steps into $g(x) = -(1/2)f(x+3) - 1$

## Exercises and Problems for Section 6.2

### Skill Refresher

**S1.** If $f(x) = x^2$, evaluate at $x = 6$:

**(a)** $2f(x)$ **(b)** $-\frac{1}{2}f(x)$
**(c)** $5f(x) - 3$ **(d)** $\frac{1}{4}f(x-1)$

**S2.** Evaluate at $x = -2$ if $g(x) = \frac{1}{x}$:

**(a)** $-4g(x)$ **(b)** $\dfrac{2}{3}g(x)$
**(c)** $2g(x+5)$ **(d)** $-\frac{1}{5}g(-x)$

**S3.** Write a formula for the transformations of $f(x) = \sqrt{x}$:

**(a)** $y = -\frac{1}{3}f(x)$ **(b)** $y = 5f(-x)$
**(c)** $y = 6f(x-8)$ **(d)** $y = \frac{1}{4}f(2-x)$

**S4.** Write a formula for the transformations of $p(x) = 3x^2 - 6$:

**(a)** $y = 5p(x)$ **(b)** $y = -\frac{1}{3}p(-x)$
**(c)** $y = -2p(x+3)$ **(d)** $y = \frac{5}{3}p(x-1)$

### Exercises

**1.** Let $y = f(x)$. Write a formula for the transformation that both increases the $y$-value by a factor of 10 and shifts the graph to the right by 2 units.

**2.** Using Figure 6.25, match the functions (i)–(v) with a graph (a)–(i).

(i) $y = 2f(x)$ (ii) $y = \frac{1}{3}f(x)$
(iii) $y = -f(x+1)$ (iv) $y = f(x+2) + 1$
(v) $y = f(-x)$

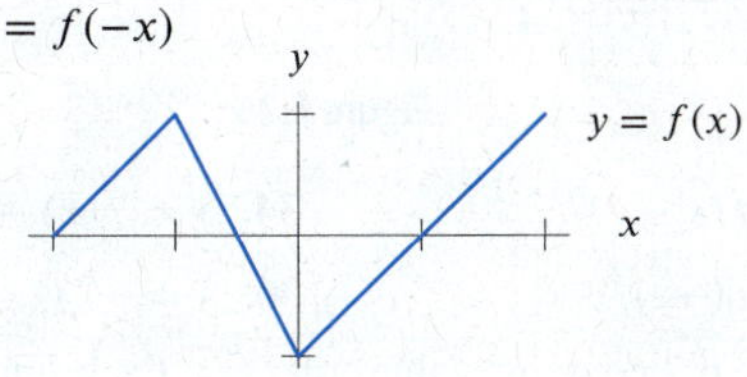


Figure 6.25

(a)
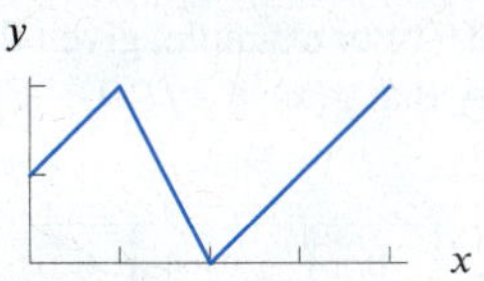

(b)
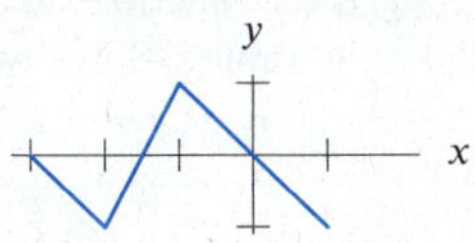

(c)
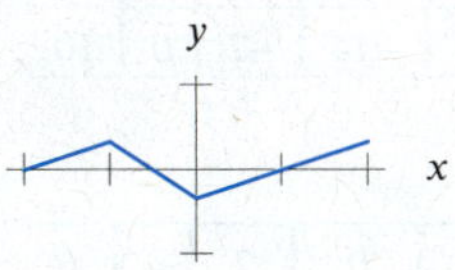

(d)
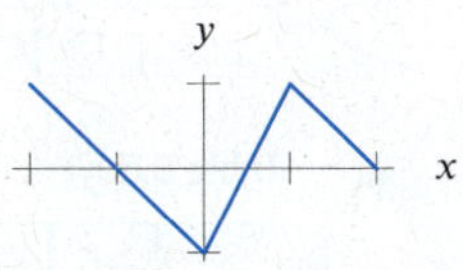

(e)
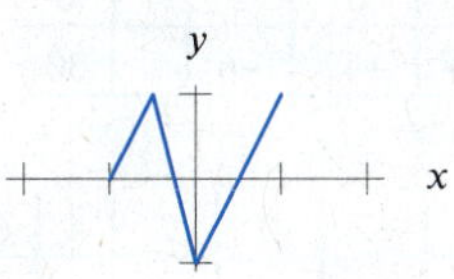

(f)
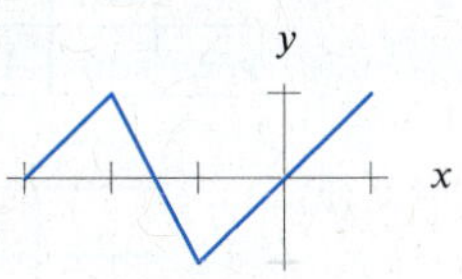

(g)
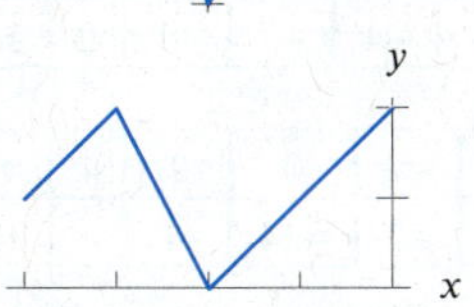

(h)
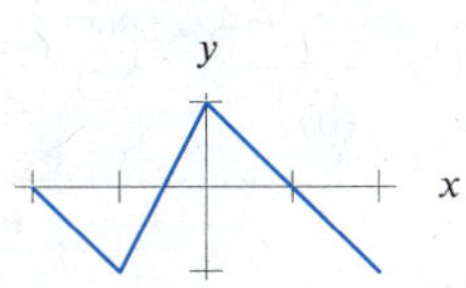

(i)
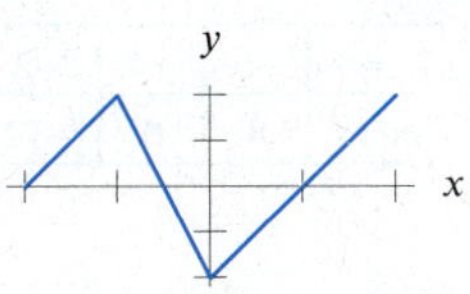

In Exercises 3–6, graph and label $f(x)$, $4f(x)$, $-\frac{1}{2}f(x)$, and $-5f(x)$ on the same axes.

**3.** $f(x) = \sqrt{x}$ **4.** $f(x) = -x^2 + 7x$
**5.** $f(x) = e^x$ **6.** $f(x) = \ln x$

**7.** Using Table 6.5, make tables for the following transformations of $f$ on an appropriate domain.

**(a)** $\frac{1}{2}f(x)$ **(b)** $-2f(x+1)$ **(c)** $f(x) + 5$
**(d)** $f(x-2)$ **(e)** $f(-x)$ **(f)** $-f(x)$

**Table 6.5**

| $x$ | −3 | −2 | −1 | 0 | 1 | 2 | 3 |
|---|---|---|---|---|---|---|---|
| $f(x)$ | 2 | 3 | 7 | −1 | −3 | 4 | 8 |

**8.** Using Table 6.6, create a table of values for

**(a)** $f(-x)$ **(b)** $-f(x)$ **(c)** $3f(x)$

**(d)** Which of these tables from parts (a), (b), and (c) represents an even function?

**Table 6.6**

| $x$ | −4 | −3 | −2 | −1 | 0 | 1 | 2 | 3 | 4 |
|---|---|---|---|---|---|---|---|---|---|
| $f(x)$ | 13 | 6 | 1 | −2 | −3 | −2 | 1 | 6 | 13 |

Without a calculator, graph the transformations in Exercises 9–14. Label at least three points.

**9.** $y = f(x+3)$ if $f(x) = |x|$
**10.** $y = f(x) + 3$ if $f(x) = |x|$
**11.** $y = -g(x)$ if $g(x) = x^2$
**12.** $y = g(-x)$ if $g(x) = x^2$
**13.** $y = 3h(x)$ if $h(x) = 2^x$
**14.** $y = 0.5h(x)$ if $h(x) = 2^x$

## Problems

**15.** Describe the effect of the transformation $2f(x+1)$ on the graph of $y = f(x)$.

**16.** The function $s(t)$ gives the distance (miles) in terms of time (hours). If the average rate of change of $s(t)$ on $0 \le t \le 4$ is 70 mph, what is the average rate of change of $\frac{1}{2}s(t)$ on this interval?

In Problems 17–21, let $f(t) = 1/(1+t^2)$. Graph the function, labeling intercepts and asymptotes.

**17.** $y = f(t)$ **18.** $y = -f(t)$

**19.** $y = 0.5f(t)$ **20.** $y = f(t-3)$

**21.** $y = f(t+5) - 5$

**22.** The graph of the function $g(x)$ contains the point $(5, \frac{1}{3})$. What point must be on the graph of $y = 3g(x+1)$?

**23.** The range of the function $C(x)$ is $-1 \le C(x) \le 1$. What is the range of $0.25C(x)$?

Problems 24–27 refer to functions obtained by applying a transformation to $P(n)$ whose domain is $-3 < n \le 8$ and range is $-6 \le P(n) < 12$. Find a possible formula for the function in terms of $P(n)$.

**24.** The domain of $Q(n)$ is $-3 < n \le 8$ and the range is $-2 \le Q(n) < 4$.

**25.** The domain of $R(n)$ is $-3 < n \le 8$ and the range is $-60 < R(n) \le 30$.

**26.** The domain of $S(n)$ is $-8 \le n < 3$ and the range is $2 \le S(n) < 20$.

**27.** The domain of $T(n)$ is $-10 < n \le 1$ and the range is $-1.5 \le T(n) < 3$.

**28.** The number of gallons of paint, $n = f(A)$, needed to cover a house is a function of the surface area $A$, in ft$^2$. Match each story to one expression.

**(a)** I figured out how many gallons I needed and then bought two extra gallons just in case.
**(b)** I bought enough paint to cover my house twice.
**(c)** I bought enough paint to cover my house and my welcome sign, which measures 2 square feet.

(i) $2f(A)$ (ii) $f(A+2)$ (iii) $f(A)+2$

**29.** The US population in millions is $P(t)$ today and $t$ is in years. Match each statement (I)–(IV) with one of the formulas (a)–(h).

I. The population 10 years before today.
II. Today's population plus 10 million immigrants.
III. Ten percent of the population we have today.
IV. The population after 100,000 people have emigrated.

**(a)** $P(t) - 10$ **(b)** $P(t-10)$ **(c)** $0.1P(t)$
**(d)** $P(t) + 10$ **(e)** $P(t+10)$ **(f)** $P(t)/0.1$
**(g)** $P(t) + 0.1$ **(h)** $P(t) - 0.1$

**30.** Let $R = P(t)$ be the number of rabbits living in a national park in month $t$. What do the following expressions represent?

**(a)** $P(t+1)$ **(b)** $2P(t)$

**31.** Suppose $C(t)$ represents the cost in euros of the first $t$ days of a business trip to France from the United States. Find a formula for $h(t)$, the cost in US dollars for the first $t$ days. During the trip, 1 euro was equivalent to 1.3 US dollars.

**32.** Suppose that $f(x)$ is the age in years of the ice $x$ cm below the surface of a glacier in Antarctica. Find the age $g(x)$ for the same ice in centuries.

Graph the transformations of $f$ in Problems 33–36 using Figure 6.26. Label the points corresponding to $A$ and $B$.

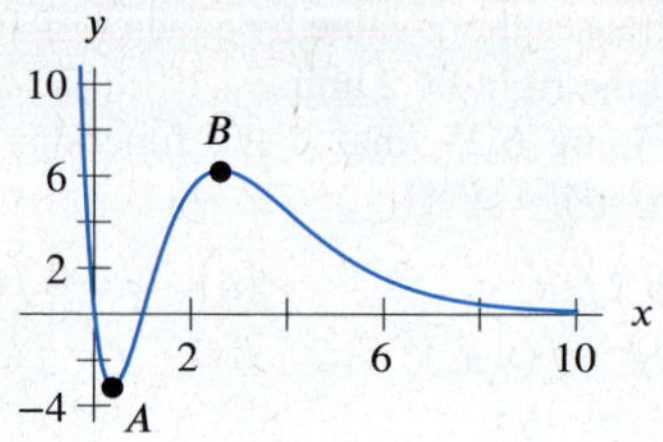


Figure 6.26

**33.** $y = f(x-3)$ **34.** $y = f(x) - 3$

**35.** $y = f(-x)/3$ **36.** $y = -2f(x)$

**37.** Table 6.7 contains values of $f(x)$. Each function in parts (a)–(c) is obtained by applying a single transformation to $f(x)$. The transformation may be a stretch, compression, shift, or reflection. Find a possible formula for each of these functions in terms of $f$. For example, given the data in Table 6.8, we would say that $g(x) = 3f(x)$.

**Table 6.7**

| $x$ | −6 | −4 | −2 | 0 | 2 | 4 | 6 |
|---|---|---|---|---|---|---|---|
| $f(x)$ | 12 | 4 | −8 | −14 | −2 | 0 | 10 |

**Table 6.8**

| $x$ | −6 | −4 | −2 | 0 | 2 | 4 | 6 |
|---|---|---|---|---|---|---|---|
| $g(x)$ | 36 | 12 | −24 | −42 | −6 | 0 | 30 |

**(a)**

| $x$ | −6 | −4 | −2 | 0 | 2 | 4 | 6 |
|---|---|---|---|---|---|---|---|
| $h(x)$ | 6 | 2 | −4 | −7 | −1 | 0 | 5 |

**(b)**

| $x$ | −6 | −4 | −2 | 0 | 2 | 4 | 6 |
|---|---|---|---|---|---|---|---|
| $k(x)$ | 10 | 0 | −2 | −14 | −8 | 4 | 12 |

**(c)**

| $x$ | −6 | −4 | −2 | 0 | 2 | 4 | 6 |
|---|---|---|---|---|---|---|---|
| $m(x)$ | 8 | 0 | −12 | −18 | −6 | −4 | 6 |

**38.** Using Figure 6.27, find formulas, in terms of $f$, for the horizontal and vertical shifts of the graph of $f$ in parts (a)–(c). What is the equation of each asymptote?

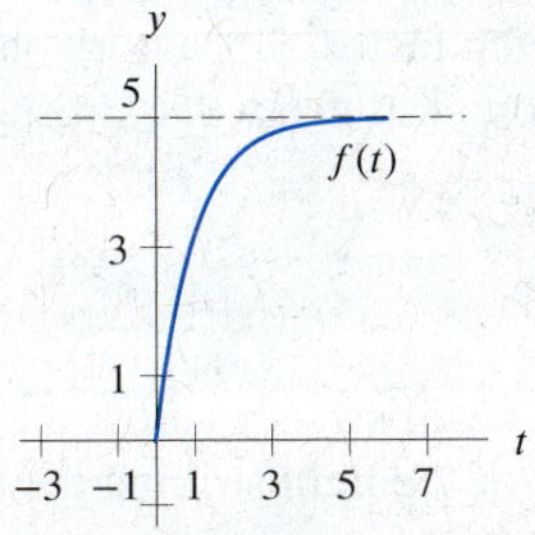


Figure 6.27

(a)

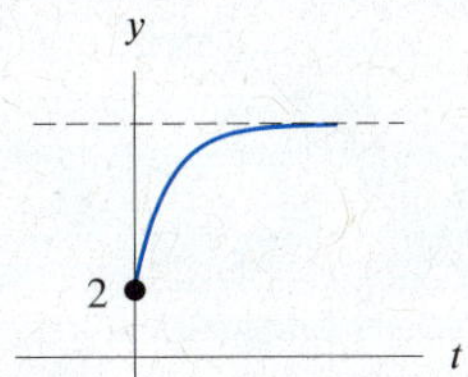


(b)

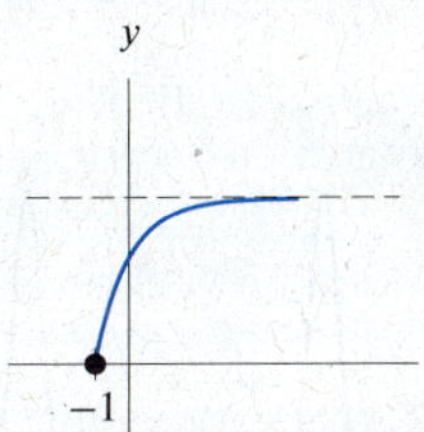


(c)

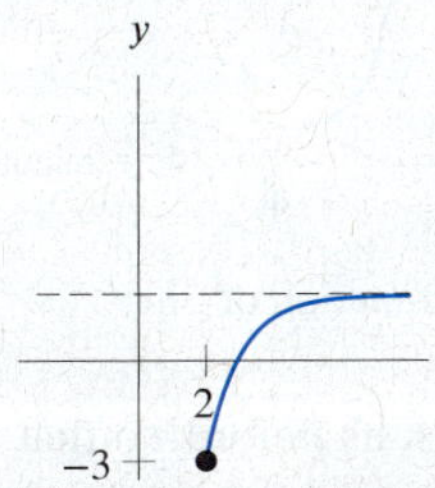


**39.** Using Figure 6.28, find formulas, in terms of $f$, for the transformations of $f$ in parts (a)–(c).

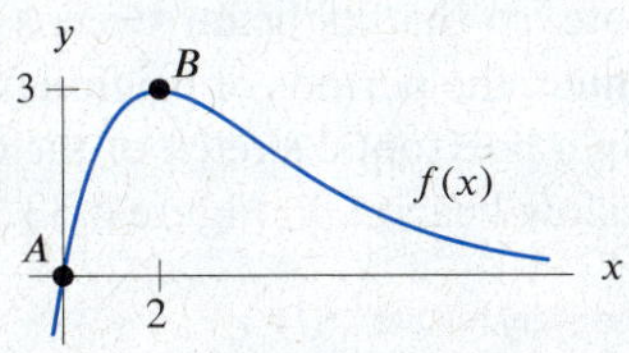


Figure 6.28

(a)

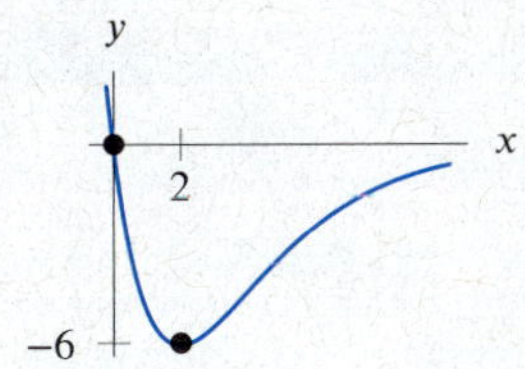


(b)

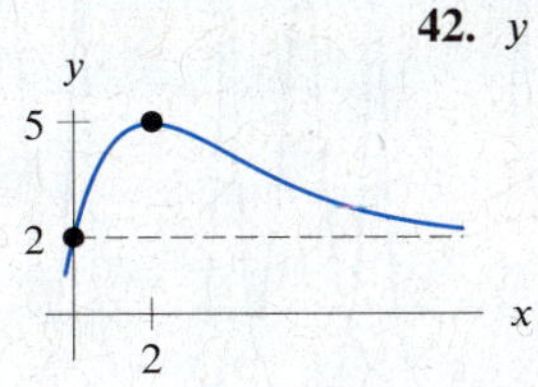


(c)

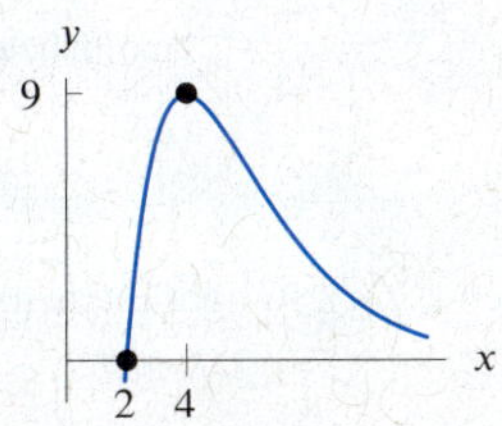


**40.** In Figure 6.29, the point $b$ is labeled on the $x$-axis. On the $y$-axis, locate and label the output values:

**(a)** $f(b)$ **(b)** $-2f(b)$ **(c)** $-2f(-b)$

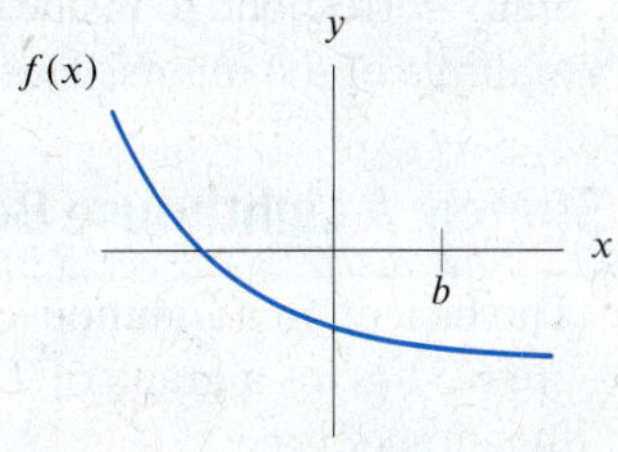


Figure 6.29

**41.** Figure 6.30 gives a graph of $y = f(x)$. Consider the transformations $y = \frac{1}{2}f(x)$ and $y = 2f(x)$. Which points on the graph of $y = f(x)$ stay fixed under these transformations? Compare the intervals on which all three functions are increasing and decreasing.

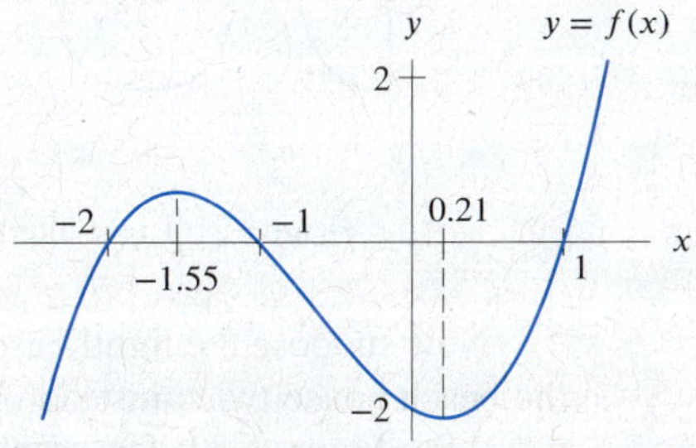


Figure 6.30

The average rate of change of $f(x)$ over the interval $a \le x \le b$ is 28, where $a$ and $b$ denote constants. In Problems 42–44 give the average rate of change over the interval $a \le x \le b$ for each transformation of $f(x)$.

**42.** $y = 3f(x)$ **43.** $y = -\frac{1}{4}f(x)$ **44.** $y = f(x) + 3$

# 6.3 HORIZONTAL STRETCHES AND COMBINATIONS OF TRANSFORMATIONS

In Section 6.2, we observed that a vertical stretch of a function's graph corresponds to an outside change in its formula, specifically, multiplication by a stretch factor. Since horizontal changes generally correspond to inside changes, we expect that a horizontal stretch will correspond to a constant multiple of the inputs. This turns out to be the case.

## Horizontal Stretch: A Lighthouse Beacon

The beacon in a lighthouse turns once per minute, and its beam sweeps across a beach house. Figure 6.31 gives a graph of $L(t)$, the intensity, or brightness, of the light striking the beach house as a function of time.

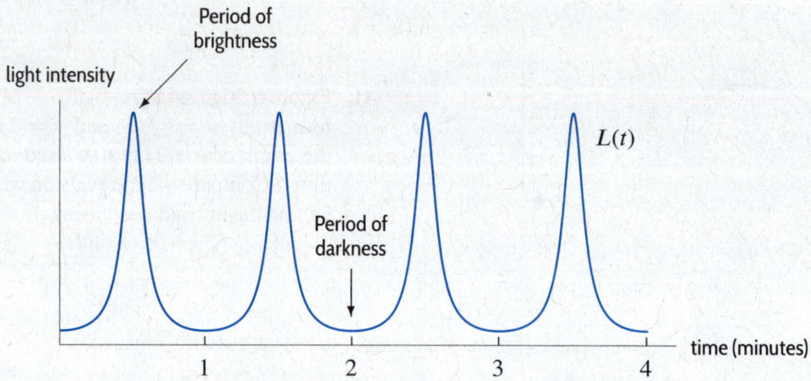


**Figure 6.31**: Light intensity or brightness, $L(t)$, as a function of time

Now suppose the lighthouse beacon turns twice as fast as before, so that its beam sweeps past the beach house twice instead of once each minute. The periods of brightness now occur twice as often. See Figure 6.32. The graph of $f(t)$, the intensity of light from this faster beacon, is a horizontal squeezing or compression of the original graph of $L(t)$.

If the lighthouse beacon turns at half its original rate, so that its beam sweeps past the beach house once every two minutes instead of once every minute, the periods of brightness occur half as often as originally. Slowing the beacon's speed results in a horizontal stretch of the original graph, illustrated by the graph of $s(t)$, the light intensity of the slow beacon, in Figure 6.33.

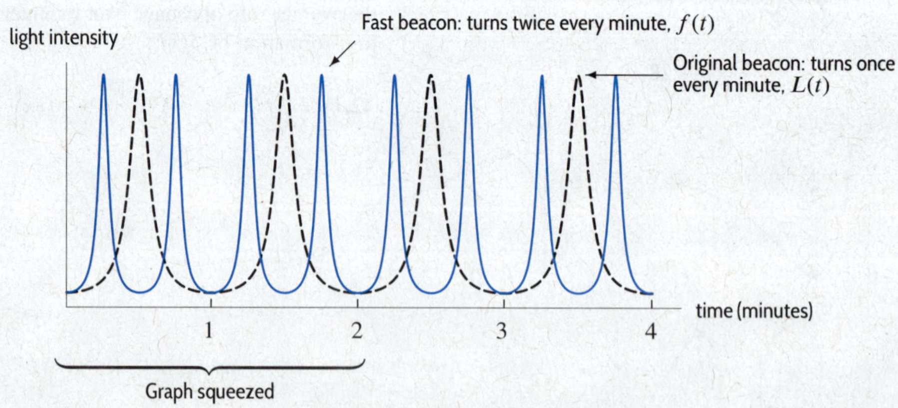


**Figure 6.32**: Comparing light intensity from the fast beacon, $f(t)$, to light intensity from the original beacon, $L(t)$

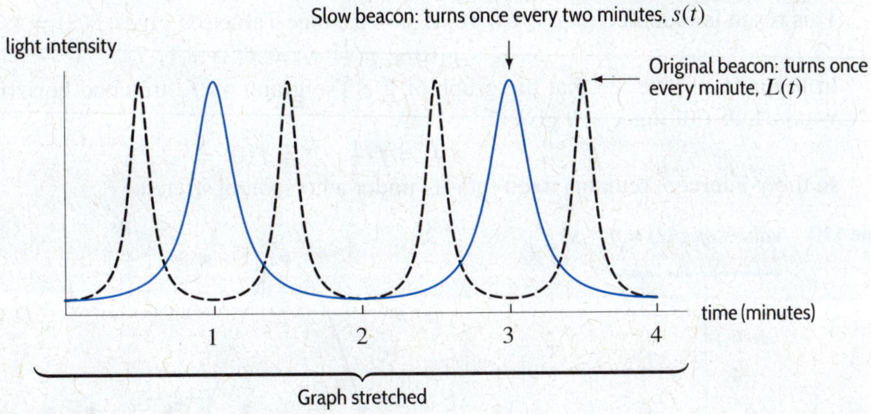


**Figure 6.33:** Comparing light intensity from the slow beacon, $s(t)$, to light intensity from the original beacon, $L(t)$

## Formula for Horizontal Stretch or Compression

How are the formulas for the three light functions related? We expect that multiplying the function's input by a constant will horizontally stretch or compress its graph. The fast beacon corresponds to speeding up by a factor of 2, or multiplying the input by 2. Thus

$$f(t) = L(2t).$$

Similarly for the slow beacon, the input times are multiplied by $1/2$, so

$$s(t) = L(\tfrac{1}{2}t).$$

Generalizing the lighthouse example gives the following result:

> If $f$ is a function and $k$ is a constant, then the graph of $y = f(kx)$ is the graph of $y = f(x)$
> - Horizontally compressed by a factor of $1/k$ if $k > 1$.
> - Horizontally stretched by a factor of $1/k$ if $0 < k < 1$.
> - Horizontally stretched or compressed by a factor $1/|k|$ and reflected across $y$-axis, if $k < 0$.

**Example 1** Values of the function $f(x)$ are in Table 6.9 and its graph is in Figure 6.34. Make a table and a graph of the function $g(x) = f(\frac{1}{2}x)$.

**Table 6.9** *Values of $f(x)$*

| $x$ | $f(x)$ |
|---|---|
| −3 | 0 |
| −2 | 2 |
| −1 | 0 |
| 0 | −1 |
| 1 | 0 |
| 2 | −1 |
| 3 | 1 |

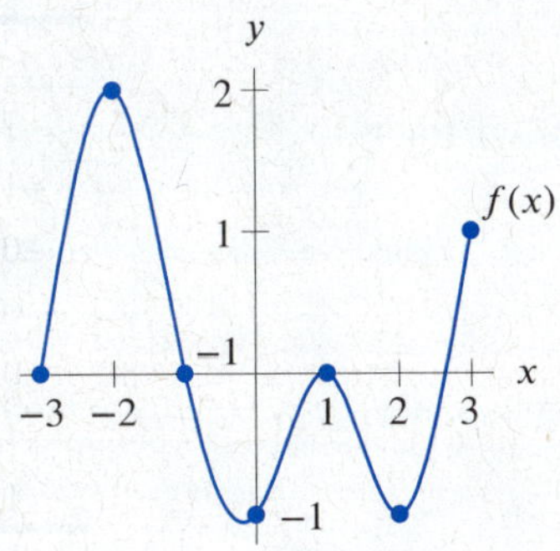


**Figure 6.34**

**Solution** To make a table for $g(x) = f(\frac{1}{2}x)$, we substitute values for $x$. For example, if $x = 4$, then

$$g(4) = f(\tfrac{1}{2} \cdot 4) = f(2).$$

Table 6.9 shows that $f(2) = -1$, so

$$g(4) = f(2) = -1.$$

This result is recorded in Table 6.10. If $x = 6$, since Table 6.9 gives $f(3) = 1$, we have

$$g(6) = f(\tfrac{1}{2} \cdot 6) = f(3) = 1.$$

In Figure 6.35, we see that the graph of $g$ is the graph of $f$ stretched horizontally away from the $y$-axis. Substituting $x = 0$ gives

$$g(0) = f(\tfrac{1}{2} \cdot 0) = f(0) = -1,$$

so the $y$-intercept remains fixed (at $-1$) under a horizontal stretch.

**Table 6.10** *Values of* $g(x) = f(\frac{1}{2}x)$

| $x$ | $g(x)$ |
|---|---|
| −6 | 0 |
| −4 | 2 |
| −2 | 0 |
| 0 | −1 |
| 2 | 0 |
| 4 | −1 |
| 6 | 1 |

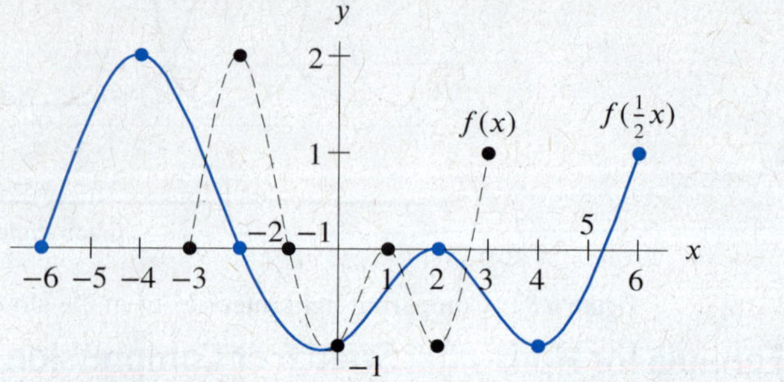


**Figure 6.35**: The graph of $g(x) = f(\frac{1}{2}x)$ is the graph of $y = f(x)$ stretched away from the $y$-axis by a factor of 2

Example 1 shows the effect of an inside multiple of $1/2$. We see that the graph of $f$ is stretched by a factor of 2 horizontally from the $y$-axis. The next example shows the effect on the graph of an inside multiple of 2.

**Example 2** Let $f(x)$ be the function in Example 1. Make a table and a graph for the function $h(x) = f(2x)$.

**Solution** We use Table 6.9 and the formula $h(x) = f(2x)$ to evaluate $h(x)$ at several values of $x$. For example, if $x = 1$, then

$$h(1) = f(2 \cdot 1) = f(2).$$

Table 6.9 shows that $f(2) = -1$, so $h(1) = -1$. These values are recorded in Table 6.11. Similarly, substituting $x = 1.5$ gives

$$h(1.5) = f(2 \cdot 1.5) = f(3) = 1.$$

Since $h(0) = f(2 \cdot 0) = f(0)$, the $y$-intercept remains fixed (at $-1$). In Figure 6.36 we see that the graph of $h$ is the graph of $f$ compressed by a factor of 1/2 horizontally toward the $y$-axis.

**Table 6.11** *Values of* $h(x) = f(2x)$

| $x$ | $h(x)$ |
|---|---|
| −1.5 | 0 |
| −1.0 | 2 |
| −0.5 | 0 |
| 0.0 | −1 |
| 0.5 | 0 |
| 1.0 | −1 |
| 1.5 | 1 |

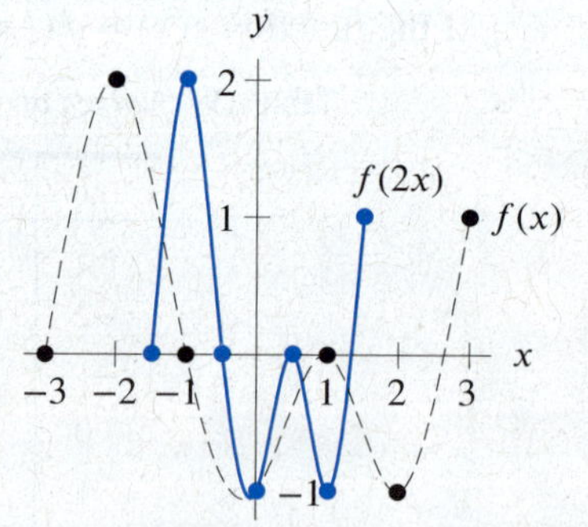


**Figure 6.36**: The graph of $h(x) = f(2x)$ is the graph of $y = f(x)$ compressed horizontally by a factor of 1/2

If $t$ is in years since 2014 and we use the function $P = 317e^{0.007t}$ to model the US population in millions,[2] this function is a transformation of the exponential function $f(t) = e^t$, since

$$P = 317e^{0.007t} = 317f(0.007t).$$

[2]www.census.gov/newsroom/releases/archives/population/cb13-tps112.html. Accessed June 18, 2014.

Thus, the US population is $f(t) = e^t$ stretched vertically by a factor of 317 and stretched horizontally by a factor of $1/0.007 \approx 143$.

**Example 3** Match the functions $f(t) = e^t$, $g(t) = e^{0.5t}$, $h(t) = e^{0.8t}$, $j(t) = e^{2t}$ with the graphs in Figure 6.37.

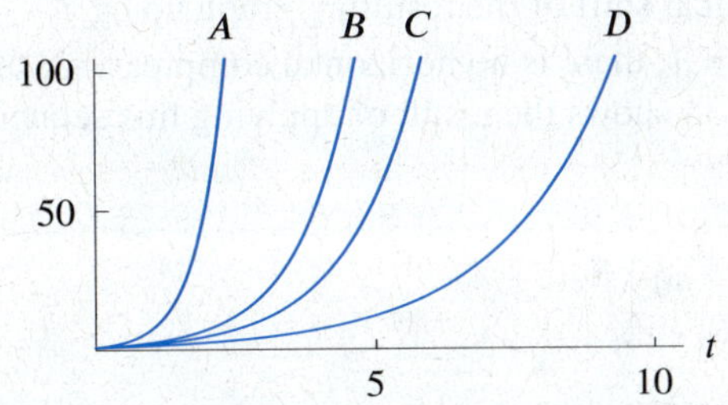


Figure 6.37: Matching transformations of $e^t$ with formulas

**Solution** Since the function $j(t) = e^{2t}$ climbs fastest of the four and $g(t) = e^{0.5t}$ climbs slowest, graph $A$ must be $j$ and graph $D$ must be $g$. Similarly, graph $B$ is $f$ and graph $C$ is $h$.

## Combining Transformations

Now we have seen several types of transformations, what happens if we combine them? We know that if there are several inside changes, the order we do them matters. Similarly, the ordering of the outside changes matters. However, it does not actually matter if we do the inside changes or the outside changes first.

### Ordering Horizontal and Vertical Transformations

We can follow the effect of a sequence of transformations on the graph by writing the function in the following form:

> For nonzero constants $A$, $B$, $h$ and $k$, the graph of the function
>
> $$y = Af(B(x-h)) + k$$
>
> is obtained by applying the transformations to the graph of $f(x)$ in the following order:
> - Horizontal stretch/compression by a factor of $1/|B|$
> - Horizontal shift by $h$ units
> - Vertical stretch/compression by a factor of $|A|$
> - Vertical shift by $k$ units
>
> If $A < 0$, follow the vertical stretch/compression by a reflection about the $x$-axis.
> If $B < 0$, follow the horizontal stretch/compression by a reflection about the $y$-axis.

Example 4 shows these steps applied to a familiar function, $f(x) = x^2$. In Section 3.2 we expressed a quadratic function in vertex form. We now see the vertex in the context of transformations.

**Example 4**
(a) Let $y = 5(x+2)^2 + 7$. Determine the values of $A$, $B$, $h$, and $k$ when $y$ is put in the form $y = Af(B(x-h)) + k$ with $f(x) = x^2$. List the transformations applied to $f(x) = x^2$ to give $y = 5(x+2)^2 + 7$.
(b) Using your answers in part (a), sketch a graph of $y = 5(x+2)^2 + 7$, labeling the vertex.

**Solution**

(a) If $f(x) = x^2$, we see $y = 5(x+2)^2 + 7 = 5f(x+2) + 7$, so $A = 5$, $B = 1$, $h = -2$, and $k = 7$. Based on the values of these constants, we carry out these steps:

- Horizontal shift 2 to the left of the graph of $f(x) = x^2$.
- Vertical stretch of the resulting graph by a factor of 5.
- Vertical shift of the resulting graph up by 7.

Since $B = 1$, there is no horizontal compression, stretch, or reflection.

(b) Figure 6.38 shows the result of applying this sequence of transformations to the graph of $f$.

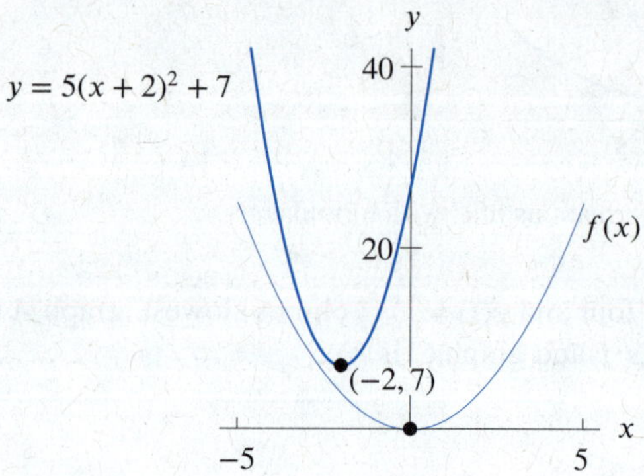


**Figure 6.38:** Graphs of $y = x^2$ and $y = 5(x+2)^2 + 7$

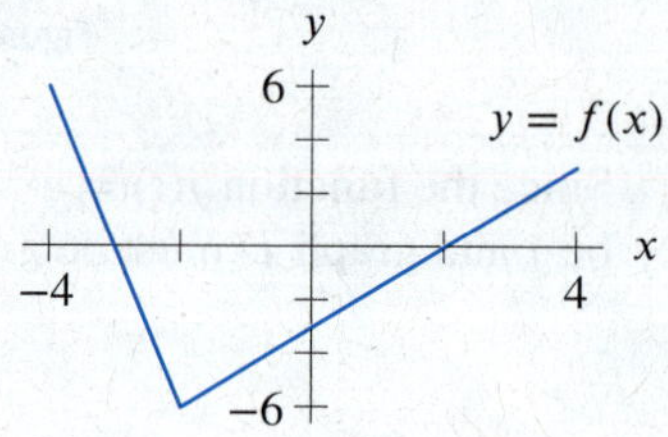


**Figure 6.39:** Graph of $y = f(x)$

Example 5 illustrates the use of these steps to graph multiple inside and outside changes.

**Example 5**

Figure 6.39 shows the graph of a function $f$.

(a) Put the function $r(x) = -\frac{1}{3}(f(2x-8)+3)$ in the form $y = Af(B(x-h)) + k$ and determine the values for constants $A$, $B$, $h$, and $k$.

(b) Express $r$ as the result of applying a list of transformations to $f$, specifying the order in which the transformations are applied. Sketch a graph of $r$ using the graph of $f$.

**Solution**

(a) Expanding on the outside and factoring on the inside, we have

$$r(x) = -\frac{1}{3}f(2x-8) - 1 = -\frac{1}{3}f\,(2(x-4)) - 1,$$

so $A = -1/3$, $B = 2$, $h = 4$, and $k = -1$.

(b) See Figure 6.40. To find the graph of $r$, we carry out these steps:

- Horizontally compress the graph of $f$ by $1/|B| = 1/2$.
- Horizontally shift the resulting graph to the right by $h = 4$.
- Vertically compress the resulting graph by $|A| = 1/3$ and reflect the graph about the $x$-axis.
- Vertically shift the resulting graph by $k = -1$.

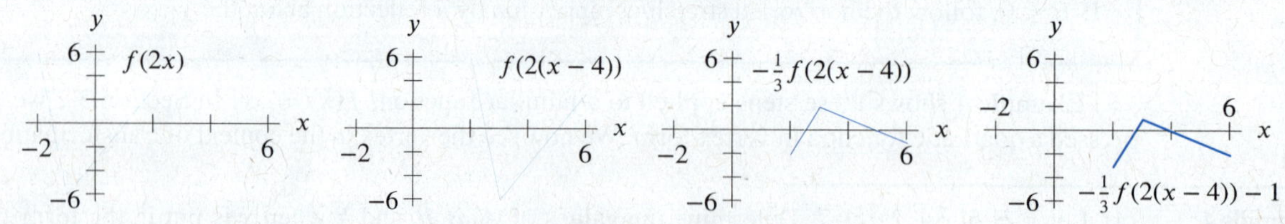


**Figure 6.40:** Graph of $y = -\frac{1}{3}(f(2x-8)+3)$

## Exercises and Problems for Section 6.3

### Skill Refresher

In Exercises S1–S4, write a formula for each of the transformations of $f(x) = x^3 - 5$.

**S1.** $y = f(2x)$ **S2.** $y = 2f(x)$

**S3.** $y = f\left(-\frac{1}{3}x\right)$ **S4.** $y = \frac{1}{5}f(3x)$

In Exercises S5–S8, write a formula for each of the transformations of $Q(t) = 4e^{6t}$.

**S5.** $y = Q\left(\frac{1}{3}t\right)$ **S6.** $y = \frac{1}{3}Q(t)$

**S7.** $y = Q(2t) + 11$ **S8.** $y = 7Q(t-3)$

In Exercises S9–S12, solve for the constant $h$.

**S9.** $4x + 12 = 4(x - h)$ **S10.** $\frac{1}{5}t - 10 = \frac{1}{5}(t - h)$

**S11.** $-3z + 10 = -3(z - h)$ **S12.** $-\frac{1}{9}x - 4 = -\frac{1}{9}(x - h)$

In Exercises S13–S16, rewrite as $y = Af(B(x - h)) + k$ and give values for $A$, $B$, $h$, and $k$.

**S13.** $y = f(-2x) + 9$ **S14.** $y = -f(2x - 6) + 9$

**S15.** $y = 6f\left(-\frac{1}{3}x - 9\right)$

**S16.** $y = -5(f(-x - 7) + 2)$

### Exercises

**1.** Using Figure 6.41, graph the following functions.

(a) $y = -f(x) + 2$ (b) $y = 2f(x)$

(c) $y = f(x - 3)$ (d) $y = -\frac{1}{2}f(x + 1) - 3$

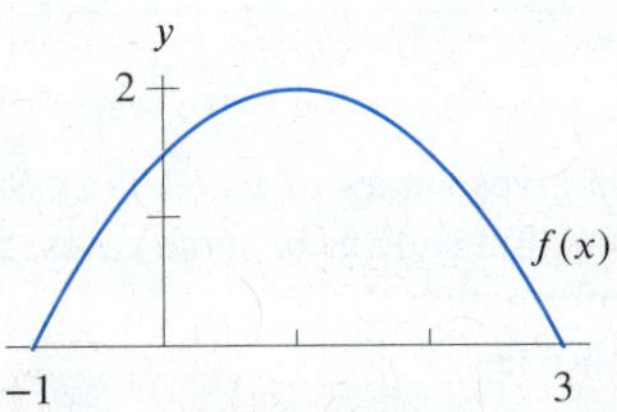


Figure 6.41

**2.** Table 6.12 gives values of function $v$. Create a table giving five values of the function $w(t) = 40 - 2v(-0.5t)$.

Table 6.12

| $t$ | 0 | 1 | 2 | 3 | 4 |
|---|---|---|---|---|---|
| $v(t)$ | 20 | 17 | 16 | 19 | 23 |

**3.** Graph $m(x) = e^x$, $n(x) = e^{2x}$, and $p(x) = 2e^x$ on the same axes and describe how the graphs of $n(x)$ and $p(x)$ compare with that of $m(x)$.

**4.** Graph $y = h(3x)$ if $h(x) = 2^x$.

In Exercises 5–7, graph and label $f(x)$, $f(\frac{1}{2}x)$, and $f(-3x)$ on the same axes between $x = -2$ and $x = 2$.

**5.** $f(x) = e^x + x^3 - 4x^2$

**6.** $f(x) = 5e^x + x^2$

**7.** $f(x) = \ln(x^4 + 3x^2 + 4)$

**8.** Using Figure 6.42, match each function to a graph (if any) that represents it:

(i) $y = f(2x)$ (ii) $y = 2f(2x)$ (iii) $y = f(\frac{1}{2}x)$

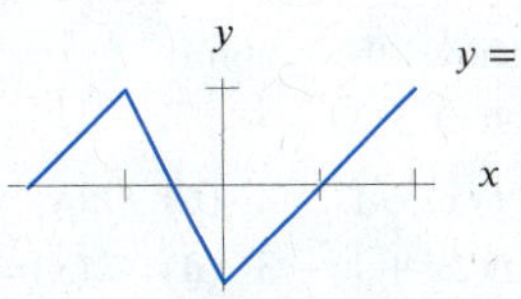


Figure 6.42

(a)
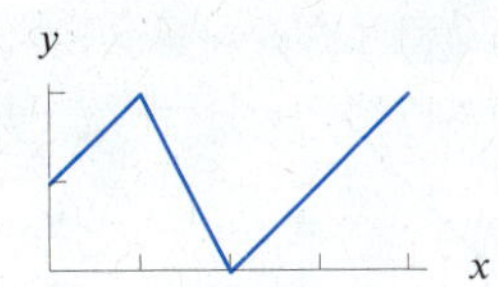


(b)
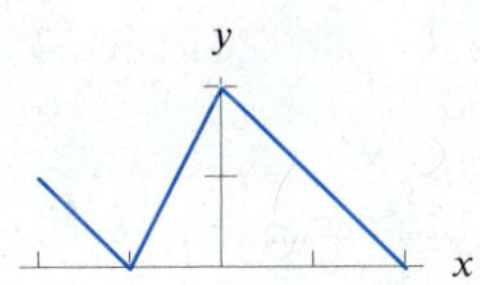


(c)
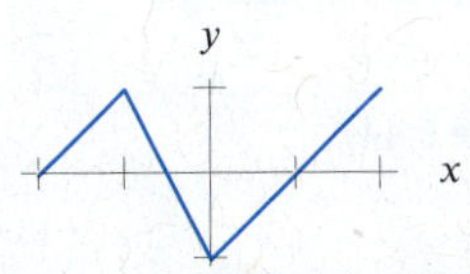


(d)
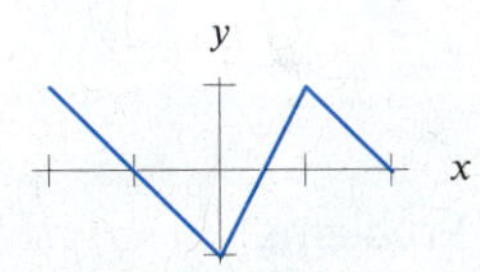


(e)
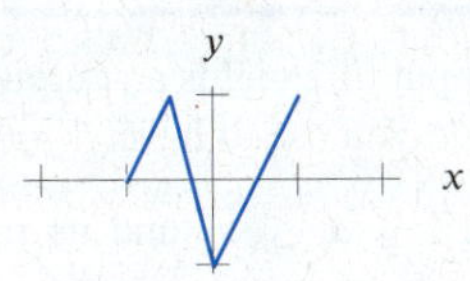


(f)
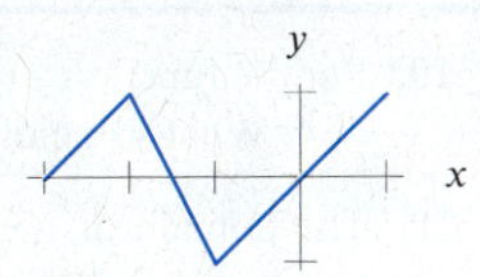


(g)
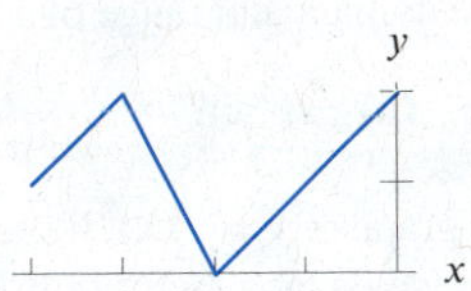


(h)
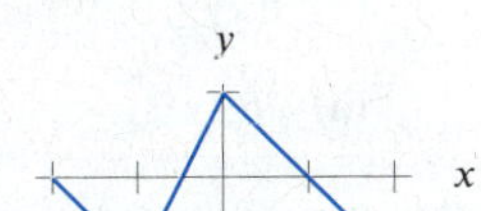


(i)
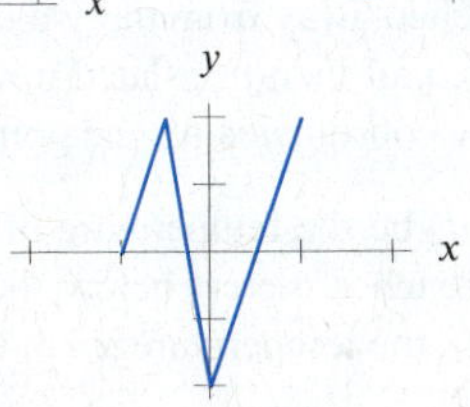

**9.** Using Figure 6.43, graph the following functions on separate sets of axes, together with the graph of the original function. Label any intercepts or special points.

**(a)** $y = f(3x)$ **(b)** $y = f(-2x)$ **(c)** $y = f(\frac{1}{2}x)$

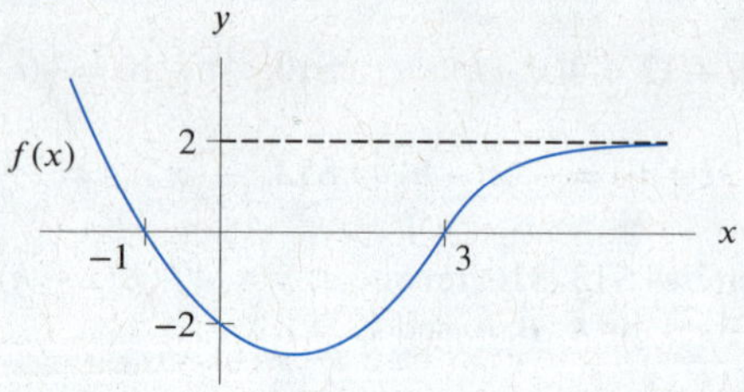


Figure 6.43

**10.** Describe the effect of the transformation $10f(\frac{1}{10}x)$ on the graph of $f(x)$.

**11.** Describe $y = f(3x - 2)$ as the result of first applying a stretch or compression and then applying a shift to $f$.

**12.** Describe $y = 5(g(x) - 8)$ as the result of first applying a stretch or compression and then applying a shift to $g$.

**13.** Figure 6.44 shows the graph of $fx) = |x|$. Graph each transformation of $f(x)$:

**(a)** $g(x) = f(x) + 1$ **(b)** $h(x) = f(x + 1)$
**(c)** $j(x) = f(2x + 1) - 3$ **(d)** $k(x) = \frac{1}{2}f(2x-4)+1$
**(e)** $m(x) = -\frac{1}{2}f(4x+12)-3$

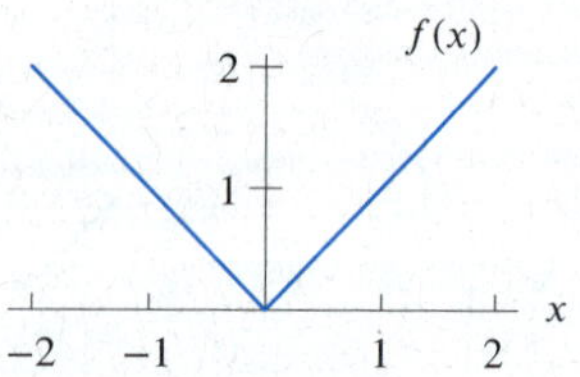


Figure 6.44

**14.** The point (2, 3) lies on the graph of $g(x)$. What point must lie on the graph of $g(2x)$?

**15.** The graph of $g(x)$ contains the point (6, −9). For each of the following functions, find a point on its graph:

**(a)** $g(2x) - 5$ **(b)** $3g(x) + 1$
**(c)** $-(g(\frac{1}{3}(x + 4)) - 8)$ **(d)** $\frac{1}{2}g(-5x - 15) - 8$

**16.** Using Table 6.13, make a table of values for $f(\frac{1}{2}x)$ for an appropriate domain.

Table 6.13

| $x$ | −3 | −2 | −1 | 0 | 1 | 2 | 3 |
|---|---|---|---|---|---|---|---|
| $f(x)$ | 2 | 3 | 7 | −1 | −3 | 4 | 8 |

**17.** Fill in all the blanks in Table 6.14 for which you have sufficient information.

Table 6.14

| $x$ | −3 | −2 | −1 | 0 | 1 | 2 | 3 |
|---|---|---|---|---|---|---|---|
| $f(x)$ | −4 | −1 | 2 | 3 | 0 | −3 | −6 |
| $f(\frac{1}{2}x)$ | | | | | | | |
| $f(2x)$ | | | | | | | |

**18.** Table 6.15 gives values of $x$ and $f(x)$. Supply the values of each function shown. In some cases, there may not be enough information to fill in a box.

Table 6.15

| $x$ | −2 | −1 | 0 | 1 | 2 |
|---|---|---|---|---|---|
| $f(x)$ | −3 | −4 | 2 | 0 | 5 |
| $2f(x) + 3$ | | | | | |
| $f(x - 1) + 1$ | | | | | |
| $f(x + 2) - 1$ | | | | | |
| $3f(2x + 2) - 1$ | | | | | |

## Problems

**19.** For the function $f(p)$ an input of 2 yields an output value of 4. What value of $p$ would you use to have $f(3p) = 4$?

**20.** The domain of $l(x)$ is $-12 \leq x \leq 12$ and its range is $0 \leq l(x) \leq 3$. What are the domain and range of

**(a)** $l(2x)$? **(b)** $l(\frac{1}{2}x)$?

**21.** The point $(a, b)$ lies on the graph of $y = f(x)$. If the graph is stretched away from the $y$-axis by a factor of $d$ (where $d > 1$), and then translated upward by $c$ units, what are the new coordinates for the point?

**22.** Let $T(d)$ be the temperature of the seawater in the Mariana Trench $d$ meters below the surface. Find a formula for $p(x)$, the temperature $x$ km below the surface.

**23.** Let $f(t)$ be a man's height in meters when he turns $t$ months old. Find a formula for his height $g(n)$ in centimeters when he turns $n$ years old.

**24.** Let $A(t)$ be the altitude in meters of a balloonist $t$ minutes after liftoff. Find a formula for

**(a)** $f(s)$, the altitude $s$ seconds after liftoff.
**(b)** $g(h)$, the altitude $h$ hours after liftoff.

**25.** Every day I take the same taxi over the same route from home to the train station. The trip is $x$ miles, so the cost for the trip is $f(x)$. Match each story in (a)–(d) to a function in (i)–(iv) representing the amount paid to the driver.

**(a)** I received a raise yesterday, so today I gave my driver a five-dollar tip.

**(b)** I had a new driver today and he got lost. He drove five extra miles and charged me for it.
**(c)** I haven't paid my driver all week. Today is Friday and I'll pay what I owe for the week.
**(d)** The meter in the taxi went crazy and showed five times the number of miles I actually traveled.

(i) $5f(x)$ (ii) $f(x)+5$
(iii) $f(5x)$ (iv) $f(x+5)$

**26.** A company projects a total profit, $P(t)$ dollars, in year $t$. Explain the economic meaning of $r(t) = 0.5P(t)$ and $s(t) = P(0.5t)$.

**27.** Let $A = f(r)$ be the area of a circle of radius $r$.

**(a)** Write a formula for $f(r)$.
**(b)** Which expression represents the area of a circle whose radius is increased by 10%? Explain.

(i) $0.10f(r)$ (ii) $f(r+0.10)$ (iii) $f(0.10r)$
(iv) $f(1.1r)$ (v) $f(r)+0.10$

**(c)** By what percent does the area increase if the radius is increased by 10%?

In Problems 28–29, state which graph represents

**(a)** $f(x)$ **(b)** $f(-2x)$ **(c)** $f(-\frac{1}{2}x)$ **(d)** $f(2x)$

**28.**

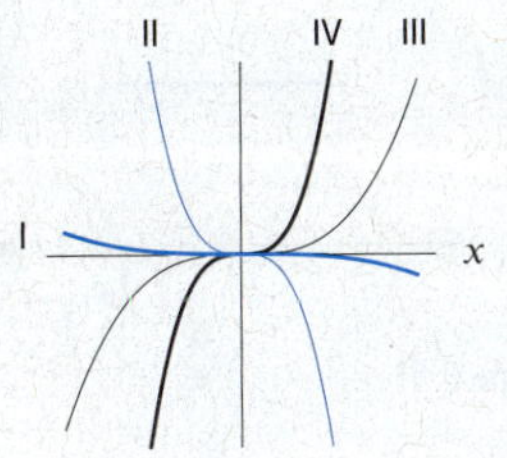


**29.**

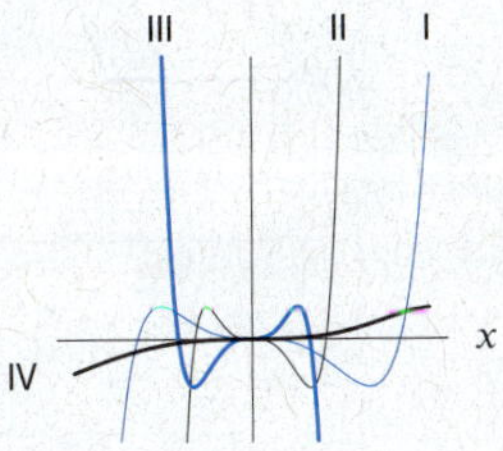


In Problems 30–31, graph the transformation of $f$, the function in Figure 6.45.

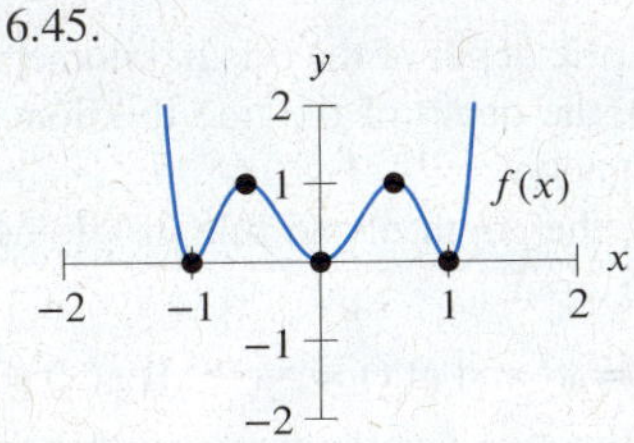


Figure 6.45

**30.** $y = -2f(x-1)$ **31.** $y = f(x/2) - 1$

In Problems 32–37, let $f$ is the function in Figure 6.46. Find a formula and sketch the graph of the function obtained by performing the given transformations on $f$.

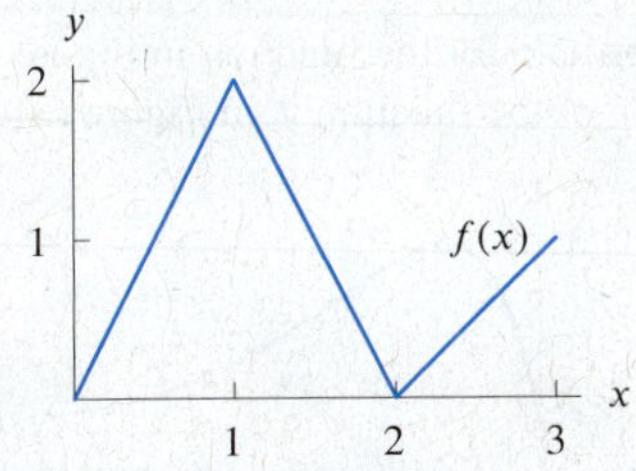


Figure 6.46

**32.** Shift $f$ up by 2, then reflect it about the $x$-axis.

**33.** Reflect $f$ about the $x$-axis, then shift it up by 2.

**34.** Shift $f$ right by 1, then reflect it about the $y$-axis.

**35.** Reflect $f$ about the $y$-axis, then shift it right by 1.

**36.** Stretch $f$ vertically by a factor of 2, then shift down 2.

**37.** Shift $f$ down 2, then stretch vertically by a factor of 2.

**38.** This problem investigates the effect of a horizontal stretch on the zeros of a function.

**(a)** Graph $f(x) = 4 - x^2$. Mark the zeros of $f$ on the graph.
**(b)** Graph and find a formula for $g(x) = f(0.5x)$. What are the zeros of $g(x)$?
**(c)** Graph and find a formula for $h(x) = f(2x)$. What are the zeros of $h(x)$?
**(d)** Without graphing, what are the zeros of $f(10x)$?

**39.** In Figure 6.47, the point $c$ is labeled on the $x$-axis. On the $y$-axis, locate and label output values:

**(a)** $g(c)$ **(b)** $2g(c)$ **(c)** $g(2c)$

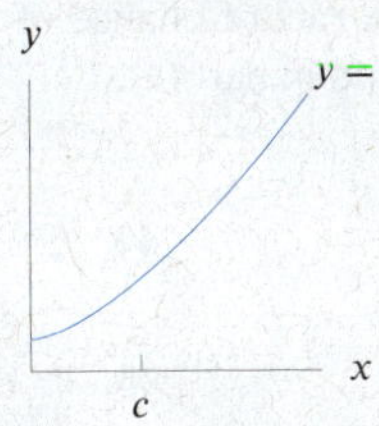


Figure 6.47

**40.** Figure 6.48 shows the graphs of $f(x)$ and $g(x)$. It is claimed that $f(x)$ is a horizontal stretch of $g(x)$. If that could be true, find the stretch. If that could not be true, explain.

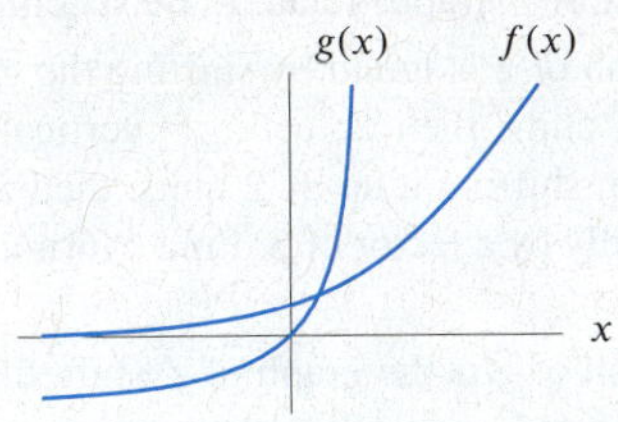


Figure 6.48

**41.** Find a formula for the function in Figure 6.50 as a transformation of the function $f$ in Figure 6.49.

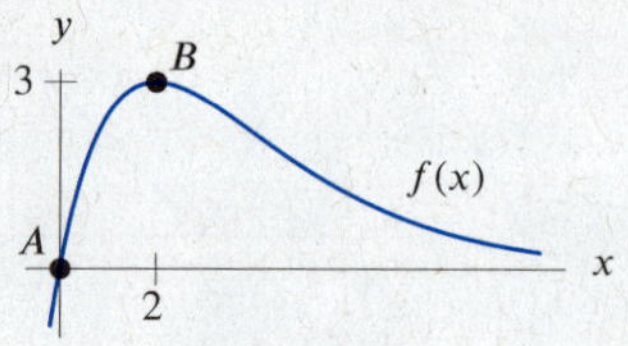


Figure 6.49

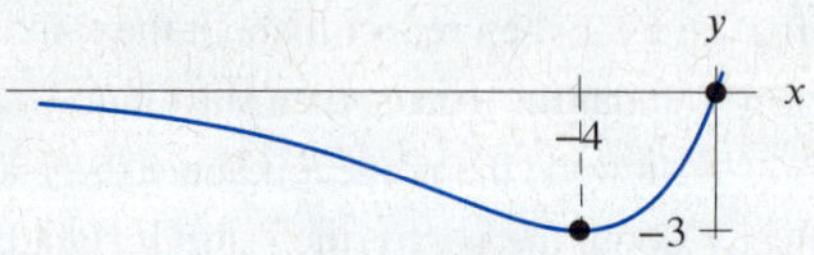


Figure 6.50

**42.** The *Heaviside step function*, $H(x)$, is defined as follows:

$$H(x) = 1 \text{ for } x \geq 0$$
$$H(x) = 0 \text{ for } x < 0.$$

Graph the Heaviside function, then sketch graphs of the following transformations of $H$:

**(a)** $y = H(x) - 2$ **(b)** $y = H(x + 2)$
**(c)** $y = -3H(-x) + 4$

The function $f(x)$ has domain $-6 \leq x \leq 2$, and the average rate of change of $f(x)$ over the interval $-6 \leq x \leq 2$ is $-3$. For each transformation of $f(x)$ in Problems 43–46,

**(a)** State the domain of the function.
**(b)** Give the average rate of change of the function over its new domain found in part (a).

**43.** $g(x) = f(2x)$ **44.** $h(x) = f\left(\frac{1}{4}x\right)$

**45.** $k(x) = f(x + 2)$ **46.** $m(x) = f(-x)$

**47.** Use shifts, reflections, and vertical stretches to graph each parabola without a calculator. Then write each equation in standard form:

**(a)** $y = (x + 3)^2 - 4$ **(b)** $y = -2(x + 1)^2 + 3$

**48.** **(a)** Graph $h(x) = -2x^2 - 8x - 8$.
**(b)** Compare the graphs of $h(x)$ and $f(x) = x^2$. How are these two graphs related? Be specific.

**49.** The graph of $g$ is found by shifting the graph of $f$ to the left by 4 units, then reflecting it vertically across the $x$-axis, then shifting it up by 2 units, then finally stretching it vertically by a factor of 3. Find a formula for $g$ in terms of $f$.

**50.** The graph of $g$ is the graph of $f$ shifted right by 3 units, then reflected vertically across the $x$-axis, then shifted down by 2 units, then stretched vertically by a factor of 2. Find a formula for $g$ in terms of $f$.

**51.** The graph of $g$ is found by shifting the graph of $f$ up by 2 units and left by 3 units. The graph of $h$ is found by reflecting the graph of $g$ horizontally, then reflecting it vertically and stretching it vertically by a factor of 2. Find a formula for $h$ in terms of $f$.

**52.** Which (if any) of the following transformations of $f$ is not the same as the others?

**(a)** Shift $f$ up by 3, then stretch it vertically by a factor of 2, then reflect it vertically.
**(b)** Reflect $f$ vertically, then shift it down by 3, then stretch it vertically by a factor of 2.
**(c)** Stretch $f$ vertically by a factor of 2, then shift it up by 6, then reflect it vertically.
**(d)** Reflect $f$ vertically, then stretch it vertically by a factor of 2, then shift it down by 6.

**53.** The function $d(t)$ graphed in Figure 6.51 gives the winter temperature in °F at a high school, $t$ hours after midnight.

**(a)** Describe in words the heating schedule for this building during the winter months.
**(b)** Graph $c(t) = 142 - d(t)$.
**(c)** Explain why $c$ might describe the cooling schedule for summer months.

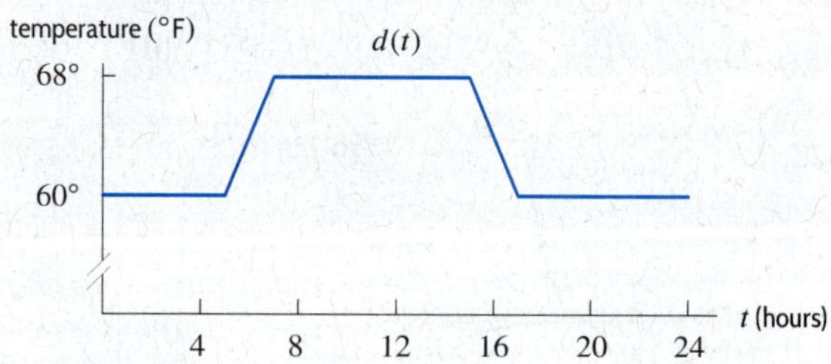


Figure 6.51

**54.** Let $D(t)$ denote the depth, in meters, of the water at a fixed location in a bay $t$ hours from noon. Find a formula for each function below by applying transformation(s) to $D(t)$.

**(a)** $f(t)$, the depth of the tide in kilometers.
**(b)** $g(x)$, the depth of the tide in kilometers $x$ minutes from noon.
**(c)** $j(x)$, the depth of the tide in kilometers $x$ minutes from 2 pm.

**55.** Let $f(x) = e^x$ and $g(x) = 5e^{x-2}$. If $g(x) = kf(x)$, find $k$.

**56.** The log function has the property that the graph resulting from a horizontal stretch can also be obtained by a vertical shift.

**(a)** Graph $f(x) = \log x$ and $g(x) = \log(10x)$ and determine the vertical shift.
**(b)** Explain how you could have predicted the answer to part (a) from the properties of logarithms.
**(c)** If $h(x) = \log(ax)$, what is the vertical shift $k$ making $h(x) = \log(x) + k$?

## CHAPTER SUMMARY

- **Vertical and Horizontal Shifts**
  Vertical: $y = g(x) + k$.
  Upward if $k > 0$; downward if $k < 0$.
  Horizontal: $y = g(x + k)$.
  Left if $k > 0$; right if $k < 0$.
- **Reflections**
  Across $x$-axis: $y = -f(x)$.
  Across $y$-axis: $y = f(-x)$.
- **Symmetry**
  About $y$-axis: $f(-x) = f(x)$; even function.
  About the origin: $f(-x) = -f(x)$; odd function.
- **Stretches and Compressions**
  Vertical: $y = kf(x)$. Stretch if $|k| > 0$; compress if $0 < |k| < 1$; reflect across $x$-axis if $k < 0$.
  Horizontal: $y = f(kx)$. Compress if $|k| > 0$; stretch if $0 < |k| < 1$; reflect across $y$-axis if $k < 0$.
- **Combining Transformations**
  Standard form: $y = Af(B(x - h)) + k$.
  Order of changes to graph of $f$:
  Horizontal stretch/compression by a factor of $1/|B|$
  Horizontal shift by $h$ units
  Vertical stretch/compression by a factor of $|A|$
  Vertical shift by $k$ units
  If $A < 0$, reflect about the $x$-axis.
  If $B < 0$, reflect about the $y$-axis.

## REVIEW EXERCISES AND PROBLEMS FOR CHAPTER SIX

### Exercises

**1.** Suppose $x = 2$. Determine the value of the input of the function $f$ in each of the following expressions:

**(a)** $f(2x)$ **(b)** $f(\frac{1}{2}x)$ **(c)** $f(x+3)$ **(d)** $f(-x)$

**2.** Determine the value of $x$ in each of the following expressions that leads to an input of 2 to the function $f$:

**(a)** $f(2x)$ **(b)** $f(\frac{1}{2}x)$ **(c)** $f(x+3)$ **(d)** $f(-x)$

**3.** The point $(2, 5)$ is on the graph of $y = f(x)$. Give the coordinates of one point on the graph of each of the following functions.

**(a)** $y = f(x - 4)$ **(b)** $y = f(x) - 4$
**(c)** $y = f(4x)$ **(d)** $y = 4f(x)$

**4.** The point $(-3, 4)$ is on the graph of $y = g(x)$. Give the coordinates of one point on the graph of each of the following functions.

**(a)** $y = g(\frac{1}{3}x)$ **(b)** $y = \frac{1}{3}g(x)$
**(c)** $y = g(-3x)$ **(d)** $y = -g(3x)$

Are the functions in Exercises 5–10 even, odd, or neither?

**5.** $a(x) = \dfrac{1}{x}$ **6.** $m(x) = \dfrac{1}{x^2}$

**7.** $e(x) = x + 3$ **8.** $p(x) = x^2 + 2x$

**9.** $b(x) = |x|$ **10.** $q(x) = 2^{x+1}$

**11.** Let $f(x) = 1 - x$. Evaluate and simplify:

**(a)** $f(2x)$ **(b)** $f(x+1)$ **(c)** $f(1-x)$
**(d)** $f(x^2)$ **(e)** $f(1/x)$ **(f)** $f(\sqrt{x})$

**12.** Fill in all the blanks in Table 6.16 for which you have sufficient information.

**Table 6.16**

| $x$ | −3 | −2 | −1 | 0 | 1 | 2 | 3 |
|---|---|---|---|---|---|---|---|
| $f(x)$ | −4 | −1 | 2 | 3 | 0 | −3 | −6 |
| $f(-x)$ | | | | | | | |
| $-f(x)$ | | | | | | | |
| $f(x) - 2$ | | | | | | | |
| $f(x - 2)$ | | | | | | | |
| $f(x) + 2$ | | | | | | | |
| $f(x + 2)$ | | | | | | | |
| $2f(x)$ | | | | | | | |
| $-f(x)/3$ | | | | | | | |

## Problems

In Problems 13–14, use Figure 6.52 to sketch the function.

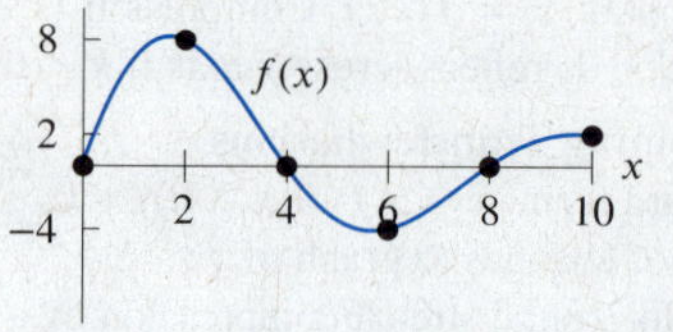


Figure 6.52

**13.** $y = f(x+2)+2$

**14.** $y = -2f(-x)$

In Problems 15–16, use Figure 6.52 to find a possible formula for the transformation of $f$ shown.

**15.**

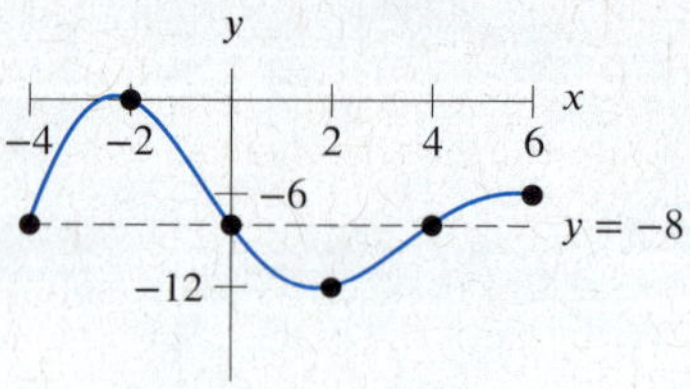


**16.**

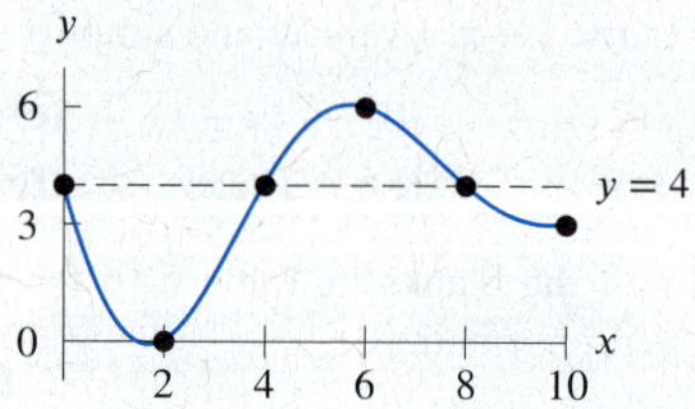


**17.** The function $f(x)$ contains the point $(-3, 1)$. The function $g(x)$ is obtained by applying a single transformation to the graph of $f(x)$ such that $g(x)$ contains the point $(3, 1)$. Describe the transformation that is applied to $f(x)$ in order to get the function $g(x)$ if

**(a)** $g(x)$ is obtained by applying only one reflection to the graph of $f(x)$.

**(b)** $g(x)$ is obtained by applying only one shift to the graph of $f(x)$.

**18.** Let $D(p)$ be the number of iced cappuccinos sold each week by a coffeehouse when the price is $p$ cents.

**(a)** What does the expression $D(225)$ represent?

**(b)** Do you think that $D(p)$ is an increasing function or a decreasing function? Why?

**(c)** What does the following equation tell you about $p$? $D(p) = 180$

**(d)** The coffeehouse sells $n$ iced cappuccinos when they charge the average price in their area, $t$ cents. Thus, $D(t) = n$. What is the meaning of the following expressions: $D(1.5t)$, $1.5D(t)$, $D(t+50)$, $D(t)+50$?

**19.** Without a calculator, match each of the functions (a)–(f) with one of the graphs (I)–(VI).

**(a)** $y = e^x$ **(b)** $y = e^{5x}$ **(c)** $y = 5e^x$
**(d)** $y = e^{x+5}$ **(e)** $y = e^{-x}$ **(f)** $y = e^x + 5$

(I)

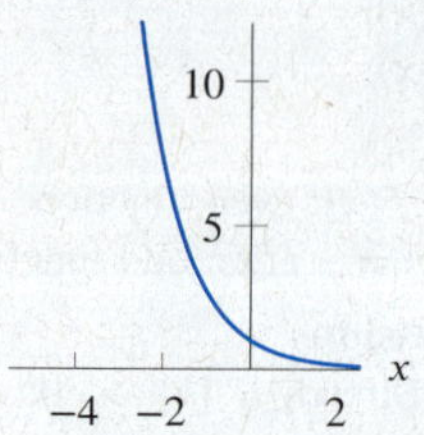


(II)

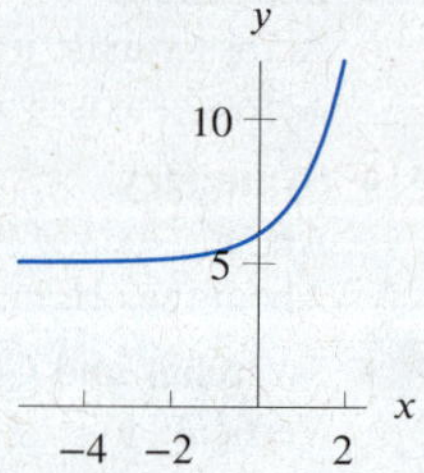


(III)

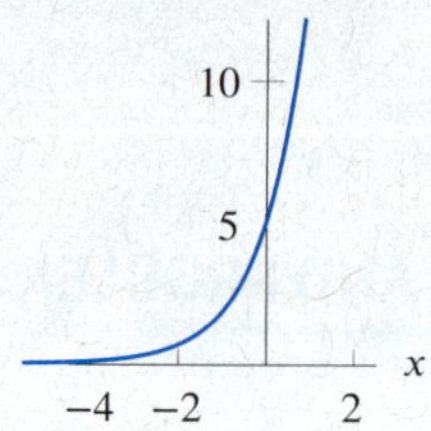


(IV)

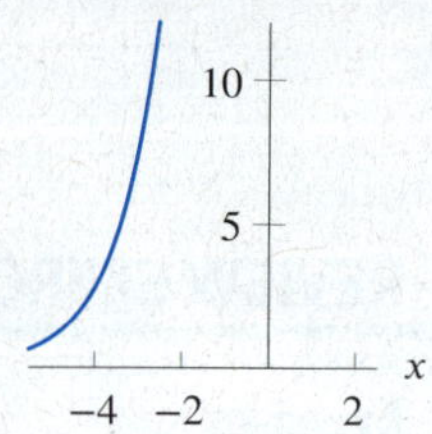


(V)

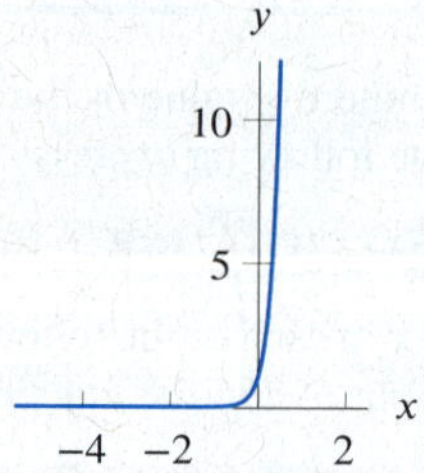


(VI)

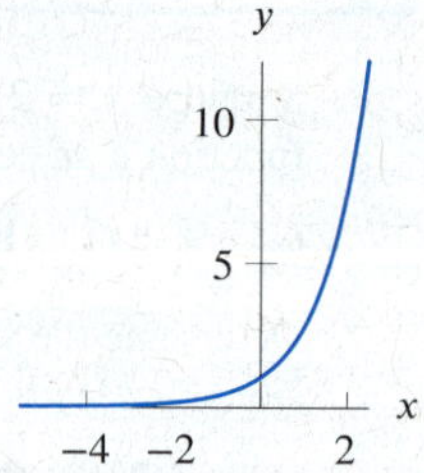


**20.** The graph in Figure 6.53 gives the number of hours of daylight in Charlotte, North Carolina on day $d$ of the year, where $d = 0$ is January 1. Graph the number of hours of daylight in Buenos Aires, Argentina, which is as far south of the equator as Charlotte is north. [Hint: When it is summer in the Northern Hemisphere, it is winter in the Southern Hemisphere.]

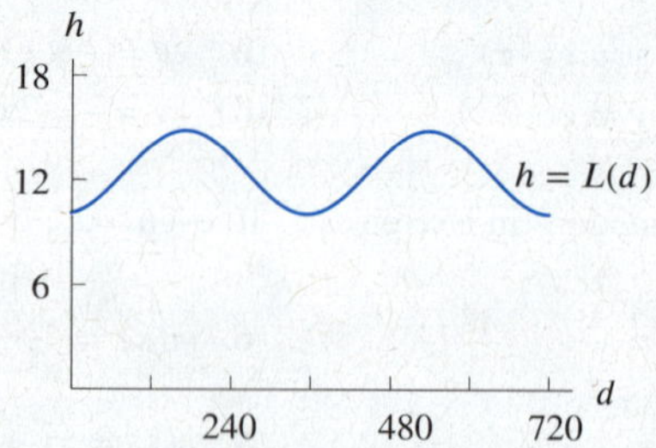


Figure 6.53

**21.** During a hurricane, a brick breaks loose from the top of a chimney, 38 feet above the ground. As the brick falls, its distance from the ground after $t$ seconds is given by:

$$d(t) = -16t^2 + 38.$$

**(a)** Find formulas for $d(t) - 15$ and $d(t - 1.5)$.
**(b)** On the same axes, graph $d(t)$, $d(t) - 15$, $d(t - 1.5)$.
**(c)** Suppose $d(t)$ represents the height of a brick which began to fall at noon. What might $d(t) - 15$ represent? $d(t - 1.5)$?
**(d)** Using algebra, determine when the brick hits the ground:
(i) If $d(t)$ represents the distance of the brick from the ground,
(ii) If $d(t) - 15$ represents the distance of the brick from the ground.
**(e)** Use one of your answers in part (d) to determine when the brick hits the ground if $d(t - 1.5)$ represents its distance above the ground at time $t$.

The functions graphed in Problems 22–23 are transformations of some basic function. Give a possible formula for each one.

**22.**

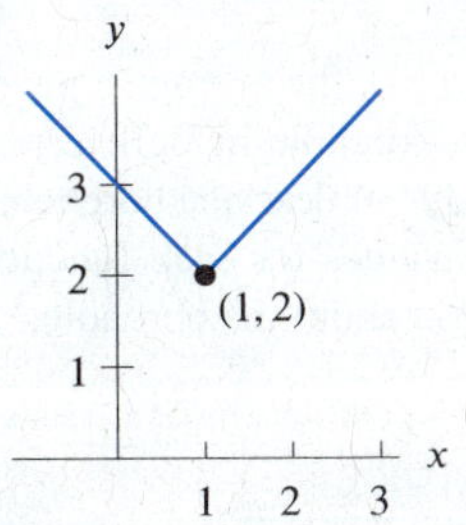


**23.**

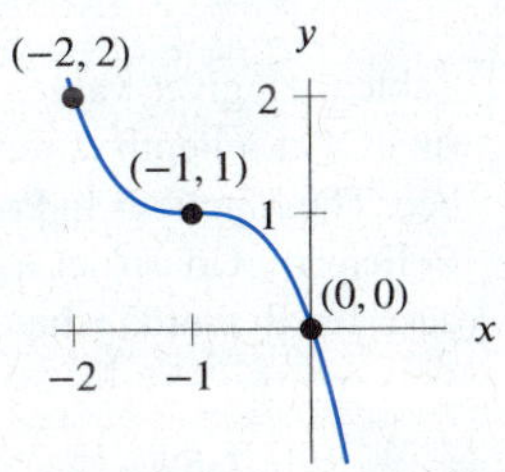


In Problems 24–25, use Figure 6.54 to find a formula for the transformations of $h(x)$.

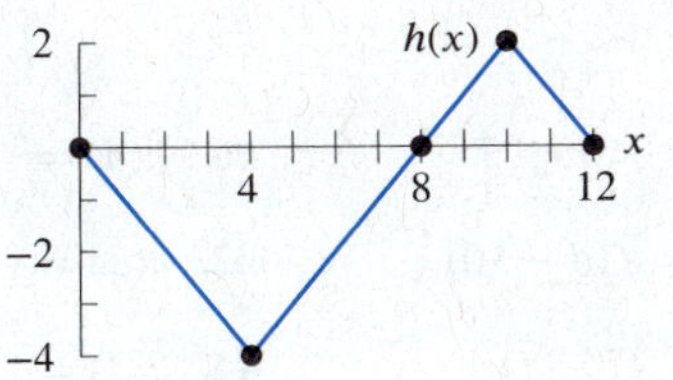


Figure 6.54

**24.**

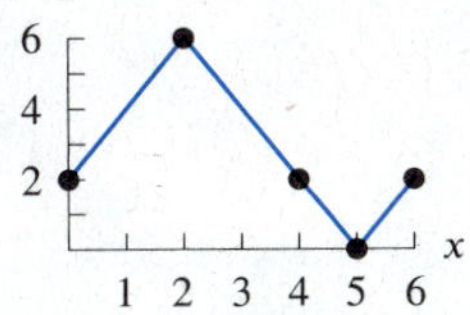


**25.**

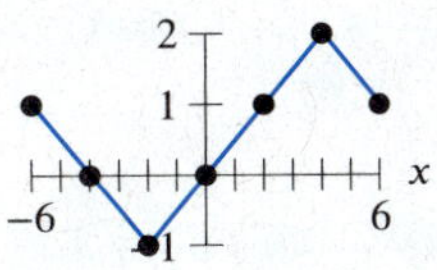


Problems 26–30 use Table 6.17, which gives the total cost, $C = f(n)$, for a carpenter to build $n$ wooden chairs.

Table 6.17

| $n$ | 0 | 10 | 20 | 30 | 40 | 50 |
|---|---|---|---|---|---|---|
| $f(n)$ | 5000 | 6000 | 6800 | 7450 | 8000 | 8500 |

**26.** Evaluate the following expressions. Explain in everyday terms what they mean.

**(a)** $f(10)$ **(b)** $f(x)$ if $x = 30$
**(c)** $z$ if $f(z) = 8000$ **(d)** $f(0)$

**27.** Find approximate values for $p$ and $q$ if $f(p) = 6400$ and $q = f(26)$.

**28.** Let $d_1 = f(30) - f(20)$, $d_2 = f(40) - f(30)$, and $d_3 = f(50) - f(40)$.

**(a)** Evaluate $d_1$, $d_2$ and $d_3$.
**(b)** What do these numbers tell you about the carpenter's cost of building chairs?

**29.** Graph $f(n)$. Label the quantities you found in Problems 26–28 on your graph.

**30.** The carpenter currently builds $k$ chairs per week.

**(a)** What do the following expressions represent?
(i) $f(k + 10)$ (ii) $f(k) + 10$
(iii) $f(2k)$ (iv) $2f(k)$

**(b)** If the carpenter sells his chairs at 80% above cost, plus an additional 5% sales tax, write an expression for his gross income (including sales tax) each week.

**31.** In Figure 6.55, the value of $d$ is labeled on the $x$-axis. Locate the following quantities on the $y$-axis:

**(a)** $g(d)$ **(b)** $g(-d)$ **(c)** $-g(-d)$

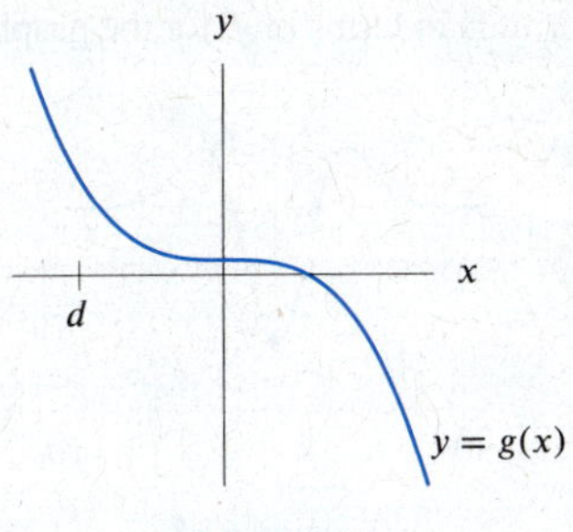


Figure 6.55

In Problems 32–34, use Figure 6.56 to find a formula for the graphs in terms of $h$.

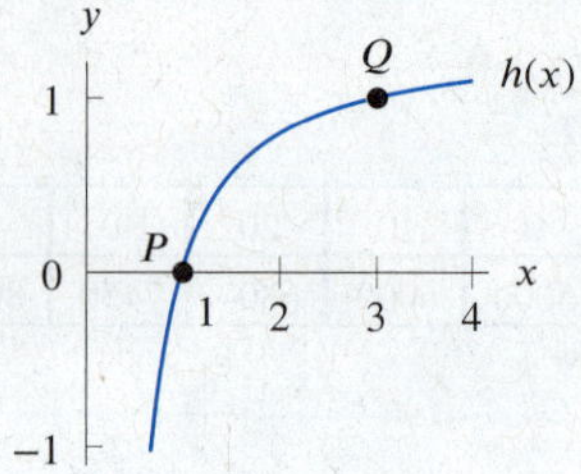


Figure 6.56

**32.**

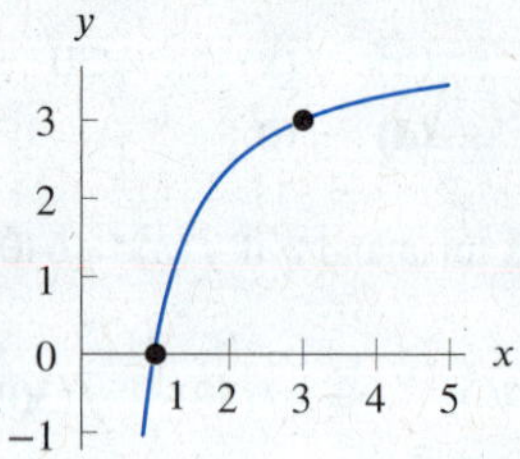

**33.**

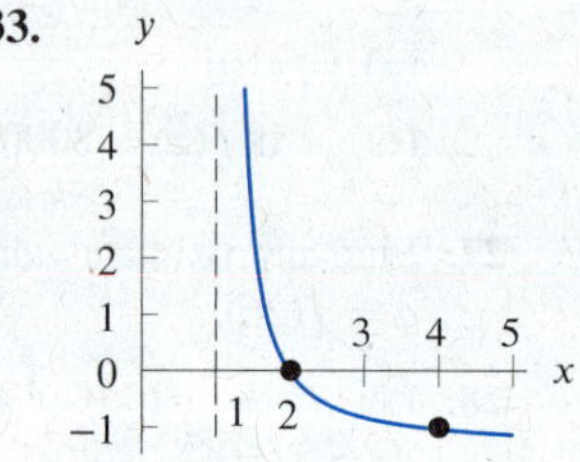

**34.**

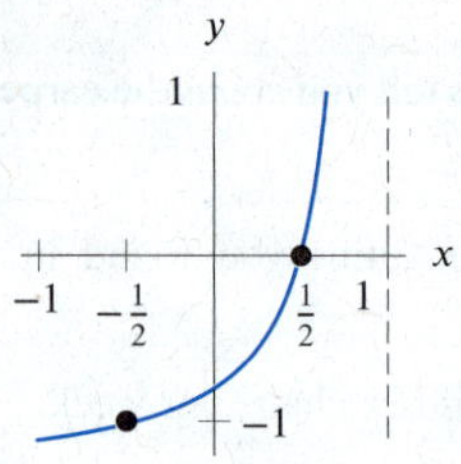

For Problems 35–36, use the graph of $y = f(x)$ in Figure 6.57.

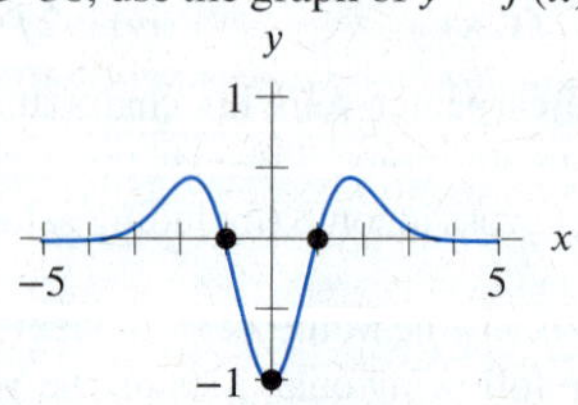

Figure 6.57

**35.** Graph $y = 2 - f(x - 2)$.

**36.** Find a formula in terms of $f$ for the graph in Figure 6.58.

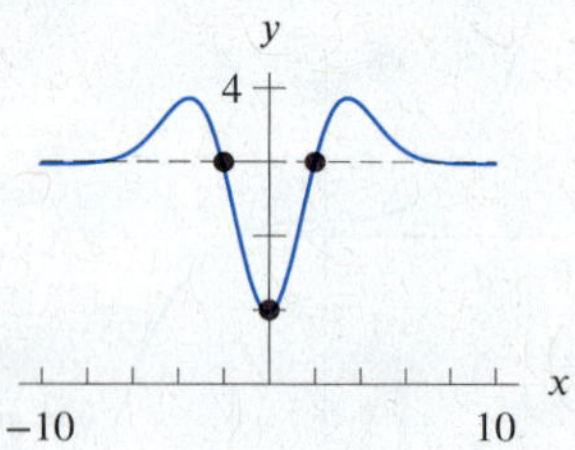

Figure 6.58

**37.** Gwendolyn, a pleasant parabola, was taking a peaceful nap when her dream turned into a nightmare: she dreamt that a low-flying pterodactyl was swooping toward her. Startled, she flipped over the horizontal axis, darted up (vertically) by three units, and to the left (horizontally) by two units. Finally she woke up and realized that her equation was $y = (x - 1)^2 + 3$. What was her equation before she had the bad dream?

**38.** Suppose $w = j(x)$ is the average daily quantity of water (in gallons) required by an oak tree of height $x$ feet.

**(a)** What does the expression $j(25)$ represent? What about $j^{-1}(25)$?

**(b)** What does the following equation tell you about $v$: $j(v) = 50$? Rewrite this statement in terms of $j^{-1}$.

**(c)** Oak trees are on average $z$ feet high and a tree of average height requires $p$ gallons of water. Represent this fact in terms of $j$ and then in terms of $j^{-1}$.

**(d)** Using the definitions of $z$ and $p$ from part (c), what do the following expressions represent?

$$j(2z),\quad 2j(z),\quad j(z+10),\quad j(z)+10,$$
$$j^{-1}(2p),\quad j^{-1}(p+10),\quad j^{-1}(p)+10.$$

Table 6.18 gives values of $T = f(d)$, the average temperature (in °C) at a depth $d$ meters in a borehole in Belleterre, Quebec. The functions in Problems 39–44 describe boreholes near Belleterre. Construct a table of values for each function and describe in words what it tells you about the borehole.[3]

Table 6.18

| $d$, depth (m) | 25 | 50 | 75 | 100 |
|---|---|---|---|---|
| $T$, temp (°C) | 5.5 | 5.2 | 5.1 | 5.1 |
| $d$, depth (m) | 125 | 150 | 175 | 200 |
| $T$, temp (°C) | 5.3 | 5.5 | 5.75 | 6 |

**39.** $g(d) = f(d) - 3$

**40.** $h(d) = f(d + 5)$

**41.** $m(d) = f(d - 10)$

**42.** $n(d) = 1.5f(d)$

**43.** $p(d) = f(0.8d)$

**44.** $q(d) = 1.5f(d) + 2$

[3]Hugo Beltrami of St. Francis Xavier University and David Chapman of the University of Utah posted this data at http://esrc.stfx.ca/borehole/node3.html, accessed December 20, 2005.

## STRENGTHEN YOUR UNDERSTANDING

Are the statements in Problems 1–19 true or false? Give an explanation for your answer.

1. If $g(x) = f(x) + k$ and $k$ is negative, the graph of $g(x)$ is the same as the graph of $f$, but shifted down.
2. Vertical and horizontal shifts are called translations.
3. The reflection of $y = x^2$ across the $x$-axis is $y = -x^2$.
4. If $f(x)$ is an odd function, then $f(x) = f(-x)$.
5. The graphs of odd functions are symmetric about the $y$-axis.
6. The graph of $y = -f(x)$ is the reflection of the graph of $y = f(x)$ across the $x$-axis.
7. The graph of $y = f(-x)$ is the reflection of the graph of $y = f(x)$ across the $y$-axis.
8. If the graph of a function $f$ is symmetric about the $y$-axis then $f(x) = f(-x)$.
9. If $g(x) = x^2 + 4$ then $g(x - 2) = x^2$.
10. For any function $f$, we have $f(x + k) = f(x) + k$.
11. Let $f(x) = 3^x$. If the graph of $f(x)$ is reflected across the $x$-axis and then shifted up four units, the new graph has the equation $y = -3^x + 4$.
12. If $q(p) = p^2 + 2p + 4$ then $-q(-p) = p^2 - 2p + 4$.
13. Multiplying a function by a constant $k$, with $k > 1$, vertically stretches its graph.
14. If $g(x) = kf(x)$, then on any interval the average rate of change of $g$ is $k$ times the average rate of change of $f$.
15. Figure 6.59 suggests that $g(x) = -2f(x + 1) + 3$.

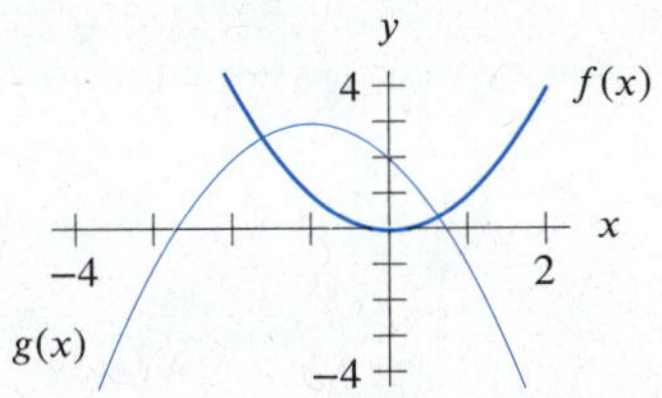

Figure 6.59

16. Using Table 6.19, we can conclude that if $g(x) = -\frac{1}{2}f(x + 1) - 3$, then $g(-2) = -10$.

Table 6.19

| $x$ | −3 | −2 | −1 | 0 | 1 | 2 | 3 |
|---|---|---|---|---|---|---|---|
| $f(x)$ | 10 | 6 | 4 | 1 | −2 | −4 | −10 |

17. Shifting the graph of a function up by one unit and then compressing it vertically by a factor of $\frac{1}{2}$ produces the same result as first compressing the graph by a factor of $\frac{1}{2}$ and then shifting it up by one unit.
18. Figure 6.60 suggests that $g(x) = 3f(\frac{1}{2}x)$.

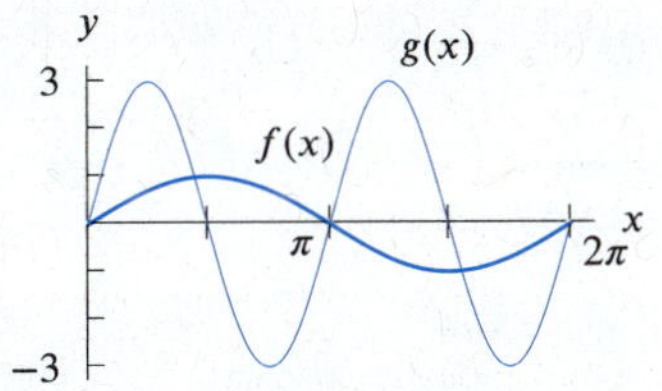

Figure 6.60

19. For the function given in Table 6.20, for $x = -2$, we have $3f(2x) + 1 = -2f(\frac{1}{2}x)$.

Table 6.20

| $x$ | −4 | −3 | −2 | −1 | 0 | 1 | 2 |
|---|---|---|---|---|---|---|---|
| $f(x)$ | 1 | 4 | 0 | −2 | 0 | 0 | −2 |

Chapter Seven

# TRIGONOMETRY AND PERIODIC FUNCTIONS

© Patrick Zephyr/Patrick Zephyr Nature Photography

## Contents

# 7.1 INTRODUCTION TO PERIODIC FUNCTIONS

## The London Eye Ferris Wheel

To celebrate the millennium, British Airways funded construction of the "London Eye," at that time the world's largest Ferris wheel.[1] The wheel is located on the south bank of the river Thames, in London, England, measures 400 feet in diameter, and carries up to 800 passengers in 32 capsules. It turns continuously, completing a single rotation once every 30 minutes. This is slow enough for people to hop on and off while it turns. The wheel rests on a boarding platform that is 50 feet above the ground.

## Ferris Wheel Height as a Function of Time

Suppose you hop on this Ferris wheel at time $t = 0$ and ride it for two full turns. Let $f(t)$ be your height above the platform, measured in feet as a function of $t$, the number of minutes you have been riding. We can find some values of $f(t)$.

Viewed from the boarding platform, the wheel turns in a counterclockwise direction. At time $t = 0$ you have just boarded the wheel, so your height is 0 ft above the platform (not counting the height of your seat). Thus, $f(0) = 0$. Since the wheel turns all the way around once every 30 minutes, after 7.5 minutes the wheel has turned one-quarter of the way around. Thinking of the wheel as a giant clock, this means you have been carried from the 6 o'clock position to the 3 o'clock position, as shown in Figure 7.1. You are now halfway up the wheel, or 200 feet above the platform, so $f(7.5) = 200$.

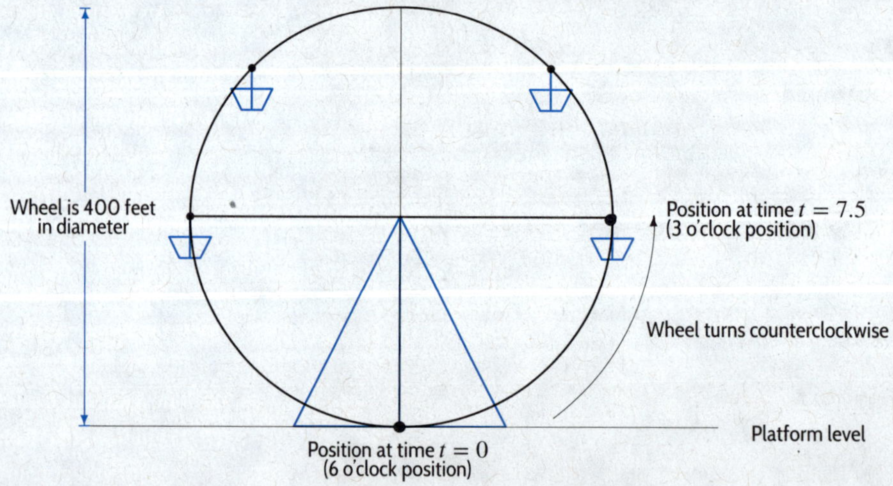


**Figure 7.1:** The London Eye Ferris wheel is 400 ft in diameter and turns around once every 30 minutes. Seats not drawn to scale

After 15 minutes, the wheel has turned halfway around, so you are now at the top, in the 12 o'clock position. Thus, $f(15) = 400$. And after 22.5 minutes, the wheel has turned three quarters of the way around, bringing you into the 9 o'clock position. You have descended from the top of the wheel halfway down to the platform, and you are once again 200 feet above the platform. Thus, $f(22.5) = 200$. (See Figures 7.2 and 7.3.) Finally, after 30 minutes, the wheel has turned all the way around, bringing you back to platform level, so $f(30) = 0$.

[1] http://en.wikipedia.org/wiki/London_Eye, accessed June 1, 2014.

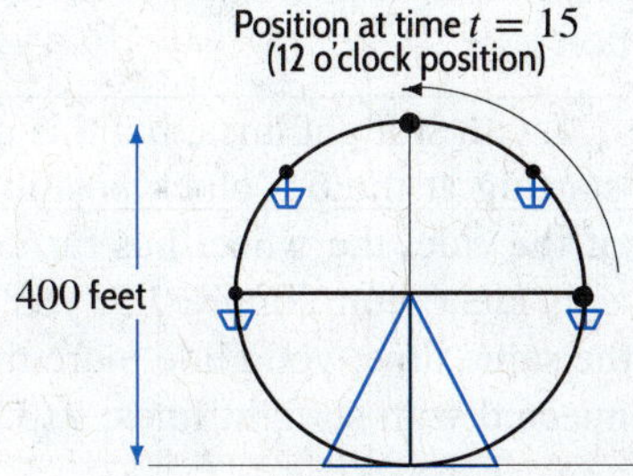


Figure 7.2: At time $t = 15$, the wheel has turned halfway around

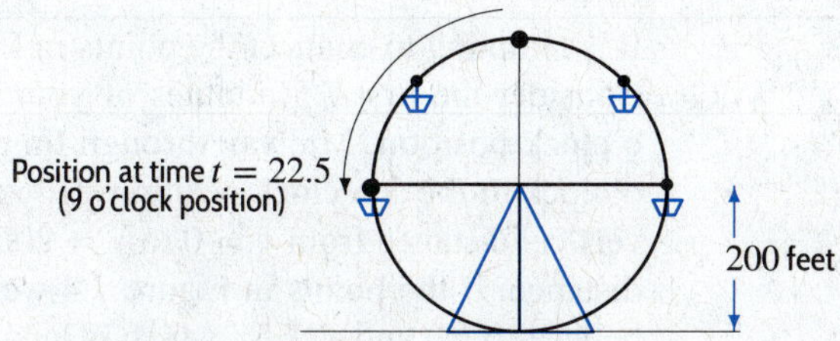


Figure 7.3: At time $t = 22.5$, the wheel has turned three quarters of the way around

### The Second Time Around on the Wheel and Later

Since the wheel turns without stopping, at time $t = 30$ it begins its second turn. Thus, at time $t = 37.5$, the wheel has again turned one quarter of the way around, and $f(37.5) = 200$. Similarly, at time $t = 45$ the wheel has again turned halfway around, bringing you back to the top. Likewise, at time $t = 52.5$, the wheel has carried you halfway back down to the platform. This means that $f(45) = 400$ and $f(52.5) = 200$. Finally, at time $t = 60$, the wheel has completed its second full turn and you are back at platform level, so $f(60) = 0$.

**Table 7.1** *Values of $f(t)$, your height above the platform t minutes after boarding the wheel*

| $t$ (minutes) | 0 | 7.5 | 15 | 22.5 | 30 | 37.5 | 45 | 52.5 | |
|---|---|---|---|---|---|---|---|---|---|
| $f(t)$ (feet) | 0 | 200 | 400 | 200 | 0 | 200 | 400 | 200 | |
| $t$ (minutes) | 60 | 67.5 | 75 | 82.5 | 90 | 97.5 | 105 | 112.5 | 120 |
| $f(t)$ (feet) | 0 | 200 | 400 | 200 | 0 | 200 | 400 | 200 | 0 |

### Repeating Values of the Ferris Wheel Function

Notice that the values of $f(t)$ in Table 7.1 begin repeating after 30 minutes. This is because the second turn is just like the first turn, except that it happens 30 minutes later. If you ride the wheel for more full turns, the values of $f(t)$ continue to repeat at 30-minute intervals.

## Graphing the Ferris Wheel Function

The data from Table 7.1 are plotted in Figure 7.4. The graph begins at $y = 0$ (platform level), rises to $y = 200$ (halfway up the wheel) and then to $y = 400$ (the top of the wheel). The graph then falls to $y = 200$ and then down to $y = 0$. This cycle then repeats itself three more times, once for each rotation of the wheel.

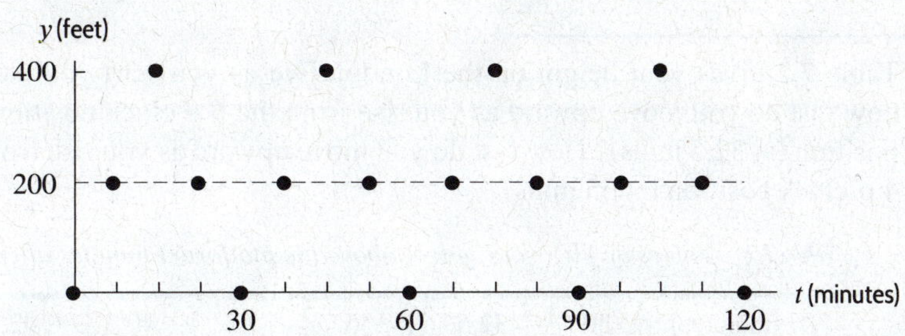


Figure 7.4: Values of $f(t)$, the Ferris wheel height function, at 7.5-minute intervals

### Filling in the Graph of the Ferris Wheel Function

It is tempting to connect the points in Figure 7.4 with straight lines, but this does not reflect reality. Consider the first 7.5 minutes of your ride, starting at the 6 o'clock position and ending at the 3 o'clock position. Midway through this part of the ride, the wheel has turned halfway from the 6 o'clock to the 3 o'clock position. However, as is clear from Figure 7.5, you rise less than half the vertical distance from $y = 0$ to $y = 200$. At the same time, you glide more than half the horizontal distance. If the points in Figure 7.4 were connected with straight lines, $f(3.75)$ would be halfway between $f(0)$ and $f(7.5)$, which is incorrect.

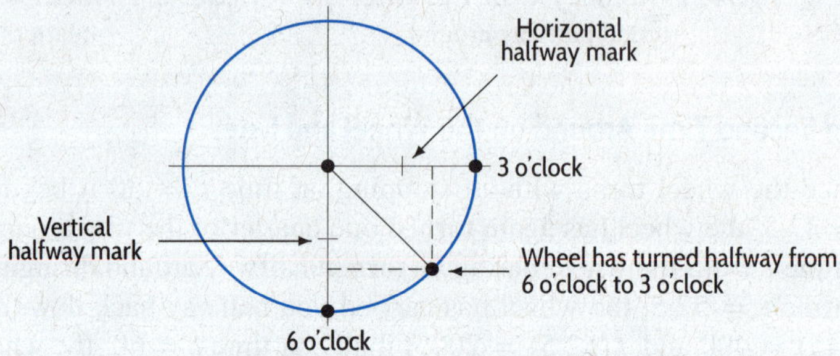


Figure 7.5: As the wheel turns half the way from 6 o'clock to 3 o'clock, you rise less than half the vertical distance but glide more than half the horizontal distance

The graph of $f(t)$ in Figure 7.6 is a smooth curve that more accurately reflects how the vertical distance changes over time. It looks the same from $t = 0$ to $t = 30$ as from $t = 30$ to $t = 60$, or from $t = 60$ to $t = 90$, or from $t = 90$ to $t = 120$.

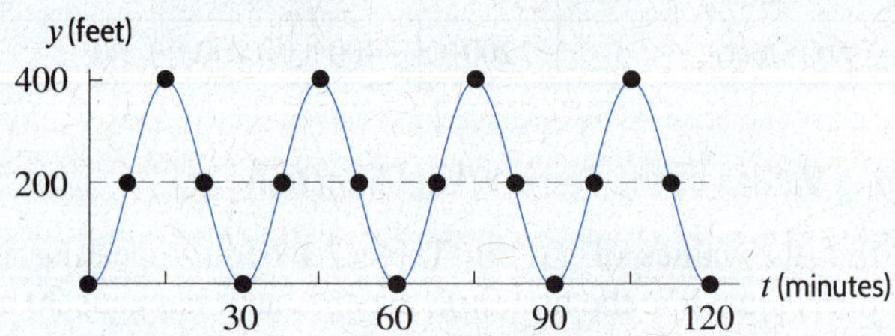


Figure 7.6: The graph of $y = f(t)$ is a smooth wave-shaped curve

## Average Rate of Change: How Fast We Move Up Depends on Where We Are

If the dots in Figure 7.4 were joined with straight lines, between two consecutive dots we would be moving the same distance upward (or downward) per unit time. But our upward (or downward) speed is not constant; it changes depending on where we are on our ride.

**Example 1** Table 7.2 gives your height on the London Eye as you start your second go-around. On average, how fast do you move upward as you rise from the 6 o'clock position (at 30 mins) to the 5 o'clock position (at 32.5 mins)? How fast do you move upward as you rise from the 5 o'clock position to the 4 o'clock position (at 35 mins)?

**Table 7.2** *Values of $f(t)$, your height above the platform t minutes after boarding the London Eye*

| $t$ (minutes) | 30 | 32.5 | 35 | 37.5 | 40 | 42.5 | 45 | 47.5 |
|---|---|---|---|---|---|---|---|---|
| $f(t)$ (feet) | 0 | 26.795 | 100 | 200 | 300 | 373.205 | 400 | 373.205 |

**Solution** To figure out how fast we move up over a certain time interval, we calculate the average rate of change in height, in feet per minute, over that interval. As we move near the platform,

$$\text{Average height increase from 6 to 5 o'clock} = \frac{f(32.5) - f(30)}{32.5 - 30} = \frac{26.795 - 0}{2.5} = 10.718 \text{ ft/min.}$$

As we move farther up,

$$\text{Average height increase from 5 to 4 o'clock} = \frac{f(35) - f(32.5)}{35 - 32.5} = \frac{100 - 26.795}{2.5} = 29.282 \text{ ft/min.}$$

The calculations in Example 1 illustrate that we rise more slowly on the Ferris wheel when we are near the platform than when we are slightly higher up. The upward concavity of the lower half of the graph in Figure 7.6 confirms this: Our vertical speed is increasing as we first begin to move up.

## Periodic Functions: Period, Midline, and Amplitude

The Ferris wheel function, $f$, is said to be *periodic*, because its values repeat on a regular interval or *period*. (Periodic functions are more formally defined in Section 7.5.) In Figure 7.7, the period is indicated by the horizontal gap between the first two peaks. The period represents the time elapsed in a full revolution of the wheel.

The dashed horizontal line in Figure 7.7 is the *midline* of the graph of $f$. In the context of the Ferris wheel, the midline represents the distance from the platform to the hub (the center) of the wheel. The vertical distance shown in Figure 7.7 between the first peak and the midline is called the *amplitude*. The amplitude represents the radius of the Ferris wheel.

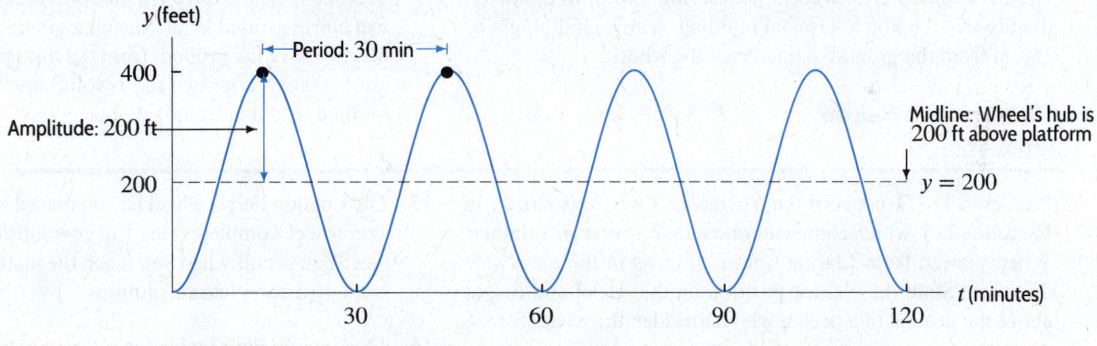


Figure 7.7: The graph of $y = f(t)$ showing the amplitude, period, and midline

Wave-shaped graphs such as Figure 7.7 arise from modeling many natural phenomena such as the length of a bobbing spring; see Problem 30. For such functions, we define:

- The **period** is the distance between consecutive peaks
- The **midline** is the horizontal line midway between the maximum and minimum values
- The **amplitude** is the vertical distance from midline to peak

## Exercises and Problems for Section 7.1

### Exercises

**1.** Which of the graphs in Figure 7.8 might represent periodic functions? You should list only those functions for which more than one full period is shown.

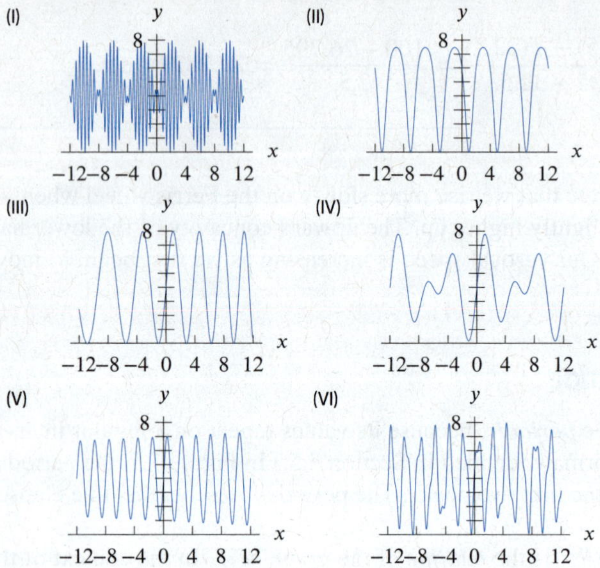


Figure 7.8

In Exercises 2–5, state the height above the ground of a person in the given position on the Singapore Flyer, until recently the world's largest Ferris wheel.[2] Measuring 150 m in diameter, the Flyer is set atop a terminal building, with a total height of 165 m from the ground to the top of the wheel.

**2.** 12 o'clock position

**3.** 3 o'clock position

**4.** 6 o'clock position

**5.** 9 o'clock position

In Exercises 6–9, estimate the period of the periodic functions.

**6.**

| $t$ | 0 | 1 | 2 | 3 | 4 | 5 | 6 |
|---|---|---|---|---|---|---|---|
| $f(t)$ | 12 | 13 | 14 | 12 | 13 | 14 | 12 |

**7.**

| $z$ | 1 | 11 | 21 | 31 | 41 | 51 | 61 | 71 | 81 |
|---|---|---|---|---|---|---|---|---|---|
| $g(z)$ | 5 | 3 | 2 | 3 | 5 | 3 | 2 | 3 | 5 |

**8.**

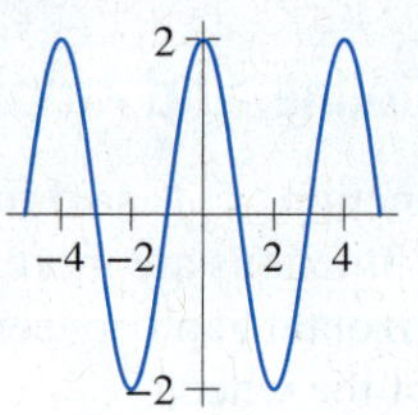


**9.**

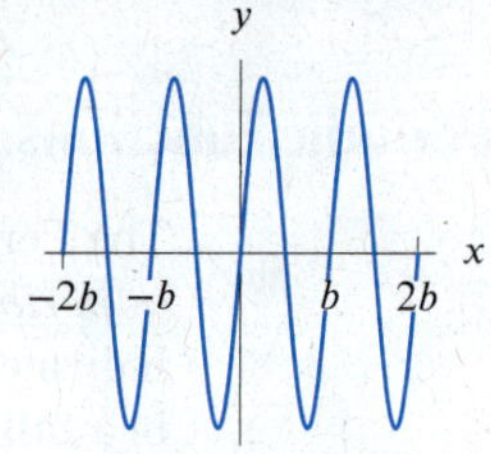


**10.** The Vegas High Roller in Las Vegas, Nevada, currently the world's largest Ferris wheel, is 520 feet in diameter.[3] The wheel completes one full revolution every 30 minutes and is boarded from a platform at its lowest point, 30 feet above ground level. Sketch a graph of $h = f(t)$, your height above the ground, from the time you board at $t = 0$ until you exit after two full revolutions. Show the period, midline, and amplitude on your graph.

### Problems

Problems 11–14 concern the Singapore Flyer, introduced in Exercises 2–5, which completes one rotation every 32 minutes. When viewed from Marina Centre, it turns in the clockwise direction. State the o'clock position on the wheel and height above the ground of a person who has ridden the wheel for the given time.

**11.** 8 minutes

**12.** 16 minutes

**13.** 24 minutes

**14.** 32 minutes

In Problems 15–17, you board the London Ferris wheel described in this section. Graph $h = f(t)$, your height in feet above the platform $t$ minutes after the wheel begins to turn. Label the period, the amplitude, and the midline of each graph, as well as both axes. In each case, first determine an appropriate interval for $t$, with $t \geq 0$.

**15.** The London Ferris wheel has increased its rotation speed. The wheel completes one full revolution every ten minutes. You get off when you reach the platform after having made two complete revolutions.

**16.** The London Ferris wheel is rotating at twice the speed as the wheel in Problem 15.

**17.** Everything is the same as Problem 15 (including the rotation speed) except the wheel has a 600-foot diameter.

In Problems 18–22, the table gives the height above the ground, $h$, of a passenger traveling on the Vegas High Roller, currently the largest Ferris wheel in the world with a 520-foot diameter.[4] Assume passengers board the wheel at its lowest

[2] http://en.wikipedia.org/wiki/List_of_Ferris_wheels, accessed February 12, 2014.
[3] http://en.wikipedia.org/wiki/High_Roller_(Ferris_wheel), accessed February 11, 2014.
[4] http://en.wikipedia.org/wiki/High_Roller_(Ferris_wheel), accessed February 11, 2014.

point (30 feet off the ground), time is measured in minutes since boarding, and a full ride takes 30 minutes. Write a difference quotient that represents the quantity asked, then calculate its value using data from the table. (Recall that heights on the way down are equal to heights on the way up.)

| $t$ | 0 | 2.5 | 5 | 7.5 | 10 | 12.5 | 15 |
|---|---|---|---|---|---|---|---|
| $h = f(t)$ | 30 | 64.8 | 160 | 290 | 420 | 515.2 | 550 |

18. The passenger's average vertical speed during her trip from the bottom to the top of the wheel.
19. The passenger's average vertical speed during the last 5 minutes of her trip up to the top of the wheel.
20. The passenger's average vertical speed during her trip from the top to the halfway-down point on the wheel.
21. The passenger's average vertical speed during the last 5 minutes of her ride down.
22. The passenger's average vertical speed during the last 2.5 minutes of her ride down.

Problems 23–25 involve different Ferris wheels. Graph $h = f(t)$ where $h$ is the height above ground (in meters) at time $t$, in minutes. Label the period, the amplitude, and the midline for each graph. In each case, first determine an appropriate interval for $t$, with $t \geq 0$.

23. A Ferris wheel is 20 meters in diameter and boarded from a platform that is 4 meters above the ground. The six o'clock position on the Ferris wheel is level with the loading platform. The wheel completes one full revolution every 2 minutes. At $t = 0$ you are in the twelve o'clock position. You then make two complete revolutions and any additional part of a revolution needed to return to the boarding platform.
24. A Ferris wheel is 35 meters in diameter and boarded at ground level. The wheel completes one full revolution every 5 minutes. At $t = 0$ you are in the three o'clock position and ascending. You then make two complete revolutions and return to the boarding platform.
25. A Ferris wheel is 50 meters in diameter and boarded from a platform that is 5 meters above the ground. The six o'clock position on the Ferris wheel is level with the loading platform. The wheel completes one full revolution every 8 minutes. You make two complete revolutions on the wheel, starting at $t = 0$.

The graphs in Problems 26–29 describe your height, $h = f(t)$, above the ground on different Ferris wheels, where $h$ is in meters and $t$ is time in minutes. You boarded the wheel before $t = 0$. For each graph, determine the following: your position and direction at $t = 0$, how long it takes the wheel to complete one full revolution, the diameter of the wheel, at what height above the ground you board the wheel, and the length of time the graph shows you riding the wheel. The boarding platform is level with the bottom of the wheel.

26.

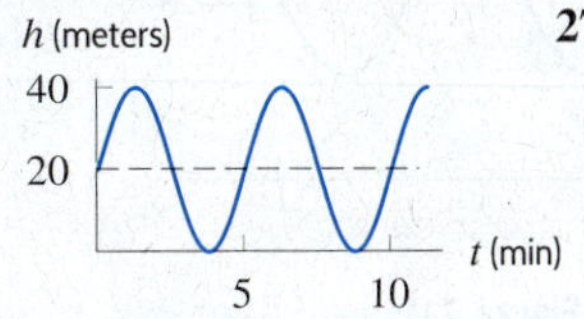


27.

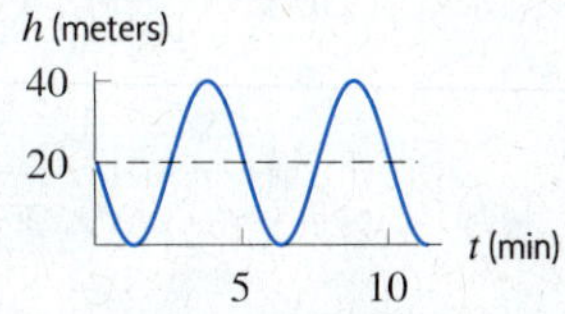


28.

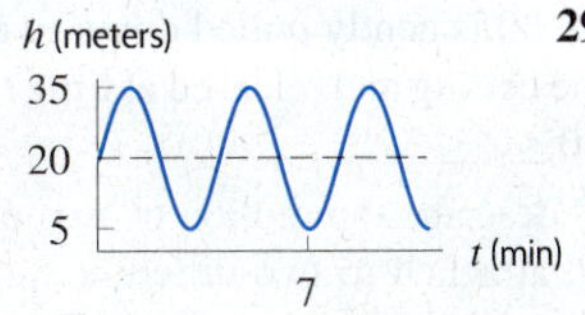


29.

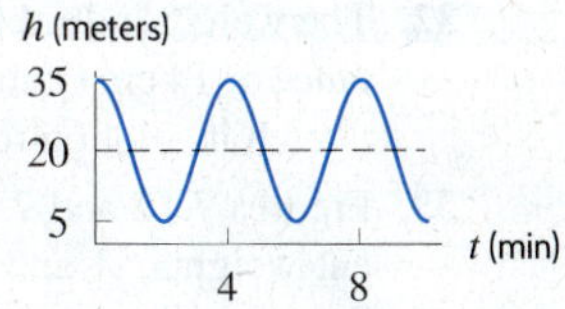


Problems 30–33 concern a weight suspended from the ceiling by a spring. (See Figure 7.9.) Let $d$ be the distance in centimeters from the ceiling to the weight. When the weight is motionless, $d = 10$. If the weight is disturbed, it begins to bob up and down, or *oscillate*. Then $d$ is a periodic function of $t$, time in seconds, so $d = f(t)$.

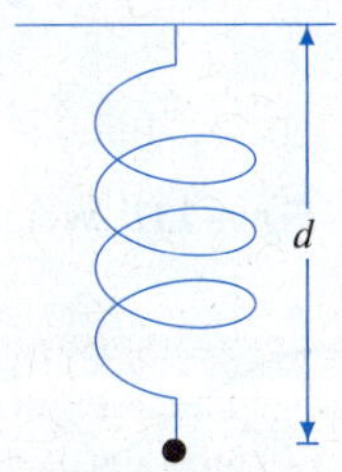


Figure 7.9

30. Determine the midline, period, amplitude, and the minimum and maximum values of $f$ from the graph in Figure 7.10. Interpret these quantities physically; that is, use them to describe the motion of the weight.

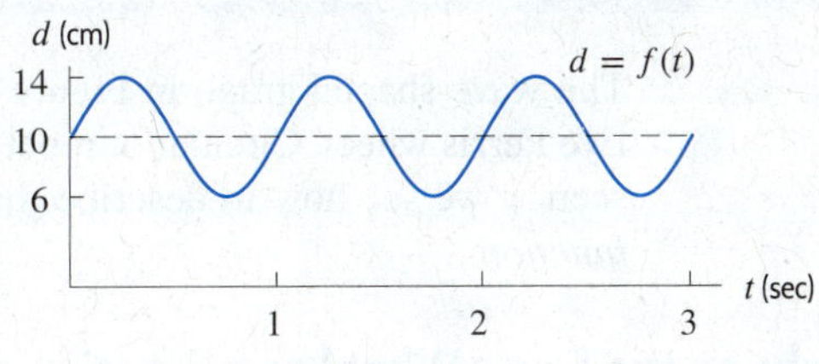


Figure 7.10

31. A new experiment with the same weight and spring is represented by Figure 7.11. Compare Figure 7.11 to Figure 7.10. How do the oscillations differ? For both figures, the weight was disturbed at time $t = -0.25$ and then left to move naturally; determine the nature of the initial disturbances.

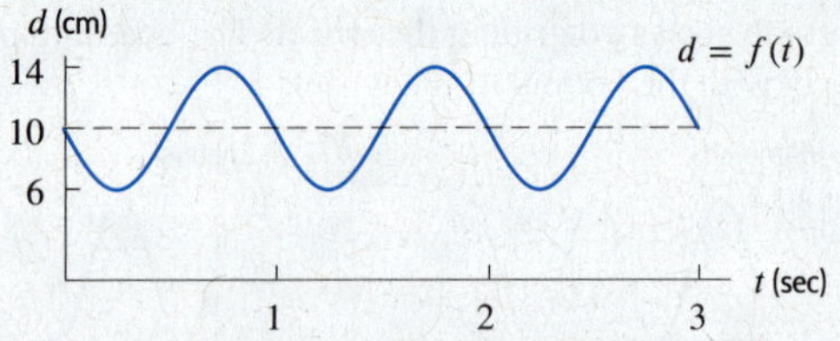


Figure 7.11

**32.** The weight in Problem 30 is gently pulled down to a distance of 14 cm from the ceiling and released at time $t = 0$. Sketch its motion for $0 \le t \le 3$.

**33.** Figures 7.12 and 7.13 describe the motion of two different weights, $A$ and $B$, attached to two different springs. Based on these graphs, which weight:

**(a)** Is closest to the ceiling when not in motion?
**(b)** Makes the largest oscillations?
**(c)** Makes the fastest oscillations?

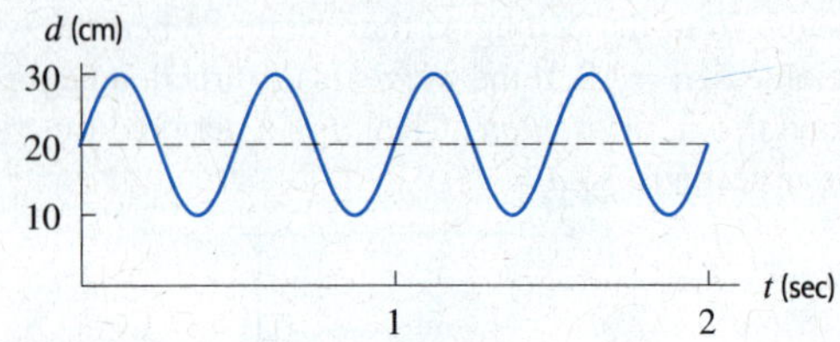


Figure 7.12: Weight $A$

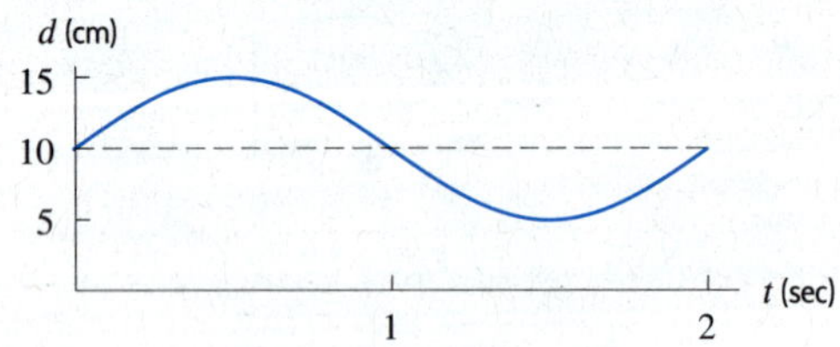


Figure 7.13: Weight $B$

**34.** The temperature of a chemical reaction oscillates between a low of 30°C and a high of 110°C. The temperature is at its lowest point when $t = 0$ and completes one cycle over a five-hour period.

**(a)** Sketch the temperature, $T$, against the elapsed time, $t$, over a ten-hour period.
**(b)** Find the period, the amplitude, and the midline of the graph you drew in part (a).

**35.** Table 7.3 gives the number of white blood cells (in 10,000s) in a patient with chronic myelogenous leukemia with nearly periodic relapses. Plot these data and estimate the midline, amplitude and period.

Table 7.3

| Day | 0 | 10 | 40 | 50 | 60 | 70 | 75 | 80 | 90 |
|---|---|---|---|---|---|---|---|---|---|
| WBC | 0.9 | 1.2 | 10 | 9.2 | 7.0 | 3.0 | 0.9 | 0.8 | 0.4 |
| Day | 100 | 110 | 115 | 120 | 130 | 140 | 145 | 150 | 160 |
| WBC | 1.5 | 2.0 | 5.7 | 10.7 | 9.5 | 5.0 | 2.7 | 0.6 | 1.0 |
| Day | 170 | 175 | 185 | 195 | 210 | 225 | 230 | 240 | 255 |
| WBC | 2.0 | 6.0 | 9.5 | 8.2 | 4.5 | 1.8 | 2.5 | 6.0 | 10.0 |

**36.** Use the height values in Table 7.2 on page 268 to compare each pair of quantities. Explain what each comparison tells you about a ride on the London Eye.

**(a)** $\dfrac{f(35) - f(32.5)}{35 - 32.5}$ and $\dfrac{f(37.5) - f(35)}{37.5 - 35}$

**(b)** $\dfrac{f(32.5) - f(30)}{32.5 - 30}$ and $\dfrac{f(47.5) - f(45)}{47.5 - 45}$

# 7.2 THE SINE AND COSINE FUNCTIONS

The wave-shaped graph in Figure 7.7 on page 269 arises from the circular motion of the London Eye Ferris wheel. Circular or repetitive motion often results in this sort of wavelike behavior. In this section we see how to describe this behavior using a new kind of function called a *trigonometric function*.

## Height on the Ferris Wheel as a Function of Position

The graph of $y = f(t)$ in Figure 7.7 shows your height above the platform as a function of the time spent riding the wheel. But we can also think of your height as a function of position on the wheel, not of time. For instance, if you are in the 3 o'clock position, we know your height is 200 ft.

### Using Angles to Measure Position on a Circle

The standard way to describe a position on a circle is to use angles, not clock positions. When working with angles, we adopt the following standard conventions:

### Conventions for Working with Angles

- We measure angles with respect to the horizontal, not the vertical, so that 0° describes the 3 o'clock position (the positive $x$-axis if the circle is centered at the origin).
- Positive angles are measured in the counterclockwise direction, negative angles in the clockwise direction.
- Large angles (greater than 360° or less than −360°) wrap around a circle more than once.

Note also that we often name angles using Greek letters like $\theta$ ("theta") and $\phi$ ("phi").

**Example 1** Sketch the following angles on the Ferris wheel: (a) 90° (b) −90° (c) 720° What clock position does each represent?

**Solution**
(a) This is a 90° counterclockwise turn from the 3 o'clock position, so it describes the 12 o'clock position. See Figure 7.14.
(b) This is a 90° clockwise turn from the 3 o'clock position, so it describes the 6 o'clock position. See Figure 7.15.
(c) This is a counterclockwise turn that wraps twice around the circle starting from the 3 o'clock position, so it describes the 3 o'clock position. See Figure 7.16.

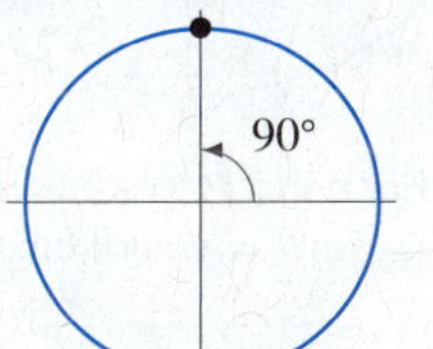


Figure 7.14: The angle $\theta = 90°$ specifies the 12 o'clock position

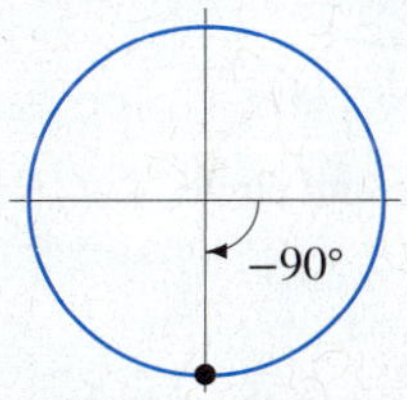


Figure 7.15: The angle $\theta = -90°$ specifies the 6 o'clock position

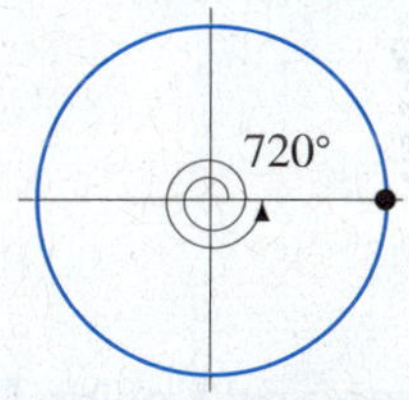


Figure 7.16: The angle $\theta = 720°$ specifies the 3 o'clock position

## Height on the Ferris Wheel as a Function of Angle

Since we can measure position on the Ferris wheel using angles, we see that:

$$\begin{pmatrix}\text{Height above platform}\\ \text{on Ferris wheel}\end{pmatrix} \text{ is a function of } \begin{pmatrix}\text{Angle position}\\ \text{on the wheel}\end{pmatrix}.$$

For example, in Table 7.1 on page 267, at time $t = 0$ you are in the 6 o'clock position, or at angle $\theta = -90°$. Likewise, after 7.5 minutes you are in the 3 o'clock position, or at angle $\theta = 0°$. Table 7.4 shows the heights in Table 7.1 rewritten as a function of angle, instead of time.

**Table 7.4** *Your height above the platform, y, as a function of θ, the angle turned through by the wheel*

| $\theta$ (degrees) | −90° | 0° | 90° | 180° | 270° | 360° | 450° | 540° | |
|---|---|---|---|---|---|---|---|---|---|
| $y$ (feet) | 0 | 200 | 400 | 200 | 0 | 200 | 400 | 200 | |
| $\theta$ (degrees) | 630° | 720° | 810° | 900° | 990° | 1080° | 1170° | 1260° | 1350° |
| $y$ (feet) | 0 | 200 | 400 | 200 | 0 | 200 | 400 | 200 | 0 |

In Table 7.1, the $y$-values repeat every 30 minutes. Similarly, in Table 7.4, the values of $y$ repeat every 360°. In both cases, the $y$-values repeat every time the wheel completes one full revolution.

## The Unit Circle

The heights in Table 7.4 depend on the radius of the Ferris wheel, 200 feet. Thus, Ferris wheels of different radii give rise to different periodic functions. Trigonometric functions, like the sine and cosine functions, are defined using a particular circle called the *unit circle*. The unit circle is the circle of radius one centered at the origin. (See Figure 7.17.) This circle gets its name from the fact that its radius measures exactly one unit.

We measure position on the unit circle just as we do on any other circle. In Figure 7.17, the angle $\theta = 0°$ determines the point $(1, 0)$, and the angle $\theta = 90°$ determines the point $(0, 1)$. The angle $\theta$ corresponds to the point $P$ with coordinates $(x, y)$.

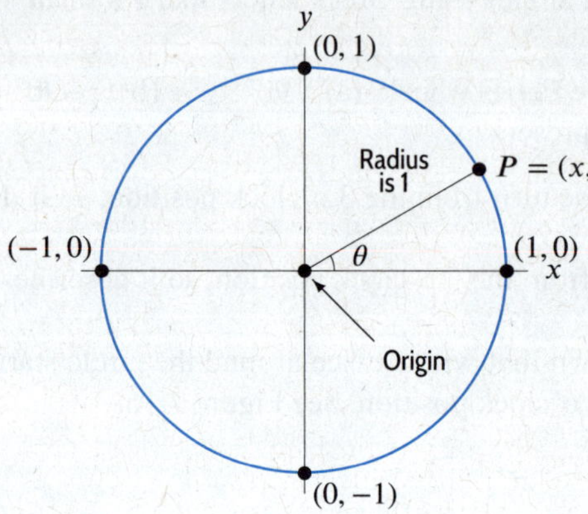


Figure 7.17: The unit circle

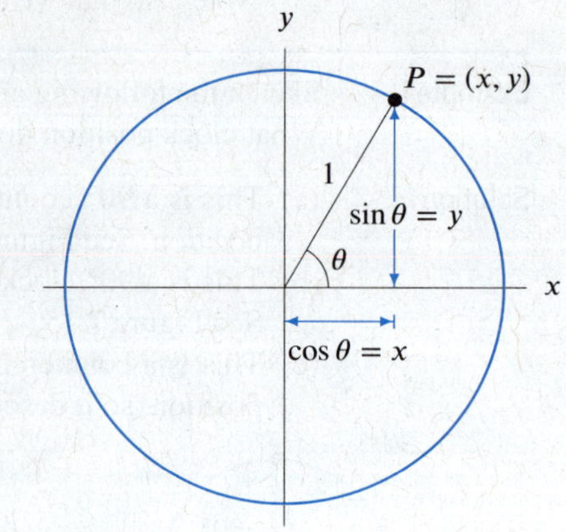


Figure 7.18: For point $P$, $\cos\theta = x$-coordinate and $\sin\theta = y$-coordinate

## The Sine and Cosine Functions

On the Ferris wheel the $y$-coordinate of your position is a function of the angle $\theta$. In general, we can think of the $x$- and $y$-coordinates of a point on a circle as functions of a corresponding angle $\theta$. With this in mind, we define the *sine* and *cosine functions* as follows: Given an angle $\theta$ that determines a point on the unit circle,

$$\text{Sine of } \theta = \text{the } y\text{-coordinate of the point on the unit circle}$$
$$\text{Cosine of } \theta = \text{the } x\text{-coordinate of the point on the unit circle.}$$

> Suppose $P = (x, y)$ in Figure 7.18 is the point on the unit circle specified by the angle $\theta$. We define the functions, **cosine** of $\theta$, or $\cos\theta$, and **sine** of $\theta$, or $\sin\theta$, by
>
> $$\cos\theta = x \qquad \text{and} \qquad \sin\theta = y.$$
>
> In other words, $\cos\theta$ is the $x$-coordinate of the point $P$ and $\sin\theta$ is the $y$-coordinate.

We sometimes put parentheses around the independent variable: $\cos(\theta)$ or $\sin(\theta + 1)$.

**Example 2** Find the values of $\sin\theta$ and $\cos\theta$ for $\theta = 0°, 90°, 180°, 270°$.

**Solution** We know that the angle $\theta = 0°$ specifies the point $P = (1, 0)$. Since $\sin 0°$ is the $y$-coordinate of this point and $\cos 0°$ is the $x$-coordinate, this means $\sin 0° = 0$ and $\cos 0° = 1$. Likewise, we know that $\theta = 90°$ specifies the point $P = (0, 1)$, so $\sin 90° = 1$ and $\cos 90° = 0$. Continuing, we can find the values of $\sin\theta$ and $\cos\theta$ for the other angles as follows:

| | | | | | |
|---|---|---|---|---|---|
| At $\theta = 0°$, | $P = (1, 0)$ | so | $\cos 0° = 1$ | and | $\sin 0° = 0$ |
| At $\theta = 90°$, | $P = (0, 1)$ | so | $\cos 90° = 0$ | and | $\sin 90° = 1$ |
| At $\theta = 180°$, | $P = (-1, 0)$ | so | $\cos 180° = -1$ | and | $\sin 180° = 0$ |
| At $\theta = 270°$, | $P = (0, -1)$ | so | $\cos 270° = 0$ | and | $\sin 270° = -1$. |

In principle, we can find values of $\sin\theta$ and $\cos\theta$ for any value of $\theta$. In practice, we rely on tables of values, or on calculators and computers, as in the next example.

**Example 3** (a) In Figure 7.19, find the coordinates of the point $Q$ on the unit circle.
(b) Find the lengths of the line segments labeled $m$ and $n$ in Figure 7.19.

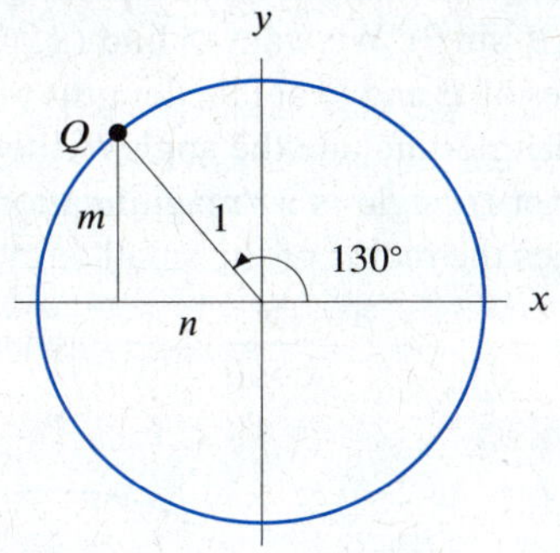


Figure 7.19: The point $Q$ designated by 130° on the unit circle

**Solution** (a) The coordinates of the point $Q$ are $(\cos 130°, \sin 130°)$. A calculator[5] gives, approximately,

$$\cos 130° = -0.643 \quad \text{and} \quad \sin 130° = 0.766.$$

(b) The length of line segment $m$ is the same as the $y$-coordinate of point $Q$, so length of $m = \sin 130° = 0.766$. The $x$-coordinate of $Q$ is negative because $Q$ is in the second quadrant. The length of line segment $n$ has the same magnitude as the $x$-coordinate of $Q$ but is positive. Thus, the length of $n = -\cos 130° = 0.643$.

## Symmetry and Reference Angles

In Figure 7.20, point $Q$ is in the second quadrant, so the symmetry of the unit circle gives

$$\cos 130° = -\cos 50° \quad \text{and} \quad \sin 130° = \sin 50°.$$

Similarly, in Figure 7.21, where point $R$ is in the third quadrant,

$$\cos 230° = -\cos 50° \quad \text{and} \quad \sin 230° = -\sin 50°.$$

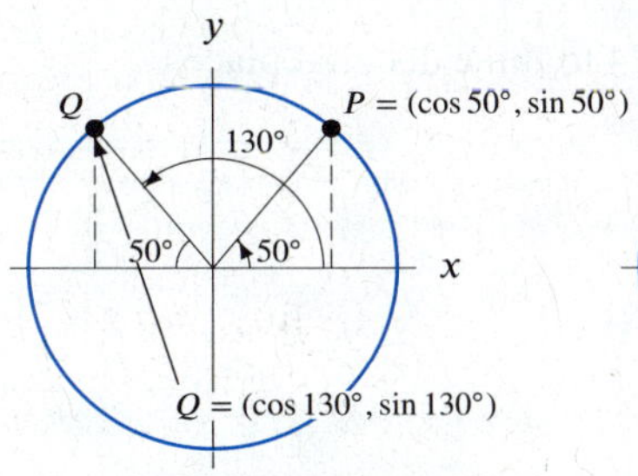


Figure 7.20: Angle 130° has a reference angle of 50°

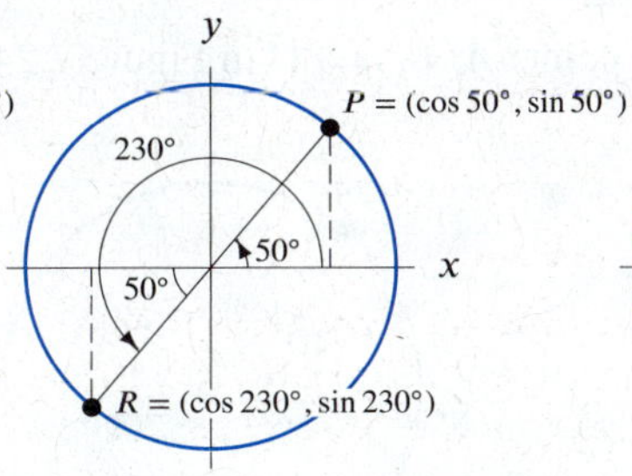


Figure 7.21: Angle 230° has a reference angle of 50°

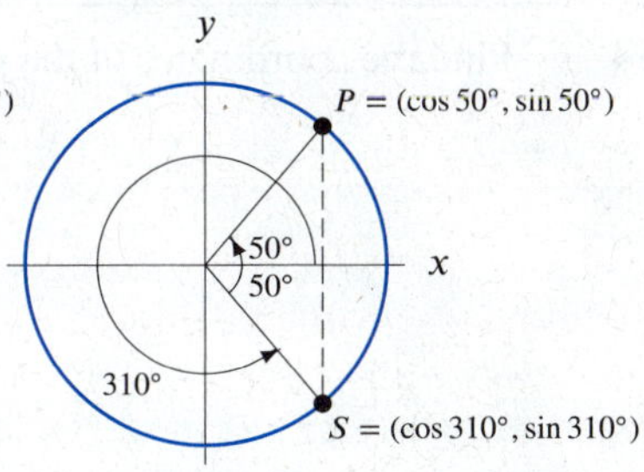


Figure 7.22: Angle 310° has a reference angle of 50°

[5] Many calculators can measure angles in more than one way, so be sure your calculator is in "degree mode."

In Figure 7.22, where point $S$ is in the fourth quadrant,

$$\cos 310^\circ = \cos 50^\circ \quad \text{and} \quad \sin 310^\circ = -\sin 50^\circ.$$

The angle 50° is called the *reference angle* for each of the angles 130°, 230°, 310°. Reference angles are measured to the nearest part of the $x$-axis and are always between 0 and 90 degrees. The sine and cosine of the original angle have the same magnitude as the sine or cosine of the reference angle, with the sign determined by the quadrant.

## Coordinates of a Point on a Circle of Radius $r$

Using the sine and cosine, we can find the coordinates of points on circles of any size. Figure 7.23 shows two concentric circles: the inner circle has radius 1 and the outer circle has radius $r$. The angle $\theta$ designates point $P$ on the unit circle and point $Q$ on the larger circle. We know that the coordinates of point $P$ are $(\cos\theta, \sin\theta)$. We want to find $(x, y)$, the coordinates of $Q$.

The coordinates of $P$ and $Q$ are the lengths of the sides of the two right triangles in Figure 7.23. Since both right triangles include the angle $\theta$, they are similar and their sides are proportional. This means that the larger triangle is a "magnification" of the smaller triangle. Since the radius of the large circle is $r$ times the radius of the small circle, we have

$$\frac{x}{\cos\theta} = \frac{r}{1} \quad \text{and} \quad \frac{y}{\sin\theta} = \frac{r}{1}.$$

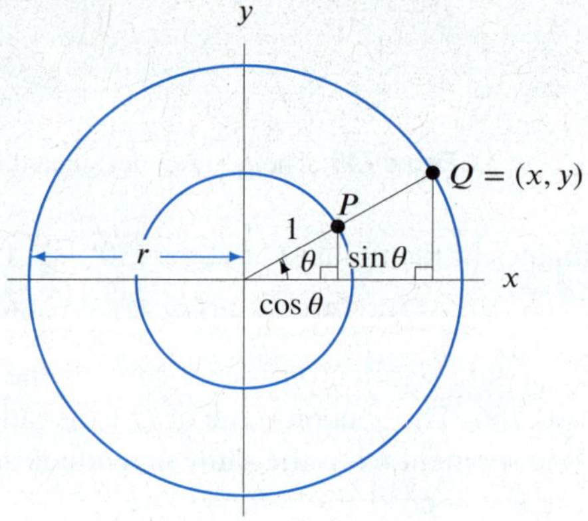


Figure 7.23: Points $P$ and $Q$ on circles of different radii specify the same angle $\theta$

Solving for $x$ and $y$ gives us the following result:

> The coordinates $(x, y)$ of the point $Q$ in Figure 7.23 are given by
>
> $$x = r\cos\theta \quad \text{and} \quad y = r\sin\theta.$$

**Example 4** Find the coordinates of the points $A$, $B$, and $C$ in Figure 7.24 to three decimal places.

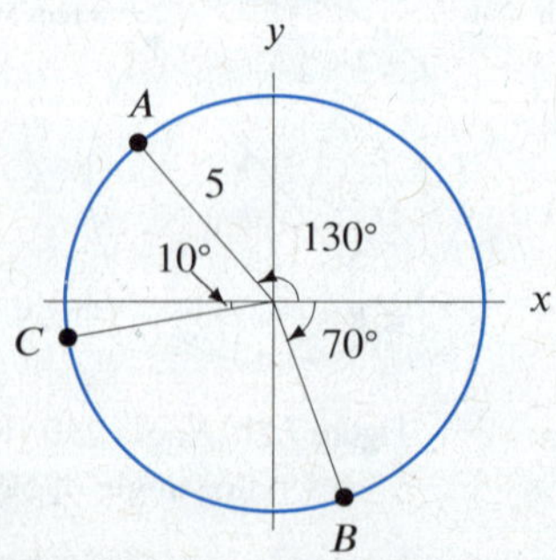


Figure 7.24: Finding coordinates of points on a circle of radius $r = 5$

**Solution** Since the circle has radius 5, the coordinates of point $A$ are given by

$$x = 5\cos 130° = 5(-0.6427) = -3.214,$$
$$y = 5\sin 130° = 5(0.766) = 3.830.$$

Point $B$ corresponds to an angle of $-70°$ (because the angle is measured clockwise), so $B$ has coordinates

$$x = 5\cos(-70°) = 5(0.342) = 1.710,$$
$$y = 5\sin(-70°) = 5(-0.93969) = -4.698.$$

For point $C$, we are given a reference angle, $10°$, measured from the negative $x$-axis. Since the point is in the third quadrant, the angle we want is $180° + 10° = 190°$, so

$$x = 5\cos(190°) = 5(-0.9848) = -4.924,$$
$$y = 5\sin(190°) = 5(-0.1736) = -0.868.$$

## Exercises and Problems for Section 7.2

### Exercises

**1.** Mark the following angles on a unit circle and give the coordinates of the point determined by each angle.

(a) 100° (b) 200° (c) −200°
(d) −45° (e) 1000° (f) −720°

In Exercises 2–4, what angle (in degrees) corresponds to the given number of rotations around the unit circle?

**2.** 4 **3.** −2 **4.** 16.4

In Exercises 5–8, find the reference angle for the given angle.

**5.** 100° **6.** 200° **7.** 50° **8.** 320°

**9.** Sketch the angles $\phi = 420°$ and $\theta = -150°$ as a displacement on a Ferris wheel, starting from the 3 o'clock position. What position on the wheel do these angles indicate?

In Exercises 10–18, find the coordinates of the point at the given angle on a circle of radius 3.8 centered at the origin.

**10.** 90° **11.** 180° **12.** −90°

**13.** −270° **14.** −540° **15.** 1426°

**16.** 1786° **17.** −10° **18.** −20°

In Exercises 19–20, evaluate $\sin\theta$ and $\cos\theta$ for the angle $\theta$.

**19.**

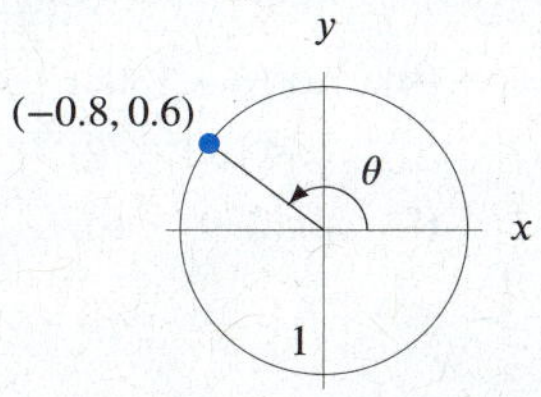


**20.**

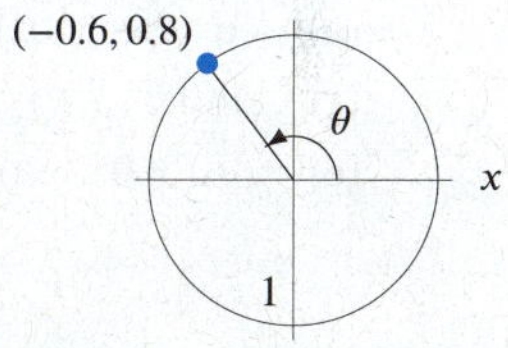


### Problems

In Problems 21–22, find four angles in different quadrants with the given reference angle.

**21.** 70° **22.** 15°

**23.** Find angles between 0° and 360° (but not 53°) that have the same

(a) Cosine as 53° (b) Sine as 53°

**24.** Find angles between 0° and 360° (but not 240°) that have the same

(a) Cosine as 240° (b) Sine as 240°

**25.** **(a)** Given that $P \approx (0.707, 0.707)$ is a point on the unit circle with angle 45°, estimate sin 135° and cos 135° without a calculator.

**(b)** Given that $Q \approx (0.259, 0.966)$ is a point on the unit circle with angle 75°, estimate sin 285° and cos 285° without a calculator.

**26.** Without using a calculator, rank the following numbers in order from smallest to largest.

$$\cos 20°,\ \cos 145°,\ \cos 35°,\ \cos 70°,\ \cos 150°$$

**27.** Without using a calculator, rank the following numbers in order from smallest to largest.

$$\sin 20°,\ \sin 145°,\ \sin 35°,\ \sin 70°,\ \sin 150°$$

**28.** For the angle $\phi$ shown in Figure 7.25, sketch each of the following angles.

**(a)** $180° + \phi$ **(b)** $180° - \phi$
**(c)** $90° - \phi$ **(d)** $360° - \phi$

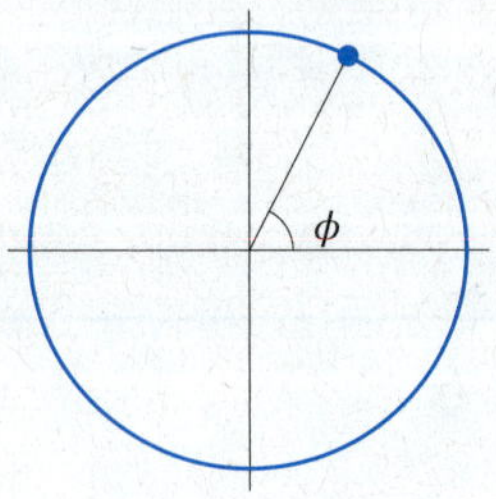


Figure 7.25

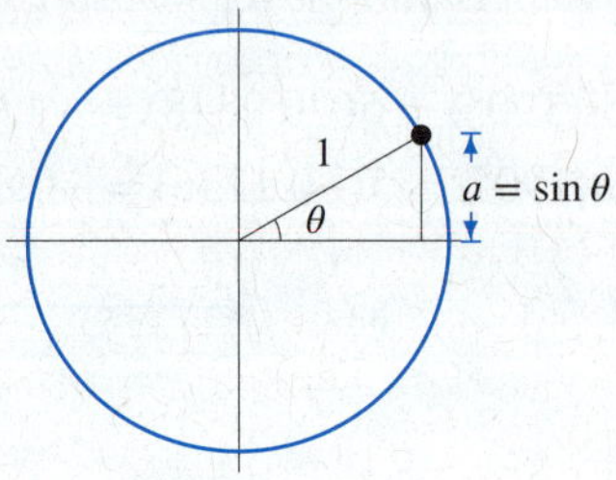


Figure 7.26

**29.** Let $\theta$ be an angle in the first quadrant with $\sin\theta = a$. (See Figure 7.26.) Evaluate the following in terms of $a$.

**(a)** $\sin(\theta + 360°)$ **(b)** $\sin(\theta + 180°)$
**(c)** $\cos(90° - \theta)$ **(d)** $\sin(180° - \theta)$
**(e)** $\sin(360° - \theta)$ **(f)** $\cos(270° - \theta)$

**30.** Let $\theta$ be an angle in the first quadrant with $\cos\theta = a$. (See Figure 7.27.) Evaluate the following in terms of $a$.

**(a)** $\cos(\theta + 360°)$ **(b)** $\cos(\theta + 180°)$
**(c)** $\sin(90° - \theta)$ **(d)** $\cos(180° - \theta)$
**(e)** $\cos(360° - \theta)$ **(f)** $\sin(270° - \theta)$

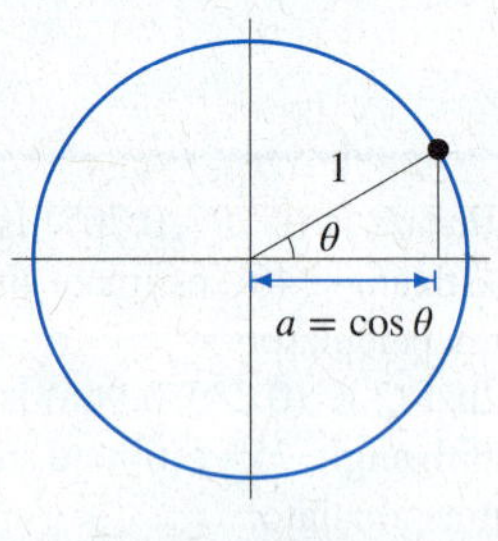


Figure 7.27

**31.** Explain in your own words the definition of $\sin\theta$ on the unit circle ($\theta$ in degrees).

**32.** The revolving door in Figure 7.28 rotates counterclockwise and has four equally spaced panels.

**(a)** What is the angle between two adjacent panels?
**(b)** What is the angle created by a panel rotating from $B$ to $A$?
**(c)** When the door is as shown in Figure 7.28, a person going outside rotates the door from $D$ to $B$. What is this angle of rotation?
**(d)** If the door is initially as shown in Figure 7.28, a person coming inside rotates the door from $B$ to $D$. What is this angle of rotation?
**(e)** The door starts in the position shown in Figure 7.28. Where is the panel at $A$ after three people enter and five people exit? Assume that people going in and going out do so in the manner described in parts (c) and (d), and that each person goes completely through the door before the next enters.

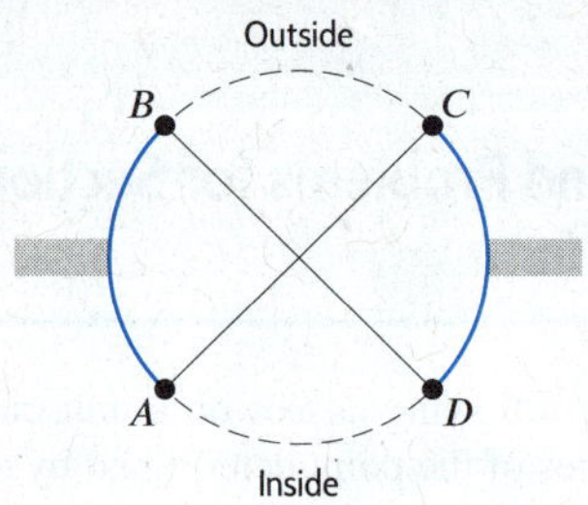


Figure 7.28

**33.** A revolving door (that rotates counterclockwise in Figure 7.29) was designed with five equally spaced panels for the entrance to the Pentagon. The arcs $BC$ and $AD$ have equal length.

**(a)** What is the angle between two adjacent panels?
**(b)** A four-star general enters by pushing on the panel at point $B$, and leaves the panel at point $D$. What is the angle of rotation?
**(c)** With the door in the position shown in Figure 7.29, an admiral leaves the Pentagon by pushing the panel between $A$ and $D$ to point $B$. What is the angle of rotation?

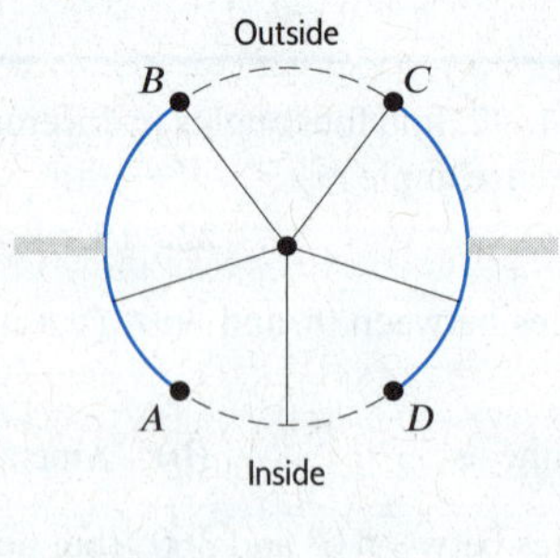


Figure 7.29

**34.** Calculate sin 45° and cos 45° exactly. Use the fact that the point $P$ corresponding to 45° on the unit circle, $x^2 + y^2 = 1$, lies on the line $y = x$.

**35.** **(a)** In Figure 7.30, what can be said about the lengths of the three sides of triangle $KOL$?
**(b)** What is the distance from $K$ to $P$?
**(c)** Using the Pythagorean theorem, find the distance from $O$ to $P$.
**(d)** What are the coordinates of point $K$?
**(e)** Using your results, determine sin 30° and cos 30°.
**(f)** Using the three sides in triangle $KOP$, find sin 60° and cos 60°.

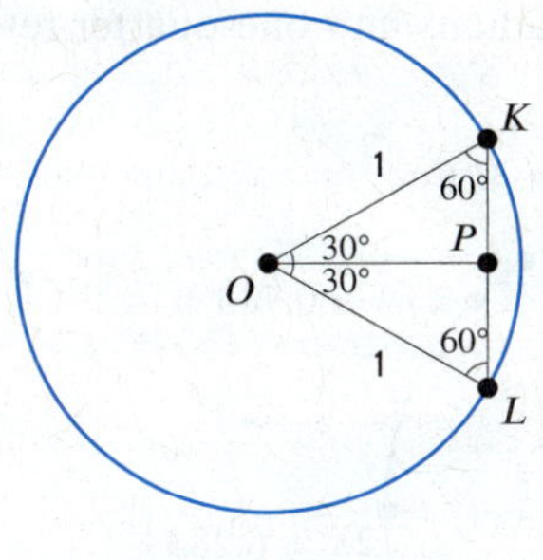


Figure 7.30

**36.** A circle of radius 5 is centered at the point $(-6, 7)$. Find a formula for $f(\theta)$, the $x$-coordinate of the point $P$ in Figure 7.31.

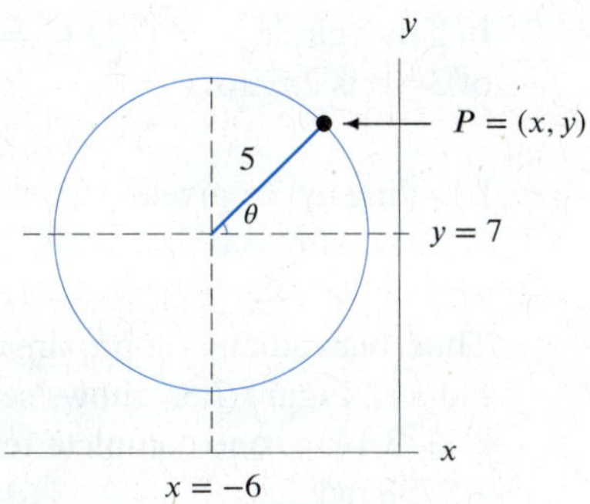


Figure 7.31

**37.** A DVD is 120 millimeters across with a center hole of diameter 15 millimeters. The center of the disc is at the origin. What are the coordinates of the points at which the inner and outer edge intersect the positive $x$-axis? What are the coordinates of the points at which the inner and outer edges cut a line making an angle $\theta$ with the positive $x$-axis?

# 7.3 RADIANS AND ARC LENGTH

In Section 7.2, we measured angles in degrees. Another way to measure angles, which is surprisingly helpful in calculus, is called *radian measure* and uses the arc length of a circle.

## Definition of a Radian

The arc length *spanned*, or cut off, by an angle is shown in Figure 7.32. If the radius of a circle is fixed (say the radius is 1), the arc length is completely determined by the angle $\theta$. (The radius and arc length must be measured in the same units.) Figure 7.33 illustrates radian measure:

> An angle of **1 radian** is defined to be the angle, in the counterclockwise direction, at the center of a unit circle, which spans an arc of length 1.

An angle of 2 radians cuts off an arc of length 2 (in the same units) in a unit circle; an angle of $-0.6$ radians is measured clockwise and cuts off an arc of length 0.6. See Figure 7.34. Note that radians are dimensionless units of measurement for angles (radians are not units of length).

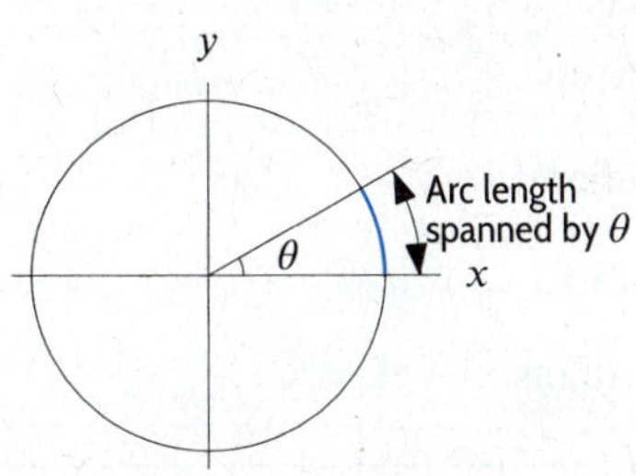


Figure 7.32: Arc length spanned by the angle $\theta$

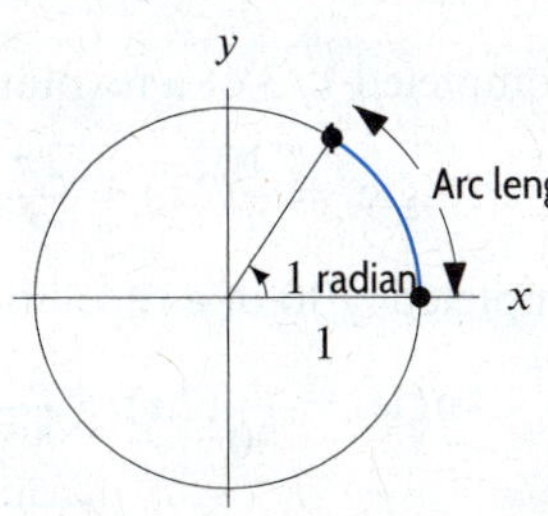


Figure 7.33: One radian cuts off an arc length of one in a unit circle

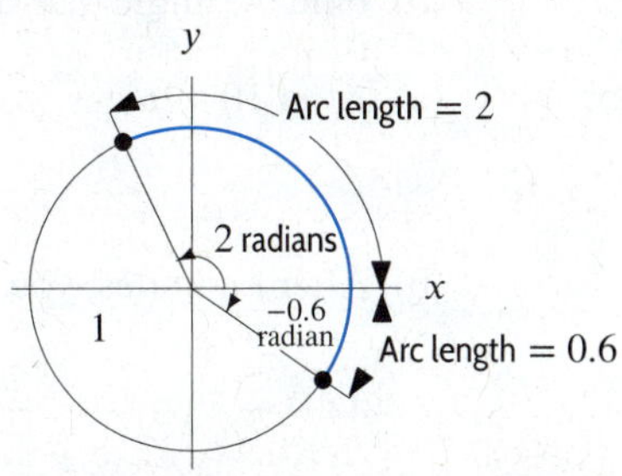


Figure 7.34: Angles of 2 radians and −0.6 radian

## Relationship Between Radians and Degrees

The circumference, $C$, of a circle of radius $r$ is given by

$$C = 2\pi r.$$

In a unit circle, $r = 1$, so $C = 2\pi$. This means that the arc length spanned by a complete revolution of 360° is $2\pi$, so

$$360^\circ = 2\pi \text{ radians}.$$

Dividing by $2\pi$ gives

$$1 \text{ radian} = \frac{360^\circ}{2\pi} \approx 57.296^\circ.$$

Thus, one radian is approximately 57.296°. One-quarter revolution, or 90°, is equal to $\frac{1}{4}(2\pi)$ or $\pi/2$ radians. Figure 7.35 shows several positions on the unit circle described in radian measure. Since $\pi \approx 3.1416$, one complete revolution is about 6.2832 radians and one-quarter revolution is about 1.5708 radians.

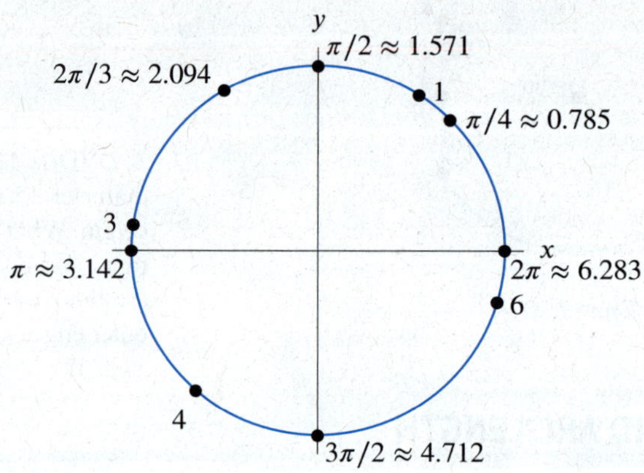


Figure 7.35: Angles in radian measure marked on circle

---

**Example 1** In which quadrant is an angle of 2 radians? An angle of 5 radians?

**Solution** Refer to Figure 7.35. The second quadrant includes angles between $\pi/2$ and $\pi$ (that is, between 1.571 and 3.142 radians), so an angle of 2 radians lies in the second quadrant. An angle of 5 radians is between 4.71 and 6.283, that is, between $3\pi/2$ and $2\pi$ radians, so an angle of 5 radians lies in the fourth quadrant.

---

**Example 2** The Ferris wheel described in Section 7.1 makes one rotation every 30 minutes.

(a) If you start at the 3 o'clock position, find the angle in radians that specifies your position after 10 minutes.

(b) Find the angle that specifies your position after $t$ minutes.

**Solution** (a) After 10 minutes, you have completed $1/3$ of a revolution, so

$$\text{Angle} = \frac{1}{3}(2\pi) = \frac{2\pi}{3} \text{ radians}.$$

(b) After $t$ minutes, you have completed $t/30$ of a revolution, so

$$\text{Angle} = \frac{t}{30}(2\pi) = \frac{\pi t}{15} \text{ radians}.$$

### Converting Between Degrees and Radians

To convert degrees to radians, or vice versa, we use the fact that $2\pi$ radians $= 360°$. So

$$1 \text{ radian} = \frac{180°}{\pi} \approx 57.296° \quad \text{and} \quad 1° = \frac{\pi}{180} \approx 0.01745 \text{ radians.}$$

Thus, to convert from radians to degrees, multiply the radian measure by $180°/\pi$ radians. To convert from degrees to radians, multiply the degree measure by $\pi$ radians$/180°$.

**Example 3** (a) Convert 3 radians to degrees. (b) Convert 3 degrees to radians.

**Solution**

(a) $3 \text{ radians} \cdot \dfrac{180°}{\pi \text{ radians}} = \dfrac{540°}{\pi} \approx 171.887°.$

(b) $3° \cdot \dfrac{\pi \text{ radians}}{180°} = \dfrac{\pi \text{ radians}}{60} \approx 0.052 \text{ radians.}$

The word "radians" is often dropped, so if an angle or rotation is referred to without units, it is understood to be in radians. When referring to an angle, we may see, for instance, $90° = \pi/2$ and $\pi = 180°$.

**Example 4** Find the arc length spanned by an angle of 30° in a circle of radius 1 meter.

**Solution** Converting 30° to radians, we have:

$$30° = 30° \cdot \frac{\pi}{180°} = \frac{\pi}{6}.$$

The definition of a radian tells us that the arc length spanned by an angle of $\pi/6$ in a circle of radius 1 meter is $\pi/6 \approx 0.524$ meter. See Figure 7.36.

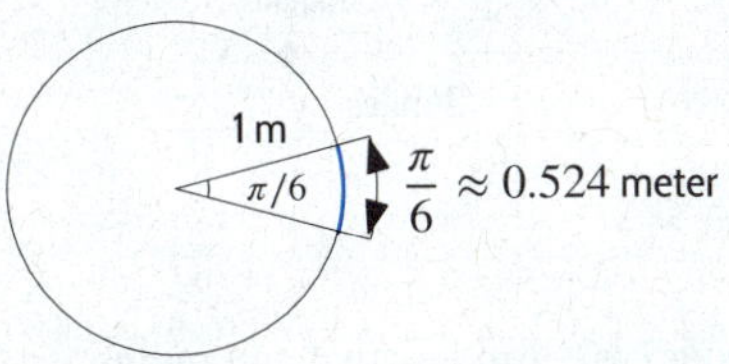


Figure 7.36: Arc length spanned by $30° = \pi/6$ radian in a unit circle

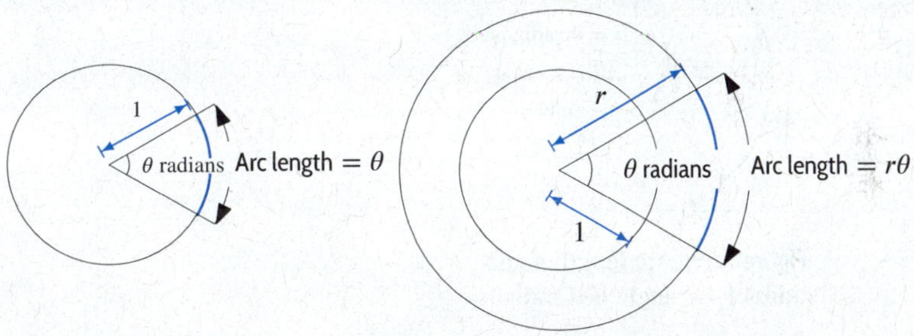


Figure 7.37: Arc length spanned by angle $\theta$ is proportional to the radius of the circle

## Arc Length in Circle of Radius $r$

Radians were defined using a unit circle, but they can be used to calculate arc length in circles of any radius. Figure 7.37 suggests that the arc length spanned by a fixed angle $\theta$ is proportional to the radius of the circle. If we go no more than once around the circle, it can be shown that

> The **arc length**, $s$, spanned in a circle of radius $r$ by an angle of $\theta$ in radians is
>
> $$s = r\theta.$$

**Example 5** What length of arc is cut off by an angle of 120° on a circle of radius 12 cm?

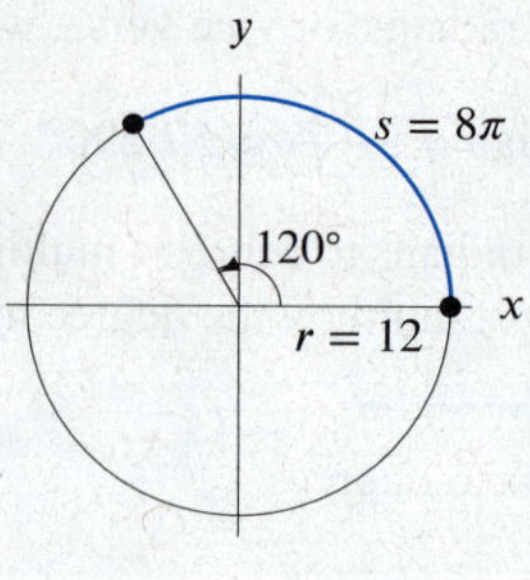


Figure 7.38: An arc cut off by 120°

**Solution** See Figure 7.38. Converting 120° to radians gives

$$\text{Angle} = 120° \cdot \frac{\pi}{180°} = \frac{2}{3}\pi \text{ radians,}$$

so

$$\text{Arc length} = 12 \cdot \frac{2}{3}\pi = 8\pi \text{ cm.}$$

**Example 6** You walk 4 miles around a circular lake. Give an angle in radians which represents your final position relative to your starting point if the radius of the lake is: (a) 1 mile (b) 3 miles

**Solution** (a) The path around the lake is a unit circle, so you have traversed an angle in radians equal to the arc length traveled, 4 miles. An angle of 4 radians is in the middle of the third quadrant relative to your starting point. (See Figure 7.39.)

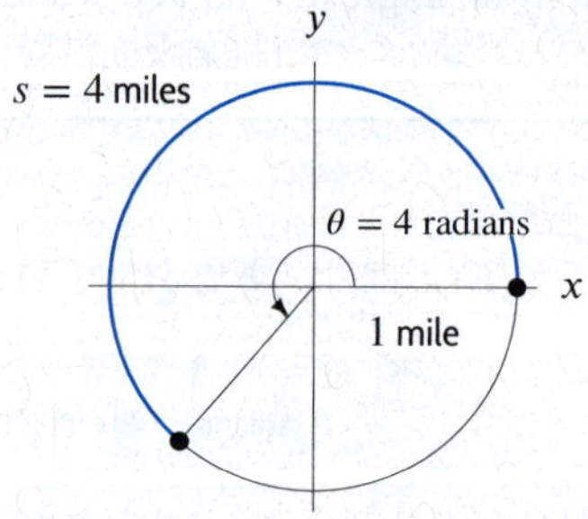


Figure 7.39: Arc length 4 and radius 1, so angle is 4 radians

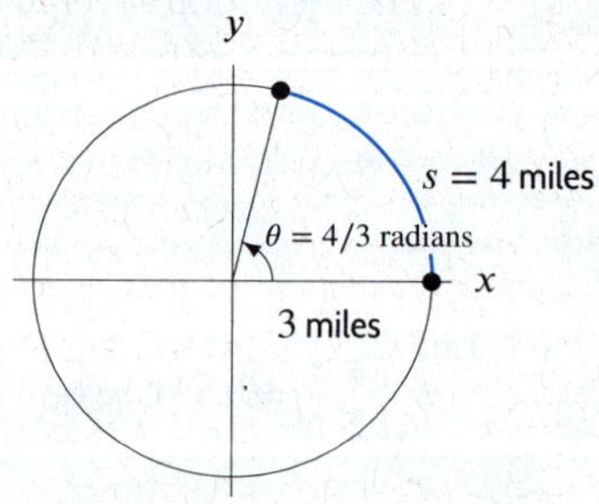


Figure 7.40: Arc length 4 and radius 3, so angle is 4/3 radians

(b) See Figure 7.40. To find $\theta$, use $s = r\theta$, giving

$$\theta = \frac{s}{r}.$$

The angle you have traversed is

$$\frac{s}{r} = \frac{4}{3} \approx 1.33 \text{ radians.}$$

One-quarter revolution is $\pi/2 \approx 1.571$ radians, so you are nearly one-quarter of the way around the lake.

## Sine and Cosine of a Number

We have defined the sine and cosine of an angle measured in degrees. For any real number $t$, we define $\cos t$ and $\sin t$ by interpreting $t$ as an angle of $t$ radians.

**Example 7** Evaluate: (a) $\cos 3.14°$ (b) $\cos 3.14$

**Solution** (a) Using a calculator in degree mode, we have $\cos 3.14^\circ = 0.9985$. This is reasonable, because a $3.14^\circ$ angle is quite close to a $0^\circ$ angle, so

$$\cos 3.14^\circ \approx \cos 0^\circ = 1.$$

(b) Here, 3.14 is not an angle measured in degrees; instead we interpret it as an angle of 3.14 radians. Using a calculator in radian mode, we have $\cos 3.14 = -0.99999873$. This is reasonable, because 3.14 radians is extremely close to $\pi$ radians or $180^\circ$, so

$$\cos 3.14 \approx \cos \pi = -1.$$

## Exact Values of the Sine and Cosine

In Example 2 on page 274, we found sine and cosine of $90^\circ$ and $180^\circ$. Because $90^\circ = \pi/2$ and $180^\circ = \pi$ we have

$$\cos \frac{\pi}{2} = 0, \quad \sin \frac{\pi}{2} = 1, \quad \cos \pi = -1, \quad \sin \pi = 0.$$

**Example 8** Evaluate $\sin \theta$ and $\cos \theta$ for $\theta = 0, 3\pi/2$, and $2\pi$.

**Solution** Figure 7.41 gives the coordinates of the points on the unit circle specified by $0, 3\pi/2$, and $2\pi$. We use these coordinates to evaluate the sines and cosines of these angles.

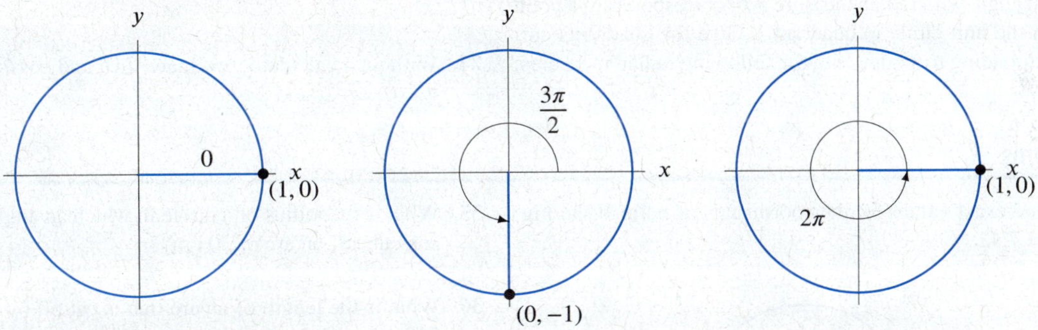


**Figure 7.41:** The coordinates of the points on the unit circle specified by the angles $0, 3\pi/2$, and $2\pi$

The definition of the cosine function tells us that $\cos 0$ is the $x$-coordinate of the point on the unit circle specified by the angle 0. Since the $x$-coordinate of this point is $x = 1$, we have

$$\cos 0 = 1.$$

Similarly, the $y$-coordinate of the point $(1, 0)$ is $y = 0$, which gives

$$\sin 0 = 0.$$

From Figure 7.41 we see that the coordinates of the remaining points are $(0, -1)$, and $(1, 0)$, so

$$\cos \frac{3\pi}{2} = 0, \quad \sin \frac{3\pi}{2} = -1, \quad \text{and} \quad \cos 2\pi = 1, \quad \sin 2\pi = 0.$$

In the skills section at the end of this chapter, triangles are used to compute the exact values of the sine and cosine of $30^\circ = \pi/6$, $45^\circ = \pi/4$, $60^\circ = \pi/3$. The results are:

$$\cos 30^\circ = \cos \frac{\pi}{6} = \frac{\sqrt{3}}{2} \qquad \cos 45^\circ = \cos \frac{\pi}{4} = \frac{1}{\sqrt{2}} \qquad \cos 60^\circ = \cos \frac{\pi}{3} = \frac{1}{2}$$

$$\sin 30^\circ = \sin \frac{\pi}{6} = \frac{1}{2} \qquad \sin 45^\circ = \sin \frac{\pi}{4} = \frac{1}{\sqrt{2}} \qquad \sin 60^\circ = \sin \frac{\pi}{3} = \frac{\sqrt{3}}{2}$$

# Exercises and Problems for Section 7.3

## Exercises

In Exercises 1–8 determine the radian measure of the angles.

**1.** 60° **2.** 45° **3.** 100° **4.** 17°

**5.** 150° **6.** 120° **7.** −270° **8.** $\pi$°

In Exercises 9–13, convert the angle given in radians to degrees.

**9.** $\frac{7}{2}\pi$ **10.** $5\pi$ **11.** 90 **12.** 2 **13.** 45

**14.** Convert each of the following angles to radians in two forms: as a multiple of $\pi$ and as a decimal approximation rounded to two decimal places.

**(a)** 30° **(b)** 120° **(c)** 200° **(d)** 315°

**15.** An angle with radian measure $\pi/4$ corresponds to a point on the unit circle in quadrant I. Give the quadrants corresponding to angles with the following radian measures:

**(a)** 1 radian **(b)** 2 radians **(c)** 3 radians
**(d)** 4 radians **(e)** 5 radians **(f)** 6 radians
**(g)** 7 radians **(h)** 8 radians **(i)** 9 radians
**(j)** 10 radians

In Exercises 16–19, what angle in radians corresponds to the given number of rotations around the unit circle?

**16.** 1 **17.** −2 **18.** 0.75 **19.** 4.27

In Exercises 20–23, find the arc length corresponding to the given angle on a circle of radius 6.2.

**20.** −180° **21.** 45° **22.** $\left(\frac{180}{\pi}\right)^{\circ}$ **23.** $a$°

**24.** Without a calculator, evaluate $\sin\theta$ and $\cos\theta$ for $\theta = 0, 3\pi/2$, and $2\pi$.

## Problems

**25.** Find exact values for the coordinates of point $W$ in Figure 7.42.

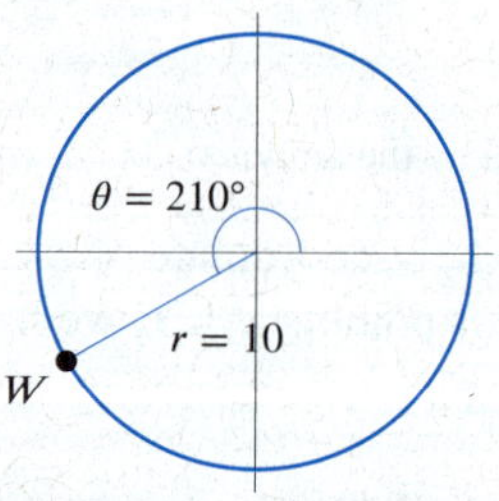


Figure 7.42

In Problems 26–31, find the exact coordinates of the point at the given angle on a circle of radius 5 centered at the origin.

**26.** 0° **27.** −180° **28.** 45°

**29.** 135° **30.** 225° **31.** 315°

In Problems 32–34, find exact values without a calculator.

**32.** $\sin\left(\frac{2\pi}{3}\right)$ **33.** $\cos\left(\frac{3\pi}{4}\right)$ **34.** $\cos\left(\frac{11\pi}{6}\right)$

**35.** What is the radius of a circle in which an angle of 3 radians cuts off an arc of 30 cm?

**36.** What is the length of an arc that is cut off by an angle of 225° in a circle of radius 4 feet?

**37.** What is the length of an arc cut off by an angle of 2 radians on a circle of radius 8 inches?

**38.** What is the angle determined by an arc of length $2\pi$ meters on a circle of radius 18 meters?

In Problems 39–44, where possible give the radius $r$, the measure of $\theta$ in both degrees and radians, the arc length $s$, and the coordinates of point $P$.

**39.**

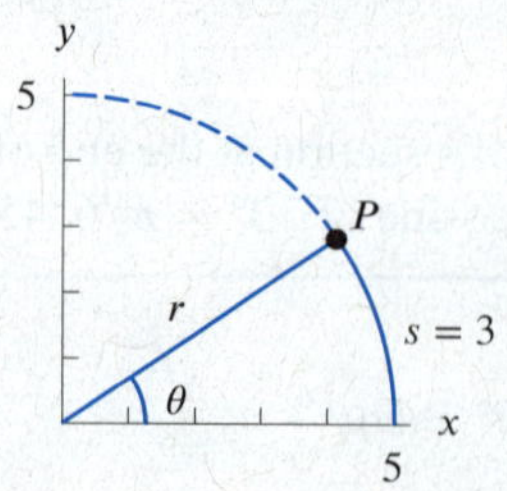


**40.**

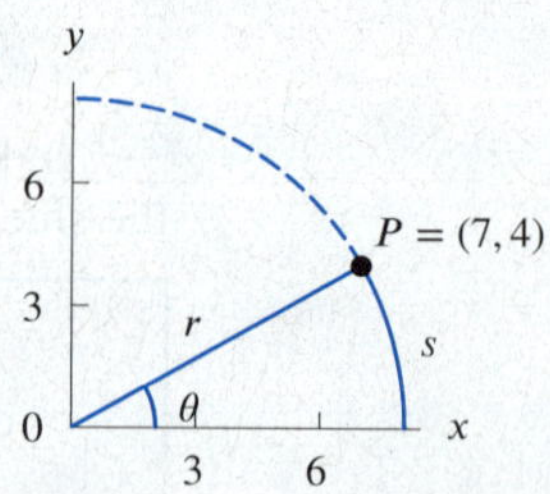

41.

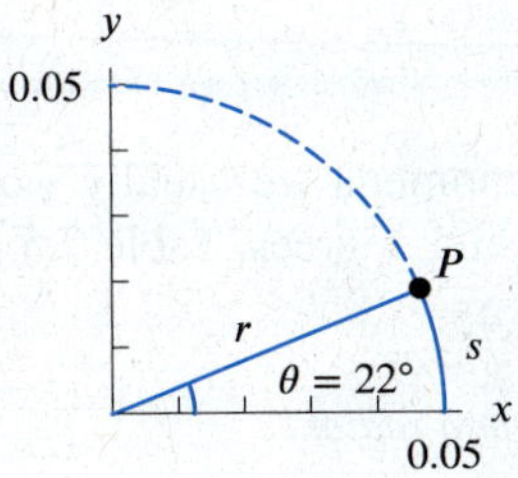


42.

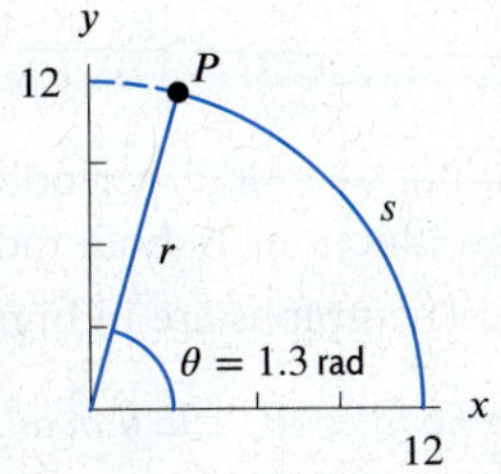


43.

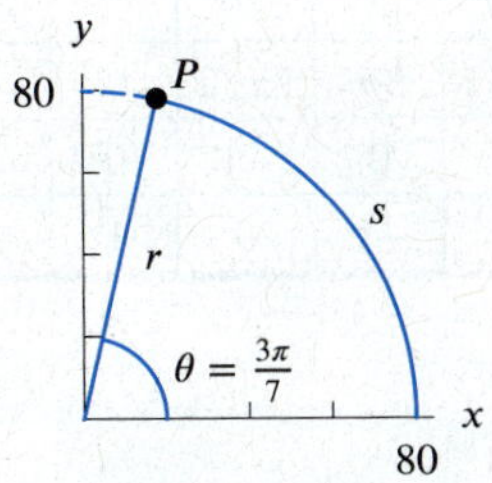


44.

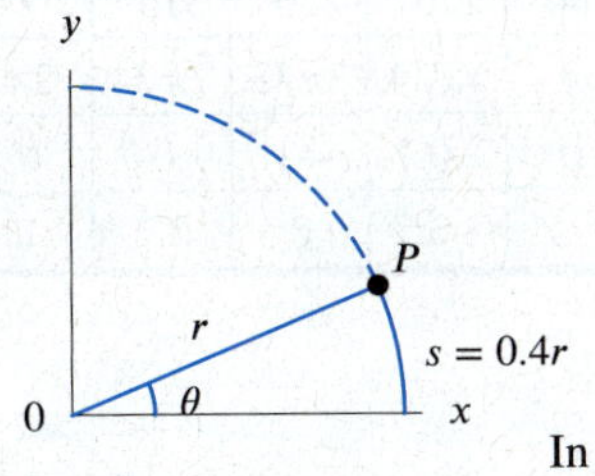


45. A radian can be defined to be the angle, at the center of a circle, that cuts off an arc of length equal to the radius of the circle. This is because the arc length formula in radians, $s = r \cdot \theta$, becomes just $s = r$ when $\theta = 1$. Figure 7.43 shows circles of radii 2 cm and 3 cm.

**(a)** Does the angle 1 radian appear to be the same in both circles?

**(b)** Estimate the number of arcs of length 2 cm that fit in the circumference of the circle of radius 2 cm.

Figure 7.43

46. Without using a calculator, give the sign of each of the following numbers:

**(a)** $\cos 3$ **(b)** $\sin 4$ **(c)** $\sin(-4)$ **(d)** $\cos 7$

47. Without using a calculator, rank the following in order from smallest to largest.

**(a)** The angles: $\frac{2\pi}{3}, 2.3, \frac{2}{3}, -\frac{2\pi}{3}$

**(b)** The numbers:

$$\cos\left(\frac{2\pi}{3}\right), \quad \cos(2.3), \quad \cos\left(\frac{2}{3}\right), \quad \cos\left(-\frac{2\pi}{3}\right)$$

48. How many miles on the surface of the earth correspond to one degree of latitude? (The earth's radius is 3960 miles.)

49. For $\phi$ in Figure 7.44, sketch the following angles.

**(a)** $\pi + \phi$ **(b)** $\pi - \phi$

**(c)** $\pi/2 - \phi$ **(d)** $2\pi - \phi$

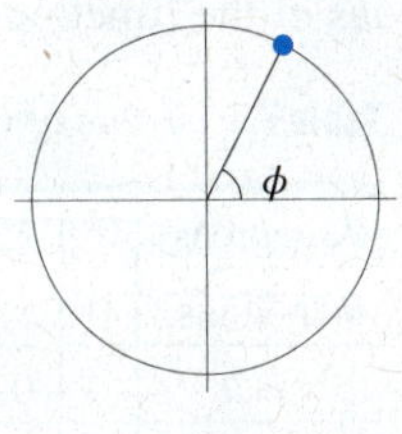


Figure 7.44

In Problems 50–51, find exact values for the lengths of the labeled segments. ($s$ is the arc length.)

50.

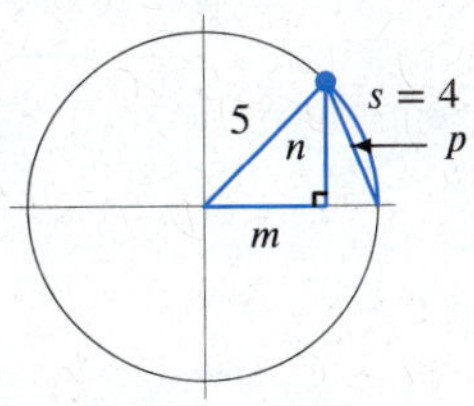


51.

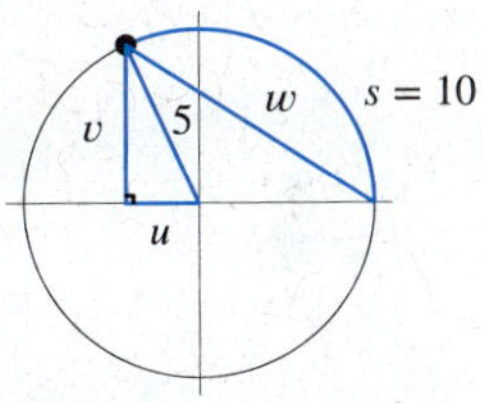


52. If you have the wrong degree/radian mode set on your calculator, you can get unexpected results.

**(a)** If you intend to enter an angle of 30 degrees but you are in radian mode, the angle value would be considered 30 radians. How many degrees is a 30 radian angle?

**(b)** If you intend to enter an angle of $\pi/6$ radians but you are in degree mode, the angle value would be considered $\pi/6$ degrees. How many radians is a $\pi/6$ degree angle?

53. Without a calculator, decide which is bigger, $t$ or $\sin t$, for $0 < t < \pi/2$. Illustrate your answer with a sketch.

54. A person on earth is observing the moon, which is 238,860 miles away. The moon has a diameter of 2160 miles. What is the angle in degrees spanned by the moon in the eye of the beholder?

55. How far does the tip of the minute hand of a clock move in 35 minutes if the hand is 6 inches long?

56. A weather satellite orbits the earth in a circular orbit 500 miles above the earth's surface. What is the radian measure of the angle (measured at the center of the earth) through which the satellite moves in traveling 600 miles along its orbit? (The radius of the earth is 3960 miles.)

## 7.4 GRAPHS OF THE SINE AND COSINE

To visualize angles we often use degrees, but to model periodic phenomena we usually work in radians. Thus, we graph the sine and cosine functions in both radians and degrees. Table 7.5 gives values of the functions, using $1/\sqrt{2} \approx 0.7$. The graphs are in Figure 7.45.

**Table 7.5** *Table of values (rounded) for* $\sin\theta$ *and* $\cos\theta$*, with* $\theta$ *in degrees and radians*

| $\theta$ (degrees) | 0° | 45° | 90° | 135° | 180° | 225° | 270° | 315° | 360° | 450° | 540° | 630° | 720° | ⋯ |
|---|---|---|---|---|---|---|---|---|---|---|---|---|---|---|
| $\theta$ (radians) | 0 | $\pi/4$ | $\pi/2$ | $3\pi/4$ | $\pi$ | $5\pi/4$ | $3\pi/2$ | $7\pi/4$ | $2\pi$ | $5\pi/2$ | $3\pi$ | $7\pi/2$ | $4\pi$ | ⋯ |
| $\sin\theta$ | 0 | 0.7 | 1 | 0.7 | 0 | −0.7 | −1 | −0.7 | 0 | 1 | 0 | −1 | 0 | ⋯ |
| $\cos\theta$ | 1 | 0.7 | 0 | −0.7 | −1 | −0.7 | 0 | 0.7 | 1 | 0 | −1 | 0 | 1 | ⋯ |

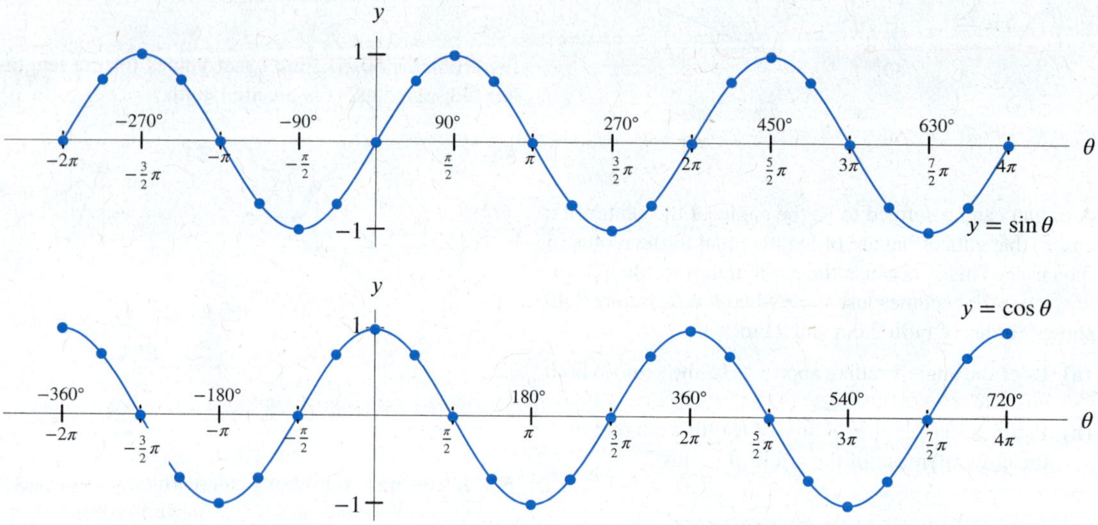


**Figure 7.45:** The graphs of $\sin\theta$ and $\cos\theta$

Properties of the sine and cosine functions that are apparent from their graphs include:

- **Domain**: All values of $\theta$, since any angle, positive or negative, specifies a point on the unit circle.
- **Range**: Since values of the sine and cosine are coordinates of points on the unit circle, they lie between −1 and 1. So the ranges of the sine and cosine are
$$-1 \le \sin\theta \le 1 \quad \text{and} \quad -1 \le \cos\theta \le 1.$$
- **Odd/Even Symmetry**: The sine function is odd, since it is symmetric with respect to the origin. The cosine function is even, since it is symmetric with respect to the $y$-axis:
$$\sin(-\theta) = -\sin\theta \quad \text{and} \quad \cos(-\theta) = \cos\theta.$$
- **Midline**: Both sine and cosine have midline $y = 0$, because they both have $y = 1$ and $y = -1$ as their maximum and minimum values.
- **Amplitude**: Both sine and cosine have an amplitude of 1, because 1 is the vertical distance from the midline to the maximum (or minimum) value of the functions. Note that this distance is also the radius of the unit circle.
- **Period**: Both sine and cosine are periodic functions with a period of $2\pi = 360°$, because their values repeat every $2\pi$ units.

## Amplitude and Midline: Vertical Stretches and Shifts of the Sine and Cosine

The midline of a function is the horizontal line midway between the function's maximum and minimum values. The amplitude is the vertical distance between the maximum (or minimum) value and the midline. By stretching and shifting the graphs of sine and cosine, which have midlines at $y = 0$ and amplitudes of 1, we can obtain new periodic functions with different amplitudes and midlines.

We have seen how to find the coordinates of a point on a circle of radius $r$. If we let the point vary, its $x$ and $y$ coordinates become functions of the angle giving the point's position on the circle. The resulting functions are periodic with amplitude $r$.

**Example 1** Using Figure 7.46, write the height of the point $P$ above the $x$-axis as a function of the angle $\theta$ and graph the function.

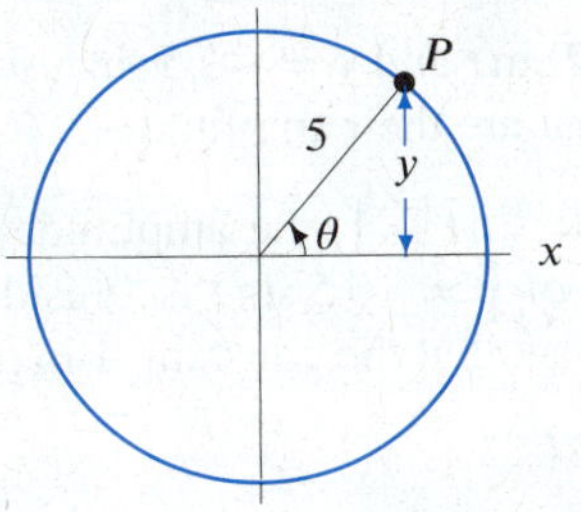

Figure 7.46: Height of point $P$ above $x$-axis

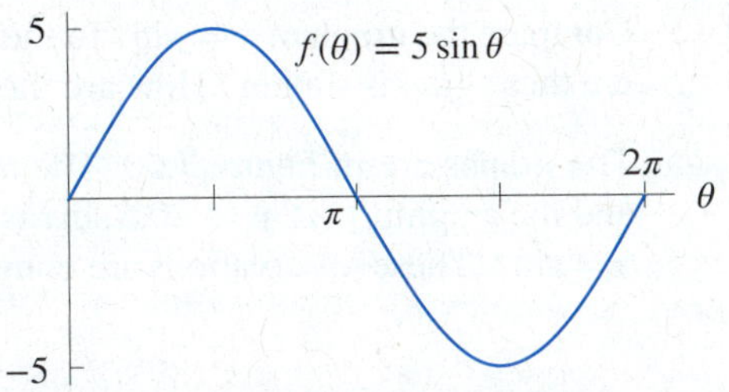

Figure 7.47: Graph of height of point $P$ above $x$-axis

**Solution** We want $y$ as a function of the angle $\theta$. Since the radius of the circle is 5, we have

$$y = f(\theta) = 5\sin\theta.$$

Figure 7.47 shows the graph of this function. Notice that the amplitude, 5, is the radius of the circle and the period, $2\pi$, is the angle of one rotation.

A periodic function resulting from motion on a circle may be a vertical stretch of the sine or cosine, as Example 1 shows. When this is the case, the radius, $A$, of the circle is the amplitude of the function.

**Example 2** In Figure 7.48, the circle from Example 1 has been moved upward so that the center of the circle is at the point $(0, 6)$. Write the height of the point $P$ above the $x$-axis as a function of the angle $\theta$ and graph the function.

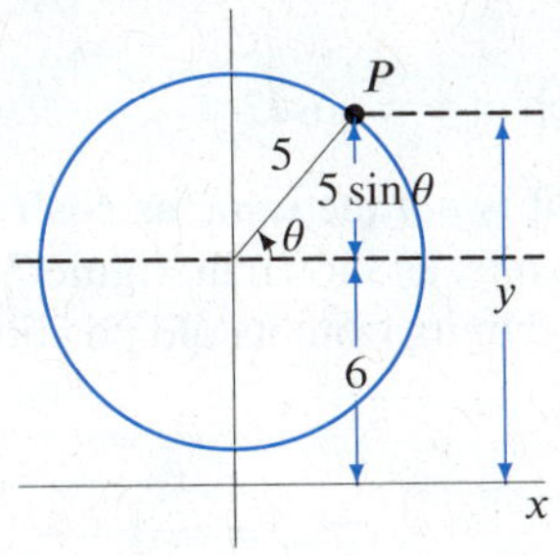

Figure 7.48: Height of point $P$ above $x$-axis

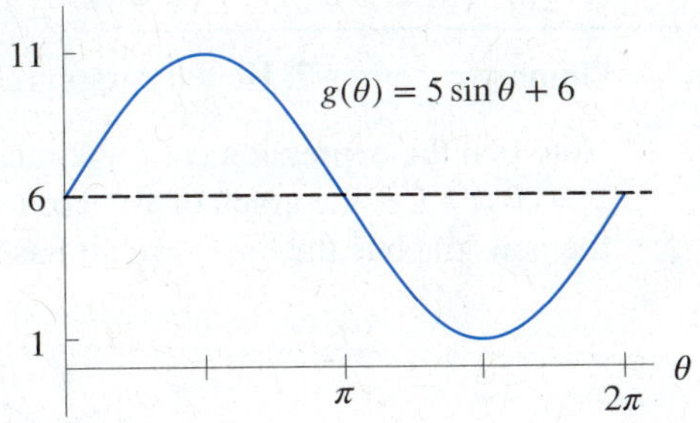

Figure 7.49: Graph of height of point $P$ above $x$-axis

**Solution** The new function, $g$, is the function from Example 1 shifted vertically upward by 6 units, so $g(\theta) = f(\theta) + 6$. Thus,

$$y = g(\theta) = 5\sin\theta + 6.$$

Figure 7.49 shows the graph of this function. Notice that again the amplitude is the radius of the circle and the period, $2\pi$, is the angle of one rotation. In the graphs of both Example 1 and Example 2 the midline corresponds to the $y$-coordinate of the center of the circle.

As Example 2 shows, a periodic function may be both a stretch and a vertical shift of the sine or the cosine. The vertical shift depends on the location of the circle's center and determines the midline of the periodic function.

In general, we have:

> The functions $y = A \sin t + k$ and $y = A \cos t + k$ have **amplitude** $|A|$, and their **midline** is the horizontal line $y = k$.

**Example 3** Compare the graph of $y = \sin t$ to the graphs of $y = 2\sin t$ and $y = -0.5 \sin t$, for $0 \le t \le 2\pi$. How are these graphs similar? How are they different? What are their amplitudes?

**Solution** The graphs are in Figure 7.50. The amplitude of $y = \sin t$ is 1, the amplitude of $y = 2 \sin t$ is 2, and the amplitude of $y = -0.5 \sin t$ is 0.5. The graph of $y = -0.5 \sin t$ is "upside-down" relative to $y = \sin t$. These observations are consistent with the fact that the constant $A$ in the equation

$$y = A \sin t$$

stretches or shrinks the graph vertically, and reflects it about the $t$-axis if $A$ is negative. The amplitude of the function is $|A|$.

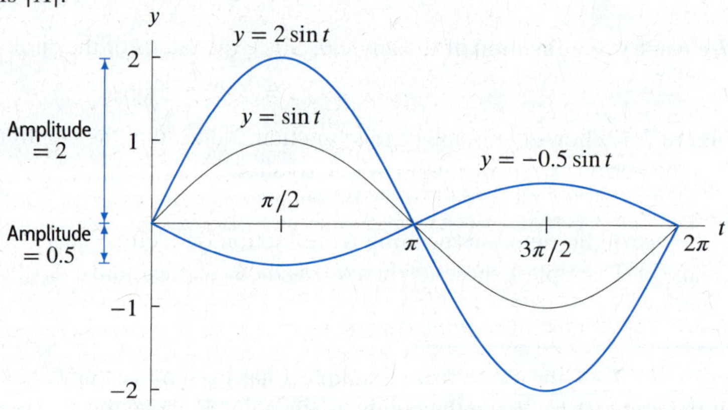


Figure 7.50: The graphs of $y = \sin t$, $y = 2\sin t$, and $y = -0.5 \sin t$ all have different amplitudes

**Example 4** Graph $y = \cos t + 2$. How its graph related to the graph of $y = \cos t$?

**Solution** Note that the expression $\cos t + 2$ means $(\cos t) + 2$; it is *not* the same as $\cos(t + 2)$. The graph of $y = \cos t + 2$ is the graph of $y = \cos t$ shifted up by 2 units, as shown in Figure 7.51. The midline of the new graph is the line $y = 2$; it has been shifted up 2 units from its old position at $y = 0$.

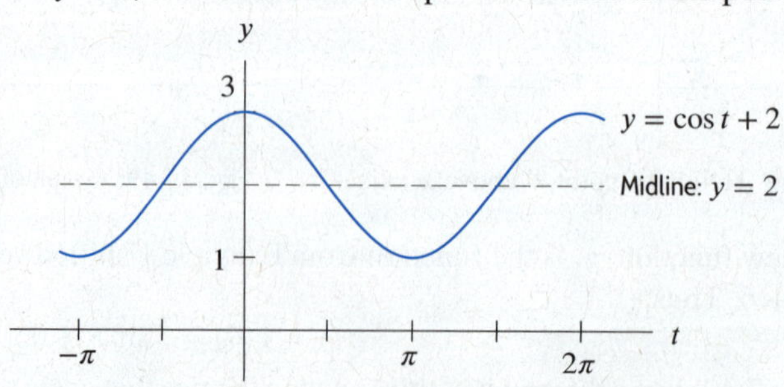


Figure 7.51: The graph of $y = \cos t + 2$ and its midline $y = 2$

**Example 5** Give the midline and amplitude of the following functions and show them on graphs:

(a) $y = 3\sin t + 5$ (b) $y = \dfrac{4 - 3\cos t}{20}$.

**Solution** (a) Here, $A = 3$ and $k = 5$, so the amplitude is 3 and the midline is $y = 5$. See Figure 7.52.

(b) Writing this as

$$y = \frac{4 - 3\cos t}{20}$$
$$y = \frac{4}{20} - \frac{3\cos t}{20}$$
$$y = 0.2 - 0.15\cos t,$$

we have $A = -0.15$ and $k = 0.2$. Thus, the amplitude is $|-0.15| = 0.15$ and the midline is $y = 0.2$. See Figure 7.53.

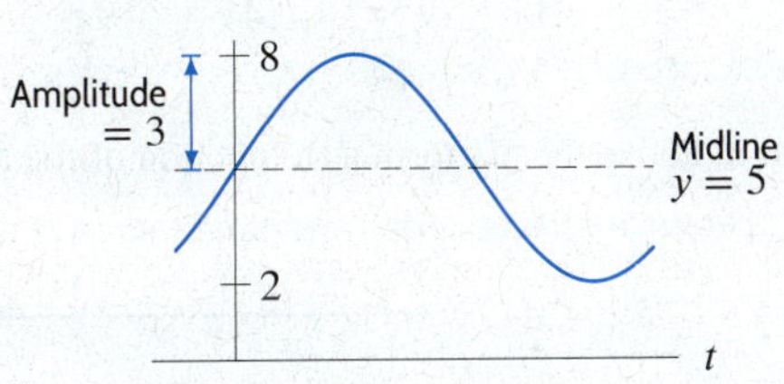


Figure 7.52: Graph of $y = 3\sin t + 5$

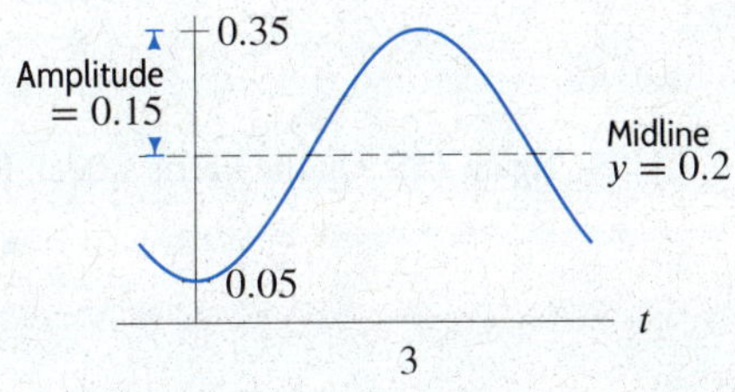


Figure 7.53: Graph of $y = 0.2 - 0.15\cos t$

**Example 6** The London Eye Ferris wheel has a radius, $r$, of 200 feet. Find your height above the platform as a function of the angle $\theta$ measured from the 3 o'clock position. Graph the height function and find the amplitude and midline.

**Solution** Position the origin at the center of the Ferris wheel, so that the positive $x$-axis extends horizontally to the right, through the 3 o'clock position (see Figure 7.54). Since $r = 200$, the $y$-coordinate of point $P$ is given by

$$y = r\sin\theta = 200\sin\theta.$$

Your height above the platform is given by $200 + y$, so the formula for your height in terms of $\theta$ is

$$\text{Height} = 200 + y = 200 + 200\sin\theta.$$

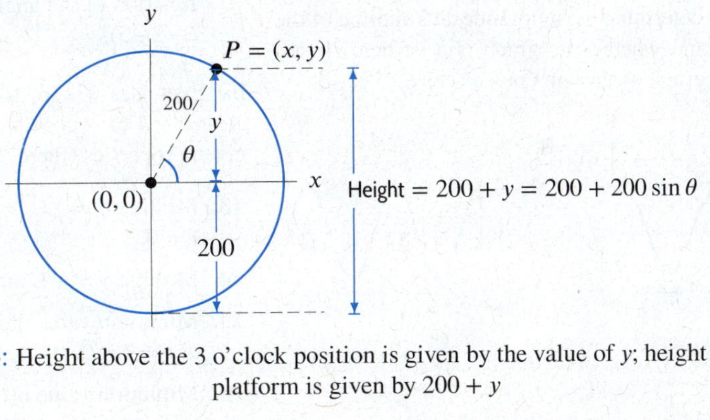


Figure 7.54: Height above the 3 o'clock position is given by the value of $y$; height above the platform is given by $200 + y$

Notice that the midline of the height function, $y = 200$, is the height of the center of the Ferris wheel above the platform. Also, as in Example 1, the radius, $r$, of the wheel is the amplitude of the height function. The angle of a full revolution, $2\pi$, is the period.

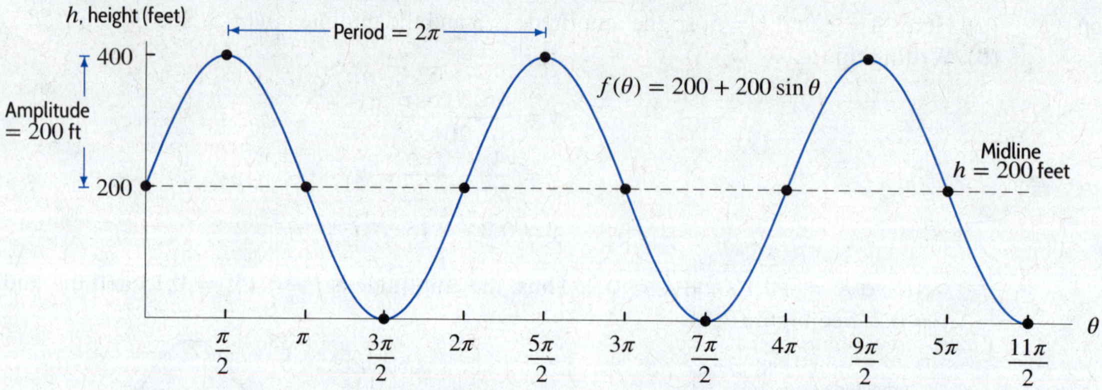


**Figure 7.55**: On the Ferris wheel: Height, $h$, above the platform as a function of the angle, $\theta$

## Exercises and Problems for Section 7.4

### Exercises

In Exercises 1–4, find the amplitude and midline.

**1.** $y = 2\sin t + 3$

**2.** $y = 4 - 2\cos t$

**3.** $y = \dfrac{3\sin t}{20}$.

**4.** $y = \dfrac{5 - 2\sin t}{5}$.

In Exercises 5–6, find the range of the function.

**5.** $f(t) = -2\cos t + 4$.

**6.** $f(t) = 3\sin t - \pi$.

In Exercises 7–10, determine the amplitude and midline of the periodic function and whether its graph is a vertical stretch and/or shift of the graph of $y = \sin t$ or $y = \cos t$.

**7.**

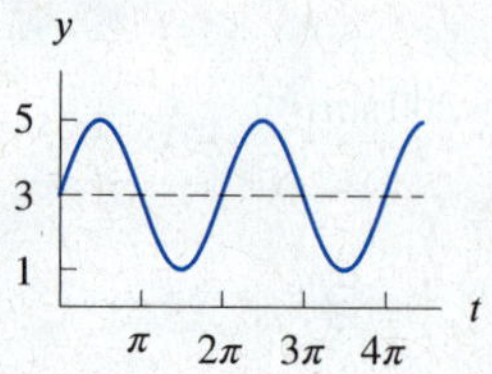


**8.**

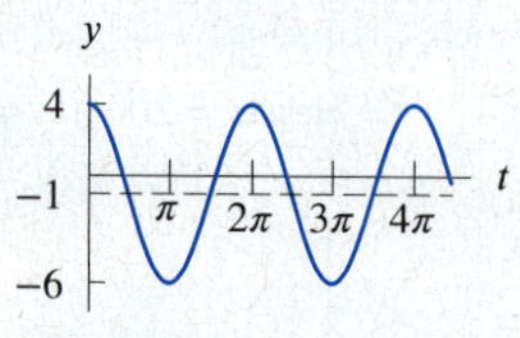


**9.**

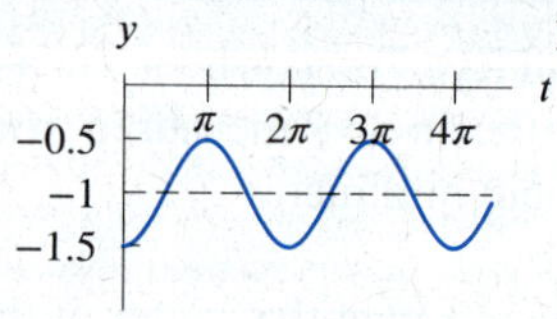


**10.**

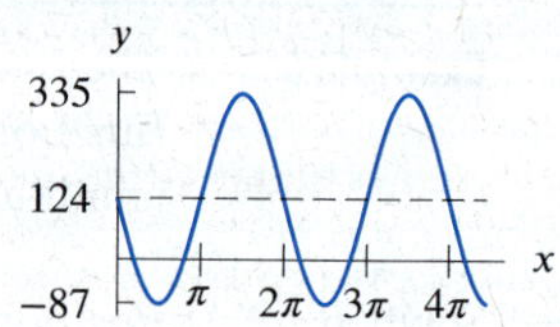


In Exercises 11–12, find functions that describe the following.

**11.** The height of a point $P$ above the $x$-axis as a function of the angle $\theta$ for a circle of radius 3 and center $(3, 3)$.

**12.** The horizontal distance of a point $P$ from the $y$-axis as a function of the angle $\theta$ for a circle of radius 7 and center $(-2, -5)$.

In Exercises 13–16, find all possible formulas of the form $A\sin t + k$ or $A\cos t + k$ for the periodic functions with the given properties (there may be more than one).

**13.** Midline $y = 3$, amplitude 2 with a maximum value at $t = 0$.

**14.** Midline $y = -1$, amplitude 0.5 with $g(\pi) = -1$.

**15.** Minimum value of 2 at $t = \pi/2$ and maximum value of 8 at $t = 3\pi/2$.

**16.** Minimum value of $-6$ and maximum value of 10.

## Problems

**17.** Figure 7.56 shows the graphs of $y = (\sin x) - 1$ and $y = (\sin x) + 1$. Identify which is which.

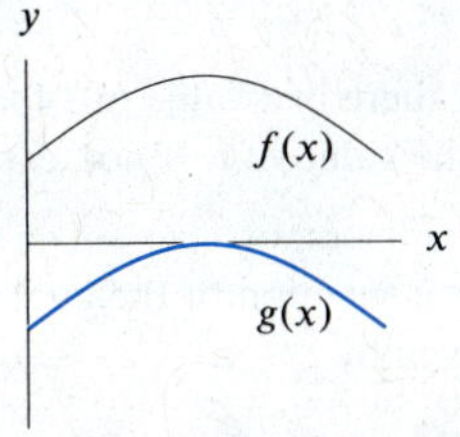


Figure 7.56

**18.** Figure 7.57 shows $y = \sin x$ and $y = \cos x$ starting at $x = 0$. Which is $y = \cos x$? Find values for $a$ and $b$.

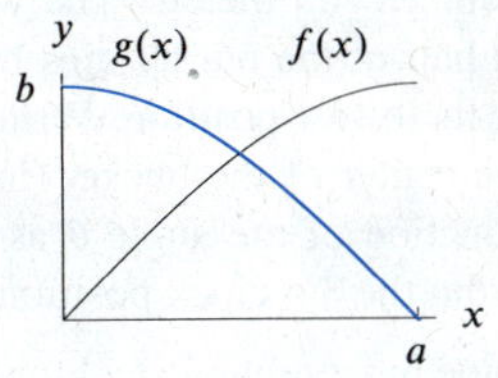


Figure 7.57

**19.** Figure 7.58 shows $y = \sin x$ and $y = 2\sin x$. Which function is $y = 2\sin x$? Find values for $a$ and $b$.

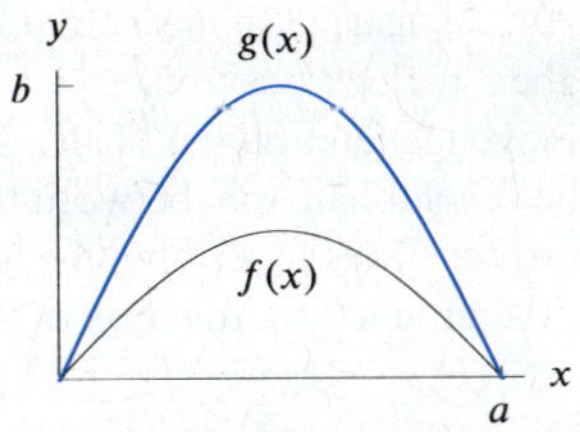


Figure 7.58

**20.** Figure 7.59 is $y = \sin(x - 90°)$ and $y = \sin(x + 90°)$ starting at $x = 0$. Identify which is which.

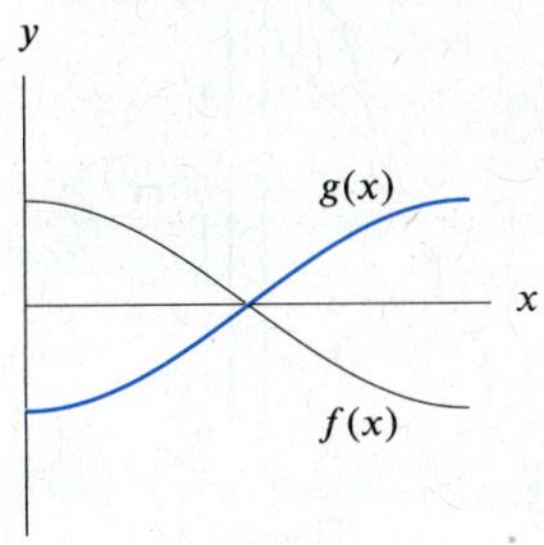


Figure 7.59

**21.** Compare the graph of $y = \sin\theta$ to the graphs of $y = 0.5\sin\theta$ and $y = -2\sin\theta$ for $0 \le \theta \le 360°$. How are these graphs similar? How are they different?

**22.** The graph of $y = \sin\theta$ never goes higher than 1. Explain why this is true by using a sketch of the unit circle and the definition of the $\sin\theta$ function.

**23.** **(a)** Graph the following functions and use the graphs to determine which are odd functions.

(i) $f(\theta) = 2\sin\theta$

(ii) $g(\theta) = \sin\theta + 1$

(iii) $h(\theta) = -3\sin\theta$

(iv) $l(\theta) = 2\sin\theta - 1$

**(b)** Use your answers to decide on conditions for $A$ and $k$ that ensure that $f(t) = A\sin\theta + k$ is odd.

**24.** **(a)** Graph the following functions and use the graphs to determine which are even functions.

(i) $f(\theta) = 3\cos\theta$

(ii) $g(\theta) = 2\cos\theta + 1$

(iii) $h(\theta) = -3\sin\theta - 4$

**(b)** Use your answers to decide conditions on $A$ and $k$ that ensure that $f(t) = A\cos\theta + k$ is even.

**25.** Find a possible formula of the form $A\sin x + k$ or $B\cos x + k$ for the trigonometric function whose values are in the following table.

| $x$ | $-\frac{\pi}{2}$ | $0$ | $\frac{\pi}{2}$ |
|---|---|---|---|
| $g(x)$ | $-4.8$ | $-3.3$ | $-1.5$ |

**26.** Figure 7.60 shows $y = \frac{1}{2}\sin x$ and $y = \sin x + \frac{1}{2}$. Which graph is $y = \frac{1}{2}\sin x$? Find values for $a$, $b$, $c$, $d$.

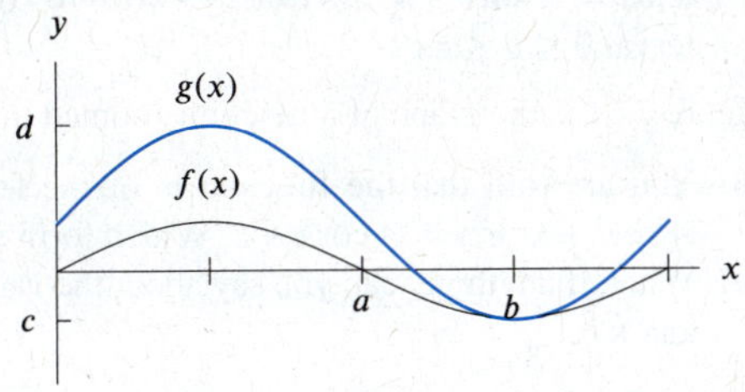


Figure 7.60

In Problems 27–30, find a formula for the graph in terms of $f(x) = \sin x$ or $g(x) = \cos x$.

**27.**

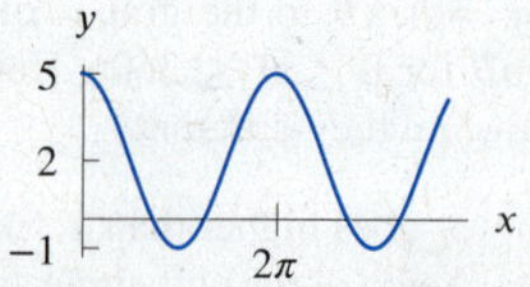


**28.**

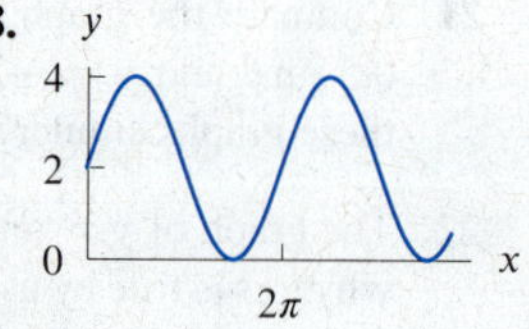


**29.**

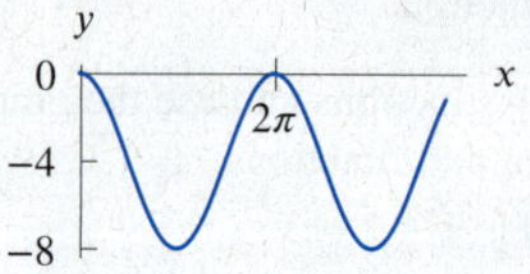


**30.**

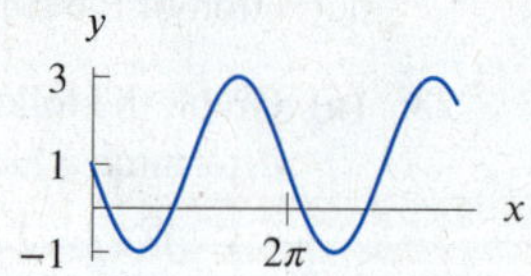


**31.** A stunt pilot flies her plane in two complete loops. At the bottom of the loop (an angle of $\theta = -90°$) she is $s = 300$ meters above the ground; at the top of the loop ($\theta = 90°$) she is $s = 1200$ meters above the ground. Find a formula for $s = f(\theta)$. Sketch a graph showing her height $s$ above ground for $-90° \le \theta \le 630°$.

**32.** A small frog hops onto the Laxey wheel, a waterwheel in the Isle of Man, the largest in the world. Clinging to the wheel, the frog rides from its base, at a height of $s = 0$ ft, to its top at a height of 72.5 ft, then back down. It rides the wheel once more before hopping off. Find a formula for $s = f(\theta)$, the frog's height above ground as a function of $\theta$, the angle (in degrees), $-90° \le \theta \le 630°$, and sketch its graph.

**33.** A spy pushes a thumbtack into the bicycle tire of his enemy. The wheel has diameter 740 mm. When the bike begins to roll, the tack is at an angle of $\theta = 0°$, at the height of the wheel's hub, or $s = 370$ mm above the ground. Find a formula for $s = f(\theta)$. Sketch a graph showing the tack's height above ground for $0° \le \theta \le 720°$.

**34.** **(a)** Use graphs of the following to find the number of zeros the function has on the interval $0 \le \theta \le 4\pi$.

(i) $f(\theta) = 2\sin\theta + 1$

(ii) $g(\theta) = 2\sin\theta + 3$

(iii) $h(\theta) = -2\sin\theta + 2$

**(b)** How must $A$ and $k$ be related so that the function $f(\theta) = A\sin\theta + k$ has no zeros on this interval?

**(c)** How must $A$ and $k$ be related so that the function $f(\theta) = A\sin\theta + k$ has exactly two zeros on the interval $0 \le \theta \le 4\pi$?

**35.** Figure 7.61 is the graph of a function zoomed in at $t = 0$.

**(a)** You are told that the function is of the form $y = A\sin t + k$ or $y = A\cos t + k$. Which form is it?

**(b)** What, if anything, can you say about the values of $A$ and $k$?

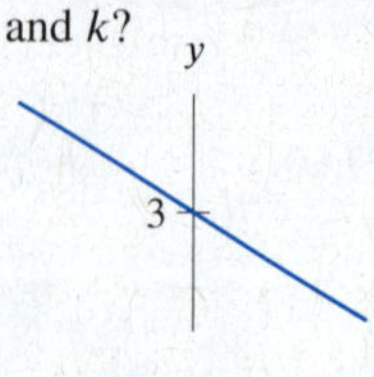


Figure 7.61

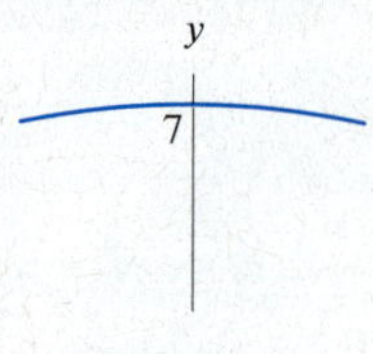


Figure 7.62

**36.** Figure 7.62 is the graph of a function zoomed in at $t = 0$.

**(a)** You are told that the function is of the form $y = A\sin t + k$ or $y = A\cos t + k$. Which form is it?

**(b)** What, if anything, can you say about the values of $A$ and $k$?

**37.** Do you think there is a value of $t$ for which $\cos t = t$? If so, estimate the value of $t$. If not, explain why not.

**38.** The height (in meters) of a person on a Ferris wheel above the ground as a function of the angle from the 3 o'clock position is given by

$$f(\theta) = 50 + 45\sin\theta.$$

**(a)** What are the highest and lowest points on the ride?

**(b)** Describe the dimensions of the Ferris wheel.

**39.** The top of a bucket 0.5 meter high is attached to a water wheel of diameter 4.5 meters. The wheel sits above the river so that half of the bucket dips below the surface of the water at its lowest position. Write a function for the height of the center of the bucket (in meters) above the river as a function of the angle $\theta$ as measured counterclockwise from the 3 o'clock position.

**40.** A wind turbine has a tubular steel tower that is 60 meters high. It has three 30-meter-long blades attached to the top of the tower, as in Figure 7.63.

**(a)** Blade 1 starts at the 3 o'clock position and rotates counterclockwise. Find a formula for the height $h$, in meters, of the tip of blade 1 as a function of the angle $\theta°$. As usual, we measure $\theta$ counterclockwise from the 3 o'clock position.

**(b)** Two ladybugs land on the blade, the first on the tip and the second half way between the tip and the rotor. Find functions describing the height of each bug above the ground as a function of the angle $\theta$.

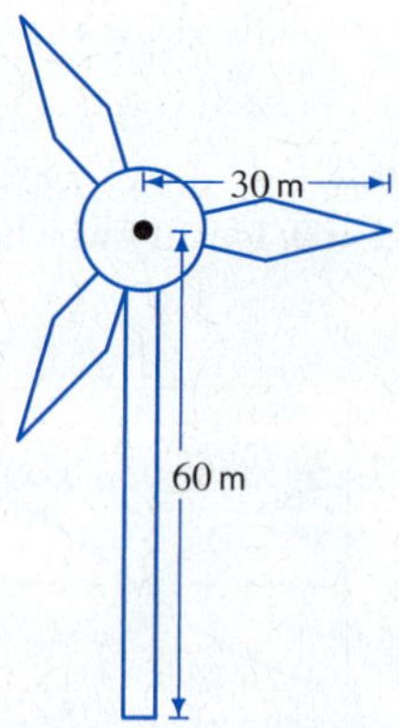


Figure 7.63

# 7.5 SINUSOIDAL FUNCTIONS

Periodic functions, like height on a Ferris wheel, tides, human energy levels, and the motion of a spring, have graphs showing regular oscillations. They are called *sinusoidal functions.* and can be represented as shifts and stretches of the sine and cosine functions.

## Period

The sine and cosine are examples of periodic functions. We now say more precisely what is meant by a periodic functions in terms of shifts, as follows:

> A function $f$ is **periodic** if its values repeat at regular intervals. Then if the graph of $f$ is shifted horizontally by $c$ units, for some constant $c$, the new graph is identical to the original graph. In function notation, periodic means that, for all $t$ in the domain of $f$,
>
> $$f(t+c) = f(t).$$
>
> The smallest positive constant $c$ for which this relationship holds for all values of $t$ is called the **period** of $f$.

Note that, for the sine and cosine functions, we have

$$\sin(\theta + 2\pi) = \sin\theta \quad \text{and} \quad \cos(\theta + 2\pi) = \cos\theta.$$

The constant $2\pi$, the period of sine and cosine, is the smallest constant for which these relationships hold.

### Changing the Period: Horizontal Stretches and Compressions

We now study functions that are transformation of the sine and cosine and calculate their periods.

**Example 1** Graph $y = \sin t$ and $y = \sin 2t$ for $0 \le t \le 2\pi$. Describe any similarities and differences. What are their periods?

**Solution** The graphs are in Figure 7.64. The two functions have the same amplitude and midline, but their periods are different. The period of $y = \sin t$ is $2\pi$, but the period of $y = \sin 2t$ is $\pi$. This is because the factor of 2 causes a horizontal compression, squeezing the graph twice as close to the $y$-axis.

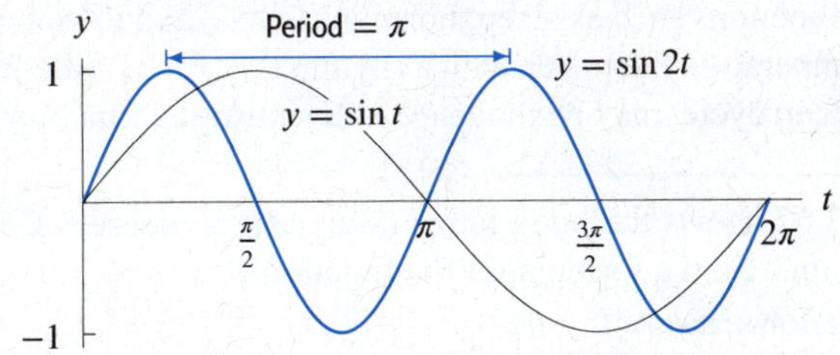


Figure 7.64: The functions $y = \sin t$ and $y = \sin 2t$ have different periods

If $B > 0$ the function $y = \sin(Bt)$ resembles the function $y = \sin t$ except that it is stretched or compressed horizontally. The constant $B$ determines how many cycles the function completes on an interval of length $2\pi$. For example, we see from Figure 7.64 that the function $y = \sin 2t$ completes two cycles on the interval $0 \le t \le 2\pi$.

More generally, for $B > 0$, the graph of $y = \sin(Bt)$ completes $B$ cycles on the interval $0 \le t \le 2\pi$, so each cycle has length $2\pi/B$. The period is thus $2\pi/B$. In general, for $B$ of any sign, we have:

Functions $y = \sin(Bt)$ and $y = \cos(Bt)$ have **period** $P = 2\pi/|B|$. Generally, $B$ is positive.

The number of cycles in one unit of time is called the *frequency*, $f = 1/\text{period} = |B|/2\pi$.

**Example 2** Find possible formulas for the functions $f$ and $g$ shown in Figures 7.65 and 7.66.

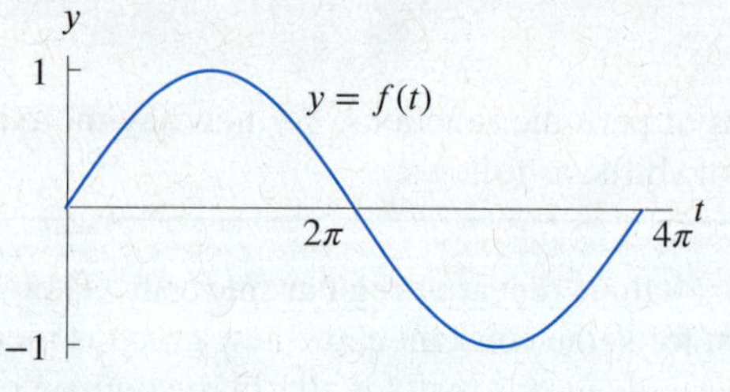


Figure 7.65: This function has period $4\pi$

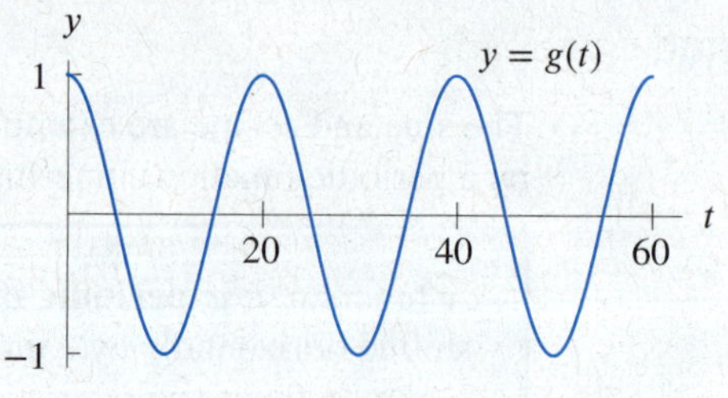


Figure 7.66: This function has period 20

**Solution** The graph of $f$ resembles the graph of $y = \sin t$ except that its period is $P = 4\pi$. Using $P = 2\pi/B$ gives

$$4\pi = \frac{2\pi}{B} \quad \text{so} \quad B = \frac{1}{2}.$$

Thus, $f(t) = \sin\left(\frac{1}{2}t\right)$.

The function $g$ resembles the function $y = \cos t$ except that its period is $P = 20$. This gives

$$20 = \frac{2\pi}{B} \quad \text{so} \quad B = \frac{\pi}{10}.$$

Thus, $g(t) = \cos\left(\frac{\pi}{10}t\right)$.

Sinusoidal functions of the following form represent oscillating quantities:

$$y = A\sin(Bt) + k \quad \text{and} \quad y = A\cos(Bt) + k.$$

The amplitude is $|A|$, the midline is $k$, and the period is $2\pi/|B|$.

Plants and animals have natural internal "clocks", called circadian rhythms, that regulate biological functions such as sleep, hormone levels, and the movement of flower petals. Normal human body temperature fluctuates with a rhythm that varies little from person to person. This rhythm, tied to our sleep cycle, may be modeled by a sinusoidal function.

**Example 3** Figure 7.67 shows the body temperature, $H$, in degrees Celsius, as a function of time $t$, in hours since 9 am.[6] Find a formula for the function.

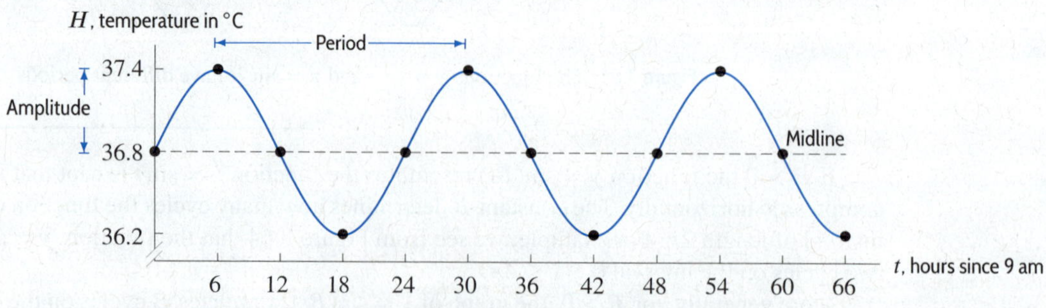


Figure 7.67: Human body temperature, $H$, as a function of time, $t$

[6]Model based on data from circadian.org, retrieved January, 2014.

**Solution** At 9 am the body temperature is at the midline and rising, so we choose to model the temperature with a sine function of the form $H = A\sin(Bt) + k$.

The temperature cycles between a high of 37.4°C and a low of 36.2°C. At 9 am the body temperature is 36.8°C. Thus, the midline and amplitude are given by

$$\text{Midline} = k = \frac{(37.4 + 36.2)}{2} = 36.8°\text{C}.$$
$$\text{Amplitude} = A = 37.4 - 36.8 = 0.6°\text{C}.$$

This tells us that the body temperature varies by 0.6°C above and below the average of 36.8°C, so we have

$$H = 0.6\sin(Bt) + 36.8.$$

The temperature repeats every 24 hours, so the period is 24 hours. This is expected, as the word circadian refers to internal clocks that reset daily. To find $B$, we use

$$24 = \frac{2\pi}{B}, \quad \text{so} \quad B = \frac{2\pi}{24} = \frac{\pi}{12}.$$

Thus, we have

$$H = 0.6\sin\left(\frac{\pi t}{12}\right) + 36.8.$$

Note that the frequency of this function, 1/period, is 1/24. This makes sense, as in one hour the function has completed $1/24^{\text{th}}$ of its cycle.

## Horizontal Shift

Figure 7.68 shows the graphs of two trigonometric functions, $f$ and $g$, each with period $P = 12$. The graph of $f$ resembles a sine function, so a possible formula for $f$ is $f(t) = \sin Bt$. Since the period of $f$ is 12, we have $12 = 2\pi/B$, so $B = 2\pi/12$, so $f(t) = \sin(\pi t/6)$.

The graph of $g$ looks like the graph of $f$ shifted to the right by 2 units. Thus, a possible formula for $g$ is

$$g(t) = f(t-2),$$

or

$$g(t) = \sin\left(\frac{\pi}{6}(t-2)\right).$$

Notice that we can also write the formula for $g(t)$ as

$$g(t) = \sin\left(\frac{\pi}{6}t - \frac{\pi}{3}\right),$$

but $\pi/3$ is *not* the horizontal shift in the graph! To pick out the horizontal shift from the formula, we must write the formula in factored form, that is, as $\sin(B(t-h))$.

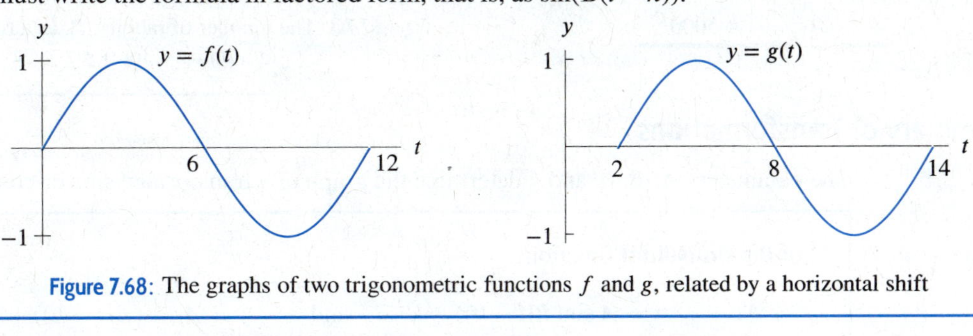


Figure 7.68: The graphs of two trigonometric functions $f$ and $g$, related by a horizontal shift

> The graphs of $y = \sin(B(t-h))$ and $y = \cos(B(t-h))$ are the graphs of $y = \sin Bt$ and $y = \cos Bt$ **shifted horizontally** by $h$ units.

**Example 4** Describe in words the graph of the function $g(t) = \cos(3t - \pi/4)$.

**Solution** Write the formula for $g$ in the form $\cos(B(t-h))$ by factoring 3 out from the expression $3t - \pi/4$ to get $g(t) = \cos(3(t - \pi/12))$. The period of $g$ is $2\pi/3$ and its graph is the graph of $f = \cos 3t$ shifted $\pi/12$ units to the right, as shown in Figure 7.69.

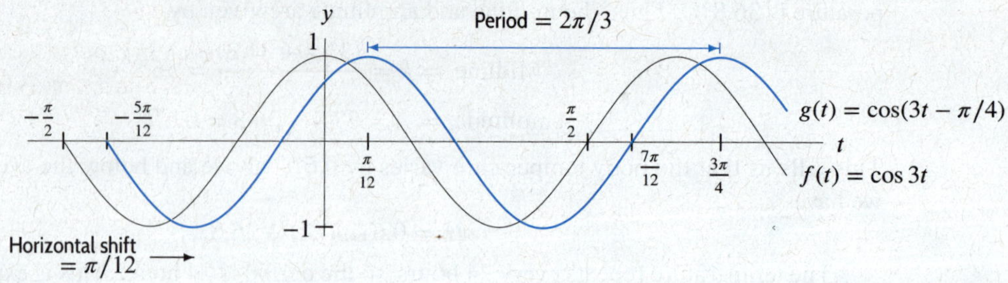


Figure 7.69: The graphs of $g(t) = \cos\left(3t - \frac{\pi}{4}\right)$ and $f(t) = \cos 3t$

**Example 5** A rabbit population in a national park rises and falls each year. In January, the population is at its minimum of 5000 rabbits. By July, with warm weather and abundant food, the population has tripled in size. By the following January, the population has again fallen to 5000 rabbits, completing the annual cycle. Use a sinusoidal function to find a possible formula for $R$, the size of the rabbit population, as a function of $t$, the number of months since January.

**Solution** Notice that January is month 0, so July is month 6. The five points in Table 7.6 have been plotted in Figure 7.70 and a curve drawn in. This curve has midline $k = 10{,}000$, amplitude $|A| = 5000$, and period $= 12$ so $B = 2\pi/12 = \pi/6$. It resembles a cosine function reflected across its midline. Thus, a possible formula for this curve is

$$R = -5000\cos\left(\frac{\pi}{6}t\right) + 10{,}000.$$

There are other possible formulas. For example, we could use a sine function and write

$$R = 5000\sin\left(\frac{\pi}{6}(t-3)\right) + 10{,}000.$$

**Table 7.6** *Rabbit population over time*

| $t$ (month) | $R$ |
|---|---|
| 0 | 5000 |
| 6 | 15,000 |
| 12 | 5000 |
| 18 | 15,000 |
| 24 | 5000 |

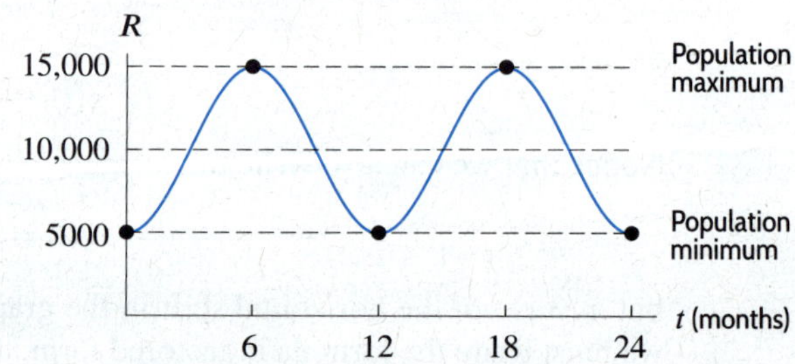


Figure 7.70: The number of rabbits, $R$, as a function of time in months, $t$, for $0 \le t \le 24$

## Summary of Transformations

The parameters $A$, $B$, $h$, and $k$ determine the graph of a transformed sine or cosine function.

For the **sinusoidal** functions

$$y = A\sin(B(t-h)) + k \quad \text{and} \quad y = A\cos(B(t-h)) + k,$$

- $|A|$ is the amplitude
- $2\pi/|B|$ is the period
- $h$ is the horizontal shift
- $y = k$ is the midline
- $|B|/2\pi$ is the frequency, that is, the number of cycles completed in one unit of time.

## Optional: Phase Shift

In Example 4, we factored $(3t - \pi/4)$ to rewrite the function as

$$g(t) = \cos(3t - \pi/4) = \cos\left(3(t - \pi/12)\right).$$

This allowed us to recognize the horizontal shift, $\pi/12$. However, in most physical applications, the quantity $\pi/4$, known as the *phase shift*, is more important than the horizontal shift. We define

$$\text{Phase shift} = \text{Fraction of period shifted} \times 2\pi.$$

The phase shift enables us to calculate the fraction of a full period that the curve has been shifted. For instance, in Example 4, the wave has been shifted

$$\frac{\text{Phase shift}}{2\pi} = \frac{\pi/4}{2\pi} = \frac{1}{8} \text{ of a full period,}$$

and the graph of $g$ in Figure 7.71 is the graph of $f(t) = \cos 3t$ shifted $1/8$ of its period to the right.

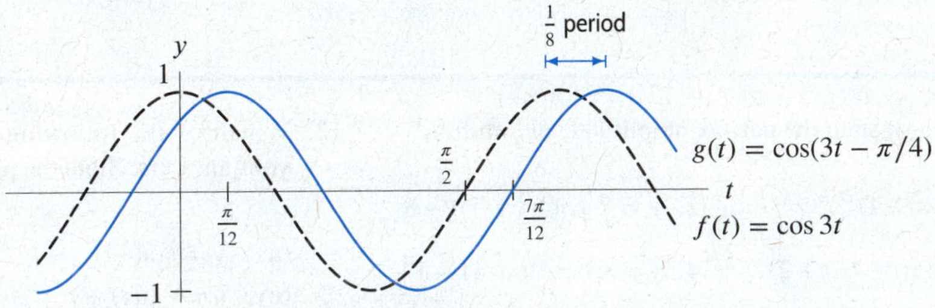


Figure 7.71: The graph of $g(t) = \cos(3t - \frac{\pi}{4})$ has phase shift $\frac{\pi}{4}$ relative to $f(t) = \cos 3t$

Phase shift is significant because in many applications, such as optical interference, we want to know if two waves reinforce or cancel each other. For two waves of the same period, a phase shift of 0 or $2\pi$ tells us that the two waves reinforce each other; a phase shift of $\pi$ tells us that the two waves cancel. Thus, the phase shift tells us the relative positions of two waves of the same period.[7]

> For sinusoidal functions written in the following form, the magnitude of $\phi$ is the **phase shift**:
>
> $$y = A\sin(Bt + \phi) \quad \text{and} \quad y = A\cos(Bt + \phi).$$
>
> If $B > 0$, a positive value of $\phi$ represents a shift to the left; a negative value of $\phi$ represents a shift to the right.

**Example 6** (a) In Figure 7.72, by what fraction of a period is the graph of $g(t)$ shifted from the graph of $f(t)$?
(b) What is the phase shift?

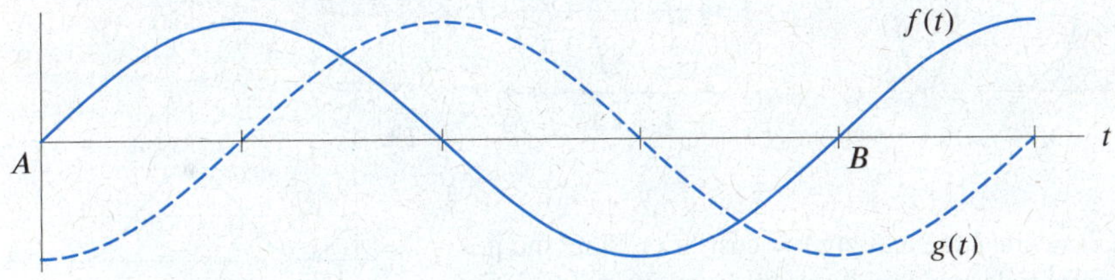


Figure 7.72

[7]The phase shift that tells us that two waves cancel is independent of their period. The horizontal shift that gives the same information is not independent of period.

**Solution** (a) The period of $f(t)$ is the length of the interval from $A$ to $B$. The graph of $g(t)$ appears to be shifted $1/4$ period to the right.

(b) The phase shift is $\frac{1}{4}(2\pi) = \pi/2$.

**Example 7** What is the phase shift of the rabbit population function in Example 5 given by

$$R = 5000\sin\left(\frac{\pi}{6}(t-3)\right) + 10{,}000?$$

**Solution** We can expand the formula for $R$, giving

$$R = 5000\sin\left(\frac{\pi t}{6} - \frac{\pi}{2}\right) + 10{,}000.$$

Thus, the phase shift is $\pi/2$, corresponding to the fact that we reach $R = 10{,}000$ after a quarter of a cycle ($\pi/2$ is a quarter of $2\pi$).

## Exercises and Problems for Section 7.5

### Exercises

In Exercises 1–4, state the period, amplitude, and midline.

**1.** $y = 6\sin(t+4)$

**2.** $y = 7\sin(4(t+7)) - 8$

**3.** $2y = \cos(8(t-6)) + 2$

**4.** $y = \pi\cos(2t+4) - 1$

In Exercises 5–10. estimate the period, midline, and amplitude of the periodic function.

**5.**

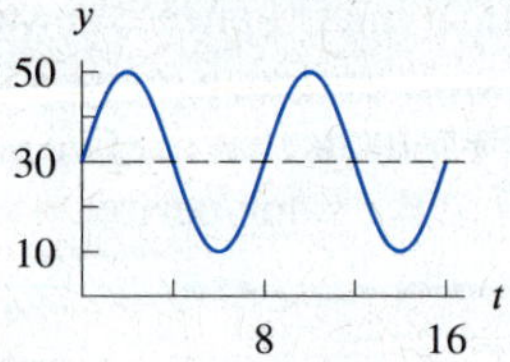


**6.**

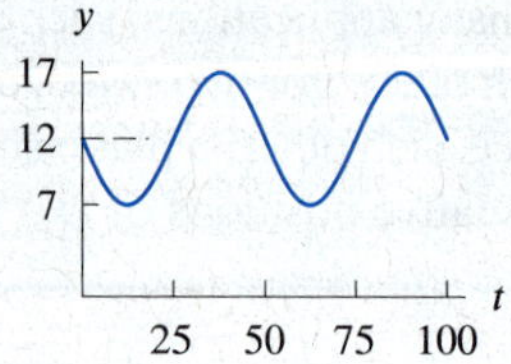


**7.**

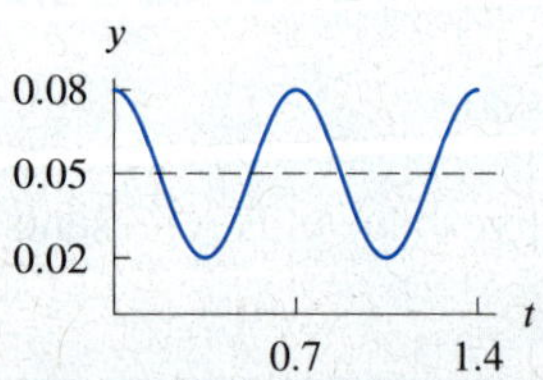


**8.**

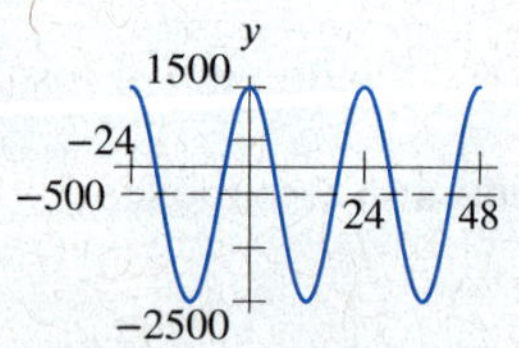


**9.**

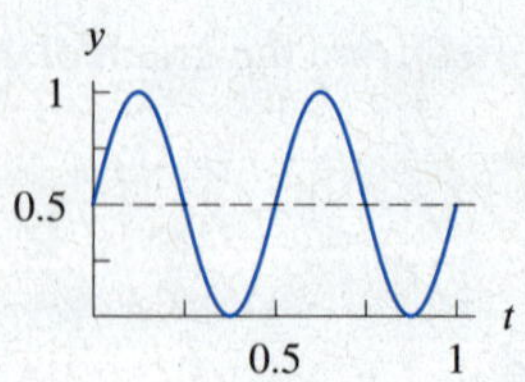


**10.**

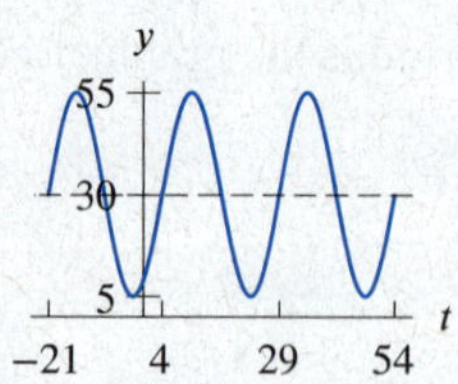


**11.** Let $f(x) = \sin(2\pi x)$ and $g(x) = \cos(2\pi x)$. State the periods, amplitudes, and midlines of $f$ and $g$.

**12.** Which of the following functions are periodic? Justify your answers. State the periods of those that are periodic.

**(a)** $y = \sin(-t)$

**(b)** $y = 4\cos(\pi t)$

**(c)** $y = (\sin t) + t$

**(d)** $y = \sin(t/2) + 1$

In Exercises 13–20, find a formula for the trigonometric function.

**13.**

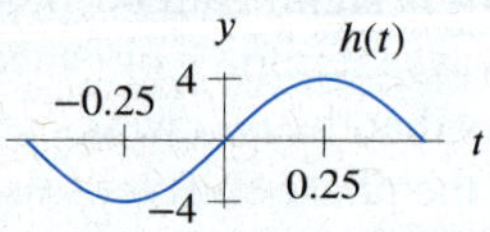


**14.**

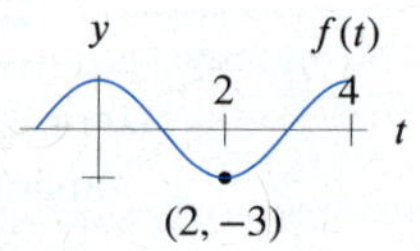


**15.**

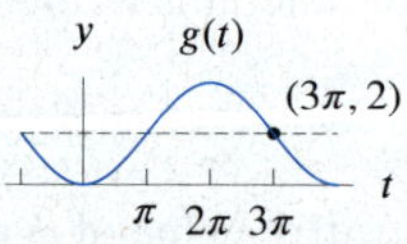


**16.**

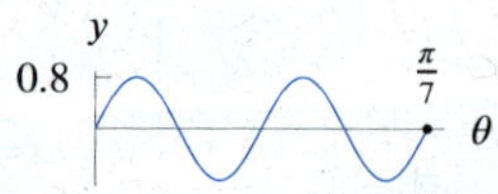


**17.**

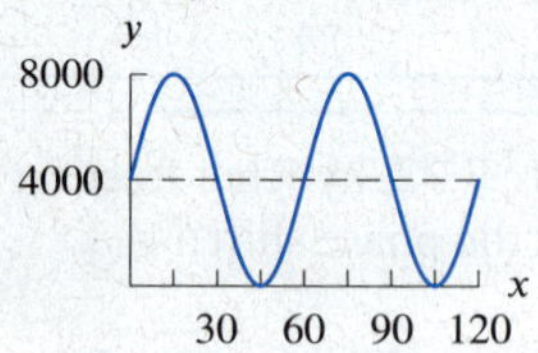


**18.**

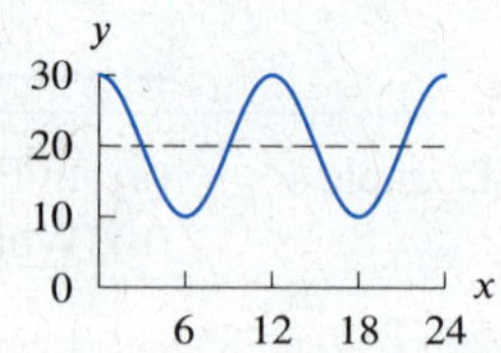


**19.**

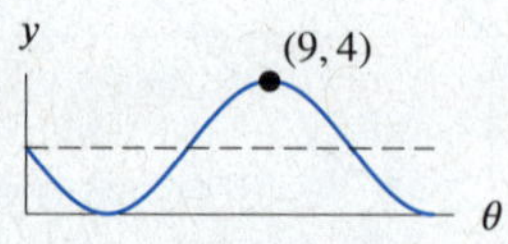


**20.**

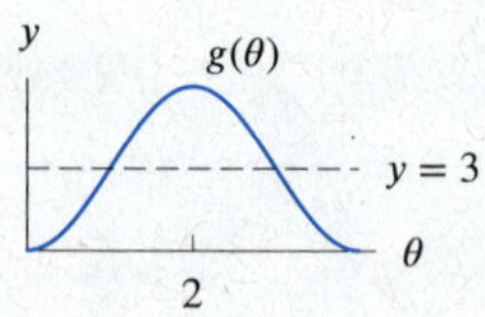

## Problems

In Problems 21–24, graph two periods of the function without a calculator.

**21.** $y = 40\sin\left(\frac{2\pi t}{7}\right) + 60$

**22.** $y = 3 - 2\cos(50\pi t)$

**23.** $y = 0.001 - 0.003\sin(0.002\pi t)$

**24.** $y = 12\cos\left(\frac{2\pi}{50}t\right) - 24$

**25.** Figure 7.73 shows $y = \sin x$ and $y = \sin 2x$. Which graph is $y = \sin x$? Identify the points $a$ to $e$.

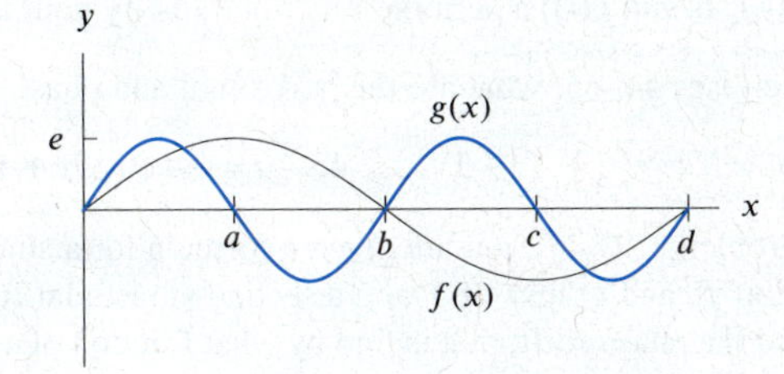


Figure 7.73

**26.** A person's blood pressure, $P$ (in millimeters of mercury, abbreviated mm Hg), is given, for time, $t$, in seconds, by

$$P = 100 - 20\cos\left(\frac{8\pi}{3}t\right),$$

Graph this function. State the period and amplitude and explain the practical significance of these quantities.

In Problems 27–28, find a formula, using the sine function, for your height above ground after $t$ minutes on the Ferris wheel. Graph the function to check that it is correct.

**27.** A Ferris wheel is 35 meters in diameter and boarded at ground level. The wheel completes one full revolution every 5 minutes. At $t = 0$ you are in the three o'clock position and ascending.

**28.** A Ferris wheel is 20 meters in diameter and boarded in the six o'clock position from a platform that is 4 meters above the ground. The wheel completes one full revolution every 2 minutes. At $t = 0$ you are in the twelve o'clock position.

**29.** The following formulas give animal populations as functions of time, $t$, in years. Describe the growth of each population in words.

**(a)** $P = 1500 + 200t$ **(b)** $P = 2700 - 80t$
**(c)** $P = 1800(1.03)^t$ **(d)** $P = 800e^{-0.04t}$
**(e)** $P = 230\sin\left(\frac{2\pi}{7}t\right) + 3800$

**30.** A population of animals oscillates between a low of 1300 on January 1 ($t = 0$) and a high of 2200 on July 1 ($t = 6$).

**(a)** Find a formula for the population, $P$, in terms of the time, $t$, in months.
**(b)** Interpret the amplitude, period, and midline of the function $P = f(t)$.
**(c)** Use a graph to estimate when $P = 1500$.

**31.** Household electrical power in the US is provided in the form of alternating current. Typically the voltage cycles smoothly between +155.6 volts and −155.6 volts 60 times per second. Use a sinusoidal function to model voltage, $V$, as a function of time, $t$, in seconds since the voltage was at its peak.

**32.** Find a possible formula for the trigonometric function whose values are in the following table.

| $x$ | 0 | .1 | .2 | .3 | .4 | .5 | .6 | .7 | .8 | .9 | 1 |
|---|---|---|---|---|---|---|---|---|---|---|---|
| $g(x)$ | 2 | 2.6 | 3 | 3 | 2.6 | 2 | 1.4 | 1 | 1 | 1.4 | 2 |

In Problems 33–36, find a formula for the graph in terms of $f(x) = \sin(2\pi x)$ and $g(x) = \cos(2\pi x)$.

**33.**

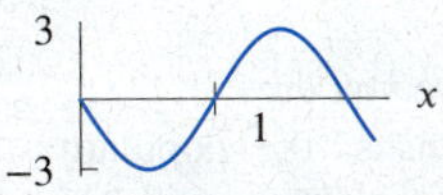


**34.**

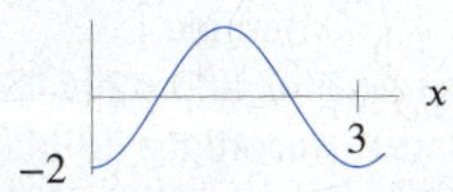


**35.**

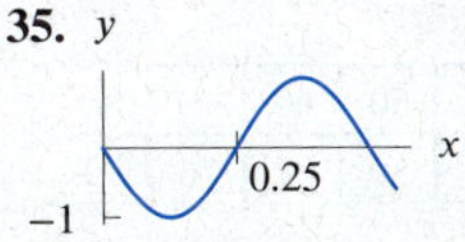


**36.**

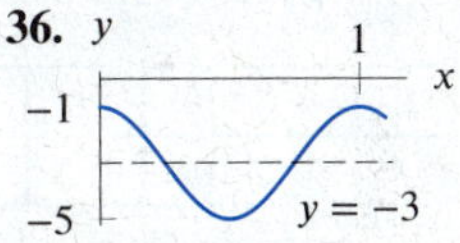


**37.** A company sells $S(t)$ thousand electric blankets in month $t$ (with $t = 0$ being January), where

$$S(t) \approx 72.25 + 41.5\sin\left(\frac{\pi t}{6} + \frac{\pi}{2}\right).$$

Graph this function over one year. Find its period and amplitude and explain their practical significance.

**38.** The pressure, $P$ (in lbs/ft$^2$), in a pipe varies over time. Five times an hour, the pressure oscillates from a low of 90 to a high of 230 and then back to a low 90. The pressure at $t = 0$ is 90.

**(a)** Graph $P = f(t)$, where $t$ is time in minutes. Label your axes.
**(b)** Find a possible formula for $P = f(t)$.
**(c)** By graphing $P = f(t)$ for $0 \le t \le 2$, estimate when the pressure first equals 115 lbs/ft$^2$.

**39.** A flight from La Guardia Airport in New York City to Logan Airport in Boston has to circle Boston several times before landing. Figure 7.74 shows the graph of the distance, $d$, of the plane from La Guardia as a function of time, $t$. Construct a function $f(t)$ whose graph approximates this one. Do this by using different formulas on the intervals $0 \le t \le 1$ and $1 \le t \le 2$.

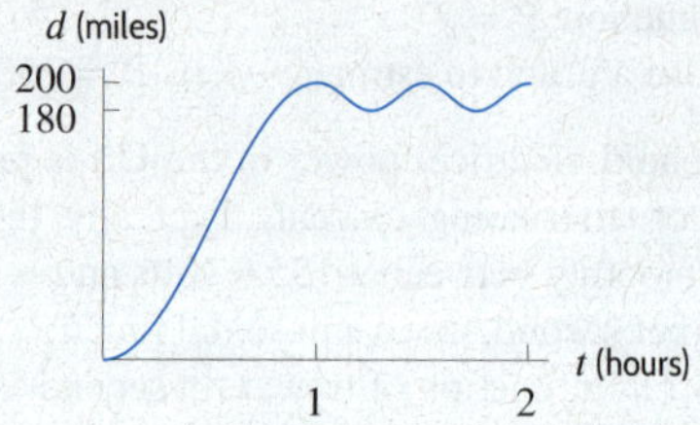


**Figure 7.74**

**40.** **(a)** Graph the data in Table 7.7, which gives the population, $P$, in thousands, of Somerville, MA, with $t$ in years since 1920.[8]

**(b)** Based on the data, a researcher decides the population varies in an approximately periodic way with time. Do you agree?

**(c)** On your graph, sketch in a sine curve that fits your data as closely as possible. Your sketch should capture the overall trend of the data but need not pass through all the data points. [Hint: Start by choosing a midline.]

**(d)** Find a formula for the curve you drew in part (c).

**(e)** According to the US census, the population of Somerville in 1910 was 77,236. How well does this agree with the value given by your formula?

**Table 7.7**

| $t$ | 0 | 10 | 20 | 30 | 40 | 50 | 60 | 70 | 80 |
|---|---|---|---|---|---|---|---|---|---|
| $P$ | 93 | 104 | 102 | 102 | 95 | 89 | 77 | 76 | 77 |

**41.** Table 7.8 shows the average daily maximum temperature in degrees Fahrenheit in Boston each month.[9]

**(a)** Plot the average daily maximum temperature as a function of the number of months past January.

**(b)** What are the amplitude and period of the function?

**(c)** Find a trigonometric approximation of this function.

**(d)** Use your formula to estimate the daily maximum temperature for October. How well does this estimate agree with the data?

**Table 7.8**

| Month | Jan | Feb | Mar | Apr | May | Jun |
|---|---|---|---|---|---|---|
| Temperature | 36.4 | 37.7 | 45.0 | 56.6 | 67.0 | 76.6 |
| **Month** | **Jul** | **Aug** | **Sep** | **Oct** | **Nov** | **Dec** |
| Temperature | 81.8 | 79.8 | 72.3 | 62.5 | 47.6 | 35.4 |

**42.** Table 7.9 gives US petroleum imports, $P$, in quadrillion BTUs, for $t$ in years since 1900. Plot the data on $73 \le t \le 92$ and find a trigonometric function $f(t)$ that approximates US petroleum imports as a function of $t$.

**Table 7.9**

| $t$ | 73 | 74 | 75 | 76 | 77 | 78 | 79 | 80 | 81 | 82 |
|---|---|---|---|---|---|---|---|---|---|---|
| $P$ | 14.2 | 14.5 | 14 | 16.5 | 18 | 17 | 17 | 16 | 15 | 14 |
| $t$ | 83 | 84 | 85 | 86 | 87 | 88 | 89 | 90 | 91 | 92 |
| $P$ | 13 | 15 | 12 | 14 | 16 | 17 | 18 | 17 | 16 | 16 |

**43.** Let $f$ be a function defined for all real numbers.

**(a)** Is $f(\sin t)$ a periodic function? Justify your answer.

**(b)** Is $\sin(f(t))$ a periodic function? Justify your answer.

In Exercises 44–45, what are the horizontal and phase shifts?

**44.** $y = 2\cos(3t + 4) - 5$

**45.** $y = -4\cos(7t + 13) - 5$

For Problems 46–47, you are given a formula for a sinusoidal function $f$, and graphs of $f$ and a second sinusoidal function $g$. Find the phase shift; that is find by what fraction of a period $f$ must be shifted to obtain $g$. Using this, find a formula for $g$.

**46.** $f(x) = 3\sin((\pi/4)x)$

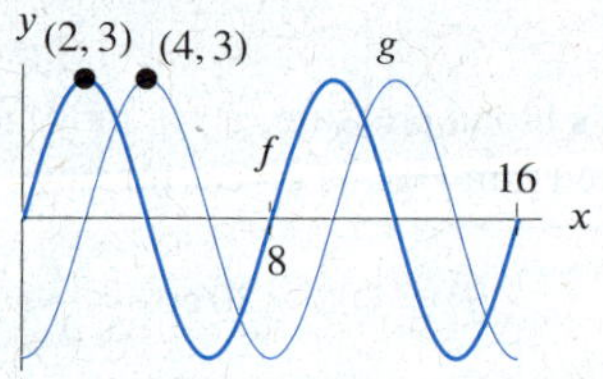


**47.** $f(x) = 10\sin((\pi/5)x)$

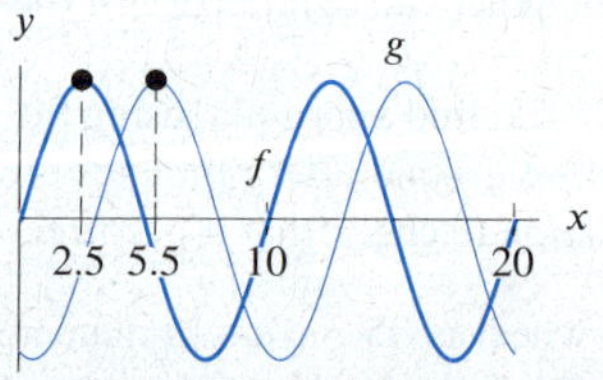


Problems 48–49 show your height $h$ meters above ground after $t$ minutes on a Ferris wheel. Using $h = A\sin(Bt + \phi) + k$, find a formula for $h$ as a function of $t$.

**48.**

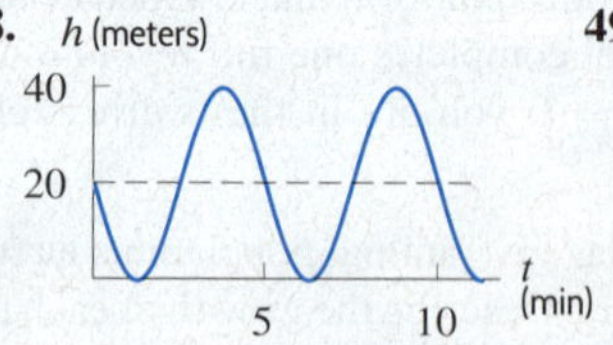


**49.**

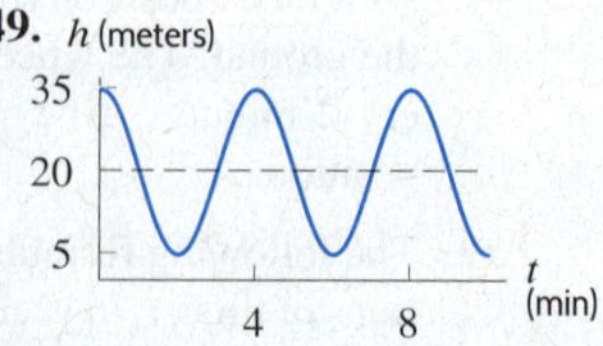


**50.** Describe in words how you can obtain the graph of $y = \cos\left(5t + \frac{\pi}{4}\right)$ from the graph of $y = \cos(5t)$.

[8]http://en.wikipedia.org/wiki/Somerville,_Massachusetts, accessed June 24, 2014.
[9]Statistical Abstract of the United States.

# 7.6 THE TANGENT FUNCTION

## The Tangent Function in the Unit Circle

Another trigonometric function, the tangent function, is defined in terms of the coordinates $(x, y)$ of the point $P$ on the unit circle in Figure 7.75 as follows:

$$\tan\theta = \frac{y}{x} \qquad \text{for } x \neq 0.$$

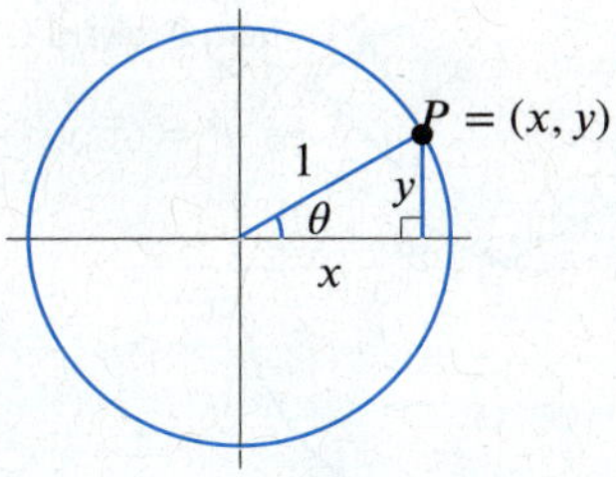


Figure 7.75: Point $P$ on unit circle

Since $x = \cos\theta$ and $y = \sin\theta$, we see that

$$\tan\theta = \frac{\sin\theta}{\cos\theta} \qquad \text{for } \cos\theta \neq 0.$$

## Interpreting the Tangent Function as a Slope

The graphical interpretation of $\tan\theta$ is as the slope of a line through the origin. In Figure 7.76, the line passing through $(0, 0)$ and another point $(x, y)$ has

$$\text{Slope} = \frac{\Delta y}{\Delta x} = \frac{y - 0}{x - 0} = \frac{y}{x}.$$

Choosing $(x, y) = (\cos\theta, \sin\theta)$ we see that

$$\text{Slope} = \frac{y}{x} = \frac{\sin\theta}{\cos\theta} = \tan\theta.$$

In words, $\tan\theta$ is the slope of the line passing through the origin and a point $(x, y)$.

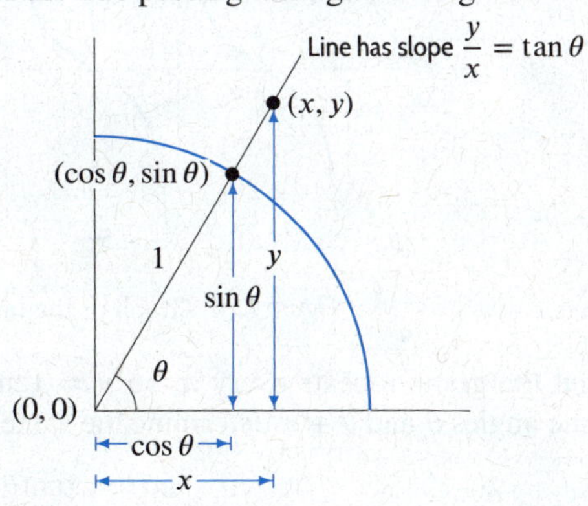


Figure 7.76: The slope of the line passing through the origin and a point $(x, y)$ is $\tan\theta$

**Example 1** Find the slope of the line passing through the origin at an angle of (a) $30^\circ$ (b) $\pi/4$

**Solution** (a) A calculator set in degree mode gives

$$\tan 30^\circ = 0.577.$$

Thus, the slope of the line is about 0.577.

(b) A calculator set in radian mode gives $\tan(\pi/4) = 1$. This makes sense because an angle of $\pi/4$ describes a ray halfway between the $x$- and $y$-axes, or the line $y = x$, which has slope of 1.

## Graph of the Tangent Function

Figure 7.77 illustrates the connection between an angle $\theta$ on the unit circle and the graph of the tangent function.

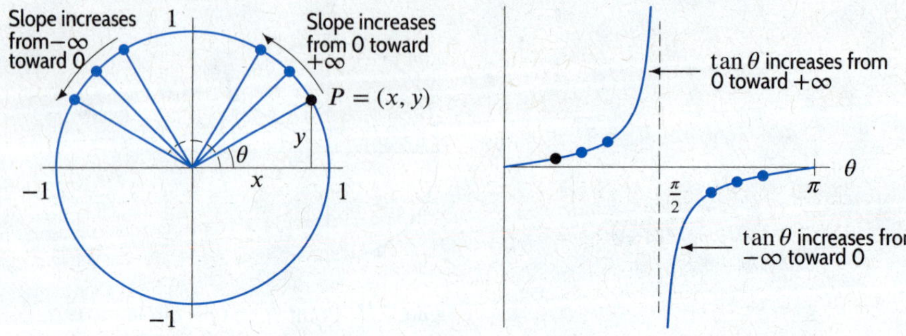


Figure 7.77: The connection between the unit circle and the graph of the tangent function

- As the angle $\theta$ increases from 0 to $\pi/2$, the line passing through $P$ and the origin gets steeper, so the slope increases from 0 to $+\infty$.
- At $\pi/2$, the line is vertical, and its slope is undefined. Thus, $\tan(\pi/2)$ is undefined and the tangent function has a vertical asymptote there.
- For $\theta$ between $\pi/2$ and $\pi$, the slopes are negative. The line is very steep near $\pi/2$, but becomes less steep as $\theta$ approaches $\pi$, where it is horizontal. Thus, $\tan\theta$ becomes less negative as $\theta$ increases in the second quadrant and reaches 0 at $\theta = \pi$.

Figure 7.78 shows the graph of the tangent function on a larger interval, $-2\pi < \theta < 2\pi$.

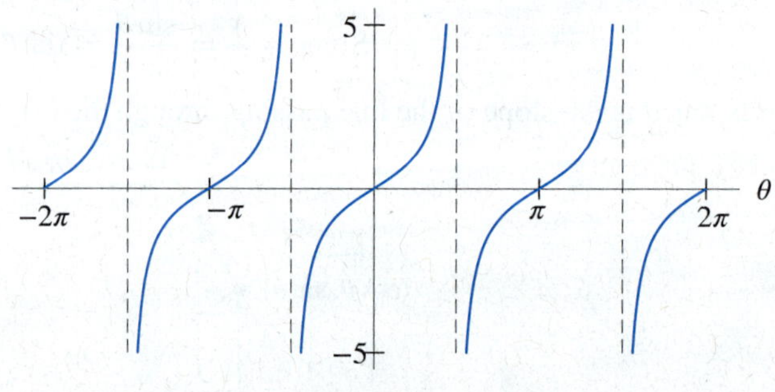


Figure 7.78: Graph of the tangent function

- Notice that the graph repeats every $\pi$, so $y = \tan\theta$ is periodic with period $\pi$. This happens because the angles $\theta$ and $\theta + \pi$ determine the same line through the origin. Therefore

$$\tan\theta = \tan(\theta + \pi).$$

- The tangent does not have an amplitude or midline because it does not have a maximum or minimum value.
- Figure 7.78 shows that the tangent is not defined at $\theta = \pm\pi/2, \pm3\pi/2, \ldots$; the graph has vertical asymptotes there. These are points where the $x$-coordinate of $P$ in Figure 7.77 is zero; that is, where $\cos\theta = 0$.

**Example 2** Give a possible formula for the function $f$ in Figure 7.79.

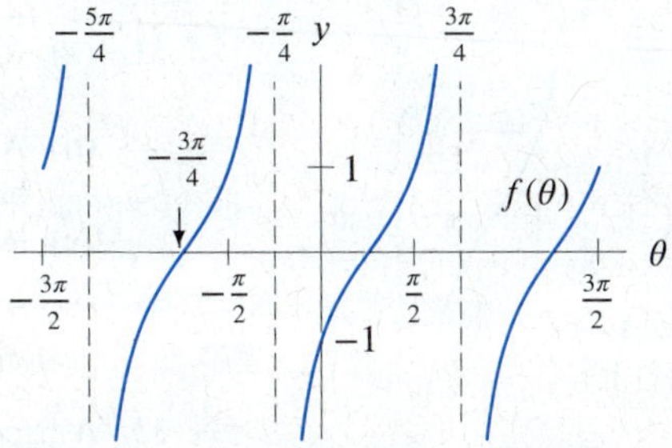


Figure 7.79: Graph of $y = f(\theta)$

**Solution** This looks like a transformation of the tangent function. The function is increasing and the zeros are halfway between the asymptotes, so there is no reflection and no vertical shift. The vertical asymptotes are a distance $\pi$ apart, so the function has period $\pi$ and is a horizontal shift of $\tan\theta$. Thus, for some positive constant $h$, the function graphed is of the form

$$f(\theta) = \tan(\theta - h),$$

Since the smallest positive zero is at $\theta = \pi/4$, a possible formula is

$$f(\theta) = \tan(\theta - \pi/4).$$

## Exercises and Problems for Section 7.6

### Exercises

**1.** Find exact values for $\sin 0°$, $\cos 0°$ and $\tan 0°$.

In Exercises 2–7, evaluate without using a calculator.

**2.** $\cos 90°$ **3.** $\tan 90°$ **4.** $\cos 540°$

**5.** $\tan 540°$ **6.** $\tan 120°$ **7.** $\tan(-30°)$

In Exercises 8–10, find exact values without a calculator.

**8.** $\tan\frac{5\pi}{4}$ **9.** $\tan\frac{\pi}{3}$ **10.** $\tan\left(-\frac{3\pi}{4}\right)$

In Exercises 11–14, find the vertical asymptotes of the function in the interval $0 \le \theta \le 2\pi$.

**11.** $f(\theta) = 2\tan\theta$ **12.** $g(\theta) = \tan(2\theta)$

**13.** $h(\theta) = \tan\theta + 2$ **14.** $r(\theta) = \tan(\theta + 2)$

In Exercises 15–20, graph the function for $-\pi \le \theta \le 2\pi$. How is the graph related to the graph of $y = \tan\theta$?

**15.** $f(\theta) = -\tan\theta$ **16.** $g(\theta) = 2\tan\theta$

**17.** $h(\theta) = 1 + \tan\theta$ **18.** $j(\theta) = \tan(-\theta)$

**19.** $k(\theta) = \tan 2\theta$ **20.** $m(\theta) = \tan(\theta - \pi/2)$

### Problems

In Exercises 21–23, find the slope of the line passing through the origin which satisfies the given additional condition.

**21.** The line forms an angle of $6\pi/7$ with the positive $x$-axis.

**22.** The line is perpendicular to a second line, $l$, passing through the origin at an angle of $5\pi/8$.

**23.** The line is perpendicular to a second line, $l$, passing through the origin at an angle of $35°$.

In Problems 24–26, give a possible formula for the function.

**24.**

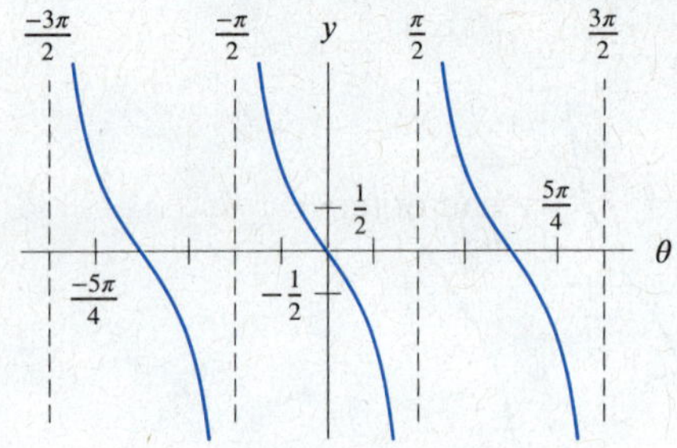


**25.**

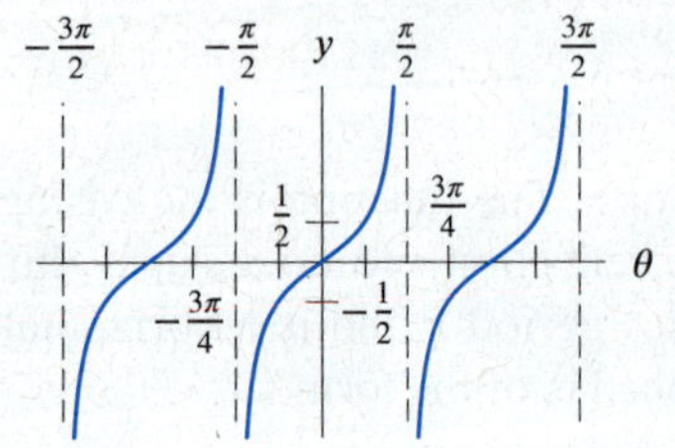


**26.**

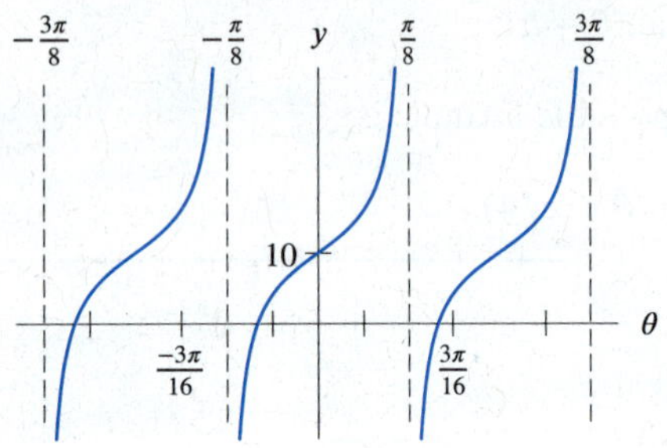


In Problems 27–30,

**(a)** Sketch the graph of a transformed tangent function with the given properties.

**(b)** Use your graph to find a possible formula of the form $f(t) = A\tan(Bt) + C$.

**27.** Vertical asymptotes at every odd integer multiple of $\pi$ and passing through the points $(0, 0)$ and $(\pi/2, 1)$

**28.** Vertical asymptotes at every odd integer multiple of $\frac{\pi}{2}$ and passing through the points $(0, 0)$ and $(-\frac{\pi}{4}, 1)$.

**29.** A vertical asymptote at every odd integer and passing through the points $(0, 0)$ and $(0.5, 1)$.

**30.** Vertical asymptotes at every odd integer multiple of $\frac{\pi}{2}$ and passing through the points $(0, 1)$ and $(-\frac{\pi}{4}, 2)$.

**31.** **(a)** At what values of $t$ does the graph of $y = \tan(t/2)$ have vertical asymptotes? Say how you know.

**(b)** At what values of $t$ does the graph of $y = \tan(t/2)$ have $t$-intercepts? Say how you know.

**(c)** What is the period of $\tan(t/2)$? How do you know?

**32.** Use the fact that sine is an odd function and cosine is an even function to show that tangent is an odd function.

**33.** A friend tells you that a periodic function that increases on one interval must decrease on another interval, else it cannot possibly repeat. Is your friend correct? Explain your answer.

**34.** Use Figure 7.80 to find an equation for the line $l$ in terms of $x_0$, $y_0$, and $\theta$.

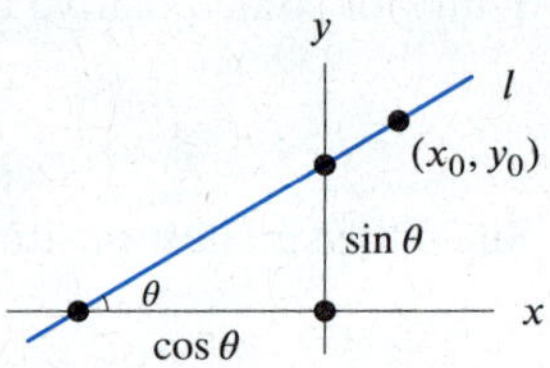


Figure 7.80

**35.** **(a)** Find an equation for the line $l$ in Figure 7.81.

**(b)** Find the $x$-intercept of the line.

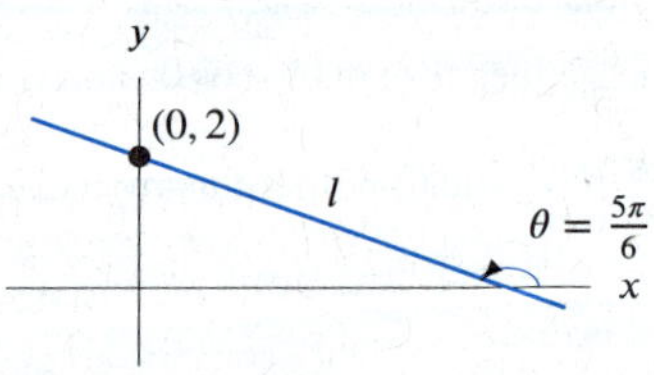


Figure 7.81

**36.** Graph $y = \cos x \cdot \tan x$. Is this function exactly the same as $y = \sin x$? Why or why not?

## 7.7 TRIGONOMETRIC RELATIONSHIPS AND IDENTITIES

In this section we identify relationships among the sine, cosine, and tangent functions. Many of these relationships can be used to solve equations, model phenomena, and explain the shapes of graphs. We also define three new functions: the secant, cosecant, and cotangent.

### The Pythagorean Identity

We now see a central relationship between sine and cosine, the *Pythagorean identity*. Figure 7.82 suggests that no matter what the value of $\theta$, the coordinates of the corresponding point $P$ satisfy the following condition:

$$x^2 + y^2 = 1.$$

But since $x = \cos\theta$ and $y = \sin\theta$, this means

$$\underbrace{(\cos\theta)^2}_{x^2} + \underbrace{(\sin\theta)^2}_{y^2} = 1.$$

To spare parentheses, we usually write $\cos^2\theta$ and $\sin^2\theta$ instead of $(\cos\theta)^2$ and $(\sin\theta)^2$. This gives the Pythagorean identity:

$$\boxed{\cos^2\theta + \sin^2\theta = 1.}$$

In Problems 33 and 34 we consider two other Pythagorean identities relating trigonometric functions.

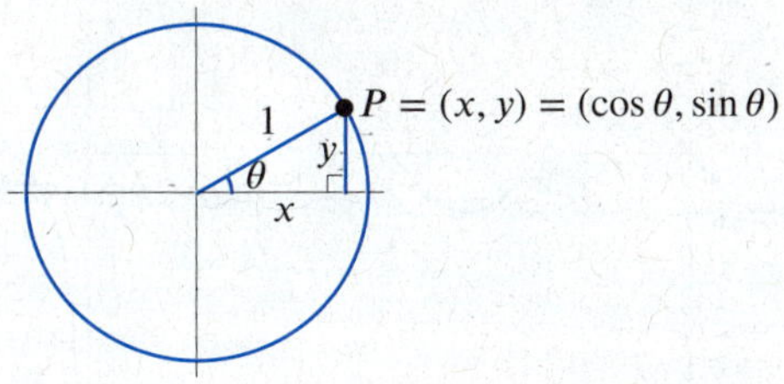


**Figure 7.82:** A right triangle shown with the unit circle

**Example 1** Check the Pythagorean identity for $\theta = 40°$.

**Solution** To three decimal places, a calculator (in degree mode) gives $\sin 40° = 0.643$ and $\cos 40° = 0.766$. We find that

$$(0.643)^2 + (0.766)^2 = 1,$$

confirming, to the accuracy with which we are working, that

$$\sin^2 40° + \cos^2 40° = 1.$$

**Example 2** Represent the Pythagorean identity graphically.

**Solution** The dashed curves in Figure 7.83 show graphs of $y = \sin^2\theta$ and $y = \cos^2\theta$. Notice that where the first graph is near $y = 0$, the second graph is near $y = 1$, and vice versa. The graphs cross at $y = 0.5$. The solid line in the figure shows the graph of $y = \sin^2\theta + \cos^2\theta$, which appears to be the horizontal line $y = 1$. This is what we should expect, since according to the Pythagorean identity, the values of $\sin^2\theta$ and $\cos^2\theta$ add to 1 for all $\theta$.

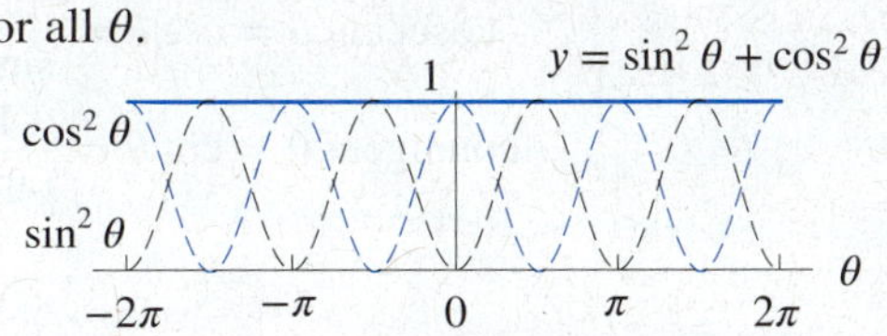


**Figure 7.83:** Graphing the Pythagorean identity

## Relationships Between the Graphs of the Sine, Cosine, and Tangent

As we saw in Section 7.6, if $P = (x, y)$ is the point on the unit circle corresponding to an angle $\theta$, then

$$\cos\theta = x \quad \sin\theta = y \quad \tan\theta = \frac{y}{x}$$

so

$$\tan\theta = \frac{\sin\theta}{\cos\theta}.$$

Many of the properties of the tangent function are explained by this relationship to the sine and the cosine. In Figure 7.84, we see that the zeros of the tangent function are the zeros of its numerator, $\sin\theta$, and the asymptotes of the tangent are the zeros of its denominator, $\cos\theta$. Also, $\tan\theta = 1$ when its numerator and denominator are equal, that is, when $\cos\theta = \sin\theta$.

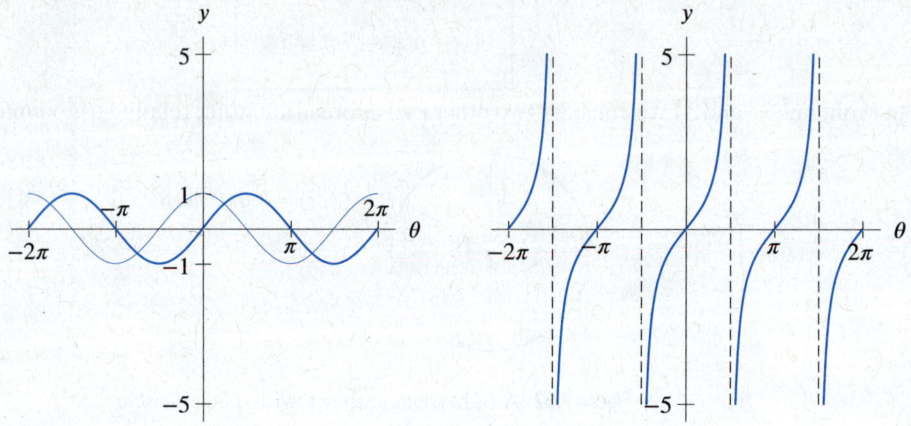


Figure 7.84: Relationships among the sine, cosine and tangent functions can be understood graphically.

## Relationships Involving Reciprocals of the Trigonometric Functions

The reciprocals of the trigonometric functions are given special names. Where the denominators are not equal to zero, we define

$$\text{secant } \theta = \sec\theta = \frac{1}{\cos\theta}.$$

$$\text{cosecant } \theta = \csc\theta = \frac{1}{\sin\theta}.$$

$$\text{cotangent } \theta = \cot\theta = \frac{1}{\tan\theta} = \frac{\cos\theta}{\sin\theta}.$$

**Example 3** Use a graph of $g(\theta) = \cos\theta$ to explain the shape of the graph of $f(\theta) = \sec\theta$.

**Solution** Figure 7.85 shows the graphs of $\cos\theta$ and $\sec\theta$. In the first quadrant $\cos\theta$ decreases from 1 to 0, so $\sec\theta$ increases from 1 toward $+\infty$. The values of $\cos\theta$ are negative in the second quadrant and decrease from 0 to $-1$, so the values of $\sec\theta$ increase from $-\infty$ to $-1$. The graph of $y = \cos\theta$ is symmetric about the vertical line $\theta = \pi$, so the graph of $f(\theta) = \sec\theta$ is symmetric about the same line. Since $\sec\theta$ is undefined wherever $\cos\theta = 0$, the graph of $f(\theta) = \sec\theta$ has vertical asymptotes at $\theta = \pi/2$ and $\theta = 3\pi/2$.

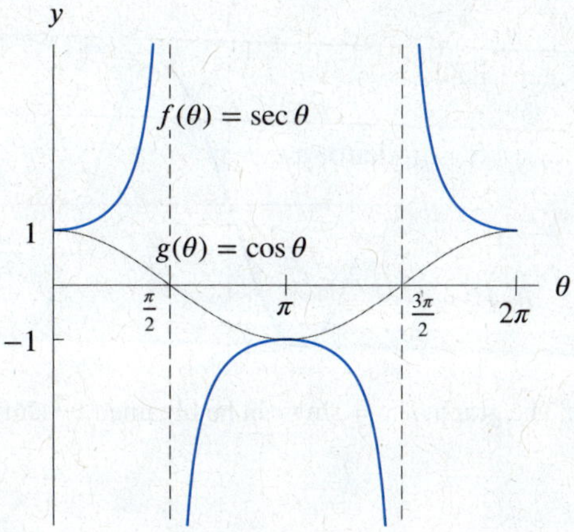


Figure 7.85: Graph of secant

The graphs of $y = \csc\theta$ and $y = \cot\theta$ are obtained in a similar fashion from the graphs of $y = \sin\theta$ and $y = \tan\theta$, respectively. See Figure 7.86 and 7.87.

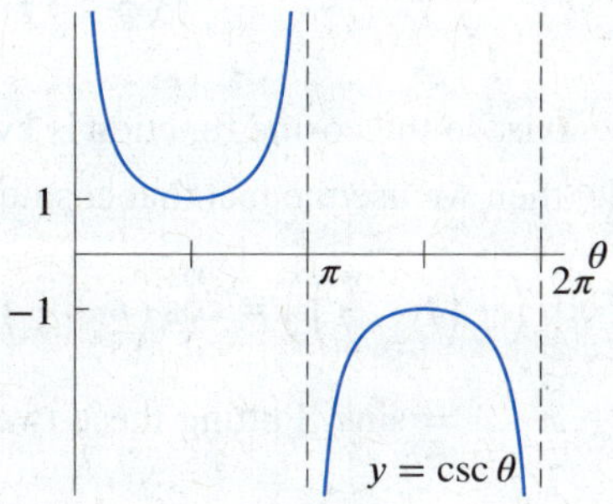


Figure 7.86: Graph of cosecant

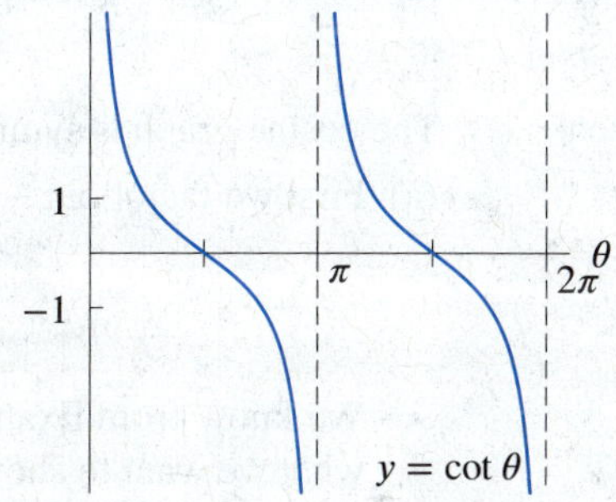


Figure 7.87: Graph of cotangent

## Relationships Between the Graphs of Sine and Cosine

You may have noticed that the graphs of sine and cosine are the same shape; this realization leads to other identities relating the two functions.

**Example 4** Use the fact that the graphs of sine and cosine are horizontal shifts of each other to find relationships between the sine and cosine functions.

Solution Figure 7.88 suggests that the graph of $y = \sin t$ is the graph of $y = \cos t$ shifted right by $\pi/2$ radians (or 90°). Likewise, the graph of $y = \cos t$ is the graph of $y = \sin t$ shifted left by $\pi/2$ radians. Thus,

$$\sin t = \cos\left(t - \frac{\pi}{2}\right), \quad \text{because } \sin t \text{ is } \cos t \text{ shifted right by } \tfrac{\pi}{2}$$

$$\cos t = \sin\left(t + \frac{\pi}{2}\right), \quad \text{because } \cos t \text{ is } \sin t \text{ shifted left by } \tfrac{\pi}{2}.$$

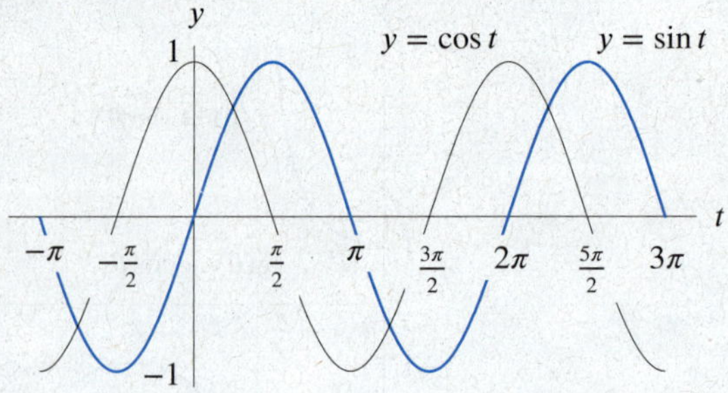


**Figure 7.88:** The graph of $y = \sin t$ can be obtained by shifting the graph of $y = \cos t$ to the right by $\pi/2$

Example 4 tells us that we can replace any sine function with a shifted cosine function, and vice versa. Example 5 shows how we can use known relationships between the graphs of sine and cosine to obtain new relationships.

**Example 5** Use the symmetry of the graph of cosine to obtain the following relationships.

(a) $\sin t = \cos\left(\frac{\pi}{2} - t\right)$ (b) $\cos t = \sin\left(\frac{\pi}{2} - t\right)$

**Solution** The cosine graph is symmetric about the $y$-axis, so the cosine function is even: $\cos(-t) = \cos t$.

(a) First we factor out $-1$ on the right side; then we use the fact that cosine is an even function:

$$\cos\left(\frac{\pi}{2} - t\right) = \cos\left(-\left(t - \frac{\pi}{2}\right)\right) = \cos\left(t - \frac{\pi}{2}\right).$$

We know from Example 4 that $\cos(t - \pi/2) = \sin t$. Putting these two facts together gives us what we want to show:

$$\sin t = \cos\left(\frac{\pi}{2} - t\right).$$

(b) Again from Example 4, we know $\cos t = \sin(t + \pi/2)$. But since cosine is an even function, we can replace $t$ with $-t$, giving us what we want to show:

$$\cos t = \cos(-t) = \sin\left(-t + \frac{\pi}{2}\right) = \sin\left(\frac{\pi}{2} - t\right).$$

We used the symmetry of the graph of cosine to obtain the results of Example 5. We can interpret these results graphically as well. Starting with the graph of $y = \cos t$, we see that:

$$f(t) = \cos\left(t + \frac{\pi}{2}\right)$$ Graph of $f(t)$ is graph of $y = \cos t$ shifted left by $\pi/2$

$$f(-t) = \cos\left(-t + \frac{\pi}{2}\right)$$ Graph of $f(-t)$ is graph of $y = f(t)$ flipped about $y$-axis

$$= \cos\left(\frac{\pi}{2} - t\right).$$

From Figure 7.88, we see that by carrying out this procedure—shifting cosine left by $\pi/2$, then flipping it about the $y$-axis—we obtain the graph of sine, as required.

## Summarizing the Trigonometric Relationships

- Pythagorean Identities

$$\cos^2 t + \sin^2 t = 1$$
$$1 + \tan^2 t = \sec^2 t$$
$$1 + \cot^2 t = \csc^2 t$$

- Tangent and Cotangent

$$\tan t = \frac{\sin t}{\cos t} = \frac{1}{\cot t}$$
$$\cot t = \frac{\cos t}{\sin t} = \frac{1}{\tan t}$$

- Secant and Cosecant

$$\sec t = \frac{1}{\cos t}$$
$$\csc t = \frac{1}{\sin t}$$

- Sine and Cosine Functions

$$\sin t = \cos\left(t - \frac{\pi}{2}\right) = \cos\left(\frac{\pi}{2} - t\right) = -\sin(-t)$$
$$\cos t = \sin\left(t + \frac{\pi}{2}\right) = \sin\left(\frac{\pi}{2} - t\right) = \cos(-t)$$

## Exercises and Problems for Section 7.7

### Exercises

In Exercises 1–7, find exact values without a calculator.

**1.** $\tan \frac{2\pi}{3}$

**2.** $\cot \frac{\pi}{4}$

**3.** $\tan \frac{11\pi}{6}$

**4.** $\csc \frac{5\pi}{4}$

**5.** $\cot \frac{5\pi}{3}$

**6.** $\sec \frac{11\pi}{6}$

**7.** $\sec\left(-\frac{\pi}{6}\right)$

In Exercises 8–15, simplify the expression (for values of the variable for which it is defined).

**8.** $\tan x \cos x$

**9.** $\cos^2(2\theta) + \sin^2(2\theta)$

**10.** $3\sin t - 2(1 - 2\sin t)$

**11.** $\sec t \cos t$

**12.** $\dfrac{\sin\theta}{\tan\theta}$

**13.** $\dfrac{1 - \sin^2 t}{\cos t}$

**14.** $5\cos t - (2\cos t - 3\sin t)$

**15.** $(\sin x - \cos x)^2 + 2\sin x \cos x$

## Problems

Problems 16–18 refer to the unit circle in Figure 7.89. The point $P$ corresponds to an angle of $t$ radians.

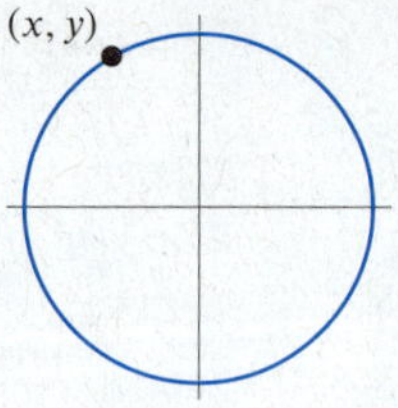


Figure 7.89

**16.** **(a)** What are the coordinates of the point $Q$ corresponding to angle $t + 2\pi$?
**(b)** Show that $\cos(t + 2\pi) = \cos t$.
**(c)** Show that $\sin(t + 2\pi) = \sin t$.

**17.** **(a)** What are the coordinates of the point $Q$ corresponding to angle $t + \pi$?
**(b)** Show that $\cos(t + \pi) = -\cos t$.
**(c)** Show that $\sin(t + \pi) = -\sin t$.
**(d)** Show that $\tan(t + \pi) = \tan t$.

**18.** **(a)** What are the coordinates of the point $Q$ corresponding to angle $-t$?
**(b)** Show that $\cos(-t) = \cos t$.
**(c)** Show that $\sin(-t) = -\sin t$
**(d)** Show that $\tan(-t) = -\tan t$.

In Problems 19–22, give exact answers for $0 \le \theta \le \pi/2$.

**19.** If $\cos\theta = \frac{1}{2}$, what is $\sec\theta$? $\tan\theta$?

**20.** If $\cos\theta = \frac{1}{2}$, what is $\csc\theta$? $\cot\theta$?

**21.** If $\sin\theta = \frac{1}{3}$, what is $\sec\theta$? $\tan\theta$?

**22.** If $\sec\theta = 17$, what is $\sin\theta$? $\tan\theta$?

**23.** If $y = \sin\theta$ for $0° < \theta < 90°$, find $\cos\theta$ in terms of $y$.

**24.** **(a)** If $\cos\alpha = -\sqrt{3}/5$ and $\alpha$ is in the third quadrant, find exact values for $\sin\alpha$ and $\tan\alpha$.
**(b)** If $\tan\beta = 4/3$ and $\beta$ is in the third quadrant, find exact values for $\sin\beta$ and $\cos\beta$.

**25.** **(a)** If $\cos\phi = 0.4626$ and $3\pi/2 < \phi < 2\pi$, find decimal approximations for $\sin\phi$ and $\tan\phi$.
**(b)** If $\sin\theta = -0.5917$ and $\pi < \theta < 3\pi/2$, find decimal approximations for $\cos\theta$ and $\tan\theta$.

Problems 26–29 give an expression for one of the three functions $\sin\theta$, $\cos\theta$, or $\tan\theta$, with $\theta$ in the first quadrant. Find expressions for the other two functions. Your answers will be algebraic expressions in terms of $x$.

**26.** $\sin\theta = x/3$

**27.** $\cos\theta = 4/x$

**28.** $x = 2\cos\theta$

**29.** $x = 9\tan\theta$

**30.** **(a)** Use graphs to decide if each of the following is an identity:
(i) $\cos\left(t - \frac{\pi}{2}\right) = \sin t$
(ii) $\sin 2t = 2\sin t$
**(b)** For each equation that is not an identity, say how many solutions it has on the interval $0 \le t \le 2\pi$.

**31.** **(a)** Graph $y = 2\sin t$ and $y = \sin t + \sin(1.01t)$ on the interval $-2\pi \le t \le 2\pi$. Compare the graphs.
**(b)** Is $\sin t + \sin(1.01t) = 2\sin t$ an identity? Explain.

**32.** Using their graphs, explain the behavior of each the functions $y = \sec x$ and $y = \csc x$ and $y = \cot x$.

**33.** Show how to obtain the identity $1 + \tan^2\theta = \sec^2\theta$ from the identity $\cos^2\theta + \sin^2\theta = 1$.

**34.** Show how to obtain the identity $\cot^2\theta + 1 = \csc^2\theta$ from the identity $\cos^2\theta + \sin^2\theta = 1$.

# 7.8 INVERSE TRIGONOMETRIC FUNCTIONS

What angle has a cosine of $-0.4$? What angle does a line of slope 2 make with the $x$-axis? Since the slope of the line is the tangent of the angle, we need to find an angle $\theta$ such that $\tan\theta = 2$. These and similar questions involving sine, cosine and tangent ask for an input value of a trig function corresponding to a given output value. They can be answered using an *inverse trig function.*

## The Inverse Cosine Function

The inverse cosine function, labeled $\boxed{\cos^{-1}}$ on a calculator, finds one of many possible angles whose cosine is a given value. Consider, for example, the equation, $\cos\theta = -0.4$, which has many solutions. Using the inverse cosine key, on the other hand, gives only one solution:

$$\theta = \cos^{-1}(-0.4) = 1.982.$$

How does the calculator select this particular solution?

In Figure 7.90, notice that, on the part of the graph where $0 \leq \theta \leq \pi$ (shaded blue), all possible cosine values from $-1$ to 1 occur once and once only. Therefore, we can define a new function that assigns just one input value between 0 and $\pi$ to each possible output value of the cosine.[10]

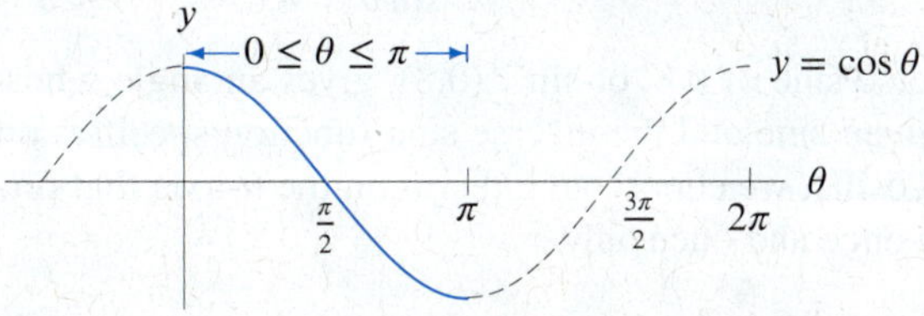


**Figure 7.90**: The solid portion of this graph, for $0 \leq \theta \leq \pi$, represents a function that has only one input value for each output value

The calculator uses the following rule:

$$\cos^{-1} y \text{ is the angle between 0 and } \pi \text{ whose cosine is } y.$$

Because an angle in radians determines an arc of the same measure on a unit circle, the inverse cosine of $y$ is sometimes called the *arccosine* of $y$.

We summarize:

> The **inverse cosine** function, also called the **arccosine** function, is written $\cos^{-1} y$ or $\arccos y$. We define
>
> $$\cos^{-1} y \text{ is the angle between 0 and } \pi \text{ whose cosine is } y.$$
>
> More formally, we say that
>
> $$\theta = \cos^{-1} y \quad \text{provided that} \quad y = \cos\theta \quad \text{and} \quad 0 \leq \theta \leq \pi.$$
>
> Note that the domain of the inverse cosine is $-1 \leq y \leq 1$ and its range is $0 \leq \theta \leq \pi$.

**Warning!** It is important to realize that the inverse function notation $\cos^{-1} y$ does *not* indicate the reciprocal of $\cos y$. In other words, $\cos^{-1} y$ is not the same as $(\cos y)^{-1}$. For example,

$$\cos^{-1}(0) = \frac{\pi}{2} \quad \text{because} \quad \cos\left(\frac{\pi}{2}\right) = 0,$$

but the value of the reciprocal is

$$(\cos 0)^{-1} = \frac{1}{\cos 0} = \frac{1}{1} = 1.$$

**Example 1** Evaluate (a) $\cos^{-1}(0)$ (b) $\arccos(1)$ (c) $\cos^{-1}(-1)$ (d) $(\cos(-1))^{-1}$

**Solution**

(a) The expression $\cos^{-1}(0)$ means the angle between 0 and $\pi$ whose cosine is 0. Since $\cos(\pi/2) = 0$, we have $\cos^{-1}(0) = \pi/2$.

(b) The expression $\arccos(1)$ means the angle between 0 and $\pi$ whose cosine is 1. Since $\cos(0) = 1$, we have $\arccos(1) = 0$.

(c) The expression $\cos^{-1}(-1)$ means the angle between 0 and $\pi$ whose cosine is $-1$. Since $\cos(\pi) = -1$, we have $\cos^{-1}(-1) = \pi$.

(d) This expression, $(\cos(-1))^{-1}$, refers to the reciprocal of the cosine of $-1$. Since (using a calculator) $\cos(-1) = 0.5403$, we have $(\cos(-1))^{-1} = (0.5403)^{-1} = 1.8508$.

[10]Other intervals, such as $-\pi \leq \theta \leq 0$, could also be used. The interval $0 \leq \theta \leq \pi$ has become the agreed-upon choice.

## The Inverse Sine and Inverse Tangent Functions

To solve equations involving sine or tangent, we need an *inverse sine* and an *inverse tangent* function. For example, suppose we want to solve the equation

$$\sin\theta = 0.8, \qquad 0 \le \theta \le 2\pi.$$

The inverse sine of 0.8, or $\sin^{-1}(0.8)$, gives an angle whose sine is 0.8. But there are many angles with a given sine and the inverse sine function specifies only one of them. Just as we did with the inverse cosine, we choose an interval on the $\theta$-axis that produces all values in the range of the sine function once and once only.

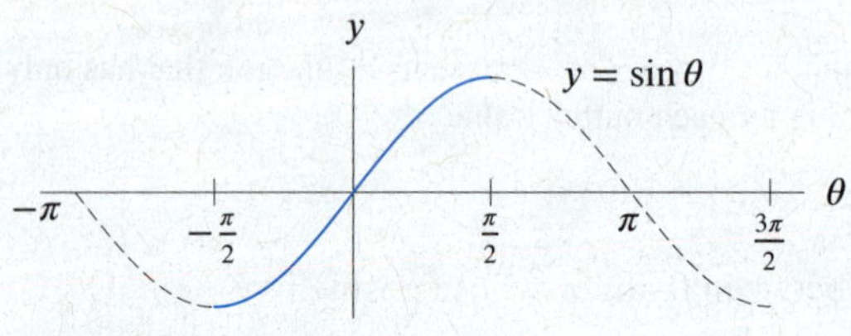


Figure 7.91: The graph of $y = \sin\theta$, $-\frac{\pi}{2} \le \theta \le \frac{\pi}{2}$

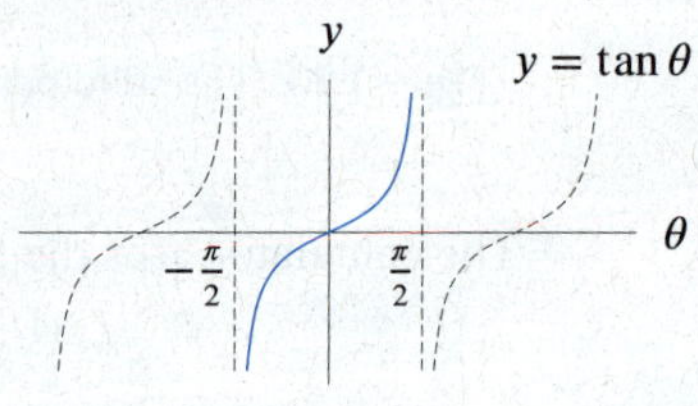


Figure 7.92: The graph of $y = \tan\theta$, $-\frac{\pi}{2} < \theta < \frac{\pi}{2}$

Figure 7.91 shows that values of the sine function repeat on the interval $0 \le \theta \le \pi$ that we chose for $\cos\theta$. However, the interval $-\pi/2 \le \theta \le \pi/2$ includes a unique input for each value of $\sin\theta$. This interval is chosen because it is the smallest interval around $\theta = 0$ that includes all values of $\sin\theta$. This same interval, except for the endpoints, is also used to define the inverse tangent function; see Figure 7.92.

The **inverse sine** function, also called the **arcsine**, is denoted by $\sin^{-1} y$ or $\arcsin y$:

$$\theta = \sin^{-1} y \quad \text{provided that} \quad y = \sin\theta \text{ and } -\frac{\pi}{2} \le \theta \le \frac{\pi}{2}.$$

The domain of the inverse sine is $-1 \le y \le 1$ and the range is $-\pi/2 \le \theta \le \pi/2$.
The **inverse tangent** function, also called the **arctangent**, is denoted by $\tan^{-1} y$ or $\arctan y$:

$$\theta = \tan^{-1} y \quad \text{provided that} \quad y = \tan\theta \text{ and } -\frac{\pi}{2} < \theta < \frac{\pi}{2}.$$

The domain of the inverse tangent is $-\infty < y < \infty$ and the range is $-\pi/2 < \theta < \pi/2$.

**Example 2** Evaluate (a) $\sin^{-1}(1)$ (b) $\arcsin(-1)$ (c) $\tan^{-1}(0)$ (d) $\arctan(1)$.

**Solution**

(a) The expression $\sin^{-1}(1)$ means the angle between $-\pi/2$ and $\pi/2$ whose sine is 1. Since $\sin(\pi/2) = 1$, we have $\sin^{-1}(1) = \pi/2$.
(b) We have, $\arcsin(-1) = -\pi/2$ because $\sin(-\pi/2) = -1$.
(c) We have, $\tan^{-1}(0) = 0$ because $\tan 0 = 0$.
(d) We have, $\arctan(1) = \pi/4$ since $\tan(\pi/4) = 1$.

## Solving Trigonometric Equations Using Reference Angles and the Unit Circle

Because of the symmetry of the unit circle, acute angles (angles in the first quadrant) can be used to find angles with known sine, cosine, or tangent in any quadrant. Acute angles used this way are reference angles, introduced on page 275.

For an angle $\theta$ with point $P$ on the unit circle, the **reference angle**, $\alpha$, is the angle between the line joining $P$ to the origin and the nearest part of the $x$-axis. See Figure 7.93. A reference angle is between 0 and $\pi/2$.

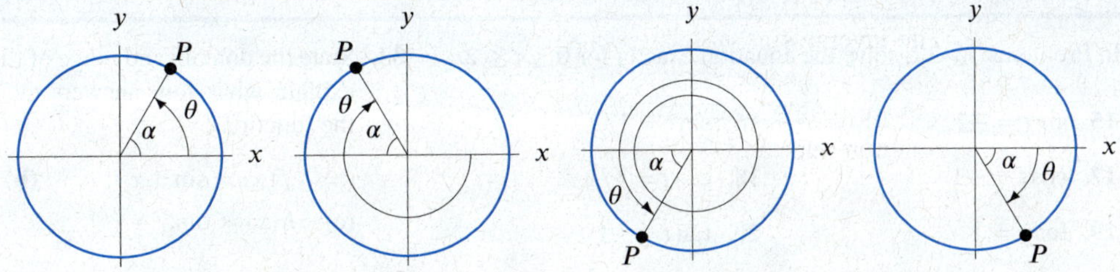


Figure 7.93: An angle $\theta$ in each quadrant whose reference angle is $\alpha$

Reference angles are useful when solving trigonometric equations using the unit circle.

**Example 3** Solve $\sin\theta = 0.9063$ for $0° \le \theta \le 360°$.

**Solution** Using our calculator, we see that one solution is given by $\theta = \sin^{-1}(0.9063) = 65°$. Referring to the unit circle in Figure 7.94, we see that another angle on the interval $0° \le \theta \le 360°$ with sine of 0.9063 is the angle in the second quadrant with a reference angle of 65°. By symmetry, we see that this second angle is $180° - 65° = 115°$.

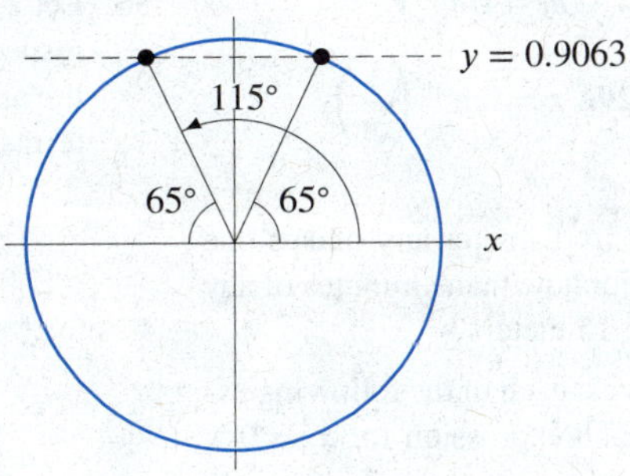


Figure 7.94: Two different angles on the interval $0° \le \theta \le 360°$ have sines equaling 0.9063

## Exercises and Problems for Section 7.8

### Exercises

**1.** Evaluate each expression.

**(a)** $\sin x$ if $x = \left(\frac{1}{2}\right)^\circ$ **(b)** $\sin^{-1} x$ if $x = \frac{1}{2}$

**(c)** $(\sin x)^{-1}$ if $x = \left(\frac{1}{2}\right)^\circ$

**2.** Evaluate each expression.

**(a)** $\tan x$ if $x = 10°$ **(b)** $\tan^{-1} x$ if $x = 10$

**(c)** $(\tan x)^{-1}$ if $x = 10°$

**3.** Evaluate each expression.

**(a)** $\sin x + \cos x + \tan x$ if $x = 45°$

**(b)** $(\sin x)^{-1} + (\cos x)^{-1} + (\tan x)^{-1}$ if $x = 45°$

**(c)** $\sin^{-1} x + \cos^{-1} x + \tan^{-1} x$ if $x = 0.45$

In Exercises 4–7, solve the equation for a value of $\theta$ in the first quadrant. Give your answer in radians and degrees.

**4.** $4\sin\theta = 1$

**5.** $6\cos\theta - 2 = 3$

**6.** $10\tan\theta - 5 = 15$

**7.** $\cos\theta + 4 = 5\cos\theta + 1$

In Exercises 8–11, find a solution with $\theta$ in radians (if possible).

**8.** $\tan\theta = 947$

**9.** $\sin\theta = 4/7$

**10.** $2\cos\theta + 6 = 9$

**11.** $3\cos\theta = 0.714$

**12.** Find the reference angles for

**(a)** 170° **(b)** $7\pi/6$

**13.** Find two angles in different quadrants with a negative cosine value and whose reference angle is $\pi/3$.

**14.** Evaluate $\cos\left(\cos^{-1}(1/2)\right)$ and $\cos^{-1}(\cos(5\pi/3))$ exactly.

## Problems

In Problems 15–22, solve the equation exactly for $0 \le t \le 2\pi$.

**15.** $\sin t = -1$ **16.** $\sin t = 1/2$

**17.** $\cos t = -1$ **18.** $\cos t = 1/2$

**19.** $\tan t = 1$ **20.** $\tan t = -1$

**21.** $\tan t = \sqrt{3}$ **22.** $\tan t = 0$

**23.** Find the angle $\theta$, in radians, in the second quadrant whose tangent is $-3$.

In Problems 24–29, give expressions for the angle and the value of the trigonometric function.

**24.** $\sin k = a$ **25.** $\cos^{-1} a = z$

**26.** $(\tan c)^{-1} = d$ **27.** $m = \sin^{-1} y$

**28.** $p = \cos n$ **29.** $z = \tan^{-1}\left(\frac{1}{y}\right)$

**30.** A Ferris wheel is 20 meters in diameter and makes one revolution every 4 minutes. For how many minutes of any revolution is your seat above 15 meters?

**31.** In your own words, explain what each of the following expressions means. Evaluate each expression for $x = 0.5$. Give an exact answer if possible.

**(a)** $\sin^{-1} x$ **(b)** $\sin(x^{-1})$ **(c)** $(\sin x)^{-1}$.

**32.** Evaluate the following expressions in radians. Give an exact answer if possible.

**(a)** arccos(0.5) **(b)** arccos(−1) **(c)** arcsin(0.1)

**33.** State the domain and range of the following functions and explain what your answers mean in terms of evaluating the functions.

**(a)** $f(x) = \sin^{-1} x$ **(b)** $g(x) = \cos^{-1} x$

**(c)** $h(x) = \tan^{-1} x$

**34.** Without a calculator, evaluate the following exactly.

**(a)** $\cos^{-1}(1/2)$ **(b)** $\cos^{-1}(-1/2)$

**(c)** $\cos(\cos^{-1}(1/2))$ **(d)** $\cos^{-1}(\cos(5\pi/3))$

**35.** One of the following statements is always true; the other is true for some values of $x$ and not for others. Which is which? Justify your answer with an example.

I. $\arcsin(\sin x) = x$ II. $\sin(\arcsin x) = x$

**36.** Let $\ell_1$ and $\ell_2$ be non-parallel lines of slope $m_1$ and $m_2$, respectively, where $m_1 > m_2 > 0$. Find a formula for $\theta$, the angle formed by the intersection of these two lines, in terms of $m_1$ and $m_2$. See Figure 7.95.

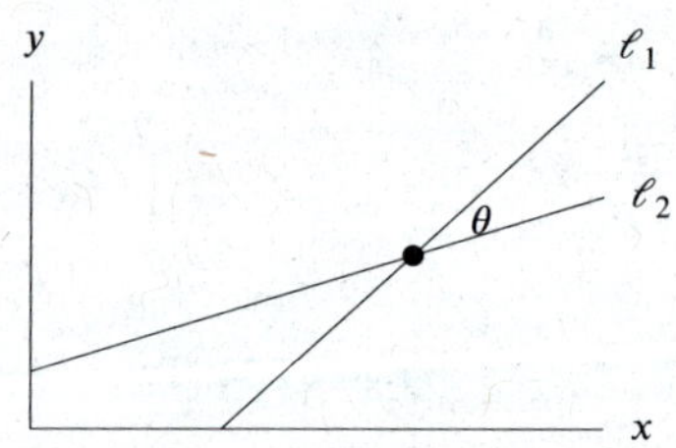


Figure 7.95

## CHAPTER SUMMARY

- **Periodic Functions**
  Values repeat on a regular interval or *period*.
  Circular motion often results in wavelike behavior.
- **Sine and Cosine Functions**
  Unit circle: circle of radius one centered at the origin.
  For $P = (x, y)$ a point on unit circle specified by $\theta$:
  $$\cos\theta = x \quad \text{and} \quad \sin\theta = y$$
  Coordinates of $P = (x, y)$, a point specified by $\theta$ on a circle of radius $r$, are
  $$x = r\cos\theta \quad \text{and} \quad y = r\sin\theta.$$
- **Radians and Arc Length**
  Definition. Relationship to degrees: $2\pi$ radians $= 360°$.
  Arc length: $s = r\theta$.
- **Graphs of Sine and Cosine**
  Domain is all values of $\theta$. Range is $-1 \le y \le 1$.
  Both functions have a period of 360° or $2\pi$:
  $$\sin(\theta + 2\pi) = \sin\theta \quad \text{and} \quad \cos(\theta + 2\pi) = \cos\theta.$$
  Amplitude of $y = A\sin t$ or $y = A\cos t$ is $|A|$.
  Midline of $y = \sin t + k$ or $y = \cos t + k$ is $y = k$.
  Exact values of Sine and Cosine for special angles.

- **Sinusoidal Functions**
General formulas:
$$y = A\sin(B(t-h)) + k,$$
$$y = A\cos(B(t-h)) + k.$$
Amplitude $= |A|$; period $= 2\pi/|B|$; midline $y = k$.
Phase shift is $|\phi|$ for:
$$y = A\sin(Bt+\phi) + k$$
or $$y = A\cos(Bt+\phi) + k.$$

- **Tangent Function**
For $P = (x, y)$ a point on the unit circle specified by $\theta$:
$$\tan\theta = \frac{y}{x}.$$
Has period 180° or $\pi$.
Graph has vertical asymptotes at
$$\theta = \ldots, -\frac{3\pi}{2}, -\frac{\pi}{2}, \frac{\pi}{2}, \frac{3\pi}{2}, \ldots.$$
Interpretation of $\tan\theta$ is the slope of line intersecting $x$-axis at angle $\theta$.

- **Trigonometric Functions and Relationships**
Tangent, secant, cosecant, and cotangent.

- **Identities:**
Relating sine and cosine:
$$\sin t = \cos\left(\frac{\pi}{2} - t\right) = \cos\left(t - \frac{\pi}{2}\right) = -\sin(-t)$$
$$\cos t = \sin\left(\frac{\pi}{2} - t\right) = \sin\left(t + \frac{\pi}{2}\right) = \cos(-t)$$
Pythagorean identity:
$$\cos^2\theta + \sin^2\theta = 1$$
$$1 + \tan^2\theta = \sec^2\theta$$
$$1 + \cot^2\theta = \csc^2\theta$$
Tangent and cotangent:
$$\tan\theta = \frac{\sin\theta}{\cos\theta} = \frac{1}{\cot\theta}.$$
Secant and cosecant:
$$\sec\theta = \frac{1}{\cos\theta}$$
$$\csc\theta = \frac{1}{\sin\theta}$$

- **Inverse Trigonometric Functions**
Inverse cosine:
$$\cos^{-1} y = t \text{ means } y = \cos t \text{ for } 0 \le t \le \pi.$$
Inverse sine:
$$\sin^{-1} y = t \text{ means } y = \sin t \text{ for } -\pi/2 \le t \le \pi/2.$$
Inverse tangent:
$$\tan^{-1} y = t \text{ means } y = \tan t \text{ for } -\pi/2 < t < \pi/2.$$
Reference angles; solving equations.

## REVIEW EXERCISES AND PROBLEMS FOR CHAPTER SEVEN

### Exercises

In Exercises 1–8, do the functions appear to be periodic with period less than 4?

1.
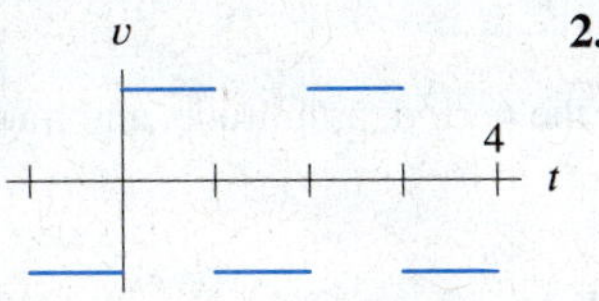


2.
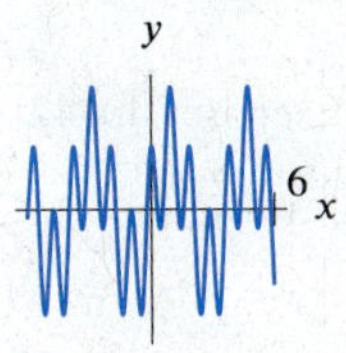


3.
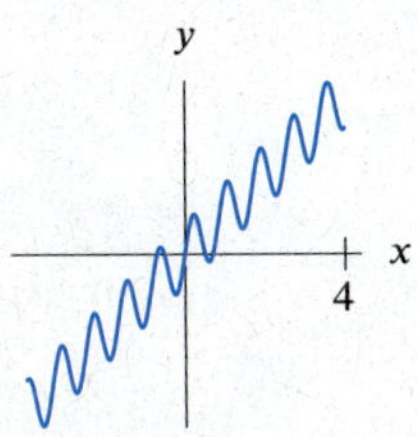


4.
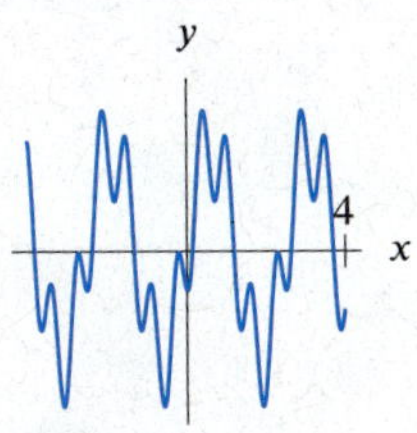


5.
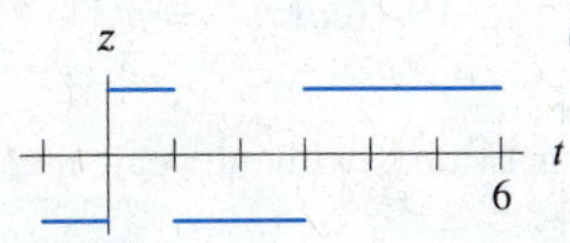


6.
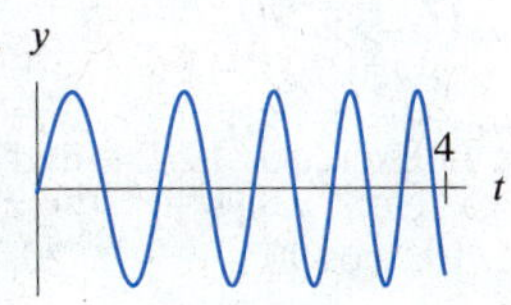


7.

| $t$ | 0 | 1 | 2 | 3 | 4 | 5 | 6 |
|---|---|---|---|---|---|---|---|
| $f(t)$ | 1 | 5 | 7 | 1 | 5 | 7 | 1 |

8.

| $r$ | 0° | 90° | 180° | 270° |
|---|---|---|---|---|
| $q(r)$ | 0 | 1 | 0 | −1 |
| $r$ | 360° | 450° | 540° | 630° |
| $q(r)$ | 1 | 0 | 1 | 0 |

For Exercises 9–10, sketch and find the coordinates of the point corresponding to each angle on the unit circle.

**9.** $S$ is at 225°, $T$ is at 270°, and $U$ is at 330°

**10.** $A$ is at 390°, $B$ is at 495°, and $C$ is at 690°

**11.** The angles in Exercise 9 are on a circle of radius 5. Evaluate the coordinates of $S$, $T$ and $U$.

**12.** The angles in Exercise 10 are on a circle of radius 3. Evaluate the coordinates of $A$, $B$ and $C$.

**13.** Find approximations to two decimal places for the coordinates of point $Z$ in Figure 7.96.

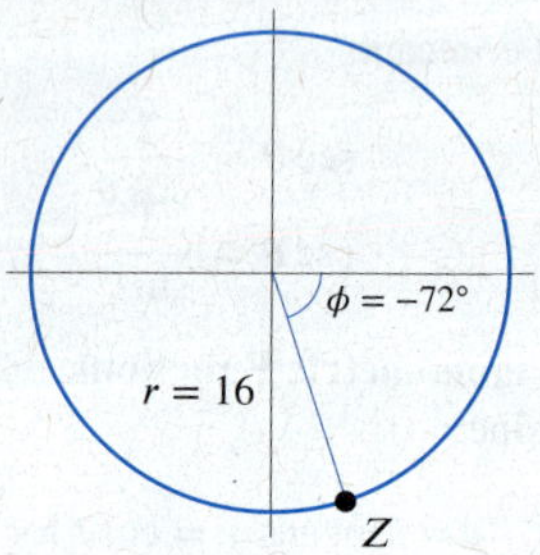


Figure 7.96

**14.** In which quadrants do the following statements hold?

**(a)** $\sin\theta > 0$ and $\cos\theta > 0$
**(b)** $\tan\theta > 0$
**(c)** $\tan\theta < 0$
**(d)** $\sin\theta < 0$ and $\cos\theta > 0$
**(e)** $\cos\theta < 0$ and $\tan\theta > 0$

In Exercises 15–18, give expressions for the angle and the value of the trigonometric function.

**15.** $\cos^{-1} x = y$ **16.** $g = (\sin w)^{-1}$

**17.** $\tan^{-1}\left(c^{-1}\right) = d$ **18.** $(\cos t)^{-1} = p^{-1}$

In Exercises 19–24, find exact values without a calculator.

**19.** $\cos 90°$ **20.** $\sin 90°$ **21.** $\sin 270°$

**22.** $\tan 225°$ **23.** $\tan 135°$ **24.** $\tan 540°$

In Exercises 25–28, convert the angle to radians.

**25.** 330° **26.** 315° **27.** −225° **28.** $6\pi°$

In Exercises 29–31, convert the angle from radians to degrees.

**29.** $\frac{3}{2}\pi$ **30.** 180 **31.** $5\pi/\pi$

In Exercises 32–34, what angle in radians corresponds to the given number of rotations around the unit circle?

**32.** 4 **33.** −6 **34.** 16.4

**35.** If you start at the point $(1, 0)$ on the unit circle and travel counterclockwise through the given angle (in radians), in which quadrant will you be?

**(a)** 2 **(b)** 4 **(c)** 6 **(d)** 1.5 **(e)** 3.2

In Exercises 36–38, find the arc length corresponding to the given angle on a circle of radius 6.2.

**36.** 17° **37.** −585° **38.** $-\dfrac{360°}{\pi}$

**39.** Without a calculator, match the graphs in Figure 7.97 to the following functions:

**(a)** $y = \sin(2t)$ **(b)** $y = (\sin t) + 2$
**(c)** $y = 2\sin t$ **(d)** $y = \sin(t + 2)$

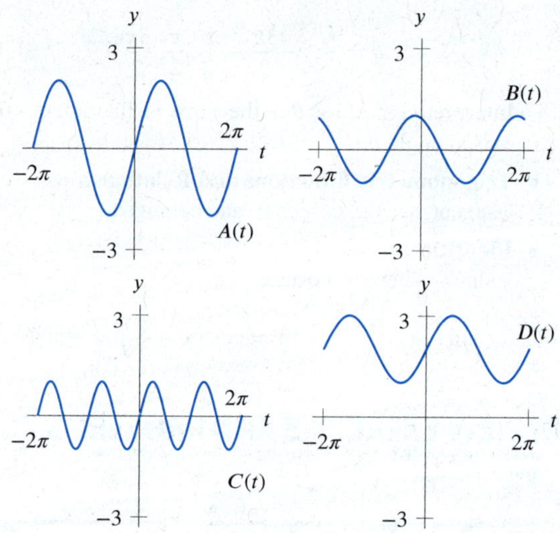


Figure 7.97

In Exercises 40–42, state the period, amplitude, and midline.

**40.** $y = \cos t + 3$

**41.** $y = \sin(t + 3) + 7$

**42.** $6y = 12\sin(\pi t - 7) + 42$

## Problems

**43.** Give an angle that specifies the following positions on the Ferris wheel:

**(a)** two o'clock **(b)** four o'clock
**(c)** ten o'clock

**44.** Use the values of the sine and cosine of 65° to find the sine and cosine of −65°, 245°, and 785°.

In Problems 45–49, use Figure 7.98 to find an interval with $-2 \leq t \leq b$ where $\Delta y/\Delta t$ has the given value. If it is not possible to find such an interval, say why.

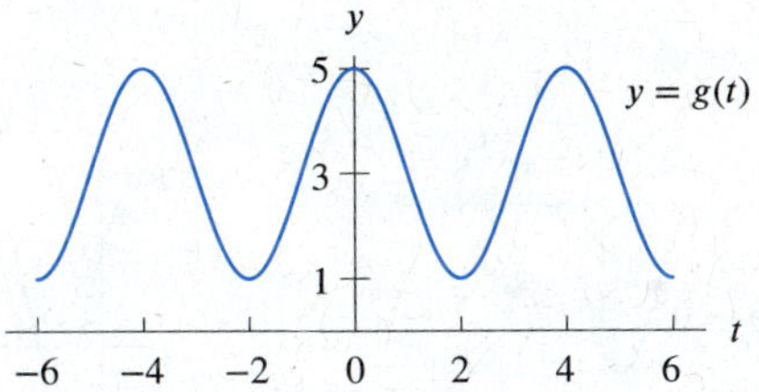


Figure 7.98

**45.** Positive

**46.** Zero

**47.** Negative

**48.** Between 0 and 1

**49.** Equal to 2

**50.** The sinusoidal curves in Figure 7.99 model three different traffic patterns: (i) moderately heavy traffic with some slowdowns; (ii) stop-and-go rush-hour traffic; and (iii) light, fast-moving traffic. Which is which?

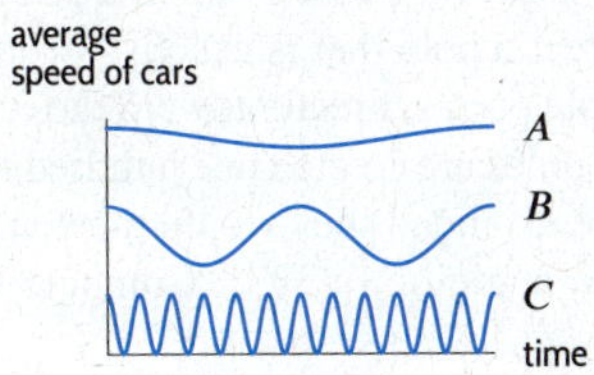


Figure 7.99

In Problems 51–52, estimate the period, amplitude, and midline.

**51.**

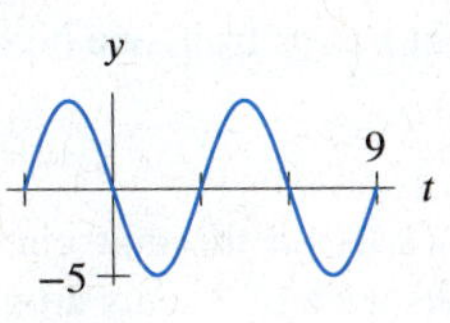


**52.**

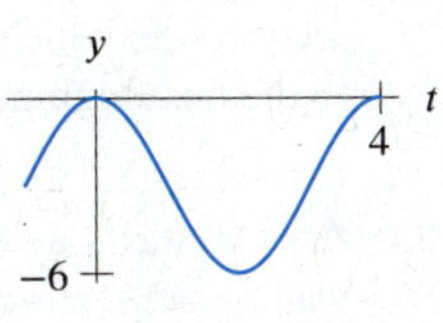


**53.** Table 7.10 gives data from a vibrating spring experiment, with time, $t$, in seconds, and height, $h = f(t)$, in centimeters. Find the midline, amplitude, and period of $f$.

Table 7.10

| $t$ | 0 | .1 | .2 | .3 | .4 | .5 | .6 | .7 | .8 | .9 | 1 |
|---|---|---|---|---|---|---|---|---|---|---|---|
| $h$ | 2 | 2.6 | 3 | 3 | 2.6 | 2 | 1.4 | 1 | 1 | 1.4 | 2 |

**54.** Table 7.11 gives the height $h = f(t)$ in feet of a weight on a spring where $t$ is time in seconds. Find the midline, amplitude, and period of the function $f$.

Table 7.11

| $t$ | 0 | 1 | 2 | 3 | 4 | 5 | 6 | 7 |
|---|---|---|---|---|---|---|---|---|
| $h$ | 4.0 | 5.2 | 6.2 | 6.5 | 6.2 | 5.2 | 4.0 | 2.8 |
| $t$ | 8 | 9 | 10 | 11 | 12 | 13 | 14 | 15 |
| $h$ | 1.8 | 1.5 | 1.8 | 2.8 | 4.0 | 5.2 | 6.2 | 6.5 |

Simplify the expressions in Exercises 55–56.

**55.** $\dfrac{\cos^2 A}{1+\sin A}$

**56.** $\dfrac{\sin 2A}{\tan 2A} + 2\cos 2A$

**57.** You have been riding on the London Eye Ferris wheel (see Section 7.1) for 17 minutes and 30 seconds. What is your height above the platform?

**58.** In the US, household electricity is in the form of *alternating current* (AC) at 155.6 volts and 60 hertz. This means that the voltage cycles from −155.6 volts to +155.6 volts and back to −155.6 volts, and that 60 cycles occur each second. Suppose that at $t = 0$ the voltage at a given outlet is at 0 volts.

**(a)** Sketch $V = f(t)$, the voltage as a function of time, for the first 0.1 seconds.
**(b)** State the period, the amplitude, and the midline of the graph you made in part (a). Describe the physical significance of these quantities.

In Problems 59–62, estimate the amplitude, midline, and period of the sinusoidal functions.

**59.**

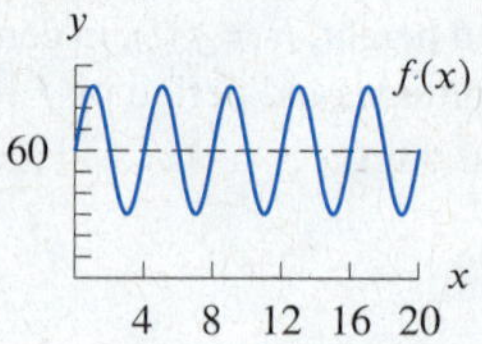


**60.**

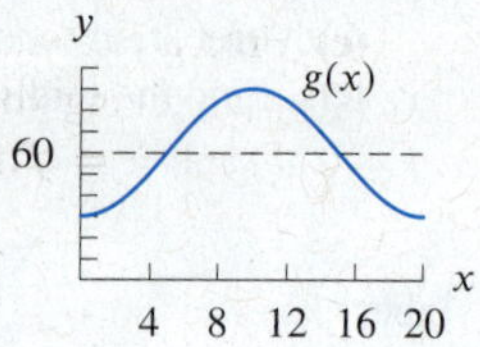


**61.**

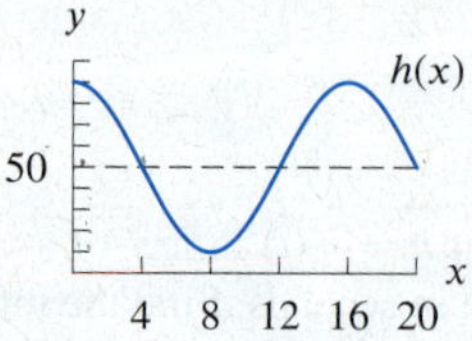


**62.**

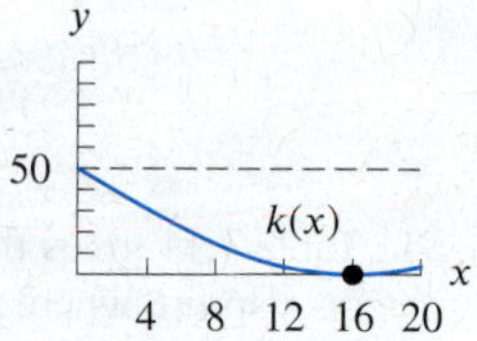


**63.** Using only vertical shifts, stretches, and flips, and horizontal stretches (but no horizontal shifts or flips), would we start with $y = \sin x$ or $y = \cos x$ if we wanted to find a formula for the functions shown in Problems 59–62?

Without a calculator, graph two periods of the functions in Problems 64–67.

**64.** $y = \sin(\frac{1}{2}t)$

**65.** $y = 4\cos(t + \frac{\pi}{4})$

**66.** $y = 5 - \sin t$

**67.** $y = \cos(2t) + 4$

**68.** Suppose $\sin\theta = 1/7$ for $\pi/2 < \theta < \pi$.

**(a)** Use the Pythagorean identity to find $\cos\theta$.
**(b)** What is $\theta$?

In Problems 69–71, find exact values without a calculator.

**69.** $\cos 540°$

**70.** $\sin\dfrac{7\pi}{6}$

**71.** $\tan\left(-\dfrac{2\pi}{3}\right)$

In Problems 72–73, find a solution with $\theta$ in radians.

**72.** $\sin\theta = 2/5$

**73.** $\tan(\theta - 1) = 0.17$

**74.** **(a)** Find a sinusoidal formula for $f$ in Figure 7.100.
**(b)** Find the $x$-coordinates of the three indicated points.

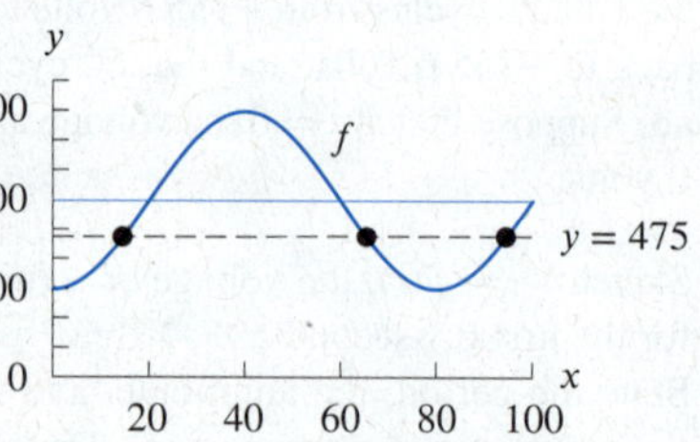


Figure 7.100

For the functions graphed in Problems 75–78, find four possible formulas for the function: $f_1(x) = A\cos(B(x-h)) + k$, where $A > 0$, $f_2(x) = A\cos(B(x-h)) + k$, where $A < 0$, $f_3(x) = A\sin(B(x-h)) + k$, where $A > 0$, and $f_4(x) = A\sin(B(x-h)) + k$, where $A < 0$.

**75.**

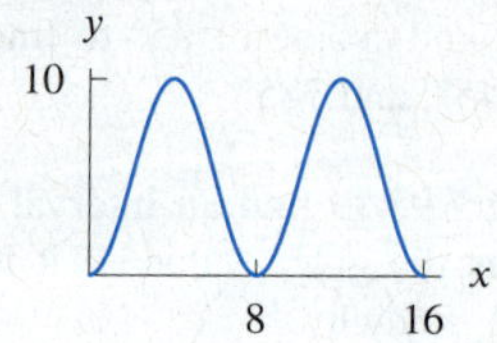


**76.**

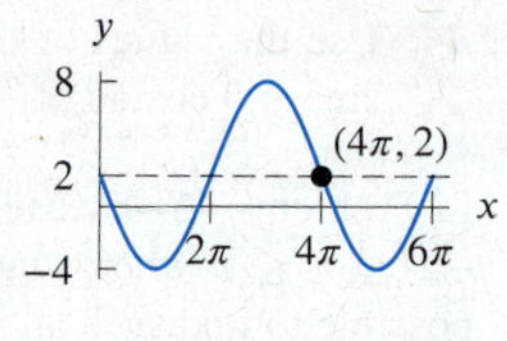


**77.**

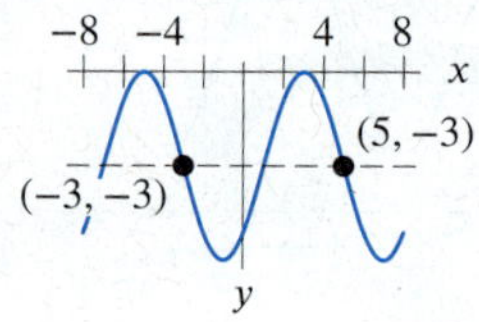


**78.**

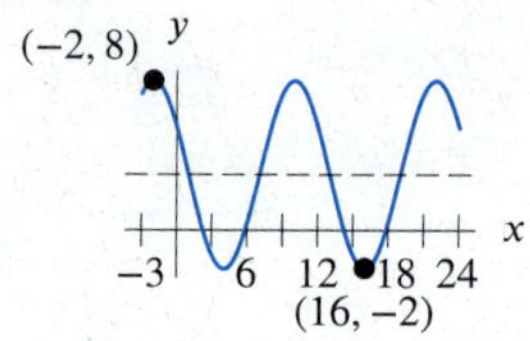


**79.** **(a)** What is the length, $s$, of the arc on a circle of diameter 38 cm cut off by an angle of $\theta = 3.83$ radians?
**(b)** Find the angle, $\varphi$, in radians which cuts off an arc of length 3.83 cm on the circumference of a circle of diameter 38 cm.

**80.** Using a weight on a string called a plumb bob, it is possible to erect a pole that is exactly vertical, which means that the pole points directly toward the center of the earth. Two such poles are erected one hundred miles apart. If the poles were extended they would meet at the center of the earth at an angle of 1.4333°. Compute the radius of the earth.

**81.** How far does the tip of the minute hand of a clock move in 1 hour and 27 minutes if the hand is 2 inches long?

**82.** If a weight hanging on a string of length 3 feet swings through 5° on either side of the vertical, how long is the arc through which the weight moves from one high point to the next high point?

**83.** An ant starts at the point (1, 0) on the unit circle and walks counterclockwise a distance of 3 units around the circle. Find the $x$ and $y$ coordinates (accurate to 2 decimal places) of the final location of the ant.

**84.** A compact disk is 12 cm in diameter and rotates at 100 rpm (revolutions per minute) when being played. The hole in the center is 1.5 cm in diameter. Find the speed in cm/min of a point on the outer edge of the disk and the speed of a point on the inner edge.

**85.** An animal population increases from a low of 1200 in year $t = 0$, to a high of 3000 four years later and then decreases back to 1200 over the next four years. Model this behavior by a sinusoidal function.

State the amplitude, period, phase shift, and horizontal shifts for the function in Problems 86–89. Without a calculator, graph the function on the given interval.

**86.** $y = -4\sin t, \quad -2\pi \le t \le 2\pi$

**87.** $y = -20\cos(4\pi t), \quad -\frac{3}{4} \le t \le 1$

**88.** $y = \cos\left(2t + \frac{\pi}{2}\right), \quad -\pi \le t \le 2\pi$

**89.** $y = 3\sin(4\pi t + 6\pi), \quad -\frac{3}{2} \le t \le \frac{1}{2}$

**90.** The data in Table 7.12 gives the height above the floor of a weight bobbing on a spring attached to the ceiling. Fit a function of the form $y = A\sin(Bt + \theta) + k$ to this data.

**Table 7.12**

| $t$, sec | 0.0 | 0.1 | 0.2 | 0.3 | 0.4 | 0.5 |
|---|---|---|---|---|---|---|
| $y$, cm | 120 | 136 | 165 | 180 | 166 | 133 |
| $t$, sec | 0.6 | 0.7 | 0.8 | 0.9 | 1.0 | 1.1 |
| $y$, cm | 120 | 135 | 164 | 179 | 165 | 133 |

**91.** The London Ferris wheel has diameter 400 feet and one complete revolution takes 30 minutes.

**(a)** Find the rate (in degrees per minute) that the London Ferris wheel is rotating.

**(b)** Let $t$, the time in minutes, be 0 when you are in the 6 o'clock position. Write $\theta$, measured from the 3 o'clock position, as a function of $t$.

**(c)** Find a formula for the Ferris wheel function, $h = f(t)$, giving your height in feet above the platform.

**(d)** Graph $h = f(t)$. What are the period, midline, and amplitude?

**92.** The website arXiv.org is used by scientists to share research papers. Figure 7.101 shows usage[11] of this site, $n = f(t)$, the number of new connections on day $t$, where $t = 0$ is Monday, August 5, 2002.

**(a)** The graph suggests that $f$ is approximately periodic with a period of 7 days. However, $f$ is not exactly periodic. How can you tell?

**(b)** Why might the usage of the arXiv.org website lend itself to being modeled by a periodic function?

**(c)** Find a trigonometric function that gives a reasonable approximation for $f$. Explain what the period, amplitude, and midline of your function tell you about the usage of the arXiv.org website.

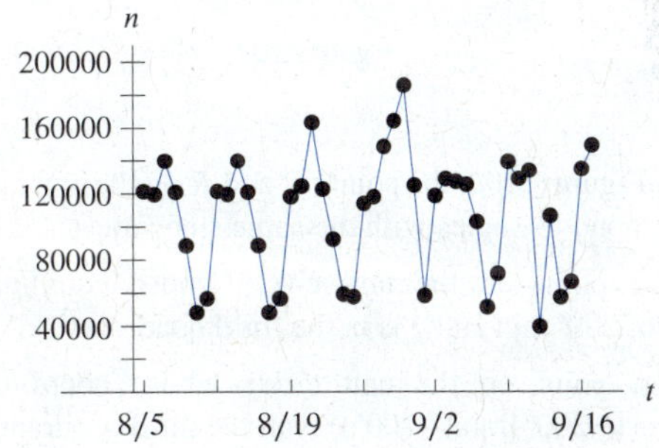


Figure 7.101

## STRENGTHEN YOUR UNDERSTANDING

Are the statements in Problems 1–4 true or false for all values of $x$? Give an explanation for your answer.

**1.** $\sin(-x) = -\sin x$

**2.** $\cos(-x) = -\cos x$

**3.** $\sin(-x) = \sin x$

**4.** $\cos(-x) = \cos x$

Are the statements in Problems 5–72 true or false? Give an explanation for your answer.

**5.** The tangent function has a period of $\pi$.

**6.** The function $\sin x$ has period $2\pi$.

**7.** The point $(1, 0)$ on the unit circle corresponds to $\theta = 0°$.

**8.** If $P = (x, y)$ is a point on the unit circle and $\theta$ is the corresponding angle, then $\sin\theta = y$.

**9.** A unit circle may have a radius of 3.

**10.** If a point $(x, y)$ is on the circumference of a circle of radius $r$ and the corresponding angle is $\theta$, then $x = r\cos\theta$.

**11.** An angle of one radian is about equal to an angle of one degree.

[11]Data obtained from usage statistics available at the arXiv.org site, accessed September, 2002. Usage statistics for August 28, 2002, are unavailable and have been estimated.

**12.** In a unit circle, one complete revolution about the circumference is about 6.28 radians.

**13.** The radian measure of an angle is the length of the arc spanned by the angle in a unit circle.

**14.** An angle of three radians corresponds to a point in the third quadrant.

**15.** The coordinates of the point of intersection of the terminal ray of a 240° angle with a circle of radius 2 are $(-\sqrt{3}, -1)$.

**16.** In Figure 7.102, the points $P$ and $Q$ on the unit circle correspond to angles with the same cosine values.

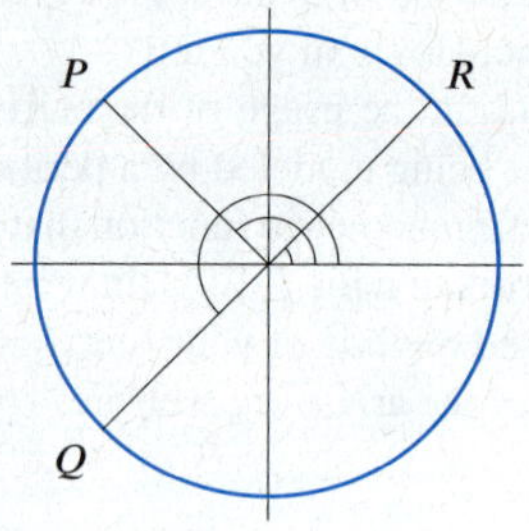


Figure 7.102

**17.** In Figure 7.102, the points $P$ and $R$ on the unit circle correspond to angles with the same sine values.

**18.** The point on the unit circle whose coordinates are $(\cos 200°, \sin 200°)$ is in the third quadrant.

**19.** The point on the unit circle whose coordinates are $(\cos(-200°), \sin(-200°))$ is in the third quadrant.

**20.** To convert an angle from degrees to radians you multiply the angle by $\dfrac{180°}{\pi}$.

**21.** The length of an arc $s$ spanned in a circle of radius 3 by an angle of $\dfrac{\pi}{3}$ is 180.

**22.** An angle of $2\pi/3$ radians corresponds to a point in the second quadrant.

**23.** The value of $\cos 315°$ is $-\sqrt{2}/2$.

**24.** The smallest positive constant $c$ for which $f(x+c) = f(x)$ is called the period of $f$.

**25.** The amplitude of a periodic function is the difference between its maximum and minimum values.

**26.** The midline of a periodic function is the horizontal line $y = \dfrac{\text{Maximum} + \text{Minimum}}{2}$.

**27.** A parabola could represent a periodic function.

**28.** If $f$ is a periodic function, then there exists a constant $c$ such that $f(x+c) = f(x)$ for all $x$ in the domain of $f$.

**29.** If $f(t)$ and $g(t)$ are periodic functions with period $A$ and $f(t) = g(t)$ for $0 \le t < A$, then $f(t) = g(t)$ for all $t$.

**30.** The cosine of 30° is the same as $\sin(\pi/3)$.

**31.** $-\sin(\pi/3) = \sin(-\pi/3)$.

**32.** In Figure 7.103, the function graphed appears to be periodic with period less than 5.

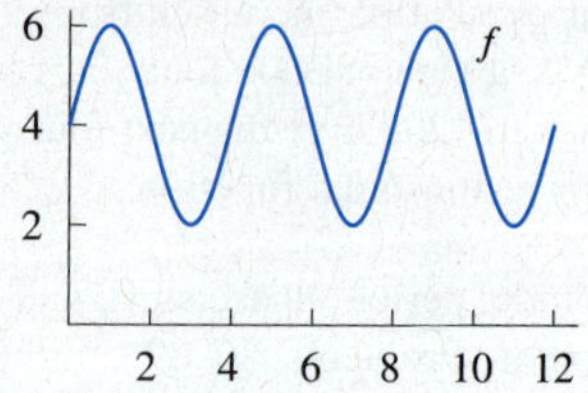


Figure 7.103

**33.** In Figure 7.103, the amplitude of the function is 6.

**34.** In Figure 7.103, the period of the function is 4.

**35.** In Figure 7.103, the midline has equation $y = 4$.

**36.** The amplitude of $y = -3\sin(2x) + 4$ is $-3$.

**37.** The amplitude of $y = 25 + 10\cos x$ is 25.

**38.** The period of $y = 25 + 10\cos x$ is $2\pi$.

**39.** The maximum $y$-value of $y = 25 + 10\cos x$ is 10.

**40.** The minimum $y$-value of $y = 25 + 10\cos x$ is 15.

**41.** The midline equation for $y = 25 + 10\cos x$ is $y = 35$.

**42.** The function $\cos x$ is a sinusoidal function.

**43.** The function $\sin(\pi x)$ has period $\pi$.

**44.** The function $y = -2\sin x + k$ has amplitude $-2$.

**45.** The graph of the function $y = 3\cos x - 4$ is the graph of the function $y = \cos x$ reflected across the $x$-axis.

**46.** The function $f(t) = \sin(2t)$ has period $\pi$.

**47.** The function $f(x) = \cos(3x)$ has a period three times as large as the function $g(x) = \cos x$.

**48.** Changing the value of $B$ in the function $y = A\sin(Bx) + k$ changes the period of the function.

**49.** A sinusoidal function that has a midline of $y = 5$, an amplitude of 3, and completes 4 cycles in the interval $0 \le x \le 2\pi$ could have the equation $y = -3\cos(4x) + 5$.

**50.** In Figure 7.104, the function graphed could have the equation $y = \frac{1}{2}\sin(2x) + 1$.

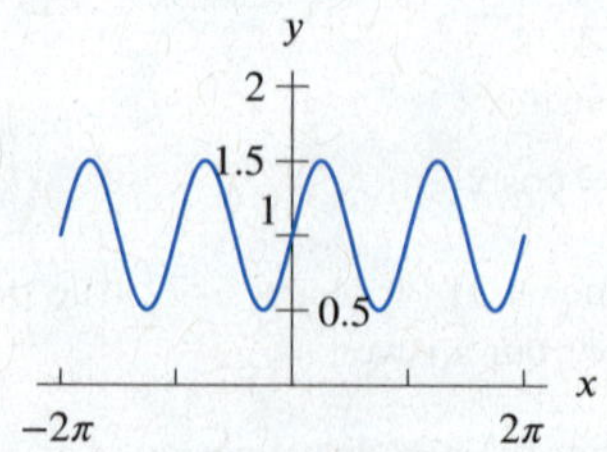


Figure 7.104

Note that $x$ is positive because $Q$ is in the first quadrant. Since $x$ and $y$ are the coordinates of $Q$, this means that $\cos 30° = \sqrt{3}/2$ and $\sin 30° = 1/2$.

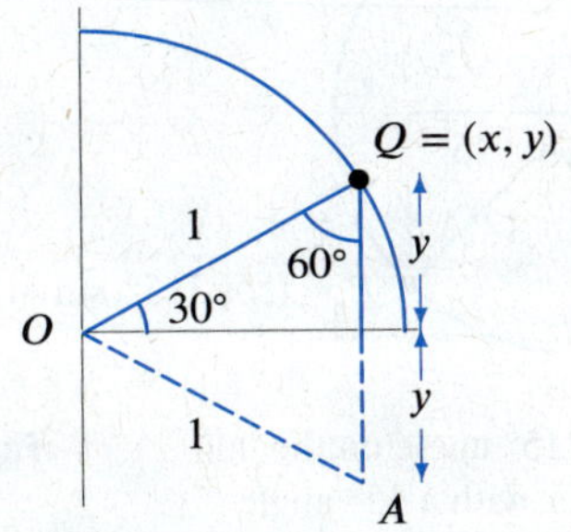


Figure 7.106: The triangle $\Delta OQA$ has three equal angles and three equal sides, so $2y = 1$

A similar argument shows that $\cos 60° = 1/2$ and $\sin 60° = \sqrt{3}/2$.

It is worth memorizing[12] the values of sine and cosine for these special angles as well as for the angles $\theta = 0°$ and $\theta = 90°$, which we found in Example 2 on page 274. See Table 7.13.

Table 7.13 *Trigonometric functions of special angles*

| $\theta$ (degrees) | $\theta$ (radians) | $\cos\theta$ | $\sin\theta$ |
|---|---|---|---|
| 0° | 0 | 1 | 0 |
| 30° | $\pi/6$ | $\sqrt{3}/2$ | $1/2$ |
| 45° | $\pi/4$ | $\sqrt{2}/2$ | $\sqrt{2}/2$ |
| 60° | $\pi/3$ | $1/2$ | $\sqrt{3}/2$ |
| 90° | $\pi/2$ | 0 | 1 |

## Other Values of the Sine and Cosine Functions

We can use the symmetry of the unit circle to find other values of the sine and cosine functions, as illustrated by Examples 3 and 4.

**Example 3** Find $\sin 315°$ and $\cos 315°$.

**Solution** Figure 7.107 shows the angle $\theta = 315°$ on the unit circle, together with a reference angle of 45°. From the figure, we see that point $P$ has coordinates $P = (\cos 315°, \sin 315°)$. But from the triangle in the figure, we see that $P = (\sqrt{2}/2, -\sqrt{2}/2)$. We conclude that

$$\cos 315° = \frac{\sqrt{2}}{2} \quad \text{and} \quad \sin 315° = -\frac{\sqrt{2}}{2}.$$

[12]A useful trick is to notice that for $\theta = 0°, 30°, 45°, 60°, 90°$, the value of sine follows the pattern $\sqrt{0}/2$, $\sqrt{1}/2$, $\sqrt{2}/2$, $\sqrt{3}/2$, $\sqrt{4}/2$. The value of cosine reverses this pattern.

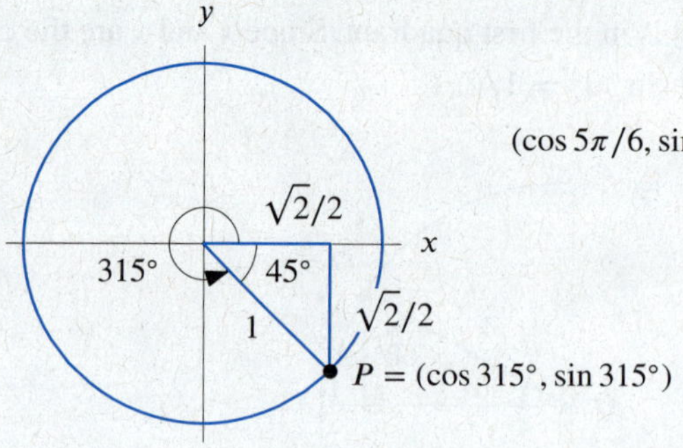


**Figure 7.107:** A 315° angle on the unit circle, together with a 45° angle

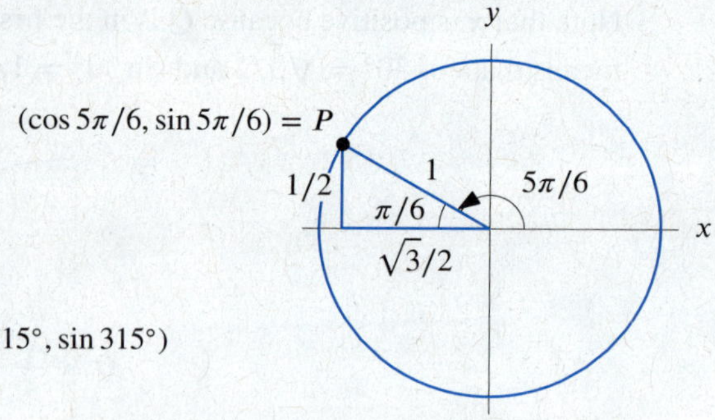


**Figure 7.108:** An angle of $5\pi/6$ on the unit circle, together with an angle of $\pi/6$

**Example 4** Find $\sin 5\pi/6$ and $\cos 5\pi/6$.

**Solution** Figure 7.108 shows the angle of $5\pi/6$ on the unit circle, together with a reference angle of $\pi/6$. The coordinates of $P$ are $(\sqrt{3}/2, 1/2)$ so

$$\cos 5\pi/6 = -\frac{\sqrt{3}}{2} \quad \text{and} \quad \sin 5\pi/6 = \frac{1}{2}.$$

## Exercises to Skills for Special Angles

### Exercises

Find exact values for the expressions in Exercises 1–17.

**1.** $\sin 30°$

**2.** $\sin 150°$

**3.** $\sin 300°$

**4.** $\sin(-60°)$

**5.** $\cos(-60°)$

**6.** $\sin 120°$

**7.** $\cos(-30°)$

**8.** $\sin 210°$

**9.** $\sin 135°$

**10.** $\cos 300°$

**11.** $\cos 210°$

**12.** $\sin 330°$

**13.** $\sin 405°$

**14.** $\cos 120°$

**15.** $\cos 135°$

**16.** $\cos 225°$

**17.** $\sin 300°$

### Problems

**18.** Compare the values of $\sqrt{1/2}$ and $\sqrt{2}/2$ using a calculator. Explain your observations.

**19.** Compare the values of $\sqrt{3/4}$ and $\sqrt{3}/2$ using your calculator. Explain your observations.

In Exercises 20–24, use a trigonometric function to find the exact value of $x$.

**20.**

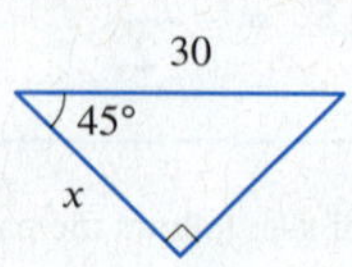


**21.**

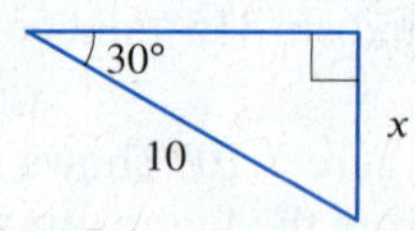


**22.**

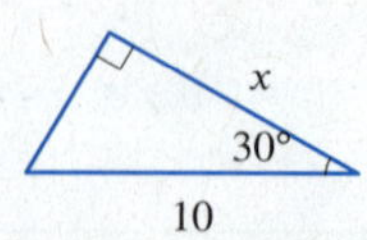

**23.**

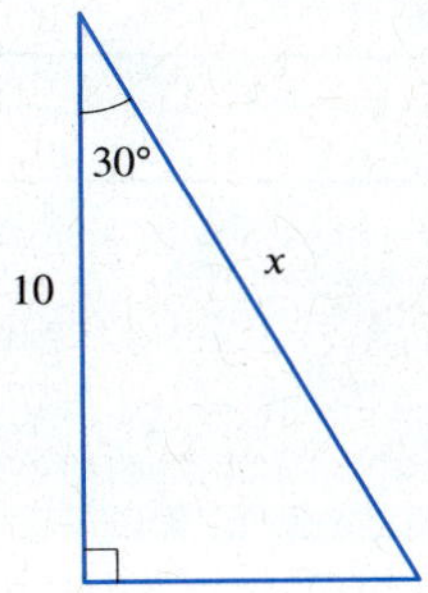


**24.**

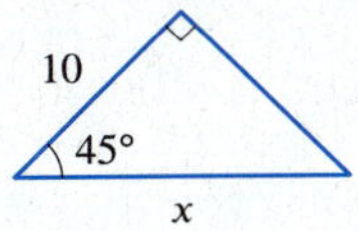


In Exercises 25–28, find exact values for the missing lengths in the triangle. Use the relationships between the sides in a right triangle with special angles.

**25.**

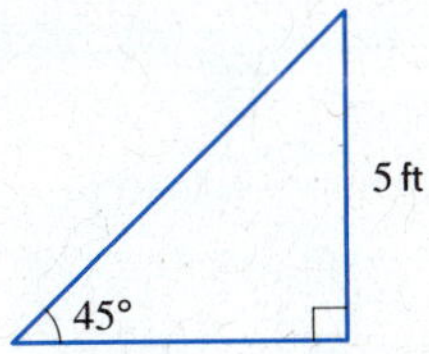


**26.**

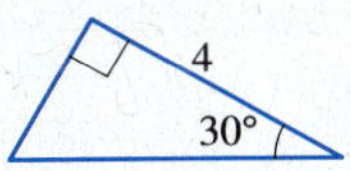


**27.**

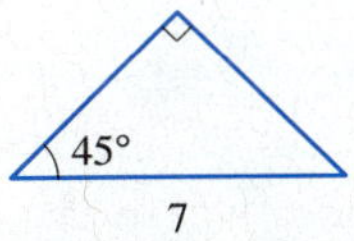


**28.**

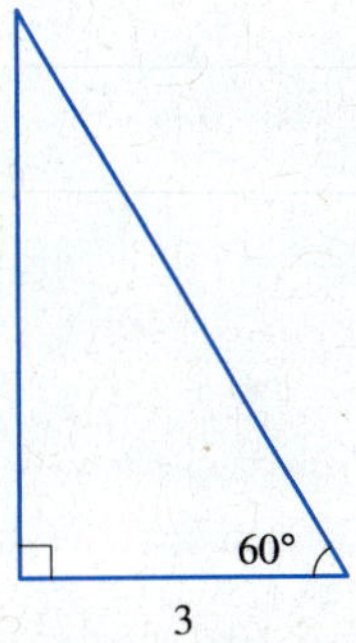


**29.** A right triangle has two sides of length 4. What are the angles? What is the exact length of the third side?

**30.** A right triangle has one side of length 7 and hypotenuse of 14. What are its angles? What is the exact length of the third side?

**31.** Find the exact coordinates of a point $B$ designated by 315° on a circle of radius 6.

**32.** A revolving door is shown in Figure 7.109. Each of the four panels is one meter wide. What is $d$, the width of the opening from $B$ to $C$?

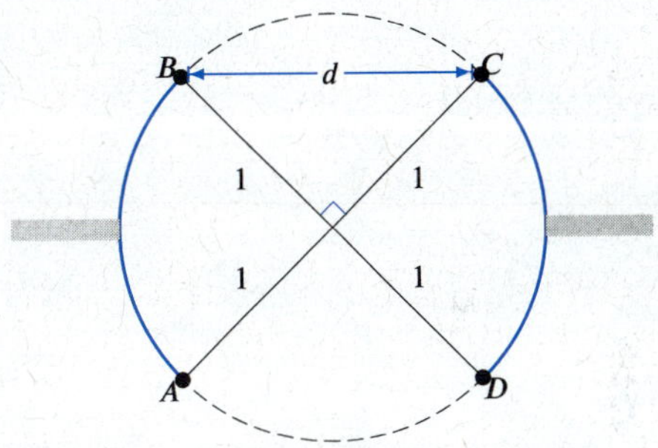


Figure 7.109

Chapter Eight

# TRIANGLE TRIGONOMETRY AND POLAR COORDINATES

© Patrick Zephyr/Patrick Zephyr Nature Photography

## Contents

# 8.1 TRIG FUNCTIONS AND RIGHT TRIANGLES

## The Sine and Cosine Functions in Right Triangles

The sine and cosine functions from Section 7.2 are called trigonometric functions because they can be defined in terms of triangles. Relationships between the sides and angles in a right triangle are used to calculate unknown lengths or angles.

When the cloud ceiling is too low, for example, it can be dangerous for planes to fly. To determine how low it is, a scientist shines a laser vertically upward from the ground, making a white spot on a cloud. A second scientist at a different location measures the angle above the horizon where she sees the spot. The second scientist is 2 km away from the first on level ground and sees the white spot 35° above the horizon. See Figure 8.1. Can a pilot who needs the cloud ceiling to be above 800 meters fly safely?

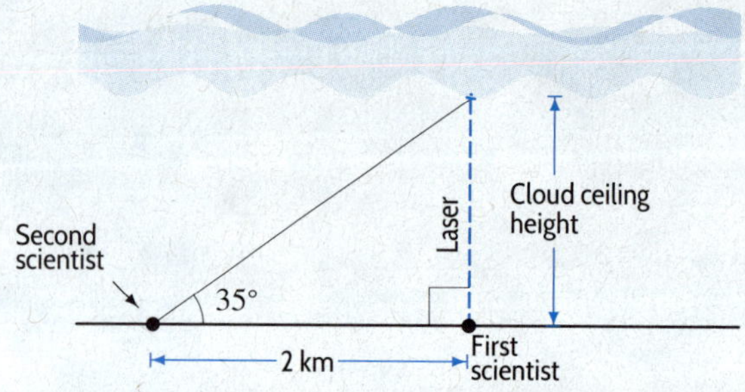


Figure 8.1: How high is the cloud ceiling?

Notice that the triangle in Figure 8.1 is a right triangle. Before finding the height of the cloud ceiling in Example 4, we look at the relationship between the sides and angles in right triangles. See Figure 8.2.

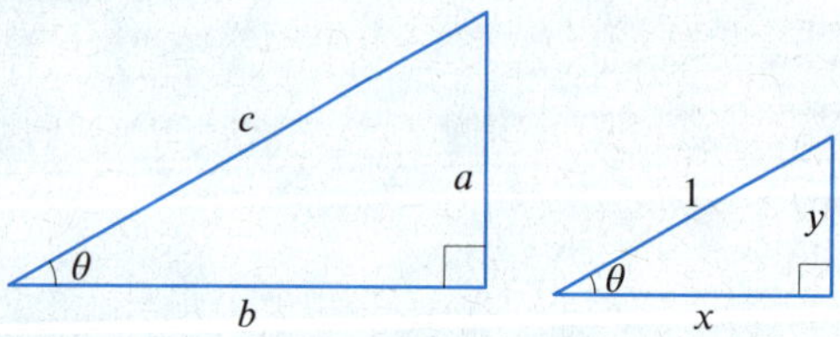


Figure 8.2: Similar triangles: Corresponding sides have equal ratios

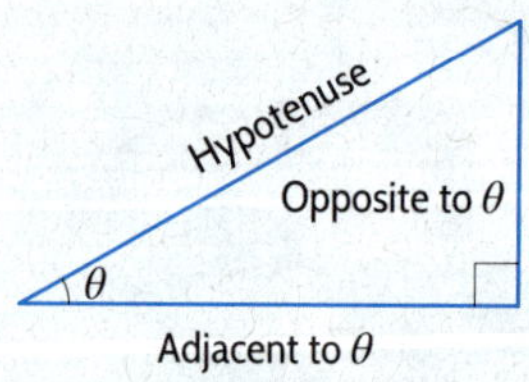


Figure 8.3: Adjacent and opposite sides to $\theta$

Both right triangles in Figure 8.2 have an angle $\theta$, so they have all the same angles; that is, they are similar. Thus the ratios of the lengths of corresponding sides of these triangles are equal:

$$\frac{a}{c} = \frac{y}{1} \quad \text{and} \quad \frac{b}{c} = \frac{x}{1}.$$

Since these equal ratios depend only on the angle $\theta$, we can define:

> If $\theta$ is an angle in a right triangle (other than the right angle),
>
> $$\sin\theta = \frac{\text{Opposite}}{\text{Hypotenuse}}, \qquad \cos\theta = \frac{\text{Adjacent}}{\text{Hypotenuse}}.$$

Note that the term *opposite* refers to the side of a right triangle directly across from the angle $\theta$, and the term *adjacent* refers to the other side (that is not the hypotenuse). See Figure 8.3. For $\theta$ between 0° and 90°, these definitions of $\sin\theta$ and $\cos\theta$ agree with our earlier definitions using the circle. See Problem 23.

**Example 1** Using Figure 8.4, find $\sin\theta$ and $\cos\theta$.

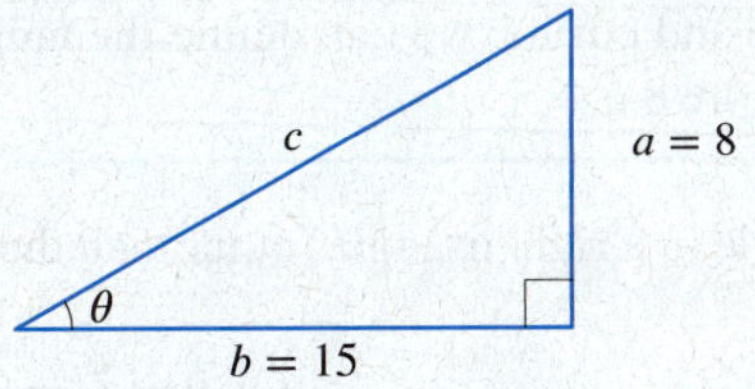


Figure 8.4: Find $\sin\theta$ and $\cos\theta$

**Solution** The hypotenuse of the triangle is $c = \sqrt{8^2 + 15^2} = \sqrt{289} = 17$. We have:

$$\sin\theta = \frac{\text{Opposite}}{\text{Hypotenuse}} = \frac{a}{c} = \frac{8}{17} = 0.4706$$
$$\cos\theta = \frac{\text{Adjacent}}{\text{Hypotenuse}} = \frac{b}{c} = \frac{15}{17} = 0.8824.$$

## Solving for Angles Using the Inverse Sine and Cosine

Now we see how to find angles in a right triangle when we know the sides. We use the inverse trigonometric functions introduced in Section 7.8 which tell us a value[1] of $\theta$ corresponding to a value of $\sin\theta$ or $\cos\theta$. Recall that if $\theta$ is an angle in a right triangle (other than the right angle):

- Inverse cosine: $\theta = \cos^{-1} x$ is an angle $\theta$ whose cosine is $x$
- Inverse sine: $\theta = \sin^{-1} x$ is an angle $\theta$ whose sine is $x$

Using the $\cos^{-1}$ and $\sin^{-1}$ keys on a calculator gives these angles.

**Example 2** The triangle in Figure 8.5, called a 3-4-5 right triangle, is special because its sides are whole numbers.[2] Find the angles $\theta$ and $\phi$.

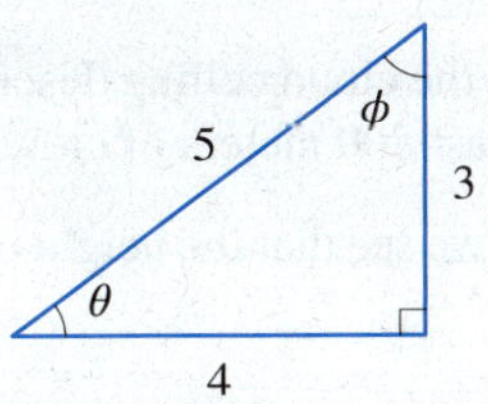


Figure 8.5: A 3-4-5 triangle

**Solution** From Figure 8.5, we see that

$$\sin\theta = \frac{\text{Opposite}}{\text{Hypotenuse}} = \frac{3}{5} = 0.6.$$

Using our calculator's inverse sine function:

$$\sin\theta = \frac{3}{5} = 0.6$$
$$\text{so} \quad \theta = \sin^{-1}(0.6) = 36.87^\circ \quad \text{using a calculator}$$
$$\sin\phi = \frac{4}{5} = 0.8$$
$$\text{so} \quad \phi = \sin^{-1}(0.8) = 53.13^\circ. \quad \text{using a calculator}$$

Notice that we could have found $\phi$ from $\theta$ using the relation $\phi = 90^\circ - \theta$.

[1] Calculators can measure angles in more than one way, so if you are working in degrees, be sure your calculator is in "degree mode."

[2] The numbers 3, 4, 5 are an example of a *Pythagorean triple*, because $3^2 + 4^2 = 5^2$. Other Pythagorean triples include 5, 12, 13 and 8, 15, 17.

## The Tangent Function in Right Triangles

Just as with sine and cosine, we can define the tangent function in terms of side lengths in a right triangle. See Figure 8.6.

> If $\theta$ is an angle in a right triangle (other than the right angle),
>
> $$\tan\theta = \frac{a}{b} = \frac{\text{Opposite}}{\text{Adjacent}}.$$

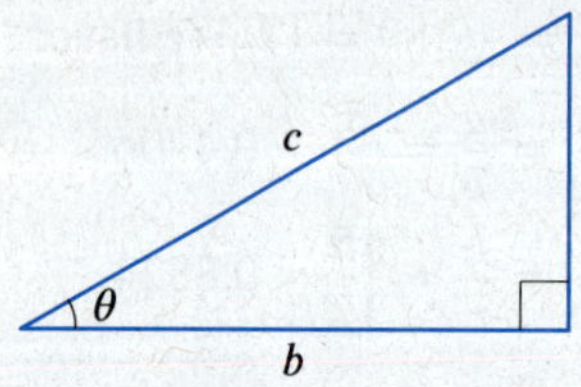


Figure 8.6: $\tan\theta = a/b$

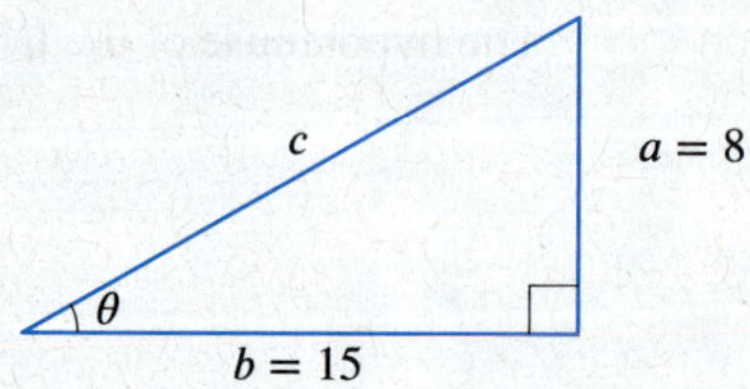


Figure 8.7: Find $\tan\theta$

Problem 24 asks you to explain why the definition of $\tan\theta$ in terms of sides of a right triangle agrees with our earlier definition of the tangent function.

**Example 3** Find $\tan\theta$ using Figure 8.7.

**Solution** We have

$$\tan\theta = \frac{a}{b} = \frac{8}{15} = 0.533.$$

**Example 4** Calculate the height of the cloud ceiling described in Figure 8.1 at the beginning of this section. Can a pilot who needs at least 800 meters of cloud ceiling safely fly?

**Solution** Looking at Figure 8.8, we see that the height of the cloud ceiling is equal to the length of the opposite side of $\theta = 35°$.

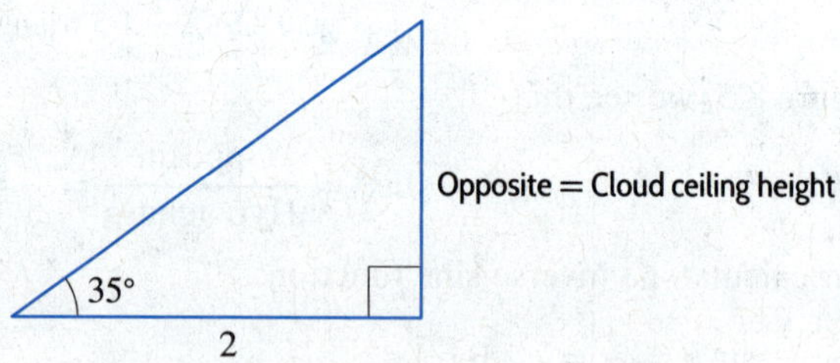


Figure 8.8: Height of the cloud ceiling

Since we know the adjacent side of the triangle but not its hypotenuse, we use the formula for $\tan\theta$:

$$\tan(35°) = \frac{\text{Opposite}}{\text{Adjacent}} = \frac{\text{Opposite}}{2}.$$

We have:

$$\text{Height of cloud ceiling} = \text{Opposite} = 2\tan(35°) = 1.4004 \text{ km}.$$

Thus, the lowest clouds are 1400 meters high, so the pilot can safely fly.

**Example 5** The grade of a road is calculated from its vertical rise per 100 feet. For instance, a road that rises 8 ft in every one hundred feet has a grade of

$$\text{Grade} = \frac{8 \text{ ft}}{100 \text{ ft}} = 8\%.$$

Suppose a road climbs at an angle of 6° to the horizontal. What is its grade?

**Solution** From Figure 8.9, we see that

$$\tan 6^\circ = \frac{x}{100}$$
$$\text{so} \quad x = 100 \tan 6^\circ$$
$$= 10.510. \quad \text{using a calculator}$$

Thus, the road rises 10.51 ft every 100 feet, so its grade is 10.51/100 = 10.51%.

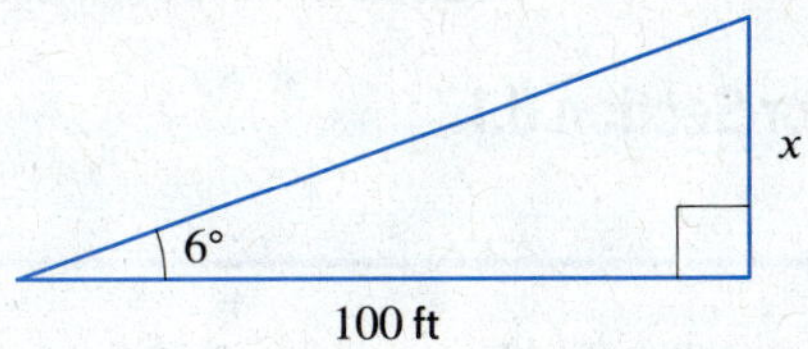


Figure 8.9: A road rising at an angle of 6°

## Solving for Angles Using the Inverse Tangent

To find an angle, $\theta$, corresponding to a given value of $\tan\theta$ we use the *inverse tangent* function introduced in Section 7.8. Recall that if $\theta$ is an angle in a right triangle (other than the right angle),

- Inverse tangent: $\theta = \tan^{-1} x$ is an angle $\theta$ whose tangent is $x$

The $\tan^{-1}$ key on a calculator gives this angle.

**Example 6** The grade of a road is 5.8%. What angle does the road make with the horizontal? (See Example 5.)

**Solution** Since the grade is 5.8%, the road climbs 5.8 feet for 100 feet; see Figure 8.10. We see that

$$\tan\theta = \frac{5.8}{100} = 0.058$$
$$\text{so} \quad \theta = \tan^{-1}(0.058)$$
$$= 3.319^\circ. \quad \text{using a calculator}$$

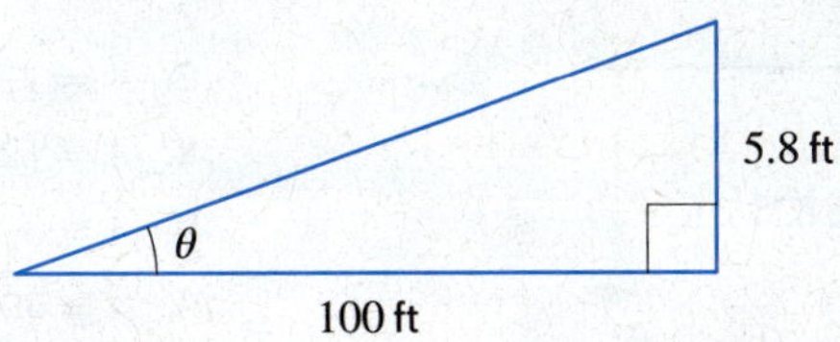


Figure 8.10: A road of 5.8% grade (not to scale)

**Example 7** A hiker heads across a plain in a direction she believes to be due east, but, her compass being faulty, her course is off by an angle $\theta$ to the north. She reaches a north-south road after traveling 12 miles. Had she been heading due east the hike would have taken only 11 miles. (See Figure 8.11.) Find the value of $\theta$.

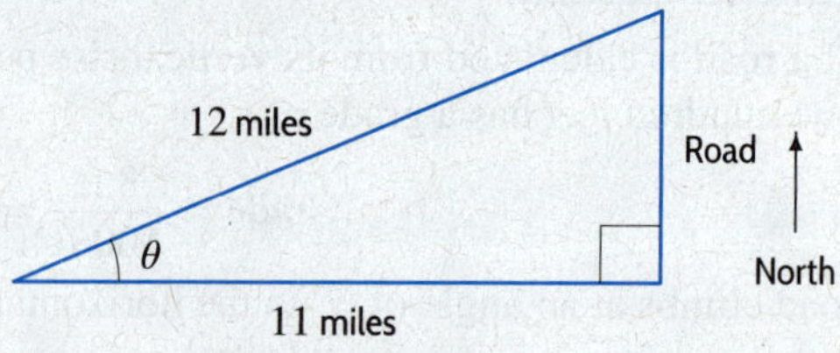


Figure 8.11: A hiker's path

**Solution** From Figure 8.11, we see that

$$\cos\theta = \frac{11}{12}$$
$$\text{so}\quad \theta = \cos^{-1}\left(\frac{11}{12}\right)$$
$$= 23.5565^\circ. \quad \text{using a calculator}$$

## Exercises and Problems for Section 8.1

### Exercises

**1.** Use Figure 8.12 to find the following exactly:

**(a)** $\tan\theta$ **(b)** $\sin\theta$ **(c)** $\cos\theta$

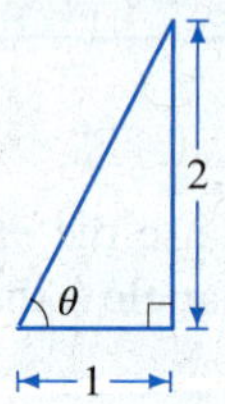


Figure 8.12

**2.** Using Figure 8.13, find exactly:

**(a)** $\sin\theta$ **(b)** $\sin\phi$ **(c)** $\cos\theta$
**(d)** $\cos\phi$ **(e)** $\tan\theta$ **(f)** $\tan\phi$

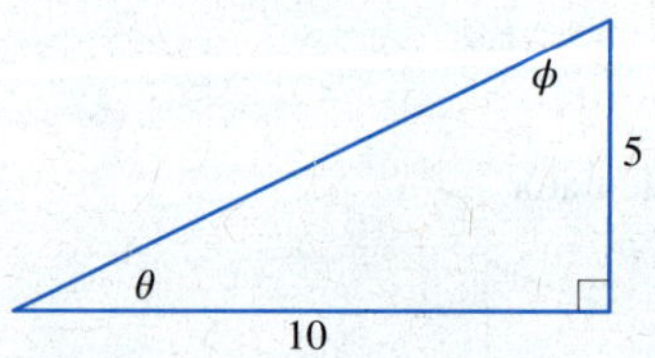


Figure 8.13

In Exercises 3–6, find (a) $\sin\theta$ (b) $\cos\theta$

**3.**

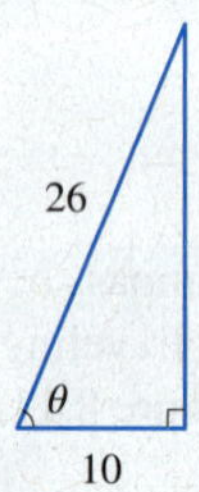


**4.**

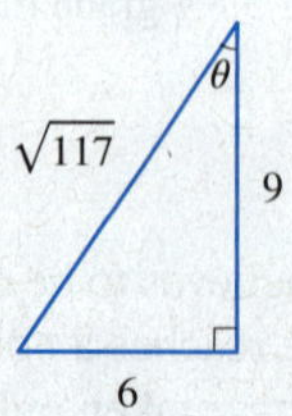


**5.**

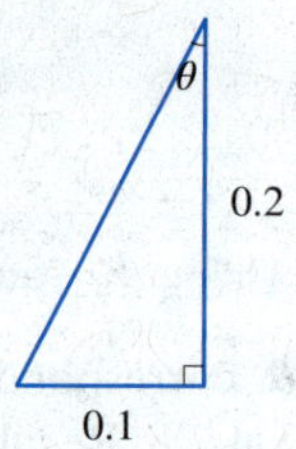


**6.**

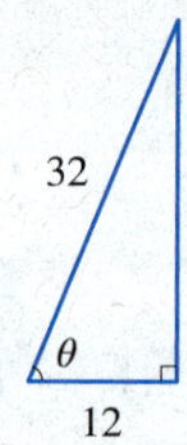


In Exercises 7–12, one of the sides $x$, $y$ and $r$ of the triangle in Figure 8.14 is given. Find exact values of the other two sides.

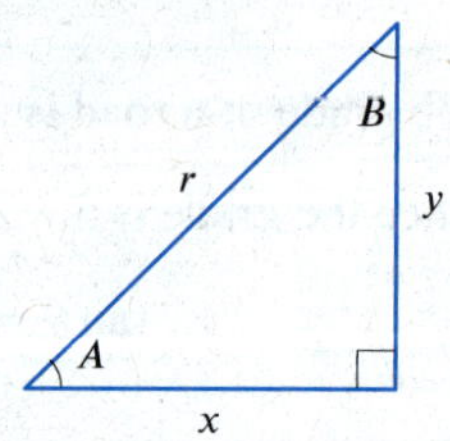


Figure 8.14

**7.** $A = 17^\circ,\ B = 73^\circ,\ r = 7$

**8.** $A = 12^\circ,\ B = 78^\circ,\ y = 4$

**9.** $A = 37^\circ,\ B = 53^\circ,\ x = 6$

**10.** $A = 40^\circ,\ r = 15$

**11.** $B = 77^\circ,\ x = 9$

**12.** $B = 22^\circ,\ x = \lambda$

For Exercises 13–18, find $\theta$, an angle in a right triangle.

**13.** $\sin\theta = 0.876$

**14.** $\cos\theta = 0.016$

**15.** $\tan\theta = 0.123$

**16.** $\tan\theta = 54.169$

**17.** $\sin\theta = 0.999$

**18.** $\cos\theta = 0.999$

In Exercises 19–22, find the missing sides and angles in the right triangle, where $a$ is the side across from angle $A$, $b$ across from $B$, and $c$ across from the right angle.

**19.** $a = 20, b = 28$

**20.** $a = 20, c = 28$

**21.** $c = 20, A = 28^\circ$

**22.** $a = 20, B = 28^\circ$

## Problems

**23.** Use Figure 8.15 to explain why, for $\theta$ between $0^\circ$ and $90^\circ$, the unit circle definition of $\sin\theta$ and $\cos\theta$ in Section 7.2 agree with the right triangle definitions in this section.

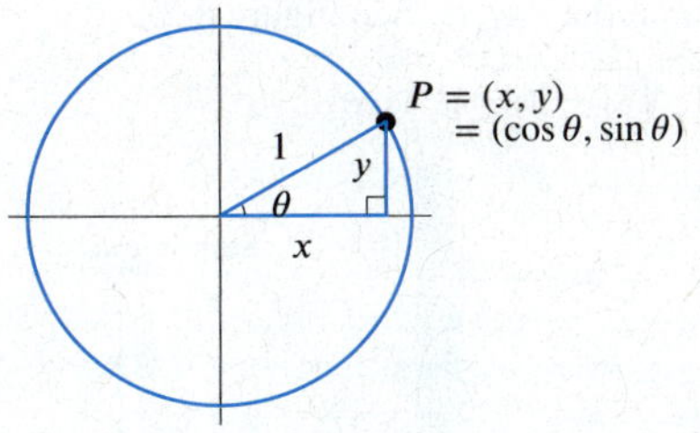


Figure 8.15

**24.** Using the definition of the trigonometric functions in terms of the lengths of the sides of a right triangle, explain why

$$\tan\theta = \frac{\sin\theta}{\cos\theta}.$$

**25.** Figure 8.16 shows a 45-45-90 right triangle with side $a$ and hypotenuse $c = \sqrt{a^2 + a^2}$. Find exact values for

**(a)** $\sin 45^\circ$ **(b)** $\cos 45^\circ$ **(c)** $\tan 45^\circ$

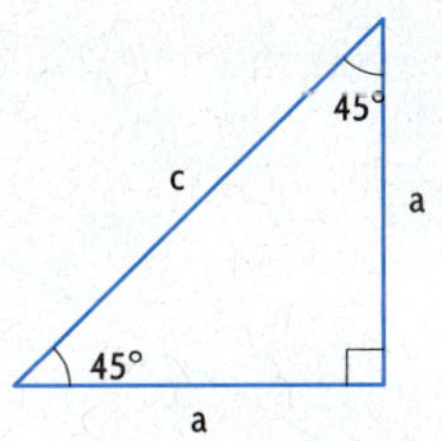


Figure 8.16

**26.** Figure 8.17 shows an equilateral triangle $PQR$ with side 2 and altitude $a$. Use right triangle $PQS$ to find exact values for

**(a)** $\sin 60^\circ$ **(b)** $\cos 60^\circ$ **(c)** $\tan 60^\circ$

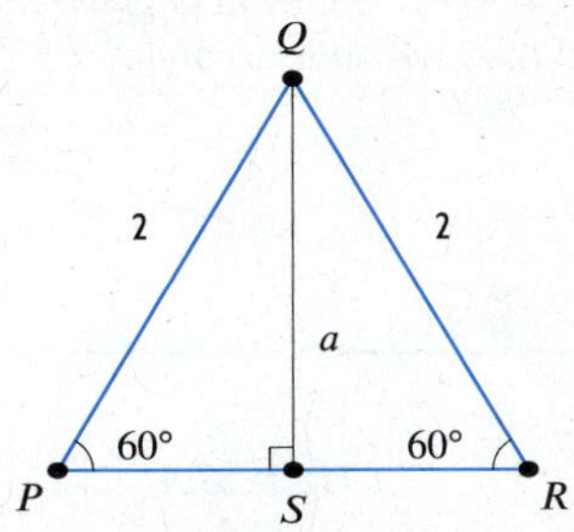


Figure 8.17

**27.** The top of a 200-foot vertical tower is to be anchored by cables that make an angle of $30^\circ$ with the ground. How long must the cables be? How far from the base of the tower should anchors be placed?

**28.** A surveyor must measure the distance between the two banks of a straight river. (See Figure 8.18.) She sights a tree at point $T$ on the opposite bank of the river and drives a stake into the ground (at point $P$) directly across from the tree. Then she walks 50 meters upstream and places a stake at point $Q$. She measures angle $PQT$ and finds that it is $58^\circ$. Find the width of the river.

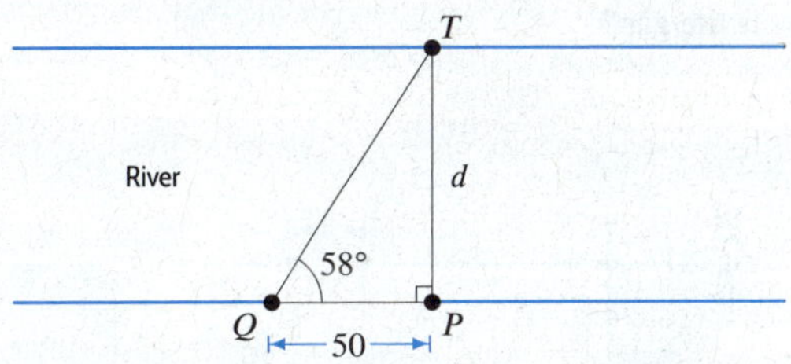


Figure 8.18

**29.** A search and rescue volunteer leaves a rendezvous point in the Arizona desert walking 75 degrees north of east. She reaches a river that runs east-west and is located 1.3 miles directly north of the rendezvous point. See Figure 8.19. How far east of the rendezvous point is the volunteer when she reaches the river?

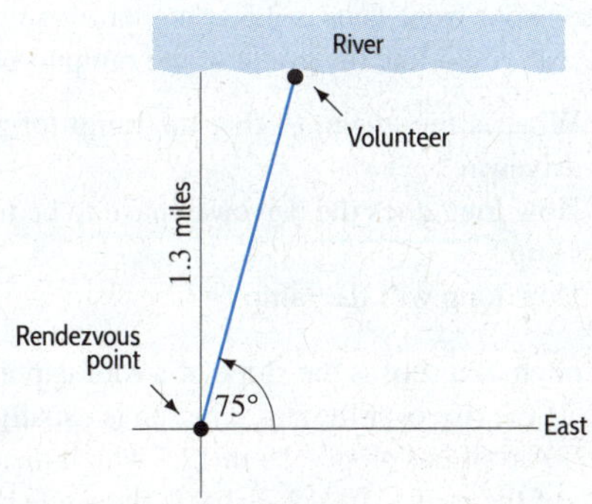


Figure 8.19

In Problems 30–33, a line passing through the origin and the given point $P$ forms the given angle $\theta$ with the $x$-axis. Find the missing coordinate of $P$.

**30.** $P = (16, b), \theta = 50^\circ$.

**31.** $P = (p, 5), \theta = 15^\circ$.

**32.** $P = (-3, q), \theta = 3\pi/8$.

**33.** $P = (p, -1), \theta = 5\pi/6$.

**34.** A 240-ft tree casts a 130-foot shadow on horizontal ground. A girl lying on the grass at the tip of the shadow is looking at a bird nesting at the top of the tree. At what angle is the girl looking up?

**35.** A plane is flying at an elevation of 35,000 feet when the Gateway Arch in St. Louis, Missouri comes into view. The pilot wants to estimate her horizontal distance from the arch, so she notes the angle of depression, $\theta$, between the horizontal and a line joining her eye to a point on the ground directly below the arch. Make a sketch. Express her horizontal distance to that point as a function of $\theta$.

**36.** A bridge over a river was damaged in an earthquake and you are called in to determine the length, $d$, of the steel beam needed to fill the gap. (See Figure 8.20.) You cannot be on the bridge, but you are able to drop a line from $T$, the beginning of the bridge, and measure a distance of 50 ft to the point $P$. From $P$ you find the angles of elevation to the two ends of the gap to be 42° and 35°. How wide is the gap?

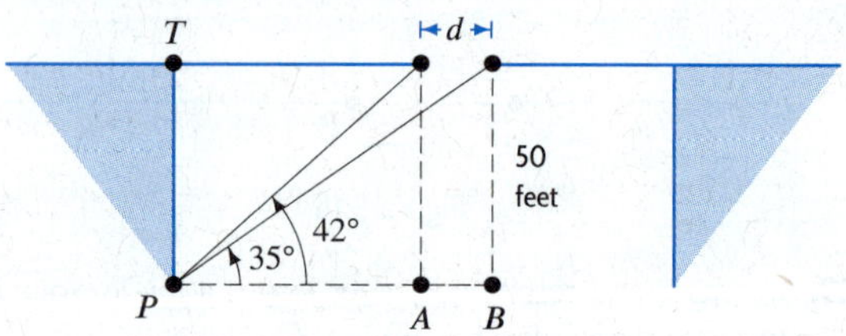


Figure 8.20: Gap in damaged bridge

**37.** You want to build a wheelchair ramp leading up to your house. Your front door is 2 feet higher than the driveway and you would like the grade of the ramp to be 7%.

**(a)** What is the angle, $\theta$, that the ramp forms with the driveway?
**(b)** How long does the driveway have to be to build this ramp?
**(c)** How long will the ramp be?

**38.** The pitch of a roof is the slope of a roof expressed as the ratio of the rise over the run. The run is usually expressed as 12. A roof has slope "10 in 12," which means that the slope of the roof is 10/12. What is the angle between the roof and the horizontal?

**39.** Find approximately the acute angle formed by the line $y = -2x + 5$ and the $x$-axis.

**40.** **(a)** Find expressions in terms of $a, b$, and $c$ for the sine, cosine, and tangent of the angle $\phi$ in Figure 8.21.
**(b)** Using your answers in part (a), show that $\sin\phi = \cos\theta$ and $\cos\phi = \sin\theta$.

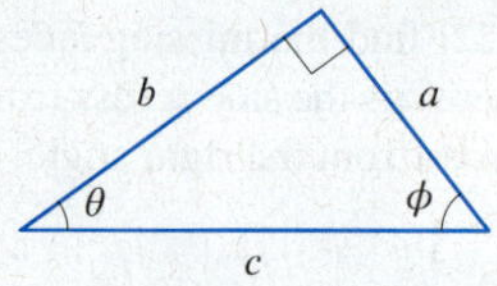


Figure 8.21

**41.** Knowing the height of the Columbia Tower in Seattle, determine the height of the Seafirst Tower and the distance between the towers. See Figure 8.22.

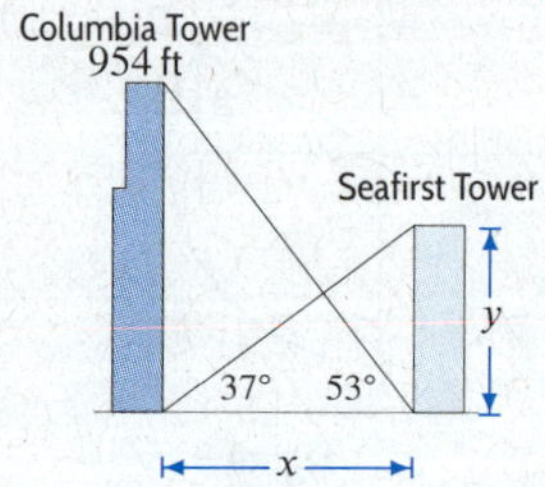


Figure 8.22

**42.** A tree 50 feet tall casts a shadow 60 feet long. Find the angle of elevation $\theta$ of the sun.

**43.** A staircase is to rise 17.3 feet over a horizontal distance of 10 feet. At approximately what angle with respect to the floor should it be built?

**44.** **(a)** In the right triangle in Figure 8.23, angle $A = 30°$ and $b = 2\sqrt{3}$. Find the lengths of the other sides and the other angles.
**(b)** Repeat part (a), this time assuming only that $a = 25$ and $c = 24$.

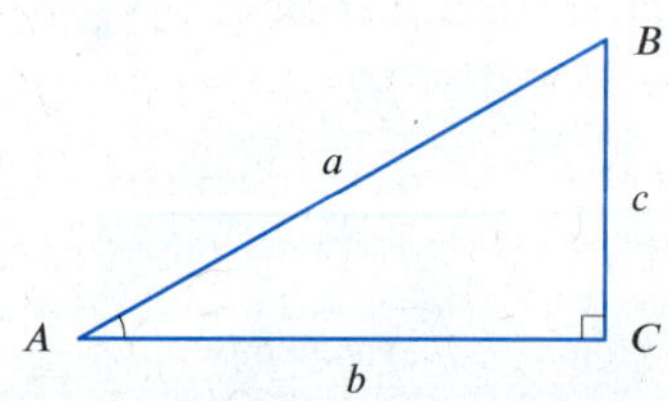


Figure 8.23

**45.** To check the calibration of their transit (an instrument to measure angles), two student surveyors used the setup in Figure 8.24. What angles in degrees for $\alpha$ and $\beta$ should they get if their transit is accurate?

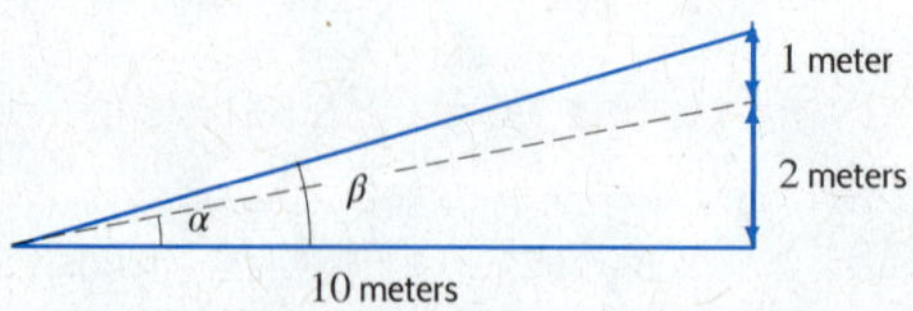


Figure 8.24

**46.** A ladder 3 meters long leans against a house, making an angle $\alpha$ with the ground. How far is the base of the ladder from the base of the wall, in terms of $\alpha$?

**47.** You are parasailing on a rope that is 125 feet long behind a boat. (Disregard sag in the rope.) See Figure 8.25.

**(a)** At first, you stabilize at a height that forms a 45° angle with the water. What is that height?

**(b)** After enjoying the scenery, you encounter a strong wind that blows you down to a height that forms a 30° angle with the water. At what height are you now?

**(c)** Find $c$ and $d$, that is, the horizontal distances between you and the boat, in parts (a) and (b).

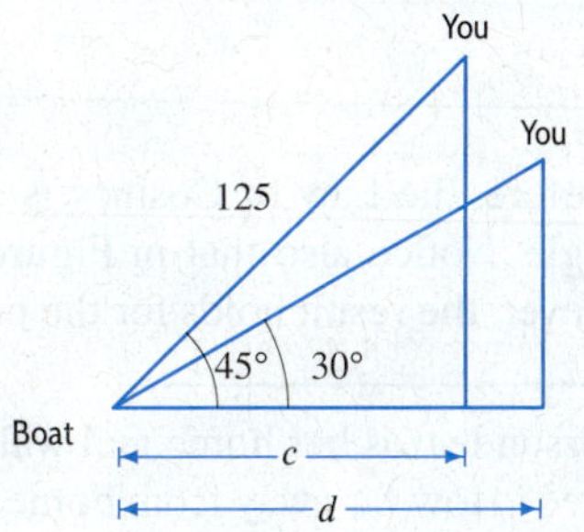

Figure 8.25

## 8.2 NON-RIGHT TRIANGLES

Sines and cosines relate the angles of a right triangle to its sides. Similar, although more complicated, relationships exist for all triangles, not just right triangles.

### The Law of Cosines

The Pythagorean theorem relates the three sides of a right triangle. The *Law of Cosines* relates the three sides of any triangle.

> **Law of Cosines:** For a triangle with sides $a,\ b,\ c$, and angle $C$ opposite side $c$, we have
>
> $$c^2 = a^2 + b^2 - 2ab\cos C$$

We use Figure 8.26 to derive the Law of Cosines. The dashed line of length $h$ is at right angles to side $a$ and it divides this side into two pieces, one of length $x$ and one of length $a - x$.

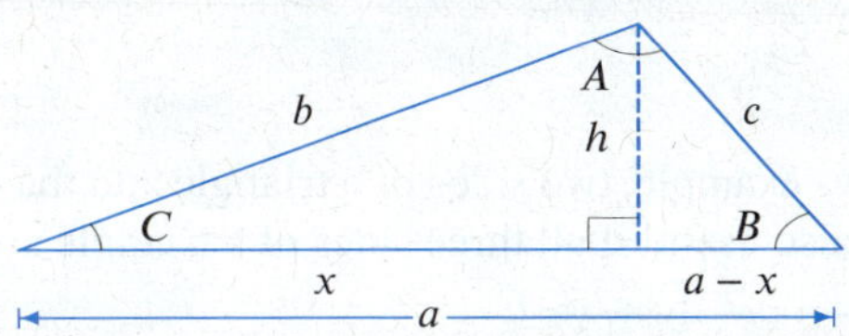

Figure 8.26: Triangle used to derive the Law of Cosines

Applying the Pythagorean theorem to the right-hand right triangle, we get

$$(a-x)^2 + h^2 = c^2$$
$$a^2 - 2ax + x^2 + h^2 = c^2.$$

Applying the Pythagorean theorem to the left-hand triangle, we get $x^2 + h^2 = b^2$. Substituting into the previous equation gives

$$a^2 - 2ax + \underbrace{x^2 + h^2}_{b^2} = c^2$$
$$a^2 + b^2 - 2ax = c^2.$$

But $\cos C = x/b$, so $x = b\cos C$. This gives the Law of Cosines:

$$a^2 + b^2 - 2ab\cos C = c^2.$$

Notice that if $C$ happens to be a right angle, that is, if $C = 90°$, then $\cos C = 0$. In this case, the Law of Cosines reduces to the Pythagorean theorem:

$$a^2 + b^2 - 2ab \cdot 0 = c^2$$
$$a^2 + b^2 = c^2.$$

Therefore, the Law of Cosines is a generalization of the Pythagorean theorem that works for any triangle. Notice also that in Figure 8.26 we assumed that angle $C$ is acute, that is, less than $90°$. However, the result holds for the case where $90° < C < 180°$. (See Problem 40 for the derivation.)

**Example 1** A person leaves her home and walks 5 miles due east and then 3 miles northeast. How far has she walked? How far away from home is she?

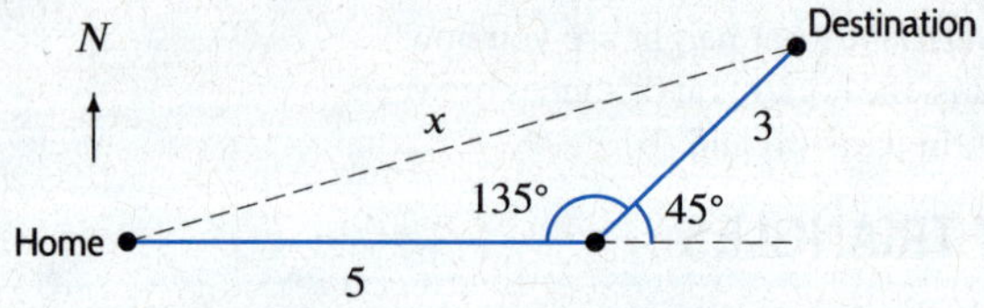


Figure 8.27: A person walks 5 miles east and then 3 miles northeast

**Solution** She has walked $5+3 = 8$ miles in total. One side of the triangle in Figure 8.27 is 5 miles long, while the second side is 3 miles long and forms an angle of 135° with the first. This is because when the person turns northeast, she turns through an angle of 45°. Thus, we know two sides of this triangle, 5 and 3, and the angle between them, which is 135°. To find her distance from home, we find the third side $x$, using the Law of Cosines:

$$\begin{aligned} x^2 &= 5^2 + 3^2 - 2 \cdot 5 \cdot 3 \cos 135° \\ &= 34 - 30\left(-\frac{\sqrt{2}}{2}\right) \\ &= 55.213. \end{aligned}$$

This gives $x = \sqrt{55.213} = 7.431$ miles. Notice that this is less than 8 miles, the total distance she walked.

In the previous example, two sides of a triangle and the angle between them were known. The Law of Cosines is also useful if all three sides of a triangle are known.

**Example 2** At what angle must the person from Example 1 walk to go directly home?

**Solution** According to Figure 8.28, if the person faces due west and then turns south through an angle of $\theta$, she heads directly home. This same angle $\theta$ is opposite the side of length 3 in the triangle. The Law of Cosines tells us that

$$\begin{aligned} 5^2 + 7.431^2 - 2 \cdot 5 \cdot 7.431 \cos\theta &= 3^2 \\ -74.31 \cos\theta &= -71.220 \\ \cos\theta &= 0.958 \\ \theta &= \cos^{-1}(0.958) = 16.582°. \end{aligned}$$

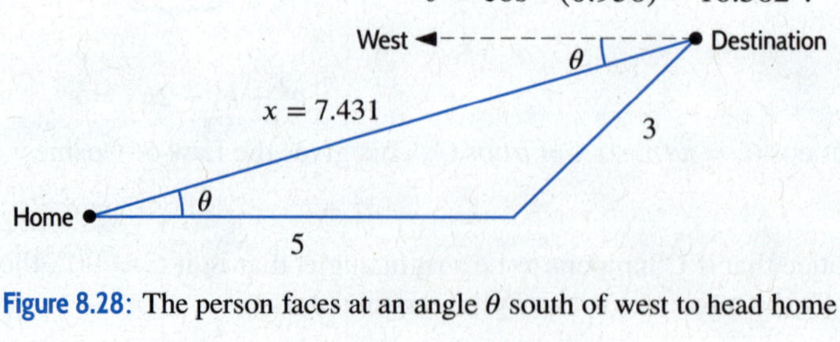


Figure 8.28: The person faces at an angle $\theta$ south of west to head home

In Examples 1 and 2, notice that we used the Law of Cosines in two different ways for the same triangle.

## The Law of Sines

The *Law of Sines* relates the sides and angles of any triangle.

> **Law of Sines:** For a triangle with sides $a$, $b$, $c$ opposite angles $A$, $B$, $C$, respectively:
>
> $$\frac{\sin A}{a} = \frac{\sin B}{b} = \frac{\sin C}{c}.$$

We derive the Law of Sines using Figure 8.29 , which shows the same triangle as in Figure 8.26. Since $\sin C = h/b$ and $\sin B = h/c$, we have $h = b \sin C$ and $h = c \sin B$. This means that

$$b \sin C = c \sin B$$

so

$$\frac{\sin B}{b} = \frac{\sin C}{c}.$$

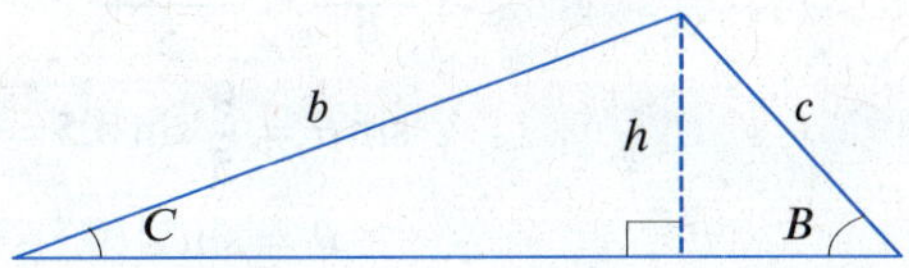


Figure 8.29: Triangle used to derive the Law of Sines

A similar argument (see Problem 41) shows that

$$\frac{\sin A}{a} = \frac{\sin B}{b}.$$

The Law of Sines is useful when we know a side and the angle opposite it.

**Example 3** An aerial tram starts at a point one half mile from the base of a mountain whose face has a 60° angle of elevation. (See Figure 8.30.) The tram ascends at an angle of 20°. What is the length of the cable from $T$ to $A$?

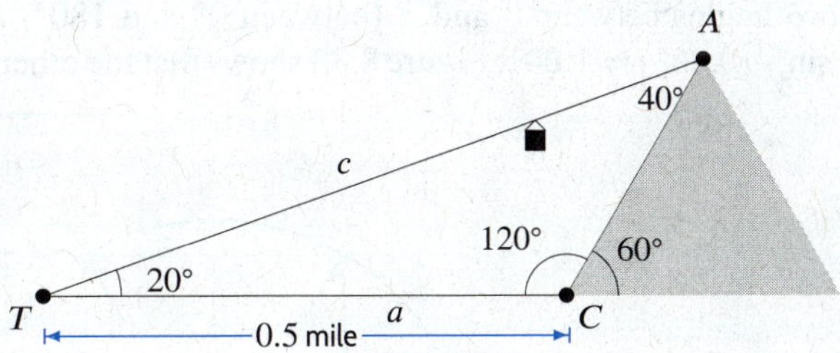


Figure 8.30

**Solution** The Law of Cosines does not help us here because we only know the length of one side of the triangle. We do, however, know two angles in this diagram. Thus, we can use the Law of Sines:

$$\frac{\sin A}{a} = \frac{\sin C}{c}$$

$$\frac{\sin 40°}{0.5} = \frac{\sin 120°}{c}$$

so

$$c = 0.5\left(\frac{\sin 120°}{\sin 40°}\right) = 0.674.$$

Therefore, the cable from $T$ to $A$ is 0.674 miles.

## The Ambiguous Case

There is a drawback to using the Law of Sines for finding angles. The problem is that the Law of Sines does not tell us the angle, but only its sine, and there are two angles between 0° and 180° with a given sine. For example, if the sine of an angle is 1/2, the angle may be either 30° or 150°.

**Example 4** Solve the following triangles for $\theta$ and $\phi$, using radians.

(a)

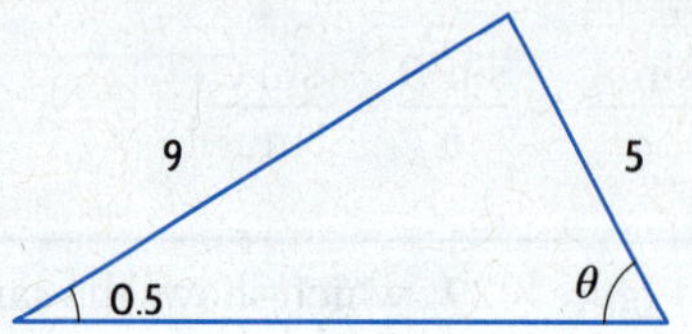


Figure 8.31

(b)

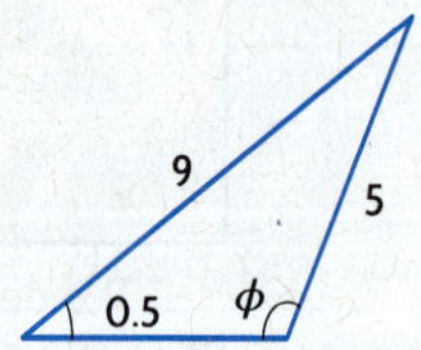


Figure 8.32

**Solution** (a) Using the Law of Sines in Figure 8.31 and a calculator set in radian mode, we have

$$\frac{\sin\theta}{9} = \frac{\sin 0.5}{5}$$
$$\sin\theta = \frac{9}{5}\sin 0.5 = 0.863$$
$$\theta = \sin^{-1}(0.863) = 1.041.$$

Note that 1.041 is the radian measure of an angle whose sine is 0.863. An angle of 1.041 radians measures approximately 59.6°.

(b) From Figure 8.32, we get

$$\frac{\sin\phi}{9} = \frac{\sin 0.5}{5}$$
$$\sin\phi = \frac{9}{5}\sin 0.5 = 0.863.$$

This is the same equation we had for $\theta$ in part (a). However, judging from the figures, $\phi$ is not equal to $\theta$. Knowing the sine of an angle is not enough to tell us the angle. In fact, there are *two* angles between 0 and $\pi$ (between 0° and 180°) whose sine is 0.863. One of them is $\theta = \sin^{-1}(0.863) = 1.041$. Figure 8.33 shows that the other is $\phi = \pi - \theta = 2.100$, or $\phi = 120.4°$.

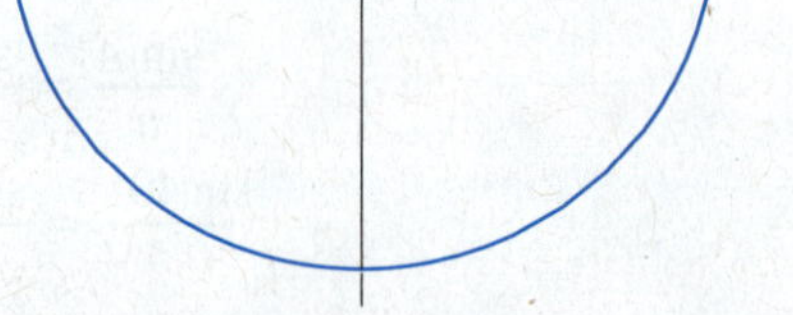


Figure 8.33: Two angles, $\theta = 1.041$ and $\phi = 2.100$, whose sine is 0.863

## Exercises and Problems for Section 8.2

### Exercises

In Exercises 1–2, solve for $x$.

**1.**

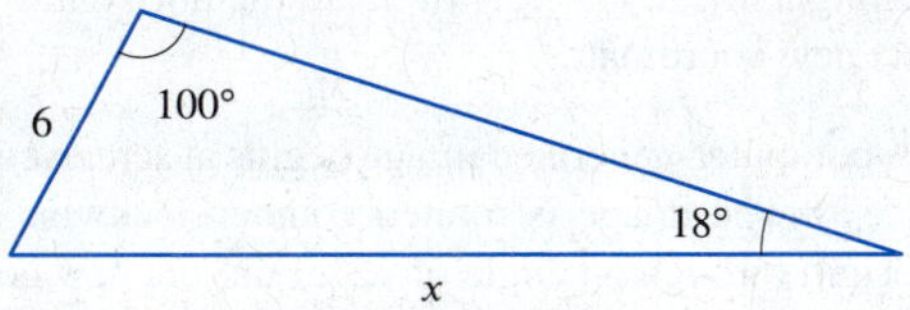


**2.**

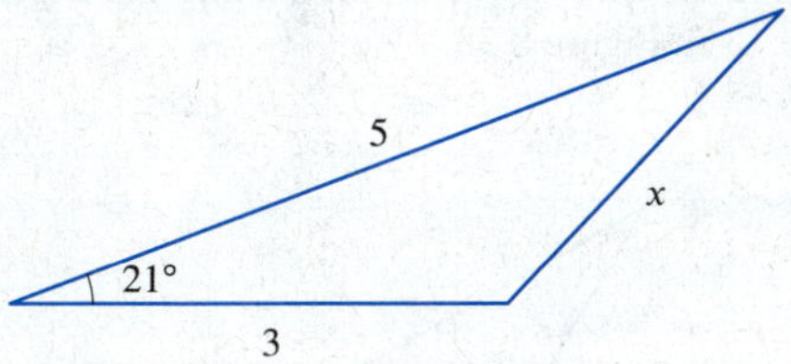


Find all sides and angles of the triangles in Problems 3–8. (Sides and angles are not necessarily to scale.)

**3.**

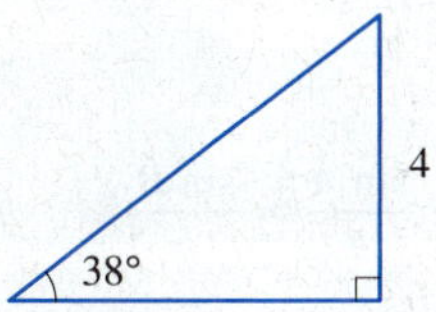


**4.**

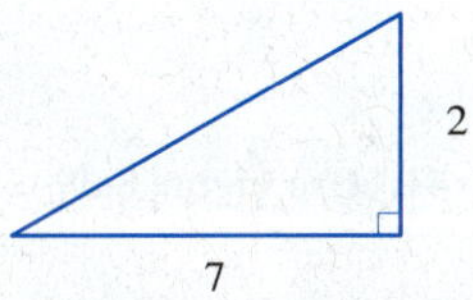


**5.**

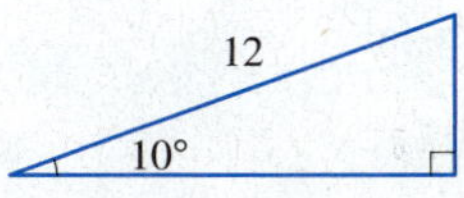


**6.**

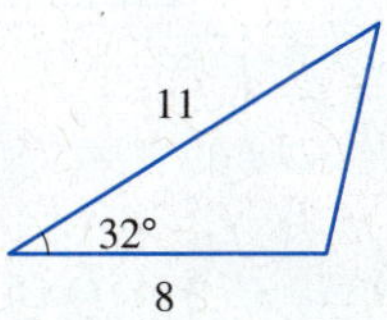


**7.**

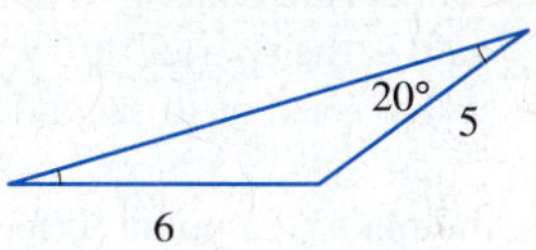


**8.**

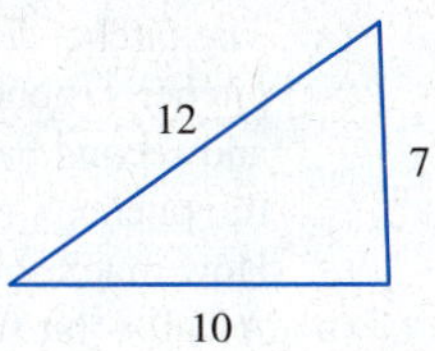


In Exercises 9–27, use Figure 8.34 to find the missing sides, $a, b, c$, and angles, $A, B, C$ (if possible). If there are two solutions, find both.

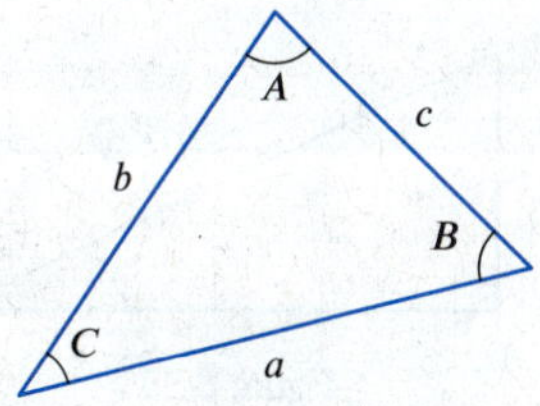


Figure 8.34

**9.** $a = 20,\ b = 28,\ c = 41$

**10.** $a = 14,\ b = 12,\ C = 23°$

**11.** $a = 20,\ B = 81°,\ c = 28$

**12.** $a = 20,\ b = 28,\ C = 12°$

**13.** $a = 9,\ b = 8,\ C = 80°$

**14.** $a = 8,\ b = 11,\ C = 114°$

**15.** $a = 5$, $b = 11$, and $C = 32°$

**16.** $A = 13°,\ B = 25°, c = 4$

**17.** $a = 8$, and $C = 98°$, $c = 17$

**18.** $A = 12°,\ C = 150°,\ c = 5$

**19.** $A = 92°,\ C = 35°,\ c = 9$

**20.** $A = 5°,\ C = 9°,\ c = 3$

**21.** $B = 95°,\ b = 5,\ c = 10$

**22.** $B = 72°,\ b = 13,\ c = 4$

**23.** $B = 75°,\ b = 7,\ c = 2$

**24.** $B = 17°,\ b = 5,\ c = 8$

**25.** $A = 3\pi/5,\ B = \pi/20,\ c = 15$

**26.** $A = 2\pi/3,\ B = \pi/9,\ c = 7$

**27.** $A = 3\pi/8,\ B = \pi/4,\ c = 9$

Find all sides and angles of the triangles in Exercises 28–31. Sketch each triangle. If there is more than one possible triangle, solve and sketch both. Note that $\alpha$ is the angle opposite side $a$, and $\beta$ is the angle opposite side $b$, and $\gamma$ is the angle opposite side $c$.

**28.** $a = 18.7$ cm, $c = 21.0$ cm, $\beta = 22°$

**29.** $a = 2.00$ m, $\alpha = 25.80°$, $\beta = 10.50°$

**30.** $b = 510.0$ ft, $c = 259.0$ ft, $\gamma = 30.0°$

**31.** $a = 16.0$ m, $b = 24.0$ m, $c = 20.0$ m

## Problems

**32.** **(a)** Find an expression for $\sin\theta$ in Figure 8.35 and $\sin\phi$ in Figure 8.36.
**(b)** Explain how you can find $\theta$ and $\phi$ by using the inverse sine function. In what way does the method used for $\theta$ differ from the method used for $\phi$?

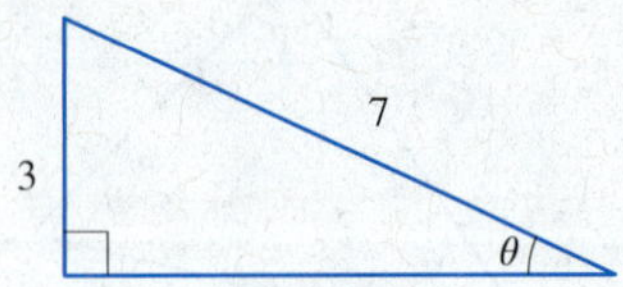


Figure 8.35

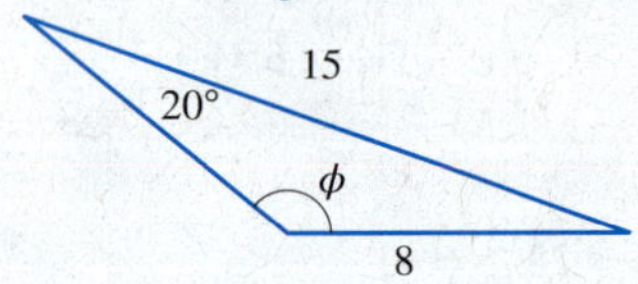


Figure 8.36

**33.** In Figure 8.37: **(a)** Find $\sin\theta$ **(b)** Solve for $\theta$
**(c)** Find the area of the triangle.

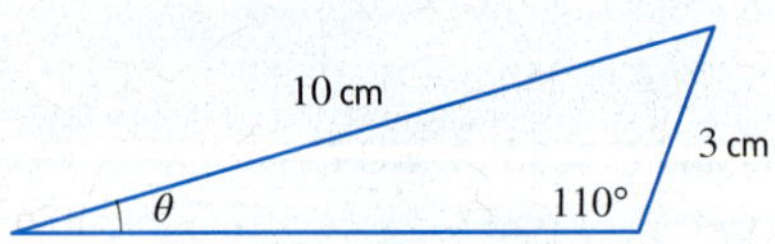


Figure 8.37

**34.** Two fire stations are located 56.7 miles apart, at points $A$ and $B$. There is a forest fire at point $C$. If $\angle CAB = 54°$ and $\angle CBA = 58°$, which fire station is closer? How much closer?

**35.** To measure the height of the Eiffel Tower in Paris, a person stands away from the base and measures the angle of elevation to the top of the tower to be 60°. Moving 210 feet closer, the angle of elevation to the top of the tower is 70°. How tall is the Eiffel Tower?

**36.** Two airplanes leave Kennedy airport in New York at 11 am. The air traffic controller reports that they are traveling away from each other at an angle of 103°. The A380 travels 509 mph and the 787 travels at 503 mph. How far apart are they at 11:30 am?

**37.** A parcel of land is in the shape of an isosceles triangle. The base has length 425 feet; the other sides, which are of equal length, meet at an angle of 39°. How long are they?

**38.** In video games, images are drawn on the screen using $xy$-coordinates. The origin, $(0, 0)$, is the lower-left corner of the screen. An image of an animated character moves from its position at $(8, 5)$ through a distance of 12 units along a line at an angle of 25° to the horizontal. What are its new coordinates?

**39.** A computer-generated image begins at screen coordinates $(5, 3)$. The image is rotated counterclockwise about the origin through an angle of 42°. Find the new coordinates of the image.

**40.** Derive the Law of Cosines assuming the angle $C$ is obtuse, as in Figure 8.38.

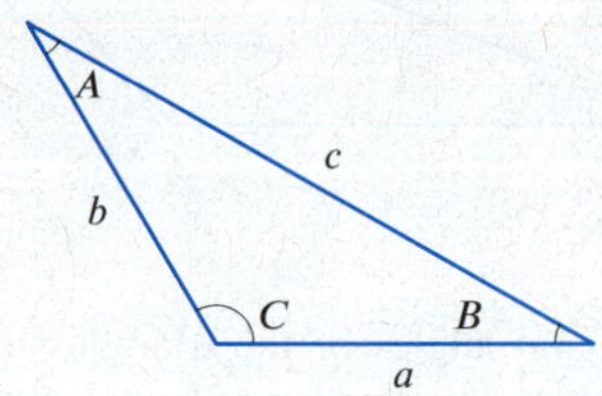


Figure 8.38

**41.** Use Figure 8.39 to show that $\dfrac{\sin A}{a} = \dfrac{\sin B}{b}$.

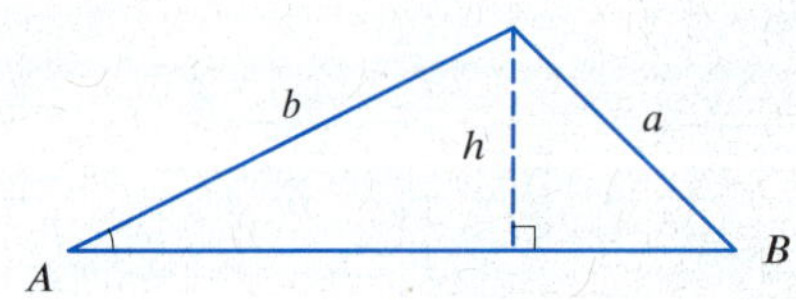


Figure 8.39

**42.** In baseball, the four bases form a square. The pitcher stands near the center on the pitcher's rubber. See Figure 8.40.[3]

**(a)** The pitcher fields a ground ball without leaving the pitcher's rubber, and runners are coming to both first and second bases. Which is the shorter throw: From the pitcher's rubber to first base or to second base? How much shorter?
**(b)** A ball is hit from home plate to a point 30 feet past second base. An outfielder comes in to catch the ball. How far is the throw to home plate? To third base?

[3] www.majorleaguebaseball.com, accessed June 29, 2014.

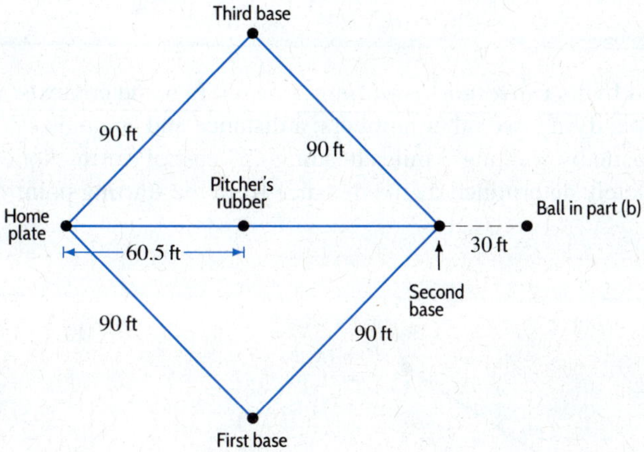


Figure 8.40

**43.** A park director wants to build a bridge across a river to a bird sanctuary. He hires a surveyor to determine the length of the bridge, represented by $AB$ in Figure 8.41. The surveyor places a transit (an instrument for measuring vertical and horizontal angles) at point $A$ and measures angle $BAC$ to be $93°$. The surveyor then moves the transit 102 feet to point $C$ and measures angle $BCA$ to be $49°$. How long should the bridge be?

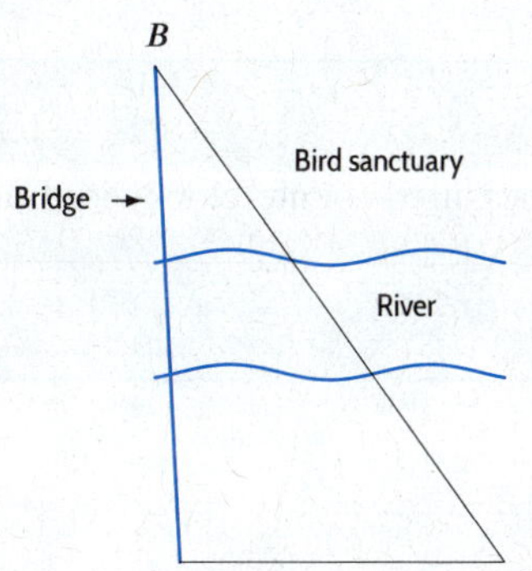


Figure 8.41

**44.** To estimate the width of an archaeological mound, archaeologists place two stakes on opposite ends of the widest point. See Figure 8.42. They set a third stake 82 feet from one stake and 97 feet from the other stake. The angle formed is $125°$. Find the width of the mound.

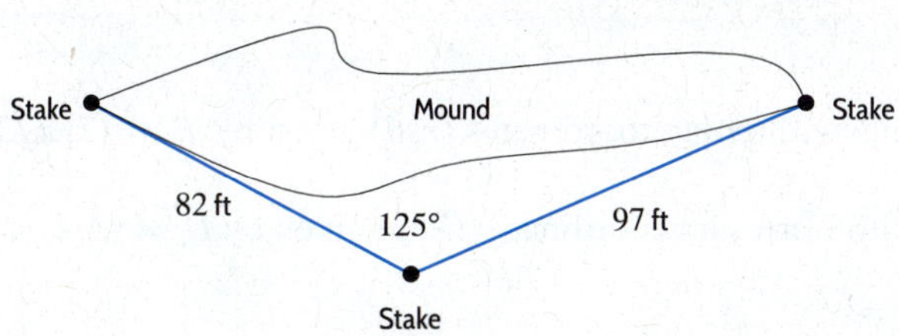


Figure 8.42

**45.** Every triangle has three sides and three angles. Make a chart showing the set of all possible triangle configurations where three of these six measures are known, and the other three measures can be deduced (or partially deduced) from the three known measures.

**46.** The telephone company needs to run a wire from the telephone pole across a street to a new house. See Figure 8.43. Since they cannot measure across the busy street, they hire a surveyor to determine the amount of wire needed. The surveyor measures a distance of 23.5 feet from the telephone pole to a stake, $D$, which she sets in the ground. She measures 145.3 feet from the house to a second stake, $E$. She sets a third stake, $F$, at a distance of 105.2 feet from the second stake. The surveyor measures angle $DEF$ to be $83°$ and angle $EFD$ to be $68°$. A total of 20 feet of wire is needed to make connections at the house and pole, and wire is sold in 100-foot rolls. How many rolls of wire are needed?

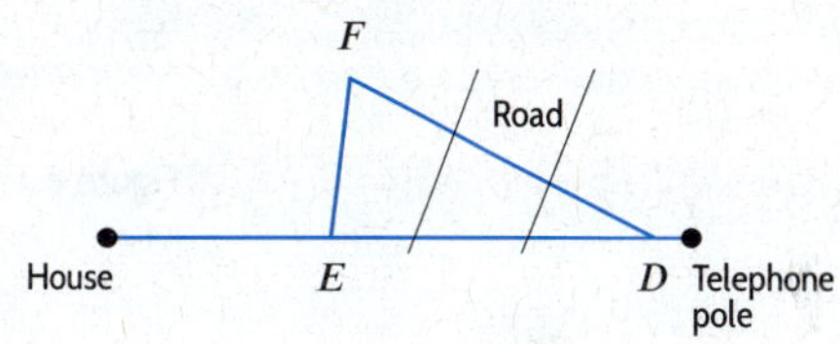


Figure 8.43

**47.** A buyer is interested in purchasing the triangular lot with vertices $LOT$ in Figure 8.44, but unfortunately, the marker at point $L$ has been lost. The deed indicates that side $TO$ is 435 ft and side $LO$ is 112 ft and that the angle at $L$ is $82.6°$.

**(a)** What is the distance from $L$ to $T$?
**(b)** What is the area of the lot?

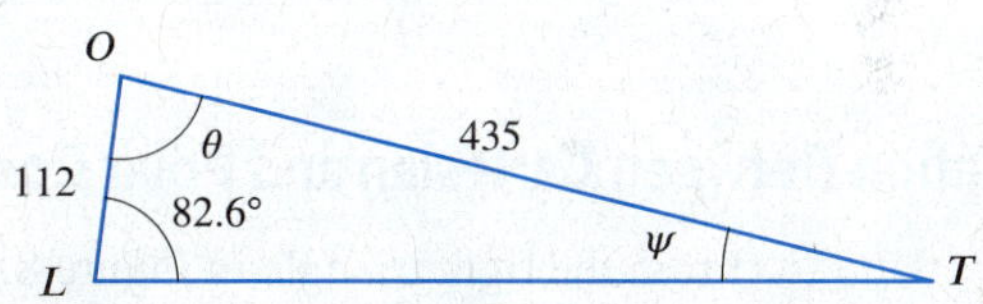


Figure 8.44

**48.** Solve the following triangles for $\theta$ and $\phi$.

(a)

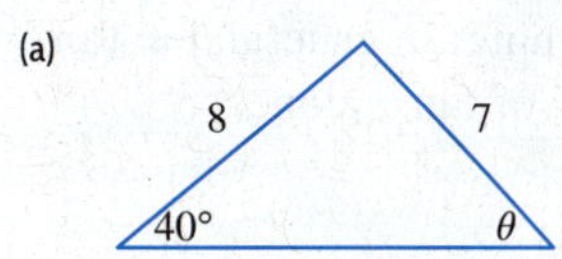


Figure 8.45

(b)

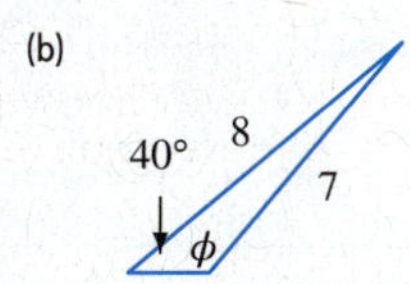


Figure 8.46

**49.** A triangle has angles $2\pi/3$, $\pi/5$, and $2\pi/15$. The length of the side opposite the $2\pi/3$ angle is 8.

**(a)** Find the lengths of the two remaining sides.
**(b)** Calculate the area of the triangle.

# 8.3 POLAR COORDINATES

A point, $P$, in the plane is often identified by its *Cartesian*[4] *coordinates* $(x, y)$. In some contexts, it is more natural to identify a point $P$ by specifying two other numbers: a distance and an angle.

A hiker in Arizona can reach a campsite by walking 3 miles heading 45° east of north. Notice that the location of the campsite is completely determined by the distance from the starting point, 3 miles, and the angle, 45°.

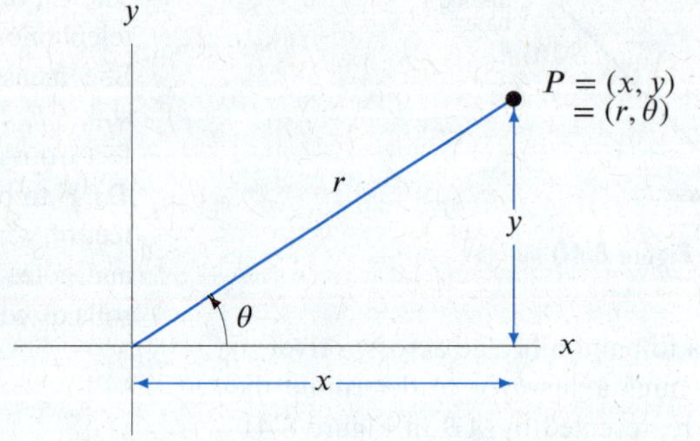


Figure 8.47: Cartesian and polar coordinates for the point $P$

This example suggests we can identify any point $P$ in the plane by specifying the following:

- Distance, $r$, from the origin, and
- Angle, $\theta$, shown in Figure 8.47. The angle $\theta$ is measured counterclockwise from the positive $x$-axis to the line joining the origin to $P$.

We call $(r, \theta)$ the **polar coordinates** of point $P$.

## Relation Between Cartesian and Polar Coordinates

From the right triangle in Figure 8.47, we see that

- $x = r\cos\theta$ and $y = r\sin\theta$
- $r = \sqrt{x^2 + y^2}$ and $\tan\theta = \dfrac{y}{x}$, $x \neq 0$

The angle $\theta$ is determined by the equations $\cos\theta = x/\sqrt{x^2+y^2}$ and $\sin\theta = y/\sqrt{x^2+y^2}$.

Warning: In general $\theta \neq \tan^{-1}(y/x)$. It is not possible to determine which quadrant $\theta$ is in from the value of $\tan\theta$ alone.

**Example 1** (a) Give Cartesian coordinates for the points with polar coordinates $(r, \theta)$ given by $P = (7, \pi/3)$, $Q = (5, 0)$, $R = (5, \pi)$.

(b) Give polar coordinates for the points with Cartesian coordinates $(x, y)$ given by $U = (3, 4)$ and $V = (0, -5)$.

[4]Cartesian coordinates can also be called rectangular coordinates.

**Solution** (a) See Figure 8.48. Point $P$ is a distance of 7 from the origin. The angle $\theta = \pi/3$ radians (60°).

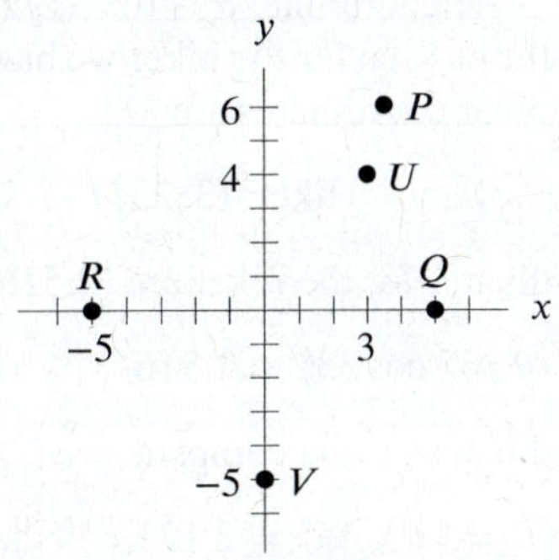


Figure 8.48: Points in the Cartesian plane

The Cartesian coordinates of $P$ are

$$x = r\cos\theta = 7\cos\frac{\pi}{3} = \frac{7}{2} \quad \text{and} \quad y = r\sin\theta = 7\sin\frac{\pi}{3} = \frac{7\sqrt{3}}{2}.$$

Point $Q$ is located a distance of 5 units along the positive $x$-axis with Cartesian coordinates

$$x = r\cos\theta = 5\cos 0 = 5 \quad \text{and} \quad y = r\sin\theta = 5\sin 0 = 0.$$

For point $R$, which is on the negative $x$-axis,

$$x = r\cos\theta = 5\cos\pi = -5 \quad \text{and} \quad y = r\sin\theta = 5\sin\pi = 0.$$

(b) For $U = (3, 4)$, we have $r = \sqrt{3^2 + 4^2} = 5$ and $\tan\theta = 4/3$. A possible value for $\theta$ is $\theta = \arctan 4/3 = 0.927$ radians, or about 53°. The polar coordinates of $U$ are $(5, 0.927)$. The point $V$ falls on the $y$-axis, so we can choose $r = 5$, $\theta = 3\pi/2$ for its polar coordinates. In this case, we cannot use $\tan\theta = y/x$ to find $\theta$, because $\tan\theta = y/x = -5/0$ is undefined.

**Example 2** The hiker looking for the campsite never reached it. She had walked 3 miles at 45° east of north using a compass. However, the campsite was actually at a heading of 45° east of true north. The hiker had not realized that compass north (magnetic north) is not the same as true north. In this part of Arizona, compass north is 12° east of true north. We choose the coordinate axes as in the figure, with east corresponding to the positive $x$-axis. Find polar and Cartesian coordinates for the hiker and campsite in Figure 8.49.

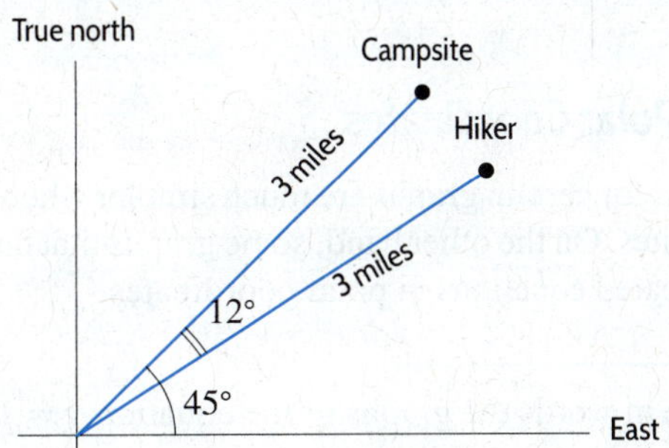


Figure 8.49: Locations of hiker and campsite

**Solution** We find the polar coordinates first. The hiker and the campsite are both 3 miles from the origin, so both have $r = 3$. The coordinate $\theta$ is the angle measured counterclockwise from the horizontal $x$-axis, that is from the east, so for the hiker we have $\theta = 90° - 45° - 12° = 33°$ and for the campsite, $\theta = 45°$. Thus, the polar coordinates are

$$\text{Hiker: } (3, 33°) \qquad \text{Campsite: } (3, 45°).$$

The Cartesian coordinates for the hiker are (2.516, 1.634) since

$$x = r\cos\theta = 3\cos 33° = 2.516 \quad \text{and} \quad y = r\sin\theta = 3\sin 33° = 1.634.$$

The Cartesian coordinates for the campsite are (2.121, 2.121) since

$$x = r\cos\theta = 3\cos 45° = 2.121 \quad \text{and} \quad y = r\sin\theta = 3\sin 45° = 2.121.$$

Because the angle $\theta$ can be allowed to wrap around the origin more than once, there are many possibilities for the polar coordinates of a point. For the point $V$ in Example 1, we can also choose $\theta = -\pi/2$ or $\theta = 7\pi/2$, so that $(5, -\pi/2)$, $(5, 7\pi/2)$, and $(5, 3\pi/2)$ are all polar coordinates for $V$. However, we usually choose $\theta$ between 0 and $2\pi$.

In some situations, negative values of $r$ are used, but we shall not do so.

**Example 3** Give three possible sets of polar coordinates for the point $P$ in Figure 8.50.

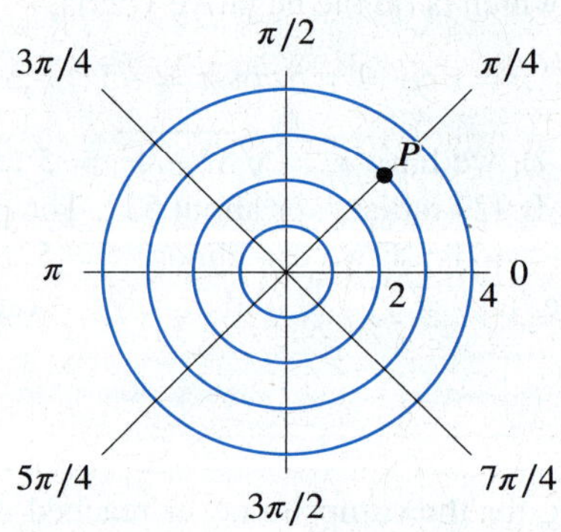


Figure 8.50: Polar coordinates for point $P$

**Solution** Because $r = 3$ and $\theta = \pi/4$, one set of polar coordinates for $P$ is $(3, \pi/4)$. We can also use $\theta = \pi/4 + 2\pi = 9\pi/4$ and $\theta = \pi/4 - 2\pi = -7\pi/4$, to get $(3, 9\pi/4)$ and $(3, -7\pi/4)$.

## Graphing Equations in Polar Coordinates

The equations for certain graphs are much simpler when expressed in polar coordinates than in Cartesian coordinates. On the other hand, some graphs that have simple equations in Cartesian coordinates have complicated equations in polar coordinates.

**Example 4**

(a) Describe in words the graphs of the equation $y = 1$ (in Cartesian coordinates) and the equation $r = 1$ (in polar coordinates).

(b) Write the equation $r = 1$ using Cartesian coordinates. Write the equation $y = 1$ using polar coordinates.

**Solution** (a) The equation $y = 1$ describes a horizontal line. Since the equation $y = 1$ places no restrictions on the value of $x$, it describes every point having a $y$-value of 1, no matter what the value of its $x$-coordinate. Similarly, the equation $r = 1$ places no restrictions on the value of $\theta$. Thus, it describes every point having an $r$-value of 1, that is, having a distance of 1 from the origin. This set of points is the unit circle. See Figure 8.51.

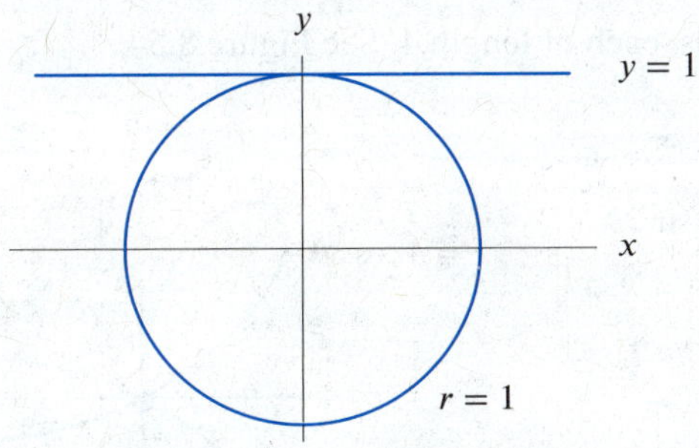


**Figure 8.51:** The graph of the equation $r = 1$ is the unit circle because $r = 1$ for every point regardless of the value of $\theta$. The graph of $y = 1$ is a horizontal line since $y = 1$ for any $x$

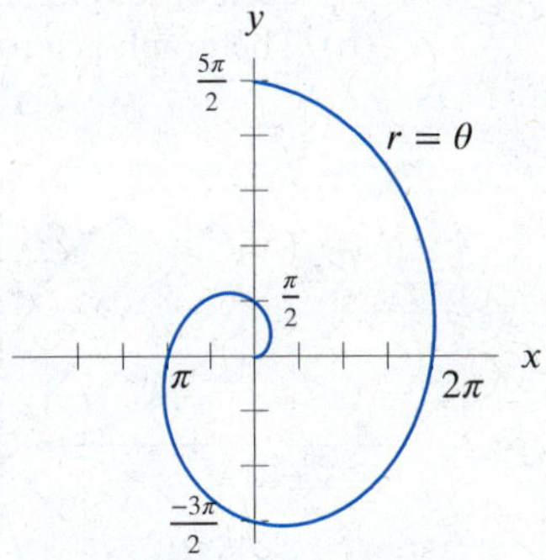


**Figure 8.52:** A graph of the Archimedean spiral $r = \theta$

(b) Since $r = \sqrt{x^2 + y^2}$, we rewrite the equation $r = 1$ using Cartesian coordinates as $\sqrt{x^2 + y^2} = 1$, or, squaring both sides, as $x^2 + y^2 = 1$. We see that the equation for the unit circle is simpler in polar coordinates than it is in Cartesian coordinates.

On the other hand, since $y = r \sin\theta$, we can rewrite the equation $y = 1$ in polar coordinates as $r \sin\theta = 1$, or, dividing both sides by $\sin\theta$, as $r = 1/\sin\theta$. We see that the equation for this horizontal line is simpler in Cartesian coordinates than in polar coordinates.

---

**Example 5** Graph the equation $r = \theta$. The graph is called an *Archimedean spiral* after the Greek mathematician Archimedes who described its properties (although not using polar coordinates).

**Solution** To construct this graph, use the values in Table 8.1. To help us visualize the shape of the spiral, we convert the angles in Table 8.1 to degrees and the $r$-values to decimals. See Table 8.2.

**Table 8.1** *Points on the Archimedean spiral $r = \theta$, with $\theta$ in radians*

| $\theta$ | 0 | $\frac{\pi}{6}$ | $\frac{\pi}{3}$ | $\frac{\pi}{2}$ | $\frac{2\pi}{3}$ | $\frac{5\pi}{6}$ | $\pi$ | $\frac{7\pi}{6}$ | $\frac{4\pi}{3}$ | $\frac{3\pi}{2}$ |
|---|---|---|---|---|---|---|---|---|---|---|
| $r$ | 0 | $\frac{\pi}{6}$ | $\frac{\pi}{3}$ | $\frac{\pi}{2}$ | $\frac{2\pi}{3}$ | $\frac{5\pi}{6}$ | $\pi$ | $\frac{7\pi}{6}$ | $\frac{4\pi}{3}$ | $\frac{3\pi}{2}$ |

**Table 8.2** *Points on the Archimedean spiral $r = \theta$, with $\theta$ in degrees and a decimal approximation to $r$*

| $\theta$ | 0 | 30° | 60° | 90° | 120° | 150° | 180° | 210° | 240° | 270° |
|---|---|---|---|---|---|---|---|---|---|---|
| $r$ | 0.00 | 0.52 | 1.05 | 1.57 | 2.09 | 2.62 | 3.14 | 3.67 | 4.19 | 4.71 |

Notice that as the angle $\theta$ increases, points on the curve move farther from the origin. At 0°, the point is at the origin. At 30°, it is 0.52 units away from the origin, at 60° it is 1.05 units away, and at 90° it is 1.57 units away. As the angle winds around, the point traces out a curve that moves away from the origin, giving a spiral. (See Figure 8.52.)

**Example 6** For an integer $n$, curves of the form $r = a\sin n\theta$ or $r = a\cos n\theta$ are called *roses*. Graph the roses

(a) $r = 3\sin 2\theta$ (b) $r = 4\cos 3\theta$

**Solution**

(a) Using a calculator or making a table of values, we see that the graph is a rose with four petals, each of length 3. See Figure 8.53.

(b) The graph is a rose with three petals, each of length 4. See Figure 8.54.

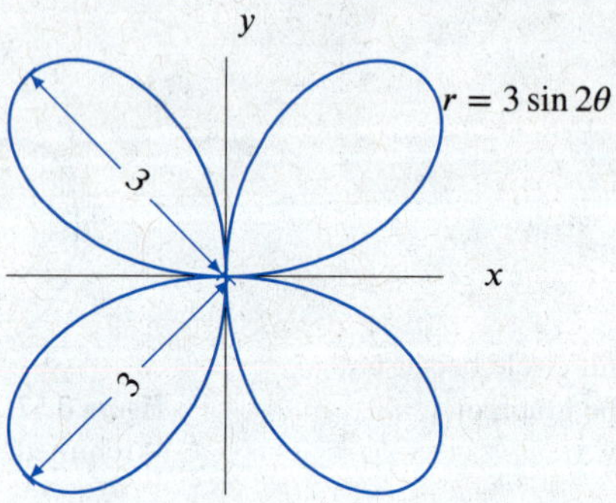


**Figure 8.53:** Graph of $r = 3\sin 2\theta$

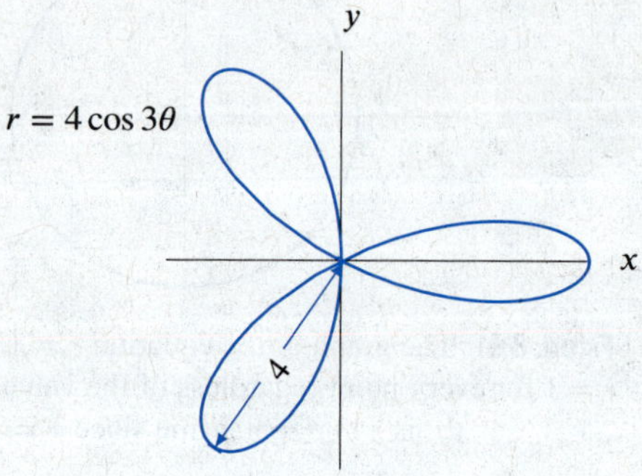


**Figure 8.54:** Graph of $r = 4\cos 3\theta$

**Example 7** Curves of the form $r = a + n\sin\theta$ or $r = a + b\cos\theta$ are called limaçons. Graph $r = 1 + 2\cos\theta$ and $r = 3 + 2\cos\theta$.

**Solution** See Figures 8.55 and 8.56.

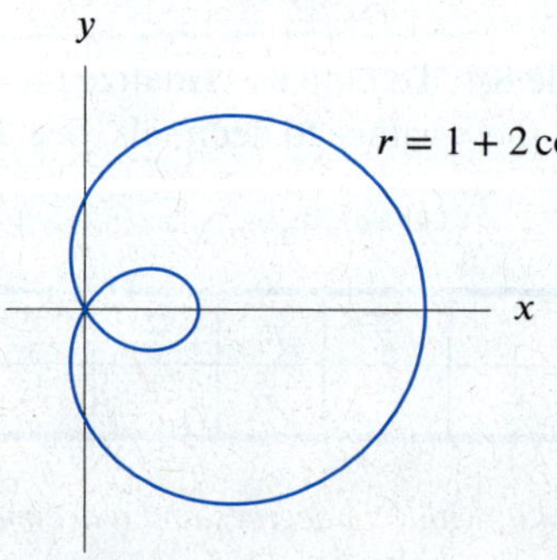


**Figure 8.55:** Graph of $r = 1 + 2\cos\theta$

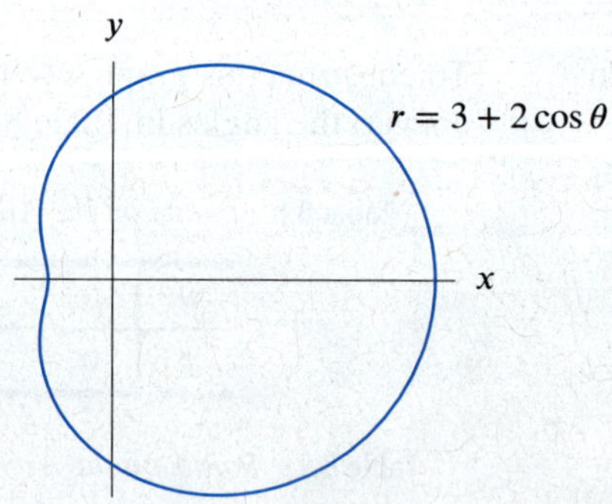


**Figure 8.56:** Graph of $r = 3 + 2\cos\theta$

## Graphing Inequalities in Polar Coordinates

Polar coordinates can be used with inequalities to describe regions that are obtained from circles. Such regions are often much harder to represent in Cartesian coordinates.

**Example 8** Using inequalities, describe a Blu-ray disk with outer diameter 120 mm and inner diameter 15 mm.

**Solution** The disk lies between two circles of radius 7.5 mm and 60 mm. See Figure 8.57. Thus, if the origin is at the center, the disk is represented by

$$7.5 \leq r \leq 60 \quad \text{and} \quad 0 \leq \theta \leq 2\pi.$$

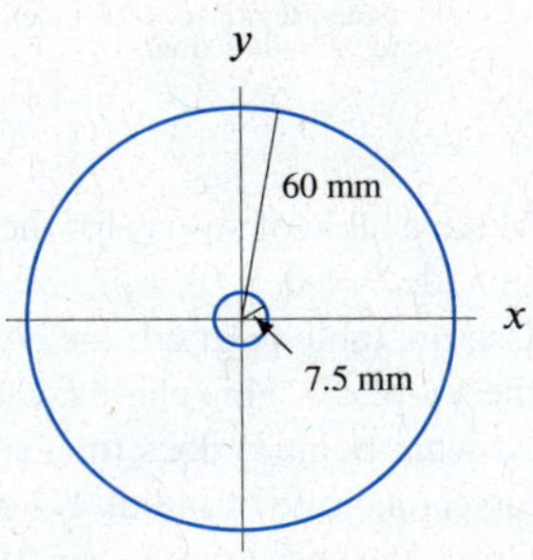


Figure 8.57: Compact disk

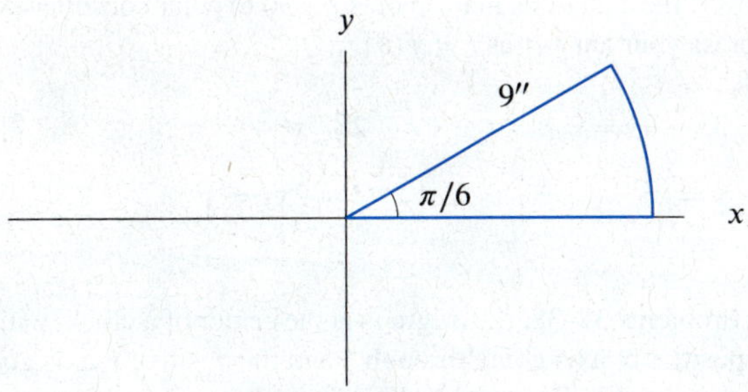


Figure 8.58: Pizza slice

**Example 9** An 18-inch pizza is cut into 12 slices. Use inequalities to describe one of the slices.

**Solution** The pizza has radius 9 inches; the angle at the center is $2\pi/12 = \pi/6$. See Figure 8.58. Thus, if the origin is at the center of the original pizza, the slice is represented by

$$0 \leq r \leq 9 \quad \text{and} \quad 0 \leq \theta \leq \frac{\pi}{6}.$$

## Exercises and Problems for Section 8.3

### Exercises

In Exercises 1–9, in which quadrant is a point with the polar coordinate $\theta$?

**1.** $\theta = 315°$ **2.** $\theta = -290°$ **3.** $\theta = 470°$

**4.** $\theta = 2.4\pi$ **5.** $\theta = 3.2\pi$ **6.** $\theta = -2.9\pi$

**7.** $\theta = 10.3\pi$ **8.** $\theta = -13.4\pi$ **9.** $\theta = -7$

In Exercises 10–13 mark the point on the $xy$-plane. In which range, 0° to 90°, 90° to 180°, 180° to 270°, or 270° to 360°, is its polar coordinate $\theta$?

**10.** $(2, 3)$ **11.** $(-1, 7)$

**12.** $(2, -3)$ **13.** $(-1.1, -3.2)$

Convert the Cartesian coordinates in Problems 14–18 to polar coordinates.

**14.** $(1, 1)$ **15.** $(-1, 0)$

**16.** $(\sqrt{6}, -\sqrt{2})$ **17.** $(-\sqrt{3}, 1)$

**18.** $(-3, 10)$

Convert the polar coordinates in Exercises 19–22 to Cartesian coordinates. Give exact answers.

**19.** $(1, 2\pi/3)$ **20.** $(\sqrt{3}, -3\pi/4)$

**21.** $(2\sqrt{3}, -\pi/6)$ **22.** $(2, 5\pi/6)$

## Problems

Convert the equations in Problems 23–26 to rectangular coordinates.

**23.** $r = 2$ **24.** $r = 6\cos\theta$

**25.** $\theta = \pi/4$ **26.** $\tan\theta = r\cos\theta - 2$

Convert the equations in Problems 27–30 to polar coordinates. Express your answer as $r = f(\theta)$.

**27.** $3x - 4y = 2$ **28.** $x^2 + y^2 = 5$

**29.** $y = x^2$ **30.** $2xy = 1$

For Problems 31–38, the origin is at the center of a clock, with the positive $x$-axis going through 3 and the positive $y$-axis going through 12. The hour hand is 3 cm long and the minute hand is 4 cm long. What are the Cartesian coordinates and polar coordinates of the tips of the hour hand and minute hand, $H$ and $M$, respectively, at the following times?

**31.** 12 noon **32.** 3 pm **33.** 9 am **34.** 1 pm

**35.** 1:30 pm **36.** 7 am **37.** 3:30 pm **38.** 9:15 am

In Problems 39–41, give inequalities for $r$ and $\theta$ that describe the following regions in polar coordinates.

**39.**

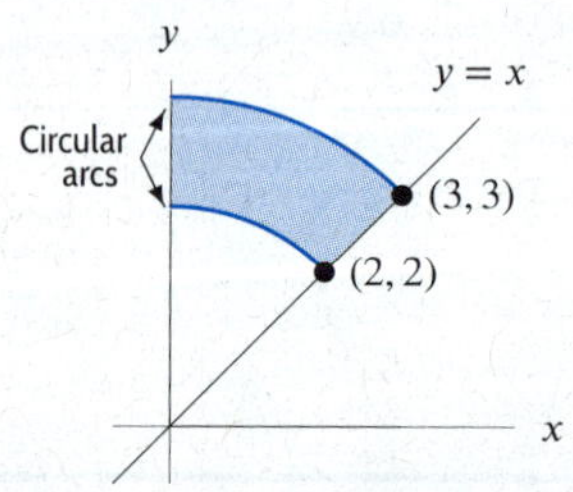


**40.**

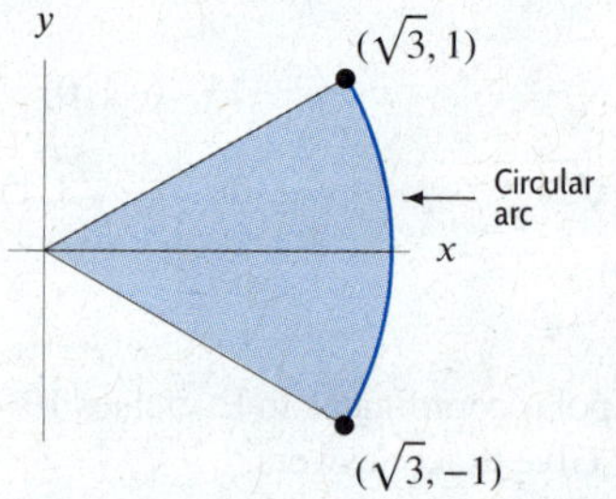


**41.**

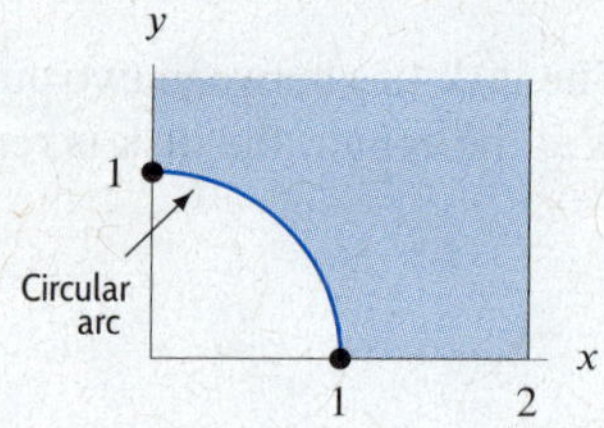


Note: Region extends indefinitely in the $y$-direction.

**42.** **(a)** Make a table of values for the equation $r = 1 - \sin\theta$. Include $\theta = 0, \pi/3, \pi/2, 2\pi/3, \pi, \cdots$.
**(b)** Use the table to graph the equation $r = 1 - \sin\theta$ in the $xy$-plane. This curve is called a *cardioid*.
**(c)** At what point(s) does the cardioid $r = 1 - \sin\theta$ intersect a circle of radius 1/2 centered at the origin?
**(d)** Graph the curve $r = 1 - \sin 2\theta$ in the $xy$-plane. Compare this graph to the cardioid $r = 1 - \sin\theta$.

**43.** Graph the equation $r = 1 - \sin(n\theta)$, for $n = 1, 2, 3, 4$. What is the relationship between the value of $n$ and the shape of the graph?

**44.** Graph the equation $r = 1 - \sin\theta$, with $0 \le \theta \le n\pi$, for $n = 2, 3, 4$. What is the relationship between the value of $n$ and the shape of the graph?

**45.** Graph the equation $r = 1 - n\sin\theta$, for $n = 2, 3, 4$. What is the relationship between the value of $n$ and the shape of the graph?

**46.** Graph the equation $r = 1 - \cos\theta$. Describe its relationship to $r = 1 - \sin\theta$.

**47.** Give inequalities that describe the flat surface of a washer that is one inch in diameter and has an inner hole with a diameter of 3/8 inch.

**48.** Graph the equation $r = 1 - \sin(2\theta)$ for $0 \le \theta \le 2\pi$. There are two loops. For each loop, give a restriction on $\theta$ that shows all of that loop and none of the other loop.

**49.** A slice of pizza is one eighth of a circle of radius 1 foot. The slice is in the first quadrant, with one edge along the $x$-axis, and the center of the pizza at the origin. Give inequalities describing this region using:
**(a)** Polar coordinates **(b)** Rectangular coordinates

**50.** A radio tower located on the coast can be configured to transmit mostly over land and not over the ocean. This way, the available energy is more efficiently used to reach farther inland. The coverage area of such a tower is described by a polar curve called a *cardioid*[5]: $r = 15(1 + \cos\theta)$. (Distances are given in miles.) The coverage area

[5] www.ictregulationtoolkit.org/Documents/Document/Document/3643, accessed May 23, 2014, or "The Electronics Handbook", Jerry C. Whitaker, CRC Press, 1996.

of the signal is shown in Figure 8.59, where the radio tower is located at the origin.

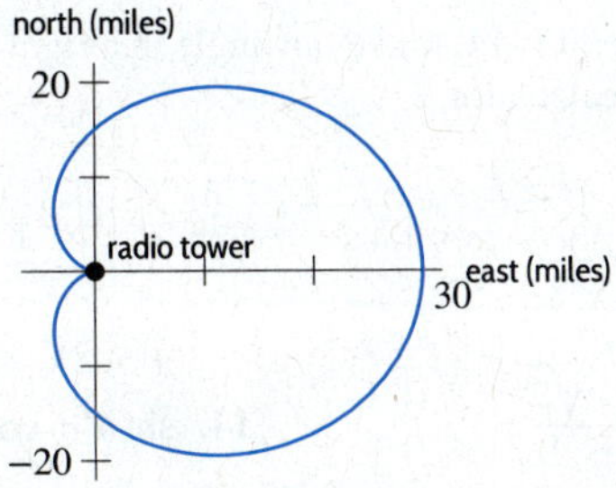


Figure 8.59

**(a)** How far does the signal reach directly north of the tower?

**(b)** Find the polar coordinates of the point directly east of the tower where you lose the signal.

**(c)** A house is located 25 miles east and 14 miles north of the radio tower. Does the signal reach the house?

**51.** A super-directional microphone, also called shotgun microphone,[6] is designed to pick up sound from directly in front of it. The pickup area of such a microphone can be described using the polar curve $r = \theta^2$ for $-\pi \le \theta \le \pi$, where the microphone is located at the origin and distances are given in feet. See Figure 8.60.

Several guests on a radio show that uses a shotgun microphone are seated at positions described by the polar coordinates $(6, \pi)$, $(6, 3\pi/4)$, $(6, -3\pi/4)$. Can all of them be picked up by the microphone? If not, suggest a place where they should move.

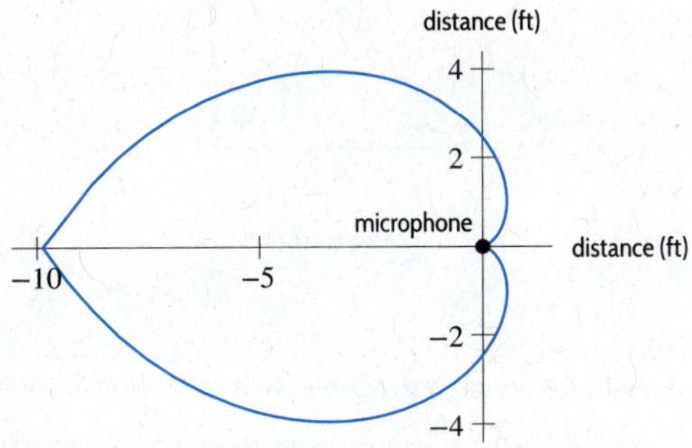


Figure 8.60

## CHAPTER SUMMARY

- **Right Triangle Trigonometry**
  For an angle $\theta$ in a right triangle (not the right angle):
  $$\sin\theta = \frac{\text{Opposite}}{\text{Hypotenuse}}, \qquad \cos\theta = \frac{\text{Adjacent}}{\text{Hypotenuse}}.$$
  For $\theta$ an angle in a right triangle (not the right angle):
  $$\tan\theta = \frac{\text{Opposite}}{\text{Adjacent}}.$$
- **Non-right Triangles**
  Law of Cosines: For a triangle with sides $a$, $b$, $c$, and angle $C$ opposite side $c$, we have
  $$c^2 = a^2 + b^2 - 2ab\cos C.$$
  Law of Sines: For a triangle with sides $a$, $b$, $c$ opposite angles $A$, $B$, $C$ respectively:
  $$\frac{\sin A}{a} = \frac{\sin B}{b} = \frac{\sin C}{c}.$$
- **Polar Coordinates**
  $$x = r\cos\theta \qquad y = r\sin\theta$$
  $$r = \sqrt{x^2 + y^2} \qquad \tan\theta = \frac{y}{x}.$$
  Graphing with polar coordinates.

[6]http://www.intmath.com/blog/polar-coordinates-and-cardioid-microphones/2496, accessed May 23, 2014.

## REVIEW EXERCISES AND PROBLEMS FOR CHAPTER EIGHT

### Exercises

In Exercises 1–6, use Figure 8.61 to find exactly

**(a)** $\sin\theta$ **(b)** $\cos\theta$ **(c)** $\tan\theta$

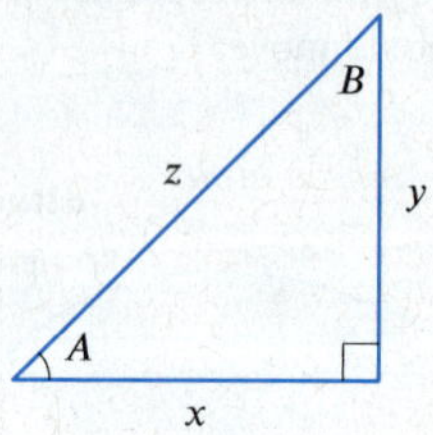


Figure 8.61

**1.** $x = 2,\ z = 7,\ A = \theta$

**2.** $x = 9,\ y = 5,\ A = \theta$

**3.** $y = 8,\ z = 12,\ A = \theta$

**4.** $z = 17,\ A = \theta,\ B = \theta$

**5.** $y = 2,\ z = 11,\ B = \theta$

**6.** $x = a,\ y = b,\ B = \theta$

For Exercises 7–10, find $\theta$, an angle in a right triangle.

**7.** $\tan\theta = 0.999$

**8.** $\sin\theta = \frac{3}{5}$

**9.** $\tan\theta = \frac{5}{3}$

**10.** $\cos\theta = \frac{5}{3}$

For Exercises 11–14, find $\theta$, an angle in a right triangle, without using a calculator.

**11.** $\cos\theta = \dfrac{\sqrt{3}}{2}$

**12.** $\sin\theta = \dfrac{1}{2}$

**13.** $\cos\theta = \dfrac{\sqrt{2}}{2}$

**14.** $\sin\theta = \cos\theta$

In Exercises 15–18, in which quadrant is a point with the polar coordinate $\theta$?

**15.** $\theta = -299^\circ$

**16.** $\theta = 730^\circ$

**17.** $\theta = -7.7\pi$

**18.** $\theta = 14.4\pi$

Convert each of the polar coordinates in Exercises 19–22 to Cartesian coordinates. Use decimal approximations where appropriate.

**19.** $(\pi/2, 0)$ **20.** $(2, 2)$ **21.** $(0, \pi/2)$ **22.** $(3, 40^\circ)$

### Problems

Find all sides and angles of the triangles in Exercises 23–26. (Not necessarily drawn to scale.)

**23.**

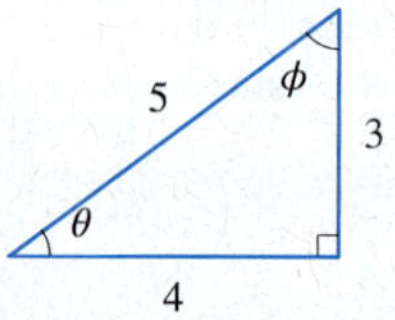


**24.**

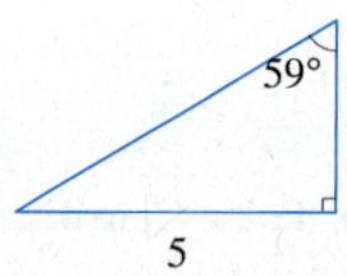


**25.**

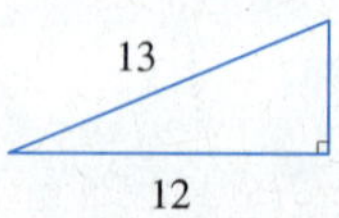


**26.**

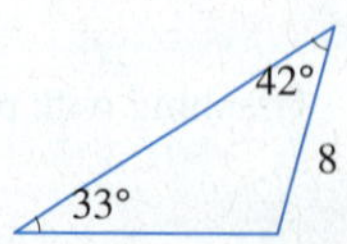


**27.** You have been asked to build a ramp for Dan's Daredevil Motorcycle Jump. The dimensions you are given are indicated in Figure 8.62. Find all the other dimensions.

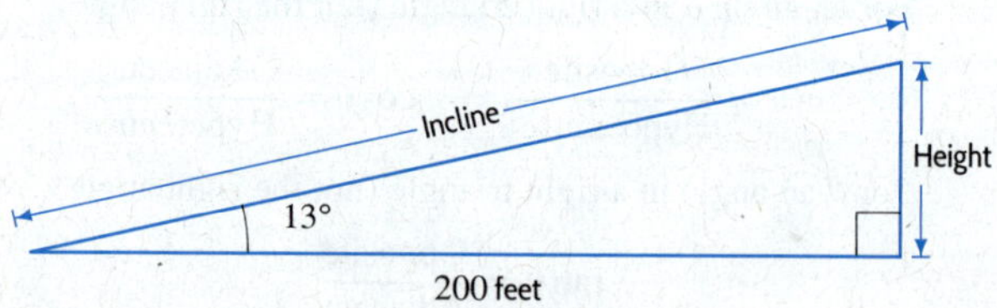


Figure 8.62

**28.** Find the value of the angle $\theta$ in Figure 8.63.

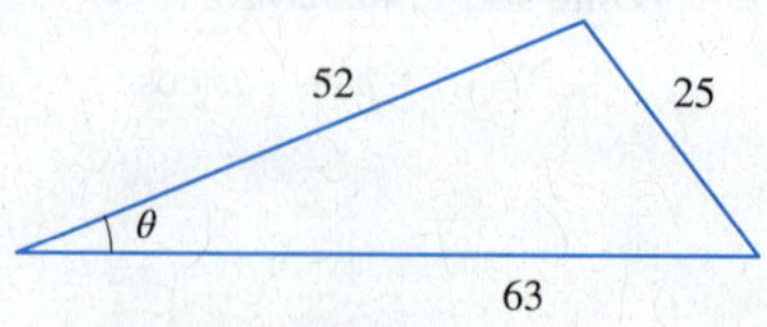


Figure 8.63

**29.** A ship spots a second ship to its east at a distance of four miles. A third ship is at an unknown distance to the north of the second ship and at an angle of 28° to the first ship. See Figure 8.64. Find $x$, the distance between the second and third ships.

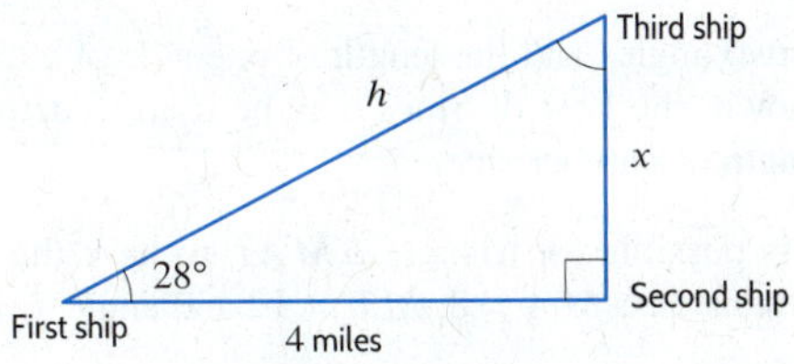


Figure 8.64

**30.** The ground crew for a hot air balloon is positioned 200 meters from the point of liftoff and monitors the ascent of the balloon. Express the height of the balloon as a function of the ground crew's angle of observation.

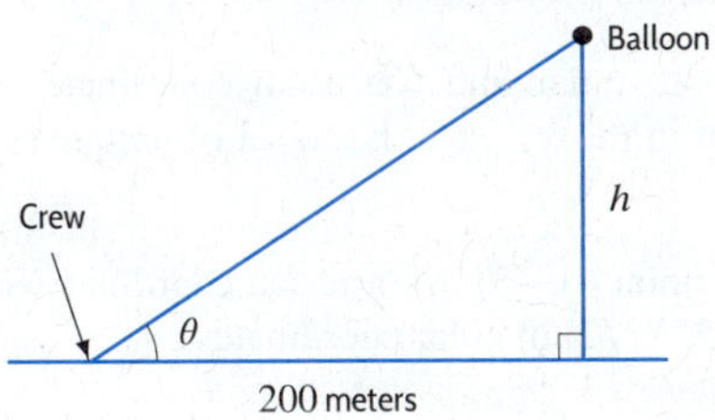


Figure 8.65

**31.** A UFO is first sighted at a point $P_1$ due east from an observer at an angle of 25° from the ground and at an altitude (vertical distance above ground) of 200 m. (See Figure 8.66.) The UFO is next sighted at a point $P_2$ due east at an angle of 50° and an altitude of 400 m. What is the distance from $P_1$ to $P_2$?

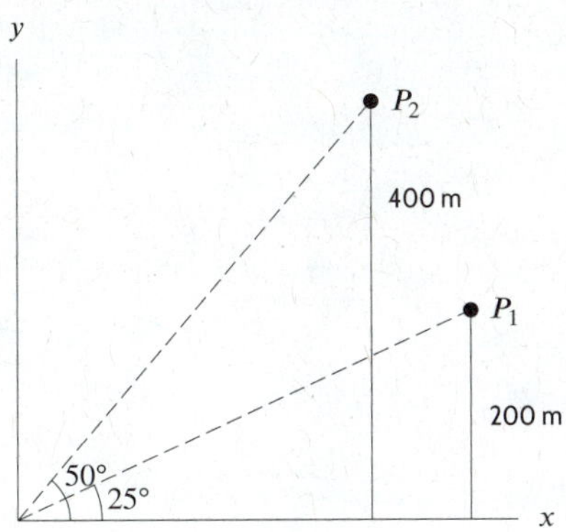


Figure 8.66

**32.** A kite flier wondered how high her kite was flying. She used a protractor to measure an angle of 38° from level ground to the kite string. If she used a full 100-yard spool of string, how high, in feet, was the kite? (Disregard the string sag and the height of the string reel above the ground.)

**33.** Hampton is a small town on a straight stretch of coastline running north and south. A lighthouse is located 3 miles offshore directly east of Hampton. The light house has a revolving searchlight that makes two revolutions per minute. The angle that the beam makes with the east-west line through Hampton is called $\phi$. Find the distance from Hampton to the point where the beam strikes the shore, as a function of $\phi$. Include a sketch.

**34.** For each of the following expressions, find a line segment in Figure 8.67 with length equal to the value of the expression.

**(a)** $\sin\theta$ **(b)** $\cos\theta$ **(c)** $\tan\theta$

**(d)** $\dfrac{1}{\sin\theta}$ **(e)** $\dfrac{1}{\cos\theta}$ **(f)** $\dfrac{1}{\tan\theta}$

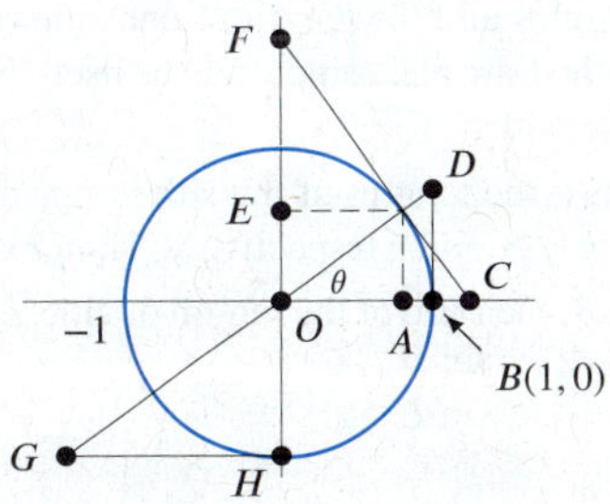


Figure 8.67

**35.** Find $\tan\theta$ exactly if $\sin\theta = -3/5$, and $\theta$ is in the fourth quadrant.

**36.** Find the radian value of $x$ in Figure 8.68.

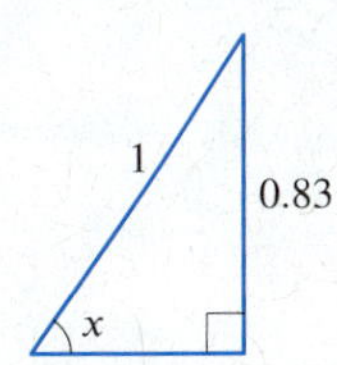


Figure 8.68

**37.** **(a)** Explain why $r = 2\cos\theta$ and $(x-1)^2 + y^2 = 1$ are equations for the same circle.

**(b)** Give Cartesian and polar coordinates for the 12, 3, 6, and 9 o'clock positions on the circle in part (a).

## STRENGTHEN YOUR UNDERSTANDING

Are the statements in Problems 1–12 true or false? Give an explanation for your answer.

1. If $\Delta ABC$ is an isosceles right triangle with angle $BAC = 90°$, then $\sin B = \sqrt{3}/2$.
2. A 16-foot ladder leans against a wall. The angle formed by the ladder and the ground is 60°. Then the top of the ladder meets the wall at a height of 8 feet above the ground.
3. The Law of Cosines may be applied only to right triangles.
4. If the lengths of three sides of a triangle are known, the Law of Cosines may be used to determine any of the angles.
5. In $\Delta NPR$, the lengths of the sides opposite angles $N$, $P$ and $R$ are $n$, $p$ and $r$ respectively. Angle $P$ is given by $\cos^{-1}\left(\frac{n^2+r^2-p^2}{2rn}\right)$.
6. When used to find the length of the side of a triangle opposite a right angle, the Law of Cosines reduces to the Pythagorean Theorem.
7. If two angles and the length of one side of a triangle are known, the Law of Cosines may be used to find the length of another side.
8. In $\Delta QXR$, the lengths of the sides opposite angles $Q$, $X$ and $R$ are $q$, $x$ and $r$ respectively. Then $r \sin Q = q \sin R$.
9. In $\Delta LAB$, the ratio of the length of side $LA$ to the length of side $BA$ is $\frac{\sin B}{\sin L}$.
10. If the lengths of two sides of a triangle and the angle included between them are known, the Law of Cosines may be used to determine the length of the third side of the triangle.
11. If two angles and the length of one side of a triangle are known, the Law of Sines may be used to determine the length of another side.
12. It is possible for triangle $\Delta MAT$ to have the following dimensions: $MA = 8$, $MT = 12$ and angle $T = 60°$.

Are the statements in Problems 13–18 true or false? Give an explanation for your answer.

13. The graph of $r = 1$ is a straight line.
14. The graph of $\theta = 1$ is the unit circle.
15. The polar coordinates $(3, \pi)$ correspond to $(-3, \pi)$ in Cartesian coordinates.
16. In both polar and Cartesian coordinate systems, each point in the $xy$-plane has a set of unique coordinate values.
17. The point $(0, -3)$ in Cartesian coordinates is represented by $(3, 3\pi/2)$ in polar coordinates.
18. The region defined in polar coordinates by $0 \le r \le 2$ and $0 \le \theta \le 2$ looks like a two-by-two square.

© Patrick Zephyr/Patrick Zephyr Nature Photography

Chapter Nine

# TRIGONOMETRIC IDENTITIES, MODELS, AND COMPLEX NUMBERS

## Contents

# 9.1 TRIGONOMETRIC EQUATIONS

A trigonometric equation is an equation involving trigonometric functions such as the sine and cosine functions. Solving a trigonometric equation means finding values that satisfy it.

## Finding Solutions Graphically

Consider the rabbit population, $R$, of Example 5 on page 296, given as a function of time, $t$, in months since January:

$$R = -5000\cos\left(\frac{\pi}{6}t\right) + 10{,}000.$$

If we want to know when the population reaches 12,000, we solve the trigonometric equation

$$-5000\cos\left(\frac{\pi}{6}t\right) + 10{,}000 = 12{,}000.$$

We can use a graph to find approximate solutions to this equation. From Figure 9.1, we see that two solutions occur near $t = 4$ and near $t = 8$. Estimating with a calculator, we obtain the solutions:

$$t = 3.79 \qquad \text{and} \qquad t = 8.21.$$

This means the rabbit population reaches 12,000 towards the end of month 3, April (since month 0 is January), and again near the start of month 8 (September).

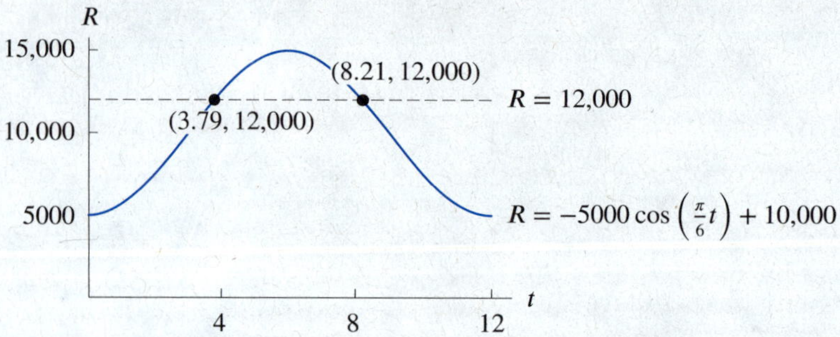


**Figure 9.1:** Using a graph to solve a trigonometric equation

## Finding Solutions Algebraically

We can also use algebra to find when the rabbit population reaches 12,000:

$$\begin{aligned}
-5000\cos\left(\frac{\pi}{6}t\right) + 10{,}000 &= 12{,}000 \\
-5000\cos\left(\frac{\pi}{6}t\right) &= 2000 && \text{subtract } 10{,}000 \\
\cos\left(\frac{\pi}{6}t\right) &= -0.4. && \text{divide by } -5000 \\
\frac{\pi}{6}t &= \cos^{-1}(-0.4) && \text{one possibility} \\
t &= \frac{6}{\pi}\cos^{-1}(-0.4) && \text{multiply by } 6/\pi \\
t &= 3.786.
\end{aligned}$$

Since the period of this function is 12, by symmetry the second solution is $12 - 3.786 = 8.214$ months. These solutions agree with the solutions we found graphically up to two decimal places.

## Solving Trigonometric Equations

An advantage of using a graph to solve a trigonometric equation on a specific interval is that you can see all possible solutions. Using algebra provides more accurate answers. In Example 1 we use a trigonometric equation to analyze the motion of the London Eye Ferris wheel.

**Example 1** On the London Eye Ferris wheel, how much time during the first turn do you spend more than 350 feet above the platform?

**Solution** Your first turn takes 30 minutes. Since you start at the bottom and your height is given by

$$f(t) = 200\sin\left(\frac{\pi}{15}(t-7.5)\right) + 200,$$

we find when you first reach 350 feet:

$$\begin{aligned}
200\sin\left(\frac{\pi}{15}(t-7.5)\right) + 200 &= 350 \\
200\sin\left(\frac{\pi}{15}(t-7.5)\right) &= 150 && \text{subtract 200} \\
\sin\left(\frac{\pi}{15}(t-7.5)\right) &= \frac{150}{200} && \text{divide by 200} \\
\frac{\pi}{15}(t-7.5) &= \sin^{-1}\left(\frac{150}{200}\right) && \text{one possibility} \\
t-7.5 &= \frac{15}{\pi}\sin^{-1}\left(\frac{150}{200}\right) && \text{multiply by } 15/\pi \\
t &= 11.55.
\end{aligned}$$

In Figure 9.2, we see there are two points on the interval $0 \le t \le 30$ where your height is 350 ft. Since the period of this function is 30, by symmetry the second solution is $30 - 11.55 = 18.45$ minutes. This means that you spend $18.45 - 11.55 = 6.9$ minutes at a height of 350 feet or more above the platform.

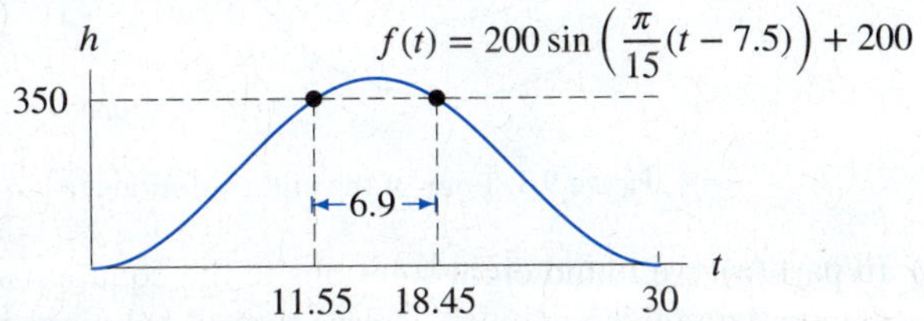


Figure 9.2: On the London Eye, you are above 350 feet between $t = 11.55$ minutes and $t = 18.45$ minutes

In Example 1, we analyzed the motion of the Ferris wheel using the graph of a sinusoidal function. We could instead have used the unit circle to find all possible solutions.

## Finding More, or All, Solutions to a Trigonometric Equation

In the next example, we see that if we want more, or all, solutions to the equation $\sin\theta = a$, we need only add multiples of $2\pi$ to the solutions between 0 and $2\pi$.

**Example 2**
(a) Solve $\sin\theta = 0.6678$ for $-2\pi \le \theta \le 6\pi$.
(b) Find all solutions to $\sin\theta = 0.6678$.

**Solution** (a) Using our calculator, we see that one solution is given by $\theta = \sin^{-1}(0.6678) = 0.7312$. Since 0.7312 radians are less than $\pi/2$ radians, a second solution is given by the angle on quadrant II with a reference angle of 0.7312 radians. We have

$$\pi - 0.7312 = 2.4103,$$

so a second angle between 0 and $2\pi$ whose sine is 0.6678 is $\theta = 2.4103$. Thus, the only two solutions between 0 and $2\pi$ are given by $\theta = 0.7312, 2.4103$.

We also know that every time $\theta$ wraps completely around the circle (in either direction), we obtain another solution. Since each time we wrap around we increase the angle by $2\pi$, we get these new solutions:

$$\begin{aligned}
0.7312 + 1\cdot 2\pi &= 7.0144 && \text{wrap once around circle} \\
0.7312 + 2\cdot 2\pi &= 13.2976 && \text{wrap twice around circle} \\
0.7312 + (-1)\cdot 2\pi &= -5.5520. && \text{wrap once around the other way}
\end{aligned}$$

Note that if we wrap three times around the unit circle, we obtain an angle larger than $6\pi$, which is outside the solution interval we wanted:

$$0.7312 + 3 \cdot 2\pi = 0.7312 + 6\pi > 6\pi.$$

If we wrap twice around the other way, we also get an angle outside our solution interval.

Reasoning the same way for $\theta = 2.4103$, we find the following solutions:

$$2.4103 + 1 \cdot 2\pi = 8.6935 \quad \text{wrap once around circle}$$
$$2.4103 + 2 \cdot 2\pi = 14.9767 \quad \text{wrap twice around circle}$$
$$2.4103 + (-1) \cdot 2\pi = -3.8728. \quad \text{wrap once around the other way}$$

Thus, the solutions on the interval $-2\pi \le \theta \le 6\pi$ are:

$$\theta = -5.5520, -3.8728, 0.7312, 2.4103, 7.0144, 8.6935, 13.2976, 14.9767.$$

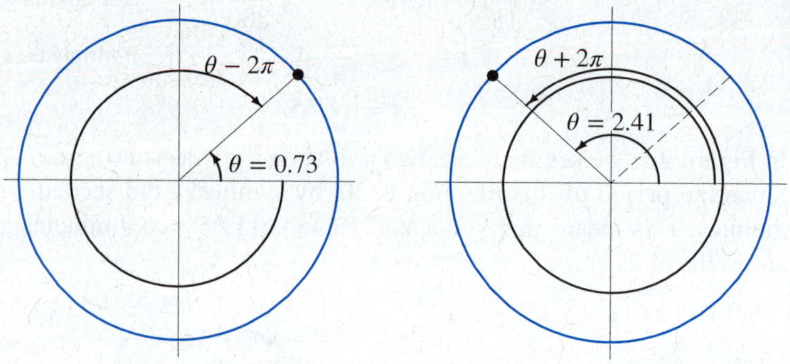


**Figure 9.3:** Four of the eight solutions to $\sin\theta = 0.6678$ on the interval

(b) In part (a), we found eight solutions to the equation $\sin\theta = 0.6678$ by adding multiples of $2\pi$. Generalizing this approach, we see that all solutions can be generated as follows:

$$\theta = 0.7312 + \left(\begin{array}{c}\text{Number of times the angle}\\ \text{wraps around the circle}\end{array}\right) \cdot 2\pi$$
$$\text{or} \quad \theta = 2.4103 + \left(\begin{array}{c}\text{Number of times the angle}\\ \text{wraps around the circle}\end{array}\right) \cdot 2\pi.$$

We can write these statements more concisely as:

$$\theta = 0.7312 + n \cdot 2\pi$$
$$\text{or} \quad \theta = 2.4103 + n \cdot 2\pi, \quad \text{where } n = 0, \pm 1, \pm 2, \ldots.$$

As Example 2 illustrates, it is possible to describe infinitely many solutions of a trigonometric equation using an algebraic expression. In this expression, the variable $n$ stands for the number of times the angle wraps around the unit circle. Alternatively, the variable $n$ stands for the number of periods by which we have shifted our original solution.

## Exercises and Problems for Section 9.1

### Exercises

**1.** Use a graph to approximate solutions to the equation $\cos t = 0.4$ on $-\pi \le t \le 3\pi$.

**2.** **(a)** Use a graph of $y = \cos t$ to estimate two solutions to the equation $\cos t = -0.3$ for $0 \le t \le 2\pi$.

**(b)** Solve the same equation using the inverse cosine.

In Exercises 3–5, use a graph to estimate all the solutions of the equations between 0 and $2\pi$.

**3.** $\sin\theta = 0.65$ **4.** $\tan x = 2.8$ **5.** $\cos t = -0.24$

In Exercises 6–11, solve the equation for a value of $\theta$ in the first quadrant. Give your answer in radians and degrees.

**6.** $5\sin(3\theta) = 4$ **7.** $9\tan(5\theta) + 1 = 10$

**8.** $2\sqrt{3}\tan(2\theta) + 1 = 3$ **9.** $3\sin\theta + 3 = 5\sin\theta + 2$

**10.** $6\cos(3\theta) + 3 = 4\cos(3\theta) + 4$

**11.** $5\tan(4\theta) + 4 = 2(\tan(4\theta) + 5)$

In Exercises 12–14, find a solution for $\theta$, if possible. Give your answer in radians.

**12.** $\cos(\theta - 2) = -50$ **13.** $\tan(5\theta + 7) = -0.241$

**14.** $2\sin(4\theta) = 0.667$

## Problems

**15.** **(a)** Find exact solutions to the equation $\cos t = 1/2$ with $-2\pi \le t \le 2\pi$. Plot them on a graph of $y = \cos t$.
**(b)** Using a unit circle, explain how many solutions to the equation $\cos t = 1/2$ you would expect in the interval $0 \le t \le 2\pi$.

**16.** Using the unit circle, find all solutions to $\cos t = 0.4$ on the interval $-2\pi \le t \le 6\pi$.

In Problems 17–22, find all solutions to the equation for $0 \le x \le 2\pi$.

**17.** $\cos x = 0.6$ **18.** $2\sin x = 1 - \sin x$

**19.** $5\cos x = 1/\cos x$ **20.** $\sin 2x = 0.3$

**21.** $\sin(x - 1) = 0.25$ **22.** $5\cos(x + 3) = 1$

**23.** Find all solutions of $\cos\theta = -0.4226$ for $0° \le \theta \le 720°$.

**24.** Solve the equations

**(a)** $\cos\theta = 0.4226$ for $0° \le \theta \le 360°$
**(b)** $\tan\theta = 2.145$ for $0° \le \theta \le 720°$

In Problems 25–28 use inverse trigonometric functions to find a solution to the equation in the given interval. Then, use a graph to find all other solutions in this interval.

**25.** $\cos x = 0.6, \quad 0 \le x \le 4\pi$

**26.** $\sin x = 0.3, \quad 0 \le x \le 2\pi$

**27.** $\cos x = -0.7, \quad 0 \le x \le 2\pi$

**28.** $\sin x = -0.8, \quad 0 \le x \le 4\pi$

In Problems 29–31, find exact values for all the solutions to the equation.

**29.** $\sin\theta = -\sqrt{2}/2$ **30.** $\cos\theta = \sqrt{3}/2$

**31.** $\tan\theta = -\sqrt{3}/3$

**32.** Solve for $\alpha$ exactly with $0 \le \alpha < 2\pi$:
$\sec^2\alpha + 3\tan\alpha = \tan\alpha$.

In Problems 33–36, solve the equation for $0 \le t \le 2\pi$. First estimate answers from a graph; then find exact answers.

**33.** $\cos(2t) = \dfrac{1}{2}$ **34.** $\tan t = \dfrac{1}{\tan t}$

**35.** $2\sin t\cos t - \cos t = 0$ **36.** $3\cos^2 t = \sin^2 t$

**37.** Use a graph to find all the solutions to the equation $12 - 4\cos 3t = 14$ between 0 and $2\pi/3$ (one cycle). How many solutions are there between 0 and $2\pi$?

**38.** Approximate the zero(s) of $f(t) = 3 - 5\sin(4t)$ for $0 \le t < \pi/2$
**(a)** Graphically. **(b)** Using the arcsine function.

**39.** A company's sales are seasonal with the peak in mid-December and the lowest point in mid-June. The company makes \$100,000 in sales in December and only \$20,000 in June.

**(a)** Find a trigonometric function, $s = f(t)$, representing sales at time $t$ months after mid-January.
**(b)** What would you expect the sales to be for mid-April?
**(c)** Find the $t$-values for which $s = 60{,}000$. Interpret your answer.

**40.** In a tidal river, the time between high tide and low tide is 6.2 hours. At high tide the depth of the water is 17.2 feet, while at low tide the depth is 5.6 feet. Assume the water depth is a trigonometric function of time.

**(a)** Graph the depth of the water over time if there is a high tide at 12:00 noon. Label your graph, indicating the high and low tide.
**(b)** Write an equation for the curve you drew in part (a).
**(c)** A boat requires a depth of 8 feet to sail, and is docked at 12:00 noon. What is the latest time in the afternoon it can set sail? Your answer should be accurate to the nearest minute.

**41.** Approximate the $x$-coordinates of points $P$ and $Q$ shown in Figure 9.4, assuming that the curve is a sine curve. [Hint: Find a formula for the curve.]

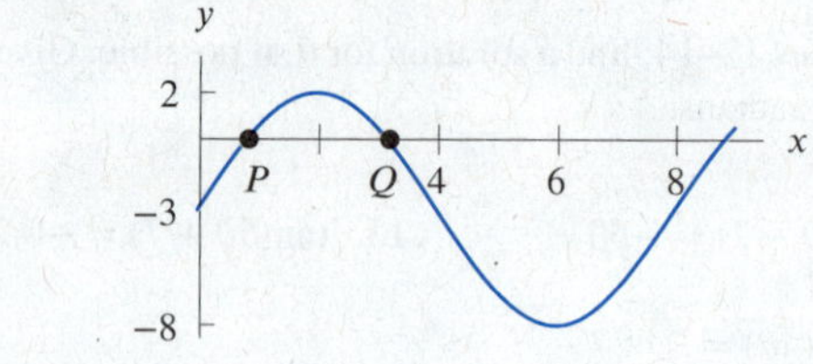


Figure 9.4

**42.** Let $k$ be a positive constant and $t$ be an angle measured in radians. Consider the equation for $t$,

$$k \sin t = t^2.$$

**(a)** Explain why any solution to the equation must be between $-\sqrt{k}$ and $\sqrt{k}$, inclusive.

**(b)** Approximate every solution to the equation when $k = 2$.

**(c)** Explain why the equation has more solutions for larger values of $k$ than it does for small values.

**(d)** Approximate the least value of $k$, if any, for which the equation has a negative solution.

**43.** You are perched in the crow's nest, $C$, on top of the mast of a ship, $S$. See Figure 9.5. You will calculate how far you can see when you are $x$ meters above the surface of the ocean.

**(a)** Find formulas for $d$, the distance you can see to the horizon, $H$, and $l$, the distance to the horizon along the earth's surface, in terms of $x$, the height of the ship's mast, and $r$, the radius of the earth.

**(b)** How far is the horizon from the top of a 50-meter mast? How far, measured along the earth's surface, is the horizon from the ship's position on the ocean? Use $r = 6{,}370{,}000$ meters.

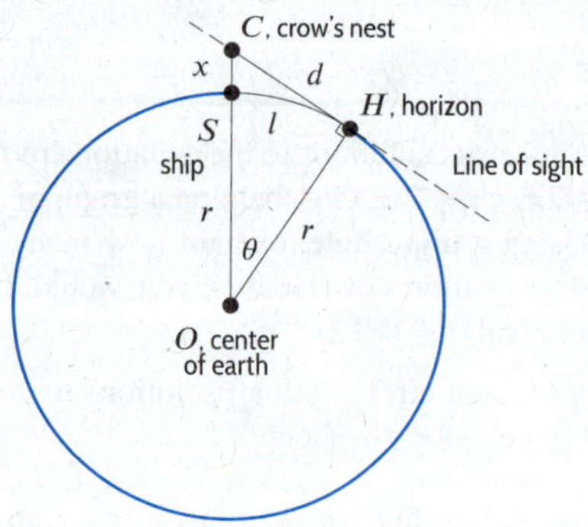


Figure 9.5

# 9.2 IDENTITIES, EXPRESSIONS, AND EQUATIONS

## Equations Versus Identities

In Chapter 7, we saw the statement $\sin^2\theta + \cos^2\theta = 1$, which we referred to as the Pythagorean identity. An *identity* is an equation that is true for all values of the variable. For example, the equation

$$2(x-1) = x$$

has only one solution, $x = 2$, since this value of $x$ makes the statement true, whereas all other values make the statement false. In contrast, the following equation is true for all values of $x$, so it is an identity.

$$2(x-1) = 2x - 2.$$

Trigonometric identities allow us to rewrite expressions so we can easily identify the behavior of trigonometric models.

### Using Identities To Simplify Expressions

Trigonometric identities can help us simplify expressions.

**Example 1** Use the identities $\sin(-t) = -\sin t$ and $\cos(\pi/2 - t) = \sin t$ to rewrite the following as a sinusoidal function and give its amplitude, midline, and period:

$$y = 2\sin t - 3\sin(-t) + 4\cos\left(\frac{\pi}{2} - t\right).$$

Solution We have

$$\begin{aligned} y &= 2\sin t - 3\underbrace{\sin(-t)}_{-\sin t} + 4\underbrace{\cos\left(\frac{\pi}{2} - t\right)}_{\sin t} \\ &= 2\sin t - 3(-\sin t) + 4\sin t \\ &= 2\sin t + 3\sin t + 4\sin t \\ &= 9\sin t. \end{aligned}$$

So this is a sinusoidal function with amplitude $A = 9$, midline $k = 0$, and period $p = 2\pi$.

**Example 2** Simplify the expression $(2\cos t + 3\sin t)(3\cos t + 2\sin t) - 13\sin t\cos t$.

Solution To make the calculations easier, let $r = \cos t$ and $s = \sin t$. Our expression becomes

$$\begin{aligned} (2r + 3s)(3r + 2s) - 13rs &= 6r^2 + 4rs + 9rs + 6s^2 - 13rs && \text{multiply out} \\ &= 6r^2 + 6s^2 + 13rs - 13rs && \text{regroup} \\ &= 6\left(r^2 + s^2\right) && \text{simplify and factor} \\ &= 6\left(\cos^2 t + \sin^2 t\right) && \text{because } r = \cos t, s = \sin t \\ &= 6. && \text{because } \cos^2 t + \sin^2 t = 1 \end{aligned}$$

We see that this complicated expression equals 6!

## Using Identities to Evaluate Expressions

If we know which quadrant an angle is in and we know the value of any one of its trigonometric functions, we can evaluate the other two by using the Pythagorean identity.

**Example 3** Suppose that $\cos\theta = 2/3$ and $3\pi/2 \le \theta \le 2\pi$. Find $\sin\theta$ and $\tan\theta$.

Solution Use the relationship $\cos^2\theta + \sin^2\theta = 1$ to find $\sin\theta$. Substitute $\cos\theta = 2/3$:

$$\begin{aligned} \left(\frac{2}{3}\right)^2 + \sin^2\theta &= 1 \\ \frac{4}{9} + \sin^2\theta &= 1 \\ \sin^2\theta &= 1 - \frac{4}{9} = \frac{5}{9} \\ \sin\theta &= \pm\sqrt{\frac{5}{9}} = \pm\frac{\sqrt{5}}{3}. \end{aligned}$$

Because $\theta$ is in the fourth quadrant, $\sin\theta$ is negative, so $\sin\theta = -\sqrt{5}/3$. To find $\tan\theta$, use the relationship

$$\tan\theta = \frac{\sin\theta}{\cos\theta} = \frac{-\sqrt{5}/3}{2/3} = -\frac{\sqrt{5}}{2}.$$

## Using Identities to Solve Equations

Identities can help us solve equations involving trigonometric functions. An equation involving several trigonometric functions can often be rewritten in terms of a single function.

**Example 4** Solve $3\sin\theta = \sqrt{3}\cos\theta$, for $0 \le \theta \le 2\pi$.

Solution We start by visualizing the solutions in Figure 9.6. There appear to be two solutions on $0 \le \theta \le 2\pi$. How can we find them exactly?

Notice that the equation involves both the sine and the cosine. But if we divide both sides by $\cos\theta$ (assuming $\cos\theta \neq 0$) and replace $\sin\theta/\cos\theta$ by $\tan\theta$, the equation is just in terms of tangent:

$$\begin{aligned} 3\cdot\frac{\sin\theta}{\cos\theta} &= \sqrt{3} \\ 3\tan\theta &= \sqrt{3} \\ \tan\theta &= \frac{\sqrt{3}}{3} = \frac{1}{\sqrt{3}}. \end{aligned}$$

Now, since we know that

$$\tan\frac{\pi}{6} = \frac{\sin(\pi/6)}{\cos(\pi/6)} = \frac{1/2}{\sqrt{3}/2} = \frac{1}{\sqrt{3}},$$

we can see that one solution to this equation is

$$\theta = \frac{\pi}{6}.$$

The tangent function has period $\pi$, so the second solution on $0 \le \theta \le 2\pi$ is

$$\theta = \frac{\pi}{6} + \pi = \frac{7\pi}{6}.$$

Dividing by $\cos\theta$ is not valid if $\cos\theta = 0$. We must check separately whether values of $\theta$ that make $\cos\theta = 0$, namely $\theta = \pi/2$ and $\theta = 3\pi/2$, are solutions. Neither of these values satisfies the original equation, as the graphs in Figure 9.6 do not cross there.

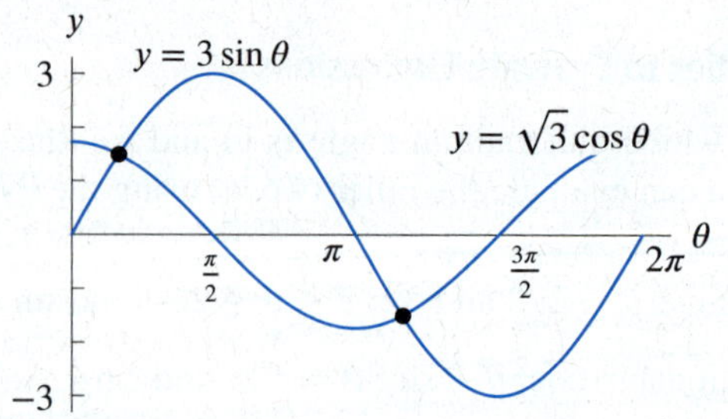


**Figure 9.6**: Graphs to show solution to $3\sin\theta = \sqrt{3}\cos\theta$

**Example 5** Solve $2\sin^2 t = 3 - 3\cos t$ for $0 \le t \le \pi$.

**Solution** This equation looks difficult to solve because it contains both sines and cosines. We use the identity $\sin^2 t = 1 - \cos^2 t$ to rewrite it entirely in terms of $\cos t$:

$$\begin{aligned} 2\underbrace{\left(1-\cos^2 t\right)}_{\sin^2 t} &= 3 - 3\cos t \\ 2 - 2\cos^2 t &= 3 - 3\cos t \\ 2\cos^2 t - 3\cos t + 1 &= 0. \end{aligned}$$

Writing $\cos t = x$, so $x^2 = (\cos t)^2 = \cos^2 t$, makes it easier to see that this equation is quadratic:

$$\begin{aligned} 2\underbrace{\cos^2 t}_{x^2} - 3\underbrace{\cos t}_{x} + 1 = 2x^2 - 3x + 1 &= 0 \\ (2x-1)(x-1) &= 0 \\ \text{so}\quad x = \frac{1}{2} \text{ or } x &= 1. \end{aligned}$$

Since $x = \cos t$, we know either $\cos t = 1/2$ with solution $t = \pi/3$, or $\cos t = 1$ with solution $t = 0$.

## Double-Angle Formula for Sine

Thinking graphically, we know that $y = \sin 2\theta$ is a different function than $y = 2\sin\theta$. For one thing, they have different amplitudes, and for another they have different periods.

Thinking algebraically, we would like to find a formula for $\sin 2\theta$ in terms of $\sin\theta$ and $\cos\theta$. We derive our formula by using Figure 9.7. The lengths of $OA$ and $OC$ are 1; the length of $AC$ is $2\sin\theta$. Writing $\alpha$ for the angle at $A$ and applying the Law of Sines to triangle $OAC$ gives

$$\frac{\sin 2\theta}{2\sin\theta} = \frac{\sin\alpha}{1}.$$

In triangle $OAB$, the length of side $OB$ is $\cos\theta$, and the hypotenuse is 1, so

$$\sin\alpha = \frac{\text{Opposite}}{\text{Hypotenuse}} = \frac{\cos\theta}{1} = \cos\theta.$$

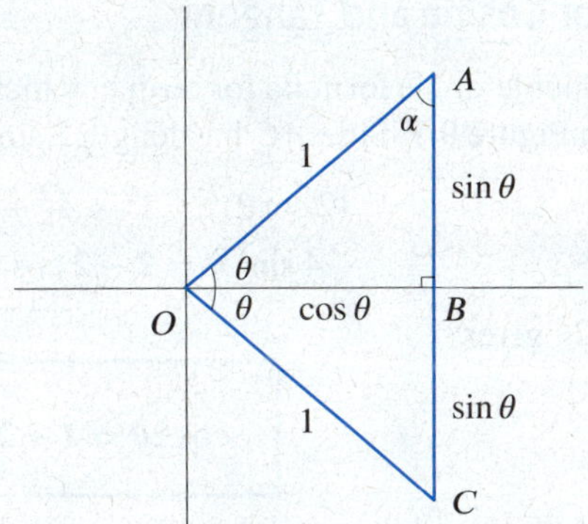


Figure 9.7: Triangle used to derive the double-angle formula for sine

Thus, substituting $\cos\theta$ for $\sin\alpha$, we have

$$\frac{\sin 2\theta}{2\sin\theta} = \cos\theta$$

so

$$\boxed{\sin 2\theta = 2\sin\theta\cos\theta.}$$

This identity is known as the *double-angle formula for sine*. We have shown the double-angle formula is true for $0 \le \theta \le \pi/2$. Problems 62 and 63 extend the result to all $\theta$.

---

**Example 6** Find all solutions to the equation $\sin 2t = 2\sin t$ on the interval $0 \le t \le 2\pi$.

**Solution** Rewriting the equation using the double-angle formula $\sin 2t = 2\sin t\cos t$, we have

$$\begin{aligned} 2\sin t\cos t &= 2\sin t \\ 2\sin t\cos t - 2\sin t &= 0 \\ 2\sin t(\cos t - 1) &= 0. \quad \text{Factoring out } 2\sin t \end{aligned}$$

Thus,

$$2\sin t = 0 \quad \text{or} \quad \cos t = 1.$$

Now, $2\sin t = 0$ for $t = 0, \pi$, and $2\pi$, and $\cos t = 1$ for $t = 0$ and $2\pi$. Thus there are three solutions to the original equation on the interval $0 \le t \le 2\pi$: $t = 0, \pi$, and $2\pi$. Figure 9.8 illustrates these solutions graphically as the points where the graphs of $y = \sin 2t$ and $y = 2\sin t$ intersect.

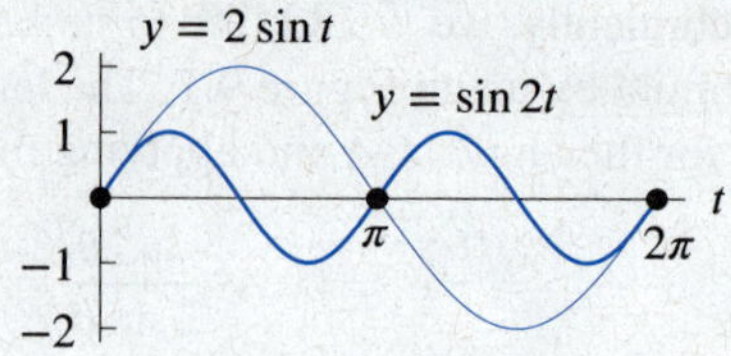


Figure 9.8: There are three solutions to the equation $\sin 2t = 2\sin t$ for $0 \le t \le 2\pi$

## Double-Angle Formulas for Cosine and Tangent

There is also a double-angle formula for cosine, which we derive by applying the Law of Cosines to triangle $OAC$ in Figure 9.7. Side $AC$ has length $2\sin\theta$, so

$$(2\sin\theta)^2 = 1^2 + 1^2 - 2\cdot 1\cdot 1\cos 2\theta$$
$$4\sin^2\theta = 2 - 2\cos 2\theta.$$

Solving for $\cos 2\theta$ gives

$$\boxed{\cos 2\theta = 1 - 2\sin^2\theta.}$$

Problems 64 and 65 extend the result to all $\theta$.

This is the double-angle formula for cosine. In Problem 28, we use the Pythagorean identity to write the formula for $\cos 2\theta$ in two other ways:

$$\boxed{\cos 2\theta = 2\cos^2\theta - 1} \quad \text{and} \quad \boxed{\cos 2\theta = \cos^2\theta - \sin^2\theta.}$$

Finally, Problem 29 asks you to derive the double-angle formula for tangent:

$$\boxed{\tan 2\theta = \frac{2\tan\theta}{1-\tan^2\theta}.}$$

Sometimes we need to use more than one identity to solve an equation.

**Example 7** Solve $\sin 2t = \sqrt{2}\sin(t + \pi/2)$ for $0 \le t < 2\pi$.

**Solution** Use the double-angle formula for sine to replace $\sin 2t$ by $2\sin t\cos t$. We also replace $\sin(t + \pi/2)$ by $\cos t$, giving

$$2\sin t\cos t = \sqrt{2}\cos t.$$

Now all of the trigonometric functions are functions of the same argument, $t$. Next we write the equation as

$$2\sin t\cos t - \sqrt{2}\cos t = 0$$

and factor the left side to get

$$\cos t(2\sin t - \sqrt{2}) = 0,$$

so

$$\cos t = 0 \quad \text{or} \quad 2\sin t - \sqrt{2} = 0.$$

Now each equation involves only one trigonometric function. The solutions of $\cos t = 0$ are $t = \pi/2$ and $t = 3\pi/2$. The solutions to $2\sin t - \sqrt{2} = 0$ are given by:

$$2\sin t = \sqrt{2}$$
$$\sin t = \frac{\sqrt{2}}{2}.$$

We know that $\sin(\pi/4) = \sqrt{2}/2$, so solutions are $t = \pi/4$ and $t = 3\pi/4$.

## Exercises and Problems for Section 9.2

### Exercises

In Exercises 1–4, are the two expressions equal for all values of $t$? If so, show this algebraically. If not, find a value of $t$ for which they are not equal.

**1.** $\sin(3t)$ and $3\sin t$

**2.** $\cos^2 t$ and $\sin^2 t - 1$

**3.** $(\cos t - \sin t)(\cos t + \sin t)$ and $\cos^2 t - \sin^2 t$

**4.** $\sec t \sin t$ and $\tan t$

Exercises 5–8 give the quadrant of an angle and the value of its sine, or cosine, or tangent. Without a calculator, find the values of the other two functions.

**5.** $\sin\theta = 1/4$ and $0 \le \theta \le \pi/2$

**6.** $\sin\theta = 3/5$ and $\pi/2 \le \theta \le \pi$

**7.** $\cos\theta = -2/5$ and $\pi \le \theta \le 3\pi/2$

**8.** $\cos\theta = 1/3$ and $3\pi/2 \le \theta \le 2\pi$

In Exercises 9–12, simplify the expression. Your answer should involve only $\sin t$ and $\cos t$.

**9.** $\tan t \cos t - \dfrac{\sin t}{\tan t}$

**10.** $2\cos t\,(3 - 7\tan t)$

**11.** $2\cos t - \cos t\,(1 - 3\tan t)$

**12.** $2\cos t\,(3\sin t - 4\tan t)$

In Exercises 13–18, simplify the expression.

**13.** $\dfrac{\sin 2\alpha}{\cos\alpha}$

**14.** $\dfrac{\cos^2\theta - 1}{\sin\theta}$

**15.** $\dfrac{\cos 2t}{\cos t + \sin t}$

**16.** $\dfrac{1}{1-\sin\theta} + \dfrac{1}{1+\sin\theta}$

**17.** $\dfrac{\cos\phi - 1}{\sin\phi} + \dfrac{\sin\phi}{\cos\phi + 1}$

**18.** $\dfrac{1}{\sin t\cos t} - \dfrac{1}{\tan t}$

In Exercises 19–24, write the expression in terms of the tangent function.

**19.** $\dfrac{\sin\sqrt{\theta}}{\cos\sqrt{\theta}}$

**20.** $\dfrac{2\sin\frac{\alpha}{2}}{\cos\frac{\alpha}{2}}$

**21.** $\dfrac{3\sin(\phi+1)}{4\cos(\phi+1)}$

**22.** $\dfrac{1}{\left(\dfrac{\cos(r^2-s^2)}{\sin(r^2-s^2)}\right)}$

**23.** $2\sin\left(\dfrac{2}{k+3}\right)\cdot\dfrac{5}{3\cos\left(\dfrac{2}{k+3}\right)}$

**24.** $\dfrac{2}{\cos\left(1-\frac{1}{z}\right)}\cdot\dfrac{\sin\left(1-\frac{1}{z}\right)}{3}$

**25.** Using the Pythagorean identity, give an expression for $\sin\theta$ in terms of $\cos\theta$.

**26.** Complete the following table, using exact values where possible. State the trigonometric identities that relate the quantities in the table.

| $\theta$ (radians) | $\sin^2\theta$ | $\cos^2\theta$ | $\sin 2\theta$ | $\cos 2\theta$ |
|---|---|---|---|---|
| 1 | | | | |
| $\pi/2$ | | | | |
| 2 | | | | |
| $5\pi/6$ | | | | |

## Problems

**27.** Use graphs to check that $\cos 2t$ and $1 - 2\sin^2 t$ have the same sign for all values of $t$.

**28.** Use the Pythagorean identity to write the double-angle formula for cosine in these two alternate forms:

**(a)** $\cos 2\theta = 2(\cos\theta)^2 - 1$
**(b)** $\cos 2\theta = (\cos\theta)^2 - (\sin\theta)^2$

**29.** Use $\tan 2\theta = \dfrac{\sin 2\theta}{\cos 2\theta}$ to derive a double-angle formula for tangent. [Hint: Use Problem 28(b).]

**30.** If $\sin\theta = 3/5$ and $\theta$ is in the second quadrant, find $\sin(2\theta)$, $\cos(2\theta)$, and $\tan(2\theta)$ exactly.

For Problems 31–33, use algebra to prove the identity.

**31.** $\dfrac{\sin t}{1 - \cos t} = \dfrac{1 + \cos t}{\sin t}$

**32.** $\dfrac{\cos x}{1 - \sin x} - \tan x = \dfrac{1}{\cos x}$

**33.** $\dfrac{\sin x \cos y + \cos x \sin y}{\cos x \cos y - \sin x \sin y} = \dfrac{\tan x + \tan y}{1 - \tan x \tan y}$

In Problems 34–37, Use trigonometric identities to solve the trigonometric equation exactly for $0 \le \theta \le 2\pi$.

**34.** $\sin^2\theta - \cos^2\theta = \sin\theta$

**35.** $\sin(2\theta) - \cos\theta = 0$

**36.** $\sec^2\theta = 1 - \tan\theta$

**37.** $\tan(2\theta) + \tan\theta = 0$

For Problems 38–53, use a graph to decide whether the equation is an identity. If the equation is an identity, prove it algebraically. If it is not an identity, find a value of the variable for which the equation is false.

**38.** $\dfrac{1}{2+x} = \dfrac{1}{2} + \dfrac{1}{x}$

**39.** $\sin\left(\dfrac{1}{x}\right) = \dfrac{1}{\sin x}$

**40.** $\sqrt{64 - x^2} = 8 - x$

**41.** $\sin(2x) = 2\sin x$

**42.** $\cos(x^2) = (\cos x)^2$

**43.** $\sin(x^2) = 2\sin x$

**44.** $\tan x = \dfrac{\sin(2x)}{1 + \cos(2x)}$

**45.** $\dfrac{\sin(2A)}{\cos(2A)} = 2\tan A$

**46.** $\dfrac{\sin^2\theta - 1}{\cos\theta} = -\cos\theta$

**47.** $\cos\left(x + \dfrac{\pi}{3}\right) = \cos x + \cos\dfrac{\pi}{3}$

**48.** $\sin x \tan x = \dfrac{1 - (\cos x)^2}{\cos x}$

**49.** $\tan t + \dfrac{1}{\tan t} = \dfrac{1}{\sin t \cos t}$

**50.** $\sin\left(\dfrac{1}{x}\right) = \sin 1 - \sin x$

**51.** $\sin(2x) = \dfrac{2\tan x}{1 + (\tan x)^2}$

**52.** $\dfrac{\sin\theta}{\cos\theta} - \dfrac{\cos\theta}{\sin\theta} = \dfrac{\cos(2\theta)}{\sin(2\theta)}$

**53.** $\cos(2x) = \dfrac{1 - (\tan x)^2}{1 + (\tan x)^2}$

**54.** Find exactly all solutions to the equations.

**(a)** $\cos 2\theta + \cos\theta = 0, \quad 0 \le \theta < 360°$
**(b)** $2\cos^2\theta = 3\sin\theta + 3, \quad 0 \le \theta \le 2\pi$

**55.** Use Figure 9.9 to express the following in terms of $\theta$.

**(a)** $y$ **(b)** $\cos\varphi$
**(c)** $1 + y^2$ **(d)** The triangle's area

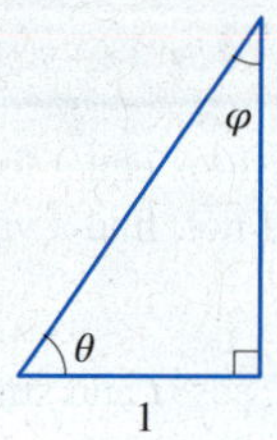


Figure 9.9

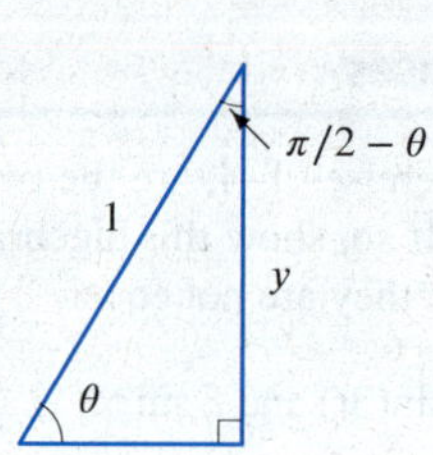


Figure 9.10

**56.** Let $y$ be the side opposite to the angle $\theta$ in a right triangle whose hypotenuse is 1. (See Figure 9.10.) Without trigonometric functions, express the following expressions in terms of $y$.

**(a)** $\cos\theta$ **(b)** $\tan\theta$ **(c)** $\cos(2\theta)$
**(d)** $\sin(\pi - \theta)$ **(e)** $\sin^2(\cos^{-1}(y))$

**57.** If $x = 3\cos\theta$, $0 < \theta < \pi/2$, express $\sin(2\theta)$ in terms of $x$.

**58.** If $x + 1 = 5\sin\theta$, $0 < \theta < \pi/2$, express $\cos(2\theta)$ in terms of $x$.

**59.** Express in terms of $x$ without trigonometric functions. [Hint: Let $\theta = \cos^{-1} x$ in part (a).]

**(a)** $\tan(2\cos^{-1} x)$ **(b)** $\sin(2\tan^{-1} x)$

**60.** Find an identity for $\cos(4\theta)$ in terms of $\cos\theta$. (You need not simplify your answer.)

**61.** Use trigonometric identities to find an identity for $\sin(4\theta)$ in terms of $\sin\theta$ and $\cos\theta$.

**62.** For $0 \le t \le \pi/2$, we know that $\sin 2t = 2\sin t\cos t$. Assume now that $\pi/2 < t \le \pi$.

**(a)** Show that $\sin 2(\pi - t) = 2\sin(\pi - t)\cos(\pi - t)$.
**(b)** Use the periodicity and the oddness of the sine function to show that $\sin 2(\pi - t) = -\sin 2t$.
**(c)** Use the fact that $\cos(t + \pi) = -\cos t$ and $\sin(t + \pi) = -\sin t$ to show that $\cos(\pi - t) = -\cos t$ and $\sin(\pi - t) = \sin t$.
**(d)** Show that $\sin 2t = 2\sin t\cos t$.

**63.** In Problem 62, we showed that $\sin 2t = 2\sin t\cos t$ for $0 \le t \le \pi$. Now assume that $-\pi \le t < 0$. Show that

**(a)** $\sin(-2t) = 2\sin(-t)\cos(-t)$.
**(b)** $\sin 2t = 2\sin t\cos t$.

**64.** For $0 \le t \le \pi/2$, we know that $\cos 2t = 1 - 2\sin^2 t$. Assume now that $\pi/2 < t \le \pi$.

**(a)** Show that $\cos 2(\pi - t) = 1 - 2\sin^2(\pi - t)$.
**(b)** Use the periodicity and the evenness of the cosine function to show that $\cos 2(\pi - t) = \cos 2t$.
**(c)** Use the fact that $\sin(t + \pi) = -\sin t$ to show that $1 - 2\sin^2(\pi - t) = 1 - 2\sin^2 t$.
**(d)** Show that $\cos 2t = 1 - 2\sin^2 t$.

**65.** In Problem 64, we showed that $\cos 2t = 1 - 2\sin^2 t$ for $0 \le t \le \pi$. Now assume that $-\pi \le t < 0$. Show that

**(a)** $\cos(-2t) = 1 - 2\sin^2(-t)$.
**(b)** $\cos 2t = 1 - 2\sin^2 t$.

**66.** The expressions (a)-(m) contain several pairs that are identically equal. Use graphs to find five such pairs.

**(a)** $2\cos^2 t + \sin t + 1$ **(b)** $\cos^2 t$
**(c)** $\cos(2t)$ **(d)** $1 - 2\sin^2 t$
**(e)** $2\sin t\cos t$ **(f)** $\cos^2 t - \sin^2 t$
**(g)** $\sin(2t)$ **(h)** $\sin(3t)$
**(i)** $-2\sin^2 t + \sin t + 3$ **(j)** $\sin(2t)\cos t + \cos(2t)\sin t$
**(k)** $\dfrac{1 - \sin t}{\sin t}$ **(l)** $\dfrac{1 + \cos(2t)}{2}$
**(m)** $\dfrac{\left(\dfrac{1}{\cos t} - 1\right)}{1 - \cos t}$

## 9.3 SUM AND DIFFERENCE FORMULAS FOR SINE AND COSINE

To help us build realistic models using trigonometric functions, we now look at some trigonometric identities that generalize the identities in the last section.

### Sums and Differences of Angles

The double-angle formulas introduced in Section 9.2 are special cases of a group of more general trigonometric identities which allow us to find the sine and cosine of the sum and difference of two angles. For example, knowing the exact values for the sine and cosine of 45° and 30°, we can find the exact value for $\cos(45° - 30°) = \cos 15°$.

Though it is tempting to speculate that $\cos 45° - \cos 30°$ equals $\cos(45° - 30°)$, you can use your calculator to see that this is not the case. Instead, we can use one of the following formulas, which are justified on pages 366–367.

**Sum and Difference Formulas**

$$\begin{aligned}\sin(\theta + \phi) &= \sin\theta\cos\phi + \cos\theta\sin\phi\\ \sin(\theta - \phi) &= \sin\theta\cos\phi - \cos\theta\sin\phi\\ \cos(\theta + \phi) &= \cos\theta\cos\phi - \sin\theta\sin\phi\\ \cos(\theta - \phi) &= \cos\theta\cos\phi + \sin\theta\sin\phi.\end{aligned}$$

**Example 1** Find an exact value for $\cos 15°$.

**Solution** Let $\theta = 45°$ and $\phi = 15°$. Then

$$\cos 15° = \underbrace{\cos(45° - 30°)}_{\cos(\theta-\phi)} = \underbrace{\cos 45°\cos 30° + \sin 45°\sin 30°}_{\cos\theta\cos\phi + \sin\theta\sin\phi}$$
$$= \frac{\sqrt{2}}{2}\cdot\frac{\sqrt{3}}{2} + \frac{\sqrt{2}}{2}\cdot\frac{1}{2} = \frac{\sqrt{6}+\sqrt{2}}{4}.$$

We can check our answer using a calculator:

$$\frac{\sqrt{6}+\sqrt{2}}{4} = 0.9659 \quad \text{and} \quad \cos 15^\circ = 0.9659.$$

**Example 2** Find an identity for $\sin 3\theta$ in terms of $\sin\theta$ and $\cos\theta$.

**Solution** We can write $\sin 3\theta = \sin(2\theta+\theta)$ and then use the sum-of-angles identity for sine and the double-angle formulas:

$$\begin{aligned}
\sin(2\theta+\theta) &= \sin(2\theta)\cos\theta + \cos(2\theta)\sin\theta && \text{sum-of-angles identity for sine}\\
&= \underbrace{2\sin\theta\cos\theta}_{\sin 2\theta}\cos\theta + \underbrace{\left(\cos^2\theta-\sin^2\theta\right)}_{\cos 2\theta}\sin\theta && \text{double-angle identities}\\
&= 2\sin\theta\cos^2\theta + \cos^2\theta\sin\theta - \sin^3\theta\\
&= 3\sin\theta\cos^2\theta - \sin^3\theta.
\end{aligned}$$

**Example 3** Using the sum and difference formulas, check the results of Example 4 on page 307,

$$\sin t = \cos\left(t-\frac{\pi}{2}\right) \quad \text{and} \quad \cos t = \sin\left(t+\frac{\pi}{2}\right).$$

**Solution** Letting $\theta = t$ and $\phi = \pi/2$ in the identity $\cos(\theta-\phi) = \cos\theta\cos\phi + \sin\theta\sin\phi$, we get the first result:

$$\cos\left(t-\frac{\pi}{2}\right) = \cos t\underbrace{\cos\frac{\pi}{2}}_{0} + \sin t\underbrace{\sin\frac{\pi}{2}}_{1} = \sin t.$$

Likewise, the identity $\sin(\theta+\phi) = \sin\theta\cos\phi + \cos\theta\sin\phi$ gives the second result:

$$\sin\left(t+\frac{\pi}{2}\right) = \sin t\underbrace{\cos\frac{\pi}{2}}_{0} + \cos t\underbrace{\sin\frac{\pi}{2}}_{1} = \cos t.$$

## Justification of $\cos(\theta-\phi) = \cos\theta\cos\phi + \sin\theta\sin\phi$

To derive the formula for $\cos(\theta-\phi)$, we find the distance between points $A$ and $B$ in Figure 9.11 in two different ways. These points correspond to the angles $\theta$ and $\phi$ on the unit circle, so their coordinates are $A = (\cos\theta, \sin\theta)$ and $B = (\cos\phi, \sin\phi)$. The angle $AOB$ is $(\theta-\phi)$, so if we use the Law of Cosines, the distance $AB$ is given by

$$AB^2 = 1^2 + 1^2 - 2\cdot 1\cdot 1\cos(\theta-\phi) = 2 - 2\cos(\theta-\phi).$$

Next, we find the distance between $A$ and $B$ using the distance formula and multiply out:

$$\begin{aligned}
AB^2 &= (\cos\theta-\cos\phi)^2 + (\sin\theta-\sin\phi)^2 && \text{distance formula}\\
&= \cos^2\theta - 2\cos\theta\cos\phi + \cos^2\phi + \sin^2\theta - 2\sin\theta\sin\phi + \sin^2\phi && \text{multiply out}\\
&= \cos^2\theta + \sin^2\theta + \cos^2\phi + \sin^2\phi - 2\cos\theta\cos\phi - 2\sin\theta\sin\phi && \text{collect like terms}\\
&= 2 - 2(\cos\theta\cos\phi + \sin\theta\sin\phi).
\end{aligned}$$

Setting the two expressions for the distance equal gives what we wanted:

$$2 - 2\cos(\theta - \phi) = 2 - 2(\cos\theta\cos\phi + \sin\theta\sin\phi)$$
$$\text{so} \quad \cos(\theta - \phi) = \cos\theta\cos\phi + \sin\theta\sin\phi.$$

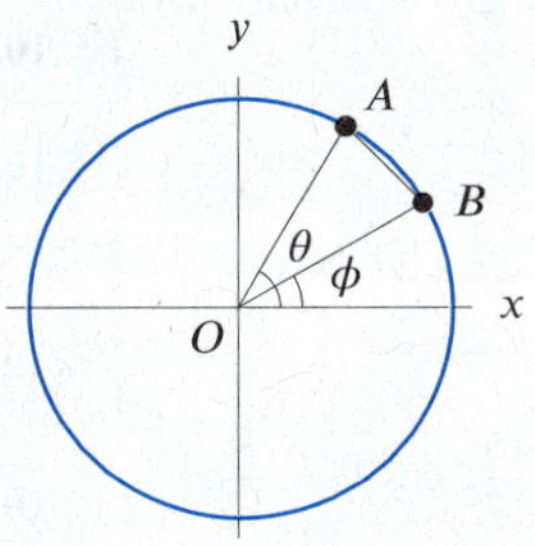


**Figure 9.11:** To justify the identity for $\cos(\theta - \phi)$, find the distance between $A$ and $B$ two ways

## Justification of Other Sum and Difference Formulas

Using $\sin\theta = \cos(\theta - \pi/2)$ with $\theta$ replaced by $(\theta + \phi)$ gives

$$\sin(\theta + \phi) = \cos\left(\theta + \phi - \frac{\pi}{2}\right).$$

Rewriting $(\theta + \phi - \pi/2)$ as $(\theta - \pi/2 - (-\phi))$, we use the identity for $\cos(\theta - \phi)$ with $\theta$ replaced by $(\theta - \pi/2)$ and $\phi$ replaced by $-\phi$:

$$\cos\left(\theta + \phi - \frac{\pi}{2}\right) = \cos\left(\theta - \frac{\pi}{2} - (-\phi)\right)$$
$$= \cos\left(\theta - \frac{\pi}{2}\right)\cos(-\phi) + \sin\left(\theta - \frac{\pi}{2}\right)\sin(-\phi).$$

Since $\cos(-\phi) = \cos\phi$ and $\sin(-\phi) = -\sin\phi$ and $\cos(\theta - \pi/2) = \sin\theta$ and $\sin(\theta - \pi/2) = -\cos\theta$, we have

$$\cos\left(\theta + \phi - \frac{\pi}{2}\right) = \sin\theta\cos\phi + (-\cos\theta)(-\sin\phi).$$

So

$$\boxed{\sin(\theta + \phi) = \sin\theta\cos\phi + \cos\theta\sin\phi.}$$

Replacing $\phi$ by $-\phi$ in the formula for $\sin(\theta + \phi)$ gives

$$\sin(\theta - \phi) = \sin(\theta + (-\phi)) = \sin\theta\cos(-\phi) + \cos\theta\sin(-\phi),$$

and since $\cos(-\phi) = \cos\phi$ and $\sin(-\phi) = -\sin\phi$,

$$\boxed{\sin(\theta - \phi) = \sin\theta\cos\phi - \cos\theta\sin\phi.}$$

Replacing $\phi$ by $-\phi$ in the formula for $\cos(\theta - \phi)$ gives

$$\cos(\theta + \phi) = \cos(\theta - (-\phi)) = \cos\theta\cos(-\phi) + \sin\theta\sin(-\phi).$$

$$\boxed{\cos(\theta + \phi) = \cos\theta\cos\phi - \sin\theta\sin\phi.}$$

We can use these formulas to obtain the double-angle formulas for sine and cosine in Section 9.2.

# Exercises and Problems for Section 9.3

## Exercises

In Exercises 1–4, assume that $\sin A = 0.84$, $\cos A = 0.54$, $\sin B = 0.92$, and $\cos B = 0.39$. Find the given quantity without finding $A$ and $B$.

**1.** $\sin(A+B)$
**2.** $\sin(A-B)$
**3.** $\cos(A+B)$
**4.** $\cos(A-B)$

In Exercises 5–8, assume that

$$\sin S = \frac{7}{25},\quad \cos S = \frac{24}{25},\quad \sin T = \frac{15}{17},\quad \cos T = \frac{8}{17}.$$

Find the given quantity without using a calculator.

**5.** $\sin(S+T)$
**6.** $\sin(S-T)$
**7.** $\cos(S+T)$
**8.** $\cos(S-T)$

**9.** Use identities and the exact values of $\sin\theta$ and $\cos\theta$ for $\theta = 30°, 45°, 60°$ to find exact values of $\sin\theta$ and $\cos\theta$ for $\theta = 15°$ and $\theta = 75°$.

In Exercises 10–12, find exact values.

**10.** $\sin 345°$
**11.** $\sin 105°$
**12.** $\cos 285°$

In Exercises 13–16,

**(a)** Evaluate both sides of the identity when $u = 15°$ and $v = 42°$ to see that they are numerically equal.

**(b)** Let $u = x$ and $v = 20°$ and compare the graphs of each side of the identity as functions of $x$ (in degrees).

**13.** $\sin(u+v) = \sin u\cos v + \sin v\cos u$

**14.** $\sin(u-v) = \sin u\cos v - \sin v\cos u$

**15.** $\cos(u+v) = \cos u\cos v - \sin v\sin u$

**16.** $\cos(u-v) = \cos u\cos v + \sin v\sin u$

## Problems

**17.** Use Figure 9.12 to find a simplified expression for the following in terms of $y$.

**(a)** $\cos(\theta-\phi)$
**(b)** $\sin(\theta-\phi)$
**(c)** $\cos\theta - \cos\phi$
**(d)** $\sin(\theta+\phi)$

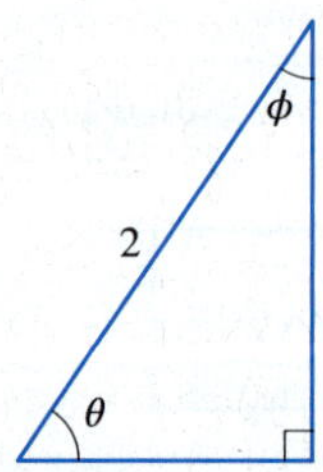


Figure 9.12

**18.** Starting from the addition formulas for sine and cosine, derive the following identities:

**(a)** $\sin t = \cos(t-\pi/2)$
**(b)** $\cos t = \sin(t+\pi/2)$

**19.** Starting from the addition formulas for sine and cosine, derive the double-angle formulas for sine and cosine.

**20.** Assuming that $\cos A = 0.74$, $\sin A = 0.67$, $\cos B = 0.48$, $\sin B = 0.87$,

**(a)** Find $\sin(A+B)$.
**(b)** Find $\cos(A+B)$.
**(c)** Use parts (a) and (b) to find $\tan(A+B)$.

**21.** Show that $\cos 3t = 4\cos^3 t - 3\cos t$.

**22.** We have calculated exact, rational values for $\sin r\pi$ when $r = 0, \pm 1, \pm\frac{1}{2}, \pm\frac{1}{6}$. Surprisingly, these are the only rational values of $\sin r\pi$ when $r$ is rational. Prove the following identity, which is used to obtain the result:[1]

$$2\cos((n+1)\theta) = (2\cos\theta)(2\cos(n\theta)) - 2\cos((n-1)\theta).$$

**23.** Use the addition formula $\cos(u+v) = \cos u\cos v - \sin u\sin v$ to derive the following identity for the average rate of change of the cosine function:

$$\frac{\cos(x+h)-\cos x}{h} = \cos x\left(\frac{\cos h - 1}{h}\right) - \sin x\left(\frac{\sin h}{h}\right).$$

**24.** Use the addition formula $\sin(u+v) = \sin u\cos v + \sin v\cos u$ to derive the following identity for the average rate of change of the sine function:

$$\frac{\sin(x+h)-\sin x}{h} = \sin x\left(\frac{\cos h - 1}{h}\right) + \cos x\left(\frac{\sin h}{h}\right).$$

**25.** Use the addition formula $\tan(u+v) = \dfrac{\tan u + \tan v}{1 - \tan u\tan v}$ to derive the following identity for the average rate of change of the tangent function:

$$\frac{\tan(x+h)-\tan x}{h} = \frac{1}{\cos^2 x}\left(\frac{\sin h}{h}\right)\frac{1}{\cos h - \sin h\tan x}.$$

[1] I. Niven and H. Zuckerman, *An Introduction to Number Theory*, 3rd ed., p. 143 (Wiley & Sons: New York, 1972).

**26.** We can obtain formulas involving half-angles from the double-angle formulas. To find an expression for $\sin \frac{1}{2}v$, we solve $\cos 2u = 1 - 2\sin^2 u$ for $\sin^2 u$ to obtain

$$\sin^2 u = \frac{1 - \cos 2u}{2}.$$

Taking the square root gives:

$$\sin u = \pm\sqrt{\frac{1 - \cos 2u}{2}},$$

and finally we make the substitution $u = \frac{1}{2}v$ to get:

$$\sin \frac{1}{2}v = \pm\sqrt{\frac{1 - \cos v}{2}}.$$

**(a)** Show that $\cos \frac{1}{2}v = \pm\sqrt{\frac{1 + \cos v}{2}}$.

**(b)** Show that $\tan \frac{1}{2}v = \pm\sqrt{\frac{1 - \cos v}{1 + \cos v}}$.

The sign in these formulas depends on the value of $v/2$. If $v/2$ is in quadrant I then the sign of $\sin v/2$ is $+$, the sign of $\cos v/2$ is $+$, and the sign of $\tan v/2$ is $+$.

**(c)** If $v/2$ is in quadrant II find the signs for $\sin v/2$, for $\cos v/2$, and for $\tan v/2$.

**(d)** If $v/2$ is in quadrant III find the signs for $\sin v/2$, for $\cos v/2$, and for $\tan v/2$.

**(e)** If $v/2$ is in quadrant IV find the signs for $\sin v/2$, for $\cos v/2$, and for $\tan v/2$.

**27.** We can find the area of triangle $ABC$ in Figure 9.13 by adding the areas of the two right triangles formed by the perpendicular from vertex $C$ to the opposite side.

**(a)** Find values for $\sin\theta$, $\cos\theta$, $\sin\phi$, and $\cos\phi$ in terms of $a$, $b$, $c_1$, $c_2$, and $h$.

**(b)** Show that

$$\text{Area } \Delta CAD = (1/2)ab \sin\theta \cos\phi$$

$$\text{Area } \Delta CDB = (1/2)ab \cos\theta \sin\phi.$$

**(c)** Use the sum formula to explain why

$$\text{Area } \Delta ABC = (1/2)ab \sin C.$$

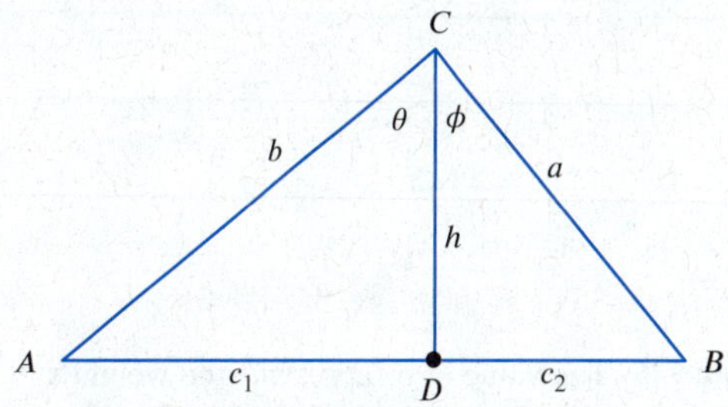

Figure 9.13

**28.** **(a)** Show that for any triangle $ABC$, we have $a = b\cos C + c\cos B$.

**(b)** Use part (a) and the Law of Sines to obtain the sum formula for sine. [Hint: Why is $\sin A = \sin(B + C)$?]

**29.** In this problem, you will derive the formula $\cos(\theta + \phi) = \cos\theta\cos\phi - \sin\theta\sin\phi$ directly.

**(a)** Find the coordinates of points $P_1$, $P_2$, $P_3$, and $P_4$ in the unit circle in Figure 9.14.

**(b)** Draw the line segments $P_1P_2$ and $P_3P_4$ and explain why $P_1P_2 = P_3P_4$, using facts from geometry.

**(c)** Use the distance formula to calculate the lengths $P_1P_2$ and $P_3P_4$.

**(d)** Write an equation and simplify to obtain the sum formula for cosine.

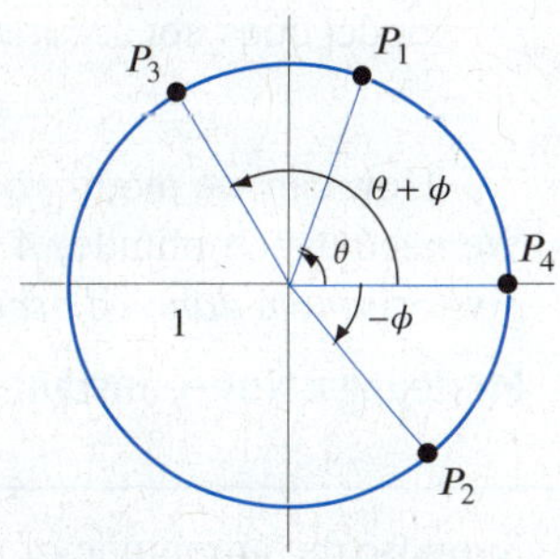

Figure 9.14

# 9.4 TRIGONOMETRIC MODELS AND SUM IDENTITIES

In Section 7.5 we used sinusoidal functions to model periodic phenomena. In this section we study other trigonometric models that allow us to more accurately describe phenomena such as spring motion and environmental fluctuations.

## Damped Oscillation

Imagine a weight attached to the ceiling by a spring. If the weight is disturbed, it begins bobbing up and down. We can model the weight's motion using a periodic trigonometric function. But is this the best model of the spring's motion? If not, how could it be improved?

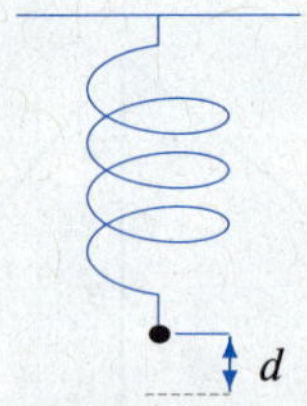


Figure 9.15: The value of $d$ represents the weight's displacement from its at-rest position, $d = 0$

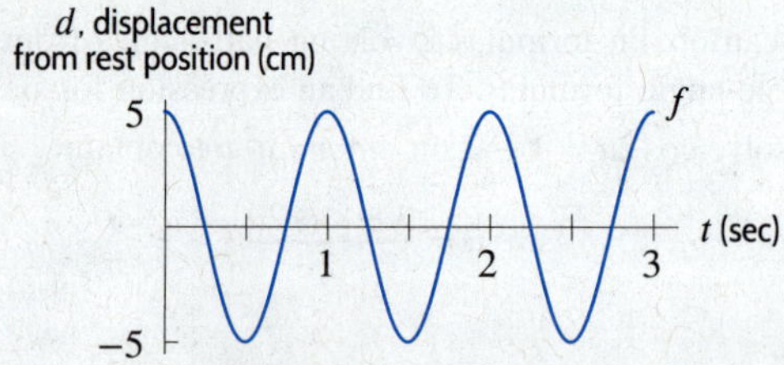


Figure 9.16: The predicted motion of the weight for the first 3 seconds

**Example 1** The weight starts at rest, and $d$ is the displacement, in centimeters, from the weight's rest position. See Figure 9.15. We raise the weight 5 cm above its rest position and release it at time $t = 0$, where $t$ is in seconds. The weight bobs up and down once every second for the first few seconds.

(a) Give a sinusoidal model for the motion.
(b) Does the sinusoidal formula accurately model the spring's behavior? Are there limitations?

**Solution** (a) From Section 7.5, we know a sinusoidal model for this motion has the form

$$d = f(t) = A\cos(B\pi t) + k.$$

The midline, $k = 0$, corresponds to the at-rest position of the weight, and the amplitude, $A = 5$, gives the maximum displacement of the weight from its rest position. Since one full cycle is completed each second, the period of the motion is 1 and $B = 2\pi$, giving the formula:

$$d = f(t) = 5\cos(2\pi t).$$

Figure 9.16 gives a graph of $f$ for the first three seconds of the weight's motion.

(b) According to this trigonometric model, the spring bobs up and down forever, which we know is not accurate. We know that, as time passes, the amplitude of the bobbing shrinks until the weight comes to a stop; the amplitude of the motion does not remain constant. Hence the sinusoidal model does not accurately represent the spring's motion past the initial moments of motion.

How can we modify our formula to model a bobbing motion that shrinks and eventually stops? We need the amplitude, $A$, to decrease over time instead of remaining constant. Such an amplitude gives rise to a *damped oscillation*.

### Modeling a Non-Constant Amplitude

**Example 2** Suppose the amplitude of the spring's motion from Example 1 decreases at a constant rate, so that after 5 seconds, the weight stops moving.

(a) Give a formula for the amplitude, $A$, of the motion as a function of time, $t$.
(b) Give a model for the spring's motion.
(c) Does this formula accurately model the spring's behavior? Are there limitations?

**Solution** (a) Since the amplitude shrinks at a constant rate, $A$ is a decreasing linear function of time. We have $A = 5$ when $t = 0$ and $A = 0$ when $t = 5$; thus:

$$A(t) = 5 - t.$$

(b) We replace the constant amplitude in the linear model with the linearly decreasing amplitude. So, instead of writing

$$d = f(t) = \underbrace{\text{Constant amplitude}}_{5} \cdot \cos 2\pi t,$$

we write

$$d = f(t) = \underbrace{\text{Decreasing amplitude}}_{A(t)=(5-t)} \cdot \cos 2\pi t$$

so that the trigonometric model becomes

$$d = f(t) = (5 - t) \cdot \cos 2\pi t.$$

Figure 9.17 shows a graph of this function.

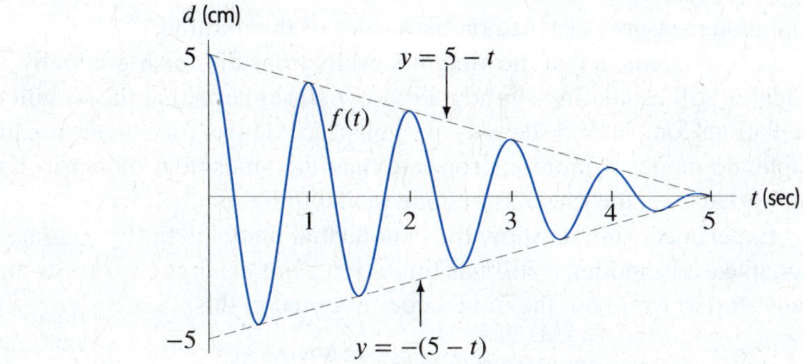


**Figure 9.17**: Graph of the weight's displacement assuming amplitude is a decreasing linear function of time

(c) While the function $d = f(t) = (5 - t)\cos 2\pi t$ is a better model for the behavior of the spring for $t < 5$, Figure 9.18 shows that this model does not work for $t > 5$. The breakdown in the model occurs because the magnitude of $A(t) = 5 - t$ starts to increase when $t$ grows larger than 5. So, this decreasing amplitude model is very inaccurate past the first 5 seconds of motion.

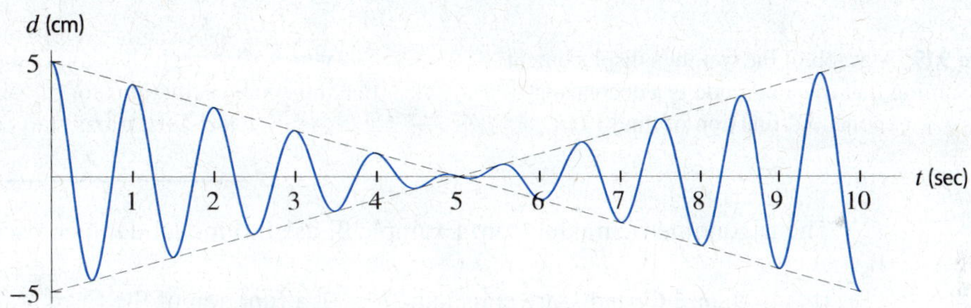


**Figure 9.18**: The formula $d = f(t) = (5 - t)\cos 2\pi t$ makes inaccurate predictions for values of $t$ larger than 5

We improve the spring motion model in Example 2 by picking a decreasing function to represent the amplitude as a function of time that does not become negative. Since a decreasing exponential function is always positive, we try an exponentially decaying amplitude.

**Example 3** Suppose that the amplitude of the spring's motion from Example 1 shrinks by half every second.

(a) Give a formula for the amplitude, $A$, of the motion as a function of time, $t$.

(b) Give a model for the spring's motion.

(c) Does this formula accurately model the spring's behavior? Are there limitations?

**Solution** (a) Since the amplitude shrinks by 50% every second, the amplitude $A$ decreases by a constant percentage rate per unit time, so is a decaying exponential function. We have, at $t = 0$ the amplitude is 5, and, at time $t$, the amplitude is given by

$$A(t) = 5\left(\frac{1}{2}\right)^t.$$

(b) A trigonometric model for the spring's motion is

$$d = f(t) = \underbrace{5\left(\frac{1}{2}\right)^t}_{\text{Decreasing amplitude}} \cdot \cos 2\pi t.$$

Figure 9.19 shows a graph of this function. The dashed curves in Figure 9.19 show the decreasing exponential function. There are two dashed curves $\left(y = 5\left(\frac{1}{2}\right)^t \text{ and } y = -5\left(\frac{1}{2}\right)^t\right)$ because the amplitude measures distance on both sides of the midline.

(c) Figure 9.19 predicts that the weight's oscillations diminish gradually, so that at time $t = 5$ the weight is still oscillating slightly. Figure 9.20 suggests that the weight continues to make small oscillations long after 5 seconds have elapsed. Hence, for longer time intervals this exponential amplitude model is more appropriate than the sinusoidal model in Example 1 and also more appropriate than the linear amplitude model in Example 2.

Experiments in fact show the exponential model is fairly accurate for many springs. However, like every model, it still has limitations. For example, we know an actual spring will eventually stop moving, but the model does not predict this.

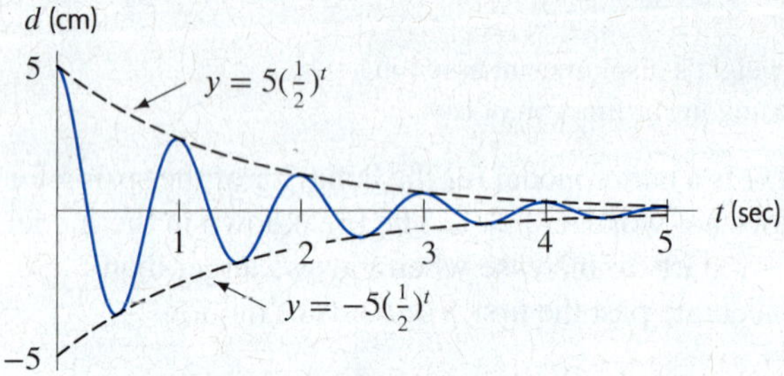


**Figure 9.19:** A graph of the weight's displacement assuming that the amplitude is a decreasing exponential function of time

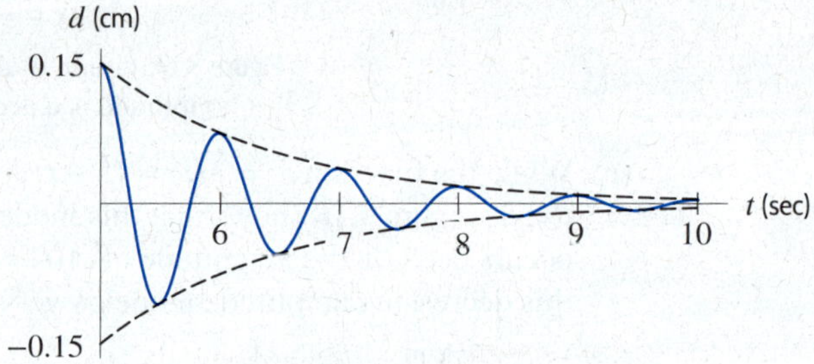


**Figure 9.20:** Graph showing that the weight is still making small oscillations for $t \geq 5$. Note reduced scale on $d$-axis; $t$-axis starts at $t = 5$

The trigonometric model from Example 3, used to model damped oscillations, is summarized below.

If $A_0$, $B$, and $C$, and $k$ are constants, $k > 0$, a function of the form

$$y = A_0 e^{-kt}\cos(Bt) + C \qquad \text{or} \qquad y = A_0 e^{-kt}\sin(Bt) + C$$

can be used to model an oscillating quantity whose amplitude, $A(t)$, decreases exponentially according to

$$A(t) = A_0 e^{-kt},$$

where $A_0$ is the initial amplitude.

Our trigonometric model in Example 3 is in this form, with $k = \ln 2$.

## Oscillation With a Rising Midline

In the next example, we consider an oscillating quantity that does not have a horizontal midline, but whose amplitude of oscillation is, in some sense, constant.

**Example 4** In Section 7.5, Example 5, we modeled a rabbit population undergoing seasonal fluctuations by the function

$$R = f(t) = 10{,}000 - 5000\cos\left(\frac{\pi}{6}t\right),$$

where $R$ is the size of the rabbit population $t$ months after January 1. See Figure 9.21. Give the meaning of the two terms in the formula.

**Solution** The rabbit population varies periodically about the midline, $R = 10{,}000$. Thus, in this model, the average number of rabbits is 10,000, but, depending on the time of year, the actual number may be above or below the average.

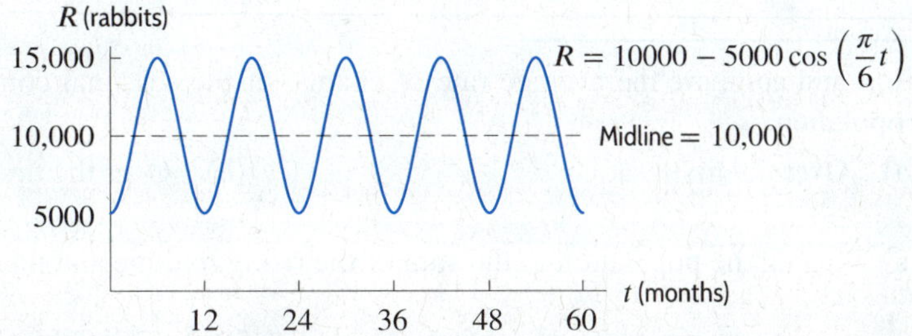


**Figure 9.21:** The rabbit population over a 5-year (60-month) period

This suggests we can think of the formula for $R$, the number of rabbits, as the sum of the average population and a seasonal component giving the deviation from the average:

$$R = \underbrace{10{,}000}_{\substack{\text{Midline:}\\ \text{average population}}} + \underbrace{-5000\cos\left(\frac{\pi}{6}t\right)}_{\text{Seasonal component}}.$$

Notice that to say the average number of rabbits is 10,000 we need to look at the rabbit population during an entire year. For example, if we looked at the population over the first two months, an interval on which the number of rabbits is always below 10,000, then the average population would be less than 10,000.

Now let us imagine a different situation. What if the average population, even over long periods of time, does not remain constant?

**Example 5** Suppose that, due to conservation efforts, there is a steady increase of 50 rabbits per month in the average rabbit population. Find a function that models both the oscillation and the average increase.

**Solution** Instead of writing

$$R = f(t) = \underbrace{10{,}000}_{\substack{\text{Constant midline:}\\ \text{average population}\\ \text{is constant.}}} + \underbrace{-5000\cos\left(\frac{\pi}{6}t\right)}_{\text{Seasonal component}},$$

we could write

$$R = f(t) = \underbrace{10{,}000 + 50t}_{\substack{\text{Rising midline:}\\ \text{average population}\\ \text{grows by 50 per month}}} + \underbrace{-5000\cos\left(\frac{\pi}{6}t\right)}_{\text{Seasonal component}}.$$

Figure 9.22 gives a graph of this new function over a five-year period.

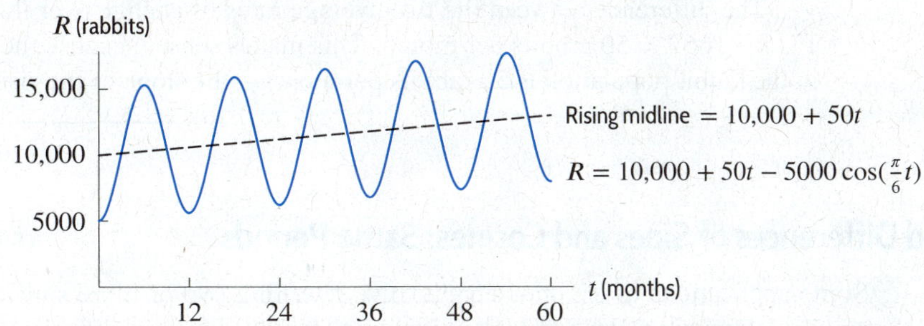


**Figure 9.22:** A graph of the population showing a gradual increase in the average number of rabbits

The number of rabbits changes both with the seasons and in the long term. What can we say about the average rate of change of the rabbit population due to the seasons? What can we say about the average rate of change in the population in the long term?

**Example 6** Find and compare the average rate of change of the seasonal component, and of the total rabbit population

(a) Over the first year (b) Over the first 6 months.

**Solution** (a) Writing the population as the sum of the rising midline and the seasonal component, $S(t)$:

$$R = f(t) = \underbrace{10{,}000 + 50t}_{\text{Rising midline}} + \underbrace{-5000\cos\left(\frac{\pi}{6}t\right)}_{\text{Seasonal component, } S(t)}.$$

The seasonal component, $S(t)$, is periodic with period 12 months, or 1 year. Thus $S(0) = S(12)$ and the average rate of change of the seasonal component for the first year is 0 rabbits per month.

The total rabbit population is not periodic; it is the sum of the periodic seasonal component and the rising midline. We have:

$$\begin{aligned}\text{Average population over first 12 months} &= \frac{\Delta R}{\Delta t} = \frac{f(12) - f(0)}{12}\\ &= \frac{5600 - 5000}{12} = \frac{600}{12} = 50 \text{ rabbits/month.}\end{aligned}$$

We observe that the difference between the two average rates of change is 50 rabbits per month—the slope of the rising midline.

(b) The seasonal component $S(t)$ increases during the first 6 months of the year, so we expect a positive average rate of change. We have

$$\begin{aligned}\text{Average seasonal variation over first 6 months} &= \frac{S(6) - S(0)}{6}\\ &= \frac{5000 - (-5000)}{6} = 1667 \text{ rabbits/month.}\end{aligned}$$

In Figure 9.22 we see that the rabbit population increases during the first 6 months of the year, so we expect a positive average rate of change for the total number of rabits. We have

$$\begin{aligned}\text{Average change in population over first 6 months} &= \frac{R(6) - R(0)}{12}\\ &= \frac{15300 - 5000}{6} = \frac{10300}{6} = 1717 \text{ rabbits/month.}\end{aligned}$$

The difference between the two average rates of change over this 6-month period is again $1717 - 1667 = 50$ rabbits per month. This makes sense, because the average monthly increase in the rabbit population is 50 rabbits per month—the slope of the rising midline.

## Sums and Differences of Sines and Cosines: Same Periods

Some applications of trigonometry involve rewriting two or more sinusoidal functions as a single sinusoidal function. We first look at identities for the sums and differences of sines and cosines.

We start with two sinusiodal functions with the same period. For instance, what can we say about the function $f(t) = \sin t + \cos t$? The graph of $f$, in Figure 9.23, looks remarkably like the graph of the sine, having the same period but a larger amplitude (somewhere between 1 and 2).

Is $f$ actually sinusoidal? That is, can $f$ be written in the form

$$f(t) = A\sin(B(t-h)) + k?$$

If so, Figure 9.23 suggests that the period is $2\pi$, so $B = 1$, and that the midline is $k = 0$.

The leftward horizontal shift appears to be about $1/8$ of the period of $2\pi$, so

$$h = -\frac{2\pi}{8} = -\frac{\pi}{4}.$$

So far, we have:

$$f(t) = A\sin\left(t + \frac{\pi}{4}\right).$$

What about the value of $A$?

Assuming $h = -\pi/4$, let's use our identity for $\sin(\theta + \phi)$ to rewrite $\sin(t + \pi/4)$. Substituting $\theta = t$ and $\phi = \pi/4$, we obtain

$$\begin{aligned}\sin\left(t + \frac{\pi}{4}\right) &= \cos\left(\frac{\pi}{4}\right)\sin t + \sin\left(\frac{\pi}{4}\right)\cos t \quad \text{because } \sin(\theta+\phi) = \sin\theta\cos\phi + \sin\phi\cos\theta \\ &= \underbrace{\left(\frac{1}{\sqrt{2}}\right)}_{\cos\frac{\pi}{4}}\sin t + \underbrace{\left(\frac{1}{\sqrt{2}}\right)}_{\sin\frac{\pi}{4}}\cos t.\end{aligned}$$

Multiplying both sides of this equation by $\sqrt{2}$ gives

$$\sqrt{2}\sin\left(t + \frac{\pi}{4}\right) = \sin t + \cos t.$$

Thus, since $f(t) = \sin t + \cos t$, we see that

$$f(t) = \sqrt{2}\sin\left(t + \frac{\pi}{4}\right).$$

We have shown that $f$, the sum of two sinusoidal functions, is indeed a sinusoidal function with period $2\pi$, amplitude $A = \sqrt{2} \approx 1.414$, and horizontal shift $h = -\pi/4$. See Figure 9.23.

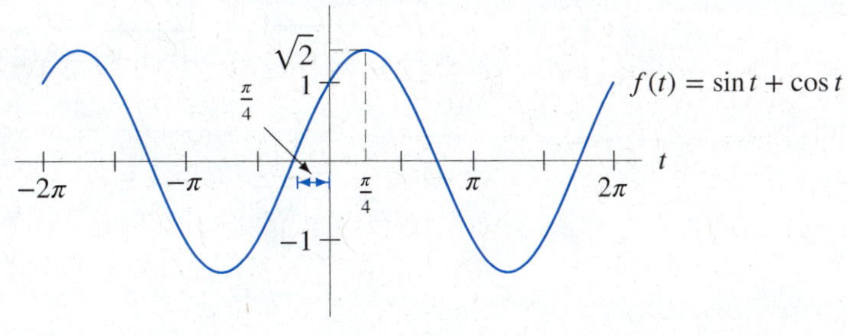


**Figure 9.23:** The graph of $f(t) = \sin t + \cos t$ is a sinusoidal function with amplitude $\sqrt{2}$ and leftward horizontal shift $\pi/4$: that is, $f(t) = \sqrt{2}\sin(t + \pi/4)$

## Rewriting $a_1 \sin Bt + a_2 \cos Bt$

We have rewritten $f(t) = \sin t + \cos t$ as a sinusoidal function. We now show that this result can be extended: The sum of any two cosine and sine functions having the same periods can be written as a single sinusoidal function.

We start with

$$a_1 \sin(Bt) + a_2 \cos(Bt),$$

where $a_1, a_2$, and $B$ are constants. We want to write this expression in the form $A\sin(Bt+\phi)$, which is equivalent to the formula $A\sin(B(t-h))$ with $\phi = -Bh$.

To do this, we show that we can find the constants $A$ and $\phi$ for any $a_1$ and $a_2$. We use the identity $\sin(\theta+\phi) = \sin\theta\cos\phi + \sin\phi\cos\theta$ with $\theta = Bt$:

$$\sin(Bt+\phi) = \sin(Bt)\cos\phi + \sin\phi\cos(Bt).$$

Multiplying by $A$, we have

$$A\sin(Bt+\phi) = A\cos\phi\sin(Bt) + A\sin\phi\cos(Bt).$$

Letting $a_1 = A\cos\phi$ and $a_2 = A\sin\phi$, we have

$$A\sin(Bt+\phi) = \underbrace{A\cos\phi}_{a_1}\sin(Bt) + \underbrace{A\sin\phi}_{a_2}\cos(Bt) = a_1\sin(Bt) + a_2\cos(Bt).$$

We want to write $A$ and $\phi$ in terms of $a_1$ and $a_2$; so far, we know $a_1$ and $a_2$ in terms of $A$ and $\phi$. Since

$$a_1^2 + a_2^2 = A^2\cos^2\phi + A^2\sin^2\phi = A^2,$$

we have

$$A = \sqrt{a_1^2 + a_2^2}.$$

From $a_1 = A\cos\phi$ and $a_2 = A\sin\phi$, we get $\cos\phi = a_1/A$ and $\sin\phi = a_2/A$. If $a_1 \neq 0$, then

$$\frac{a_2}{a_1} = \frac{A\sin\phi}{A\cos\phi} = \frac{\sin\phi}{\cos\phi} = \tan\phi.$$

These formulas allow $A$ and $\phi$ to be determined from $a_1$ and $a_2$.

**Example 7** Check algebraically that $\sin t + \cos t = \sqrt{2}\sin\left(t + \frac{\pi}{4}\right)$.

**Solution** Here, $a_1 = 1$ and $a_2 = 1$, so

$$A = \sqrt{a_1^2 + a_2^2} = \sqrt{1^2 + 1^2} = \sqrt{2}$$

and

$$\cos\phi = \frac{a_1}{A} = \frac{1}{\sqrt{2}} \qquad \sin\phi = \frac{a_2}{A} = \frac{1}{\sqrt{2}}$$

so

$$\phi = \frac{\pi}{4}.$$

This confirms what we already knew, namely that $\sin t + \cos t = \sqrt{2}\sin(t + \pi/4)$.

In summary:

**Sum and sine and cosine with the same period**

Provided their periods are equal, the sum of a sine function and a cosine function can be written as a single sinusoidal function. We have

$$a_1 \sin(Bt) + a_2 \cos(Bt) = A \sin(Bt + \phi)$$

where

$$A = \sqrt{a_1^2 + a_2^2} \quad \text{and} \quad \tan\phi = \frac{a_2}{a_1}.$$

The angle $\phi$ is determined by the equations $\cos\phi = a_1/A$ and $\sin\phi = a_2/A$.

To find the angle $\phi$, find its quadrant from the signs of $\cos\phi = a_1/A$ and $\sin\phi = a_2/A$. Then find the angle $\phi$ in that quadrant with $\tan\phi = a_2/a_1$.

**Example 8** If $g(t) = 2\sin 3t + 5\cos 3t$, write $g(t)$ as a single sinusoidal function.

**Solution** We have $a_1 = 2$, $a_2 = 5$, and $B = 3$. Thus,

$$A = \sqrt{2^2 + 5^2} = \sqrt{29} \quad \text{and} \quad \tan\phi = \frac{5}{2}.$$

Since $\cos\phi = 2/\sqrt{29}$ and $\sin\phi = 5/\sqrt{29}$ are both positive, $\phi$ is in the first quadrant. Since $\arctan(5/2) = 1.190$ is in the first quadrant, we take $\phi = 1.190$. Therefore,

$$g(t) = \sqrt{29}\sin(3t + 1.190).$$

## Modeling with Sums of Trigonometric Functions with the Same Period

In this example, we see how the identities for sums of sinusoidal functions enable us to analyze phenomena.

**Example 9** A utility company serves two different cities. Let $P_1$ be the power requirement in megawatts (mw) for City 1 and $P_2$ be the requirement for City 2. Both $P_1$ and $P_2$ are functions of $t$, the number of hours elapsed since midnight. Suppose $P_1$ and $P_2$ are given by the following formulas:

$$P_1 = 40 - 15\cos\left(\frac{\pi}{12}t\right) \quad \text{and} \quad P_2 = 50 + 10\sin\left(\frac{\pi}{12}t\right).$$

(a) Describe the power requirements of each city in words.
(b) Find a sinusoidal function giving the total power, $P = P_1 + P_2$.
(c) What is the maximum total power the utility company must be prepared to provide? Explain why it is not the sum of the maximum power provided in each city.

**Solution** (a) The power requirement of City 1 is at a minimum of $40 - 15 = 25$ mw at $t = 0$, or midnight. It rises to a maximum of $40 + 15 = 55$ mw at noon and falls back to 25 mw by the following midnight. The power requirement of City 2 is at a maximum of 60 mw at 6 am. It falls to a minimum of 40 mw by 6 pm but by the following morning has climbed back to 60 mw, again at 6 am. Figure 9.24 shows $P_1$ and $P_2$ over a two-day period.

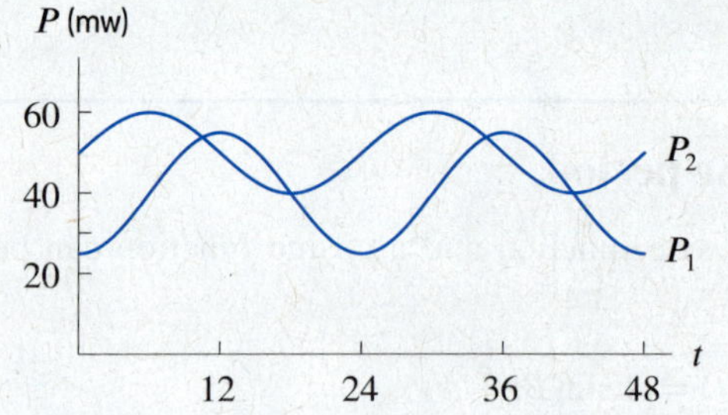


Figure 9.24: Power requirements for cities 1 and 2

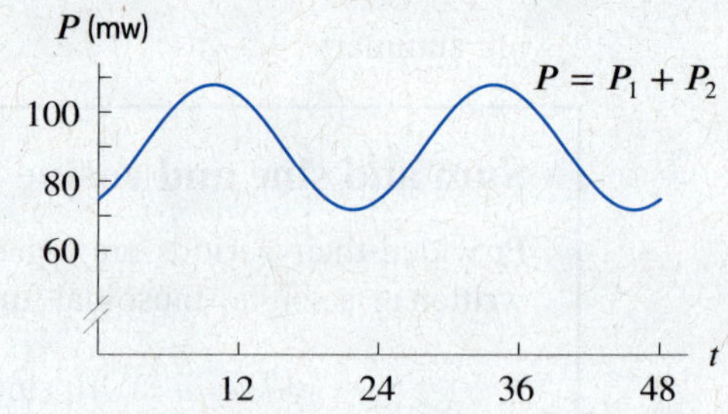


Figure 9.25: Total power demand for both cities combined

(b) The utility company must provide enough power to satisfy the needs of both cities. The total power required is given by

$$P = P_1 + P_2 = 90 + 10\sin\left(\frac{\pi}{12}t\right) - 15\cos\left(\frac{\pi}{12}t\right).$$

The graph of total power in Figure 9.25 looks like a sinusoidal function. It varies between about 108 mw and 72 mw, giving it an amplitude of roughly 18 mw. The period of $P$ is 24 hours because the values of both $P_1$ and $P_2$ begin repeating after 24 hours, so their sum repeats with that same frequency as well.

The box on page 377 leads to a formula for the sinusoidal function for the total power:

$$P = P_1 + P_2 = 90 + 10\sin\left(\frac{\pi}{12}t\right) - 15\cos\left(\frac{\pi}{12}t\right) = 90 + A\sin(Bt + \phi),$$

where

$$A = \sqrt{10^2 + (-15)^2} = 18.028 \text{ mw}.$$

Since $\cos\phi = 10/18.028 = 0.555$ and $\sin\phi = -15/18.028 = -0.832$, we know $\phi$ must be in the fourth quadrant. Also,

$$\tan\phi = -\frac{15}{10} = -1.5 \quad \text{and} \quad \tan^{-1}(-1.5) = -0.983,$$

and since $-0.983$ is in the fourth quadrant, we take $\phi = -0.983$. Thus,

$$P = P_1 + P_2 = 90 + 18.028\sin\left(\frac{\pi}{12}t - 0.983\right).$$

(c) The maximum value of $P$ is $90 + 18.028 = 108.028$ mw. The utility company must be prepared to provide at least this much power at all times.

Because the maximum value of $P_1$ is 55 and the maximum value of $P_2$ is 60, we might have expected that the maximum value of $P$ would be $55 + 60 = 115$ mw. The reason that this is not true is that the maximum values of $P_1$ and $P_2$ occur at different times, so they never combine to give a value as high as 115 mw.

However, the midline of $P$ is a horizontal line at 90 mw, and does equal the sum of the midlines of $P_1$ and $P_2$.

## Exercises and Problems for Section 9.4

### Exercises

In Exercises 1–4, write in the form $A\sin(Bt+\phi)$ using sum or difference formulas.

**1.** $8\sin t - 6\cos t$

**2.** $8\sin t + 6\cos t$

**3.** $-\sin t + \cos t$

**4.** $-2\sin 3t + 5\cos 3t$

**5.** Graph $y = t + 5\sin t$ and $y = t$ on $0 \le t \le 2\pi$. Where do the two graphs intersect if $t$ is not restricted to $0 \le t \le 2\pi$?

**6.** Graph the function $f(x) = e^{-x}(2\cos x + \sin x)$ on the interval $0 \le x \le 12$. Use a calculator or computer to find the maximum and minimum values of the function.

In Exercises 7–10, match the function to a graph, (I)–(IV).

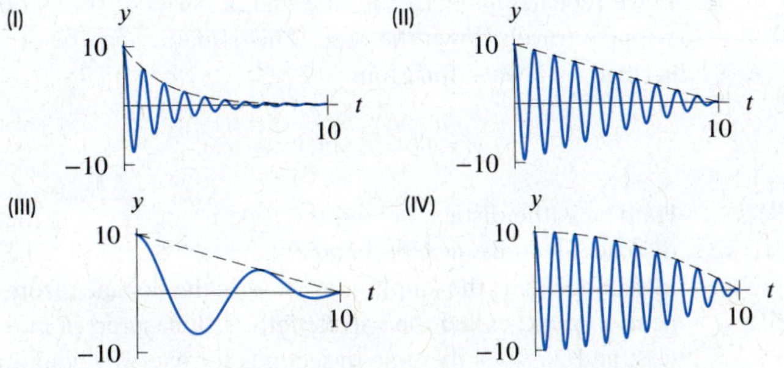


7. $y = 10e^{-t/2}\cos 2\pi t$

8. $y = 10(1 - 0.01t^2)\cos 2\pi t$

9. $y = (10 - t)\cos 2\pi t$

10. $y = (10 - t)\cos t$

11. For time $t$ in months since January 1, a national park rabbit population is modeled by

$$R = f(t) = -5000\cos(\pi t/6) + 10{,}000.$$

(a) Evaluate and interpret $f(3) - f(2)$.
(b) Find and interpret all solutions to the equation

$$f(t) = 12{,}000 \qquad 0 \le t \le 12.$$

## Problems

12. Write a possible formula for the graph in Figure 9.26.

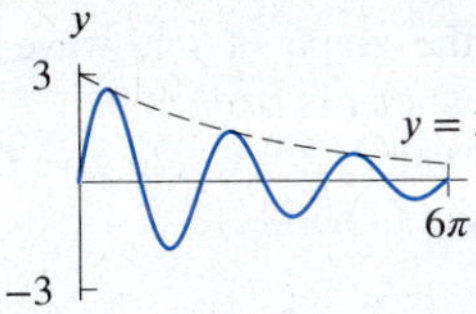


**Figure 9.26**

13. Find formulas for each of the following populations.

(a) In year $t = 0$, this population of 5000 began growing at the rate of 300 per year.
(b) In year $t = 0$, this population of 3200 began growing at an annual rate of 4% per year.
(c) This population oscillates between a low of 1200 in year $t = 0$ and a high of 3000. The period of oscillation is 5 years.

14. From its past behavior, John knows that the value of a stock has a cyclical component that increases for the first three months of each year, falls for the next six, and rises again for the last three. In addition, inflation adds a linear component to the stock's price. John is seeking a model of the form

$$f(t) = mt + b + A\sin\frac{\pi t}{6}.$$

with $t$ in months since Jan 1. He has the following data:

| Date | Jan 1 | Apr 1 | Jul 1 | Oct 1 | Jan 1 |
|---|---|---|---|---|---|
| Price | \$20.00 | \$37.50 | \$35.00 | \$32.50 | \$50.00 |

(a) Find values of $m$, $b$, and $A$ so that $f$ fits the data.
(b) During which month(s) does this stock appreciate the most?
(c) During what period each year is this stock actually losing value?

15. A power company serves two different cities, City $A$ and City $B$. The power requirements of both cities vary in a predictable fashion over the course of a typical day.

(a) At midnight, the power requirement of City $A$ is at a minimum of 40 megawatts. (A megawatt is a unit of power.) By noon the city has reached its maximum power consumption of 90 megawatts and by midnight it once again requires only 40 megawatts. This pattern repeats every day. Find a possible formula for $f(t)$, the power, in megawatts, required by City $A$ as a function of $t$, in hours since midnight.
(b) The power requirements, $g(t)$ megawatts, of City $B$ differ from those of City $A$. For $t$, in hours since midnight,

$$g(t) = 80 - 30\sin\left(\frac{\pi}{12}t\right).$$

Give the amplitude and the period of $g(t)$, and a physical interpretation of these quantities.
(c) Graph and find all $t$ such that

$$f(t) = g(t), \qquad 0 \le t < 24.$$

Interpret your solution(s) in terms of power usage.
(d) Why should the power company be interested in the maximum value of the function

$$h(t) = f(t) + g(t), \qquad 0 \le t < 24?$$

What is the approximate maximum of this function, and approximately when is it attained?
(e) Find a formula for $h(t)$ as a single sine function. What is the exact maximum of this function?

16. Let $f(t) = \cos(e^t)$, where $t$ is measured in radians.

(a) Note that $f(t)$ has a horizontal asymptote as $t \to -\infty$. Find its equation, and explain why $f$ has this asymptote.
(b) Describe the behavior of $f$ as $t \to \infty$. Explain why $f$ behaves this way.
(c) Find the vertical intercept of $f$.
(d) Let $t_1$ be the least zero of $f$. Find $t_1$ exactly. [Hint: What is the smallest positive zero of the cosine function?]
(e) Find an expression for $t_2$, the least zero of $f$ greater than $t_1$.

**17.** Let $f(x) = \sin\left(\frac{1}{x}\right)$ for $x > 0$, $x$ in radians.

**(a)** $f(x)$ has a horizontal asymptote as $x \to \infty$. Find the equation for the asymptote and explain carefully why $f(x)$ has this asymptote.
**(b)** Describe the behavior of $f(x)$ as $x \to 0$. Explain why $f(x)$ behaves in this way.
**(c)** Is $f(x)$ a periodic function?
**(d)** Let $z_1$ be the greatest zero of $f(x)$. Find the exact value of $z_1$.
**(e)** How many zeros do you think the function $f(x)$ has?
**(f)** Suppose $a$ is a zero of $f$. Find a formula for $b$, the largest zero of $f$ less than $a$.

**18.** Derive the following identity used in an electrical engineering text[2] to represent the received AM signal function. Note that $A, \omega_c, \omega_d$ are constants and $M(t)$ is a function of time, $t$.

$$\begin{aligned} r(t) &= (A + M(t)) \cos \omega_c t + I \cos(\omega_c + \omega_d)t \\ &= ((A + M(t)) + I \cos \omega_d t) \cos \omega_c t - I \sin \omega_d t \sin \omega_c t \end{aligned}$$

**19.** In the July 1993 issue of *Standard and Poor's Industry Surveys,* the editors stated:[3]

> The strength (of sales) of video games, seven years after the current fad began, is amazing .... What will happen next year is anything but clear. While video sales ended on a strong note last year, the toy industry is nothing if not cyclical.

**(a)** Why might sales of video games be cyclical?
**(b)** Does $s(t) = a \sin(bt)$, where $t$ is time, serve as a reasonable model for the sales graph in Figure 9.27? What about $s(t) = a \cos(bt)$?
**(c)** Modify your choice in part (b) to provide for the higher amplitude in the years 1985–1992 as compared to 1979–1982.
**(d)** Graph the function created in part (c) and compare your results to Figure 9.27. Modify your function to improve your approximation.
**(e)** Use your function to predict sales for 1993.

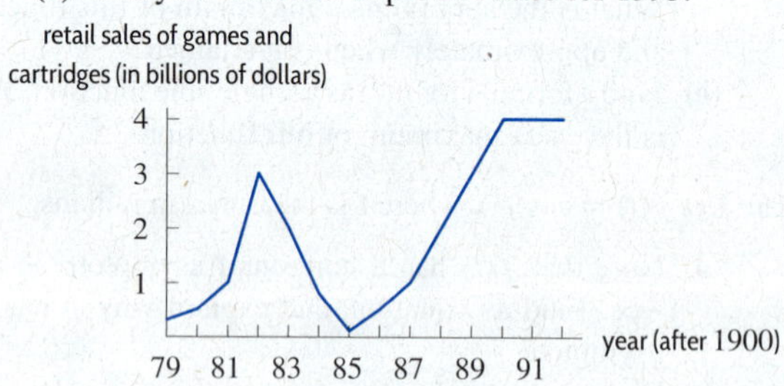


Figure 9.27

**20.** A rope has one free end. If we give the free end a small upward shake, a wiggle travels down the length of the rope. If we repeatedly shake the free end, a periodic series of wiggles travels down the rope. This situation can be described by a wave function:

$$y(x, t) = A \sin(kx - \omega t).$$

Here, $x$ is the distance along the rope in meters; $y$ is the displacement distance perpendicular to the rope; $t$ is time in seconds; $A$ is the amplitude; $2\pi/k$ is the distance from peak to peak, called the wavelength, $\lambda$, measured in meters; and $2\pi/\omega$ is the time in seconds for one wavelength to pass by. Suppose $A = 0.06$, $k = 2\pi$, and $\omega = 4\pi$.

**(a)** What is the wavelength of the motion?
**(b)** How many peaks of the wave pass by a given point each second?
**(c)** Construct the graph of this wave from $x = 0$ to $x = 1.5$ m when $t$ is fixed at 0.
**(d)** What other values of $t$ would give the same graph as the one found in part (c)?

**21.** The concentration of carbon dioxide[4] in the air, $y$, measured regularly at Mauna Loa Observatory in Hawaii, may be modeled by

$$y = C(t) = 3.5 \sin(\pi t/6) + 381 + t/6,$$

where $y$ is measured in parts per million, ppm, and $t$ is measured in months since December 2005. See Figure 9.28.

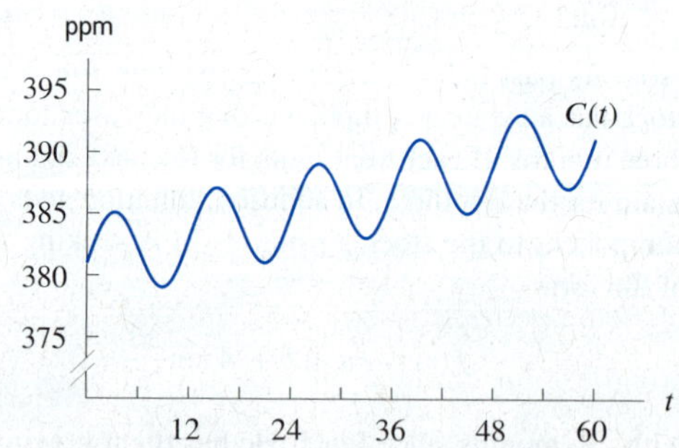


Figure 9.28

**(a)** On average, how much did $C(t)$ increase between December 2005 and June 2006? How about between December 2010 and June 2011? How about between March and September 2012?

---

[2]B. P. Lathi, *Modern Digital and Analog Communication Systems*, 2nd ed. (OUP, 1989), page 269.
[3]Source: Nintendo of America.
[4]http://www.esrl.noaa.gov/gmd/ccgg/trends/. Accessed March 2011. NOAA data reports one concentration value per month; $y$ is the monthly average concentration. Our model uses a continuous function to represent the trend in the monthly average concentration.

**(b)** The concentration of carbon dioxide model may be written as

$$y = C(t) = S(t)+381+t/6, \quad \text{where } S(t) = 3.5\sin(\pi t/6).$$

What is the practical meaning of the sinusoidal function $S(t)$ in terms of carbon dioxide?

**(c)** On average, how much did the function $S(t)$ increase between December 2005 and June 2006? How about between December 2010 and June 2011? How about between March and September 2012?

**(d)** Calculate the difference between the average rate of change of $C(t)$ and $S(t)$ over each of the time periods in parts (a) and (c).

**(e)** What is the practical significance of the difference(s) you calculated in part (d) in terms of carbon dioxide?

## 9.5 SUMS WITH DIFFERENT PERIODS AND ACCOUSTIC BEATS

When our eardrums vibrate because of changes in air pressure, we hear a sound. We hear a high pitch when the oscillation is sinusoidal with a short period, and a low pitch when the period is long. But when we hear two tones at the same time, what kind of oscillation is behind the sound we hear? Is the oscillation sinusoidal?

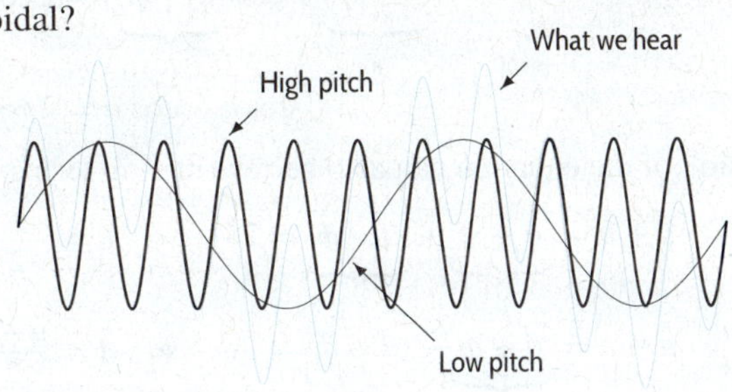


**Figure 9.29**: When high- and low-pitch tones sound at the same time, what do we hear?

In Section 9.4 we used trigonometric identities to see that the sum of a sine and a cosine with the same period is sinusoidal. In this section, we use more trigonometric identities to show that the sum of sine or cosine functions with different periods is not always sinusoidal.

**Example 1** Sketch and describe the graph of $y = \sin 2x + \sin 3x$.

**Solution** Figure 9.30 shows that the function $y = \sin 2x + \sin 3x$ is not sinusoidal. It is, however, periodic. Its period seems to be $2\pi$, since it repeats twice on the interval of length $4\pi$ shown in the figure.

We can see that the function $y = \sin 2x + \sin 3x$ has period $2\pi$ by looking at the periods of $\sin 2x$ and $\cos 3x$. Since the period of $\sin 2x$ is $\pi$ and the period of $\sin 3x$ is $2\pi/3$, on any interval of length $2\pi$, the function $y = \sin 2x$ completes two cycles and the function $y = \sin 3x$ completes three cycles. Both functions are at the beginning of a new cycle after an interval of $2\pi$, so their sum begins to repeat at this point. See Figure 9.31.

Notice that even though the maximum value of each of the functions $\sin 2x$ and $\sin 3x$ is 1, the maximum value of their sum is not 2; it is less than 2. This is because $\sin 2x$ and $\sin 3x$ achieve their maximum values for different $x$ values, so they never combine to reach the value 2.

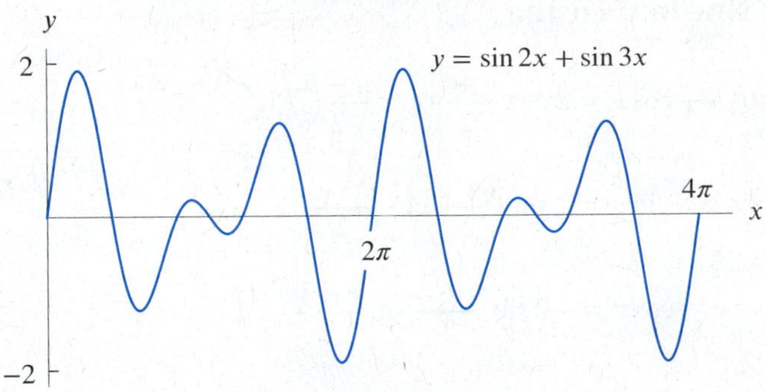


**Figure 9.30**: A graph of the sum $y = \sin 2x + \sin 3x$

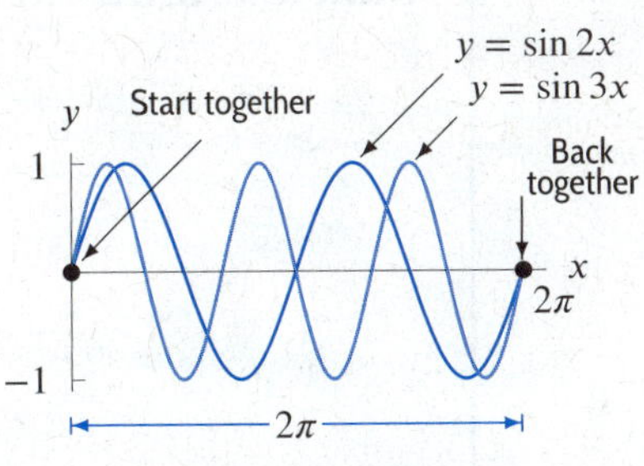


**Figure 9.31**: Graphs of $y = \sin 2x$ and $y = \sin 3x$ on an interval of $2\pi$

The two sines in Example 1 have different periods but equal amplitudes. Next, we derive an identity for the sum of two cosine functions having different periods and equal amplitudes.

## Sums and Differences of Sines and Cosines: Same Amplitudes

We start with two identities we know for the cosine of the sum and of the difference of two angles:

$$\cos(\theta + \phi) = \cos\theta\cos\phi - \sin\theta\sin\phi$$
$$\cos(\theta - \phi) = \cos\theta\cos\phi + \sin\theta\sin\phi.$$

If we add these two identities, the terms involving the sine function cancel out, giving

$$\cos(\theta + \phi) + \cos(\theta - \phi) = 2\cos\theta\cos\phi.$$

The left side of this equation can be rewritten by substituting $u = \theta + \phi$ and $v = \theta - \phi$:

$$\cos\underbrace{(\theta + \phi)}_{u} + \cos(\underbrace{\theta - \phi}_{v}) = 2\cos\theta\cos\phi$$
$$\cos u + \cos v = 2\cos\theta\cos\phi.$$

The right side of the equation can also be rewritten in terms of $u$ and $v$:

$$u + v = \underbrace{\theta + \phi}_{u} + \underbrace{\theta - \phi}_{v} = 2\theta \quad \text{so} \quad \theta = \frac{u+v}{2}$$
$$u - v = \underbrace{\theta + \phi}_{u} - \underbrace{(\theta - \phi)}_{v} = 2\phi \quad \text{so} \quad \phi = \frac{u-v}{2}.$$

With these substitutions, we have

$$\cos u + \cos v = 2\cos\frac{u+v}{2}\cos\frac{u-v}{2}.$$

This identity relates the sum of two cosine functions to the product of two new cosine functions.

**Example 2** Write $\cos(30t) + \cos(28t)$ as the product of two cosine functions.

**Solution** Letting $u = 30t$, $v = 28t$, we have

$$\cos(30t) + \cos(28t) = 2\cos\frac{30t + 28t}{2}\cos\frac{30t - 28t}{2} = 2\cos(29t)\cos t.$$

There are similar formulas for the sum of two sine functions and the difference between sine functions and cosine functions, all having the same amplitude. These are summarized below.

### Summary of Sums and Differences of Sine and Cosine: Same Amplitudes

**Sum and difference of sine and cosine**

$$\cos u + \cos v = 2\cos\frac{u+v}{2}\cos\frac{u-v}{2}$$

$$\sin u + \sin v = 2\sin\frac{u+v}{2}\cos\frac{u-v}{2}$$

$$\cos u - \cos v = -2\sin\frac{u+v}{2}\sin\frac{u-v}{2}$$

$$\sin u - \sin v = 2\cos\frac{u+v}{2}\sin\frac{u-v}{2}$$

These formulas can be derived using the four formulas for the sines and cosines of the sums and differences of two angles from Section 9.3. See Problems 11, 12, and 13.

**Example 3** Write the function $y = \sin 2x + \sin 3x$ from Example 1 as a product of sines and cosines.

**Solution** Letting $u = 2x$, $v = 3x$ and using the formula for the sum of two sines, we have:

$$\begin{aligned}\sin(2x) + \sin(3x) &= 2\sin\frac{2x+3x}{2}\cos\frac{2x-3x}{2} \\ &= 2\sin\left(\frac{5}{2}x\right)\cos\left(-\frac{1}{2}x\right) \\ &= 2\sin(2.5x)\cos(0.5x), \quad \text{since the cosine function is even.}\end{aligned}$$

Next we apply these identities to tones of different pitch but equal loudness, which combine to give tones of varying loudness. (See Figure 9.29 at the beginning of this section.)

## Acoustic Beats

Acoustic beats arise in the process of tuning musical instruments. Tuning is done according to an international agreement. The agreement specifies the pitch that a particular note must have on a perfectly tuned instrument. In the current agreement, the musical note $A$ above[5] middle $C$ must have a frequency of 440 cycles per second, also written 440 hertz (hz).

The next example describes how acoustic beats arise and are used in the tuning process.

**Example 4** To tune a piano, we strike a note on its keyboard and, at the same time, we strike the desired frequency for this note on a tuning fork. A low pitched note on the piano has frequency 55 hz.

(a) Suppose the pitch of this key on an out-of-tune piano is 61 hz instead of the desired 55 hz. The sound wave $W$ that results from striking a frequency of $f$ hz is given by

$$W = \cos(2\pi f t), \quad \text{with } t \text{ in seconds.}$$

Give formulas for the sound waves emitted by the piano and the tuning fork.

(b) If both notes are struck at the same time, then their combined sound wave is the sum of their separate waves, sketched in Figure 9.32. Describe the combined sound wave.

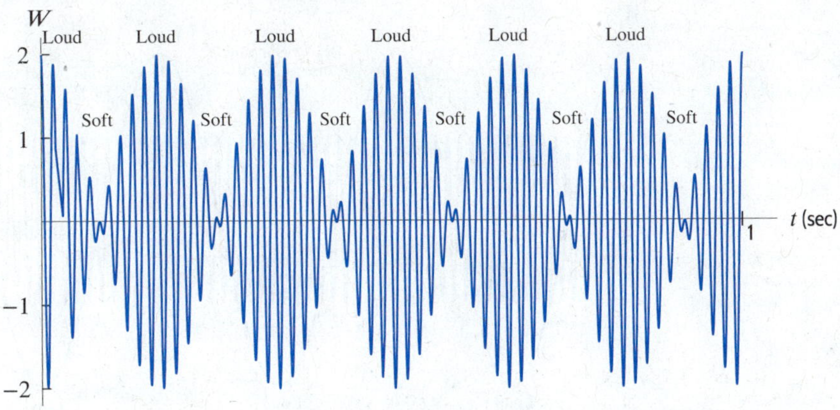


Figure 9.32: A graph of $W = \cos(2\pi 61t) + \cos(2\pi 55t)$ during one second of sound, showing acoustic beats

[5] On a piano, these notes correspond to two specific keys on the keyboard.

**Solution** (a) The sound wave, $W_1$, of the tone produced by this piano's key is

$$W_1 = \cos(2\pi \cdot 61t);$$

the sound wave, $W_2$, of the tone from the tuning fork is

$$W_2 = \cos(2\pi \cdot 55t).$$

(b) The combined sound wave is given by

$$W = W_1 + W_2 = \cos(2\pi \cdot 61t) + \cos(2\pi \cdot 55t).$$

The graph of the combined sound wave resembles a rapidly varying sinusoidal function except that its amplitude is not constant. We note that each original sound wave has constant amplitude, but the combined wave does not: its maximum and minimum values vary.

The ear perceives this variation in amplitude as a variation in loudness, so the tone we hear appears to waver (or *beat*) in a regular way. This is an example of *acoustic beats*.

How can we see the beats of the combined sound wave in its formula? We do this next, using trigonometric identities to rewrite the wave formula and reinterpret it.

**Example 5** Write the formula for the acoustic beats function in Example 4 as a product of two cosine functions.

**Solution** Using the identity for a sum of cosines on page 382, we rewrite the combined sound function as

$$\begin{aligned} W = \cos(2\pi \cdot 61t) + \cos(2\pi \cdot 55t) &= 2\cos\frac{2\pi \cdot 61t + 2\pi \cdot 55t}{2} \cdot \cos\frac{2\pi \cdot 55t - 2\pi \cdot 61t}{2} \\ &= 2\cos\frac{2\pi \cdot (61+55)t}{2} \cdot \cos\frac{2\pi \cdot (61-55)t}{2} \\ &= 2\cos(2\pi \cdot 58t)\cos(2\pi \cdot 3t). \end{aligned}$$

The formula for the acoustic beats in Example 4 can be interpreted as the product of a varying amplitude, $A(t)$, and a pure tone, $W_3(t)$, as follows:

$$W = 2\cos(2\pi \cdot 3t) \cdot \cos(2\pi \cdot 58t) = A(t)W_3(t).$$

Since $W_3(t) = \cos(2\pi \cdot 58t)$, the tone $W$ has a pitch of 58 hz, midway between the tones sounded by the tuning fork and the out-of-tune piano. Since the amplitude $A(t) = 2\cos(2\pi \cdot 3t)$ rises and falls, the tone $W$ grows louder and softer, but its pitch remains 58 hz. (See Figure 9.33.)

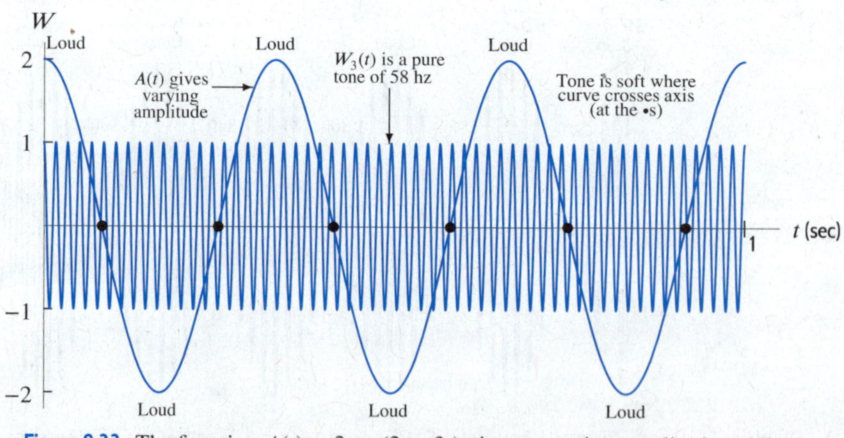


**Figure 9.33**: The function $A(t) = 2\cos(2\pi \cdot 3t)$ gives a varying amplitude, while the function $W_3(t) = \cos(2\pi \cdot 58t)$ gives a pure tone of 58 hz

Notice in Figure 9.33 that the function $A(t)$ completes three full cycles on the interval $0 \le t \le 1$. The tone is loudest when $A(t) = 2$ or $A(t) = -2$, as we see in Figure 9.32. Hence, the tone is at its loudest six times every second.

## Exercises and Problems for Section 9.5

### Exercises

In Exercises 1–2, find exact values.

**1.** $\cos 165° - \cos 75°$

**2.** $\cos 75° + \cos 15°$

In Exercises 3–6, write the sum or difference as a product of two trigonometric functions.

**3.** $\cos(5t) + \cos(3t)$

**4.** $\cos(4t) + \cos(6t)$

**5.** $\sin(7t) + \sin(3t)$

**6.** $\sin(7t) - \sin(3t)$

In Exercises 7–10,

**(a)** Evaluate both sides of the identity when $u = 35°$ and $v = 40°$ to see that they are numerically equal.

**(b)** Let $u = x$ and $v = 25°$ and compare the graphs of each side of the identity as functions of $x$ (in degrees).

**7.** $\cos u + \cos v = 2\cos\left(\frac{u+v}{2}\right)\cos\left(\frac{u-v}{2}\right)$

**8.** $\cos u - \cos v = -2\sin\left(\frac{u+v}{2}\right)\sin\left(\frac{u-v}{2}\right)$

**9.** $\sin u + \sin v = 2\sin\left(\frac{u+v}{2}\right)\cos\left(\frac{u-v}{2}\right)$

**10.** $\sin u - \sin v = 2\cos\left(\frac{u+v}{2}\right)\sin\left(\frac{u-v}{2}\right)$

### Problems

**11.** Prove: $\cos u - \cos v = -2\sin\left(\frac{u+v}{2}\right)\sin\left(\frac{u-v}{2}\right)$.

**12.** Prove: $\sin u + \sin v = 2\sin\left(\frac{u+v}{2}\right)\cos\left(\frac{u-v}{2}\right)$.

**13.** Prove: $\sin u - \sin v = 2\cos\left(\frac{u+v}{2}\right)\sin\left(\frac{u-v}{2}\right)$.

**14.** Solve the equation $\sin 4x + \sin x = 0$ for values of $x$ in the interval $0 < x < 2\pi$ by applying an identity to the left side of the equation.

**15.** An amusement park has a giant double Ferris wheel as in Figure 9.34. The double Ferris wheel has a 30-meter rotating arm attached at its center to a 25-meter main support. At each end of the rotating arm is attached a Ferris wheel measuring 20 meters in diameter, rotating in the direction shown in Figure 9.34. The rotating arm takes 6 minutes to complete one full revolution, and each wheel takes 4 minutes to complete a revolution about that wheel's hub. At time $t = 0$ the rotating arm is parallel to the ground and your seat is at the 3 o'clock position of the rightmost wheel.

**(a)** Find a formula for $h = f(t)$, your height above the ground in meters, as a function of time in minutes. [Hint: Your height above ground equals the height of your wheel hub above ground plus your height above that hub.]

**(b)** Graph $f(t)$. Is $f(t)$ periodic? If so, what is its period?

**(c)** Approximate the least value of $t$ such that $h$ is at a maximum value. What is this maximum value?

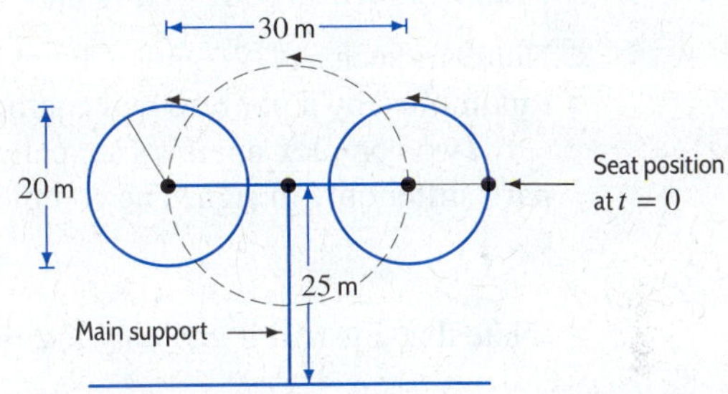


Figure 9.34

# 9.6 COMPLEX NUMBERS AND DE MOIVRE'S THEOREM

In this section, we expand our idea of number to include *complex numbers*. These numbers make it possible to interpret expressions like $\sqrt{-4}$ in a meaningful way.

## Using Complex Numbers to Solve Equations

The quadratic equation

$$x^2 - 2x + 2 = 0$$

is not satisfied by any real number $x$. Applying the quadratic formula gives

$$x = \frac{2 \pm \sqrt{4-8}}{2} = 1 \pm \frac{\sqrt{-4}}{2}.$$

But $-4$ does not have a square root that is a real number. To overcome this problem, we define the imaginary number $i$ such that:

$$i^2 = -1.$$

Using $i$, we see that $(2i)^2 = -4$. Now we can write our solution of the quadratic as:

$$x = 1 \pm \frac{\sqrt{-4}}{2} = 1 \pm \frac{2i}{2} = 1 \pm i.$$

The numbers $1 + i$ and $1 - i$ are examples of complex numbers.

> A **complex number** is defined as any number that can be written in the form
>
> $$z = a + bi,$$
>
> where $a$ and $b$ are real numbers and $i = \sqrt{-1}$.
> The *real part* of $z$ is the number $a$; the *imaginary part* is the number $b$.

Calling the number $i$ imaginary makes it sound like $i$ does not exist in the same way as real numbers. In some cases, it is useful to make such a distinction between real and imaginary numbers. If we measure mass or position, we want our answers to be real. But imaginary numbers are just as legitimate mathematically as real numbers. For example, complex numbers are used in studying wave motion in electric circuits.

## Algebra of Complex Numbers

Numbers such as $0$, $1$, $\frac{1}{2}$, $\pi$, $e$ and $\sqrt{2}$ are called *purely real*, because their imaginary part is zero. Numbers such as $i$, $2i$, and $\sqrt{2}i$ are called *purely imaginary*, because they contain only the number $i$ multiplied by a nonzero real coefficient.

Two complex numbers are called *conjugates* if their real parts are equal and if their imaginary parts differ only in sign. The complex conjugate of the complex number $z = a + bi$ is denoted $\overline{z}$, so

$$\overline{z} = a - bi.$$

(Note that $z$ is real if and only if $z = \overline{z}$.)

- Two complex numbers, $z = a + bi$ and $w = c + di$, are equal only if $a = c$ and $b = d$.
- Adding two complex numbers is done by adding real and imaginary parts separately: $(a + bi) + (c + di) = (a + c) + (b + d)i$.
- Subtracting is similar: $(a + bi) - (c + di) = (a - c) + (b - d)i$.
- Multiplication works just like for polynomials, using $i^2 = -1$ to simplify:

$$\begin{aligned}(a + bi)(c + di) &= a(c + di) + bi(c + di) = ac + adi + bci + bdi^2\\ &= ac + adi + bci - bd = (ac - bd) + (ad + bc)i.\end{aligned}$$

- Powers of $i$: We know that $i^2 = -1$; then, $i^3 = i \cdot i^2 = -i$, and $i^4 = (i^2)^2 = (-1)^2 = 1$. Then $i^5 = i \cdot i^4 = i$, and so on. Thus we have

$$i^n = \begin{cases} i & \text{for } n = 1, 5, 9, 13, \ldots \\ -1 & \text{for } n = 2, 6, 10, 14, \ldots \\ -i & \text{for } n = 3, 7, 11, 15, \ldots \\ 1 & \text{for } n = 4, 8, 12, 16, \ldots. \end{cases}$$

- The product of a number and its conjugate is always real and nonnegative:

$$z \cdot \overline{z} = (a + bi)(a - bi) = a^2 - abi + abi - b^2i^2 = a^2 + b^2.$$

- Dividing is done by multiplying the numerator and denominator by the conjugate of the denominator, thereby making the denominator real:

$$\frac{a+bi}{c+di} = \frac{a+bi}{c+di} \cdot \frac{c-di}{c-di} = \frac{ac - adi + bci - bdi^2}{c^2+d^2} = \frac{ac+bd}{c^2+d^2} + \frac{bc-ad}{c^2+d^2}i.$$

**Example 1** Compute the sum $(3 + 6i) + (5 - 2i)$.

**Solution** Adding real and imaginary parts gives $(3 + 6i) + (5 - 2i) = (3 + 5) + (6i - 2i) = 8 + 4i$.

**Example 2** Simplify $(2 + 7i)(4 - 6i) - i$.

**Solution** Multiplying out gives $(2 + 7i)(4 - 6i) - i = 8 + 28i - 12i - 42i^2 - i = 8 + 15i + 42 = 50 + 15i$.

**Example 3** Compute $\frac{2+7i}{4-6i}$.

**Solution** The conjugate of the denominator is $4 + 6i$, so we multiply by $(4 + 6i)/(4 + 6i)$.

$$\frac{2+7i}{4-6i} = \frac{2+7i}{4-6i} \cdot \frac{4+6i}{4+6i} = \frac{8 + 12i + 28i + 42i^2}{4^2+6^2} = \frac{-34+40i}{52} = -\frac{17}{26} + \frac{10}{13}i.$$

## The Complex Plane and Polar Coordinates

It is often useful to picture a complex number $z = x + iy$ in the plane, with $x$ along the horizontal axis and $y$ along the vertical. The $xy$-plane is then called the *complex plane*. Figure 9.35 shows the complex numbers $3i$, $-2 + 3i$, $-3$, $-2i$, 2, and $1 + i$.

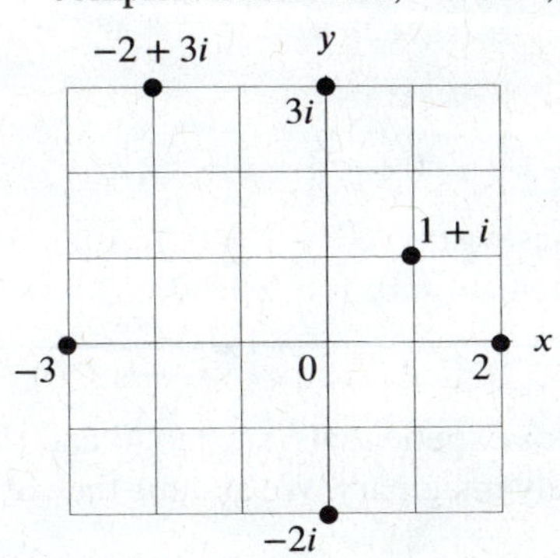


Figure 9.35: Points in the complex plane

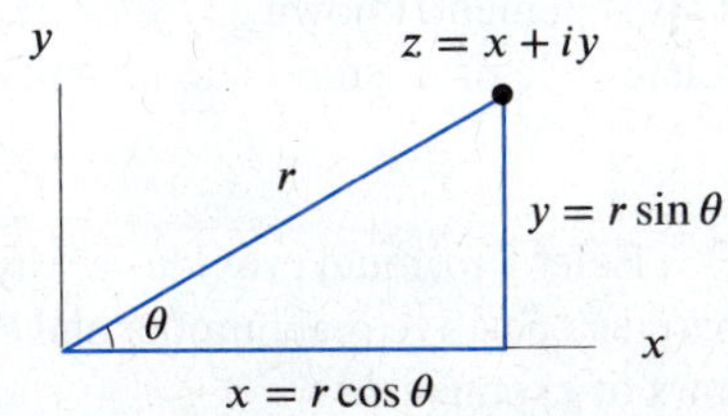


Figure 9.36: The point $z = x + iy$ in the complex plane, showing polar coordinates

Figure 9.36 shows that a complex number can be written using polar coordinates as follows:

$$z = x + iy = r\cos\theta + ir\sin\theta.$$

**Example 4** Express $z = -2i$ and $z = -2 + 3i$ using polar coordinates. (See Figure 9.35.)

**Solution** For $z = -2i$, the distance of $z$ from the origin is 2, so $r = 2$. Also, one value for $\theta$ is $\theta = 3\pi/2$. Thus, using polar coordinates,

$$z = -2i = 2\cos(3\pi/2) + i\,2(\sin 3\pi/2).$$

For $z = -2 + 3i$, we have $x = -2$, $y = 3$, so $r = \sqrt{(-2)^2 + 3^2} = 3.606$ and $\tan\theta = 3/(-2)$, so since $z$ is in quadrant II, we can take $\theta = \pi + \tan^{-1}(-1.5) = 2.159$. Thus, using polar coordinates,

$$z = -2 + 3i = 3.606\cos(2.159) + i\,3.606\sin(2.159).$$

**Example 5** What complex number is represented by the point with polar coordinates $r = 5$ and $\theta = 3\pi/4$?

**Solution** Since $x = r\cos\theta$ and $y = r\sin\theta$ we see that $z = x + iy = 5\cos(3\pi/4) + i(5\sin(3\pi/4)) = -5/\sqrt{2} + i\,5/\sqrt{2}$.

## Euler's Formula

Based on what we have seen so far, there is no reason to expect a connection between the exponential function $e^x$ and the trigonometric functions $\sin x$ and $\cos x$. However, in the eighteenth century, the Swiss mathematician Leonhard Euler[6] discovered a surprising connection between these functions that involves complex numbers. This result, called *Euler's formula*, states that, for real $\theta$ in radians,

$$e^{i\theta} = \cos\theta + i\sin\theta$$

**Example 6** Evaluate $e^{i\pi}$.

**Solution** Using Euler's formula,

$$e^{i\pi} = \cos\pi + i\sin\pi = -1 + i\cdot 0 = -1.$$

This statement, known as *Euler's identity*, is sometimes written $e^{i\pi} + 1 = 0$. It is famous because it relates five of the most fundamental constants in mathematics: 0, 1, $e$, $\pi$, and $i$.

Euler's formula provides a way to evaluate complex exponentials, such as $4^i$ and $e^{i\theta}$, where the base is a positive real number and the exponent is purely imaginary. We assume they obey the usual laws of exponents.

**Example 7** Convert $4^i$ to the form $a + bi$.

**Solution** Using Euler's formula and the exponent rules, we can evaluate the expression $4^i$. Since $4 = e^{\ln 4}$,

$$\begin{aligned} 4^i &= \left(e^{\ln 4}\right)^i \\ &= e^{i(\ln 4)} && \text{using an exponent rule} \\ &= \cos(\ln 4) + i\sin(\ln 4) && \text{using Euler's formula} \\ &= 0.183 + 0.983i. && \text{using a calculator} \end{aligned}$$

[6]Pronounced "Oiler."

## Polar Form of a Complex Number

Euler's formula allows us to write the complex number $z$ in *polar form*:

> If the point representing $z$ has polar coordinates $(r, \theta)$, then
> $$z = r(\cos\theta + i\sin\theta) = re^{i\theta}.$$
> The expression $re^{i\theta}$ is called the **polar form** of the complex number $z$.

Similarly, since $\cos(-\theta) = \cos\theta$ and $\sin(-\theta) = -\sin\theta$, we have

$$re^{-i\theta} = r(\cos(-\theta) + i\sin(-\theta)) = r(\cos\theta - i\sin\theta).$$

Thus, $re^{-i\theta}$ is the conjugate of $re^{i\theta}$.

**Example 8** Express the complex number represented by the point with polar coordinates $r = 8$ and $\theta = 3\pi/4$ in Cartesian form, $a + bi$, and in polar form, $z = re^{i\theta}$. See Figure 9.37.

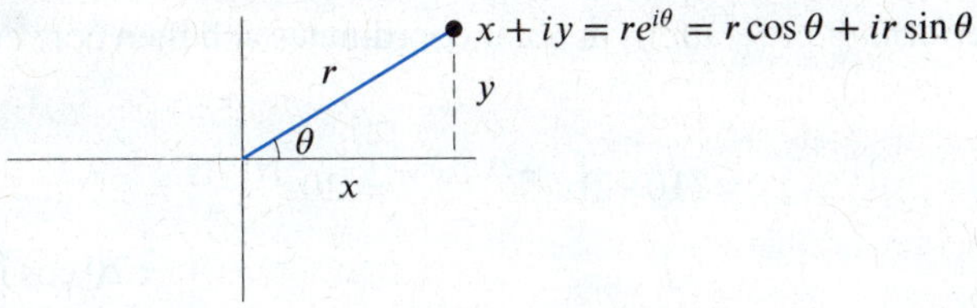


Figure 9.37: Cartesian and polar forms of a complex number

**Solution** Using Cartesian coordinates, the complex number is

$$z = 8\left(\cos\left(\frac{3\pi}{4}\right) + i\sin\left(\frac{3\pi}{4}\right)\right) = \frac{-8}{\sqrt{2}} + i\frac{8}{\sqrt{2}}.$$

The polar form is

$$z = 8e^{i3\pi/4}.$$

Since increasing an angle by $2\pi$ does not change the point on the unit circle to which it corresponds, a complex number can have more than one polar form.

**Example 9** Show that the polar coordinates $(8, 3\pi/4), (8, 11\pi/4), (8, 19\pi/4)$ all represent the same complex number.

**Solution** Since $11\pi/4 = 3\pi/4 + 2\pi$ and $19\pi/4 = 3\pi/4 + 4\pi$, these polar coordinates all represent the point $P$ in Figure 9.38 and the same complex number.

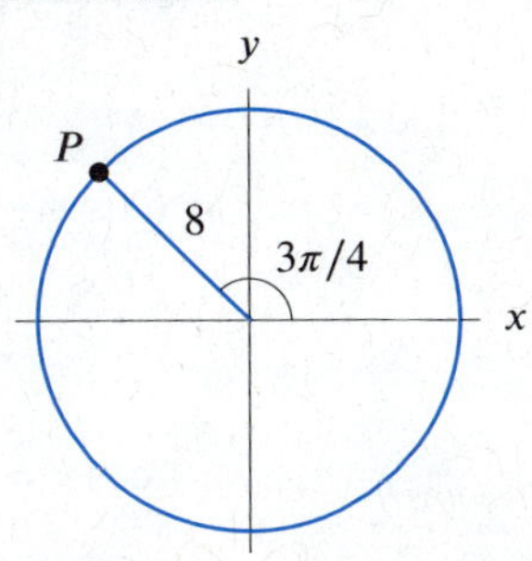


Figure 9.38: Point $P$ with $r = 8$ and $\theta = 3\pi/4$

## Products of Complex Numbers in Polar Form

It is easy to add and subtract complex numbers written in Cartesian form. The polar form makes multiplication and division of complex numbers easy. If

$$z_1 = r_1 e^{i\theta_1} \text{ and } z_2 = r_2 e^{i\theta_2}$$

then

$$z_1 z_2 = (r_1 r_2) e^{i(\theta_1+\theta_2)} \text{ and } \frac{z_1}{z_2} = \frac{r_1}{r_2} e^{i(\theta_1-\theta_2)}.$$

In words,

- To multiply two complex numbers, multiply the $r$ coordinates and add the $\theta$ coordinates.
- To divide two complex numbers, divide the $r$ coordinates and subtract the $\theta$ coordinates.

**Example 10** Let $z = 10e^{i\pi/3}$ and $z_2 = 2e^{i\pi/4}$. Find the Cartesian form of $z_1 z_2$ and $z_1/z_2$.

**Solution** We perform the operations in polar coordinates and then convert to Cartesian coordinates. The product is

$$\begin{aligned} z_1 z_2 = (10 \cdot 2)e^{i(\pi/3+\pi/4)} = 20e^{i7\pi/12} &= 20\left(\cos\left(\frac{7\pi}{12}\right) + i\sin\left(\frac{7\pi}{12}\right)\right) \\ &= 20\cos\left(\frac{7\pi}{12}\right) + i20\sin\left(\frac{7\pi}{12}\right). \end{aligned}$$

The quotient is

$$\frac{z_1}{z_2} = \frac{10}{2}e^{i(\pi/3-\pi/4)} = \frac{10}{2}e^{i\pi/12} = 5\left(\cos\left(\frac{\pi}{12}\right) + i\sin\left(\frac{\pi}{12}\right)\right) = 5\cos\left(\frac{\pi}{12}\right) + i5\sin\left(\frac{\pi}{12}\right).$$

**Example 11** Show that $\cos x = \dfrac{e^{ix} + e^{-ix}}{2}$.

**Solution** We know from Euler's formula that

$$e^{ix} = \cos x + i\sin x.$$

Also, we have

$$e^{-ix} = \cos x - i\sin x.$$

Then

$$\begin{aligned} e^{ix} + e^{-ix} &= (\cos x + i\sin x) + (\cos x - i\sin x) \\ &= 2\cos x, \end{aligned}$$

so

$$\cos x = \frac{e^{ix} + e^{-ix}}{2}.$$

## Powers of Complex Numbers in Polar Form

We can use the polar form to find any integer power $p$ of $z$ as follows:

$$z^p = (re^{i\theta})^p = r^p e^{ip\theta}.$$

The special case where $r = 1$ gives

$$z^p = \left(e^{i\theta}\right)^p = e^{ip\theta}.$$

Since $e^{i\theta} = \cos\theta + i\sin\theta$, we have

**de Moivre's formula**[7]

For $p$ an integer:

$$(\cos\theta + i\sin\theta)^p = \cos p\theta + i\sin p\theta.$$

In fact, the formula $z^p = (re^{i\theta})^p = r^p e^{ip\theta}$ holds for any real number $p$. We can use it to find the roots of a complex number by choosing $p$ to be a fraction, as in the following example.

**Example 12** Find a cube root of the complex number represented by the point with polar coordinates $(8, 3\pi/4)$.

**Solution** From Example 8, we have $z = 8e^{i3\pi/4}$. Thus,

$$\sqrt[3]{z} = \left(8e^{i\,3\pi/4}\right)^{1/3} = 8^{1/3}e^{i(3\pi/4)\cdot(1/3)} = 2e^{i\pi/4} = 2\left(\cos\left(\frac{\pi}{4}\right) + i\sin\left(\frac{\pi}{4}\right)\right)$$
$$= 2\left(\frac{1}{\sqrt{2}} + \frac{i}{\sqrt{2}}\right) = \frac{2}{\sqrt{2}} + \frac{2i}{\sqrt{2}} = 1.414 + 1.414i.$$

Every nonzero complex number has three cube roots. To find them all, apply the method of the preceding example using different polar forms of the complex number.

**Example 13** Find the Cartesian form of all cube roots of the complex number $z = 8e^{i3\pi/4}$. Plot the cube roots on the complex plane.

**Solution** The point with polar coordinates $(8, 3\pi/4)$ also has polar coordinates $(8, 3\pi/4 + 2\pi) = (8, 11\pi/4)$ and $(8, 3\pi/4 + 4\pi) = (8, 19\pi/4)$. The three cube roots of $z$, in polar form, are

$$\begin{aligned}
\left(8e^{i3\pi/4}\right)^{1/3} &= 8^{1/3}e^{i(3\pi/4)\cdot(1/3)} &&= 2e^{i\pi/4} &&= 1.414 + 1.414i\\
\left(8e^{i11\pi/4}\right)^{1/3} &= 8^{1/3}e^{i(11\pi/4)\cdot(1/3)} &&= 2e^{i11\pi/12} &&= -1.932 + 0.518i\\
\left(8e^{i19\pi/4}\right)^{1/3} &= 8^{1/3}e^{i(19\pi/4)\cdot(1/3)} &&= 2e^{i19\pi/12} &&= 0.518 - 1.932i.
\end{aligned}$$

Using more polar forms of $z$ does not produce more cube roots. For example, computing with $(r, \theta) = (8, 3\pi/4 + 6\pi) = (8, 27\pi/4)$ leads to a cube root that we have already found:

$$\left(8e^{i27\pi/4}\right)^{1/3} = 8^{1/3}e^{i(27\pi/4)\cdot(1/3)} = 2e^{i9\pi/4} = 2e^{i\pi/4}.$$

[7] Abraham de Moivre, 1667–1754, was a French mathematician.

The polar forms of the cube roots all have $r = 2$. Thus the roots are on the circle of radius 2 centered at the origin of the complex plane. They lie at angles $\theta = \pi/4, 11\pi/12$, and $19\pi/12$. See Figure 9.39.

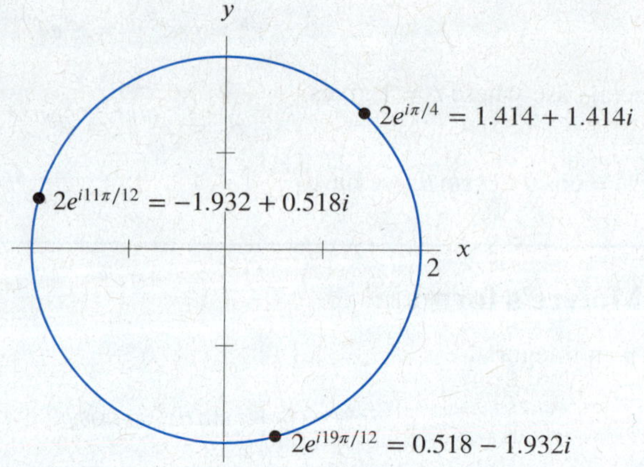


**Figure 9.39**: The three cube roots of $8e^{i3\pi/4}$ are evenly spaced on a circle

## Exercises and Problems for Section 9.6

### Exercises

For Exercises 1–6, express the complex number in polar form, $z = re^{i\theta}$.

**1.** $-5$ **2.** $-i$ **3.** $0$

**4.** $2i$ **5.** $-3-4i$ **6.** $-1+3i$

In Exercises 7–14. give a value in Cartesian form, $z = x + iy$.

**7.** $(2+3i)^2$ **8.** $(2+3i)(5+7i)$

**9.** $(2+3i)+(-5-7i)$ **10.** $(0.5-i)(1-i/4)$

**11.** $(e^{i\pi/3})^2$ **12.** $(2i)^3-(2i)^2+2i-1$

**13.** $\sqrt{e^{i\pi/3}}$ **14.** $\sqrt[4]{10e^{i\pi/2}}$

### Problems

In Problems 15–23, use the polar form of the complex number to find a value in Cartesian form, $z = x + iy$.

**15.** $\sqrt{4i}$ **16.** $\sqrt{-i}$ **17.** $\sqrt[3]{i}$

**18.** $\sqrt{7i}$ **19.** $\sqrt[4]{-1}$ **20.** $(1+i)^{2/3}$

**21.** $(\sqrt{3}+i)^{1/2}$ **22.** $(\sqrt{3}+i)^{-1/2}$ **23.** $(\sqrt{5}+2i)^{\sqrt{2}}$

In Problems 24–25, solve the simultaneous equations for the complex numbers $A_1$ and $A_2$.

**24.** $A_1 + A_2 = 2$
$(1-i)A_1 + (1+i)A_2 = 0$

**25.** $A_1 + A_2 = i$
$iA_1 - A_2 = 3$

**26.** If the roots of the equation $x^2 + 2bx + c = 0$ are the complex numbers $p \pm iq$, find expressions for $p$ and $q$ in terms of $b$ and $c$.

**27.** **(a)** Find the polar form of the complex number $i$.
**(b)** If $z = re^{i\theta}$ is any point in the complex plane, show that the point $iz$ can by found by rotating the segment from 0 to $z$ by 90° counterclockwise, as in Figure 9.40.

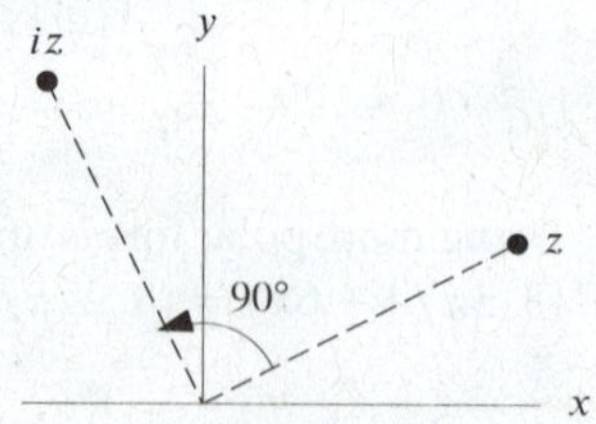


**Figure 9.40**

For Problems 28–31, use Euler's formula to derive the identity. (Note that if $a, b, c, d$ are real numbers, $a + bi = c + di$ means that $a = c$ and $b = d$.)

**28.** $\sin 2\theta = 2\sin\theta\cos\theta$

**29.** $\cos 2\theta = \cos^2\theta - \sin^2\theta$

**30.** $\cos(-\theta) = \cos\theta$

**31.** $\sin(-\theta) = -\sin\theta$

In Problems 32–35, find polar and Cartesian forms for all three cube roots of the complex number.

**32.** $e^{2i}$

**33.** $-8$

**34.** $8$

**35.** $1 + i$

In Problems 36–39, use de Moivre's formula to evaluate the power in Cartesian form.

**36.** $(\cos\pi/4 + i\sin\pi/4)^4$

**37.** $(\cos 2\pi/3 + i\sin 2\pi/3)^3$

**38.** $(\cos 2 + i\sin 2)^{-1}$

**39.** $(\cos\pi/4 + i\sin\pi/4)^{-7}$

**40.** Show the Pythagorean identity, $\cos^2\theta + \sin^2\theta = 1$, follows from Euler's formula and the fact that $e^{i\theta}e^{-i\theta} = 1$.

**41.** Show that the identities for $\cos(\theta + \phi)$ and $\sin(\theta + \phi)$ follow from Euler's formula.

## CHAPTER SUMMARY

- **Solving Trigonometric Equations**
- **Identities and Formulas** (see inside cover)
  Double-Angle Formulas:

$$\sin 2\theta = 2\sin\theta\cos\theta$$
$$\cos 2\theta = \cos^2\theta - \sin^2\theta = 2\cos^2\theta - 1 = 1 - 2\sin^2\theta$$
$$\tan 2\theta = \frac{2\tan\theta}{1 - \tan^2\theta}.$$

  Sums and differences of angles:

$$\sin(\theta + \phi) = \sin\theta\cos\phi + \cos\theta\sin\phi$$
$$\sin(\theta - \phi) = \sin\theta\cos\phi - \cos\theta\sin\phi$$
$$\cos(\theta + \phi) = \cos\theta\cos\phi - \sin\theta\sin\phi$$
$$\cos(\theta - \phi) = \cos\theta\cos\phi + \sin\theta\sin\phi.$$

  Sums and differences: Same periods:

$$a_1\sin(Bt) + a_2\cos(Bt) = A\sin(Bt + \phi)$$

  where $A = \sqrt{a_1^2 + a_2^2}$ and $\tan\phi = \dfrac{a_2}{a_1}$.

  Sums and differences: Same amplitudes:

$$\cos u + \cos v = 2\cos\frac{u+v}{2}\cos\frac{u-v}{2}$$
$$\sin u + \sin v = 2\sin\frac{u+v}{2}\cos\frac{u-v}{2}$$
$$\cos u - \cos v = -2\sin\frac{u+v}{2}\sin\frac{u-v}{2}$$
$$\sin u - \sin v = 2\cos\frac{u+v}{2}\sin\frac{u-v}{2}$$

- **Modeling with Trigonometric Functions**
  Damped oscillations; rising midline; acoustic beats.
- **Complex Numbers**
  Operations and graphing with $z = a + bi$.
  Finding roots of complex numbers.
- **Euler's Formula**
  $e^{i\theta} = \cos\theta + i\sin\theta$.
  Polar form: $z = r(\cos\theta + i\sin\theta) = re^{i\theta}$.
- **de Moivre's formula**
  $(\cos\theta + i\sin\theta)^p = \cos(p\theta) + i\sin(p\theta)$

## REVIEW EXERCISES AND PROBLEMS FOR CHAPTER NINE

### Exercises

In Exercises 1–2, simplify the expression to contain only $\sin t$ and $\cos t$.

**1.** $(1 - \sin t)(1 - \cos t) - \cos t\sin t$

**2.** $2\cos t - 3\sin t - (2\sin t - 3\cos t)$

In Exercises 3–8, simplify the expression to contain only $\sin t$, $\cos t$, and $\tan t$.

**3.** $\sec t\sin t + 3\tan t$

**4.** $\cot t\tan t\sin t$

**5.** $\dfrac{\sec t}{\csc t}$

**6.** $\dfrac{\cot t}{\csc t}$

**7.** $(\sec t\cot t - \csc t\tan t)\sin t\cos t$

**8.** $\dfrac{1}{2\csc t\cos t - 3\cot t}$

In Exercises 9–12, simplify the expression.

**9.** $\dfrac{1 - \cos^2\theta}{\sin\theta}$

**10.** $\dfrac{\cot x}{\csc x}$

**11.** $\dfrac{\cos 2\phi + 1}{\cos\phi}$

**12.** $\cos^2\theta(1 + \tan\theta)(1 - \tan\theta)$

In Exercises 13–14, perform the calculation. Give your answer in Cartesian form, $z = x + iy$.

**13.** $(5 + 4i) + (3 - 9i)$

**14.** $(2 + 3i)(3 + 5i)$

For Exercises 15–16, express the complex number in polar form, $z = re^{i\theta}$.

**15.** $2 - 4i$

**16.** $3 + 5i$

## Problems

In Problems 17–20, solve the equation with $0 \le \alpha < 2\pi$. Give exact answers if possible.

**17.** $2\cos\alpha = 1$

**18.** $\tan\alpha = \sqrt{3} - 2\tan\alpha$

**19.** $4\tan\alpha + 3 = 2$

**20.** $3\sin^2\alpha + 4 = 5$

**21.** **(a)** Find a sinusoidal formula for the graph in Figure 9.41.
**(b)** Find $x_1$ and $x_2$.

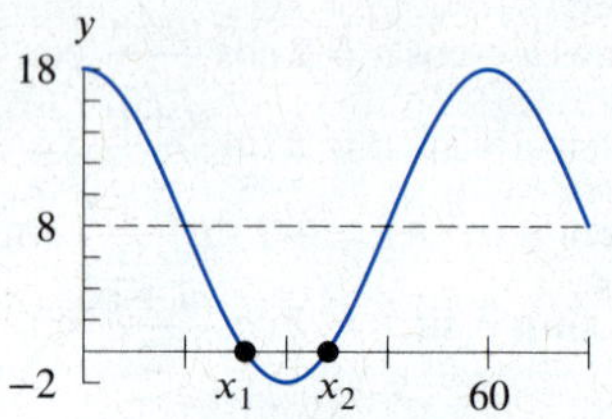


Figure 9.41

**22.** A weight is suspended from the ceiling by a spring. Figure 9.42 shows a graph of the distance from the ceiling to the weight, $d = f(t)$, as a function of time.

**(a)** Find a possible formula for $f(t)$.
**(b)** Solve $f(t) = 12$ exactly. Interpret your results.

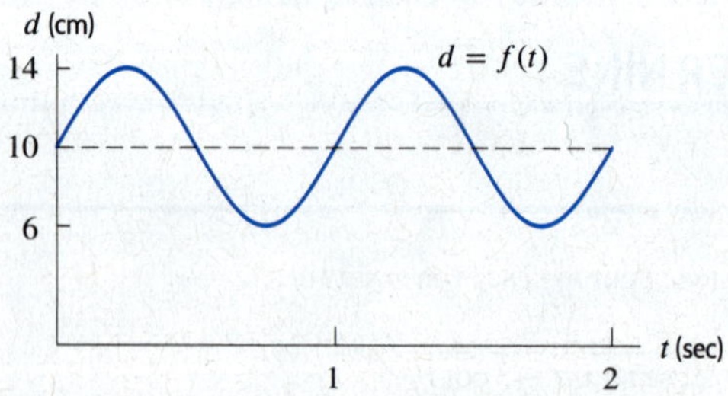


Figure 9.42

**23.** Table 9.1 gives the average monthly temperature, $y$, in degrees Fahrenheit for the city of Fairbanks, Alaska, as a function of $t$, the month, where $t = 0$ indicates January.

**(a)** Plot the data points. On the same graph, draw a curve which fits the data.
**(b)** What kind of function best fits the data? Be specific.
**(c)** Find a formula for a function, $f(t)$, that models the temperature. [Note: There are many correct answers.]
**(d)** Use your answer to part (c) to solve $f(t) = 32$. Interpret your results.
**(e)** Check your results from part (d) graphically.
**(f)** In the southern hemisphere, the times at which summer and winter occur are reversed relative to the northern hemisphere. Modify your formula from part (c) so that it represents the average monthly temperature for a southern-hemisphere city whose summer and winter temperatures are similar to Fairbanks.

Table 9.1

| $t$ | 0 | 1 | 2 | 3 | 4 | 5 |
|---|---|---|---|---|---|---|
| $y$ | −11.5 | −9.5 | 0.5 | 18.0 | 36.9 | 53.1 |
| $t$ | 6 | 7 | 8 | 9 | 10 | 11 |
| $y$ | 61.3 | 59.9 | 48.8 | 31.7 | 12.2 | −3.3 |

**24.** One student said that $\sin 2\theta \ne 2\sin\theta$, but another student said let $\theta = \pi$ and then $\sin 2\theta = 2\sin\theta$. Who is right?

**25.** Graph $\tan^2 x + \sin^2 x$, $(\tan^2 x)(\sin^2 x)$, $\tan^2 x - \sin^2 x$, $\tan^2 x/\sin^2 x$ to see if any two expressions appear to be identical. Prove any identities you find.

**26.** For $y$ and $\theta$ and $\phi$ in Figure 9.43, evaluate the following in terms of $y$.

**(a)** $\tan\theta$ **(b)** $\cos\theta$ **(c)** $\theta$
**(d)** $\cos(2\phi)$ for $0 < \phi < \pi/4$

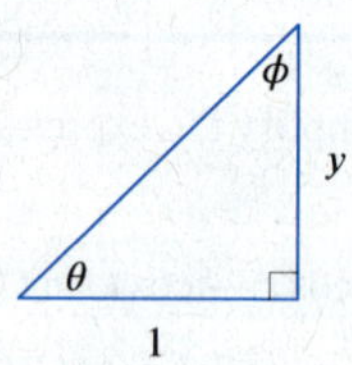


Figure 9.43

In Problems 27–28, give exact answers for $0 \le \theta \le \pi/2$.

**27.** If $\sin\theta = \frac{8}{11}$, what is $\csc\theta$? $\tan\theta$?

**28.** If $\csc\theta = 94$, what is $\cos\theta$? $\tan\theta$?

**29.** Let $\cos\theta = 0.27$. Find one possible value for $\sin\theta$ and for $\tan\theta$.

**30.** The angle $\theta$ is in the first quadrant and $\tan\theta = 3/4$. Since $\tan\theta = (\sin\theta)/(\cos\theta)$, does this mean that $\sin\theta = 3$ and $\cos\theta = 4$?

**31.** If $\cos(2\theta) = 2/7$ and $\theta$ is in the first quadrant, find $\cos\theta$ exactly.

**32.** Use a trigonometric identity to find exactly all solutions: $\cos 2\theta = \sin\theta, \quad 0 \le \theta < 2\pi$.

**33.** Solve exactly: $\cos(2\alpha) = -\sin\alpha$ with $0 \le \alpha < 2\pi$.

**34.** Simplify the expression $\sin\left(2\cos^{-1}\left(\frac{5}{13}\right)\right)$ to a rational number.

Prove the identities in Problems 35–38.

**35.** $\tan\theta = \dfrac{1-\cos 2\theta}{2\cos\theta\sin\theta}$

**36.** $(\sin^2 2\pi t + \cos^2 2\pi t)^3 = 1$

**37.** $\sin^4 x - \cos^4 x = \sin^2 x - \cos^2 x$

**38.** $\dfrac{1+\sin\theta}{\cos\theta} = \dfrac{\cos\theta}{1-\sin\theta}$

**39.** With $x$ and $\theta$ as in Figure 9.44 and with $0 < \theta < \pi/4$, express the following in terms of $x$ without using trigonometric functions:

**(a)** $\cos\theta$ **(b)** $\cos(\frac{\pi}{2}-\theta)$ **(c)** $\tan^2\theta$

**(d)** $\sin(2\theta)$ **(e)** $\cos(4\theta)$ **(f)** $\sin(\cos^{-1} x)$

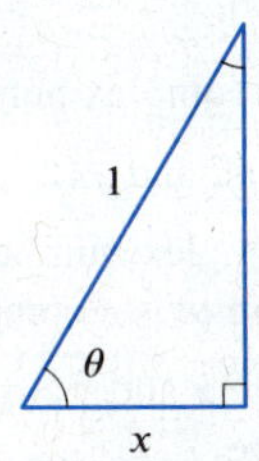

Figure 9.44

In Problems 40–41, solve for $\alpha$ for $0 \le \alpha \le 2\pi$.

**40.** $3\cos^2\alpha + 2 = 3 - 2\cos\alpha$

**41.** $3\sin^2\alpha + 3\sin\alpha + 4 = 3 - 2\sin\alpha$

**42.** Use the cosine addition formula and other identities to find a formula for $\cos 3\theta$ in terms of $\cos\theta$. Your formula should contain no trigonometric functions other than $\cos\theta$ and its powers.

**43.** Use the identity $\cos 2x = 2\cos^2 x - 1$ to find an expression for $\cos(\theta/2)$, assuming $0 \le \theta \le \pi/2$. Do this by letting $x = \theta/2$, and then finding an expression for $\cos x$ in terms of $\cos(2x)$.

**44.** Suppose that $\sin(\ln x) = \frac{1}{3}$, and that $\sin(\ln y) = \frac{1}{5}$. If $0 < \ln x < \frac{\pi}{2}$ and $0 < \ln y < \frac{\pi}{2}$, use the sum-of-angle sine formula to evaluate

$$\sin(\ln(xy)).$$

**45.** **(a)** Graph $g(\theta) = \sin\theta - \cos\theta$.

**(b)** Write $g(\theta)$ as a sine function without using cosine. Write $g(\theta)$ as a cosine function without using sine.

**46.** For positive constants $a, b$ and $t$ in years, the sizes of two populations are given by

$$P_1 = f(t) = a\sin\left(\frac{2\pi}{b}\cdot t\right) + 7a,$$
$$P_2 = g(t) = 2a\sin\left(\frac{2\pi}{b+2}\cdot t\right) + 3a.$$

Identify which of the two populations:

**(a)** Has the larger swings from its minimum to maximum size.

**(b)** Has the slower swings from its minimum to maximum size.

**(c)** At its smallest, is larger than the other population.

**(d)** Is more vulnerable to sudden extinction.

**47.** Graph $f(t) = \cos t$ and $g(t) = \sin(t + \pi/2)$. Explain what you see.

**48.** Let $z_1 = -3 - i\sqrt{3}$ and $z_2 = -1 + i\sqrt{3}$.

**(a)** Find $z_1 z_2$ and $z_1/z_2$. Give your answer in Cartesian form, $z = x + iy$.

**(b)** Put $z_1$ and $z_2$ into polar form, $z = re^{i\theta}$. Find $z_1 z_2$ and $z_1/z_2$ using the polar form, and check that you get the same answer as in part (a).

**49.** **(a)** Let $z = 3 + 2i$. Plot $z$ and $iz$ on the complex plane.

**(b)** Show that the line from the origin to $z$ is perpendicular to the line from the origin to $iz$.

## STRENGTHEN YOUR UNDERSTANDING

Are the statements in Problems 1–30 true or false? Give an explanation for your answer.

1. The function $f(t) = e^{-t}\cos t$ is a damped oscillation.
2. For all values of $\phi$ for which the expression $\tan\phi$ is defined, $\tan\phi \cdot \cos\phi = \sin\phi$.
3. If the left side of an equation and the right side of the equation are equal for all values of the variable for which the equation is defined, then the equation is an identity.
4. The function $f(\theta) = \sin\theta\tan\theta$ is an odd function.
5. For all values of $\theta$ for which the expressions are defined, $1 - 2\tan^2\theta\cos^2\theta = 2\cos^2\theta - 1$
6. The equation $\sin\beta^2 + \cos\beta^2 = 1$ holds for all values of $\beta$.
7. If $\cos\alpha = -\frac{2}{3}$ and the point corresponding to $\alpha$ is in the third quadrant, then $\sin(2\alpha) = \dfrac{4\sqrt{5}}{9}$.
8. Since $\sin 45^\circ = \cos 45^\circ$, we see that $\sin\theta = \cos\theta$ is an identity.
9. For all values of $\theta$, we have $\sin^2\theta = \frac{1}{2}(1 - \cos 2\theta)$.
10. For all values of $\theta$, we have $\cos(4\theta) = -\cos(-4\theta)$.
11. For all values of $\theta$ and $\phi$, we have $\sin(\theta + \phi) = \sin\theta + \sin\phi$.
12. For all values of $\theta$, we have $\sin 2\theta = 2\sin\theta$.
13. For all value of $\theta$, we have $\sin(\theta + \frac{\pi}{2}) = \cos\theta$.
14. The function $f(t) = \sin t\cos t$ is periodic with period $2\pi$.
15. For all values of $t$, we have $\sin(t - \pi) = \sin(t + \pi)$.
16. For all values of $t$, we have $\cos(t - \pi/2) = \cos(t + \pi/2)$.
17. The graph of $A\cos(Bt)$ is a horizontal shift of the graph of $A\sin(Bt)$.
18. The function $f(x) = \sin 2x + \sin 3x$ is a periodic function.
19. The function $f(x) = \sin 2x + \sin 3x$ is a sinusoidal function.
20. The sum of two sinusoidal functions is sinusoidal.
21. For $a_i > 0$, when writing $a_1\sin(Bt) + a_2\cos(Bt)$ as a single sine function, the amplitude of the resulting single sine function is larger than that of either function in the sum.
22. The function $f(t) = 3\sin(2\pi t) + 4\cos(2\pi t)$ is sinusoidal.
23. The function $g(t) = (1/2)e^t\sin(t/2)$ has a decreasing amplitude.
24. If a company has a power requirement of $g(t) = 55 - 10\sin(\pi t/12)$, then the peak requirement is 65 units.
25. A CRT screen that has a refresh rating of 60 hertz will refresh every 1/60th of a second.
26. The function $f(t) = \cos(\pi t) + 1$ has all of its zeros at odd integer values.
27. The parameter $B$ in $g(x) = A\sin(Bx) + C$ affects the amplitude of the function.
28. One solution of $3e^t\sin t - 2e^t\cos t = 0$ is $t = \arctan(\frac{2}{3})$.
29. For for all $x$ values, $-1 \le x \le 1$, we have $0 \le \cos^{-1}(\sin(\cos^{-1}x)) \le \frac{\pi}{2}$.
30. If $a > b > 0$, then the domain of $f(t) = \ln(a + b\sin t)$ is all real numbers $t$.

Are the statements in Problems 31–39 true or false? Give an explanation for your answer.

31. Every nonnegative real number has a real square root.
32. For any complex number $z$, the product $z \cdot \bar{z}$ is a real number.
33. The square of any complex number is a real number.
34. If $f$ is a polynomial, and $f(z) = i$, then $f(\bar{z}) = i$.
35. Every nonzero complex number $z$ can be written in the form $z = e^w$, where $w$ is a complex number.
36. If $z = x + iy$, where $x$ and $y$ are positive, then $z^2 = a + ib$ has $a$ and $b$ positive.
37. For all values of $\theta$ we have $e^{i\theta} = \cos\theta + i\sin\theta$.
38. $(1 + i)^2 = 2i$.
39. $i^{101} = i$.

Chapter Ten

# COMPOSITIONS, INVERSES, AND COMBINATIONS OF FUNCTIONS

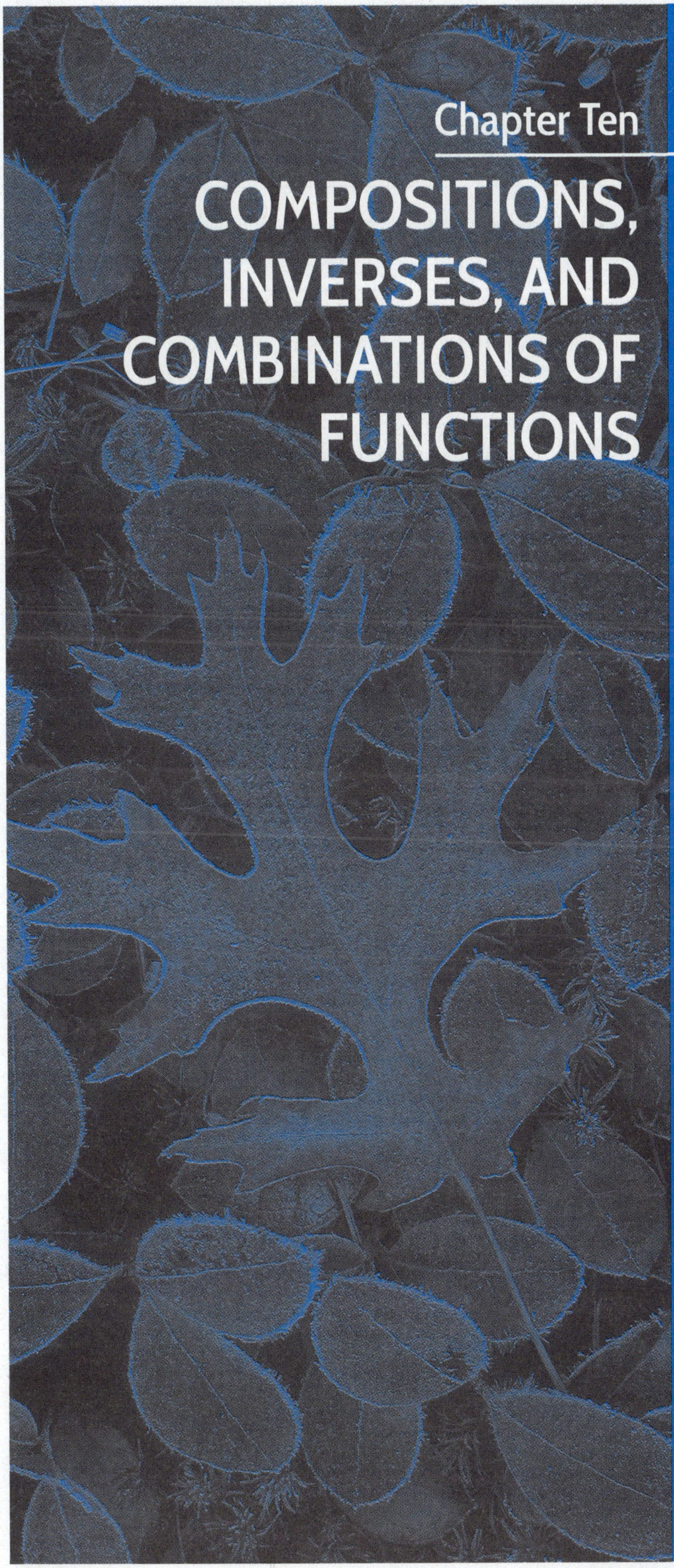

© Patrick Zephyr/Patrick Zephyr Nature Photography

## Contents

# 10.1 COMPOSITION OF FUNCTIONS

In Chapter 2, we introduced the composition of functions; we now look at extensions of this idea. In Section 2.5, for two functions $f$ and $g$, we have the following definition:

> The **composite function** $f(g(t))$ uses the output of the $g$ as the input to $f$. It is defined for all values of $t$ in the domain of $g$ whose $g(t)$ values are in the domain of $f$.

For example, recall the London Eye Ferris wheel example from Section 7.1. We wished to find the height on the Ferris wheel at a particular time. If $H = f(\theta)$ represents the height on the Ferris wheel (in feet) as a function of angular position (in degrees) and $\theta = g(t)$ represents the angular position as a function of time (in minutes), then the composition of $f$ with $g$ gives a function $H = f(g(t))$ that represents the height on the Ferris wheel as a function of time.

In Section 7.1 we composed functions defined by formulas. We can also compose functions defined by graphs.

**Example 1** Let $u$ and $v$ be two functions defined by the graphs in Figure 10.1. Evaluate:

(a) $v(u(-1))$ (b) $u(v(5))$

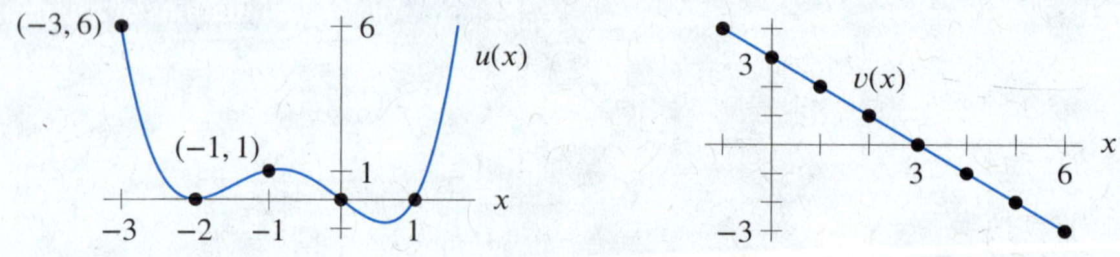


Figure 10.1: Evaluate the composition of functions $u$ and $v$ defined by their graphs

**Solution** (a) To evaluate $v(u(-1))$, start with $u(-1)$. From Figure 10.1, we see that $u(-1) = 1$. Thus,

$$v(u(-1)) = v(1).$$

From the graph we see that $v(1) = 2$, so

$$v(u(-1)) = 2.$$

(b) Since $v(5) = -2$, we have $u(v(5)) = u(-2) = 0$.

Sometimes we do not have a formula or a graph, but we have a table of values for a function. In the next example, we compose functions given by tables.

**Example 2** Complete the table given that $h(x) = f(g(x))$.

| $x$ | $f(x)$ | $g(x)$ | $h(x)$ |
|---|---|---|---|
| 0 | 1 | 2 | 5 |
| 1 | 9 | 0 | |
| 2 | | 1 | |

**Solution** We have $h(1) = f(g(1))$. From the table, we see $g(1) = 0$, so

$$h(1) = f(g(1)) = f(0) = 1.$$

| $x$ | $f(x)$ | $g(x)$ | $h(x)$ |
|---|---|---|---|
| 0 | 1 | 2 | 5 |
| 1 | 9 | 0 | 1 |
| 2 | 5 | 1 | 9 |

Similarly, we have

$$h(2) = f(g(2)) = f(1) = 9.$$

The value for $f(2)$ is still missing in the table. Since $h(x) = f(g(x))$, to find a value for $f(2)$, we first find a value of $x$ such that $g(x) = 2$. From the table, we see that $g(0) = 2$, so

$$h(0) = f(g(0)) = f(2).$$

Since $h(0) = 5$ we find that $f(2) = 5$.

So far we have considered examples of two functions composed together, but there is no limit on the number of functions that can be composed.

**Example 3** Let $p(x) = \sin x + 1$ and $q(x) = x^2 - 3$. Find a formula in terms of $x$ for $w(x) = p(p(q(x)))$.

**Solution** We work from inside the parentheses outward. First we find $p(q(x))$, and then input the result to $p$.

$$\begin{aligned} w(x) &= p(p(q(x))) \\ &= p(p(x^2 - 3)) \\ &= p(\underbrace{\sin(x^2 - 3) + 1}_{\text{Input for } p}) \\ &= \sin(\sin(x^2 - 3) + 1) + 1. \qquad \text{because } p(\text{Input}) = \sin(\text{Input}) + 1 \end{aligned}$$

## Decomposition of Functions

Sometimes we reason backward to find the functions that went into a composition. This process is called *decomposition*.

**Example 4** Let $h(x) = f(g(x)) = e^{x^2+1}$. Find possible formulas for $f(x)$ and $g(x)$.

**Solution** In the formula $h(x) = e^{x^2+1}$, the expression $x^2+1$ is in the exponent. We can take the inside function to be $g(x) = x^2 + 1$. This means that we can write

$$h(x) = e^{\overbrace{x^2+1}^{g(x)}} = e^{g(x)}.$$

Thus the outside function is $f(x) = e^x$. We check that composing $f$ and $g$ gives $h$:

$$f(g(x)) = f(x^2 + 1) = e^{x^2+1} = h(x).$$

There are many possible solutions to Example 4. For example, we could choose $f(x) = e^{x+1}$ and $g(x) = x^2$. Then

$$f(g(x)) = e^{g(x)+1} = e^{x^2+1} = h(x).$$

Alternatively, we might choose $f(x) = e^{x^2+1}$ and $g(x) = x$. Although this satisfies the condition that $h(x) = f(g(x))$, it is not very useful, because $f$ is the same as $h$. This kind of decomposition is referred to as *trivial*. Another example of a trivial decomposition of $h(x)$ is $f(x) = x$ and $g(x) = e^{x^2+1}$.

We saw in Example 3 that we can compose more than two functions. Similarly, it may be possible to decompose a function into more than two simpler functions.

**Example 5** Let $p(z) = \sin^2(\ln z)$. Decompose $p(z)$ into three simpler functions by giving formulas for $f(z)$, $g(z)$, and $h(z)$ where

$$p(z) = f(g(h(z))).$$

**Solution** Note that $p(z) = \sin^2(\ln z) = (\sin(\ln z))^2$. In words, to find $p(z)$ we first find $\ln z$, then take the sin of $\ln z$, and then square the result. In the formula $p(z) = f(g(h(z)))$, the innermost function is $h(z)$. Since the first operation corresponds to the innermost function, we let

$$h(z) = \ln z.$$

Next we want to take the sin of $h(z)$. Thus, we let

$$g(\text{Input}) = \sin(\text{Input})$$
$$g(z) = \sin z.$$

So we have

$$g(h(z)) = g(\ln z) = \sin(\ln z).$$

Finally, since $p(z) = \sin^2(\ln z) = (\sin(\ln z))^2$, we square the expression $\sin(\ln z)$ and let

$$f(\text{Input}) = \text{Input}^2$$
$$f(z) = z^2.$$

To check, we compute

$$\begin{aligned} f(g(h(z))) &= f(g(\ln z)) && \text{since } h(z) = \ln z \\ &= f(\underbrace{\sin(\ln z)}_{\text{Input for } f}) && \text{since } g(\ln z) = \sin(\ln z) \\ &= \sin^2(\ln z). && \text{since } f(\sin(\ln z)) = (\sin(\ln z))^2 \end{aligned}$$

## Exercises and Problems for Section 10.1

### Exercises

**1.** Use Figures 10.2 and 10.3 to calculate the following:

**(a)** $f(g(-1))$ **(b)** $g(f(2))$
**(c)** $f(g(-2))$ **(d)** $g(f(0))$
**(e)** $f(f(1))$ **(f)** $g(g(g(-1)))$

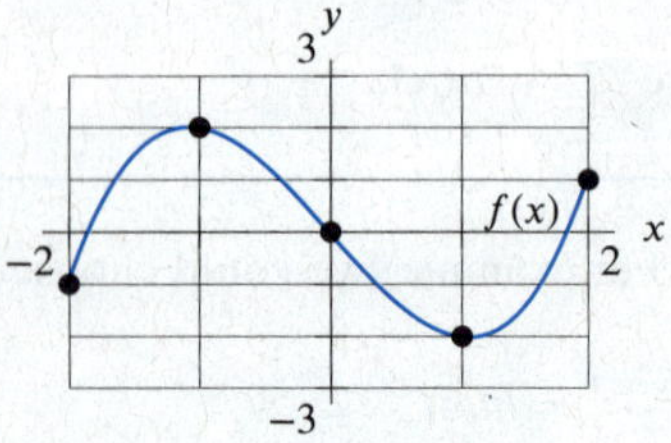

Figure 10.2

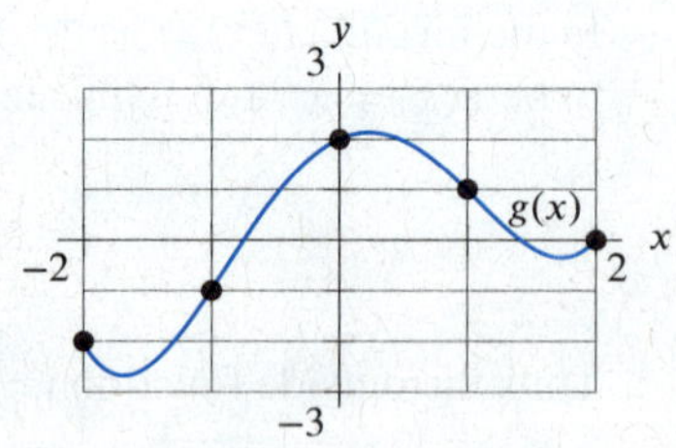

Figure 10.3

**2.** Use Table 10.1 to construct a table of values for $r(x) = p(q(x))$.

**Table 10.1**

| $x$ | 0 | 1 | 2 | 3 | 4 | 5 |
|---|---|---|---|---|---|---|
| $p(x)$ | 1 | 0 | 5 | 2 | 3 | 4 |
| $q(x)$ | 5 | 2 | 3 | 1 | 4 | 8 |

**3.** Let $p$ and $q$ be the functions in Exercise 2. Construct a table of values for $s(x) = q(p(x))$.

In Exercises 4–5, find and simplify for $f(x) = 2^x$ and $g(x) = \frac{x}{x+1}$.

**4.** $f(g(x))$ **5.** $g(f(x))$

**6.** Let $f(x) = \sin 4x$ and $g(x) = \sqrt{x}$. Find formulas for $f(g(x))$ and $g(f(x))$.

**7.** Let $m(x) = 3 + x^2$ and $n(x) = \tan x$. Find formulas for $m(n(x))$ and $n(m(x))$.

**8.** Find a formula in terms of $x$ for the function $w(x) = p(p(x))$, where $p(x) = 2x + 1$.

In Exercises 9–14, let $f(x) = 3x^2$, $g(x) = 9x - 2$, $m(x) = 4x$, and $r(x) = \sqrt{3x}$. Find and simplify the composite function.

**9.** $r(g(x))$ **10.** $f(r(x))$

**11.** $r(f(x))$ **12.** $g(f(x))$

**13.** $g(m(f(x)))$ **14.** $f(m(g(x)))$

**15.** Write $f(g(x))$ in the form $a^b$ given that

$$f(x) = e^{0.5(1-x)}, \quad g(x) = x^2 + 1.$$

**16.** Write $p(t) = q(r(t))$ in the form $\sqrt{a}$ given that

$$r(t) = 1 - \sqrt{t}, \quad q(t) = \sqrt{2-t}.$$

In Exercises 17–21, identify the function $f(x)$.

**17.** $h(x) = e^{f(x)} = e^{\sin x}$

**18.** $j(x) = \sqrt{f(x)} = \sqrt{\ln(x^2 + 4)}$

**19.** $k(x) = \sin(f(x)) = \sin(x^3 + 3x + 1)$

**20.** $l(x) = (f(x))^2 = \cos^2 2x$

**21.** $m(x) = \ln f(x) = \ln(5 + 1/x)$

## Problems

In Problems 22–25, give a practical interpretation in words of the function.

**22.** $f(h(t))$, where $A = f(r)$ is the area of a circle of radius $r$ and $r = h(t)$ is the radius of the circle at time $t$.

**23.** $k(g(t))$, where $L = k(H)$ is the length of a steel bar at temperature $H$ and $H = g(t)$ is temperature at time $t$.

**24.** $R(Y(q))$, where $R$ gives a farmer's revenue as a function of corn yield per acre, and $Y$ gives the corn yield as a function of the quantity, $q$, of fertilizer.

**25.** $t(f(H))$, where $t(v)$ is the time of a trip at velocity $v$, and $v = f(H)$ is velocity at temperature $H$.

**26.** Using Figure 10.4, estimate the following:

**(a)** $f(g(2))$ **(b)** $g(f(2))$ **(c)** $f(f(3))$ **(d)** $g(g(3))$

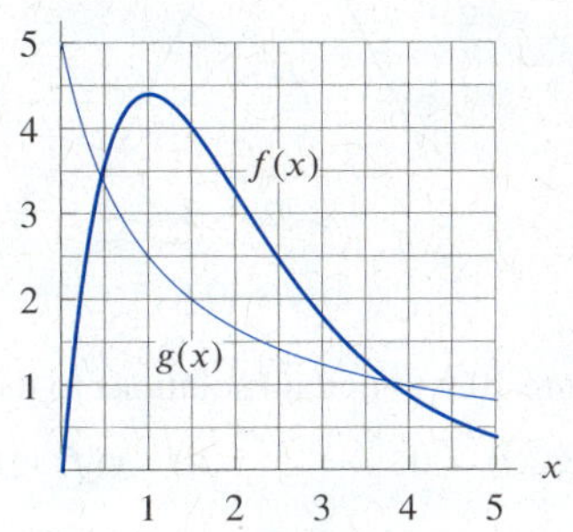


Figure 10.4

**27.** Complete Table 10.2.

**Table 10.2**

| $x$ | $f(x)$ | $g(x)$ | $g(f(x))$ |
|---|---|---|---|
| 0 | 2 | 3 | 2 |
| 1 | 3 | 1 | 3 |
| 2 | 1 | | |

**28.** Complete Table 10.3 given that $h(x) = g(f(x))$.

**Table 10.3**

| $x$ | $f(x)$ | $g(x)$ | $h(x)$ |
|---|---|---|---|
| 0 | 1 | 2 | 8 |
| 1 | 2 | | |
| 2 | 1 | 6 | |

**29.** Complete the table given $h(x) = g(f(x))$.

| $x$ | $f(x)$ | $g(x)$ | $h(x)$ |
|---|---|---|---|
| 0 | 2 | | |
| 1 | | 0 | 0 |
| 2 | | 3 | 2 |
| 3 | 0 | | 1 |
| 4 | 3 | 2 | 4 |

**30.** Complete Table 10.4 given that $w(t) = v(u(t))$.

**Table 10.4**

| $t$ | $u$ | $v$ | $w$ |
|---|---|---|---|
| 0 | 2 | 3 | |
| 1 | | | 2 |
| 2 | 1 | 1 | 4 |
| 3 | | 2 | 0 |
| 4 | 0 | 0 | |

In Problems 31–32, find a simplified formula for $g$ given that $f(x) = \dfrac{x+1}{x+3}$.

**31.** $g(x) = f(h(x))$ where $h(x) = 3x^2 - 1$.

**32.** $f(x) = g(x) + \dfrac{2x+4}{x+3}$.

In Problems 33–36, find a simplified formula for the difference quotient

$$\frac{f(x+h) - f(x)}{h}.$$

**33.** $f(x) = x^2 + x$

**34.** $f(x) = \sqrt{x}$

**35.** $f(x) = \dfrac{1}{x}$

**36.** $f(x) = 2^x$

In Problems 37–40, let $x > 0$ and $k(x) = e^x$. Find a possible formula for $f(x)$.

**37.** $k(f(x)) = e^{2x}$

**38.** $f(k(x)) = 1 - e^x$

**39.** $f(k(x)) = e^{2x}$

**40.** $k(f(x)) = x$

In Problems 41–44, let $f(x) = g(h(x))$. Find possible formulas for $g(x)$ and $h(x)$ (There may be more than one possible answer. Assume $g(x) \neq x$ and $h(x) \neq x$.)

**41.** $f(x) = (x+3)^2$

**42.** $f(x) = \sqrt{1 + \sqrt{x}}$

**43.** $f(x) = 9x^2 + 3x$

**44.** $f(x) = \dfrac{1}{x^2 + 8x + 16}$

In Problems 45–50, decompose the functions into $u(v(x))$ for each given $u$ or $v$.

**45.** $y = \dfrac{1+x^2}{2+x^2}$ given that

**(a)** $v(x) = x^2$ **(b)** $v(x) = x^2 + 1$

**46.** $y = e^{-\sqrt{x}}$ given that

**(a)** $u(x) = e^x$ **(b)** $v(x) = \sqrt{x}$

**47.** $y = \sqrt{1 - x^3}$ given that

**(a)** $u(x) = \sqrt{1 + x^3}$ **(b)** $v(x) = x^3$

**48.** $y = 2^{x+1}$ given that

**(a)** $u(x) = 2x$ **(b)** $v(x) = -x$

**49.** $y = \sin^2 x$ given that

**(a)** $u(x) = x^2$ **(b)** $v(x) = x^2$

**50.** $y = e^{2\cos x}$ given that

**(a)** $u(x) = e^x$ **(b)** $v(x) = \cos x$

**51.** **(a)** Use Table 10.5 and Figure 10.5 to calculate:

(i) $f(g(4))$ (ii) $g(f(4))$
(iii) $f(f(0))$ (iv) $g(g(0))$

**(b)** Solve $g(g(x)) = 1$ for $x$.

**Table 10.5**

| $x$ | 0 | 1 | 2 | 3 | 4 | 5 |
|---|---|---|---|---|---|---|
| $f(x)$ | 2 | 5 | 3 | 4 | 1 | 0 |

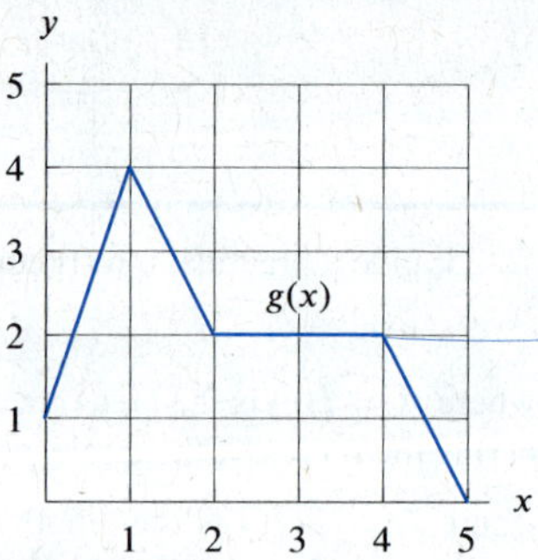


**Figure 10.5**

**52.** Use Figure 10.6 to calculate the following:

**(a)** $f(f(1))$ **(b)** $g(g(1))$
**(c)** $f(g(2))$ **(d)** $g(f(2))$

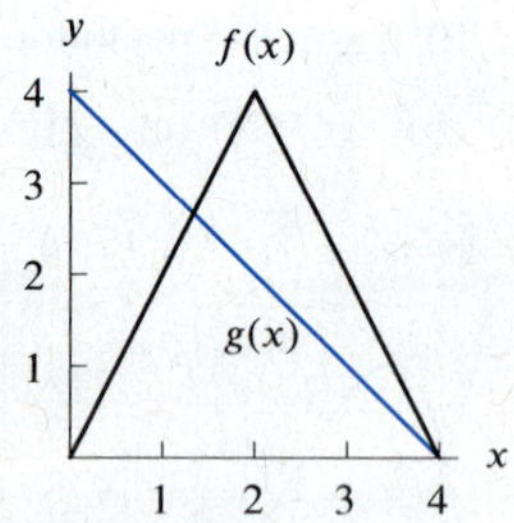


**Figure 10.6**

**53.** Use Figure 10.6 to find all solutions to the equations:

**(a)** $f(g(x)) = 0$ **(b)** $g(f(x)) = 0$

**54.** Let $f(x)$ and $g(x)$ be the functions in Figure 10.6.

**(a)** Graph the functions $f(g(x))$ and $g(f(x))$.
**(b)** On what interval(s) is $f(g(x))$ increasing?
**(c)** On what interval(s) is $g(f(x))$ increasing?

In Problems 55–58, use the information from Figures 10.7 and 10.8 to graph the functions.

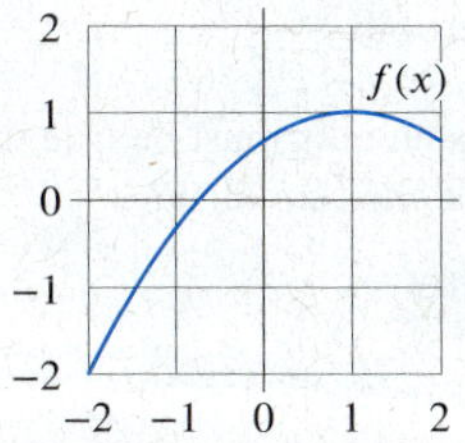


Figure 10.7

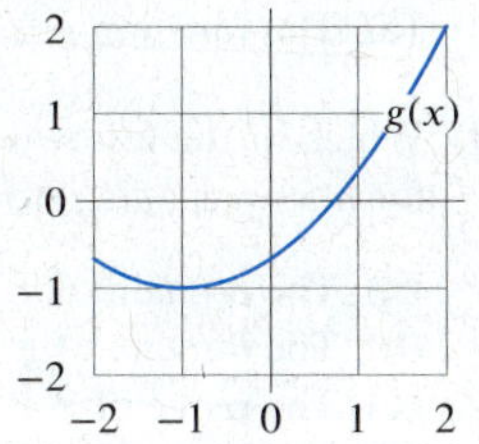


Figure 10.8

**55.** $f(g(x))$ **56.** $g(f(x))$ **57.** $f(f(x))$ **58.** $g(g(x))$

In Problems 59–62, the graph of a function $g(x)$ is shown. Let $f(x) = x^2$.

**(a)** Graph $f(g(x))$ **(b)** Graph $g(f(x))$

**59.**

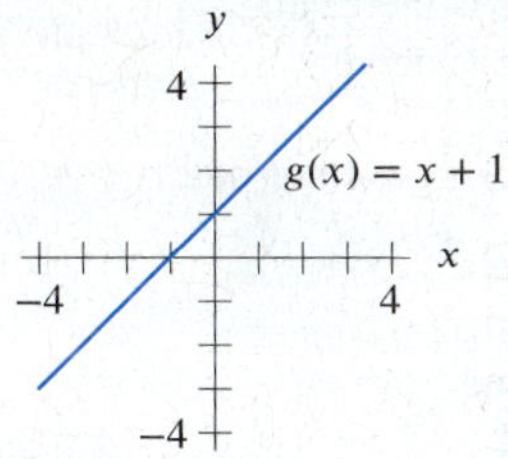


**60.**

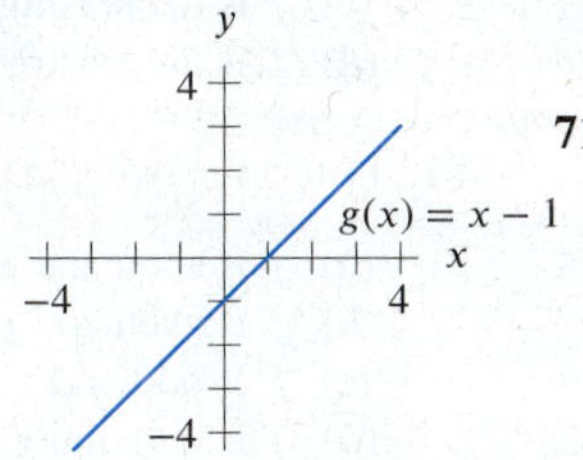


**61.**

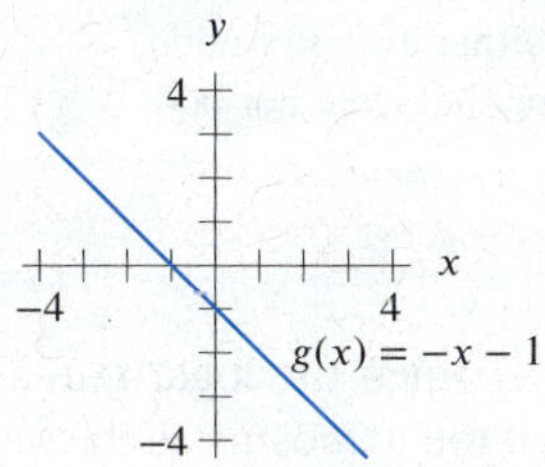


**62.**

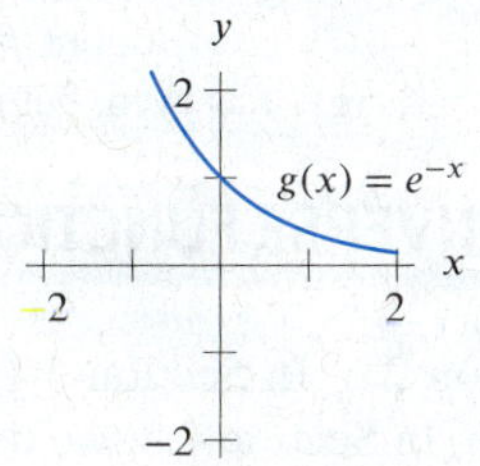


**63.** Find $f(f(1))$ for

$$f(x) = \begin{cases} 2 & \text{if } x \le 0 \\ 3x+1 & \text{if } 0 < x < 2 \\ x^2 - 3 & \text{if } x \ge 2 \end{cases}$$

**64.** Suppose $u(v(x)) = \dfrac{1}{x^2-1}$ and $v(u(x)) = \dfrac{1}{(x-1)^2}$. Find possible formulas for $u(x)$ and $v(x)$.

**65.** If $s(x) = 5 + \dfrac{1}{x+5} + x$, $k(x) = x + 5$, and $s(x) = v(k(x))$, what is $v(x)$?

**66.** Let $f(x) = 12 - 4x$, $g(x) = 1/x$, and $h(x) = \sqrt{x-4}$. Find the domain of the functions:

**(a)** $g(f(x))$ **(b)** $h(f(x))$

**67.** Find possible formulas for $f$ and $g$ given that

$$h(x) = x^2$$
$$h(f(x)) = e^{-2x}$$
$$h(g(x)) = x^2 - 4x + 4.$$

**68.** Find possible formulas for $v$ and $w$ given that

$$u(x) = \sqrt{x}$$
$$v(u(x)) = 3 + \sqrt{2x}$$
$$w(u(x)) = 3 + \sin\left(3\sqrt{x}\right).$$

**69.** Letting $r(x) = 3x^3 - 4x^2$, find $q(x)$ given that $q(r(x)) = \dfrac{8^{x^3}}{16^{x^2}}$.

**70.** You have two money machines, both of which increase any money inserted into them. The first machine doubles your money. The second adds five dollars. The money that comes out is described by $d(x) = 2x$ in the first case and $a(x) = x + 5$ in the second, where $x$ is the number of dollars inserted. The machines can be hooked up so that the money coming out of one machine goes into the other. Find formulas for each of the two possible composition machines. Is one machine more profitable than the other?

**71.** A side effect of a therapeutic drug is raising the patient's heart rate. The relation between $Q$, the amount of drug in the patient's body (in milligrams), and $r$, the patient's heart rate (in beats per minute), is

$$r = f(Q) = 60 + 0.2Q.$$

Over time, the level of the drug in the patient's bloodstream falls. The drug level as a function of time $t$, in hours since the initial injection, is

$$Q = g(t) = 250(0.8)^t.$$

Find a formula for the heart rate $r$ as a function of time $t$.

**72.** A software company issues a bug patch. After $t$ days, the company estimates that the number of installed patches (in thousands) is

$$g(t) = \frac{60}{10 + 40\,(1/2)^{t/10}}.$$

For $a = 10$, $h = 20$, evaluate $\dfrac{g(a+h) - g(a)}{h}$. What does it tell you about the adoption of the patch?

**73.** A car company issues a recall for a model. For every thousand cars, the company estimates that after $t$ weeks,

$$f(t) = 1000\left(1 - 2^{-t/5}\right).$$

have received the repairs. For $a = 5, b = 10$, evaluate $\dfrac{f(b) - f(a)}{b - a}$. What does it tell you about the recall?

**74.** Let $p(t) = 10(0.01)^t$ and $q(t) = \log t^2$. Solve the equation $q(p(t)) = 0$ for $t$.

**75.** Let $f(t) = \sin t$ and $g(t) = 3t - \pi/4$. Solve the equation $f(g(t)) = 1$ for $t$ in the interval $0 \le t \le 2\pi/3$.

**76.** Let $f(x) = \sqrt{x}$ and $g(x) = x^2$. Calculate the domain of $f(g(x))$ and the domain of $g(f(x))$.

**77.** The graphs for $y = f(x)$ and $y = g(x)$ are shown in Figure 10.9.

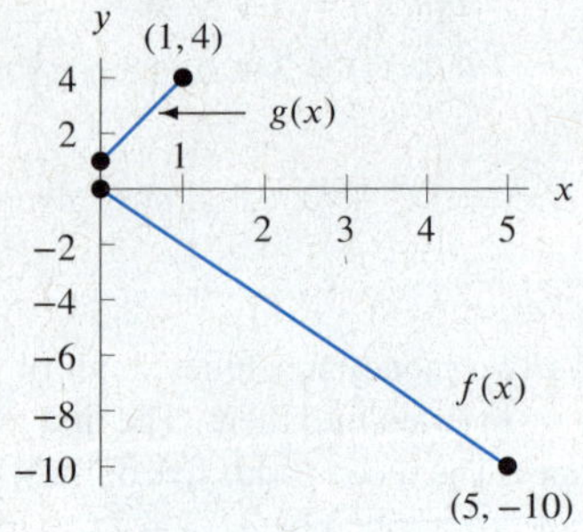


Figure 10.9

(a) Write an equation for $f(x)$.
(b) State the domain and range of $f(x)$.
(c) Write an equation for $g(x)$.
(d) State the domain and range of $g(x)$.
(e) Write an equation for $h(x) = f(g(x))$.
(f) State the domain and range of $h(x)$.
(g) Draw the graph of $h(x)$.

**78.** Which of the following statements must be true in order for the point (2, 5) to be on the graph of $h(x) = g(f(x))$?

(a) The domain of $f$ includes 2.
(b) The domain of $g$ includes 2.
(c) The range of $f$ includes 5.
(d) The range of $g$ includes 5.
(e) The function $g$ is invertible.

**79.** Which of the following statements must be true in order for $h(x) = g(f(x))$ to be defined for all $x$?

(a) The domain of $f$ is all $x$.
(b) The domain of $g$ is all $x$.
(c) The range of $f$ is all $y$.
(d) The range of $g$ is all $y$.
(e) The range of $f$ is within the domain of $g$.

**80.** Determine whether $f(g(x))$ is increasing or decreasing for all $x$.

(a) $f$ is increasing and $g$ is increasing.
(b) $f$ is increasing and $g$ is decreasing.
(c) $f$ is decreasing and $g$ is increasing.
(d) $f$ is decreasing and $g$ is decreasing.

**81.** Determine whether $f(g(x))$ is even or odd.

(a) $f$ is even and $g$ is even.
(b) $f$ is even and $g$ is odd.
(c) $f$ is odd and $g$ is even.
(d) $f$ is odd and $g$ is odd.
(e) $g$ is even, but $f$ is neither even nor odd.
(f) $g$ is odd, but $f$ is neither even nor odd.
(g) $f$ is even, but $g$ is neither even nor odd.

# 10.2 INVERTIBILITY AND PROPERTIES OF INVERSE FUNCTIONS

Inverse functions were introduced in Section 2.5. In Section 5.1, we defined the logarithm as the inverse function of the exponential function. In Section 7.8, we defined the arccosine as the inverse of cosine.

## Definition of Inverse Function

Recall that the statement $f^{-1}(50) = 20$ means that $f(20) = 50$. In fact, the values of $f^{-1}$ are determined in just this way. In general,

> Suppose $Q = f(t)$ is a function with the property that each value of $Q$ determines exactly one value of $t$. Then $f$ has an **inverse function**, $f^{-1}$, and
>
> $$f^{-1}(Q) = t \quad \text{if and only if} \quad Q = f(t).$$
>
> If a function has an inverse, it is said to be **invertible**.

The definitions of the logarithm and of the inverse cosine have the same form as the definition of $f^{-1}$. Since $y = \log x$ is the inverse function of $y = 10^x$, we have

$$x = \log y \quad \text{if and only if} \quad y = 10^x,$$

and since $y = \cos^{-1} t$ is the inverse function of $y = \cos t$,

$$t = \cos^{-1} y \quad \text{if and only if} \quad y = \cos t.$$

**Example 1** Find a solution to the equation $\sin x = 0.8$ using an inverse function.

**Solution** A solution is $x = \sin^{-1}(0.8)$. A calculator (set in radians) gives $x = \sin^{-1}(0.8) \approx 0.927$.

## Notation and Formulas for Inverse Functions

As we saw in Section 2.5, it is sometimes possible to find a formula for an inverse function $f^{-1}$ from a formula for $f$. In abstract mathematical examples, a function $y = f(x)$ often has its inverse function written with $x$ as the independent variable.

**Example 2** Find the inverse of the function.

$$f(x) = \frac{3x}{2x+1}.$$

**Solution** First, we solve the equation $y = f(x)$ for $x$:

$$\begin{aligned} y &= \frac{3x}{2x+1} \\ 2xy + y &= 3x \\ 2xy - 3x &= -y \\ x(2y-3) &= -y \\ x &= \frac{-y}{2y-3} = \frac{y}{3-2y}. \end{aligned}$$

As before, we write $x = f^{-1}(y) = \dfrac{y}{3-2y}$.

Since $y$ is now the independent variable, by convention we rewrite the inverse function with $x$ as the independent variable. We have

$$f^{-1}(x) = \frac{x}{3-2x}.$$

## Non-invertible Functions: Horizontal Line Test

Not every function has an inverse function. A function $Q = f(t)$ has no inverse if it returns the same $Q$-value for two different $t$-values. When that happens, the value of $t$ cannot be uniquely determined from the value of $Q$.

For example, if $q(x) = x^2$ then $q(-3) = 9$ and $q(+3) = 9$. This means that we cannot say what the value $q^{-1}(9)$ would be. (Is it $+3$ or $-3$?) Thus, $q$ is not invertible. In Figure 10.10, notice that the horizontal line $y = 9$ intersects the graph of $q(x) = x^2$ at two different points: $(-3, 9)$ and $(3, 9)$. This corresponds to the fact that the function $q$ returns $y = 9$ for two different $x$-values, $x = +3$ and $x = -3$.

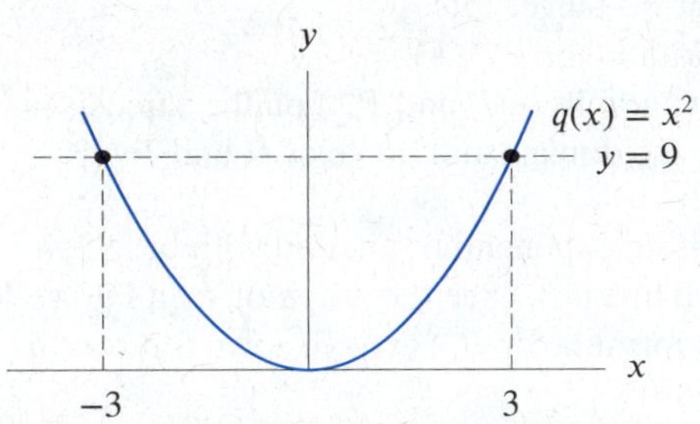


Figure 10.10: The graph of $q(x) = x^2$ fails the horizontal line test

We have the following general result:

**The Horizontal Line Test** If there is a horizontal line that intersects a function's graph in more than one point, then the function does not have an inverse. If every horizontal line intersects a function's graph at most once, then the function has an inverse.

## Evaluating an Inverse Function Graphically

Finding a formula for an inverse function can be difficult. However, this does not mean that the inverse function does not exist. Even without a formula, we may be able to find values of the inverse function.

**Example 3** Let $u(x) = x^3 + x + 1$. Explain why a graph suggests the function $u$ is invertible. Assuming $u$ has an inverse, estimate $u^{-1}(4)$.

**Solution** To show that $u$ is invertible, we could try to find a formula for $u^{-1}$. To do this, we would solve the equation $y = x^3 + x + 1$ for $x$. Unfortunately, this is difficult. However, the graph in Figure 10.11 suggests that $u$ passes the horizontal line test and therefore that $u$ is invertible. To estimate $u^{-1}(4)$, we find an $x$-value such that

$$x^3 + x + 1 = 4.$$

In Figure 10.11, the graph of $y = u(x)$ and the horizontal line $y = 4$ intersect at the point $x \approx 1.213$. Thus, tracing along the graph, we estimate $u^{-1}(4) \approx 1.213$.

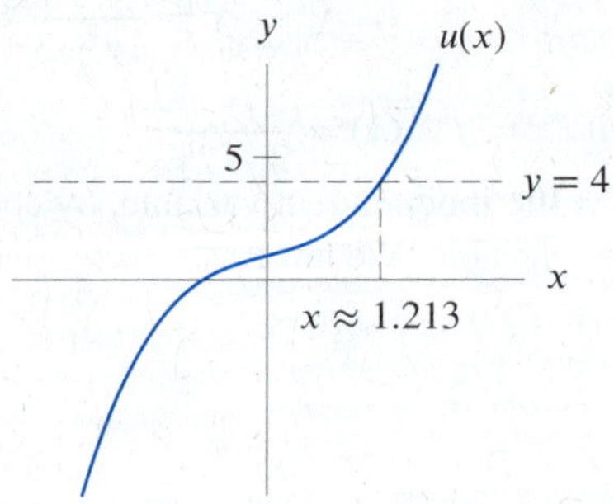


**Figure 10.11**: The graph of $u(x)$ passes the horizontal line test.
Since $u(1.213) \approx 4$, we have $u^{-1}(4) \approx 1.213$

In Example 3, we can approximate $u^{-1}(a)$ for any value of $a$ even though we do not have a formula for $u^{-1}$.

**Example 4** Let $P(x) = 2^x$.

(a) Show that $P$ is invertible.
(b) Find a formula for $P^{-1}(x)$.
(c) Sketch the graphs of $P$ and $P^{-1}$ on the same axes.
(d) What are the domain and range of $P$ and $P^{-1}$?

**Solution**

(a) Since $P$ is an exponential function with base 2, it is always increasing, and therefore passes the horizontal line test. (See the graph of $P$ in Figure 10.12.) Thus, $P$ has an inverse function.
(b) To find a formula for $P^{-1}(x)$, we solve for $x$ in the equation

$$2^x = y.$$

We take the log of both sides to get

$$\log 2^x = \log y$$
$$x \log 2 = \log y$$
$$x = P^{-1}(y) = \frac{\log y}{\log 2}.$$

Thus, we have a formula for $P^{-1}$ with $y$ as the input. To graph $P$ and $P^{-1}$ on the same axes, we write $P^{-1}$ as a function of $x$:

$$P^{-1}(x) = \frac{\log x}{\log 2} = \frac{1}{\log 2} \cdot \log x = 3.322 \log x.$$

(c) Table 10.6 gives values of $P(x)$ for $x = -3, -2, \ldots, 3$. Interchanging the columns of Table 10.6 gives Table 10.7 for $P^{-1}(x)$. We use these tables to sketch Figure 10.12.

**Table 10.6** *Values of* $P(x) = 2^x$

| $x$ | $P(x) = 2^x$ |
|---|---|
| −3 | 0.125 |
| −2 | 0.25 |
| −1 | 0.5 |
| 0 | 1 |
| 1 | 2 |
| 2 | 4 |
| 3 | 8 |

**Table 10.7** *Values of* $P^{-1}(x)$

| $x$ | $P^{-1}(x)$ |
|---|---|
| 0.125 | −3 |
| 0.25 | −2 |
| 0.5 | −1 |
| 1 | 0 |
| 2 | 1 |
| 4 | 2 |
| 8 | 3 |

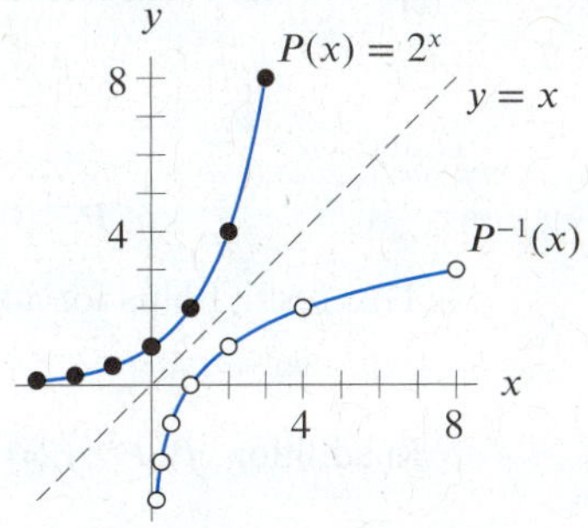


**Figure 10.12:** The graphs of $P(x) = 2^x$ and its inverse are symmetrical across the line $y = x$

(d) The domain of $P$, an exponential function, is all real numbers, and its range is all positive numbers. The domain of $P^{-1}$, a logarithmic function, is all positive numbers and its range is all real numbers.

## The Graph, Domain, and Range of an Inverse Function

In Figure 10.12, we see that the graph of $P^{-1}$ is the mirror image of the graph of $P$ across the line $y = x$. In general, this is true if the $x$- and $y$-axes have the same scale. To understand why this occurs, consider how a function is related to its inverse.

If $f$ is an invertible function with, for example, $f(2) = 5$, then $f^{-1}(5) = 2$. Thus, the point $(2, 5)$ is on the graph of $f$ and the point $(5, 2)$ is on the graph of $f^{-1}$. Generalizing, if $(a, b)$ is any point on the graph of $f$, then $(b, a)$ is a point on the graph of $f^{-1}$. Figure 10.13 shows how reflecting the point $(a, b)$ across the line $y = x$ gives the point $(b, a)$. Consequently, the graph of $f^{-1}$ is the reflection of the graph of $f$ across the line $y = x$.

Notice that outputs from a function are inputs to its inverse function. Similarly, outputs from the inverse function are inputs to the original function. This is expressed in the statement $f^{-1}(b) = a$ if and only if $f(a) = b$, and also in the fact that we can obtain a table for $f^{-1}$ by interchanging the columns of a table for $f$. Consequently, the domain and range for $f^{-1}$ are obtained by interchanging the domain and range of $f$. In other words,

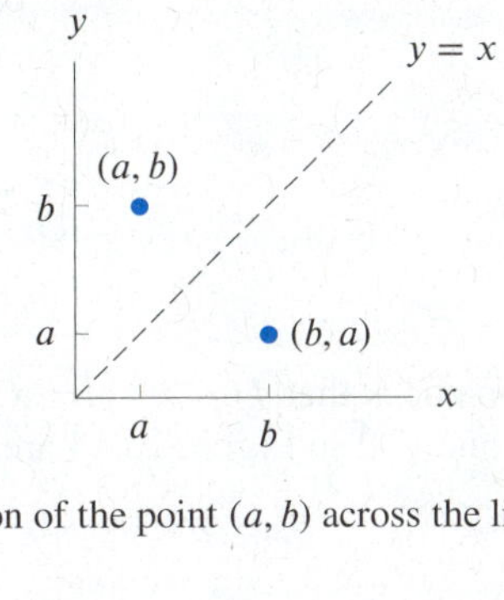


**Figure 10.13:** The reflection of the point $(a, b)$ across the line $y = x$ is the point $(b, a)$

Graph of $f^{-1}$ is reflection of graph of $f$ across the line $y = x$.

Domain of $f^{-1}$ = Range of $f$ and Range of $f^{-1}$ = Domain of $f$.

As we saw in Example 4, the function $P$ has all real numbers as its domain and all positive numbers as its range; the function $P^{-1}$ has all positive numbers as its domain and all real numbers as its range.

## A Property of Inverse Functions

The fact that Tables 10.6 and 10.7 contain the same values, but with the columns switched, reflects the special relationship between the values of $P(x)$ and $P^{-1}(x)$. For the function in Example 4,

$$P^{-1}(2) = 1 \quad \text{and} \quad P(1) = 2 \quad \text{so} \quad P^{-1}(P(1)) = 1,$$

and

$$P^{-1}(0.25) = -2 \quad \text{and} \quad P(-2) = 0.25 \quad \text{so} \quad P^{-1}(P(-2)) = -2.$$

This result holds for any input $x$, so in general,

$$P^{-1}(P(x)) = x.$$

In addition, $P(P^{-1}(2)) = 2$ and $P(P^{-1}(0.25)) = 0.25$, and for any $x$

$$P(P^{-1}(x)) = x.$$

Similar reasoning holds for any other invertible function, suggesting the general result:

If $y = f(x)$ is an invertible function and $y = f^{-1}(x)$ is its inverse, then
- $f^{-1}(f(x)) = x$ for all values of $x$ for which $f(x)$ is defined,
- $f(f^{-1}(x)) = x$ for all values of $x$ for which $f^{-1}(x)$ is defined.

This property tell us that composing a function and its inverse function returns the original value as the end result. We can use this property to decide whether two functions are inverses.

**Example 5** (a) Check that $f(x) = \dfrac{x}{2x+1}$ and $f^{-1}(x) = \dfrac{x}{1-2x}$ are inverse functions of each other.
(b) Graph $f$ and $f^{-1}$ on axes with the same scale. What are the domains and ranges of $f$ and $f^{-1}$?

**Solution** (a) To check that these functions are inverses, we compose

$$f^{-1}(f(x)) = \frac{f(x)}{1-2f(x)} = \frac{\frac{x}{2x+1}}{1-2\left(\frac{x}{2x+1}\right)} = \frac{\frac{x}{2x+1}}{\frac{2x+1}{2x+1} - \frac{2x}{2x+1}} = \frac{\frac{x}{2x+1}}{\frac{1}{2x+1}} = x.$$

Similarly, you can check that $f(f^{-1}(x)) = x$.

(b) The graphs of $f$ and $f^{-1}$ in Figure 10.14 are symmetric about the line $y = x$.

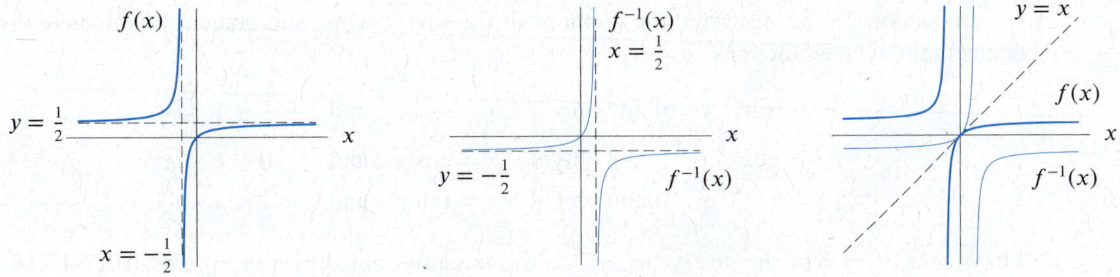


**Figure 10.14:** The graph of $f(x) = x/(2x+1)$ and the inverse $f^{-1}(x) = x/(1-2x)$

The function $f(x) = x/(2x+1)$ is undefined at $x = -1/2$, so its domain consists of all real numbers except $-1/2$. Figure 10.14 suggests that $f$ has a horizontal asymptote at $y = 1/2$ which it does not cross and that its range is all real numbers except $1/2$.

Because the inverse function $f^{-1}(x) = x/(1-2x)$ is undefined at $x = 1/2$, its domain is all real numbers except $1/2$. Note that this is the same as the range of $f$. The graph of $f^{-1}$ appears to have a horizontal asymptote which it does not cross at $y = -1/2$, suggesting that its range is all real numbers except $-1/2$. Note that this is the same as the domain of $f$.

The ranges of the functions $f$ and $f^{-1}$ can be confirmed algebraically.

## Restricting the Domain to Create an Inverse

A function that fails the horizontal line test is not invertible. For this reason, the function $f(x) = x^2$ does not have an inverse function. However, by considering only part of the graph of $f$, we can eliminate the duplication of $y$-values. Suppose we consider the half of the parabola with $x \geq 0$. See Figure 10.15. This part of the graph does pass the horizontal line test because there is only one (positive) $x$-value for each $y$-value in the range of $f$.

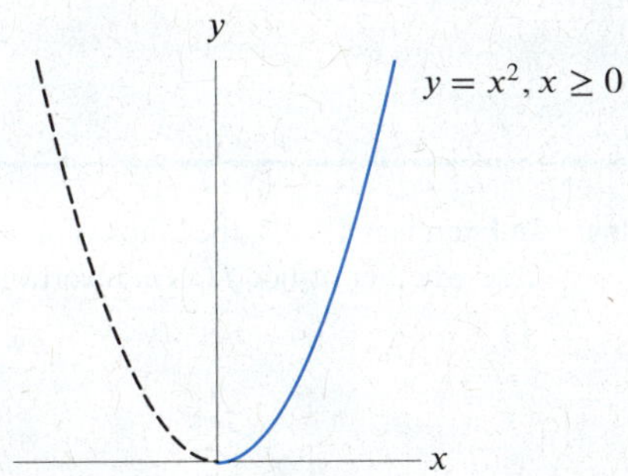


**Figure 10.15**

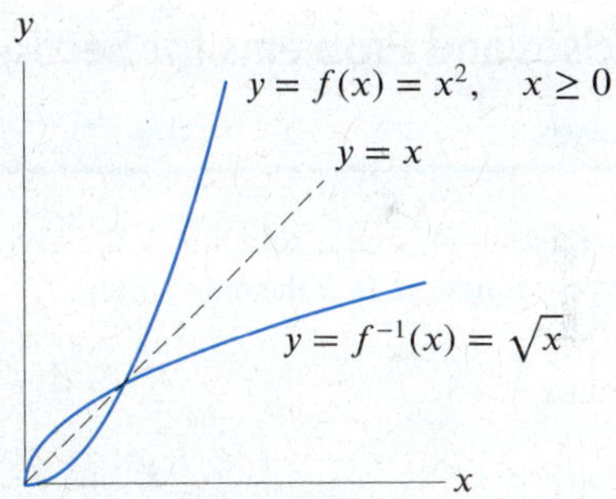


**Figure 10.16**

We can find an inverse for $f(x) = x^2$ on its restricted domain,[1] $x \geq 0$. Using the fact that $x \geq 0$ and solving $y = x^2$ for $x$ gives

$$x = \sqrt{y}.$$

Thus a formula for the inverse function is

$$x = f^{-1}(y) = \sqrt{y}.$$

Rewriting the formula for $f^{-1}$ with $x$ as the input, we have

$$f^{-1}(x) = \sqrt{x}.$$

The graphs of $f$ and $f^{-1}$ are shown in Figure 10.16. Note that the domain of $f$ is the the range of $f^{-1}$, and the domain of $f^{-1}$ $(x \geq 0)$ is the range of $f$.

[1] Technically, changing the domain results in a new function, but we will continue to call it $f(x)$.

In Section 7.8 we restricted the domains of the sine, cosine, and tangent functions in order to define their inverse functions:

$$y = \sin^{-1} x \quad \text{if and only if} \quad x = \sin y \quad \text{and} \quad -\frac{\pi}{2} \le y \le \frac{\pi}{2}$$
$$y = \cos^{-1} x \quad \text{if and only if} \quad x = \cos y \quad \text{and} \quad 0 \le y \le \pi$$
$$y = \tan^{-1} x \quad \text{if and only if} \quad x = \tan y \quad \text{and} \quad -\frac{\pi}{2} < y < \frac{\pi}{2}.$$

The graphs of each of the inverse trigonometric functions are shown in Figures 10.17-10.19. Note the symmetry about the line $y = x$ for each trigonometric function and its inverse.

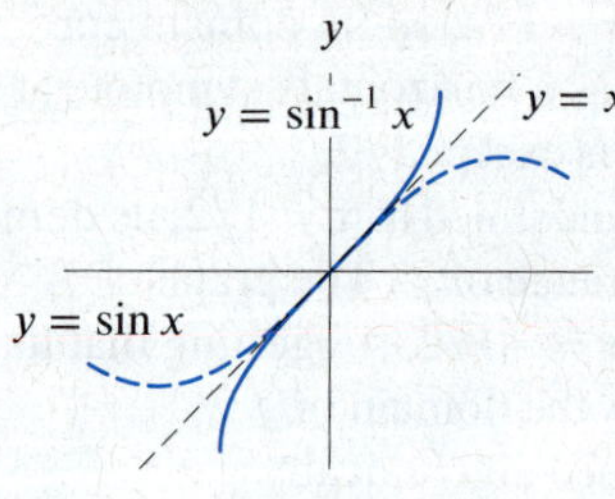


Figure 10.17

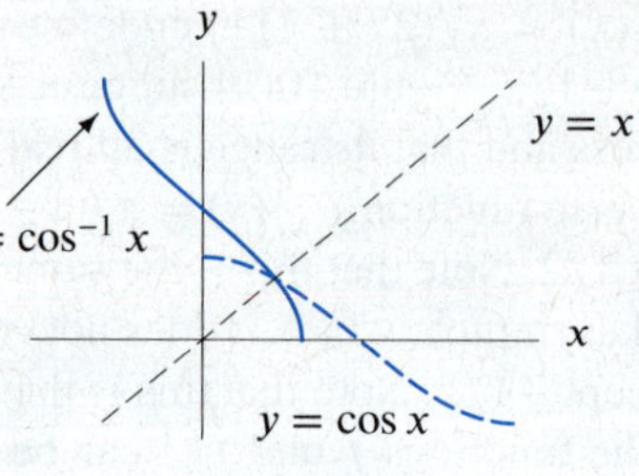


Figure 10.18

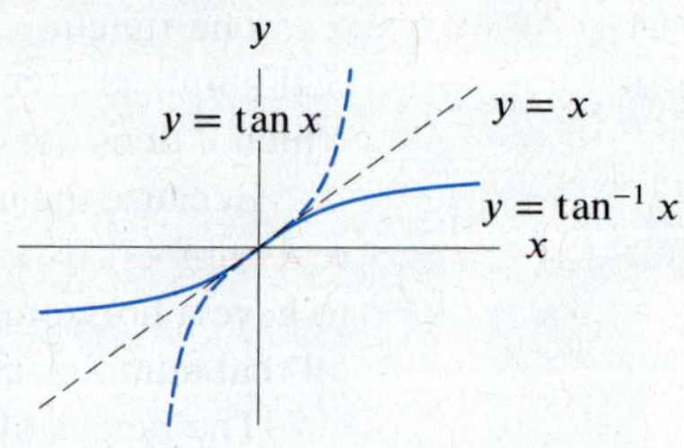


Figure 10.19

In each case, the restricted domain of the function is the range of the inverse function. In addition, the domain of the inverse is the range of the original function. For example,

$$y = \sin x \quad \text{has restricted domain } -\frac{\pi}{2} \le x \le \frac{\pi}{2} \quad \text{and} \quad \text{range} \quad -1 \le y \le 1$$
$$y = \sin^{-1} x \quad \text{has domain} \quad -1 \le x \le 1 \quad \text{and} \quad \text{range} \quad -\frac{\pi}{2} \le y \le \frac{\pi}{2}.$$

## Exercises and Problems for Section 10.2

### Exercises

In Exercises 1–4, find a solution to the given equation using an inverse function. Is it the only solution?

**1.** $\tan x = -5$

**2.** $e^x = 3$

**3.** $\log x = 1.2$

**4.** $\cos x = -0.2$

In Exercises 5–12, find a formula for the inverse function.

**5.** $f(x) = x + 5$

**6.** $g(x) = 1 - x$

**7.** $h(x) = \sqrt{x}$

**8.** $j(x) = \dfrac{1}{x}$

**9.** $f(x) = 3x - 7$

**10.** $k(x) = \dfrac{x}{x-1}$

**11.** $m(x) = \dfrac{1}{x} - x$

**12.** $h(x) = \dfrac{2x+1}{3x-2}$

In Exercises 13–16, the graph of a function $f(x)$ is shown. Decide whether or not $f(x)$ is invertible.

**13.**

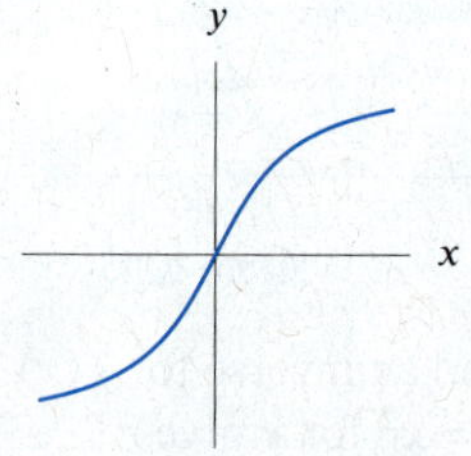


**14.**

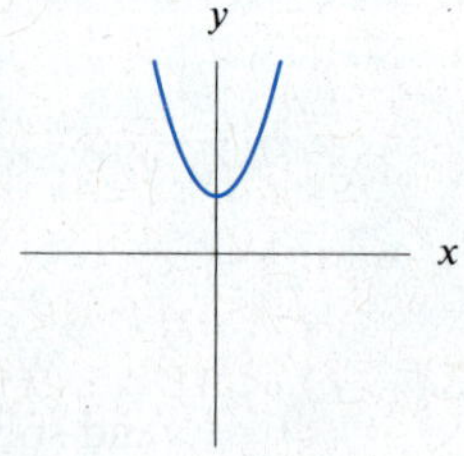


**15.**

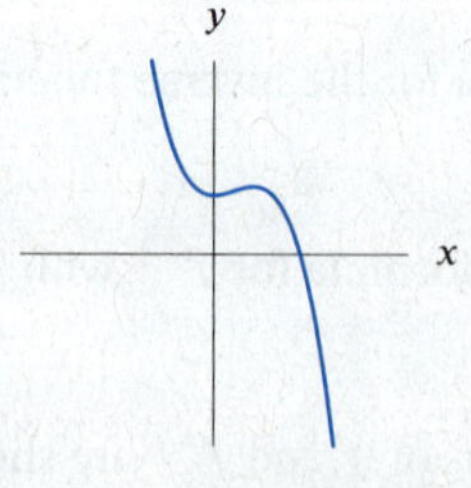


**16.**

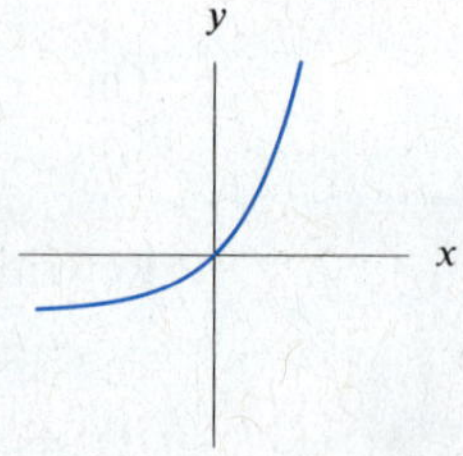

In Exercises 17–22, use a graph to decide whether or not the function is invertible.

**17.** $y = x^6 + 2x^2 - 10$

**18.** $y = x^4 - 6x + 12$

**19.** $y = e^{x^2}$

**20.** $y = \cos(x^3)$

**21.** $y = |x|$

**22.** $y = x + \ln x$

**23.** In Figure 10.20 are the graphs of three functions and their inverse functions. Match each function to its inverse.

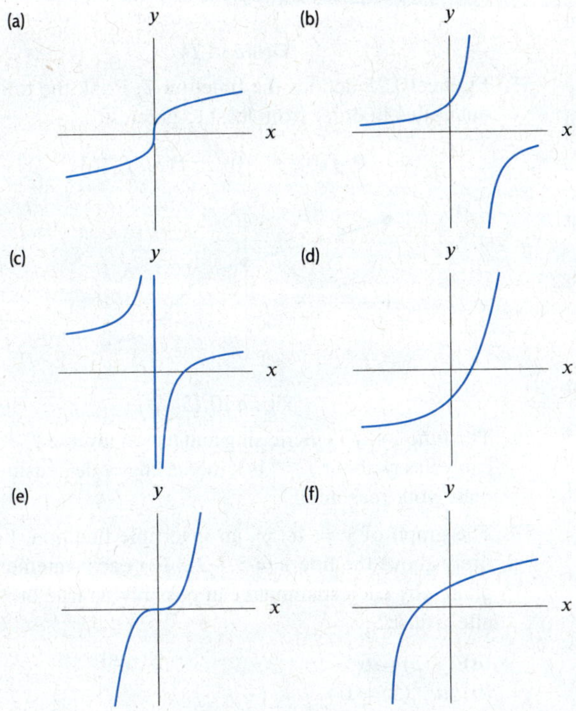


Figure 10.20

In Exercises 24–25, sketch a graph of the inverse function.

**24.**

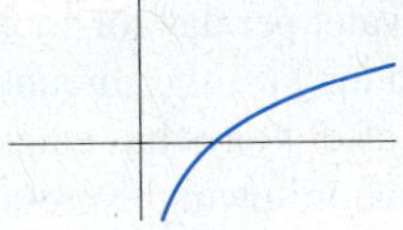

Figure 10.21

**25.**

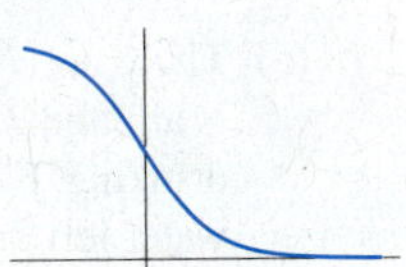

Figure 10.22

**26.** Check that the functions $R = f(T) = 150 + 5T$ and $T = f^{-1}(R) = \frac{1}{5}R - 30$ satisfy the identities

$$f^{-1}(f(T)) = T \quad \text{and} \quad f(f^{-1}(R)) = R.$$

In Exercises 27–34, check that the functions are inverses.

**27.** $f(x) = \dfrac{x}{4} - \dfrac{3}{2}$ and $g(t) = 4\left(t + \dfrac{3}{2}\right)$

**28.** $f(x) = 1 + 7x^3$ and $f^{-1}(x) = \sqrt[3]{\dfrac{x-1}{7}}$

**29.** $g(x) = 1 - \dfrac{1}{x-1}$ and $g^{-1}(x) = 1 + \dfrac{1}{1-x}$

**30.** $h(x) = \sqrt{2x}$ and $k(t) = \dfrac{t^2}{2}$, for $x, t \geq 0$

**31.** $f(x) = e^{x+1}$ and $f^{-1}(x) = \ln x - 1$

**32.** $f(x) = e^{2x}$ and $f^{-1}(x) = \dfrac{\ln x}{2}$

**33.** $f(x) = e^{x/2}$ and $f^{-1}(x) = 2\ln x$

**34.** $f(x) = \ln(x/2)$ and $f^{-1}(x) = 2e^x$

In Exercises 35–43, find a formula for the inverse function. Assume these functions are defined on domains on which they are invertible.

**35.** $l(x) = \sqrt{1 - 2x^2}$

**36.** $n(x) = (1 + x^2)^2$

**37.** $o(x) = 1/(1 + \dfrac{1}{x})$

**38.** $j(x) = \sqrt{1 + \sqrt{x}}$

**39.** $k(x) = \dfrac{3 - \sqrt{x}}{\sqrt{x} + 2}$

**40.** $g(x) = \dfrac{\ln x - 5}{2\ln x + 7}$

**41.** $h(x) = \log\left(\dfrac{x+5}{x-4}\right)$

**42.** $f(x) = \cos\sqrt{x}$

**43.** $g(x) = 2^{\sin x}$

## Problems

**44.** If $C = f(q)$ gives the cost, in dollars, to manufacture $q$ items, what does $f^{-1}(C)$ represent? What are its units?

**45.** If $P = f(t)$ gives the population of a city, in thousands, as a function of time, $t$, in years, what does $f^{-1}(P)$ represent? What are its units?

**46.** The balance $B$, in dollars, in an account after $t$ years with an initial deposit of \$500 that pays 4% compounded annually is $B = f(t) = 500(1.04)^t$.

**(a)** Find a formula for $t = f^{-1}(B)$.

**(b)** What does the $f^{-1}(B)$ represent in terms of the account?

**47.** Let $R = f(T) = 150 + 5T$ give the resistance of a circuit element as a function of temperature.

**(a)** Find a formula for $T = f^{-1}(R)$.
**(b)** Make a table of values showing values of $f$. Make another table showing values of $f^{-1}$. Explain the relationship between the tables.

**48.** Let $C = f(q) = 200+0.1q$ give the cost in dollars to manufacture $q$ kg of a chemical. Find and interpret $f^{-1}(C)$.

**49.** Let $P = f(t) = 10e^{0.02t}$ give the population in millions at time $t$ in years. Find and interpret $f^{-1}(P)$.

**50.** The police can determine the speed a car was traveling from the length of the skid marks it leaves. The function they use is $S = f(L) = 2\sqrt{(5L)}$ where $S$ is speed (mph) and $L$ is the length of the skid marks (feet).

**(a)** If skid marks of length 125 feet are measured, what was the speed of the car?
**(b)** Find a formula for the inverse function.
**(c)** If the car had been traveling at 80 mph, how long would the skid marks be?

**51.** Let $P = f(t) = 37.8(1.044)^t$ be the population of a town (in thousands) in year $t$.

**(a)** Describe the town's population in words.
**(b)** Evaluate $f(50)$. What does this quantity tell you about the population?
**(c)** Find a formula for $f^{-1}(P)$ in terms of $P$.
**(d)** Evaluate $f^{-1}(50)$. What does this quantity tell you about the population?

**52.** Let $f(x) = 5^x$.

**(a)** Evaluate $f(3)$ and $f^{-1}(1/25)$ exactly.
**(b)** Approximate the value of $f^{-1}(10)$.

**53.** Figure 10.23 is the graph of a function $f(x)$. Use it to find the following:

**(a)** $f^{-1}(-1)$ **(b)** $f^{-1}(0)$
**(c)** $f^{-1}(3)$ **(d)** $f^{-1}(-2)$
**(e)** $f^{-1}(f(\frac{1}{2}))$ **(f)** $f(f^{-1}(-0.05))$

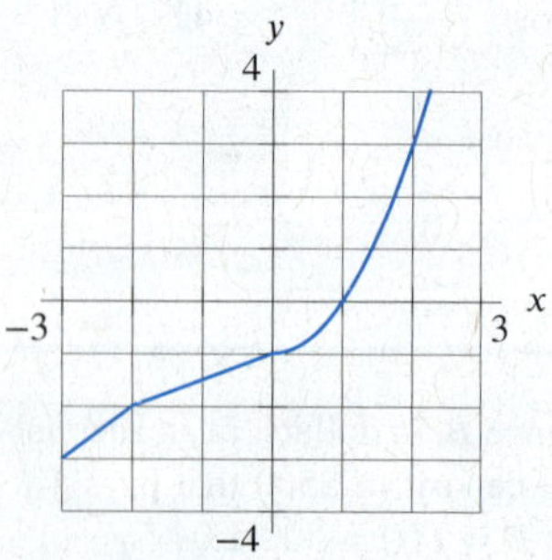


Figure 10.23

**54.** Figure 10.24 shows a graph of the function $f$. Sketch a graph of $f^{-1}$.

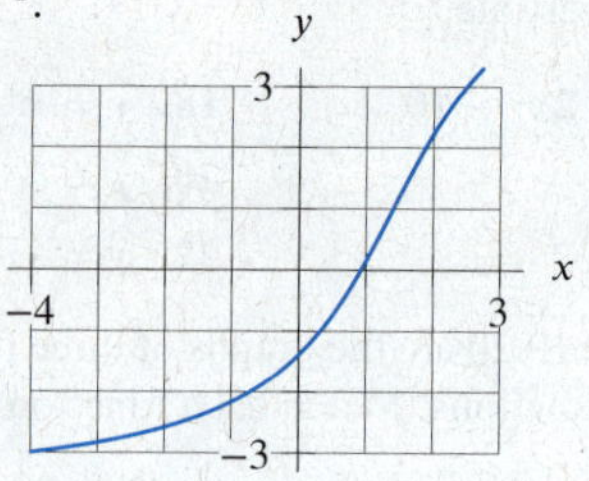


Figure 10.24

**55.** Figure 10.25 defines the function $f$. Rank the following quantities in order from least to greatest:

$$0, f(0), f^{-1}(0), 3, f(3), f^{-1}(3).$$

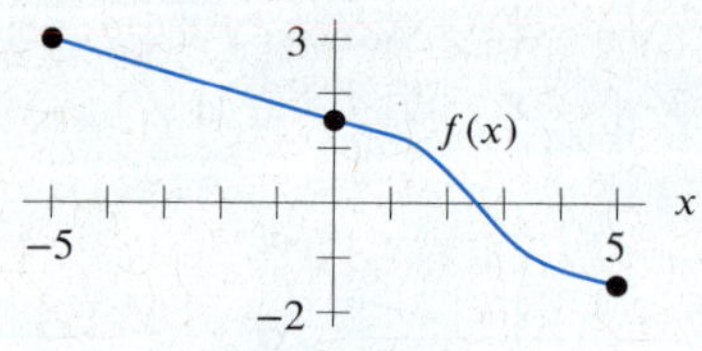


Figure 10.25

**56.** The function $f$ is increasing and has an inverse $f^{-1}$. What can you say about $f^{-1}$? Is it increasing or decreasing? Explain your reasoning.

**57.** The graph of $y = w(x)$, an invertible function, lies entirely above the line $y = 3 + 2x$. For each statement, explain why each statement can possibly be true or is definitely false?

**(a)** $w(2) = 6$
**(b)** $w^{-1}(5) = 0$
**(c)** $\left(w^{-1}(-6)\right)^{-1} = -2$
**(d)** $\Delta w/\Delta x = 4$ on $-2 \leq x \leq 3$

**58.** A research facility on the Isle of Shoals has 800 gallons of fresh water for a two-month period.

**(a)** There are 7 members of the research team and each is allotted 2 gallons of water per day for cooking and drinking. Find a formula for $f(t)$, the amount of fresh water left on the island after $t$ days has elapsed.
**(b)** Evaluate and interpret the following expressions:
(i) $f(0)$ (ii) $f^{-1}(0)$
(iii) $t$ if $f(t) = \frac{1}{2}f(0)$ (iv) $800 - f(t)$

**59.** For a T-shirt company, $q = f(p)$ represents the number of T-shirts sold at a if the price is $p$ dollars per shirt, and $R = g(n)$ represents the revenue generated, in hundreds of dollars, from selling $n$ T-shirts. What is a practical interpretation of each statement, (a)–(f)?

**(a)** $f(20)$ **(b)** $f^{-1}(20)$
**(c)** $g(10) = 2$ **(d)** $g^{-1}(10) = 50$
**(e)** $f(10) + 5$ **(f)** $f^{-1}(10) + 5$

**60.** The noise level, $N$, of a sound in decibels is given by

$$N = f(I) = 10\log\left(\frac{I}{I_0}\right),$$

where $I$ is the intensity of the sound and $I_0$ is a constant. Find and interpret $f^{-1}(N)$.

**61.** The rating, $r$, of an earthquake of intensity $I$ is given by $r = f(I) = \log(I/I_0)$, where $I_0$ is a constant. Find and interpret $f^{-1}(r)$.

**62.** Let $f(x) = e^x$ and $g(x) = \ln x$.

**(a)** Find $f(g(x))$ and $g(f(x))$. What can you conclude about the relationship between these two functions?
**(b)** Graph $f(x)$ and $g(x)$ together on axes with the same scales. What is the line of symmetry of the graph?

**63.** Let $p(t) = 10^t$ and $q(t) = \log t$.

**(a)** Find $p(q(t))$ and $q(p(t))$. What can you conclude about the relationship between these two functions?
**(b)** Graph $p(t)$ and $q(t)$ together on axes with the same scales. What is the line of symmetry of the graph?

**64.** Table 10.8 gives the number $N$ of cows in a herd.

**(a)** Find an exponential function that approximates the data.
**(b)** Find the inverse function of the function in part (a).
**(c)** When do you predict that the herd will contain 400 cows?

**Table 10.8**

| $t$ (years) | 0 | 1 | 2 |
|---|---|---|---|
| $N = P(t)$ (cows) | 150 | 165 | 182 |

**65.** Values of $f$ and $g$ are in Table 10.9. Based on this table:

**(a)** Is $f(x)$ invertible? If not, explain why; if so, construct a table of values of $f^{-1}(x)$ for all values of $x$ for which $f^{-1}(x)$ is defined.
**(b)** Answer the same question as in part (a) for $g(x)$.
**(c)** Make a table of values for $h(x) = f(g(x))$, with $x = -3, -2, -1, 0, 1, 2, 3$.
**(d)** Explain why you cannot define a function $j(x)$ by the formula $j(x) = g(f(x))$.

**Table 10.9**

| $x$ | −3 | −2 | −1 | 0 | 1 | 2 | 3 |
|---|---|---|---|---|---|---|---|
| $f(x)$ | 9 | 7 | 6 | −4 | −5 | −8 | −9 |
| $g(x)$ | 3 | 1 | 3 | 2 | −3 | −1 | 3 |

**66.** There is a linear relationship between the number of units, $N(x)$, of a product that a company sells and the amount of money, $x$, spent on advertising. If the company spends \$25,000 on advertising, it sells 400 units, and for each additional \$5,000 spent, it sells 20 units more.

**(a)** Calculate and interpret $N(20{,}000)$.
**(b)** Find a formula for $N(x)$ in terms of $x$.
**(c)** Give interpretations of the slope and the $x$- and $y$-intercepts of $N(x)$ if possible.
**(d)** Calculate and interpret $N^{-1}(500)$.
**(e)** An internal audit reveals that the profit made by the company on the sale of 10 units of its product, before advertising costs have been accounted for, is \$2,000. What are the implications regarding the company's advertising campaign? Discuss.

**67.** A hot brick is removed from a kiln at 200°C above room temperature. Over time, the brick cools off. After 2 hours have elapsed, the brick is 20°C above room temperature. Let $t$ be the time in hours since the brick was removed from the kiln. Let $y = H(t)$ be the difference between the brick's and the room's temperature at time $t$. Assume that $H(t)$ is an exponential function.

**(a)** Find a formula for $H(t)$.
**(b)** How many degrees does the brick's temperature drop during the first quarter hour? During the next quarter hour?
**(c)** Find and interpret $H^{-1}(y)$.
**(d)** How much time elapses before the brick's temperature is 5°C above room temperature?
**(e)** Interpret the physical meaning of the horizontal asymptote of $H(t)$.

**68.** The *quadratic mean* (or RMS) of $x$ and the constant $A$ is given by $f(x) = 0.5(x^2 + A^2)^{0.5}$. Assuming $x \geq 0$, find a formula for $f^{-1}(x)$.

**69.** The *harmonic mean* of $x$ and the constant $A$ is given by $f(x) = 0.5(x^{-1} + A^{-1})^{-1}$. Find a formula for $f^{-1}(x)$.

Problems 70–71 involve the *Lambert W function*,[2] defined by $W(x) = f^{-1}(x)$ where $f(x) = xe^x$ with $x \geq -1$. Since $f\left(f^{-1}(x)\right) = x$, this means $W(x)e^{W(x)} = x$.

**70.** Find three points on the graph of $W$ by finding $f(-1)$, $f(0)$, and $f(1)$.

**71.** Given that $W(10) = 1.746$, find a solution to the equation $te^{2t} = 5$.

**72.** Suppose $P = f(t)$ is the population (in thousands) in year $t$, and that $f(7) = 13$ and $f(12) = 20$.

**(a)** Find a formula for $f(t)$ assuming $f$ is exponential.
**(b)** Find a formula for $f^{-1}(P)$.
**(c)** Evaluate $f(25)$ and $f^{-1}(25)$. Explain what these expressions mean in terms of population.

[2] See http://en.wikipedia.org/wiki/Lambert%27s_W_function, accessed April 30, 2008.

**73.** A 100-ml solution contains 99% alcohol and 1% water. Let $y = C(x)$ be the concentration of alcohol in the solution after $x$ ml of alcohol are removed, so

$$C(x) = \frac{\text{Amount of alcohol}}{\text{Amount of solution}}.$$

**(a)** What is $C(0)$?
**(b)** Find a formula in terms of $x$ for $C(x)$.
**(c)** Find a formula in terms of $y$ for $C^{-1}(y)$.
**(d)** Explain the physical significance of $C^{-1}(y)$.

**74.** **(a)** How much alcohol do you think should be removed from the 99% solution in Problem 73 in order to obtain a 98% solution? (Make a guess.)
**(b)** Express the exact answer to part (a) using the function $C^{-1}$ you found in Problem 73.
**(c)** Determine the exact answer to part (a). Are you surprised by your result?

**75.** A gymnast at Ringling Bros. and Barnum & Bailey Circus is fired straight up in the air from a cannon. While she is in the air, a trampoline is moved into the spot where the cannon was. Figure 10.26 is a graph of the gymnast's height, $h$, as a function of time, $t$.

**(a)** Approximately what is her maximum height?
**(b)** Approximately when does she land on the trampoline?
**(c)** Restrict the domain of $h(t)$ so that $h(t)$ has an inverse. That is, pick a piece of the graph on which $h(t)$ does have an inverse. Graph this new restricted function.
**(d)** Change the story to go with your graph in part (c).
**(e)** Graph the inverse of the function in part (c). Explain in your story why it makes sense that the inverse is a function.

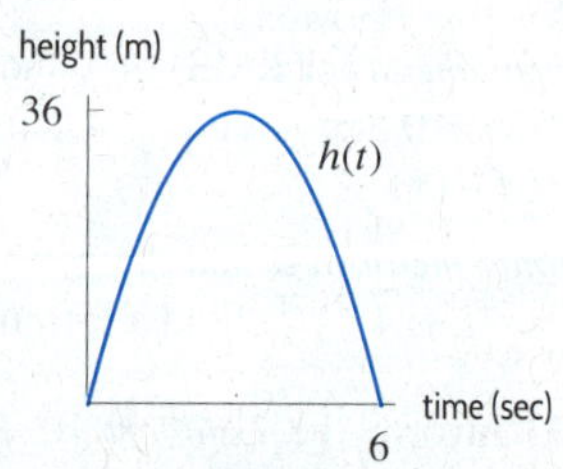


Figure 10.26

In Problems 76–81, what does the statement tell you about caffeine and a person's pulse? The pulse in beats per minute (bpm) of a healthy person fifteen minutes after consuming $q$ milligrams of caffeine is given by $r = f(q)$. The amount of caffeine in a cup of coffee is $q_c$, and $r_c = f(q_c)$. Assume that $f$ is an increasing function for nontoxic levels of caffeine.

**76.** $f(2q_c)$

**77.** $f^{-1}(r_c + 20)$

**78.** $2f^{-1}(r_c) + 20$

**79.** $f(q_c) - f(0)$

**80.** $f^{-1}(r_c + 20) - q_c$

**81.** $f^{-1}(1.1f(q_c))$

**82.** Figure 10.27 is the graph of a function $f(x)$.

**(a)** Assuming this is the whole graph, what is the largest domain on which you can restrict $f(x)$ so it has an inverse?
**(b)** Sketch the graph of $f^{-1}(x)$ with $f(x)$ restricted to the domain you specified in (a).

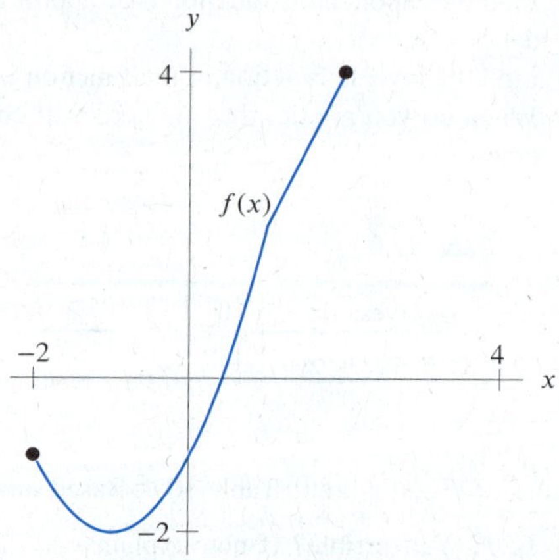


Figure 10.27

# 10.3 COMBINATIONS OF FUNCTIONS

Like numbers, functions can be combined using addition, subtraction, multiplication, and division.

## The Difference of Two Functions Defined by Formulas: A Measure of Prosperity

We can define new functions as the sum or difference of two functions. In Chapter 4, we discussed Thomas Malthus, who predicted widespread food shortages because he believed that human populations increase exponentially, whereas the food supply increases linearly. We considered a country with population $P(t)$ million in year $t$. The population is initially 2 million and grows at the rate of 4% per year, so

$$P(t) = 2(1.04)^t.$$

Let $N(t)$ be the number of people (in millions) that the country can feed in year $t$. The annual food supply is initially adequate for 4 million people and it increases by enough for an additional 0.5 million people every year. Thus

$$N(t) = 4 + 0.5t.$$

This country first experiences shortages in about 78 years. (See Figure 10.28.) When is it most prosperous?

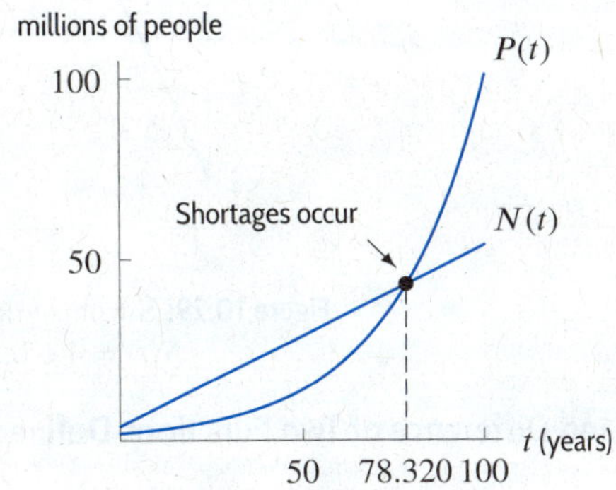


**Figure 10.28:** Predicted population, $P(t)$, and number of people who can be fed, $N(t)$, over a 100-year period

The answer depends on how we decide to measure prosperity. We could measure prosperity in one of the following ways:

- By the food surplus—that is, the amount of food the country has over and above its needs. This surplus food could be warehoused or exported in trade.
- By the per-capita food supply—that is, how much food there is per person. (The term *per capita* means per person, or literally, "per head.") This indicates the portion of the country's wealth each person might enjoy.

First, we choose to measure prosperity in terms of food surplus, $S(t)$, in year $t$, where

$$S(t) = \underbrace{\text{Number of people that can be fed}}_{N(t)} - \underbrace{\text{Number of people living in the country}}_{P(t)}$$

so

$$S(t) = N(t) - P(t).$$

For example, to determine the surplus in year $t = 25$, we evaluate

$$S(25) = N(25) - P(25).$$

Since $N(25) = 4 + 0.5(25) = 16.5$ and $P(25) = 2(1.04)^{25} \approx 5.332$, we have

$$S(25) \approx 16.5 - 5.332 = 11.168.$$

Thus, in year 25 the food surplus could feed 11.168 million additional people.

We use the formulas for $N$ and $P$ to find a formula for $S$:

$$S(t) = \underbrace{N(t)}_{4+0.5t} - \underbrace{P(t)}_{2(1.04)^t},$$

so

$$S(t) = 4 + 0.5t - 2(1.04)^t.$$

A graph of $S$ is shown in Figure 10.29. The maximum surplus occurs sometime during the 48th year. In that year, there is surplus food sufficient for an additional 14.865 million people.

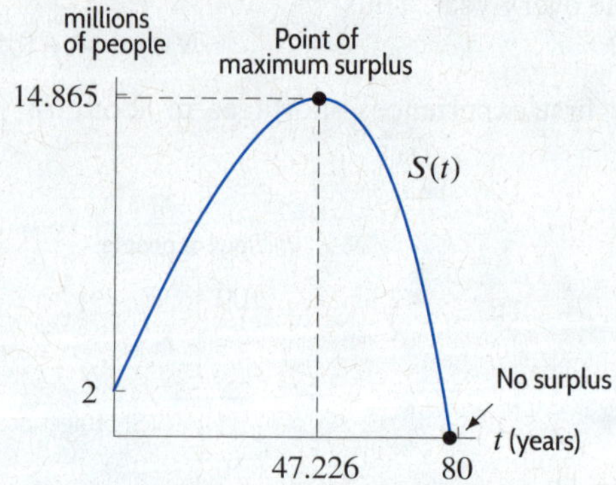


Figure 10.29: Surplus graphed using the formula $S(t) = 4 + 0.5t - 2(1.04)^t$

## The Sum and Difference of Two Functions Defined by Graphs

How does the graph of the surplus function $S$, shown in Figure 10.29, relate to the graphs of $N$ and $P$ in Figure 10.28? Since

$$S(t) = N(t) - P(t),$$

the value of $S(t)$ is represented graphically as the vertical distance between the graphs of $N(t)$ and $P(t)$. See Figure 10.30. Figure 10.31 shows the surplus plotted against time.

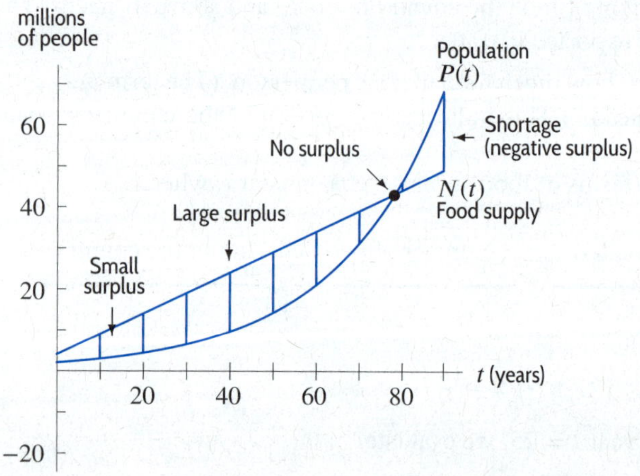


Figure 10.30: Surplus, $S(t) = N(t) - P(t)$, as vertical distance between $N(t)$ and $P(t)$ graphs

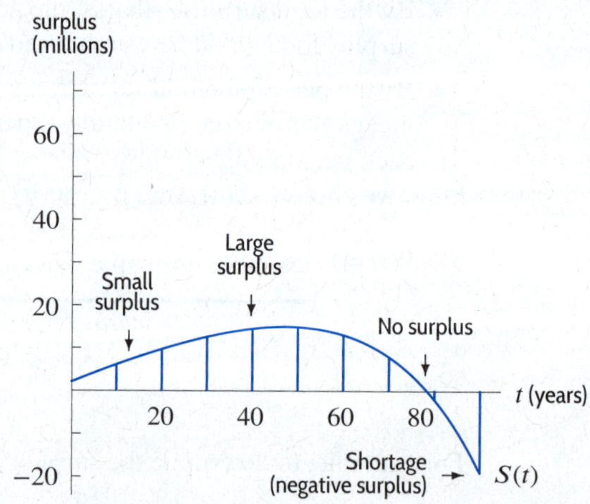


Figure 10.31: Surplus as a function of time

From year $t = 0$ to $t \approx 78.320$, the food supply is more than the population needs. Therefore the surplus, $S(t)$, is positive on this time interval. At time $t = 78.320$, the food supply is exactly sufficient for the population, so $S(t) = 0$, resulting in the horizontal intercept $t = 78.320$ on the graph of $S(t)$ in Figure 10.31. For times $t > 78.320$, the food supply is less than the population needs. Therefore the surplus is negative, representing a food shortage.

In the next example we consider a sum of two functions.

**Example 1** Let $f(x) = x$ and $g(x) = \dfrac{1}{x}$. By adding vertical distances on the graphs of $f$ and $g$, sketch

$$h(x) = f(x) + g(x) \quad \text{for } x > 0.$$

**Solution** The graphs of $f$ and $g$ are shown in Figure 10.32. Since both functions are positive, for each value of $x$, we can add the vertical distances that represent $f(x)$ and $g(x)$ to get a point on the graph of $h(x)$. Compare the graph of $h(x)$ to the values shown in Table 10.10.

**Table 10.10** *Adding function values*

| $x$ | $\frac{1}{4}$ | $\frac{1}{2}$ | 1 | 2 | 4 |
|---|---|---|---|---|---|
| $f(x) = x$ | $\frac{1}{4}$ | $\frac{1}{2}$ | 1 | 2 | 4 |
| $g(x) = 1/x$ | 4 | 2 | 1 | $\frac{1}{2}$ | $\frac{1}{4}$ |
| $h(x) = f(x) + g(x)$ | $4\frac{1}{4}$ | $2\frac{1}{2}$ | 2 | $2\frac{1}{2}$ | $4\frac{1}{4}$ |

Note that as $x$ increases, $g(x)$ decreases toward zero, so the values of $h(x)$ get closer to the values of $f(x)$. On the other hand, as $x$ approaches zero, $h(x)$ gets closer to $g(x)$.

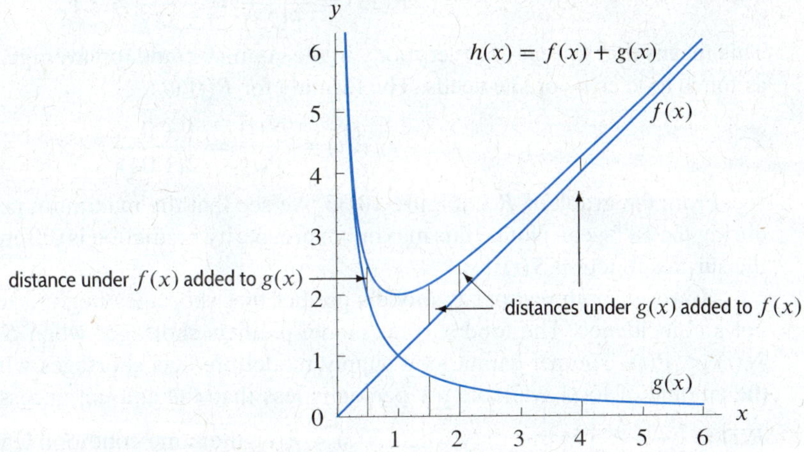


**Figure 10.32**: Graph of $h(x) = f(x) + g(x)$ constructed by adding vertical distances under $f$ and $g$

## Factoring a Function's Formula into a Product

It is often useful to be able to express a given function as a product of functions.

**Example 2** Find exactly all the zeros of the function

$$p(x) = 2^x \cdot 6x^2 - 2^x \cdot x - 2^{x+1}.$$

**Solution** We could approximate the zeros by finding the points where the graph of the function $p$ crosses the $x$-axis. Unfortunately, these solutions are not exact. Alternatively, we can express $p$ as a product. Using the fact that

$$2^{x+1} = 2^x \cdot 2^1 = 2 \cdot 2^x,$$

we rewrite the formula for $p$ as

$$\begin{aligned} p(x) &= 2^x \cdot 6x^2 - 2^x x - 2 \cdot 2^x \\ &= 2^x(6x^2 - x - 2) \qquad \text{Factoring out } 2^x \\ &= 2^x(2x+1)(3x-2). \qquad \text{Factoring the quadratic} \end{aligned}$$

Thus $p$ is the product of the exponential function $2^x$ and two linear functions. Since $p$ is a product, it equals zero if one or more of its factors equals zero. But $2^x$ is never equal to 0, so $p(x)$ equals zero if and only if one of the linear factors is zero:

$$(2x+1)=0 \qquad \text{or } (3x-2)=0$$
$$x=-\frac{1}{2} \qquad\qquad x=\frac{2}{3}.$$

## The Quotient of Functions Defined by Formulas and Graphs: Prosperity

Now let's think about our second proposed measure of prosperity, the per-capita food supply, $R(t)$. With this definition of prosperity

$$R(t)=\frac{\text{Number of people that can be fed}}{\text{Number of people living in the country}}=\frac{N(t)}{P(t)}.$$

For example,

$$R(25)=\frac{N(25)}{P(25)}=\frac{16.5}{5.332}\approx 3.1.$$

This means that in year 25, everybody in the country could, on average, have more than three times as much food as he or she needs. The formula for $R(t)$ is

$$R(t)=\frac{N(t)}{P(t)}=\frac{4+0.5t}{2(1.04)^t}.$$

From the graph of $R$ in Figure 10.33, we see that the maximum per-capita food supply occurs during the $18^{\text{th}}$ year. Notice this maximum prosperity prediction is different from the one made using the surplus function $S(t)$.

However, both prosperity models predict that shortages begin after time $t = 78.320$. This is not a coincidence. The food surplus model predicts shortages when $S(t) = N(t) - P(t) < 0$, or $N(t) < P(t)$. The per-capita food supply model predicts shortages when $R(t) < 1$, meaning that the amount of food available per person is less than the amount necessary to feed 1 person. Since $R(t) = \dfrac{N(t)}{P(t)} < 1$ is true only when $N(t) < P(t)$, the same condition leads to shortages.

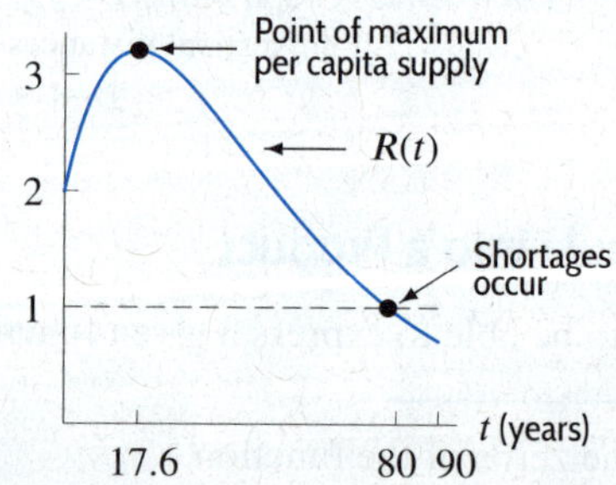


Figure 10.33: Per-capita food supply, $R(t) = \frac{N(t)}{P(t)}$

## The Quotient of Functions Defined by Tables: Per-Capita Crime Rate

Table 10.11 gives the number of violent crimes committed in two cities between 2005 and 2010. It appears that crime in both cities is on the rise and that there is less crime in City $B$ than in City $A$.

**Table 10.11** *Number of violent crimes committed each year in two cities*

| Year | 2005 | 2006 | 2007 | 2008 | 2009 | 2010 |
|---|---|---|---|---|---|---|
| $t$, years since 2005 | 0 | 1 | 2 | 3 | 4 | 5 |
| Crimes in City $A$ | 793 | 795 | 807 | 818 | 825 | 831 |
| Crimes in City $B$ | 448 | 500 | 525 | 566 | 593 | 652 |

Table 10.12 gives the population for these two cities from 2005 to 2010. The population of City $A$ is larger than that of City $B$ and both cities are growing.

**Table 10.12** *Population of the two cities*

| Year | 2005 | 2006 | 2007 | 2008 | 2009 | 2010 |
|---|---|---|---|---|---|---|
| $t$, years since 2000 | 0 | 1 | 2 | 3 | 4 | 5 |
| Population of City $A$ | 61,000 | 62,100 | 63,220 | 64,350 | 65,510 | 66,690 |
| Population of City $B$ | 28,000 | 28,588 | 29,188 | 29,801 | 30,427 | 31,066 |

Can we attribute the growth in crime in both cities to the population growth? Can we attribute the larger number of crimes in City $A$ to its larger population? To answer these questions, we consider the per-capita crime rate in each city.

Let's define $N_A(t)$ to be the number of crimes in City $A$ during year $t$ (where $t = 0$ means 2005). Similarly, let's define $P_A(t)$ to be the population of City $A$ in year $t$. Then the per-capita crime rate in City $A$, $r_A(t)$, is given by

$$r_A(t) = \frac{\text{Number of crimes in year } t}{\text{Number of people in year } t} = \frac{N_A(t)}{P_A(t)}.$$

We have defined a new function, $r_A(t)$, as the quotient of $N_A(t)$ and $P_A(t)$. For example, the data in Tables 10.11 and 10.12 shows that the per-capita crime rate for City $A$ in year $t = 0$ is

$$r_A(0) = \frac{N_A(0)}{P_A(0)} = \frac{793}{61{,}000} = 0.0130 \text{ crimes per person.}$$

Similarly, the per-capita crime rate for the year $t = 1$ is

$$r_A(1) = \frac{N_A(1)}{P_A(1)} = \frac{795}{62{,}100} = 0.0128 \text{ crimes per person.}$$

Thus, the per-capita crime rate in City $A$ actually decreased from 0.0130 crimes per person in 2005 to 0.0128 crimes per person in 2006.

**Example 3**

(a) Make a table of values for $r_A(t)$ and $r_B(t)$, the per-capita crime rates of Cities $A$ and $B$.
(b) Use the table to decide which city is more dangerous.

**Solution**

(a) Table 10.13 gives values of $r_A(t)$ for $t = 0, 1, \ldots, 5$. The per-capita crime rate in City $A$ declined between 2005 and 2010 despite the fact that the total number of crimes rose during this period. Table 10.13 also gives values of $r_B(t)$, the per-capita crime rate of City $B$, defined by

$$r_B(t) = \frac{N_B(t)}{P_B(t)},$$

where $N_B(t)$ is the number of crimes in City $B$ in year $t$ and $P_B(t)$ is the population of City $B$ in year $t$. For example, the per-capita crime rate in City $B$ in year $t = 0$ is

$$r_B(0) = \frac{N_B(0)}{P_B(0)} = \frac{448}{28{,}000} = 0.016 \text{ crimes per person.}$$

**Table 10.13** *Values of $r_A(t)$ and $r_B(t)$, the per-capita violent crime rates of Cities A and B*

| Year | 2005 | 2006 | 2007 | 2008 | 2009 | 2010 |
|---|---|---|---|---|---|---|
| $t$, years since 2005 | 0 | 1 | 2 | 3 | 4 | 5 |
| $r_A(t) = N_A(t)/P_A(t)$ | 0.0130 | 0.0128 | 0.01276 | 0.01271 | 0.01259 | 0.01246 |
| $r_B(t) = N_B(t)/P_B(t)$ | 0.01600 | 0.01749 | 0.01799 | 0.01899 | 0.01949 | 0.02099 |

(b) From Table 10.13, we see that between 2005 and 2010, City $A$ had a lower per-capita crime rate than City $B$. The crime rate of City $A$ is decreasing, whereas the crime rate of City $B$ is increasing. Thus, even though Table 10.11 indicates that there are more crimes committed in City $A$, Table 10.13 tells us that City $B$ is, in some sense, more dangerous. Table 10.13 also tells us that, even though the number of crimes is rising in both cities, City $A$ is getting safer, while City $B$ is getting more dangerous.

## Exercises and Problems for Section 10.3

### Exercises

In Exercises 1–6, find the following functions.

(a) $f(x)+g(x)$ (b) $f(x)-g(x)$
(c) $f(x)g(x)$ (d) $f(x)/g(x)$

1. $f(x)=x+5 \quad g(x)=x-5$
2. $f(x)=x^3 \quad g(x)=x^2$
3. $f(x)=x+1 \quad g(x)=3x^2$
4. $f(x)=x^2+4 \quad g(x)=x+2$
5. $f(x)=x^2+4 \quad g(x)=x^2+2$
6. $f(x)=\sqrt{x} \quad g(x)=x^2+2$

In Exercises 7–12, find simplified formulas using

$$u(x)=2x-1, \quad v(x)=1-x, \quad \text{and} \quad w(x)=\frac{1}{x}.$$

7. $f(x)=u(x)+v(x)$
8. $g(x)=v(x)w(x)$
9. $h(x)=2u(x)-3v(x)$
10. $j(x)=\dfrac{u(x)}{w(x)}$
11. $k(x)=(v(x))^2$
12. $l(x)=u(x)-v(x)-w(x)$

In Exercises 13–16, let $u(x)=e^x$ and $v(x)=2x+1$. Find a simplified formula for the function.

13. $f(x)=u(x)v(x)$
14. $g(x)=u(x)^2+v(x)^2$
15. $h(x)=(v(u(x)))^2$
16. $k(x)=v(u(x)^2)$

17. Use Table 10.14 to make tables of values for $x=-1, 0, 1, 2, 3, 4$ for the following functions.

(a) $h(x)=f(x)+g(x)$ (b) $j(x)=2f(x)$
(c) $k(x)=(g(x))^2$ (d) $m(x)=g(x)/f(x)$

Table 10.14

| $x$ | −1 | 0 | 1 | 2 | 3 | 4 |
|---|---|---|---|---|---|---|
| $f(x)$ | −4 | −1 | 2 | 5 | 8 | 11 |
| $g(x)$ | 4 | 1 | 0 | 1 | 4 | 9 |

18. Figures 10.34 and 10.35 show the graphs of $f(x)$ and $g(x)$ respectively. Let $a(x)=f(x)+g(x)$, $d(x)=f(x)-g(x)$, $s(x)=f(x)g(x)$ and $q(x)=f(x)/g(x)$. Find the following.

(a) $a(1)$ (b) $d(2)$
(c) $s(-1)$ (d) $q(0)$

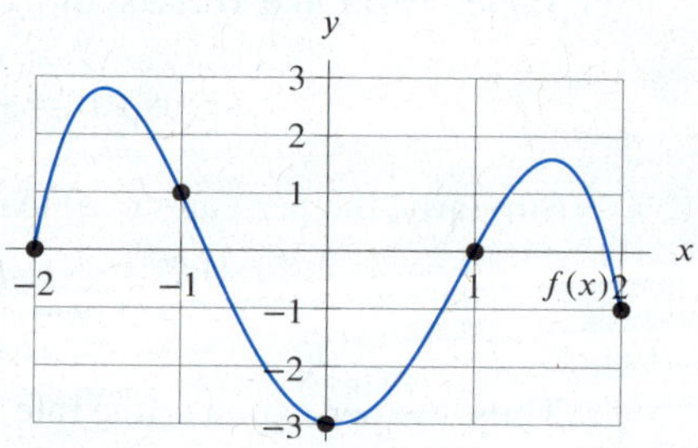


Figure 10.34

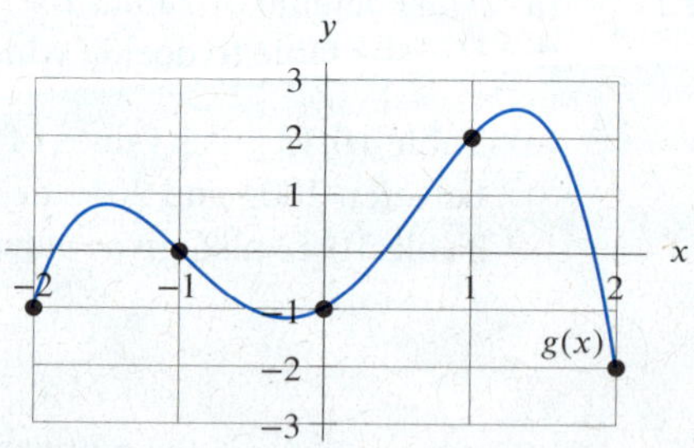


Figure 10.35

Find formulas for the functions in Exercises 19–24. Let $f(x)=\sin x$ and $g(x)=x^2$.

19. $f(x)+g(x)$
20. $g(x)f(x)$
21. $f(x)/g(x)$
22. $f(g(x))$
23. $g(f(x))$
24. $1-(f(x))^2$

## Problems

**25.** Let $f(x) = x+1$ and $g(x) = x^2 - 1$. In parts (a)–(e), write a formula in terms of $f(x)$ and $g(x)$ for the function. Then evaluate the formula for $x = 3$. Write a formula in terms of $x$ for each function. Check your formulas for $x = 3$.

**(a)** $h(x)$ is the sum of $f(x)$ and $g(x)$.
**(b)** $j(x)$ is the difference between $g(x)$ and twice $f(x)$.
**(c)** $k(x)$ is the product of $f(x)$ and $g(x)$.
**(d)** $m(x)$ is the ratio of $g(x)$ to $f(x)$.
**(e)** $n(x)$ is defined by $n(x) = (f(x))^2 - g(x)$.

**26.** Let $f(x) = \dfrac{1}{x+1}$. Find and simplify $f\left(\dfrac{1}{x}\right) + \dfrac{1}{f(x)}$.

**27.** Use Table 10.15 to make tables of values for the following functions.

**(a)** $f(x) = r(x) + t(x)$ **(b)** $g(x) = 4 - 2s(x)$
**(c)** $h(x) = r(x)t(x)$ **(d)** $j(x) = \dfrac{r(x) - t(x)}{s(x)}$
**(e)** $k(x) = r(x)^2$ **(f)** $l(x) = r(x) + s(x)t(x)$

**Table 10.15**

| $x$ | −2 | −1 | 0 | 1 | 2 | 3 |
|---|---|---|---|---|---|---|
| $r(x)$ | 4 | 5 | 6 | 7 | 8 | 9 |
| $s(x)$ | −2 | 2 | −2 | 2 | −2 | 2 |
| $t(x)$ | 8 | 5 | 7 | −3 | 2 | 13 |

Create tables for the functions in Problems 28–29 based on Table 10.16.

**Table 10.16**

| $x$ | 0 | 1 | 2 | 3 | 4 | 5 |
|---|---|---|---|---|---|---|
| $v(x)$ | 4 | 3 | 3 | 5 | 4 | 4 |
| $w(x)$ | 2 | 1 | 3 | 4 | 0 | 5 |

**28.** $p(x) = v(x)w(x)$ **29.** $q(x) = w^{-1}(v(x))$

Evaluate the expressions in Problems 30–31 using Figures 10.36 and 10.37, giving estimates if necessary.

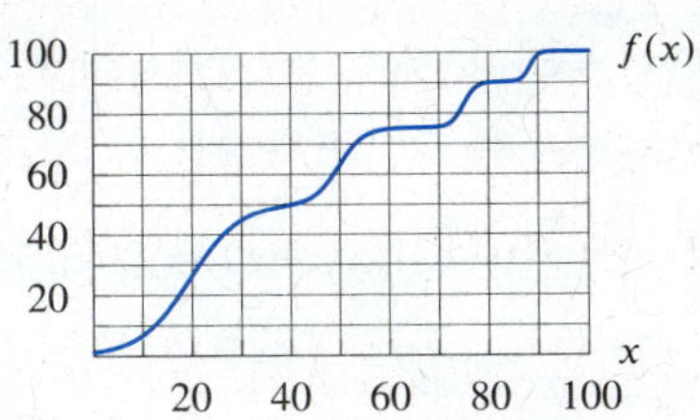


Figure 10.36

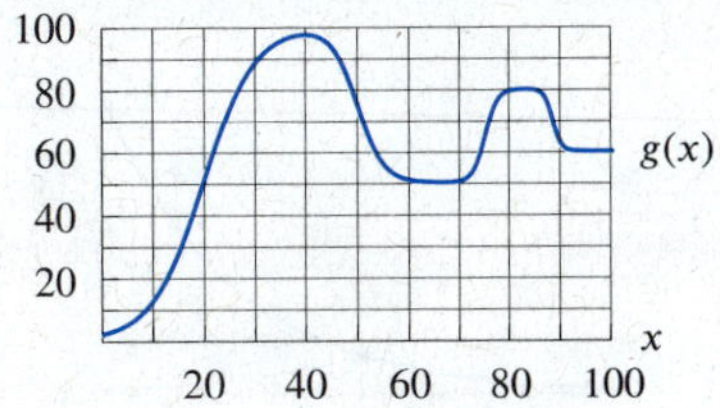


Figure 10.37

**30.** $f(g(65))$

**31.** $v(50)$ where $v(x) = g(x)f(x)$

**32.** Using Figure 10.38, sketch a graph of $h(x) = f(x)g(x)$. [Hint: Note where $g(x) = 0$ and $g(x) = 1$.]

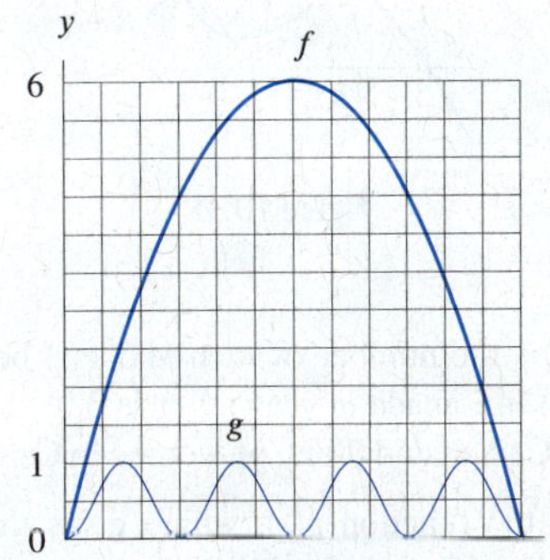


Figure 10.38

**33.** Using Figure 10.39, sketch a graph of $h(x) = g(x) - f(x)$. (Hint: Note where $g(x) = f(x)$.)

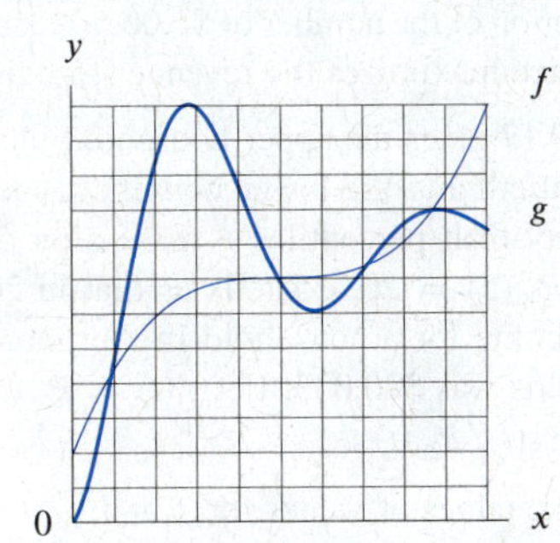


Figure 10.39

**34.** Graph $h(x) = f(x) + g(x)$ using Figure 10.40.

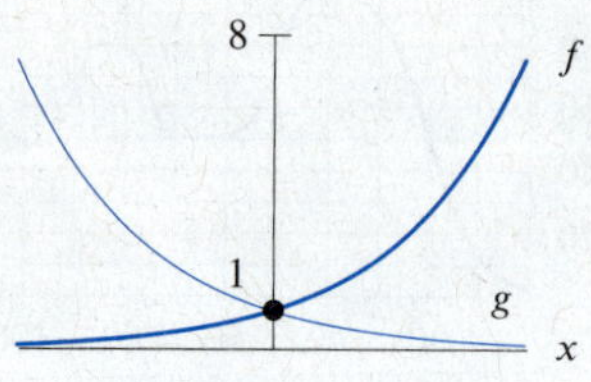


Figure 10.40

**35.** Use Figure 10.41 to graph $h(x) = g(x) - f(x)$. On the graph of $h(x)$, label the points whose $x$-coordinates are $x = a$, $x = b$, and $x = c$. Label the $y$-intercept.

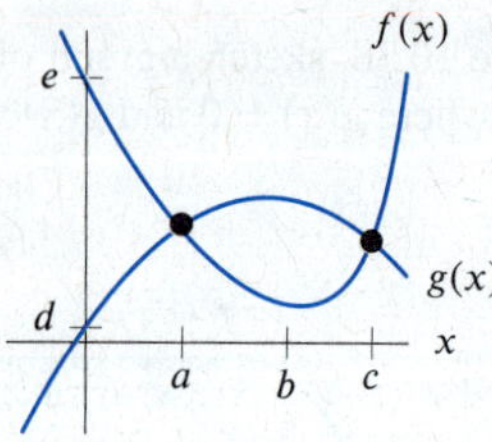


Figure 10.41

**36.** Let $f(t)$ be the number of men and $g(t)$ be the number of women in Canada in year $t$. Let $h(t)$ be the average income, in Canadian dollars, of women in Canada in year $t$.

**(a)** Find the function $p(t)$, which gives the number of people in Canada in year $t$.

**(b)** Find the total amount of money $m(t)$ earned by Canadian women in year $t$.

**37.** An average of 50,000 people visit Riverside Park each day in the summer. The park charges \$15.00 for admission. Consultants predict that for each \$1.00 increase in the entrance price, the park would lose an average of 2500 customers per day. Express the daily revenue from ticket sales as a function of the number of \$1.00 price increases. What ticket price maximizes the revenue from ticket sales?

**38.** Table 10.17 gives the upper household-income limits for the twentieth and the lower household-income limits of the ninety-fifth percentiles $t$ years after 2006.[3] For instance, $P_{20}(5) = 20{,}035$ tells us that in 2011 the maximum income for a household in the poorest 20% of all households was \$20,035. Let $f(t) = P_{95}(t) - P_{20}(t)$ and $g(t) = P_{95}(t)/P_{20}(t)$.

**(a)** Make tables of values for $f$ and $g$.

**(b)** Describe in words what $f$ and $g$ tell you about household income.

Table 10.17

| $t$ (yrs) | 0 | 1 | 2 |
|---|---|---|---|
| $P_{20}(t)$(\$) | 20,035 | 20,291 | 20,712 |
| $P_{95}(t)$(\$) | 174,012 | 177,000 | 180,000 |
| $t$ (yrs) | 3 | 4 | 5 |
| $P_{20}(t)$ (\$) | 20,453 | 20,000 | 20,262 |
| $P_{95}(t)$ (\$) | 180,001 | 180,485 | 186,000 |

**39.** In 1961 and 1962, large amounts of the radioactive isotope carbon-14 were produced by tests of nuclear bombs.[4] If $t$ is years since 1963, the amount of carbon-14, as a percent in excess of the normal level, is given by

$$C(t) = 108(e^{-0.1t} - e^{-0.7t}).$$

**(a)** Graph the function.

**(b)** Approximately when was the level of carbon-14 the highest?

**(c)** What happens to the level of carbon-14 in the long run?

**40.** Sketch two linear functions whose product is the function $f$ graphed in Figure 10.42(a). Explain why this is not possible for the function $q$ graphed in Figure 10.42(b).

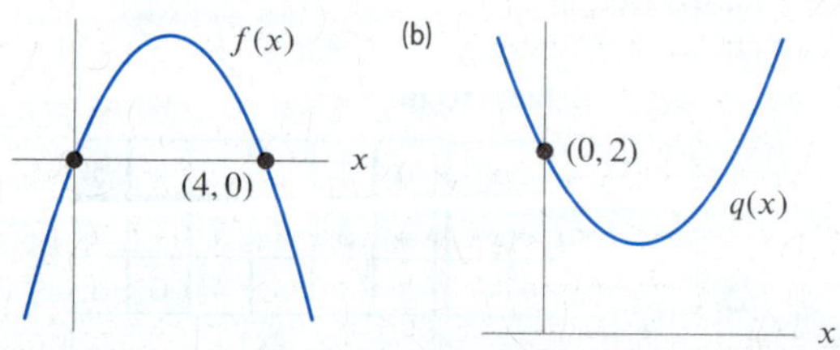


Figure 10.42

**41.** Find formulas for $H$ and $h$ given that:

$$F(x) = \cos x \qquad f(x) = -\sin x$$
$$G(x) = \sqrt{x} \qquad g(x) = \frac{1}{2\sqrt{x}}$$
$$H(x) = F(G(x))\,; \quad h(x) = f(G(x)) \cdot g(x).$$

[3] US Census Bureau, http://www.census.gov/hhes/www/income/data/historical/household/, accessed July 27, 2013.
[4] Adapted from William Bolton, *Patterns in Physics* (New York: McGraw-Hill, 1974).

**42.** Find simplified formulas for $H$ and $h$ given that:

$$F(x) = e^{-x^2} \qquad f(x) = -2xe^{-x^2}$$
$$G(x) = x^4 \qquad g(x) = 4x^3$$
$$H(x) = \frac{F(x)}{G(x)} \qquad h(x) = \frac{f(x)G(x) - F(x)g(x)}{(G(x))^2}.$$

Problems 43–44 involve the following formulas, which arise in calculus:

$$H(x) = F(x)G(x)$$
$$h(x) = f(x)G(x) + F(x)g(x)$$
$$j(x) = \frac{f(x)G(x) - F(x)g(x)}{(G(x))^2}.$$

Find a formula for the function requested, given that

$$F(x) = x^3, f(x) = 3x^2,$$
$$h(x) = 3x^2 \sin x + x^3 \cos x.$$

**43.** $H(x)$ **44.** $j(x)$

**45.** At the Mauna Loa Observatory, measurements of $CO_2$ levels (ppm) in the atmosphere reveal a slow exponential increase due to deforestation and burning of fossil fuels, and a periodic seasonal variation.

**(a)** In 1960, the average $CO_2$ level was 316.75 parts per million and was rising by 0.4% per year. Write the average annual $CO_2$ level as an exponential function of time, $t$, in years since 1960.

**(b)** Each year, the $CO_2$ level oscillates once between 3.25 ppm above and 3.25 ppm below the average level. Write a sinusoidal function for the seasonal variation in $CO_2$ levels in terms of time, $t$.

**(c)** Graph the sum of your functions in parts (a) and (b).

**46.** Table 10.18 gives data on strawberry production from 2008 through 2012,[5] where $t$ is in years since 2008. Let $f_{CA}(t)$, $f_{FL}(t)$, and $f_{US}(t)$ be the harvested area in year $t$ for strawberries grown in California, Florida, and the US overall, respectively. Likewise, let $g_{CA}(t)$, $g_{FL}(t)$, and $g_{US}(t)$ give the yield in thousands of pounds per acre for these three regions.

**(a)** Let $h_{CA}(t) = f_{CA}(t) \cdot g_{CA}(t)$. Create a table of values for $h_{CA}(t)$ for $0 \le t \le 4$. Describe in words what $h_{CA}(t)$ tells you about strawberry production.

**(b)** Let $p(t)$ be the fraction of all US strawberries (by weight) grown in Florida and California in year $t$. Find a formula for $p(t)$ in terms of $f_{CA}, f_{FL}, f_{US}$. Use Table 10.18 to make a table of values for $p(t)$.

**Table 10.18**

| | Harvested area (acres) | | | Yield (1000 lbs per acre) | | |
|---|---|---|---|---|---|---|
| $t$ | CA | FL | US total | CA | FL | US total |
| 0 | 37,600 | 7100 | 54,700 | 60.5 | 26.0 | 46.5 |
| 1 | 39,800 | 8800 | 58,080 | 62.0 | 27.0 | 48.2 |
| 2 | 38,600 | 8800 | 56,900 | 67.0 | 22.0 | 50.1 |
| 3 | 38,000 | 9900 | 57,420 | 68.0 | 25.0 | 50.8 |
| 4 | 38,500 | 8700 | 56,140 | 72.0 | 21.0 | 53.7 |

In Problems 47–53, you hire either Ace Construction or Space Contractors to build office space. Let $f(x)$ be the average total cost in dollars of building $x$ square feet of office space, as estimated by Ace. Let $h(x)$ be the total number of square feet of office space you can build with $x$ dollars, as estimated by Space.

**47.** Describe in words what the following statement tells you: $f(2000) = 200{,}000$.

**48.** Let $g(x) = f(x)/x$. Using the information from Problem 47, evaluate $g(2000)$, and describe in words what $g(2000)$ represents. [Hint: Think about the units.]

**49.** Ace tells you that, due to the economies of scale, "Building twice as much office space always costs less than twice as much." Express this statement symbolically, in terms of $f$ and $x$. [Hint: If you are building $x$ square feet, how do you represent the cost? How would you represent twice the cost? How do you represent the cost of building twice as many square feet?]

**50.** Suppose that $q > p$ and $p > 1$. Assuming that the contractor's statement in Problem 49 is correct, rank the following in increasing order, using inequality signs: $f(p)$, $g(p)$, $f(q)$, $g(q)$.

**51.** What does the statement $h(200{,}000) = 1500$ tell you?

**52.** Let $j(x)$ be the average cost in dollars per square foot of office space as estimated by Space Contractors. Give a formula for $j(x)$. (Your formula will have $h(x)$ in it.)

**53.** Research reveals that $h(f(x)) < x$ for every value of $x$ you check. Explain the implications of this statement. [Hint: Which company seems more economical?]

**54.** In Figure 10.43, the line $l_2$ is fixed and the point $P$ moves along $l_2$. Define $f(\theta)$ as the $y$-coordinate of $P$.

**(a)** Find a formula for $f(\theta)$ if $0 < \theta < \pi/2$. [Hint: Use the equation for $l_2$.]

**(b)** Graph $y = f(\theta)$ on the interval $-\pi \le \theta \le \pi$.

**(c)** How does the $y$-coordinate of $P$ change as $\theta$ changes? Is $y = f(\theta)$ periodic?

[5] http://usda.mannlib.cornell.edu/MannUsda/viewDocumentInfo.do?documentID=1381, accessed July 27, 2013.

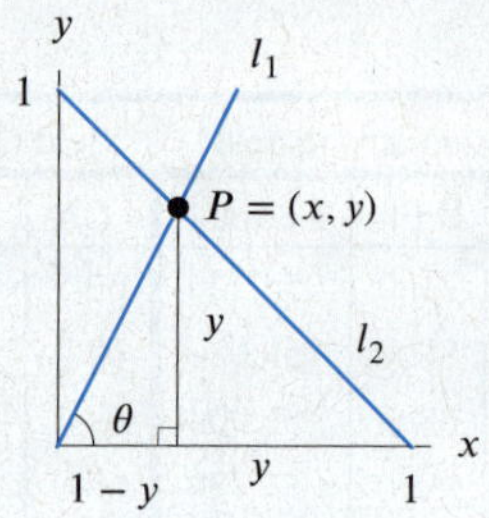


Figure 10.43

**55.** Let $f(x) = kx^2 + B$ and $g(x) = C^{2x}$ and

$$h(x) = kx^2C^{2x} + BC^{2x} + C^{2x}.$$

Suppose $f(3) = 7$ and $g(3) = 5$. Evaluate $h(3)$.

## CHAPTER SUMMARY

- **Composition of Functions**
  Notation: $h(t) = f(g(t))$.
  Domain and range; Decomposition.
- **Inverse Functions**
  Definition: $f^{-1}(Q) = t$ if and only if $Q = f(t)$.
  Invertibility; horizontal line test.
  Domain and range of an inverse function.
  Restricting domain of a function to construct an inverse.
  Graph of an inverse function.
  $f^{-1}(f(x)) = x$ and $f(f^{-1}(x)) = x$.
- **Combinations of Functions**
  Sums, differences, products, quotients of functions.

## REVIEW EXERCISES AND PROBLEMS FOR CHAPTER TEN

### Exercises

**1.** Let $h(x) = 2^x$ and $k(x) = x^2$. Find formulas for $h(k(x))$ and $k(h(x))$.

Find formulas for the functions in Exercises 2–7 and simplify. Let $f(x) = x^2 + 1$, $g(x) = \dfrac{1}{x-3}$, and $h(x) = \sqrt{x}$.

**2.** $f(g(x))$ **3.** $g(f(x))$ **4.** $f(h(x))$

**5.** $h(f(x))$ **6.** $g(g(x))$ **7.** $g(f(h(x)))$

**8.** Using Tables 10.19 and 10.20, complete Table 10.21:

Table 10.19

| $x$ | $f(x)$ |
|---|---|
| 0 | 0 |
| $\pi/6$ | $1/2$ |
| $\pi/4$ | $\sqrt{2}/2$ |
| $\pi/3$ | $\sqrt{3}/2$ |
| $\pi/2$ | 1 |

Table 10.20

| $y$ | $g(y)$ |
|---|---|
| 0 | $\pi/2$ |
| $1/4$ | $\pi$ |
| $\sqrt{2}/4$ | 0 |
| $1/2$ | $\pi/3$ |
| $\sqrt{2}/2$ | $\pi/4$ |
| $3/4$ | 0 |
| $\sqrt{3}/2$ | $\pi/6$ |
| 1 | 0 |

Table 10.21

| $x$ | $g(f(x))$ |
|---|---|
| 0 | |
| $\pi/6$ | |
| $\pi/4$ | |
| $\pi/3$ | |
| $\pi/2$ | |

**9.** For each of the following functions, use a graph to decide whether or not the function is invertible.

**(a)** $y = e^{x^2}$ **(b)** $y = \cos(x^3)$ **(c)** $y = x + \ln x$

In Exercises 10–22, find a formula for the inverse function. Assume these functions are defined on domains on which they are invertible.

**10.** $h(x) = 12x^3$

**11.** $h(x) = \dfrac{x}{2x+1}$

**12.** $k(x) = 3 \cdot e^{2x}$

**13.** $g(x) = e^{3x+1}$

**14.** $n(x) = \log(x - 3)$

**15.** $h(x) = \ln(1 - 2x)$

**16.** $h(x) = \dfrac{\sqrt{x}}{\sqrt{x}+1}$

**17.** $g(x) = \dfrac{x-2}{2x+3}$

**18.** $f(x) = \sqrt{\dfrac{4-7x}{4-x}}$

**19.** $f(x) = \dfrac{\sqrt{x}+3}{11-\sqrt{x}}$

**20.** $f(x) = \ln\left(1 + \dfrac{1}{x}\right)$

**21.** $s(x) = \dfrac{3}{2 + \log x}$

**22.** $q(x) = \ln(x + 3) - \ln(x - 5)$

In Exercises 23–28, state whether the function is invertible. If so, give a formula for the inverse function.

**23.** $f(x) = 6x^2 - 2x - 3$

**24.** $q(x) = 3x^3 - 2$

**25.** $m(x) = 3x^3 - 2x^2 + 5x$

**26.** $p(x) = 5x^5 + 4$

**27.** $r(x) = e^x - 7$

**28.** $b(x) = \sqrt[3]{x} - 2$

**29.** Check that $h^{-1}(h(x)) = x$ if $h(x) = \sqrt{\frac{1-x}{x}}$ for $0 < x \le 1$ and $h^{-1}(x) = \frac{1}{x^2+1}$ for $x \ge 0$.

**30.** Suppose $f(x) = \frac{x}{2x+1}$. Find a formula for $f^{-1}(x)$.

In Exercises 31–36, find simplified formulas if $f(x) = e^x$, $g(x) = 2x - 1$, and $h(x) = \sqrt{x}$.

**31.** $g(f(x))$ **32.** $g(x)f(x)$ **33.** $g(g(x))$

**34.** $f(g(h(x)))$ **35.** $f(g(x))h(x)$ **36.** $(f(h(x)))^2$

**37.** Find formulas for the following functions, given that

$$f(x) = x^2 + x, \quad g(x) = 2x - 3, \quad h(x) = \frac{x}{1-x}.$$

**(a)** $f(2x)$ **(b)** $g(x^2)$ **(c)** $h(1-x)$
**(d)** $(f(x))^2$ **(e)** $g^{-1}(x)$ **(f)** $(h(x))^{-1}$
**(g)** $f(x) \cdot g(x)$ **(h)** $h(f(x))$

In Exercises 38–40, find simplified formulas if $u(x) = \frac{1}{1+x^2}$, $v(x) = e^x$, and $w(x) = \ln x$.

**38.** $v(x)/u(x)$ **39.** $u(v(x)) \cdot w(v(x))$

**40.** $\frac{w(2+h) - w(2)}{h}$

In Exercises 41–46, find a simplified formula for the function. Let $m(x) = 3x^2 - x$, $n(x) = 2x$, and $o(x) = \sqrt{x+2}$.

**41.** $f(x) = m(x) + n(x)$ **42.** $g(x) = (o(x))^2$

**43.** $h(x) = n(x)o(x)$ **44.** $i(x) = m(o(x))n(x)$

**45.** $j(x) = (m(x))/n(x)$

**46.** $k(x) = m(x) - n(x) - o(x)$

In Exercises 47–50, find simplified formulas if $f(x) = x^{3/2}$, $g(x) = \frac{(3x-1)^2}{4}$, and $h(x) = \tan 2x$.

**47.** $f(x)h(x)$ **48.** $\frac{h(x)}{f(g(x))}$

**49.** $h(g(x)) - f(9x)$ **50.** $h(x/2)\cos x$

## Problems

**51.** Complete Table 10.22 if $r(t) = q(p(t))$.

**Table 10.22**

| $t$ | $p(t)$ | $q(t)$ | $r(t)$ |
|---|---|---|---|
| 0 | 4 | | |
| 1 | | 2 | 1 |
| 2 | | | 0 |
| 3 | 2 | 0 | 4 |
| 4 | 1 | 5 | |
| 5 | 0 | 1 | 3 |

**52.** Complete Table 10.23, Table 10.24, and Table 10.25 given that $h(x) = g(f(x))$. Assume that different values of $x$ lead to different values of $f(x)$.

**Table 10.23**

| $x$ | $f(x)$ |
|---|---|
| −2 | 4 |
| −1 | |
| 0 | |
| 1 | 5 |
| 2 | 1 |

**Table 10.24**

| $x$ | $g(x)$ |
|---|---|
| 1 | |
| 2 | 1 |
| 3 | 2 |
| 4 | 0 |
| 5 | −1 |

**Table 10.25**

| $x$ | $h(x)$ |
|---|---|
| −2 | |
| −1 | 1 |
| 0 | 2 |
| 1 | |
| 2 | −2 |

Decompose the functions in Problems 53–60 into two new functions, $u$ and $v$, where $v$ is the inside function, $u(x) \neq x$, and $v(x) \neq x$.

**53.** $f(x) = \sqrt{3-5x}$ **54.** $g(x) = \sin(x^2)$

**55.** $h(x) = \sin^2 x$ **56.** $k(x) = e^{\sin x} + \sin x$

**57.** $F(x) = (2x+5)^3$ **58.** $G(x) = \frac{2}{1+\sqrt{x}}$

**59.** $H(x) = 3^{2x-1}$ **60.** $J(x) = 8 - 2|x|$

**61.** Using your knowledge of the absolute value function, explain in a few sentences the relationship between the graph of $y = |\sin x|$ and the graph of $y = \sin x$.

**62.** Graph the following functions for $-2\pi \le x \le 2\pi$.

**(a)** $f(x) = \sin x$ **(b)** $g(x) = |\sin x|$
**(c)** $h(x) = \sin|x|$ **(d)** $i(x) = |\sin|x||$
**(e)** Do any two of these functions have identical graphs? If so, explain why this makes sense.

**63.** Suppose $p(x) = (1/x) + 1$ and $q(x) = x - 2$.

**(a)** Let $r(x) = p(q(x))$. Find a formula for $r(x)$ and simplify it.
**(b)** Write formulas for $s(x)$ and $t(x)$ such that $p(x) = s(t(x))$, where $s(x) \neq x$ and $t(x) \neq x$.
**(c)** Let $a$ be different from 0 and $-1$. Find a simplified expression for $p(p(a))$.

**64.** Let $f(x) = \frac{1}{x}$ and $g(x) = \sin x$. Calculate the domain of $f(g(x))$ and the domain of $g(f(x))$.

**65.** Suppose that $h(x) = f(g(x))$, and that $f$ is invertible. Complete the following table.

| $x$ | $f(x)$ | $g(x)$ | $h(x)$ |
|---|---|---|---|
| 0 | 9 | 1 | |
| 1 | 0 | | 1 |
| 2 | 1 | 0 | |

In Problems 66–67, a population is given by the formula $P = f(t) = 20 + 0.4t$ where $P$ is the number of people (in thousands) and $t$ is the number of years since 1985.

**66.** Evaluate the following quantities. Explain in words what each tells you about the population.

**(a)** $f(25)$ **(b)** $f^{-1}(25)$
**(c)** Show how to estimate $f^{-1}(25)$ from a graph of $f$.

**67.** **(a)** Find a formula for $t = f^{-1}(P)$.
**(b)** Construct tables of values for both functions for values of $t$ from $t = 0$ years to $t = 20$ years, using five-year intervals.

**68.** Let $P = f(t) = 14 \cdot 2^{t/12}$ give the size in 1000s of an animal population in year $t$.

**(a)** Find $f^{-1}(P)$.
**(b)** Evaluate $f^{-1}(24)$. Say what this tells you about the population.

**69.** If $t = g(v)$ represents the time in hours it takes to drive to the next town at velocity $v$ mph, what does $g^{-1}(t)$ represent? What are its units?

**70.** Let $f(x) = e^x$. Solve each of the following equations exactly for $x$.

**(a)** $(f(x))^{-1} = 2$ **(b)** $f^{-1}(x) = 2$
**(c)** $f(x^{-1}) = 2$

**71.** Simplify the expression $\cos^2(\arcsin t)$, using the property that inverses "undo" each other.

Solve the equations in Problems 72–77 exactly. Use an inverse function when appropriate.

**72.** $7\sin(3x) = 2$ **73.** $2^{x+5} = 3$

**74.** $x^{1.05} = 1.09$ **75.** $\ln(x + 3) = 1.8$

**76.** $\dfrac{2x+3}{x+3} = 8$ **77.** $\sqrt{x + \sqrt{x}} = 3$

**78.** Let $Q = f(t) = 20(0.96)^{t/3}$ be the number of grams of a radioactive substance remaining after $t$ years.

**(a)** Describe the behavior of the radioactive substance as a function of time.
**(b)** Evaluate $f(8)$. Explain the meaning of this quantity in practical terms.
**(c)** Find a formula for $f^{-1}(Q)$ in terms of $Q$.
**(d)** Evaluate $f^{-1}(8)$. Explain the meaning of this quantity in practical terms.

**79.** **(a)** What is the formula for the area of a circle in terms of its radius?
**(b)** Graph this function for the domain all real numbers.
**(c)** What domain actually applies in this situation? On separate axes, graph the function for this domain.
**(d)** Find the inverse of the function in part (c).
**(e)** Graph the inverse function on the domain you gave in part (c) on the same axes used in part (c).
**(f)** If area is a function of the radius, is radius a function of area? Explain carefully.

**80.** A company believes there is a linear relationship between the consumer demand for its products and the price charged. When the price was \$3 per unit, the quantity demanded was 500 units per week. When the unit price was raised to \$4, the quantity demanded dropped to 300 units per week. Let $D(p)$ be the quantity per week demanded by consumers at a unit price of \$$p$.

**(a)** Estimate and interpret $D(5)$.
**(b)** Find a formula for $D(p)$ in terms of $p$.
**(c)** Calculate and interpret $D^{-1}(5)$.
**(d)** Give an interpretation of the slope of $D(p)$ in terms of demand.
**(e)** Currently, the company can produce 400 units every week. What should the price of the product be if the company wants to sell all 400 units?
**(f)** If the company produced 500 units per week instead of 400 units per week, would its weekly revenues increase, and if so, by how much?

**81.** Use Figure 10.44.

**(a)** Evaluate $f(g(a))$.
**(b)** Evaluate $g(f(c))$.
**(c)** Evaluate $f^{-1}(b) - g^{-1}(b)$.
**(d)** For what positive value(s) of $x$ is $f(x) \leq g(x)$?

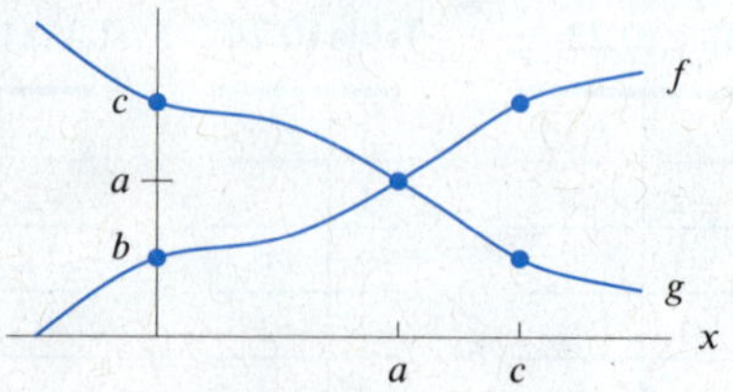


Figure 10.44

**82.** Use Table 10.26 to make tables of values for the following functions.

**(a)** $n(x) = f(x) + g(x)$
**(b)** $p(x) = 2f(x)g(x) - f(x)$
**(c)** $q(x) = g(x)/f(x)$

Table 10.26

| $x$ | 1 | 2 | 3 | 4 |
|---|---|---|---|---|
| $f(x)$ | 3 | 4 | 1 | 2 |
| $g(x)$ | 2 | 1 | 4 | 3 |

For Problems 83–87, let

$$p(x) = 2x - 3 \qquad q(x) = \sqrt{x} - 3$$
$$r(x) = \frac{2x-1}{2x+1} \qquad s(x) = (x-1)^2$$

**83.** Find and simplify $p(q(x))$.

**84.** Find and simplify $r^{-1}(x)$.

**85.** Solve for $x$ exactly: $p(x) = r(x)$.

**86.** Find $u(x)$ given that $q(x) = p(u(x))$.

**87.** Graph and label $p(s(x))$.

**88.** Let $f(x) = e^x$. For each of the following, use the rules of logarithms and exponents (if possible) to find the exact solution to each equation. If this is not possible, use a graph to find an approximate solution.

**(a)** $f(3x) + f(3x) = 1$ **(b)** $f(x) + f(3x) = 1$
**(c)** $f(3x)f(3x) = 2$ **(d)** $f(x)f(3x) = 2$

**89.** Let $u(x) = \sin x$, $v(x) = \cos x$, and $w(x) = x^2$.

**(a)** Let $f(x) = \sin^2 x$. Is $f(x)$ equal to $u(u(x))$ or to $(u(x))^2$? Justify your answer.
**(b)** Let $p(x) = \sin(\cos^2 x)$. Identify the functions (i)–(iv) that are equal to $p(x)$.
 (i) $u(x)(v(x))^2$
 (ii) $u((v(x))^2)$
 (iii) $u(w(v(x)))$
 (iv) $u(x)w(v(x))$
**(c)** Calculate and simplify as far as possible:
 (i) $(u(x) + v(x))^2$
 (ii) $(u(x))^2 + (v(x))^2$
 (iii) $u(x^2) + v(x^2)$

Using Figures 10.45 and 10.46, graph the functions in Problems 90–93.

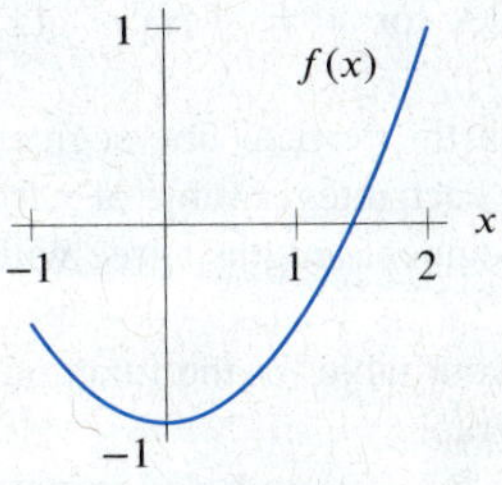

Figure 10.45

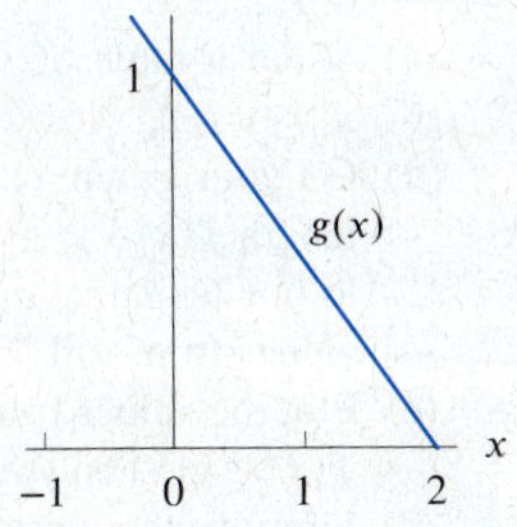

Figure 10.46

**90.** $f(x) - g(x)$ **91.** $f(g(x))$

**92.** $g(f(x))$ **93.** $g(f(x-2))$

**94.** Using Figure 10.47, match the functions (a)–(g) and graphs (I)–(IV). There may be some functions whose graphs are not shown.

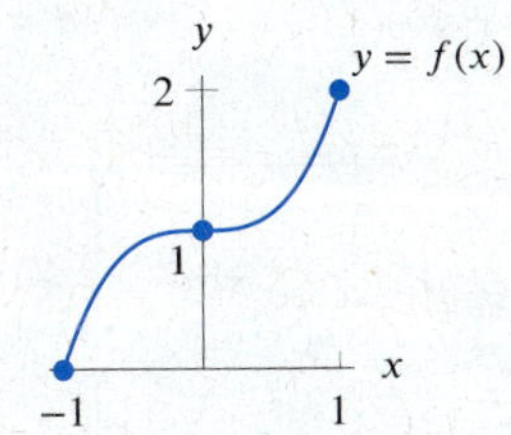

Figure 10.47

**(a)** $y = -f(x)$ **(b)** $y = f(-x)$
**(c)** $y = f(-x) - 2$ **(d)** $y = f^{-1}(x)$
**(e)** $y = -f^{-1}(x)$ **(f)** $y = f(x+1)$
**(g)** $y = -(f(x) - 2)$

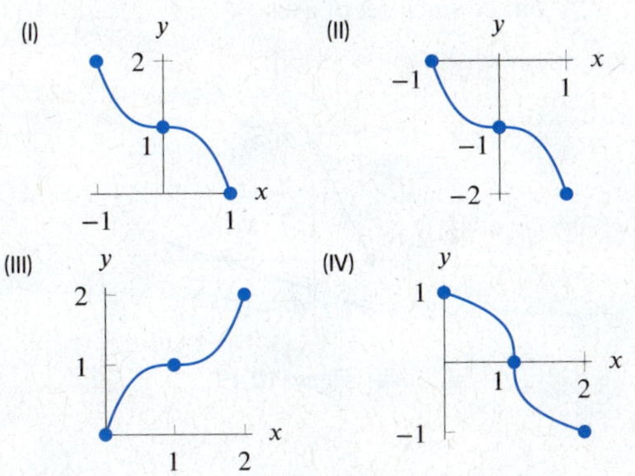

**95.** **(a)** On the same set of axes, graph $f(x) = (x-4)^2 - 2$ and $g(x) = -(x-2)^2 + 8$.
**(b)** Make a table of values for $f$ and $g$ for $x = 0, 1, 2, ..., 6$.
**(c)** Make a table of values for $y = f(x) - g(x)$ for $x = 0, 1, 2, ..., 6$.
**(d)** On your graph, sketch the vertical line segment of length $f(x) - g(x)$ for each integer value of $x$ from 0 to 6. Check that the segment lengths agree with the values from part (c).
**(e)** Plot the values from your table for the function $y = f(x) - g(x)$ on your graph.
**(f)** Simplify the formulas for $f(x)$ and $g(x)$ in part (a). Find a formula for $y = f(x) - g(x)$.
**(g)** Use part (f) to graph $y = f(x) - g(x)$ on the same axes as $f$ and $g$. Does the graph pass through the points you plotted in part (e)?

**96.** Figure 10.48 shows a weight attached to the end of a spring that is hanging from the ceiling. The weight is pulled down from the ceiling and then released. The weight oscillates up and down, but over time, friction decreases the magnitude of the vertical oscillations. Which of the following functions could describe the distance of the weight from the ceiling, $d$, as a function of time, $t$?

(i) $d = 2 + \cos(10t)$ (ii) $d = 2 + e^{-t}\cos t$
(iii) $d = 2 + \cos(e^t)$ (iv) $d = 2 + e^{\cos t}$

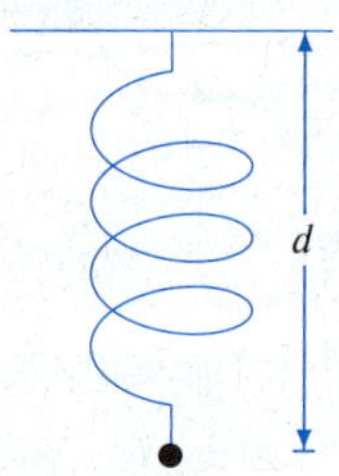


Figure 10.48

**97.** **(a)** Find possible formulas for the functions in Figure 10.49.
**(b)** Let $h(x) = f(x) \cdot g(x)$. Graph $f(x)$, $g(x)$ and $h(x)$ on the same set of axes.

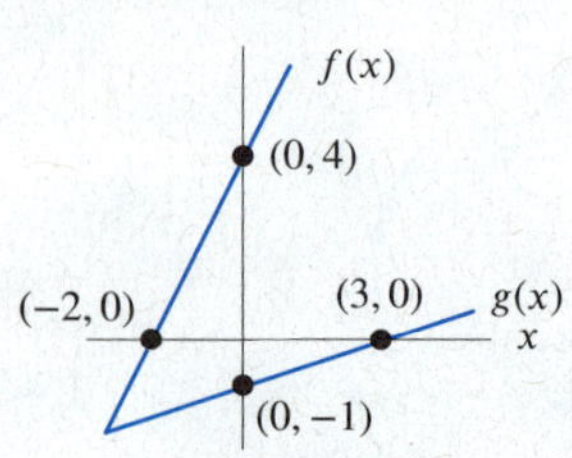


Figure 10.49

**98.** Use Figure 10.50 to graph $c(x) = a(x) \cdot b(x)$. [Hint: There is not enough information to determine formulas for $a$ and $b$ but you can use the method of Problem 97.]

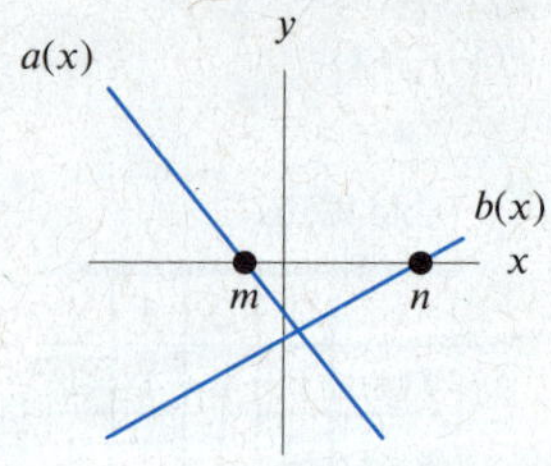


Figure 10.50

**99.** Use Figure 10.51 to graph the following functions.

**(a)** $y = g(x) - 3$ **(b)** $y = g(x) + x$

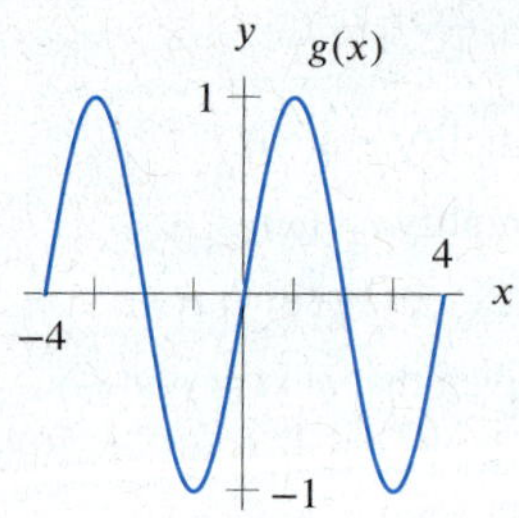


Figure 10.51

**100.** Describe the similarities and differences between the graphs of $y = \sin(1/x)$ and $y = 1/\sin x$.

**101.** Is the following statement true or false? If $f(x) \cdot g(x)$ is an odd function, then both $f(x)$ and $g(x)$ are odd functions. Explain your answer.

In Problems 102–103, find a simplified formula for $g$ given that $f(x) = x^2 + 5x + 6$.

**102.** $g(f(x)) = 64 \cdot 2^{x^2} \cdot 32^x$

**103.** $f(x) = g(x)h(x)$ where $h(x) = 2x + 6$.

**104.** Assume that $f(x) = 3 \cdot 9^x$ and that $g(x) = 3^x$.

**(a)** If $f(x) = h(g(x))$, find a formula for $h(x)$.
**(b)** If $f(x) = g(j(x))$, find a formula for $j(x)$.

**105.** In calculus, it is often necessary to write functions in the form $y = f(G(x)) \cdot g(x)$. Do this for the following functions, giving formulas for $f$, $g$, and $G$. There may be more than one possible answer.

**(a)** $y = 6xe^{3x^2}$ **(b)** $y = -\dfrac{\sin\left(\sqrt{x}\right)}{2\sqrt{x}}$

**106.** Currency traders often move investments from one country to another in order to make a profit. Table 10.27 gives exchange rates for US dollars, Japanese yen, and the European euro.[6] In November, 2009, for example, 1 US dollar purchased 89.1823 Japanese yen or 0.6686 European euro. Similarly, 1 European euro purchases 133.39 Japanese yen or 1.4957 US dollars. Let

$f(x)$ = Number of yen one can buy with $x$ dollars
$g(x)$ = Number of euros one can buy with $x$ dollars
$h(x)$ = Number of euros one can buy with $x$ yen

**(a)** Find formulas for $f$, $g$, and $h$.
**(b)** Evaluate $h(f(1000))$ and interpret it in terms of currency.

**Table 10.27** *Exchange rate for US dollars, Japanese yen and euros, November 18, 2009*

| Amount invested | Dollars purchased | Yen purchased | Euros purchased |
|---|---|---|---|
| 1 dollar | 1.0000 | 89.1823 | 0.6686 |
| 1 yen | 0.01121 | 1.0000 | 0.007496 |
| 1 euro | 1.4957 | 133.39 | 1.0000 |

**107.** Letting $h(x) = f(x) + g(x)$, say which of the following statements must be true. Briefly explain your reasoning.

**(a)** If $f$ and $g$ have the same domain, so does $h$.
**(b)** If $f$ and $g$ have the same range, so does $h$.
**(c)** If $f$ and $g$ lack zeros, so does $h$.
**(d)** Where defined, $\dfrac{\Delta h}{\Delta x} = \dfrac{\Delta f}{\Delta x} + \dfrac{\Delta g}{\Delta x}$.

**108.** **(a)** Is the sum of two even functions even, odd, or neither? Justify your answer.
**(b)** Is the sum of two odd functions even, odd, or neither? Justify your answer.
**(c)** Is the sum of an even and an odd function even, odd, or neither? Justify your answer.

In Problems 109–112, let $f(x)$ be an increasing function and let $g(x)$ be a decreasing function. Are the following functions increasing, decreasing, or is it impossible to tell? Explain.

**109.** $f(f(x))$

**110.** $g(f(x))$

**111.** $f(x) + g(x)$

**112.** $f(x) - g(x)$

**113.** For a positive integer $x$, let $f(x)$ be the remainder obtained by dividing $x$ by 3. For example, $f(6) = 0$, because 6 divided by 3 equals 2 with a remainder of 0. Likewise, $f(7) = 1$, because 7 divided by 3 equals 2 with a remainder of 1.

**(a)** Evaluate $f(8)$, $f(17)$, $f(29)$, $f(99)$.
**(b)** Find a formula for $f(3x)$.
**(c)** Is $f(x)$ invertible?
**(d)** Find a formula in terms of $f(x)$ for $f(f(x))$.
**(e)** Does $f(x+y)$ necessarily equal $f(x) + f(y)$?

**114.** Suppose that $f$, $g$, and $h$ are invertible and that

$$f(x) = g(h(x)).$$

Find a formula for $f^{-1}(x)$ in terms of $g^{-1}$ and $h^{-1}$.

**115.** The von Bertalanffy growth model predicts the mean length $L$ of fish of age $t$ (in years):[7]

$$L = f(t) = L_\infty\left(1 - e^{-k(t+t_0)}\right), \text{ for constant } L_\infty, k, t_0.$$

Find a formula for $f^{-1}$. Describe in words what $f^{-1}$ tells you about fish. What is the domain of $f^{-1}$?

[6] www.x-rates.com, accessed November 18, 2009. Currency exchange rates fluctuate constantly.

[7] Per Sparre, *Introduction to Tropical Fish Stock Assessment*, Danish Institute for Fisheries Research, and Siebren C. Venema, FAO Fisheries Department, available at www.fao.org/docrep/W5449E/w5449e00.htm., accessed November 2001.

## STRENGTHEN YOUR UNDERSTANDING

Let $f(x) = \frac{1}{x}$, $g(x) = \sqrt{x}$, and $h(x) = x - 5$. Are the statements in Problems 1–10 true or false? Give an explanation for your answer.

1. $f(4) + g(4) = (f + g)(8)$.
2. $\frac{h(x)}{f(x)} = \frac{x-5}{x}$.
3. $f(4) + g(4) = 2\frac{1}{4}$.
4. $f(g(x))$ is defined for all $x$.
5. $g(f(x)) = \sqrt{\frac{1}{x}}$.
6. $f(x)g(x) = f(g(x))$.
7. $2f(2) = g(1)$.
8. $f(1)g(1)h(1) = -4$.
9. $\frac{f(3) + g(3)}{h(3)} = \frac{\frac{1}{3} + \sqrt{3}}{-2}$.
10. $4h(2) = h(8)$.

Are the statements in Problems 11–37 true or false? Give an explanation for your answer.

11. If $f(x) = x^2$ and $g(x) = \sqrt{x+3}$. Then $f(g(x))$ is defined for all $x$.
12. If $f(x) = x^2$ and $g(x) = \sqrt{x+3}$. Then $f(g(6)) = 9$.
13. In general, $f(g(x)) = g(f(x))$.
14. The formula for the area of a circle is $A = \pi r^2$ and the formula for the circumference of a circle is $C = 2\pi r$. Then the area of a circle as a function of the circumference is $A = \frac{C^2}{2\pi}$.
15. If $f(x) = x^2 + 2$ then $f(f(1)) = 11$.
16. If $h(x) = f(g(x))$, $h(2) = 1\frac{1}{4}$ and $g(2) = \frac{1}{2}$ then $f(x)$ might be equal to $x^2 + 1$.
17. If $f(x) = \frac{1}{x}$ then $f(x+h) = \frac{1}{x} + \frac{1}{h}$.
18. If $f(x) = x^2 + x$, then $\frac{f(x+h) - f(x)}{h} = 2x + h$.
19. If $f(x) = x^2$ and $g(x) = \sin x$ then $f(g(x)) = x^2 \sin x$.
20. If $f(x)$ and $g(x)$ are linear, then $f(g(x))$ is linear.
21. If $f(x)$ and $g(x)$ are quadratic, then $f(g(x))$ is quadratic.
22. There is more than one way to write $h(x) = 3(x^2+1)^3$ as a composition $h(x) = f(g(x))$.
23. The composition $f(g(x))$ is never the same as the composite $g(f(x))$.
24. If $f$ and $g$ are both increasing, then $h(x) = f(g(x))$ is also increasing.
25. The functions $f$ and $g$ given in Tables 10.28 and 10.29 satisfy $g(f(2)) = f(g(3))$.

**Table 10.28**

| $t$ | 1 | 2 | 3 | 4 |
|---|---|---|---|---|
| $f(t)$ | 3 | 1 | 2 | 4 |

**Table 10.29**

| $x$ | 1 | 2 | 3 | 4 |
|---|---|---|---|---|
| $g(x)$ | 3 | 4 | 1 | 2 |

26. If $f$ is increasing and invertible, then $f^{-1}$ is decreasing.
27. If there is a vertical line that intersects a graph in more than one point, then the graph does not represent a function.
28. Every function has an inverse.
29. If $g(3) = g(5)$, then $g$ is not invertible.
30. If no horizontal line intersects the graph of a function in more than one point, then the function has an inverse.
31. Most quadratic functions have an inverse.
32. All linear functions of the form $f(x) = mx + b, m \neq 0$ have inverses.
33. The following table describes $y$ as a function of $x$.

**Table 10.30**

| $x$ | 1 | 2 | 3 | 4 | 5 | 6 |
|---|---|---|---|---|---|---|
| $y$ | 3 | 4 | 7 | 8 | 5 | 3 |

34. The table in Problem 33 describes a function from $x$ to $y$ that is invertible.
35. The graph in Figure 10.52 is the graph of a function.

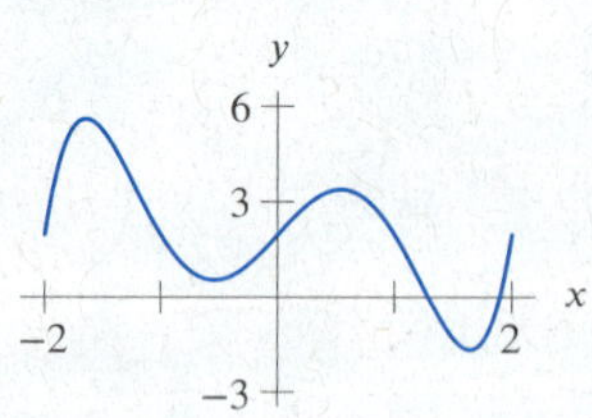


Figure 10.52

36. The function graphed in Figure 10.52 is invertible.
37. For an invertible function $g$, it is always true that $g^{-1}(g(x)) = x$.

Chapter Eleven

# POLYNOMIAL AND RATIONAL FUNCTIONS

© Patrick Zephyr/Patrick Zephyr Nature Photography

## Contents

# 11.1 POWER FUNCTIONS AND PROPORTIONALITY

The goal of this chapter is to learn about polynomial and rational functions such as

$$y = x^3 + 3x^2 - 7 \qquad \text{and} \qquad y = \frac{x^3}{4x^5 + 2}.$$

Both types of functions are created from constant multiples of a power of $x$, so we start by studying these. We define:

> A **power function** is a function of the form
>
> $$f(x) = kx^p, \qquad \text{where } k \text{ and } p \text{ are constants.}$$

## Graphs of Power Functions

We know that varying $k$ stretches or compresses the graph of the function vertically but does not change its basic shape. If $k < 0$, the graph is reflected across the $x$-axis. To isolate the effect of the power $p$, we first study functions with $k = 1$.

## The Effect of the Power $p$

### Positive Integer Powers: $y = x^2, x^3, x^4, x^5 \ldots$

In the next example, we see that positive integer powers lead to two characteristic shapes.

**Example 1** Graph $y = x^2$, $y = x^3$, $y = x^4$, $y = x^5$. Describe and explain the similarities and differences in the graphs.

**Solution** The graphs of $y = x^2$ and $y = x^4$ in Figure 11.1 are both ⋃-shaped and symmetric about the $y$-axis. Since the power, 2 or 4, is even, the function takes only nonnegative function values. Thus, the graph lies above the $x$-axis, touching the $x$-axis only at the origin. As $x$ increases in either the positive and or the negative direction, the $y$-values become infinitely large. That is, $y \to \infty$ as $x \to \pm\infty$.

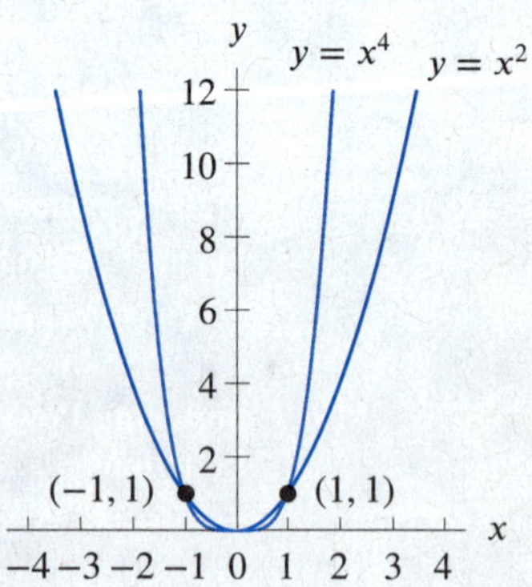


**Figure 11.1:** Graphs of positive even powers of $x$ are ⋃-shaped

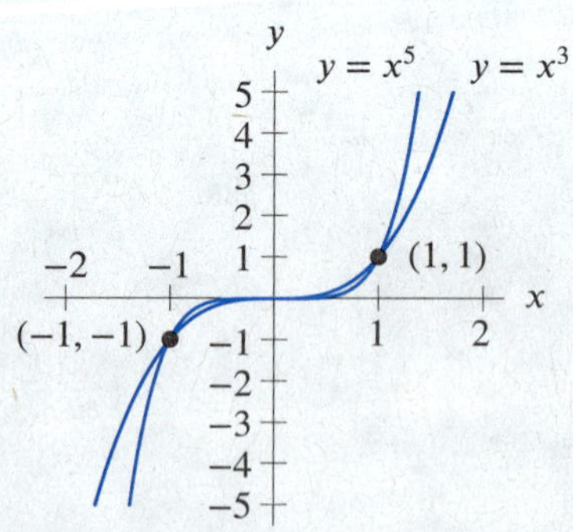


**Figure 11.2:** Graphs of positive odd powers of $x$ are "chair-shaped"

The graphs of $y = x^3$ and $y = x^5$ in Figure 11.2 resemble the side view of a chair and are symmetric about the origin. Since the power, 3 or 5, is odd, the function values are positive for positive $x$ and negative for negative $x$. Thus the graph lies below the $x$-axis for $x < 0$ and above the $x$-axis for $x > 0$. Functions with an odd power are increasing everywhere and pass through the origin. For $x \to \infty$, the long-run behavior is the same as for even powers because the function values get infinitely large. But for $x \to -\infty$, the function values go toward $-\infty$.

Generalizing, power functions with positive integer powers have the shapes in Figures 11.3 and 11.4:

- For $p$ a positive even integer, the graph of $y = x^p$ is $\bigcup$-shaped.
- For $p$ a positive odd integer, the graph of $y = x^p$ is "chair-shaped."

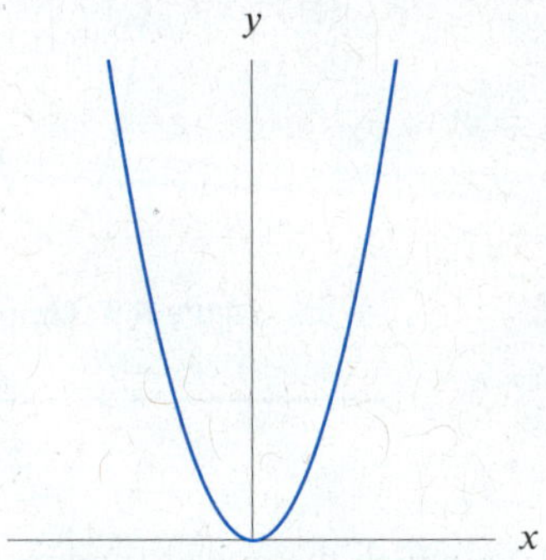


**Figure 11.3**: Positive even powers: Graph of $y = x^p$ is $\bigcup$-shaped

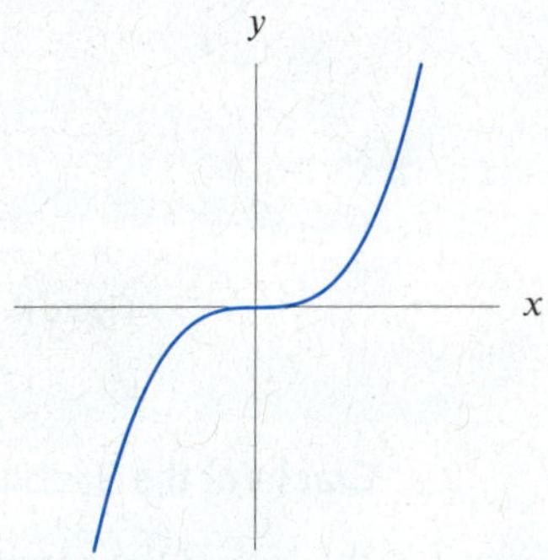


**Figure 11.4**: Positive odd powers: Graph of $y = x^p$ is "chair-shaped"

## Negative Integer Powers: $y = x^{-1}, x^{-2}, x^{-3}, x^{-4}, x^{-5}, \ldots$

The graphs of power functions with negative integer exponents also have two distinct shapes, one for odd and one for even powers. See Figures Figure 11.6 and 11.5.

Since $y = x^{-p} = 1/x^p$ for $x \neq 0$, we can use our knowledge of functions with positive powers to understand the graphs of their reciprocals, which have negative powers.

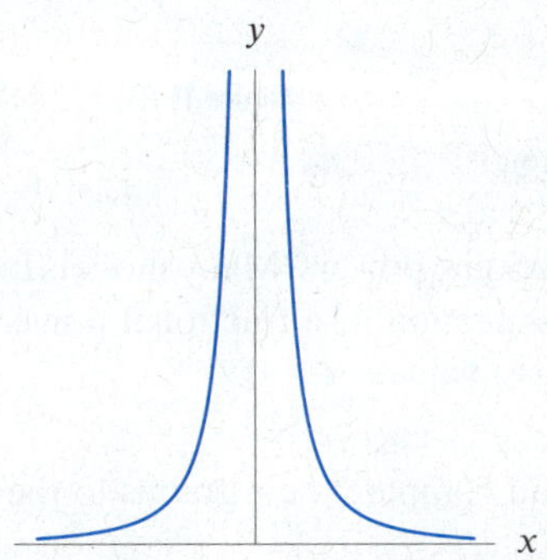


**Figure 11.5**: Negative even power: Graph of $y = x^{-2} = 1/x^2$

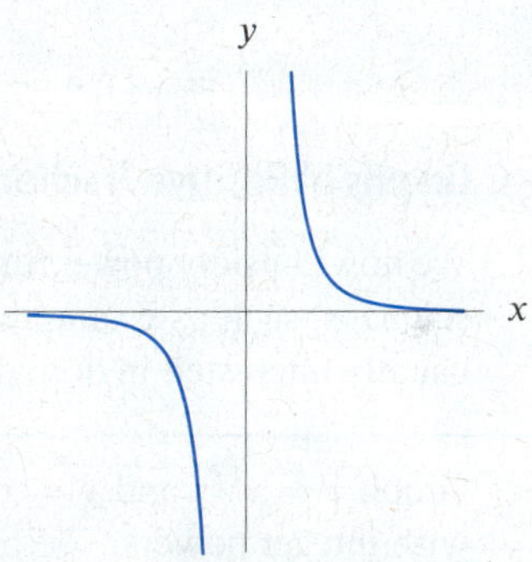


**Figure 11.6**: Negative odd power: Graph of $y = x^{-3} = 1/x^3$

**Example 2** Describe and explain the features of the graph of $y = x^{-2}$ in terms of the behavior of $y = x^2$.

**Solution** Figures 11.7 and 11.8 show the graphs of $y = x^{-2}$ and $y = x^2$ side by side. The two functions are reciprocals of each other. Since $y = x^2$ is nonnegative for all $x$, the function

$$y = x^{-2} = \frac{1}{x^2},$$

is also nonnegative for all $x$ and its graph lies above the $x$-axis. Since the graph of $y = x^2$ goes through the origin, $y = x^{-2}$ is undefined at $x = 0$ and the $y$-axis is a vertical asymptote.

As $x$ becomes large in the positive and negative direction, we saw that $x^2$ becomes increasingly large, so $x^{-2} = 1/x^2$ approaches zero. As a result, $y = 0$ is a horizontal asymptote of $y = x^{-2}$.

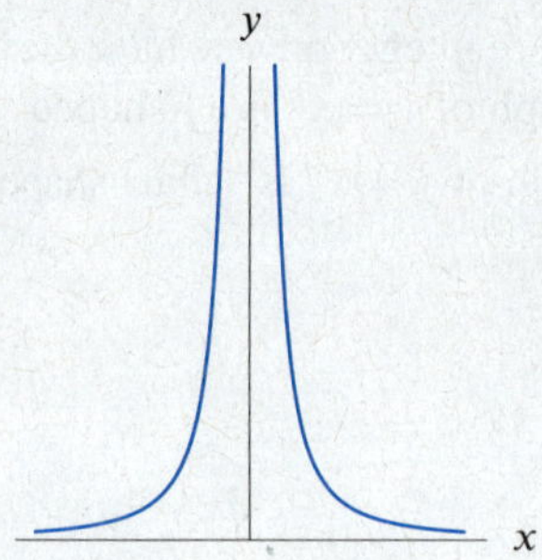


Figure 11.7: Graph of $y = x^{-2} = 1/x^2$

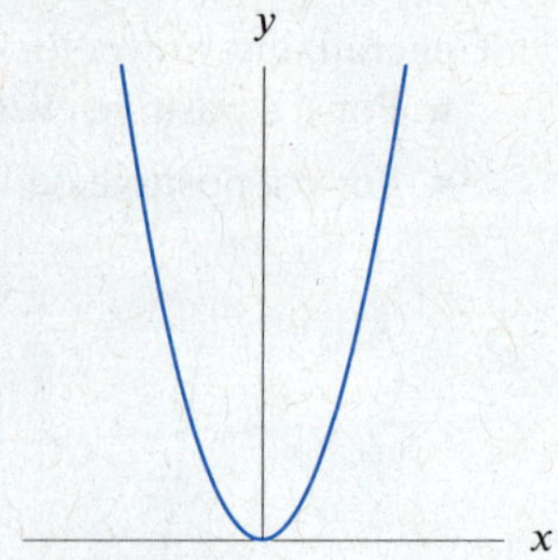


Figure 11.8: Graph of $y = x^2$

## Graphs of the Special Cases $y = x^0$ and $y = x^1$

We look at the special cases $p = 0$ and $p = 1$, in which the power function is linear. The graph of $y = x^0 = 1$ is a horizontal line through the point (1, 1). The graph of $y = x^1 = x$ is a line through the origin with slope 1.

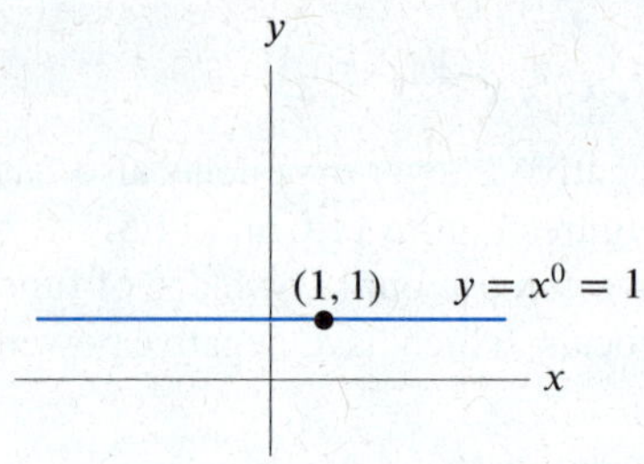


Figure 11.9: Graph of $y = x^0 = 1$

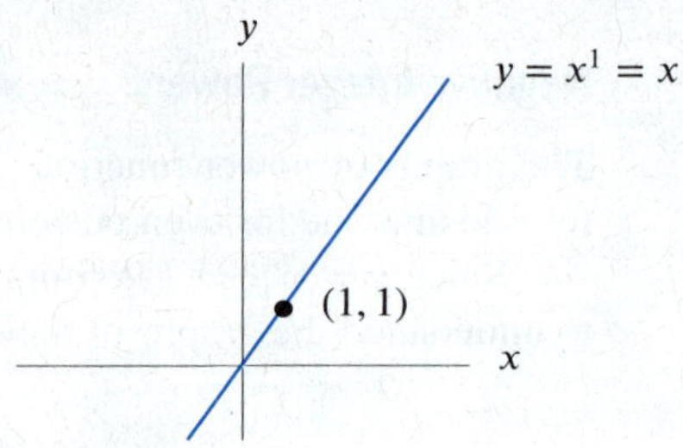


Figure 11.10: Graph of $y = x^1 = x$

## Graphs of Positive Fractional Powers

We now consider power functions with fractional powers. Many models in the biological and physical sciences, such as Examples 8 and 9 in this section, use fractional powers. For applications, we are usually interested in domains of nonnegative values, $x \geq 0$.

**Example 3** Graph $y = x^{1/2}$ and $y = x^{2.5}$ for $x \geq 0$ and compare their graphs to the graphs of power functions with integer powers.

**Solution** Figure 11.11 shows that the graph of $y = x^{1/2}$ bends in a direction opposite to that of the graph of $y = x^2$. The graph of $y = x^2$ is concave up, whereas the graph of $y = x^{1/2}$ is concave down. (Problem 29 on page 440 analyzes the concavity using rates of change.) Any power function with $p$ between 0 and 1 has a graph that is concave down.

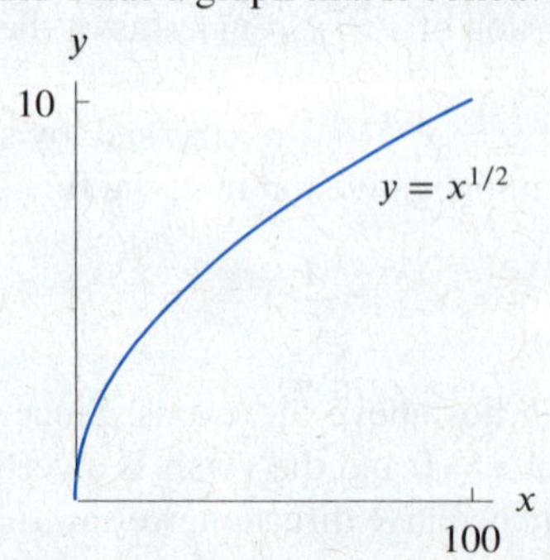


Figure 11.11: Fractional power $0 < p < 1$: Graph of $y = x^{1/2}$

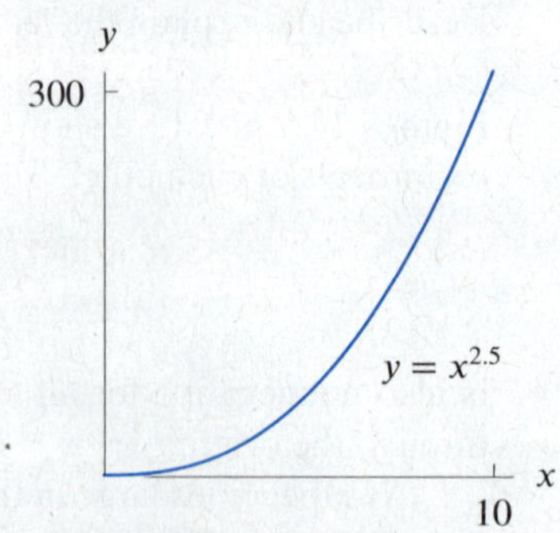


Figure 11.12: Fractional power $p > 1$: Graph of $y = x^{2.5}$

Figure 11.12, on the other hand, shows that the graph of $y = x^{2.5}$ has the same shape as $y = x^2$. Any power function with $p > 1$ has a graph which is concave up.

As $x$ increases, both $y = x^{1/2}$ and $y = x^{2.5}$ become infinitely large. All power functions with $p > 0$ have the same long-run behavior: As $x \to \infty$, the value of $x^p \to \infty$.

## Dominance

All power functions $y = x^p$ with a positive power, $p$, grow infinitely large for large values of $x$. However, they grow large at different rates. The larger the value of $p$, the steeper the graph and the faster the function values grow in the long run. See Figure 11.13. We say that $y = x^2$ *dominates* $y = x$ because $x^2$ is greater in the long run. For the same reason, $y = x$ dominates $y = x^{1/2}$. In general, if $0 < p < q$, the function $y = x^q$ dominates $y = x^p$.

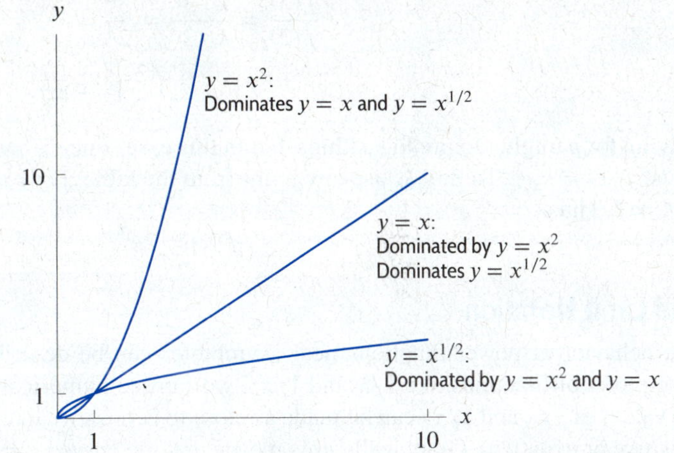


**Figure 11.13:** Comparing growth: $x^2$ dominates $x$ and $x$ dominates $x^{1/2}$

So far we have considered only power functions $y = kx^p$ with $k = 1$. The next example shows that even for large positive $k$, the function with the higher power eventually dominates ones with smaller powers.

**Example 4** Explain why the power function $y = x^3$ dominates $y = 100x^2$. Find a window that shows this behavior and the intersection points of the two graphs.

**Solution** We find the intersection points of the two graphs by setting the $y$-values equal:

$$x^3 = 100x^2$$
$$x^2(x - 100) = 0$$
$$x = 0 \quad \text{and} \quad x = 100.$$

Note that $x^3 = x \cdot x^2$. Therefore, for $x > 100$, we see that $x \cdot x^2 > 100x^2$, so $x^3$ dominates $100x^2$.

To show the points of intersection, the viewing window must include $x = 0$ and $x = 100$. Suppose we choose a window with $0 \le x \le 120$. Since $x = 120$ gives output values $y = 120^3 = 1{,}728{,}000$ and $y = 100 \cdot 120^2 = 1{,}440{,}000$, we can choose the window with $0 \le y \le 1{,}500{,}000$. See Figure 11.14.

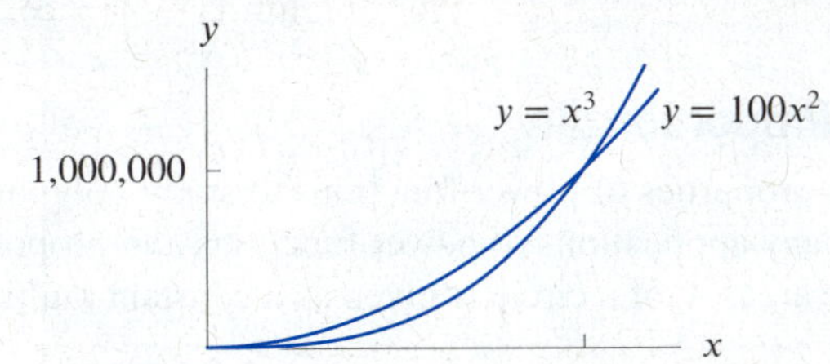


**Figure 11.14:** Dominance of $y = x^3$ over $y = 100x^2$

## Finding the Formula for a Power Function

As for linear and exponential functions, the formula of a power function can be found using two points on its graph.

**Example 5** Find a possible formula for the power function $f$ whose values are given in the table:

| $x$ | 2 | 3 | 4 | 5 |
|---|---|---|---|---|
| $f(x)$ | 16 | 54 | 128 | 250 |

**Solution** We need to solve $f(x) = kx^p$ for $p$ and $k$. Since $f(3) = 54$ we have $54 = k3^p$ and since $f(2) = 16$ we have $16 = k2^p$. Taking the ratio enables us to cancel $k$, giving an equation for $p$:

$$\frac{f(3)}{f(2)} = \frac{k \cdot 3^p}{k \cdot 2^p} = \frac{3^p}{2^p} = \frac{54}{16}$$

$$\left(\frac{3}{2}\right)^p = \frac{27}{8}.$$

Solving for $p$ might require logarithms, but in this case, since $3^3 = 27$ and $2^3 = 8$, we can see $p = 3$. Thus, $f(x) = kx^3$. To find $k$, use any point from the table. For example, $f(2) = k \cdot 2^3 = 8k = 16$, so $k = 2$. Thus,

$$f(x) = 2x^3.$$

## Asymptotes and Limit Notation

The behavior of power functions near asymptotes can be described using limit notation. For the asymptotes of functions like $1/x$ and $1/x^2$, we can see numerically (Problem 27 on page 440) that the values of $1/x$ and $1/x^2$ can be made as close to zero as we like by choosing a sufficiently large $x$ (positive or negative). Graphically, this means that the curves $y = 1/x$ and $y = 1/x^2$ get closer and closer to the $x$-axis for large values of $x$. We write

$$\frac{1}{x} \to 0 \text{ as } x \to \pm\infty \quad \text{and} \quad \frac{1}{x^2} \to 0 \text{ as } x \to \pm\infty.$$

Using limit notation, we say

$$\lim_{x\to\infty}\left(\frac{1}{x}\right) = 0, \quad \lim_{x\to-\infty}\left(\frac{1}{x}\right) = 0 \quad \text{and} \quad \lim_{x\to\infty}\left(\frac{1}{x^2}\right) = 0, \quad \lim_{x\to-\infty}\left(\frac{1}{x^2}\right) = 0.$$

On the other hand, as $x$ gets close to zero, the values of $1/x$ and $1/x^2$ get very large in magnitude. (See Problem 28 on page 440.) The $y$-axis is a vertical asymptote for each graph. We write

$$\frac{1}{x} \to \infty \text{ as } x \to 0^+ \quad \text{and} \quad \frac{1}{x} \to -\infty \text{ as } x \to 0^-.$$

Using limit[1] notation, we write

$$\lim_{x\to0^+}\left(\frac{1}{x}\right) = \infty \quad \text{and} \quad \lim_{x\to0^-}\left(\frac{1}{x}\right) = -\infty.$$

We see that $\frac{1}{x^2} \to \infty$ as $x \to 0$, namely

$$\lim_{x\to0}\left(\frac{1}{x^2}\right) = \infty.$$

## Power Functions and Proportionality

We use the properties of power functions to study polynomials and rational functions. In addition, there are many applications of power functions and proportionality. An example of proportionality is when the area, $A$, of a circle is given as a constant multiple of the square of the radius, $r$:

$$A = \pi r^2.$$

In this case, the constant is $\pi$. We make the following definition:

[1] Some authors say that these limits do not exist.

> A quantity $y$ is **(directly) proportional** to $x^n$ if
> $$y = kx^n, \qquad \text{where } k \text{ and } n \text{ are constants.}$$
> $k$ is called the **proportionality constant.** Thus, $y$ is a **power function** of $x$.

Notice that if $n$ is positive, then $x$ and $y$ increase together. If $n$ is negative, the behavior is different.

For example, the weight, $w$, of an object[2] is proportional to the object's distance, $d$, from the earth's center to the power of negative 2. That is,

$$w = kd^{-2}.$$

Because $d^{-2} = 1/d^2$, we can also write this relationship in the another form:

$$w = \frac{k}{d^2},$$

which is an example of *inverse proportionality*. We make the following new definition:

> A quantity $y$ is **inversely proportional** to $x^n$ if
> $$y = \frac{k}{x^n}, \qquad \text{where } k \text{ and } n \text{ are constants.}$$
> $k$ is the proportionality constant.

Thus, if $y$ is inversely proportional to $x^n$, then $y$ is also (directly) proportional to $x^{-n}$. In Example 6 we see that if variables are inversely related, when one increases, the other decreases.

**Example 6** The weight, $w$, of an object is inversely proportional to the square of the object's distance, $d$, from the earth's center. An object weighs 44 pounds on the surface of the earth, which is 3960 miles from the earth's center. Make a table of values and a graph of $w = f(d)$ for $4000 \le d \le 8000$.

**Solution** We have

$$w = \frac{k}{d^2}.$$

Substituting $w = 44$ when $d = 3960$ enables us to find $k$:

$$44 = \frac{k}{3960^2} \qquad \text{so} \qquad k = 44 \cdot 3960^2 = 6.900 \cdot 10^8.$$

Using this value of $k$, we get the data in Table 11.1 and the graph in Figure 11.15.

**Table 11.1** *Weight of an object, w, is inversely proportional to the square of the distance, d, from the earth's center*

| $d$, miles | $w = f(d)$, lbs |
|---|---|
| 4000 | 43.1 |
| 5000 | 27.6 |
| 6000 | 19.2 |
| 7000 | 14.1 |
| 8000 | 10.8 |

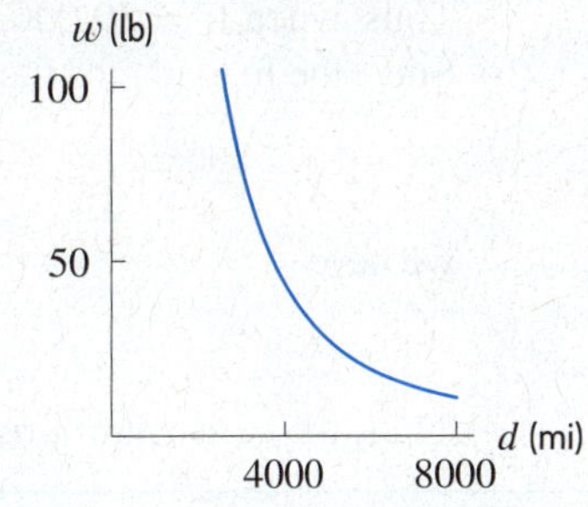


**Figure 11.15:** Weight, $w$, is inversely proportional to the square of the object's distance, $d$, from the earth's center

[2]There is a distinction between mass and weight. For example, astronauts in orbit may be weightless, but still have mass.

**Example 7** Which of the following functions are power functions? For each power function, give the value of the constant of proportionality, $k$, and the power, $p$, in the formula $y = kx^p$.

(a) $f(x) = 13\sqrt[3]{x}$ (b) $g(x) = 2(x+5)^3$ (c) $u(x) = \sqrt{\dfrac{25}{x^3}}$ (d) $v(x) = 6 \cdot 3^x$

**Solution** The functions $f$ and $u$ are power functions; the functions $g$ and $v$ are not.

(a) The function $f(x) = 13\sqrt[3]{x}$ is a power function because we can write its formula as

$$f(x) = 13x^{1/3}.$$

Here, $k = 13$ and $p = 1/3$.

(b) Although the value of $g(x) = 2(x+5)^3$ is proportional to the cube of $x+5$, it is *not* proportional to a power of $x$. We see this by rewiting $(x+5)^3 = x^3 + 15x^2 + 75x + 125$. Since there are terms involving $x^2$, $x$, and a constant, we cannot write $g(x)$ in the form $g(x) = kx^p$. Thus, $g$ is not a power function.

(c) We can rewrite the formula for $u(x) = \sqrt{25/x^3}$ as

$$u(x) = \frac{\sqrt{25}}{\sqrt{x^3}} = \frac{5}{(x^3)^{1/2}} = \frac{5}{x^{3/2}} = 5x^{-3/2}.$$

Thus, $u$ is a power function. Here, $k = 5$ and $p = -3/2$.

(d) Although the value of $v(x) = 6 \cdot 3^x$ is proportional to a power of 3, the power is not a constant—it is the variable $x$. In fact, $v(x) = 6 \cdot 3^x$ is an exponential function, not a power function. Notice that $y = 6 \cdot x^3$ is a power function. However, $6 \cdot x^3$ and $6 \cdot 3^x$ are quite different.

**Example 8** Geometry tells us that the radius of a sphere is directly proportional to the cube root of its volume.

(a) Use this proportionality relationship and the fact that a sphere of radius 18.2 cm has a volume of 25,252.4 cm$^3$ to find the radius of a sphere whose volume is 30,000 cm$^3$.

(b) Confirm that the value of the proportionality constant found in part (a) is the same as that obtained from the formula for the volume of a sphere.

**Solution** (a) Since the radius of the sphere is proportional to the cube root of its volume, we know that

$$r = kV^{1/3}, \quad \text{for } k \text{ constant}.$$

We also know that $r = 18.2$ cm when $V = 25{,}252.4$ cm$^3$, so we have:

$$18.2 = k(25{,}252.4)^{1/3},$$

giving the value of $k$:

$$k = \frac{18.2}{(25{,}252.4)^{1/3}} = 0.620.$$

Thus, when $V = 30{,}000$, the radius of the sphere is $r = 0.620(30{,}000)^{1/3} = 19.265$ cm.

(b) Since the formula for the volume of a sphere is

$$V = \frac{4}{3}\pi r^3,$$

we have

$$r = \left(\frac{3}{4\pi}V\right)^{1/3} = \left(\frac{3}{4\pi}\right)^{1/3} V^{1/3}.$$

Thus, the constant of proportionality is $(3/(4\pi))^{1/3} = 0.620$, as in part (a).

**Example 9** Water is leaking out of a container with a hole in the bottom. Torricelli's Law states that at any instant, the velocity $v$ with which water escapes from the container is a power function of $d$, the depth of the water at that moment. When $d = 9$ feet, then $v = 24$ ft/sec; when $d = 1/4$ foot, then $v = 4$ ft/sec. Express $v$ as a function of $d$.

**Solution** Torricelli's Law tells us that $v = kd^p$, where $k$ and $p$ are constants. Since $v = 24$ when $d = 9$ we have $24 = k9^p$ and since $v = 4$ when $d = 1/4$ we have $4 = k(1/4)^p$. Taking the ratio gives

$$\frac{24}{4} = \frac{k9^p}{k(1/4)^p}$$
$$6 = 36^p.$$

Since $36^{1/2} = 6$, we must have $p = 1/2$, so $v = kd^{1/2}$. Substituting $v = 24$ and $d = 9$ gives

$$24 = k9^{1/2}, \qquad \text{so } k = 8.$$

Therefore we have

$$v = 8d^{1/2}.$$

*Note*: Torricelli's Law is often written in the form $v = \sqrt{2gd}$, where $g = 32$ ft/sec$^2$ is the acceleration due to gravity. Since $\sqrt{2 \cdot 32d} = \sqrt{64d} = 8\sqrt{d}$, these two forms are equivalent.

## Exercises and Problems for Section 11.1

### Skill Refresher

In Exercises S1–S4, simplify the expression.

**S1.** $\sqrt{36t^2}$

**S2.** $\left(3x\sqrt{x^3}\right)^2$

**S3.** $(0.1)^2\left(4xy^2\right)^2$

**S4.** $7\left(5w^{1/2}\right)\left(2w^{1/3}\right)$

In Exercises S5–S6, solve for $x$.

**S5.** $\dfrac{10x^5}{x^2} = 2$

**S6.** $5x^{-2} = 500$

In Exercises S7–S10, is the statement true or false?

**S7.** $t^3t^4 = t^{12}$

**S8.** $(p^3)^8 = p^{11}$

**S9.** $\dfrac{m^8}{2m^2} = \dfrac{1}{2}m^4$

**S10.** $5z^{-4} = \dfrac{1}{5z^4}$

### Exercises

**1.** Write $y = 3\left(\dfrac{2}{5\sqrt{7x}}\right)^4$ in the form $y = ax^p$ and state the values of $a$ and $p$.

**2.** Write $y = \sqrt{\pi(2x)^3}$ in the form $y = ax^p$ and state the values of $a$ and $p$.

In Exercises 3–8, is the function a power function? If so, write it in the form $f(x) = kx^p$.

**3.** $g(x) = \dfrac{(-x^3)^3}{6}$

**4.** $R(t) = \dfrac{4}{\sqrt{16t}}$

**5.** $f(x) = 4(x+7)^2$

**6.** $T(s) = (6s^{-2})(es^{-3})$

**7.** $h(x) = 22(7^x)^2$

**8.** $K(w) = \dfrac{w^4}{4\sqrt{w^3}}$

In Exercises 9–12, does the power function appear to have an odd, even, or fractional power?

**9.**

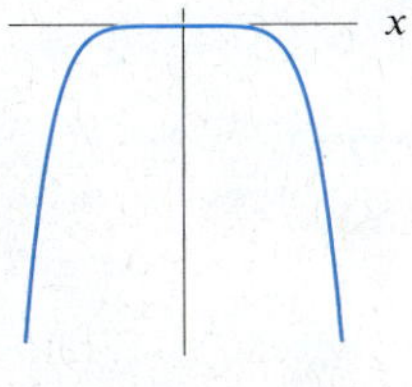


**10.**

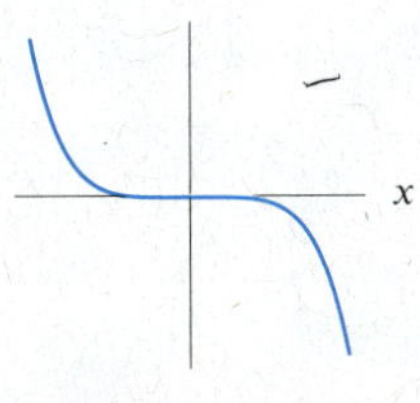


**11.**

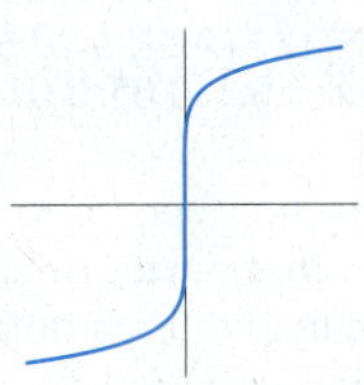


**12.**

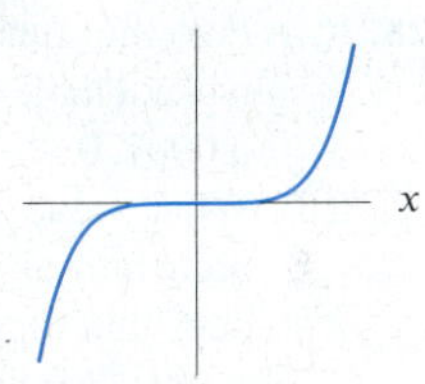


In Exercises 13–14, find a possible formula for the power function with the values given.

**13.** $f(1) = \frac{3}{2}$ and $f(2) = \frac{3}{8}$

**14.** $g\left(-\frac{1}{5}\right) = 25$ and $g(2) = -\frac{1}{40}$

In Exercises 15–16, find a power function through the two points.

**15.** (1, 3) (4, 13)

**16.** (7, 8) (1, 0.7)

In Exercises 17–20, find a possible formula for the power function.

**17.**

| $x$ | 0 | 1 | 2 | 3 |
|---|---|---|---|---|
| $j(x)$ | 0 | 2 | 16 | 54 |

**18.**

| $x$ | 2 | 3 | 4 | 5 |
|---|---|---|---|---|
| $f(x)$ | 12 | 27 | 48 | 75 |

**19.**

| $x$ | −6 | −2 | 3 | 4 |
|---|---|---|---|---|
| $g(x)$ | 36 | 4/3 | −9/2 | −32/3 |

**20.**

| $x$ | −2 | −1/2 | 1/4 | 4 |
|---|---|---|---|---|
| $h(x)$ | −1/2 | −8 | −32 | −1/8 |

**21.** Suppose $c$ is directly proportional to the square of $d$. If $c = 45$ when $d = 3$, find the constant of proportionality and write the formula for $c$ as a function of $d$. Use your formula to find $c$ when $d = 5$.

**22.** Suppose $c$ is inversely proportional to the square of $d$. If $c = 45$ when $d = 3$, find the constant of proportionality and write the formula for $c$ as a function of $d$. Use your formula to find $c$ when $d = 5$.

**23.** If $y$ is directly proportional to $x$, and $y = 6$ when $x = 4$, find the constant of proportionality, write a formula for $y$ in terms of $x$, and find $x$ when $y = 8$.

**24.** If $y$ is inversely proportional to $x$, and $y = 6$ when $x = 4$, find the constant of proportionality, write a formula for $y$ in terms of $x$, and find $x$ when $y = 8$.

**25.** Find **(a)** $\lim_{x\to\infty} x^{-4}$ **(b)** $\lim_{x\to-\infty} 2x^{-1}$

**26.** Find **(a)** $\lim_{t\to\infty}(t^{-3} + 2)$ **(b)** $\lim_{y\to-\infty} (5 - 7y^{-2})$

## Problems

**27.** **(a)** For the functions $f(x) = 1/x$ and $g(x) = 1/x^2$, make a table of values for $x = 0, 10, 20, 30, 40, 50$.

**(b)** What value does each function appear to be approaching as $x$ increases? What does this suggest about the limit of each function as $x \to \infty$? What about the limit as $x \to -\infty$?

**(c)** Which function appears to be approaching its limit more rapidly as $x$ increases?

**28.** **(a)** For the functions $f(x) = 1/x$ and $g(x) = 1/x^2$, make a table of values for $x = 0.1, 0.05, 0.01, 0.001, 0.0001, 0$.

**(b)** What value does each function appear to be approaching as $x$ approaches zero from the right? What does this suggest about the limit of each function as $x \to 0^+$? What about the limit as $x \to 0^-$?

**(c)** Which function appears to be approaching the value more rapidly as $x$ goes to 0 from the right?

**29.** **(a)** Find the average rate of change of $f(x) = x^{1/2}$ between successive values of $x$ for $x = 0, 2, 4, 6, 8$. Do the same for $g(x) = x^2$.

**(b)** What do these rates of change suggest about the concavity of the graphs of $f(x) = x^{1/2}$ and $g(x) = x^2$?

**30.** Compare the graphs of $y = x^2$, $y = x^4$, and $y = x^6$. Describe the similarities and differences.

**31.** Describe the behavior of $y = x^{-3}$ and $y = x^{1/3}$ as

**(a)** $x \to 0$ from the right **(b)** $x \to \infty$

**32.** Describe the behavior of $y = x^{-10}$ and $y = -x^{10}$ as

**(a)** $x \to 0$ **(b)** $x \to \infty$ **(c)** $x \to -\infty$

**33.** Figure 11.16 shows the power function $y = c(t)$. Is $c(t) = 1/t$ the only possible formula for $c$? Could there be others?

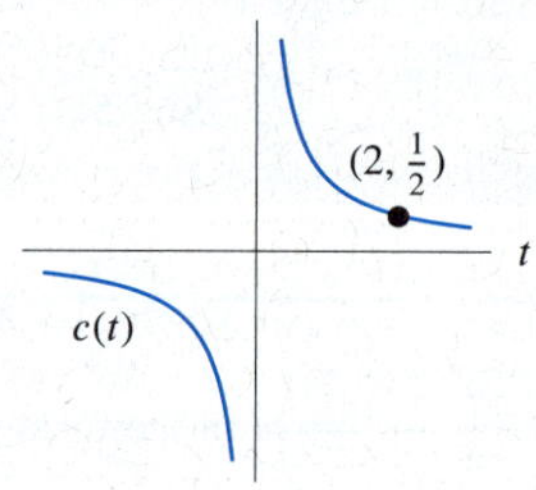


Figure 11.16

**34.** **(a)** Figure 11.17 shows $g(x)$, a mystery power function. If you learn that the point $(-1, 3)$ lies on its graph, do you have enough information to write a formula for $g(x)$?

**(b)** If you are told that the point $(1, -3)$ also lies on the graph, what new deductions can you make?

**(c)** If the point $(2, -96)$ lies on the graph $g$, in addition to the points already given, state three other points that also lie on it.

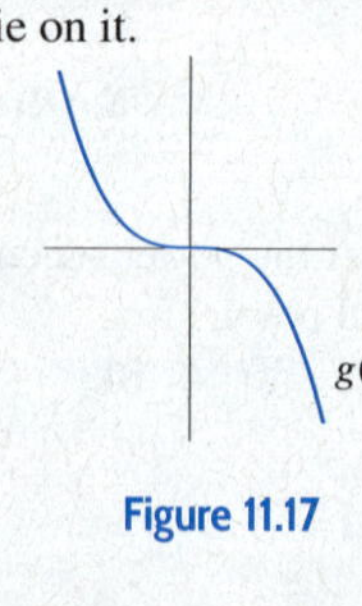


Figure 11.17

**35.** **(a)** Match the functions $x$, $x^2$, $x^3$, $x^{1/2}$, $x^{1/3}$, $x^{3/2}$ with the graphs in Figure 11.18. Justify your choice.

**(b)** What is the relationship between the concavity of $y = x^2$ and $y = x^{1/2}$? Between the concavity of $y = x^3$ and $y = x^{1/3}$ for $x > 0$? Explain why this happens.

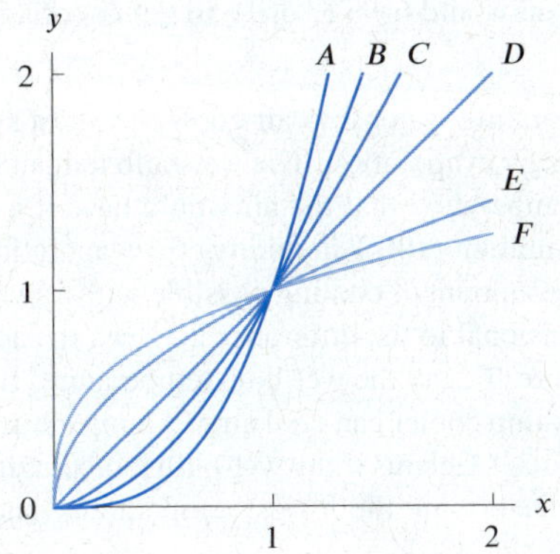


Figure 11.18

**36.** Figures 11.19 and 11.20 show the graphs of $y = x^{1/2}$, $y = x^{1/3}$, $y = x^{1/4}$, $y = x^{1/5}$.

**(a)** What is the domain of each of these four functions?

**(b)** Using your answer to part (a), match the functions to the graphs.

**(c)** What are the points of intersection in the graphs?

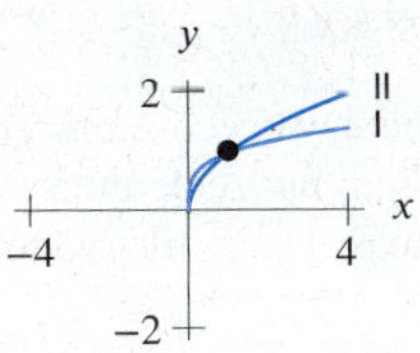


Figure 11.19

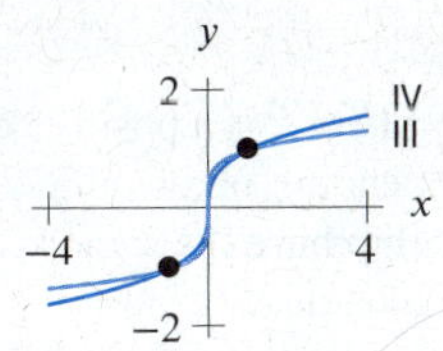


Figure 11.20

**37.** Figure 11.21 shows graphs of two power functions, $f(x) = x^r$ and $g = x^s$. Given that $s < r < 0$, is $x_0$ larger or smaller than $x_1$?

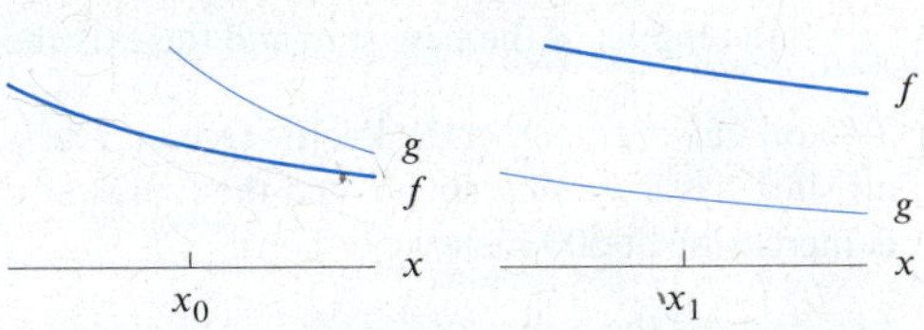


Figure 11.21

**38.** Figure 11.22 shows graphs of two power functions, $p(x) = x^m$ and $q(x) = x^n$. Given that $0 < x_0 < x_1$, rank 0, $m$, and $n$ in order from smallest to largest.

[3] *The World Almanac Book of Facts*, 1999,p. 718.

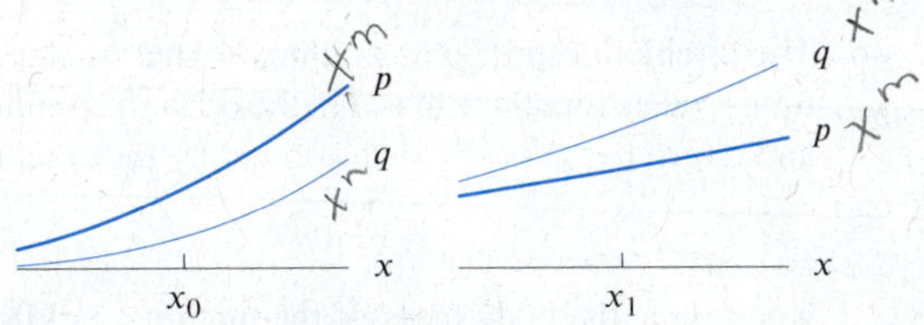


Figure 11.22

Problems 39–40 refer to the power functions $f, g, w, v$ in Figure 11.23. All four graphs contain the point (1, 1).

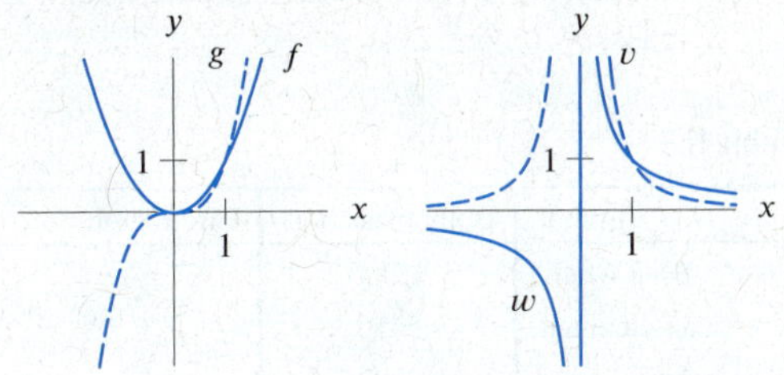


Figure 11.23

**39.** Rank the four functions in order of $p$, the power, from least to greatest.

**40.** For which (if any) of the four functions is the power $p$ odd?

**41.** For the power function $F(x) = kx^n$, let $f(x) = nkx^{n-1}$. Find $f$ given that $F(x) = \dfrac{1}{\sqrt[3]{7x}}$.

**42.** For the power function $f(x) = kx^n$, let $F(x) = \dfrac{kx^{n+1}}{n+1}$. Find $F$ given that $f(x) = \dfrac{\sqrt[3]{x^2}}{4}$.

**43.** The cost of denim fabric is directly proportional to the amount that you buy. Let $C(x)$ be the cost, in dollars, of $x$ yards of denim fabric.

**(a)** Write a formula for the cost, $C(x)$, in terms of $x$. Your answer will contain a constant, $k$.

**(b)** A particular type of denim costs \$28.50 for 3 yards. Find $k$ and rewrite the formula for $C(x)$ using it.

**(c)** Graph $C(x)$.

**(d)** How much does it cost to buy 5.5 yards of denim?

**44.** The musical pitch, $P$, of a guitar string can be varied by changing its density, $\rho$. The pitch is inversely proportional to the square root of the density, so a string that is denser produces a lower pitch. Write an expression for the pitch in terms of the density.

**45.** Do you expect the number of calories in a serving of broiled ground beef to be proportional, or inversely proportional, to the numbers of ounces of beef? Which? Explain. If three ounces of broiled ground beef contains 245 calories,[3] write a formula for the relationship and find the number of calories in 4 ounces of broiled hamburger.

**46.** The circulation time of a mammal—that is, the average time it takes for all the blood in the body to circulate once and return to the heart—is governed by the equation

$$t = 17.4m^{1/4},$$

where $m$ is the body mass of the mammal in kilograms, and $t$ is the circulation time in seconds.[4]

**(a)** Complete Table 11.2, which shows typical body masses in kilograms for various mammals.[5]

**(b)** If the circulation time of one mammal is twice that of another, what is the relationship between their body masses?

**Table 11.2**

| Animal | Body mass (kg) | Circulation time (sec) |
|---|---|---|
| Blue whale | 91000 | |
| African elephant | 5450 | |
| White rhinoceros | 3000 | |
| Hippopotamus | 2520 | |
| Black rhinoceros | 1170 | |
| Horse | 700 | |
| Lion | 180 | |
| Human | 70 | |

**47.** In 2004, Agutter and Wheatley[6] reported

> The relationship between body mass, $M$, and standard metabolic rate, $B$, among living organisms remains controversial, though it is widely accepted that in many cases $B$ is approximately proportional to the three-quarters power of $M$.

**(a)** Write a function that represents this relationship.

**(b)** The average mass of an African forest elephant is 4.6 metric tons[7] and that of a typical mouse is 20 grams. Use part (a) to determine how many times greater the metabolic rate of an elephant is than that of a mouse. (1 metric ton = 1,000,000 grams.)

**48.** A 30-second commercial during Super Bowl XLIV in 2010 cost advertisers up to \$3 million. For the first Super Bowl in 1967, an advertiser could have purchased approximately 34.439 minutes of advertising time for the same amount of money.[8]

**(a)** Assuming that cost is proportional to time, find the cost of advertising, in dollars/second, during the 1967 and 2010 Super Bowls.

**(b)** How many times more expensive was Super Bowl advertising in 2010 than in 1967?

**49.** At a constant speed, is the time to drive from Long Island to Albany, NY, proportional, or inversely proportional, to the speed? Explain. At 55 mph, it takes 3.5 hours to make the trip. Write a formula for the relationship and find how fast you would have to drive to get to Albany in 3 hours.

**50.** Evaporating water into air cools the air; a swamp cooler works by evaporation. The wet-bulb temperature of air is the temperature that the air would have if it were cooled to saturation (100% humidity) by evaporating water into it. The amount of cooling possible with a swamp cooler is proportional to the difference between the actual air temperature, $T$, and the wet-bulb temperature, $W$. A particular swamp cooler can cool air at a temperature of 40° Celsius to 23° Celsius if the wet-bulb temperature is 20° Celsius. Determine the lowest temperature that this swamp cooler could achieve:

**(a)** In Phoenix, where the actual temperature is 48°C and the wet-bulb temperature is 22°C.

**(b)** In Miami, where the actual temperature is 32°C and the wet-bulb temperature is 26°C.

**51.** Ship designers usually construct scale models before building a real ship. The formula that relates the speed $u$ to the hull length $l$ of a ship is

$$u = k\sqrt{l},$$

where $k$ is a positive constant. This constant $k$ varies depending on the ship's design, but scale models of a real ship have the same $k$ as the real ship after which they are modeled.[9]

**(a)** How fast should a scale model with hull length 4 meters travel to simulate a real ship with hull length 225 meters traveling 9 meters/sec?

**(b)** A new ship is to be built whose speed is to be 10% greater than the speed of an existing ship with the same design. What is the relationship between the hull lengths of the new ship and the existing ship?

**52.** Two oil tankers crash in the Pacific Ocean. The spreading oil slick has a circular shape, and the radius of the circle is increasing at 200 meters per hour.

**(a)** Express the radius of the spill, $r$, as a power function of time, $t$, in hours since the crash.

---

[4] K. Schmidt-Nielsen, *Scaling, Why is Animal Size So Important?* (Cambridge: CUP, 1984).

[5] R. McNeill Alexander, *Dynamics of Dinosaurs and Other Extinct Giants* (New York: Columbia University Press, 1989).

[6] P. S. Agutter and D. N. Wheatley (2004), "Metabolic scaling: consensus or controversy?" *Theor. Biol. Med. Mod.* 1:13, PMID: 15546492

[7] www.pbs.org/wnet/nature/elephants/life.html, accessed June 14, 2010.

[8] www.dailyfinance.com/2009/09/20/cbs-super-bowl-ad-sales-slip-behind-last-years-pace, accessed November 27, 2009.

[9] R. McNeill Alexander, *Dynamics of Dinosaurs and Other Extinct Giants* (New York: Columbia University Press, 1989).

(b) Express the area of the spill, $A$, as a power function of time, $t$.

(c) Cleanup efforts begin 7 hours after the spill. How large an area is covered by oil at that time?

53. A volcano erupts in a powerful explosion. The sound from the explosion is heard in all directions for many hundreds of kilometers. The speed of sound is about 340 meters per second.

(a) Fill in Table 11.3 showing the distance, $d$, that the sound of the explosion has traveled at time $t$. Write a formula for $d$ as a function of $t$.

(b) How long after the explosion will a person living 200 km away hear the explosion?

(c) Fill in Table 11.3 showing the land area, $A$, over which the explosion can be heard as a function of time. Write a formula for $A$ as a function of $t$.

(d) The average population density around the volcano is 31 people per square kilometer. Write a formula for $P$ as function of $t$, where $P$ is the number of people who have heard the explosion at time $t$.

(e) Graph the function $P = f(t)$. How long will it take until 1 million people have heard the explosion?

**Table 11.3**

| Time, $t$ | 5 sec | 10 sec | 1 min | 5 min |
|---|---|---|---|---|
| Distance, $d$ (km) | | | | |
| Area, $A$ ($\text{km}^2$) | | | | |

54. In a microwave oven, cooking time is inversely proportional to the amount of power used. It takes 6.5 minutes to heat a frozen dinner at 750 watts.

(a) Write a formula for the cooking time, $t$, as a function of power level, $w$.

(b) Fill in Table 11.4 with the cooking times needed to heat the frozen dinner at various power levels.

(c) Graph the function $t = f(w)$.

(d) If it takes 2 minutes to heat a rhubarb crumble at 250 watts, how long will it take at 500 watts?

**Table 11.4**

| Power, $w$ (watts) | 250 | 300 | 500 | 650 |
|---|---|---|---|---|
| Time, $t$ (mins) | | | | |

55. An average hailstone is a sphere of radius 0.3 centimeter. Severe thunderstorms can produce hailstones of radius 0.95 centimeter. The largest hailstone found in the US had radius 7.05 centimeters. Table 11.5 gives the masses of these hailstones in grams.[10]

(a) Using the data given, check that mass, $m$, is proportional to the cube of the radius, $r$.

(b) Find the constant of proportionality and write $m$ as a function of $r$.

(c) The largest recorded hailstone was found in India in 1939 and weighed 3.4 kilograms. What was its radius?

(d) Calculate the density of ice in grams per cubic centimeter. [Hint: Density is mass per unit volume.]

**Table 11.5**

| Radius, $r$ (cm) | 0.3 | 0.95 | 7.05 |
|---|---|---|---|
| Mass, $m$ (gm) | 0.058 | 1.835 | 750 |

56. The following questions involve the behavior of the power function $y = x^{-p}$, for $p$ a positive integer. If a distinction between even and odd values of $p$ is significant, the significance should be indicated.

(a) What is the domain of $y = x^{-p}$? What is the range?

(b) What symmetries does the graph of $y = x^{-p}$ have?

(c) What is the behavior of $y = x^{-p}$ as $x \to 0$?

(d) What is the behavior of $y = x^{-p}$ for large positive values of $x$? For large negative values of $x$?

57. Let $f(x) = 16x^4$ and $g(x) = 4x^2$.

(a) If $f(x) = g(h(x))$, find a possible formula for $h(x)$, assuming $h(x) \leq 0$ for all $x$.

(b) If $f(x) = j(2g(x))$, find a possible formula for $j(x)$, assuming $j(x)$ is a power function.

58. Consider the power function $y = t(x) = k \cdot x^{p/3}$ where $p$ is any integer, $p \neq 0$.

(a) For what values of $p$ does $t(x)$ have domain restrictions? What are those restrictions?

(b) What is the range of $t(x)$ if $p$ is even? If $p$ is odd?

(c) What symmetry does the graph of $t(x)$ exhibit if $p$ is even? If $p$ is odd?

## 11.2 POLYNOMIAL FUNCTIONS

A *polynomial function* is a sum of power functions whose exponents are nonnegative integers. We use what we learned about power functions to study polynomials.

[10]C. Donald Ahrens, *Essentials of Meteorology* (Wadsworth: Belmont, CA, 1998).

**Example 1** You make five separate deposits of \$1000 each into a savings account, one deposit per year, beginning today. What annual interest rate gives a balance in the account of \$6000 five years from today? (Assume the interest rate is constant over these five years.)

**Solution** Let $r$ be the annual interest rate. Our goal is to determine what value of $r$ gives you \$6000 in five years. In year $t = 0$, you make a \$1000 deposit. One year later, you have \$1000 plus the interest earned on that amount. At that time, you add another \$1000.

To picture how this works, imagine the account pays $r = 5\%$ annual interest, compounded annually. Then, after one year, your balance would be

$$\begin{aligned}\text{Balance} &= (100\% \text{ of Initial deposit}) + (5\% \text{ of Initial deposit}) + \text{Second deposit}\\ &= 105\% \text{ of } \underbrace{\text{Initial deposit}}_{\$1000} + \underbrace{\text{Second deposit}}_{\$1000}\\ &= 1.05(1000) + 1000.\end{aligned}$$

Here $1.05 = 1 + 0.05$ is the annual growth factor. Let $x$ represent the annual growth factor, $1 + r$. Then we write the balance after one year in terms of $x$:

$$\text{Balance after one year} = 1000x + 1000.$$

After two years, you would have earned interest on the first-year balance. This gives

$$\text{Balance after earning interest} = \underbrace{(1000x + 1000)}_{\text{First-year balance}}x = 1000x^2 + 1000x.$$

The third \$1000 deposit brings your balance to

$$\text{Balance after two years} = 1000x^2 + 1000x + \underbrace{1000.}_{\text{Third deposit}}$$

A year's worth of interest on this amount, plus the fourth \$1000 deposit, brings your balance to

$$\begin{aligned}\text{Balance after three years} &= \underbrace{(1000x^2 + 1000x + 1000)x}_{\text{Second-year balance}} + \underbrace{1000}_{\text{Fourth deposit}}\\ &= 1000x^3 + 1000x^2 + 1000x + 1000.\end{aligned}$$

The pattern is this: Each of the \$1000 deposits grows to $\$1000x^n$ by the end of its $n^{\text{th}}$ year in the bank. Thus,

$$\text{Balance after five years} = 1000x^5 + 1000x^4 + 1000x^3 + 1000x^2 + 1000x.$$

If the interest rate is chosen correctly, then the balance will be \$6000 in five years. This gives us

$$1000x^5 + 1000x^4 + 1000x^3 + 1000x^2 + 1000x = 6000.$$

Dividing by 1000 and moving the 6 to the left side, we have the equation

$$x^5 + x^4 + x^3 + x^2 + x - 6 = 0.$$

Solving this equation for $x$ determines how much interest we must earn. Using a computer or calculator, we find where the graph of $Q(x) = x^5 + x^4 + x^3 + x^2 + x - 6$ crosses the $x$-axis. Figure 11.24

shows that this occurs at $x \approx 1.0614$. Since $x = 1 + r$, this means $r = 0.0614$. So the account must earn 6.14% annual interest[11] for the balance to be \$6000 at the end of five years.

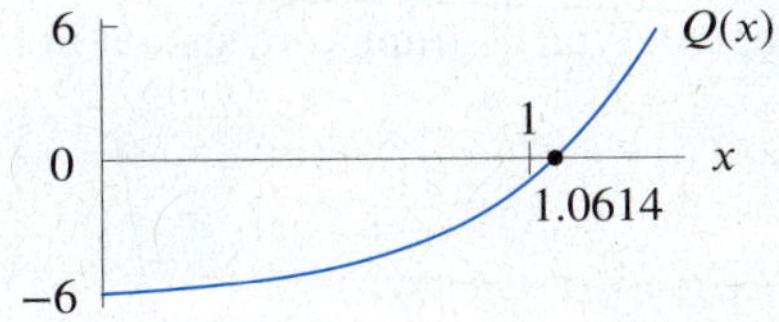


**Figure 11.24**: Finding where $Q(x)$ crosses the $x$-axis, for $x \geq 0$

You may wonder if $Q$ crosses the $x$-axis more than once. For $x \geq 0$, graphing $Q$ on a larger scale suggests that $Q$ increases for all values of $x$ and crosses the $x$-axis only once. For $x > 1$, we expect $Q$ to be an increasing function, because larger values of $x$ indicate higher interest rates and therefore larger values of $Q(x)$. Having crossed the axis once, the graph of $Q$ does not "turn around" to cross it again.

The function $Q(x) = x^5 + x^4 + x^3 + x^2 + x - 6$ is the sum of power functions; $Q$ is called a *polynomial*. (Note that the expression $-6$ can be written as $-6x^0$, so it, too, is a power function.)

## A General Formula for the Family of Polynomial Functions

The general formula for a polynomial function can be written as

$$p(x) = a_n x^n + a_{n-1} x^{n-1} + \dots + a_1 x + a_0,$$

where $n$ is called the *degree* of the polynomial and $a_n$ is the *leading coefficient*. For example, the function

$$g(x) = 3x^2 + 4x^5 + x - x^3 + 1,$$

is a polynomial of degree 5 because the term with the highest power is $4x^5$. It is customary to write a polynomial with the powers in decreasing order from left to right:

$$g(x) = 4x^5 - x^3 + 3x^2 + x + 1.$$

The function $g$ has one other term, $0 \cdot x^4$, which we don't bother to write down. The values of $g$'s coefficients are $a_5 = 4, a_4 = 0, a_3 = -1, a_2 = 3, a_1 = 1$, and $a_0 = 1$. In summary:

> The general formula for the family of polynomial functions can be written as
>
> $$p(x) = a_n x^n + a_{n-1} x^{n-1} + \dots + a_1 x + a_0,$$
>
> where $n$ is a positive integer called the **degree** of $p$ and where $a_n \neq 0$.
>
> - Each power function $a_i x^i$ in this sum is called a **term**.
> - The constants $a_n, a_{n-1}, \dots, a_0$ are called **coefficients**.
> - The term $a_0$ is called the **constant term**. The term with the highest power, $a_n x^n$, is called the **leading term**, and the term $a_n$ is called the **leading coefficient**.
> - To write a polynomial in **standard form**, we arrange its terms from highest power to lowest power, going from left to right.

[11] This is 6.14% interest per year, compounded annually.

Like the power functions from which they are built, polynomials are defined for all values of $x$. Except for polynomials of degree zero (whose graphs are horizontal lines), the graphs of polynomials do not have horizontal or vertical asymptotes; they are smooth and unbroken. The shape of the graph depends on its degree; typical graphs are shown in Figure 11.25.

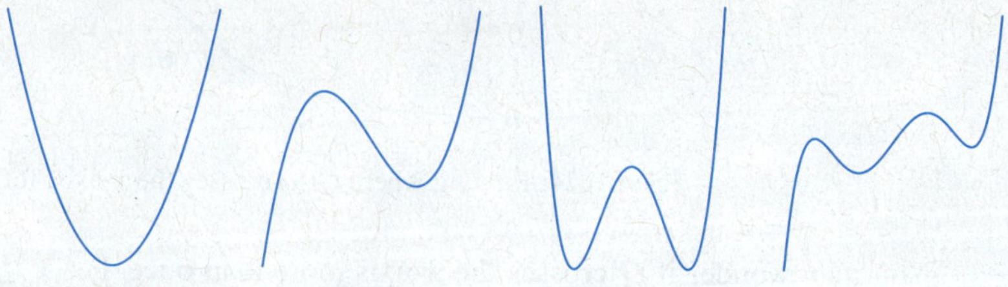


**Figure 11.25:** Graphs of typical polynomials of degree $n$

**Example 2** Which of the following functions are polynomials? For each polynomial, state the degree, the leading coefficient, and the constant term.

(a) $f(x) = 3x^5 + 4x^3 + 2x^{-1}$ (b) $g(x) = x^2 + \sqrt{2}x^3 + 1$ (c) $h(x) = 2(x+5)(x-1)^2$

**Solution** The functions $g$ and $h$ are polynomials; the function $f$ is not.

(a) The function $f(x) = 3x^5 + 4x^3 + 2x^{-1}$ is not a polynomial since the power of the last term is $-1$.

(b) The function $g(x) = x^2 + \sqrt{2}x^3 + 1$ is a polynomial with degree 3 because the term with the highest power is $\sqrt{2}x^3$. The leading coefficient is $\sqrt{2}$ and the constant term is 1.

(c) The function $h(x)$ is a polynomial since $h(x) = 2(x+5)(x-1)^2 = 2x^3 + 6x^2 - 18x + 10$. Hence its degree is 3, the leading coefficient is 2, and the constant term is 10. Note that to determine the leading term of a polynomial given in its factored form we do not necessarily have to multiply out each term. The leading term of $h(x)$ is the product of the leading terms of each factor, namely $2 \cdot x \cdot x^2 = 2x^3$. Similarly, the constant term of $h(x)$ is the product of the constant terms of each factor, namely $2 \cdot 5 \cdot (-1)^2 = 10$. The constant term can also be obtained by setting $x = 0$.

## The Long-Run Behavior of Polynomial Functions

We have seen that, as $x$ grows large, $y = x^2$ increases fast, $y = x^3$ increases faster, and $y = x^4$ increases faster still. In general, power functions with larger positive powers eventually grow much faster than those with smaller powers. This tells us about the behavior of polynomials for large $x$. For instance, consider the polynomial $g(x) = 4x^5 - x^3 + 3x^2 + x + 1$. Provided $x$ is large enough, the value of the term $4x^5$ is much larger than the value of the other terms combined. For example, if $x = 100$,

$$4x^5 = 4(100)^5 = 40{,}000{,}000{,}000,$$

and the other terms in $g(x)$ are

$$\begin{aligned} -x^3 + 3x^2 + x + 1 &= -(100)^3 + 3(100)^2 + 100 + 1 \\ &= -1{,}000{,}000 + 30{,}000 + 100 + 1 = -969{,}899. \end{aligned}$$

Therefore $g(100) = 39{,}999{,}030{,}101$, which is approximately equal to the value of the $4x^5$ term. In general, if $x$ is large enough, the most important contribution to the value of a polynomial is made by the leading term; we can ignore the lower-power terms.

When viewed on a large enough scale, the graph of the polynomial

$$p(x) = a_n x^n + a_{n-1} x^{n-1} + \cdots + a_1 x + a_0$$

looks like the graph of the power function

$$y = a_n x^n.$$

This behavior is called the **long-run behavior** of the polynomial. To determine the values that $p(x)$ takes on as $x$ grows large (either positive or negative) we check the values of its leading term, $a_n x^n$, as $x \to \infty$ and as $x \to -\infty$, respectively.

Using limit notation, the long-run behavior of a polynomial, $p(x)$, can be written as:

$$\lim_{x\to\infty} p(x) = \lim_{x\to\infty} a_n x^n \quad \text{and} \quad \lim_{x\to-\infty} p(x) = \lim_{x\to-\infty} a_n x^n.$$

**Example 3** Find a window in which the graph of $f(x) = x^3 + x^2$ resembles the power function $y = x^3$.

**Solution** Figure 11.26 gives the graphs of $f(x) = x^3 + x^2$ and $y = x^3$. On this scale, $f$ does not look like a power function. On the larger scale in Figure 11.27, the graph of $f$ resembles the graph of $y = x^3$. On this larger scale, the "bumps" in the graph of $f$ are too small to be seen. On an even larger scale, as in Figure 11.28, the graph of $f$ is indistinguishable from the graph of $y = x^3$.

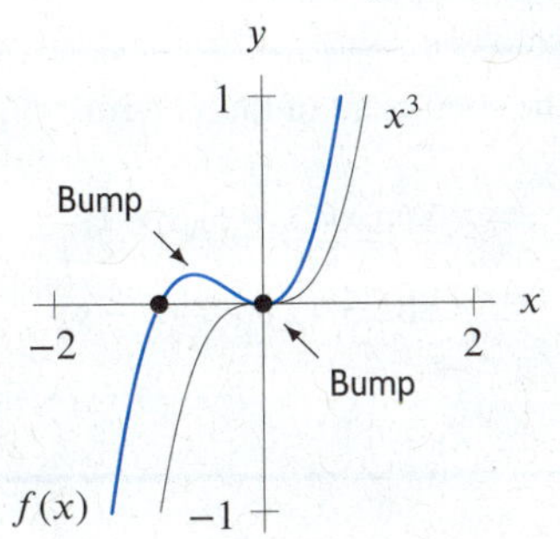


Figure 11.26: On this scale, $f(x) = x^3 + x^2$ does not look like a power function

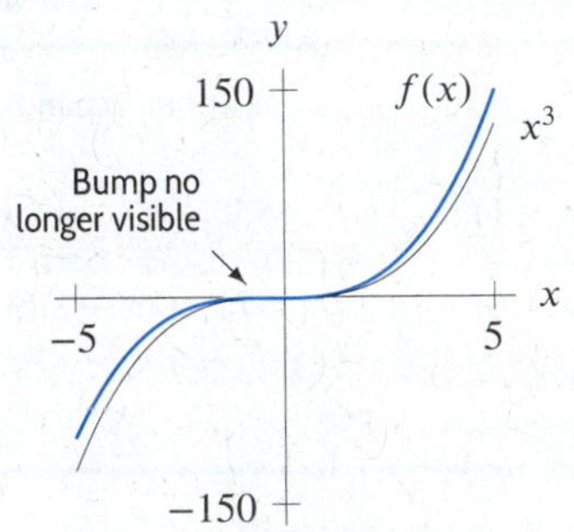


Figure 11.27: On this scale, $f(x) = x^3 + x^2$ resembles the power function $y = x^3$

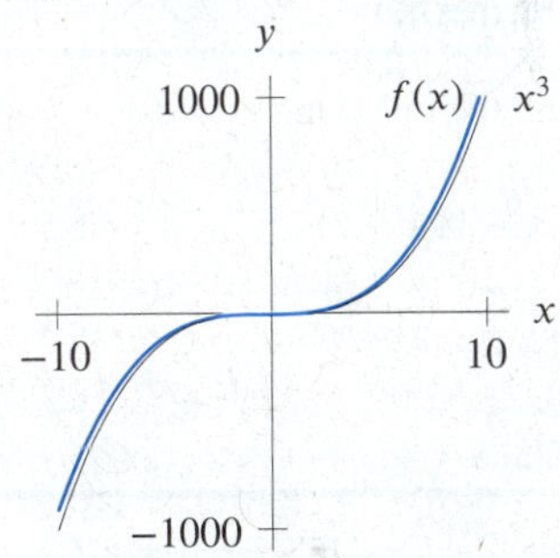


Figure 11.28: On this scale, $f(x) = x^3 + x^2$ is nearly indistinguishable from $y = x^3$

## Zeros of Polynomials

The *zeros* of a polynomial $p$ are values of $x$ for which $p(x) = 0$. The zeros are also the $x$-intercepts, because they tell us where the graph of $p$ crosses the $x$-axis. Factoring can sometimes be used to find the zeros of a polynomial; however, the graphical method of Example 1 can always be used to estimate the zeros. In addition, the long-run behavior of the polynomial can give us clues as to how many zeros (if any) there may be.

**Example 4** Given the polynomial

$$q(x) = 3x^6 - 2x^5 + 4x^2 - 1,$$

where $q(0) = -1$, is there a reason to expect a solution to the equation $q(x) = 0$? If not, explain why not. If so, how do you know?

**Solution** The equation $q(x) = 0$ must have at least two solutions. We know this because on a large scale, $q$ looks like the power function $y = 3x^6$. (See Figure 11.29.) The function $y = 3x^6$ takes on large positive values as $x$ grows large (either positive or negative). Since the graph of $q$ is smooth and unbroken, it must cross the $x$-axis at least twice to get from $q(0) = -1$ to the positive values it attains as $x \to \infty$ and $x \to -\infty$.

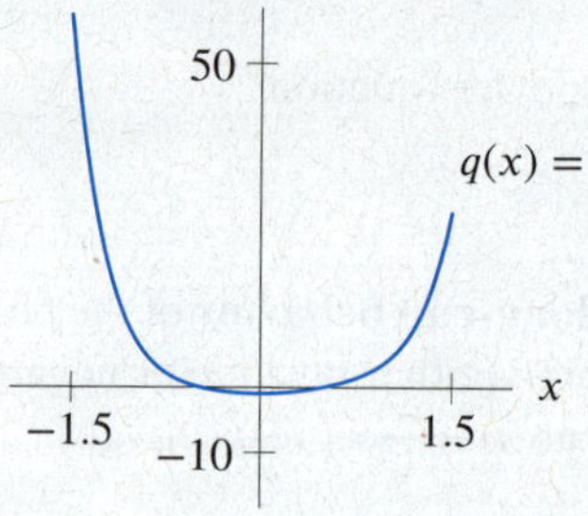


**Figure 11.29:** Graph must cross $x$-axis at least twice since $q(0) = -1$ and $q(x)$ looks like $3x^6$ for large $x$

A sixth-degree polynomial such as $q$ in Example 4 can have as many as six real zeros. We consider the zeros of a polynomial in more detail in Section 11.3.

## Exercises and Problems for Section 11.2

### Skill Refresher

Expand the following expressions.

**S1.** $(x-1)(x+2)$

**S2.** $x(x+1)(x-1)$

**S3.** $x(2x+1)(3x-1)$

**S4.** $x(x+3)(x-4)$

Without expanding, find the coefficient of the $x^3$ term.

**S5.** $\frac{1}{3}x(x+1)(x-1)$

**S6.** $x(2x+1)(x-1)$

**S7.** $x(2x+1)(3x-1)$

**S8.** $3x(2x+1)(2-x)$

### Exercises

Are the functions in Exercises 1–6 polynomials? If so, of what degree?

**1.** $y = 5^x - 2$

**2.** $y = 5 + x$

**3.** $y = 4x^2 + 2$

**4.** $y = 7t^6 - 8t + 7.2$

**5.** $y = 4x^4 - 3x^3 + 2e^x$

**6.** $y = 4x^2 - 7\sqrt{x^9} + 10$

In Exercises 7–12, state the leading term.

**7.** $x^3 + 2x^2 - 5x^4 - 7$

**8.** $x^2 - 3x^3 + 5x + 7x^3$

**9.** $5x^4 - 3x^2 - 2x^4 + 1 - 3x^4$

**10.** $(x-1)(2x+1)(3x-5)$

**11.** $\left(3x^2+3\right)^2 + \left(2x^3+5\right)^3$

**12.** $(4x-7)(x+1)(2x+1)(3x+2) + 2x^5$

In Exercises 13–14, give the leading term assuming that $n$ is a positive integer.

**13.** $4x^{n+3} + 4x^{n-1} - 7x^2$

**14.** $3x^4 + 5x^{n+2} - 6x^{n+5} - 8x^3$

In Exercises 15–17, determine the leading term and state which terms from each factor contribute to it. For example, the leading term of $\left(x^2+x+1\right)\left(4x^3+2x+3\right)$ is $x^2 \cdot 4x^3 = 4x^5$, so the contributing terms are $x^2$ and $4x^3$.

**15.** $(2x+5)(5x+3)(4x+2)$

**16.** $\left(2x^2+4x+1\right)\left(3x+5x^2+3\right)$

**17.** $\left(x^2+3x+1\right)\left(x^2+2x^3+1\right)$

For the polynomials in Exercises 18–20, state the degree and the number of terms, and describe the long-run behavior.

**18.** $y = 2x^3 - 3x + 7$

**19.** $y = 1 - 2x^4 + x^3$

**20.** $y = (x+4)(2x-3)(5-x)$

**21.** Find

**(a)** $\lim_{x\to\infty}(3x^2 - 5x + 7)$ **(b)** $\lim_{x\to-\infty}(7x^2 - 9x^3)$

## Problems

22. Estimate the zeros of $f(x) = x^4 - 3x^2 - x + 2$.

23. Estimate the minimum value of $g(x) = x^4 - 3x^3 - 8$.

24. Find the equation of the line through the $y$-intercept of $y = x^4 - 3x^5 - 1 + x^2$ and the $x$-intercept of $y = 2x - 4$.

25. Let $u(x) = -\frac{1}{5}(x-3)(x+1)(x+5)$ and $v(x) = -\frac{1}{5}x^2(x-5)$.
    **(a)** Graph $u$ and $v$ for $-10 \le x \le 10$, $-10 \le y \le 10$. How are the graphs similar? How are they different?
    **(b)** Compare the graphs of $u$ and $v$ on the window $-20 \le x \le 20$, $-1600 \le y \le 1600$, the window $-50 \le x \le 50$, $-25{,}000 \le y \le 25{,}000$, and the window $-500 \le x \le 500$, $-25{,}000{,}000 \le y \le 25{,}000{,}000$. Discuss.

26. Compare the graphs of $f(x) = x^3 + 5x^2 - x - 5$ and $g(x) = -2x^3 - 10x^2 + 2x + 10$ on a window that shows all intercepts. How are the graphs similar? Different? Discuss.

27. Let $f(x) = \left(\frac{1}{50{,}000}\right)x^3 + \left(\frac{1}{2}\right)x$.
    **(a)** For small values of $x$, which term of $f$ is more important? Explain your answer.
    **(b)** Graph $y = f(x)$ for $-10 \le x \le 10$, $-10 \le y \le 10$. Is this graph linear? How does the appearance of this graph agree with your answer to part (a)?
    **(c)** How large a value of $x$ is required for the cubic term of $f$ to be equal to the linear term?

In Problems 28–31, find a viewing window on which the graph of $f(x) = x^3 + x^2$ resembles the plot.

28.
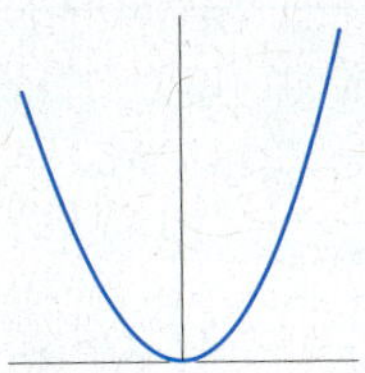

29.
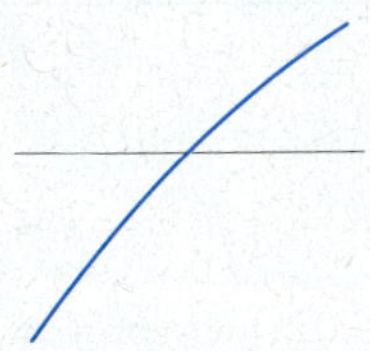

30.
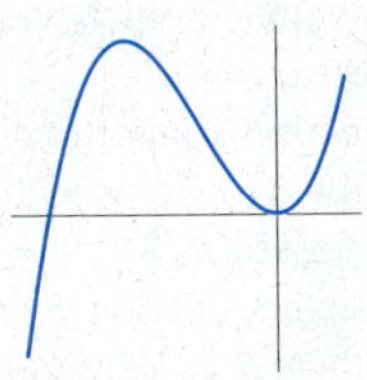

31.
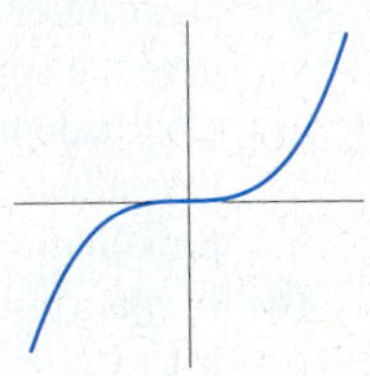

32. Find four different viewing windows on which $f(x) = (x+2)(x-1)(x-3)^2$ resembles graphs (a)–(d).

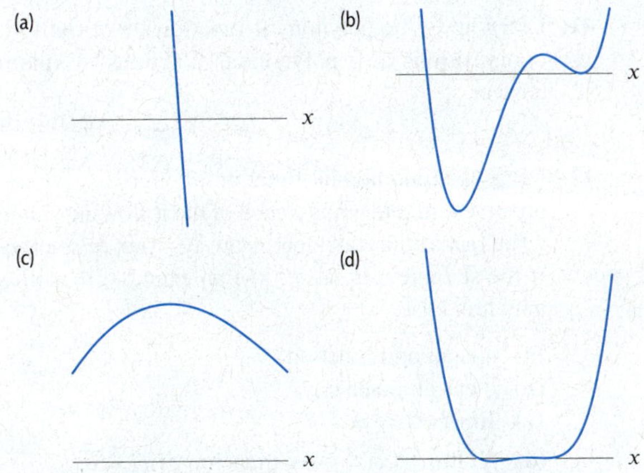


In Problems 33–36, using what you know about the long-run behavior of a polynomial, match the given function to one of the graphs in Figures (I)–(IV). Explain your reasoning.

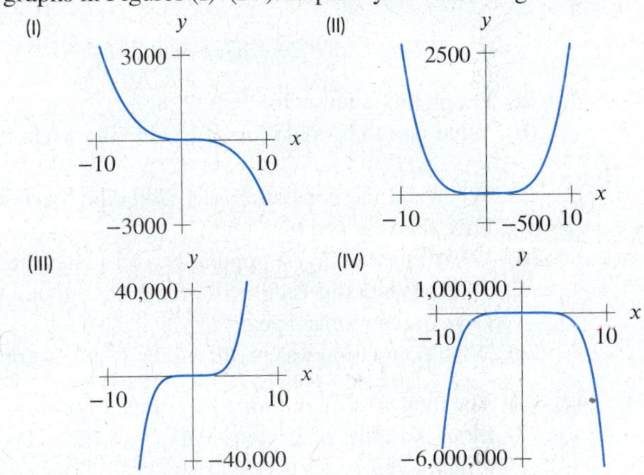


33. $0.3x^4 - 2x^2 + 17$

34. $-3x^3 + 7x^2 - 2x + 5$

35. $4x^5 + 2x^3 - 7x - 5$

36. $-8x^6 - 4x^3 + 12x^2$

37. The polynomial function $f(x) = x^3 + x + 1$ is invertible—that is, this function has an inverse.
    **(a)** Graph $y = f(x)$. Explain how you can tell from the graph that $f$ is invertible.
    **(b)** Find $f(0.5)$ and an approximate value for $f^{-1}(0.5)$.

38. If $f(x) = x^2$ and $g(x) = (x+2)(x-1)(x-3)$, find all $x$ for which $f(x) < g(x)$.

39. In calculus, we often consider pairs of polynomials. If $F(x) = 3x^4 - 4x^3 + 5x - 4$, find $a_4, a_3, a_2, a_1, a_0$. Then use these values to construct the cubic polynomial

$$f(x) = 4a_4x^3 + 3a_3x^2 + 2a_2x + a_1.$$

**40.** **(a)** Suppose $f(x) = ax^2 + bx + c$. What must be true about the coefficients if $f$ is an even function?

**(b)** Suppose $g(x) = ax^3 + bx^2 + cx + d$. What must be true about the coefficients if $g$ is an odd function?

**41.** Let $f$ and $g$ be polynomial functions. Are the following compositions also polynomial functions? Explain your answer.

$$f(g(x)) \quad \text{and} \quad g(f(x))$$

**42.** Let $g$ be a polynomial function of degree $n$, where $n$ is a positive odd integer. For each of the following statements, write *true* if the statement is always true, *false* otherwise. If the statement is false, give an example that illustrates why it is false.

**(a)** $g$ is an odd function.

**(b)** $g$ has an inverse.

**(c)** $\lim_{x\to\infty} g(x) = \infty$.

**(d)** If $\lim_{x\to-\infty} g(x) = -\infty$, then $\lim_{x\to\infty} g(x) = \infty$.

**43.** The town of Liddleville was founded in 1890. Its population $y$ (in hundreds) was modeled, for $t$ in years since 1890, by

$$y = 1 - 0.58t + 4.89t^2 - 1.872t^3 + 0.247t^4 - 0.0111t^5.$$

**(a)** Graph the function for $0 \le t \le 8$, $-2 \le y \le 12$.

**(b)** What was the population of Liddleville when it was founded?

**(c)** When did the population of Liddleville reach zero? Give the year and the month.

**(d)** What was the largest population of Liddleville after 1890? When did Liddleville reach that population? Give the month and year.

**(e)** What population was predicted for 1898? Comment.

**44.** **(a)** The total cost, in millions of dollars, of producing $x$ thousand units of an item is $C(x) = 4(x-1)^2 + 4$. Graph $C(x)$.

**(b)** The revenue (in millions of dollars) from selling $x$ thousand units of the item is $R(x) = 10x$. What does this tell you about the price of each unit?

**(c)** Profit is revenue minus cost. For what values of $x$ does the firm make a profit? Break even? Lose money?

**45.** Let $V$ represent the volume in liters of air in the lungs during a 5-second respiratory cycle. If $t$ is time in seconds, $V$ is given by

$$V = 0.1729t + 0.1522t^2 - 0.0374t^3.$$

**(a)** Graph this function for $0 \le t \le 5$.

**(b)** What is the maximum value of $V$ on this interval? What is its practical significance?

**(c)** Explain the practical significance of the $t$- and $V$-intercepts on the interval $0 \le t \le 5$.

**46.** The volume, $V$, in milliliters, of 1 kg of water as a function of temperature $T$ is given, for $0 \le T \le 30°\text{C}$, by:

$$V = 999.87 - 0.06426T + 0.0085143T^2 - 0.0000679T^3.$$

**(a)** Graph $V$.

**(b)** Describe the shape of your graph. Does $V$ increase or decrease as $T$ increases? Does the graph curve upward or downward? What does the graph tell us about how the volume varies with temperature?

**(c)** At what temperature does water have the maximum density? How does that appear on your graph? (Density = Mass/Volume. In this problem, the mass of the water is 1 kg.)

**47.** Table 11.6 gives $v$, the speed of sound (in m/sec) in water as a function of the temperature $T$ (in °C).[12]

**(a)** An approximate linear formula for $v$ is given by $v = 1402.385 + 5.038813T$. Over what temperature range does this formula agree with the values in Table 11.6 to within 1°C?

**(b)** The formula in part (a) can be improved by adding the quadratic term $-5.799136 \cdot 10^{-2}T^2$. Repeat part (a) using this adjusted formula.

**(c)** The formula in part (b) can be further improved by adding the cubic term $3.287156 \cdot 10^{-4}T^3$. Repeat part (a) using this adjusted formula.

**(d)** The speed of sound in water at 50°C is 1542.6 m/s. If we want to improve our formula still further by adding a quartic (fourth-degree) term, should this term be positive or negative?

**Table 11.6**

| $T$ | 0 | 5 | 10 | 15 | 20 | 25 | 30 |
|---|---|---|---|---|---|---|---|
| $v$ | 1402.4 | 1426.2 | 1447.3 | 1466.0 | 1482.4 | 1496.7 | 1509.2 |

**48.** Let $f(x) = x - \dfrac{x^3}{6} + \dfrac{x^5}{120}$.

**(a)** Graph $y = f(x)$ and $y = \sin x$ for $-2\pi \le x \le 2\pi$, $-3 \le y \le 3$.

**(b)** The graph of $f$ resembles the graph of $\sin x$ on a small interval. Based on your graphs from part (a), give the approximate interval.

**(c)** Your calculator uses a function similar to $f$ in order to evaluate the sine function. How reasonable an approximation does $f$ give for $\sin(\pi/8)$?

**(d)** Explain how you could use the function $f$ to approximate the value of $\sin\theta$, where $\theta = 18$ radians. [Hint: Use the fact that the sine function is periodic.]

[12]Data from the Marczak formula at the UK National Physical Laboratory: www.npl.co.uk, accessed 2001.

**49.** For certain $x$-values, the function $f(x) = 1/(1+x)$ can be well approximated by the polynomial

$$p(x) = 1 - x + x^2 - x^3 + x^4 - x^5.$$

**(a)** Show that $p(0.5) \approx f(0.5) = 2/3$. To how many decimal places do $p(0.5)$ and $f(0.5)$ agree?

**(b)** Calculate $p(1)$. How well does $p(1)$ approximate $f(1)$?

**(c)** Graph $p(x)$ and $f(x)$ together on the same set of axes for $-1 \leq x \leq 1$. Based on your graph, for what range of values of $x$ does $p(x)$ appear to give a good estimate for $f(x)$?

# 11.3 THE SHORT-RUN BEHAVIOR OF POLYNOMIALS

The long-run behavior of a polynomial is determined by its leading term. However, polynomials with the same leading term and the same long-run behavior may have very different short-run behaviors.

**Example 1** Compare the graphs of the polynomials $f$, $g$, and $h$ given by

$$f(x) = x^4 - 4x^3 + 16x - 16, \quad g(x) = x^4 - 4x^3 - 4x^2 + 16x, \quad h(x) = x^4 + x^3 - 8x^2 - 12x.$$

**Solution** Each of these functions is a fourth-degree polynomial, and each has $x^4$ as its leading term. Thus, all their graphs resemble the graph of $x^4$ on a large scale. See Figure 11.30.

However, on a smaller scale, the functions look different. See Figure 11.31. Two of the graphs go through the origin while the third does not. The graphs also differ from one another in the number of bumps each one has and in the number of times each one crosses the $x$-axis.

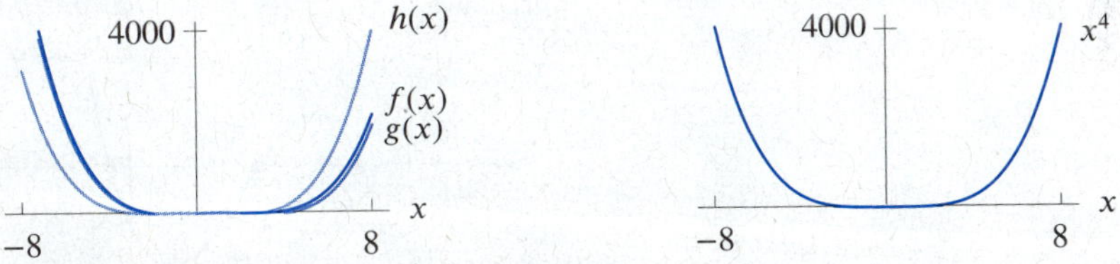


**Figure 11.30**: On a large scale, the polynomials $f$, $g$, and $h$ resemble the power function $y = x^4$

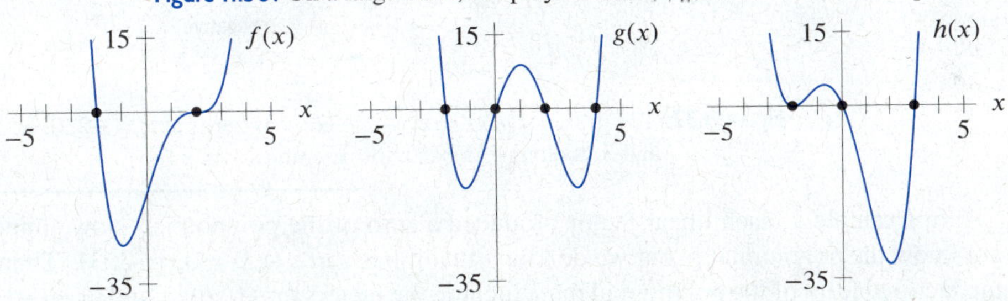


**Figure 11.31**: On a smaller scale, the polynomials $f$, $g$, and $h$ look quite different from one another

## Factored Form, Zeros, and the Short-Run Behavior of a Polynomial

To predict the long-run behavior of a polynomial, we use the highest-power term. To determine the zeros and the short-run behavior of a polynomial, we write it in factored form with as many linear factors as possible.

**Example 2** Investigate the short-run behavior of the third-degree polynomial $u(x) = x^3 - x^2 - 6x$.

(a) Rewrite $u(x)$ as a product of linear factors.

(b) Find the zeros of $u(x)$.

(c) Use the factored form to describe the graph of $u(x)$. Where does it cross the $x$-axis? The $y$-axis? Where is $u(x)$ positive? Negative?

**Solution**

(a) By factoring out an $x$ and then factoring the quadratic, $x^2 - x - 6$, we rewrite $u(x)$ as

$$u(x) = x^3 - x^2 - 6x = x(x^2 - x - 6) = x(x-3)(x+2).$$

Thus, we have expressed $u(x)$ as the product of three linear factors, $x$, $x-3$, and $x+2$.

(b) The polynomial equals zero if and only if at least one of its factors is zero. We solve the equation

$$x(x-3)(x+2) = 0,$$

giving

$$x = 0, \quad \text{or} \quad x - 3 = 0, \quad \text{or} \quad x + 2 = 0,$$

so

$$x = 0, \quad \text{or} \quad x = 3, \quad \text{or} \quad x = -2.$$

These are the zeros, or $x$-intercepts, of $u$. To check, evaluate $u(x)$ for these $x$-values; you should get 0. There are no other zeros.

(c) To describe the graph of $u$, we give the $x$- and $y$-intercepts, and the long-run behavior.

The factored form, $u(x) = x(x-3)(x+2)$, shows that the graph crosses the $x$-axis at $x = 0$, $3, -2$. The graph of $u$ crosses the $y$-axis at $u(0) = 0^3 - 0^2 - 6 \cdot 0 = 0$; that is, at $y = 0$. For large values of $x$, the graph of $y = u(x)$ resembles the graph of its leading term, $y = x^3$. Figure 11.32 shows where $u$ is positive and where $u$ is negative.

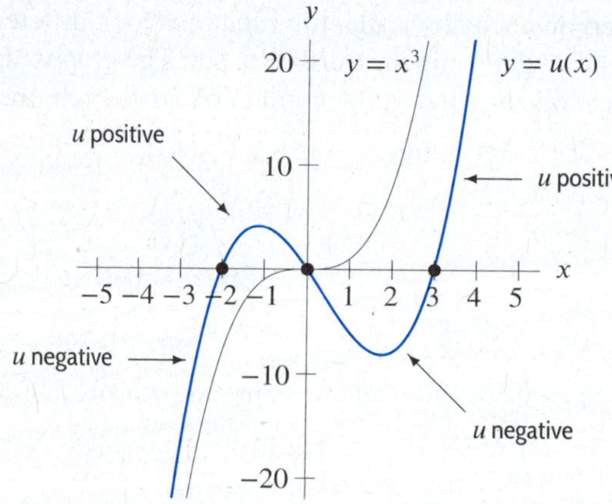


**Figure 11.32:** The graph of $u(x) = x^3 - x^2 - 6x$ has zeros at $x = -2, 0,$ and 3. Its long-run behavior resembles $y = x^3$

In Example 2, each linear factor produced a zero of the polynomial. Now suppose that we do not know the polynomial $p$, but we do know that it has zeros at $x = 0, -12, 31$. Then we know that the factored form of the polynomial must include the factors $(x-0)$ or $x$, and $(x-(-12))$ or $(x+12)$, and $(x-31)$. It may include other factors too. In summary:

> Suppose $p$ is a polynomial. If the formula for $p$ has a **linear factor**, that is, a factor of the form $(x-k)$, then $p$ has a zero at $x = k$.
> Conversely, if $p$ has a **zero** at $x = k$, then $p$ has a linear factor of the form $(x-k)$.

## The Number of Factors, Zeros, and Turning Points

The number of linear factors is always less than or equal to the degree of a polynomial. For example, a fourth-degree polynomial can have no more than four linear factors. This makes sense because if we had another factor in the product and multiplied out, the highest power of $x$ would be greater than four. Since each zero corresponds to a linear factor, the number of zeros is less than or equal to the degree of the polynomial.

Between any two consecutive zeros of a polynomial, there is at least one "bump," or *turning point*. For example, in Figure 11.32, the function is zero at $x = 0$ and negative at $x = 1$, and must change direction to come back up and cross the $x$-axis at $x = 3$. Using calculus, it can be shown that any third-degree polynomial has no more than two bumps. In general:

> The graph of an $n^{\text{th}}$-degree polynomial has at most $n$ zeros and turns at most $(n-1)$ times.

## Multiple Zeros

The functions $s(x) = (x-4)^2$ and $t(x) = (x+1)^3$ are both polynomials in factored form. Each is a horizontal shift of a power function. We refer to the zeros of $s$ and $t$ as *multiple zeros*, because in each case the factor contributing the value of $y = 0$ occurs more than once. For instance, we say that $x = 4$ is a *double zero* of $s$, since

$$s(x) = (x-4)^2 = \underbrace{(x-4)(x-4)}_{\text{Occurs twice}}.$$

Likewise, we say that $x = -1$ is a *triple zero* of $t$, since

$$t(x) = (x+1)^3 = \underbrace{(x+1)(x+1)(x+1)}_{\text{Occurs three times}}.$$

The graphs of $s$ and $t$ in Figures 11.33 and 11.34 show typical behavior near multiple zeros.

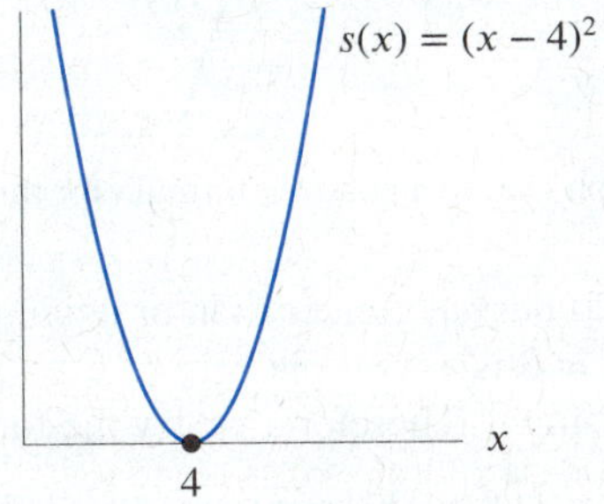


**Figure 11.33:** Double zero at $x = 4$

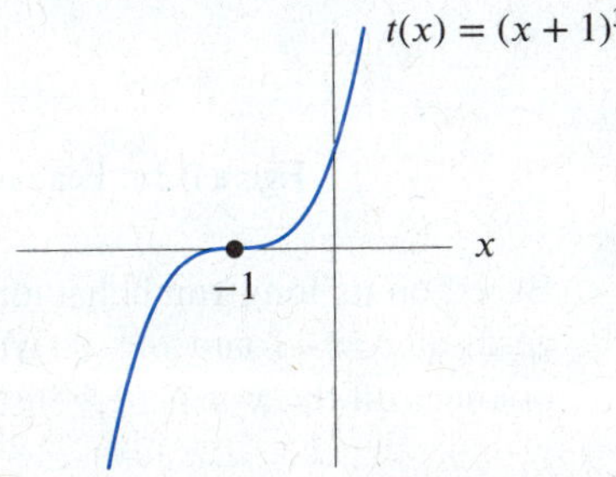


**Figure 11.34:** Triple zero at $x = -1$

In general:

> If $p$ is a polynomial with a repeated linear factor, then $p$ has a **multiple zero**.
> - If the factor $(x-k)$ occurs an even number of times, the graph of $y = p(x)$ does not cross the $x$-axis at $x = k$, but instead "bounces" off the $x$-axis at $x = k$. (See Figure 11.33.)
> - If the factor $(x-k)$ occurs an odd number of times, the graph of $y = p(x)$ crosses the $x$-axis at $x = k$, but it looks flattened there. (See Figure 11.34.)

**Example 3** Describe in words the zeros of the $4^{\text{th}}$-degree polynomials $f(x)$, $g(x)$, and $h(x)$, in Figure 11.35.

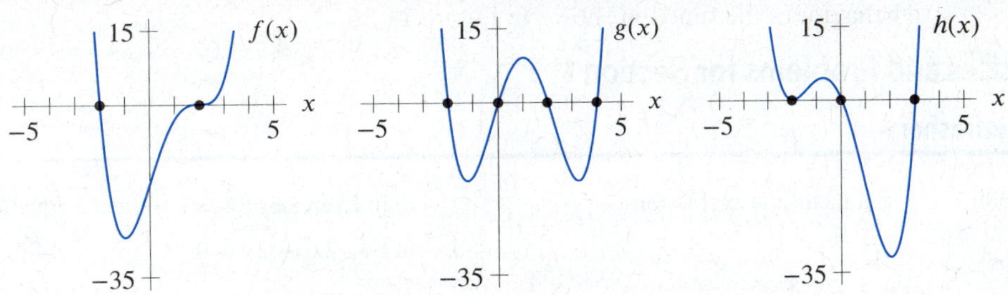


**Figure 11.35:** Three $4^{\text{th}}$-degree polynomials

**Solution** The graph suggests that $f$ has a single zero at $x = -2$. The flattened appearance near $x = 2$ suggests that $f$ has a multiple zero there. Since the graph crosses the $x$-axis at $x = 2$ (instead of bouncing off it), this zero must occur an odd number of times. Since $f$ is $4^{\text{th}}$ degree, $f$ has at most 4 factors, so there must be a triple zero at $x = 2$.

The graph of $g$ has four single zeros. The graph of $h$ has two single zeros (at $x = 0$ and $x = 3$) and a double zero at $x = -2$. The multiplicity of the zero at $x = -2$ is not higher than two because $h$ is of degree $n = 4$.

## Finding the Formula for a Polynomial from its Graph

The graph of a polynomial often enables us to find a possible formula for the polynomial.

**Example 4** Find a possible formula for the polynomial function $f$ graphed in Figure 11.36.

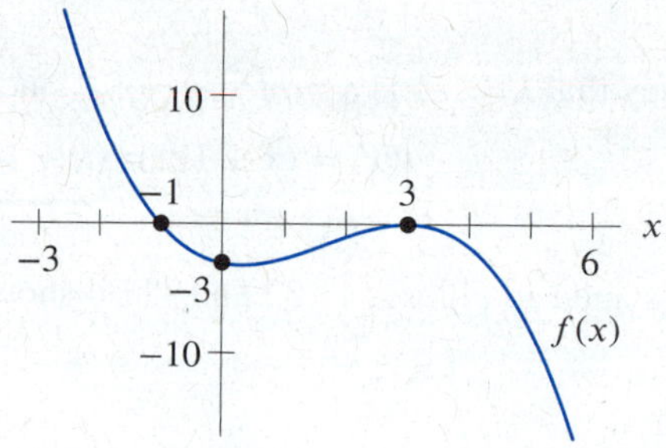


**Figure 11.36:** Features of the graph lead to a possible formula for this polynomial

**Solution** Based on its long-run behavior, $f$ is of odd degree greater than or equal to 3. The polynomial has zeros at $x = -1$ and $x = 3$. We see that $x = 3$ is a multiple zero of even power, because the graph bounces off the $x$-axis here instead of crossing it. Therefore, we try the formula

$$f(x) = k(x+1)(x-3)^2$$

where $k$ represents a stretch factor. The shape of the graph shows that $k$ must be negative.

To find $k$, we use the fact that $f(0) = -3$, so

$$f(0) = k(0+1)(0-3)^2 = -3,$$

which gives

$$9k = -3 \qquad \text{so} \qquad k = -\frac{1}{3}.$$

Thus, $f(x) = -\frac{1}{3}(x+1)(x-3)^2$ is a possible formula for this polynomial.

The formula for $f$ we found in Example 4 is the polynomial of least degree we could have chosen. However, there are other polynomials, such as $y = -\frac{1}{27}(x+1)(x-3)^4$, with the same overall behavior as the function shown in Figure 11.36.

## Exercises and Problems for Section 11.3

### Skill Refresher

In Exercises S1–S3, factor the expressions.

**S1.** $x^6 - x^4$

**S2.** $x^3 - 6x^2 + 5x$

**S3.** $16x - x^3$

In Exercises S4–S6, solve the equations.

**S4.** $2x^3 - 2x = 0$

**S5.** $(2x+4)(x-1)^2 = 0$

**S6.** $-x^3 - x^2 + 4x + 4 = 0$

**S7.** Let $f(x) = k(x-1)(x+1)^2$. Assume $f(x) = 10$, when $x = 3$. Use this information to find the value of $k$.

## Exercises

In Exercises 1–4, find the zeros of the functions.

1. $y = 7(x+3)(x-2)(x+7)$
2. $y = x^3 + 7x^2 + 12x$
3. $y = (x^2 + 2x - 7)(x^3 + 4x^2 - 21x)$
4. $y = a(x+2)(x-b)$, where $a, b$ are nonzero constants
5. Factor $f(x) = 8x^3 - 4x^2 - 60x$ completely, and determine the zeros of $f$.
6. Use the graph of $g(x)$ in Figure 11.31 on page 451 to determine the factored form of
$$g(x) = x^4 - 4x^3 - 4x^2 + 16x.$$
7. Use the graph of $h(x)$ in Figure 11.31 on page 451 to determine the factored form of
$$h(x) = x^4 + x^3 - 8x^2 - 12x.$$
8. Use the graph of $f(x)$ in Figure 11.31 on page 451 to determine the factored form of
$$f(x) = x^4 - 4x^3 + 16x - 16.$$
9. Find a possible formula for a polynomial with zeros at (and only at) $x = -2, 2, 5$, a $y$-intercept at $y = 5$, and long-run behavior of $y \to -\infty$ as $x \to \pm\infty$.

In Exercises 10–12, graph the polynomials without a calculator. Label all $x$-intercepts and $y$-intercepts.

10. $f(x) = -5(x^2 - 4)(25 - x^2)$
11. $g(x) = 5(x-4)(x^2 - 25)$
12. $y = 0.5(x-2)^2(x+3)(x+2)$

## Problems

13. Without using a calculator, decide which of the equations A–E best describes the polynomial in Figure 11.37.
    - A $y = (x+2)(x+1)(x-2)(x-3)$
    - B $y = x(x+2)(x+1)(x-2)(x-3)$
    - C $y = -\frac{1}{2}(x+2)(x+1)(x-2)(x-3)$
    - D $y = \frac{1}{2}(x+2)(x+1)(x-2)(x-3)$
    - E $y = -(x+2)(x+1)(x-2)(x-3)$

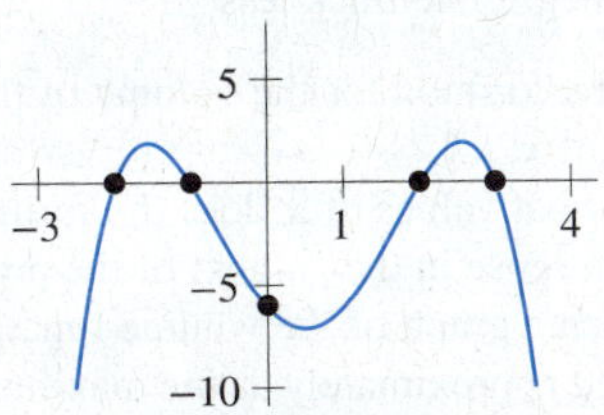

Figure 11.37

14. (a) Let $f(x) = (2x-1)(3x-1)(x-7)(x-9)$. What are the zeros of this polynomial?
    (b) Is it possible to find a viewing window that shows all of the zeros and all of the the turning points of $f$?
    (c) Find two separate viewing windows that together show all the zeros and all the turning points of $f$.
15. Let $p(x) = x^4 + 10x^3 - 68x^2 + 102x - 45$. By experimenting with various viewing windows, determine the zeros of $p$ and use this information to write $p(x)$ in factored form.
16. (a) Experiment with various viewing windows to determine the zeros of $f(x) = 2x^4 + 9x^3 - 7x^2 - 9x + 5$. Then write $f$ in factored form.
    (b) Find a single viewing window that clearly shows all of the turning points of $f$.

In Problems 17–22, find a possible formula for each polynomial with the given properties.

17. $f$ has degree $\leq 2$, $f(0) = 0$ and $f(1) = 1$.
18. $f$ has degree $\leq 2$, $f(0) = f(1) = f(2) = 1$.
19. $f$ has degree $\leq 2$, $f(0) = f(2) = 0$ and $f(3) = 3$.
20. $f$ is third degree with $f(-3) = 0$, $f(1) = 0$, $f(4) = 0$, and $f(2) = 5$.
21. $g$ is fourth degree, $g$ has a double zero at $x = 3$, $g(5) = 0$, $g(-1) = 0$, and $g(0) = 3$.
22. Least possible degree through the points $(-3, 0)$, $(1, 0)$, and $(0, -3)$.
23. Find at least two different third-degree polynomials having zeros at $x = -1$ and $x = 2$ (and nowhere else), and $y$-intercept at $y = 3$.

For Problems 24–29, find the real zeros (if any) of the polynomials.

24. $y = 4x^2 - 1$
25. $y = x^4 + 6x^2 + 9$
26. $y = (x^2 - 8x + 12)(x - 3)$
27. $y = x^2 + 5x + 6$
28. $y = 4x^2 + 1$
29. $y = ax^2(x^2 + 4)(x + 3)$, where $a$ is a nonzero constant.

In Problems 30–39, give a possible formula for the polynomial.

**30.**

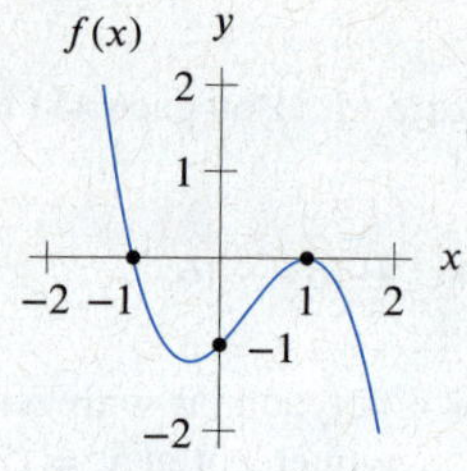


**31.**

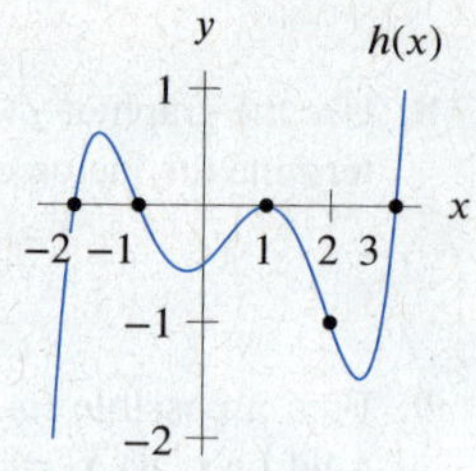


**32.**

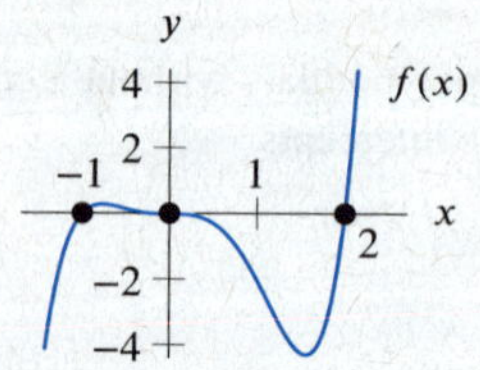


Note appearance near origin

**33.**

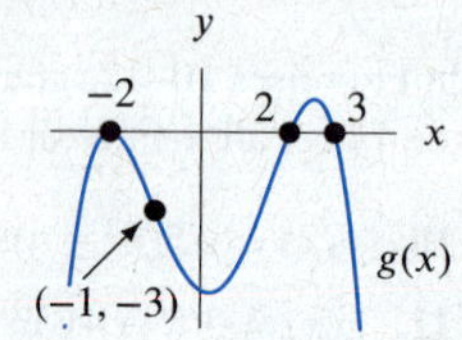


**34.**

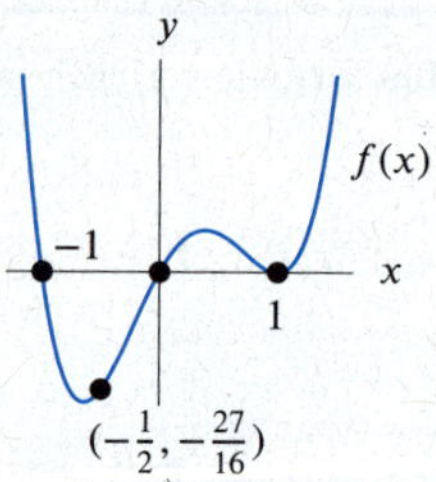


**35.**

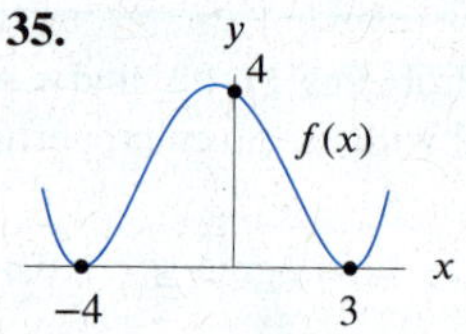


**36.**

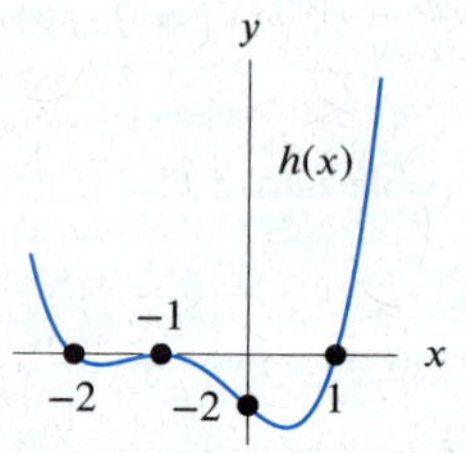


**37.**

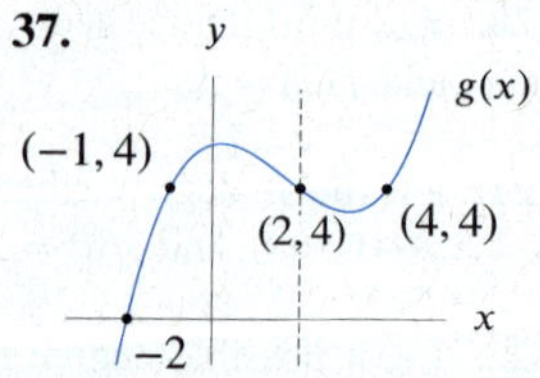


**38.**

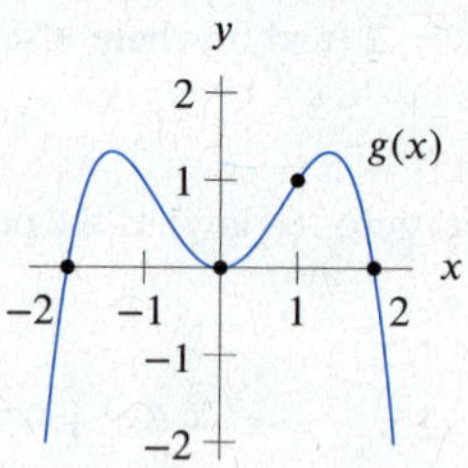


**39.**

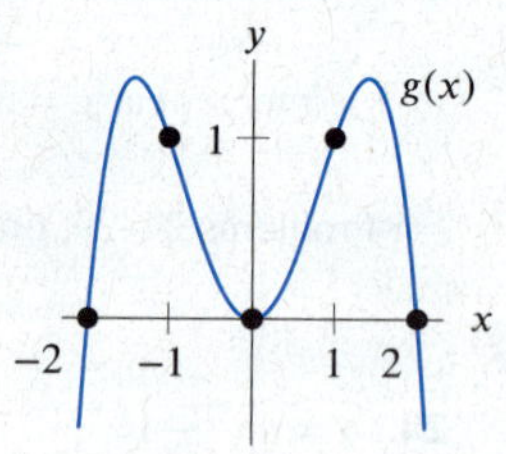


**40.** Which of these functions have inverses that are functions? Discuss.

**(a)** $f(x) = (x-2)^3 + 4$.
**(b)** $g(x) = x^3 - 4x^2 + 2$.

**41.** The polynomial

$$f(x) = (x-5)^2(x-3)^2(x-1)(x-r)(x+3)^s \cdot g(x)$$

is an even function. What can you say about the constants $r$, $s$ and the second-degree polynomial function $g(x)$?

**42.** In each of the following cases, find a possible formula for the polynomial $f$.

**(a)** Suppose $f$ has zeros at $x = -2$, $x = 3$, $x = 5$ and a $y$-intercept of 4.
**(b)** In addition to the properties in part (a), suppose $f$ has the following long-run behavior: As $x \to \pm\infty$, $y \to -\infty$. [Hint: Assume $f$ has a double zero.]
**(c)** In addition to the properties in part (a), suppose $f$ has the following long-run behavior: As $x \to \pm\infty$, $y \to +\infty$.

**43.** You wish to pack a cardboard box inside a wooden crate. In order to have room for the packing materials, you need to leave a 0.5-ft space around the front, back, and sides of the box, and a 1-ft space around the top and bottom of the box. If the cardboard box is $x$ feet long, $(x+2)$ feet wide, and $(x-1)$ feet deep, find a formula in terms of $x$ for the amount of packing material needed.

**44.** An open-top box is to be constructed from a 6-in by 8-in rectangular sheet of tin by cutting out squares of equal size at each corner, then folding up the resulting flaps. Let $x$ denote the length of the side of each cut-out square. Assume negligible thickness.

**(a)** Find a formula for the volume of the box as a function of $x$.
**(b)** For what values of $x$ does the formula from part (a) make sense in the context of the problem?
**(c)** Sketch a graph of the volume function.
**(d)** What, approximately, is the maximum volume of the box?

**45.** Take an 8.5- by 11-inch piece of paper and cut out four equal squares from the corners. Fold up the sides to create an open box. Find the dimensions of the box that has maximum volume.

**46.** Consider the function $a(x) = x^5 + 2x^3 - 4x$.

**(a)** Without using a calculator or computer, what can you say about the graph of $a$?
**(b)** Use a calculator or a computer to determine the zeros of this function to three decimal places.
**(c)** Explain why you think that you have all the possible zeros.
**(d)** What are the zeros of $b(x) = 2x^5 + 4x^3 - 8x$? Does your answer surprise you?

**47.** Give the domain for $g(x) = \ln\left((x-3)^2(x+2)\right)$.

**48.** Given that $a, b$, and $c$ are constants, $a < b < c$, state the domain of

$$y = \sqrt{(x-a)(x-b)(x-c)}.$$

[Hint: Graph $y = (x-a)(x-b)(x-c)$.]

**49.** The following statements about $f(x)$ are true:

- $f(x)$ is a polynomial function
- $f(x) = 0$ at exactly four different values of $x$
- $f(x) \to -\infty$ as $x \to \pm\infty$

For each of the following statements, write *true* if the statement must be true, *never true* if the statement is never true, or *sometimes true* if it is sometimes true and sometimes not true.

**(a)** $f(x)$ is an odd function
**(b)** $f(x)$ is an even function
**(c)** $f(x)$ is a fourth-degree polynomial
**(d)** $f(x)$ is a fifth-degree polynomial
**(e)** $f(-x) \to -\infty$ as $x \to \pm\infty$
**(f)** $f(x)$ is invertible

# 11.4 RATIONAL FUNCTIONS

## The Average Cost of Producing a Therapeutic Drug

A pharmaceutical company wants to begin production of a new drug. The total cost $C$, in dollars, of making $q$ grams of the drug is given by the linear function

$$C(q) = 2{,}500{,}000 + 2000q.$$

The fact that $C(0) = 2{,}500{,}000$ tells us that the company spends \$2,500,000 before it starts making the drug. This quantity is known as the *fixed cost* because it does not depend on how much of the drug is made. It represents the cost for research, testing, and equipment. In addition, the slope of $C$ tells us that each gram of the drug costs an extra \$2000 to make. This quantity is known as the *variable cost* per unit. It represents the additional cost, in labor and materials, to make an additional gram of the drug.

The fixed cost of \$2.5 million is large compared to the variable cost of \$2000 per gram. This means that it is impractical for the company to make a small amount of the drug. For instance, the total cost for 10 grams is

$$C(10) = 2{,}500{,}000 + 2000 \cdot 10 = 2{,}520{,}000,$$

which works out to an average cost of \$252,000 per gram. The company would probably never sell such an expensive drug.

However, as larger quantities of the drug are manufactured, the initial expenditure of \$2.5 million seems less significant. The fixed cost averages out over a large quantity. For example, if the company makes 10,000 grams of the drug,

$$\text{Average cost} = \frac{\text{Cost of producing 10,000 grams}}{10{,}000} = \frac{2{,}500{,}000 + 2000 \cdot 10{,}000}{10{,}000} = 2250,$$

or \$2250 per gram of drug produced.

We define the average cost, $a(q)$, as the cost per gram to produce $q$ grams of the drug:

$$a(q) = \begin{array}{c}\text{Average cost of}\\ \text{producing } q \text{ grams}\end{array} = \frac{\text{Total cost}}{\text{Number of grams}} = \frac{C(q)}{q} = \frac{2{,}500{,}000 + 2000q}{q}.$$

Figure 11.38 gives a graph of $y = a(q)$ for $q > 0$. The horizontal asymptote reflects the fact that for large values of $q$, the value of $a(q)$ is close to 2000. This is because, as more of the drug is produced, the average cost gets closer to \$2000 per gram. See Table 11.7.

The vertical asymptote of $y = a(q)$ is the $y$-axis, which tells us that the average cost per gram is very large if a small amount of the drug is made. This is because the initial \$2.5 million expenditure is averaged over very few units. We saw that producing only 10 grams costs a staggering \$252,000 per gram.

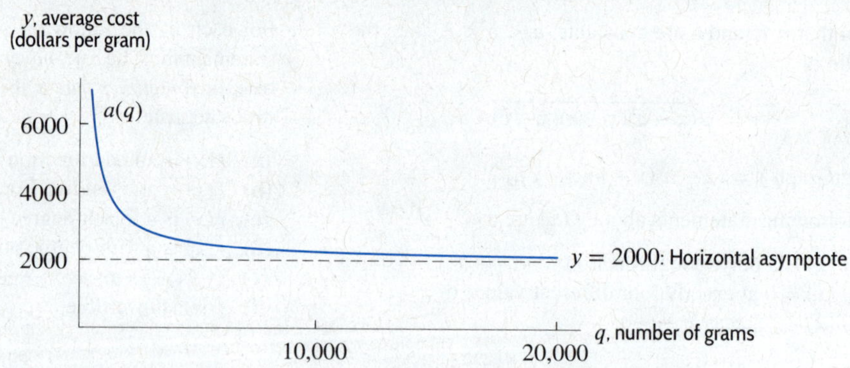


**Figure 11.38:** The graph of $y = a(q)$, a rational function, has a horizontal asymptote at $y = 2000$ and a vertical asymptote at $q = 0$

**Table 11.7** *As quantity q increases, the average cost a(q) draws closer to* \$2000 *per gram*

| Quantity, $q$ | Total cost, $C(q) = 2{,}500{,}000 + 2000q$ | Average cost, $a(q) = C(q)/q$ |
|---|---|---|
| 10,000 | $2{,}500{,}000 + 20{,}000{,}000 = 22{,}500{,}000$ | 2250 |
| 50,000 | $2{,}500{,}000 + 100{,}000{,}000 = 102{,}500{,}000$ | 2050 |
| 100,000 | $2{,}500{,}000 + 200{,}000{,}000 = 202{,}500{,}000$ | 2025 |
| 500,000 | $2{,}500{,}000 + 1{,}000{,}000{,}000 = 1{,}002{,}500{,}000$ | 2005 |

## What Is a Rational Function?

The formula for $a(q)$ is the ratio of the polynomial $2{,}500{,}000 + 2000q$ and the polynomial $q$. Since $a(q)$ is given by the ratio of two polynomials, $a(q)$ is an example of a *rational function*. In general:

> A function $r$ is called a **rational function** if $r$ can be written as the ratio of two polynomial functions $p(x)$ and $q(x)$, that is, if
> $$r(x) = \frac{p(x)}{q(x)}.$$
> We assume that $q(x)$ is not the constant polynomial $q(x) = 0$.

## The Long-Run Behavior of Rational Functions

In the long run, every rational function behaves like a power function. For example, consider

$$f(x) = \frac{6x^4 + x^3 + 1}{-5x + 2x^2}.$$

Since the long-run behavior of a polynomial is determined by its highest-power term, for large $x$ the numerator behaves like $6x^4$ and the denominator behaves like $2x^2$. The long-run behavior of $f$ is

$$f(x) = \frac{6x^4 + x^3 + 1}{-5x + 2x^2} \approx \frac{6x^4}{2x^2} = 3x^2,$$

so $f(x) \to \infty$ as $x \to \pm\infty$.

Using limit notation, the information about the long-run behavior of $f(x)$ is given as follows:

$$\lim_{x\to\pm\infty} f(x) = \lim_{x\to\pm\infty} (3x^2) = \infty.$$

See Figure 11.39.

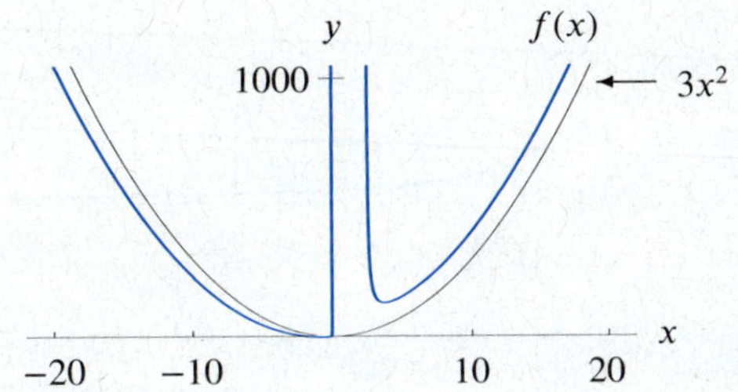


**Figure 11.39:** In the long run, the graph of $f(x)$ looks like the graph of $3x^2$

In general, if $r$ is any rational function, then for large enough values of $x$,

$$r(x) = \frac{a_n x^n + a_{n-1}x^{n-1} + \cdots + a_0}{b_m x^m + b_{m-1}x^{m-1} + \cdots + b_0} \approx \frac{a_n x^n}{b_m x^m} = \frac{a_n}{b_m} x^{n-m}.$$

This means that on a large scale $r$ resembles the function $y = \left(\frac{a_n}{b_m}\right) x^{n-m}$, which is a power function of the form $y = kx^p$, where $k = a_n/b_m$ and $p = n - m$. In summary:

> For $x$ of large enough magnitude (either positive or negative), the graph of the rational function $r$ looks like the graph of a power function. If $r(x) = p(x)/q(x)$, then the **long-run behavior** of $y = r(x)$ is given by
>
> $$y = \frac{\text{Leading term of } p}{\text{Leading term of } q}.$$

Using limits, the long-run behavior of a rational function, $r(x) = p(x)/q(x)$, is given as:

$$\lim_{x\to\pm\infty} \frac{p(x)}{q(x)} = \lim_{x\to\pm\infty} \frac{\text{Leading term of } p}{\text{Leading term of } q}.$$

Note that this limit, if it exists, gives the horizontal asymptote of $r(x)$.

**Example 1** For positive $x$, describe the long-run behavior of the rational function

$$r(x) = \frac{x+3}{x+2}.$$

**Solution** If $x$ is a large positive number, then

$$r(x) = \frac{\text{Big number} + 3}{\text{Same big number} + 2} \approx \frac{\text{Big number}}{\text{Same big number}} = 1.$$

For example, if $x = 100$, we have

$$r(x) = \frac{103}{102} = 1.0098\ldots \approx 1.$$

If $x = 10{,}000$, we have

$$r(x) = \frac{10{,}003}{10{,}002} = 1.00009998\ldots \approx 1,$$

For large positive $x$-values, $r(x) \approx 1$. Thus, for large enough values of $x$, the graph of $y = r(x)$ looks like the line $y = 1$, its horizontal asymptote. We write $\lim_{x\to\infty} r(x) = 1$. See Figure 11.40. However, for $x > 0$, the graph of $r$ is above the line since the numerator is larger than the denominator.

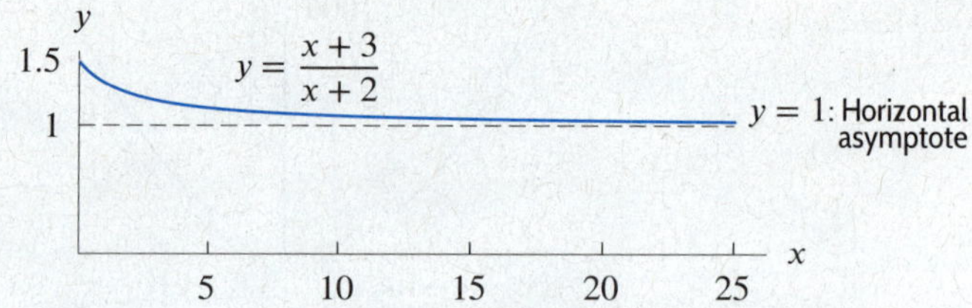


Figure 11.40: For large positive values of $x$, the graph of $r(x) = (x+3)/(x+2)$ looks like the horizontal line $y = 1$

**Example 2** For positive $x$, describe the positive long-run behavior of the rational function

$$g(x) = \frac{3x+1}{x^2+x-2}.$$

**Solution** The leading term in the numerator is $3x$ and the leading term in the denominator is $x^2$. Thus for large enough values of $x$,

$$g(x) \approx \frac{3x}{x^2} = \frac{3}{x},$$

so since $\frac{3}{x} \to 0$ as $x \to \infty$ we see that $g(x) \to 0$ as $x \to \infty$.

In limit notation,

$$\lim_{x\to\infty} g(x) = \lim_{x\to\infty} \left(\frac{3}{x}\right) = 0.$$

Figure 11.41 shows the graphs of $y = g(x)$ and $y = 3/x$. For large values of $x$, the two graphs are nearly indistinguishable. Both graphs have a horizontal asymptote at $y = 0$.

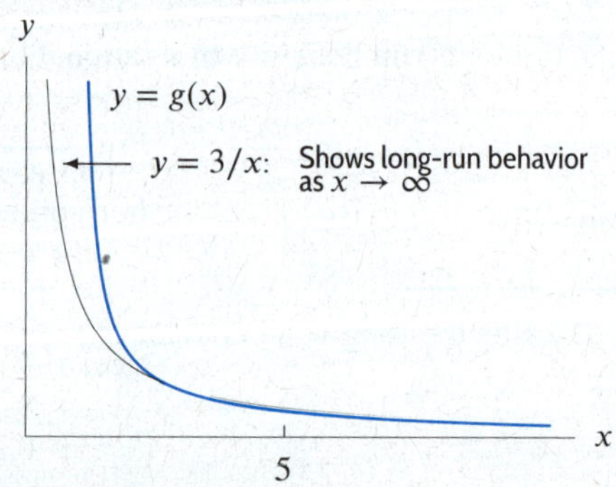


Figure 11.41: For large enough values of $x$, the function $g$ looks like the function $y = 3x^{-1}$

## What Causes Asymptotes?

The graphs of rational functions often behave differently from the graphs of polynomials. Polynomial graphs (except constant functions) cannot level off to a horizontal line as the graphs of rational functions can. In Example 1, the numerator and denominator are approximately equal for large $x$, producing the horizontal asymptote $y = 1$. In Example 2, the denominator grows faster than the numerator, driving the quotient toward zero.

The rapid rise (or fall) of the graph of a rational function near its vertical asymptote is due to the denominator becoming small (close to zero). It is tempting to assume that any function that has a denominator has a vertical asymptote. However, this is not true. To have a vertical asymptote, the denominator must equal zero. For example, suppose that

$$r(x) = \frac{1}{x^2+3}.$$

The denominator is always greater than 3; it is never 0. We see from Figure 11.42 that $r$ does not have a vertical asymptote.

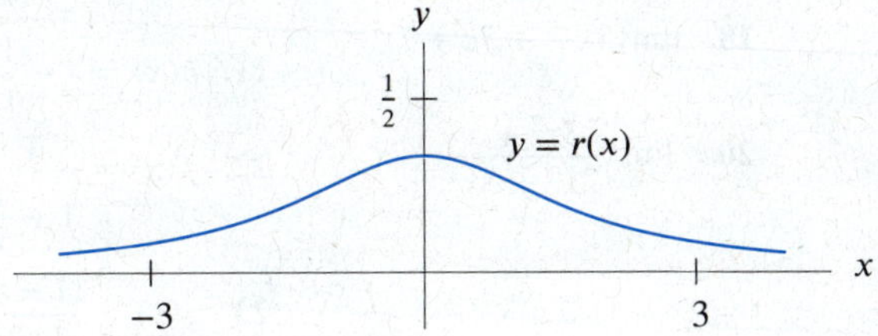


Figure 11.42: The rational function $r(x) = 1/(x^2 + 3)$ has no vertical asymptote

## Exercises and Problems for Section 11.4

### Skill Refresher

For Exercises S1–S5, perform the operations. Express answers in reduced form.

**S1.** $\dfrac{6}{y} + \dfrac{7}{y^3}$

**S2.** $\dfrac{13}{x-1} + \dfrac{14}{2x-2}$

**S3.** $\dfrac{\frac{1}{x} - \frac{2}{x^2}}{\frac{2x-4}{x^5}}$

**S4.** $\dfrac{9}{x^2+5x+6} + \dfrac{12}{x+3}$

**S5.** $\dfrac{5}{(x-2)^2(x+1)} - \dfrac{18}{(x-2)}$

In Exercises S6–S9, simplify, if possible.

**S6.** $\dfrac{1/(x+y)}{x+y}$

**S7.** $\dfrac{(w+2)/2}{w+2}$

**S8.** $\dfrac{a^2-b^2}{a^2+b^2}$

**S9.** $\dfrac{x^{-1}+x^{-2}}{1-x^{-2}}$.

### Exercises

Are the functions in Exercises 1–7 rational functions? If so, write them in the form $p(x)/q(x)$, the ratio of polynomials.

**1.** $f(x) = \dfrac{x+2}{x^2-1}$

**2.** $f(x) = \dfrac{4^x+3}{3^x-1}$

**3.** $f(x) = \dfrac{x^2}{2} + \dfrac{1}{x}$

**4.** $f(x) = \dfrac{x^4+3^x-x^2}{x^3-2}$

**5.** $f(x) = \dfrac{\sqrt{x}+1}{x+1}$

**6.** $f(x) = \dfrac{x^3}{2x^2} + \dfrac{1}{6}$

**7.** $f(x) = \dfrac{9x-1}{4\sqrt{x}+7} + \dfrac{5x^3}{x^2-1}$

**8.** Compare and discuss the long-run behaviors of the following functions:

$$f(x) = \frac{x^2+1}{x^2+5}, \quad g(x) = \frac{x^3+1}{x^2+5}, \quad h(x) = \frac{x+1}{x^2+5}.$$

In Exercises 9–12, determine the long-run behavior of the rational function.

**9.** $\dfrac{1}{x+2} + \dfrac{x}{x+1}$

**10.** $\dfrac{-x}{x+1} + \dfrac{x}{x+2}$

**11.** $\dfrac{(x+3)(2x+1)}{(3x+1)(2x-5)}$

**12.** $y = x + \dfrac{2x+1}{4x-3}$

In Exercises 13–16, using what you know about the long-run behavior of a rational function, match the function to one of the graphs in Figures (I)–(IV). Explain your reasoning.

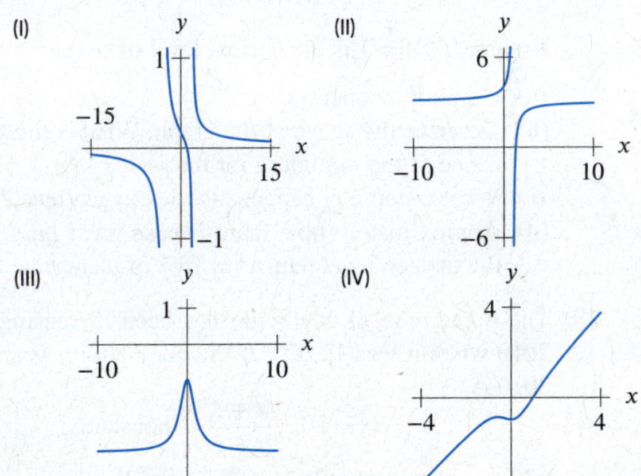


**13.** $y = -\dfrac{3x^2+1}{x^2+1}$

**14.** $y = \dfrac{x-1}{(x+3)(x-2)}$

**15.** $y = \dfrac{x^3-1}{x^2+1}$

**16.** $y = \dfrac{3x-4}{x-1}$

Evaluate the limits in Exercises 17–20.

**17.** $\lim_{x\to\infty}(2x^{-3}+4)$

**18.** $\lim_{x\to\infty}(3x^{-2}+5x+7)$

**19.** $\lim_{x\to\infty}\dfrac{4x+3x^2}{4x^2+3x}$

**20.** $\lim_{x\to-\infty}\dfrac{3x^2+x}{2x^2+5x^3}$

Find the horizontal asymptote, if it exists, of the functions in Exercises 21–23.

**21.** $h(x)=3-\dfrac{1}{x}+\dfrac{x}{x+1}$

**22.** $f(x)=\dfrac{1}{1+\dfrac{1}{x}}$

**23.** $g(x)=\dfrac{(1-x)(2+3x)}{2x^2+1}$

## Problems

**24.** Let $r(x)=p(x)/q(x)$, where $p$ and $q$ are polynomials of degrees $m$ and $n$, respectively. What conditions on $m$ and $n$ ensure that the following statements are true?

**(a)** $\lim_{x\to\infty} r(x)=0$
**(b)** $\lim_{x\to\infty} r(x)=k$, with $k\neq 0$.

**25.** Find a formula for $f^{-1}(x)$ given that

$$f(x)=\frac{4-3x}{5x-4}.$$

**26.** Give examples of rational functions with even symmetry, odd symmetry, and neither. How does the symmetry of $f(x)=p(x)/q(x)$ depend on the symmetry of $p(x)$ and $q(x)$?

**27.** Let $t$ be the time in weeks. At time $t=0$, organic waste is dumped into a pond. The oxygen level in the pond at time $t$ is given by

$$f(t)=\frac{t^2-t+1}{t^2+1}.$$

Assume $f(0)=1$ is the normal level of oxygen.

**(a)** Graph this function.
**(b)** Describe the shape of the graph. What is the significance of the minimum for the pond?
**(c)** What eventually happens to the oxygen level?
**(d)** Approximately how many weeks must pass before the oxygen level returns to 75% of its normal level?

**28.** The population of Mathville has been increasing since 2010 when it was 12,000. If the population $t$ years after 2010 is

$$P(t)=20\left(\frac{4t+3}{2t+5}\right)\text{ thousand},$$

when will the population of Mathville reach 20,000? When will it reach 50,000?

**29.** A car is driven for 60 miles. The first 10 miles are through a large city at a speed of 40 mph; once out of the city, a speed of $V$ mph is maintained.

**(a)** Calculate the average speed for the 60-mile trip.
**(b)** If you want to average 60 mph for the trip, how fast need you go during the last 50 miles?

**30.** A small printing house agrees to publish a book of poems illustrated by its author. The printing house plans to recover its investment of \$80,000 and make a profit of \$40,000. The price of the book will depend on the number of copies they expect to sell.

**(a)** Fill in the table with the price per copy for each projected sales figure.

| Number of copies sold | 1000 | 2000 | 4000 | 6000 |
|---|---|---|---|---|
| Price per copy | | | | |

**(b)** Give a formula for the price per copy, $p$, as a function of projected sales, $s$.
**(c)** Graph the function $p=f(s)$.

**31.** A chemist is studying the properties of a bronze alloy (mixture) of copper and tin. She begins with 2 kg of an alloy that is one-half tin. Keeping the amount of copper constant, she adds small amounts of tin to the alloy. Letting $x$ be the total amount of tin added, define

$$C(x)=\text{Concentration of tin}=\frac{\text{Total amount of tin}}{\text{Total amount of alloy}}.$$

**(a)** Find a formula for $C(x)$.
**(b)** Evaluate $C(0.5)$ and $C(-0.5)$. Explain the physical significance of these quantities.
**(c)** Graph $y=C(x)$, labeling all interesting features. Describe the physical significance of the features you have labeled.

**32.** Bronze is an alloy, or mixture, of copper and tin. The alloy initially contains 3 kg copper and 9 kg tin. You add $x$ kg of copper to this 12 kg of alloy. The concentration of copper in the alloy is a function of $x$:

$$f(x)=\text{Concentration of copper}=\frac{\text{Total amount of copper}}{\text{Total amount of alloy}}.$$

**(a)** Find a formula for $f$ in terms of $x$, the amount of copper added.

**(b)** Evaluate the following expressions and explain their significance for the alloy:

(i) $f(\frac{1}{2})$ (ii) $f(0)$ (iii) $f(-1)$
(iv) $f^{-1}(\frac{1}{2})$ (v) $f^{-1}(0)$

**(c)** Graph $f(x)$ for $-5 \leq x \leq 5, -0.25 \leq y \leq 0.5$. Interpret the intercepts in the context of the alloy.

**(d)** Graph $f(x)$ for $-3 \leq x \leq 100, 0 \leq y \leq 1$. Describe the appearance of your graph for large $x$-values. Does the appearance agree with what you expect to happen when large amounts of copper are added to the alloy?

**33.** An alcohol solution consists of 5 gallons of pure water and $x$ gallons of alcohol, $x > 0$. Let $f(x)$ be the ratio of the volume of alcohol to the total volume of liquid. [Note that $f(x)$ is the concentration of the alcohol in the solution.]

**(a)** Find a possible formula for $f(x)$.

**(b)** Evaluate and interpret $f(7)$ in the context of the mixture.

**(c)** What is the zero of $f$? Interpret your result in the context of the mixture.

**(d)** Find an equation for the horizontal asymptote of $f$. Explain its significance in the context of the mixture.

**34.** The total cost $C(n)$ for a producer to manufacture $n$ units of a good is given by

$$C(n) = 5000 + 50n.$$

The average cost of producing $n$ units is $a(n) = C(n)/n$.

**(a)** Evaluate and interpret the economic significance of:

(i) $C(1)$ (ii) $C(100)$
(iii) $C(1000)$ (iv) $C(10000)$

**(b)** Evaluate and interpret the economic significance of:

(i) $a(1)$ (ii) $a(100)$
(iii) $a(1000)$ (iv) $a(10000)$

**(c)** Based on part (b), what trend do you notice in the values of $a(n)$ as $n$ gets large? Explain this trend in economic terms.

**35.** It costs a company \$30,000 to begin production of a good, plus \$3 for every unit of the good produced. Let $x$ be the number of units produced by the company.

**(a)** Find a formula for $C(x)$, the total cost for the production of $x$ units of the good.

**(b)** Find a formula for the company's average cost per unit, $a(x)$.

**(c)** Graph $y = a(x)$ for $0 < x \leq 50{,}000$, $0 \leq y \leq 10$. Label the horizontal asymptote.

**(d)** Explain in economic terms why the graph of $a$ has the long-run behavior that it does.

**(e)** Explain in economic terms why the graph of $a$ has the vertical asymptote that it does.

**(f)** Find a formula for $a^{-1}(y)$. Give an economic interpretation of $a^{-1}(y)$.

**(g)** The company makes a profit if the average cost of its good is less than \$5 per unit. Find the minimum number of units the company can produce and make a profit.

**36.** Figure 11.43 shows the cost function, $C(n)$, from Problem 34, and a line, $l$, that passes through the origin.

**(a)** What is the slope of line $l$?

**(b)** How does line $l$ relate to $a(n_0)$, the average cost of producing $n_0$ units (as defined in Problem 34)?

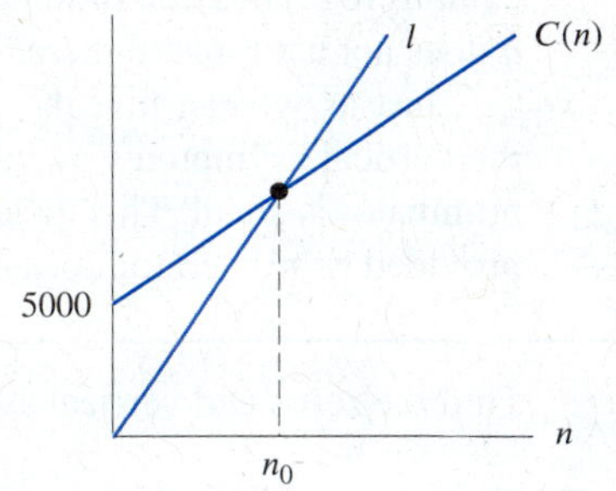


Figure 11.43

**37.** Typically, the average cost of production (as defined in Problem 34) decreases as the level of production increases. Is this always the case for the goods whose total cost function is graphed in Figure 11.44? Use the result of Problem 36 and explain your reasoning.

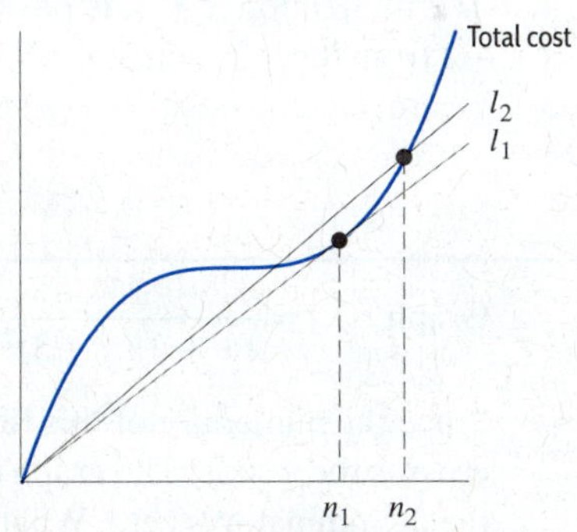


Figure 11.44

**38.** Find a rational function of the form

$$R(x) = \frac{1}{1 + cx}$$

whose values equal those of $f(x) = e^x$ at $x = 0$ and $x = 1$. Plot and compare your approximation on the interval $(0, 1)$ and comment on the result.

# 11.5 THE SHORT-RUN BEHAVIOR OF RATIONAL FUNCTIONS

The short-run behavior of a polynomial can often be determined from its factored form. The same is true of rational functions. If $r$ is a rational function given by

$$r(x) = \frac{p(x)}{q(x)}, \qquad p, q \text{ polynomials,}$$

then the short-run behaviors of $p$ and $q$ tell us about the short-run behavior of $r$.

## The Zeros and Vertical Asymptotes of a Rational Function

A fraction is equal to zero if and only if its numerator equals zero (and its denominator does not equal zero). Thus, the rational function $r(x) = p(x)/q(x)$ has a zero wherever $p$ has a zero, provided $q$ does not have a zero there.

Just as we can find the zeros of a rational function by looking at its numerator, we can find the vertical asymptotes by looking at its denominator. A rational function is large wherever its denominator is small. This means that $r$ has a vertical asymptote wherever its denominator has a zero, provided its numerator does not also have a zero there.

**Example 1** Find the zeros and vertical asymptotes of the rational function $r(x) = \dfrac{x+3}{x+2}$.

**Solution** We see that $r(x) = 0$ if

$$\frac{x+3}{x+2} = 0.$$

This ratio equals zero only if the numerator is zero (and the denominator is not zero), so

$$x + 3 = 0$$
$$x = -3.$$

The only zero of $r$ is $x = -3$. To check, note that $r(-3) = 0/(-1) = 0$.

The denominator, $x+2$, has a zero at $x = -2$, so the graph of $r(x)$ has a vertical asymptote there. The numerator, $x + 3$, is positive both to the left and to the right of $x = -2$. Thus, as $x$ approaches $-2$ from the left, where $x+2$ is negative, $r(x)$ tends toward $-\infty$. As $x$ approaches $-2$ from the right, where $x + 2$ is positive, $r(x)$ tends toward $\infty$.

**Example 2** Graph $r(x) = \dfrac{25}{(x+2)(x-3)^2}$, showing all the important features.

**Solution** Since the numerator of this function is never zero, $r$ has no zeros, meaning that the graph of $r$ never crosses the $x$-axis. The graph of $r$ has vertical asymptotes at $x = -2$ and $x = 3$ because this is where the denominator is zero. What does the graph of $r$ look like near its asymptote at $x = -2$? At $x = -2$, the numerator is 25 and the value of the factor $(x-3)^2$ is $(-2-3)^2 = 25$. Thus, near $x = -2$,

$$r(x) = \frac{25}{(x+2)(x-3)^2} \approx \frac{25}{(x+2)(25)} = \frac{1}{x+2}.$$

So, near $x = -2$, the graph of $r$ looks like the graph of $y = 1/(x+2)$. Note that the graph of $y = 1/(x+2)$ is the graph of $y = 1/x$ shifted to the left by 2 units. We see that as $x$ approaches $-2$ from the left $r(x)$ tends toward $-\infty$ and as $x$ approaches $-2$ from the right $r(x)$ tends toward $\infty$.

Alternatively, we can write this using limit notation as

$$\lim_{x\to-2^-} r(x) = -\infty \quad \text{and} \quad \lim_{x\to-2^+} r(x) = \infty.$$

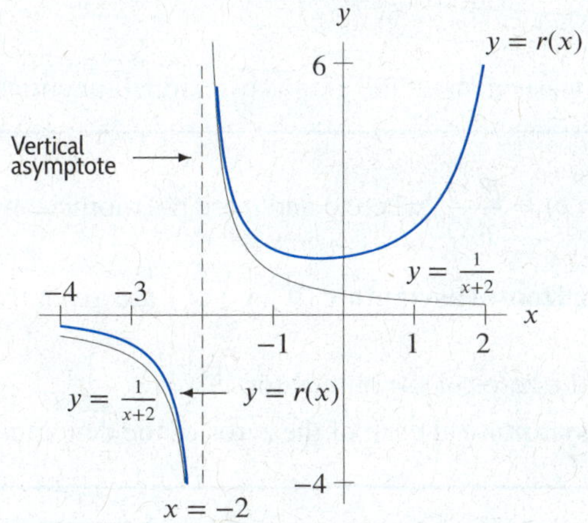


Figure 11.45: The rational function $r$ resembles the shifted power function $1/(x+2)$ near the asymptote at $x=-2$

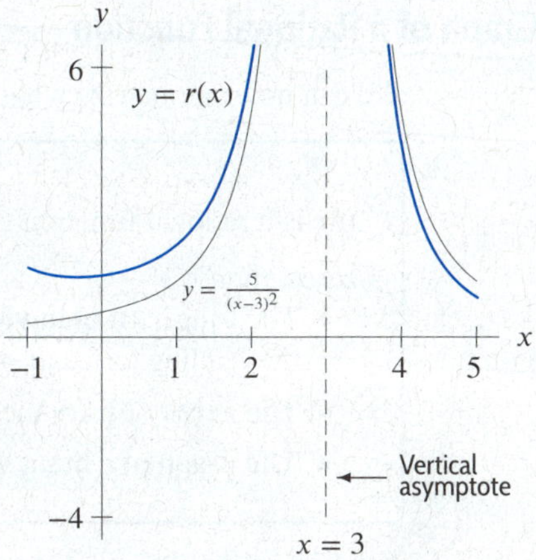


Figure 11.46: The rational function $r$ resembles the shifted power function $5/(x-3)^2$ near the asymptote at $x=3$

What does the graph of $r$ look like near its vertical asymptote at $x=3$? Near $x=3$, the numerator is 25 and value of the factor $(x+2)$ is approximately $(3+2)=5$. Thus, near $x=3$,

$$r(x) \approx \frac{25}{(5)(x-3)^2} = \frac{5}{(x-3)^2}.$$

Near $x=3$, the graph of $r$ looks like the the graph of $y=5/(x-3)^2$. We see that $r(x)\to\infty$ as $x\to 3$, in other words

$$\lim_{x\to 3} r(x) = \infty.$$

The graph of $y=5/(x-3)^2$ is the graph of $y=5/x^2$ shifted to the right 3 units. Since

$$r(0) = \frac{25}{(0+2)(0-3)^2} = \frac{25}{18} \approx 1.4,$$

the graph of $r$ crosses the $y$-axis at $25/18$. The long-run behavior of $r$ is given by the ratio of the leading term in the numerator to the leading term in the denominator. The numerator is 25, and if we multiply out the denominator, we see that its leading term is $x^3$. Thus, the long-run behavior of $r$ is given by $y=25/x^3$, which has a horizontal asymptote at $y=0$. See Figure 11.47. We see that $r(x)\to 0$ as $x\to\pm\infty$. Using limit notation, we have

$$\lim_{x\to\pm\infty} r(x) = 0.$$

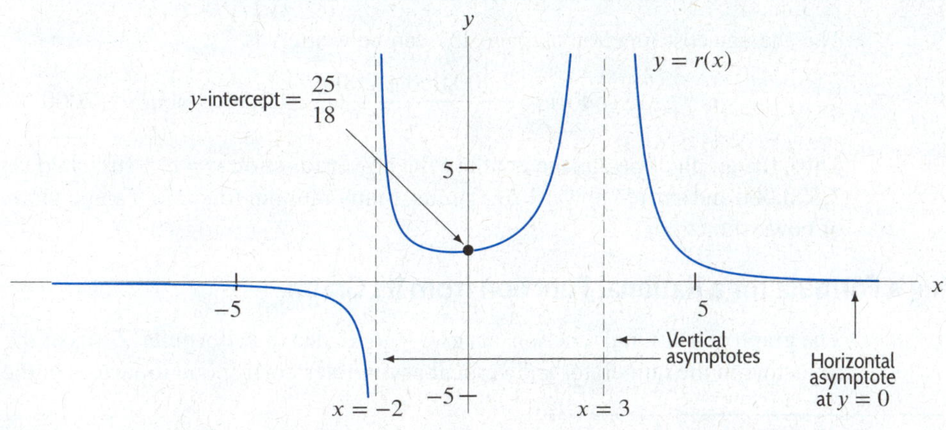


Figure 11.47: A graph of the rational function $r(x) = \dfrac{25}{(x+2)(x-3)^2}$, showing intercepts and asymptotes

## The Graph of a Rational Function

We can now summarize what we have learned about the graphs of rational functions.

If $r$ is a rational function given by $r(x) = \dfrac{p(x)}{q(x)}$, where $p$ and $q$ are polynomials with different zeros, then:

- The **long-run behavior** and **horizontal asymptote** (if any) of $r$ are given by the ratio of the leading terms of $p$ and $q$.
- The **zeros** of $r$ are the same as the zeros of the numerator, $p$.
- The graph of $r$ has a **vertical asymptote** at each of the zeros of the denominator, $q$.

If $p$ and $q$ have zeros at the same $x$-values, the rational function may behave differently. See the discussion about holes in graphs on page 467.

### Can a Graph Cross an Asymptote?

The graph of a rational function never crosses a vertical asymptote. However, the graphs of some rational functions cross their horizontal asymptotes. The difference is that a vertical asymptote occurs where the function is undefined, so there can be no $y$-value there, whereas a horizontal asymptote represents the limiting value of the function as $x \to \pm\infty$. There is no reason that the function cannot take on this limiting $y$-value for some finite $x$-value. For example, the graph of $r(x) = \dfrac{x^2 + 2x - 3}{x^2}$ crosses the line $y = 1$, its horizontal asymptote; the graph does not cross the vertical asymptote, the $y$-axis. See Figure 11.48.

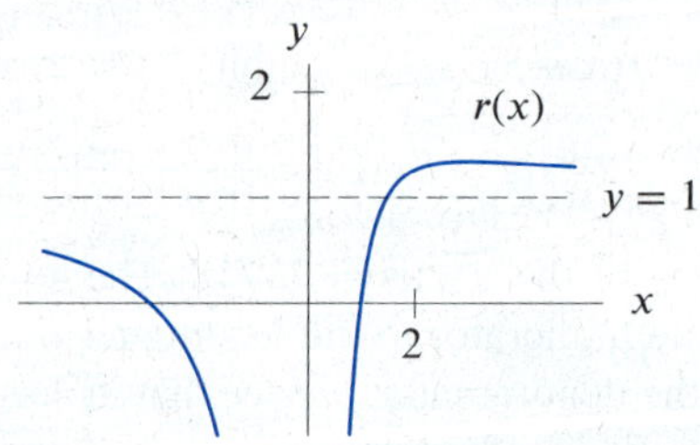


**Figure 11.48:** A rational function can cross its horizontal asymptote

## Rational Functions as Transformations of Power Functions

The average cost function on page 457 can be written as

$$a(q) = \frac{2{,}500{,}000 + 2000q}{q} = 2{,}500{,}000q^{-1} + 2000.$$

Thus, the graph of $a(q)$ is the graph of the power function $y = q^{-1}$ stretched vertically by a factor of 2,500,000 and shifted upward 2000 units. Many rational functions can be viewed as transformations of power functions.

## Finding a Formula for a Rational Function from its Graph

The graph of a rational function can give a good idea of its formula. Zeros of the function correspond to factors in the numerator and vertical asymptotes correspond to factors in the denominator.

**Example 3** Find a possible formula for the rational function, $g(x)$, graphed in Figure 11.49.

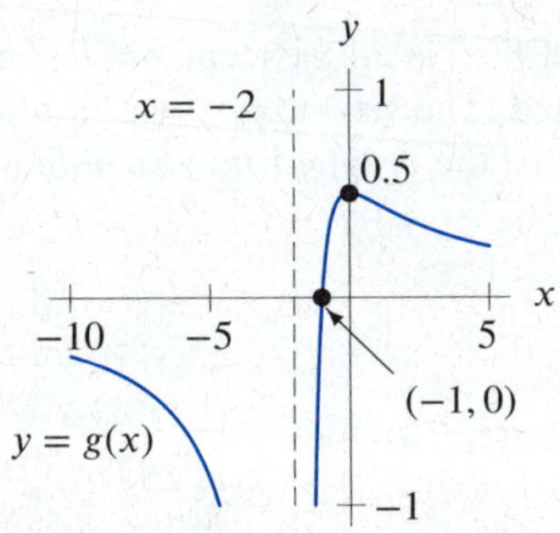


**Figure 11.49:** The graph of $y = g(x)$, a rational function

**Solution** From the graph, we see that $g$ has a zero at $x = -1$ and a vertical asymptote at $x = -2$. This means that the numerator of $g$ has a zero at $x = -1$ and the denominator of $g$ has a zero at $x = -2$. The zero of $g$ does not seem to be a multiple zero because the graph crosses the $x$-axis instead of bouncing and does not have a flattened appearance. Thus, we conclude that the numerator of $g$ has one factor of $(x + 1)$.

The values of $g(x)$ have the same sign on both sides of the vertical asymptote. Thus, the behavior of $g$ near its vertical asymptote is more like the behavior of $y = 1/(x+2)^2$ than like $y = 1/(x+2)$. We conclude that the denominator of $g$ has a factor of $(x + 2)^2$. This suggests

$$g(x) = k \cdot \frac{x+1}{(x+2)^2},$$

where $k$ is a stretch factor. To find the value of $k$, use the fact that $g(0) = 0.5$. So

$$\begin{aligned} 0.5 &= k \cdot \frac{0+1}{(0+2)^2} \\ 0.5 &= k \cdot \frac{1}{4} \\ k &= 2. \end{aligned}$$

Thus, a possible formula for $g$ is $g(x) = \dfrac{2(x+1)}{(x+2)^2}$.

## When Numerator and Denominator Have the Same Zeros: Holes

The rational function $h(x) = \dfrac{x^2 + x - 2}{x - 1}$ is undefined at $x = 1$ because the denominator equals zero at $x = 1$. However, the graph of $h$ may not have a vertical asymptote at $x = 1$ because the numerator of $h$ also equals zero at $x = 1$. At $x = 1$, this function is undefined since

$$h(1) = \frac{x^2 + x - 2}{x - 1} = \frac{1^2 + 1 - 2}{1 - 1} = \frac{0}{0},$$

What does the graph of $h$ look like? Factoring the numerator of $h$ gives

$$h(x) = \frac{(x-1)(x+2)}{x-1} = \frac{x-1}{x-1}(x+2).$$

For any $x \neq 1$, we can cancel $(x - 1)$ top and bottom and rewrite the formula for $h$ as

$$h(x) = x + 2, \qquad \text{provided } x \neq 1.$$

Thus, the domain of $h(x)$ is all $x$ except $x = 1$. The graph of $h$ is the line $y = x + 2$ except at $x = 1$, where $h$ is undefined. The line $y = x + 2$ contains the point $(1, 3)$, but the graph of $h$ does not. Therefore, we say that the graph of $h$ has a *hole* in it at the point $(1, 3)$. See Figure 11.50.

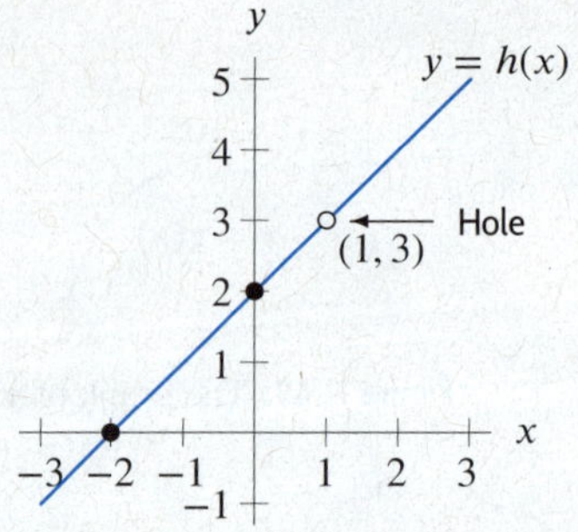


Figure 11.50: The graph of $y = h(x)$ is the line $y = x + 2$, except at the point $(1, 3)$, where it has a hole

## Exercises and Problems for Section 11.5

### Exercises

**1.** Without a calculator, match the functions (a)–(f) with their graphs in (i)–(vi) by finding the zeros, asymptotes, and end behavior for each function.

**(a)** $y = \dfrac{-1}{(x-5)^2} - 1$ **(b)** $y = \dfrac{x-2}{(x+1)(x-3)}$

**(c)** $y = \dfrac{2x+4}{x-1}$ **(d)** $y = \dfrac{1}{x+1} + \dfrac{1}{x-3}$

**(e)** $y = \dfrac{1-x^2}{x-2}$ **(f)** $y = \dfrac{1-4x}{2x+2}$

(i)

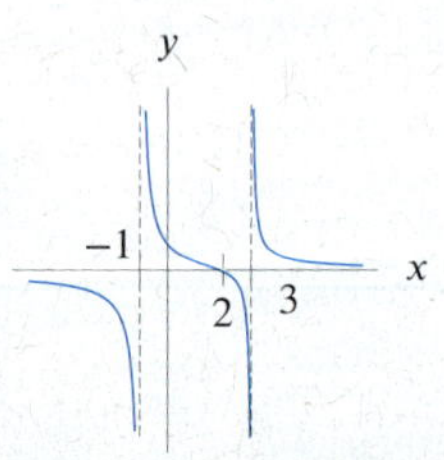


(ii)

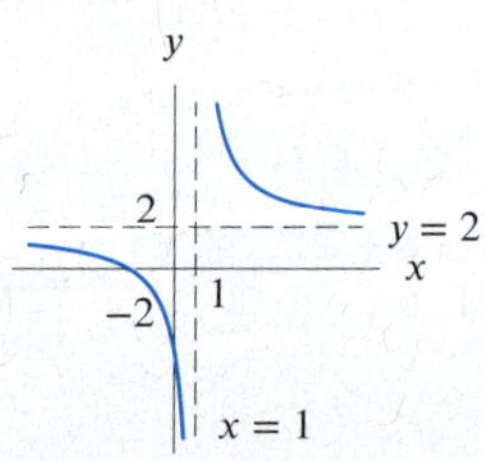


(iii)

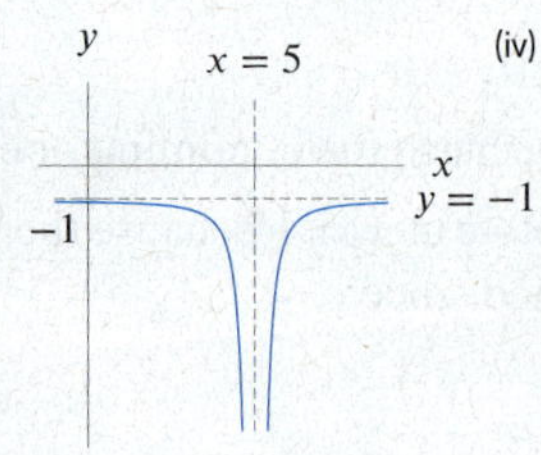


(iv)

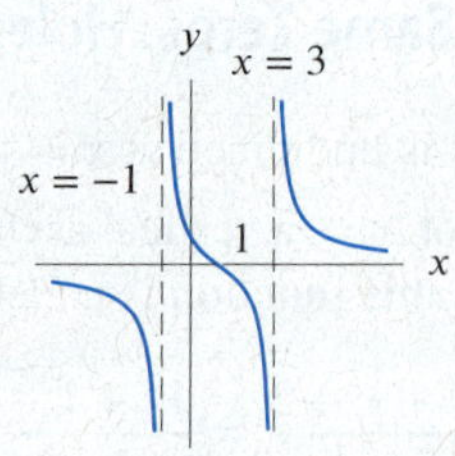


(v)

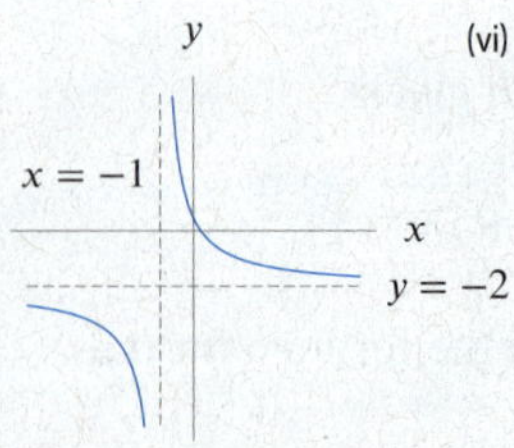


(vi)

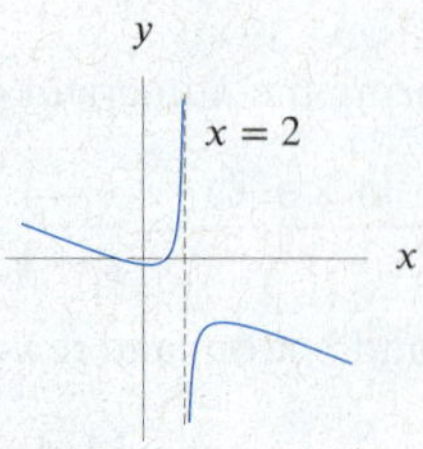


**2.** Let $G(x) = \dfrac{2x}{x+4}$.

**(a)** Complete Table 11.8 for $x$-values close to $-4$. What happens to the values of $G(x)$ as $x$ approaches $-4$ from the left? From the right?

**Table 11.8**

| $x$ | $-5$ | $-4.1$ | $-4.01$ | $-4$ | $-3.99$ | $-3.9$ | $-3$ |
|---|---|---|---|---|---|---|---|
| $G(x)$ | | | | | | | |

**(b)** Complete Tables 11.9 and 11.10. What happens to the values of $G(x)$ as $x$ takes very large positive values? As $x$ takes very large negative values?

**Table 11.9**

| $x$ | 5 | 10 | 100 | 1000 |
|---|---|---|---|---|
| $G(x)$ | | | | |

**Table 11.10**

| $x$ | $-5$ | $-10$ | $-100$ | $-1000$ |
|---|---|---|---|---|
| $G(x)$ | | | | |

**(c)** Without a calculator, graph $y = G(x)$. Give equations for the horizontal and vertical asymptotes.

**3.** Let $g(x) = \dfrac{1}{(x+2)^2}$.

**(a)** Complete Table 11.11 for $x$-values close to $-2$. What happens to the values of $g(x)$ as $x$ approaches $-2$ from the left? From the right?

**Table 11.11**

| $x$ | $-3$ | $-2.1$ | $-2.01$ | $-2$ | $-1.99$ | $-1.9$ | $-1$ |
|---|---|---|---|---|---|---|---|
| $g(x)$ | | | | | | | |

**(b)** Complete Tables 11.12 and 11.13. What happens to the values of $g(x)$ as $x$ takes very large positive values? As $x$ takes very large negative values?

**Table 11.12**

| $x$ | 5 | 10 | 100 | 1000 |
|---|---|---|---|---|
| $g(x)$ | | | | |

**Table 11.13**

| $x$ | $-5$ | $-10$ | $-100$ | $-1000$ |
|---|---|---|---|---|
| $g(x)$ | | | | |

**(c)** Without a calculator, graph $y = g(x)$. Give equations for the horizontal and vertical asymptotes.

For the rational functions in Exercises 4–7, find all zeros and vertical asymptotes and describe the long-run behavior, then graph the function.

**4.** $y = \dfrac{x-4}{x^2-9}$ **5.** $y = \dfrac{x^2-4}{x-9}$

**6.** $y = \dfrac{x+3}{x+5}$ **7.** $y = \dfrac{x+3}{(x+5)^2}$

In Exercises 8–11, what are the $x$-intercepts, $y$-intercepts, and horizontal and vertical asymptotes (if any)?

**8.** $h(x) = \dfrac{x^2-4}{x^3+4x^2}$ **9.** $k(x) = \dfrac{x(4-x)}{x^2-6x+5}$

**10.** $f(x) = \dfrac{x-2}{x-4}$ **11.** $g(x) = \dfrac{x^2-9}{x^2+9}$

## Problems

In Problems 12–13, estimate the one-sided limits:

**(a)** $\lim\limits_{x\to a^+} f(x)$ **(b)** $\lim\limits_{x\to a^-} f(x)$

**12.** $f(x) = \dfrac{x}{5-x}$ with $a = 5$

**13.** $f(x) = \dfrac{5-x}{(x-2)^2}$ with $a = 2$

In Problems 14–15,

**(a)** Estimate $\lim\limits_{x\to\infty} f(x)$ and $\lim\limits_{x\to-\infty} f(x)$.

**(b)** What does the vertical asymptote tell you about limits?

**14.**

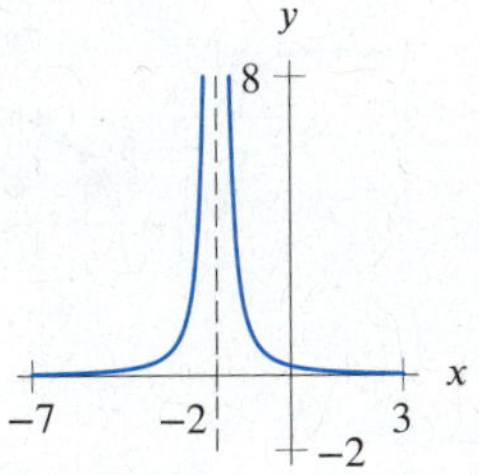


**15.**

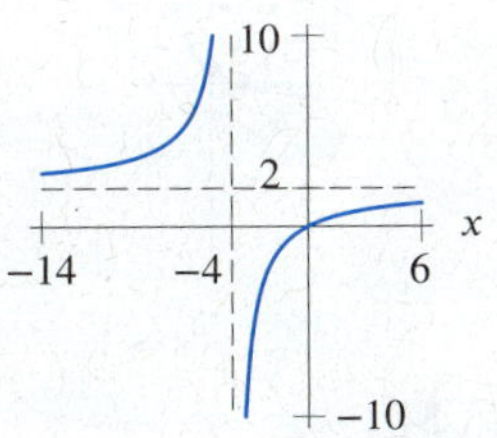


**16.** Let $f(x) = x^2-4$, $g(x) = x^2+4$, and $h(x) = x+5$. Without a calculator, match the functions in (a)–(f) to the descriptions in (i)–(viii). Some of the functions may match none of the descriptions.

**(a)** $y = \dfrac{f(x)}{g(x)}$ **(b)** $y = \dfrac{g(x)}{f(x)}$ **(c)** $y = \dfrac{h(x)}{f(x)}$

**(d)** $y = f(\frac{1}{x})$ **(e)** $y = \dfrac{g(x)}{h(x)}$ **(f)** $y = \dfrac{h(x^2)}{h(x)}$

**(g)** $y = \dfrac{1}{g(x)}$ **(h)** $y = f(x)\cdot g(x)$

(i) Horizontal asymptote at $y = 0$ and one zero at $x = -5$.

(ii) No horizontal asymptote, no zeros, and a vertical asymptote at $x = -5$.

(iii) Zeros at $x = -5$, $x = -2$, and $x = 2$.

(iv) No zeros, a horizontal asymptote at $y = 0$, and a vertical asymptote at $x = -5$.

(v) Two zeros, no vertical asymptotes, and a horizontal asymptote at $y = 1$.

(vi) No zeros, no vertical asymptotes, and a horizontal asymptote at $y = 1$.

(vii) Horizontal asymptote at $y = -4$.

(viii) No horizontal asymptotes, two zeros, and a vertical asymptote at $x = -5$.

**17.** Suppose that $n$ is a constant and that $f(x)$ is a function defined when $x = n$. Complete the following sentences.

**(a)** If $f(n)$ is large, then $\dfrac{1}{f(n)}$ is ...

**(b)** If $f(n)$ is small, then $\dfrac{1}{f(n)}$ is ...

**(c)** If $f(n) = 0$, then $\dfrac{1}{f(n)}$ is ...

**(d)** If $f(n)$ is positive, then $\dfrac{1}{f(n)}$ is ...

**(e)** If $f(n)$ is negative, then $\dfrac{1}{f(n)}$ is ...

**18. (a)** Use the results of Problem 17 to graph $y = 1/f(x)$ given the graph of $y = f(x)$ in Figure 11.51.

**(b)** Find a possible formula for the function in Figure 11.51. Use this formula to check your graph for part (a).

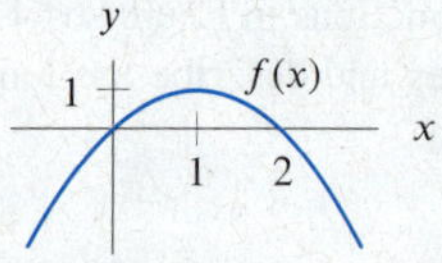


Figure 11.51

19. Use the graph of $f$ in Figure 11.52 to graph

(a) $y=-f(-x)+2$ (b) $y=\dfrac{1}{f(x)}$

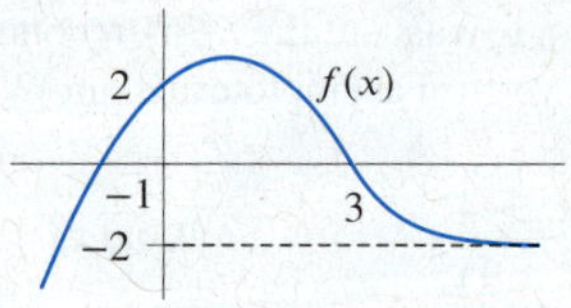


Figure 11.52

Problems 20–22 show a transformation of $y=1/x$.

**(a)** Find a possible formula for the graph.

**(b)** Write the formula from part (a) as the ratio of two linear polynomials.

**(c)** Find the coordinates of the intercepts of the graph.

**20.**

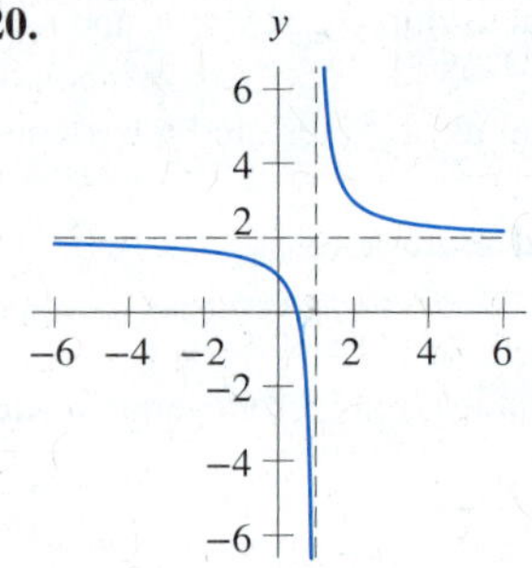

**21.**

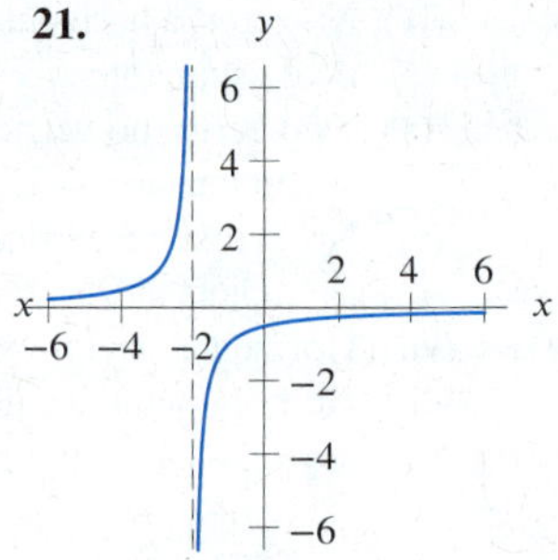

**22.**

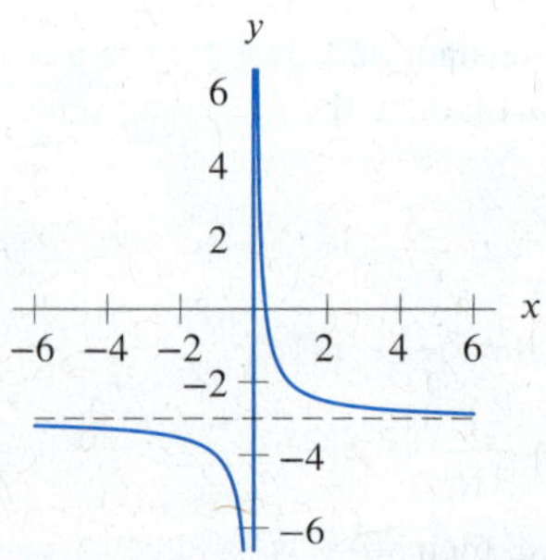

Each of the functions in Problems 23–25 is a transformation of $y=1/x^p$. For each function, determine $p$, describe the transformation in words, and graph the function, labeling any intercepts and asymptotes.

**23.** $f(x)=\dfrac{1}{x-3}+4$ **24.** $g(x)=-\dfrac{1}{(x-2)^2}-3$

**25.** $h(x)=\dfrac{1}{x-1}+\dfrac{2}{1-x}+2$

Problems 26–29 give values of transformations of either $y=1/x$ or $y=1/x^2$. In each case

**(a)** Determine if the values are from a transformation of $y=1/x$ or $y=1/x^2$. Explain your reasoning.

**(b)** Find a possible formula for the function.

**26.**

| $x$ | $y$ |
|---|---|
| −1000 | 1.000001 |
| −100 | 1.00001 |
| −10 | 1.01 |
| 10 | 1.01 |
| 100 | 1.0001 |
| 1000 | 1.000001 |

**27.**

| $x$ | $y$ |
|---|---|
| 1.5 | −1.5 |
| 1.9 | −9.5 |
| 1.95 | −19.5 |
| 2 | Undefined |
| 2.05 | 20.5 |
| 2.1 | 10.5 |
| 2.5 | 2.5 |

**28.**

| $x$ | $y$ |
|---|---|
| 2.7 | 12.1 |
| 2.9 | 101 |
| 2.95 | 401 |
| 3 | Undefined |
| 3.05 | 401 |
| 3.1 | 101 |
| 3.3 | 12.1 |

**29.**

| $x$ | $y$ |
|---|---|
| −1000 | 0.499 |
| −100 | 0.490 |
| −10 | 0.400 |
| 10 | 0.600 |
| 100 | 0.510 |
| 1000 | 0.501 |

Graph the functions in Problems 30–31 without a calculator.

**30.** $y=2+\dfrac{1}{x}$ **31.** $y=\dfrac{2x^2-10x+12}{x^2-16}$

Find possible formulas for the functions in Problems 32–37.

**32.**

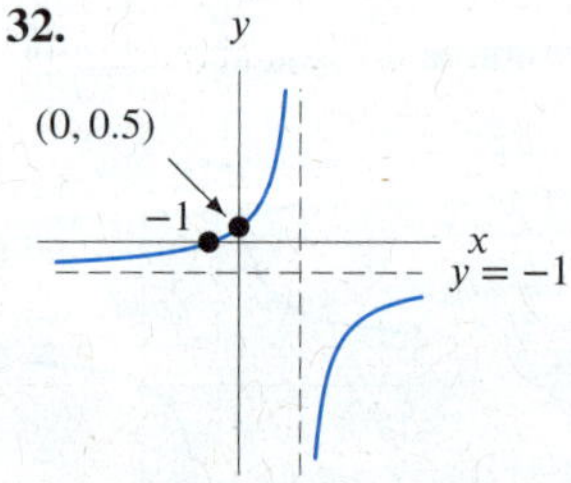


**33.**

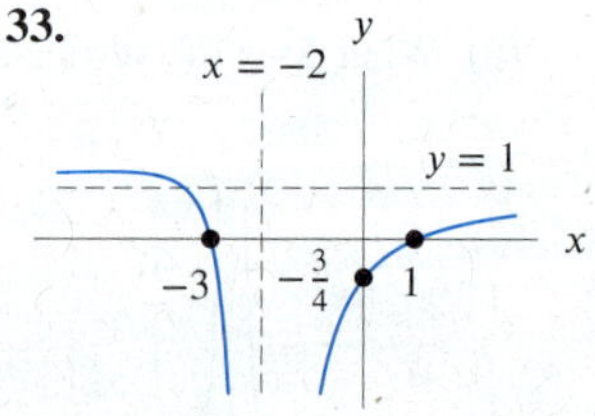


**34.**

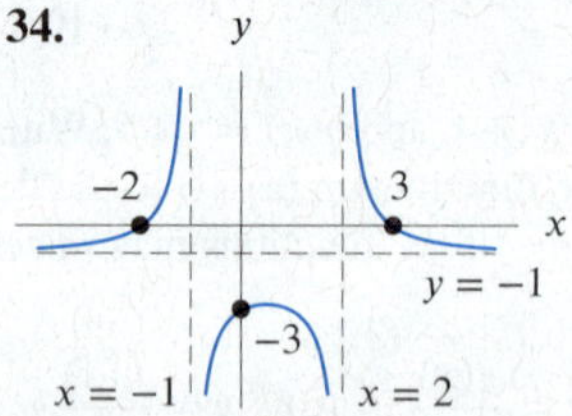


**35.**

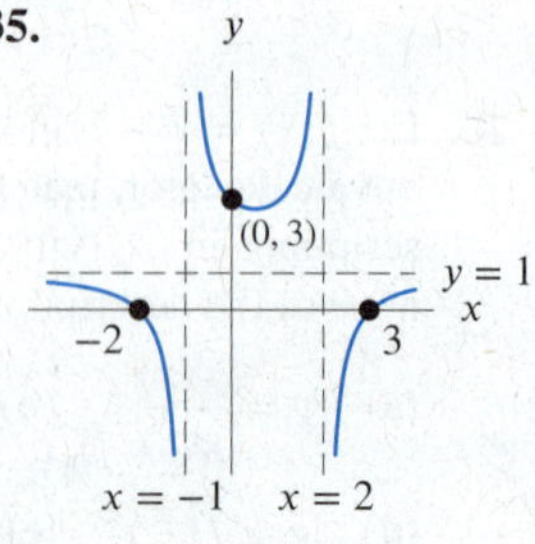


**36.**

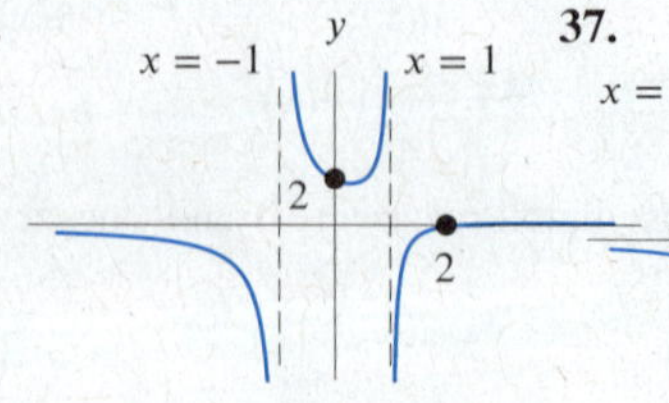


**37.**

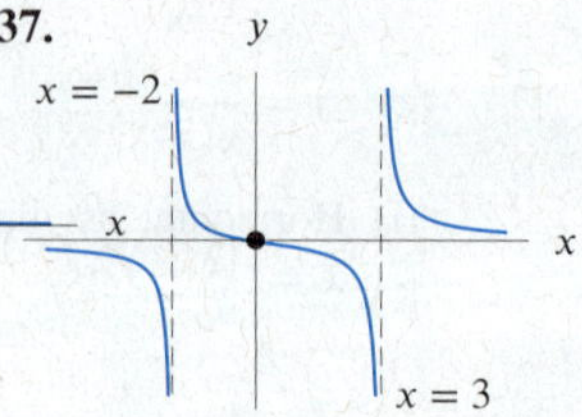

**38.** The graph of $f(x) = \dfrac{18 - 11x + x^2}{x - 2}$ is a line with a hole in it. What is the equation of the line? What are the coordinates of the hole?

**39.** The graph of $g(x) = \dfrac{x^3 + 5x^2 + x + 5}{x + 5}$ is a parabola with a hole in it. What is the equation of the parabola? What are the coordinates of the hole?

**40.** Write a formula for a function, $h(x)$, whose graph is identical to the graph of $y = x^3$, except that the graph of $h$ has a hole at $(2, 8)$. Express the formula as a ratio of two polynomials.

In Problems 41–44, the numerator and denominator share common zeros. Does the graph have a vertical asymptote or a hole at these values of $x$?

**41.** $y = \dfrac{x - 1}{x^2 - 4x + 3}$

**42.** $y = \dfrac{x^2 + 4x + 4}{x^2 + 3x + 2}$

**43.** $y = \dfrac{x^2 + 5x + 6}{(x + 2)^2}$

**44.** $y = \dfrac{x - 3}{x^2 - 6x + 9}$

In Problems 45–47, find a possible formula for the rational functions.

**45.** This function has zeros at $x = 2$ and $x = 3$. It has a vertical asymptote at $x = 5$. It has a horizontal asymptote of $y = -3$.

**46.** The graph of $y = g(x)$ has two vertical asymptotes: one at $x = -2$ and one at $x = 3$. It has a horizontal asymptote of $y = 0$. The graph of $g$ crosses the $x$-axis once, at $x = 5$.

**47.** The graph of $y = h(x)$ has two vertical asymptotes: one at $x = -2$ and one at $x = 3$. It has a horizontal asymptote of $y = 1$. The graph of $h$ touches the $x$-axis once, at $x = 5$.

**48.** Cut four equal squares from the corners of a 8.5″ × 11″ piece of paper. Fold up the sides to create an open box. Find the dimensions of the box with the maximum volume per surface area.

## 11.6 COMPARING POWER, EXPONENTIAL, AND LOG FUNCTIONS

In preceding chapters, we encountered exponential and logarithmic functions. In this section, we compare the long- and short-run behaviors of these functions and power functions.

### Comparing Power Functions

For power functions $y = kx^p$ with $p$ a positive integer for large $x$, the higher the power of $x$, the faster the function climbs. See Figure 11.53. Not only are the higher powers larger, but they are *much* larger. This is because if $x = 100$, for example, $100^5$ is one hundred times as big as $100^4$, which is one hundred times as big as $100^3$. As $x$ gets larger, any positive power of $x$ grows much faster than all lower powers of $x$. We say that, as $x \to \infty$, higher powers of $x$ *dominate* lower powers.

As $x$ approaches zero, the situation is reversed. Figure 11.54 is a close-up view near the origin. For $x$ between 0 and 1, $x^3$ is bigger than $x^4$, which is bigger than $x^5$. (Try $x = 0.1$ to confirm this.) For values of $x$ near zero, smaller powers dominate.

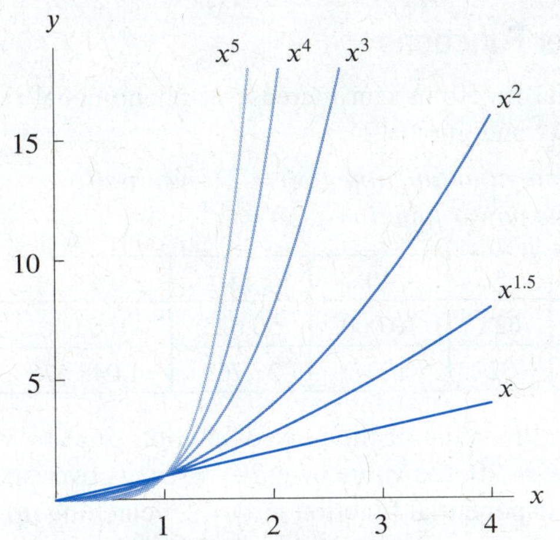


**Figure 11.53**: For large $x$: Large powers of $x$ dominate

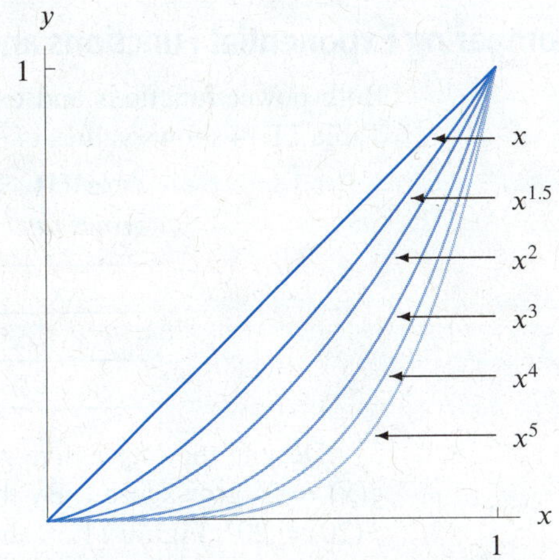


**Figure 11.54**: For $0 \leq x \leq 1$: Small powers of $x$ dominate

In Chapter 6 we saw the effect of $k$ on the graph of $f(x) = kx^p$. The coefficient $k$ stretches or compresses the graph vertically; if $k$ is negative, the graph is reflected across the $x$-axis. How does the value of $k$ affect the long-term growth rate of $f(x) = kx^p$? Is the growth of a power function affected more by the size of the coefficient or by the size of the power?

**Example 1** Let $f(x) = 100x^3$ and $g(x) = x^4$ for $x > 0$. Compare the long-term behavior of these two functions using graphs.

**Solution** For $x < 10$, Figure 11.55 suggests that $f$ is growing faster than $g$ and that $f$ dominates $g$. Eventually, however, the fact that $g$ has a higher power than $f$ asserts itself. In Figure 11.56, we see that $g(x)$ has caught up to $f(x)$ at $x = 100$. In Figure 11.57, we see that for $x > 100$, values of $g$ are larger than values of $f$.

Could the graphs of $f$ and $g$ intersect again for some value of $x > 100$? To show that this cannot be the case, solve the equation $g(x) = f(x)$:

$$x^4 = 100x^3$$
$$x^4 - 100x^3 = 0$$
$$x^3(x - 100) = 0.$$

Since the only solutions to this equation are $x = 0$ and $x = 100$, the graphs of $f$ and $g$ do not cross for $x > 100$.

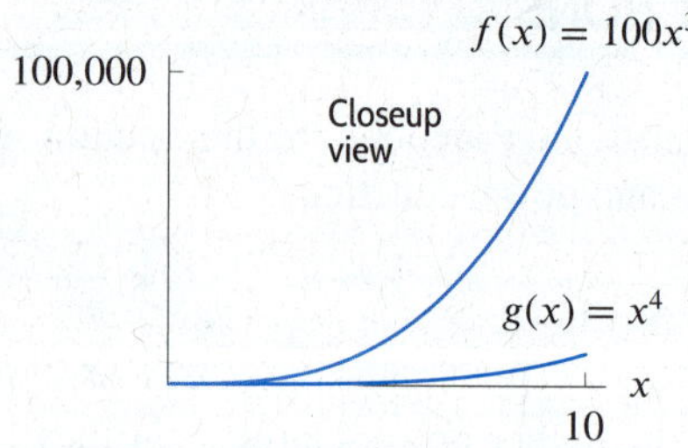


**Figure 11.55:** On this interval, $f$ climbs faster than $g$

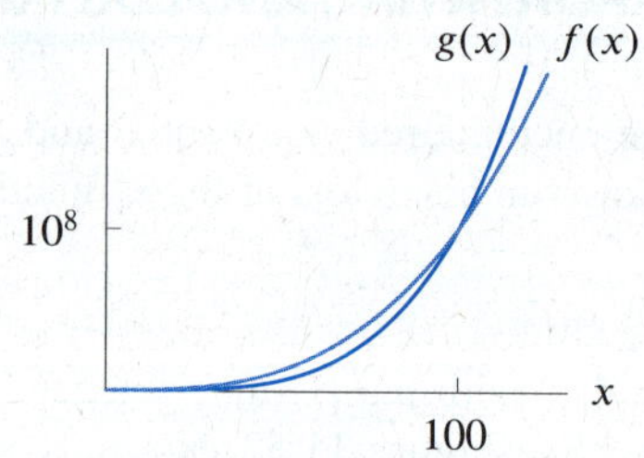


**Figure 11.56:** On this interval, $g$ catches up to $f$

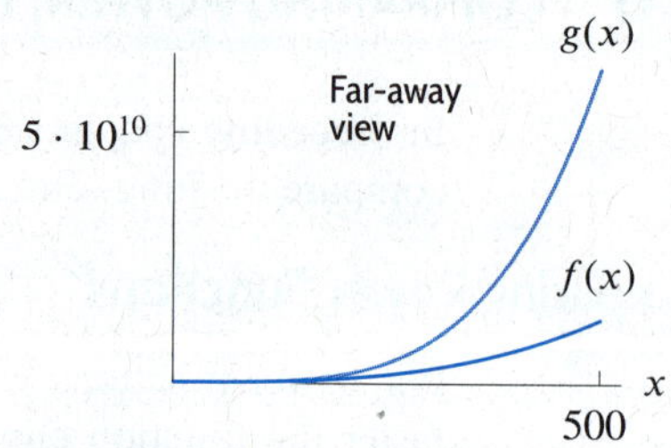


**Figure 11.57:** On this interval, $g$ ends up far ahead of $f$

> When comparing power functions with positive coefficients, higher powers dominate.

## Comparing Exponential Functions and Power Functions

Both power functions and exponential functions can increase at phenomenal rates. For example, Table 11.14 shows values of $f(x) = x^4$ and $g(x) = 2^x$.

**Table 11.14** *The exponential function $g(x) = 2^x$ eventually grows faster than the power function $f(x) = x^4$*

| $x$ | 0 | 5 | 10 | 15 | 20 |
|---|---|---|---|---|---|
| $f(x) = x^4$ | 0 | 625 | 10,000 | 50,625 | 160,000 |
| $g(x) = 2^x$ | 1 | 32 | 1024 | 32,768 | 1,048,576 |

Despite the impressive growth in the value of the power function $f(x) = x^4$, in the long run $g(x) = 2^x$ grows faster. By the time $x = 20$, the value of $g(20) = 2^{20}$ is over six times as large as $f(20) = 20^4$. Figure 11.58 shows the exponential function $g(x) = 2^x$ catching up to $f(x) = x^4$.

But what about a more slowly growing exponential function? After all, $y = 2^x$ increases at a 100% growth rate. Figure 11.59 compares $y = x^4$ to the exponential function $y = 1.005^x$. Despite the fact that this exponential function creeps along at a 0.5% growth rate, at around $x = 7000$, it overtakes the power function. In summary,

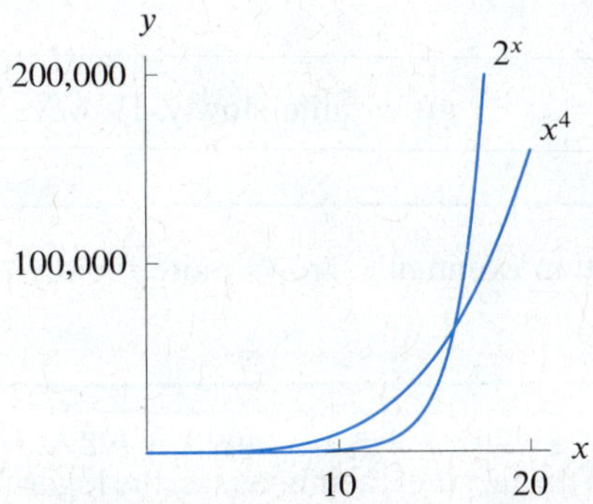


Figure 11.58: The exponential function $y = 2^x$ dominates the power function $y = x^4$

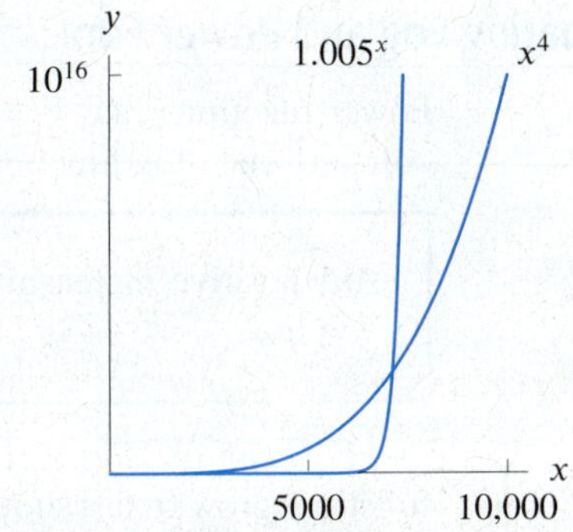


Figure 11.59: The exponential function $y = 1.005^x$ dominates the power function $y = x^4$

> *Any* positive increasing exponential function eventually grows faster than *any* power function.

## Decreasing Exponential Functions and Decreasing Power Functions

Just as an increasing exponential function eventually outpaces any increasing power function, an exponential decay function wins the race toward the $x$-axis. In general:

> *Any* positive decreasing exponential function eventually approaches the horizontal axis faster than any positive decreasing power function.

For example, let's compare the long-term behavior of the decreasing exponential function $y = 0.5^x$ with the decreasing power function $y = x^{-2}$. By rewriting

$$y = 0.5^x = \left(\frac{1}{2}\right)^x = \frac{1}{2^x} \quad \text{and} \quad y = x^{-2} = \frac{1}{x^2}$$

we can see the comparison more easily. In the long run, the smallest of these two fractions is the one with the largest denominator. The fact that $2^x$ is eventually larger than $x^2$ means that $1/2^x$ is eventually smaller than $1/x^2$.

Figure 11.60 shows $y = 0.5^x$ and $y = x^{-2}$. Both graphs have the $x$-axis as a horizontal asymptote. As $x$ increases, the exponential function $y = 0.5^x$ approaches the $x$-axis faster than the power function $y = x^{-2}$. Figure 11.61 shows what happens for large values of $x$. The exponential function approaches the $x$-axis so rapidly that it becomes invisible compared to $y = x^{-2}$.

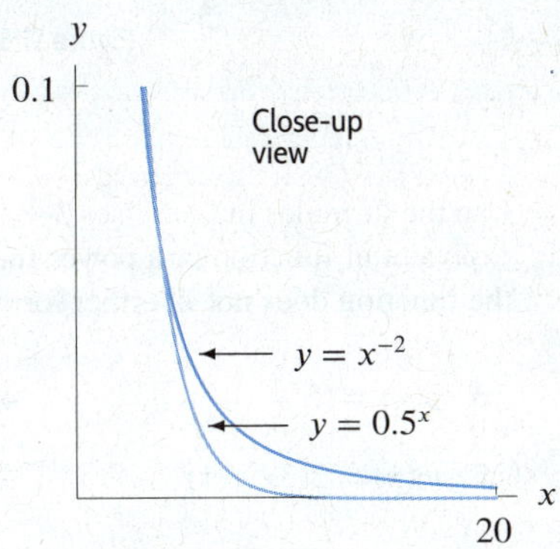


Figure 11.60: Graphs of $y = x^{-2}$ and $y = 0.5^x$

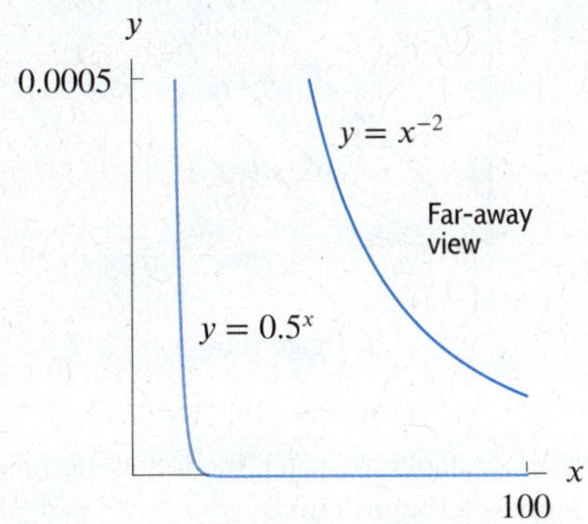


Figure 11.61: Graphs of $y = x^{-2}$ and $y = 0.5^x$

## Comparing Log and Power Functions

Power functions like $y = x^{1/2}$ and $y = x^{1/3}$ grow quite slowly. However, they grow rapidly in comparison to log functions. In fact:

> *Any* positive increasing power function eventually grows more rapidly than $y = \log x$ and $y = \ln x$.

For example, Figure 11.62 shows the graphs of $y = x^{1/2}$ and $y = \log x$. The fact that exponential functions grow so fast should alert you to the fact that their inverses, the logarithms, grow very slowly. See Figure 11.63.

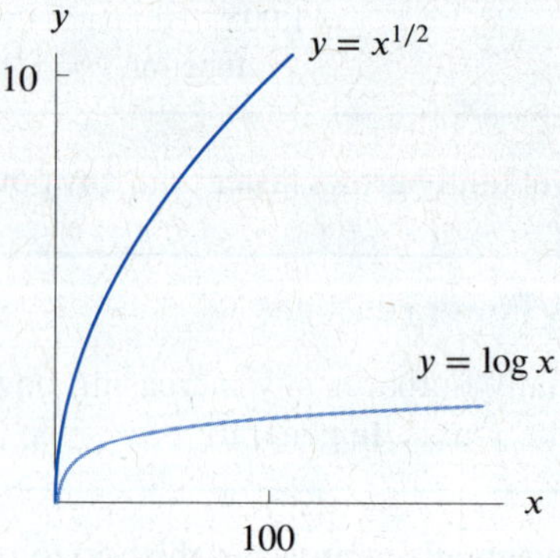


Figure 11.62: Graphs of $y = x^{1/2}$ and $y = \log x$

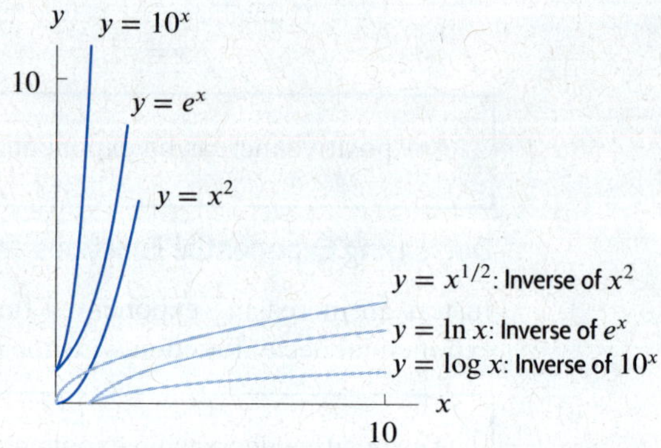


Figure 11.63: Graphs of $y = 10^x$, $y = e^x$, $y = x^2$, $y = x^{1/2}$, $y = \ln x$, and $y = \log x$

## Exercises and Problems for Section 11.6

### Exercises

**1.** Without a calculator, match the following functions with the graphs in Figure 11.64.

(i) $y = x^5$ (ii) $y = x^2$ (iii) $y = x$ (iv) $y = x^3$

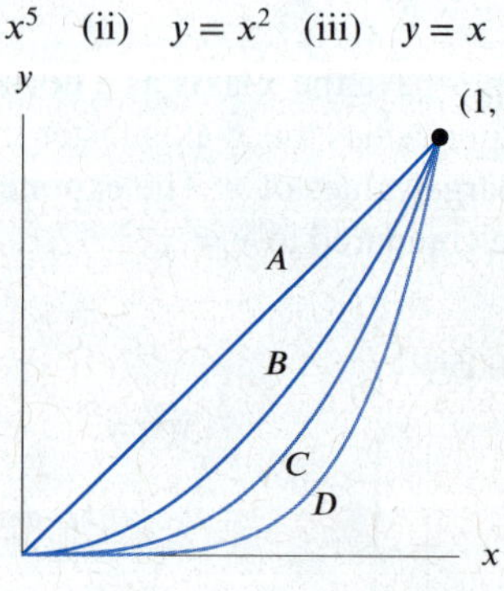


Figure 11.64

**2.** Without a calculator, match the following functions with the graphs in Figure 11.65.

(i) $y = x^5$ (ii) $y = x^2$ (iii) $y = x$ (iv) $y = x^3$

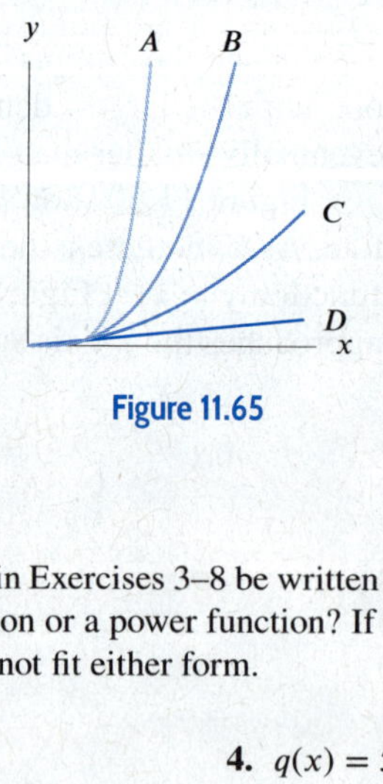


Figure 11.65

Can the formulas in Exercises 3–8 be written in the form of an exponential function or a power function? If not, explain why the function does not fit either form.

**3.** $p(x) = (5^x)^2$

**4.** $q(x) = 5^{(x^2)}$

**5.** $m(x) = 3(3x + 1)^2$

**6.** $n(x) = 3 \cdot 2^{3x+1}$

**7.** $r(x) = 2 \cdot 3^{-2x}$

**8.** $s(x) = \dfrac{4}{5x^{-3}}$

**9.** Let $f(x) = 3^x$ and $g(x) = x^3$.

**(a)** Complete the following table of values:

| $x$ | $-3$ | $-2$ | $-1$ | 0 | 1 | 2 | 3 |
|---|---|---|---|---|---|---|---|
| $f(x)$ | | | | | | | |
| $g(x)$ | | | | | | | |

**(b)** Describe the long-run behaviors of $f$ and $g$ as $x \to -\infty$ and as $x \to +\infty$.

In Exercises 10–13, which function dominates as $x \to \infty$?

**10.** $y = ax^3$, $y = bx^2$, $a, b > 0$

**11.** $y = 7(0.99)^x$, $y = 6x^{35}$

**12.** $y = 4e^x$, $y = 2x^{50}$

**13.** $y = 50x^{1.1}$, $y = 1000x^{1.08}$

## Problems

**14.** Match the graphs in Figure 11.66 with the functions $y = kx^{9/16}$, $y = kx^{3/8}$, $y = kx^{5/7}$, $y = kx^{3/11}$.

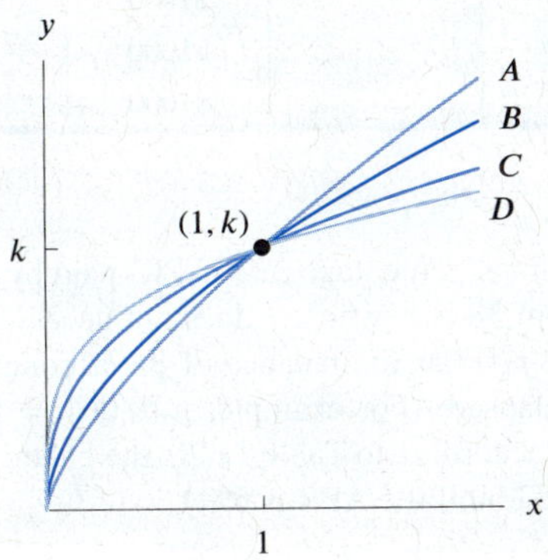


Figure 11.66

**15.** **(a)** Match the functions $f(x) = x^2, g(x) = 2x^2$, and $h(x) = x^3$ to their graphs in Figure 11.67.

**(b)** Do graphs $A$ and $B$ intersect for $x > 0$? If so, for what value(s) of $x$? If not, explain how you know.

**(c)** Do graphs $C$ and $A$ intersect for $x > 0$? If so, for what value(s) of $x$? If not, explain how you know.

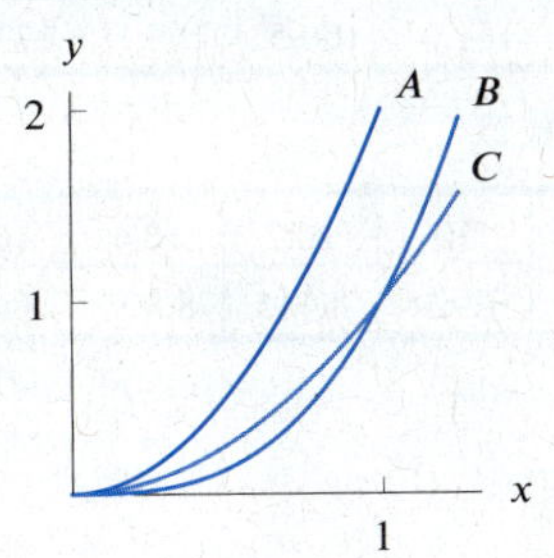


Figure 11.67

**16.** In Figure 11.68, find the values of $m$, $t$, and $k$.

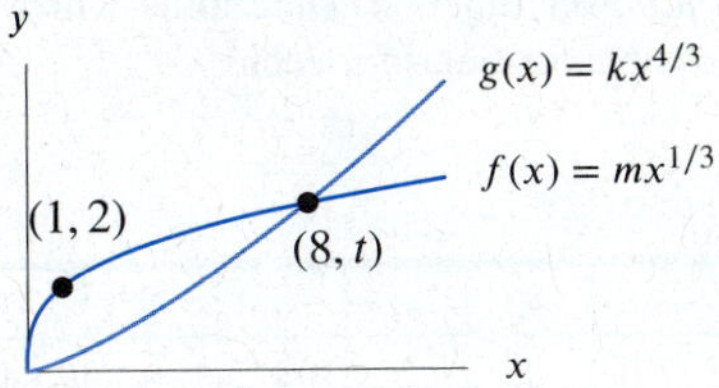


Figure 11.68

**17.** The functions $y = x^{-3}$ and $y = 3^{-x}$ both approach zero as $x \to \infty$. Which function approaches zero faster? Support your conclusion numerically.

**18.** The functions $y = x^{-3}$ and $y = e^{-x}$ both approach zero as $x \to \infty$. Which function approaches zero faster? Support your conclusion numerically.

**19.** Let $f(x) = x^x$. Is $f$ a power function, an exponential function, both, or neither? Discuss.

In Problems 20–22, find a possible formula for $f$ if $f$ is
**(a)** Linear **(b)** Exponential **(c)** Power function.

**20.** $f(1) = 18$ and $f(3) = 1458$

**21.** $f(1) = 16$ and $f(2) = 128$

**22.** $f(-1) = \frac{3}{4}$ and $f(2) = 48$

**23.** Data from four functions are in Tables 11.15–11.18. One function is linear, one is logarithmic, and the other two are power functions (one cubic and one quadratic). Find a formula for each function, and explain how you made your choices.

**Table 11.15**

| $x$ | $j(x)$ |
|---|---|
| 1.8 | 1.75 |
| 2.2 | 3.19 |
| 2.6 | 5.27 |
| 3.0 | 8.10 |
| 3.4 | 11.79 |

**Table 11.16**

| $x$ | $k(x)$ |
|---|---|
| 2.8 | 0.36 |
| 3.0 | 0.60 |
| 3.2 | 0.84 |
| 3.4 | 1.08 |
| 3.6 | 1.32 |

Table 11.17

| $x$ | $m(x)$ |
|---|---|
| 0.2 | −0.699 |
| 0.6 | −0.222 |
| 1.0 | 0.00 |
| 1.4 | 0.146 |
| 1.8 | 0.255 |

Table 11.18

| $x$ | $z(x)$ |
|---|---|
| −4.0 | 6.4 |
| −2 | 1.6 |
| 0 | 0 |
| 2 | 1.6 |
| 4 | 6.4 |

**24.** Table 11.19 gives approximate values for three functions, $f$, $g$, and $h$. One is exponential, one is trigonometric, and one is a power function. Determine which is which and find possible formulas for each.

Table 11.19

| $x$ | −2 | −1 | 0 | 1 | 2 |
|---|---|---|---|---|---|
| $f(x)$ | 4 | 2 | 4 | 6 | 4 |
| $g(x)$ | 20.0 | 2.5 | 0.0 | −2.5 | −20.0 |
| $h(x)$ | 1.33 | 0.67 | 0.33 | 0.17 | 0.08 |

**25.** **(a)** Given $t(x) = x^{-2}$ and $r(x) = 40x^{-3}$, find $v$ such that $t(v) = r(v)$.
**(b)** For $0 < x < v$, which is greater, $t(x)$ or $r(x)$?
**(c)** For $x > v$, which is greater, $t(x)$ or $r(x)$?

What is the long-run behavior of the functions in Problems 26–37?

**26.** $y = \dfrac{x^2+5}{x^8}$

**27.** $y = \dfrac{5-t^2}{(7+t+\sqrt{t})t^5}$

**28.** $y = \dfrac{2^t+7}{5^t+9}$

**29.** $y = \dfrac{3^{-t}}{4^t+7}$

**30.** $y = \dfrac{x(x+5)(x-7)}{4+x^2}$

**31.** $y = \dfrac{2^x+3}{x^2+5}$

**32.** $y = \dfrac{\ln x}{\sqrt{x}+5}$

**33.** $y = \dfrac{e^t+t^2}{\ln|t|}$

**34.** $y = \dfrac{e^x-e^{-x}}{2}$

**35.** $y = \dfrac{e^x-e^{-x}}{e^x+e^{-x}}$

**36.** $y = \dfrac{e^x+5}{x^{100}+50}$

**37.** $y = \dfrac{e^{2t}}{e^{3t}+5}$

**38.** A woman opens a bank account with an initial deposit of \$1000. At the end of each year thereafter, she deposits an additional \$1000.

**(a)** The account earns 6% annual interest, compounded annually. Complete Table 11.20.
**(b)** Does the balance of this account grow linearly, exponentially, or neither? Justify your answer.

Table 11.20

| Years elapsed | Start-of-year balance | End-of-year deposit | End-of-year interest |
|---|---|---|---|
| 0 | \$1000.00 | \$1000 | \$60.00 |
| 1 | \$2060.00 | \$1000 | \$123.60 |
| 2 | \$3183.60 | \$1000 | |
| 3 | | \$1000 | |
| 4 | | \$1000 | |
| 5 | | \$1000 | |

**39.** The annual percentage rate (APR) paid by the account in Problem 38 is $r$, where $r$ does not necessarily equal 6%. Define $p_n(r)$ as the balance of the account after $n$ years have elapsed. (For example, $p_2(0.06) = \$3183.60$, because, according to Table 11.20, the balance after 2 years is \$3183.60 if the APR is 6%.)

**(a)** Find formulas for $p_5(r)$ and $p_{10}(r)$.
**(b)** What is APR if the woman in Problem 38 has \$10,000 in 5 years?

**40.** Values of $f$ and $g$ are in Tables 11.21 and 11.22. One function is of the form $y = a \cdot d^{p/q}$ with $p > q$; the other is of the form $y = b \cdot d^{p/q}$ with $p < q$. Which is which? How can you tell?

Table 11.21

| $d$ | 2 | 2.2 | 2.4 | 2.6 | 2.8 |
|---|---|---|---|---|---|
| $f(d)$ | 151.6 | 160.5 | 169.1 | 177.4 | 185.5 |

Table 11.22

| $d$ | 10 | 10.2 | 10.4 | 10.6 | 10.8 |
|---|---|---|---|---|---|
| $g(d)$ | 7.924 | 8.115 | 8.306 | 8.498 | 8.691 |

## 11.7 FITTING EXPONENTIALS AND POLYNOMIALS TO DATA

In Section 1.6 we used linear regression to find the equation for a line of best fit for a set of data. In this section, we fit an exponential or a power function to a set of data.

## The Spread of AIDS

The data in Table 11.23 give the total number of deaths in the US from AIDS from 1981 to 1996.[13] Figure 11.69 suggests that a linear function may not give the best possible fit for these data.

**Table 11.23** *US deaths from AIDS, 1981–96*

| $t$ | $N$ | $t$ | $N$ |
|---|---|---|---|
| 1 | 159 | 9 | 90039 |
| 2 | 622 | 10 | 121577 |
| 3 | 2130 | 11 | 158193 |
| 4 | 5635 | 12 | 199287 |
| 5 | 12607 | 13 | 244885 |
| 6 | 24717 | 14 | 295303 |
| 7 | 41129 | 15 | 346420 |
| 8 | 62248 | 16 | 384445 |

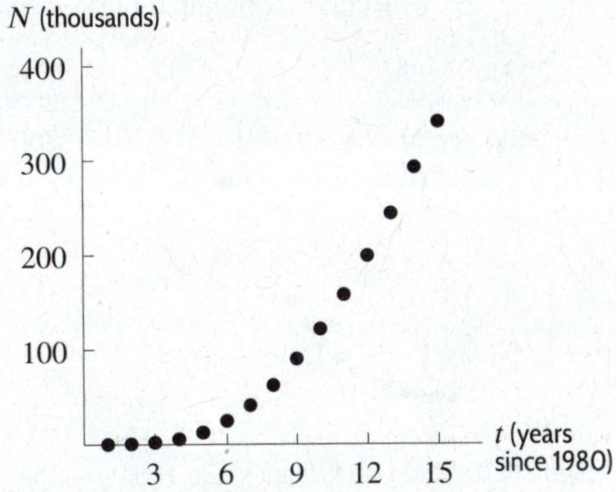


**Figure 11.69:** US deaths from AIDS, 1981–96

### Fitting an Exponential

We first fit an exponential function to the data in Table 11.23:

$$N = ae^{kt},$$

where $N$ is the total number of deaths $t$ years after 1980. Using exponential regression on a calculator or computer, we obtain[14]

$$N \approx 626e^{0.47t}.$$

Figure 11.70 shows how the graph of this formula fits the data points.

### Fitting a Power Function

Now we fit the AIDS data with a power function of the form

$$N = at^{p},$$

where $a$ and $p$ are constants. Some scientists have suggested that a power function may be a better model for the growth of AIDS than an exponential function.[15] Using power function regression on a calculator or a computer, we obtain

$$N \approx 106t^{3.009}.$$

Figure 11.70 shows the graph of this power function with the data.

### Which Function Best Fits the Data?

Both the exponential function

$$N = 626e^{0.47t}$$

and the power function

$$N = 106t^{3.009}$$

[13]www.cdc.gov. Accessed May 27, 2014.

[14]Calculator and computer answers for exponential regression may vary slightly due to different algorithms used.

[15]Stirling A. Colgate, E. Ann Stanley, James M. Hyman, Scott P. Layne, and Clifford Qualls, "Risk behavior-based model of the cubic growth of acquired immunodeficiency syndrome in the United States," *Proc. Natl. Acad. Sci. USA*, Vol. 86, June 1989, Population Biology.

fit the AIDS data reasonably well. By visual inspection alone, the power function arguably provides the better fit. If we fit a linear function to the original data we get

$$N = -99176 + 26296t.$$

Even this linear function gives a possible fit for $t \geq 4$, that is, for 1984 to 1996. (See Figure 11.70.)

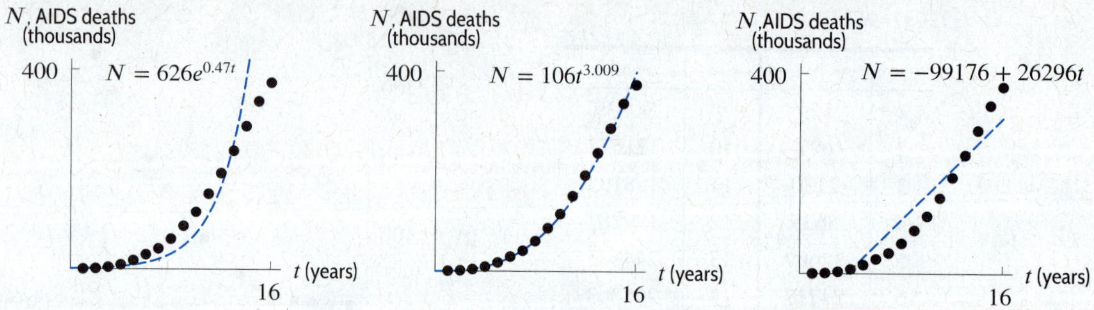


**Figure 11.70:** The AIDS data since 1981 together with an exponential model, a power-function model, and a linear model

Despite the fact that all three functions fit the data reasonably well up to 1996, it's important to realize that they give wildly different predictions. If we use each model to estimate the total number of AIDS deaths by the year 2010 (when $t = 30$), the exponential model gives

$$N = 626e^{(0.47)30} \approx 832{,}006{,}134, \quad \text{about triple the US population;}$$

the power model gives

$$N = 106(30)^{3.009} \approx 2{,}961{,}017, \quad \text{or about 1\% of the US population;}$$

and the linear model gives

$$N = -99176 + 26296 \cdot 30 = 689{,}704, \quad \text{or about 0.22\% of the US population.}$$

Which function is the best predictor of the future? To explore this question, let us add some more recent data to our previous data on AIDS deaths. See Table 11.24.

**Table 11.24** *US deaths from AIDS, 1997–2010*

| $t$ | $N$ | $t$ | $N$ |
|---|---|---|---|
| 17 | 406,444 | 24 | 527,954 |
| 18 | 424,841 | 25 | 544,222 |
| 19 | 442,013 | 26 | 558,954 |
| 20 | 459,152 | 27 | 576,613 |
| 21 | 476,763 | 28 | 595,704 |
| 22 | 494,307 | 29 | 615,685 |
| 23 | 512,156 | 30 | 635,028 |

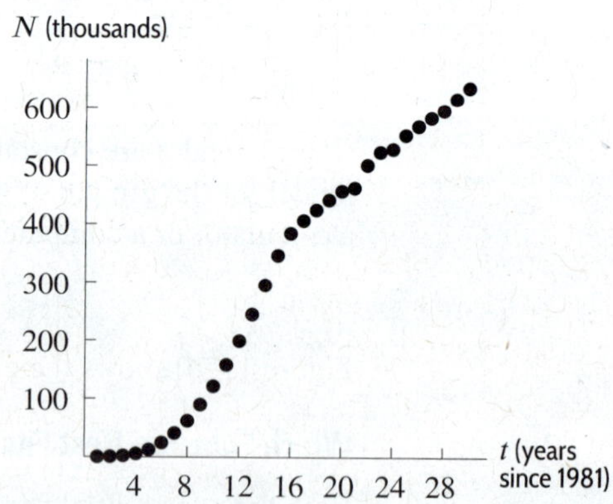


**Figure 11.71:** US deaths from AIDS, 1981–2010

When data from the entire period from 1981 to 2010 are plotted together in Figure 11.71, we see that the rate of increase of AIDS deaths reached a peak sometime around 1995 and then began to taper off. Since none of the functions we have used exhibit this behavior, another type of function is needed to describe the number of AIDS deaths accurately over the entire 30 years.

This example illustrates that while a function may fit a set of data over a short period of time, care must be taken when using a mathematical model to make predictions about the future.

## Exercises and Problems for Section 11.7

### Exercises

In Exercises 1–4, what sort of function might best model the graphed data—linear, exponential or logarithmic?

**1.**

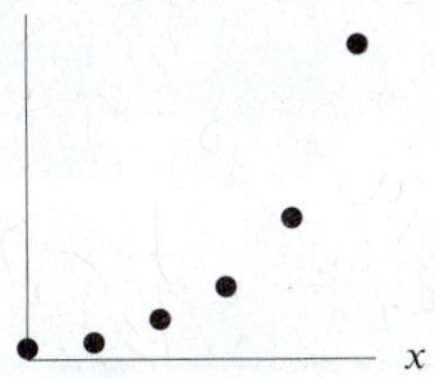


**2.**

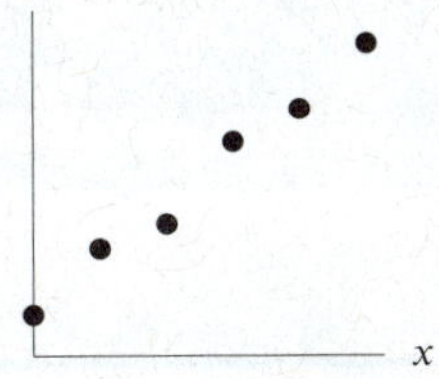


**3.**

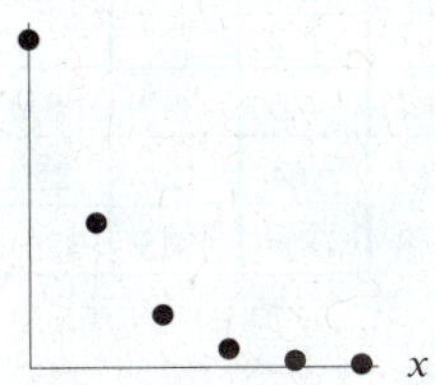


**4.**

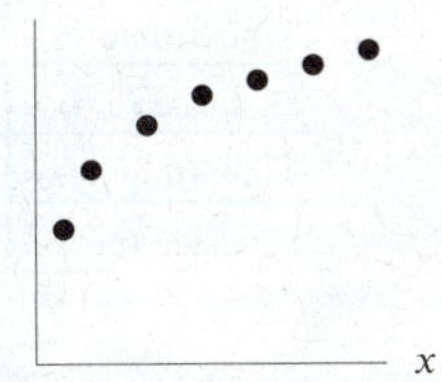


**5.** Find a formula for the power function $f(x)$ such that $f(1) = 1$ and $f(2) = c$.

**6.** Find a formula for an exponential function $h(x)$.

| $x$ | 2 | 3 | 4 | 5 |
|---|---|---|---|---|
| $h(x)$ | 4.5948 | 7.4744 | 10.5561 | 13.7973 |

**7.** Find a formula for the power function $g(x)$.

| $x$ | 2 | 3 | 4 | 5 |
|---|---|---|---|---|
| $g(x)$ | 4.5948 | 7.4744 | 10.5561 | 13.7973 |

**8.** Students in the School of Forestry & Environmental Studies at Yale University collected data measuring sassafras trees. Table 11.25 lists the diameter at breast height (dbh, in cm) and the total dry weight ($w$, in gm) of different trees.[16]

**(a)** Find a power function that fits the data.
**(b)** Predict the total weight of a tree with a dbh of 20 cm.
**(c)** If a tree has a total dry weight of 100,000 gm, what is its expected dbh?

**Table 11.25**

| dbh | 5 | 23.4 | 11.8 | 16.7 | 4.2 | 5.6 |
|---|---|---|---|---|---|---|
| $w$ | 5353 | 169,290 | 30,696 | 76,730 | 3436 | 5636 |
| dbh | 3.8 | 4.3 | 6.5 | 21.9 | 17.7 | 25.5 |
| $w$ | 14,983 | 2098 | 7364 | 177,596 | 100,848 | 171,598 |

**9.** Table 11.26 shows the Maine lobster catch[17] (in millions of pounds) from 2002 to 2012.

**(a)** With $t$ in years since 2000, use a calculator or computer to fit the data with
  (i) A power function of the form $y = at^b$.
  (ii) A quadratic function of the form $y = at^2+bt+c$.
**(b)** Discuss which function is a better fit.

**Table 11.26**

| Year | 2002 | 2004 | 2006 | 2008 | 2010 | 2012 |
|---|---|---|---|---|---|---|
| $t$ | 2 | 4 | 6 | 8 | 10 | 12 |
| Lobster | 62.3 | 70.9 | 75.32 | 69.7 | 93 | 123.3 |

**10.** Anthropologists suggest that the relationship between the body weight and brain weight of primates can be modeled with a power function. Table 11.27 lists various body weights and the corresponding brain weights of different primates.[18]

**(a)** Using Table 11.27, find a power function that gives the brain weight, $Q$ (in mg), as a function of the body weight, $b$ (in gm).
**(b)** The erythrocebus (Patas monkey) has a body weight of 7800 gm. Estimate its brain weight.

**Table 11.27**

| $b$ | 6667 | 960 | 6800 | 9500 | 1088 |
|---|---|---|---|---|---|
| $Q$ | 56,567 | 18,200 | 110,525 | 120,100 | 20,700 |
| $b$ | 2733 | 3000 | 6300 | 1500 | 665 |
| $Q$ | 78,250 | 58,200 | 96,400 | 31,700 | 25,050 |

In Exercises 11–16, find an equation for $y$ in terms of $x$.

**11.**

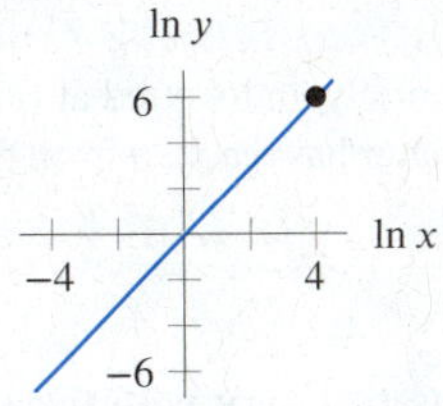


**12.**

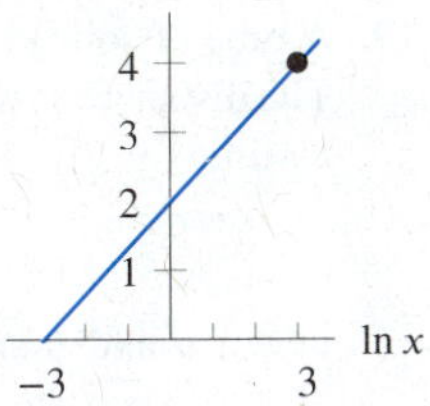


[16] www.yale.edu/fes519b/totoket/allom/allom.htm, accessed December 15, 2002.
[17] www.maine.gov/dmr/rm/lobster/?, accessed May 17, 2013.
[18] mac-huwis.lut.ac.uk/˜wis/lectures/primate-adaptation/10PrimateBrains.pdf, accessed December 15, 2002.

13.

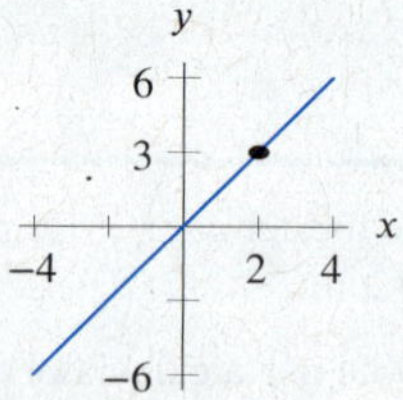


14.

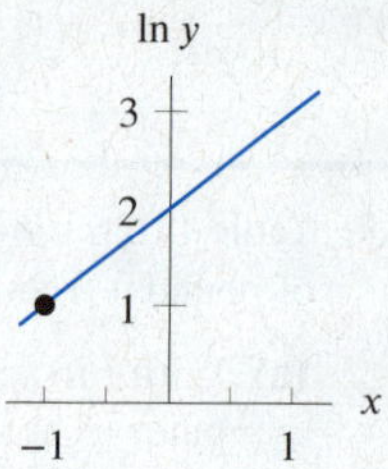


15.

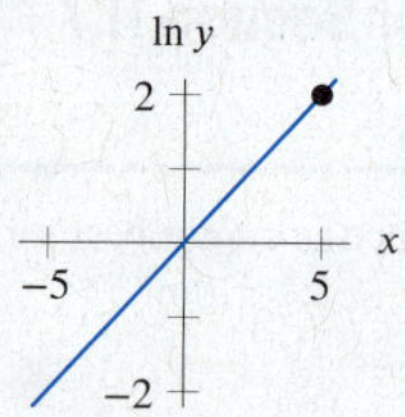


16.

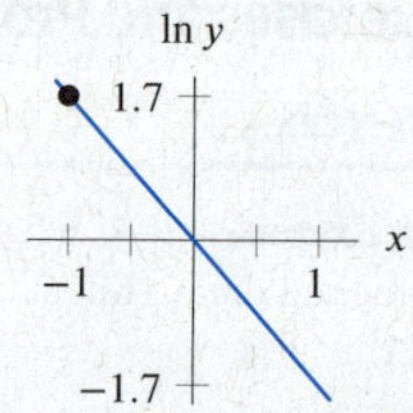


## Problems

**17.** **(a)** Find a linear function that fits the data in Table 11.28. How good is the fit?

**(b)** The data in the table was generated using the power function $y = 5x^3$. Explain why (in this case) a linear function gives such a good fit to a power function. Does the fit remain good for other values of $x$?

**Table 11.28**

| $x$ | 2.00 | 2.01 | 2.02 | 2.03 | 2.04 | 2.05 |
|---|---|---|---|---|---|---|
| $y$ | 40.000 | 40.603 | 41.212 | 41.827 | 42.448 | 43.076 |

**18.** The managers of a furniture store have compiled data showing the weekly demand for recliners at various prices.

**(a)** In Table 11.29, fill in the revenue generated by selling the number of recliners at the corresponding price.

**(b)** Find the quadratic function that best fits the data.

**(c)** According to the function you found, what price should the store charge for their recliners to maximize revenue? What is the maximum revenue?

**Table 11.29**

| Recliner price ($) | 399 | 499 | 599 | 699 | 799 |
|---|---|---|---|---|---|
| Demand (recliners) | 62 | 55 | 47 | 40 | 34 |
| Revenue ($) | | | | | |

**19.** A tube of soil is held horizontally and wetted at one end. The distance, $x$, which the water has reached from the end by time $t$ is given by

$$x = at^b,$$

where $a$ and $b$ are constants that vary with the sample of soil. The data for such an experiment is given in Table 11.30.[19] For this soil it is known that $b \approx 0.4$. What value for $a$ makes this a reasonable model?

**Table 11.30** *Movement of wetting front*

| Time (minutes) | 0 | 1 | 2 | 4 | 8 |
|---|---|---|---|---|---|
| Distance (cm) | 0 | 3.7 | 4.5 | 6.0 | 8.0 |
| Time (minutes) | 16 | 32 | 64 | 128 | 256 |
| Distance (cm) | 10.6 | 13.8 | 18.6 | 24.3 | 32.0 |

**20.** An analog radio dial can be measured in millimeters from left to right. Although the scale of the dial can be different from radio to radio, Table 11.31 gives typical measurements.

**(a)** Which radio band data appear linear? Graph and connect the data points for each band.

**(b)** Which radio band data appear exponential?

**(c)** Find a possible formula for the FM station number in terms of $x$.

**(d)** Find a possible formula for the AM station number in terms of $x$.

**Table 11.31**

| $x$, millimeters | 5 | 15 | 25 | 35 | 45 | 55 |
|---|---|---|---|---|---|---|
| FM (mhz) | 88 | 92 | 96 | 100 | 104 | 108 |
| AM (khz/10) | 53 | 65 | 80 | 100 | 130 | 160 |

**21.** In this problem you will fit a quartic polynomial to the AIDS data.

**(a)** With $N$ as the total number of AIDS deaths in the US $t$ years after 1980, use a calculator or computer to fit the data in Table 11.23 on page 477 with a polynomial of the form

$$N = at^4 + bt^3 + ct^2 + dt + e.$$

**(b)** Graph the data and your quartic for $0 \leq t \leq 16$. Comment on the fit.

**(c)** Graph the data and your quartic for $0 \leq t \leq 30$. Comment on the predictions made by this model.

[19] Adapted from I. A. Guerrini, "An example of motion in a course of physics for agriculture," *The Physics Teacher*, February 1984, 102–103.

22. The population of Armenia,[20] the smallest of the former Soviet republics, is shown from 2009 to 2013 in Table 11.32.

(a) Does the function giving the population as a function of time appear to be increasing or decreasing? Concave up or concave down?
(b) Find a power function that approximates this data with time, $t$, in years since 2008.
(c) Using your function, estimate the population in 2015. Have you used interpolation or extrapolation?

**Table 11.32**

| Year | 2009 | 2010 | 2011 | 2012 | 2013 |
|---|---|---|---|---|---|
| Pop. (millions) | 2.976 | 2.976 | 2.968 | 2.970 | 2.974 |

23. Cellular telephone use has increased over the past two decades. Table 11.33 gives the number of cellular telephone subscriptions in the US, in millions, from 2005 to 2010.[21]

(a) Fit an exponential function to this data with time in years since 2005.
(b) Based on your model, by what percent was the number of cell subscribers increasing each year?
(c) Based on common sense, what do you expect of the rate of growth in the long run? What does this mean in terms of the shape of the graph?

**Table 11.33**

| Year | 2005 | 2006 | 2007 | 2008 | 2009 | 2010 |
|---|---|---|---|---|---|---|
| Subscriptions | 207 | 233 | 255 | 263 | 277 | 301 |

24. The use of one-way pagers declined as cell phones became more popular.[22] The number of users is given in Table 11.34 and plotted in Figure 11.72, along with a quadratic regression function.

(a) How well does the graph of the quadratic function fit the data?
(b) Find a cubic regression function. Does it fit better?

**Table 11.34**

| Year | 1990 | 1991 | 1992 | 1993 | 1994 | 1995 |
|---|---|---|---|---|---|---|
| Users, millions | 10 | 12 | 15 | 19 | 25 | 32 |
| Year | 1996 | 1997 | 1998 | 1999 | 2000 | |
| Users, millions | 38 | 43 | 44 | 43 | 37 | |

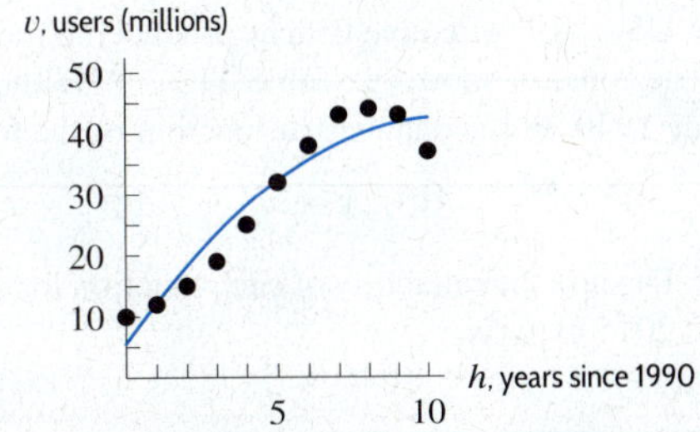


**Figure 11.72**

25. Table 11.35 gives the estimated population, in thousands, of the American colonies from 1650 to 1770.[23]

(a) Make a scatterplot of the data using $t = 0$ to represent the year 1650.
(b) Fit an exponential function to the data.
(c) Explain the meaning of the parameters in your model.
(d) Use your function to predict the population in 1750. Is it high or low?
(e) According to the US Census Bureau,[24] the US population in 1800 was 5,308,483. Use your function to make a prediction for 1800. Is it high or low?

**Table 11.35**

| Year | 1650 | 1670 | 1690 | 1700 |
|---|---|---|---|---|
| Population | 50.4 | 111.9 | 210.4 | 250.9 |
| Year | 1720 | 1740 | 1750 | 1770 |
| Population | 466.2 | 905.6 | 1170.8 | 2148.1 |

26. The US Census Bureau began recording census data in 1790. Table 11.36 gives the population of the US in millions from 1790 to 1860.[25]

(a) With t = 0 representing the number of years since 1790, fit an exponential function to the data.
(b) The 1800 census value is 5.3 million. Find the population predicted by your function for 1800. Problem 25 gave a prediction of 5.5 million using the data for 1650 to 1750. Using the two data sets, explain the difference in predicted values.
(c) Use your function to project the population of the US in 2010. Is this prediction reasonable?

**Table 11.36**

| Year | 1790 | 1800 | 1810 | 1820 |
|---|---|---|---|---|
| Population | 3.929 | 5.308 | 7.240 | 9.638 |
| Year | 1830 | 1840 | 1850 | 1860 |
| Population | 12.861 | 17.063 | 23.192 | 31.443 |

[20] http://www.census.gov/ipc/www/idb/country.php, accessed May 17, 2013.
[21] http://www.infoplease.com/ipa/A0933563.html, accessed May, 2013.
[22] *The New York Times*, p. 16, April 11, 2002.
[23] *The World Almanac and Book of Facts, 2002*, New York, NY, p. 376.
[24] http://www.census.gov/, accessed January 15, 2003.
[25] http://www.census.gov/, accessed January 15, 2003.

**27.** The US export of edible fishery produce, in thousands of metric tons, is shown in Table 11.37.[26] With $t$ in years since 1940, fit the data with a function of the form

**(a)** $y = at^b$ **(b)** $y = ab^t$ **(c)** $y = at^2+bt+c$

**(d)** Discuss the reliability of each function for estimating 2015 exports.

**Table 11.37**

| Year | 1945 | 1950 | 1955 | 1960 | 1965 | 1970 | 1975 |
|---|---|---|---|---|---|---|---|
| Fish export | 62 | 55 | 50 | 31 | 50 | 73 | 109 |
| Year | 1980 | 1985 | 1990 | 1995 | 2000 | 2005 | 2010 |
| Fish export | 275 | 305 | 883 | 929 | 982 | 1329 | 1239 |

**28.** The data in Table 11.37 show a big jump in fish exports between 1985 and 1990. This suggests fitting a piecewise defined function. With $t$ in years since 1935, fit a quadratic function to the data from

**(a)** 1940 to 1985 **(b)** 1990 to 2005

**(c)** Write a piecewise defined function using parts (a) and (b). Graph the function and the data.

**29.** Table 11.38 gives $N$, the number of transistors per integrated circuit chip, $t$ years after 1970.[27]

**(a)** Plot $N$ vs $t$ and fit an exponential curve to the data.
**(b)** According to the formula of your curve of best fit, approximately how often does the number of transistors double?

**Table 11.38**

| Chip name | $t$ | $N$ |
|---|---|---|
| 4004 | 1 | 2,300 |
| 8008 | 2 | 2,500 |
| 8080 | 4 | 4,500 |
| 8086 | 8 | 29,000 |
| Intel286 | 12 | 134,000 |
| Intel386 | 15 | 275,000 |
| Intel486 | 19 | 1,200,000 |
| Pentium | 23 | 3,100,000 |
| Pentium II | 27 | 7,500,000 |
| Pentium III | 29 | 9,500,000 |
| Pentium 4 | 30 | 42,000,000 |
| Itanium | 31 | 25,000,000 |
| Itanium 2 | 33 | 220,000,000 |
| Itanium 2 (9MB cache) | 34 | 592,000,000 |
| Dual Core Itanium | 36 | 1,720,000,000 |
| Xeon 7460 | 38 | 1,900,000,000 |
| Xeon 7500 | 39 | 2,300,000,000 |

**30.** **(a)** Using the data in Table 11.23 on page 477, plot $\ln N$ against $t$. If the original data were exponential, the points would lie on a line.
**(b)** Fit a line to the graph from part (a).
**(c)** From the equation of the line, obtain the formula for $N$ as an exponential function of $t$.

**31.** **(a)** Using the data in Table 11.23 on page 477, plot $\ln N$ against $\ln t$. If a power function fitted the original data, the points would lie on a line.
**(b)** Fit a line to the graph from part (a).
**(c)** From the equation of the line, obtain the formula for $N$ as a power function of $t$.

**32.** **(a)** Let $N = at^p$, with $a, p$ constant. Explain why if you plot $\ln N$ against $\ln t$, you get a line.
**(b)** To decide if a function of the form $N = at^p$ fits some data, you plot $\ln N$ against $\ln t$. Explain why this plot is useful.

**33.** According to the US Census Bureau, the 2008 mean income by age is as given in Table 11.39.[28]

**(a)** Choose the best type of function to fit the data: linear, exponential, power, or quadratic.
**(b)** Using a mid-range age value for each interval, find an equation to fit the data.
**(c)** Interpolation estimates incomes for ages within the range of the data. Predict the income of a 37-year-old.
**(d)** Extrapolation estimates incomes outside the range of data. Use your function to predict the income of a 10-year-old. Is it reasonable?

**Table 11.39**

| Age | Mean income, dollars |
|---|---|
| 15 to 24 | 14,268 |
| 25 to 34 | 36,146 |
| 35 to 44 | 47,520 |
| 45 to 54 | 49,570 |
| 55 to 64 | 46,408 |
| 65 to 74 | 33,286 |
| 75+ years | 24,396 |

**34.** Table 11.40 gives the development time $t$ (in days) for eggs of the pea weevil (*Bruchus pisorum*) at temperature $H$ (°C).[29]

**(a)** Plot these data and fit a power function.
**(b)** Ecologists define the development rate $r = 1/t$ where $t$ is the development time. Plot $r$ against $H$, and fit a linear function.
**(c)** At a certain temperature, the value of $r$ drops to 0 and pea weevil eggs will not develop. What is this tem-

[26] www.st.nmfs.noaa.gov/st1/fus/fus11/FUS_2011.pdf, accessed May 2013.
[27] The Intel Corporation, www.intel.com/museum/archives/history_docs/mooreslaw.htm, intel.com, www.geek.com.
[28] www.census.gov/hhes/www/income/histinc/p10ar.html, accessed November 27, 2009.
[29] From website created by A. Sharov, http://www.ento.vt.edu/~sharov/PopEcol/lec8/quest8.html, accessed November, 2001. The site attributes the data to Smith, A. M., 1992, *Environ. Entomol.* 21:314-321.

perature according to the model from part (a)? Part (b)? Which model's prediction do you think is more reasonable?

**Table 11.40**

| $H$, °C | 10.7 | 14.4 | 16.2 | 18.1 | 21.4 | 23.7 | 24.7 | 26.9 |
|---|---|---|---|---|---|---|---|---|
| $t$, days | 38.0 | 19.5 | 15.6 | 9.6 | 9.5 | 7.3 | 4.5 | 4.5 |

**35.** German physicist Arnd Leike of the University of Munich won the 2002 Ig Nobel prize in Physics for experiments with beer foam conducted with his students.[30] The data in Table 11.41 give the height (in cm) of beer foam after $t$ seconds for three different types of beer, Erdinger Weissbier, Augustinerbräu München, and Budweiser Budvar. The heights are denoted $h_e$, $h_a$, and $h_b$, respectively.

**(a)** Plot these points and fit exponential functions to them. Give the equations in the form $h = h_0 e^{-t/\tau}$.

**(b)** What does the value of $h_0$ tell you for each type of beer? What does the value of $\tau$ tell you for each type of beer?

**Table 11.41**

| $t$ | $h_e$ | $h_a$ | $h_b$ | $t$ | $h_e$ | $h_a$ | $h_b$ |
|---|---|---|---|---|---|---|---|
| 0 | 17.0 | 14.0 | 14.0 | 120 | 10.7 | 6.0 | 7.0 |
| 15 | 16.1 | 11.8 | 12.1 | 150 | 9.7 | 5.3 | 6.2 |
| 30 | 14.9 | 10.5 | 10.9 | 180 | 8.9 | 4.4 | 5.5 |
| 45 | 14.0 | 9.3 | 10.0 | 210 | 8.3 | 3.5 | 4.5 |
| 60 | 13.2 | 8.5 | 9.3 | 240 | 7.5 | 2.9 | 3.5 |
| 75 | 12.5 | 7.7 | 8.6 | 300 | 6.3 | 1.3 | 2.0 |
| 90 | 11.9 | 7.1 | 8.0 | 360 | 5.2 | 0.7 | 0.9 |
| 105 | 11.2 | 6.5 | 7.5 | | | | |

**36.** In 1619, Kepler published his third law, which relates $D$, the distance of a planet from the Sun, to $P$, the period of the planet—the time it takes for the planet to orbit the Sun. Kepler conjectured that $P = kD^{3/2}$, where $k$ is a constant that he determined empirically from experimental data. Table 11.42 represents modern observational data. Does this model fit the data? What is your estimate for $k$? What is the final form of Kepler's law?

**Table 11.42** *The period and distance of a planet from the Sun*

| Planet | Distance (kms $\times 10^6$) | Period (days) |
|---|---|---|
| Mercury | 57.9 | 88 |
| Venus | 108.2 | 225 |
| Earth | 149.6 | 365 |
| Mars | 227.9 | 687 |
| Jupiter | 778.3 | 4329 |
| Saturn | 1427.0 | 10753 |
| Uranus | 2870.0 | 30660 |
| Neptune | 4497.0 | 60150 |
| Pluto | 5900.0 | 90670 |

## CHAPTER SUMMARY

- **Proportionality**
  Direct and indirect.
- **Power Functions**
  $y = kx^p$.
- **Polynomials**
  General formula:
  $$p(x) = a_n x^n + a_{n-1}x^{n-1} + \cdots + a_1 x + a_0.$$
  All terms have nonnegative, integer exponents. Leading term $a_n x^n$; coefficients $a_0, \ldots, a_n$; degree $n$.
  Long-run behavior: Like $y = a_n x^n$.
  Short-run behavior: Zeros corresponding to each factor; multiple zeros.
- **Rational Functions**
  Ratio of polynomials: $r(x) = \dfrac{p(x)}{q(x)}$.
  Long-run behavior: Horizontal asymptote of $r(x)$:
  Given by ratio of highest-degree terms.
  Short-run behavior: Vertical asymptote of $r(x)$:
  At zeros of $q(x)$ (if $p(x) \neq 0$).
  Short-run behavior: Zeros of $r(x)$:
  At zeros of $p(x)$ (if $q(x) \neq 0$).
  Using limits to understand short- and long-run behavior.
- **Comparing Functions**
  Exponential functions eventually dominate power functions. Power functions eventually dominate logs.
- **Fitting Exponentials and Polynomials to Data**

[30] http://ignobel.com/ig/ig-pastwinners.html. The Ig Nobel prize is a spoof of the Nobel prize and honors researchers whose achievements "cannot or should not be reproduced." The data here is taken from *Demonstration of the Exponential Decay Law Using Beer Froth*, Arnd Leike, *European Journal of Physics*, vol. 23, January 2002, pp. 21-26.

# REVIEW EXERCISES AND PROBLEMS FOR CHAPTER ELEVEN

## Exercises

In Exercises 1–4, does the function represent proportionality to a power of the independent variable? That is, can the function be written in the form $y = kx^p$ for the variables given in the problem? If so, identify the constant, $k$, and the power, $p$.

**1.** $y = \dfrac{\frac{1}{3}}{2x^7}$ **2.** $y = \dfrac{6}{-2/x^5}$

**3.** $z = 5(3)^x$ **4.** $C = 2q^3 - 5$

In Exercises 5–8, is $y$ a power function of $x$? If so, write it in the form $y = kx^p$.

**5.** $y = 6x^3 + 2$ **6.** $3y = 9x^2$

**7.** $y - 9 = (x + 3)(x - 3)$ **8.** $y = 4(x-2)(x+2) + 16$

Does the power function in Exercises 9–14 appear to have an odd power, an even power, or a power between 0 and 1?

**9.**

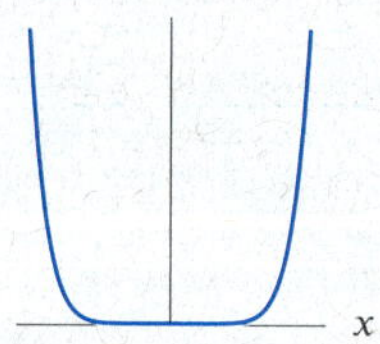

**10.**

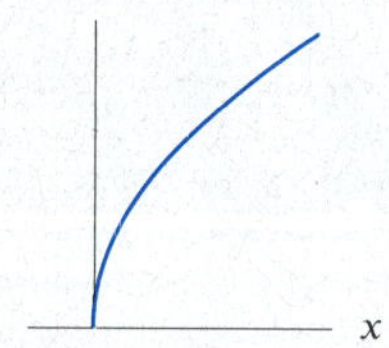

**11.**

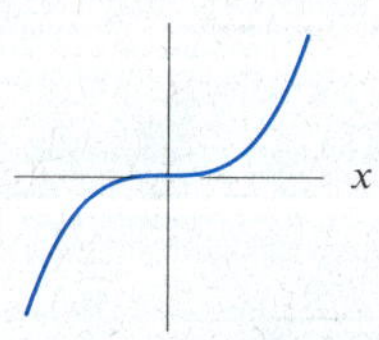

**12.**

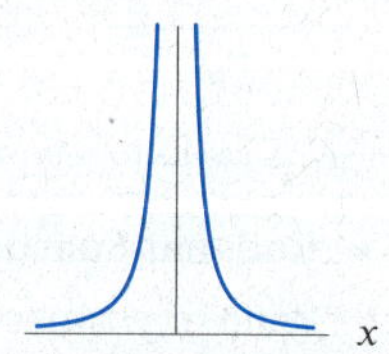

**13.**

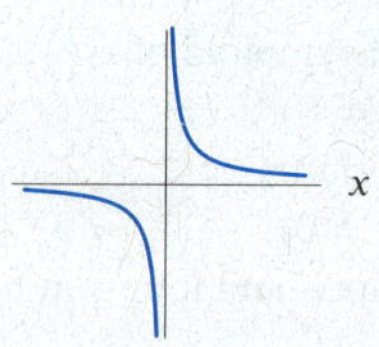

**14.**

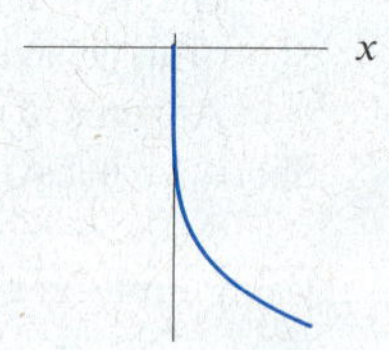

**15.** State the values of $k$ and $p$ if $r(x) = 2\sqrt[3]{7x}\sqrt[5]{x^2}$ is written in the form $kx^p$.

**16.** Find a possible formula for the power function $f(t)$ given that $f(3) = 5$ and $f(5) = 3$.

**17.** Show that the function $y = (x^2 - 4)(x^2 - 2x - 3)$ is a polynomial. What is its degree?

Describe in words the long-run behavior as $x \to \infty$ of the functions in Exercises 18–21. What power function does each resemble?

**18.** $y = 16x^3 - 4023x^2 - 2$

**19.** $y = 4x^4 - 2x^2 + 3$

**20.** $y = 5x^2/x^{3/2} + 2$

**21.** $y = 3x^3 + 2x^2/x^{-7} - 7x^5 + 2$

In Exercises 22–23, find the zeros of the functions.

**22.** $y = 3x^5 + 7x + 1$ **23.** $y = 2x^2 - 3x - 3$

Are the functions in Exercises 24–25 rational functions? If so, write them in the form $p(x)/q(x)$, the ratio of polynomials.

**24.** $f(x) = \dfrac{x^2}{x-3} - \dfrac{5}{x-3}$ **25.** $f(x) = \dfrac{x^2+4}{e^x}$

In Exercises 26–27, which function dominates as $x \to \infty$?

**26.** $y = 12x^3$, $y = 7/x^{-4}$

**27.** $y = 4/e^{-x}$, $y = 17x^{43}$

**28.** Find

**(a)** $\displaystyle\lim_{x\to\infty} \frac{x(x^2-4)}{5+5x^3}$ **(b)** $\displaystyle\lim_{x\to-\infty} \frac{3x(x-1)(x-2)}{5-6x^4}$

**29.** Find

**(a)** $\displaystyle\lim_{x\to\infty} \frac{2x+1}{x-5}$ **(b)** $\displaystyle\lim_{x\to-\infty} \frac{2+5x}{6x+3}$

**30.** For each of the following functions, state whether it is even, odd, or neither.

**(a)** $f(x) = x^2 + 3$ **(b)** $g(x) = x^3 + 3$
**(c)** $h(x) = 5/x$ **(d)** $j(x) = |x - 4|$
**(e)** $k(x) = \log x$ **(f)** $l(x) = \log(x^2)$
**(g)** $m(x) = 2^x + 2$ **(h)** $n(x) = \cos x + 2$

**31.** A function $f$ satisfies $f(1) = 2$ and $f(3) = 18$. Find $f(2)$ if

**(a)** $f$ is linear.
**(b)** $f$ is exponential.
**(c)** $f$ is a power function.

## Problems

**32.** It is claimed that Figure 11.73 is the graph of a power function $kx^p$. If it is, estimate $k$ and $p$. If it is not, explain why.

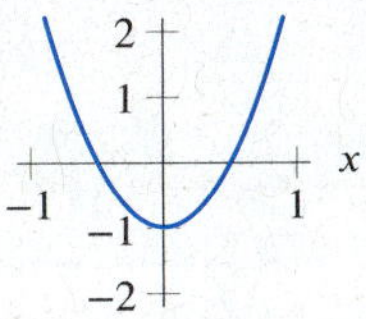


Figure 11.73

**33.** **(a)** One of the graphs in Figure 11.74 is $y = x^n$ and the other is $y = x^{1/n}$, where $n$ is a positive integer. Which is which? How do you know?
**(b)** What are the coordinates of point $A$?

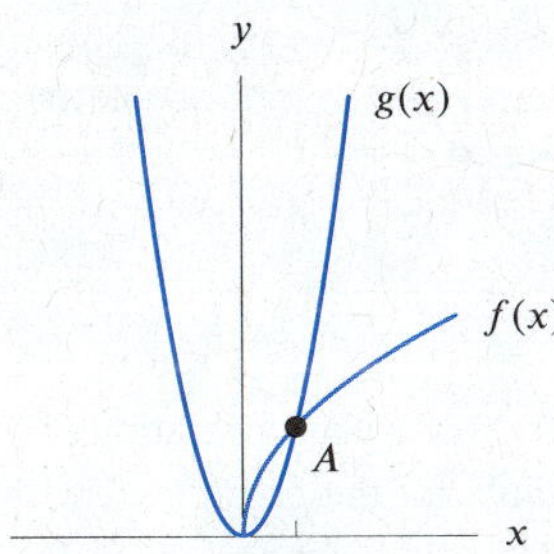


Figure 11.74

**34.** Without a calculator, match each graph (i)–(iv) with a function in Table 11.43.

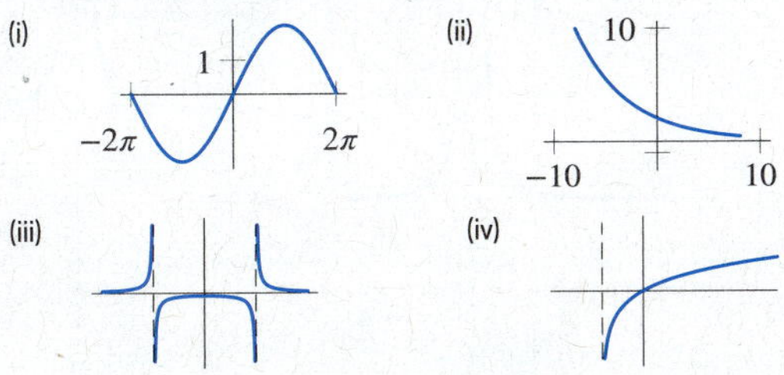


Table 11.43

| | |
|---|---|
| (A) $y = 0.5\sin(2x)$ | (J) $y = 2\sin(0.5x)$ |
| (B) $y = -\ln x$ | (K) $y = \ln(x-1)$ |
| (C) $y = 10(0.6)^x$ | (L) $y = 2e^{-0.2x}$ |
| (D) $y = 2\sin(2x)$ | (M) $y = 1/(x-6)$ |
| (E) $y = \ln(-x)$ | (N) $y = (x-2)/(x^2-9)$ |
| (F) $y = -15(3.1)^x$ | (O) $y = 1/(x^2-4)$ |
| (G) $y = 0.5\sin(0.5x)$ | (P) $y = x/(x-3)$ |
| (H) $y = \ln(x+1)$ | (Q) $y = (x-1)/(x+3)$ |
| (I) $y = 7(2.5)^x$ | (R) $y = 1/(x^2+4)$ |

**35.** Without a calculator, match each graph (i)–(viii) with a function in Table 11.44.

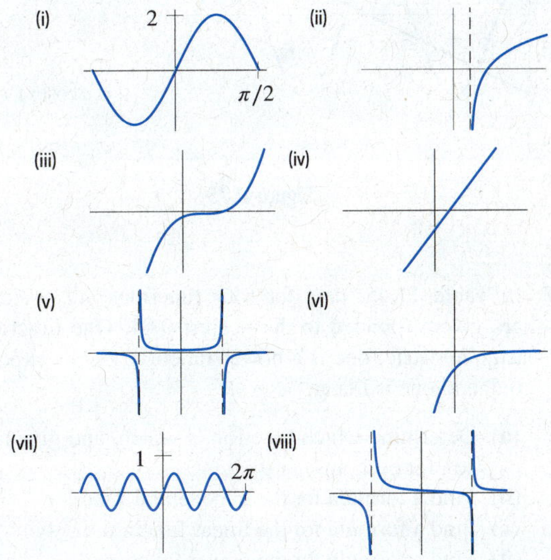


Table 11.44

| | |
|---|---|
| (A) $y = 0.5\sin(2x)$ | (M) $y = (x+3)/(x^2-4)$ |
| (B) $y = 2\sin(2x)$ | (N) $y = (x^2-4)/(x^2-1)$ |
| (C) $y = 0.5\sin(0.5x)$ | (O) $y = (x+1)^3 - 1$ |
| (D) $y = 2\sin(0.5x)$ | (P) $y = -2x - 4$ |
| (E) $y = (x-2)/(x^2-9)$ | (Q) $y = 3e^{-x}$ |
| (F) $y = (x-3)/(x^2-1)$ | (R) $y = -3e^{x}$ |
| (G) $y = (x-1)^3 - 1$ | (S) $y = -3e^{-x}$ |
| (H) $y = 2x - 4$ | (T) $y = 3e^{-x^2}$ |
| (I) $y = -\ln x$ | (U) $y = 1/(4-x^2)$ |
| (J) $y = \ln(-x)$ | (V) $y = 1/(x^2+4)$ |
| (K) $y = \ln(x+1)$ | (W) $y = (x+1)^3 + 1$ |
| (L) $y = \ln(x-1)$ | (X) $y = 2(x+2)$ |

**36.** In Figure 11.75, the graphs of four different functions are shown, defined in terms of eight constants: $a, b, c, k, m, p, q$, and $r$. Use the graphs to answer the following questions. Justify your answers with appropriate explanations.

**(a)** Which of the eight constants are definitely negative?
**(b)** Which of the eight constants are definitely positive?
**(c)** Which of the eight constants are definitely greater than zero but less than one?
**(d)** Which of the eight constants are definitely greater than one?

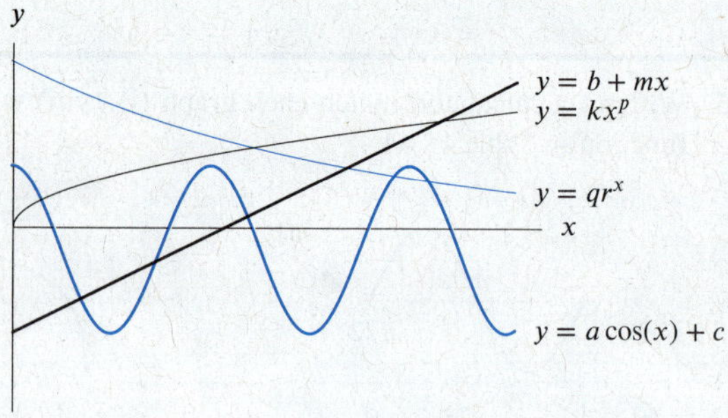


Figure 11.75

**37.** In Table 11.45, data for four functions $u, v, w$, and $z$ are given, rounded to the nearest 0.01. One function is trigonometric, one is a power function, one is exponential, and one is linear.

**(a)** Determine which function is which, and give a reason for each answer.
**(b)** Find a formula for the exponential function.
**(c)** Find a formula for the linear function.
**(d)** Find a formula for the power function.
**(e)** Find a formula for the trigonometric function.

Table 11.45

| $x$ | 1.00 | 2.00 | 3.00 | 4.00 | 5.00 | 6.00 |
|---|---|---|---|---|---|---|
| $u(x)$ | 0.50 | 1.41 | 2.60 | 4.00 | 5.59 | 7.35 |
| $v(x)$ | 0.25 | 0.50 | 0.75 | 0.50 | 0.25 | 0.50 |
| $w(x)$ | −1.34 | −0.94 | −0.54 | −0.14 | 0.26 | 0.66 |
| $z(x)$ | 64.0 | 16.0 | 4.00 | 1.00 | 0.25 | 0.06 |

Find possible polynomial formulas in Problems 38–47.

**38.**

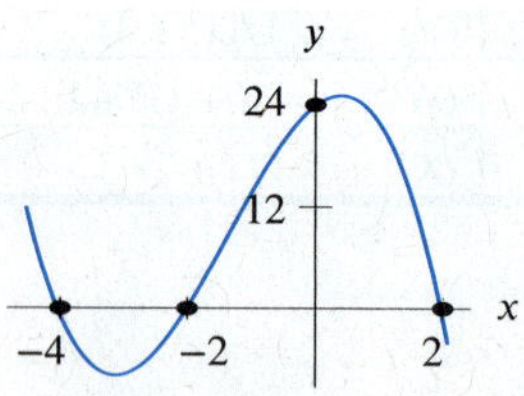


**39.**

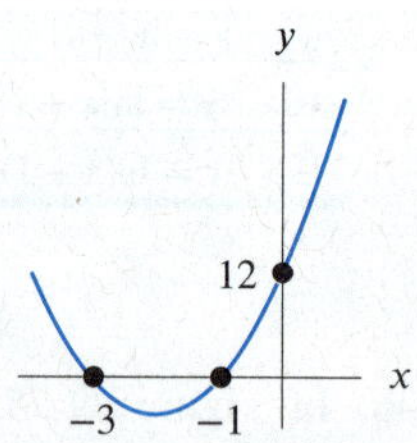


**40.**

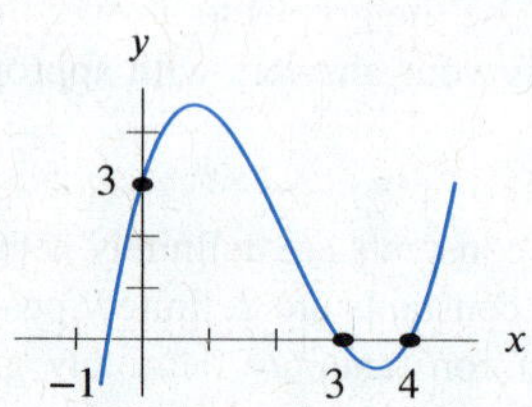


**41.**

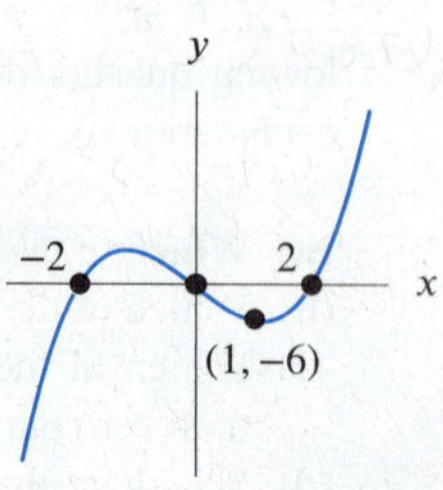


**42.**

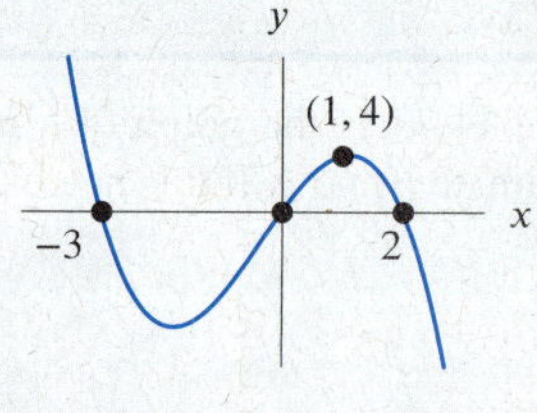


**43.**

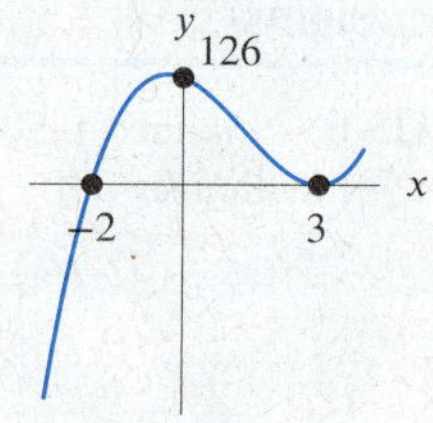


**44.**

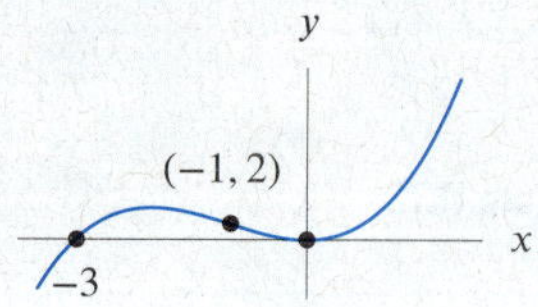


**45.**

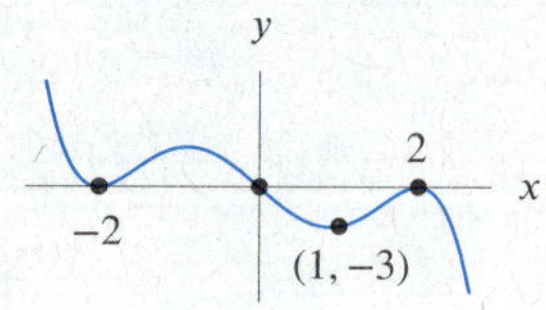


**46.**

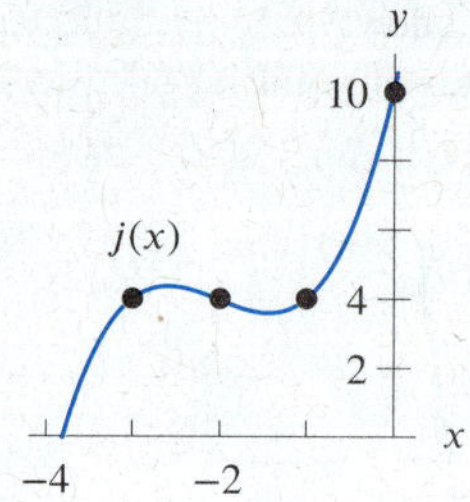


**47.**

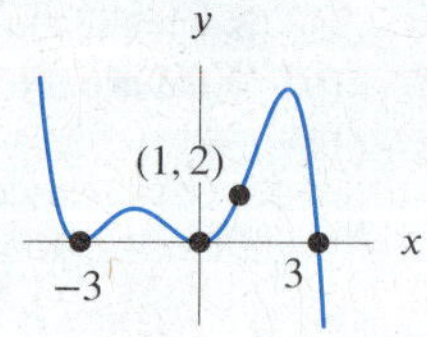


Problems 48–50 show a transformation of $y = 1/x^2$.

**(a)** Find a formula for the graph.
**(b)** Write the formula from part (a) as the ratio of two polynomials.
**(c)** Find the coordinates of any intercepts of the graph.

**48.**

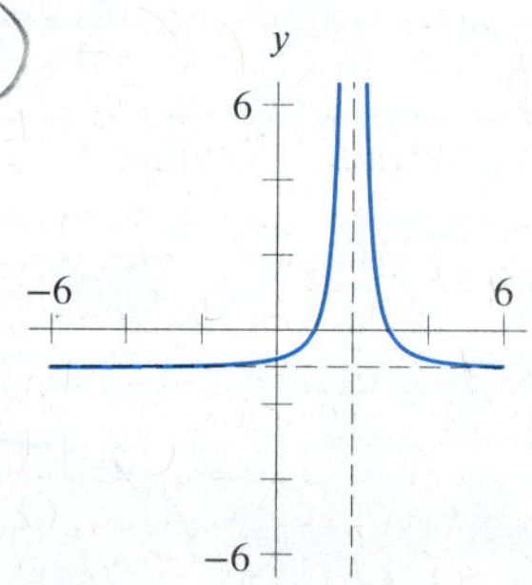


**49.**

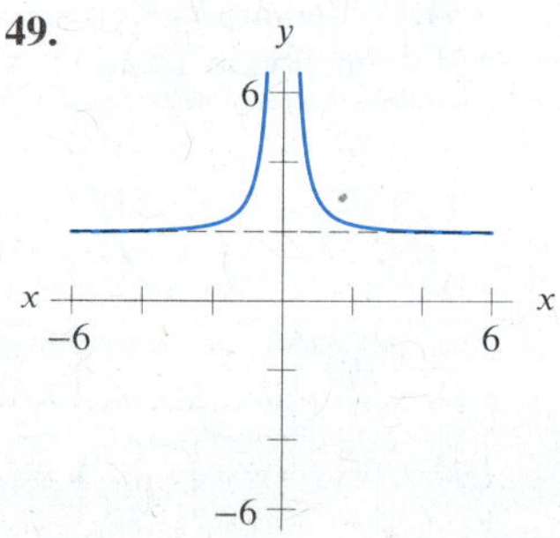


**50.**

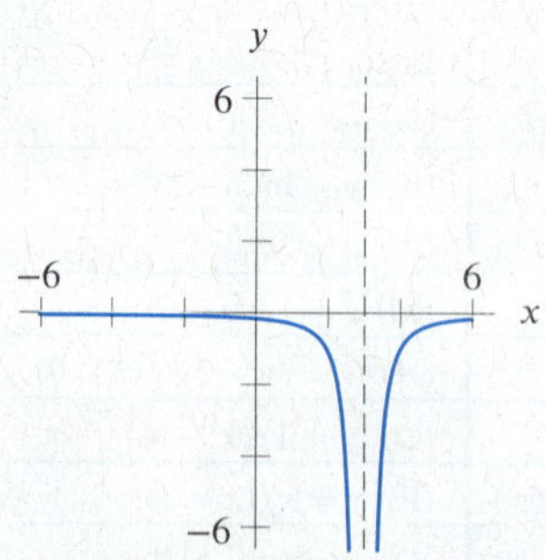

**51.** Suppose that $g(2) = 24$ and $g(4) = 96$. Find a formula for $g$, assuming $g$ is:

**(a)** A power function.
**(b)** A linear function.
**(c)** An exponential function.

**52.** Let $f(x) = x^2 + 5x + 6$ and $g(x) = x^2 + 1$.

**(a)** What are the zeros of $f$ and $g$?
**(b)** Let $r(x) = f(x)/g(x)$. Graph $r$. Does $r$ have zeros? Vertical asymptotes? What is its long-run behavior as $x \to \pm\infty$?
**(c)** Let $s(x) = g(x)/f(x)$. If you graph $s$ in the window $-10 \le x \le 10$, $-10 \le y \le 10$, it appears to have a zero near the origin. Does it? Does $s$ have a vertical asymptote? What is its long-run behavior?

**53.** Let $f(x) = (x-3)^2$, $g(x) = x^2 - 4$, $h(x) = x + 1$, and $j(x) = x^2 + 1$. Without a calculator, match the functions described in (a)–(f) to the functions in (i)–(vi). Some of the descriptions may have no matching function or more than one matching function.

(i) $p(x) = \dfrac{f(x)}{g(x)}$ (ii) $q(x) = \dfrac{h(x)}{g(x)}$

(iii) $r(x) = f(x)h(x)$ (iv) $s(x) = \dfrac{g(x)}{j(x)}$

(v) $t(x) = \dfrac{1}{h(x)}$ (vi) $v(x) = \dfrac{j(x)}{f(x)}$

**(a)** Two zeros, no vertical asymptotes, and a horizontal asymptote.
**(b)** Two zeros, no vertical asymptote, and no horizontal asymptote.
**(c)** One zero, one vertical asymptote, and a horizontal asymptote.
**(d)** One zero, two vertical asymptotes, and a horizontal asymptote.
**(e)** No zeros, one vertical asymptote, and a horizontal asymptote at $y = 1$.
**(f)** No zeros, one vertical asymptote, and a horizontal asymptote at $y = 0$.

**54.** Suppose $f$ is a polynomial function of degree $n$, where $n$ is a positive even integer. For each of the following statements, write *true* if the statement is always true, *false* otherwise. If the statement is false, give an example that illustrates why it is false.

**(a)** $f$ is an even function.
**(b)** $f$ has an inverse.
**(c)** $f$ cannot be an odd function.
**(d)** If $f(x) \to +\infty$ as $x \to +\infty$, then $f(x) \to -\infty$ as $x \to -\infty$.

**55. (a)** Sketch a graph of $f(x) = x^4 - 17x^2 + 36x - 20$ for $-10 \le x \le 10$, $-10 \le y \le 10$.
**(b)** Your graph should appear to have a vertical asymptote at $x = -5$. Does $f$ actually have a vertical asymptote here? Explain.
**(c)** How many zeros does $f$ have? Can you find a window in which all of the zeros of $f$ are clearly visible?
**(d)** Write the formula of $f$ in factored form.
**(e)** How many turning points does the graph of $f$ have? Can you find a window in which all the turning points of $f$ are clearly visible? Explain.

In Problems 56–59, find a possible formula for the rational functions.

**56.** This function has zeros at $x = -3$ and $x = 2$, and vertical asymptotes at $x = -5$ and $x = 7$. It has a horizontal asymptote of $y = 1$.

**57.** The graph of $y = f(x)$ has one vertical asymptote, at $x = -1$, and a horizontal asymptote at $y = 1$. The graph of $f$ crosses the $y$-axis at $y = 3$ and crosses the $x$-axis once, at $x = -3$.

**58.**

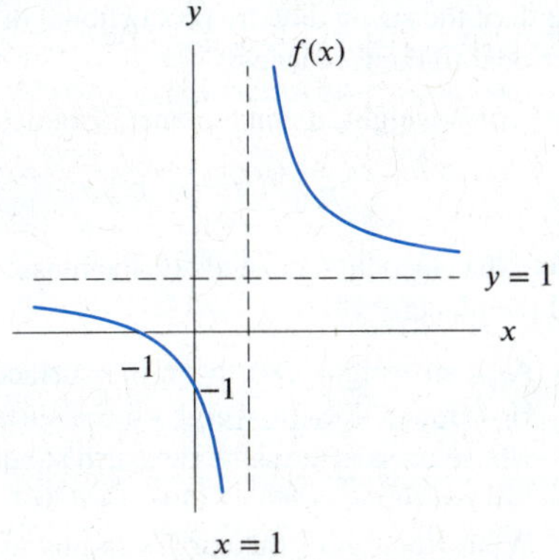


**59.**

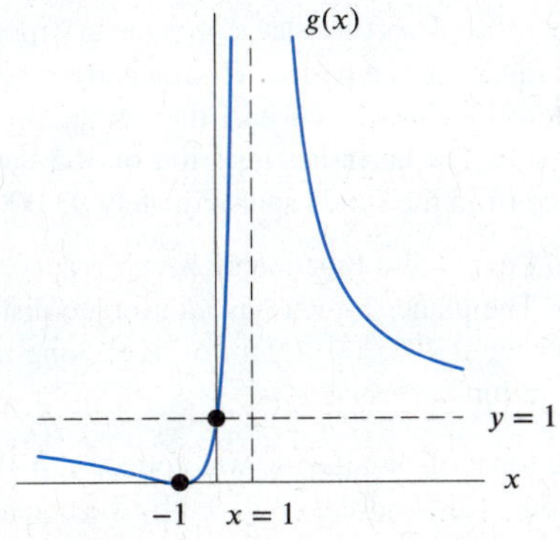


Find possible formulas for the polynomials and rational functions in Problems 60–63.

**60.** The zeros of $f$ are $x = -3$, $x = 2$, and $x = 5$, and the $y$-intercept is $y = -6$.

**61.** This function has zeros at $x = -3$, $x = 2$, $x = 5$, and a double zero at $x = 6$. It has a $y$-intercept of 7.

**62.** The polynomial $h(x) = 7$ at $x = -5, -1, 4$, and the $y$-intercept is 3. [Hint: Visualize $h$ as a vertically shifted version of another polynomial.]

**63.** The graph of $w$ intercepts the graph of $v(x) = 2x + 5$ at $x = -4, 1, 3$ and has a $y$-intercept of 2. [Hint: Let $w(x) = p(x) + v(x)$ where $p$ is another polynomial.]

**64.** On a map, 1/2 inch represents 5 miles. Is the map distance between two locations directly or inversely proportional to the actual distance that separates the two locations? Explain your reasoning and write a formula for the proportion. How far apart are two towns if the distance between these two towns on the map is 3.25 inches?

**65.** When a guitar string is plucked, the frequency of the note produced can be adjusted by varying the length of the string. The frequency is inversely proportional to the length, so a longer string results in a lower frequency, while a shorter string results in a higher frequency. Is the length of the string directly proportional or inversely proportional to the frequency?

**66.** A person's weight, $w$, on a planet of radius $d$ is given by

$$w = kd^{-2}, \quad k > 0,$$

where the constant $k$ depends on the masses of the person and the planet.

**(a)** A man weighs 180 lb on the surface of the earth. How much does he weigh on the surface of a planet whose mass is the same the earth's, but whose radius is three times as large? One-third as large?

**(b)** What fraction of the earth's radius must an equally massive planet have if, on this planet, the weight of the man in part (a) is one ton?

**67.** One of Kepler's three laws of planetary motion states that the square of the period, $P$, of a body orbiting the sun is proportional to the cube of its average distance, $d$, from the sun. The earth has a period of 365 days and its distance from the sun is approximately 93,000,000 miles.

**(a)** Find $P$ as a function of $d$.

**(b)** The planet Jupiter has an average distance from the sun of 483,000,000 miles. How long in earth days is a Jupiter year?

**68.** The town of Smallsville was founded in 1900. Its population $y$ (in hundreds) is given by the equation

$$y = -0.1x^4 + 1.7x^3 - 9x^2 + 14.4x + 5,$$

where $x$ is the number of years since 1900. Use a the graph in the window $0 \le x \le 10$, $-2 \le y \le 13$.

**(a)** What was the population of Smallsville when it was founded?

**(b)** When did Smallsville become a ghost town (nobody lived there anymore)? Give the year and the month.

**(c)** What was the largest population of Smallsville after 1905? When did Smallsville reach that population? Again, include the month and year. Explain your method.

**69.** Let $C(x)$ be a firm's total cost, in millions of dollars, for producing a quantity of $x$ thousand units of an item.

**(a)** Graph $C(x) = (x - 1)^3 + 1$.

**(b)** Let $R(x)$ be the revenue to the firm (in millions of dollars) for selling a quantity $x$ thousand units of the good. Suppose $R(x) = x$. What does this tell you about the price of each unit?

**(c)** Profit equals revenue minus cost. For what values of $x$ does the firm make a profit? Break even? Lose money?

**70.** Allometry is the study of the relative size of different parts of a body as a consequence of growth.[31] The simplest model of allometry is one in which it is assumed that the sizes of two parts $x$ and $y$ are related by a power law of the form

$$y = kx^p,$$

where $k$ and $p$ are positive constants. This equation is often called the *allometric equation.*

**(a)** Let $x$ be the length of a fish and $y$ be its weight. If $L$ is a typical unit of length, then $L^3$ is a typical unit of volume. Show that the assumptions that the length $x$ of a fish is proportional to $L$, and that its weight $y$ is proportional to its volume and therefore to $L^3$, lead to the equation $y = kx^3$.

**(b)** Table 11.46 relates the weight $y$ of plaice[32] to its length $x$. (Plaice is a type of fish.) If $y = kx^3$, what do you expect to happen if you calculate $x^3/y$ for each of the entries in the table? Do this, and estimate the proportionality constant, $k$.

**(c)** With this choice of $k$, plot the function $y = kx^3$ and the data in Table 11.46. Is the function a reasonable model?

**Table 11.46**

| Length (cm) | 33.5 | 34.5 | 35.5 | 36.5 | 37.5 | 38.5 |
|---|---|---|---|---|---|---|
| Weight (gm) | 332 | 363 | 391 | 419 | 455 | 500 |
| Length (cm) | 39.5 | 40.5 | 41.5 | 42.5 | 43.5 | |
| Weight (gm) | 538 | 574 | 623 | 674 | 724 | |

[31] J. S. Huxley, *Problems of Relative Growth* (Dover, 1972).

[32] Adapted from R. J. H. Beverton and S. J. Holt, "On the Dynamics of Exploited Fish Populations," *Fishery Investigations*, Series II, 19, 1957.

**71.** The thrust, $T$, delivered by a ship's propeller is proportional[33] to the square of the propeller rotation speed, $R$, times the fourth power of the propeller diameter, $D$.

**(a)** Write a formula for $T$ in terms of $R$ and $D$.
**(b)** What happens to the thrust if the propeller speed is doubled?
**(c)** What happens to the thrust if the propeller diameter is doubled?
**(d)** If the propeller diameter is increased by 50%, by how much can the propeller speed be reduced to deliver the same thrust?

**72.** A function that is not a polynomial can often be approximated by a polynomial. For example, for certain $x$-values, the function $f(x) = e^x$ can be approximated by the fifth-degree polynomial

$$p(x) = 1 + x + \frac{x^2}{2} + \frac{x^3}{6} + \frac{x^4}{24} + \frac{x^5}{120}.$$

**(a)** Show that $p(1) \approx f(1) = e$. How good is the estimate?
**(b)** Calculate $p(5)$. How well does $p(5)$ approximate $f(5)$?
**(c)** Graph $p(x)$ and $f(x)$ together on the same set of axes. Based on your graph, for what range of values of $x$ does $p(x)$ give a good estimate for $f(x)$?

**73.** The resolution, $r\%$, of a gamma ray telescope depends on the energy $v$ (in millions of electron volts, or MeVs) of the detected gamma rays.[34] The smaller the value of $r$, the better the telescope is at distinguishing two gamma ray photons of slightly different energies, and the more detailed observations that can be made. Table 11.47 gives values of $r$ for gamma rays at different energies.

**(a)** Plot the data in Table 11.47, with $r$ on the vertical axis.
**(b)** Based on this data, is the telescope better able to distinguish between high-energy photons or low-energy photons?
**(c)** Fit both power and exponential functions to the data, and give their formulas. Which appears to give the better fit?
**(d)** The telescope is predicted to grow rapidly worse and worse at distinguishing photons as the energy level drops toward 0 MeV. Which curve, power or exponential, is most consistent with this prediction?

**Table 11.47**

| $v$, MeV | 0.5 | 0.7 | 0.9 | 1.3 | 1.8 | 4.0 | 4.4 |
|---|---|---|---|---|---|---|---|
| $r$, % | 16.0 | 13.5 | 12.0 | 8.5 | 7.0 | 4.5 | 4.0 |

## STRENGTHEN YOUR UNDERSTANDING

Are the statements in Problems 1–47 true or false? Give an explanation for your answer.

**1.** All quadratic functions are power functions.

**2.** The function $y = 3 \cdot 2^x$ is a power function.

**3.** Let $g(x) = x^p$. If $p$ is a positive, even integer, then the graph of $g$ passes through the point $(-1, 1)$.

**4.** Let $g(x) = x^p$. If $p$ is a positive, even integer, then the graph of $g$ is symmetric about the $y$-axis.

**5.** Let $g(x) = x^p$. If $p$ is a positive, even integer, then the graph of $g$ is concave up.

**6.** The graph of $f(x) = x^{-1}$ passes through the origin.

**7.** The graph of $f(x) = x^{-2}$ has the $x$-axis as its only asymptote.

**8.** If $f(x) = x^{-1}$ then $f(x)$ approaches $+\infty$ as $x$ approaches zero.

**9.** As $x$ grows very large, the values of $f(x) = x^{-1}$ approach zero.

**10.** The function $2^x$ eventually grows faster than $x^b$ for any $b$.

**11.** The function $f(x) = x^{0.5}$ eventually grows faster than $g(x) = \ln x$.

**12.** We have $2^x \geq x^2$ on the interval $0 \leq x \leq 4$.

**13.** The function $f(x) = x^{-3}$ approaches the $x$-axis faster than $g(x) = e^{-x}$ as $x$ grows very large.

**14.** The function $f(x) = 3^x$ is an example of a power function.

**15.** The function $y = 3x$ is an example of a power function.

**16.** Every quadratic function is a polynomial function.

**17.** The power of the first term of a polynomial is its degree.

**18.** Far from the origin, the graph of a polynomial looks like the graph of its highest-power term.

**19.** A zero of a polynomial $p$ is the value $p(0)$.

**20.** The zeros of a polynomial are the $x$-coordinates where its graph intersects the $x$-axis.

**21.** The $y$-intercept of a polynomial $y = p(x)$ can be found by evaluating $p(0)$.

[33] Thomas C. Gillner, *Modern Ship Design* (US Naval Institute Press, 1972).

[34] E. Aprile et al., *The LXeGRIT Compton Status and Future Prospects*, posted at http://arxiv.org as arXiv:astro-ph/0212005v2, accessed December 4, 2002.

**22.** For very large $x$-values $f(x) = 1000x^3 + 345x^2 + 17x + 394$ is less than $g(x) = 0.01x^4$.

**23.** If $y = f(x)$ is a polynomial of degree $n$, where $n$ is a positive even number, then $f$ has an inverse.

**24.** If $y = f(x)$ is a polynomial of degree $n$, where $n$ is a positive odd number, then $f$ has an inverse.

**25.** If $p(x)$ is a polynomial and $x - a$ is a factor of $p$, then $x = a$ is a zero of $p$.

**26.** A polynomial of degree $n$ cannot have more than $n$ zeros.

**27.** The polynomial in Figure 11.76 has a multiple zero at $x = -2$.

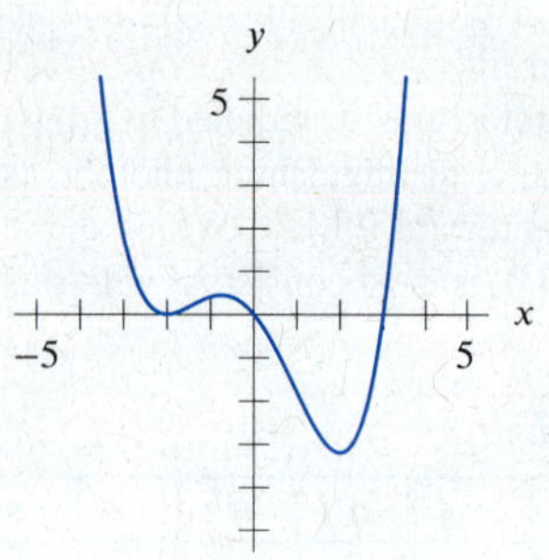


Figure 11.76

**28.** The polynomial in Figure 11.76 has a multiple zero at $x = 0$.

**29.** A rational function is the quotient of two polynomials. We assume the denominator is not equal to zero.

**30.** The function $f(x) = \dfrac{1}{x}$ is a rational function.

**31.** In order to determine the long-run behavior of a rational function, it is sufficient to consider only the ratio of the highest-power term in the numerator to the highest-power term in the denominator.

**32.** As $x$ grows through large positive values, $y = \dfrac{x+18}{x+9}$ approaches $y = 2$.

**33.** As $x$ grows through large positive values, $y = \dfrac{2x+125}{x^2-1}$ approaches $y = 0$.

**34.** As $x$ grows through large positive values, $y = \dfrac{x^3+4x^2-16x+12}{4x^3-16x+1}$ has an asymptote at $y = 4$.

**35.** As $x$ grows through large positive values, $y = \dfrac{1-4x^2}{x^2+1}$ approaches $y = 0$.

**36.** As $x$ grows through large positive values, $y = \dfrac{5x}{x+1}$ approaches $y = -5$.

**37.** As $x$ grows through large positive values, $y = \dfrac{3x^4-6x^3+10x^2-16x+7}{-3x+x^2}$ behaves like $y = -x^3$.

**38.** As $x$ decreases through large negative values, $f(x) = \dfrac{x^3-7x^2+28x+76}{-x^2-101x+72}$ approaches positive infinity.

**39.** A fraction is equal to zero if and only if its numerator equals zero and its denominator does not.

**40.** The zeros of a function $y = f(x)$ are the values of $x$ that make $y = 0$.

**41.** The function $f(x) = \dfrac{x+4}{x-3}$ has a zero at $x = -4$.

**42.** The rational function $y = \dfrac{x+2}{x^2-4}$ has a zero at $x = -2$.

**43.** The rational function $g(w) = \dfrac{12}{(w-2)(w+3)}$ has exactly two zeros.

**44.** If $p(x)$ and $q(x)$ have no zeros in common, then the rational function $r(x) = \dfrac{p(x)}{q(x)}$ has an asymptote at each of the zeros of $p(x)$.

**45.** In general, the rational function $r(x) = \dfrac{p(x)}{q(x)}$ must have at least one zero.

**46.** Rational functions can never cross an asymptote.

**47.** The rational function $g(w) = \dfrac{3w-3}{(w-12)(w+4)}$ has a vertical asymptote at $w = 1$.

# SKILLS REFRESHER FOR CHAPTER 11: ALGEBRAIC FRACTIONS

Algebraic fractions are combined in the same way as numeric fractions according to the following rules:

Add numerators when denominators are equal: $\dfrac{a}{c} + \dfrac{b}{c} = \dfrac{a+b}{c}$

Find a common denominator: $\dfrac{a}{b} + \dfrac{c}{d} = \dfrac{a \cdot d}{b \cdot d} + \dfrac{b \cdot c}{b \cdot d} = \dfrac{ad+bc}{bd}$

Multiply numerators and denominators for a product: $\dfrac{a}{b} \cdot \dfrac{c}{d} = \dfrac{ac}{bd}$

To divide by a fraction, multiply by its reciprocal: $\dfrac{a/b}{c/d} = \dfrac{a}{b} \cdot \dfrac{d}{c} = \dfrac{ad}{bc}$

The sign of a fraction is changed by changing the sign of the numerator or the denominator (but not both):

$$-\frac{a}{b} = \frac{-a}{b} = \frac{a}{-b}$$

We assume that no denominators are zero, since we cannot divide by zero; that is, $a/0$ is not defined.

We can simplify a fraction in which either the numerator or denominator is itself a fraction as follows:

$$\frac{a/b}{c} = \frac{a/b}{c/1} = \frac{a}{b} \cdot \frac{1}{c} = \frac{a}{bc} \quad \text{and} \quad \frac{a}{b/c} = \frac{a/1}{b/c} = \frac{a}{1} \cdot \frac{c}{b} = \frac{ac}{b}.$$

**Example 1** Perform the indicated operations and express the answers as a single fraction.

(a) $\dfrac{4}{x^2+1} - \dfrac{1-x}{x^2+1}$ (b) $\dfrac{M}{M^2-2M-3} + \dfrac{1}{M^2-2M-3}$

(c) $\dfrac{-H^2P}{17} \cdot \dfrac{\left(PH^{1/3}\right)^2}{K^{-1}}$ (d) $\dfrac{2z/w}{w(w-3z)}$

**Solution**

(a) $\dfrac{4}{x^2+1} - \dfrac{1-x}{x^2+1} = \dfrac{4-(1-x)}{x^2+1} = \dfrac{3+x}{x^2+1}.$

(b) $\dfrac{M}{M^2-2M-3} + \dfrac{1}{M^2-2M-3} = \dfrac{M+1}{(M^2-2M-3)} = \dfrac{M+1}{(M+1)(M-3)} = \dfrac{1}{M-3}$ if $M \neq -1$.

(c) $\dfrac{-H^2P}{17} \cdot \dfrac{\left(PH^{1/3}\right)^2}{K^{-1}} = \dfrac{-H^2P\left(P^2H^{2/3}\right)}{17K^{-1}} = -\dfrac{H^{8/3}P^3K}{17}.$

(d) $\dfrac{2z/w}{w(w-3z)} = \dfrac{2z}{w} \cdot \dfrac{1}{w(w-3z)} = \dfrac{2z}{w^2(w-3z)}.$

**Example 2** Simplify the following expressions, giving your answer as a single fraction.

(a) $2x^{-1/2} + \dfrac{\sqrt{x}}{3}$ (b) $2\sqrt{t+3} + \dfrac{1-2t}{\sqrt{t+3}}$

**Solution**

(a) $2x^{-1/2} + \dfrac{\sqrt{x}}{3} = \dfrac{2}{\sqrt{x}} + \dfrac{\sqrt{x}}{3} = \dfrac{2 \cdot 3 + \sqrt{x}\sqrt{x}}{3\sqrt{x}} = \dfrac{6+x}{3\sqrt{x}} = \dfrac{6+x}{3x^{1/2}}.$

(b)
$$\begin{aligned} 2\sqrt{t+3} + \frac{1-2t}{\sqrt{t+3}} = \frac{2\sqrt{t+3}}{1} + \frac{1-2t}{\sqrt{t+3}} &= \frac{2\sqrt{t+3}\sqrt{t+3}+1-2t}{\sqrt{t+3}} \\ &= \frac{2(t+3)+1-2t}{\sqrt{t+3}} \\ &= \frac{7}{\sqrt{t+3}} = \frac{7}{(t+3)^{1/2}}. \end{aligned}$$

## Finding a Common Denominator

We can multiply (or divide) both the numerator and denominator of a fraction by the same nonzero number without changing the fraction's value. This is equivalent to multiplying by a factor of +1. We are using this rule when we add or subtract fractions with different denominators. For example, to add $\frac{x}{3a} + \frac{1}{a}$, we multiply $\frac{1}{a} \cdot \frac{3}{3} = \frac{3}{3a}$. Then

$$\frac{x}{3a} + \frac{1}{a} = \frac{x}{3a} + \frac{3}{3a} = \frac{x+3}{3a}.$$

**Example 3** Perform the indicated operations:

(a) $3 - \frac{1}{x-1}$ (b) $\frac{2}{x^2+x} + \frac{x}{x+1}$

**Solution** (a) $3 - \frac{1}{x-1} = 3\frac{(x-1)}{(x-1)} - \frac{1}{x-1} = \frac{3(x-1)-1}{x-1} = \frac{3x-3-1}{x-1} = \frac{3x-4}{x-1}.$

(b) $\frac{2}{x^2+x} + \frac{x}{x+1} = \frac{2}{x(x+1)} + \frac{x}{x+1} = \frac{2}{x(x+1)} + \frac{x(x)}{(x+1)(x)} = \frac{2+x^2}{x(x+1)}.$

**Note**: We can multiply (or divide) the numerator and denominator by the same nonzero number because this is the same as multiplying by a factor of +1, and multiplying by a factor of 1 does not change the value of the expression. However, we cannot perform any other operation that would change the value of the expression. For example, we cannot add the same number to the numerator and denominator of a fraction nor can we square both, take the logarithm of both, etc., without changing the fraction.

## Reducing Fractions: Canceling

We can reduce a fraction when we have the same (nonzero) factor in both the numerator and the denominator. For example,

$$\frac{ac}{bc} = \frac{a}{b} \cdot \frac{c}{c} = \frac{a}{b} \cdot 1 = \frac{a}{b}.$$

**Example 4** Reduce the following fractions (if possible).

(a) $\frac{2x}{4y}$ (b) $\frac{2+x}{2+y}$

(c) $\frac{5n-5}{1-n}$ (d) $\frac{x^2(4-2x)-(4x-x^2)2x}{x^4}$

**Solution** (a) $\frac{2x}{4y} = \frac{2}{2} \cdot \frac{x}{2y} = \frac{x}{2y}$

(b) $\frac{2+x}{2+y}$ cannot be reduced further.

(c) $\frac{5n-5}{1-n} = \frac{5(n-1)}{(-1)(n-1)} = -5$

(d)

$$\begin{aligned}\frac{x^2(4-2x)-\left(4x-x^2\right)2x}{x^4} &= \frac{x^2(4-2x)-(4-x)2x^2}{x^4}\\ &= \frac{(4-2x)-2(4-x)}{x^2}\left(\frac{x^2}{x^2}\right)\\ &= \frac{4-2x-8+2x}{x^2} = \frac{-4}{x^2}.\end{aligned}$$

## Complex Fractions

A *complex fraction* is a fraction whose numerator or denominator (or both) contains one or more fractions. To simplify a complex fraction, we change the numerator and denominator to single fractions and then divide.

**Example 5** Write the following as simple fractions in reduced form.

(a) $\dfrac{\dfrac{1}{x+h}-\dfrac{1}{x}}{h}$. (b) $\dfrac{a+b}{a^{-2}-b^{-2}}$.

**Solution** (a) $\dfrac{\dfrac{1}{x+h}-\dfrac{1}{x}}{h}=\dfrac{\dfrac{x-(x+h)}{x(x+h)}}{h}=\dfrac{\dfrac{-h}{x(x+h)}}{\dfrac{h}{1}}=\dfrac{-h}{x(x+h)}\cdot\dfrac{1}{h}=\dfrac{-1}{x(x+h)}\dfrac{(h)}{(h)}=\dfrac{-1}{x(x+h)}$.

(b) $\dfrac{a+b}{a^{-2}-b^{-2}}=\dfrac{a+b}{\dfrac{1}{a^2}-\dfrac{1}{b^2}}=\dfrac{a+b}{\dfrac{b^2-a^2}{a^2b^2}}=\dfrac{a+b}{1}\cdot\dfrac{a^2b^2}{b^2-a^2}=\dfrac{(a+b)(a^2b^2)}{(b+a)(b-a)}=\dfrac{a^2b^2}{b-a}$.

## Splitting Expressions

We can reverse the rule for adding fractions to split up an expression into two fractions,

$$\frac{a+b}{c}=\frac{a}{c}+\frac{b}{c}.$$

**Example 6** Split $\dfrac{3x^2+2}{x^3}$ into two reduced fractions.

**Solution** $\dfrac{3x^2+2}{x^3}=\dfrac{3x^2}{x^3}+\dfrac{2}{x^3}=\dfrac{3}{x}+\dfrac{2}{x^3}$

Sometimes we can alter the form of the fraction even further if we can create a duplicate of the denominator within the numerator. This technique is useful when graphing some rational functions. For example, we may rewrite the fraction $\dfrac{x+3}{x-1}$ by creating a factor of $(x-1)$ within the numerator. To do this, we write

$$\frac{x+3}{x-1}=\frac{x-1+1+3}{x-1}$$

which can be written as

$$\frac{(x-1)+4}{x-1}.$$

Then, splitting this fraction, we have

$$\frac{x+3}{x-1}=\frac{x-1}{x-1}+\frac{4}{x-1}=1+\frac{4}{x-1}.$$

**Note**: It is not possible to split a sum that occurs in the denominator of a fraction. For example,

$$\frac{a}{b+c}\text{ does not equal }\frac{a}{b}+\frac{a}{c}.$$

# Exercises to Skills Review for Chapter 11

For Exercises 1–30, perform the operations. Express answers in reduced form.

**1.** $\frac{3}{5}+\frac{4}{7}$

**2.** $\frac{7}{10}-\frac{2}{15}$

**3.** $\frac{1}{2x}-\frac{2}{3}$

**4.** $\frac{6}{7y}+\frac{9}{y}$

**5.** $\frac{-2}{yz}+\frac{4}{z}$

**6.** $\frac{-2z}{y}+\frac{4}{y}$

**7.** $\frac{2}{x^2}-\frac{3}{x}$

**8.** $\frac{3/4}{7/20}$

**9.** $\frac{5/6}{15}$

**10.** $\frac{3/x}{x^2/6}$

**11.** $\frac{3/x}{6/x^2}$

**12.** $\frac{14}{x-1}+\frac{13}{2x-2}$

**13.** $\frac{4z}{x^2y}-\frac{3w}{xy^4}$

**14.** $\frac{10}{y-2}+\frac{3}{2-y}$

**15.** $\frac{8y}{y-4}+\frac{32}{y-4}$

**16.** $\frac{8y}{y-4}+\frac{32}{4-y}$

**17.** $\frac{8}{3x^2-x-4}-\frac{9}{x+1}$

**18.** $\frac{15}{(x-3)^2(x+5)}+\frac{7}{(x-3)(x+5)^2}$

**19.** $\frac{3}{x-4}-\frac{2}{x+4}$

**20.** $\frac{x^2}{x-1}-\frac{1}{1-x}$

**21.** $\frac{1}{2r+3}+\frac{3}{4r^2+6r}$

**22.** $u+a+\frac{u}{u+a}$

**23.** $\frac{1}{\sqrt{x}}-\frac{1}{(\sqrt{x})^3}$

**24.** $\frac{1}{e^{2x}}+\frac{1}{e^x}$

**25.** $\frac{a+b}{2}\cdot\frac{8x+2}{b^2-a^2}$

**26.** $\frac{0.07}{M}+\frac{3}{4}M^2$

**27.** $\frac{1}{r_1}+\frac{1}{r_2}+\frac{1}{r_3}$

**28.** $\frac{8y}{y-4}-\frac{32}{y-4}$

**29.** $\frac{a}{a^2-9}+\frac{1}{a-3}$

**30.** $\frac{x^3}{x-4}\Big/\frac{x^2}{x^2-2x-8}$

In Exercises 31–40, simplify, if possible.

**31.** $\dfrac{\frac{1}{(x+h)^2}-\frac{1}{x^2}}{h}$

**32.** $\frac{a^{-2}+b^{-2}}{a^2+b^2}$

**33.** $\frac{4-(x+h)^2-(4-x^2)}{h}$

**34.** $\frac{b^{-1}(b-b^{-1})}{b+1}$.

**35.** $\frac{1-a^{-2}}{1+a^{-1}}$.

**36.** $p-\dfrac{q}{\frac{p}{q}+\frac{q}{p}}$

**37.** $\dfrac{\frac{3}{xy}-\frac{5}{x^2y}}{\frac{6x^2-7x-5}{x^4y^2}}$

**38.** $\dfrac{\frac{1}{x}\left(3x^2\right)-(\ln x)(6x)}{\left(3x^2\right)^2}$

**39.** $\frac{2x(x^3+1)^2-x^2(2)(x^3+1)(3x^2)}{[(x^3+1)^2]^2}$

**40.** $\frac{\frac{1}{2}(2x-1)^{-1/2}(2)-(2x-1)^{1/2}(2x)}{(x^2)^2}$

In Exercises 41–46, split into a sum or difference of reduced fractions.

**41.** $\frac{26x+1}{2x^3}$

**42.** $\frac{\sqrt{x}+3}{3\sqrt{x}}$

**43.** $\frac{6l^2+3l-4}{3l^4}$

**44.** $\frac{7+p}{p^2+11}$

**45.** $\dfrac{\frac{1}{3}x-\frac{1}{2}}{2x}$

**46.** $\frac{t^{-1/2}+t^{1/2}}{t^2}$

In Exercises 47–52, rewrite in the form $1+(A/B)$.

**47.** $\frac{x-2}{x+5}$

**48.** $\frac{q-1}{q-4}$

**49.** $\frac{R+1}{R}$

**50.** $\frac{3+2u}{2u+1}$

**51.** $\frac{\cos x+\sin x}{\cos x}$

**52.** $\frac{1+e^x}{e^x}$

Are the statements in Exercises 53–58 true or false?

**53.** $\frac{a+c}{a}=1+c$

**54.** $\frac{rs-s}{s}=r-1$

**55.** $\frac{y}{y+z}=1+\frac{y}{z}$

**56.** $\frac{2u^2-w}{u^2-w}=2$

**57.** $\frac{x^2yz}{2x^2y}=\frac{z}{2}$

**58.** $x^{5/3}-3x^{2/3}=\frac{x^2-3x}{x^{1/3}}$

Chapter Twelve

# VECTORS AND MATRICES

© Patrick Zephyr/Patrick Zephyr Nature Photography

## Contents

# 12.1 VECTORS

## Distance Versus Displacement

If you start at home and walk 3 miles east and then 4 miles north, you have walked a total of 7 miles. However, your distance from home is not 7 miles, but $5 = \sqrt{3^2 + 4^2}$ miles, by the Pythagorean Theorem. See Figure 12.1. As this example illustrates, there is a distinction between *distance* and *displacement*. Distance is a number that measures separation, while displacement consists of separation and direction. For example, walking 3 miles north and walking 3 miles east give different displacements, but both correspond to a distance of 3 miles.

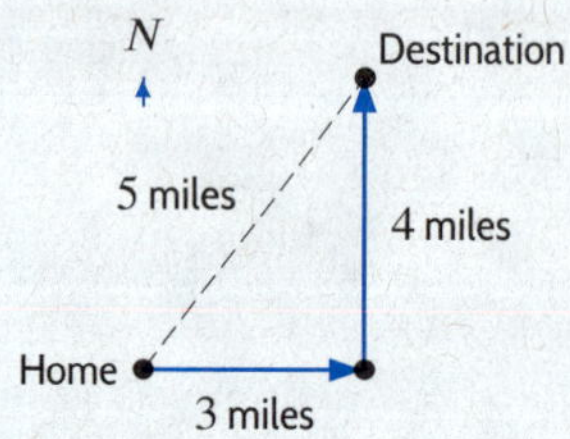


**Figure 12.1:** After walking 3 miles east then 4 miles north, your distance from home is 5 miles

## Adding Displacements Using Triangles

Displacements are added by treating them as arrows called *vectors*. The length of each arrow indicates the magnitude of the displacement, while the orientation of the arrow (which way the arrow points) indicates the direction. To add two displacements, join the tail of the second arrow to the head of the first arrow. Together, the arrows form two sides of a triangle[1] whose third side represents their sum. The sum is the *net displacement* resulting from the two displacements.

**Example 1** A person leaves her home and walks 5 miles due east and then 3 miles northeast. How far has she walked? How far away from home is she? What is her net displacement?

**Solution** In Figure 12.2, the first arrow is 5 units long because she walks 5 miles, while the second arrow is 3 units long and at an angle of 45° to the first. The length of the third side, $x$, of the triangle can be found by the Law of Cosines:

$$\begin{aligned} x^2 &= 5^2 + 3^2 - 2\cdot 5\cdot 3\cos 135^\circ \\ &= 34 - 30\left(-\frac{\sqrt{2}}{2}\right) \\ &= 55.213. \end{aligned}$$

This gives $x = \sqrt{55.213} = 7.431$ miles for the distance from home to destination, which is less than the total distance walked ($5 + 3 = 8$ miles). To specify her net displacement we use the angle $\theta$ in Figure 12.2. We find $\theta$ using the Law of Sines:

$$\frac{\sin\theta}{3} = \frac{\sin 135^\circ}{7.431},$$

giving

$$\sin\theta = 3\cdot\frac{\sqrt{2}/2}{7.431} = 0.285.$$

We know $\theta$ is less than 90°. Taking the arcsin of both sides, we obtain $\theta = \sin^{-1}(0.285) = 16.588^\circ$. Thus, the net displacement is 7.431 miles in a direction 16.588° north of east.

[1] If the vectors are parallel, the triangle becomes a line.

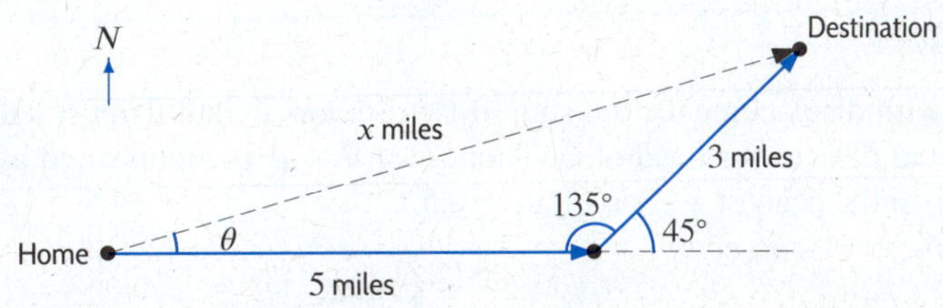


**Figure 12.2**: A person walks 5 miles east and then 3 miles northeast

### The Order of Addition Does Not Affect the Sum

When two displacements are added, the sum is independent of the order in which the two arrows are joined. For example, in Figure 12.3 the 5-mile displacement followed by the 3-mile displacement leads to the same destination as the 3-mile displacement followed by the 5-mile displacement.

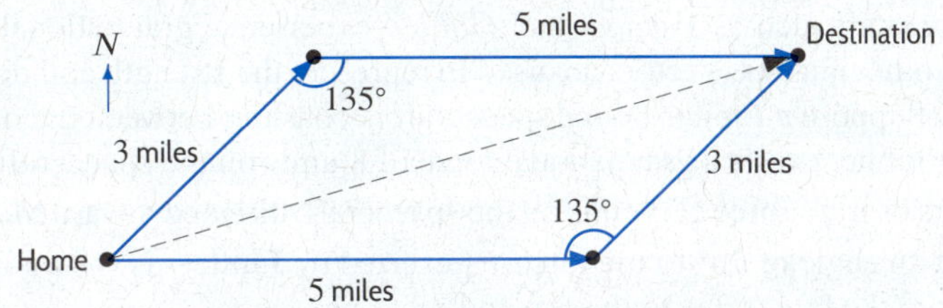


**Figure 12.3**: No matter which order the arrows are joined, the resulting net displacement is the same

## Vectors

Many physical quantities add in the same way as displacements. Such quantities are known as *vectors*. Like displacements, vectors are often represented as arrows. A vector has *magnitude* (the arrow's length) and *direction* (which way the arrow points). Examples of vectors include:

- **Velocity** This is the speed and direction of travel.
- **Force** This is, loosely speaking, the strength and direction of a push or a pull.
- **Magnetic fields** A vector gives the direction and intensity of a magnetic field at a point.
- **Vectors in economics** Vectors are used to keep track of prices and quantities.
- **Vectors in computer animation** Computers generate animations by performing enormous numbers of vector-based calculations.
- **Population vectors** In biology, populations of different animals or age groups can be represented using vectors.

## Vector Notation

In this book we write vectors as variables with arrows over them: $\vec{v}$. The notation is intended to ensure that a vector $\vec{v}$ is not mistaken for a *scalar*, which is another name for an ordinary number.

### Notation for the Magnitude of a Vector

The length or magnitude of a vector $\vec{v}$ is written $||\vec{v}\,||$. This notation looks like the absolute value notation, $|x|$, used for scalars. The absolute value of a number is its size without regard to sign, so $|-10| = |+10| = 10$. Similarly, the vectors $\vec{s}$ and $\vec{t}$ in Figure 12.4 are both of length 5, even though they point in different directions. Therefore, $||\vec{s}\,|| = ||\vec{t}\,|| = 5$ (although $\vec{s} \neq \vec{t}$).

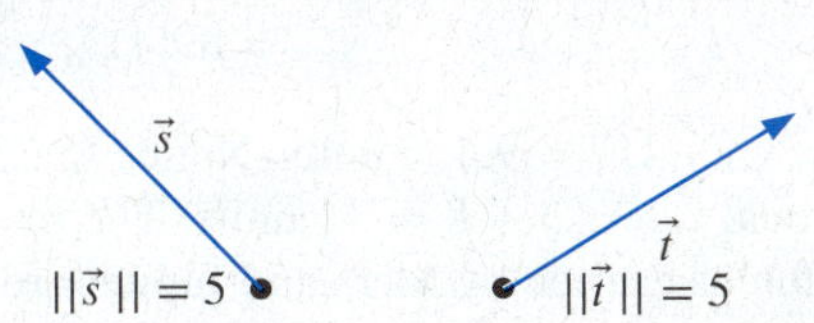


**Figure 12.4**: The vectors $\vec{s}$ and $\vec{t}$ are both of magnitude 5, but they point in different directions

## Addition of Vectors

As with displacements, the sum of two vectors $\vec{u}$ and $\vec{v}$ represented by arrows is found by joining the tail of $\vec{v}$ to the head of $\vec{u}$. Then $\vec{w} = \vec{u} + \vec{v}$ is represented by the arrow drawn from the tail of of $\vec{u}$ to the head of $\vec{v}$. See Figure 12.5.[2]

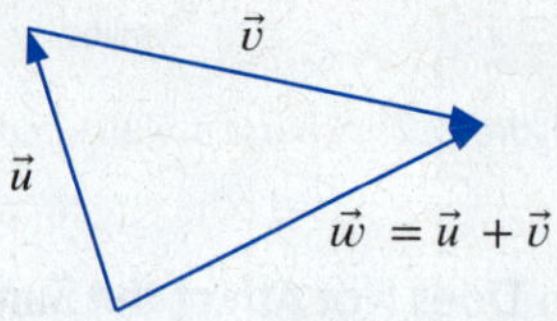


**Figure 12.5:** Constructing the vector sum $\vec{w} = \vec{u} + \vec{v}$

---

**Example 2** Spacecraft such as *Voyager* and *Galileo* experience gravitational forces exerted by the sun, earth, and other planets. Vectors are used to represent the strength and direction of the gravitational forces.

Suppose a Jupiter-bound spacecraft is coasting between the orbits of Mars and Jupiter. Assume that Jupiter exerts a gravitational force of 8 units on the spacecraft, directed toward Jupiter, and that Mars exerts a force of 3 units on the spacecraft, directed toward Mars. The force $\vec{F}_{\text{M}}$ exerted by Mars is at an angle of 70° to the force $\vec{F}_{\text{J}}$ exerted by Jupiter, as shown in Figure 12.6. (For simplicity, we ignore the forces due to the sun and other planets.)

(a) Find $||\vec{F}_{\text{MJ}}||$, the magnitude of the net force on the spacecraft due to Mars and Jupiter.

(b) What is the direction of $\vec{F}_{\text{MJ}}$?

(c) Without engine power, will the spacecraft stay on course?

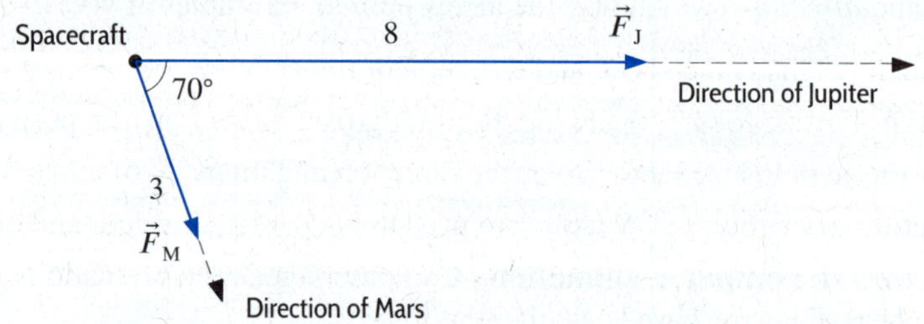


**Figure 12.6:** Forces $\vec{F}_{\text{M}}$ and $\vec{F}_{\text{J}}$ exerted by Mars and Jupiter, respectively

**Solution** (a) The net force, $\vec{F}_{\text{MJ}}$, is the sum of $\vec{F}_{\text{M}}$ and $\vec{F}_{\text{J}}$, the forces due to Mars and Jupiter, so

$$\vec{F}_{\text{MJ}} = \vec{F}_{\text{M}} + \vec{F}_{\text{J}}.$$

Figure 12.7 shows $\vec{F}_{\text{MJ}}$, obtained by joining the tail of $\vec{F}_{\text{J}}$ to the head of $\vec{F}_{\text{M}}$. The Law of Cosines tells us that $||\vec{F}_{\text{MJ}}||$, the length of the resulting vector, is given by

$$\begin{aligned} ||\vec{F}_{\text{MJ}}||^2 &= 8^2 + 3^2 - 2 \cdot 8 \cdot 3 \cos 110° \\ &= 73 - 48 \cos 110° \\ &= 89.417, \end{aligned}$$

so $||\vec{F}_{\text{MJ}}|| = \sqrt{89.417} = 9.456$ units. Notice that 9.456 is less than the sum of the magnitudes of the individual forces $3 + 8 = 11$ units. The magnitude of the net force is less than 11 units because the forces exerted by Mars and Jupiter are not perfectly aligned.

[2]If $\vec{u}$ and $\vec{v}$ are parallel, the vectors are added by the same method. However, in this case the figure is not a triangle.

(b) The direction of $\vec{F}_{MJ}$, denoted by $\theta$ in Figure 12.7, can be found using the Law of Sines:

$$\frac{\sin\theta}{3} = \frac{\sin 110^\circ}{||\vec{F}_{MJ}||}$$

$$\sin\theta = 3\left(\frac{\sin 110^\circ}{||\vec{F}_{MJ}||}\right) = 3\left(\frac{0.939}{9.456}\right) = 0.298.$$

The angle $\theta$ is less than 90°, so

$$\theta = \arcsin(0.298) = 17.345^\circ.$$

(c) The gravity of Mars is pulling the spacecraft off its Jupiter-bound course by about 17.345°. The spacecraft will veer off course if it does not use its engines.

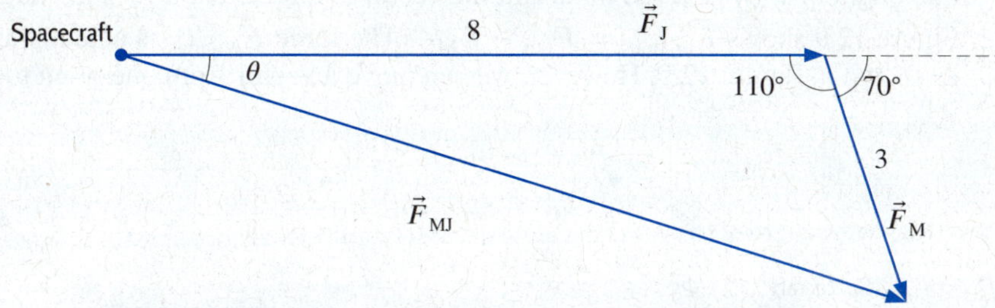


Figure 12.7: The net force, $\vec{F}_{MJ}$, on the spacecraft

## Subtraction of Vectors

We have seen how vectors are added. Now we see how to subtract one vector from another. If

$$\vec{w} = \vec{u} - \vec{v},$$

then it is reasonable to assume that by adding $\vec{v}$ to both sides we obtain

$$\vec{w} + \vec{v} = \vec{u}.$$

In other words, $\vec{w}$ is the vector that when added to $\vec{v}$ gives $\vec{u}$. This is illustrated in Figure 12.8.

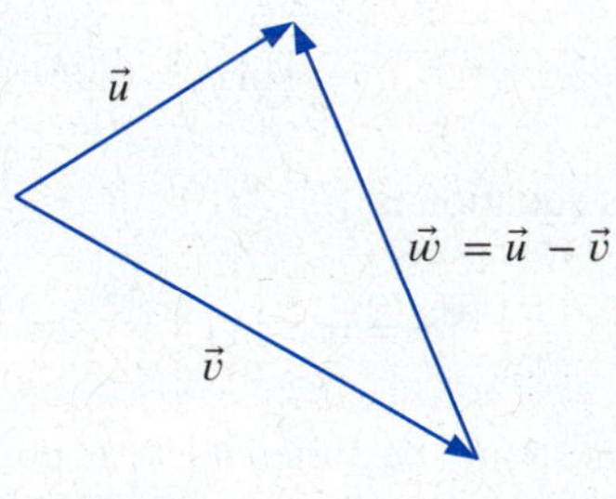


Figure 12.8: If $\vec{w} = \vec{u} - \vec{v}$, then $\vec{w}$ is the vector that when added to $\vec{v}$ gives $\vec{u}$

Figure 12.8 suggests the following rule: To find $\vec{w} = \vec{u} - \vec{v}$, join the tails of $\vec{v}$ and $\vec{u}$, and then $\vec{w}$ is the vector drawn from the head of $\vec{v}$ to the head of $\vec{u}$.

**Example 3** Suppose the spacecraft in Example 2 can fire its engine thrusters in any direction, but that the strength of the thrust is always 4 units. In what direction should the spacecraft aim its engine thrusters in order to stay on course toward Jupiter?

**Solution** The engine-thrust vector $\vec{F}_{\text{engines}}$ must be chosen so as to push the spacecraft back on course. The combined force on the craft due to Mars and Jupiter is given by $\vec{F}_{\text{MJ}}$. When the engines are on, the net force due to Mars, Jupiter, and the engines is given by

$$\vec{F}_{\text{net}} = \vec{F}_{\text{MJ}} + \vec{F}_{\text{engines}}.$$

We choose $\vec{F}_{\text{engines}}$ so that $\vec{F}_{\text{net}}$ points directly toward Jupiter. Solving for $\vec{F}_{\text{engines}}$ gives

$$\vec{F}_{\text{engines}} = \vec{F}_{\text{net}} - \vec{F}_{\text{MJ}}.$$

We do not know the length of $\vec{F}_{\text{net}}$, but we do know that it points toward Jupiter. From Example 2, the length of $\vec{F}_{\text{MJ}}$ is 9.456 units and the vector is directed at a 17.345° angle clockwise from $\vec{F}_{\text{net}}$. Figure 12.9 shows $\vec{F}_{\text{engines}} = \vec{F}_{\text{net}} - \vec{F}_{\text{MJ}}$. (The force $\vec{F}_{\text{engines}}$ is actually applied to the spacecraft, as shown in Figure 12.9. However, we can move a vector from one point to another when adding.)

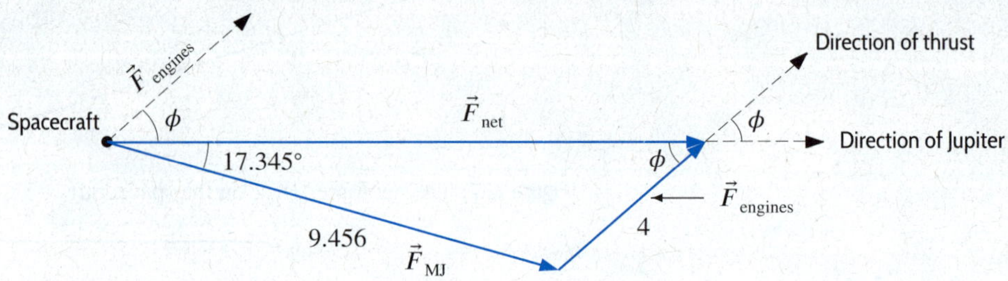


**Figure 12.9:** The engine-thrust vector is given by $\vec{F}_{\text{engines}} = \vec{F}_{\text{net}} - \vec{F}_{\text{MJ}}$, where $\vec{F}_{\text{net}}$ points directly toward Jupiter and $\vec{F}_{\text{MJ}}$ is as shown in Figure 12.7

We want to find the direction of thrust, labeled $\phi$ in Figure 12.9. By the Law of Sines, we have

$$\frac{\sin\phi}{9.456} = \frac{\sin 17.345^\circ}{4}$$
$$\sin\phi = 9.456\left(\frac{\sin 17.345^\circ}{4}\right).$$

One solution to this equation is

$$\phi = \arcsin\left(\frac{9.456}{4}\sin 17.345^\circ\right) = 44.811^\circ.$$

The engine thrusters should be aimed 44.811° counterclockwise from the direction of Jupiter.

There is another possible value for $\phi$, the direction of the engine thrust in Example 3. This is because there are two solutions to the equation

$$\sin\phi = \frac{9.456}{4}\sin 17.345^\circ.$$

The second solution, $180^\circ - 44.811^\circ = 135.189^\circ$, is shown in Figure 12.10. Both values of $\phi$ serve to counteract the gravity of Mars, but the 44.811° direction helps push the craft toward Jupiter, whereas the 135.189° direction helps push it away from Jupiter. Thus, the first solution is the one we need.

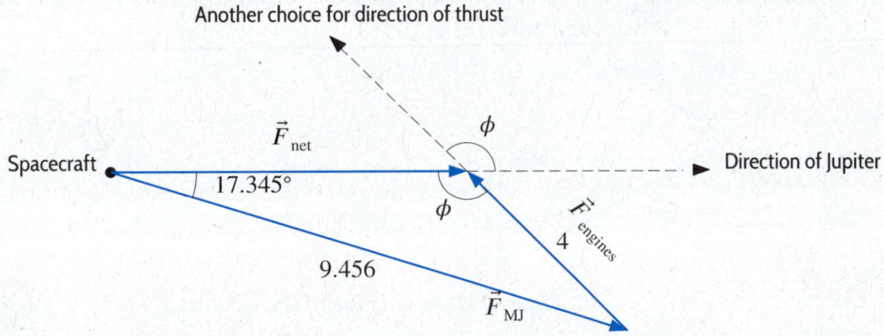


**Figure 12.10**: There are two possible choices for the direction of thrust. One has the effect of pushing the spacecraft toward Jupiter; the other (shown here) pushes the spacecraft away from Jupiter

## Scalar Multiplication

A vector $\vec{u}$ can be added to itself. In Figure 12.11, we see that the resulting vector, $\vec{u}+\vec{u}$, points in the same direction as $\vec{u}$ but is twice as long. The vector $\vec{u}+\vec{u}+\vec{u}$ also points in the same direction as $\vec{u}$, but is three times as long. We write

$$2\vec{u} \quad \text{for} \quad \vec{u}+\vec{u} \qquad \text{and} \qquad 3\vec{u} \quad \text{for} \quad \vec{u}+\vec{u}+\vec{u}.$$

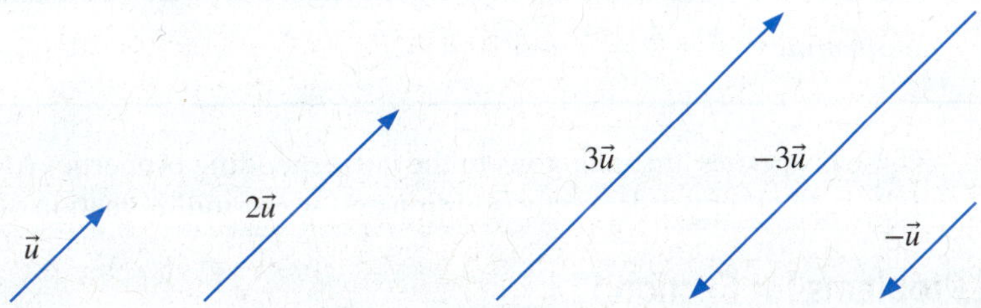

**Figure 12.11**: The vector $2\vec{u}=\vec{u}+\vec{u}$ points in the same direction as $\vec{u}$ but is twice as long as $\vec{u}$. Similarly, $3\vec{u}=\vec{u}+\vec{u}+\vec{u}$ points in the same direction and is 3 times as long as $\vec{u}$, and $-\vec{u}$ is in the opposite direction to $\vec{u}$

Similarly, $0.5\vec{u}$ points in the same direction as $\vec{u}$ but is only half as long. In general, if $\vec{v}$ is a vector and $k$ a positive number (a positive scalar), then $k\vec{v}$ is a vector that is $k$ times as long as $\vec{v}$ and pointing in the same direction as $\vec{v}$. The vector $k\vec{v}$ is called a *scalar multiple* of $\vec{v}$.

If $k$ is a negative number, the vector $k\vec{v}$ has length $|k|$ times the length of $\vec{v}$, but points in the opposite direction to $\vec{v}$. For example, $-3\vec{u}$ is the vector that is three times as long as $\vec{u}$ but pointing in the opposite direction. See Figure 12.11.

### The Zero Vector

The *zero vector*, denoted by the symbol $\vec{0}$, represents no displacement, which is the displacement that leaves you where you started. The zero vector has no direction.

The following rules summarize scalar multiplication.

- If $k>0$, then $k\vec{v}$ points in the same direction as $\vec{v}$ and is $k$ times as long.
- If $k<0$, then $k\vec{v}$ points in the opposite direction as $\vec{v}$ and is $|k|$ times as long.
- If $k=0$, then $k\vec{v}=\vec{0}$, the zero vector.

Since they either point in the same direction or opposite directions, two nonzero vectors $\vec{u}$ and $\vec{v}$ are *parallel* if there is a nonzero scalar $k$ with $\vec{u}=k\vec{v}$. For example, all the vectors in Figure 12.11 are parallel. The zero vector is not considered parallel to any other vector.

### Alternate View of Subtraction

Since the vector $-\vec{v}$ points in the opposite direction to $\vec{v}$, the difference $\vec{u} - \vec{v}$ is the same as the sum $\vec{u} + (-\vec{v})$. See Figure 12.12.

**Figure 12.12**: Illustration of the fact that $\vec{u} - \vec{v} = \vec{u} + (-\vec{v})$

## Properties of Vector Addition and Scalar Multiplication

The following properties hold true for any three vectors $\vec{u}$, $\vec{v}$, $\vec{w}$ and any two scalars $a$ and $b$:

1. *Commutativity of addition:* $\vec{u} + \vec{v} = \vec{v} + \vec{u}$
2. *Associativity of addition:* $(\vec{u} + \vec{v}) + \vec{w} = \vec{u} + (\vec{v} + \vec{w})$
3. *Associativity of scalar multiplication:* $a(b\vec{v}) = (ab)\vec{v}$
4. *Distributivity of scalar multiplication:* $(a + b)\vec{v} = a\vec{v} + b\vec{v}$ and $a(\vec{u} + \vec{v}) = a\vec{u} + a\vec{v}$
5. *Identities:* $\vec{v} + \vec{0} = \vec{v}$ and $1 \cdot \vec{v} = \vec{v}$

These properties are analogous to the corresponding properties for addition and multiplication of numbers. In other words, vector addition and scalar multiplication of vectors behave as expected.

## Exercises and Problems for Section 12.1

### Exercises

We use scalars to describe quantities like weight because these require only a single number (for example, 50 lbs). We use vectors to describe quantities like velocity because these require more than one number (for example, 20 mph at a heading of 80°). In Exercises 1–8, should we describe the given quantity with a vector or a scalar?

**1.** The elevation of a national monument.

**2.** The temperature at a point on the earth's surface.

**3.** The wind velocity at a point on the earth's surface.

**4.** The distance from Seattle to St. Louis.

**5.** The population of the US.

**6.** The population of a city.

**7.** The location in space of a satellite.

**8.** The number of men and women living in a city.

**9.** The vectors $\vec{w}$ and $\vec{u}$ are in Figure 12.13. Match the vectors $\vec{p}, \vec{q}, \vec{r}, \vec{s}, \vec{t}$ with five of the following vectors: $\vec{u} + \vec{w}$, $\vec{u} - \vec{w}$, $\vec{w} - \vec{u}$, $2\vec{w} - \vec{u}$, $\vec{u} - 2\vec{w}$, $2\vec{w}$, $-2\vec{w}$, $2\vec{u}$, $-2\vec{u}$, $-\vec{w}$, $-\vec{u}$.

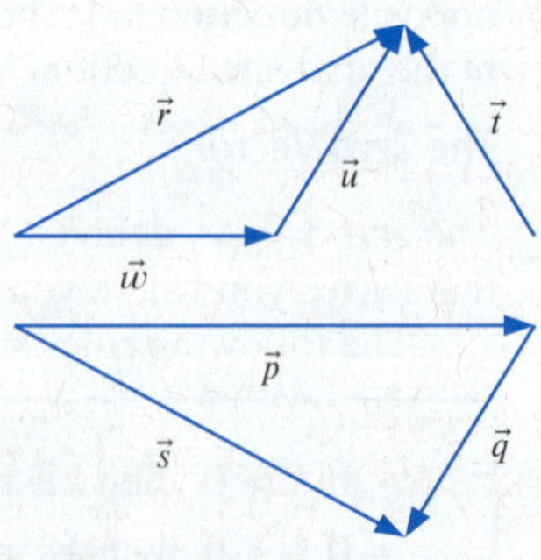


**Figure 12.13**

## Problems

Given the displacement vectors $\vec{v}$ and $\vec{w}$ in Figure 12.14, draw the vectors in Problems 10–15.

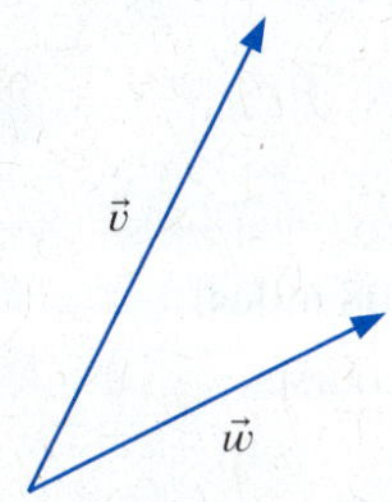


Figure 12.14

**10.** $\vec{v} + \vec{w}$  **11.** $\vec{v} - \vec{w}$  **12.** $2\vec{v}$

**13.** $\vec{w} + \vec{v}/2$  **14.** $2\vec{v} + \vec{w}$  **15.** $\vec{v} - 2\vec{w}$

**16.** **(a)** A kite on a 50-foot string is flying at an angle of 20° to flat ground. What are the magnitude and the direction of the vector from the kite to the ground?
**(b)** A wind blows the kite upward so that its angle with the ground is 40°. How far is the new position from the original position?

**17.** Oracle Road heads due north from its intersection with Route 10, which heads 20° west of north.

**(a)** If you travel 5 miles up Route 10 from the intersection, how far are you from Oracle Road?
**(b)** How far do you have to travel up Route 10 from the intersection to be 2 miles from Oracle Road?

**18.** A person leaves home and walks 2 miles due west. She then walks 3 miles southwest. How far away from home is she? In what direction must she walk to head directly home? (Your answer should be an angle in degrees and should include a sketch.)

**19.** The person from Problem 18 next walks 4 miles southeast. How far away from home is she? In what direction must she walk to head directly home?

**20.** A hockey puck starts on the edge of the rink and slides with a constant velocity $\vec{v}$, at a speed of 7 ft/sec and an angle of 35° with the edge. After 2 seconds, how far has the puck traveled? How far is it from the edge?

**21.** A helicopter is hovering at 3000 meters directly over the eastern perimeter of a secret Air Force installation. The eastern perimeter is 5000 meters straight east from the installation headquarters. The helicopter receives a transmission from headquarters that a UFO has been sighted at 7500 meters directly over the installation's western perimeter, which is 9000 meters straight west from the headquarters. How far must the helicopter travel to intercept the UFO? In what direction must it head? Be specific.

**22.** Suppose instead the UFO in Problem 21 is sighted directly over the installation's northern perimeter, which is 12,000 meters from headquarters. How far must the helicopter travel to intercept the UFO? In what direction must it head? Be specific.

**23.** A spacecraft is traveling directly toward Saturn, which exerts a gravitational force of 7 units on it. The moon Titan exerts a force of 10 units at a 40° angle to the right of the line joining the spacecraft to Saturn. The moon Rhea exerts a force of 5 units 80° to the left of this line.

**(a)** Find $||\vec{F}_{STR}||$, the magnitude of the sum of the forces acting on the spacecraft.
**(b)** Does the craft remain on course or veer to the left or to the right?
**(c)** If the engines have a thrust of 1 unit, is it possible to aim them in such a way as to correct the spacecraft's course?

**24.** A ball is thrown horizontally at 5 feet per second relative to still air. At the same time a wind blows at 3 feet per second at an angle of 45° to the ball's path. What is the velocity of the ball relative to the ground? (There are two answers.)

Use the definitions of addition and scalar multiplication to explain each of the properties in Problems 25–31.

**25.** $\vec{w} + \vec{v} = \vec{v} + \vec{w}$  **26.** $(a + b)\vec{v} = a\vec{v} + b\vec{v}$

**27.** $a(\vec{v} + \vec{w}) = a\vec{v} + a\vec{w}$  **28.** $\vec{v} + \vec{0} = \vec{v}$

**29.** $1\vec{v} = \vec{v}$  **30.** $\vec{v} + (-1)\vec{w} = \vec{v} - \vec{w}$

**31.** $(\vec{u} + \vec{v}) + \vec{w} = \vec{u} + (\vec{v} + \vec{w})$

# 12.2 THE COMPONENTS OF A VECTOR

So far, vectors have been described in terms of length and direction. There is another way to think about a vector, and that is in terms of *components*. We begin with an example illustrating the use of components; we define this term precisely later.

**Example 1** A ship travels 200 miles in a direction that its compass says is due east. The captain then discovers that the compass is faulty and that the ship has actually been heading in a direction that is 17.4° to the north of due east. How far north is the ship from its intended course?

**Solution** In Figure 12.15, a vector $\vec{v}$ has been drawn for the ship's actual displacement. We know that $\|\vec{v}\| = 200$ miles and the direction of $\vec{v}$ is 17.4° to the north of due east. We see that $\vec{v}$ can be considered the sum of two vectors, one pointing due east and the other pointing due north, so

$$\vec{v} = \vec{v}_{\text{north}} + \vec{v}_{\text{east}}.$$

From Figure 12.15, we have

$$\sin 17.4° = \frac{\|\vec{v}_{\text{north}}\|}{200},$$

so

$$\|\vec{v}_{\text{north}}\| = 200 \sin 17.4° = 59.808 \text{ miles}.$$

Therefore, the ship is 59.808 miles due north of its intended course.

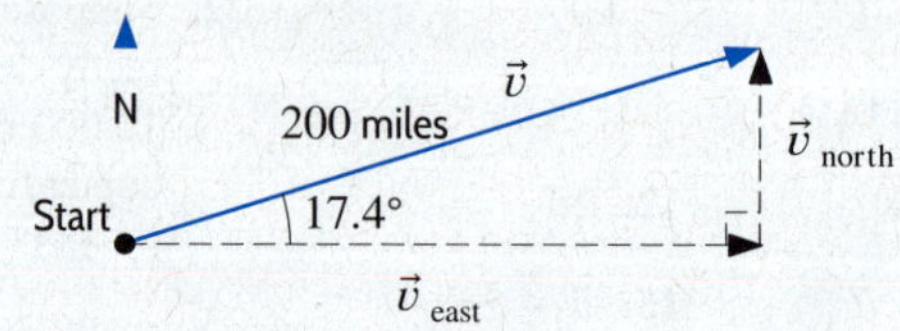


**Figure 12.15:** How far north is the ship from its intended eastward course?

## Unit Vectors

A *unit vector* is a vector of length 1 unit. The unit vectors in the directions of the positive $x$- and $y$-axes are called $\vec{i}$ and $\vec{j}$, respectively. See Figure 12.16. These two vectors are important because any vector in the plane can be expressed in terms of $\vec{i}$ and $\vec{j}$.

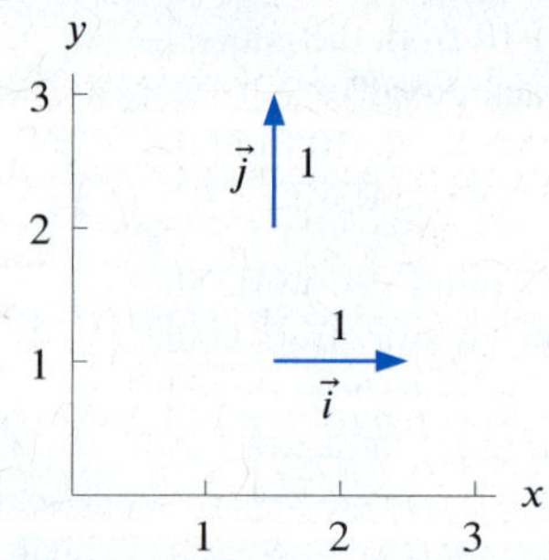


**Figure 12.16:** The unit vectors $\vec{i}$ and $\vec{j}$

**Example 2** Let the $x$-axis point east and the $y$-axis point north. A person walks 3 miles east and then 4 miles north. Express her displacement in terms of the unit vectors $\vec{i}$ and $\vec{j}$.

**Solution** The unit of distance is miles, so the unit vector $\vec{i}$ represents a displacement of 1 mile east, and the unit vector $\vec{j}$ represents a displacement of 1 mile north. The person's displacement can be written

$$\text{Displacement} = 3 \text{ miles east} + 4 \text{ miles north} = 3\vec{i} + 4\vec{j}.$$

**Example 3** Let the $x$-axis point east and the $y$-axis point north. Describe the displacement $5\vec{i} - 7\vec{j}$ in words.

**Solution** Since $\vec{i}$ represents a 1-mile displacement to the east, $5\vec{i}$ represents a 5-mile displacement to the east. Since $\vec{j}$ represents a 1-mile displacement to the north, $-7\vec{j}$ represents a 7-mile displacement to the south. Thus, $5\vec{i} - 7\vec{j}$ represents a 5-mile displacement to the east followed by a 7-mile displacement to the south.

## Resolving a Vector in the Plane into Components

In this book, we focus chiefly on vectors that lie flat in the plane. Such vectors can always be broken (or *resolved*) into sum of two vectors,[3] one parallel to the $x$-axis and the other to the $y$-axis. These two vectors are called the *components* parallel to the axes.

For the vector $\vec{v}$ in Figure 12.17, we can write

$$\vec{v} = \vec{v}_1 + \vec{v}_2,$$

where $\vec{v}_1$ is called the $x$-*component* of $\vec{v}$ and $\vec{v}_2$ is the $y$-*component* of $\vec{v}$. What can we say about the components $\vec{v}_1$ and $\vec{v}_2$?

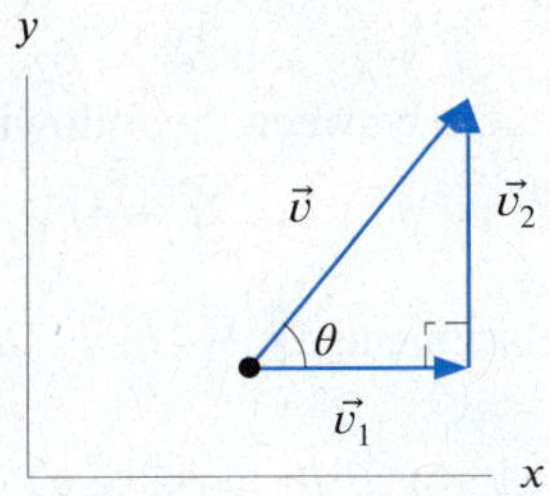


**Figure 12.17:** A vector $\vec{v}$ can be written as the sum of components, $\vec{v}_1$ and $\vec{v}_2$, parallel to the $x$- and $y$-axes

In Figure 12.17, the vector $\vec{v}$ has length $||\vec{v}\,||$ and makes an angle $\theta$ with the positive $x$-axis, with $0 \le \theta < 2\pi$. Thus, we see that

$$\cos\theta = \frac{||\vec{v}_1||}{||\vec{v}\,||} \quad \text{and} \quad \sin\theta = \frac{||\vec{v}_2||}{||\vec{v}\,||}.$$

Therefore

$$||\vec{v}_1|| = ||\vec{v}\,||\cos\theta \quad \text{and} \quad ||\vec{v}_2|| = ||\vec{v}\,||\sin\theta.$$

We know that $\vec{v}_1$ is parallel to the $x$-axis, so it is a scalar multiple of the unit vector $\vec{i}$. Similarly, $\vec{v}_2$ is a scalar multiple of $\vec{j}$. After checking that the signs of $\cos\theta$ and $\sin\theta$ give the correct directions for $\vec{v}_1$ and $\vec{v}_2$, we have

$$\vec{v}_1 = \left(||\vec{v}\,||\cos\theta\right)\vec{i} \quad \text{and} \quad \vec{v}_2 = \left(||\vec{v}\,||\sin\theta\right)\vec{j}.$$

These are the $x$- and $y$- components of $\vec{v}$. We have the following result:

> In the plane, a vector $\vec{v}$ of length $||\vec{v}\,||$ that makes an angle $\theta$ with the positive $x$-axis can be written in terms of its components:
>
> $$\vec{v} = \left(||\vec{v}\,||\cos\theta\right)\vec{i} + \left(||\vec{v}\,||\sin\theta\right)\vec{j}.$$

**Example 4** In Example 1, suppose the $x$-axis points east and the $y$-axis points north. Resolve the displacement vector $\vec{v}$ into components parallel to the axes.

**Solution** The vector $\vec{v}_{\text{north}}$ is the $y$-component of $\vec{v}$ and $\vec{v}_{\text{east}}$ is the $x$-component. We already know that $||\vec{v}_{\text{north}}|| = 59.808$ miles. From Figure 12.15, we see that $||\vec{v}_{\text{east}}|| = 200\cos 17.4^\circ = 190.848$ miles. (This last piece of information tells us that although the ship traveled a total of 200 miles, its eastward progress amounted to only 190.848 miles.) The easterly and northerly components are

$$\vec{v}_{\text{east}} = 190.848\vec{i} \qquad \text{and} \qquad \vec{v}_{\text{north}} = 59.808\vec{j}.$$

[3]However, one or both of the vectors may be of zero length.

Thus, the course taken by the ship can be written as follows:

$$\vec{v} = \vec{v}_{\text{east}} + \vec{v}_{\text{north}} = 190.848\vec{i} + 59.808\vec{j}.$$

To check the calculation, notice that

$$||\vec{v}_{\text{east}}||^2 + ||\vec{v}_{\text{north}}||^2 = (190.848)^2 + (59.808)^2 \approx (200)^2 = ||\vec{v}||^2.$$

Sums and scalar multiples are easy to compute if vectors are written in terms of components. For a sum, add corresponding components; for a multiple, multiply each component by the scalar.

**Example 5** Let $\vec{u} = 5\vec{i} + 7\vec{j}$, $\vec{v} = \vec{i} + \vec{j}$, and $\vec{w} = 6\vec{j} - 3\vec{i} + 2\vec{j} - 2\vec{i} - 4\vec{j} + 7\vec{i}$. Describe in words the similarities and differences between the following displacements:

$$2\vec{i} + 4\vec{j}, \quad \vec{u} - 3\vec{v}, \quad \vec{w}.$$

**Solution** We show that the displacements $2\vec{i} + 4\vec{j}$, $\vec{u} - 3\vec{v}$, and $\vec{w}$ are all the same. Using the properties on page 502, we have

$$\begin{aligned}\vec{u} - 3\vec{v} &= (5\vec{i} + 7\vec{j}) - 3(\vec{i} + \vec{j}) \\ &= (5\vec{i} + 7\vec{j}) - (3\vec{i} + 3\vec{j}) \\ &= (5\vec{i} - 3\vec{i}) + (7\vec{j} - 3\vec{j}) = 2\vec{i} + 4\vec{j}.\end{aligned}$$

The vector $\vec{w}$ can be simplified as

$$\vec{w} = (-3\vec{i} - 2\vec{i} + 7\vec{i}) + (6\vec{j} + 2\vec{j} - 4\vec{j}) = 2\vec{i} + 4\vec{j}.$$

Alternatively, we can see that all three displacements are the same by doing head-to-tail addition for each case, with the same starting point. Figure 12.18 shows that all three end points are the same, although the path followed is different in each case.

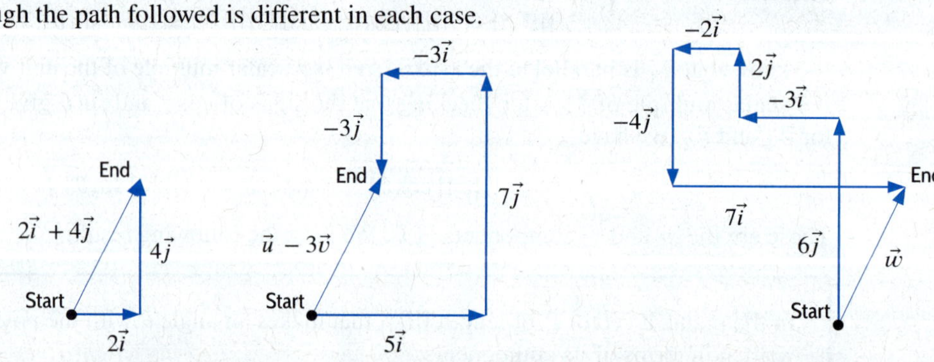


**Figure 12.18:** These three displacements, $2\vec{i} + 4\vec{j}$, $\vec{u} - 3\vec{v}$, $\vec{w}$, are all the same

## Displacement Vectors

Vectors that indicate displacement are often designated by giving two points that specify the initial and final positions. For example, the vector $\vec{v} = \overrightarrow{PQ}$ describes a displacement or motion from point $P$ to point $Q$. If we know the coordinates of $P$ and $Q$, we can find the components of $\vec{v}$.

**Example 6** If $P$ is the point $(-3, 1)$ and $Q$ is the point $(2, 4)$, find the components of the vector $\overrightarrow{PQ}$.

**Solution** Figure 12.19 shows the vector $\overrightarrow{PQ}$ and its components $\vec{v}_1$ and $\vec{v}_2$. The length of $\vec{v}_1$ is the horizontal distance from $P$ to $Q$, so

$$||\vec{v}_1|| = 2 - (-3) = 5.$$

The length of $\vec{v}_2$ is the vertical distance from $P$ to $Q$, so

$$||\vec{v}_2|| = 4 - 1 = 3.$$

Thus, the components are $\vec{v}_1 = 5\vec{i}$ and $\vec{v}_2 = 3\vec{j}$. We write $\vec{v} = \overrightarrow{PQ} = 5\vec{i} + 3\vec{j}$.

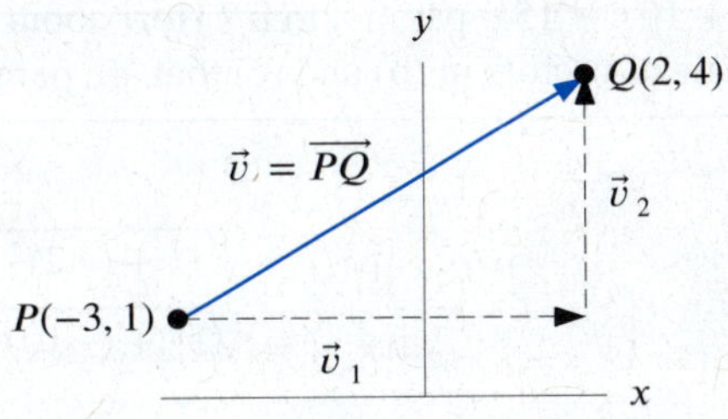


**Figure 12.19:** Finding components $\vec{v}_1$ and $\vec{v}_2$ from the coordinates of the points $P$ and $Q$

In general, we have the following result:

> If $P = (x_1, y_1)$ and $Q = (x_2, y_2)$, then the vector $\vec{v} = \overrightarrow{PQ}$ from $P$ to $Q$ is given by
>
> $$\vec{v} = (x_2 - x_1)\vec{i} + (y_2 - y_1)\vec{j}.$$

## Vectors in $n$ Dimensions

So far we have considered 2-dimensional vectors which have two components. Sometimes, we want to analyze vectors in space, which have three components. In addition, there are *n-dimensional* vectors, which have $n$ components, where $n$ is a positive integer.

**Example 7** A balloon rises vertically a distance of 2 miles, floats west a distance of 3 miles, and floats north a distance of 4 miles. The $x$-axis points eastward, the $y$-axis points northward, and the $z$-axis points upward. Write the balloon's displacement vector.

**Solution** The displacement vector has three components: a vertical component, an eastward component, and a northward component. Therefore, this displacement is a 3-dimensional vector. Writing $\vec{i}, \vec{j}, \vec{k}$ for unit vectors in the three directions, we have

$$\text{Displacement} = -3\vec{i} + 4\vec{j} + 2\vec{k}.$$

## Lengths of Vectors and Deciding Whether Vectors Are Parallel

When a 2-dimensional vector is written in terms of its components, we can use Pythagoras' Theorem to find its length. Similarly, we can use the components of a vector in space to determine its *length*: If $\vec{v} = v_1\vec{i} + v_2\vec{j} + v_3\vec{k}$, then

$$\text{Length of } \vec{v} = \|\vec{v}\| = \sqrt{(v_1)^2 + (v_2)^2 + (v_3)^2}.$$

The direction of vector is also determined by its components—a large component in one direction means the vector points roughly in that direction. We can use components to decide whether two vectors are *parallel*. Two nonzero vectors $\vec{v} = v_1\vec{i} + v_2\vec{j} + v_3\vec{k}$ and $\vec{w} = w_1\vec{i} + w_2\vec{j} + w_3\vec{k}$ are parallel when there exists a nonzero scalar $k$ with $\vec{w} = k\vec{v}$, so that for this value of $k$, their components satisfy:

$$w_1 = kv_1 \qquad \text{and} \qquad w_2 = kv_2 \qquad \text{and} \qquad w_3 = kv_3.$$

**Example 8** Let $\vec{v} = \vec{i} - 2\vec{j} + 3\vec{k}$ and $\vec{w} = 5\vec{i} - 10\vec{j} + 15\vec{k}$.

(a) Find $\|\vec{v}\|$ and $\|\vec{w}\|$.

(b) Are $\vec{v}$ and $\vec{w}$ parallel?
(c) Is $\vec{a} = 5\vec{i} + 10\vec{j} + 15\vec{k}$ parallel to $\vec{v}$? How about $\vec{b} = -5\vec{i} + 10\vec{j} - 15\vec{k}$?
(d) For each pair of vectors in (b) and (c) that are parallel, interpret the value of $k$.

**Solution** (a) We have

$$||\vec{v}|| = \sqrt{1^2 + (-2)^2 + 3^2} = \sqrt{14},$$
$$||\vec{w}|| = \sqrt{5^2 + (-10)^2 + 15^2} = \sqrt{350}.$$

(b) We have

$$w_1 = 5 = 5 \cdot 1 = 5v_1 \quad \text{and} \quad w_2 = -10 = 5 \cdot (-2) = 5v_2 \quad \text{and} \quad w_3 = 15 = 5 \cdot 3 = 5v_3.$$

Thus $\vec{v}$ and $\vec{w}$ are parallel, with $k = 5$. We have $\vec{w} = 5\vec{v}$.

(c) For $\vec{a}$ and $\vec{v}$, we have

$$a_1 = 5 = 5 \cdot 1 = 5v_1 \text{ and } a_2 = 10 = -5 \cdot (-2) = -5v_2 \text{ and } a_3 = 15 = 5 \cdot 3 = 5v_3,$$

The three equations do not all have the same value of $k$, so $\vec{v}$ and $\vec{a}$ are *not* parallel. For $\vec{b}$ and $\vec{v}$, we have

$$b_1 = -5 = -5 \cdot 1 = -5v_1 \text{ and } b_2 = 10 = -5 \cdot (-2) = -5v_2 \text{ and } w_3 = -15 = -5 \cdot 3 = -5v_3.$$

We see that $\vec{v}$ and $\vec{b}$ are parallel, with $k = -5$. We have $\vec{b} = -5\vec{v}$.

(d) We have $\vec{w} = 5\vec{v}$ and $\vec{b} = -5\vec{v}$, so $\vec{w}$ and $\vec{b}$ are both 5 times as long as $\vec{v}$. All three vectors are parallel, but $\vec{w}$ points in the same direction as $\vec{v}$, while $\vec{b}$ points in the opposite direction.

In general, for two parallel vectors, $\vec{w}$ and $\vec{v}$, with $\vec{w} = k\vec{v}$, we interpret $k$ as follows:

- The sign of $k$ tells us whether they point in the same or opposite directions.
- The magnitude of $k$ tells us the ratio of their lengths.

## Exercises and Problems for Section 12.2

### Exercises

Resolve the vectors in Exercises 1–3 into components.

**1.** A vector starting at the point $Q = (4, 6)$ and ending at the point $P = (1, 2)$.

**2.** A vector starting at the point $P = (1, 2)$ and ending at the point $Q = (4, 6)$.

**3.** The vector shown in Figure 12.20, with components expressed in inches.

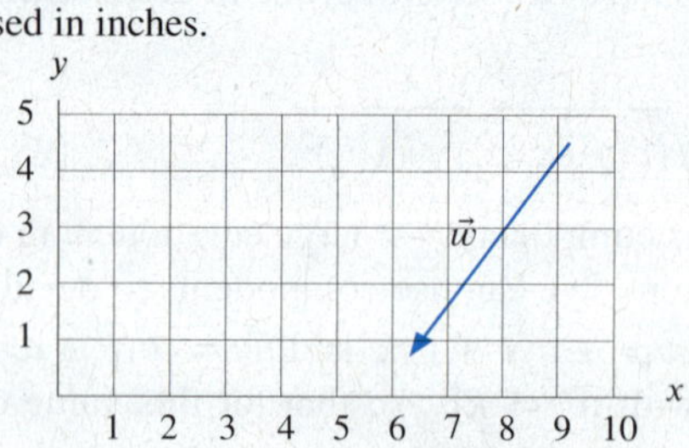


**Figure 12.20**: Scale: 1 unit on the $x$- and $y$-axes is 0.25 inches

For Exercises 4–7, perform the indicated computations.

**4.** $5(2\vec{i} - \vec{j}) + \vec{j}$

**5.** $(4\vec{i} + 2\vec{j}) - (3\vec{i} - \vec{j})$

**6.** $-(\vec{i} + \vec{j}) + 2(2\vec{i} - 3\vec{j})$

**7.** $2(0.45\vec{i} - 0.9\vec{j} - 0.01\vec{k}) - 0.5(1.2\vec{i} - 0.1\vec{k})$

**8.** On the graph in Figure 12.21, draw the vector $\vec{v} = 4\vec{i} + \vec{j}$ twice, once with its tail at the origin and once with its tail at the point $(3, 2)$.

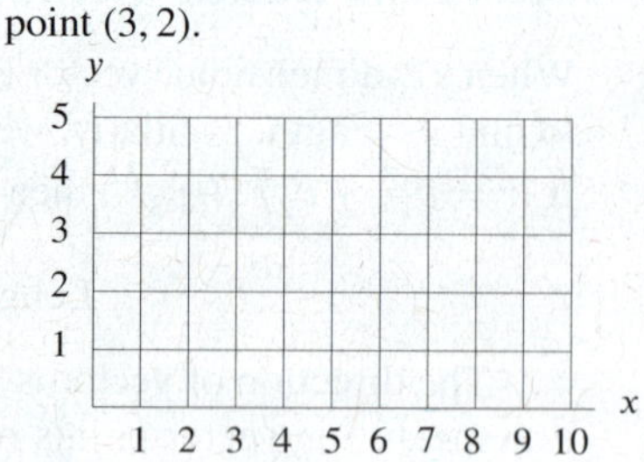


**Figure 12.21**

Find the length of the vectors in Exercises 9–12.

**9.** $\vec{v} = \vec{i} - \vec{j} + 3\vec{k}$

**10.** $\vec{v} = \vec{i} - \vec{j} + 2\vec{k}$

**11.** $\vec{v} = 7.2\vec{i} - 1.5\vec{j} + 2.1\vec{k}$

**12.** $\vec{v} = 1.2\vec{i} - 3.6\vec{j} + 4.1\vec{k}$

## Problems

**13.** A ship steams due north at 10 knots but experiences an ocean current that moves east to west at 5 knots.[4] Resolve into components the velocity of the ship in a coordinate system in which the $x$-axis points east and the $y$ axis points north.

In Problems 14–17, use the information in Figure 12.22. Each grid square is 1 unit along each side.

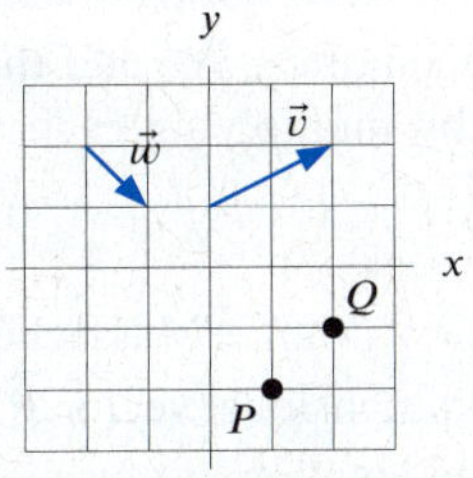


Figure 12.22

**14.** Write the following vectors in component form.

**(a)** $\vec v$ **(b)** $2\vec w$ **(c)** $\vec v + \vec w$ **(d)** $\vec w - \vec v$

**(e)** The displacement vector $\overrightarrow{PQ}$.
**(f)** The vector from the point $P$ to the point $(2, 0)$.
**(g)** A vector perpendicular to the $y$-axis.
**(h)** A vector perpendicular to the $x$-axis.

**15.** What is the angle between the vector $\vec w$ and the negative $y$-axis?

**16.** What is the angle between the vectors $\vec v$ and $\vec w$?

**17.** What is the angle between the vector $\vec w$ and the displacement vector $\overrightarrow{PQ}$?

**18.** A car is traveling at a speed of 50 km/hr. The positive $y$-axis is north and the positive $x$-axis is east. Resolve the car's velocity vector into two components if the car is traveling in each of the following directions:

**(a)** East **(b)** South
**(c)** Southeast **(d)** Northwest

**19.** Shortly after takeoff, a plane is climbing northwest through still air at an airspeed of 200 km/hr and rising at a rate of 300 m/min. Resolve into components its velocity vector in a coordinate system in which the $x$-axis points east, the $y$-axis points north, and the $z$-axis points up.

**20.** **(a)** Find a unit vector from the point $P = (1, 2)$ toward the point $Q = (4, 6)$.
**(b)** Find a vector of length 10 pointing in the same direction.

**21.** Which is traveling faster, a car whose velocity vector is $21\vec i + 35\vec j$, or a car whose velocity vector is $40\vec i$, assuming that the units are the same for both directions?

**22.** A truck is traveling due north at 30 km/hr toward a crossroad. On a perpendicular road a police car is traveling west toward the intersection at 40 km/hr. Both vehicles will reach the crossroad in exactly one hour. Find the vector currently representing the displacement of the truck with respect to the police car.

**23.** An athlete throws a shot at a 30° angle to the ground at speed 50 km/hr.

**(a)** What is the magnitude of the velocity vector of the shot?
**(b)** Resolve the velocity vector into vertical and horizontal components.

**24.** Which of the following vectors are parallel?

$$\vec u = 2\vec i + 4\vec j - 2\vec k,\quad \vec v = \vec i - \vec j + 3\vec k,$$
$$\vec w = -\vec i - 2\vec j + \vec k,\quad \vec p = \vec i + \vec j + \vec k,$$
$$\vec q = 4\vec i - 4\vec j + 12\vec k,\ \vec r = \vec i - \vec j + \vec k.$$

**25.** **(a)** A man swims northeast at 5 mph. Give a vector representing his velocity.
**(b)** He now swims in a river at the same velocity relative to still water; the river's current flows north at 1.2 mph. What is his velocity relative to the riverbed?

**26.** The coastline is along the $x$-axis and the sea is in the region $y > 0$. A sailing boat has velocity $\vec v = 2.5\vec i + 3.1\vec j$ mph. What is the speed of the boat? What angle does the boat's path make with the coastline?

**27.** The hour hand and the minute hand of a clock are represented by the vectors $\vec h$ and $\vec m$, respectively, with $\|\vec h\| = 2$ and $\|\vec m\| = 3$. The origin is at the center of the clock and the positive $x$-axis goes through the three o'clock position.

**(a)** What are the components of $\vec h$ and $\vec m$ at the following times? Illustrate with a sketch.

(i) 12 noon (ii) 3 pm
(iii) 1 pm (iv) 1:30 pm

**(b)** Sketch the displacement vector from the tip of the hour hand to the tip of the minute hand at 3 pm. What are the components of this displacement vector?
**(c)** Sketch the vector representing the sum $\vec h + \vec m$ at 1:30 pm. What are the components of this sum?

A cat on the ground at the point $(1, 4, 0)$ watches a squirrel at the top of a tree. The tree is one unit high with base at the point $(2, 4, 0)$. Find the displacement vectors in Problems 28–31.

**28.** From the origin to the cat.

**29.** From the bottom of the tree to the squirrel.

**30.** From the bottom of the tree to the cat.

**31.** From the cat to the squirrel.

[4] 1 knot is 1 nautical mile (approximately 6080 feet) per hour.

# 12.3 APPLICATIONS OF VECTORS

## Alternate Notation for the Components of a Vector

The vector $\vec{v} = 3\vec{i} + 4\vec{j}$ is sometimes written $\vec{v} = (3, 4)$, evn though this notation can be confused with the coordinate notation used for points. (For instance, $(3, 4)$ might mean the point $x = 3$, $y = 4$ or the vector $3\vec{i} + 4\vec{j}$.) This notation is particularly useful for vectors in $n$ dimensions.

## Population Vectors

Vectors in $n$ dimensions are used for keeping track of $n$ quantities. We add them by adding the corresponding components; scalar multiples are obtained by multiplying each component by the scalar.

**Example 1** The 2005 population of the six New England states (Connecticut, Maine, Massachusetts, New Hampshire, Rhode Island, and Vermont) can be thought of as a 6-dimensional vector $\vec{P}$. The components of $\vec{P}$ are the populations of the six states. Using the alternate notation, we write

$$\vec{P} = (P_{\text{CT}}, P_{\text{ME}}, P_{\text{MA}}, P_{\text{NH}}, P_{\text{RI}}, P_{\text{VT}}),$$

where $P_{\text{CT}}$ is the population of Connecticut, $P_{\text{ME}}$ is the population of Maine, and so on.

The population vector $\vec{P}$ from the previous example does not have a geometrical interpretation. We cannot draw a picture of $\vec{P}$, or interpret its length and direction as we do for displacement vectors. However, the following example shows how it is used.

**Example 2** The vectors $\vec{P}$ and $\vec{Q}$ give the populations of the six New England states in 2005 and 2010, respectively. According to the Census Bureau,[5] these vectors are given, in millions of people, by

$$\vec{P} = (3.51, 1.32, 6.40, 1.31, 1.08, 0.62)$$
$$\vec{Q} = (3.57, 1.33, 6.55, 1.32, 1.05, 0.63).$$

For instance, the population of Connecticut was 3.51 million in 2005 and 3.57 million in 2010.

(a) Find $\vec{R} = \vec{Q} - \vec{P}$. Explain its significance in terms of the population of New England.

(b) Let $\vec{S}$ be the estimated population of New England in the census year 2020. One expert predicts that $\vec{S} = \vec{Q} + 2\vec{R}$. Find $\vec{S}$ and explain the assumption this expert is making about the New England population.

(c) Another expert makes a different prediction, claiming that $\vec{T} = 1.08\vec{Q}$ will give the population of New England in 2020. Find $\vec{T}$ and explain the assumption that this expert is making.

**Solution** (a) We calculate $\vec{R} = \vec{Q} - \vec{P}$ component by component, giving

$$\begin{aligned}\vec{R} &= \vec{Q} - \vec{P}\\ &= (3.57, 1.33, 6.55, 1.32, 1.05, 0.63) - (3.51, 1.32, 6.40, 1.32, 1.08, 0.62)\\ &= (3.57 - 3.51, 1.33 - 1.32, 6.55 - 6.40, 1.32 - 1.31, 1.05 - 1.08, 0.63 - 0.62)\\ &= (0.06, 0.01, 0.15, 0.01, -0.03, 0.01).\end{aligned}$$

The components of $\vec{R}$ give the change in population for each New England state. For instance, the population of Connecticut rose from 3.51 million in 2005 to 3.57 million in 2010, a change of $3.57 - 3.51 = 0.06$ million people, so $R_{\text{CT}} = 0.06$.

[5] www.census.gov, accessed May 17, 2013.

(b) The formula $\vec{S} = \vec{Q} + 2\vec{R}$ means that

$$\text{Population in 2020} = \text{Population in 2010} + 2 \cdot \text{Change between 2005 and 2010.}$$

According to this expert, states will see their populations increase or decrease between 2010 and 2020 as they did between 2005 and 2010. Note that that the factor of 2 is required because the interval is twice as long. Algebraically, we have

$$\begin{aligned}\vec{S} &= \vec{Q} + 2\vec{R}\\ &= (3.57, 1.33, 6.55, 1.32, 1.05, 0.63) + 2 \cdot (0.06, 0.01, 0.15, 0.01, -0.03, 0.01)\\ &= (3.57, 1.33, 6.55, 1.32, 1.05, 0.63) + (0.12, 0.02, 0.30, 0.02, -0.06, 0.02)\\ &= (3.69, 1.35, 6.85, 1.34, 0.99, 0.65).\end{aligned}$$

For instance, Connecticut's population climbed by 0.06 million between 2005 and 2010. The predicted population of Connecticut in 2020 is given by

$$S_{\text{CT}} = Q_{\text{CT}} + 2 \cdot R_{\text{CT}} = 3.57 + 2 \cdot (0.06) = 3.69.$$

The 0.12 million jump in Connecticut's population between 2010 and 2020 is twice as large as the 0.06 million jump between 2005 and 2010.

(c) The formula $\vec{T} = 1.08\vec{Q}$ means that each component of $\vec{T}$ is 1.08 times as large as the corresponding component of $\vec{Q}$. In other words, the population of each state is predicted to grow by 8%. We have

$$\begin{aligned}\vec{T} = 1.08\vec{Q} &= 1.08 \cdot (3.57, 1.33, 6.55, 1.32, 1.05, 0.63)\\ &= (1.08 \cdot 3.57, 1.08 \cdot 1.33, 1.08 \cdot 6.55, 1.08 \cdot 1.32, 1.08 \cdot 1.05, 1.08 \cdot 0.63)\\ &= (3.86, 1.44, 7.07, 1.43, 1.13, 0.68).\end{aligned}$$

For instance, the population of New Hampshire is predicted to grow from $Q_{\text{NH}} = 1.32$ million in 2010 to $T_{\text{NH}} = 1.08Q_{\text{NH}} = 1.08(1.32) = 1.43$ million in the year 2020. Here uniform growth is assumed across the region, which means that Rhode Island has growth instead of a decline that was predicted in $\vec{S}$.

Notice how vector addition and subtraction were used in parts (a) and (b) of the previous example, and scalar multiplication was used in parts (b) and (c).

## Economics

In 2009, the blockbuster movie *Avatar* set a record by grossing \$761 million at domestic (US) box offices.[6] It made even more money at international box offices, where it grossed \$2023 million. Rounding out its wild success is the \$191 million for video and DVD domestic sales. The *revenue vector*, $\vec{a} = (761, 2023, 191)$, summarizes the domestic, international, and video/DVD gross revenues in millions of dollars.

**Example 3** After *Avatar*, two other major financially successful US movies were *Harry Potter and the Deathly Hallows: Part II* in 2011 and *The Avengers* in 2012. If $\vec{s}$ and $\vec{t}$, are the revenue vectors for these two movies, then

$$\vec{s} = (381, 947, 90) \qquad \vec{t} = (623, 891, 91).$$

Find $\vec{R}$, the revenue vector for all three movies combined.

Solution The revenue vector, $\vec{R}$, is the sum of $\vec{a}$, $\vec{s}$, and $\vec{t}$:

$$\vec{R} = \vec{a} + \vec{s} + \vec{t} = (761 + 381 + 623, 2023 + 947 + 891, 191 + 90 + 91) = (1765, 3888, 372).$$

[6] Data compiled from www.the-numbers.com. Accessed May 17, 2012.

### Price and Consumption Vectors

A car dealership has several different models of cars in its inventory, with different prices for each model. A *price vector*, $\vec{P}$, gives the price of each model:

$$\vec{P} = (P_1, P_2, \dots, P_n).$$

Here $P_1$ is the price of car model 1, $P_2$ is the price of model 2, and so on. A *consumption vector* $\vec{C}$ gives the number of each model of car purchased (or consumed) during a given month:

$$\vec{C} = (C_1, C_2, \dots, C_n).$$

**Example 4** Suppose $\vec{I}$ gives the number of cars of each model a car dealer currently has in inventory. Explain the meaning of the following in terms of the car dealership.

(a) $\vec{I} - \vec{C}$ (b) $\vec{I} - 2\vec{C}$ (c) $\vec{C} = 0.3\vec{I}$

**Solution**

(a) This expression represents the difference between the number of each model currently in inventory and the number of that model purchased each month. So $\vec{I} - \vec{C}$ represents the numbers of each model that remain on the lot after one month.
(b) This expression represents the number of each model the dealer has left after two months, assuming that no new cars are added to inventory and that $\vec{C}$ does not change.
(c) This equation tells us that 30% of the inventory of each model is purchased by consumers each month.

**Example 5** Let $\vec{E}$ represent expenses incurred by the dealer for acquisition, insurance, and overhead, for each model. What does the vector $\vec{P} - \vec{E}$ represent?

**Solution** This represents the dealer's profit for each model. For instance, a model that the dealer sells for \$29,000 may cost \$25,000 to acquire from the factory, insure, and maintain. The difference of \$4000 represents profit for the dealer. The vector $\vec{P} - \vec{E}$ keeps track of this profit for each model.

## Physics

Newton's law of gravitation states that the magnitude of force exerted on an object of mass $m$ by the earth is given by

$$\|\vec{F}_E\| = G\frac{mM_E}{r_E^2},$$

where $M_E$ is the mass of the earth, $r_E$ is the distance from the object to earth's center, and $G$ is a constant. Similarly, the force exerted on the object by the moon has magnitude

$$\|\vec{F}_L\| = G\frac{mM_L}{r_L^2},$$

where $M_L$ is the mass of the moon and $r_L$ is the distance to the moon's center. The force exerted by the earth is directed toward the earth; the force exerted by the moon is exerted toward the moon.

**Example 6** Figure 12.23 shows the position of a 100,000-kg spacecraft relative to the earth and moon. The vector $\vec{r}$ is the position of the moon relative to the earth, $\vec{r}_E$ is the position of the spacecraft relative to the earth, and $\vec{r}_L$ is the position of the spacecraft relative to the moon. With distances in thousands of kilometers, we have

$$\vec{r} = 384\vec{i} \quad \text{and} \quad \vec{r}_E = 280\vec{i} + 90\vec{j}.$$

Which force is stronger, the pull of the moon on the spacecraft or the pull of the earth? [Use $M_E = 5.98 \cdot 10^{24}$ kg, $M_L = 7.34 \cdot 10^{22}$ kg, and $G = 6.67 \cdot 10^{-23}$ if distance is in thousands of kilometers.]

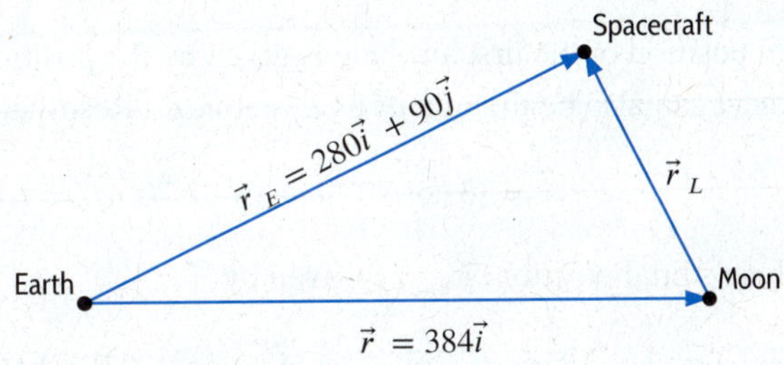


**Figure 12.23:** Gravitational forces between the earth, the moon, and a 100,000-kg spacecraft

**Solution** To use Newton's law of gravitation, we first find $||\vec{r}_E||$ and $||\vec{r}_L||$. Since $\vec{r}_E = 280\vec{i} + 90\vec{j}$,

$$||\vec{r}_E|| = \sqrt{(280)^2 + (90)^2} = \sqrt{86500} = 294.109.$$

Since $\vec{r}_L = \vec{r}_E - \vec{r}$, we have

$$\vec{r}_L = (280\vec{i} + 90\vec{j}) - 384\vec{i} = -104\vec{i} + 90\vec{j}.$$

Thus,

$$||\vec{r}_L|| = \sqrt{(-104)^2 + (90)^2} = \sqrt{18916} = 137.535.$$

Using the given values of $m$, $M_E$, $M_L$, and $G$, we calculate

$$F_E = G\frac{mM_E}{||\vec{r}_E||^2} = 461.116 \quad \text{and} \quad F_L = G\frac{mM_L}{||\vec{r}_L||^2} = 25.882.$$

Thus, the pull of the earth on this spacecraft is much stronger than the pull of the moon; in fact, it is over ten times as strong. The earth's pull is stronger, even though the spaceship is closer to the moon, because the earth has a much larger mass than the moon.

## Computer Graphics: Position Vectors

Video games usually incorporate computer-generated graphics, as do the flight simulators used by commercial and military flight-training schools. Hollywood makes ever-increasing use of computer graphics in its films. Enormous amounts of computation are involved in creating such effects and much of this computation involves vectors.

A computer screen can be thought of as an $xy$-grid with the origin at the lower left corner. The position of a point on the screen is specified by a vector pointing from the origin to the point. Such a vector is called a *position vector*. The tail of a position vector is always fixed at the origin.

**Example 7** A video game shows two airplanes on the screen at the points $(3, 5)$ and $(7, 2)$. See Figure 12.24. Both airplanes move a distance of 3 units at an angle of $70°$ counterclockwise from the $x$-axis. What are the new positions of the airplanes?

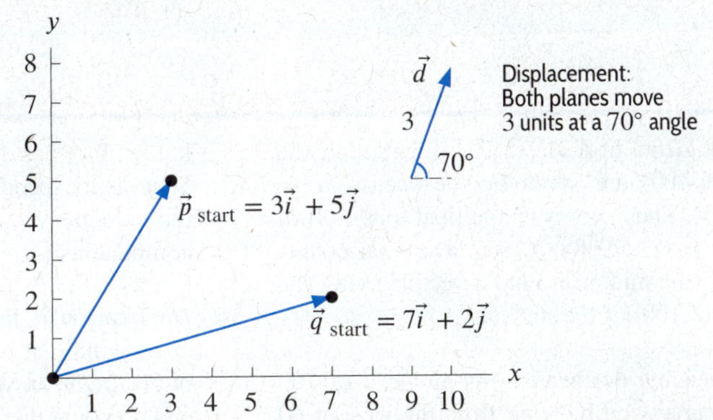


**Figure 12.24:** Computer screen showing two airplanes moving by a displacement vector $\vec{d}$

**Solution** The initial position of the first airplane is given by the position vector $\vec{p}_{\text{start}} = 3\vec{i} + 5\vec{j}$. The plane's displacement can also be thought of as a vector, $\vec{d}$. Resolving $\vec{d}$ into components gives

$$\vec{d} = (3\cos 70^\circ)\vec{i} + (3\sin 70^\circ)\vec{j} = 1.026\vec{i} + 2.819\vec{j}.$$

The airplane's final position, $\vec{p}_{\text{end}}$, is given by

$$\begin{aligned}\vec{p}_{\text{end}} = \vec{p}_{\text{start}} + \vec{d} &= 3\vec{i} + 5\vec{j} + 1.026\vec{i} + 2.819\vec{j} \\ &= 4.026\vec{i} + 7.819\vec{j}.\end{aligned}$$

The second airplane's initial position is given by the position vector $\vec{q}_{\text{start}} = 7\vec{i} + 2\vec{j}$. The second plane's displacement, however, is exactly the same as the first airplane's, $\vec{d} = 1.026\vec{i} + 2.819\vec{j}$. Therefore, the final position of the second airplane, $\vec{q}_{\text{end}}$, is given by

$$\begin{aligned}\vec{q}_{\text{end}} = \vec{q}_{\text{start}} + \vec{d} &= 7\vec{i} + 2\vec{j} + 1.026\vec{i} + 2.819\vec{j} \\ &= 8.026\vec{i} + 4.819\vec{j}.\end{aligned}$$

Example 7 illustrates the difference between position vectors and other vectors: The tail of a position vector is fixed to the origin, but the tail of the displacement vector $\vec{v}$ can be anywhere, so long as its length and orientation do not change.

## Exercises and Problems for Section 12.3

### Exercises

In Exercises 1–8, find the vector using $\vec{N} = (5, 6, 7, 8, 9, 10)$ and $\vec{M} = (1, 1, 2, 3, 5, 8)$.

**1.** $\vec{B} = 2\vec{M}$

**2.** $\vec{G} = \vec{N} + \vec{M}$

**3.** $\vec{A} = \vec{M} - \vec{N}$

**4.** $\vec{\epsilon} = 2\vec{N} - 7\vec{M}$

**5.** $\vec{K} = \dfrac{\vec{N}}{3} + \dfrac{2\vec{N}}{3}$

**6.** $\vec{\rho} = 1.067\vec{M} + 2.361\vec{N}$

**7.** $\vec{Z} = \dfrac{\vec{N}}{3} + \dfrac{\vec{M}}{2}$

**8.** $\vec{R} = \sqrt{3}\vec{M} + 4\sqrt{3}\vec{N}$

In Exercises 9–12, use $\vec{Q}$ from Example 2 on page 510.

**9.** If the population of each New England state increases by 120,000 people from 2010 to 2015, what is $\vec{S}$, the population vector for 2015?

**10.** If the population of each New England state increases by 2% from 2010 to 2015, what is $\vec{R}$, the population vector for 2015?

**11.** If the population of each New England state decreases by 43,000 people from 2010 to 2015, what is $\vec{U}$, the population vector for 2015?

**12.** If the population of each New England state decreases by 22% from 2010 to 2015, what is $\vec{T}$, the population vector for 2015?

### Problems

**13.** There are five students in a class. Their scores on the midterm (out of 100) are given by the vector $\vec{v} = (73, 80, 91, 65, 84)$. Their scores on the final (out of 100) are given by $\vec{w} = (82, 79, 88, 70, 92)$. The final counts twice as much as the midterm. Find a vector giving the total scores (out of 100) of the students.

**14.** An airplane is heading northeast at an airspeed of 700 km/hr, but there is a wind blowing from the west at 60 km/hr. In what direction does the plane end up flying? What is its speed relative to the ground? [Hint: Resolve the velocity vectors for the airplane and the wind into components.]

**15.** An airplane is flying at an airspeed of 600 km/hr in a cross-wind that is blowing from the northeast at a speed of 50 km/hr. In what direction should the plane head to end up going due east?

**16.** Two children are throwing a ball back-and-forth straight across the back seat of a car. Suppose the ball is being thrown at 10 mph relative to the car and the car is going 25 mph down the road.

**(a)** Make a sketch showing the relevant velocity vectors.

**(b)** If one child does not catch the ball and it goes out an open window, what angle does the ball's horizontal motion make with the road?

**17.** A man walks 5 miles in a direction 30° north of east. He then walks a distance $x$ miles due east. He turns around to look back at his starting point, which is at an angle of 10° south of west.

**(a)** Make a sketch. Give vectors in $\vec{i}$ and $\vec{j}$ components for each part of the man's walk.

**(b)** What is $x$?

**(c)** How far is the man from his starting point?

**18.** A clothing store sells four different proposed children's outfits of pants, shirt, shoes, belt, and jacket. The prices of the four outfits can be represented by vectors:

$$\vec{a} = (75, 30, 120, 40, 200)$$
$$\vec{b} = (25, 20, 75, 30, 90)$$
$$\vec{c} = (145, 50, 100, 40, 300)$$
$$\vec{d} = (60, 45, 200, 35, 150)$$

**(a)** Which two outfits have the same belt?

**(b)** David's children want 3 outfits $\vec{a}$, 2 outfits $\vec{c}$ and 1 outfit $\vec{d}$.

(i) Write their wishes as a sum of vectors.

(ii) Write out a single vector expressing what David would pay for each of the five items.

**(c)** David buys 5 outfits for his children and spends \$145 on shirts.

(i) Express his purchase as a sum of the vectors $\vec{a}, \vec{b}, \vec{c}, \vec{d}$,

(ii) Write out the vector $\vec{e}$ giving the sum.

**19.** Three different electric charges $q_1$, $q_2$, and $q_3$ exert forces on a test charge $Q$. The forces are, respectively, $\vec{F}_1 = (3, 6)$, $\vec{F}_2 = (-2, 5)$, and $\vec{F}_3 = (7, -4)$. The net force, $\vec{F}_{\text{net}}$ is given by $\vec{F}_{\text{net}} = \vec{F}_1 + \vec{F}_2 + \vec{F}_3$.

**(a)** Calculate $\vec{F}_{\text{net}}$.

**(b)** If a fourth charge $q_4$ is added, what force $\vec{F}_4$ must it exert on $Q$ so that $Q$ feels no net force at all, that is, so that $\vec{F}_{\text{net}} = \vec{0}$?

**20.** Let $q = 20$ be an electric charge at the origin $(0, 0)$, and let $Q = 30$ be another charge at position $\vec{r} = (r_x, r_y)$. Then $q$ exerts a force $\vec{F}$ on $Q$ given by *Coulomb's Law*,

$$||\vec{F}\,|| = \frac{qQ}{||\vec{r}\,||^2}.$$

The direction of $\vec{F}$ is the same as the direction of $\vec{r}$, so that $Q$ tends to be pushed away[7] from $q$.

**(a)** Let the position of $Q$ be $\vec{r} = (3, 5)$. Find $||\vec{r}\,||$.

**(b)** Using Coulomb's Law, find $||\vec{F}\,||$.

**(c)** Find the components of $\vec{F} = (F_x, F_y)$. [Hint: $\vec{F}$ points in the same direction as $\vec{r}$, so both vectors make the same angle with the $x$-axis.]

**21.** Figure 12.25 shows a rectangle whose four corners are the points

$$a = (2, 1), \quad b = (4, 1), \quad c = (4, 2), \quad \text{and} \quad d = (2, 2).$$

As part of a video game, this rectangle is rotated counter-clockwise through an angle of 35° about the origin. See Figure 12.26. What are the new coordinates of the corners of the rectangle?

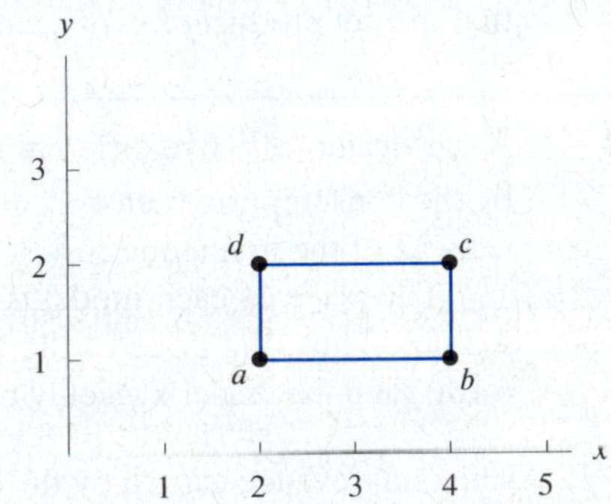


**Figure 12.25**

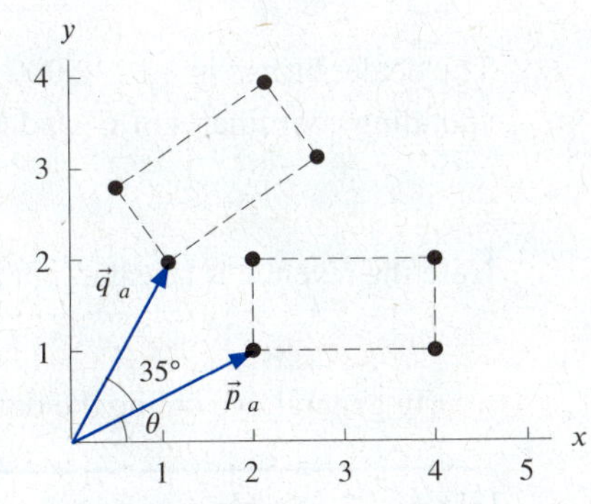


**Figure 12.26**: Rectangle rotated about the origin through an angle of 35°

[7]This is true because $q$ and $Q$ have the same sign. The force would be opposite to $\vec{r}$ if $Q$ and $q$ had opposite signs.

22. Provided that $x_1 \neq x_2$ and $y_1 \neq y_2$, the points $p_1 = (x_1, y_1)$ and $p_2 = (x_2, y_2)$ define a rectangle with $p_1$ in one corner and $p_2$ in the opposite corner. We represent this rectangle as a 4-vector $\vec{r} = (x_1, y_1, x_2, y_2)$. For instance, Figure 12.27 shows the rectangle described by $\vec{r} = (2, 2, 7, 6)$, which has $p_1 = (2, 2)$ and $p_2 = (7, 6)$.
    (a) Sketch the rectangle described by $\vec{s} = (2, 4, 5, 8)$.
    (b) Let $\vec{t}$ be a rectangle and let $\vec{u} = (1, 0, 1, 0)$, $\vec{v} = (0, 1, 0, 1)$, and $\vec{w} = \vec{u} + \vec{v}$. Describe in words how the following rectangles are related to $\vec{t}$.
        (i) $\vec{t} + \vec{u}$ (ii) $\vec{t} + \vec{v}$
        (iii) $\vec{t} - \vec{u}$ (iv) $\vec{t} + 2\vec{u} + 3\vec{v}$
        (v) $\vec{t} + \vec{w}$ (vi) $\vec{t} + k\vec{w}$, $k$ a constant.

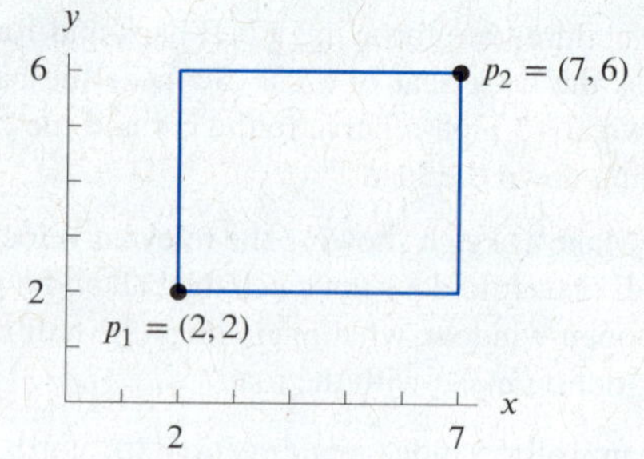

Figure 12.27

## 12.4 THE DOT PRODUCT

We have added and subtracted vectors and we have multiplied vectors by scalars. We now see how to multiply one vector by another vector.

To compute the product of $\vec{u}$ and $\vec{v}$, we multiply each coordinate of $\vec{u}$ by the corresponding coordinate of $\vec{v}$ and add these products. The result is called the *dot product*, written $\vec{u} \cdot \vec{v}$. Notice that the dot product of two vectors is a scalar, not a vector.

**Example 1** A car dealer sells five different models of car. The number of each model sold each week is given by the consumption vector $\vec{C} = (22, 14, 8, 12, 19)$. For instance, we see that in one week the dealer sells 22 of the first model, 14 of the second, and so forth.

The price of each model is given by the vector $\vec{P} = (19, 23, 40, 47, 32)$, where the units are \$1000s. Thus, the price of the first model is \$19,000, the price of the second is \$23,000, and so forth. Find the dealer's weekly revenue.

Solution The total revenue earned by the dealer each week (in 1000s of dollars) is given by

$$\begin{aligned}\text{Revenue (in \$1000s)} &= 22 \text{ cars} \cdot 19 \text{ per car} + 14 \text{ cars} \cdot 23 \text{ per car} + 8 \text{ cars} \cdot 40 \text{ per car} \\ &\quad + 12 \text{ cars} \cdot 47 \text{ per car} + 19 \text{ cars} \cdot 32 \text{ per car} \\ &= 2232.\end{aligned}$$

The dealer brings in \$2,232,000 each week. Since the revenue is obtained by multiplying the corresponding coordinates of $\vec{C}$ and $\vec{P}$ and adding, we can write revenue as a dot product:

$$\text{Revenue} = 2232 = \vec{C} \cdot \vec{P}.$$

Note the revenue is a scalar, not a vector, as the dot product always gives a number.

In general, we define the dot product of two vectors as follows:

if $\vec{u} = (u_1, u_2, \ldots, u_n)$ and $\vec{v} = (v_1, v_2, \ldots, v_n)$ are two $n$-dimensional vectors, then the **dot product**, $\vec{u} \cdot \vec{v}$, is the scalar given by

$$\vec{u} \cdot \vec{v} = u_1 v_1 + u_2 v_2 + \cdots + u_n v_n.$$

**Example 2** The population vector $\vec{P}_{\text{New England}} = (3.51, 1.32, 6.40, 1.31, 1.08, 0.62)$ gives the populations (in millions) of the six New England states (CT, ME, MA, NH, RI, VT) in 2005. The vector $\vec{r} = (1.71\%, 0.76\%, 2.34\%, 0.76\%, -2.78\%, 1.62\%)$ gives the percent change in population between 2005 and 2010 for each state. Calculate and interpret the dot product: $\vec{r} \cdot \vec{P}_{\text{New England}}$.

**Solution** We calculate the dot product:

$$\begin{aligned}\vec{r} \cdot \vec{P}_{\text{New England}} &= 0.0171 \cdot 3.51 + 0.0076 \cdot 1.32 + 0.0234 \cdot 6.40 \\ &\quad +0.0076 \cdot 1.31 - 0.02781 \cdot 1.08 + 0.0162 \cdot 0.62 \\ &= 0.209789.\end{aligned}$$

To interpret this dot product, we look at the each term separately:

$$0.0171 \cdot 3.51 = 0.060021 = \text{Change in Connecticut population,} \quad \text{and so on.}$$

So the dot product gives the total change in population of New England: it increased by 209,789 people (0.209789 million) between 2005 and 2010:

$$\Delta P_{\text{New England}} = \vec{r} \cdot \vec{P}_{\text{New England}}.$$

## Properties of the Dot Product

It can be shown that the following properties hold:

We take the angle $\theta$ between two vectors to be $0° \leq \theta \leq 180°$. Then:

- $\vec{u} \cdot \vec{v} = \vec{v} \cdot \vec{u}$ *(Commutative Law)*
- $\vec{u} \cdot (\vec{v} + \vec{w}) = \vec{u} \cdot \vec{v} + \vec{u} \cdot \vec{w}$ *(Distributive Law)*
- $\vec{v} \cdot \vec{v} = ||\vec{v}||^2$
- $\vec{u} \cdot \vec{v} = ||\vec{u}|| \cdot ||\vec{v}|| \cos\theta$, where $\theta$ is the angle between $\vec{u}$ and $\vec{v}$

The commutative law holds because multiplication of coordinates is commutative. A proof of the distributive law using coordinates is outlined in Problem 23. For the fourth property, note that

$$\vec{v} \cdot \vec{v} = (v_1, v_2) \cdot (v_1, v_2) = v_1^2 + v_2^2.$$

By the Pythagorean theorem, $||\vec{v}||^2 = {v_1}^2 + {v_2}^2$, so $\vec{v} \cdot \vec{v} = ||v||^2$. A similar argument applies to vectors of higher dimension.

### Dot Product and Angles: What Does the Dot Product Mean?

From the formula $\vec{u} \cdot \vec{v} = \|\vec{u}\| \cdot \|\vec{v}\| \cos\theta$, we see that if $0° < \theta < 90°$, then $\vec{u} \cdot \vec{v}$ is positive, and if $90° < \theta < 180°$, then $\vec{u} \cdot \vec{v}$ is negative. This property provides a useful way to calculate the dot product without using coordinates. Later in this section, we use the Law of Cosines to show why it works for 2-dimensional vectors.

The dot product can be interpreted as a measure of the alignment of two vectors. If two vectors $\vec{u}$ and $\vec{v}$ are perpendicular, then the angle between them is $\theta = 90°$. (See Figure 12.28.) In this case, $\cos\theta = \cos 90° = 0$, so

$$\vec{u} \cdot \vec{v} = ||\vec{u}|| \cdot ||\vec{v}|| \cos 90° = 0.$$

A dot product of zero tells us that the two vectors are perpendicular.

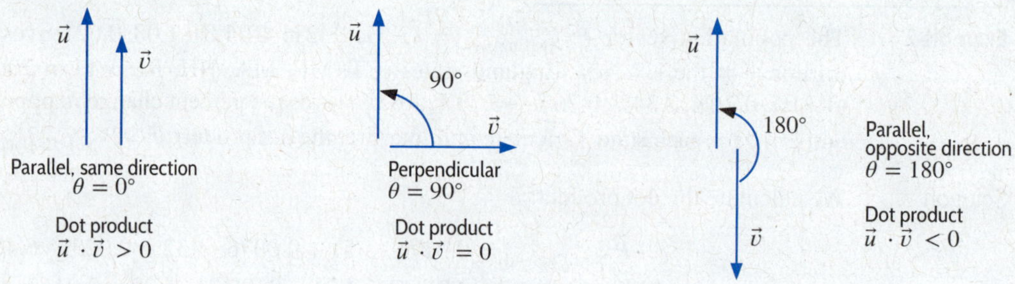


**Figure 12.28:** Dot product is positive, zero, or negative depending on alignment of vectors

We obtain the following results connecting the dot product and the alignment of two vectors:

- The largest possible value for $\vec{u} \cdot \vec{v}$ occurs if $\vec{u}$ and $\vec{v}$ are parallel and pointing in the same direction, with $\theta = 0°$.
- Perpendicularity results in $\vec{u} \cdot \vec{v} = 0$. It occurs if $\vec{u}$ and $\vec{v}$ are at angle of $\theta = 90°$.
- The most negative value for $\vec{u} \cdot \vec{v}$ occurs if $\vec{u}$ and $\vec{v}$ are parallel but pointing in opposite directions, with $\theta = 180°$.

**Example 3**

(a) Find $\vec{v} \cdot \vec{w}$ where $\vec{v} = 3\vec{i} + 4\vec{j}$ and $\vec{w} = 2\vec{i} + 5\vec{j}$.

(b) One person walks 3 miles east and then 4 miles north to point $A$. Another person walks 2 miles east and then 5 miles north to point $B$. Both started from the same spot, $O$. What is the angle of separation of these two people? (See Figure 12.29.)

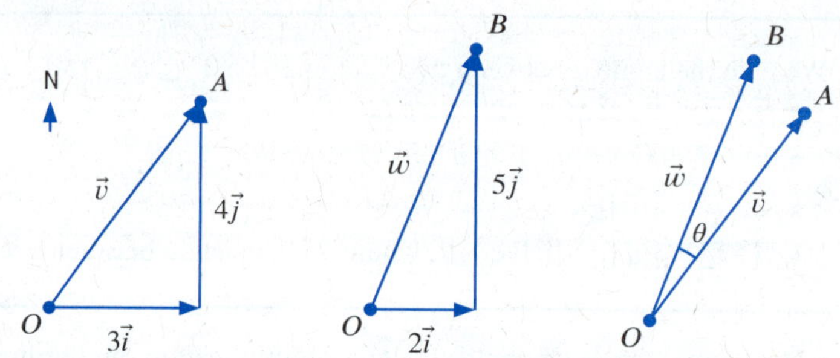


**Figure 12.29:** What is the angle of separation $\theta$ between the people at $A$ and $B$?

**Solution**

(a) We have $\vec{v} \cdot \vec{w} = (3\vec{i} + 4\vec{j}) \cdot (2\vec{i} + 5\vec{j}) = 3 \cdot 2 + 4 \cdot 5 = 26$.

(b) Assuming $\vec{i}$ points east and $\vec{j}$ points north, we see that $\vec{v}$ gives the first person's position and $\vec{w}$ gives the second person's position. The angle of separation between these two people is labeled $\theta$ in Figure 12.29. We use the formula

$$\vec{v} \cdot \vec{w} = ||\vec{v}|| \cdot ||\vec{w}|| \cos\theta.$$

Since $||\vec{v}|| = \sqrt{3^2 + 4^2} = 5$ and $||\vec{w}|| = \sqrt{2^2 + 5^2} = \sqrt{29}$ and, from part (a), $\vec{v} \cdot \vec{w} = 26$, we have

$$26 = 5\sqrt{29}\cos\theta,$$

$$\cos\theta = \frac{26}{5\sqrt{29}},$$

$$\theta = \arccos\frac{26}{5\sqrt{29}} = 15.068°.$$

### Justification of $\vec{u} \cdot \vec{v} = ||\vec{u}|| \cdot ||\vec{v}|| \cos\theta$

Figure 12.30 shows two vectors, $\vec{u}$ and $\vec{v}$, with an angle $\theta$ between them. The vectors $\vec{u}$, $\vec{v}$, and $\vec{w} = \vec{u} - \vec{v}$ form a triangle. Now, we know (from the fourth property of dot products) that $\vec{w} \cdot \vec{w} = ||\vec{w}||^2$. We can also calculate $\vec{w} \cdot \vec{w}$ using the distributive and commutative laws:

$$\begin{aligned}\vec{w} \cdot \vec{w} &= (\vec{v} - \vec{u}) \cdot (\vec{v} - \vec{u}) \\ &= \vec{v} \cdot \vec{v} - \vec{u} \cdot \vec{v} - \vec{v} \cdot \vec{u} + \vec{u} \cdot \vec{u} \\ &= ||\vec{v}||^2 + ||\vec{u}||^2 - 2\vec{u} \cdot \vec{v}.\end{aligned}$$

Since $\vec{w} \cdot \vec{w} = ||\vec{w}||^2$, we have shown that

$$||\vec{w}||^2 = ||\vec{v}||^2 + ||\vec{u}||^2 - 2\vec{u} \cdot \vec{v}.$$

But applying the Law of Cosines to the triangle in Figure 12.30 gives

$$||\vec{w}||^2 = ||\vec{u}||^2 + ||\vec{v}||^2 - 2||\vec{u}|| \cdot ||\vec{v}|| \cos\theta.$$

Thus setting these two expressions for $||\vec{w}||^2$ equal gives

$$||\vec{u}||^2 + ||\vec{v}||^2 - 2\vec{u} \cdot \vec{v} = ||\vec{u}||^2 + ||\vec{v}||^2 - 2||\vec{u}|| \cdot ||\vec{v}|| \cos\theta,$$

which simplifies to the formula we wanted:

$$\vec{u} \cdot \vec{v} = ||\vec{u}|| \cdot ||\vec{v}|| \cos\theta.$$

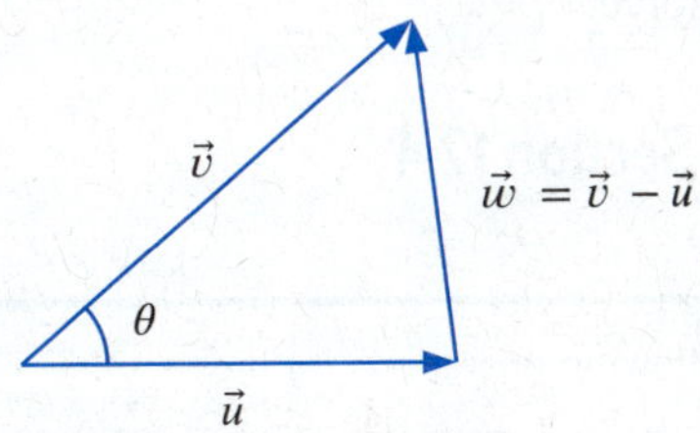


**Figure 12.30:** Triangle used to justify $\vec{u} \cdot \vec{v} = ||\vec{u}|| \cdot ||\vec{v}|| \cos\theta$

## Work

In physics, the concept of *work* is represented by the dot product. In everyday language, work means effort expended. In physics, the term has a similar, but more precise, meaning.

Suppose you load a heavy refrigerator onto a truck. The refrigerator is on casters and glides with little effort along the floor. However, to lift the refrigerator takes a lot of work. The *work* done in moving the refrigerator against the force of gravity is defined by

$$\text{Work} = \vec{F} \cdot \vec{d},$$

where $\vec{F}$ is the force exerted (assumed constant) and $\vec{d}$ is the displacement. If we measure distance in feet and force in pounds, work is measured in foot-pounds, where 1 ft-lb is the amount of work required to raise 1 pound a distance of 1 foot.

Suppose we push the refrigerator up a ramp that makes a 10° angle with the floor. If the ramp is 12 ft long and the force exerted on the refrigerator is 350 lbs vertically upward, then the displacement $\vec{d}$ has a magnitude of 12, and the angle $\theta$ between $\vec{F}$ and $\vec{d}$ is $90° - 10° = 80°$. (See Figure 12.31.) The work done by the force $\vec{F}$ is

$$\text{Work} = ||\vec{F}|| \cdot ||\vec{d}|| \cos 80° = 350 \cdot 12 \cos 80° = 729.322 \text{ ft-lbs}.$$

To push the refrigerator up the ramp, we do 729.322 ft-lbs of work.

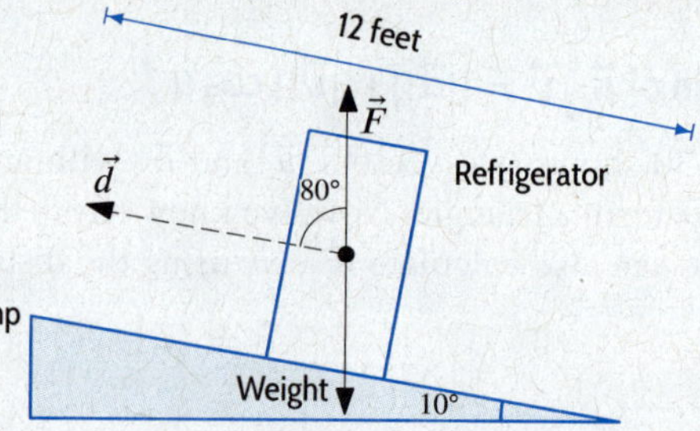


Figure 12.31: Refrigerator being pushed up ramp; the angle between $\vec{F}$ and $\vec{r}$ is $\theta = 100°$

It is informative to consider the two extreme cases: horizontal and vertical ramps. A horizontal ramp leads to a 90° angle between $\vec{F}$ and $\vec{d}$, so

$$\text{Work} = ||\vec{F}|| \cdot ||\vec{d}|| \cos 90° = 350 \cdot 12 \cdot 0 = 0 \text{ ft-lbs.}$$

Since we push the refrigerator in a direction perpendicular to its weight, we don't have to fight the weight at all (assuming frictionless casters). For a vertical ramp, we have

$$\text{Work} = ||\vec{F}|| \cdot ||\vec{d}|| \cos 0° = 350 \cdot 12 \cdot 1 = 4200 \text{ ft-lbs.}$$

In this case, we are pushing (or hoisting) the refrigerator in a direction opposite to that of its weight, so we feel its full force.

## Exercises and Problems for Section 12.4

### Exercises

For Exercises 1–12, perform the following operations on the given 3-dimensional vectors.

$$\vec{a} = 2\vec{j} + \vec{k} \qquad \vec{b} = -3\vec{i} + 5\vec{j} + 4\vec{k}$$
$$\vec{c} = \vec{i} + 6\vec{j} \qquad \vec{y} = 4\vec{i} - 7\vec{j}$$
$$\vec{z} = \vec{i} - 3\vec{j} - \vec{k}$$

**1.** $\vec{z} \cdot \vec{a}$

**2.** $\vec{a} \cdot \vec{z}$

**3.** $\vec{c} \cdot \vec{y}$

**4.** $\vec{a} \cdot \vec{y}$

**5.** $\vec{a} \cdot \vec{b}$

**6.** $\vec{b} \cdot \vec{z}$

**7.** $\vec{c} \cdot \vec{a} + \vec{a} \cdot \vec{y}$

**8.** $\vec{a} \cdot (\vec{c} + \vec{y})$

**9.** $(\vec{a} \cdot \vec{b})\vec{a}$

**10.** $((\vec{c} \cdot \vec{c})\vec{a}) \cdot \vec{a}$

**11.** $(\vec{a} \cdot \vec{y})(\vec{c} \cdot \vec{z})$

**12.** $(\vec{z} \cdot \vec{c})(\vec{y} \cdot \vec{a})$

### Problems

**13.** What is the angle between the vectors $\vec{v}$ and $\vec{w}$ in Figure 12.32?

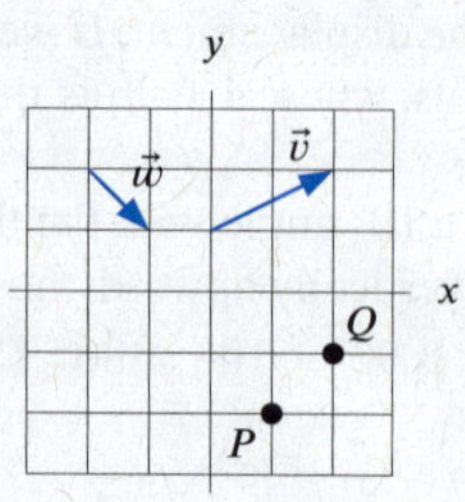


Figure 12.32

**14.** The force of gravity acting on a ball is 2 lb downward. How much work is done by gravity if the ball

**(a)** Falls 3 feet?

**(b)** Moves upward 5 feet?

**15.** How much work is done in pushing a 350-lb refrigerator up a 12-ft ramp that makes a 30° angle with the floor?

**16.** Which pairs of the vectors $\sqrt{3}\vec{i} + \vec{j}$, $3\vec{i} + \sqrt{3}\vec{j}$, $\vec{i} - \sqrt{3}\vec{j}$ are parallel and which are perpendicular?

**17.** Compute the angle between the vectors $\vec{i} + \vec{j} + \vec{k}$ and $\vec{i} - \vec{j} - \vec{k}$.

**18.** For what values of $t$ are $\vec{u} = t\vec{i} - \vec{j} + \vec{k}$ and $\vec{v} = t\vec{i} + t\vec{j} - 2\vec{k}$ perpendicular? Are there values of $t$ for which $\vec{u}$ and $\vec{v}$ are parallel?

**19.** Suppose that $||\vec{a}|| = 7$ and $||\vec{b}|| = 4$, but that $\vec{a}$ and $\vec{b}$ can point in any direction. What are the maximum and minimum possible lengths for the vectors $\vec{a} + \vec{b}$ and $\vec{a} - \vec{b}$? Illustrate your answers with sketches.

**20.** A 100-meter dash is run on a track in the direction of the vector $\vec{v} = 2\vec{i} + 6\vec{j}$. The wind velocity $\vec{w}$ is $5\vec{i} + \vec{j}$ km/hr. The rules say that a legal wind speed measured in the direction of the dash must not exceed 5 km/hr. Will the race results be disqualified due to an illegal wind? Justify your answer.

**21.** Let $A, B, C$ be the points $A = (1, 2)$; $B = (4, 1)$; $C = (2, 4)$. Is triangle $\Delta ABC$ a right triangle?

**22.** Show that the vectors $(\vec{b} \cdot \vec{c})\vec{a} - (\vec{a} \cdot \vec{c})\vec{b}$ and $\vec{c}$ are perpendicular.

**23.** In this problem, you will check the distributive law for 2-vectors. Show that if $\vec{u} = (u_1, u_2)$, $\vec{v} = (v_1, v_2)$, and $\vec{w} = (w_1, w_2)$, then

$$\vec{u} \cdot (\vec{v} + \vec{w}) = \vec{u} \cdot \vec{v} + \vec{u} \cdot \vec{w}.$$

**24.** Let $\vec{u} = u_1\vec{i} + u_2\vec{j}$ and $\vec{v} = v_1\vec{i} + v_2\vec{j}$. Using $\vec{u} \cdot \vec{v} = ||\vec{u}|| \cdot ||\vec{v}|| \cos\theta$ and the distributive law, show that $\vec{u} \cdot \vec{v} = u_1v_1 + u_2v_2$.

**25.** **(a)** Bread, eggs, and milk cost \$3.00 per loaf, \$2.00 per dozen, and \$4.00 per gallon, respectively, at Acme Store. Use a price vector $\vec{a}$ and a consumption vector $\vec{c}$ to write a vector equation that describes what may be bought for \$40.

**(b)** At Beta Mart, where the food is fresher, the price vector is $\vec{b} = (3.20, 1.80, 4.50)$. Explain the meaning of $(\vec{b} - \vec{a}) \cdot \vec{c}$ in practical terms. Is $\vec{b} - \vec{a}$ ever perpendicular to $\vec{c}$?

**(c)** Some people think Beta Mart's freshness makes each grocery item at Beta equivalent to 110% of the corresponding Acme item. What does it mean for a consumption vector to satisfy the inequality $(1/1.1)\vec{b} \cdot \vec{c} < \vec{a} \cdot \vec{c}$?

**26.** Recall that in 2 or 3 dimensions, if $\theta$ is the angle between $\vec{v}$ and $\vec{w}$, the dot product is given by

$$\vec{v} \cdot \vec{w} = ||\vec{v}||\,||\vec{w}|| \cos\theta.$$

We use this relationship to define the angle between two vectors in $n$ dimensions. If $\vec{v}, \vec{w}$ are $n$-vectors, then the dot product, $\vec{v} \cdot \vec{w} = v_1w_1 + v_2w_2 + \cdots + v_nw_n$, is used to define the angle $\theta$ by

$$\cos\theta = \frac{\vec{v} \cdot \vec{w}}{||\vec{v}||\,||\vec{w}||} \qquad \text{provided } ||\vec{v}||, ||\vec{w}|| \neq 0.$$

We now use this idea of angle to measure how close two populations are to one another genetically. Table 12.1 shows the relative frequencies of four alleles (variants of a gene) in four populations.

**Table 12.1**

| Allele | Eskimo | Bantu | English | Korean |
|---|---|---|---|---|
| $A_1$ | 0.29 | 0.10 | 0.20 | 0.22 |
| $A_2$ | 0.00 | 0.08 | 0.06 | 0.00 |
| $B$ | 0.03 | 0.12 | 0.06 | 0.20 |
| $O$ | 0.67 | 0.69 | 0.66 | 0.57 |

Let $\vec{a}_1, \vec{a}_2, \vec{a}_3, \vec{a}_4$ be the 4-vectors showing the relative frequencies of each allele in the Eskimo, Bantu, English, Korean populations, respectively. The genetic distance between two populations is defined as the angle between the corresponding vectors. Using this definition, is the English population closer genetically to the Bantus or to the Koreans? Explain.[8]

**27.** A basketball gymnasium is 25 meters high, 80 meters wide and 200 meters long. For a half-time stunt, the cheerleaders want to run two strings, one from each of the two corners of the gym above one basket to the diagonally opposite corners of the gym floor. What is the angle made by the strings as they cross?

**28.** Let $S$ be the triangle with vertices $A = (2, 2, 2)$, $B = (4, 2, 1)$, and $C = (2, 3, 1)$.

**(a)** Find the length of the shortest side of $S$.
**(b)** Find the cosine of the angle $BAC$ at vertex $A$.

**29.** We can represent a rectangle as a 4-vector $\vec{r} = (x_1, y_1, x_2, y_2)$, where $(x_1, y_1)$ and $(x_2, y_2)$ are opposite corners. (See Problem 22 on page 516.) Assume that $x_2 > x_1$ and $y_2 > y_1$. Let $\vec{w} = (-1, 0, 1, 0)$ and $\vec{h} = (0, -1, 0, 1)$. What do the following quantities tell you about the rectangle?

**(a)** $\vec{r} \cdot \vec{w}$ **(b)** $\vec{r} \cdot \vec{h}$
**(c)** $2\vec{r} \cdot (\vec{w} + \vec{h})$

[8] Adapted from L. Cavalli-Sforza and A. Edwards, "Phylogenetic Analysis: Models and Estimation Procedures," *Am. J. Hum. Genet.*, Vol. 19 (1967), pp. 223-57.

## 12.5 MATRICES

Table 12.2 shows the latest census data (rounded, in 1000s) by age group for the six New England states.[9]

**Table 12.2** *Year 2010 population (1000s) by age group for the six New England states*

| | CT | ME | MA | NH | RI | VT |
|---|---|---|---|---|---|---|
| under 15 | 665 | 223 | 1158 | 232 | 182 | 104 |
| 15–24 | 479 | 168 | 938 | 178 | 162 | 90 |
| 25–34 | 420 | 145 | 845 | 144 | 127 | 70 |
| 35–44 | 484 | 171 | 887 | 179 | 137 | 78 |
| 45–54 | 576 | 219 | 1012 | 226 | 162 | 103 |
| 55–64 | 443 | 192 | 803 | 178 | 131 | 90 |
| 65–74 | 255 | 113 | 456 | 97 | 74 | 50 |
| over 74 | 252 | 89 | 429 | 82 | 78 | 42 |

We can treat the array of numbers in the table as a mathematical object in its own right, independent of the row and column headers. This rectangular grid of numbers is called a *matrix*,[10] usually written inside parentheses. Since the numbers in the table are populations, we use **P** to denote this matrix:

$$\mathbf{P} = \begin{pmatrix} 665 & 223 & 1158 & 232 & 182 & 104 \\ 479 & 168 & 938 & 178 & 162 & 90 \\ 420 & 145 & 845 & 144 & 127 & 70 \\ 484 & 171 & 887 & 179 & 137 & 78 \\ 576 & 219 & 1012 & 226 & 162 & 103 \\ 443 & 192 & 803 & 178 & 131 & 90 \\ 255 & 113 & 456 & 97 & 74 & 50 \\ 252 & 89 & 429 & 82 & 78 & 42 \end{pmatrix}.$$

The individual entries are called *entries* in the matrix; we write $p_{ij}$ for the entry in the $i^{\text{th}}$ row of the $j^{\text{th}}$ column. Thus for example, $p_{21} = 479$ and $p_{16} = 104$.

### Addition, Subtraction, and Scalar Multiplication

Like vectors, matrices provide a convenient way of organizing information. And, as with vectors, we can manipulate matrices algebraically by adding them, subtracting them, or multiplying them by scalars.

As for vectors, to multiply a matrix by a scalar, we multiply each entry by the scalar. To add or subtract two matrices, we add or subtract the corresponding entries.[11]

**Example 1** Evaluate the following matrices. What do they tell you about the population of New England?

(a) Evaluate $1.1\mathbf{P}$.

(b) Evaluate $1.1\mathbf{P} - \mathbf{P}$.

(c) Show that $1.1\mathbf{P} - \mathbf{P} = 0.1\mathbf{P}$.

[9] The US Census Bureau, www.census.gov.

[10] In general, matrices can include objects other than numbers, such as complex numbers, functions, or even operators from calculus like $\partial/\partial x$.

[11] We can add or subtract two matrices provided that they have the same number of rows and columns.

Solution (a) To multiply a matrix by a scalar, we multiply each entry by the scalar. For instance, the first (upper-left) entry of **P** is 665, so the first entry of 1.1**P** is 1.1(665) = 731.5. This matrix tells us what the population for each age group would be for each state after a 10% increase:

$$1.1\mathbf{P} = \begin{pmatrix} 731.5 & 245.3 & 1273.8 & 255.2 & 200.2 & 114.4 \\ 526.9 & 184.8 & 1031.8 & 195.8 & 178.2 & 99.0 \\ 462.0 & 159.5 & 929.5 & 158.4 & 139.7 & 77.0 \\ 532.4 & 188.1 & 975.7 & 196.9 & 150.7 & 85.8 \\ 633.6 & 240.9 & 1113.2 & 248.6 & 178.2 & 113.3 \\ 487.3 & 211.2 & 883.3 & 195.8 & 144.1 & 99.0 \\ 280.5 & 124.3 & 501.6 & 106.7 & 81.4 & 55.0 \\ 277.2 & 97.9 & 471.9 & 90.2 & 85.8 & 46.2 \end{pmatrix}.$$

(b) To subtract one matrix from another, we subtract the corresponding entries. For instance, the first entry of 1.1**P** is 731.5, and the first entry of **P** is 665, and so the first entry of 1.1**P** − **P** = 731.5 − 665 = 66.5. This matrix tells us how much the population in part (a) increased for each age group for each state:

$$1.1\mathbf{P} - \mathbf{P} = \begin{pmatrix} 66.5 & 22.3 & 24.4 & 23.2 & 18.2 & 10.4 \\ 47.9 & 16.8 & 14.5 & 17.8 & 16.2 & 9.0 \\ 42.0 & 14.5 & 14.7 & 14.4 & 12.7 & 7.0 \\ 48.4 & 17.1 & 21.4 & 17.9 & 13.7 & 7.8 \\ 57.6 & 21.9 & 19.4 & 22.6 & 16.2 & 10.3 \\ 44.3 & 19.2 & 11.8 & 17.8 & 13.1 & 9.0 \\ 25.5 & 11.3 & 9.5 & 9.7 & 7.4 & 5.0 \\ 25.2 & 8.9 & 8.3 & 8.2 & 7.8 & 4.2 \end{pmatrix}.$$

(c) In part (b), we found the matrix 1.1**P** − **P**. We now find 0.1**P** by multiplying every entry of **P** by 0.1. Then we can

---

If **A** is a matrix having $m$ rows and $n$ columns, we say that **A** is an $m \times n$ matrix, and we let $a_{ij}$ stand for the entry at row $i$, column $j$. If $m = 1$ or $n = 1$, the matrix has only one row or one column, and is a vector. A $1 \times n$ matrix is often called a *row* vector and a $m \times 1$ matrix is often called a *column* vector.

> If **A** and **B** are $m \times n$ matrices and $k$ is a constant:
> - **C** = $k$**A** is an $m \times n$ matrix such that $c_{ij} = ka_{ij}$. This is called *scalar multiplication* of a matrix.
> - **C** = **A** + **B** is an $m \times n$ matrix such that $c_{ij} = a_{ij} + b_{ij}$. This is called *matrix addition.*
> - **C** = **A** − **B** is an $m \times n$ matrix such that $c_{ij} = a_{ij} - b_{ij}$. This is called *matrix subtraction.*

As with vector subtraction, matrix subtraction can be defined in terms of matrix addition by rewriting **A** + (−1)**B** as **A** − **B**.

## Properties of Scalar Multiplication and Matrix Addition

The properties of scalar multiplication and matrix addition are similar to the properties for vectors.

- *Commutativity of addition*: $\mathbf{A}+\mathbf{B}=\mathbf{B}+\mathbf{A}$.
- *Associativity of addition*: $(\mathbf{A}+\mathbf{B})+\mathbf{C}=\mathbf{A}+(\mathbf{B}+\mathbf{C})$.
- *Associativity of scalar multiplication*: $k_1(k_2\mathbf{A})=(k_1k_2)\mathbf{A}$.
- *Distributivity of scalar multiplication*: $(k_1+k_2)\mathbf{A}=k_1\mathbf{A}+k_2\mathbf{A}$ and $k(\mathbf{A}+\mathbf{B})=k\mathbf{A}+k\mathbf{B}$.

**Example 2** Use the following matrices to illustrate the properties of matrix addition.

$$\mathbf{A}=\begin{pmatrix}2&0\\-3&5\end{pmatrix}\quad \mathbf{B}=\begin{pmatrix}-1&4\\3&1\end{pmatrix}\quad\text{and}\quad \mathbf{C}=\begin{pmatrix}3&1\\4&5\end{pmatrix}.$$

**Solution** (a) $\mathbf{A}+\mathbf{B}=\mathbf{B}+\mathbf{A}$:

$$\mathbf{A}+\mathbf{B}=\begin{pmatrix}2&0\\-3&5\end{pmatrix}+\begin{pmatrix}-1&4\\3&1\end{pmatrix}=\begin{pmatrix}2-1&0+4\\-3+3&5+1\end{pmatrix}=\begin{pmatrix}1&4\\0&5\end{pmatrix}.$$

$$\mathbf{B}+\mathbf{A}=\begin{pmatrix}-1&4\\3&1\end{pmatrix}+\begin{pmatrix}2&0\\-3&5\end{pmatrix}=\begin{pmatrix}-1+2&4+0\\3-3&1+5\end{pmatrix}=\begin{pmatrix}1&4\\0&5\end{pmatrix}.$$

(b) $(\mathbf{A}+\mathbf{B})+\mathbf{C}=\mathbf{A}+(\mathbf{B}+\mathbf{C})$:

$$(\mathbf{A}+\mathbf{B})+\mathbf{C}=\left[\begin{pmatrix}2&0\\-3&5\end{pmatrix}+\begin{pmatrix}-1&4\\3&1\end{pmatrix}\right]+\begin{pmatrix}3&1\\4&5\end{pmatrix}$$
$$=\begin{pmatrix}1&4\\0&6\end{pmatrix}+\begin{pmatrix}3&1\\4&5\end{pmatrix}=\begin{pmatrix}4&5\\4&11\end{pmatrix}.$$

$$\mathbf{A}+(\mathbf{B}+\mathbf{C})=\begin{pmatrix}2&0\\-3&5\end{pmatrix}+\left[\begin{pmatrix}-1&4\\3&1\end{pmatrix}+\begin{pmatrix}3&1\\4&5\end{pmatrix}\right]$$
$$=\begin{pmatrix}2&0\\-3&5\end{pmatrix}+\begin{pmatrix}2&5\\7&6\end{pmatrix}=\begin{pmatrix}4&5\\4&11\end{pmatrix}.$$

**Example 3** Use the matrices in Example 2 and scalars $k=5$, $k_1=2$, and $k_2=-3$, to illustrate the properties of scalar multiplication of matrices.

**Solution** (a) $k_1(k_2\mathbf{A})=(k_1k_2\mathbf{A})$:

$$k_1(k_2\mathbf{A})=2\left[-3\begin{pmatrix}2&0\\-3&5\end{pmatrix}\right]=2\begin{pmatrix}-6&0\\9&-15\end{pmatrix}=\begin{pmatrix}-12&0\\18&-30\end{pmatrix}.$$

$$(k_1k_2)\mathbf{A}=(2\cdot-3)\begin{pmatrix}2&0\\-3&5\end{pmatrix}=-6\begin{pmatrix}2&0\\-3&5\end{pmatrix}=\begin{pmatrix}-12&0\\18&-30\end{pmatrix}.$$

(b) $(k_1 + k_2)\mathbf{A} = k_1\mathbf{A} + k_2\mathbf{A}$:

$$(k_1 + k_2)\mathbf{A} = -1\begin{pmatrix} 2 & 0 \\ -3 & 5 \end{pmatrix} = \begin{pmatrix} -2 & 0 \\ 3 & -5 \end{pmatrix}.$$

$$k_1\mathbf{A} + k_2\mathbf{A} = 2\begin{pmatrix} 2 & 0 \\ -3 & 5 \end{pmatrix} + (-3)\begin{pmatrix} 2 & 0 \\ -3 & 5 \end{pmatrix}$$

$$= \begin{pmatrix} 4 & 0 \\ -6 & 10 \end{pmatrix} + \begin{pmatrix} -6 & 0 \\ 9 & -15 \end{pmatrix} = \begin{pmatrix} -2 & 0 \\ 3 & -5 \end{pmatrix}.$$

(c) $k(\mathbf{A} + \mathbf{B}) = k\mathbf{A} + k\mathbf{B}$:

$$k(\mathbf{A} + \mathbf{B}) = 5\left[\begin{pmatrix} 2 & 0 \\ -3 & 5 \end{pmatrix} + \begin{pmatrix} -1 & 4 \\ 3 & 1 \end{pmatrix}\right] = 5\begin{pmatrix} 1 & 4 \\ 0 & 6 \end{pmatrix} = \begin{pmatrix} 5 & 20 \\ 0 & 30 \end{pmatrix}.$$

$$k\mathbf{A} + k\mathbf{B} = 5\begin{pmatrix} 2 & 0 \\ -3 & 5 \end{pmatrix} + 5\begin{pmatrix} -1 & 4 \\ 3 & 1 \end{pmatrix} = \begin{pmatrix} 10 & 0 \\ -15 & 25 \end{pmatrix} + \begin{pmatrix} -5 & 20 \\ 15 & 5 \end{pmatrix} = \begin{pmatrix} 5 & 20 \\ 0 & 30 \end{pmatrix}.$$

## Multiplication of a Matrix and a Vector

Although there are more general cases of matrix multiplication, in this text we focus on multiplying a $n \times n$ square matrix by an $n$-dimensional column vector. We introduce this topic by considering two dependent populations.

### Matrix Multiplication of 2-Dimensional Vectors

A country begins with $x_0 = 4$ million people and another country begins with $y_0 = 2$ million people. Every year, 20% of the people in country $X$ move to country $Y$, and 30% of the people in country $Y$ move to country $X$. Suppose also that no one is born and no one dies. What happens to the populations of these two countries over time? After one year, $x_1$, the number of people (in millions) in country $X$, is given by

$$x_1 = x_0 - \underbrace{0.2x_0}_{\text{20\% leave } X} + \underbrace{0.3y_0}_{\text{30\% leave } Y}$$
$$= 0.8x_0 + 0.3y_0.$$

Likewise, $y_1$, the number of people (in millions) in country $Y$, is given by

$$y_1 = y_0 + \underbrace{0.2x_0}_{\text{20\% leave } X} - \underbrace{0.3y_0}_{\text{30\% leave } Y}$$
$$= 0.2x_0 + 0.7y_0.$$

Thus, we have

$$x_1 = 0.8(4) + 0.3(2) = 3.2 + 0.6 = 3.8$$
$$y_1 = 0.2(4) + 0.7(2) = 0.8 + 1.4 = 2.2.$$

The population of country $X$ drops by 0.2 million, while the population of country $Y$ goes up by 0.2 million. The same reasoning shows that after two years,

$$x_2 = 0.8x_1 + 0.3y_1$$
$$y_2 = 0.2x_1 + 0.7y_1,$$

so

$$x_2 = 0.8(3.8) + 0.3(2.2) = 3.04 + 0.66 = 3.7$$
$$y_2 = 0.2(3.8) + 0.7(2.2) = 0.76 + 1.54 = 2.3.$$

In general, we see that

$$x_{\text{new}} = 0.8x_{\text{old}} + 0.3y_{\text{old}}$$
$$y_{\text{new}} = 0.2x_{\text{old}} + 0.7y_{\text{old}}.$$

Thinking in terms of vectors, we can write $\vec{P}_{\text{new}} = (x_{\text{new}}, y_{\text{new}})$ and $\vec{P}_{\text{old}} = (x_{\text{old}}, y_{\text{old}})$. Then, using the dot product, we have

$$x_{\text{new}} = (0.8, 0.3) \cdot \vec{P}_{\text{old}}$$
$$y_{\text{new}} = (0.2, 0.7) \cdot \vec{P}_{\text{old}}.$$

This pair of equations tells us how the components of $\vec{P}_{\text{new}}$, are related to the components of $\vec{P}_{\text{old}}$, when both are written as a column vector. Notice that each component in $\vec{P}_{\text{new}}$ is a combination of both components of $\vec{P}_{\text{old}}$. Using matrices, we write the pair of equations as a single equation:

$$\vec{P}_{\text{new}} = \begin{pmatrix} x_{\text{new}} \\ y_{\text{new}} \end{pmatrix} = \underbrace{\begin{pmatrix} 0.8 & 0.3 \\ 0.2 & 0.7 \end{pmatrix} \vec{P}_{\text{old}}}_{\text{Matrix multiplication}}.$$

We place the two coefficients from the first equation, $x_{\text{new}} = 0.8x_{\text{old}} + 0.3y_{\text{old}}$, in the first row of the matrix. We place the two coefficients from the second equation, $y_{\text{new}} = 0.2x_{\text{old}} + 0.7y_{\text{old}}$, in the second row of the matrix. In general, we use the following notation:

> If $\vec{P}_{\text{new}} = (x_{\text{new}}, y_{\text{new}})$ and $\vec{P}_{\text{old}} = (x_{\text{old}}, y_{\text{old}})$, then we can write the equations
>
> $$x_{\text{new}} = ax_{\text{old}} + by_{\text{old}}$$
> $$y_{\text{new}} = cx_{\text{old}} + dy_{\text{old}}$$
>
> in the compact form
>
> $$\vec{P}_{\text{new}} = \begin{pmatrix} a & b \\ c & d \end{pmatrix} \vec{P}_{\text{old}}.$$
>
> The coefficients in the first row of the matrix tell us how $x_{\text{new}}$ is related to $\vec{P}_{\text{old}}$ and the coefficients in the second row tell us how $y_{\text{new}}$ is related to $\vec{P}_{\text{old}}$. If we let $\mathbf{A} = \begin{pmatrix} a & b \\ c & d \end{pmatrix}$, then we can write
>
> $$\vec{P}_{\text{new}} = \mathbf{A}\vec{P}_{\text{old}}.$$

**Example 4** What does the following matrix equation tell you about the relationship between $\vec{P}_{\text{new}}$ and $\vec{P}_{\text{old}}$?

$$\vec{P}_{\text{new}} = \begin{pmatrix} 0.9 & 0.4 \\ 0.1 & 0.6 \end{pmatrix} \vec{P}_{\text{old}}.$$

**Solution** We have

$$x_{\text{new}} = (0.9, 0.4) \cdot \vec{P}_{\text{old}} = 0.9x_{\text{old}} + 0.4y_{\text{old}}$$
$$y_{\text{new}} = (0.1, 0.6) \cdot \vec{P}_{\text{old}} = 0.1x_{\text{old}} + 0.6y_{\text{old}}.$$

This tells us that each year, 10% of the population of country $X$ moves to country $Y$ and 40% of the population of country $Y$ moves to country $X$.

## Matrix Multiplication of $n$-Dimensional Vectors

Let $\vec{u} = (u_1, u_2, \ldots, u_n)$ be an $n$-dimensional vector and let $\mathbf{A}$ be an $n \times n$ square matrix with entries $a_{11}, a_{12}, \ldots, a_{nn}$. Let $\vec{v}$ be the $n$-dimensional vector with coordinates given by equations:

$$\begin{array}{rclcl} v_1 & = & (a_{11}, a_{12}, \ldots, a_{1n}) \cdot (u_1, u_2, \ldots, u_n) & = & a_{11}u_1 + \cdots + a_{1n}u_n \\ v_2 & = & (a_{21}, a_{22}, \ldots, a_{2n}) \cdot (u_1, u_2, \ldots, u_n) & = & a_{21}u_1 + \cdots + a_{2n}u_n \\ & \vdots & & \vdots & \\ v_n & = & (a_{n1}, a_{n2}, \ldots, a_{nn}) \cdot (u_1, u_2, \ldots, u_n) & = & a_{n1}u_1 + \cdots + a_{nn}u_n. \end{array}$$

Then, if

$$\mathbf{A} = \begin{pmatrix} a_{11} & \cdots & a_{1n} \\ \vdots & & \vdots \\ a_{n1} & \cdots & a_{nn} \end{pmatrix},$$

we see that $\vec{v}$ is the product of $\mathbf{A}$ and $\vec{u}$:

$$\vec{v} = \mathbf{A}\vec{u}.$$

**Example 5** Let $e$ be the number of employed people in a certain city, and $u$ be the number of unemployed people. We define $\vec{E} = (e, u)$ as the *employment vector* of this city. Suppose that each year, 10% of employed peopled become unemployed, and 20% of unemployed people become employed. Find a matrix $\mathbf{A}$ such that $\vec{E}_{\text{new}} = \mathbf{A}\vec{E}_{\text{old}}$.

**Solution** We have

$$\begin{aligned} e_{\text{new}} &= e_{\text{old}} - \underbrace{0.1e_{\text{old}}}_{\text{10\% become unemployed}} + \underbrace{0.2u_{\text{old}}}_{\text{20\% become employed}} \\ &= 0.9e_{\text{old}} + 0.2u_{\text{old}} \\ u_{\text{new}} &= u_{\text{old}} - \underbrace{0.2u_{\text{old}}}_{\text{20\% become employed}} + \underbrace{0.1e_{\text{old}}}_{\text{10\% become unemployed}} \\ &= 0.1e_{\text{old}} + 0.8u_{\text{old}}. \end{aligned}$$

We can rewrite this pair of equations using matrix notation:

$$\vec{E}_{\text{new}} = \mathbf{A}\vec{E}_{\text{old}} = \begin{pmatrix} 0.9 & 0.2 \\ 0.1 & 0.8 \end{pmatrix} \vec{E}_{\text{old}}.$$

**Example 6** A country produces agricultural products valued at \$70 billion and industrial products valued at \$50 billion. Some of these agricultural products are used up in order to produce other products (for instance, cotton is needed to produce cloth, or soybeans are needed to feed cattle), and some of these industrial products are used up in order to produce other products (for instance, tractors are needed to harvest cotton, or ball bearings are needed to build engines). The production vector of this country is $\vec{P} = (70, 50)$.

(a) Let $\mathbf{C}$ be the consumption matrix defined in the following equation. Calculate $\vec{U}$, the amount of agricultural and industrial products used during production:

$$\vec{U} = \mathbf{C}\vec{P} = \begin{pmatrix} 0.2 & 0.1 \\ 0.3 & 0.2 \end{pmatrix} \vec{P}.$$

Find $\vec{U}$ and describe what it tells you about the country's economy.

(b) Let the surplus vector $\vec{S} = \vec{P} - \mathbf{C}\vec{P}$. Find $\vec{S}$ and describe what it tells you about the country's economy.

**Solution** (a) We have

$$\vec{U} = \begin{pmatrix} 0.2 & 0.1 \\ 0.3 & 0.2 \end{pmatrix} \begin{pmatrix} 70 \\ 50 \end{pmatrix},$$

so

$$U_1 = 0.2(70) + 0.1(50) = 19$$
$$U_2 = 0.3(70) + 0.2(50) = 31.$$

Thus, $\vec{U} = (19, 31)$. This tells us that \$19 billion in agricultural products are used up during production, and \$31 billion in industrial products are used up during production.

(b) We have

$$\begin{aligned} \vec{S} &= \vec{P} - \mathbf{C}\vec{P} \\ &= \vec{P} - \vec{U} && \text{because } \vec{U} = \mathbf{C}\vec{P} \\ &= (70, 50) - (19, 31), && \text{from part (a)} \end{aligned}$$

so $\vec{S} = (51, 19)$. This tells us that, after internal consumption is accounted for, \$51 billion in agricultural products are available for general consumption, as are \$19 billion in industrial products.

---

**Example 7** Let $\vec{P}_0 = (x_0, y_0)$ be a position vector. This vector is rotated about the origin through an angle $\phi$ without changing its length. What is its new position, $\vec{P}_1$ ? Find a matrix $\mathbf{R}$ such that $\vec{P}_1 = \mathbf{R}\vec{P}_0$.

**Solution** Let $r = \|\vec{P}_0\| = \|\vec{P}_1\|$. We see from Figure 12.33 that $x_0 = r\cos\theta$, $y_0 = r\sin\theta$, $x_1 = r\cos(\theta + \phi)$, and $y_1 = r\sin(\theta + \phi)$. Notice that we can rewrite the coordinates of $\vec{P}_1$ as follows:

$$\begin{aligned} x_1 &= r\cos(\theta + \phi) \\ &= r(\cos\theta\cos\phi - \sin\theta\sin\phi) \\ &= (r\cos\theta)\cos\phi - (r\sin\theta)\sin\phi \\ &= x_0\cos\phi - y_0\sin\phi \\ y_1 &= r\sin(\theta + \phi) \\ &= r(\sin\theta\cos\phi + \cos\theta\sin\phi) \\ &= (r\sin\theta)\cos\phi + (r\cos\theta)\sin\phi \\ &= y_0\cos\phi + x_0\sin\phi. \end{aligned}$$

We can rewrite this pair of equations using a matrix $\mathbf{R}$:

$$\vec{P}_1 = \mathbf{R}\vec{P}_0 = \begin{pmatrix} \cos\phi & -\sin\phi \\ \sin\phi & \cos\phi \end{pmatrix} \vec{P}_0.$$

Notice that this equation holds for any original vector, $\vec{P}_0$, and any angle of rotation, $\phi$.

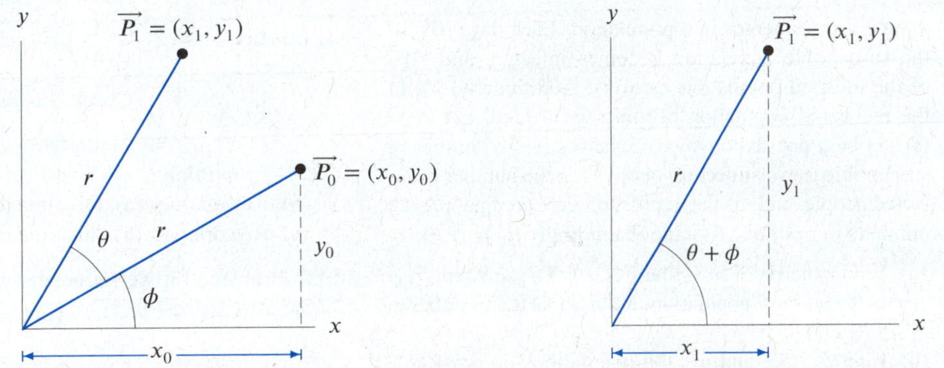


**Figure 12.33**: The vector $\vec{P_0}$ is rotated through an angle $\phi$

## Exercises and Problems for Section 12.5

### Exercises

**1.** Evaluate the following expressions given that:

$$\mathbf{R} = \begin{pmatrix} 3 & 7 \\ 2 & -1 \end{pmatrix} \quad \text{and} \quad \mathbf{S} = \begin{pmatrix} 1 & -5 \\ 0 & 8 \end{pmatrix}.$$

**(a)** $5\mathbf{R}$ **(b)** $-2\mathbf{S}$
**(c)** $\mathbf{R}+\mathbf{S}$ **(d)** $\mathbf{S}-3\mathbf{R}$
**(e)** $\mathbf{R}+2\mathbf{R}+2(\mathbf{R}-\mathbf{S})$ **(f)** $k\mathbf{S}$, $k$ constant

**2.** Evaluate the following expressions given that:

$$\mathbf{A} = \begin{pmatrix} 2 & 5 & 7 \\ 4 & -6 & 3 \\ 16 & -5 & 0 \end{pmatrix} \quad \text{and} \quad \mathbf{B} = \begin{pmatrix} 8 & -6 & 0 \\ 5 & 3 & -2 \\ 3 & 7 & 12 \end{pmatrix}.$$

**(a)** $2\mathbf{A}$ **(b)** $-3\mathbf{B}$ **(c)** $\mathbf{A}+\mathbf{B}$
**(d)** $2\mathbf{A}-3\mathbf{B}$ **(e)** $5(\mathbf{A}+\mathbf{B})$ **(f)** $\mathbf{A}+(\mathbf{A}+(\mathbf{A}+\mathbf{B}))$

**3.** Evaluate the following expressions given that:

$$\mathbf{U} = \begin{pmatrix} 3 & 2 & 5 & 1 \\ 4 & 6 & 7 & 3 \\ 1 & 9 & 5 & 8 \\ 0 & -2 & 4 & 6 \end{pmatrix} \text{ and } \mathbf{V} = \begin{pmatrix} 1 & 6 & 4 & 2 \\ 3 & 5 & -1 & 7 \\ 9 & 4 & 7 & 3 \\ 2 & 8 & 4 & 5 \end{pmatrix}.$$

**(a)** $4\mathbf{U}$ **(b)** $-2\mathbf{V}$
**(c)** $\mathbf{U}-\mathbf{V}$ **(d)** $3\mathbf{U}-3\mathbf{V}$
**(e)** $\mathbf{U}+\mathbf{U}-(\mathbf{U}-\mathbf{V})$ **(f)** $2(2\mathbf{U}-\mathbf{V})$

**4.** Let **R** and **S** be the matrices from Problem 1, and let $\vec{p} = (3, 1)$ and $\vec{q} = (-1, 5)$ be vectors. Evaluate the following expressions.

**(a)** $\mathbf{R}\vec{p}$ **(b)** $\mathbf{S}\vec{q}$ **(c)** $\mathbf{S}(\vec{q}+\vec{p})$
**(d)** $(\mathbf{R}+\mathbf{S})\vec{p}$ **(e)** $\mathbf{R}\vec{p}\cdot\mathbf{S}\vec{q}$ **(f)** $(\vec{p}\cdot\vec{q})\mathbf{S}$

**5.** Let **A** and **B** be the matrices defined in Problem 2, and let $\vec{u} = (3, 2, 5)$ and $\vec{v} = (-1, 0, 3)$ be vectors. Evaluate the following expressions.

**(a)** $\mathbf{A}\vec{u}$ **(b)** $\mathbf{B}\vec{v}$ **(c)** $\mathbf{A}(\vec{u}+\vec{v})$
**(d)** $(\mathbf{A}+\mathbf{B})\vec{v}$ **(e)** $\mathbf{A}\vec{u}\cdot\mathbf{B}\vec{v}$ **(f)** $(\vec{u}\cdot\vec{v})\mathbf{A}$

**6.** Let **U** and **V** be the matrices defined in Problem 3, and let $\vec{s} = (2, 0, -1, 7)$ and $\vec{t} = (4, 5, 1, -1)$ be vectors. Evaluate the following expressions.

**(a)** $\mathbf{U}\vec{t}$ **(b)** $\mathbf{V}\vec{s}$
**(c)** $\mathbf{U}\vec{t}\cdot\mathbf{U}\vec{s}$ **(d)** $\mathbf{U}(\vec{s}-\vec{t})$
**(e)** $(\mathbf{U}+\mathbf{V})(\vec{s}+\vec{t})$ **(f)** $(\vec{s}\cdot\vec{t})(\mathbf{U}+\mathbf{V})$

### Problems

**7.** Let **A** and **B** be the matrices given by

$$\mathbf{A} = \begin{pmatrix} 3 & 7 \\ 2 & -1 \end{pmatrix} \quad \text{and} \quad \mathbf{B} = \begin{pmatrix} 1 & -5 & 0 \\ 0 & -1 & 1 \\ 1 & 0 & 1 \end{pmatrix}$$

and $\vec{u} = (1, 1)$ and $\vec{v} = (1, 0, 3)$ be vectors. Which of the following are defined?

**(a)** $\mathbf{A}\vec{u}$ **(b)** $\mathbf{B}\vec{u}$
**(c)** $\mathbf{A}(\vec{u}+\vec{v})$ **(d)** $\mathbf{A}+\vec{u}$
**(e)** $(\mathbf{A}\vec{u})/(\vec{v}\cdot\vec{v})$ **(f)** $(\vec{u}\cdot\vec{v})\mathbf{A}$

**8.** Let

$$\mathbf{A} = \begin{pmatrix} 1 & 2 \\ 2 & -1 \end{pmatrix} \quad \text{and} \quad \vec{u} = (5, -5).$$

**(a)** Find $\vec{w}$ such that $\vec{w} = \mathbf{A}\vec{u}$.
**(b)** Find $\vec{w}$ such that $\mathbf{A}\vec{w} = \vec{u}$.

**9.** A new virus emerges in a population. Each day, 10% of the susceptible population becomes infected, and 50% of the infected population recovers. Additionally, 2% of the recovered population becomes reinfected. Let $\vec{p} = (s, i, r)$ be a population vector where $s$ is the number of susceptible (never-infected) people, $i$ is the number of infected people, and $r$ is the number of recovered people, all numbers in millions. Assume that initially $\vec{p}_0 = (2, 0, 0)$.

**(a)** Find a matrix $\mathbf{T}$ such that $\vec{p}_{\text{new}} = \mathbf{T}\vec{p}_{\text{old}}$ where $\vec{p}_{\text{old}}$ is the starting population and $\vec{p}_{\text{new}}$ is the population a day later.

**(b)** Find $\vec{p}_1$, $\vec{p}_2$, and $\vec{p}_3$, the populations on days 1, 2, and 3, respectively.

**10.** A certain species of insect has a three-year life cycle. The vector $\vec{P} = (f, s, t)$ gives the number of insects in their first, second, and third years of life, respectively. Each year, insects of all three ages lay eggs, and so the value of $f_{\text{new}}$ depends on the values of $f_{\text{old}}$, $s_{\text{old}}$, and $t_{\text{old}}$. Only some first-year insects survive until their second year, and so $s_{\text{new}}$ depends only on $f_{\text{old}}$; likewise, $t_{\text{new}}$ depends only on $s_{\text{old}}$.

**(a)** Suppose $\vec{P}_{\text{new}} = \mathbf{T}\vec{P}_{\text{old}} = \begin{pmatrix} 0.3 & 0.6 & 0.5 \\ 0.7 & 0 & 0 \\ 0 & 0.4 & 0 \end{pmatrix} \vec{P}_{\text{old}}$.

Describe in words what this tells you about the life cycle of this insect. Be specific.

**(b)** Let $\vec{P}_0 = (2000, 0, 0)$ be the initial population vector in year $t = 0$. Evaluate $\vec{P}_1$, $\vec{P}_2$, and $\vec{P}_3$.

**11.** Let $\vec{p}_{2010} = (A_{2010}, B_{2010}) = (200, 400)$ describe the populations (in 1000s) of two cities in the census year 2010. Each year, 3% of the population of city $A$ moves to the city $B$, and 5% of the population of the city $B$ moves to city $A$.

**(a)** Find a matrix $\mathbf{T}$ such that $\vec{p}_{\text{new}} = \mathbf{T}\vec{p}_{\text{old}}$, where $\vec{p}_{\text{old}}$ is the starting population and $\vec{p}_{\text{new}}$ is the population 1 year later.

**(b)** Find $\vec{p}_{2011}$ and $\vec{p}_{2012}$, the populations in 2011 and 2012, respectively.

**12.** Let $\mathbf{I}_2 = \begin{pmatrix} 1 & 0 \\ 0 & 1 \end{pmatrix}$ and $\mathbf{I}_3 = \begin{pmatrix} 1 & 0 & 0 \\ 0 & 1 & 0 \\ 0 & 0 & 1 \end{pmatrix}$.

**(a)** Evaluate $\mathbf{I}_2\vec{u}$ for (i) $\vec{u} = (3, 2)$, (ii) $\vec{u} = (0, 7)$, and (iii) $\vec{u} = (a, b)$.

**(b)** Evaluate $\mathbf{I}_3\vec{v}$ for (i) $\vec{v} = (-1, 5, 7)$, (ii) $\vec{v} = (3, 8, 1)$, and (iii) $\vec{v} = (a, b, c)$.

**(c)** Find a matrix $\mathbf{I}_4$ such that $\mathbf{I}_4\vec{w} = \vec{w}$ where $\vec{w}$ is any 4-vector.

**13.** In an algebraic equation such as $y = kx$, $k \neq 0$, we can multiply through by $k^{-1} = 1/k$ in order to solve for $x$, giving $x = k^{-1}y$. We can do a similar thing with matrices: if $\vec{v} = \mathbf{A}\vec{u}$, it is sometimes possible to find an *inverse* of $\mathbf{A}$, written $\mathbf{A}^{-1}$, such that $\vec{u} = \mathbf{A}^{-1}\vec{v}$.

**(a)** Let $\mathbf{A} = \begin{pmatrix} 2 & 1 \\ 3 & 2 \end{pmatrix}$. It can be shown that $\mathbf{A}^{-1} = \begin{pmatrix} 2 & -1 \\ -3 & 2 \end{pmatrix}$. Check this fact using the vector $\vec{u} = (3, 5)$, first by multiplying $\vec{u}$ by $\mathbf{A}$ to get $\vec{v}$, and then by multiplying $\vec{v}$ by $\mathbf{A}^{-1}$ to show that we get $\vec{u}$ back.

**(b)** Rework part (a), this time for $\vec{u} = (-1, 7)$.

**(c)** Rework part (a), this time for $\vec{u} = (a, b)$.

**14.** In Problem 13, we introduced the notion of the inverse of a matrix $\mathbf{A}$, written $\mathbf{A}^{-1}$. If $\mathbf{A} = \begin{pmatrix} a & b \\ c & d \end{pmatrix}$, then $\mathbf{A}^{-1} = \frac{1}{D}\begin{pmatrix} d & -b \\ -c & a \end{pmatrix}$ where $D = ad - bc$. Notice that $\mathbf{A}^{-1}$ is undefined if $D = 0$.

**(a)** Use this formula to show that $\mathbf{A}^{-1}$ from Problem 13 is $\begin{pmatrix} 2 & -1 \\ -3 & 2 \end{pmatrix}$.

**(b)** Let $\mathbf{B} = \begin{pmatrix} 3 & 11 \\ 1 & 7 \end{pmatrix}$. Find $\mathbf{B}^{-1}$. Verify that if $\vec{v} = \mathbf{B}\vec{u}$ then $\vec{u} = \mathbf{B}^{-1}\vec{v}$ for $\vec{u} = (a, b)$.

**(c)** Show that $\mathbf{C} = \begin{pmatrix} 2 & 8 \\ 3 & 12 \end{pmatrix}$ does not have an inverse.

**15.** Multiplying a vector by a matrix is a much more complicated process than multiplying a vector by a scalar. However, in certain special cases, matrix and scalar multiplication can lead to the same result. If $\mathbf{A}\vec{v} = \lambda\vec{v}$ where $\mathbf{A}$ is a matrix, $\vec{v}$ a nonzero vector, and $\lambda$ a scalar, then $\vec{v}$ is said to be an *eigenvector* of $\mathbf{A}$, and $\lambda$ is said to be an *eigenvalue* of $\mathbf{A}$. For instance, let $\mathbf{A} = \begin{pmatrix} -2 & -1 \\ 8 & 7 \end{pmatrix}$, $\vec{v}_1 = (1, -8)$, and $\lambda_1 = 6$. We have:

$$\mathbf{A}\vec{v}_1 = \begin{pmatrix} -2 & -1 \\ 8 & 7 \end{pmatrix} \cdot \begin{pmatrix} 1 \\ -8 \end{pmatrix} = \begin{pmatrix} 6 \\ -48 \end{pmatrix}$$

$$\lambda_1\vec{v}_1 = 6 \cdot \begin{pmatrix} 1 \\ -8 \end{pmatrix} = \begin{pmatrix} 6 \\ -48 \end{pmatrix}.$$

We see that multiplying $\vec{v}_1$ by $\lambda_1 = 6$ works out the same as multiplying $\vec{v}_1$ by $\mathbf{A}$; we say that $\vec{v}_1$ is an *eigenvector* of $\mathbf{A}$ with an *eigenvalue* of $\lambda_1 = 6$.

**(a)** Show that $\vec{v}_2 = (1, -1)$ is an eigenvector of $\mathbf{A}$. What is the eigenvalue?

**(b)** Show that $\vec{v}_3 = (-3, 3)$ is an eigenvector of $\mathbf{A}$. What is the eigenvalue?

**(c)** If $\vec{v}$ is an eigenvector of $\mathbf{A}$, explain why the vectors $\vec{v}$ and $\mathbf{A}\vec{v}$ are parallel.

**16.** In this problem, we will consider a new way to think about matrix multiplication of vectors. A vector such as $\vec{v} = (9, 8)$ can be written as a combination of unit vectors, like this:

$$\vec{v} = (9, 8) = 9\vec{i} + 8\vec{j} = 9\underbrace{(1, 0)}_{\vec{i}} + 8\underbrace{(0, 1)}_{\vec{j}}.$$

But we can also write $\vec{v}$ as a combination[12] of other vectors besides $\vec{i}$ and $\vec{j}$. For instance, if $\vec{c}_1 = (1,2)$, and $\vec{c}_2 = (3,1)$, notice that

$$\vec{v} = (9,8) = 3\vec{c}_1 + 2\vec{c}_2 = 3\underbrace{(1,2)}_{\vec{c}_1} + 2\underbrace{(3,1)}_{\vec{c}_2}.$$

This is where matrix multiplication comes in: if we think of $\vec{c}_1$ and $\vec{c}_2$ as the columns of a matrix $\mathbf{C}$, we see that

$$\begin{pmatrix} 9 \\ 8 \end{pmatrix} = \Bigg(\underbrace{\boxed{\begin{matrix} 1 \\ 2 \end{matrix}}}_{\vec{c}_1}\ \underbrace{\boxed{\begin{matrix} 3 \\ 1 \end{matrix}}}_{\vec{c}_2}\Bigg)\begin{pmatrix} 3 \\ 2 \end{pmatrix}$$

is the same as

$$\begin{pmatrix} 9 \\ 8 \end{pmatrix} = 3\underbrace{\begin{pmatrix} 1 \\ 2 \end{pmatrix}}_{\vec{c}_1} + 2\underbrace{\begin{pmatrix} 3 \\ 1 \end{pmatrix}}_{\vec{c}_2}.$$

In general, if $\vec{v} = \mathbf{C}\vec{u}$, then the coordinates of $\vec{u}$ tell us how to combine the columns of $\mathbf{C}$ in order to get $\vec{v}$.

**(a)** Let $\vec{v} = 3\vec{r}_1 + 5\vec{r}_2$ where $r_1 = (3,2)$ and $r_2 = (0,1)$. Find $\vec{v}$, $\vec{u}$, and $\mathbf{R}$ such that $\vec{v} = \mathbf{R}\vec{u}$.

**(b)** Let $\vec{q} = \mathbf{S}\vec{p} = \begin{pmatrix} 3 & 4 \\ 2 & 3 \end{pmatrix}\begin{pmatrix} 3 \\ -2 \end{pmatrix}$. Show that $\vec{q}$ can be written as a combination of $\vec{s}_1 = (3,2)$ and $\vec{s}_2 = (4,3)$, the two columns of $\mathbf{S}$.

**17.** Let $\vec{v} = (2,5)$, $\vec{c}_1 = (3,2)$, and $\vec{c}_2 = (5,4)$. In this problem we will use the results of Problem 16 to write $\vec{v}$ as a combination of $\vec{c}_1$ and $\vec{c}_2$ given by $\vec{v} = a\vec{c}_1 + b\vec{c}_2$.

**(a)** Let $\mathbf{C}$ be a matrix whose columns are given by $\vec{c}_1$ and $\vec{c}_2$, and let $\vec{u} = (a,b)$. Show that $\mathbf{C}\vec{u} = a\vec{c}_1 + b\vec{c}_2$.

**(b)** Let $\vec{v} = \mathbf{C}\vec{u}$ where $\vec{v} = (2,5)$. Referring to Problem 13 and Problem 14, solve for $\vec{u}$.

**(c)** Show that $\vec{v}$ is a combination of $\vec{c}_1$ and $\vec{c}_2$ given by $\vec{v} = a\vec{c}_1 + b\vec{c}_2$.

**18.** Following the procedure outlined in Problem 17, write $\vec{v} = (5,8)$ as a combination of $\vec{c}_1 = (6,1)$ and $\vec{c}_2 = (7,2)$ given by $\vec{v} = a\vec{c}_1 + b\vec{c}_2$.

## CHAPTER SUMMARY

- **Vectors**
  Displacement; vector notation; components.
  Length: $\|\vec{v}\| = \sqrt{v_1^2 + v_2^2 + v_3^2}$; unit vectors.
- **Addition, Subtraction, and Scalar Multiplication of Vectors**
- **Dot Product**
  $\vec{v} \cdot \vec{w} = v_1w_1 + v_2w_2 + v_3w_3 = \|\vec{v}\|\|\vec{w}\|\cos\theta.$
- **Applications**
  Economics; computer graphics; work.
- **Matrices**
  Addition; subtraction; multiplication by scalars, by vectors.

## REVIEW EXERCISES AND PROBLEMS FOR CHAPTER TWELVE

### Exercises

Calculate the vectors in Exercises 1–8, using $\vec{a} = (5,1,0)$, $\vec{b} = (2,-1,9)$, $\vec{c} = (1,1,2)$.

**1.** $3\vec{c}$

**2.** $\vec{a} + \vec{b}$

**3.** $\vec{b} - \vec{a}$

**4.** $\vec{a} + 2(\vec{b} + \vec{c})$

**5.** $2\vec{a} - 3(\vec{b} - \vec{c})$

**6.** $2(\vec{b} + 4(\vec{a} + \vec{c}))$

**7.** $\vec{a} + \vec{b} - (\vec{a} - \vec{b})$

**8.** $4(\vec{c} + 2(\vec{a} - \vec{c}) - 2\vec{a})$

Resolve the vectors in Exercises 9–10 into components.

**9.**

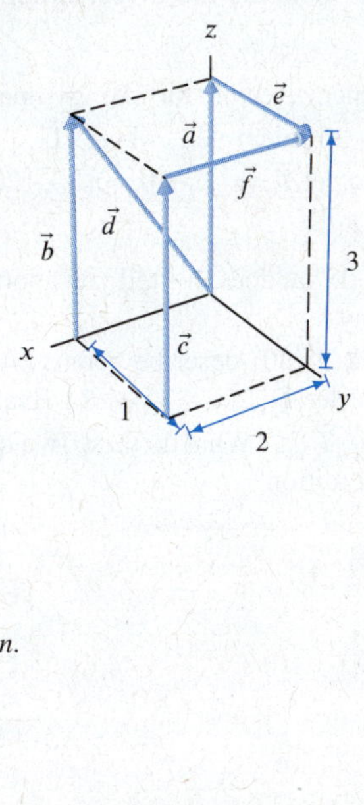


**10.**

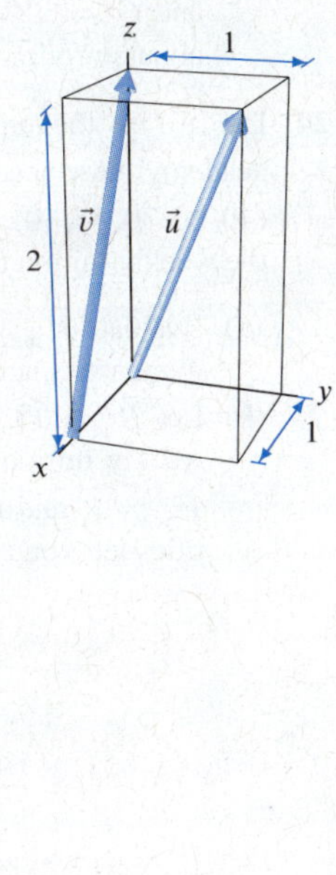


[12]This sort of combination is called a *linear combination*.

For Exercises 11–18, perform the indicated computations.

**11.** $-4(\vec{i} - 2\vec{j}) - 0.5(\vec{i} - \vec{k})$

**12.** $(\vec{i} + 2\vec{j}) + (-3)(2\vec{i} + \vec{j})$

**13.** $(3\vec{i} + \vec{j}) \cdot (5\vec{i} - 2\vec{j})$

**14.** $(3\vec{j} - 2\vec{k} + \vec{i}) \cdot (4\vec{k} - 2\vec{i} + 3\vec{j})$

**15.** $(5\vec{i} - \vec{j} - 3\vec{k}) \cdot (2\vec{i} + \vec{j} + \vec{k})$

**16.** $(2\vec{i} \cdot 5\vec{j})(\vec{i} + \vec{j} + \vec{k})$

**17.** $(2\vec{i} + 5\vec{j}) \cdot 3\vec{i}(\vec{i} + \vec{j} + \vec{k})$

**18.** $(\vec{i} + \vec{j} + \vec{k})(2\vec{i} + 3\vec{j} + \vec{k}) \cdot (3\vec{i} + \vec{j} + 4\vec{k})$

**19.** Find the length of the vectors $\vec{u}$ and $\vec{v}$ in Problem 10.

**20.** Find a vector of length 2 that points in the same direction as $\vec{i} - \vec{j} + 2\vec{k}$.

## Problems

**21.** **(a)** Are the vectors $4\vec{i} + a\vec{j} + 6\vec{k}$ and $a\vec{i} + (a-1)\vec{j} + 3\vec{k}$ parallel for any values of the constant $a$?
**(b)** Are these vectors ever perpendicular?

**22.** A point $P$ is on the rim of a moving bicycle wheel of radius 1 ft. The bicycle is moving forward at $6\pi$ ft/sec.

**(a)** Sketch the velocity of $P$ relative to the wheel's axle at the 3, 6, 9, 12 o'clock positions.
**(b)** Sketch the velocity of the axle relative to the ground.
**(c)** Sketch the velocity of $P$ relative to the ground at the positions in part (a).
**(d)** Does $P$ ever stop, relative to the ground? What is the greatest speed that $P$ moves, relative to the ground?

**23.** A gymnastics academy offers classes to children of different ages. Software used by the academy tracks enrollment data using vectors. Let $\vec{E} = (e_1, e_2, e_3, e_4, e_5)$ represent the number of children enrolled at each of five different experience levels, from beginner (1) to advanced (5), and let $\vec{E}_{\text{max}} = (40, 40, 30, 15, 10)$ be the maximum allowed enrollment at each level.

**(a)** Let $\vec{E} = (51, 47, 41, 22, 23)$ give the number of applicants at each level for the next session. After full enrollment is reached, the remaining applicants are placed on a wait list, $\vec{L}$. Find $\vec{L}$.
**(b)** At the start of the next session, $\vec{E} = \vec{E}_{\text{max}}$. Let $\vec{D} = (8, 4, 9, 7, 6)$ be the number of dropouts after the first week of the next session. Find $\vec{F} = \vec{E} - \vec{D}$ and $\vec{G} = \vec{L} - \vec{D}$. What do these vectors tell you about enrollment?

**24.** Let $\vec{E}$ be the enrollment vector for the gymnastics academy described in Problem 23, where $\vec{E}_{\text{max}} = (40, 40, 30, 15, 10)$. Also, let $\vec{T} = (30, 30, 40, 80, 120)$ be the weekly tuition (in \$).

**(a)** Evaluate $\vec{E}_{\text{max}} \cdot \vec{T}$. What does this tell you about the gymnastics academy?
**(b)** Let $\vec{R} = (5, 5, 10, 20, 30)$ describe a rate hike in weekly tuition, and let $\vec{T}_{\text{new}} = \vec{T} + \vec{R}$. Evaluate $\vec{E}_{\text{max}} \cdot \vec{R}$ and $\vec{E}_{\text{max}} \cdot \vec{T}_{\text{new}}$. What do these two quantities tell you about tuition?

In Problems 25–27, a 5-pound block sits on a plank of wood. If one end of the plank is raised, the block slides down the plank. However, friction between the block and the plank prevents it from sliding until the plank has been raised a certain height. It turns out that the *sliding force* exerted by gravity on the block is proportional to the sine of the angle made by the plank with the ground.

**25.** Find a formula for $F = f(\theta)$, the sliding force (in lbs) exerted on a block if the plank makes an angle of $\theta$ with the ground. [Hint: What is sliding force if the plank is horizontal? Vertical?]

**26.** One end of the plank is lifted at a constant rate of 2 ft per second, while the other end rests on the ground.

**(a)** Find a formula for $F = h(t)$, the sliding force exerted on the block as a function of time.
**(b)** Suppose the block begins to slide once the sliding force equals 3 lbs. At what time will the block begin to slide?

**27.** The 5-lb force exerted on the block by gravity can be resolved into two components, the sliding force $\vec{F}_S$ parallel to the ramp and the normal force $\vec{F}_N$ perpendicular to the ramp. Use this information to show that your formula in Problem 25 is correct.

**28.** A plane is heading due east and climbing at the rate of 80 km/hr. If its airspeed is 480 km/hr and there is a wind blowing 100 km/hr to the northeast, what is the ground speed of the plane?

**29.** A particle moving with speed $v$ hits a barrier at an angle of 60° and bounces off at an angle of 60° in the opposite direction with speed reduced by 20 percent, as shown in Figure 12.34. Find the velocity vector of the object after impact.

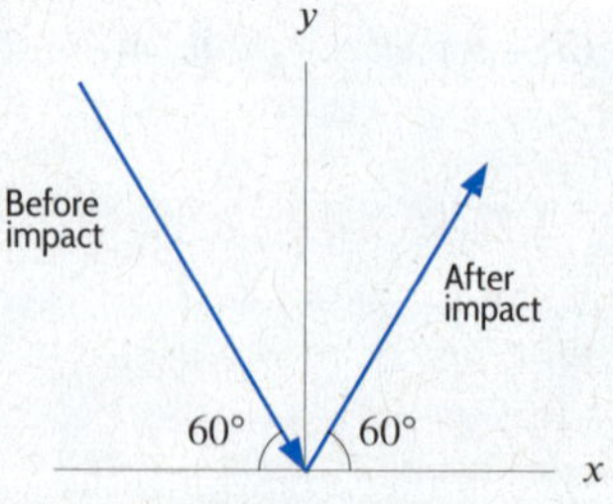


Figure 12.34

**30.** Figure 12.35 shows a molecule with four atoms at $O$, $A$, $B$ and $C$. Show that every atom in the molecule is 2 units away from every other atom.

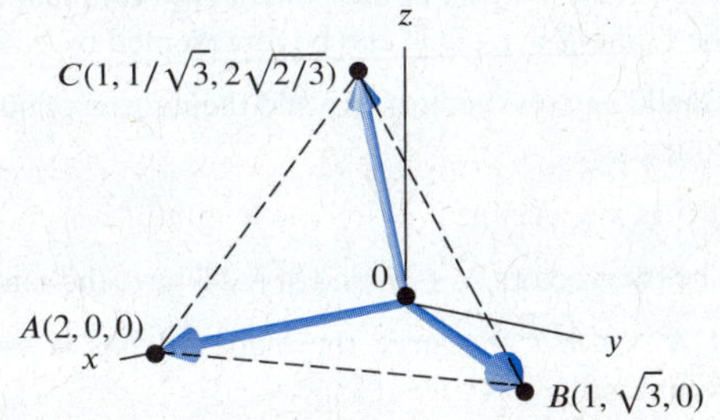


Figure 12.35

**31.** Two cylindrical cans of radius 2 and height 7 are shown in Figure 12.36. The cans touch down the side. Let $A$ be the point at the top rims where they touch. Let $B$ be the front point on the bottom rim of the left can, and $C$ be the back point on the bottom rim of the right can. The origin is at $A$ and the $z$-axis points upward; the $x$-axis points forward (out of the paper) and the $y$-axis points to the right.

**(a)** Write vectors in component form for $\overrightarrow{AB}$ and $\overrightarrow{AC}$.

**(b)** What is the angle between $\overrightarrow{AB}$ and $\overrightarrow{AC}$?

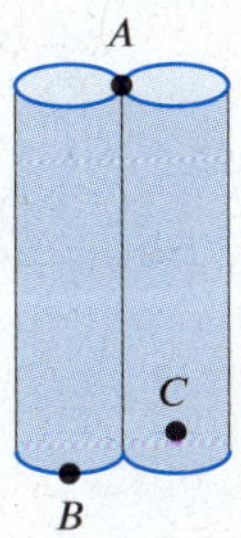


Figure 12.36

**32. (a)** Using the fact that $\vec{u} \cdot \vec{v} = ||\vec{u}|| \cdot ||\vec{v}|| \cos\theta$, show that

$$\vec{u} \cdot (-\vec{v}) = -(\vec{u} \cdot \vec{v}).$$

[Hint: What happens to the angle when you multiply $\vec{v}$ by $-1$?]

**(b)** Using the fact that $\vec{u} \cdot \vec{v} = ||\vec{u}|| \cdot ||\vec{v}|| \cos\theta$, show that for any negative scalar $\lambda$

$$\vec{u} \cdot (\lambda\vec{v}) = \lambda(\vec{u} \cdot \vec{v})$$
$$(\lambda\vec{u}) \cdot \vec{v} = \lambda(\vec{u} \cdot \vec{v}).$$

**33.** A consumption vector of three goods is given by $\vec{x} = (x_1, x_2, x_3)$, where $x_1$, $x_2$ and $x_3$ are the quantities consumed of the three goods. Consider a budget constraint represented by the equation $\vec{p} \cdot \vec{x} = k$, where $\vec{p}$ is the price vector of the three goods and $k$ is a constant. Show that the difference between two consumption vectors corresponding to points satisfying the same budget constraint is perpendicular to the price vector $\vec{p}$.

**34.** Consider the grid in Figure 12.37. Write expressions for $\overrightarrow{AB}$ and $\overrightarrow{CD}$ in terms of $\vec{i}$ and $\vec{j}$.

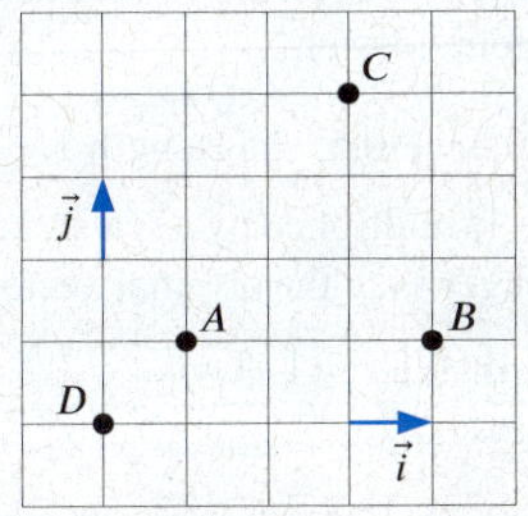


Figure 12.37

**35.** Consider the grid of equilateral triangles in Figure 12.38. Find expressions for $\overrightarrow{AB}$, $\overrightarrow{BC}$, $\overrightarrow{AC}$, and $\overrightarrow{AD}$ in terms of $\vec{u}$ and $\vec{v}$.

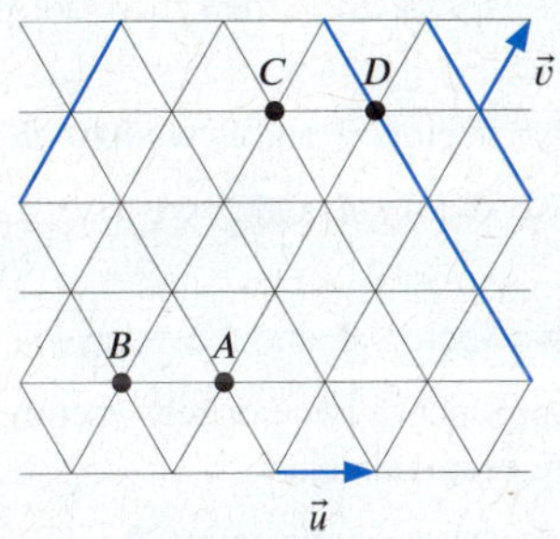


Figure 12.38

**36.** Consider the regular hexagon in Figure 12.39. Express the six sides and all three diameters in terms of $\vec{m}$ and $\vec{n}$.

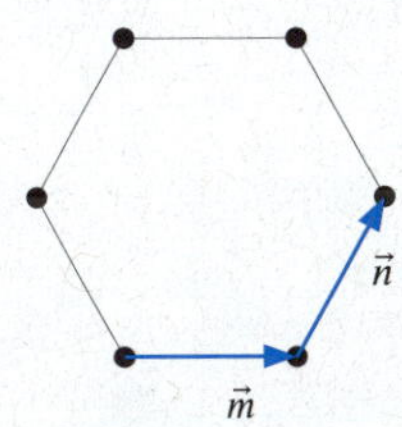


Figure 12.39

**37.** Consider the grid of regular hexagons in Figure 12.40. Express $\overrightarrow{AC}$, $\overrightarrow{AB}$, $\overrightarrow{AD}$ and $\overrightarrow{BD}$ in terms of $\vec{m}$ and $\vec{n}$.

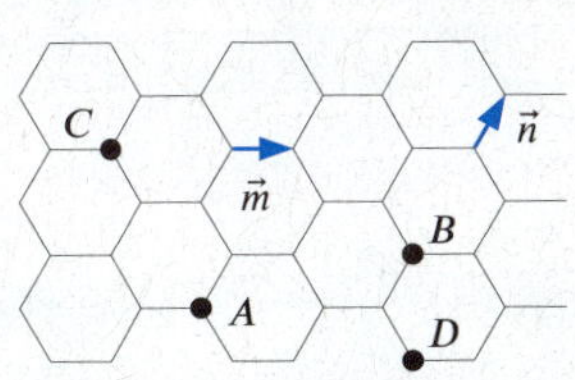


Figure 12.40

## STRENGTHEN YOUR UNDERSTANDING

Are the statements in Problems 1–24 true or false? Give an explanation for your answer.

1. The vector $0.5\vec{i} + 0.5\vec{j}$ has length 1.
2. If a vector $\vec{v}$ is multiplied by a scalar, $k$, the resultant vector $k\vec{v}$ is longer than the original vector.
3. If $\|\vec{u}\| = 5$, then $\| -\vec{u}\| = -5$.
4. The vectors $3\vec{i} - \vec{j} + \vec{k}$ and $6\vec{i} - 2\vec{j} + 2\vec{k}$ are parallel.
5. The vectors $2\vec{i} + \vec{j}$ and $2\vec{i} - \vec{j}$ are perpendicular.
6. The two vectors $2\vec{i} + 3\vec{j}$ and $3\vec{i} + 2\vec{j}$ point in opposite directions
7. If $\vec{u} \cdot \vec{v} < 0$ then $\vec{u}$ and $\vec{v}$ form an acute angle (between 0° and 90°).
8. If $\vec{u}$ and $\vec{v}$ are vectors, then their dot product is also a vector.
9. For any two vectors $\vec{u}$ and $\vec{v}$ we have $\vec{u} + \vec{v} = \vec{v} + \vec{u}$.
10. For any two vectors $\vec{u}$ and $\vec{v}$ we have $\|\vec{u} + \vec{v}\| = \|\vec{u}\| + \|\vec{v}\|$.
11. For any two vectors $\vec{u}$ and $\vec{v}$ we have $\vec{u} \cdot \vec{v} = \vec{v} \cdot \vec{u}$.
12. If the dot product of two nonzero vectors is zero, then the vectors are perpendicular.
13. The distance between the points $P = (2, 3)$ and $Q = (3, 4)$ is the vector $\vec{i} + \vec{j}$.
14. The vector starting at the point $P = (1, 2)$ and ending at the point $Q = (-3, 7)$ can be represented by $-2\vec{i} + 9\vec{j}$.
15. To add any two vectors, we add their corresponding components.
16. If $\vec{v}$ is any vector, $\vec{v} \cdot \vec{v}$ is the length of $\vec{v}$.
17. The two vectors $2\vec{i} + 3\vec{j}$ and $3\vec{i} + 2\vec{j}$ have the same length.
18. If $A$ is a $m \times n$ matrix, the matrix $A$ has $m$ rows and $n$ columns.
19. If $\mathbf{A} = \begin{pmatrix} 1 & 2 & 3 \\ 4 & 5 & 6 \\ 7 & 8 & 9 \end{pmatrix}$ then $a_{23} = 8$.
20. The matrix $\mathbf{A} = \begin{pmatrix} 1 & 2 & 3 \\ 4 & 5 & 6 \end{pmatrix}$ is a $3 \times 2$ matrix.
21. If $A$ is a $5 \times 5$ matrix, then $a_{35}$ refers to the element of $A$ that is in the third row and fifth column.
22. To rotate a vector $\vec{v}$ through an angle $\phi$, multiply $\vec{v}$ by the matrix $R = \begin{pmatrix} -\cos\phi & \sin\phi \\ \sin\phi & \cos\phi \end{pmatrix}$.
23. You cannot add a $2 \times 2$ matrix $\mathbf{A}$ and a $3 \times 3$ matrix $\mathbf{B}$.
24. If $A = \begin{pmatrix} 2 & 3 \\ -1 & 5 \end{pmatrix}$ and $\vec{v} = \begin{pmatrix} 10 \\ 5 \end{pmatrix}$, then $A\vec{v} = \begin{pmatrix} 35 \\ 15 \end{pmatrix}$.

Chapter Thirteen

# SEQUENCES AND SERIES

© Patrick Zephyr/Patrick Zephyr Nature Photography

## Contents

# 13.1 SEQUENCES

Elections for president of the United States have been held every four years since 1792, that is, in 1792, 1796, 1800, 1804, and so on up to the year 2012. Any ordered list of numbers, such as

$$1792, 1796, 1800, 1804, \ldots, 2004, 2008, 2012$$

is called a *sequence*, and the individual numbers are the *terms* of the sequence. A sequence can be a finite list, such as the sequence of past presidential election years, or it can be an infinite list, such as the sequence of positive integers

$$1, 2, 3, 4, \ldots.$$

**Example 1**

(a) $0, 1, 4, 9, 16, 25, \ldots$ is the sequence of squares of integers.
(b) $2, 4, 8, 16, 32, \ldots$ is the sequence of positive integer powers of 2.
(c) $3, 1, 4, 1, 5, 9, \ldots$ is the sequence of digits in the decimal expansion of $\pi$.
(d) $3.9, 5.3, 7.2, 9.6, 12.9, 17.1, 23.1, 38.6, 50.2$ is the sequence of U.S. population figures, in millions, for the first 10 census reports (1790 to 1880).
(e) $40.4, 60.2, 73.3, 84.4, 93.4, 101.5$ is the sequence of smartphone users in the US, in millions, from the years 2009 to 2014.[1]

## Notation for Sequences

We denote the terms of a sequence by

$$a_1, a_2, a_3, \ldots, a_n, \ldots$$

so that $a_1$ is the first term, $a_2$ is the second term, and so on. We use $a_n$ to denote the $n^{\text{th}}$ or *general term* of the sequence. If there is a pattern in the sequence, we may be able to find a formula for $a_n$.

**Example 2** Find the first three terms and the $98^{\text{th}}$ term of the sequence.

(a) $a_n = 1 + \sqrt{n}$  (b) $b_n = (-1)^n \dfrac{n}{n+1}$

**Solution**

(a) $a_1 = 1 + \sqrt{1}, a_2 = 1 + \sqrt{2} \approx 2.414, a_3 = 1 + \sqrt{3} \approx 2.732$, and $a_{98} = 1 + \sqrt{98} \approx 10.899$.
(b) $b_1 = (-1)^1 \dfrac{1}{1+1} = -\dfrac{1}{2}, b_2 = (-1)^2 \dfrac{2}{2+1} = \dfrac{2}{3}, b_3 = (-1)^3 \dfrac{3}{3+1} = -\dfrac{3}{4}$, and $b_{98} = (-1)^{98} \dfrac{98}{98+1} = \dfrac{98}{99}$.
This sequence is called *alternating* because the terms alternate in sign.

A sequence can be thought of as a function whose domain is a set of integers. Each term of the sequence is an output value for the function, so $a_n = f(n)$.

**Example 3** List the first five terms of the sequence $a_n = f(n)$, where $f(x) = 500 - 10x$.

**Solution** Evaluate $f(x)$ for $x = 1, 2, 3, 4, 5$:

$$a_1 = f(1) = 500 - 10 \cdot 1 = 490 \quad \text{and} \quad a_2 = f(2) = 500 - 10 \cdot 2 = 480.$$

Similarly, $a_3 = 470$, $a_4 = 460$, and $a_5 = 450$.

## Arithmetic Sequences

You buy a used car that has already been driven 15,000 miles and drive it 8000 miles per year. The odometer registers 23,000 miles 1 year after your purchase, 31,000 miles after 2 years, and so on. The yearly odometer readings form a sequence $a_n$ whose terms are

$$15{,}000, \; 23{,}000, \; 31{,}000, \; 39{,}000, \; \ldots.$$

[1] http://blog.signalhq.com. Accessed May 25, 2014.

Each term of the sequence is obtained from the previous term by adding 8000; that is, the difference between successive terms is 8000. A sequence in which the difference between pairs of successive terms is a fixed quantity is called an *arithmetic sequence*.

**Example 4** Which of the following sequences are arithmetic?

(a) $9, 5, 1, -3, -7$
(b) $3, 6, 12, 24, 48$
(c) $2, 2 + p, 2 + 2p, 2 + 3p$
(d) $10, 5, 0, 5, 10$

**Solution**

(a) Each term is obtained from the previous term by subtracting 4. This sequence is arithmetic.
(b) This sequence is not arithmetic: each term is twice the previous term. The differences are 3, 6, 12, 24.
(c) This sequence is arithmetic: $p$ is added to each term to obtain the next term.
(d) This is not arithmetic. The difference between the second and first terms is $-5$, but the difference between the fifth and fourth terms is 5.

We can write a formula for the general term of an arithmetic sequence. Look at the sequence $2, 6, 10, 14, 18, \ldots$ in which the terms increase by 4, and observe that

$$\begin{aligned} a_1 &= 2 \\ a_2 &= 6 = 2 + 1 \cdot 4 \\ a_3 &= 10 = 2 + 2 \cdot 4 \\ a_4 &= 14 = 2 + 3 \cdot 4. \end{aligned}$$

When we get to the $n^{\text{th}}$ term, we have added $(n-1)$ copies of 4, so that $a_n = 2 + (n-1)4$. In general:

> For $n \geq 1$, the $n^{\text{th}}$ term of an arithmetic sequence is
>
> $$a_n = a_1 + (n-1)d,$$
>
> where $a_1$ is the first term, and $d$ is the difference between consecutive terms.

**Example 5**

(a) Write a formula for the general term of the odometer sequence of the car that is driven 8000 miles per year and had gone 15,000 miles when it was bought.
(b) What is the car's odometer reading seven years after its purchase?

**Solution**

(a) The odometer reads 15,000 miles initially, so $a_1 = 15{,}000$. Each year the odometer reading increases by 8000, so $d = 8000$. Thus, $a_n = 15{,}000 + (n-1)8000$.
(b) Seven years from the date of purchase is the start of the $8^{\text{th}}$ year, so $n = 8$. The total number of miles on the car is

$$a_8 = 15{,}000 + (8-1) \cdot 8000 = 71{,}000.$$

### Arithmetic Sequences and Linear Functions

You may have noticed that the arithmetic sequence for the car's odometer reading looks like a linear function. The formula for the $n^{\text{th}}$ term, $a_n = 15{,}000 + (n-1)8000$, can be simplified to $a_n = 7000 + 8000n$, a linear function with slope $m = 8000$ and initial value $b = 7000$. However, for a sequence we consider only positive integer inputs, whereas a linear function is defined for all values of $n$. We can think of an arithmetic sequence as a linear function whose domain has been restricted to the positive integers.

## Geometric Sequences

You are offered a job at a salary of $40,000 for the first year with a 5% pay raise every year. Under this plan, your annual salaries form a sequence with terms

$$\begin{aligned} a_1 &= 40{,}000 \\ a_2 &= 40{,}000(1.05) = 42{,}000 \\ a_3 &= 42{,}000(1.05) = 40{,}000(1.05)^2 = 44{,}100 \\ a_4 &= 44{,}100(1.05) = 40{,}000(1.05)^3 = 46{,}305, \end{aligned}$$

and so on, where each term is obtained from the previous one by multiplying by 1.05. A sequence in which each term is a constant multiple of the preceding term is called a *geometric sequence*. In a geometric sequence, the ratio of successive terms is constant.

**Example 6** Which of the following sequences are geometric?

(a) $5, 25, 125, 625, \ldots$ (b) $-8, 4, -2, 1, -\frac{1}{2}, \ldots$ (c) $12, 6, 4, 3, \ldots$

**Solution**

(a) This sequence is geometric. Each term is 5 times the previous term. Note that the ratio of any term to its predecessor is 5.

(b) This sequence is geometric. The ratio of any term to the previous term is $-\frac{1}{2}$.

(c) This sequence is not geometric. The ratios of successive terms are not constant: $\frac{a_2}{a_1} = \frac{6}{12} = \frac{1}{2}$, but $\frac{a_3}{a_2} = \frac{4}{6} = \frac{2}{3}$.

As for arithmetic sequences, there is a formula for the general term of a geometric sequence. Consider the sequence $8, 2, \frac{1}{2}, \frac{1}{8}, \ldots$ in which each term is $\frac{1}{4}$ times the previous term. We have

$$\begin{aligned} a_1 &= 8 \\ a_2 &= 2 = 8\left(\frac{1}{4}\right) \\ a_3 &= \frac{1}{2} = 8\left(\frac{1}{4}\right)^2 \\ a_4 &= \frac{1}{8} = 8\left(\frac{1}{4}\right)^3. \end{aligned}$$

When we get to the $n^{\text{th}}$ term, we have multiplied 8 by $(n-1)$ factors of $\frac{1}{4}$, so that $a_n = 8\left(\frac{1}{4}\right)^{n-1}$. In general,

> For $n \geq 1$, the $n^{\text{th}}$ term of a geometric sequence is
>
> $$a_n = a_1 r^{n-1},$$
>
> where $a_1$ is the first term, and $r$ is the ratio of consecutive terms.

**Example 7**

(a) Write a formula for the general term of the salary sequence that starts at $40,000 and increases by 5% each year.

(b) What is your salary after 10 years on the job?

**Solution**

(a) Your starting salary is $40,000, so $a_1 = 40{,}000$. Each year your salary increases by 5%, so $r = 1.05$. Thus, $a_n = 40{,}000(1.05)^{n-1}$.

(b) After 10 years on the job, you are at the start of your $11^{\text{th}}$ year, so $n = 11$. Your salary is

$$a_{11} = 40{,}000(1.05)^{11-1} \approx 65{,}156 \text{ dollars.}$$

## Geometric Sequences and Exponential Functions

The formula for the salary sequence, $a_n = 40{,}000(1.05)^{n-1}$, looks like a formula for an exponential function, $f(t) = a \cdot b^t$. A geometric sequence is an exponential function whose domain is restricted to the positive integers. For many applications, this restricted domain is more realistic than an interval of real numbers. For example, salaries are usually increased once a year, rather than continuously.

# Exercises and Problems for Section 13.1

## Exercises

Are the sequences in Exercises 1–4 arithmetic?

**1.** $2, 7, 11, 14, \ldots$

**2.** $2, -5, -12, -19, \ldots$

**3.** $2, 7, 12, 17, \ldots$

**4.** $2, -5, -11, -16, \ldots$

Are the sequences in Exercises 5–7 arithmetic? For those that are, give a formula for the $n^{\text{th}}$ term.

**5.** $6, 9, 12, 15, \ldots$

**6.** $1, -1, 2, -2, \ldots$

**7.** $-1, -1.1, -1.2, -1.3, \ldots$

Are the sequences in Exercises 8–11 geometric?

**8.** $2, 6, 18, 54, \ldots$

**9.** $2, 4, 8, 12, \ldots$

**10.** $2, -1, \frac{1}{2}, -\frac{1}{4}, \frac{1}{8}, \ldots$

**11.** $2, 0.2, 0.02, 0.002, \ldots$

Are the sequences in Exercises 12–17 geometric? For those that are, give a formula for the $n^{\text{th}}$ term.

**12.** $4, 12, 36, 108, \ldots$

**13.** $4, 1, \frac{1}{4}, \frac{1}{8}, \ldots$

**14.** $2, -4, 8, -16, \ldots$

**15.** $4, 2, 1, \frac{1}{2}, \frac{1}{4}, \ldots$

**16.** $4, 0.4, 0.04, 0.004, \ldots$

**17.** $1, \frac{1}{1.2}, \frac{1}{(1.2)^2}, \frac{1}{(1.2)^3}, \ldots$

## Problems

In Problems 18–23, write out the first four terms of each sequence and state if it is geometric.

**18.** $a_n = 2^n$

**19.** $a_n = \frac{2n+1}{n+2}$

**20.** $a_n = \left(-\frac{1}{3}\right)^n$

**21.** $a_n = \cos(n\pi)$

**22.** $a_n = n^2 - n$

**23.** $a_n = \frac{1}{\sqrt{n}}$

**24.** An arithmetic sequence has first term of 10 and difference of 5; what is the tenth term?

**25.** An arithmetic sequence has first term of 5 and difference of 10. After how many terms will the sequence exceed 1000?

In Problems 26–29, find the $5^{th}$, $50^{th}$, $n^{th}$ term of the arithmetic sequences.

**26.** $3, 5, 7, \ldots$

**27.** $6, 7.2, \ldots$

**28.** $a_1 = 2.1$, $a_3 = 4.7$

**29.** $a_3 = 5.7$, $a_6 = 9$

In Problems 30–33, find the $6^{th}$ and $n^{th}$ of the geometric sequences.

**30.** $1, 2, 4, \ldots$

**31.** $7, 5.25, \ldots$

**32.** $a_1 = 3$, $a_3 = 48$

**33.** $a_2 = 6$, $a_4 = 54$

**34.** During 2012, global oil consumption grew by 0.7%, to reach 88 million barrels per day.[2] Assume that it continues to increase at this rate.

**(a)** Write the first four terms of the sequence $a_n$ giving daily oil consumption $n$ years after 2011; give a formula for the general term $a_n$.

**(b)** In what year is consumption expected to exceed 100 million barrels a day?

**35.** In 2011, US natural gas consumption was 690.1 billion cubic meters. Asian consumption was 590.6 billion cubic meters.[3] US consumption increased by 2.4% a year during 2011, while Asian consumption grew by 5.9% a year. Assume these rates continue into the future.

**(a)** Give the first four terms of the sequence, $a_n$, giving US consumption of natural gas $n$ years after 2010.

**(b)** Give the first four terms of a similar sequence $b_n$ showing Asian gas consumption.

**(c)** According to this model, when will Asian yearly gas consumption exceed US consumption?

[2]http://www.bp.com/extendedsectiongenericarticle.do?categoryId=9041228&contentId=7074736, accessed May 18, 2013.

[3]www.bp.com/downloads, Statistical Review of World Energy 2011, accessed May 18, 2013.

**36.** The population[4] of Nevada grew from 2.4 million in 2005 to 2.76 million in 2012. Assuming a constant percent growth rate:

**(a)** Find a formula for $a_n$, the population in millions $n$ years after 2012.
**(b)** When is the population predicted to reach 5 million?

**37.** Florida's population[5] was 19.042 million in 2012 and 18.934 million in 2011. Assuming the population continues to increase at the same percentage rate:

**(a)** Find the first three terms of the sequence $a_n$ giving the population, in millions, $n$ years after 2012.
**(b)** Write a formula for $a_n$.
**(c)** What is the doubling time of the population?

The graphs in Problems 38–41 represent either an arithmetic or a geometric sequence; decide which. For an arithmetic sequence, say if the common difference, $d$, is positive or negative. For a geometric sequence, say if the common ratio, $r$, is greater or smaller than 1.

**38.**

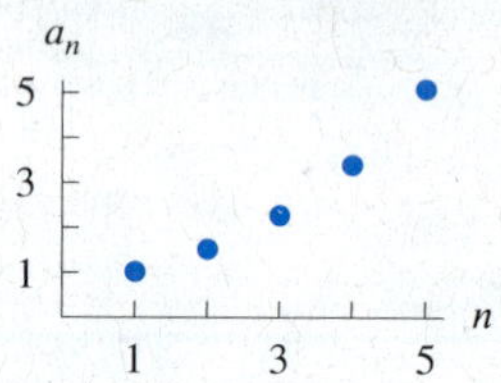

**39.**

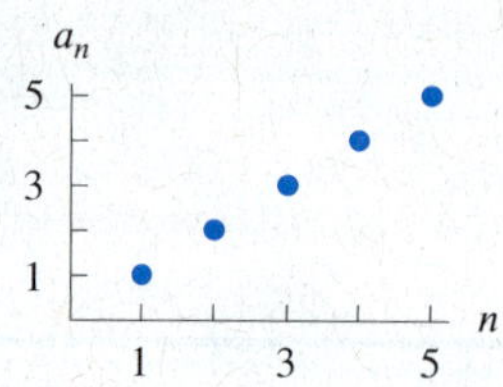

**40.**

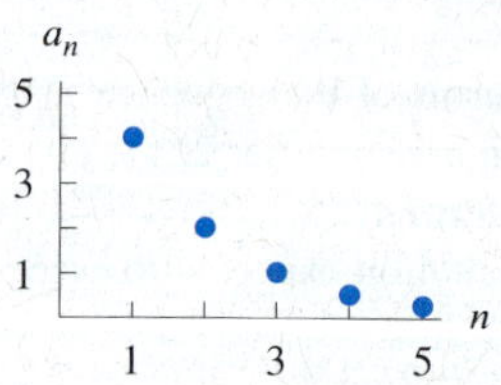

**41.**

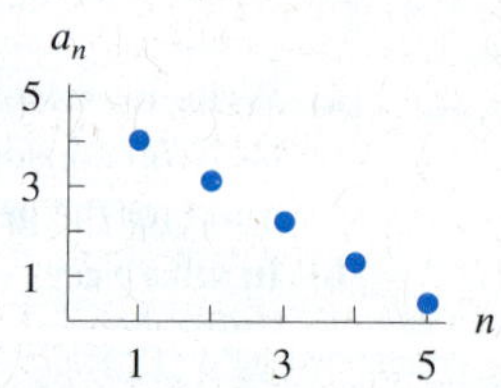

A sequence $a_n$ can be defined by a *recurrence relation*, which gives $a_n$ in terms of the previous term, $a_{n-1}$, and the first term $a_1$. In Problems 42–45, find the first four terms of the sequence and a formula for the general term.

**42.** $a_n = 2a_{n-1}$; $a_1 = 3$

**43.** $a_n = a_{n-1} + 5$; $a_1 = 2$

**44.** $a_n = -a_{n-1}$; $a_1 = 1$

**45.** $a_n = 2a_{n-1} + 1$; $a_1 = 3$

**46.** A geometric sequence has first term of 10 and ratio of $-0.2$. After how many terms will the sequence have absolute value less than $10^{-7}$?

**47.** The sequence defined by

$$p_{n+1} = 2p_n - \frac{p_n^2}{200}, \quad p_0 = 150,$$

is called a discrete logistic equation. Such a sequence is often used to model a population $p_n$, where the initial population is $p_0$.

**(a)** Compute the first 6 terms of the sequence. What do you conclude?
**(b)** Show that if $0 < p_0 < 200$, then $p_n$ increases but never exceeds 200.

**48.** The Fibonacci sequence starts with 1, 1, 2, 3, 5, …, and each term is the sum of the previous two terms.

**(a)** Write the next three terms in the sequence.
**(b)** Write an expression for $a_n$ in terms of $a_{n-1}$ and $a_{n-2}$.
**(c)** Suppose $r_n$ is the ratio $a_n/a_{n-1}$ and $r_{n-1}$ is the ratio $a_{n-1}/a_{n-2}$. Using your answer to part (b), find a formula for $r_n$ in terms of $r_{n-1}$ and without any of the $a$s.
**(d)** The terms $r_n$ form another sequence. Suppose $r_n$ tends to a fixed value, $r$ as $n$ increases without bound; that is, $r_n \to r$ as $n \to \infty$. Use your answer to part (c) to find an equation for $r$. Solve this equation. The number that you find is called the *golden ratio*.

**49.** Some people believe they can make money from a chain letter (they are usually disappointed). A chain letter works roughly like this: A letter arrives with a list of four names attached and instructions to mail a copy to four more friends and to send \$1 to the top name on the list. When you mail the four letters, you remove the top name (to whom the money was sent) and add your own name to the bottom of the list.

**(a)** If no one breaks the chain, how much money do you receive?
**(b)** Let $d_n$ be the number of dollars you receive if there are $n$ names on the list instead of 4, but you still mailed to four friends. Find a formula for $d_n$.

**50.** For a positive integer $n$, let $a_n$ be the fraction of the US population with income less than or equal to \$$n$ thousand dollars.

**(a)** Which is larger, $a_{40}$ or $a_{50}$? Why?
**(b)** What does the quantity $a_{50} - a_{40}$ represent in terms of US population?
**(c)** Is there any value of $n$ with $a_n = 0$? Explain.
**(d)** What happens to the value of $a_n$ as $n$ increases?

[4] http://www.infoplease.com/us/census/data/nevada/, accessed May, 2013.
[5] http://www.floridacharts.com/FLQUERY/Population/PopulationRpt.aspx, accessed May 18, 2013.

# 13.2 DEFINING FUNCTIONS USING SUMS: ARITHMETIC SERIES

## Domestic Deaths from AIDS $n$ years after 1980

Table 13.1 shows the number of US AIDS deaths[6] that occurred each year since 1980, where $a_n$ is the number of deaths in year $n$, and $n = 1$ corresponds to 1981. For each year, we can calculate the total number of death from AIDS since 1980. For example, in 1985

Total number of AIDS deaths (1981-1985) $= 159 + 463 + 1508 + 3505 + 6972 = 12{,}607$.

We write $S_n$ to denote the sum of the first $n$ terms of a sequence. In this example, 1985 corresponds to $n = 5$, so we have $S_5 = 12{,}607$.

**Table 13.1** *US deaths from AIDS each year from 1981 to 2010*

| $n$ | $a_n$ | $n$ | $a_n$ | $n$ | $a_n$ | $n$ | $a_n$ | $n$ | $a_n$ | $n$ | $a_n$ |
|---|---|---|---|---|---|---|---|---|---|---|---|
| 1 | 159 | 6 | 12,110 | 11 | 36,616 | 16 | 38,025 | 21 | 18,524 | 26 | 14,732 |
| 2 | 463 | 7 | 16,412 | 12 | 41,094 | 17 | 21,999 | 22 | 17,557 | 27 | 16,659 |
| 3 | 1508 | 8 | 21,119 | 13 | 45,598 | 18 | 18,397 | 23 | 18,017 | 28 | 20,091 |
| 4 | 3505 | 9 | 27,791 | 14 | 50,418 | 19 | 17,172 | 24 | 15,798 | 29 | 19,981 |
| 5 | 6972 | 10 | 31,538 | 15 | 51,117 | 20 | 15,245 | 25 | 16,268 | 30 | 19,343 |

**Example 1** Find and interpret $S_8$ for the AIDS sequence.

**Solution** Since $S_8$ is the sum of the first 8 terms of the sequence, we have.

$$\begin{aligned} S_8 &= \underbrace{a_1 + a_2 + a_3 + a_4 + a_5}_{S_5=12{,}607} + a_6 + a_7 + a_8 \\ &= 12{,}607 + 12{,}110 + 16{,}412 + 21{,}119 \\ &= 62{,}248. \end{aligned}$$

Here, $S_8$ is the number of deaths from AIDS from 1980 to 1988.

The sum of the terms of a sequence is called a *series*. We write $S_n$ for the sum of the first $n$ terms of the sequence, called the $n^{\text{th}}$ *partial sum*. We see that $S_n$ is a function of $n$, the number of terms in the partial sum. In this section we see how to evaluate functions defined by sums.

## Arithmetic Series

Landscape timbers are large beams of wood used to landscape gardens. To make the terrace in Figure 13.1, one timber is set into the slope, followed by a stack of two, then a stack of three, then a stack of four. The stacks are separated by earth.

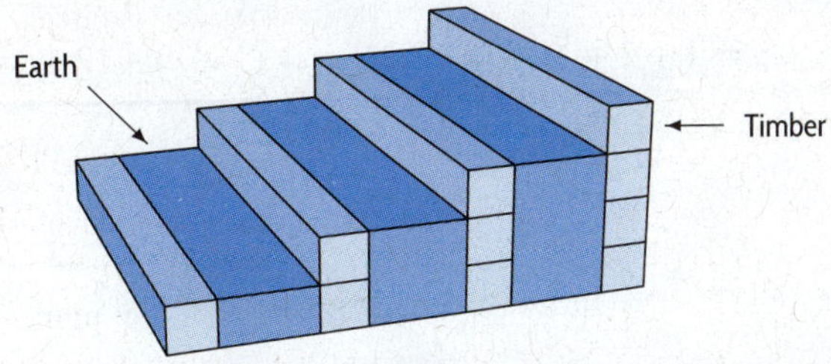


**Figure 13.1:** A slope terraced for planting using landscape timbers

[6]From the HIV/AIDS Surveillance Report, 2011, Centers for Disease Control and Prevention, Atlanta. http://www.cdc.gov/hiv/pdf/statistics_2011_HIV_Surveillance_Report_vol_23.pdf. Accessed May 25, 2014.

The total number of timbers in 4 stacks is $S_4$, the sum of the number of timbers in each stack:

$$\text{Total number of timbers} = S_4 = 1 + 2 + 3 + 4 = 10.$$

For a larger terrace using 5 stacks of timbers, the total number is given by

$$S_5 = 1 + 2 + 3 + 4 + 5 = 15.$$

For an even larger terrace using 6 stacks, the total number is given by

$$S_6 = 1 + 2 + 3 + 4 + 5 + 6 = 21.$$

For a terrace made from $n$ stacks of landscape timbers, $S_n$, the total number of timbers needed is a function of $n$, so

$$S_n = f(n) = 1 + 2 + \cdots + n.$$

The symbol $\cdots$ means that all the integers from 1 to $n$ are included in the sum.

Notice that each stack of landscape timbers contains one more timber than the previous one. Thus, the number of timbers in each stack,

$$1, 2, 3, 4, 5, \ldots$$

is an arithmetic sequence, so $S_n$ is the sum of the terms of an arithmetic sequence. Such a sum is called an *arithmetic series*.

## The Sum of an Arithmetic Series

We now find a formula for the sum of an arithmetic series. A famous story concerning this series[7] is told about the great mathematician Carl Friedrich Gauss (1777–1855), who as a young boy was asked by his teacher to add the numbers from 1 to 100. He did so almost immediately:

$$S_{100} = 1 + 2 + 3 + \cdots + 100 = 5050.$$

Of course, no one really knows how Gauss accomplished this, but he probably did not perform the calculation directly, by adding 100 terms. He might have noticed that the terms in the sum can be regrouped into pairs, as follows:

$$\begin{aligned} S_{100} = 1 + 2 + \cdots + 99 + 100 &= \underbrace{(1 + 100) + (2 + 99) + \cdots + (50 + 51)}_{50 \text{ pairs}} \\ &= \underbrace{101 + 101 + \cdots + 101}_{50 \text{ terms}} \qquad \text{Each pair adds to 101} \\ &= 50 \cdot 101 = 5050. \end{aligned}$$

The approach of pairing numbers works for the sum from 1 to $n$, no matter how large $n$ is. Provided $n$ is an even number, we can write

$$\begin{aligned} S_n = 1 + 2 + \cdots + (n-1) + n &= \underbrace{(1 + n) + (2 + (n-1)) + (3 + (n-2)) + \cdots}_{\frac{1}{2}n \text{ pairs}} \\ &= \underbrace{(1 + n) + (1 + n) + \cdots}_{\frac{1}{2}n \text{ pairs}} \qquad \text{Each pair adds to } 1 + n \end{aligned}$$

[7]As told by E.T. Bell, *The Men of Mathematics*, p. 221 (New York: Simon and Schuster, 1937), the series involved was arithmetic, but more complicated than this one.

so we have the formula:

$$S_n = 1 + 2 + \cdots + n = \frac{1}{2}n(n+1).$$

**Example 2** Check this formula for $S_n$ with $n = 100$.

**Solution** Using the formula, we get the same answer, 5050, as before:

$$S_{100} = 1 + 2 + \cdots + 100 = \frac{1}{2} \cdot 100(100+1) = 50 \cdot 101 = 5050.$$

A similar derivation shows that this formula for $S_n$ also holds for odd values of $n$.

To find a formula for the sum of a general arithmetic series, we first assume that $n$ is even and write

$$\begin{aligned} S_n &= a_1 + a_2 + \cdots + a_n \\ &= \underbrace{(a_1 + a_n) + (a_2 + a_{n-1}) + (a_3 + a_{n-2}) + \cdots}_{\frac{1}{2}n \text{ pairs}}. \end{aligned}$$

We pair the first term with the last term, the second term with the next to last term, and so on, just as Gauss may have done. Each pair of terms adds up to the same value, just as each of Gauss's pairs added to 101. Using the formula for the terms of an arithmetic sequence, $a_n = a_1 + (n-1)d$, the first pair, $a_1 + a_n$, can be written as

$$a_1 + a_n = a_1 + \underbrace{a_1 + (n-1)d}_{a_n} = 2a_1 + (n-1)d,$$

and the second pair, $a_2 + a_{n-1}$, can be written as

$$a_2 + a_{n-1} = \underbrace{a_1 + d}_{a_2} + \underbrace{a_1 + (n-2)d}_{a_{n-1}} = 2a_1 + (n-1)d.$$

Both the first two pairs have the same sum: $2a_1 + (n-1)d$. The remaining pairs also all have the same sum, so

$$S_n = \underbrace{(a_1 + a_n) + (a_2 + a_{n-1}) + (a_3 + a_{n-2}) + \cdots}_{\frac{1}{2}n \text{ pairs}} = \frac{1}{2}n\left(2a_1 + (n-1)d\right).$$

The same formula gives the sum of the series when $n$ is odd. See Problem 45 on page 547.

The **sum**, $S_n$, of the first $n$ terms of the **arithmetic series** with $a_n = a_1 + (n-1)d$ is

$$S_n = \frac{1}{2}n(a_1 + a_n) = \frac{1}{2}n(2a_1 + (n-1)d).$$

**Example 3** Calculate the sum $1 + 2 + \cdots + 100$ using the formula for $S_n$.

**Solution** Here, $a_1 = 1$, $n = 100$, and $d = 1$. We get the same answer as before:

$$S_{100} = \frac{1}{2}n\left(2a_1 + (n-1)d\right) = \frac{1}{2} \cdot 100\,(2 \cdot 1 + (100-1) \cdot 1) = 50 \cdot 101 = 5050.$$

## Summation Notation

The symbol $\Sigma$ is used to indicate addition. This symbol, pronounced *sigma*, is the Greek capital letter for $S$, which stands for sum. Using this notation, we write

$$\sum_{i=1}^{n} a_i \quad \text{to stand for the sum } a_1 + a_2 + \cdots + a_n.$$

The $\Sigma$ tells us we are adding some numbers. The $a_i$ tells us that the numbers we are adding are called $a_1$, $a_2$, and so on.[8] The sum begins with $a_1$ and ends with $a_n$ because the subscript $i$ starts at $i = 1$ (at the bottom of the $\Sigma$ sign) and ends at $i = n$ (at the top of the $\Sigma$ sign):

This tells us that the sum ends at $a_n$ ↓

$$\sum_{i=1}^{n} a_i \longleftarrow$$ This tells us that the numbers we are adding are called $a_1, a_2, a_3, \ldots$

↑ This tells us that the sum starts at $a_1$

**Example 4** Write $S_n = f(n)$, the total number of landscape timbers in $n$ stacks, using sigma notation, and give a formula for $S_n$.

**Solution** Since $S_n = f(n) = 1 + 2 + \cdots + n$, we have $a_i = i$. We start at $i = 1$ and end at $i = n$. Thus, $S_n = \sum_{i=1}^{n} i$. The formula for the sum of an arithmetic series with $a_1 = d = 1$ gives

$$S_n = f(n) = \sum_{i=1}^{n} i = \frac{1}{2}n(n+1).$$

**Example 5** Use sigma notation to write the sum of the first 20 positive odd numbers. Evaluate this sum.

**Solution** The odd numbers form an arithmetic sequence: 1, 3, 5, 7, … with $a_1 = 1$ and $d = 2$. The $i^{\text{th}}$ odd number is

$$a_i = 1 + (i-1)2 = 2i - 1.$$

(As a check: $a_1 = 1 + (1-1)2 = 1$ and $a_2 = 1 + (2-1)2 = 3$.) Thus, we have

$$\text{Sum of the first 20 odd numbers} = \sum_{i=1}^{20} a_i = \sum_{i=1}^{20} (2i-1).$$

We evaluate the sum using $n = 20$, $a_1 = 1$, and $d = 2$:

$$\text{Sum} = \frac{1}{2}n\left(2a_1 + (n-1)d\right) = \frac{1}{2} \cdot 20\,(2 \cdot 1 + (20-1) \cdot 2) = 400.$$

[8] The letter $i$ is often called a dummy index since it can be switched with any other letter without affecting the sum.

**Example 6** If air resistance is neglected, a falling object travels 16 ft during the first second, 48 ft during the next, 80 ft during the next, and so on. These distances form the arithmetic sequence 16, 48, 80, .... In this sequence, $a_1 = 16$ and $d = 32$.

(a) Find a formula for the $n^{\text{th}}$ term in the sequence of distances. Calculate the fifth and tenth terms.
(b) Calculate $S_1$, $S_2$, and $S_3$, the total distance an object falls in 1, 2, and 3 seconds, respectively.
(c) Give a formula for $S_n$, the distance fallen in $n$ seconds.

**Solution** (a) The $n^{\text{th}}$ term is $a_n = a_1 + (n-1)d = 16 + 32(n-1) = 32n - 16$. Thus, the fifth term is $a_5 = 32(5) - 16 = 144$. The value of $a_{10} = 32(10) - 16 = 304$.

(b) Since $a_1 = 16$ and $d = 32$,

$$S_1 = a_1 = 16 \text{ feet}$$
$$S_2 = a_1 + a_2 = a_1 + (a_1 + d) = 16 + 48 = 64 \text{ feet}$$
$$S_3 = a_1 + a_2 + a_3 = 16 + 48 + 80 = 144 \text{ feet.}$$

(c) The formula for $S_n$ is $S_n = \frac{1}{2}n\left(2a_1 + (n-1)d\right)$. We can check our answer to part (b) using this formula:

$$S_1 = \frac{1}{2}\cdot 1(2\cdot 16 + 0\cdot 32) = 16, \quad S_2 = \frac{1}{2}\cdot 2(2\cdot 16 + 1\cdot 32) = 64, \quad S_3 = \frac{1}{2}\cdot 3(2\cdot 16 + 2\cdot 32) = 144.$$

**Example 7** An object falls from 1000 feet starting at time $t = 0$ seconds. What is its height, $h$, in feet above the ground at $t = 1, 2, 3$ seconds? Show these values on a graph of height against time.

**Solution** At time $t = 0$, the height $h = 1000$ feet. At time $t = 1$, as in Example 6, the object has fallen $S_1 = 16$ feet, so

$$h = 1000 - 16 = 984 \text{ feet.}$$

At time $t = 2$, the object has fallen a total distance of $S_2 = 64$ feet, so

$$h = 1000 - 64 = 936 \text{ feet.}$$

At time $t = 3$, the object has fallen a total distance of $S_3 = 144$ feet, so

$$h = 1000 - 144 = 856 \text{ feet.}$$

These heights are marked on the graph in Figure 13.2.

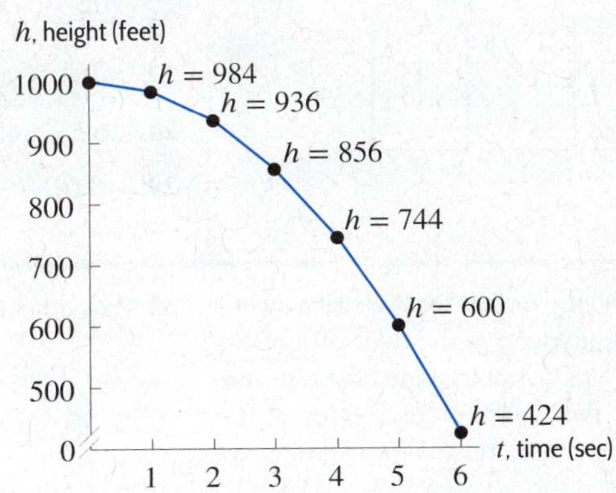


Figure 13.2: Height of a falling object

## Exercises and Problems for Section 13.2

### Exercises

Are the series in Exercises 1–4 arithmetic?

**1.** $2+4+8+16+\cdots$

**2.** $10+8+6+4+2+\cdots$

**3.** $1+2+4+5+7+8+\cdots$

**4.** $-\frac{1}{3}+\frac{2}{3}+\frac{5}{3}+\frac{8}{3}+\cdots$

Expand the sums in Exercises 5–10. (Do not evaluate.)

**5.** $\sum_{i=-1}^{5} i^2$

**6.** $\sum_{i=10}^{20} (i+1)^2$

**7.** $\sum_{k=0}^{5} 2k+1$

**8.** $\sum_{j=1}^{6} 3(j-3)$

**9.** $\sum_{j=2}^{10} (-1)^j$

**10.** $\sum_{n=1}^{7} (-1)^{n-1}2^n$

In Exercises 11–14, write the sum using sigma notation.

**11.** $3+6+9+12+15+18+21$

**12.** $10+13+16+19+22$

**13.** $1/2+1+3/2+2+5/2+3+7/2+4$

**14.** $30+25+20+15+10+5$

**15.** **(a)** Use sigma notation to write the sum $2+4+6+\cdots+20$ of the first 10 even numbers.

**(b)** Without a calculator, find the sum in part (a).

In Exercises 16–20, complete the tables with the terms of the arithmetic series $a_1, a_2, \ldots, a_n$, and the sequence of partial sums, $S_1, S_2, \ldots, S_n$. State the values of $a_1$ and $d$ where $a_n = a_1 + (n-1)d$.

**16.**

| $n$ | 1 | 2 | 3 | 4 | 5 | 6 | 7 | 8 |
|---|---|---|---|---|---|---|---|---|
| $a_n$ | 2 | 7 | 12 | 17 | | | | |
| $S_n$ | 2 | 9 | 21 | 38 | | | | |

**17.**

| $n$ | 1 | 2 | 3 | 4 | 5 | 6 | 7 | 8 |
|---|---|---|---|---|---|---|---|---|
| $a_n$ | 3 | 7 | 11 | | | | | |
| $S_n$ | 3 | 10 | | | | | | |

**18.**

| $n$ | 1 | 2 | 3 | 4 | 5 | 6 | 7 | 8 |
|---|---|---|---|---|---|---|---|---|
| $a_n$ | 7 | | | 16 | | | | |
| $S_n$ | | | | | | | | |

**19.**

| $n$ | 1 | 2 | 3 | 4 | 5 | 6 | 7 | 8 |
|---|---|---|---|---|---|---|---|---|
| $a_n$ | 2 | | | | | | | |
| $S_n$ | 2 | 13 | 33 | 62 | | | | |

**20.**

| $n$ | 1 | 2 | 3 | 4 | 5 | 6 | 7 | 8 |
|---|---|---|---|---|---|---|---|---|
| $a_n$ | | | | | | | | |
| $S_n$ | | | | | | 201 | 273 | 356 |

**21.** Find the sum of the first 1000 integers: $1+2+3+\cdots+1000$.

Without using a calculator, find the sum of the series in Problems 22–29.

**22.** $\sum_{i=1}^{50} 3i$

**23.** $\sum_{i=1}^{30} (5i+10)$

**24.** $\sum_{n=0}^{15} \left(2+\frac{1}{2}n\right)$

**25.** $\sum_{n=0}^{10} (8-4n)$

**26.** $\sqrt{2}+2\sqrt{2}+3\sqrt{2}+\cdots+25\sqrt{2}$

**27.** $-101-91-81-71-61-\cdots-11-1$

**28.** $26.5+24.5+22.5+\cdots+2.5+0.5$

**29.** $-3.01-3.02-3.03-\cdots-3.35$

### Problems

**30.** Jenny decides to raise money for her local charity by encouraging people to lay quarters on the floor of a classroom in the shape of an equilateral triangle. She puts one quarter in the first row, two in the second, three in the third, and so on. Her target is to raise \$400. How many rows does she need?

**31.** Find the thirtieth positive multiple of 5 and the sum of the first thirty positive multiples of 5.

**32.** For the AIDS data in Table 13.1 on page 541,

**(a)** Find and interpret in terms of AIDS

(i) The partial sums $S_5, S_6, S_7, S_8$.

(ii) $S_6 - S_5, S_7 - S_6, S_8 - S_7$.

**(b)** Use your answer to part (a) (ii) to explain the value of $S_{n+1} - S_n$ for any positive integer $n$.

**33.** Table 13.2 shows US Census figures,[9] in millions. Interpret these figures as partial sums, $S_n$ of a sequence, $a_n$, where $n$ is the number of decades since 1950, so $S_1 = 179.3$, $S_2 = 203.3$, $S_3 = 226.6$ and so on.

**(a)** Find and interpret in terms of population
  (i) $S_4, S_5, S_6$
  (ii) $a_2, a_5, a_6$
  (iii) $a_6/10$

**(b)** For any positive integer $n$, what are the meanings of $S_n$, $a_n$, and $a_n/10$ in terms of population?

**Table 13.2**

| Year | 1960 | 1970 | 1980 | 1990 | 2000 | 2010 |
|---|---|---|---|---|---|---|
| Population | 179.3 | 203.3 | 226.6 | 248.7 | 281.4 | 308.7 |

Simplify the expressions in Problems 34–37.

**34.** $\sum_{i=1}^{5} i^2 - \sum_{j=0}^{4}(j+1)^2$

**35.** $\sum_{i=4}^{20} i - \sum_{j=4}^{20}(-2j)$

**36.** $\sum_{i=1}^{20} 1$

**37.** $\sum_{i=1}^{15} i^3 - \sum_{j=3}^{15} j^3$

Problems 38–41 refer to the falling object of Example 6 on page 545.

**38.** Calculate the distance, $S_7$, the object falls in 7 seconds.

**39.** **(a)** Find the total distance that the object falls in 4, 5, 6 seconds.

**(b)** The object falls from 1000 feet at time $t = 0$. Calculate its height at $t = 4$, $t = 5$, $t = 6$ seconds. Explain how you can check your answer using Figure 13.2.

**40.** Find a formula for $f(n)$, the distance fallen by the object in $n$ seconds.

**41.** If the object falls from 1000 feet, how long does it take to hit the ground?

**42.** A boy is dividing M&Ms between himself and his sister. He gives one to his sister and takes one for himself. He gives another to his sister and takes two for himself. He gives a third one to his sister and takes three for himself, and so on.

**(a)** On the $n^{\text{th}}$ round, how many M&Ms does the boy give his sister? How many does he take himself?

**(b)** After $n$ rounds, how many M&Ms does his sister have? How many does the boy have?

**43.** An auditorium has 30 seats in the first row, 34 seats in the second row, 38 seats in the third row, and so on. If there are twenty rows in the auditorium, how many seats are there in the last row? How many seats are there in the auditorium?

**44.** **(a)** Show that $n^3 - (n-1)^3 = 3n^2 - 3n + 1$

**(b)** Write $n^3 = n^3 - (n-1)^3 + (n-1)^3 - (n-2)^3 + (n-2)^3 - \ldots - 2^3 + 2^3 - 1^3 + 1^3 - 0^3$ and then show

$$n^3 = \sum_{j=1}^{n}(3j^2 - 3j + 1).$$

**(c)** Use the result $\sum_{j=1}^{n} j = \frac{1}{2}n(n+1)$ to show that

$$\sum_{j=1}^{n} j^2 = \frac{n}{6}(n+1)(2n+1).$$

**45.** In the text we showed how to calculate the sum of an arithmetic series with an even number of terms. Consider the arithmetic series

$$5 + 12 + 19 + 26 + 33 + 40 + 47 + 54 + 61.$$

Here, there are $n = 9$ terms, and the difference between each term is $d = 7$. Adding these terms directly, we find that their sum is 297. In this problem we find the sum of this arithmetic sequence in two different ways. We then use our results to obtain a general formula for the sum of an arithmetic series with an odd number of terms.

**(a)** The sum of the first and last terms is $5+61 = 66$, the sum of the second and next-to-last terms is $12+54 = 66$, and so on. Find the sum of this arithmetic series by pairing off terms in this way. Notice that since the number of terms is odd, one of them will be unpaired.

**(b)** This arithmetic series can be thought of as a series of eight terms $(5+\cdots+54)$ plus an additional term (61). Use the formula we found for the sum of an arithmetic series containing an even number of terms to find the sum of the given arithmetic series.

**(c)** Find a formula for the sum of an arithmetic series with $n$ terms where $n$ is odd. Let $a_1$ be the first term in the series, and let $d$ be the difference between consecutive terms. Show that the two approaches used in parts (a) and (b) give the same result, and show that your formula is the same as the formula given for even values of $n$.

[9]http://www.census.gov, accessed June 27, 2014.

# 13.3 FINITE GEOMETRIC SERIES

In the previous section, we studied *arithmetic series*. An arithmetic series is the sum of terms in a sequence in which each term is obtained by adding a constant to the preceding term. In this section, we study another type of sum, a *geometric series*. In a geometric series, each term is a constant multiple of the preceding term.

## Bank Balance

A person deposits \$2000 every year in an IRA that pays 6% interest per year, compounded annually. After the first deposit (but before any interest has been earned), the balance in the account in dollars is

$$B_1 = 2000.$$

After 1 year has passed, the first deposit has earned interest, so the balance becomes 2000(1.06) dollars. Then the second deposit is made and the balance becomes

$$\begin{aligned} B_2 &= \underbrace{2^{\text{nd}}\text{ deposit}}_{2000} + \underbrace{1^{\text{st}}\text{ deposit with interest}}_{2000(1.06)} \\ &= 2000 + 2000(1.06) \text{ dollars.} \end{aligned}$$

After 2 years have passed, the third deposit is made, and the balance is

$$\begin{aligned} B_3 &= \underbrace{3^{\text{rd}}\text{ deposit}}_{2000} + \underbrace{2^{\text{nd}}\text{ dep. with 1 year interest}}_{2000(1.06)} + \underbrace{1^{\text{st}}\text{ deposit with 2 years interest}}_{2000(1.06)^2} \\ &= 2000 + 2000(1.06) + 2000(1.06)^2. \end{aligned}$$

Let $B_n$ be the balance in dollars after $n$ deposits. Then we see that

$$\begin{aligned} &\text{After 4 deposits} \quad B_4 = 2000 + 2000(1.06) + 2000(1.06)^2 + 2000(1.06)^3 \\ &\text{After 5 deposits} \quad B_5 = 2000 + 2000(1.06) + 2000(1.06)^2 + 2000(1.06)^3 + 2000(1.06)^4 \\ &\qquad\qquad\vdots \\ &\text{After } n \text{ deposits} \quad B_n = 2000 + 2000(1.06) + 2000(1.06)^2 + \cdots + 2000(1.06)^{n-1}. \end{aligned}$$

**Example 1** How much money is in this IRA in 5 years, right after a deposit is made? In 25 years, right after a deposit is made?

**Solution** After 5 years, we have made 6 deposits. A calculator gives

$$\begin{aligned} B_6 &= 2000 + 2000(1.06) + 2000(1.06)^2 + \cdots + 2000(1.06)^5 \\ &= \$13{,}950.64. \end{aligned}$$

In 25 years, we have made 26 deposits. Even using a calculator, it would be tedious to evaluate $B_{26}$ by adding 26 terms. Fortunately, there is a shortcut. Start with the formula for $B_{26}$:

$$B_{26} = 2000 + 2000(1.06) + 2000(1.06)^2 + \cdots + 2000(1.06)^{25}.$$

Multiply both sides of this equation by 1.06 and add 2000, giving

$$1.06B_{26} + 2000 = 1.06\left(2000 + 2000(1.06) + 2000(1.06)^2 + \cdots + 2000(1.06)^{25}\right) + 2000.$$

We simplify the right-hand side to get

$$1.06B_{26} + 2000 = 2000(1.06) + 2000(1.06)^2 + \cdots + 2000(1.06)^{25} + 2000(1.06)^{26} + 2000.$$

Notice that the right-hand side of this equation and the formula for $B_{26}$ have almost every term in common. We can rewrite this equation as

$$1.06B_{26} + 2000 = \underbrace{2000 + 2000(1.06) + 2000(1.06)^2 + \cdots + 2000(1.06)^{25}}_{B_{26}} + 2000(1.06)^{26}$$

$$= B_{26} + 2000(1.06)^{26}.$$

Solving for $B_{26}$ gives

$$\begin{aligned}1.06B_{26} - B_{26} &= 2000(1.06)^{26} - 2000\\ 0.06B_{26} &= 2000(1.06)^{26} - 2000\\ B_{26} &= \frac{2000(1.06)^{26} - 2000}{0.06}.\end{aligned}$$

Using a calculator to evaluate this expression for $B_{26}$, we find that $B_{26} = 118{,}312.77$ dollars.

## Geometric Series

The formula for the IRA balance,

$$B_{26} = 2000 + 2000(1.06) + 2000(1.06)^2 + \cdots + 2000(1.06)^{25},$$

is an example of a *geometric series*. This is a finite geometric series, because there are a finite number of terms (in this case, 26). In general, a geometric series is the sum of the terms of a geometric sequence—that is, in which each term is a constant multiple of the preceding term.

> A **finite geometric series** is a sum of the form
>
> $$S_n = a + ar + ar^2 + \cdots + ar^{n-1} = \sum_{i=0}^{n-1} ar^i.$$

Notice that $S_n$ is defined to contain exactly $n$ terms. Since the first term is $a = ar^0$, we stop at $ar^{n-1}$. For instance, the series

$$50 + 50(1.06) + 50(1.06)^2 + \cdots + 50(1.06)^{25}$$

contains 26 terms, so $n = 26$. For this series, $r = 1.06$ and $a = 50$.

### The Sum of a Geometric Series

The shortcut from Example 1 can be used to find the sum of a general geometric series. Let $S_n$ be the sum of a geometric series of $n$ terms, so that

$$S_n = a + ar + ar^2 + \cdots + ar^{n-1}.$$

Multiply both sides of this equation by $r$, giving

$$\begin{aligned}rS_n &= r\left(a + ar + ar^2 + \cdots + ar^{n-1}\right)\\ &= ar + ar^2 + ar^3 + \cdots + ar^{n-1} + ar^n.\end{aligned}$$

The right-hand side now looks like the $S_n$ with the first term, $a$, missing, and the term $ar^n$ added at the end. Subtract $S_n$ from both sides, writing $S_n$ on the left and $S_n$ in its term-by-term form on the right:

$$\begin{aligned} rS_n - S_n &= (ar + ar^2 + ar^3 + \cdots + ar^{n-1} + ar^n) - \left(\underbrace{a + ar + ar^2 + \cdots + ar^{n-1}}_{S_n}\right) \\ &= ar^n - a. \end{aligned}$$

Solving the equation $rS_n - S_n = ar^n - a$ for $S_n$ gives

$$\begin{aligned} rS_n - S_n &= ar^n - a \\ S_n(r-1) &= ar^n - a \qquad \text{factoring out } S_n \\ S_n &= \frac{ar^n - a}{r-1} \\ &= \frac{a(r^n - 1)}{r-1}. \end{aligned}$$

By multiplying the numerator and denominator by $-1$, this formula can be rewritten as follows:

> The sum of a finite **geometric series of $n$ terms** is given by
>
> $$S_n = a + ar + ar^2 + \cdots + ar^{n-1} = \frac{a(1-r^n)}{1-r}, \qquad \text{for } r \neq 1.$$

This formula is called a *closed form* of the sum.

**Example 2** Use the formula for the sum of a geometric series to solve Example 1.

**Solution** We need to find $B_6$ and $B_{26}$ where

$$B_n = 2000 + 2000(1.06) + 2000(1.06)^2 + \cdots + 2000(1.06)^{n-1}.$$

Using the formula for $S_n$ with $a = 2000$ and $r = 1.06$, we get the same answers as before:

$$\begin{aligned} B_6 &= \frac{2000(1 - (1.06)^6)}{1 - 1.06} = 13{,}950.64, \\ B_{26} &= \frac{2000(1 - (1.06)^{26})}{1 - 1.06} = 118{,}312.77. \end{aligned}$$

## Drug Levels in the Body

Geometric series arise naturally in many different contexts. The following example illustrates a geometric series with decreasing terms.

**Example 3** A patient is given a 20-mg injection of a therapeutic drug. Each day, the patient's body metabolizes 50% of the drug present, so that after 1 day only one-half of the original amount remains, after 2 days only one-fourth remains, and so on. The patient is given a 20-mg injection of the drug every day at the same time. Write a geometric series that gives the drug level in this patient's body right after the $n^{\text{th}}$ injection.

**Solution** Immediately after the $1^{\text{st}}$ injection, the drug level in the body is given by

$$Q_1 = 20.$$

One day later, the original 20 mg has fallen to $20 \cdot \frac{1}{2} = 10$ mg and the second 20-mg injection is given. Right after the second injection, the drug level is given by

$$Q_2 = \underbrace{2^{\text{nd}} \text{ injection}}_{20} + \underbrace{\text{Residue of } 1^{\text{st}} \text{ injection}}_{\frac{1}{2}\cdot 20}$$
$$= 20 + 20\left(\frac{1}{2}\right).$$

Two days later, the original 20 mg has fallen to $(20 \cdot \frac{1}{2}) \cdot \frac{1}{2} = 20(\frac{1}{2})^2 = 5$ mg, the second 20 mg injection has fallen to $20 \cdot \frac{1}{2} = 10$ mg, and the third 20 mg injection is given. Right after the third injection, the drug level is given by

$$Q_3 = \underbrace{3^{\text{rd}} \text{ injection}}_{20} + \underbrace{\text{Residue of } 2^{\text{nd}} \text{ injection}}_{20\cdot\frac{1}{2}} + \underbrace{\text{Residue of } 1^{\text{st}} \text{ injection}}_{20\cdot\frac{1}{2}\cdot\frac{1}{2}}$$
$$= 20 + 20\left(\frac{1}{2}\right) + 20\left(\frac{1}{2}\right)^2.$$

Continuing, we see that

$$\text{After } 4^{\text{th}} \text{ injection} \quad Q_4 = 20 + 20\left(\frac{1}{2}\right) + 20\left(\frac{1}{2}\right)^2 + 20\left(\frac{1}{2}\right)^3$$
$$\text{After } 5^{\text{th}} \text{ injection} \quad Q_5 = 20 + 20\left(\frac{1}{2}\right) + 20\left(\frac{1}{2}\right)^2 + \cdots + 20\left(\frac{1}{2}\right)^4$$
$$\vdots$$
$$\text{After } n^{\text{th}} \text{ injection} \quad Q_n = 20 + 20\left(\frac{1}{2}\right) + 20\left(\frac{1}{2}\right)^2 + \cdots + 20\left(\frac{1}{2}\right)^{n-1}.$$

This is another example of a geometric series. Here, $a = 20$ and $r = 1/2$ in the geometric series formula

$$Q_n = a + ar + ar^2 + \cdots + ar^{n-1}.$$

To calculate the drug level for a specific value of $n$, we use the formula for the sum of a geometric series.

**Example 4** What quantity of the drug remains in the patient's body after the $10^{\text{th}}$ injection?

**Solution** After the $10^{\text{th}}$ injection, the drug level in the patient's body is given by

$$Q_{10} = 20 + 20\left(\frac{1}{2}\right) + 20\left(\frac{1}{2}\right)^2 + \cdots + 20\left(\frac{1}{2}\right)^9.$$

Using the formula for $S_n$ with $n = 10$, $a = 20$, and $r = 1/2$, we get

$$Q_{10} = \frac{20(1 - (\frac{1}{2})^{10})}{1 - \frac{1}{2}} = 39.961 \text{ mg}.$$

# Exercises and Problems for Section 13.3

## Exercises

How many terms are there in the series in Exercises 1–2? Find the sum.

**1.** $\sum_{j=5}^{18} 3 \cdot 2^j$

**2.** $\sum_{k=2}^{20} (-1)^k 5(0.9)^k$

Find the sum of the series in Exercises 3–7.

**3.** $3 + \frac{3}{2} + \frac{3}{4} + \frac{3}{8} + \cdots + \frac{3}{2^{10}}$

**4.** $5 + 15 + 45 + 135 + \cdots + 5(3^{12})$

**5.** $1/125 + 1/25 + 1/5 + \cdots + 625$

**6.** $\sum_{n=1}^{10} 4(2^n)$

**7.** $\sum_{k=0}^{7} 2\left(\frac{3}{4}\right)^k$

In Exercises 8–11, decide which of the following are geometric series. For those which are, give the first term and the ratio between successive terms. For those which are not, explain why not.

**8.** $2 + 1 + \frac{1}{2} + \frac{1}{4} + \frac{1}{8} + \cdots + \frac{1}{128}$

**9.** $1 - \frac{1}{2} + \frac{1}{4} - \frac{1}{8} + \frac{1}{16} + \cdots + \frac{1}{256}$

**10.** $1 + \frac{1}{2} + \frac{1}{3} + \frac{1}{4} + \frac{1}{5} + \cdots + \frac{1}{100}$

**11.** $5 - 10 + 20 - 40 + 80 - \cdots - 2560$

Write each of the sums in Exercises 12–15 in sigma notation.

**12.** $1 + 4 + 16 + 64 + 256$

**13.** $3 - 9 + 27 - 81 + 243 - 729$

**14.** $2 + 10 + 50 + 250 + 1250 + 6250 + 31250$

**15.** $32 - 16 + 8 - 4 + 2 - 1$

Evaluate the sums in Exercises 16–19.

**16.** $\sum_{n=0}^{4} (0.1)^n$

**17.** $\sum_{n=0}^{5} \frac{3}{2^n}$

**18.** $\sum_{j=0}^{n-1} e^{jx}$

**19.** $\sum_{n=0}^{N} (-1)^n$

## Problems

**20.** Write $ar^3 + ar^5 + ar^7 + ar^9 + ar^{11}$ in $\sum$ notation.

**21.** Worldwide consumption of oil was about 88 billion barrels in 2012.[10] Assume that consumption continues to increase at 0.7% per year, the rate for the previous decade.

**(a)** Write a sum representing the total oil consumption for 25 years, starting with 2012.
**(b)** Evaluate this sum.

**22.** Figure 13.3 shows the quantity of the drug atenolol in the blood as a function of time, with the first dose at time $t = 0$. Atenolol is taken in 50-mg doses once a day to lower blood pressure.

**(a)** If the half-life of atenolol in the blood is 6.3 hours, what percentage of the atenolol present at the start of a 24-hour period is still there at the end?
**(b)** Find expressions for the quantities $Q_0, Q_1, Q_2, Q_3, \ldots$, and $Q_n$ shown in Figure 13.3. Write the expression for $Q_n$ in closed form.
**(c)** Find expressions for the quantities $P_1, P_2, P_3, \ldots$, and $P_n$ shown in Figure 13.3. Write the expression for $P_n$ in closed form.

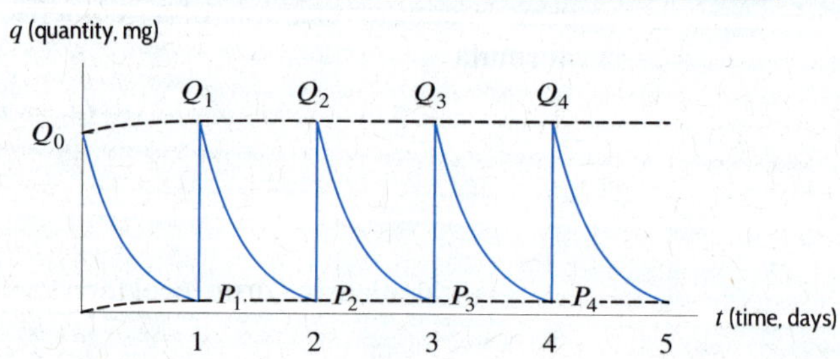


**Figure 13.3**

**23.** Annual deposits of $4000 are made into a bank account earning 2% interest per year. What is the balance in the account right after the 15th deposit if interest is calculated

**(a)** Annually
**(b)** Continuously

**24.** A deposit of $1000 is made once a year, starting today, into a bank account earning 3% interest per year, compounded annually. If 20 deposits are made, what is the balance in the account on the day of the last deposit?

[10] http://www.bp.com/extendedsectiongenericarticle.do?categoryId=9041228&contentId=7074736, accessed May 18, 2013.

25. What effect does doubling each of the following quantities (leaving other quantities the same) have on the answer to Problem 24? Is the answer doubled, more than doubled, or less than doubled?

    (a) The deposit
    (b) The interest rate
    (c) The number of deposits made

26. To save for a new car, you put \$500 a month into an account earning interest at 3% per year, compounded continuously.

    (a) How much money do you have 2 years after the first deposit, right before you make a deposit?
    (b) When does the balance first go over \$10,000?

27. A bank account in which interest is earned at 2% per year, compounded annually, starts with a balance of \$50,000. Payments of \$1000 are made out of the account once a year for 10 years, starting today. Interest is earned right before each payment is made.

    (a) What is the balance in the account right after the tenth payment is made?
    (b) Assume that the tenth payment exhausts the account. What is the largest yearly payment that can be made from this account?

## 13.4 INFINITE GEOMETRIC SERIES

In this section, we look at geometric series with an infinite number of terms and see under what circumstances they have a finite sum.

### Long-term Drug Level in the Body

Suppose the patient from Example 3 on page 550 receives injections over a long period of time. What happens to the drug level in the patient's body? To find out, we calculate the drug level after 10, 15, 20, and 25 injections:

$$Q_{10} = \frac{20(1-(\frac{1}{2})^{10})}{1-\frac{1}{2}} = 39.960938 \text{ mg},$$
$$Q_{15} = \frac{20(1-(\frac{1}{2})^{15})}{1-\frac{1}{2}} = 39.998779 \text{ mg},$$
$$Q_{20} = \frac{20(1-(\frac{1}{2})^{20})}{1-\frac{1}{2}} = 39.999962 \text{ mg},$$
$$Q_{25} = \frac{20(1-(\frac{1}{2})^{25})}{1-\frac{1}{2}} = 39.999999 \text{ mg}.$$

The drug level appears to approach 40 mg. To see why this happens, notice that if there is exactly 40 mg of drug in the body, then half of this amount is metabolized in one day, leaving 20 mg. The next 20 mg injection replaces what was lost, and the level returns to 40 mg. We say that the *equilibrium* drug level is 40 mg.

Initially the patient has less than 40 mg of the drug in the body. Then the amount metabolized in one day is less than 20 mg. Thus, after the next 20 mg injection, the drug level is higher than it was before. For instance, if there are currently 30 mg, then after one day, half of this has been metabolized, leaving 15 mg. At the next injection the level rises to 35 mg, or 5 mg higher than where it started. Eventually, the quantity levels off to 40 mg.

Another way to think about the patient's drug level over time is to consider an *infinite geometric series*. After $n$ injections, the drug level is given by the sum of the finite geometric series

$$Q_n = 20 + 20\left(\frac{1}{2}\right) + 20\left(\frac{1}{2}\right)^2 + \cdots + 20\left(\frac{1}{2}\right)^{n-1} = \frac{20(1-(\frac{1}{2})^n)}{1-\frac{1}{2}}.$$

What happens to the value of this sum as the number of terms approaches infinity? It does not seem possible to add up an infinite number of terms. However, we can look at the *partial sums*, $Q_n$, to see what happens for large values of $n$. For large values of $n$, we see that $(1/2)^n$ is very small, so that

$$Q_n = \frac{20\,(1 - \text{Small number})}{1 - \frac{1}{2}} = \frac{20\,(1 - \text{Small number})}{1/2}.$$

We write $\to$ to mean "tends toward." Thus, as $n \to \infty$, we know that $(1/2)^n \to 0$, so

$$Q_n \to \frac{20(1-0)}{1/2} = \frac{20}{1/2} = 40 \text{ mg}.$$

## The Sum of an Infinite Geometric Series

Consider the geometric series $S_n = a + ar + ar^2 + \cdots + ar^{n-1}$. In general, if $|r| < 1$, then $r^n \to 0$ as $n \to \infty$, so

$$S_n = \frac{a(1-r^n)}{1-r} \to \frac{a(1-0)}{1-r} = \frac{a}{1-r} \quad \text{as } n \to \infty.$$

Thus, if $|r| < 1$, the partial sums $S_n$ approach a finite value, $S$, as $n \to \infty$. In this case, we say that the series *converges* to $S$.

For $|r| < 1$, the **sum of the infinite geometric series** is given by

$$S = a + ar + ar^2 + \cdots + ar^n + \cdots = \sum_{i=0}^{\infty} ar^i = \frac{a}{1-r}.$$

On the other hand, if $|r| > 1$, then we say that the series does not converge. The terms in the series get larger and larger as $n \to \infty$, so adding infinitely many of them does not give a finite sum.

What happens when $r = \pm 1$? For $|r| \geq 1$, the value of $S_n$ does not approach a fixed number as $n \to \infty$, so we say that the infinite series does not converge. (To see why, consider what happens to $1^n$ and $(-1)^n$ as $n$ increases.)

**Example 1** Find the sum of the geometric series $\sum_{i=0}^{17} 7(-z)^i$ and $\sum_{i=0}^{\infty} (-z)^i$ provided $|z| < 1$.

**Solution** We have

$$\sum_{i=0}^{17} 7(-z)^i = 7(-z)^0 + 7(-z)^1 + 7(-z)^2 + \cdots + 7(-z)^{17}.$$

Since $(-z)^0 = 1$, this is a geometric series with $a = 7$ and $r = -z$. There are 18 terms, so

$$\sum_{i=0}^{17} 7(-z)^i = \frac{7(1-(-z)^{18})}{1-(-z)} = \frac{7(1-z^{18})}{1+z}.$$

As for $\sum_{i=0}^{\infty} (-z)^i$, this is an infinite geometric series. Since $|z| < 1$, we have

$$\sum_{i=0}^{\infty} (-z)^i = \frac{1}{1-(-z)} = \frac{1}{1+z}.$$

## Present Value of a Series of Payments

When baseball player Alex Rodriguez was signed by the Texas Rangers, before being traded to the New York Yankees, he was given a signing bonus for \$10 million: \$2 million a year for five years. Of course, since much of the money was to be paid in the future, the team's owners did not have to have all \$10 million available on the day of the signing. How much money would the owners have to deposit in a bank account on the day of the signing in order to cover all the future payments? Assuming the account was earning interest, the owners would have to deposit less than \$10 million. This smaller amount is called the *present value* of \$10 million. We calculate the present value of Rodriguez's bonus on the day he signed.

### Definition of Present Value

Let's consider a simplified version of this problem, with only one future payment: How much money would we need to deposit in a bank account today in order to have \$2 million in one year? At an annual interest rate of 5%, compounded annually, the deposit would grow by a factor of 1.05. Thus,

$$\begin{aligned} \text{Deposit} \cdot 1.05 &= \$2 \text{ million}, \\ \text{Deposit} &= \frac{\$2{,}000{,}000}{1.05} = \$1{,}904{,}761.91. \end{aligned}$$

We need to deposit \$1,904,761.91. This is the *present value* of the \$2 million. Similarly, if we need \$2 million in 2 years, the amount we need to deposit is given by

$$\begin{aligned} \text{Deposit} \cdot (1.05)^2 &= \$2 \text{ million}, \\ \text{Deposit} &= \frac{\$2{,}000{,}000}{(1.05)^2} = \$1{,}814{,}058.96. \end{aligned}$$

The \$1,814,058.96 is the present value of \$2 million payable two years from today. In general,

> The **present value**, $\$P$, of a future payment, $\$B$, is the amount that would have to be deposited (at some interest rate, $r$) in a bank account today to have exactly $\$B$ in the account at the relevant time in the future.

If $r$ is the yearly interest rate (compounded annually) and if $n$ is the number of years, then

$$B = P(1+r)^n, \quad \text{or equivalently,} \quad P = \frac{B}{(1+r)^n}.$$

### Calculating the Present Value of Rodriguez's Bonus

The present value of Rodriguez's bonus represents what it was worth on the day it was signed. Suppose that he receives his money in 5 payments of \$2 million each, the first payment to be made on the day the contract was signed. We calculate the present values of all 5 payments assuming that interest is compounded annually at a rate of 5% per year. Since the first payment is made the day the contract is signed, we have, in millions of dollars,

$$\text{Present value of first payment} = 2.$$

Since the second payment is made a year in the future, in millions of dollars we have:

$$\text{Present value of second payment} = \frac{2}{(1.05)^1} = \frac{2}{1.05}.$$

The third payment is made two years in the future, so in millions of dollars:

$$\text{Present value of third payment} = \frac{2}{(1.05)^2},$$

and so on. Similarly, in millions of dollars:

$$\text{Present value of fifth payment} = \frac{2}{(1.05)^4}.$$

Thus, in millions of dollars,

$$\text{Total present value} = 2 + \frac{2}{1.05} + \frac{2}{(1.05)^2} + \cdots + \frac{2}{(1.05)^4}.$$

Rewriting this expression, we see that it is a finite geometric series with $a = 2$ and $r = 1/1.05$:

$$\text{Total present value} = 2 + 2\left(\frac{1}{1.05}\right) + 2\left(\frac{1}{1.05}\right)^2 + \cdots + 2\left(\frac{1}{1.05}\right)^4.$$

The formula for the sum of a finite geometric series gives

$$\text{Total present value of contract in millions of dollars} = \frac{2\left(1 - \left(\frac{1}{1.05}\right)^5\right)}{1 - \frac{1}{1.05}} \approx 9.09.$$

Thus, the total present value of the bonus is about \$9.09 million dollars.

---

**Example 2** Suppose Alex Rodriguez's bonus with the Rangers guaranteed him and his heirs an annual payment of \$2 million *forever*. How much would the owners need to deposit in an account today in order to provide these payments?

**Solution** At 5%, the total present value of an infinite series of payments is given by the infinite series

$$\begin{aligned}\text{Total present value} &= 2 + \frac{2}{1.05} + \frac{2}{(1.05)^2} + \cdots \\ &= 2 + 2\left(\frac{1}{1.05}\right) + 2\left(\frac{1}{1.05}\right)^2 + \cdots.\end{aligned}$$

The sum of this infinite geometric series can be found using the formula:

$$\text{Total present value} = \frac{2}{1 - \frac{1}{1.05}} = 42 \text{ million dollars.}$$

To see that this answer is reasonable, suppose that \$42 million is deposited in an account today, and that a \$2 million payment is immediately made to Alex Rodriguez. Over the course of a year, the remaining \$40 million earns 5% interest, which works out to \$2 million, so the next year the account again has \$42 million. Thus, it would have cost the Texas Rangers only about \$33 million more (than the \$9.09 million) to pay Rodriguez and his heirs \$2 million a year forever.

**Example 3** Using summation notation, write the present value of Alex Rodriguez's bonus, first for the 5-payment case and then for the infinite sequence of payments.

**Solution** For the 5 payments, we have

$$\text{Total present value} = 2 + \frac{2}{1.05} + \frac{2}{(1.05)^2} + \frac{2}{(1.05)^3} + \frac{2}{(1.05)^4} = \sum_{i=0}^{4} 2\left(\frac{1}{1.05}\right)^i.$$

For the infinite sequence of payments, the series goes to ∞ instead of stopping at 4, so we have

$$\text{Total present value } = 2 + \frac{2}{1.05} + \frac{2}{(1.05)^2} + \cdots = \sum_{i=0}^{\infty} 2\left(\frac{1}{1.05}\right)^i.$$

## Exercises and Problems for Section 13.4

### Exercises

In Exercises 1–7, decide which of the following are geometric series. For those that are, give the first term and the ratio between successive terms. For those that are not, explain why not.

1. $1 - x + x^2 - x^3 + x^4 - \cdots$
2. $y^2 + y^3 + y^4 + y^5 + \cdots$
3. $1 + x + 2x^2 + 3x^3 + 4x^4 + \cdots$
4. $3 + 3z + 6z^2 + 9z^3 + 12z^4 + \cdots$
5. $e^x + e^{2x} + e^{3x} + e^{4x} + \cdots$
6. $1 - e^{-3x} + e^{-6x} - e^{-9x} + \cdots$
7. $1 + \sqrt{2} + 2 + 2\sqrt{2} + \cdots$
8. Find the sum of the series in Problem 1.
9. Find the sum of the series in Problem 2.

Find the sum of the series in Exercises 10–17.

10. $-2 + 1 - \frac{1}{2} + \frac{1}{4} - \frac{1}{8} + \frac{1}{16} - \cdots$
11. $11 - 11(0.1) + 11(0.1)^2 - \cdots$
12. $\sum_{i=2}^{\infty} (0.1)^i$
13. $\sum_{i=4}^{\infty} \left(\frac{1}{3}\right)^i$
14. $\sum_{i=0}^{\infty} \frac{3^i + 5}{4^i}$
15. $\sum_{i=0}^{\infty} \frac{3^i}{2^{2i}}$
16. $\sum_{j=1}^{\infty} 7((0.1)^j + (0.2)^{j+2})$
17. $\sum_{i=1}^{\infty} x^{2i}, \quad |x| < 1$

### Problems

18. A repeating decimal can always be expressed as a fraction. This problem shows how writing a repeating decimal as a geometric series enables you to find the fraction. Consider the decimal 0.232323 … .
    **(a)** Use the fact that 0.232323 … = 0.23 + 0.0023 + 0.000023 + ⋯ to write 0.232323… as a geometric series.
    **(b)** Use the formula for the sum of a geometric series to show that 0.232323 … = 23/99.

In Problems 19–23, use the method of Problem 18 to write each of the decimals as fractions.

19. 0.235235235 …
20. 6.19191919 …
21. 0.12222222 …
22. 0.4788888 …
23. 0.7638383838 …
24. You have an ear infection and are told to take a 250-mg tablet of ampicillin (a common antibiotic) four times a day (every six hours). It is known that at the end of six hours, about 4% of the drug is still in the body. What quantity of the drug is in the body right after the third tablet? The fortieth? Assuming you continue taking tablets, what happens to the drug level in the long run?
25. In Problem 24 we found the quantity $Q_n$, the amount (in mg) of ampicillin left in the body right after the $n^{\text{th}}$ tablet is taken.
    **(a)** Make a similar calculation for $P_n$, the quantity of ampicillin (in mg) in the body right *before* the $n^{\text{th}}$ tablet is taken.
    **(b)** Find a simplified formula for $P_n$.
    **(c)** What happens to $P_n$ in the long run? Is this the same as what happens to $Q_n$? Explain in practical terms why your answer makes sense.

**26.** Draw a graph like that in Figure 13.3 for 250 mg of ampicillin taken every 6 hours, starting at time $t = 0$. Put on the graph the values of $Q_1, Q_2, Q_3, \ldots$ calculated in Problem 24 and the values of $P_1, P_2, P_3, \ldots$ calculated in Problem 25.

**27.** Basketball player Patrick Ewing received a contract from the New York Knicks in which he was offered \$3 million a year for 10 years. Determine the present value of the contract on the date of the first payment if the interest rate is 7% per year, compounded continuously.

Problems 28–30 are about *bonds*, which are issued by a government to raise money. An individual who buys a \$1000 bond gives the government \$1000 and in return receives a fixed sum of money, called the *coupon*, every six months or every year for the life of the bond. At the time of the last coupon, the individual also gets the \$1000, or *principal*, back.

**28.** What is the present value of a \$3000 bond that pays \$50 a year for 10 years, starting one year from now? Assume interest rate is 2% per year, compounded annually.

**29.** What is the present value of a \$10,000 bond that pays \$100 a year for 10 years, starting one year from now? Assume the interest rate is 2% per year, compounded annually.

**30.** **(a)** What is the present value of a \$10,000 bond that pays \$200 a year for 10 years, starting one year from now? Assume the interest rate is 2% per year, compounded annually.

**(b)** Since \$200 is 2% of \$10,000, this bond is often called a 2% bond. What does your answer to part (a) tell you about the relationship between the principal and the present value of this bond when the interest rate is 2%?

**(c)** If the interest rate is more than 2% per year, compounded annually, which one is larger: the principal or the value of the bond? Why do you think the bond is then described as *trading at discount*?

**(d)** If the interest rate is less than 2% per year, compounded annually, why is the bond described as *trading at a premium*?

## CHAPTER SUMMARY

- **Sequences**
  Arithmetic: $a_n = a_1 + (n-1)d$.
  Geometric: $a_n = a_1 r^{n-1}$
- **Sigma Notation**
  $\sum_{i=1}^{n} a_i$
- **Arithmetic Series**
  Partial sum:
  $$S_n = a_1 + a_2 + \cdots + a_n = \frac{1}{2}n(2a_1 + (n-1)d).$$
- **Geometric Series**
  Finite:
  $$S_n = a + ar + ar^2 + \cdots + ar^{n-1} = \frac{a(1-r^n)}{1-r}, \quad r \neq 1.$$
  Infinite:
  $$S = a + ar + ar^2 + \cdots = \frac{a}{1-r},$$
  converges for $|r| < 1$, does not converge for $|r| > 1$.
- **Applications**
  Bank balance; drug levels.
  Present value: $P = \dfrac{B}{(1+r)^n}$ or $P = \dfrac{B}{e^{kt}}$.

# REVIEW EXERCISES AND PROBLEMS FOR CHAPTER THIRTEEN

## Exercises

1. Find the sum of the first 18 terms of the series: $8 + 11 + 14 + \cdots$. What is the $18^{\text{th}}$ term?

2. Write the following using sigma notation: $100+90+80+70+\cdots+0$.

3. (a) Write the sum $\sum_{n=1}^{5}(4n-3)$ in expanded form.
   (b) Compute the sum.

In Exercises 4–6, decide which of the following are geometric series. For those that are, give the first term and the ratio between successive terms. For those that are not, explain why not.

4. $1 + 2z + (2z)^2 + (2z)^3 + \cdots$

5. $1 + \frac{x}{2} + \frac{x^2}{3} + \frac{x^3}{4} + \cdots$

6. $1 - y^2 + y^4 - y^6 + \cdots$

7. Find the sum of the series in Problem 4.

8. Find the sum of the series in Problem 6.

9. Find the sum of the first nine terms of the series: $7+14+21+\cdots$.

## Problems

10. Figure 13.4 shows the first four members in a sequence of square-shaped grids. In each successive grid, new dots are shown in black (•). For instance, the second grid has 3 more dots than the first grid, the third has 5 more dots than the second, and the fourth has 7 more dots than the third.
    (a) Write down the sequence of the total number of dots in each grid (black and white). Using the fact that each grid is a square, find a formula in terms of $n$ for the number of dots in the $n^{\text{th}}$ grid.
    (b) Write down the sequence of the number of black dots in each grid. Find a formula in terms of $n$ for the number of black dots in the $n^{\text{th}}$ grid.
    (c) State the relationship between the two sequences in parts (a) and (b). Use the formulas you found in parts (a) and (b) to confirm this relationship.

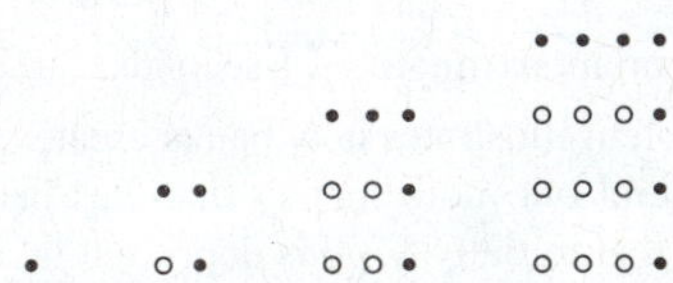

Figure 13.4

11. The numbers in the sequence 1, 4, 9, 16, … from Problem 10 are known as *square numbers* because they describe the number of dots in successive square grids. Analogously, the numbers in the sequence 1, 3, 6, 10, … are known as *triangular numbers* because they describe the number of dots in successive triangular patterns, as shown in Figure 13.5: Find a formula for the $n^{\text{th}}$ triangular number.

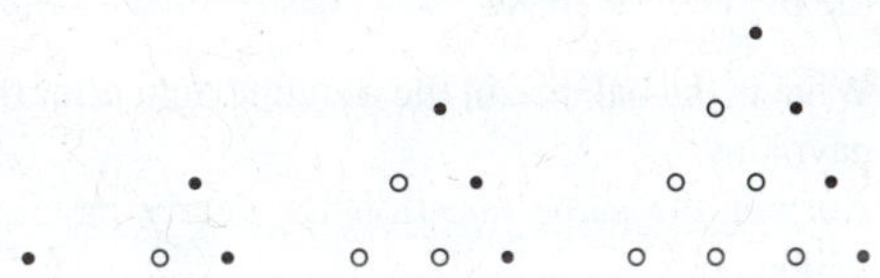

Figure 13.5

12. In a workshop, it costs \$300 to make one piece of furniture. The second piece costs a bit less, \$280. The third costs even less, \$263, and the fourth costs only \$249. The cost for each additional piece of furniture is called the *marginal cost* of production. Table 13.3 gives the marginal cost, $c$, and the change in marginal cost, $\Delta c$, in terms of the number of pierces of furniture, $n$. As the quantity produced increases, the marginal cost generally decreases and then increases again.

Table 13.3

| $n$ | 1 | | 2 | | 3 | | 4 |
|---|---|---|---|---|---|---|---|
| $c$ (\$) | 300 | | 280 | | 263 | | 249 |
| $\Delta c$ (\$) | | −20 | | −17 | | −14 | |

    (a) Assume that the arithmetic sequence −20, −17, −14, …, continues. Complete the table for $n = 5, 6, \ldots, 12$.
    (b) Find a formula for $c_n$, the marginal cost for producing the $n^{\text{th}}$ piece of furniture. Use the fact that $c_n$ is found by adding the terms in an arithmetic sequence. Using your formula, find the cost for producing the $12^{\text{th}}$ piece and the $50^{\text{th}}$ piece of furniture.
    (c) A piece of furniture can be sold at a profit if it costs less than \$400 to make. How many pieces of furniture should the workshop make each day? Discuss.

13. A store clerk has 108 cans to stack. He can fit 24 cans on the bottom row and can stack the cans 8 rows high. Use arithmetic series to determine how he can stack the cans so that each row contains fewer cans than the row beneath it and that the number of cans in each row decreases at a constant rate.

14. A university with an enrollment of 8000 students in 2007 is projected to grow by 2% in each of the next three years and by 3.5% in each of the following seven years. Find the sequence of the university's student enrollment for the next 10 years.

15. Each person in a group of 30 shakes hands with each other person exactly once. How many total handshakes take place?

16. A bank account with a \$75,000 initial deposit is used to make annual payments of \$1000, starting one year after the initial \$75,000 deposit. Interest is earned at 2% a year, compounded annually, and paid into the account right before the payment is made.

   **(a)** What is the balance in the account right after the 24$^{\text{th}}$ payment?
   **(b)** Answer the same question for yearly payments of \$3000.

17. One way of valuing a company is to calculate the present value of all its future earnings. A farm expects to sell \$1000 worth of Christmas trees once a year forever, with the first sale in the immediate future. What is the present value of this Christmas tree business? Assume that the interest rate is 4% per year, compounded continuously.

18. You inherit \$100,000 and put the money in a bank account earning 3% per year, compounded annually. You withdraw \$2000 from the account each year, right after the interest is earned. Your first \$2000 is withdrawn before any interest is earned.

   **(a)** Compare the balance in the account right after the first withdrawal and right after the second withdrawal. Which do you expect to be higher?
   **(b)** Calculate the balance in the account right after the 20$^{\text{th}}$ withdrawal is made.
   **(c)** What is the largest yearly withdrawal you can take from this account without the balance decreasing over time?

19. After breaking his leg, a patient retrains his muscles by going for walks. The first day, he manages to walk 300 yards. Each day after that he walks 50 yards farther than the day before.

   **(a)** Write a sequence that represents the distances walked each day during the first week.
   **(b)** How far is he walking after two weeks?
   **(c)** How long until he is walking at least one mile?

20. Before email made it easy to contact many people quickly, groups used telephone trees to pass news to their members. In one group, each person is in charge of calling 4 people. One person starts the tree by calling 4 people. At the second stage, each of these 4 people call another 4 people. In the third stage, each of the people in stage two calls 4 people, and so on.

   **(a)** How many people have the news by the end of the 5$^{\text{th}}$ stage?
   **(b)** Write a formula for the total number of members in a tree of 10 stages.
   **(c)** How many stages are required to cover a group with 5000 members?

21. A ball is dropped from a height of 10 feet and bounces. Each bounce is $3/4$ of the height of the bounce before. Thus after the ball hits the floor for the first time, the ball rises to a height of $10(3/4) = 7.5$ feet, and after it hits the floor for the second time, it rises to a height of $7.5(3/4) = 10(3/4)^2 = 5.625$ feet.

   **(a)** Find an expression for the height to which the ball rises after it hits the floor for the $n^{\text{th}}$ time.
   **(b)** Find an expression for the total vertical distance the ball has traveled when it hits the floor for the first, second, third, and fourth times.
   **(c)** Find an expression for the total vertical distance the ball has traveled when it hits the floor for the $n^{\text{th}}$ time. Express your answer in closed form.

22. You might think that the ball in Problem 21 keeps bouncing forever since it takes infinitely many bounces. This is not true! It can be shown that a ball dropped from a height of $h$ feet reaches the ground in $\frac{1}{4}\sqrt{h}$ seconds. It is also true that it takes a bouncing ball the same amount of time to rise $h$ feet. Use these facts to show that the ball in Problem 21 stops bouncing after

$$\frac{1}{4}\sqrt{10} + \frac{1}{2}\sqrt{10}\sqrt{\frac{3}{4}}\left(\frac{1}{1-\sqrt{3/4}}\right)$$

seconds, or approximately 11 seconds.

23. This problem illustrates how banks create credit and can thereby lend out more money than has been deposited. Suppose that initially \$100 is deposited in a bank. Experience has shown bankers that on average only 8% of the money deposited is withdrawn by the owner at any time. Consequently, bankers feel free to lend out 92% of their deposits. Thus \$92 of the original \$100 is loaned out to other customers (to start a business, for example). This \$92 will become someone else's income and, sooner or later, will be redeposited in the bank. Then 92% of \$92, or \$92(0.92) = \$84.64, is loaned out again and eventually redeposited. Of the \$84.64, the bank again loans out 92%, and so on.

   **(a)** Find the total amount of money deposited in the bank.

**(b)** The total amount of money deposited divided by the original deposit is called the *credit multiplier*. Calculate the credit multiplier for this example and explain what this number tells us.

**24.** Take a rectangle whose sides have length 1 and 2, divide it into two equal pieces, and shade one of them. Now, divide the unshaded half into two equal pieces and shade one of them. Take the unshaded area and divide it into two equal pieces and shade one of them. Continue doing this.

**(a)** Draw a picture that illustrates this process.

**(b)** Find a series that describes the shaded area after $n$ divisions.

**(c)** If you continue this process indefinitely, what would the total shaded area be? Make two arguments, one using your picture, another using the series you found in part (b).

## STRENGTHEN YOUR UNDERSTANDING

Are the statements in Problems 1–34 true or false? Give an explanation for your answer.

**1.** If $a_n = n^2 + 1$, then $a_1 = 2$.

**2.** The first term of a sequence must be a positive integer.

**3.** The sequence $1, 1, 1, 1, \ldots$ is both arithmetic and geometric.

**4.** The sequence $a, 2a, 3a, 4a$ is arithmetic.

**5.** The sequence $a, a+1, a+2, a+3$ is arithmetic.

**6.** If the difference between successive terms of an arithmetic sequence is $d$, then $a_n = a_1 + dn - d$.

**7.** The sequence $3x^2, 3x^4, 3x^6, \ldots$ is geometric.

**8.** The sequence $2, 4, 2, 4, 2, 4, \ldots$ is alternating.

**9.** For all series $S_1 = a_1$.

**10.** The accumulation of college credit is an example of a finite series.

**11.** The sum formula for a finite series with $n$ terms is $S_n = \frac{1}{2}n(n+1)$.

**12.** There are four terms in $\sum_{i=0}^{4} a_i$.

**13.** $\sum_{i=1}^{n} 3 = 3n$.

**14.** The sum of a finite arithmetic series is the average of the first and last terms, times the number of terms.

**15.** If $C_n$ denotes your college credits earned by the end of year $n$, then $C_4$ is your credits earned during your fourth year of college.

**16.** If $C_n$ denotes your college credits earned by the end of year $n$, then $C_3 - C_2$ is your credits earned in year 2.

**17.** For any series $S_n$, we must have $S_3 \le S_4$.

**18.** If $a_n = (-1)^{n+1}$ then the sequence is alternating.

**19.** The sum formula for a finite geometric series, $S_n = \dfrac{a(1-r^n)}{1-r}$, can be simplified by dividing by $(1-r)$ to get $S_n = a(1 - r^{n-1})$.

**20.** With the same interest rate, a bank balance after 5 years of \$2000 annual payments is the same as the balance after by 10 years of \$1000 annual payments.

**21.** $1 - \frac{1}{2} + \frac{1}{4} - \frac{1}{8} + \frac{1}{16} - \frac{1}{32}$ is a geometric series.

**22.** $2000(1.06)^3 + 2000(1.06)^4 + \cdots + 2000(1.06)^{10}$ is a geometric series.

**23.** $5000 + 2000(1.06) + 2000(1.06)^2 \cdots + 2000(1.06)^{10}$ is a geometric series.

**24.** $5000 + 2000(1.06) + 2000(1.06)^2 \cdots + 2000(1.06)^{10} = 3000 + \dfrac{2000(1 - 1.06^{11})}{-0.06}$.

**25.** Payments of \$2000 are made yearly into an account earning 1% interest compounded annually. If the interest rate is doubled, the balance at the end of 20 years is doubled.

**26.** The present value of a sequence of payments is the amount that you would need to invest at this moment in order to exactly make the future payments.

**27.** If the present value of a sequence of payments is 1 million dollars, then the present value of a sequence of payments which are doubled in size would be 2 million.

**28.** The series $\sum_{i=0}^{\infty} \dfrac{1}{i^2}$ is geometric.

**29.** For $Q_n = \sum_{i=1}^{n} (-1)^i$, we see $Q_2 = 0, Q_4 = 0, Q_6 = 0$, etc.

Thus, the series converges to a sum of $Q = \sum_{i=1}^{\infty} (-1)^i = 0$.

**30.** A patient takes 100 mg of a drug each day; 5% of the amount of the drug in the body decays each day. Then, the long-run level of the drug in the body is 2000 mg.

**31.** If $S = a + ar + ar^2 + \cdots$ and $|a| < 1$ then the infinite series converges.

**32.** It is possible to add an infinite number of positive terms and get a finite sum.

**33.** The sum of an infinite arithmetic series with difference $d = 0.01$ converges.

**34.** All infinite geometric series converge.

# ANSWERS TO ODD NUMBERED PROBLEMS

## Section 1.1

S1 $(3/2)c$

S2 $1.09P$

S3 $6\pi r^2$

S4 $5/3$

S5 $51/2$

S6 $-26$

S7 $1$

S8 $-12$

S9 $A = (-2, 8)$

S10 $A = (3, 21)$

1 $m = f(v)$

3

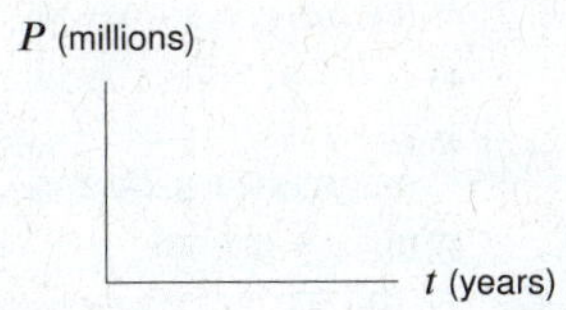


5

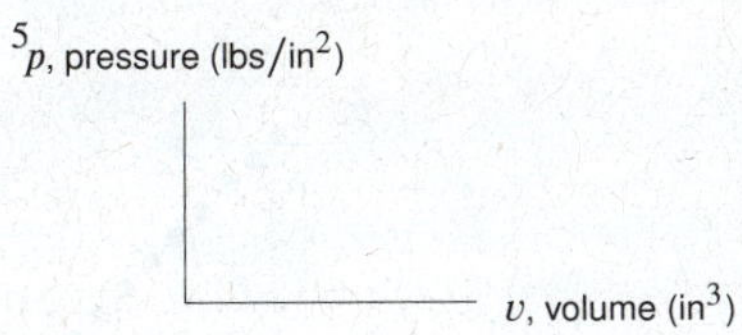


7 (a) 2
(b) 2
(c) About 12 hours

9 (a) 40
(b) 2

11 2.9

13 0, 4, 8

15 (a) $w$
(b) $(-4, 10)$
(c) $(6, 1)$

17 One ($y$ is a function of $x$)

19 Both

21 No

23 Yes

25 Yes

27

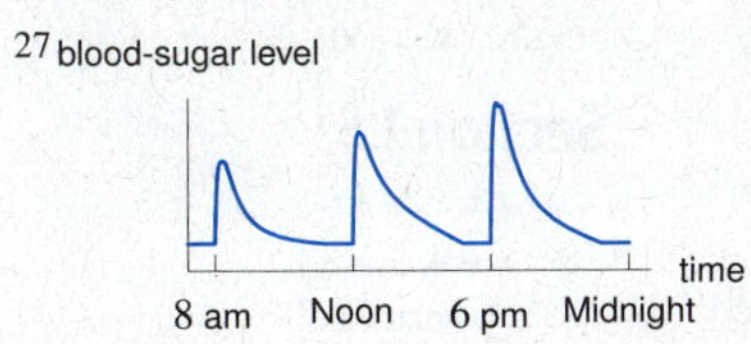


29

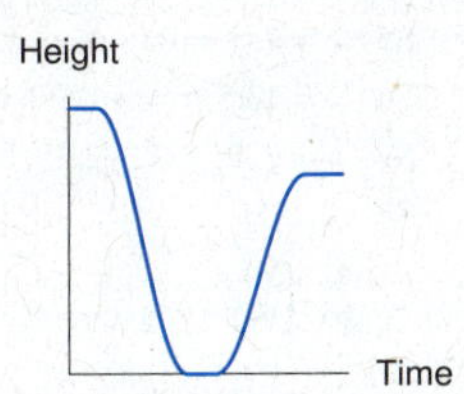


31 (a) $r_m(0) - r_h(0) = 22$
(b) $r_m(9) - r_h(9) = -2$
(c) $r_m(t) < r_a(t)$ for $t = 5$ to $t = 9$

33 (a) (ii)
(b) (i)
(c) (v)
(d) (iv)
(e) (ii)

35 (a) 100.3 m. own phones in 2000
(b) 20 m. own phones $a$ years after 1990
(c) $b$ m. own phones in 2010
(d) $n$ m. own phones $t$ years after 1990

37 (a) $72\pi$ ft$^3$
(b) $45\pi$ ft$^3$
(c) $V(h) = 9\pi h$

39 (a) 69°F
(b) July 17 and 20
(c) Yes
(d) No

41 (a) No
(b) Yes
(c) In 1981, record was 3 min 47.33 sec
(d) 1967, record of 3 min 51.1 sec

43 $0.945P$

45 (a) $x + y$
(b) $0.15x + 0.18y$
(c) $(15x + 18y)/(x + y)$

47 (a) Yes
(b) No

## Section 1.2

S1 $-2$

S2 $2/5$

S3 $-2$

S4 $-7$

S5 $-1$

S6 $-x + 2a + 3b$

S7 $-3x^2 - 4ax - a^2$

S8 $3x^2 + 2bx - b^2$

S9 $x + y$

S10 $4x + 2h$

1 Increasing $x > 0$, decreasing $x < 0$

3 Increasing $-1 < x < 1$, decreasing $x < -1$ and $x > 1$

5 (a) $-82.5, 180$
(b) $-76, 307$

7 (a) 80/3 songs per year
(b) $-20$ songs per year
(c) 0 songs per year

9 (a) (i) $\Delta D = 90$, $\Delta t = 3$
(ii) $\Delta D = 60$, $\Delta t = 2$
(iii) $\Delta D = 45$, $\Delta t = 1.5$
(b) 30 miles/hour

11 $0 \le x \le 4$, $0 \le x \le 8$, $4 \le x \le 8$,
$2.2 \le x \le 5.2$, $2.2 \le x \le 6.9$,
$5.2 \le x \le 6.9$,
other answers possible

13 Equal

15 $F(-2) > F(2)$

17 (a) $A = (10, 30)$
$B = (30, 40)$
$C = (50, 90)$
$D = (60, 40)$
$E = (90, 40)$
(b) Point $F$ is on the graph.
(c) Increasing: 6–21, 36–51, 66–81
(d) Decreasing: 22–35, 52–65, 82–96

19 (a) 2
(b) Increasing everywhere
(c)

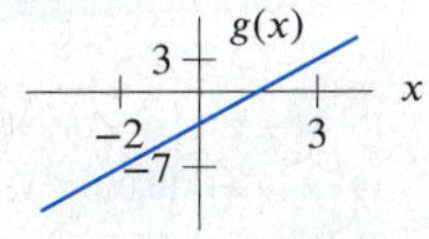


23 24.5 degrees/minute

25 (a) Year determines number
(b) $1964 < t < 1969$
$1971 < t < 1972$
$1976 < t < 1979$
$1986 < t < 1989$
$1996 < t < 2000$

27 Between 3000 and 4000 years ago

29 (a) 9
(b) $\dfrac{n-k}{m-j}$
(c) $6x + 3h$

31 (a) 10, 10, 10, 10, 5, 5
(b) 33, 30.5, 56.7, 34.2, 10.2, −2.8
(c) No; $\Delta G/\Delta t$ not constant
(d) A recycling and composting program in the US

## Section 1.3

S1 $f(0) = 5, f(3) = 7$

S2 $f(0) = 17, f(3) = 5$

S3 5

S4 $-3$

S5 $y = 3; x = 3/4$

S6 $y = -2; x = 5/4$

S7 $7/2, -2$

S8 $-8, 9$

S9 $-ab + a + 3, a - 3$

S10 $-2, 5$

1 (a) $y = -2 + 3x$
(b) $y = -1 + 2x$

3 (a) $y = (1/4)x$
(b) $y = 1 - 6x$

5 Not linear

7 Not linear

9 Could be linear
Rate of change $-10$

11 Vert int: 17.75 in; Slope: 1/250 in/yr

13 Vert int: \$29.99; Slope: \$0.05/min

15 (a) $r = f(v)$ could be linear
(b) \$5 increase/mph

(c) 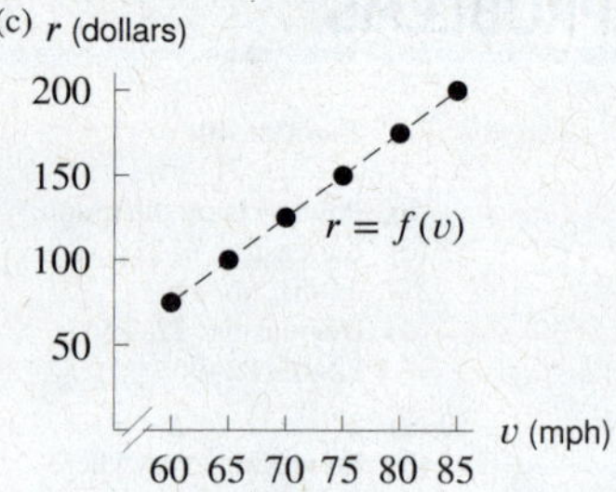


17 $V = 23{,}500 - 1559t$

19 $\pi(n) = -10{,}000 + 127n$

21 (a) 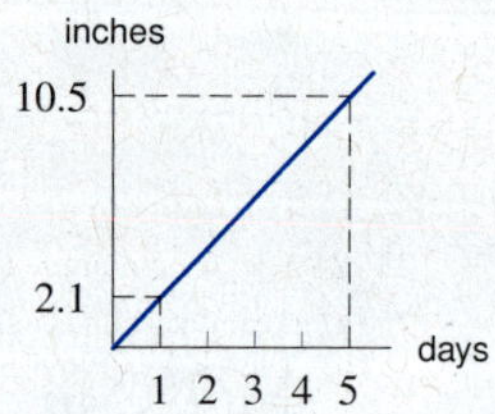


(b) 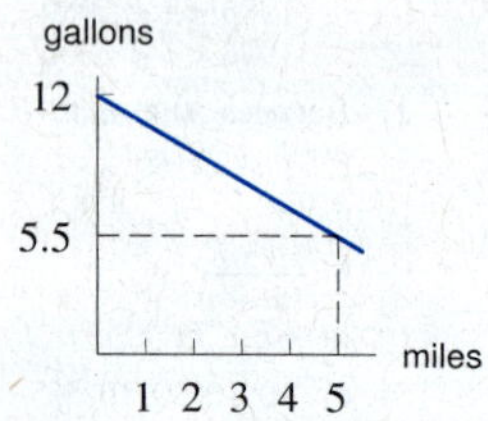


23 (a) Side length and perimeter

(b) 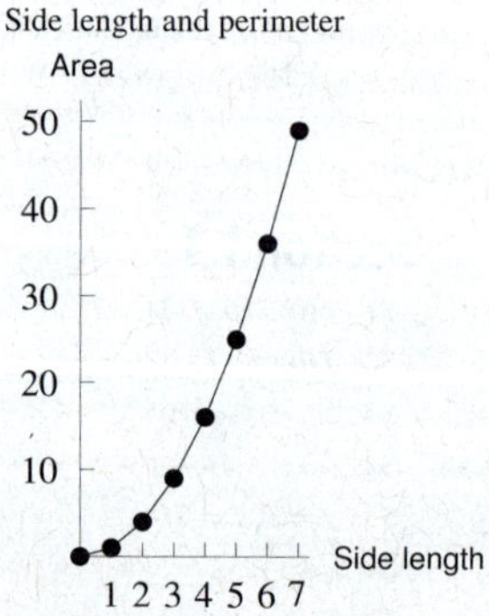


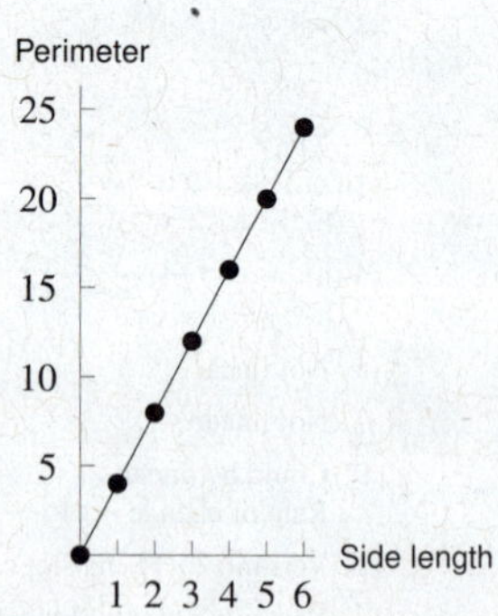


(c) 4

25 (a) Country A is Afghanistan; Country B is Sri Lanka.
(b) 0.24 million/year.
(c) 16.8 million

27 (a) No
(b) Looks linear

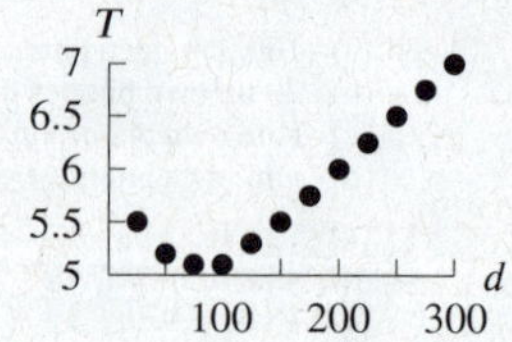


(c) $\Delta T/\Delta d = 0.01°\text{C/meter}$

29 (a) $T = \$1900$
(b) $C = 7$
(d) Twelve credits
(e) Fixed costs that do not depend on the number of credits taken

31 (a) $r = 5/2, s = 16$
(b) $k = 0.2, j = -3$

35 No

## Section 1.4

S1 $y = 26$

S2 $x = 9$

S3 $x = 2$

S4 $y = -(7/41)$

S5 $y = -17/16$

S6 $P = I/(rt)$

S7 $F = (9/5)C + 32$

S8 $r = C/(2\pi)$

S9 $x = (c - ab)/(2a)$

S10 $y = (c + d)/(b - a)$

1 $y = 28 - 4x$

3 $y = 3 - 2x$

5 $y = -5 + 5x/3$

7 $y = 1.5 - 2.4x$

9 $V = 2000 - 500t$

11 $y = -160/9 + (5/9)x$

13 $p = 125{,}000 + 5000h$

15 $s = 9 - 0.5q$

17 $y = 4 - (3/5)x$

19 $y = 2/3 + (5/3)x$

21 $y = 5 + 0x$

23 Not possible

25 Yes; $g(w) = -1/3 + 4w$

27 Not linear

29 Not linear

31 Parallel

33 Parallel

35 Neither

37 $f(x) = -12.5 - 1.5x$

39 $h(t) = 12{,}000 + 225t$

41 (a) 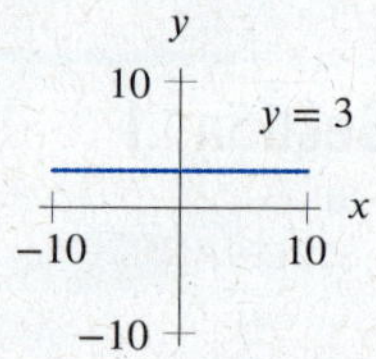


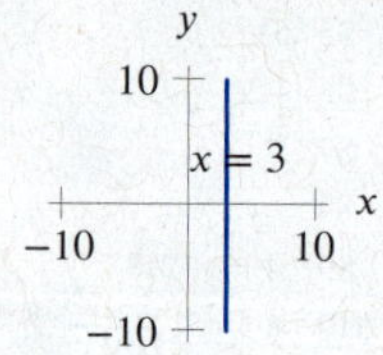


(b) Yes ($y = 3 + 0x$), No

43 $y = -\frac{1}{5}x + \frac{7}{5}$

45 (a) $f$
(b) $f(x) = 1.16x - 2.38$

47 (b) $v = 40 - 32t$

49 \$10,500

51 (b) $p = 12 - s$
(c) 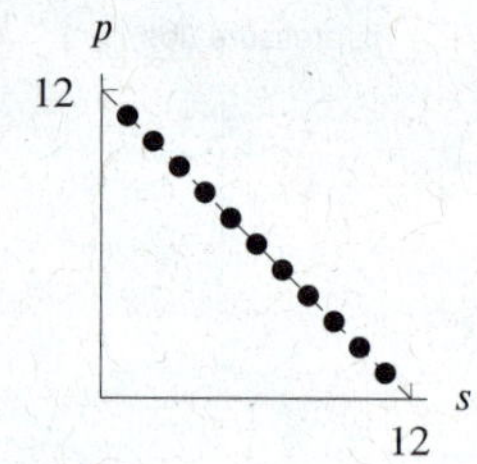


53 (a) $W = -6254.46 + 3.2368t$
(b) 282.26 million tons

55 $h(t) = 254 - 248t$

57 $y = -4 + 4x$

59 (a) $i(x) = 2.5x$
(b) $i(0) = 0$

61 (a) $p = 0.1t - 1$, and $t \geq 10$
(b) 11
(d) $t = 10p + 10$
(e) 2 hours 40 minutes

63 $w(r) = \pi x^2 - s\sqrt{x} + (-3x - 4s)r$;
$b = \pi x^2 - s\sqrt{x}; m = -3x - 4s$

65 (a) $r = 0.005H - 0.03$
(b) $S = 200$

## Section 1.5

S1 $x = -2,\ y = 5$

S2 $x = 4,\ y = -1$

S3 No solution

S4 Infinitely many solutions

S5 $x = 7, y = 4$

S6 $x = 13.5; y = -12$

S7 $x = 3/2, y = 3/2$

S8 $(3, 2.5)$

1 (a) (V)
(b) (VI)
(c) (I)
(d) (IV)
(e) (III)
(f) (II)

5 (a) (II)
(b) (IV)
(c) (I)
(d) (III)

7 Slope $= -2/5$
$y = -(2/5)x + 3$
(answers may vary)

9 (1, 0)

11 (a) Company $A$: $20 + 0.2x$
Company $B$: $35 + 0.1x$
Company $C$: 70

(b)
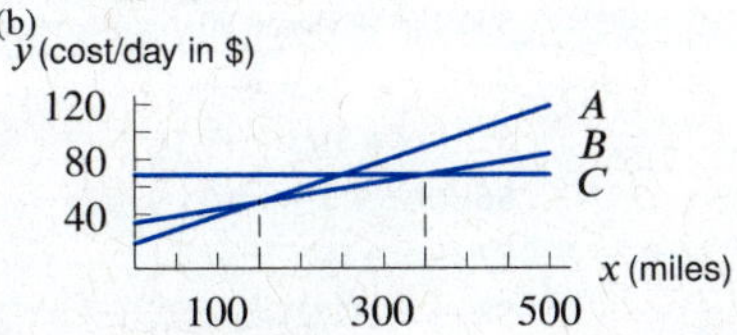


(c) Slope: mileage rate
Vertical intercept: fixed cost/day

(d) $A$ for $x < 150$
$B$ for $150 < x < 350$
$C$ for $x > 350$

13 $n < 3.5$ drinks

15 $g < 8.333$ gals

17 (b) (i)
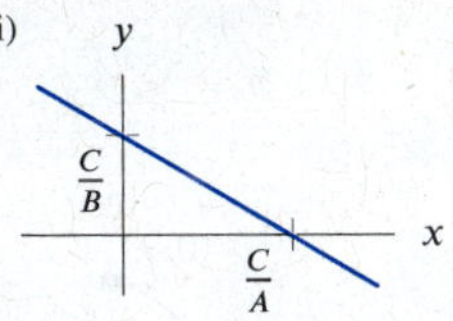


(ii)
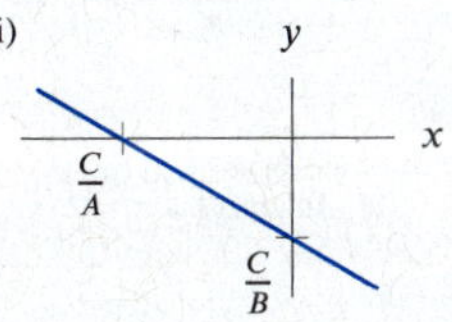


(iii)
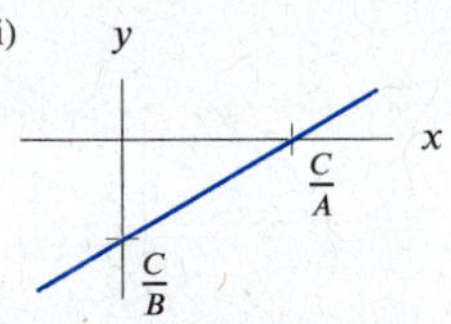


19 $y = x/3 + 2/3$

23 (a) $y = -\sqrt{3}x$
(b) $y = (1/\sqrt{3})x + 4/\sqrt{3}$

25 (a) $m_1 = m_2$ and $b_1 \neq b_2$
(b) $m_1 = m_2$ and $b_1 = b_2$
(c) $m_1 \neq m_2$
(d) Not possible

27 $\beta = 5$

29 (a) Demand decreases to 100
(b) $D = 1100 - 200p$
(c) \$5.25

31 (a)
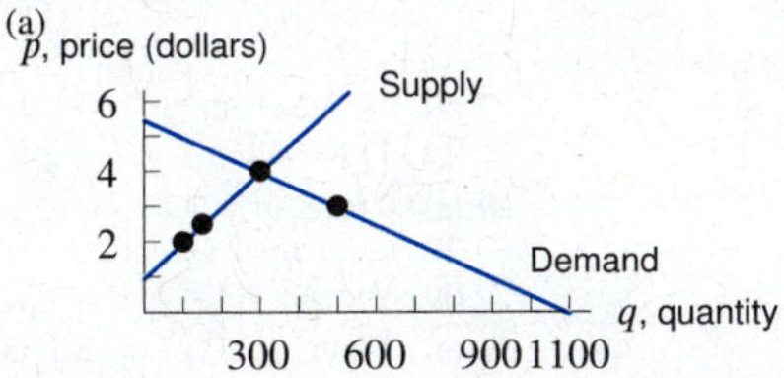


## Section 1.6

1 $r = 1$ not reasonable

3 $r = 0.9$ reasonable

5 $r = 0.9$ not reasonable

7 (a) $r = 1$
(b) $r = 0.7$
(c) $r = 0$
(d) $r = -0.98$
(e) $r = -0.25$
(f) $r = -0.5$

11 (a)
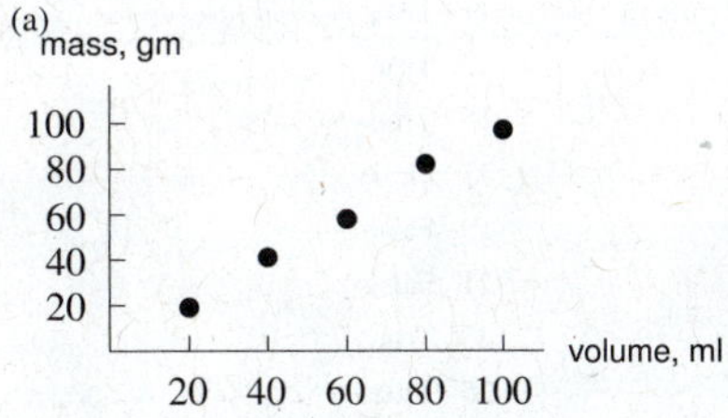


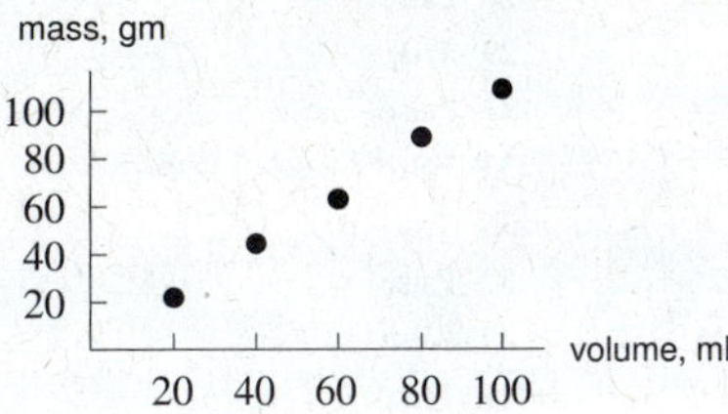


(b) $y = 0.3 + 0.985x$; $y = -0.05 + 1.0925x$

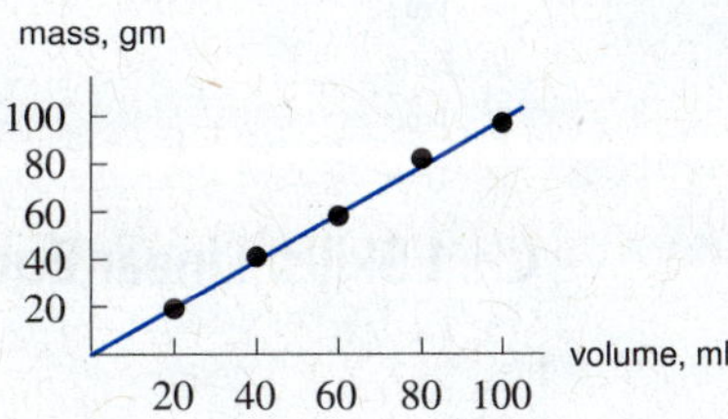


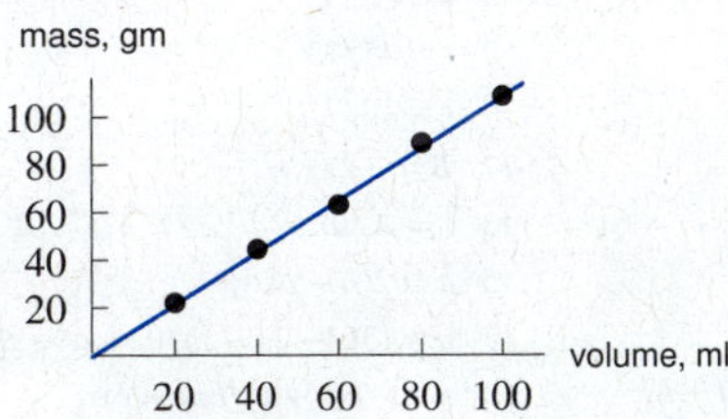


(c) gm/ml; density
(d) Salt water has higher density

13 (a)
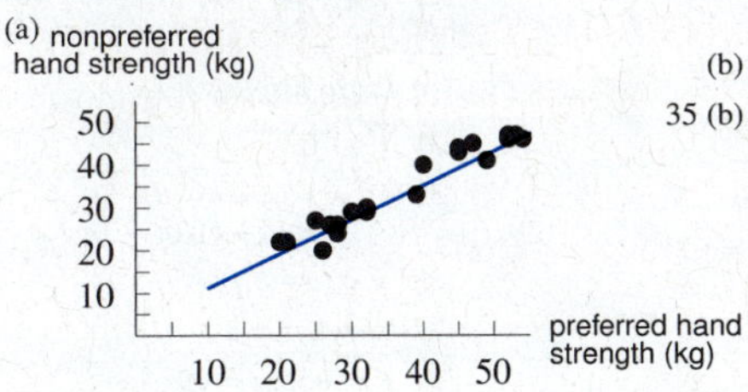


(c) $y = 3.623 + 0.825x$
(d) $\approx 34$
(e) $\approx 40$

## Chapter 1 Review

1 (a) No
(b) Yes

3 (a) (I), (III), (IV), (V)
(VII), (VIII)
(b) (i) (V) and (VI)
(ii) (VIII)
(c) (III) and (IV)

5 (a) 330, −80 people/yr
(b) 183, −75 people/yr
(c) 259, −94 people/yr

7 (b) July
(c) Increasing: Jan–July
Decreasing: July–Dec

9 No

11 $g(x) = 80 - 0.05x$

13 $f(t) = 708 - 122t$

15 (a) $y = 7 + 2x$
(b) $y = 8 - 15x$

17 Neither

19 Perpendicular

21 120

23 500 m

25
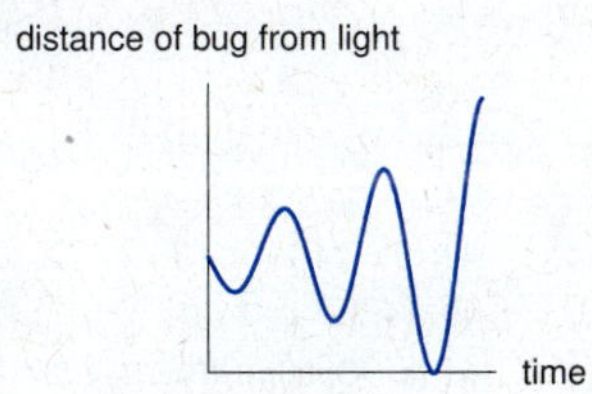


27
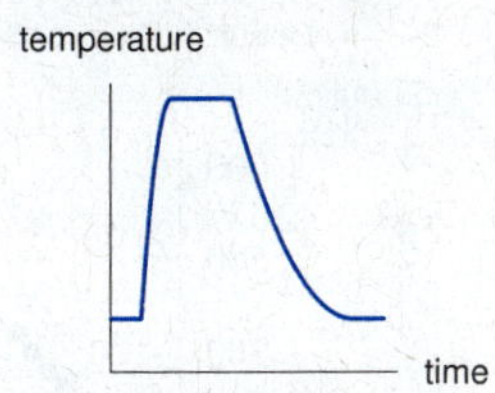


29 $T(d) = d/5 + (10 - d)/8$

31 $s = 1440 - w$

33 (a) (i) 1/2
(ii) 1/2
(iii) 1/2
(b) Always 1/2

35 (b)
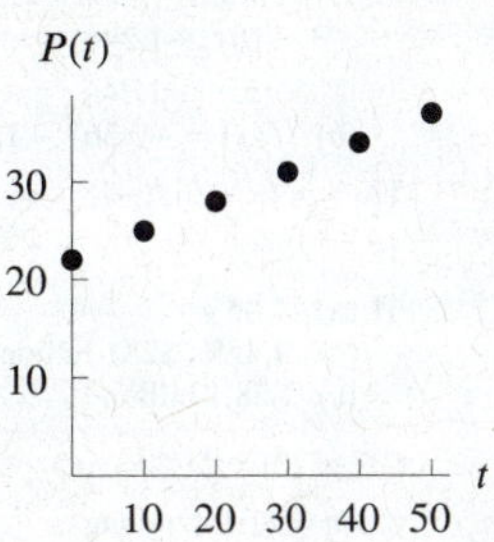

(c) 22 million
(d) 0.3 million people/year

37 (a) $F = 2C + 30$
(b) $-3°$, $-2°$, $1°$, and $4°$ respectively
(c) 10°C

39 $y = \frac{13}{\sqrt{8}} + \left(\frac{\sqrt{8}-\sqrt{13}}{\sqrt{8}}\right)x$ or
$y = \frac{13\sqrt{2}}{4} + \frac{4-\sqrt{26}}{4}x$

43 $f(r) = (x^3 + 3x^2 + 2)r + (4sx + 7s + 3)$; $m = x^3 + 3x^2 + 2$, $b = 4sx + 7s + 3$

45 (a)

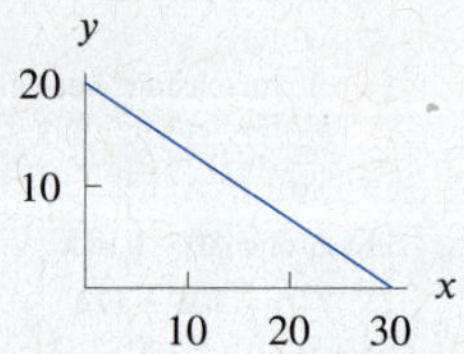


(b) $f(x) = 20 - (2/3)x$
(d)

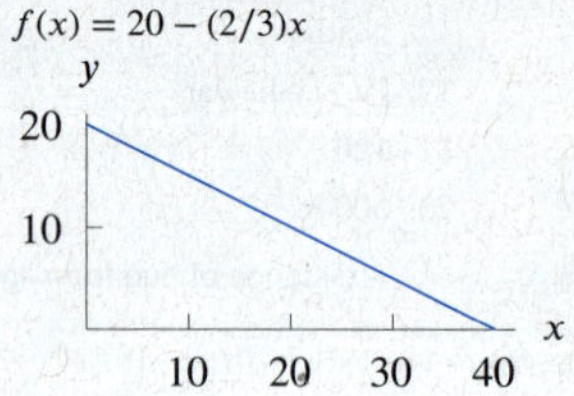


47 $y = 24 - 4x$

49 $(4, 3)$

51 (a) Option 1 if sales < \$500, Option 2 if sales > \$500
(b) Second option
(c) Second option
(d) Option 1 if sales < \$1500, Option 2 if sales > \$1500

53 (a)

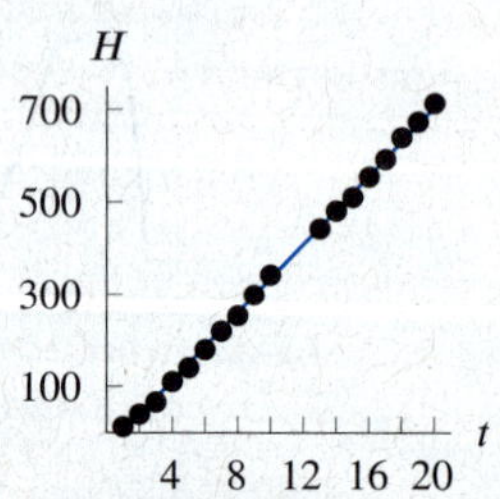


(b) Estimates will vary, e.g. $H = 37t - 37$
(c) $H = 37.26t - 39.85$, $r = 0.9995$, $r \approx 1$. The correlation between the data set and the regression line is very good.
(d) Slope gives the average number of home runs per year, about 37

55 (a) (i) $-1.246$
(ii) $-1.246$
(iii) $-1.246$
(b) $f(x) = -0.367 - 1.246x$

57 $y = (x + 6)/(-2)$; $k = 6$, $z = -2$

59 $l_1$

61 (a) 4.45%
(b) 4.45%; \$281 billion
(c) \$28.1 billion

63 $g(22) = 5^{1/4} = \sqrt[4]{5}$

65 $n = 8$; not unique

67 (a) $L = W\sqrt{2}$
(b) $w = 21.02$ cm, $L = 29.73$ cm

## Ch. 1 Understanding

1 False
3 True
5 True
7 True
9 True
11 True
13 True
15 True
17 False
19 False
21 False
23 True
25 True
27 False
29 True
31 False
33 False
35 True
37 False
39 False
41 True
43 False
45 True
47 True
49 False
51 True
53 True

## Ch. 1 Skills: Linear Equations

1 $x = 5$
3 $z = 11/2$
5 $w = -11$
7 $t = 45/13$
9 $t = 10/7$
11 $B = -2$
13 $l = A/w$
15 $a = 2(h - v_0 t)/t^2$
17 $v = (3w - 2u - z)/(u + w - z)$
19 $x = -a(b + 1)/(ad - c)$
21 $y' = 4/(y + 2x)$
23 $n > 150$
25 $x > -1$
27 $k \geq -1$
29 $-1/0.019 \geq y$
31 $t < -0.5$
33 (a) $x > (a + b)/(a - 3)$
(b) $x < (a + b)/(a - 3)$
35 $x < (a - b)/8$
37 $x = 4$, $y = 3$
39 $x = -55$, $y = 39$
41 $x = 1$, $y = a$
43 $x = 3$, $y = 6$
45 $(-8.75, -0.625)$
47 $(-0.03, -0.4)$
49 $(-52.6316, 2052.6316)$
51 $A = (-4, 7)$
53 $A = (2, 9)$, $B = (10, 1)$
55 $A = (-7, 8)$, $B = (-3, 4)$

## Section 2.1

S1 $5x - 15$
S2 $2a^2 + 5a$
S3 $4m^2 - 38m + 90$
S4 $3x^2 - 2x - 16$
S5 $(3x + 3)/3$
S6 $(3x^2 + 2 - x^3)/(x^2)$
S7 $x = \pm 3$
S8 $x = (37/2)$
S9 $(18 \pm \sqrt{285})/3$
S10 $x = 4^{2/3}$

1 3/2
3 (a) $-4$
(b) $\pm 2$
5 (a) $-1/2$
(b) $-1$
7 54
9 $(0, 2)$

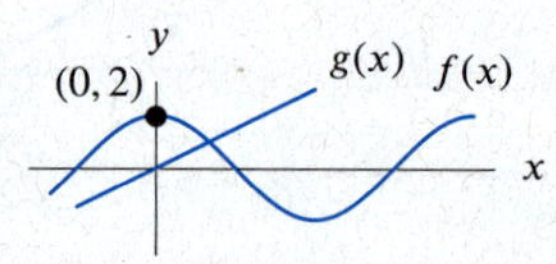


11 Intersect at $x = 2$

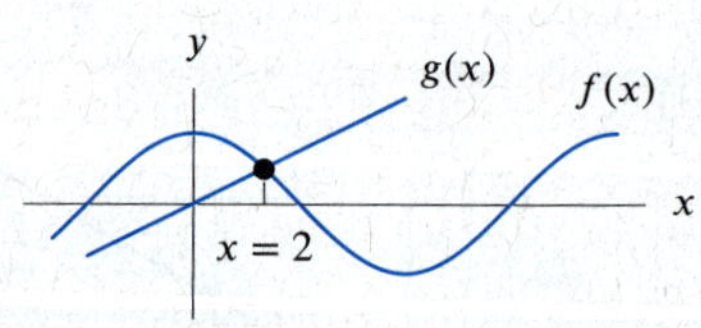


13 $x = 0$
15 $x < 1$
17 100
19 (a) Salary \$158,100 in 2004
(b) Salary \$169,300 in 2008
21 (a) 11,000
(b) 16.064
(c) 2,541,000
23 (a) (i) $1/(1 - t)$
(ii) $-1/t$
(b) $x = 3/2$
25 $\frac{2h}{h}$
27 (a) 48 ft for both
(b) 4 sec, 64 ft
29 (a) 0 ft/sec
(b) $t = 0$, $t = 2$ seconds
(c) Velocity at $t = 3$; ft/sec
31 (a) 18; the speed of the car, in feet per second, 5 seconds later than time $t$

(b) 96; a speed 5 feet per second faster than the car's speed at time $t$
(c) 2.3; the time 2 seconds prior to when the car has slowed to 40 feet per second
(d) 5; the time when the car has slowed to 30 feet per second

33 (a) \$5372.16
(b) $T(x) = 0.8x$
(c) $L(x) = 0.0532x - 479.84$
(d) \$5372.16

35 (a) Same values; same function
(b) Same values; not same function

39 (a) $h(1) = b + c + 1$
(b) $h(b+1) = 2b^2 + 3b + c + 1$

41 $a - a^2$

43

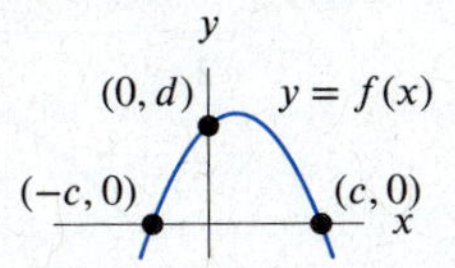


## Section 2.2

S1 $x = 3$

S2 0, 3

S3 $x < 15$

S4 $x > 15$

S5 $x > 8$

S6 $x < 5$

S7 $n < 0$

S8 $a \leq 3$

S9 $x > 5$ or $x < -5$.

S10 $-6 \leq x \leq 6$

S11 $a \geq 2$ and $a \leq -2$.

S12 $y > 3$ or $y < -3$.

1 D: $1 \leq x \leq 7$; R: $2 \leq f(x) \leq 18$

3 D: $1 \leq x \leq 5$; R: $1 \leq f(x) \leq 6$

5 $f(x) \leq -(1/2)$ or $f(x) \geq (1/2)$

7 $-4 \leq f(x) \leq 5$

9 D: all real numbers $\neq -3$

11 D: all real numbers $\neq -3$

13 D: all real numbers $> -9$

15 D: $x \geq 2$ or $x \leq -2$

17 D: all real numbers

19 D: all real numbers
R: all real numbers

21 Domain: $x > 4$
Range: $y > 0$

23 $a = 3$

25 Any $a \geq 0$

27 $y = 1/(x+2) + \sqrt{x+4}$
Other answers possible

29 D: $0 \leq t \leq 12$
R: $0 \leq f(t) \leq 200$

31 Domain: integers $0 \leq n \leq 200$
Range: $0, 4, 8, \ldots, 800$

33 D: All real numbers $\neq 0$;
R: All real numbers $\neq a$

35 (a) 9
(b) Domain: $1, 2, \ldots, N$,
where $N$ = number of listings

(c) Range: $0, 1, \ldots, 9$

37 (a) $p(0) = 50$
$p(10) \approx 131$
$p(50) \approx 911$
(c) $50 \leq p(t) < 1000$

39 (a) $A = 2\pi r^2 + 710/r$
(b)

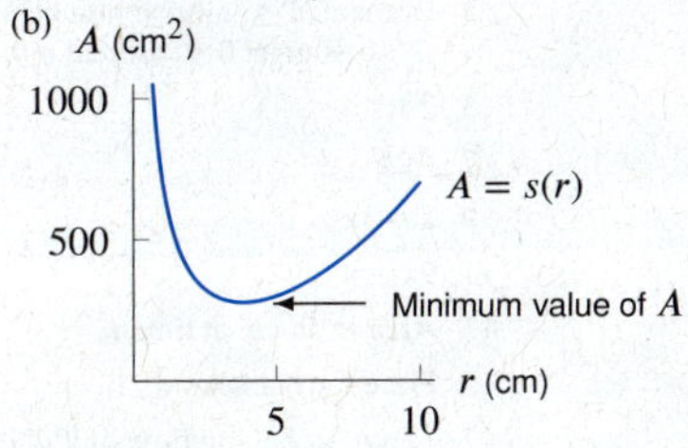


(c) Domain: $r > 0$
Approximate range: $A > 277.5$ cm$^2$
(d) $\approx 277.5$ cm$^2$
$r \approx 3.83$ cm
$h \approx 7.7$ cm

## Section 2.3

S1 $x > 0$

S2 $x \leq -3$

S3 $2 \leq x \leq 3$

S4 $0 < x \leq 4$

S5 $x \leq -1$ or $x \geq 2$

S6 $x < -1$ or $x \geq 4$

S7 Domain: $2 \leq x < 6$
Range: $3 \leq x < 5$

S8 Domain: $1 < x < 7$
Range: $-8 < x < -1$

S9 Domain: $-2 \leq x \leq 3$
Range: $-2 \leq x \leq 3$

S10 Domain: $-2 \leq x < 4$
Range: $1 \leq x \leq 5$

1

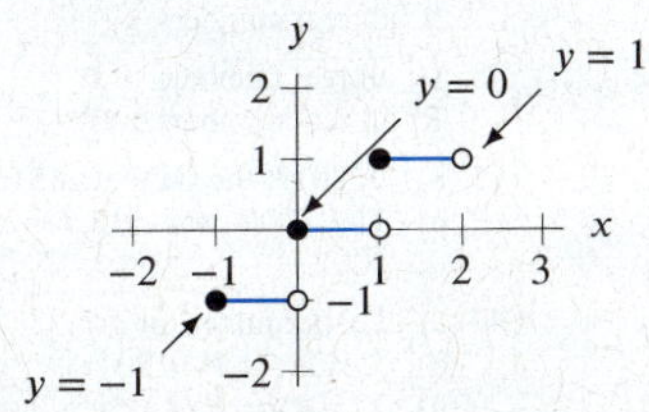


3

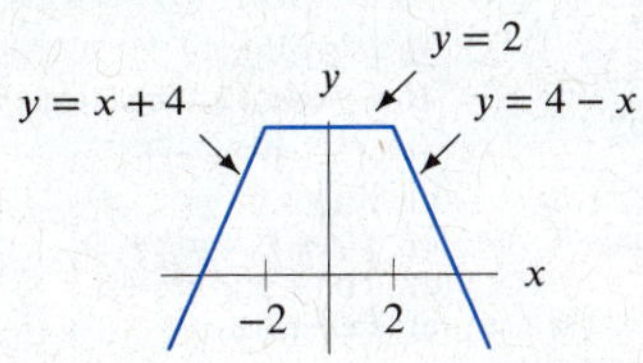


5 $y = \begin{cases} 5 - x & \text{for } x < 3 \\ -1 + (1/2)x & \text{for } x \geq 3 \end{cases}$

7 $y = \begin{cases} 4 - \frac{1}{2}x & \text{for } 1 \leq x \leq 3 \\ -9 + 2x & \text{for } 5 \leq x \leq 8 \end{cases}$

9 Domain: all reals;
Range: $G(x) < 0$ and $G(x) \geq 3$

11 $x = \pm 5$

13 $x = -4$ and $x = 10$.

15 $x = -1$ and $x = 7$.

17 (a) Yes
(b) No
(c) $y = 1, 2, 3, 4$

19 (c) Domain: all $x$, $x \neq 0$
Range: $-1$ and $1$
(d) False, $u(0)$ is undefined

21 (a) $f(0) = 0$, $f(3) = 1$
(b) Domain: $-1 \leq x \leq 5$;
Range: $-3 \leq f(x) \leq 3$.

23 $y = \begin{cases} x^2 & \text{for } x < 0 \\ x - 1 & \text{for } x \geq 0 \end{cases}$

25 (b) \$5.40
(c) 1.4 mile
(d)

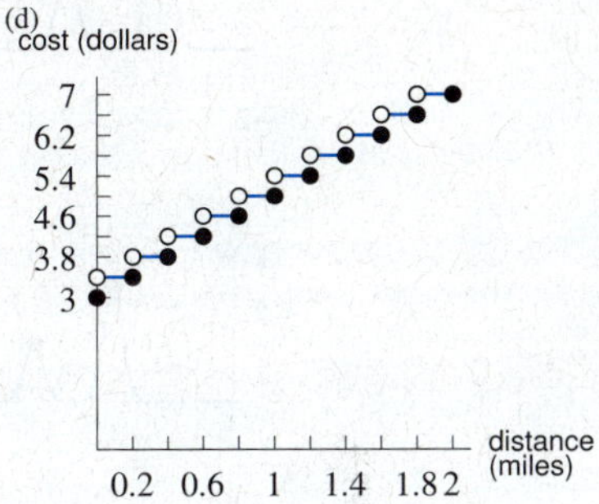


27 (a) \$1.01
(b) $y = \begin{cases} 1 + x & \text{for } 0 < x < 0.1 \\ 10x + x & \text{for } 0.1 \leq x \leq 0.5 \\ 5 + x & \text{for } x > 0.5 \end{cases}$
(c) \$4
(d)

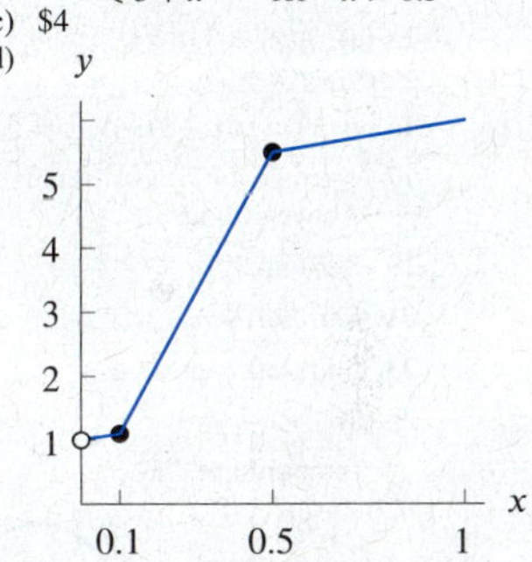


29 (b) 13-69

31 (a) $f(x) = \begin{cases} 2x - 6 & \text{for } x \geq 3 \\ 6 - 2x & \text{for } x < 3 \end{cases}$
(b)

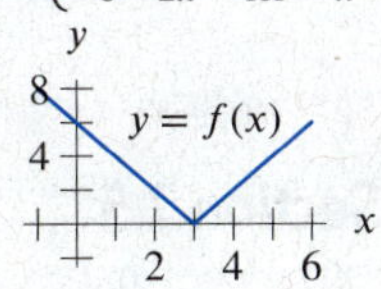


## Section 2.4

S1 2

S2 8

S3 -1

S4 3

S5 $-2, 2$

S6 $-3, 3$

S7 $-1, 3$

S8 $-3, 1$

1 (a) $-3, 0, 2, 1, -1$
One unit right
(b) $-3, 0, 2, 1, -1$
One unit left
(c) $0, 3, 5, 4, 2$
Up three units
(d) $0, 3, 5, 4, 2$
One right and three up

3

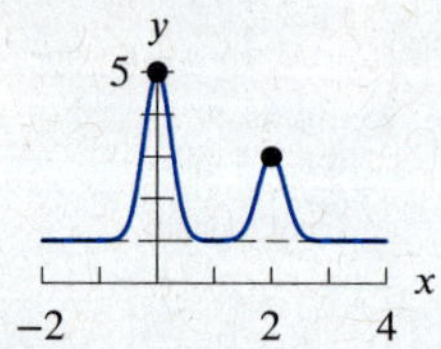


5

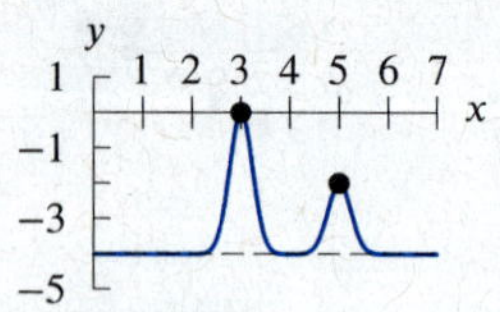


7 (a) $(3, 1)$
(b) $(-2, -4)$
(c) $(6, -6)$

9 $-50 \leq R(s) - 150 \leq 50$

11 $y = f(x-2) - 6$

13 (a) $g(x) + 3$
(b) $g(x-2)$

15 $w(x) = v(x-5) - 7; h = 5, k = -7$

17 Average for 7 mos, 10 mos
Above average

19 Shift up 3

21 Shift left 4

23 Shift left $b$, down $a$

25 (a)

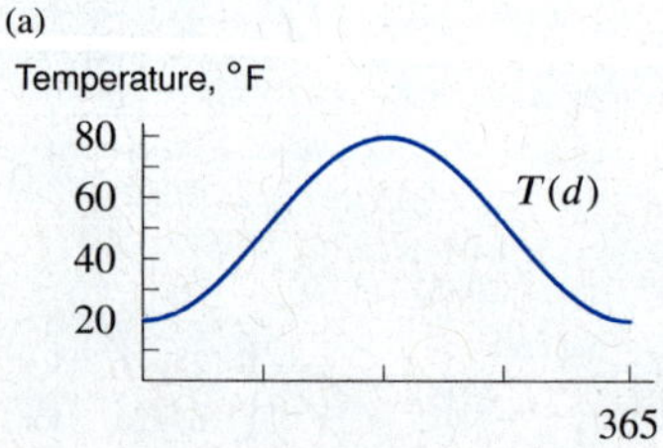


27 $H(t) - 37$

## Section 2.5

S1 $y = (x+4)/3$

S2 $(7-4x)/(-3)$

S3 $y = (2x+1)/(x-2)$

S4 $y = x/(x+4)$

S5 $y = x^2 + 4x + 4$

S6 $y = \sqrt[3]{x+4}$

S7 $y = (\sqrt[3]{5-x})/2$

S8 $x$

S9 $4x^2 + 4x - 3$

S10 $3y^2 - 12y + 5$

S11 $t^2 - t$

1 Domain: $-3 \leq$ all real numbers $\leq 2$
Range: $-2 \leq$ all real numbers $\leq 3$

3 Domain: $1 \leq$ all real numbers $\leq 4.5$
Range: $0 \leq$ all real numbers $\leq 6$

5 2

7 $-24$

9 $2 - 3x^2$

11 $9x - 4$

13 Area in sq cm at time $t$

15 Price for diameter $d$

17 Temp for baking time of $T$; °F

19 Ounces of almonds having $x$ calories

21 $f^{-1}(y) = (y-3)/2$

23 $f^{-1}(y) = (y-1)^{1/3}$

25 $f^{-1}(A) = \sqrt{A/\pi}$

27 $f^{-1}(P) = P/(3-2P)$

29 $k(3) = 2; k^{-1}(14) = 6$

31 $23/4; -2$

33 (a) $b$
(b) $a$
(c) $a$
(d) $b$

35 $A = f^{-1}(n) = 250n$

37 $f^{-1}(f(A)) = A$
$f(f^{-1}(n)) = n$

39 (a) (i) 2
(ii) 1
(iii) 1
(iv) 2
(b) $f(0) = 2$ means $f^{-1}(2) = 0$
(c) $f(1) = 0$ means $f^{-1}(0) = 1$

41 D: all real numbers
R: all real numbers

43 D: all real numbers $> 2$
R: all real numbers $> 0$

45 (a) In 2013, the GDP was \$16,011.2 billion.
(b) The GDP was \$13,776 dollars in the year 2007.

47 (a) 12, perimeter for $s = 3$
(b) 5; side for $P = 20$
(c) $f^{-1}(P) = P/4$

49 $t = f^{-1}(H) = \frac{9}{5}H + 32 = t$

51 $20 + (50/9)2^{-n}$;
$H = f(g(n))$ is temperature in °C at time $n$

53 (a) $A = f(r) = \pi r^2$
(b) $f(0) = 0$
(c) $f(r+1) = \pi(r+1)^2$
(d) $f(r) + 1 = \pi r^2 + 1$
(e) Centimeters

55 (a) $\pi(d/2)^2$
(b) $0.01196A$
(c) $0.01196 \cdot \pi(d/2)^2$;
Cost of pepperoni for $d$-in pizza
(d) Area $95.033\text{in}^2$; Pepperoni cost \$1.14

57 (a) $f(60) = 30$
(b) 40 meters
(c) $f^{-1}(70) = 100$

59 (a) Domain: $0 \leq t \leq 4$, range: $375 \leq C(t) \leq 383$
(b) $C(4)$ is the concentration in 2008; $C(4) = 383$
(c) $C^{-1}(381)$ is number of years after 2004 when the concentration was 381 ppm.

## Section 2.6

1 Concave down

3 Concave up

5 Concave up

7 Concave up

9 Rates of change: 0.043, 0.073, 0.090; Concave up

11 Possible graph:

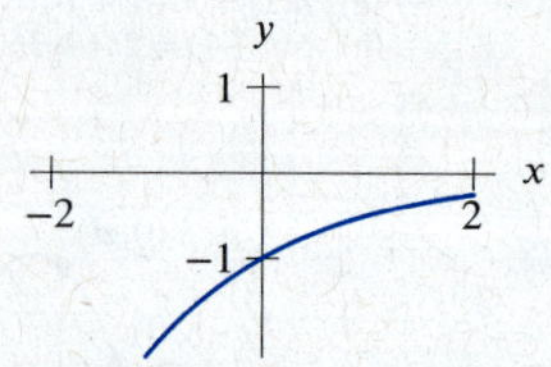


13 (a) $x < 0$
(b) Nowhere
(c) $x > 0$
(d) Nowhere

15 (a) Nowhere
(b) $x < 0$
(c) Nowhere
(d) $x > 0$

17 Increasing;
concave down

19 Increasing;
concave up then down

21 Increasing;
concave up then down

23 No

25 (a) F, IV
(b) G, I
(c) E, II
(d) H, III

27 (a) Larger swims twice as fast
(b) Increasing, concave down

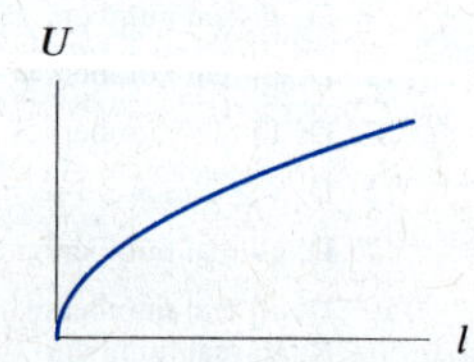


(c) $\sqrt{l}$ increasing
(d) $\sqrt{l}$ concave down;
greater

## Chapter 2 Review

1 $f(-7) = -9/2$

3 $-3/31; 1$

5 $32; \sqrt[3]{9/4}$

7 (a) 1
(b) $-1/2$

9 $-8$

11 (a) $2, 0, -2$
(b) $x = -1$

13 Domain: $x \geq 3$ or $x \leq -3$
Range: $q(x) \geq 0$

15 Domain: all real numbers
Range: all real numbers

17 D: all real numbers
R: all real numbers

21 (a) $2(1-x)$
(b) $2-x$
(c) $x$
(d) $(1-x)^2$
(e) 0
(f) $\sqrt{1-x}$

23 $(3x-7)^3+1$

25 Period in sec at time $t$

27 26

29 7

31 $2x^2+5$

33 $4x+9$

35 Interest rate for \$$I$ interest; %/year

37 D: all real numbers;
R: $h(x) \geq 6$

39 $f^{-1}(P) = (P+2)/14$.

41 Time, yrs, at which pop is $P$ mil

43 Rates of change: 4.35, 4.10, 3.80; Concave down

45 8; 81

47

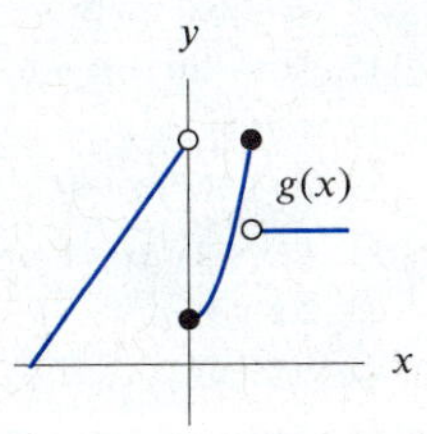


49 (a) $-22$
(b) $3-a^2$
(c) $-a^2+10a-22$
(d) $-a^2-2$
(e) $-a^2+25$

51 $g(x) = f(x+2) - 1$

53 (a) $-1$
(b) $x = \pm 3$
(c) 0
(d) $-1$
(e) $3, -3$

55 (a) $-2.9 < x < 0,\ 1.4 < x < 3$
(b) $-4 < x < -1.8,\ 0.8 < x < 3$

57

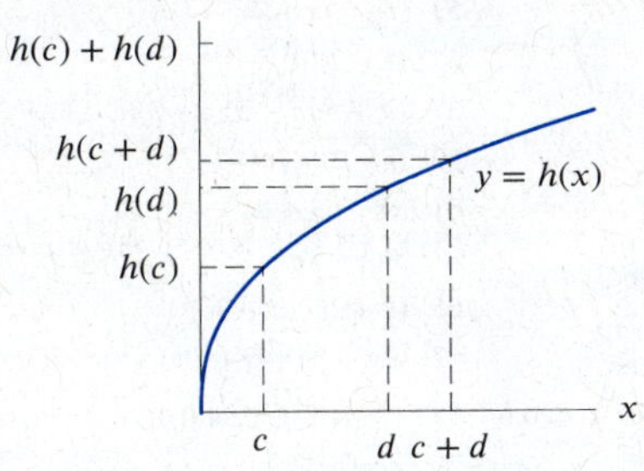


59 $f^{-1}(V) = \sqrt[3]{3V/4\pi}$

61 (a) $s^2$
(b) $0.1475A$
(c) $0.1475s^2$; cost of staining deck of side $s$ ft
(d) (i) 64 $\text{ft}^2$
(ii) \$11.80
(iii) \$14.75

63 (a) $P = f(s) = 4s$
(b) $f(s+4) = 4(s+4) = 4s+16$
(c) $f(s)+4 = 4s+4$
(d) Meters

65 (a) 2
(b) Unknown
(c) 5

67 (a) $(5, 3)$
(b) $(21, 5)$
(c) $\sqrt{7} - \sqrt{6}$

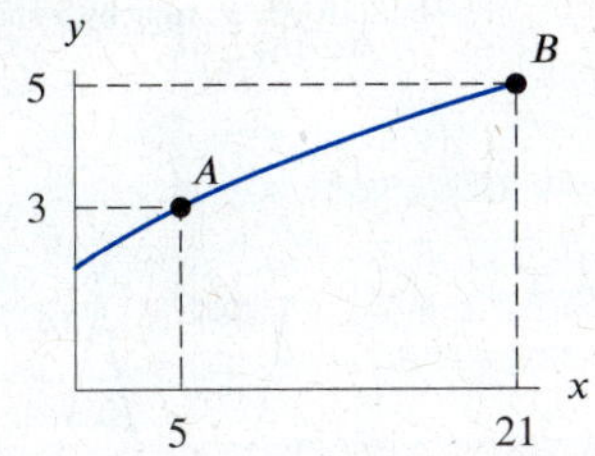


69 (b) No
(c) $f(0) = 0$
$f(-1) = 1$
$f(-2) = -1$
$f(0.5)$: undefined

73 $y = \sqrt{x-4} + 1/(x-8)$

75 (a) Increasing until year 60, then decreasing
(c) Appears concave up
(d) Greatest between 40 and 60; smallest between 60 and 70
(f) 1840; potato famine

## Ch. 2 Understanding

1 False

3 False

5 False

7 False

9 True

11 True

13 False

15 False

17 True

19 True

21 True

23 False

25 True

27 False

29 True

31 True

33 True

35 True

37 False

39 True

41 True

43 True

45 True

## Section 3.1

S1 $-200t$

S2 $A^4 - 2A^2B^2 + B^4$

S3 $u(u-2)$

S4 $(x+2)(x+1)$

S5 $(3x-4)(x+1)$

S6 $(s+2t+2p)(s+2t-2p)$

S7 $(4x-1)(4x+1)$

S8 $y(y-4)(y+3)$

S9 $x=-6$ or $x=-1$

S10 $w = (-5/2)$ or $w = 2$

1 $g(t) = 3t^2 - 12t + 19; a = 3, b = -12, c = 19$

3 $w(n) = 3n^2 + 6n + 0; a = 3, b = 6, c = 0$

5 Not quadratic

7 Yes; $T(n) = (\sqrt{3} - 1/2)n^2 + \sqrt{5}$

9 6, $-3$

11 (a) $x \approx 0.057, x \approx 1.943$
(b) No real solutions

13 $x = -1/3$

15 2 and 5

17 6 and $-3$

19 $x = 16$

21 Concave up

23 Concave up

25 $y = (7/4)(x+2)^2$

27 No

29 Impossible

31 Concave down

33

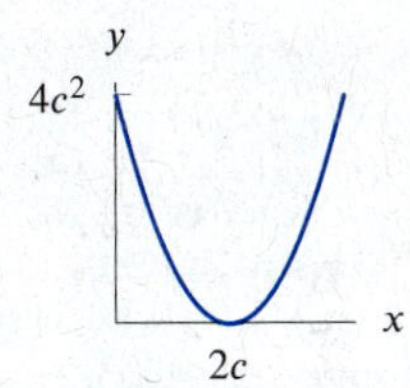


35 $y = (1/3)x^2 - (2/3)x - 1$

37 $y = (6/7)(x+1)(x-5)$

39 $k = -30, r = 8, s = 0.2$

41 (a) 5 meters below the edge
(b) 0.4184, 2.4387
(c) 2.0203 sec

43 Yes; 262.8 million; no

45 (a) 5 km
(b) 4430 m
(c) $h \approx -0.000000255d^2 + 5$

## Section 3.2

S1 $(y-6)^2 - 36$

S2 $(s+3)^2 - 17$

S3 $(c+3/2)^2 - 37/4$

S4 $4(s+1/8)^2 + 31/16$

S5 $r = 4, 2$

S6 $n = 6, -3$

S7 $q = 1/5 \pm \sqrt{41}/5$

S8 $t = (2 \pm \sqrt{31})/3$

S9 $n = -5, 1$

S10 $q = -1 \pm \sqrt{15/2}$

1 $(1, 2)$; $x = 1$; opens upward

3 Vertex: $(-11/2, -137/4)$
Axis of symmetry: $t = -11/2$

5 (a) $a = 1, b = 0, c = 3$
Axis of symmetry: $y$-axis
Vertex: $(0, 3)$
No zeros
$y$-intercept: $y = 3$

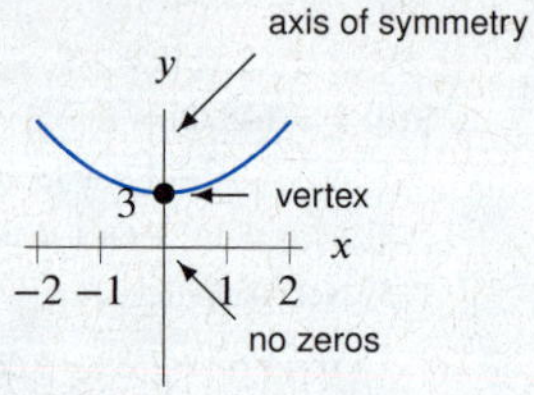


(b) $a = -2, b = 4, c = 16$
Axis of symmetry: $x = 1$
Vertex: $(1, 18)$
Zeros: $x = -2, 4$
$y$-intercept: $y = 16$

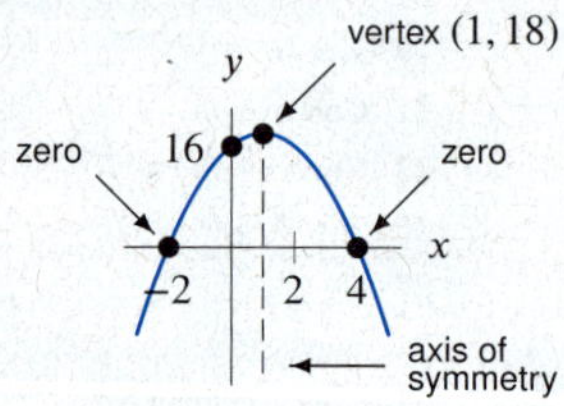


7 $y = -\frac{1}{9}(x + 6)^2 + 9$

9 $y = (3/16)(x - 6)^2 + 5$

11 $g(x) = -2(x - 3)^2 + 22$;
Vertex: $(3, 22)$; axis: $x = 3$

13 $w(z) = -3(z - 3/2)^2 + 19/4$, vertex is $(3/2, 19/4)$, axis of symmetry is $z = 3/2$

15 $(5/2)t^2 - 10$; $(5/2)(t-0)^2 - 10$; $(5/2)(t-2)(t+2)$

17 $(1/2)n^2 + 1$

19 $(1/2)n^2 - 3.7$

21 $(1/2)n^2 + \sqrt{13}$

23 $(1/2)n^2 + 3n + \frac{23}{2}$

25 $-16\left(t - \frac{47}{32}\right)^2 + \frac{2401}{64}$

27 $k = 4$

29 $y = -3(x - 2)^2 + 5$

31 $y = (-3/8)(x - 4)^2 + 2$

33 $(1/4)(x - 7)^2 + 3$

35 (a)

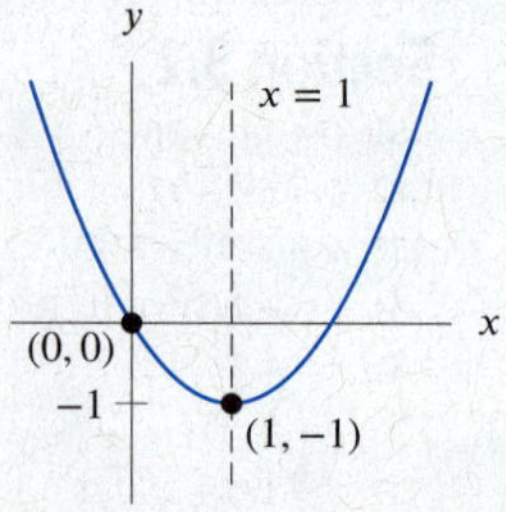


(b) $y = (x - 1)^2 - 1$ or $y = x^2 - 2x$
(c) Range: $y \geq -1$
(d) The other zero is $(2, 0)$

37 Vertex: $(-30, -25)$

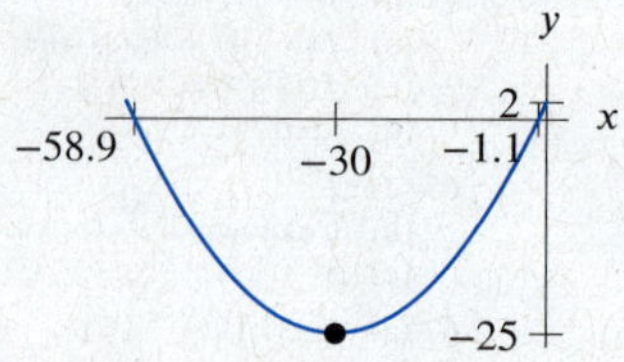


39 $g(x) = (x - 2)^2 - 1$

41 Shift $y = x^2$ right by 5 units to get $y = (x-5)^2 = x^2 - 10x + 25$

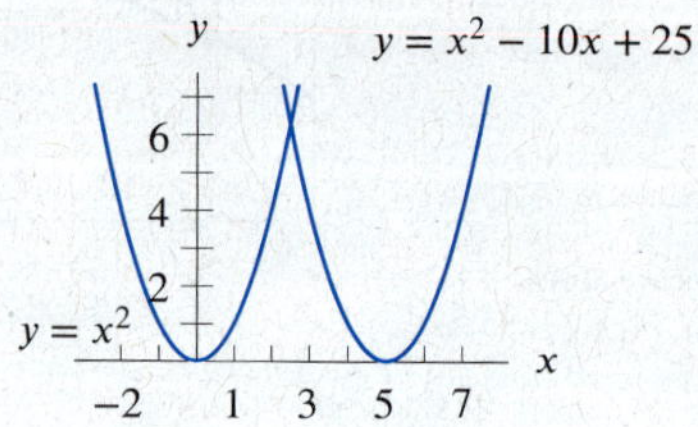


43 12.5 cm by 12.5 cm; $k/4$ by $k/4$

45 (a)

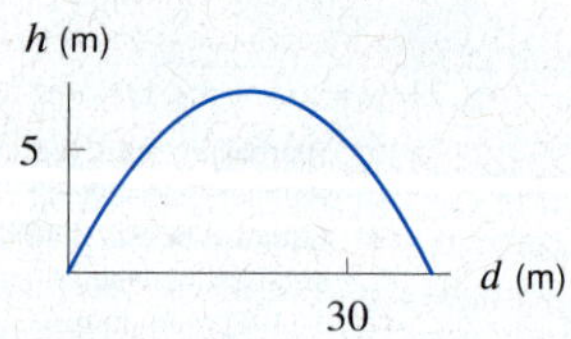


(b) 39.063 meters
(c) 7.324 meters
(d) 19.531 meters

## Chapter 3 Review

1 Yes; $f(x) = 2x^2 - 28x + 99$

3 Yes; $g(m) = -2m^2 + \sqrt{3}m + 42$

5 $x = 2, x = -1$

7 $x = (-1 \pm \sqrt{6})/5$

9 No zeros

11 Vertex is $(-5, 106)$
Axis of symmetry is $x = -5$

13 $y = (-1/16)(x - 4)^2 - 2$

15 $y = (x - 1)^2 - 1$

17 $y = -(3/16)(x - 4)^2 + 7$

19 $y = \frac{7}{9}(x - 3)^2 - 5$

21 Vertex is $(3/4, -2/3)$, axis of symmetry is $x = 3/4$, $y$-intercept $y = 11/24$, concave up

23 Vertex is $(0.6, 0)$, axis of symmetry is $x = 0.6$, $y$-intercept is $y = 0.36$, concave up

25 $y = 0.3(x-6)(x+4)$, zeros at $x = 6$ and $x = -4$, vertex at $(1, -7.5)$

27 $y = -3(x - 6)(x - 2)$, vertex is $(4, 12)$, zeros at $x = 6$ and $x = 2$

29 Rates of change: $-4, 0, 4$; Concave up

31 1/2 second

## Ch. 3 Understanding

1 True

3 False

5 False

7 False

9 False

11 True

13 False

15 True

## Ch. 3 Skills: Factoring

1 $6x - 14$

3 $12x + 12y$

5 $2x^2 + 5x$

7 $-50r^2 - 60r^2 s$

9 $5xz - 10z - 3x + 6$

11 $x^2 + 4x - 12$

13 $yz + 3y + z + 3$

15 $5xz - 10z - 3x + 6$

17 $x - 25$

19 $Pp^2 - 6Ppq + 9Pq^2$

21 $-2x - 2\sqrt{2x} - 1$

23 $2(x + 3)$

25 $5(z - 6)$

27 $5(2w - 5)$

29 $3u^2(u^5 + 4)$

31 $7rs(2r^3 s - 3t)$

33 $(x - 2)(x - 1)$

35 Cannot be factored

37 Cannot be factored

39 $(2x + 1)(x + 2)$

41 $(x + 7)(x - 4)$

43 $x(x + 3)(x - 1)$

45 $(x + 2y)(x + 3z)$

47 $(ax - b)(ax + b)$

49 $(B - 6)(B - 4)$

51 Cannot be factored.

53 $(t - 1)(t + 7)$

55 $(a - 2)(a^2 + 3)$

57 $(d + 5)(d - 5)(c + 3)(c - 3)$

59 $(r + 2)(r - s)$

61 $xe^{-3x}(x + 2)$

63 $P(1 + r)^3$

65 $(k + 2m)(d - 3e)$

67 $(2g - 3h)(4s + 5m)$

69 $x = (-3 \pm \sqrt{249})/8$

71 $x = 7/4$

73 $t = 3 \pm \sqrt{6}$

75 $x = 2, x = -4/3$

77 $N = 3, N = 1$

79 $x = 1 \pm \sqrt{2}$

81 $t = 3 \pm \sqrt{6}$

83 $a = -10, \pm 2\sqrt{5}/5$

85 $z = -7/2$

87 $L = \pm 1/2$

89 $r = \pm 5$

91 $x = 0, x = 36$

93 $x = -4/3, x = 2$

95 $b = \sqrt[5]{C/A}$

97 $x = 4m$

99 $x = 1/2$ and $y = 2$, or $x = -1/2$ and $y = -2$

101 $x = -3$ and $y = -5$, or $x = 1$ and $y = 3$

103 $(0, 0)$ and $(3, 9)$

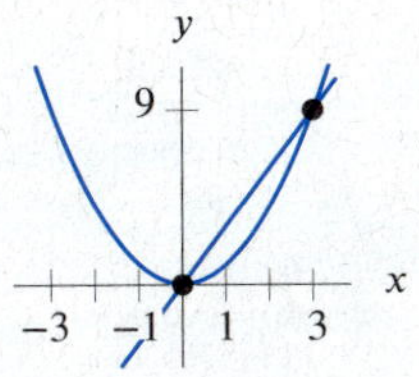


105 $(-5, 25), (3, 9)$

## Ch. 3: Completing Square

1 $(x+4)^2 - 16$

3 $2(r+5)^2 - 50$

5 $(a-1)^2 - 5$

7 $3(r+3/2)^2 - 43/4$

9 $(x-1)^2 - 4$

11 $-(x-3)^2 + 7$

13 $(-3, -6)$

15 $(-4, 18)$

17 $(1/2, -23/4)$

19 $(1, -2)$

21 $(7/4, -25/8)$

23 $g = 6, -4$

25 $d = 2, -1$

27 $s = -5/2 \pm \sqrt{27}/2$

29 $p = -9/10 \pm \sqrt{101}/10$

31 $y = -1/2, -2$

33 $w = (-1 \pm \sqrt{17})/2$

35 $q = (-3 \pm \sqrt{15})/2$

37 $s = \left(-3 \pm \sqrt{13}\right)/2$

39 $u = (3 \pm \sqrt{5})/5$

41 $y = 1 \pm \sqrt{7}$

43 $w = 3, 2, -2$

45 $m = (-5 \pm \sqrt{3})/7$

## Section 4.1

S1 0.06

S2 0.006

S3 0.12%

S4 123%

1 Yes; $g(w) = 2(1/2)^w$

3 Yes; $f(x) = (1/4)9^x$

5 Not exponential

7 Yes; $Q(t) = 2^t$

9 Yes; $K(x) = (1/3)(2/3)^x$

11 $a = 34.3; b = 0.788; r = -21.2\%$

13 $a = 0.0022; b = 0.0811; r = -91.89\%$

15 1.28 (per decade)

17 0.20 (per century)

19 (a) (i), (ii), (iv); (iii), (v), (vi)
(b) (iv); 18.5% per year
(c) (v); 22% per year
(d) (iii); (ii)

21 18.549%

23 (a) $Q = 2000(1.05)^t$
(b) 3257.789

25 (a) $Q = 112.8(0.766)^t$
(b) 7.845

27 (a) $Q = 5(2)^t$
(b) 5120

29 $P = 70(1.019)^t$

31 $Q = 726(0.94374)^t$

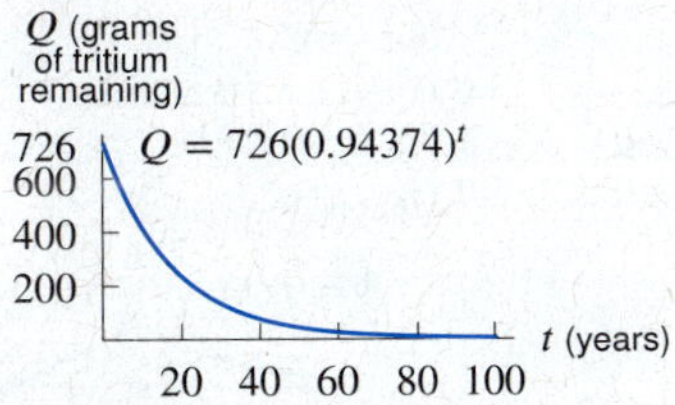


33 (a) $C = 100(0.84)^t$
(b) 41.821 mg

35 $L = 420(0.15)^n$

37 $y(t) = 300 - 300(0.97)^t$

39 (a) 134,058 megawatts; 28.4 megawatts
(b) 8.1%; 0.02%

41 6

43 20

45 (a) 1, 2, 4, 8
(b) Rate of change increases by factor of 2 over consecutive intervals
$2^{19} = 524{,}288$

47 (a) (iii) < (i) < (ii)
(b) Increasing, concave down

49 $5 \cdot 4^{-\frac{1}{6}t}; a = 5, k = -1/6$

51 $f(t) = 12(0.940)^t; a = 12, b = 0.940$

53 (a) $N(r) = 64(1/2)^r$

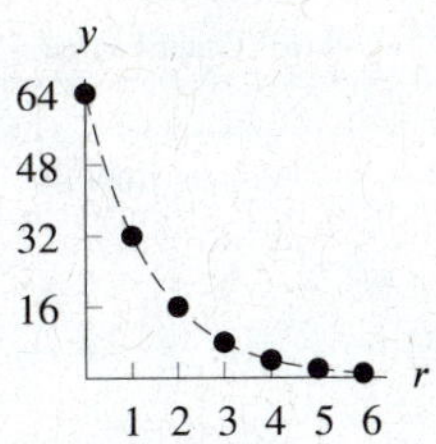


(b) 6

55 (a) 13.4%
(b) 6.489%

57 (a) (i) \$444 per month
(ii) \$286.20 per month
(iii) \$506.40 per month
(b) \$11,232
(c) \$23,112

59 366.875 miles

61 0.5

63 $a_1 > a_0; \tau_1 > \tau_0$

65 $b_0$

67 $t_0$ decreases

## Section 4.2

S1 $b^{10}$

S2 $128g^5$

S3 $3a^7b^{10}$

S4 $a^7b^{14}/2$

S5 5.6; 6.354

S6 12,837; 7608.541

S7 $x = 1.710$

S8 $x = \pm 0.2$

S9 $x = 1.393$

S10 $x = 1.842$

1 $B, C, D$ exponential

3 (a) $p = 2.50 + 0.03t$
(b) $p = 2.50 - 0.07t$
(c) $p = 2.50(1.02)^t$
(d) $p = 2.50(0.96)^t$

5 (a)

| $t$ | 0 | 1 | 2 | 3 | 4 | 5 |
|---|---|---|---|---|---|---|
| $f(t)$ | 1000 | 1200 | 1440 | 1728 | 2073.6 | 2488.32 |

(b) 1.2; 1.2; 1.2; 1.2; 1.2
(c) The growth factor is 1.2

7 (a) $g(x)$ is linear
(b) $g(x) = 2x$

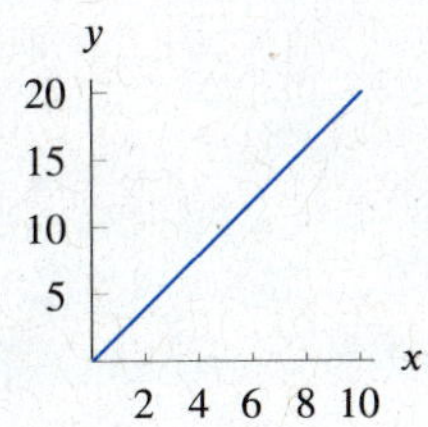


9 (a) $i(x)$ is linear
(b) $i(x) = 18 - 4x$

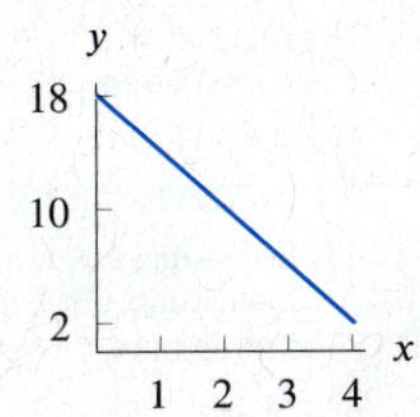

11 $f(x) = 250.4(1.0284)^x$

13 $f(x) = 3\left(\frac{1}{9}\right)^x$

15 $y = 10(1.260)^x$

17 $y = (1/2)(1/3)^x$

19 $y = 4(0.1)^x$ or $y = 4(10)^{-x}$

21 Not exponential

23 $g(t) = 5.7(0.315)^t$

25 (a) $h$
(b) $f(x) = 2.23(1.10)^{x/3} = 2.23(1.03)^x$

27 $f$ is exponential, $h$ is linear, $g$ is neither

29 Exponential,
$R(t) = 2.001(1.030)^t$

31 $y = 20 + 11x$; $y = 20(1.303)^x$

33 $x < -1.69$ and $x > 2$

35 (a) $P = 1154.160(1.20112)^t$
(b) \$1154.16
(c) 20.112%

37 (a)

$$V(t) = \begin{cases} 80 & \text{for } t < 0 \\ 80(0.8459)^t & \text{for } t \geq 0 \end{cases}$$

(b) $t \approx 39.933$ seconds

39 (a) $N = 191 + 1.9t$
(b) $N = 191(1.0095)^t$

41 (a) $W = 43.45 - 0.154t$; 39.754 seconds
(b) $W = 43.45(0.99635)^t$; 39.799 seconds

43 World: $P = 3.1(1.25)^t$
Ivory Coast: $P = 1.3 + 0.04t$

45 (a) Linear
(b) $L = 0.25t + 75.95$
(c) 88.45 years

47 (a) 0
(b) $0.35P_0$

49 (a) Neither
(b) Not possible

## Section 4.3

1 Not exponential

3 Could be exponential, $0 < b < 1$

5 (b)

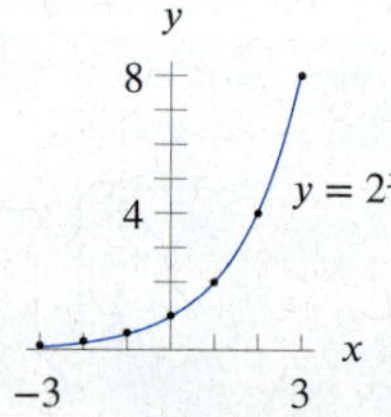


7 $h(x)$ top; $g(x)$ middle; $f(x)$ bottom

9 Yes

11 No

13 No

15 (a) 13 ft$^3$
(b) 3.2 weeks

19 $q = 5.662$

21 $t = 2.452$

23 Increasing: $b > 1, a > 0$ or $0 < b < 1, a < 0$;
Decreasing: $0 < b < 1, a > 0$ or $b > 1, a < 0$;
Concave up: $a > 0, 0 < b < 0$ or $b > 1$.

25 $D$

27 $D$

29 Zero

31

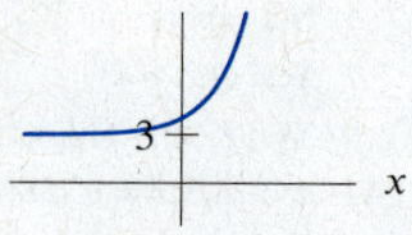


33

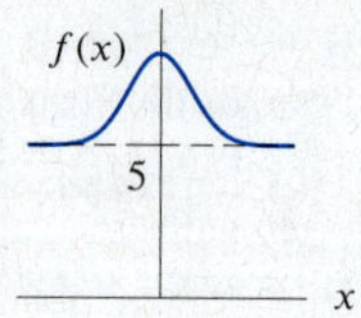


35 (a) All
(b) $b$
(c) $b, a, c, p$
(d) $a = c$
(e) $d$ and $q$

37 $y_0$ decreases, $y_0 > b$

39 (a) $P = 6.92(0.9917)^t$
(b) 6.06 million
(c) 12.42%
(d) $t = 65.8$

43 (a) $P \approx 0.538$ millibars
(b) $h \approx 0.784$ km

45 (a)

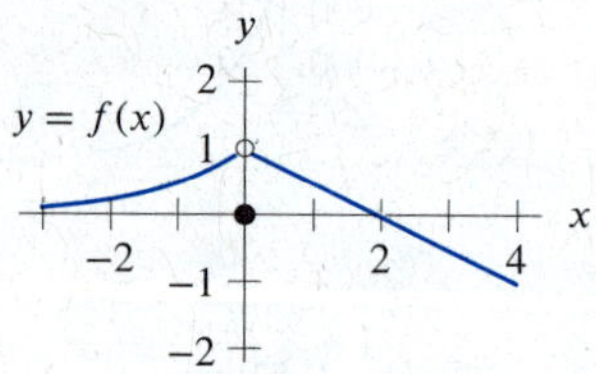


(b) $f(x) < 1$
(c) (0, 0) (2, 0)
(d) As $x \to +\infty$, $f(x) \to -\infty$
As $x \to -\infty$, $f(x) \to 0$
(e) Increasing for $x < 0$, decreasing for $x > 0$

47 (a)

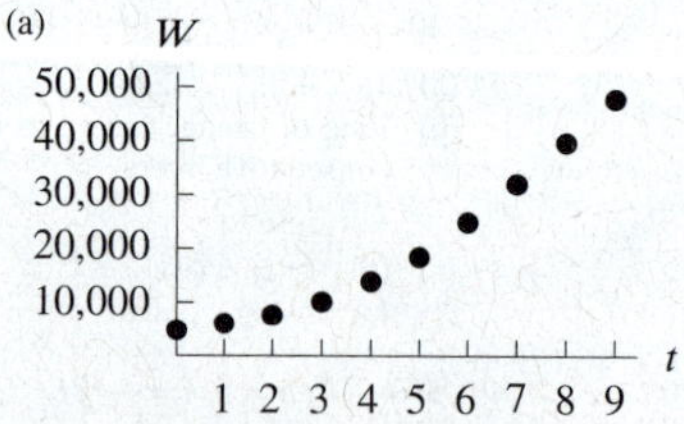


(b) $W = 4710(1.306)^t$; answers may vary
(c) 30.6%/yr

49 (a) Graphs agree

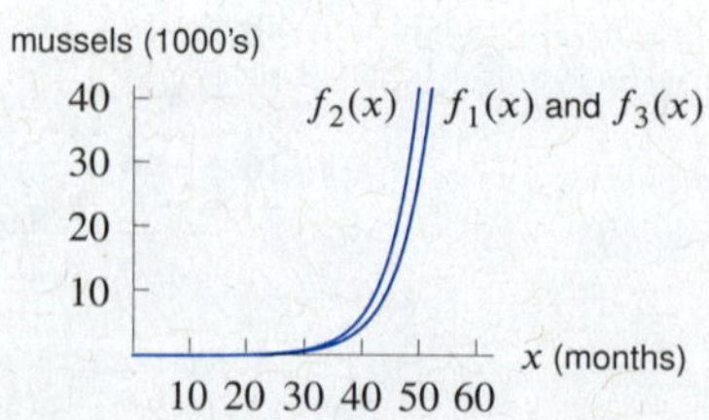


(c) Models 1 and 3 agree (approx);
Model 2 predicts larger populations

51

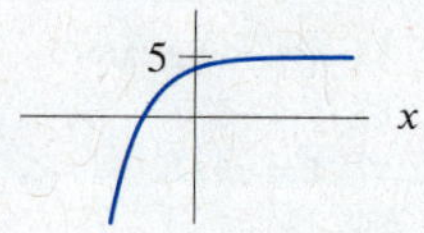


53 (a) 5
(b) −3

## Section 4.4

1 8.71%

3 (a) 8.300%
(b) 8.322%
(c) 8.328%

5 −8.3%

7 (a) \$505
(b) \$505.02
(c) \$505.03

9 (a) \$525
(b) \$525.62
(c) \$525.64

11 (a) Nom: 1% Eff: 1%
(b) Nom: 1% Eff: 1.004%
(c) Nom: 1% Eff: 1.005%

13 (a) Nom: 3% Eff: 3%
(b) Nom: 3% Eff: 3.034%
(c) Nom: 3% Eff: 3.045%

15 5.387%

17 34.392%

19 (a) \$18,655.38
(b) \$18,532.00
(c) \$18,520.84

21 (i) (b)
(ii) (a)
(iii) (c)
(iv) (b), (c) and (d)
(v) (a) and (e)

## Section 4.5

S1 1.073

S2 8.694

S3 1.433

S4 3480.202

S5 2.3; 7.636

S6 4.2; 2.599

S7 161.6; 202.027

S8 284.3; 284.334

S9 $f(t) = 27e^{0.12t}$

S10 $g(z) = 15e^{12z}$

S11 $Q = 1{,}096.633e^{-3t}$

S12 $Q = 4.482e^{3t}$

S13 $m(x) = \frac{7}{\sqrt{3}}e^{-0.3x}$

S14 $P(t) = 16e^{20/3t}$

S15 $H(r) = \frac{1}{6}e^{0.65r}$

1 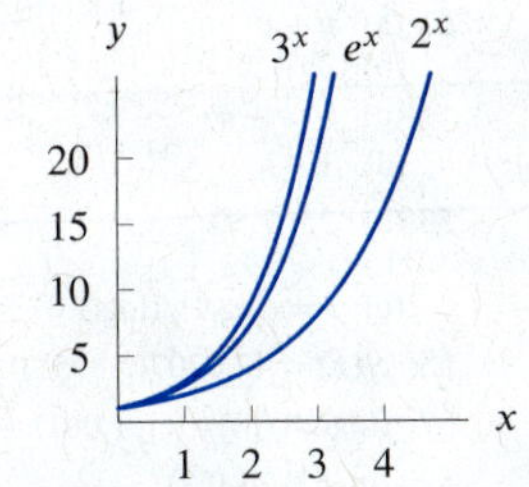


3 (a) (III)
(b) (II)
(c) (IV)
(d) (I)

5 (a)=(IV); (b)=(III); (c)=(II); (d)=(I)

7 (a)=(I); (b)=(II); (c)=(III); (d)=(IV)

9 (a) $Q_0 = 2.7$
(b) Decreasing
(c) −88%
(d) Not continuous

11 (a) $Q_0 = 158$
(b) Increasing
(c) 13.7%
(d) Not continuous

13 (a) $Q_0 = 1$
(b) Increasing
(c) 100%
(d) Not continuous

15 (a) $Q = 8(1.12)^t$; 24.847
(b) $Q = 8e^{0.12t}$; 26.561

17 (a) $Q = 500(0.93)^t$; 241.991
(b) $Q = 500e^{-0.07t}$; 248.293

19 (a) (i) 23.183
(ii) 23.645
(b) Continuous growth faster

21 (a) $P(t) = 25{,}000e^{0.075t}$
(b) 7.788%

23 (a) $P(t) = 22{,}000e^{0.071t}$
(b) ≈ 7.358%

25 (a) 272
(b) 28.4%

27 \$10,073.52.

29 (a) \$6094.97
(b) \$6107.01

31 54.931 years

33 A is continuous, B is annual
Initial deposit is \$500

35 $V = 500(1.06983)^t; a = 500, b = 1.06983, r =$ 6.983%

37 5.5% compounded annually

39 From best to worst: B, C, A

41 \$143.70

43 (a) $G = 145.8e^{0.051t}$
(b) 5.23%
(c) $G = 145.8(1.0523)^t$
(d) The two formulas have the same graph

45 (a) $P = 106e^{0.018t}$
(b) 115.98 million tons
(c) Late 2019

47 (a) $A = 50e^{-0.14t}$
(b) 12.330 mg
(c) 2025

49 (a) 13.5%
(b) 2.6%
(c) \$25.85

51 (a) Increasing, 2.7169239, 2.7181459, 2.7182682, 2.7182805
(b) Approximately $10^7$
(c) $(1 + 1/10^{16})^{10^{16}}$ appears to be 1

53 0

55 2

57 $a > 0, k > 0$

## Chapter 4 Review

1 550

3 495

5 411.8

7 $P = 2200(0.968)^t$

9 20%; 2%.

11 Linear: $p(r) = 10 + 3r$

13 Neither

15

| Year | 2013 | 2014 | 2015 | 2016 | 2017 |
|---|---|---|---|---|---|
| Cost (\$) | 126 | 132.30 | 138.92 | 145.87 | 153.15 |

17 (a) 4.2%
(b) ≈ 4.28%
(c) ≈ 4.29%

19 $h(x) = 3(5)^x$

21 $g(x) = 2(4)^x$

23 $g(x) = 14.20(0.6024)^x$

25 (a) $f(x) = \frac{31}{8}x + \frac{49}{4}$
(b) $f(x) = 5(2)^x$

27 $y = (1/2)^x$

29 $y = \frac{1}{5}(3)^x$

31 $y = 2(0.8)^x$

33 (a) $P(t) = 2.58 + 0.09t$,
increases by 90,000 people per year
(b) $P(t) = 2.68(1.026)^t$,
increases by 2.6% per year

35 $N = 10(1.13)^t$; 13%/yr

37 (a) $S = 109.5(1.138)^t$
(b) Increasing by 13.8%/yr
(c) Yes
(d) No

39 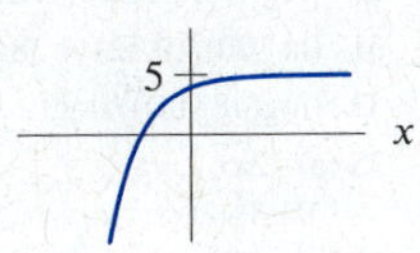


41 $f$

43 $y = 2$

47 (a) Initial balance = \$1100
Effective yield = 5%
(b) Initial balance = \$1500
Effective yield ≈ 5.13%

49 $p(x) = \frac{7\sqrt{e}}{20}e^x$

51 $s(w) = (v - 4t + kv)j^w a = v - 4t + kv, b = j$

53 $g(n) = 1000(0.7071)^n$

57 $d > b$

59 $f$ matches (ii) and (iv); $g$ matches (i) and (iii)

61 (a) 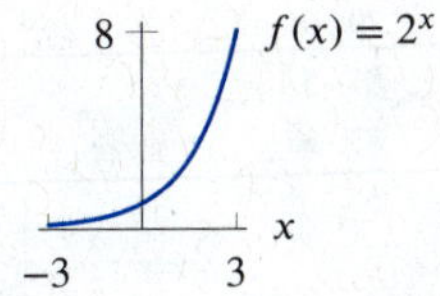


(b) 0.69
(c) 1.10
(d) $e \approx 2.72$

63 $V = 12{,}000e^{0.042t}$

65 $y = -13.1x + 2090$

67 (a) \$7401.22
(b) \$7459.12

69 (a) 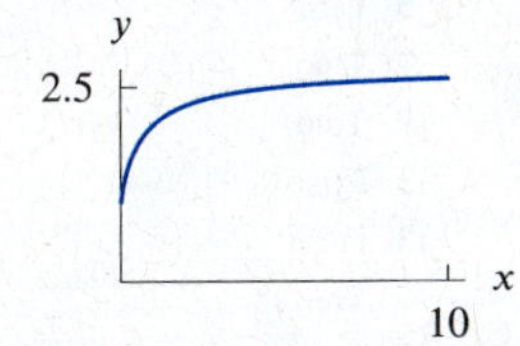


(b) Increasing
(c) Approach a limiting value
(d) Limiting value just above 2.5

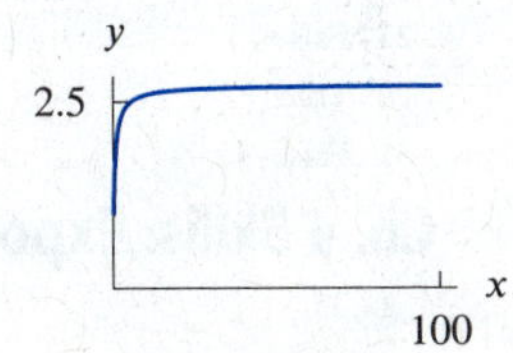


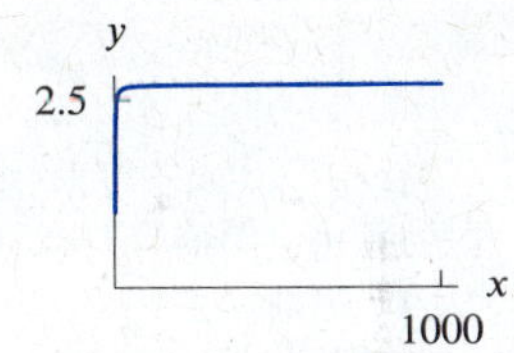


(e) $(1 + 1/x)^x$ appears to approach $e$ as $x$ gets larger
(f) 50,000

71 (a) $P(t) = 0.755\,(1.194)^t$

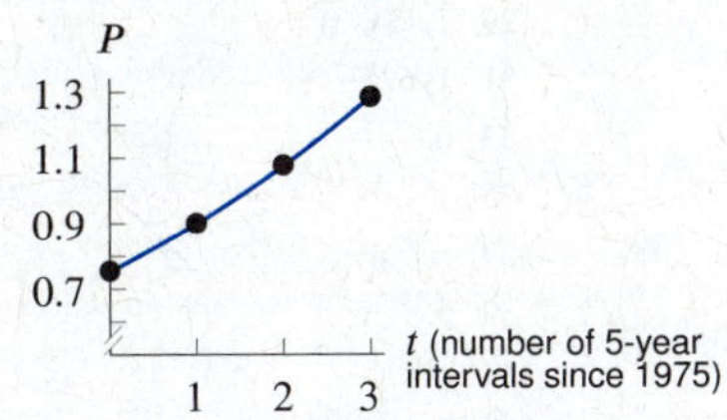


(b) 20 years; 2135

73 $t_0$ decreases

75 (a) $f(x) = p_0(2{,}087{,}372{,}982)^x$
(b) 8.55
(c) BAC of 0.051

77 97.7%

79 0

81 5.1

83 $\infty$

85

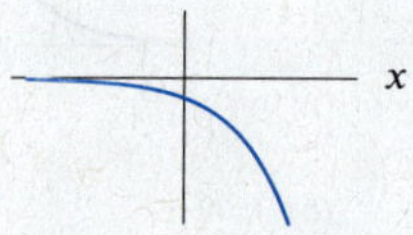


## Ch. 4 Understanding

1 True

3 True

5 False

7 True

9 True

11 False

13 False

15 True

17 True

19 False

21 True

23 False

25 True

27 False

29 True

31 False

## Ch. 4 Skills: Exponents

1 25

3 10,000

5 5

7 1

9 4

11 4

13 16

15 −121

17 2100

19 2

21 32

23 100,000

25 −6

27 4

29 $1/(3\sqrt{3})$

31 1/625

33 0.5

35 $y^4$

37 $x^{5/2}y^2$

39 $5x^{3/2}z^2$

41 $r^{3/2}$

43 $8s^{7/2}$

45 $4\sqrt{3}u^5v^6y^{5/2}$

47 $16S^2xt^2$

49 $A^3/\left(3B^3\right)$

51 $(M+2)^2$

53 $3a$

55 $25(2b+1)^{20}$

57 −8

59 Not a real number

61 1/512

63 Not a real number

65 $x = \pm 1.690$

67 (2.5, 31.25)

69 False

71 True

73 True

75 $x = r + s$

77 $x = 5/a$

79 $x = 3/a$

81 $x = b/a$

## Section 5.1

S1 $x = 6$

S2 $t = -2$

S3 $z = 3/2$

S4 $x = 0$

S5 No solution

S6 $x = -5/3$

S7 $t = 14/9$

S8 No solution

S9 $t = -1/8$

S10 $x = 5/9$

1 $19 = 10^{1.279}$

3 $26 = e^{3.258}$

5 $P = 10^t$

7 $8 = \log 100{,}000{,}000$

9 $v = \log \alpha$

11 3

13 0

15 0

17 5

19 100

21 0.01

23 0

25 1/2

27 −1/2

29 $(\log 11)/(\log 2) = 3.459$

31 $(\ln 100)/(0.12) = 38.376$

33 $(\log(48/17))/(\log(2.3)) = 1.246$

35 (a) $2x$
(b) $x^3$
(c) $-3x$

37 (a) 3, 3
(b) 5, 5
(c) −1, −1
(d) −1, −1
(e) 2, 2
(f) 3, 3
Both answers equal

39 True

41 False

43 True

45 $x = 24.990$

47 $x = (M - \ln a)/(\ln b)$

49 $x = 31.522$

51 (a) $u + v$
(b) $3u - 2$
(c) $\frac{1}{2}(v - u)$
(d) $1 - u$

53 (a) 25; 7.5%
(b) $t \approx 19$
(c) $t = (\log 4)/(\log 1.075) = 19.169$

55 $S(x) = 42.0207(0.8953)^x$

57 $(\log(14/3))/(\log 1.081)$

59 $(\ln 0.3)/(-0.2)$

61 $t = (15 \ln 6)/(\ln 2)$

63 $x = \dfrac{\ln 7 - 5}{1 - \ln 2}$

65 $(\ln 10 - 4\ln 3)/\ln 3$

67 1/5

69 $\frac{1}{4}(\ln(30/58) - 1)$

71 $(\ln Q - \ln P)/k$

73 $x = -2, \dfrac{1}{3}$, or $-\dfrac{1}{3}$

75 −1, 1

77 (a) $(\ln(0.5))/2$
(b) $(\ln(b/3))/3$

79 $B > A$

81 $q = \ln r + (\ln s)t$; $b = \ln r$, $m = \ln s$

## Section 5.2

S1 $(5x)^{-1}$

S2 $t^{-3}$

S3 $t^2/2$

S4 $100x$

S5 $x = (\log 9)/(\log 4) = 1.585$

S6 $x = \ln 8 = 2.079$

S7 $x = \ln(13/2) = 1.872$

S8 $x = (\ln 5)/4 = 0.402$

S9 $x = 93/2$

S10 $x = 10$

1 $Q = 4 \cdot 1096.633^t$

3 $Q = (14/5)1.030^t$

5 $y = 25(1.0544)^t$, 5.44%/yr, 5.3%/yr

7 $Q = 12e^{-0.105t}$

9 $Q = 14e^{-0.208t}$

11 $y = 6000e^{-0.1625t}$, −15%/yr, −16.25%/yr

13 $a = 230$, $r = 18.2\%$, $k = 16.72\%$

15 $a = 0.81$, $r = 100\%$, and $k = 69.31\%$

17 $a = 12.1$, $r = -22.38\%$, $k = -25.32\%$

19 $a = 5.4366$, $b = 0.4724$, $r = -52.76\%$, $k = -3/4$

21 About 26 years

23 About 3 years

25 About 20 days

27 About 6.3 minutes

29 (a) 4.729%
(b) 4.621%

31 23.362 years

33 $t = 30.034$

35 About 18.6 hours

37
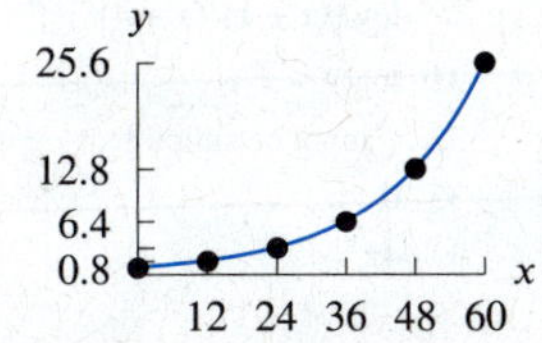


39 (a) 10; 30; and 70 yrs
(b) 14.207; 28.413; and 42.620 yrs

41 (a) 12 days
(b) 5.78% per day; $P = 50e^{0.0578t}$

43 (a) 190°F; 100°F; 77.5°F
(b) 1.292 hours; 2.292 hours

45 (a) 15.927%
(b) 7.549 years

47 (a) $P = 25{,}000(1.0466)^t$
(b) 4.66%; 4.56%
(c) 10.32 years; during the year 2020

49 (a) 2.6%
(b) 6.72 million
(c) 42.25 years
(d) 605 years ago; unreasonable

51 About 12:53 pm

53 (a) $f(x) = \frac{1}{2}(4)^x$
$g(x) = 4\left(\frac{1}{3}\right)^x$
$h(x) = x + 2$
(b) $x = \log 8/\log 12$
(c) $x = 1.378$ or $x = -1.967$

55 (a) 20, 395, 954
(b) 5.615 years, 7.2 years
(c) 1000 toads

59 $a = 9.0856, b = 1.0303, k = 0.02986, s = 23.213$

61 $t = -10\ln(-2\ln 0.5) = -3.266$

63 (a) $v = \log 1.12, w = \log 6.3$
(b) $t = w/v$; $t = 16.241$

## Section 5.3

S1 -4

S2 3

S3 $\log 100{,}000 = 5$

S4 $\ln 7.389 = 2$

S5 $x = e^{-12}$

S6 $10^2 = x + 3$

S7 $\ln x + 3\ln(7 - x)$

S8 $\ln x + 2\ln y - \ln z$

S9 $\ln x^5$

S10 $\log(2/5)$

1 D: $x > 3$
R: All $y$

3 Positive; decreases; concave down

5 Negative; decreases; concave down

7 $A$: $y = 10^x$, $B$: $y = e^x$
$C$: $y = \ln x$, $D$: $y = \log x$

9 $y = 0, y = 0, x = 0$

11
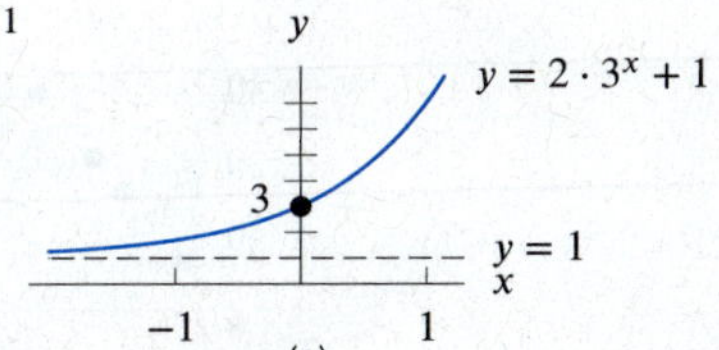

(a)

13
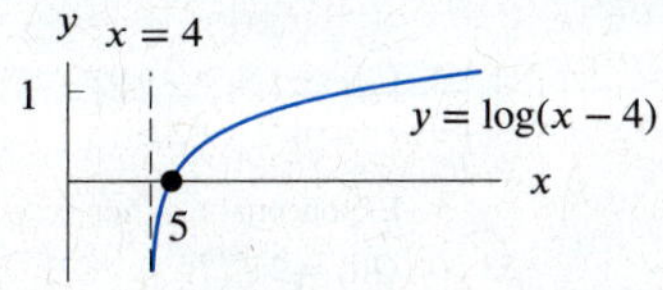

(c)

15 Vertical asymptote at 3,
Domain $(3, \infty)$

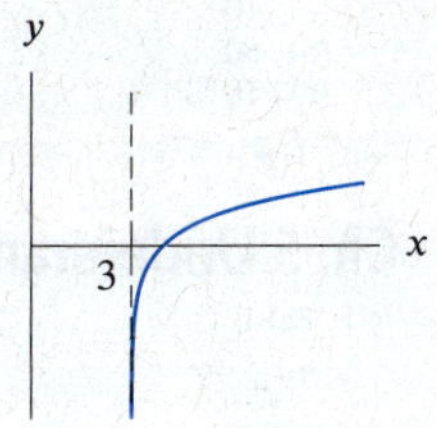


17 1 mole/l

19 $10^{-13}$ moles/l

21 $3.162 \times 10^{-5}$ moles/l

23 (a) 0
(b) $-\infty$

25 (a) $t(x)$
(b) $r(x)$
(c) $s(x)$

27 $y = b^x, 0 < b < 1$

29 $y = \ln x$

31 $y = -b^x, b > 1$

33 (a) (II)
(b) (I)
(c) (III)
(d) (I), (III) and (IV)

35 (a) (i) 0.693
(ii) 0.087
(iii) 0.01
(iv) 0.002
(b) The values appear to be approaching 0.

37 (a) $10^{-2}$, $10^{-4}$, $10^{-7}$
(b) Less

39 25.9%, −0.10018

41 (a) 0.005 moles/liter
(b) $3.3 \times 10^{-4}$ moles $H^+$ ions
$1.987 \times 10^{20}$ ions

43 100 watts/cm$^2$

45 $10^{0.306} \approx 2$ times as loud.

47 (a) $D_2 - D_1 = 10\log(I_2/I_1)$
(b) 3.01 decibels louder

49 3,162

51 32

53 (a) $-\infty$
(b) $-\infty$

## Section 5.4

S1 $1.455 \times 10^6$

S2 $4.23 \times 10^{11}$

S3 $6.47 \times 10^4$

S4 $1.231 \times 10^7$

S5 $3.6 \times 10^{-4}$

S6 $4.71 \times 10^{-3}$

S7 $10^4 < 12{,}500 < 10^5$

S8 $10^{-4} < 0.000881 < 10^{-3}$

S9 $10^{-1} < 1/3 < 10^0$

S10 $10^{11} < 3.85 \cdot 10^{11} < 10^{12}$

1 Log

3 Linear

7 3.7

9 2.2

11 0.6

13 (a) $y = -3582.145 + 236.314x$; $r \approx 0.7946$
(b) $y = 4.797(1.221)^x$; $r \approx 0.9998$
(c) Exponential is better fit

15 $10^{-3.62}$ million years ago

17 (a) Increasingly spread out
(b) Equally spaced

19 Log scale more appropriate

23 (b) Linear function
(c) $f(x)$ is exponential;
$g(x)$ is linear
(d) It is linear; yes

25 (a)
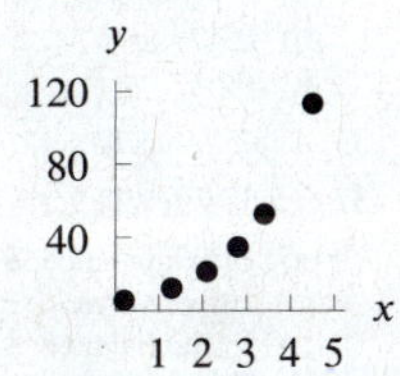


(b) Exponential
(c)
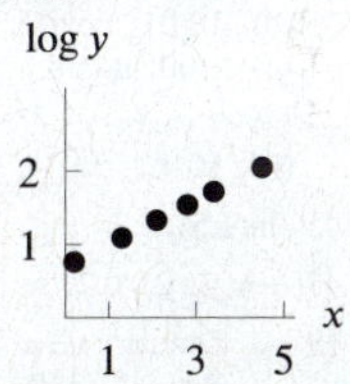


Linear

27 (a) $y = 14.227 - 0.233x$
(b) $\ln y = 2.682 - 0.0253x$
(c) $y = 14.614e^{-.0253x}$

29 (a)
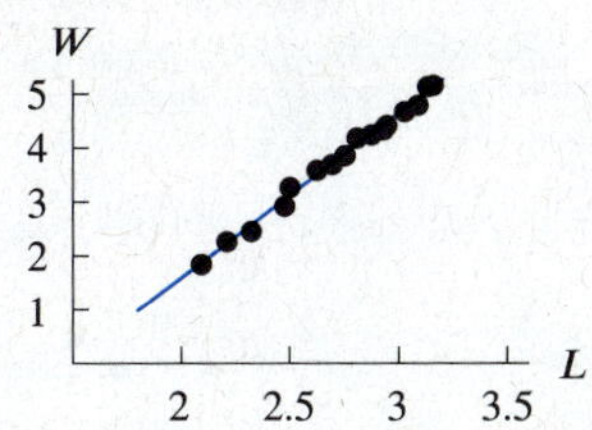


(b) $W = 3.06L - 4.54$
(c) $w = 0.011\ell^{3.06}$

## Chapter 5 Review

1 $Q = 7(0.0000454)^t$

3 $Q = 4e^{1.946t}$

5 $Q = 4e^{2.703t}$

7 $(\log 3)/(\log 1.04)$

9 $(\log(91/46))/(\log(1.1))$

11 $(\log(12/5))/(3 \log 1.014)$

13 $(\log 1.6)/(\log 1.031)$

15 47

17 $10^{\frac{\log 5}{4 \log 3}}$

19 No solution

21 $2(x+1)$

23 $\ln(AB)$

25 Domain: $x > 20$; Asymptote: $x = 20$

27 Domain: $x < 300$; Asymptote: $x = 300$

29 Domain: $x > -15$; Asymptote: $x = -15$

31 (a) $q + 4p$
(b) $-q$
(c) $p/q$
(d) $3q$

33 (a) $\ln 8 - 3$
(b) $\log 1.25/\log 1.12$
(c) $-\dfrac{\ln 4}{0.13}$
(d) 105
(e) $\frac{1}{3}e^{3/2}$
(f) $e^{1/2}/(e^{1/2} - 1)$
(g) $-1.599$ or $2.534$
(h) 2.478 or 3
(i) 0.653

35 158.5 times larger

37 15.85 times larger

39 (a) Initial balance: \$1100
Effective yield: 5%
(b) Initial balance: \$1500
Effective rate: 5.127%/yr
(c) Continuous rate: 4.879%/yr

41 (a) 1412 bacteria
(b) 10.011 hours
(c) 1.005 hours

43 (a) 7 years
(b) 10.4%

45 $\ln(1.5)/0.2$

47 $t = (\ln 2)/0.22 \approx 3.151$

49 (a) Domain: all $x$
Range: $y > 0$
Asymptote: $y = 0$
(b) Domain: all $x > 0$
Range: all $y$
Asymptote: $x = 0$

51 (a) Log function

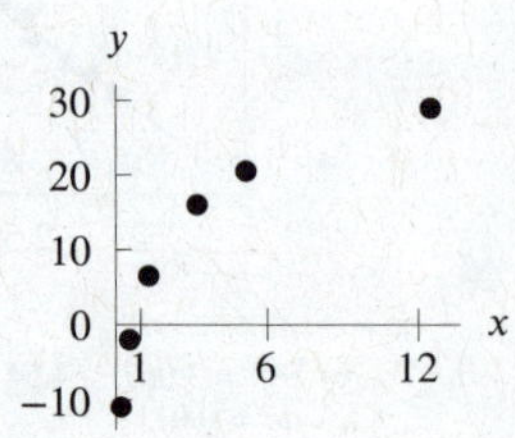


(b)

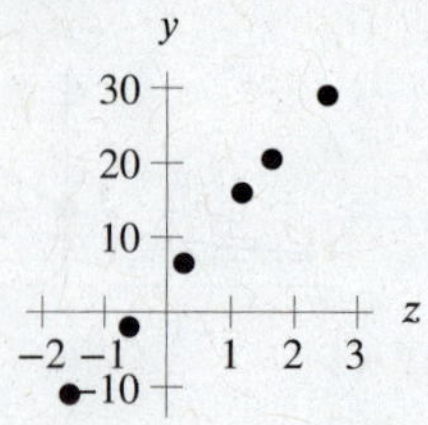


(c) Linear; $y = 4 + 9.9z$
(d) $y = 4 + 9.9 \ln x$
(e) $x = 0.67e^{0.1y}$;
Exponential function of $y$

53 (a) $Q(t) = 2e^{-0.04t}$
(b) 3.921%
(c) After 51.986 hours
(d) 54.931 hrs after second injection

55 (a) $\approx 33.517\%$
(b) $k \approx 4.082\%$, continuous hourly decay rate

57 (a) 10
(b) 50
(c) $10^{50}$

59 $\sqrt[8]{k}$

## Ch. 5 Understanding

1 False

3 True

5 True

7 True

9 True

11 False

13 False

15 False

17 True

19 False

21 False

23 True

25 True

27 True

29 True

31 True

33 False

35 False

37 False

## Ch. 5 Skills: Logs

3 8

5 0

7 2

9 $\log 0.0001 = -4$

11 $\ln 0.135 = -2$

13 $10^{-2} = 0.01$

15 $e^{x^2} = 4$

17 Cannot be rewritten

19 $\log(x^2 + 1) - 3 \log x$

21 Cannot be rewritten

23 Cannot be rewritten

25 $\log 12x$

27 $\log(\sqrt{x}y^4)$

29 $\log\left((x+1)^3(x+4)^2\right)$

31 $\log(9 - x^2)$

33 Cannot be simplified

35 0

37 $4z$

39 $-\ln(e^x + 1)$

41 $x = (\log 3)/(\log 5) \approx 0.683$

43 $x = -(\ln 9)/5 \approx -0.439$

45 $x = (\log 77)/(6 \log 19 - 4 \log 7) \approx 0.440$

47 $x = (10^{3/2} - 17)/9 \approx 1.625$

49 $x = (e+1)/6 \approx 0.620$

## Section 6.1

S1 20.086

S2 −20.086

S3 0.050

S4 −0.050

S5 (a) $f(-x) = 2x^2$
(b) $-f(x) = -2x^2$

S6 (a) $f(-x) = -1/x$
(b) $-f(x) = -1/x$

S7 (a) $f(-x) = -2x^3 - 3$
(b) $-f(x) = -2x^3 + 3$

S8 (a) $f(-x) = -4x^3 - 9x$
(b) $-f(x) = -4x^3 - 9x$

S9 (a) $f(-x) = 3x^4 + 2x$
(b) $-f(x) = -3x^4 + 2x$

S10 (a) $f(-x) = -(3x^3)/(x^2 - 1)$
(b) $-f(x) = -(3x^3)/(x^2 - 1)$

1 (a) $(-2, -3)$
(b) $(2, 3)$

3 −7

5 Domain: $t < 0$
Range: $-4 \le Q(-t) \le 7$

7 Domain: $t < 0$
Range: $-7 \le -Q(-t) \le 4$

9 $y = e^{-x}$

11 Reflected across $x$-axis;
$-g(x) = -(1/3)^x$

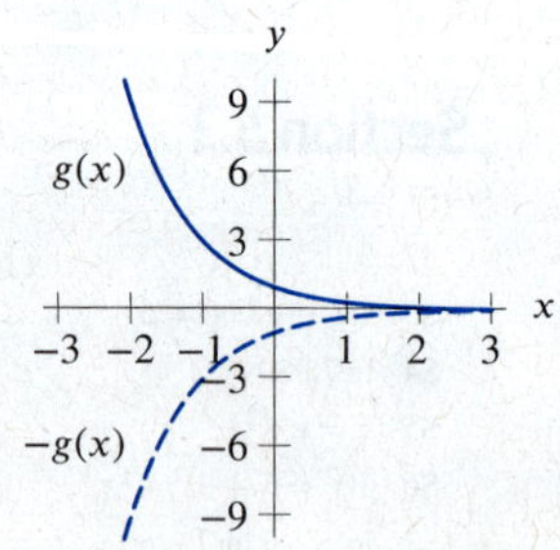


13 $m(-n) = n^2 + 4n + 5$

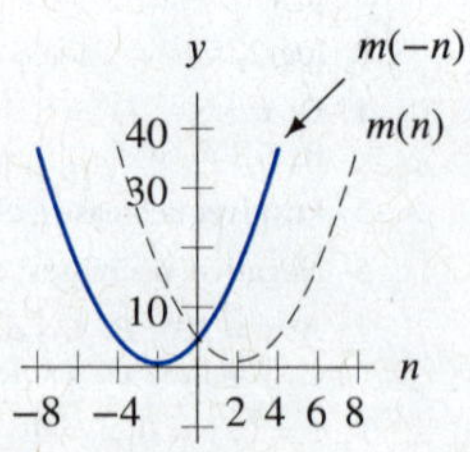

15 $-m(-n) = -n^2 - 4n - 5$

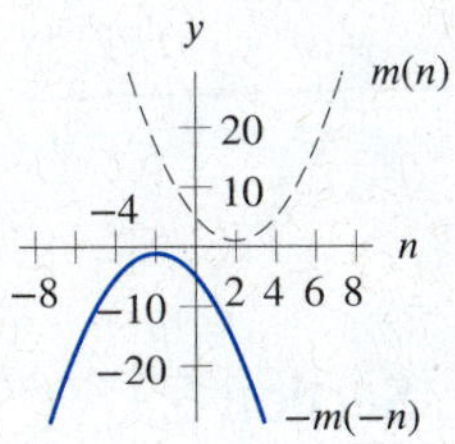


17 $k(-w) = 3^{-w}$

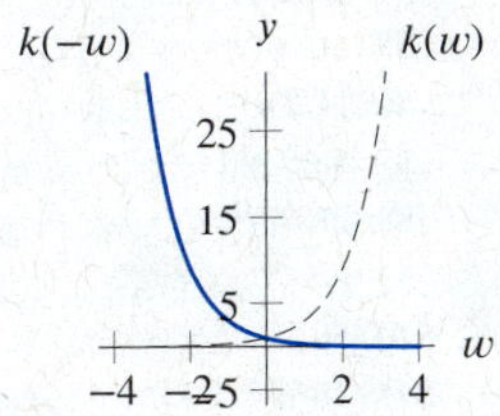


19 $-k(-w) = -3^{-w}$

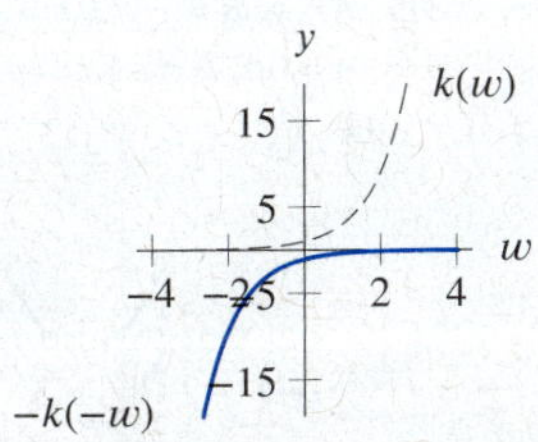


21 Neither

23 Even

25 (a) $y = -x^3 + 2$

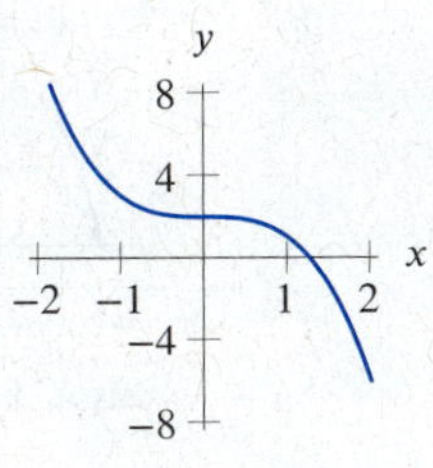


(b) $y = -(x^3 + 2)$

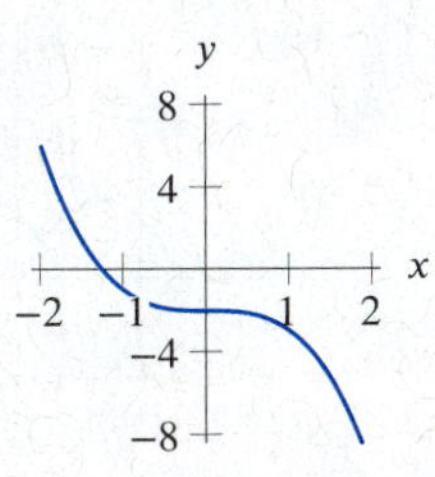


(c) No

27 (a) $-10$
(b) $-25$
(c) 10
(d) 3
(e) $-26$

29 Slope: $-m$;
$y$-intercept: $b$;
$x$-intercept: $b/m$

31 (a)

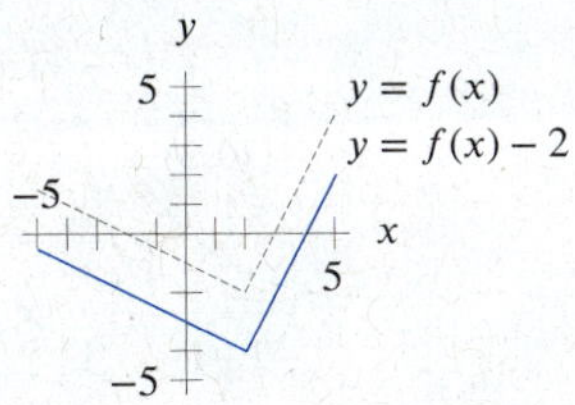


(b)

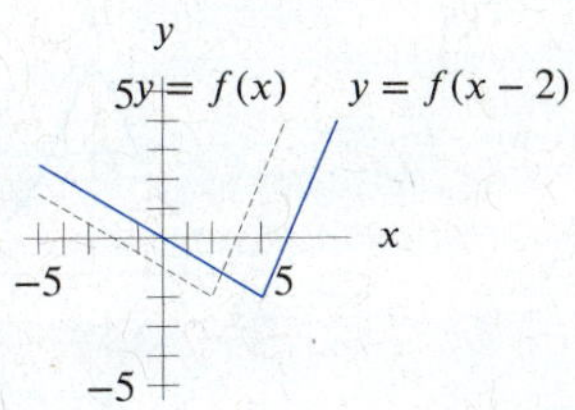


(c)

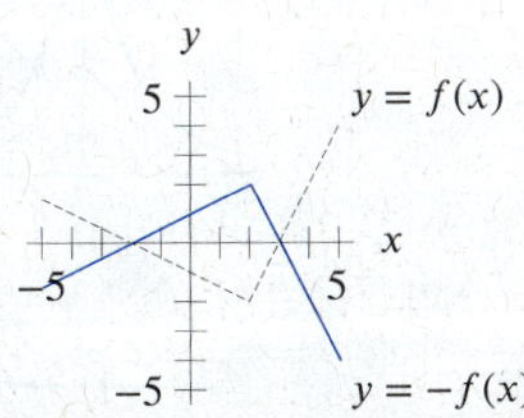


(d)

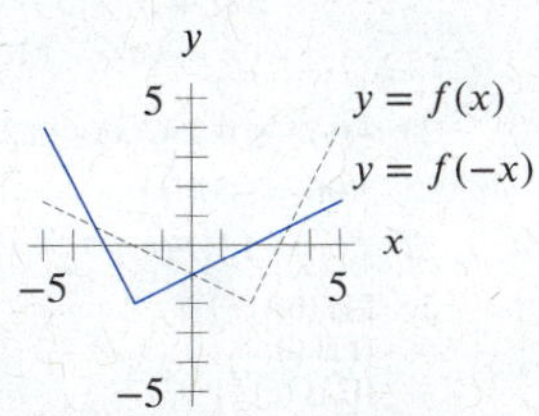


33 (a) $H(t) + k = (20 + k) + 54(0.91)^t$
(b)

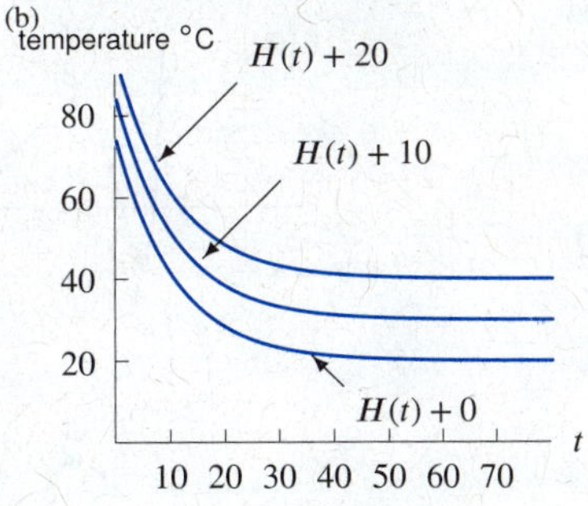


(c) $H(t) + 10$ approaches 30°C;
$H(t) + k$ approaches $(20 + k)$°C, temperature of the room

35 (a) $H(t) = 3 - 21(0.97)^t$
(b) $D(t)$: Asymptote $D = 0$, intercept $D = 21$;
$H(t)$: Asymptote $H = 3$, intercept $H = -18$

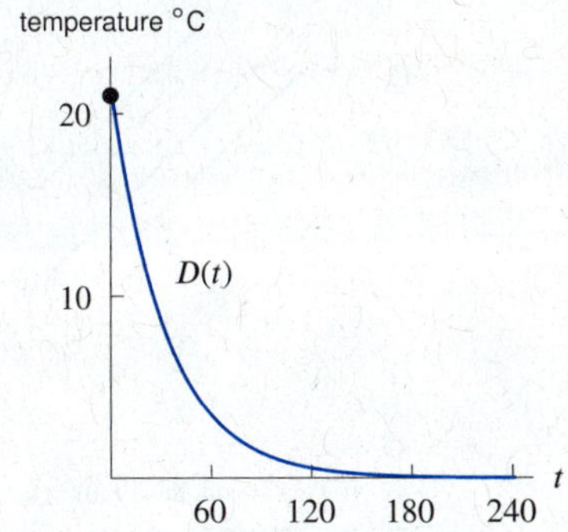


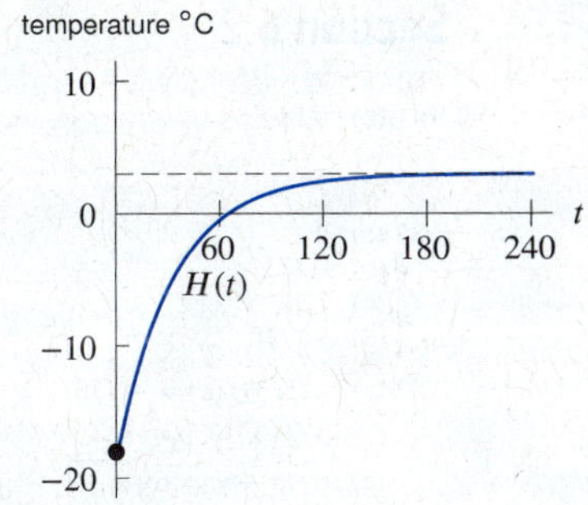


(c) $D(t)$: intercept $D = 19$, asymptote unchanged. $H(t)$: Asymptote $H = 1$, intercept unchanged.

37 (a)

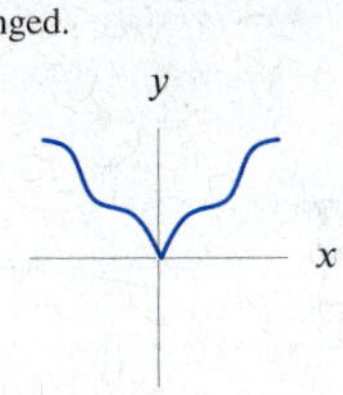


(b)

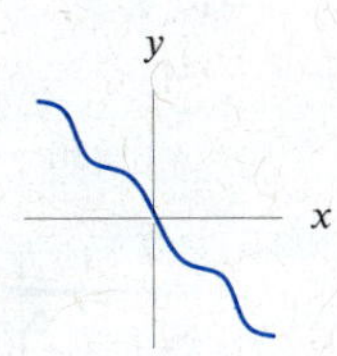


(c)

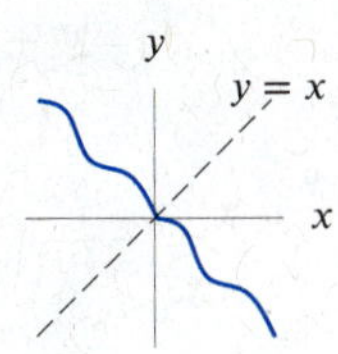


39 (a) Odd
(b) Unless $f(x) = 0$ or $g(x) = 0$, $k(x)$ is neither.
(c) Even

41 $y = x$. $y = -x + b$, where $b$ is an arbitrary constant

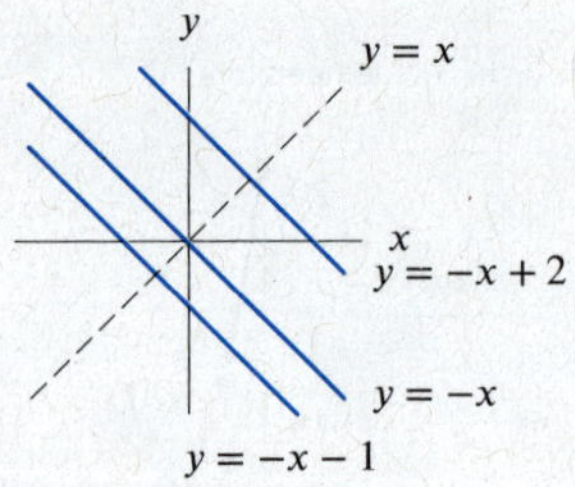


45 No

47 If $f(x)$ is odd, then $f(0) = 0$

51 Yes, $f(x) = 0$

## Section 6.2

S1 (a) 72
(b) $-18$
(c) 177
(d) $25/4$

S2 (a) 2
(b) $-1/3$
(c) $2/3$
(d) $-1/10$

S3 (a) $-(1/3)f(x) = -(1/3)\sqrt{x}$
(b) $5f(-x) = 5\sqrt{-x}$
(c) $6f(x-8) = 6\sqrt{x-8}$
(d) $(1/4)f(2-x) = (1/4)\sqrt{2-x}$

S4 (a) $5p(x) = 15x^2 - 30$
(b) $-(1/3)p(-x) = -x^2 + 2$
(c) $-2p(x+3) = -6x^2 - 36x - 42$
(d) $(5/3)p(x-1) = 5x^2 - 10x - 5$

1 $y = 10f(x-2)$

3

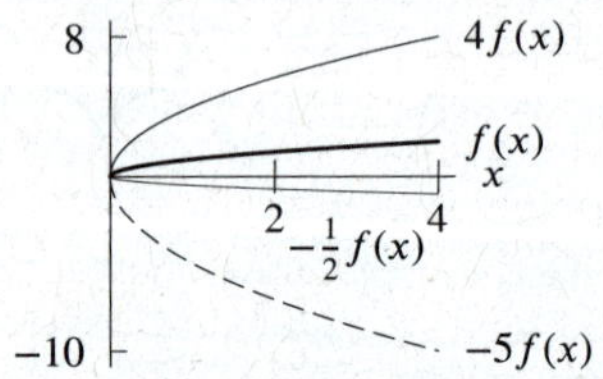


5

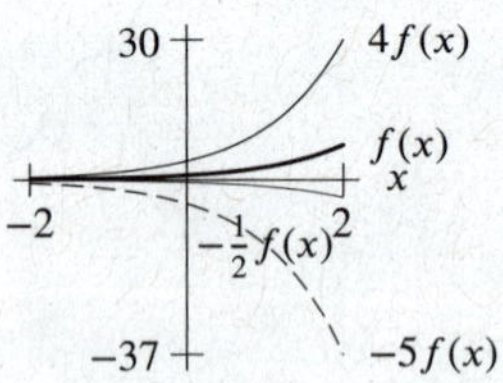


9

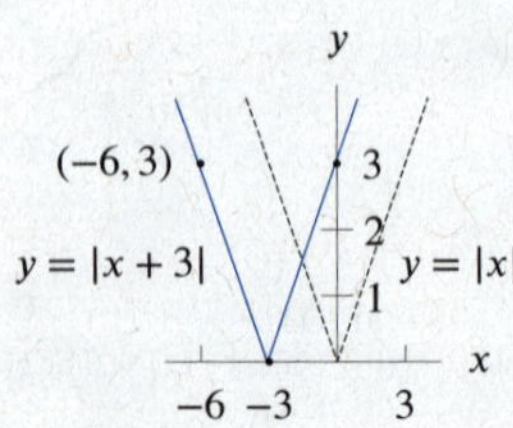


11

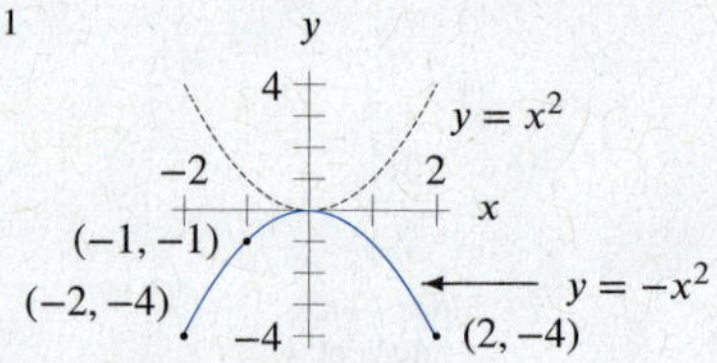


13

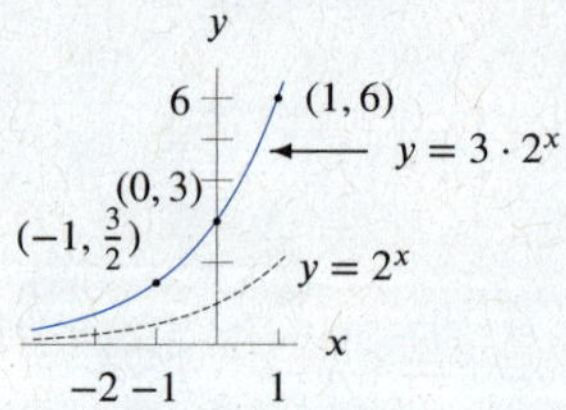


15 Stretch vertically by a factor of 2, Shift left 1 unit

17

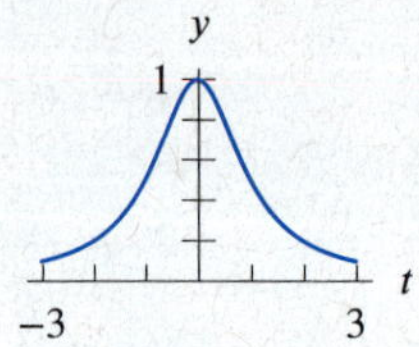


19

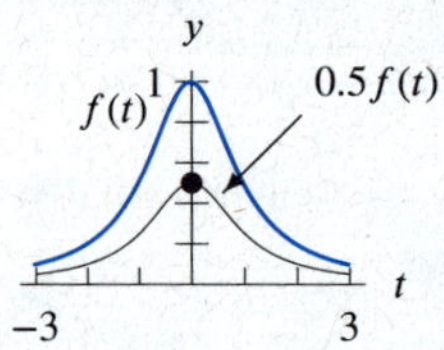


21

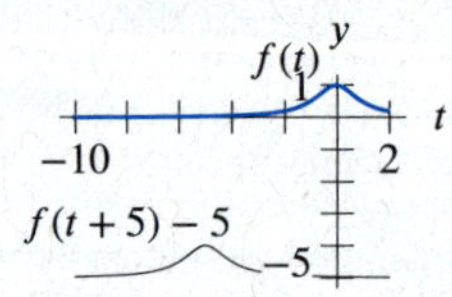


23 $-0.25 \le 0.25C(x) \le 0.25$

25 $R(n) = -5P(n)$

27 $T(n) = 1/4P(n+7)$

29 I is (b)
II is (d)
III is (c)
IV is (h)

31 $1.3C(t)$

33

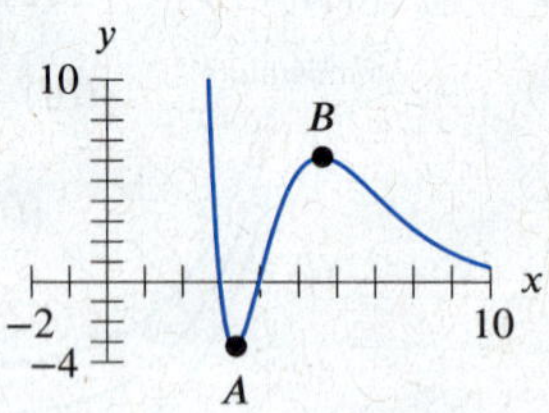


35

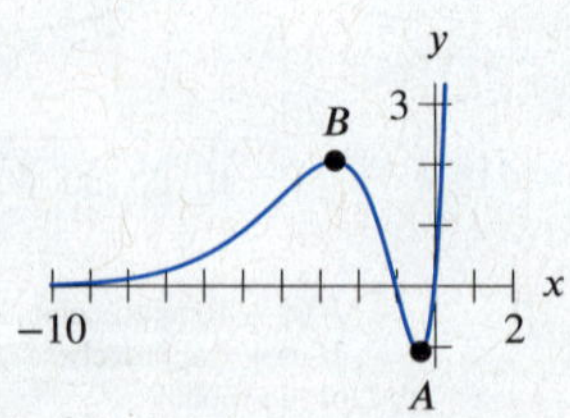


37 (a) $h(x) = 1/2f(x)$
(b) $k(x) = f(-x)$
(c) $m(x) = f(x) - 4$

39 (a) $y = -2f(x)$
(b) $y = f(x) + 2$
(c) $y = 3f(x-2)$

43 $-7$

## Section 6.3

S1 $8x^3 - 5$

S2 $2x^3 - 10$

S3 $(-x^3)/27 - 5$

S4 $(27/5)x^3 - 1$

S5 $4e^{2t}$

S6 $(4/3)e^{6t}$

S7 $4e^{12t} + 11$

S8 $28e^{6t-18}$

S9 $-3$

S10 50

S11 $10/3$

S12 $-36$

S13 $A = 1, B = -2, h = 0, k = 9$

S14 $A = -1, B = 2, h = 3, k = 9$

S15 $A = 6, B = -1/3, h = -27, k = 0$

S16 $A = -5, B = -1, h = -7, k = -10$

1 (a)

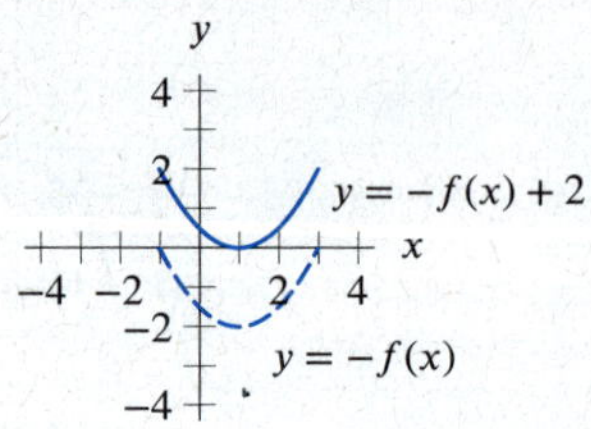


(b)

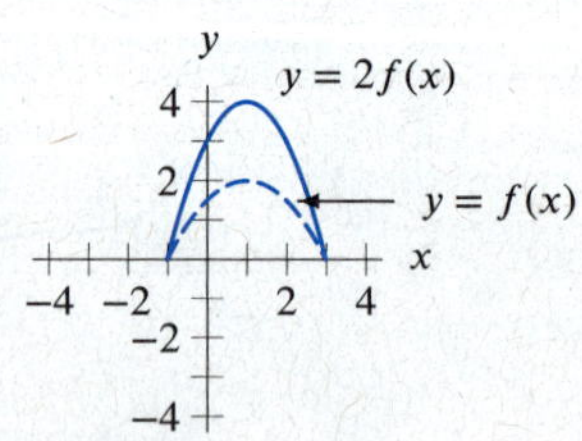


(c)

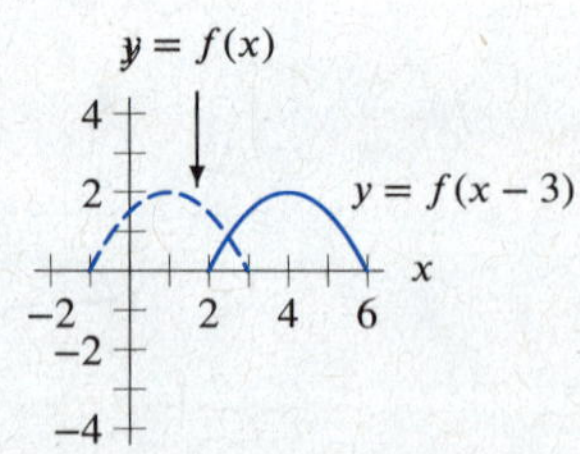

(d)

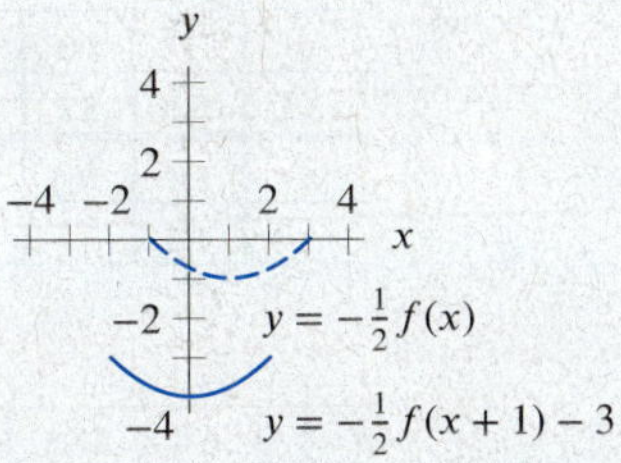


3

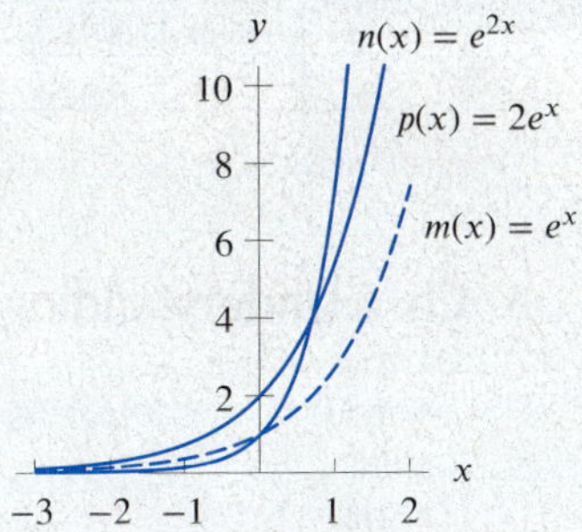


5

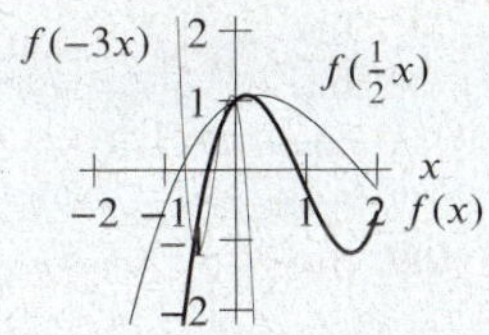


7

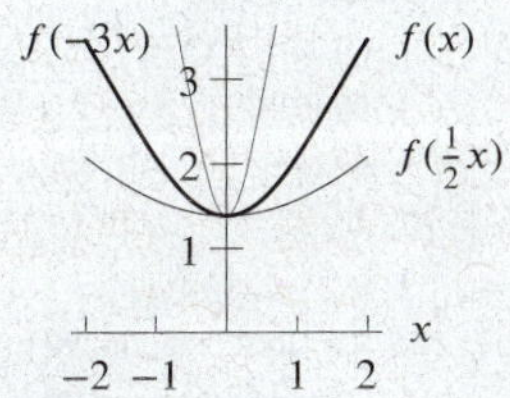


9 (a)

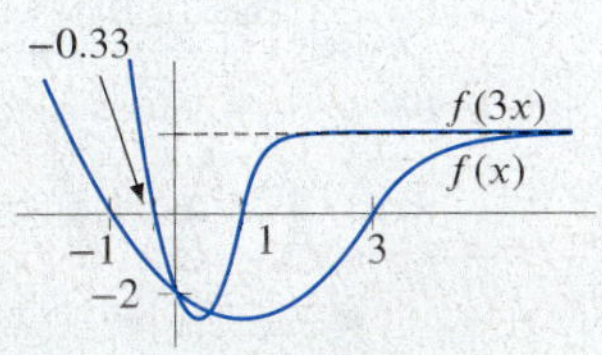


(b)

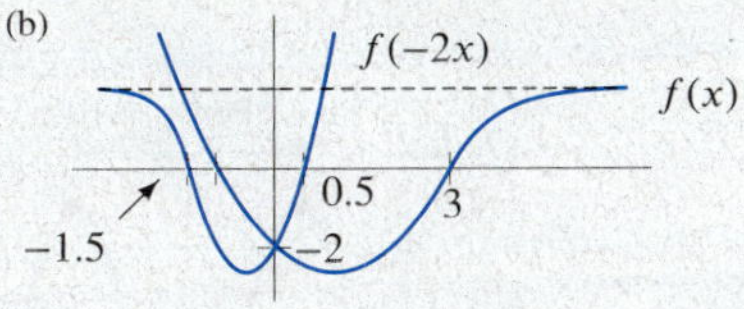


(c)

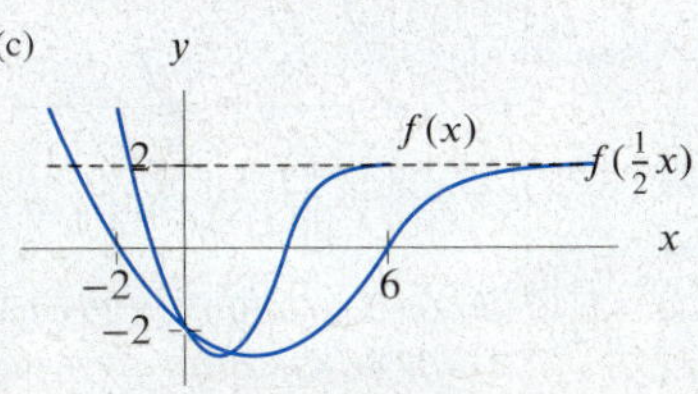


11 A horizontal compression by a factor of 1/3 and then a horizontal shift right by 2/3 units.

13 (a)

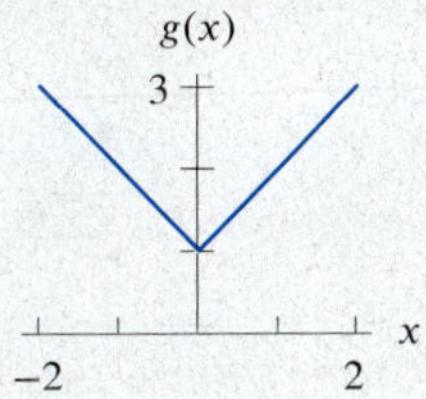


(b)

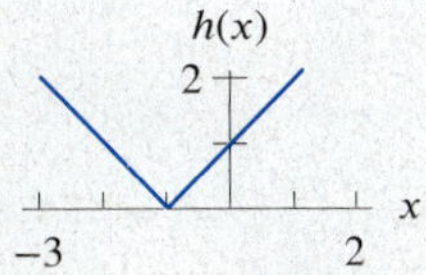


(c)

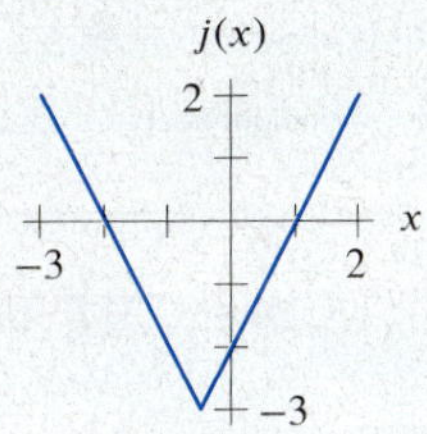


(d)

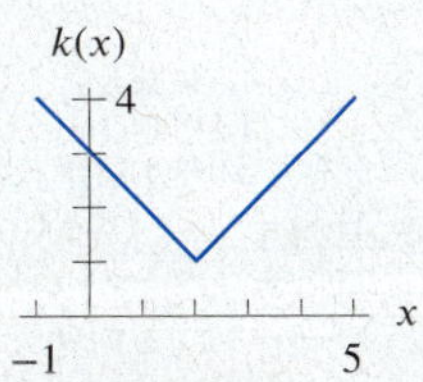


(e)

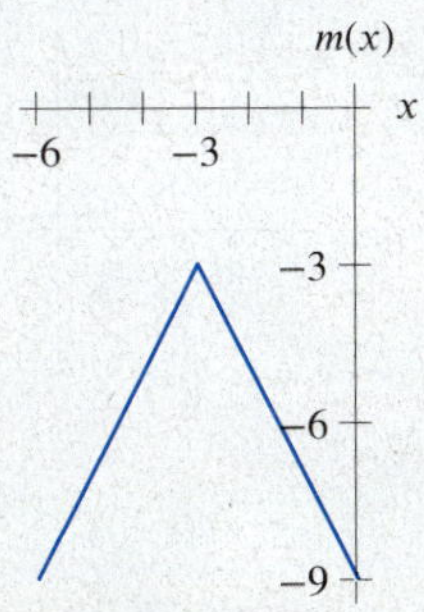


15 (a) (3, −14)
(b) (6, −26)
(c) (14, 17)
(d) (−21/5, −25/2)

19 2/3

21 $(a/d, b+c)$

23 $100f(12n)$

25 (a) (ii)
(b) (iv)
(c) (i)
(d) (iii)

27 (a) $A = f(r) = \pi r^2$
(b) $f(1.1r)$
(c) 21%

29 (a) I
(b) III
(c) IV
(d) II

31

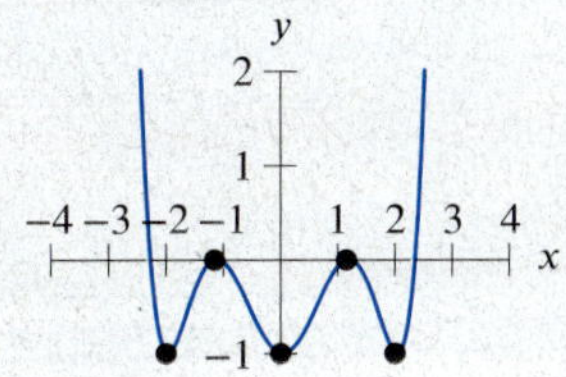


33

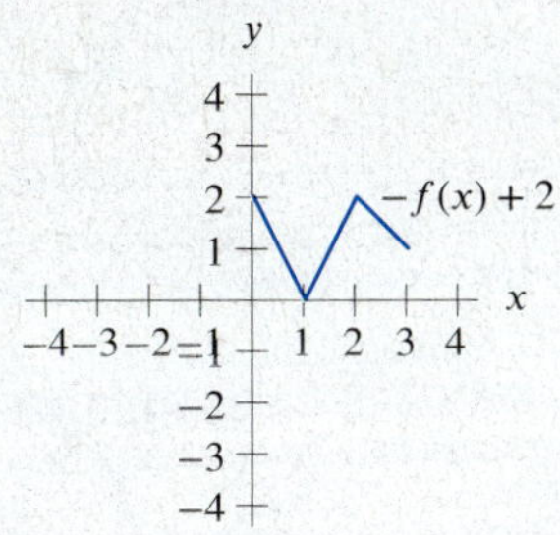


35

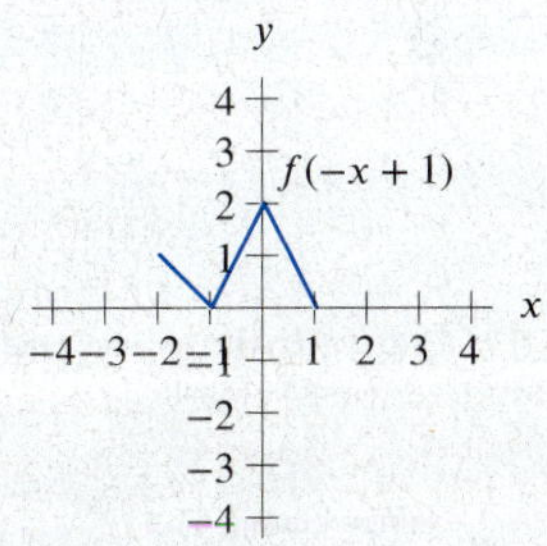


37

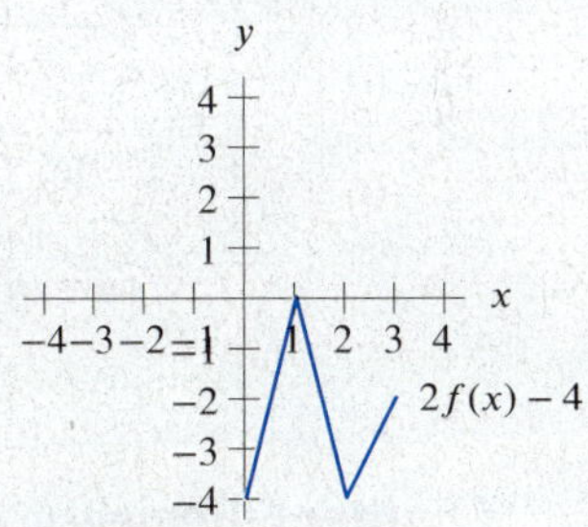


39

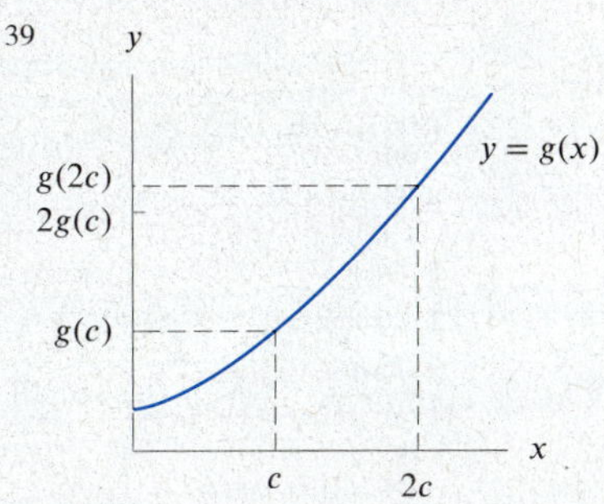

41 $y = -f(-\frac{1}{2}x)$

43 (a) $-3 \le x \le 1$
(b) $-6$

45 (a) $-8 \le x \le 0$
(b) $-3$

47 (a) $y = x^2 + 6x + 5$

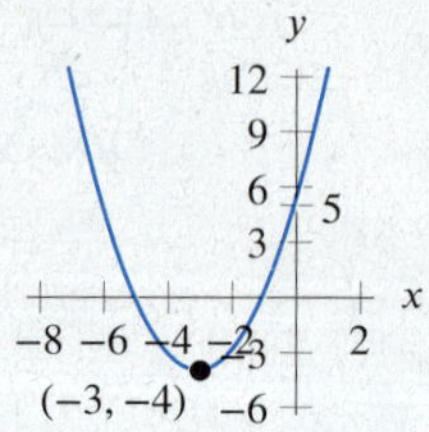


(b) $y = -2x^2 - 4x + 1$

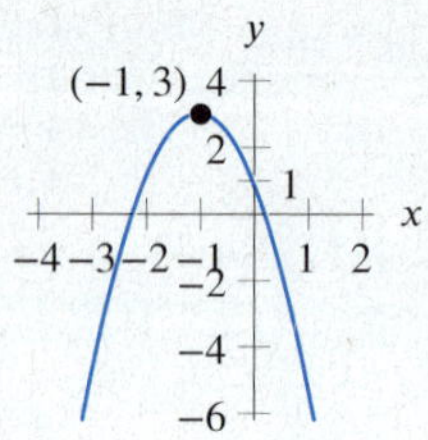


49 $g(x) = -3f(x+4) + 6$

51 $h(x) = -2f(-x+3) - 4$

53 (b) $d(t)$ reflected about the $t$-axis and then shifted up 142 units

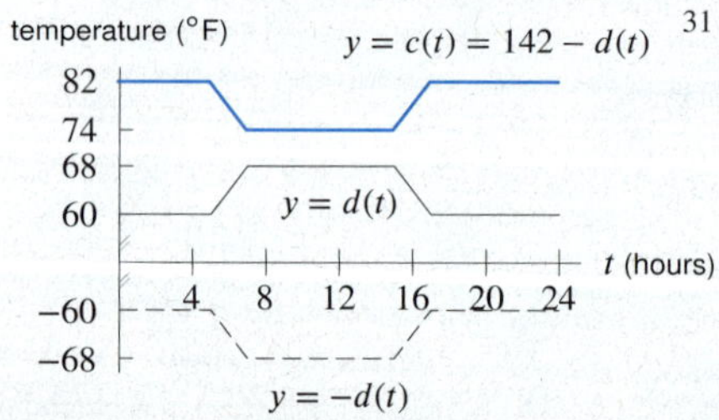


55 $5/e^2$

## Chapter 6 Review

1 (a) 4
(b) 1
(c) 5
(d) −2

3 (a) (6, 5)
(b) (2, 1)
(c) (1/2, 5)
(d) (2, 20)

5 Odd

7 Neither

9 Even

11 (a) $f(2x) = 1 - 2x$
(b) $f(x+1) = -x$
(c) $f(1-x) = x$
(d) $f(x^2) = 1 - x^2$
(e) $f(1/x) = (x-1)/x$
(f) $f(\sqrt{x}) = 1 - \sqrt{x}$

13

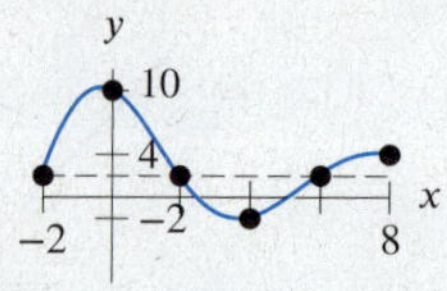


15 $y = f(t+4) - 8$

17 (a) A horizontal reflection about the $y$-axis.
(b) A horizontal shift 6 units to the right.

19 (a) VI
(b) V
(c) III
(d) IV
(e) I
(f) II

21 (a) $-16t^2 + 23$
$-16t^2 + 48t + 2$
(b)

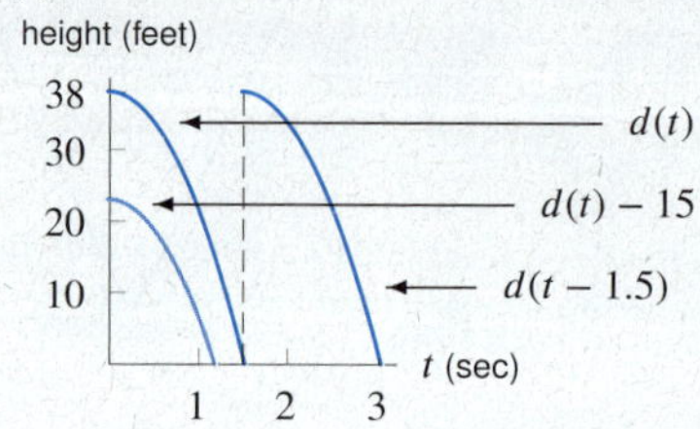


(d) 1.541 secs
1.199 secs
(e) 3.041 secs

23 $y = -(x+1)^3 + 1$

25 $y = (1/2)h(x+6) + 1$

27 $p \approx 15, q \approx 7190$

31

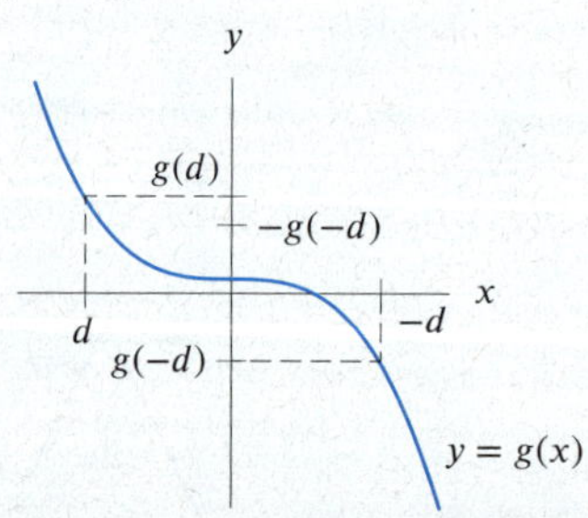


33 $y = -h(x-1)$

35

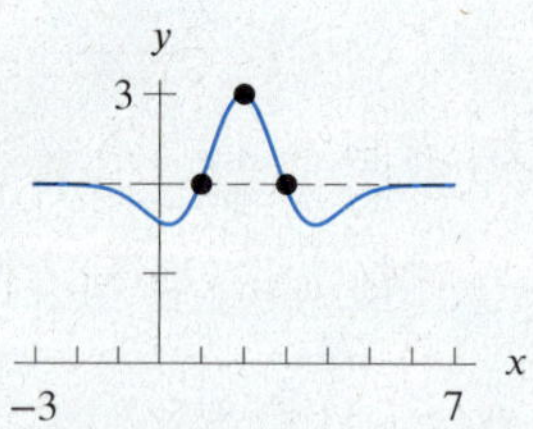


37 $y = -(x-3)^2$

39

| $d$ | 25 | 50 | 75 | 100 |
|---|---|---|---|---|
| $g(d)$ | 2.5 | 2.2 | 2.1 | 2.1 |
| $d$ | 125 | 150 | 175 | 200 |
| $g(d)$ | 2.3 | 2.5 | 2.75 | 3 |

41

| $d$ | 35 | 60 | 85 | 110 |
|---|---|---|---|---|
| $m(d)$ | 5.5 | 5.2 | 5.1 | 5.1 |
| $d$ | 135 | 160 | 185 | 210 |
| $m(d)$ | 5.3 | 5.5 | 5.75 | 6 |

43

| $d$ | 31.25 | 62.5 | 93.75 | 125 |
|---|---|---|---|---|
| $p(d)$ | 5.5 | 5.2 | 5.1 | 5.1 |
| $d$ | 156.25 | 187.5 | 218.75 | 250 |
| $p(d)$ | 5.3 | 5.5 | 5.75 | 6 |

## Ch. 6 Understanding

1 True

3 True

5 False

7 True

9 False

11 True

13 True

15 False

17 False

19 True

## Section 7.1

1 (I), (II), (IV)

3 90 m

5 90 m

7 40

9 $b$

11 9 o'clock position; 90 m

13 3 o'clock position; 90 m

15

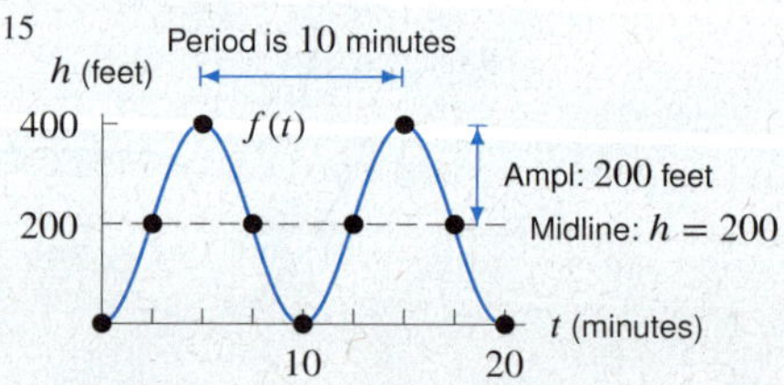


17

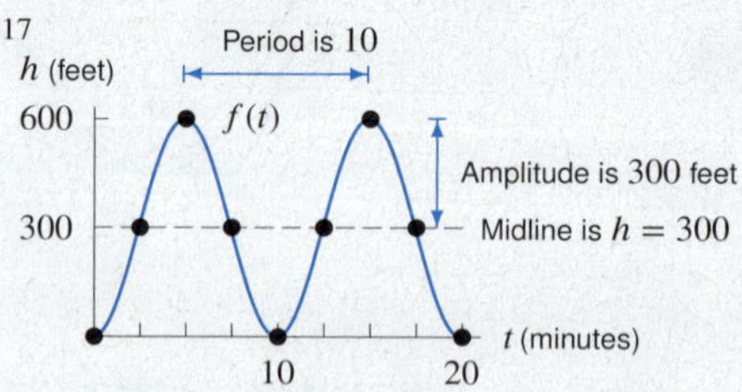


19 $(h(15) - h(10))/5 = 26$ ft/min

21 $(h(30) - h(25))/5 = -26$ ft/min

23
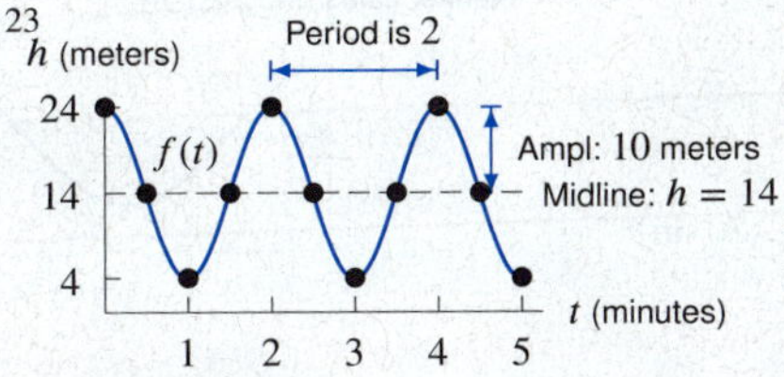


25
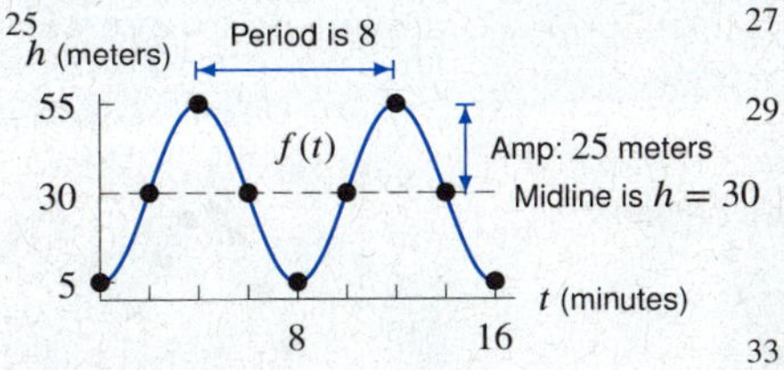


27 3 (or 9) o'clock; descending; 5 minutes; 40 meters; 0 meters; 11.25 minutes

29 12 o'clock; descending; 4 minutes; 30 meters; 5 meters; 10 minutes

33 (a) Weight B
(b) Weight A
(c) Weight A

35 Midline: $y = 5.55$;
Amplitude: 5.15 WBC $\times 10^4$/mL;
Period: 72 days

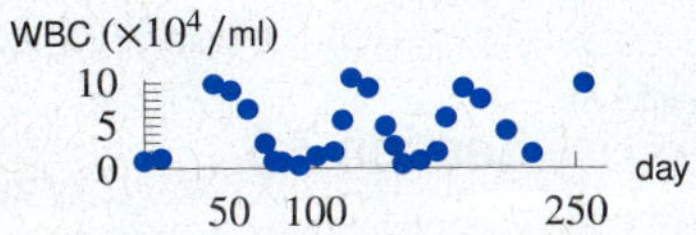


## Section 7.2

1 (a) $(-0.174, 0.985)$
(b) $(-0.940, -0.342)$
(c) $(-0.940, 0.342)$
(d) $(0.707, -0.707)$
(e) $(0.174, -0.985)$
(f) $(1, 0)$

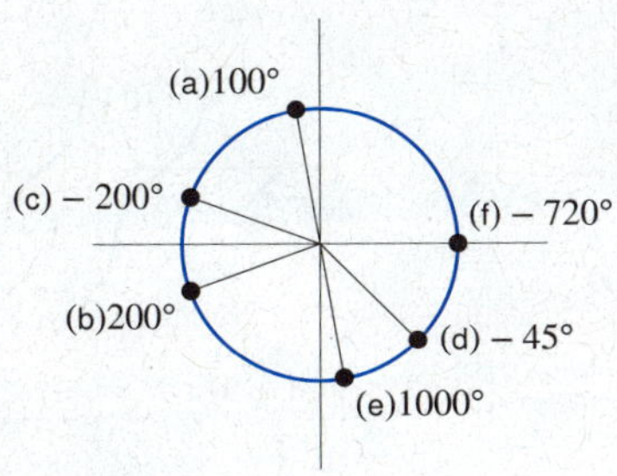


3 $-720°$

5 $80°$

7 $50°$.

11 $(-3.8, 0)$

13 $(0, 3.8)$

15 $(3.687, -0.919)$

17 $(3.742, -0.660)$

19 $\sin\theta = 0.6$;
$\cos\theta = -0.8$

21 $70°, 110°, 250°, 290°$

23 (a) $307°$
(b) $127°$

25 (a) $\sin 135° = 0.707$; $\cos 135° = -0.707$
(b) $\sin 285° = -0.966$; $\cos 285° = 0.259$

27 $\sin 20° < \sin 150° < \sin 35° = \sin 145° < \sin 70°$

29 (a) $a$
(b) $-a$
(c) $a$
(d) $a$
(e) $-a$
(f) $-a$

33 (a) $72°$
(b) $180°$
(c) $216°$

35 (a) All are equal
(b) $KP = 1/2$
(c) $OP = \sqrt{3}/2$
(d) $(\sqrt{3}/2, 1/2)$
(e) $\cos 30° = \sqrt{3}/2$;
$\sin 30° = 1/2$
(f) $\cos 60° = 1/2$;
$\sin 60° = \sqrt{3}/2$

37 $(60, 0)$, $(7.5, 0)$
$(60\cos\theta, 60\sin\theta)$
$(7.5\cos\theta, 7.5\sin\theta)$

## Section 7.3

1 $\pi/3$

3 1.7453 radians

5 $5\pi/6$

7 $-3\pi/2$

9 $630°$

11 $16{,}200/\pi \approx 5156.620°$

13 $8100/\pi \approx 2578.310°$

15 (a) I
(b) II
(c) II
(d) III
(e) IV
(f) IV
(g) I
(h) II
(i) II
(j) III

17 $-4\pi$

19 $8.54\pi$

21 $6.2\pi/4 \approx 4.869$

23 $6.2a\pi/180$

25 $(-5\sqrt{3}, -5)$

27 $(-5, 0)$

29 $(-5\sqrt{2}/2, 5\sqrt{2}/2)$

31 $(5\sqrt{2}/2, -5\sqrt{2}/2)$

33 $-1/\sqrt{2}$

35 10 cm

37 16 inches

39 $r = 5$; $\theta = 3/5$ rad $= 34.3775°$;
$s = 3$; $P = (4.1267, 2.8232)$

41 $r = 0.05$; $\theta = 0.384$ rad $= 22°$;
$s = 0.019$; $P = (0.0464, 0.0187)$

43 $r = 80$; $\theta = 3\pi/7$ rad $= 77.1429°$;
$s = 107.7117$; $P = (17.8017, 77.9942)$

45 (a) Yes
(b) Just over 6

47 (a) $-2\pi/3 < 2/3 < 2\pi/3 < 2.3$
(b) $\cos 2.3 < \cos(-2\pi/3) = \cos(2\pi/3) < \cos(2/3)$

49 (a)
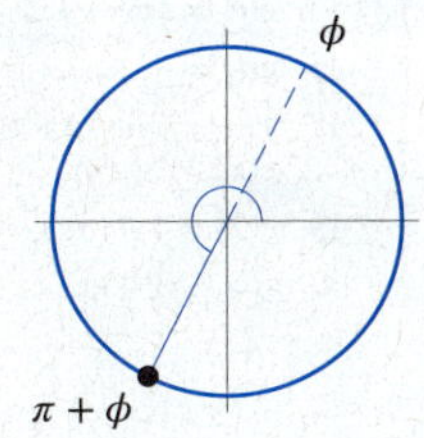


(b)
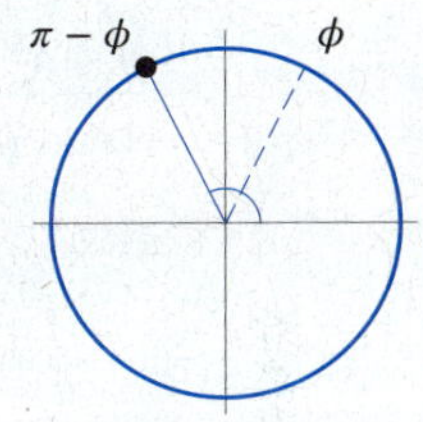


(c)
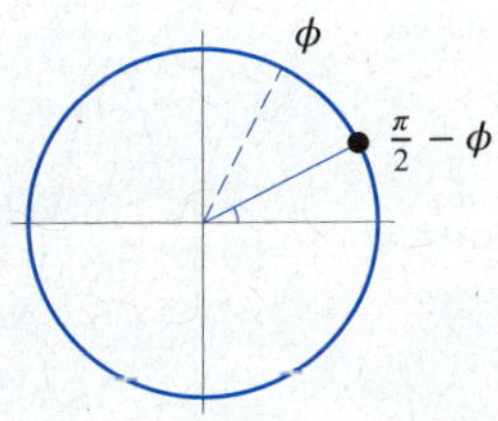


(d)
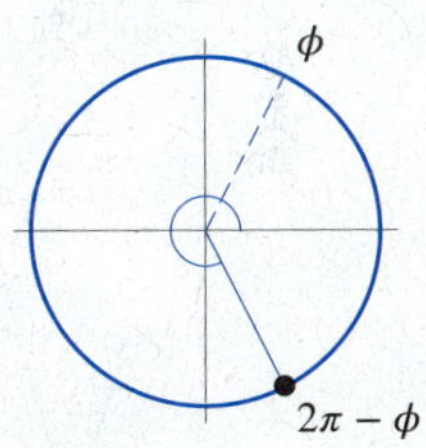


51 $u = -5\cos 2$
$v = 5\sin 2$
$w = 5\sqrt{2(1-\cos 2)}$

53 $t$

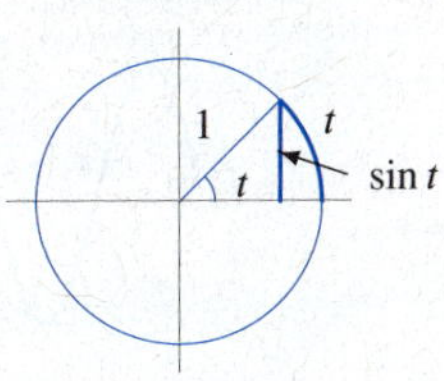


55 $7\pi$ inches

## Section 7.4

1 Mid: $y = 3$; Amp: 2

3 Mid: $y = 0$; Amp: 0.15

5 $2 \le y \le 6$

7 midline $y = 3$, amplitude 2, shift and stretch of $\sin t$

9 midline $y = -1$, amplitude 0.5, shift and stretch of $\cos t$

11 $g(\theta) = 3\sin\theta + 3$

13 $g(t) = 2\cos t + 3$

15 $g(t) = -3\sin t + 5$

17 $f(x) = (\sin x) + 1$
$g(x) = (\sin x) - 1$

19 $g(x) = 2\sin x$, $a = 180°$, $b = 2$

23 (a) (i) Odd

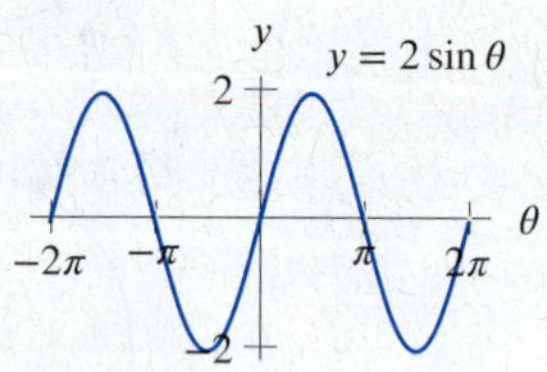


(ii) Not odd

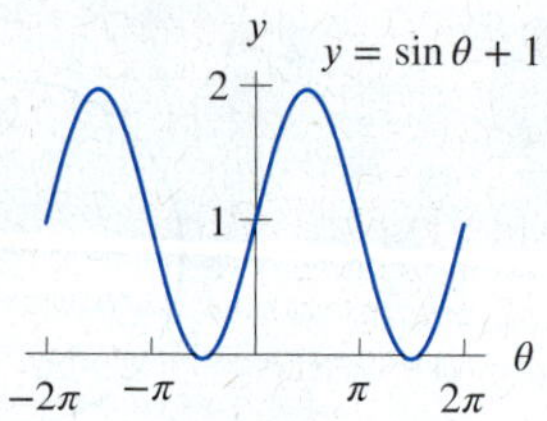


(iii) Odd

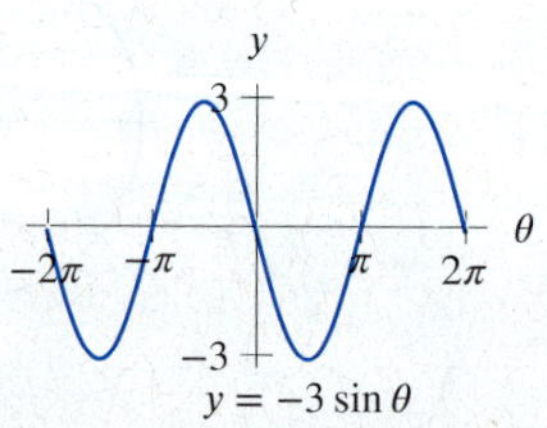


(iv) Not odd

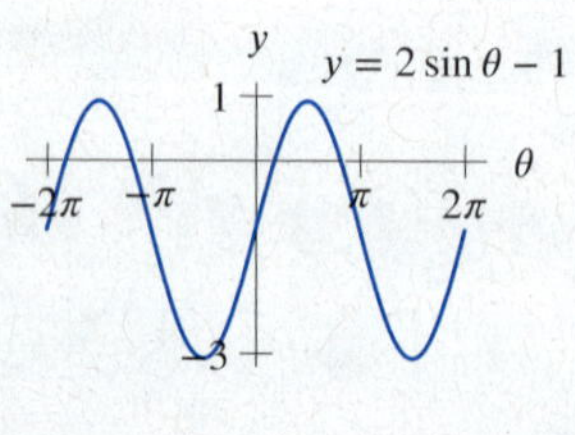


(b) Odd provided $k = 0$

25 $g(x) = 1.8\sin x - 3.3$

27 $y = 3g(x) + 2$

29 $y = 4g(x) - 4$

31

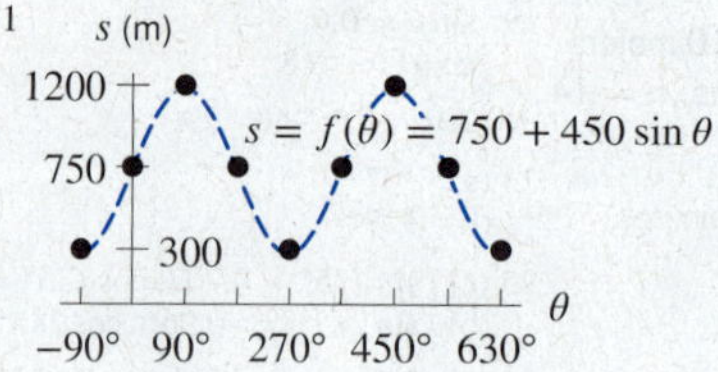


33

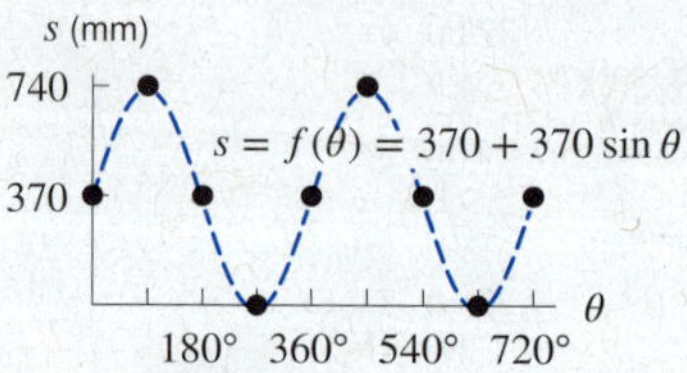


35 (a) $y = A\sin t + k$.
(b) $k = 3$ and $A < 0$

37 $t \approx 0.739$

39 $h(\theta) = 2.5 + 2.5\sin\theta$

## Section 7.5

1 Mid: $y = 0$; Amp: 6; Per: $2\pi$

3 Mid: $y = 1$; Amp: $1/2$; Per: $\pi/4$

5 period 8, midline $y = 30$, amplitude 20

7 period 0.7, midline $y = 0.05$, amplitude 0.03

9 period 0.5, midline $y = 0.5$, amplitude 0.5

11 Both $f$ and $g$ have periods of 1, amplitudes of 1, and midlines $y = 0$

13 $h(t) = 4\sin(2\pi t)$

15 $g(t) = -2\cos(t/2) + 2$

17 $y = 4000 + 4000\sin((2\pi/60)x)$

19 $y = -2\sin(\pi\theta/6) + 2$

21

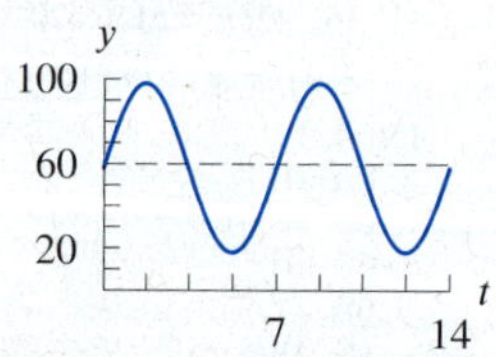


23

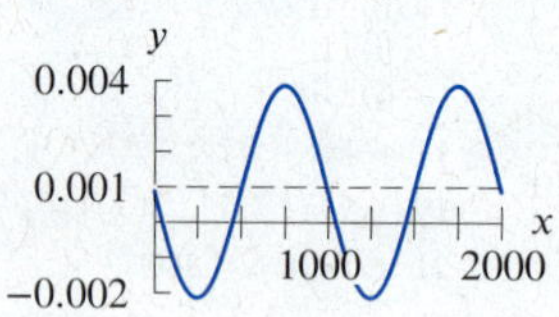


25 $f(x) = \sin x$, $a = \pi/2$, $b = \pi$, $c = 3\pi/2$, $d = 2\pi$, $e = 1$

27 $f(t) =$
$17.5 + 17.5\sin((2\pi/5)t)$

31 $V = 155.6\cos(120\pi t)$

33 $y = 3f(x)$

35 $y = -f(2x)$

37 Amplitude: 41.5;
Period: 12 months

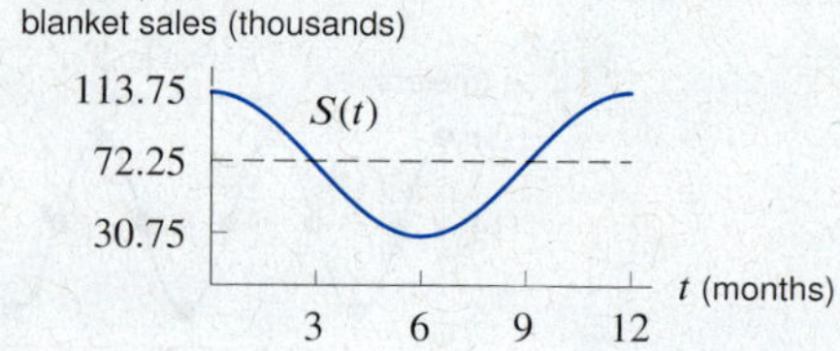


39 $f(t) =$
$-100\cos(\pi t) + 100$ (for $0 \le t \le 1$)
$10\cos(4\pi t) + 190$ (for $1 < t \le 2$)

41 (a)

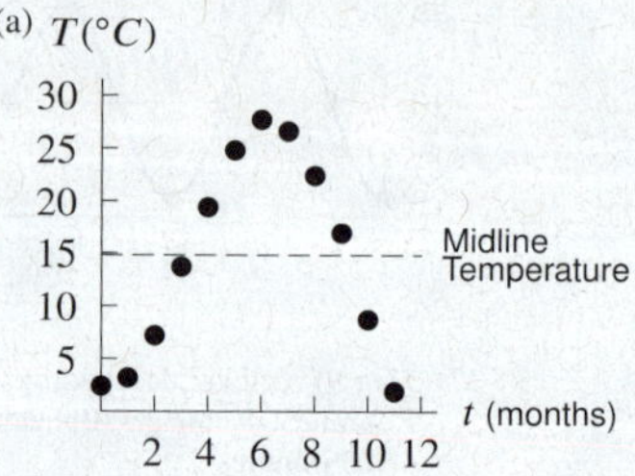


(b) 23.2°; 12 months
(c) $T = f(t) =$
$-23.2\cos((\pi/6)t) + 58.6$
(d) $T = f(9) \approx 58.6°$

43 (a) Yes
(b) No

45 Phase: 13; Horiz: $-13/7$

47 $3/10$, $g(x) = 10\sin((\pi/5)x - 3\pi/5)$

49 $h = 20 + 15\sin((\pi/2)t + \pi/2)$

## Section 7.6

1 0, 1, 0

3 Undefined

5 0

7 $-1/\sqrt{3}$

9 $\sqrt{3}$

11 $\theta = \pi/2$, $\theta = 3\pi/2$

13 $\theta = \pi/2$, $\theta = 3\pi/2$

15 Reflected about $\theta$-axis

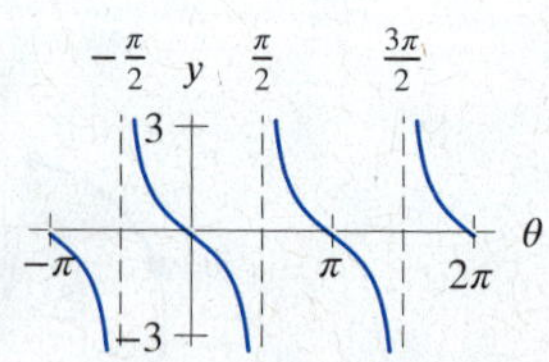


17 Shifted up by 1

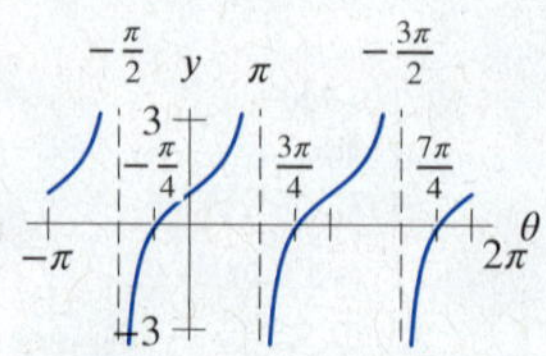

19 Compressed horizontally by 2

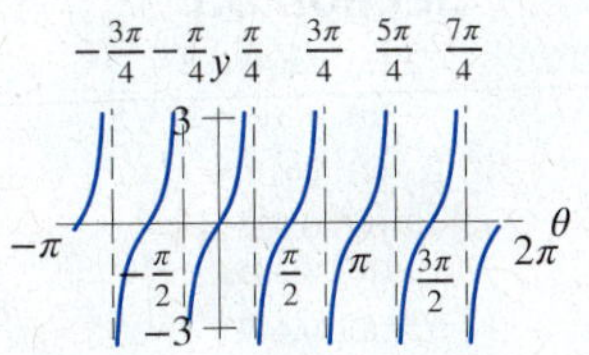

21 $-0.482$

23 $-1.428$

25 $y = (1/2)\tan\theta$

27 (a)

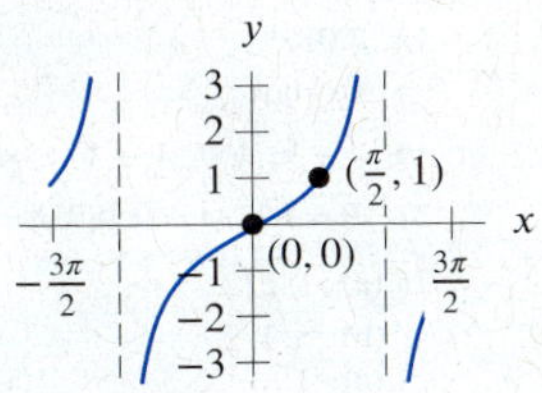

(b) $f(t) = \tan\left(\frac{t}{2}\right)$

29 (a)

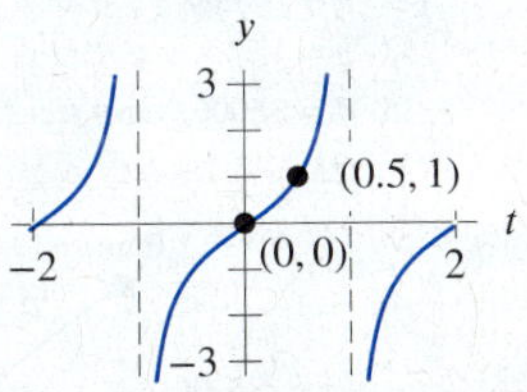

(b) $f(t) = \tan\left(\frac{\pi}{2}t\right)$

31 (a) $t = \pm\pi, \pm 3\pi, \pm 5\pi, \cdots$
(b) $t = 0, \pm 2\pi, \pm 4\pi, \pm 6\pi, \cdots$
(c) $2\pi$

33 Incorrect, $\tan t$ counterexample

35 (a) $y = (-1/\sqrt{3})x + 2$
(b) $2\sqrt{3}$

## Section 7.7

1 $-\sqrt{3}$

3 $-1/\sqrt{3}$

5 $-1/\sqrt{3}$

7 $2/\sqrt{3}$

9 $(\cos(2\theta))^2 + (\sin(2\theta))^2 = 1$

11 1

13 $\cos t$

15 1

17 (a) $(-x, -y)$

19 $\sec\theta = 2$
$\tan\theta = \sqrt{3}$

21 $\sec\theta = 3/\sqrt{8}$
$\tan\theta = 1/\sqrt{8}$

23 $\cos\theta = \sqrt{1 - y^2}$

25 (a) $\sin\phi = -0.8866$, $\tan\phi = -1.9166$
(b) $\cos\theta = -0.8062$, $\tan\theta = 0.7339$

27 $\sin\theta = \sqrt{x^2 - 16}/x$,
$\tan\theta = \sqrt{x^2 - 16}/4$

29 $\cos\theta = 9/\sqrt{x^2 + 81}$,
$\sin\theta = x/\sqrt{x^2 + 81}$

31 (a) Appear identical
(b) No

## Section 7.8

1 (a) 0.009
(b) $\pi/6$ or 30°
(c) 114.593

3 (a) $\sqrt{2} + 1$
(b) $2\sqrt{2} + 1$
(c) 114.228°

5 0.58569, 33.557°

7 0.7227, 41.4°

9 0.608

11 1.330

13 $2\pi/3, 4\pi/3$

15 $3\pi/2$

17 $\pi$

19 $\pi/4, 5\pi/4$

21 $\pi/3, 4\pi/3$

23 1.893

25 The angle is $z$; the value is $a$

27 The angle is $m$; the value is $y$

29 The angle is $z$; the value is $\frac{1}{y}$

31 (a) $\pi/6$
(b) 0.909
(c) 2.086

33 (a) $-1 \le x \le 1$; $-\pi/2 \le f(x) \le \pi/2$
(b) $-1 \le x \le 1$; $0 \le g(x) \le \pi$
(c) All real numbers; $-\pi/2 < h(x) < \pi/2$

35 Statement II is always true; statement I is not always true

## Chapter 7 Review

1 Yes

3 No

5 No

7 Yes

9 $S = (-0.707, -0.707)$, $T = (0, -1)$, $U = (0.866, -0.5)$

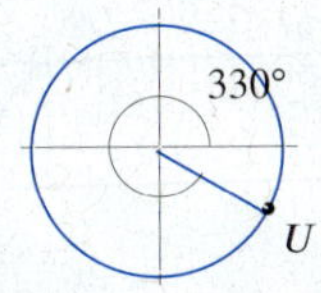

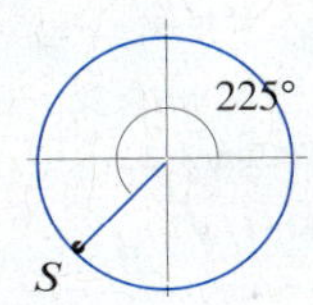

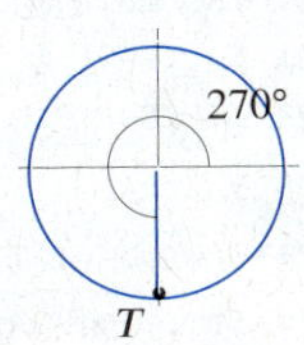

11 $S = (-3.536, -3.536)$
$T = (0, -5)$
$U = (4.330, -2.5)$

13 $(4.944, -15.217)$

15 The angle is $y$; the value is $x$

17 The angle is $d$; the value is $1/c$

19 0

21 $-1$

23 $-1$

25 $11\pi/6$

27 $-5\pi/4$

29 270°

31 $900/\pi = 286.479$

33 $-12\pi$

35 (a) II
(b) III
(c) IV
(d) I
(e) III

37 $6.2 \cdot 13\pi/4 \approx 63.303$

39 (a) $C(t)$
(b) $D(t)$
(c) $A(t)$
(d) $B(t)$

41 Mid: $y = 7$; Amp: 1; Per: $2\pi$

43 (a) 30°
(b) $-30°$
(c) 150°

45 $-2 \le t \le 1$

47 Not possible

49 $-2 \le t \le 0$

51 Period: 6; Amp: 5; Mid: $y = 0$

53 Midline: $h = 2$;
Amplitude: 1;
Period: 1

55 $1 - \sin A$

57 373.205 feet

59 Amp: 30; mid: $y = 60$; per: 4

61 Amp: 40; mid: $y = 50$; per: 16

63 $y = \sin x$ for $f(x), k(x)$; $y = \cos x$ for $g(x), h(x)$

65

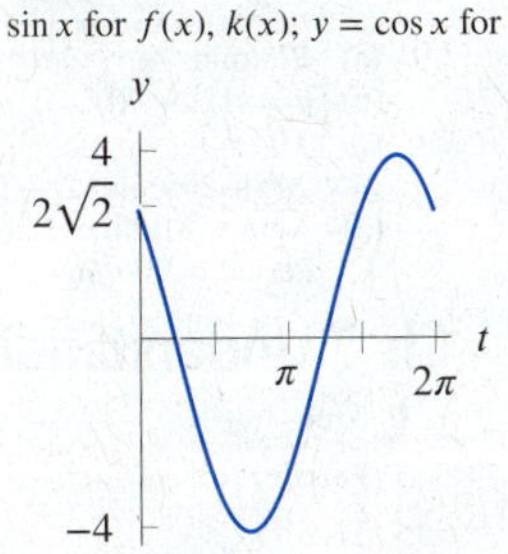

67

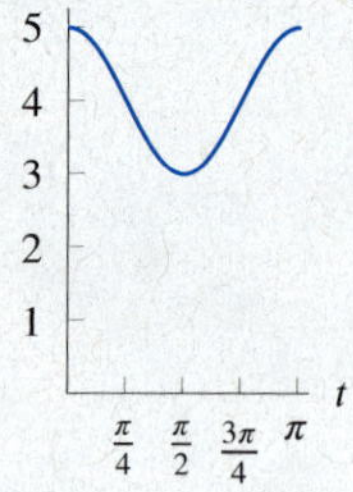


69 $-1$

71 $\sqrt{3}$

73 1.168

75 $f_1(x) = 5\cos((\pi/4)(x-4)) + 5, f_2(x) = -5\cos((\pi/4)x)+5, f_3(x) = 5\sin((\pi/4)(x-2))+5, f_4(x) = -5\sin((\pi/4)(x-6))+5$; answers may vary

77 $f_1(x) = 3\cos((\pi/4)(x-3)) - 3, f_2(x) = -3\cos((\pi/4)(x+1))-3, f_3(x) = 3\sin((\pi/4)(x-1))-3, f_4(x) = -3\sin((\pi/4)(x-5))-3$; answers may vary

79 (a) 72.77 cm
(b) 0.202 radians

81 $5.8\pi$ inches

83 $(-0.99, 0.14)$

85 $f(t) = -900\cos((\pi/4)t) + 2100$

87 Amplitude: 20
Period: 1/2
Phase shift: 0
Horizontal shift: 0

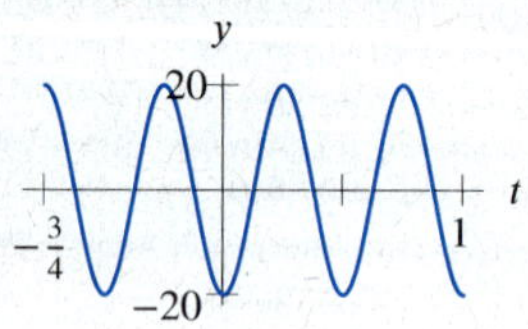


89 Amplitude: 3;
Period: 1/2;
Phase Shift: $6\pi$;
Horizontal Shift: $-3/2$ (left)

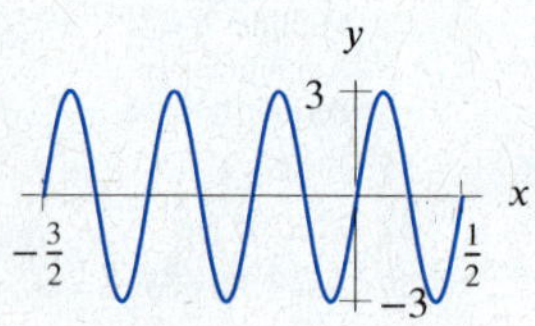


91 (a) 12°/min
(b) $\theta = (12t - 90)°$
(c) $f(t) = 200 + 200\sin(12t - 90)°$
(d) Amp = Midline = 200 feet
Period = 30 min

## Ch. 7 Understanding

1 True

3 False

5 True

7 True

9 False

11 False

13 True

15 False

17 True

19 False

21 False

23 False

25 False

27 False

29 True

31 True

33 False

35 True

37 False

39 False

41 False

43 False

45 False

47 False

49 True

51 False

53 False

55 True

57 True

59 False

61 True

63 False

65 True

67 False

69 True

71 False

73 False

75 True

77 True

79 False

## Ch. 7 Skills: Special Angles

1 $1/2$

3 $-\sqrt{3}/2$

5 $1/2$

7 $\sqrt{3}/2$

9 $1/\sqrt{2}$

11 $-\sqrt{3}/2$

13 $1/\sqrt{2}$

15 $-1/\sqrt{2}$

17 $-\sqrt{3}/2$

19 They are equal

21 5

23 $20/\sqrt{3}$

25 $5, 5, 5\sqrt{2}$

27 $7/\sqrt{2}, 7/\sqrt{2}, 7$

29 45°-45°-90°, $4\sqrt{2}$

31 $(3\sqrt{2}, -3\sqrt{2})$

## Section 8.1

1 (a) $\tan\theta = 2$
(b) $\sin\theta = 2/\sqrt{5}$
(c) $\cos\theta = 1/\sqrt{5}$

3 (a) 0.923
(b) 0.385

5 (a) 0.447
(b) 0.894

7 $y = 7\sin 17°$; $x = 7\cos 17°$

9 $r = 6/\cos 37°$; $y = 6\tan 37°$

11 $y = 9/\tan 77°$; $r = 9/\sin 77°$

13 61.164°

15 7.012°

17 89.190°

19 $c = 34.409$; $A = 35.538°$, $B = 54.462°$

21 $B = 62°$; $a = 9.389$; $b = 17.659$

25 (a) $\sqrt{2}/2$
(b) $\sqrt{2}/2$
(c) 1

27 $h = 400$ feet; $x = 346.410$ feet

29 0.35 miles

31 $p = 18.660$

33 $p = 1.732$

35 $d = 35000/\tan\theta$ feet

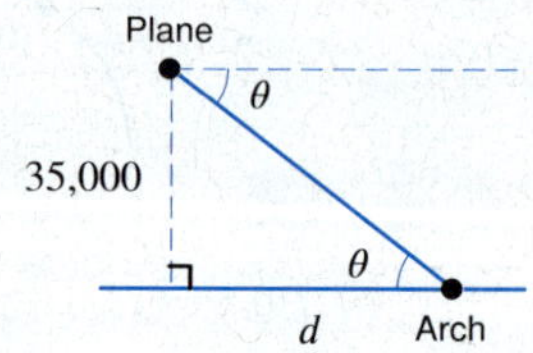


37 (a) 4.00417°
(b) 28.57 ft
(c) 28.64 ft

39 63.435°

41 Height: 541.723 ft;
Distance between: 718.891 ft

43 $\approx 59.971°$

45 $\alpha = 11.310°$; $\beta = 16.699°$

47 (a) $h \approx 88.388$ feet
(b) $h = 62.5$ feet
(c) $c \approx 88.388$ feet,
$d \approx 108.253$ feet

## Section 8.2

1 $x \approx 19.121$

3 $b \approx 5.120, c \approx 6.497; \beta = 52°$

5 $a \approx 11.818, b \approx 2.084; \theta = 80°$

7 $a = 10.450; \theta = 16.560°, \psi = 143.440°$

9 $A = 25.922°, B = 37.735°, C = 116.343°$

11 $b = 31.762, A = 38.458°, C = 60.542°$

13 $c = 10.954, A = 54.010°, B = 45.990°$

15 $c = 7.2605; A = 21.4035°; B = 126.597°$

17 $A = 27.7755°, B = 54.2245°, b = 13.9279$

19 $a = 15.681, b = 12.531, B = 53°$

21 No triangle is possible

23 $a = 7.246$, $A = 88.980°$, $C = 16.020°$

25 $a = 16.011$, $b = 2.634$, $C = 7\pi/20$ radians

27 $a = 9$, $b = 6.888$, $C = 3\pi/8$ radians

29 $b = 0.837$ m, $c = 2.720$ m; $\gamma = 143.7°$

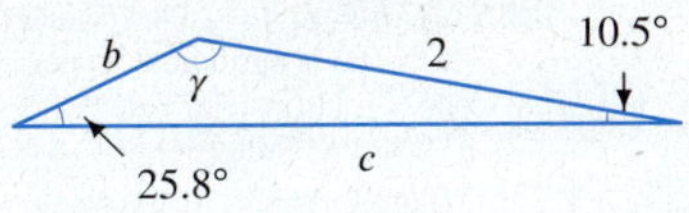


31 $\alpha \approx 41.410°, \beta \approx 82.819°, \gamma \approx 55.771°$

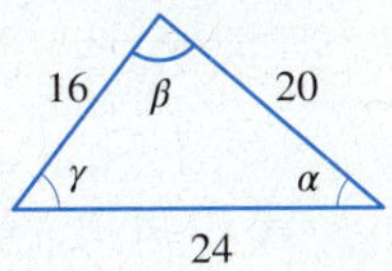


33 (a) $\sin\theta = 0.282$
(b) $\theta \approx 16.374°$
(c) $12.077\ \text{cm}^2$

35 984.160 feet

37 636.596 feet

39 (1.708, 5.575)

43 125.037 feet

47 (a) 435 feet
(b) $24{,}145.86\ \text{ft}^2$

49 (a) 5.420 and 3.757
(b) 8.833

## Section 8.3

1 IV

3 II

5 III

7 I

9 IV

11 90° to 180°

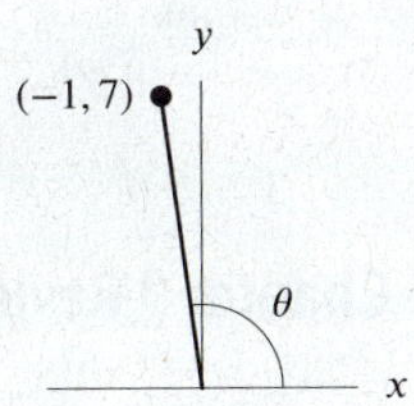


13 180° to 270°

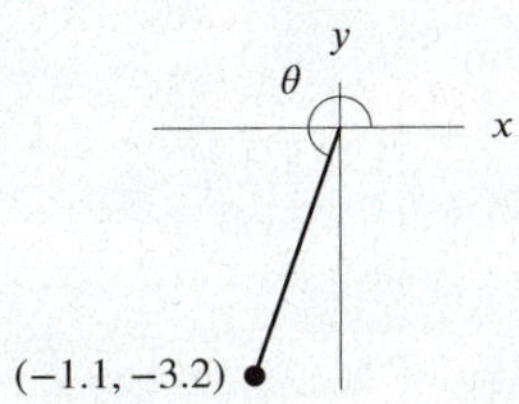


15 $(1, \pi)$

17 $(2, 5\pi/6)$

19 $(-1/2, \sqrt{3}/2)$

21 $(3, -\sqrt{3})$

23 $x^2 + y^2 = 4$

25 $y = x$

27 $r = 2/(3\cos\theta - 4\sin\theta)$

29 $r = \sin\theta/\cos^2\theta$

31 $H: x = 0,\ y = 3;\ r = 3,\ \theta = \pi/2$
$M: x = 0,\ y = 4;\ r = 4,\ \theta = \pi/2$

33 $H: x = -3,\ y = 0;\ r = 3,\ \theta = \pi$
$M: x = 0,\ y = 4;\ r = 4,\ \theta = \pi/2$

35 $H: x = 3\sqrt{2}/2,\ y = 3\sqrt{2}/2;\ r = 3,\ \theta = \pi/4$
$M: x = 0,\ y = -4;\ r = 4,\ \theta = 3\pi/2$

37 $H: x \approx 2.898,\ y \approx -0.776;\ r = 3,\ \theta = 23\pi/12$
$M: x = 0,\ y = -4;\ r = 4,\ \theta = 3\pi/2$

39 $\sqrt{8} \le r \le \sqrt{18}$ and $\pi/4 \le \theta \le \pi/2$

41 $0 \le \theta \le \pi/2$ and $1 \le r \le 2/\cos\theta$

43 $n$ loops

45 An inner and outer loop

47 $0 \le \theta \le 2\pi$ and $3/16 \le r \le 1/2$

49 (a) $0 \le \theta \le \pi/4$ and $0 \le r \le 1$
(b) Two pieces:
$0 \le x \le \sqrt{2}/2$ and $0 \le y \le x$;
$\sqrt{2}/2 \le x \le 1$ and $0 \le y \le \sqrt{1 - x^2}$

51 No; $(5, 3\pi/4)$, $(5, -3\pi/4)$

## Chapter 8 Review

1 (a) $\sqrt{45}/7$
(b) $2/7$
(c) $\sqrt{45}/2$

3 (a) $8/12$
(b) $\sqrt{80}/12$
(c) $8/\sqrt{80}$

5 (a) $\sqrt{117}/11$
(b) $2/11$
(c) $\sqrt{117}/2$

7 44.971°

9 59.036°

11 $\theta = 30°$

13 $\theta = 45°$

15 I

17 I

19 (1.571, 0)

21 (0, 0)

23 $\phi = 53.130°$; $\theta = 36.870°$

25 5; 67.380°, 22.620°

27 Height = 46.174 ft;
Incline = 205.261 ft

29 2.1268 miles

31 $d \approx 220.676$m

33 $d = 3\tan\phi$ miles

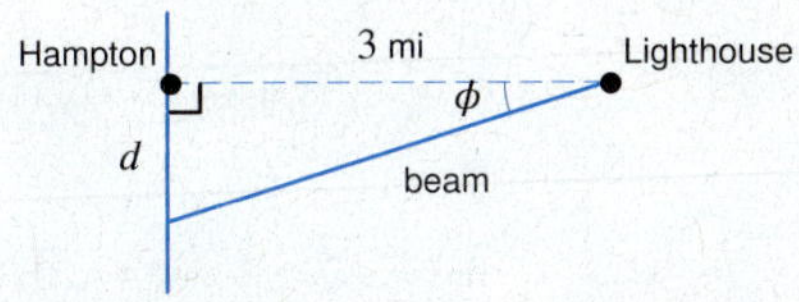


35 $\tan\theta = -3/4$

37 (b) 12 o'clock → $(x, y) = (1, 1)$ and $(r, \theta) = (\sqrt{2}, \pi/4)$,
3 o'clock → $(x, y) = (2, 0)$ and $(r, \theta) = (2, 0)$,
6 o'clock → $(x, y) = (1, -1)$ and $(r, \theta) = (\sqrt{2}, -\pi/4)$,
9 o'clock → $(x, y) = (0, 0)$ and $(r, \theta) = (0,$ any angle )

## Ch. 8 Understanding

1 False

3 False

5 True

7 False

9 True

11 True

13 False

15 False

17 True

## Section 9.1

1 −1.159, 1.159, 5.124, 7.442

3 $\theta = 0.708, 2.434$

5 $t = 1.813, 4.473$

7 0.1571 or 9°

9 0.5236 or 30°

11 0.2768 or 15.859°

13 −1.447

15 (a) $\pm\pi/3, \pm 5\pi/3$
(b) Two

17 0.927, 5.356

19 1.107, 2.034, 4.249, 5.176

21 1.253, 3.889

23 115°, 245°, 475°, 605°

25 0.927, 5.356, 7.210, 11.639

27 2.346, 3.937

29 $\theta = -\pi/4 + 2\pi k,\ 5\pi/4 + 2\pi k$, $k$ an integer

31 $\theta = -\pi/6 + \pi k$, $k$ an integer

33 $t = \pi/6,\ 5\pi/6,$
$7\pi/6,$ or $11\pi/6$

35 $t = \pi/2,\ 3\pi/2,$
$\pi/6,$ or $5\pi/6$

37 $t \approx 0.698, 1.396$; six

39 (a) $f(t) = 40{,}000\cos\left(\frac{\pi}{6}t + \frac{\pi}{6}\right) + 60{,}000$
(b) $f(3) = \$40{,}000$
(c) Mid-March and mid-September

41 $P: x \approx 0.819$;
$Q: x \approx 3.181$

43 (a) $d = \sqrt{2rx + x^2}$
(b) $d = 25{,}238.776$ meters

## Section 9.2

1 No; $t = \pi/2$

3 Yes

5 $\cos\theta = \sqrt{15}/4$, $\tan\theta = 1/\sqrt{15}$

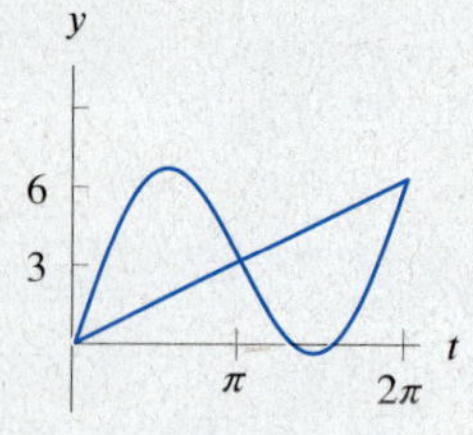


7 $\sin\theta = -\sqrt{21}/5$, $\tan\theta = \sqrt{21}/2$

9 $\sin t - \cos t$

11 $\cos t + 3\sin t$

13 $2\sin\alpha$

15 $\cos t - \sin t$

17 0

19 $\tan\sqrt{\theta}$

21 $(3/4)\tan(\phi + 1)$

23 $(10/3)\tan\left(\frac{2}{k+3}\right)$

25 $\sin\theta = \sqrt{1-\cos^2\theta}$, for $\sin\theta \geq 0$;
$\sin\theta = -\sqrt{1-\cos^2\theta}$, for $\sin\theta < 0$

35 $\pi/6, 5\pi/6, \pi/2, 3\pi/2$

37 $0, \pi/3, 2\pi/3, \pi, 4\pi/3, 5\pi/3, 2\pi$

39 Not an identity

41 Not an identity

43 Not an identity

45 Not an identity

47 Not an identity

49 Identity

51 Identity

53 Identity

55 (a) $\tan\theta$
(b) $\sin\theta$
(c) $(1/\cos\theta)^2$ or $1 + \tan^2\theta$
(d) $(1/2)\tan\theta$

57 $\sin 2\theta = \frac{2x}{9}\sqrt{9-x^2}$

59 (a) $2x\sqrt{1-x^2}/(2x^2-1)$
(b) $2x/(1+x^2)$

61 $\sin 4\theta = 4(\sin\theta\cos\theta)(2\cos^2\theta - 1)$

## Section 9.3

1 0.8244

3 −0.5622

5 416/425

7 87/425

9 $\sin 15° = \cos 75° = (\sqrt{6} - \sqrt{2})/4$
$\cos 15° = \sin 75° = (\sqrt{6} + \sqrt{2})/4$

11 $(\sqrt{6} + \sqrt{2})/4$

13 0.839

15 0.545

17 (a) $\frac{y\sqrt{4-y^2}}{2}$
(b) $\frac{y^2-2}{2}$
(c) $\frac{\sqrt{4-y^2}-y}{2}$
(d) 1

## Section 9.4

1 $10\sin(t - 0.644)$

3 $\sqrt{2}\sin(t + 3\pi/4)$

5 All integral multiples of $\pi$

7 (I)

9 (II)

11 (a) 2500; increase in population during March
(b) $t = 3.786, 8.214$ times at which pop is 12000

13 (a) $P = 5000 + 300t$
(b) $P = 3200(1.04)^t$
(c) $P(t) = -900\cos(2\pi t/5) + 2100$

15 (a) $f(t) = 65 - 25\cos(\pi t/12)$
(b) Period: 24;
Amplitude: 30
(c) ≈4 am and 1 pm

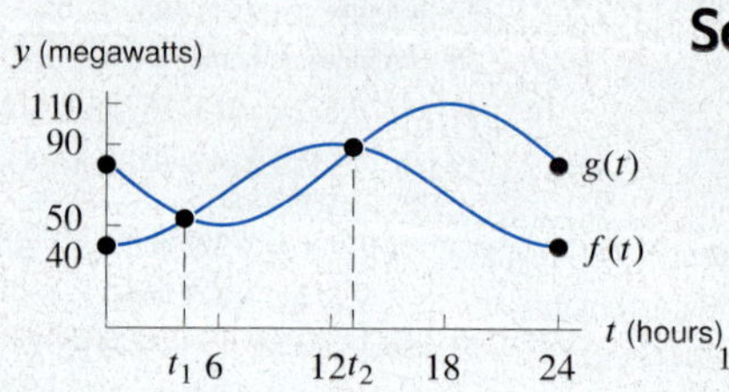


(d) 184.051 mw at 3:21 pm
(e) $h(t) = 145 - \sqrt{1525}\sin(\pi t/12 + 0.6947)$;
max $= 145 + \sqrt{1525} = 184.051$

17 (a) $y = 0$
(b) The function oscillates more and more as $x \to 0$.
(c) No
(d) $z_1 = 1/\pi$
(e) Infinitely many
(f) If $a = 1/(k\pi)$ then $b = 1/((k+1)\pi)$

19 (b) No; no, but possibly for 1979-1989
(c) Multiply by an exponential function:
$s(t) = (e^{0.05t})(-1.4\cos(\frac{2\pi}{6})t + 1.6)$
(d)

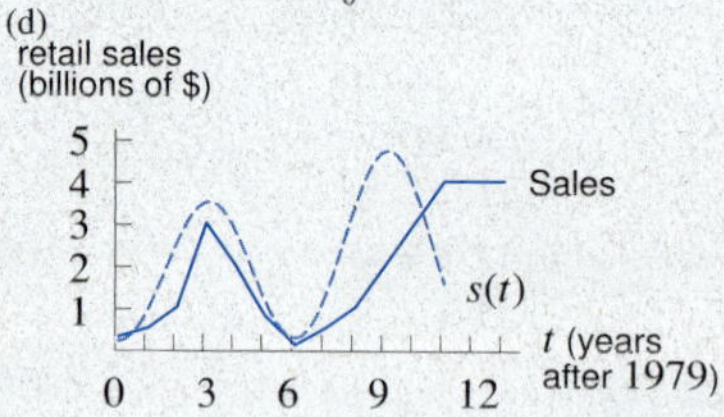


(e) 4.632 billion dollars

21 (a) 1/6 ppm/month; 1/6 ppm/month; −1 ppm/month
(b) Seasonal concentration variation
(c) 0 ppm/month; 0 ppm/month; −7/6 ppm/month
(d) 1/6 ppm/month; 1/6 ppm/month; 1/6 ppm/month
(e) Average monthly increase in concentration

## Section 9.5

1 $-\sqrt{6}/2$

3 $2\cos(4t)\cos t$

5 $2\sin(5t)\cos(2t)$

7 1.585

9 1.216

15 (a) $h = f(t) = 25 + 15\sin(\pi t/3) + 10\sin(\pi t/2)$
(b) $f(t)$ is periodic with period 12

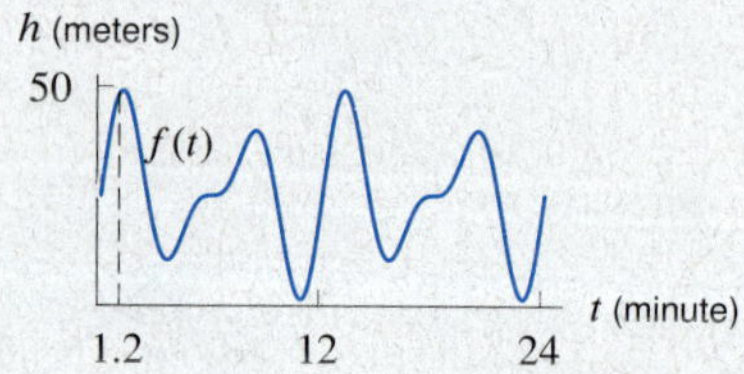


(c) $h = f(1.2) = 48.776$ m

## Section 9.6

1 $5e^{i\pi}$

3 $0e^{i\theta}$, for any $\theta$.

5 $5e^{i4.069}$

7 $-5 + 12i$

9 $-3 - 4i$

11 $-\frac{1}{2} + i\frac{\sqrt{3}}{2}$

13 $\frac{\sqrt{3}}{2} + \frac{i}{2}$

15 $\sqrt{2} + i\sqrt{2}$

17 $\frac{\sqrt{3}}{2} + \frac{i}{2}$

19 $\sqrt{2}/2 + i\sqrt{2}/2$

21 $\sqrt{2}\cos\frac{\pi}{12} + i\sqrt{2}\sin\frac{\pi}{12}$

23 $2.426 + 4.062i$

25 $A_1 = 2 - i$
$A_2 = -2 + 2i$

27 (a) $e^{i\pi/2}$

33 $2e^{\pi i/3} = 1 + 1.732i$, $2e^{\pi i} = -2$, $2e^{5\pi i/3} = 1 - 1.732i$

35 $2^{1/6}e^{\pi i/12} = 1.084 + 0.291i$, $2^{1/6}e^{3\pi i/4} = -0.794 + 0.794i$, $2^{1/6}e^{17\pi i/12} = -0.291 - 1.084i$

37 1

39 $(\sqrt{2})/2 + i(\sqrt{2})/2$

## Chapter 9 Review

1 $1 - \cos t - \sin t$

3 $4\tan t$

5 $\tan t$

7 $\cos t - \sin t$

9 $\sin\theta$

11 $2\cos\phi$

13 $8 - 5i$

15 $\sqrt{20}e^{-1.107i}$

17 $\alpha = \pi/3,\ 5\pi/3$

19 2.897, 6.038

21 (a) $y = 10\cos(2\pi x/60) + 8$
(b) $x_1 = 23.8550, x_2 = 36.1445$

23 (a)

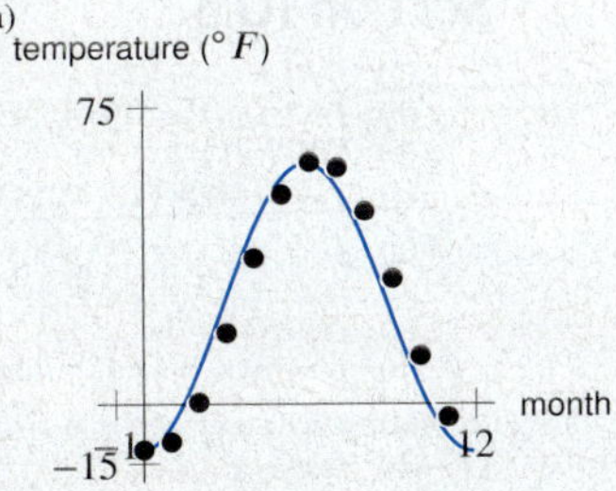


(b) Vertically reflected cosine curve
(c) $y = 24.9 - 36.4\cos(\pi t/6)$
(d) $t \approx 3.375,\ 8.625$
(f) $y = 24.9 + 36.4\cos(\pi t/6)$

25 $(\tan^2 x)(\sin^2 x) = \tan^2 x - \sin^2 x$

27 $\csc\theta = 11/8$
$\tan\theta = 8/\sqrt{57}$

29 $\sin\theta = 0.963$
$\tan\theta = 3.566$

31 $\cos\theta = 3/\sqrt{14}$

33 $7\pi/6, 11\pi/6, \pi/2$

39 (a) $\cos\theta = x$
(b) $\cos(\pi/2 - \theta) = \sqrt{1-x^2}$
(c) $\tan^2\theta = (1-x^2)/(x^2)$
(d) $\sin 2\theta = 2\sqrt{1-x^2}(x)$
(e) $\cos 4\theta = 1 - 8x^2(1-x^2)$
(f) $\sin(\cos^{-1} x) = \sqrt{1-x^2}$

41 3.376, 6.049

43 $\cos(\frac{\theta}{2}) = \sqrt{\frac{1}{2}(1+\cos\theta)}$

45 (a)

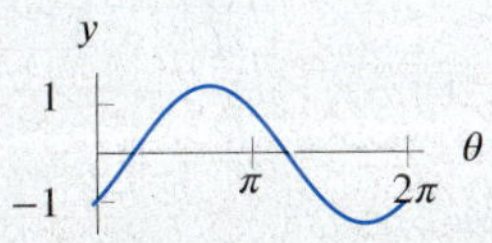


(b) $\sqrt{2}\sin(\theta - \pi/4)$
$\sqrt{2}\cos(\theta - 3\pi/4)$

47 Appear to be same

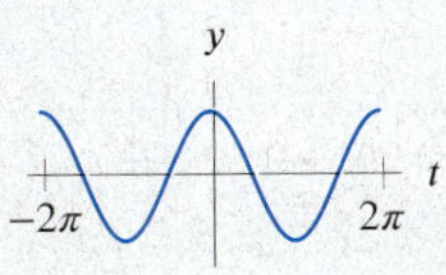


49 (a)

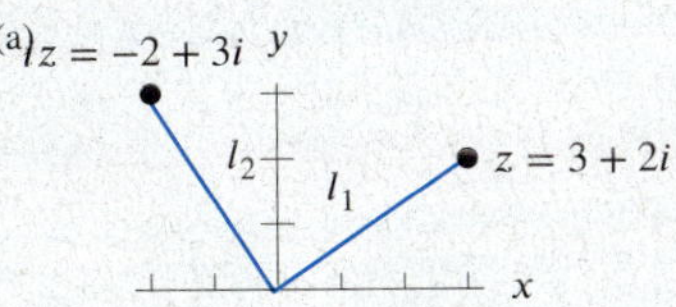


## Ch. 9 Understanding

1 True
3 True
5 True
7 True
9 True
11 False
13 True
15 True
17 True
19 False
21 True
23 False
25 True
27 False
29 True
31 True
33 False
35 True
37 True
39 True

## Section 10.1

1 (a) 2
(b) 1
(c) −1
(d) 2
(e) −1
(f) −1

3 $s(0) = 2, s(1) = 5, s(2) = 8, s(3) = 3, s(4) = 1, s(5) = 4$

5 $2^x/(2^x + 1)$

7 $3 + \tan^2 x$; $\tan(3 + x^2)$

9 $\sqrt{27x - 6}$

11 $3x$

13 $108x^2 - 2$

15 $f(g(x)) = e^{-0.5x^2}$

17 $\sin x$

19 $x^3 + 3x + 1$

21 $5 + 1/x$

23 Length in terms of time

25 Time in terms of temp

27

| $x$ | $f(x)$ | $g(x)$ | $g(f(x))$ |
|---|---|---|---|
| 0 | 2 | 3 | 2 |
| 1 | 3 | 1 | 3 |
| 2 | 1 | 2 | 1 |

29

| $x$ | $f(x)$ | $g(x)$ | $h(x)$ |
|---|---|---|---|
| 0 | 2 | 1 | 3 |
| 1 | 1 | 0 | 0 |
| 2 | 4 | 3 | 2 |
| 3 | 0 | 4 | 1 |
| 4 | 3 | 2 | 4 |

31 $g(x) = (3x^2)/(3x^2 + 2)$

33 $2x + h + 1$

35 $-1/(x(x+h))$

37 $f(x) = 2x$

39 $f(x) = x^2$

41 $g(x) = x^2$, $h(x) = x + 3$

43 $g(x) = x^2 + x$, $h(x) = 3x$

45 (a) $u(x) = (1+x)/(2+x)$
(b) $u(x) = x/(1+x)$

47 (a) $v(x) = -x$
(b) $u(x) = \sqrt{1-x}$

49 (a) $v(x) = \sin x$
(b) $u(x) = \sin^2(\sqrt{x})$

51 (a) (i) 3
(ii) 4
(iii) 3
(iv) 4
(b) 5

53 (a) 4 and 0
(b) 2

55

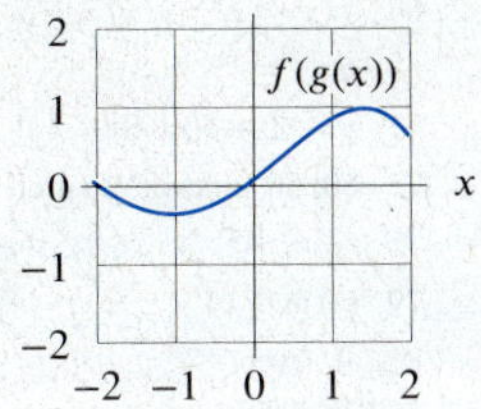


57

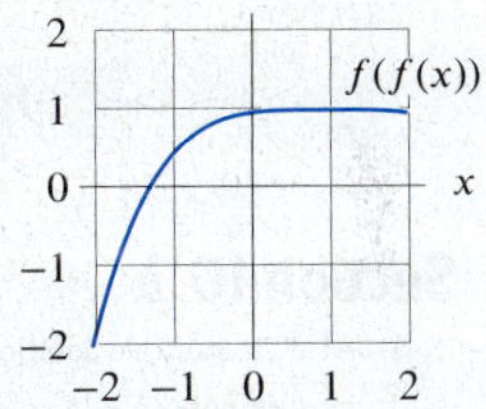


59

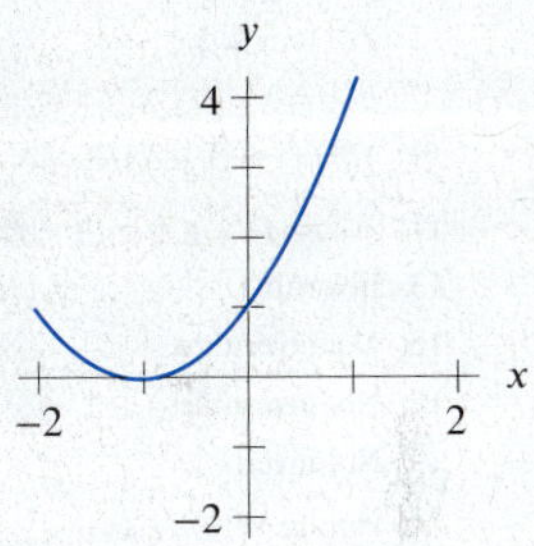


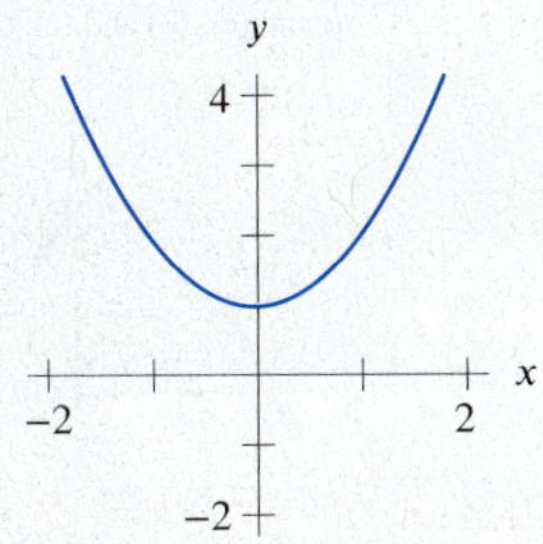


61

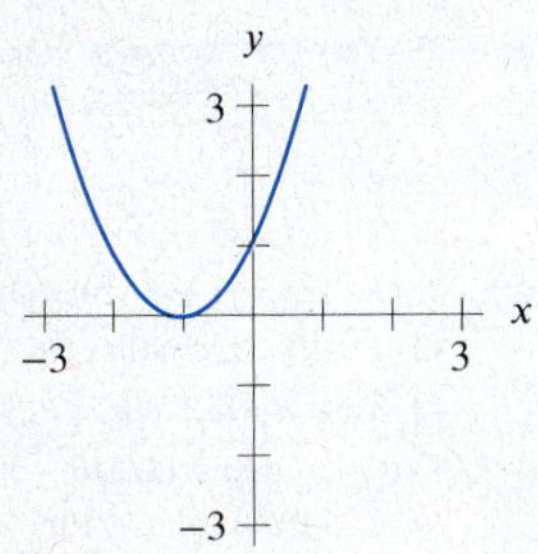

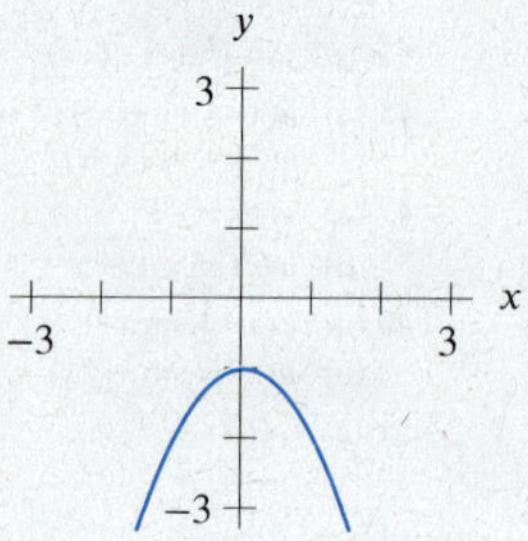


63 13

65 $v(x) = x + 1/x$

67 $f(x) = e^{-x}; g(x) = x - 2$

69 $q(x) = 2^x$

71 $r = 60 + 50(0.8)^t$

73 50; 50 cars/wk repaired on average

75 $t = \pi/4$

79 (a) and (e)

81 (a) even
(b) even
(c) even
(d) odd
(e) even
(f) neither
(g) neither

## Section 10.2

1 −1.373; multiple solutions

3 15.849; only solution.

5 $f^{-1} = x - 5$

7 $h^{-1} = x^2$

9 $f^{-1}(x) = (x + 7)/3$

11 $m^{-1} = (-x + \sqrt{x^2 + 4})/2$

13 Invertible.

15 Not invertible.

17 Not invertible

19 Not invertible

21 Not invertible

23 (a) and (e), (b) and (c), (d) and (f)

25

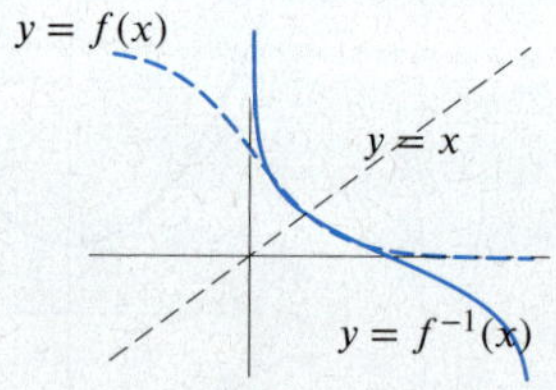


31 Yes, $f(f^{-1}(x)) = f^{-1}(f(x)) = x$

33 Yes, $f(f^{-1}(x)) = f^{-1}(f(x)) = x$

35 $l^{-1} = \sqrt{(1 - x^2)/2}$

37 $o^{-1} = x/(1 - x)$

39 $k^{-1}(x) = (3 - 2x)^2/(x + 1)^2$

41 $h^{-1}(x) = (5 + 4 \cdot 10^x)/(10^x - 1)$

43 $g^{-1}(x) = \arcsin(\ln x/\ln 2)$

45 Time at which pop is $P$; years

47 (a) $f^{-1}(R) = (1/5)R - 30$

49 $f^{-1}(P) = 50 \ln(P/10)$

51 (b) There will be approximately 325,474 people after 50 years
(c) $f^{-1}(P) =$ $(\log P - \log 37.8)/\log 1.044)$
(d) $f^{-1}(50) = 6.496$

53 (a) 0
(b) 1
(c) 2
(d) −2
(e) $\frac{1}{2}$
(f) −0.05

55 $f^{-1}(3) < f(3) < 0 < f(0) < f^{-1}(0) < 3$

57 (a) Definitely false
(b) Possibly true
(c) Definitely false
(d) Possibly true

61 $f^{-1}(r) = I_0 10^r$

63 (a) $p(q(t)) = q(p(t)) = t$; inverses
(b) Line $y = t$

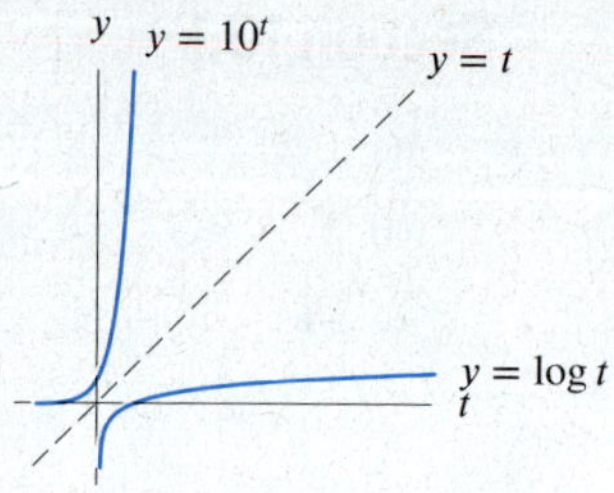


65 (a) Yes
(b) No

67 (a) $H(t) = 200e^{-1.15129t}$
(b) Dropped 50.021°C in the first 15 mins, 37.532°C in the next 15 mins
(c) $H^{-1}(y) = -\ln(y/200)/1.15129$
(d) About 3 hours and 12 minutes
(e) Brick's temperature approaches room temperature

69 $f^{-1}(x) = \left(0.5x^{-1} - A^{-1}\right)^{-1}$

71 $t = 0.5W(10) = 0.873$

73 (a) $C(0) = 99\%$
(b) $C(x) = (99 - x)/(100 - x)$
(c) $C^{-1}(y) = (99 - 100y)/(1 - y)$

75 (a) 36 m
(b) 6 seconds
(c) One possibility: $3 \le t \le 6$

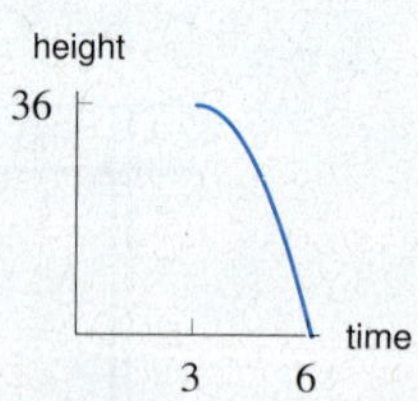


(e)

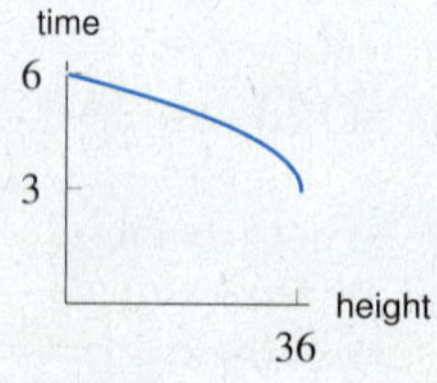


## Section 10.3

1 (a) $f(x) + g(x) = 2x$
(b) $f(x) - g(x) = 10$
(c) $f(x)g(x) = x^2 - 25$
(d) $f(x)/g(x) = (x + 5)/(x - 5)$

3 (a) $f(x) + g(x) = 3x^2 + x + 1$
(b) $f(x) - g(x) = -3x^2 + x + 1$
(c) $f(x)g(x) = 3x^3 + 3x^2$
(d) $f(x)/g(x) = (x + 1)/(3x^2)$

5 (a) $f(x) + g(x) = 2x^2 + 6$
(b) $f(x) - g(x) = 2$
(c) $f(x)g(x) = x^4 + 6x^2 + 8$
(d) $f(x)/g(x) = (x^2 + 4)/(x^2 + 2)$

7 $f(x) = x$

9 $h(x) = 7x - 5$

11 $k(x) = 1 - 2x + x^2$

13 $f(x) = e^x(2x + 1) = 2xe^x + e^x$

15 $h(x) = 4e^{2x} + 4e^x + 1$

19 $\sin x + x^2$

21 $(\sin x)/x^2$

23 $\sin^2 x$

25 (a) $h(x) = x^2 + x, h(3) = 12$
(b) $j(x) = x^2 - 2x - 3, j(3) = 0$
(c) $k(x) = x^3 + x^2 - x - 1, k(3) = 32$
(d) $m(x) = x - 1$ for $x \neq -1, m(3) = 2$
(e) $n(x) = 2x + 2, n(3) = 8$

31 4550

33

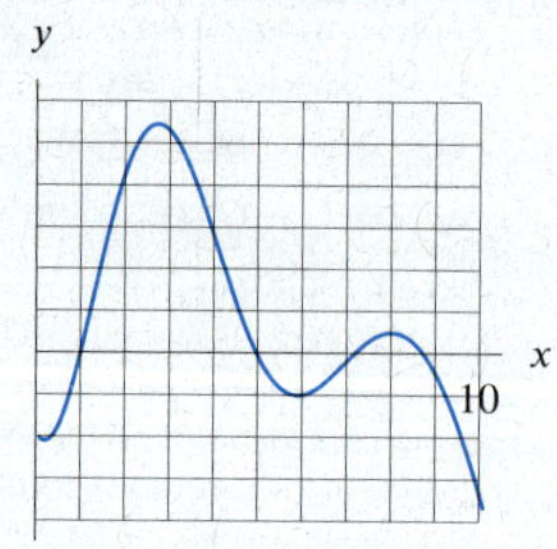


35

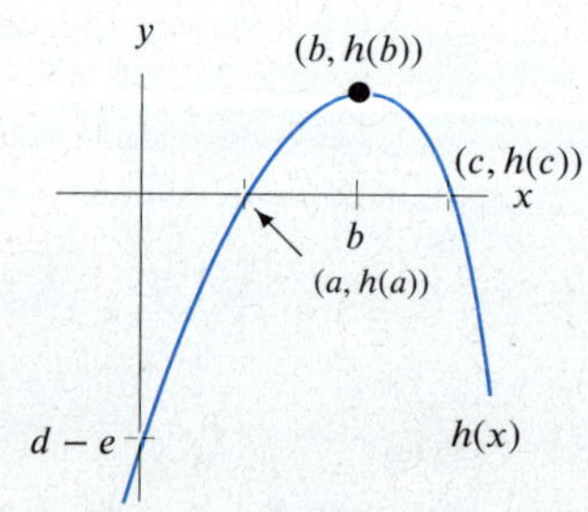


37 $17.50

39 (a)

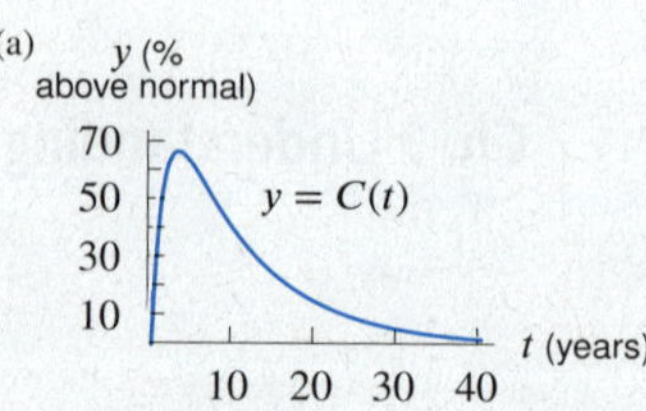

(b) $t = 3.25$ years
(c) Returns to normal

41 $H(x) = \cos\sqrt{x}, h(x) = -\sin(\sqrt{x})/(2\sqrt{x})$

43 $H(x) = F(x)G(x) = x^3 \sin x$

45 (a) $A(t) = 316.75(1.004)^t$
(b) $V(t) = 3.25 \sin 2\pi t$; other answers possible
(c)

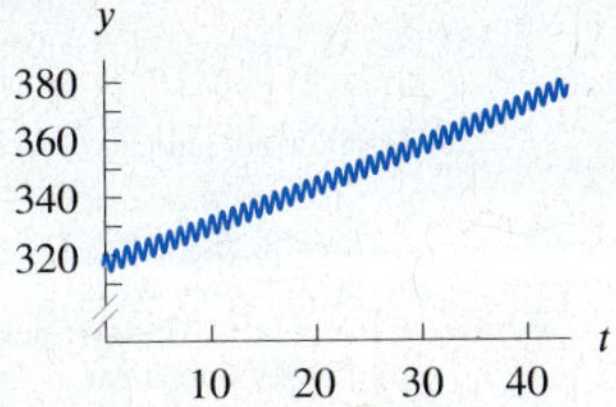


47 Ace estimates 2000 ft$^2$ of office space costs \$200,000.

49 $f(2x) < 2f(x)$

51 Space estimates 1500 ft$^2$ costs \$200,000

53 Ace seems to be a better value

55 40

## Chapter 10 Review

1 $2^{x^2}$; $4^x$

3 $1/(x^2 - 2)$

5 $\sqrt{x^2 + 1}$

7 $1/(x - 2)$

9 (a) Not invertible
(b) Not invertible
(c) Invertible

11 $h^{-1}(x) = x/(1 - 2x)$

13 $g^{-1}(x) = \frac{1}{3}\ln(x - 1)$

15 $h^{-1}(x) = \frac{1}{2}(1 - e^x)$

17 $g^{-1}(x) = (3x + 2)/(1 - 2x)$

19 $f^{-1}(x) = (11x - 3)^2/(1 + x)^2$

21 $s^{-1}(x) = 10^{(3/x)-2}$

23 Not invertible

25 Not invertible

27 $r^{-1}(y) = \ln(y + 7)$

31 $2e^x - 1$

33 $4x - 3$

35 $\sqrt{x}e^{2x-1}$

37 (a) $f(2x) = 4x^2 + 2x$
(b) $g(x^2) = 2x^2 - 3$
(c) $h(1 - x) = (1 - x)/x$
(d) $(f(x))^2 = (x^2 + x)^2$
(e) $g^{-1}(x) = (x + 3)/2$
(f) $(h(x))^{-1} = (1 - x)/x$
(g) $f(x)g(x) = (x^2 + x)(2x - 3)$
(h) $h(f(x)) = (x^2 + x)/(1 - x^2 - x)$

39 $x/(1 + e^{2x})$

41 $3x^2 + x$

43 $2x\sqrt{x + 2}$

45 $3x/2 - 1/2$

47 $x^{3/2} \tan 2x$

49 $\tan((3x - 1)^2/2) - 27x^{3/2}$

51

| $t$ | $p(t)$ | $q(t)$ | $r(t)$ |
|---|---|---|---|
| 0 | 4 | 3 | 5 |
| 1 | 5 | 2 | 1 |
| 2 | 3 | 4 | 0 |
| 3 | 2 | 0 | 4 |
| 4 | 1 | 5 | 2 |
| 5 | 0 | 1 | 3 |

53 $u(x) = \sqrt{x}, v(x) = 3 - 5x$

55 $u(x) = x^2, \quad v(x) = \sin x$

57 $u(x) = x^3, v(x) = 2x + 5$

59 $u(x) = 3^x, v(x) = 2x - 1$

63 (a) $r(x) = (x - 1)/(x - 2)$
(b) $s(x) = x + 1$ and $t(x) = 1/x$
(c) $p(p(a)) = (2a + 1)/(a + 1)$

65

| $x$ | $f(x)$ | $g(x)$ | $h(x)$ |
|---|---|---|---|
| 0 | 9 | 1 | 0 |
| 1 | 0 | 2 | 1 |
| 2 | 1 | 0 | 9 |

67 (a) $f^{-1}(P) = 2.5P - 50$
(b)

| $t$ | $P = f(t)$ |
|---|---|
| 0 | 20 |
| 5 | 22 |
| 10 | 24 |
| 15 | 26 |
| 20 | 28 |

| $P$ | $t = f^{-1}(P)$ |
|---|---|
| 20 | 0 |
| 22 | 5 |
| 24 | 10 |
| 26 | 15 |
| 28 | 20 |

69 Velocity for time $t$; mph

71 $1 - t^2$

73 $x = (\ln 3/\ln 2) - 5$

75 $x = e^{1.8} - 3$

77 $x = (19 - \sqrt{37})/2$

79 (a) $A = \pi r^2$
(b)

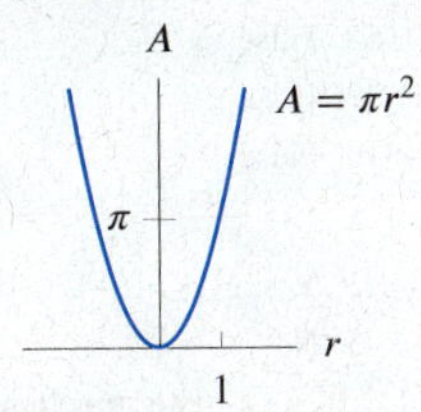


(c) $r \geq 0$

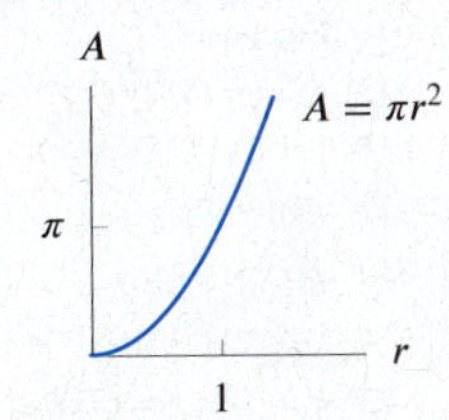


(d) $f^{-1}(A) = \sqrt{A/\pi}$
(e)

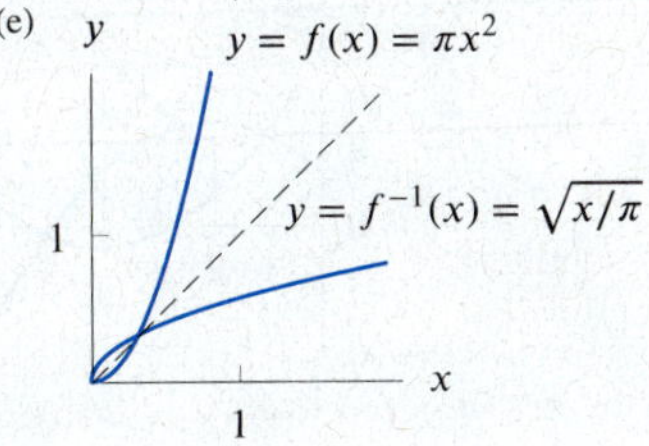


(f) Yes

81 (a) $f(g(a)) = a$
(b) $g(f(c)) = b$
(c) $f^{-1}(b) - g^{-1}(b) = -c$
(d) $0 < x \leq a$

83 $2\sqrt{x} - 9$

85 $(3 \pm \sqrt{17})/4$

87

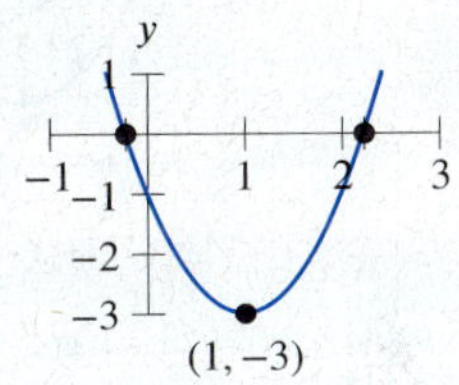


89 (a) Only to $(u(x))^2$.
(b) $u((v(x))^2)$ and $u(w(v(x)))$
(c) (i) $1 + \sin 2x$
(ii) 1
(iii) $\cos(x^2) + \sin(x^2)$

91

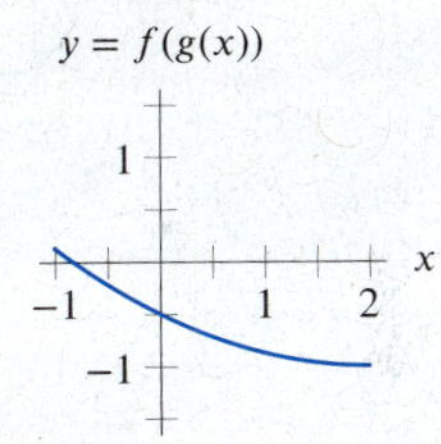


93

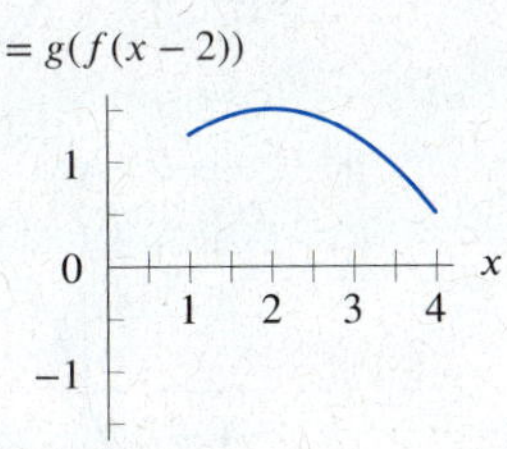


95 (a)

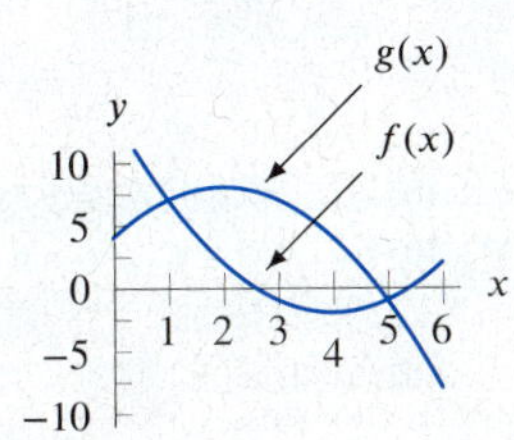

(c)

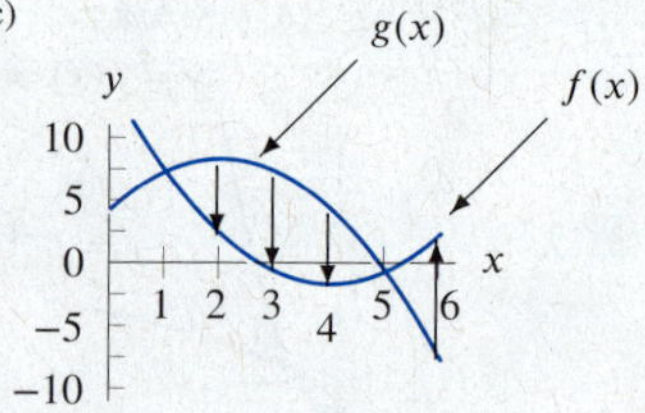


(f) $f(x) = x^2 - 8x + 14$;
$g(x) = -x^2 + 4x + 4$;
$f(x) - g(x) = 2x^2 - 12x + 10$

(g) Yes

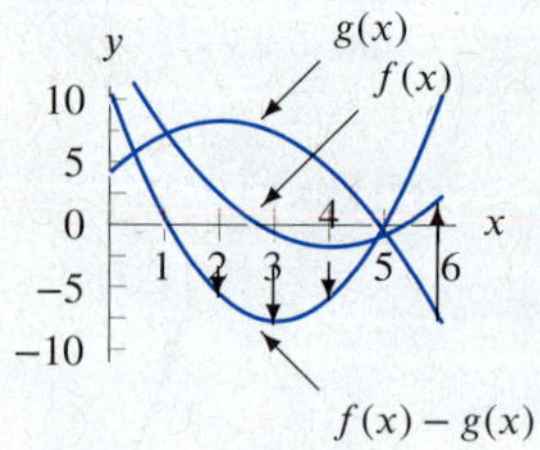


97 (a) $f(x) = 2x + 4$, $g(x) = \frac{1}{3}x - 1$

(b)

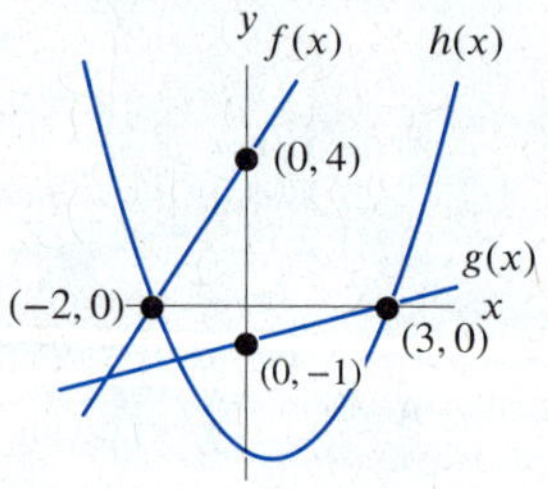


99 (a)

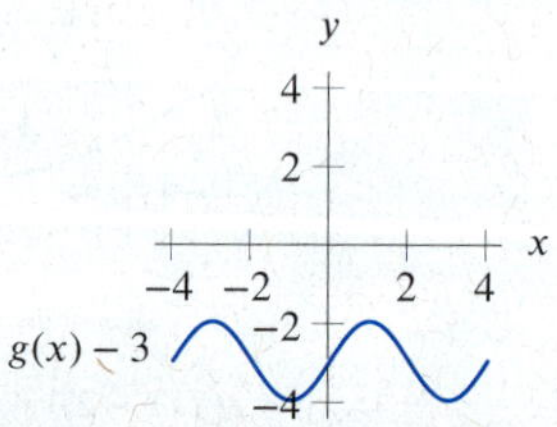


(b)

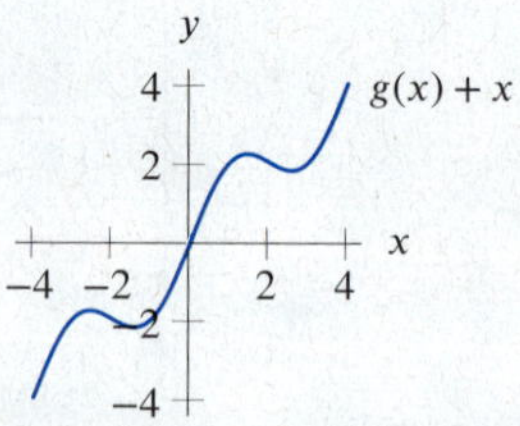


101 False

103 $g(x) = (x + 2)/2 = 0.5x + 1$

105 (a) $f(x) = e^x, g(x) = 6x, G(x) = 3x^2$
(b) $f(x) = \sin x, g(x) = -\frac{1}{2\sqrt{x}}, G(x) = \sqrt{x}$

107 (a) True
(b) False
(c) False
(d) True

109 Increasing

111 Can't tell

113 (a) $f(8) = 2, f(17) = 2,$
$f(29) = 2, f(99) = 0$
(b) $f(3x) = 0$
(c) No
(d) $f(f(x)) = f(x)$
(e) No

115 $f^{-1}(L) = -\frac{1}{k}\ln(1 - L/L_\infty)$
$f^{-1}(L) =$ Age of fish of length $L$
Domain: $0 \le L \le L_\infty$

## Ch. 10 Understanding

1 False
3 True
5 True
7 True
9 True
11 False
13 False
15 True
17 False
19 False
21 False
23 False
25 True
27 True
29 True
31 False
33 True
35 True
37 True

## Section 11.1

S1 $6|t|$
S2 $9x^5$
S3 $0.16x^2y^4$
S4 $70w^{5/6}$
S5 $x = 0.585$
S6 $x = \pm 0.1$
S7 False
S8 False
S9 False
S10 False

1 $y = \frac{48}{30625} \cdot x^{-2}, a = \frac{48}{30625}, p = -2$

3 Yes; $g(x) = (-1/6)x^9$

5 No

7 Not a power function

9 Even

11 Fractional

13 $f(x) = (3/2) \cdot x^{-2}$

15 $y = 3x^{1.058}$

17 $j(x) = 2x^3$

19 $g(x) = -\frac{1}{6}x^3$

21 $k = 5; c = 5d^2; c = 125$

23 $k = 3/2; y = (3x)/2; x = 5.33$

25 (a) 0
(b) 0

27 (b) 0
(c) $g(x)$

29 (a) For $x^{1/2}$: 0.707, 0.293, 0.225, 0.189
For $x^2$: 2, 6, 10, 14
(b) Concave down; concave up

31 (a) $x^{-3} \to +\infty, x^{1/3} \to 0$
(b) $x^{-3} \to 0, x^{1/3} \to \infty$

33 Formula not unique

35 (a) $A = x^3, B = x^2, C = x^{3/2},$
$D = x, E = x^{1/2}, F = x^{1/3}$
(b) $x^2, x^3$: Concave up
$x^{1/2}, x^{1/3}$: Concave down
Inverse functions

37 $x_1 \ge x_0$

39 $v, w, f, g$

41 $f(x) = -1/(3\sqrt[3]{7}) \cdot x^{-4/3}$

43 (a) $C(x) = kx$
(b) $k = 9.5; C(x) = 9.5x$
(c)

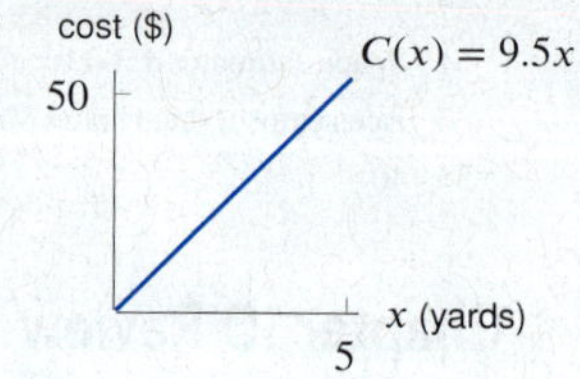


(d) $52.25

45 $c = 81.667x$; 326.667

47 (a) $B = kM^{3/4}$
(b) 10,502.6

49 $h = 192.5/v$; 64.167 mph

51 (a) 1.2 meters/sec
(b) 21% longer than the existing ship

53 (a) $d = 1.7, 3.4, 20.4, 102$
$d = 0.34t$
(b) 9.8 mins
(c) $A = 9.1, 36.3, 1307, 32685$
$A = 0.363t^2$
(d) $P = 11.25t^2$
(e) 298 sec, or approx 5 min

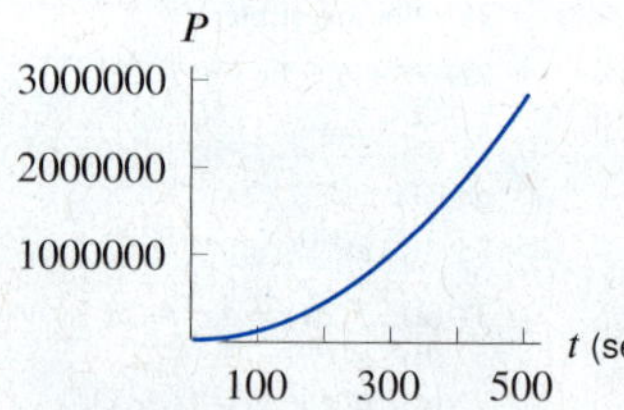


55 (b) $m = 2.1r^3$
(c) 11.7 cm
(d) 0.513 gm/cm$^3$

57 (a) $h(x) = -2x^2$
(b) $j(x) = \frac{1}{4}x^2$

## Section 11.2

S1 $x^2 + x - 2$
S2 $x^3 - x$
S3 $6x^3 + x^2 - x$

S4 $x^3 - x^2 - 12x$

S5 $\frac{1}{3}$

S6 2

S7 6

S8 $-6$

1 No

3 Yes, 2

5 No

7 $-5x^4$

9 $-3x^2$

11 $8x^9$

13 $4x^{n+3}$

15 $40x^3; 2x, 5x, 4x$

17 $2x^5; x^2, 2x^3$

19 Degree: 4; Terms: 3;
$x \to \pm\infty: y \to -\infty$

21 (a) $\infty$
(b) $\infty$

23 $-16.543$

25 (a)

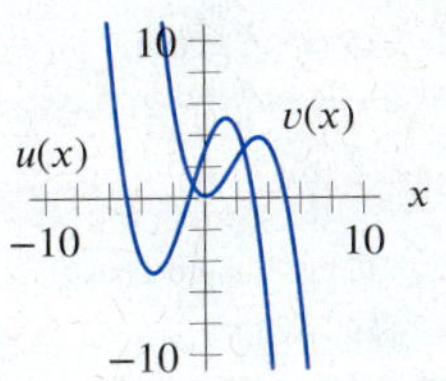


27 (a) $\frac{1}{2}x$
(b)

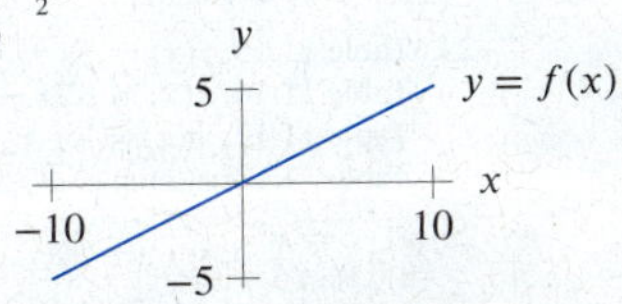


(c) $x > 158.114$ or $x < -158.114$.

29 $-1.1 \le x \le -0.9, -0.121 \le y \le 0.081$

31 $-20 \le x \le 20, -7600 \le y \le 8400$

33 (II)

35 (III)

37 (a)

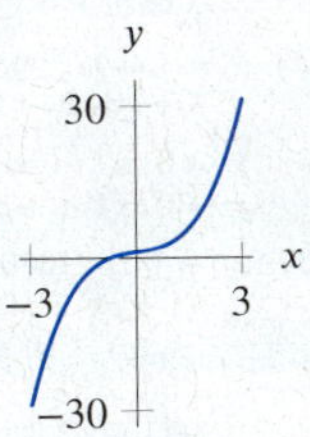


(b) $f(0.5) = 1.625$;
$f^{-1}(0.5) \approx -0.424$.

39 $a_4 = 3, a_3 = -4, a_2 = 0, a_1 = 5, a_0 = -4$
$f(x) = 12x^3 - 12x^2 + 5$

41 Yes

43 (a)

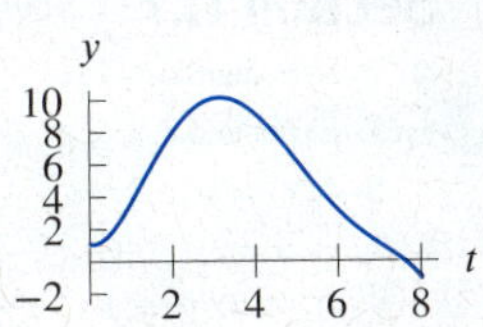


(b) 100 people
(c) July of 1897
(d) 1010; February of 1893
(e) $-115.7$; not reasonable

45 (a)

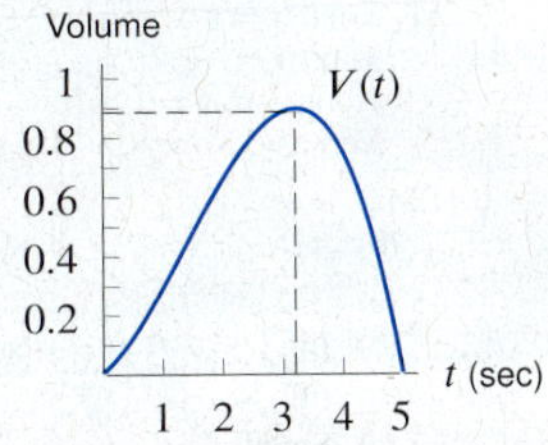


(b) $V \approx .886$ at $t \approx 3.195$
(c) (0, 0) and (5, 0);
Lungs empty at beginning and end

47 (a) $0 \le T < 5$
(b) $0 \le T < 15$
(c) $0 \le T < 30$
(d) Negative

49 (a) $p(0.5) \approx 0.65625$; 2 dec pl
(b) $p(1) = 0, f(1) = 0.5$; poor approx
(c)

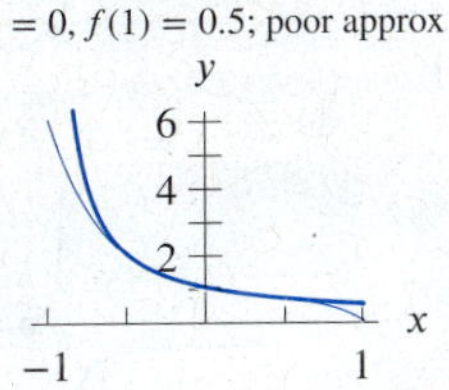


## Section 11.3

S1 $x^4(x-1)(x+1)$

S2 $x(x-5)(x-1)$

S3 $x(4-x)(4+x)$

S4 $x = 0$ or $x = 1$ or $x = -1$

S5 $x = -2$ or $x = 1$

S6 $x = -1$ or $x = 2$ or $x = -2$.

S7 $k = \dfrac{5}{16}$

1 $-3, 2, -7$

3 $0, -7, 3, -1 + \sqrt{8}, -1 - \sqrt{8}$

5 $f(x) = 4x(2x+5)(x-3)$;
Zeros: $0, -5/2, 3$

7 $h(x) = x(x+2)^2(x-3)$

9 $y = (-1/8)(x+2)(x-2)^2(x-5)$;
$y = (-1/20)(x+2)(x-2)(x-5)^2$

11

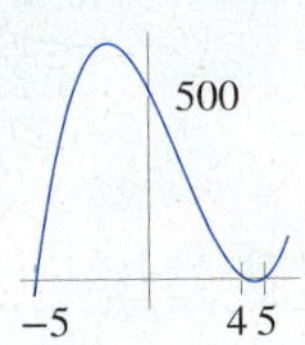


13 C

15 $p(x) = (x+15)(x-1)^2(x-3)$

17 $f(x) = x$ (and other answers)

19 $f(x) = x^2 - 2x$

21 $g(x) = -\dfrac{1}{15}(x-3)^2(x-5)(x+1)$

23 $y = \frac{3}{4}(x+1)(x-2)^2$ and $y = -\frac{3}{2}(x+1)^2(x-2)$

25 None

27 $x = -2$ or $x = -3$

29 $0, -3$

31 $h(x) = \frac{1}{12}(x+2)(x+1)(x-1)^2(x-3)$

33 $g(x) = -\dfrac{1}{4}(x+2)^2(x-2)(x-3)$

35 $f(x) = \frac{1}{36}(x+4)^2(x-3)^2$

37 $g(x) = \frac{1}{6}(x+1)(x-2)(x-4) + 4$

39 $g(x) = -\frac{1}{3}(x+2)(x-2)x^2$

41 $r = -1, s = 2, g(x) = k(x+5)^2$ or $r = -5, s = 2, g(x) = k(x+5)(x+1), k \ne 0$

43 $V = 4x^2 + 9x + 3$

45 7.83 by 5.33 by 1.585 inches

47 All $x > -2$ except $x = 3$

49 (a) Never true
(b) Sometimes true
(c) Sometimes true
(d) Never true
(e) True
(f) Never true

## Section 11.4

S1 $(6y^2 + 7)/y^3$

S2 $20/(x-1)$

S3 $x^3/2$

S4 $3(4x+11)/((x+3)(x+2))$

S5 $(-18x^2 + 18x + 41)/((x-2)^2(x+1))$

S6 $1/(x+y)^2$

S7 1/2

S8 Cannot be simplified further

S9 $1/(x-1)$

1 Rational; $(x+2)/(x^2-1)$

3 Rational; $(x^3+2)/(2x)$

5 Not rational

7 Not rational

9 $\dfrac{1}{x+2} + \dfrac{x}{x+1} \to 1$

11 $\dfrac{(x+3)(2x+1)}{(3x+1)(2x-5)} \to 1/3$

13 (III)

15 (IV)

17 4

19 3/4

21 $y = 4$

23 $y = -\frac{3}{2}$

25 $f^{-1}(x) = (4x+4)/(5x+3)$

27 (a)

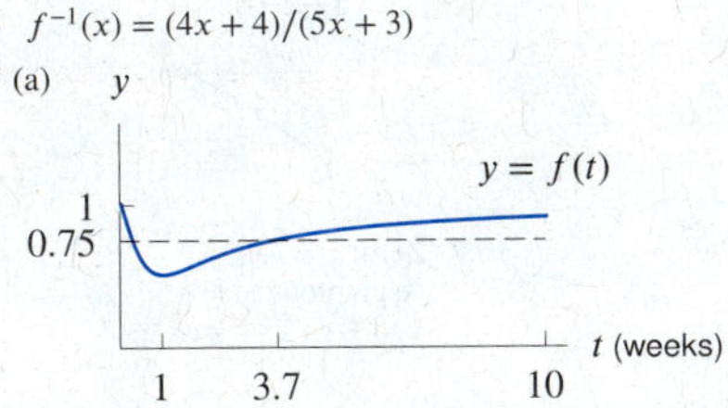


(c) Approaches 1
(d) About 3.73 weeks

29 (a) $240V/(V+200)$
(b) 200/3 mph

31 (a) $C(x) = (1+x)/(2+x)$
(b) $C(0.5) = 60\%$;
$C(-0.5) = 33.333\%$
(c)

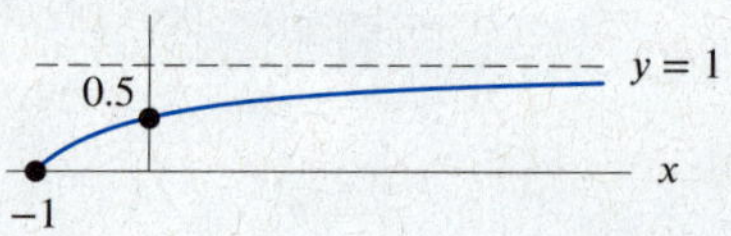


33 (a) $f(x) = x/(x+5)$
(b) $f(7) = 7/12 \approx 58.333\%$
(c) $x = 0$
(d) $y = 1$

35 (a) $C(x) = 30000 + 3x$
(b) $a(x) = 3 + 30000/x$
(c)

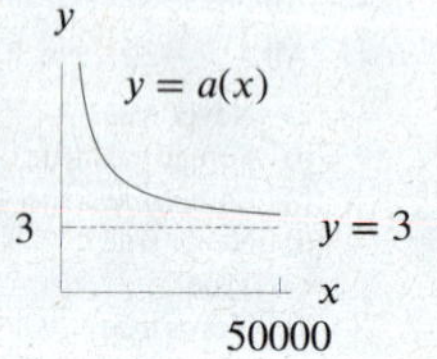


(f) $a^{-1}(y) = 30000/(y-3)$
(g) 15,000

37 No

## Section 11.5

1 (a) (iii)
(b) (i)
(c) (ii)
(d) (iv)
(e) (vi)
(f) (v)

3 (c) Horizontal: $y = 0$
Vertical: $x = -2$

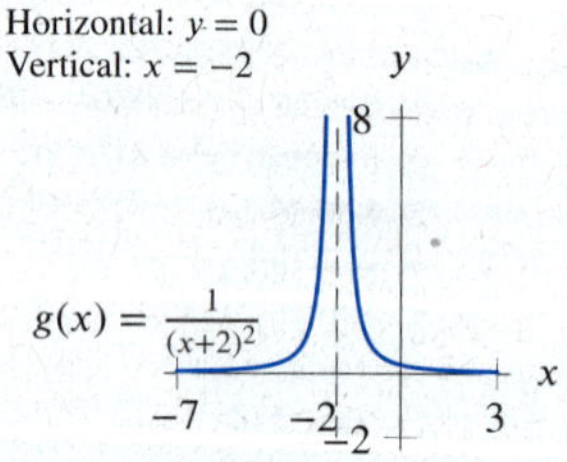


5 Zeros: $x = \pm 2$;
Asymptote: $x = 9$;
Approaches $y = x$ as
$x \to \pm\infty$

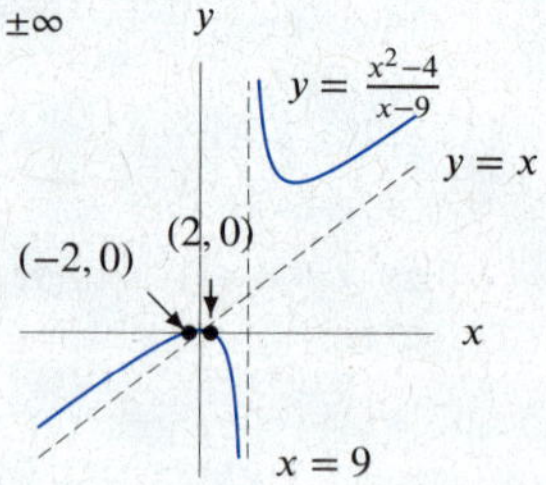


7 Zero: $x = -3$;
Asymptote: $x = -5$;
$y \to 0$ as $x \to \pm\infty$

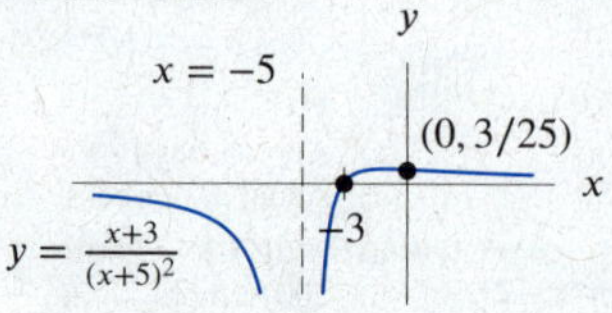


9 $x$-int: $x = 0, x = 4$
$y$-int: $y = 0$
Horiz asy: $y = -1$
Vert asy: $x = 1, x = 5$

11 $x$-int: $x = \pm 3$
$y$-int: $y = -1$
Horiz asy: $y = 1$
Vert asy: None

13 (a) $+\infty$
(b) $+\infty$

15 (a) 2, 2
(b) $\lim_{x\to -4^+} f(x) = -\infty$;
$\lim_{x\to -4^-} f(x) = \infty$

17 (a) Small
(b) Large
(c) Undefined
(d) Positive
(e) Negative

19 (a)

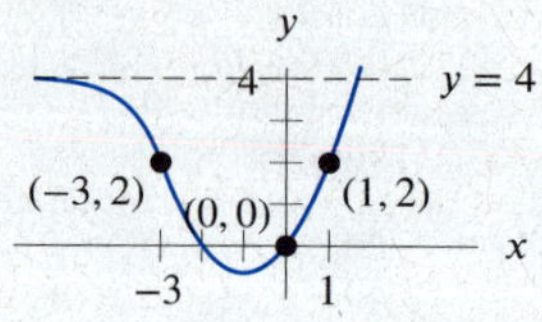


(b)

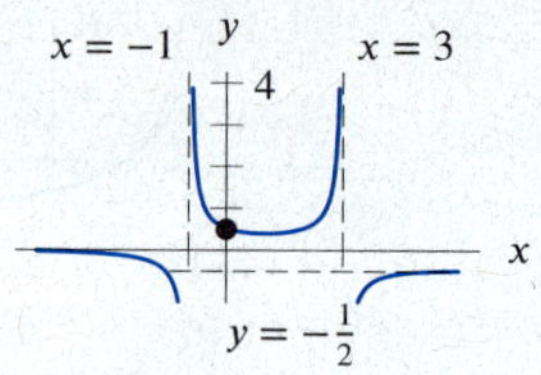


21 (a) $y = -1/(x+2)$
(b) $y = -1/(x+2)$
(c) $(0, -1/2)$

23 $p = 1, (0, 11/3), (11/4, 0)$
$x = 3, y = 4$

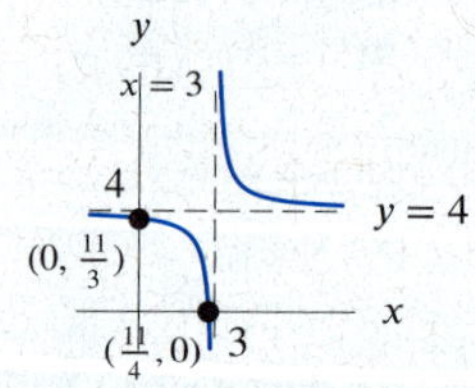


25 $p = 1, (0, 3), (3/2, 0)$
$x = 1, y = 2$

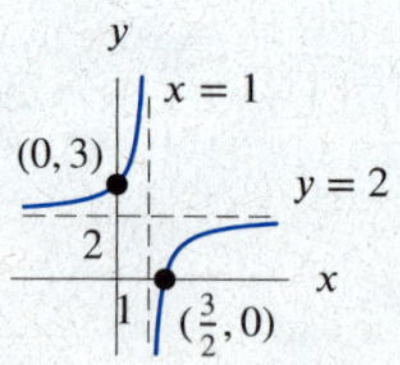


27 (a) $y = 1/x$
(b) $y = x/(2x-4)$

29 (a) $1/x$
(b) $y = (1/x) + 2$

31

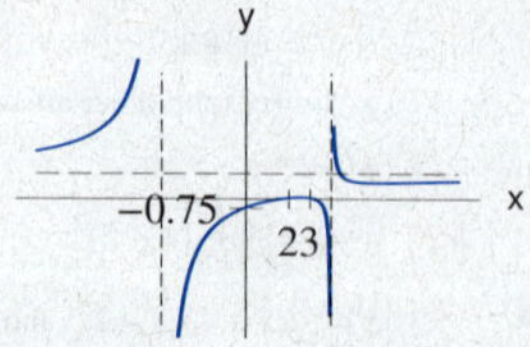


33 $y = ((x+3)(x-1))/(x+2)^2$

35 $y = (x+2)(x-3)/((x-2)(x+1))$

37 $y = x/((x+2)(x-3))$

39 $y = x^2 + 1$; $(-5, 26)$

41 Hole at $x = 1$

43 Asymptote at $x = -2$

45 $f(x) = -3(x-2)(x-3)/(x-5)^2$

47 $h(x) = (x-5)^2/((x+2)(x-3))$

## Section 11.6

1 (a) $A$ - (iii)
(b) $B$ - (ii)
(c) $C$ - (iv)
(d) $D$ - (i)

3 $p(x) = 25^x$

5 Neither

7 $r(x) = 2(\frac{1}{9})^x$

11 $y = 6x^{35}$

13 $y = 50x^{1.1}$

15 (a) $C$, $f(x)$;
$A$, $g(x)$;
$B$, $h(x)$
(b) Yes
(c) No

17 $3^{-x}$ approaches zero faster

19 Neither

21 (a) $f(x) = 112x - 96$
(b) $f(x) = 2(8)^x$
(c) $f(x) = 16x^3$

23 Table 11.15: $j(x) = .3x^3$
Table 11.16: $k(x) = 1.2x - 3$
Table 11.17: $m(x) = \log x$
Table 11.18: $z(x) = .4x^2$

25 (a) $v = 40$
(b) $r(x) > t(x)$
(c) $t(x) > r(x)$

27 $y \to 0$ as $x \to \pm\infty$

29 $y \to 0$ as $t \to \infty$
$y \to \infty$ as $t \to -\infty$

31 $y \to \infty$ as $x \to \infty$
$y \to 0$ as $x \to -\infty$

33 $y \to \infty$ as $t \to \pm\infty$

35 $y \to 1$ as $x \to \infty$
$y \to -1$ as $x \to -\infty$

37 $y \to 0$ as $t \to \infty$
$y \to 0$ as $t \to -\infty$

39 (a) $p_5(r) = 1000[(1+r)^5 + (1+r)^4 + (1+r)^3 + (1+r)^2 + (1+r) + 1]$;
$p_{10}(r) = 1000[(1+r)^{10} + (1+r)^9 + (1+r)^8 + (1+r)^7 + (1+r)^6 + (1+r)^5 + (1+r)^4 + (1+r)^3 + (1+r)^2 + (1+r) + 1]$
(b) 20.279%

## Section 11.7

1 Exponential

3 Exponential

5 $f(x) = x^{\ln c/\ln 2}$

7 $g(x) = 2x^{1.2}$

9 (a) (i) $y = 46.79t^{0.301}$

(ii) $y = 0.822t^2 - 6.282t + 76.53$
(b) Quadratic for 2002-2012

11 $y = x^{\frac{3}{2}}$

13 $y = \frac{3}{2}x$

15 $y = e^{0.4x}$

17 (a) $y = -83.039 + 61.514x$; superb fit
(b) Good only for close values

19 $a \approx 3.49$

21 (a) $N = -14t^4 + 433t^3 - 2255t^2 + 5634t - 4397$
(b) Good fit

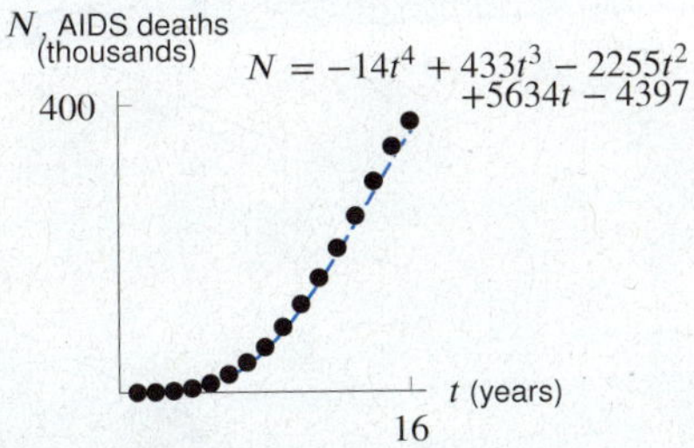


(c) Model fails for $t > 20$

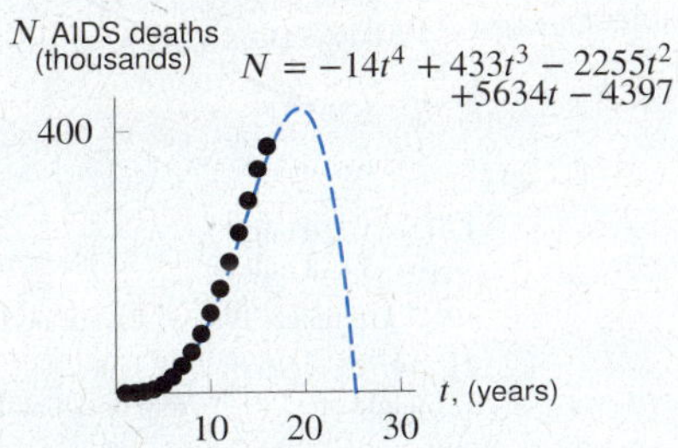


23 (a) $C(t) = 213.81(1.072)^t$
(b) 7.2% per year
(c) slower growth; concave down

25 (a)

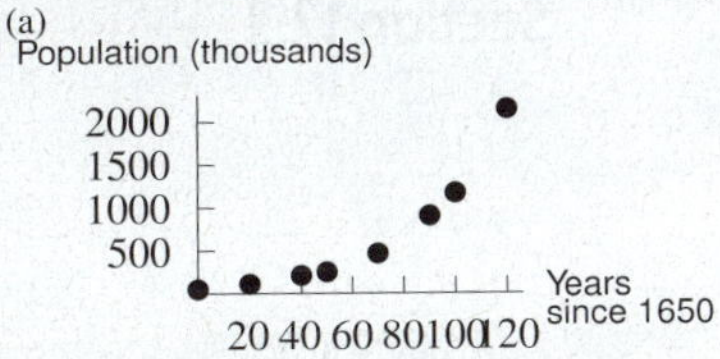


(b) $P(t) = 56.108(1.031)^t$, answers may vary
(c) 56.108 is 1650 population, 1.031 means 3.1% annual growth
(d) $P(100) = 1194.308$, Slightly higher
(e) $P(150) = 5510.118$, higher

27 (a) $y = 1.365t^{1.470}$
(b) $y = 19.065 \cdot 1.065^t$
(c) $y = 0.450t^2 - 12.1674t + 96.692$
(d) (a) low, (b) and (c) best

29 (a) $N = 1148.55e^{0.3617t}$

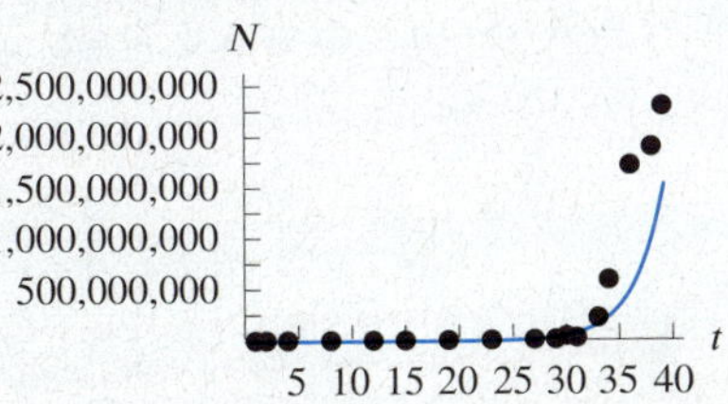


(b) About 1.92 years

31 (a)

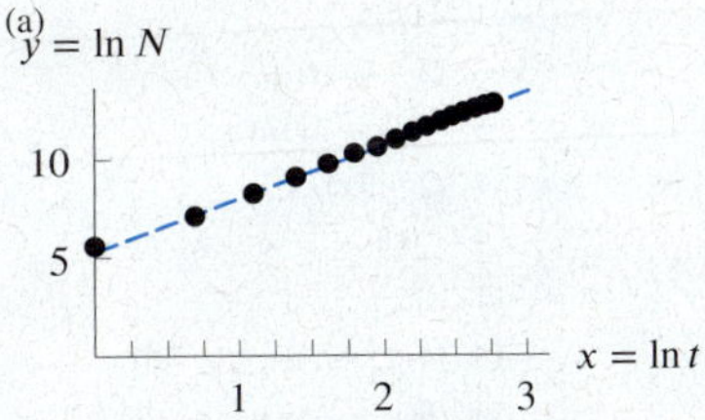


(b) $y = 4.670 + 3.005x$
(c) $N = 107t^{3.005}$

33 (a) Quadratic
(b) $y = -34.136x^2 + 3497.733x - 39{,}949.714$; answers may vary
(c) $42,734; answers may vary
(d) Age 10, −$8386, not reasonable; answers may vary

35 (a) $h_e = 16.2e^{-t/312.5}$; $h_a = 12.3e^{-t/131.6}$; $h_b = 14.7e^{-t/149.3}$

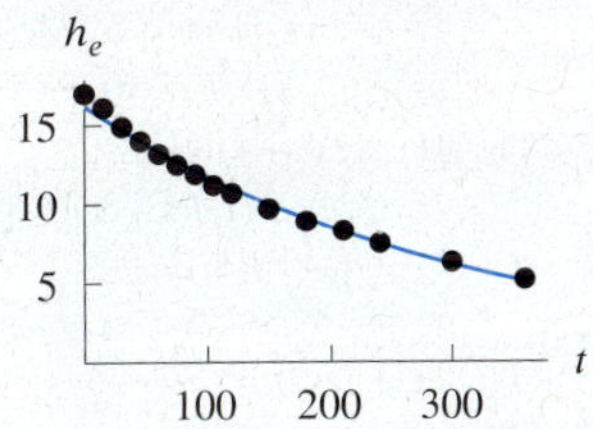


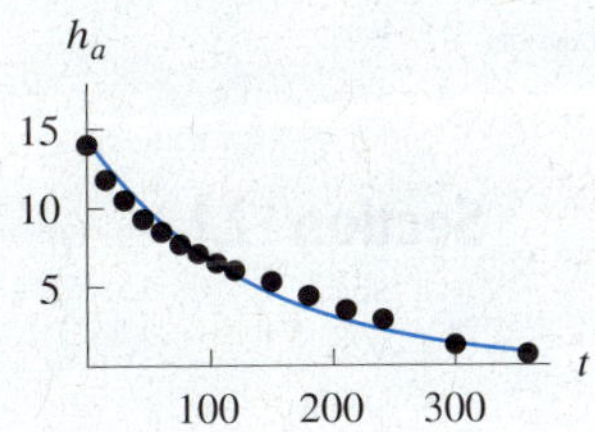


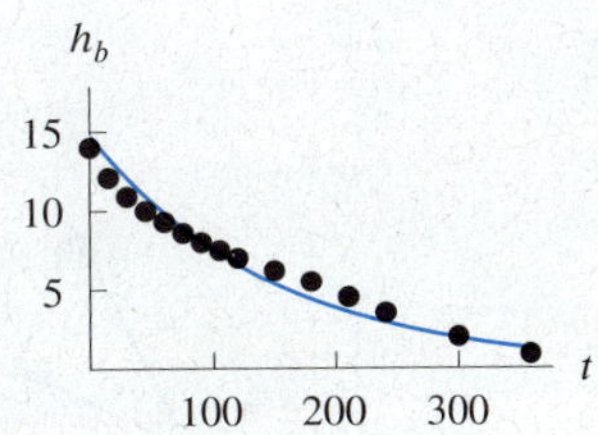


## Chapter 11 Review

1 Yes; $k = 1/6$ and $p = -7$

3 No

5 No

7 Yes; $y = x^2$

9 Even

11 Odd

13 Odd

15 $k = 2\sqrt[3]{7}$, $p = 11/15$

17 4th degree

19 $y \to \infty$; like $4x^4$

21 $y \to \infty$; like $2x^9$

23 $x = (3 \pm \sqrt{33})/4$

25 Not rational

27 $y = 4/e^{-x}$

29 (a) 2
(b) 5/6

31 (a) 8
(b) 6
(c) 8

33 (a) $f(x) = x^{1/n}$
$g(x) = x^n$
(b) (1, 1)

35 (i) B
(ii) L
(iii) G
(iv) H
(v) U
(vi) S
(vii) A
(viii) E

37 (a) $u$ is a power function, $v$ is trigonometric, $w$ is linear, $z$ is exponential
(b) $256(1/4)^x$
(c) $0.4x - 1.74$
(d) $0.5x^{1.5}$
(e) $0.25\sin((\pi/2)(x-2)) + 0.5$

39 $y = 4(x+1)(x+3)$

41 $y = 2x(x+2)(x-2)$

43 $y = 7(x+2)(x-3)^2$

45 $y = -\frac{1}{3}(x+2)^2(x)(x-2)^2$

47 $f(x) = -\frac{1}{16}(x+3)^2(x-3)x^2$

49 (a) $y = (1/x^2) + 2$
(b) $y = (2x^2+1)/x^2$
(c) None

51 (a) $g(x) = 6x^2$
(b) $g(x) = -48 + 36x$
(c) $g(x) = 6(2^x)$

53 (a) $s(x)$
(b) $r(x)$
(c) None
(d) $p(x)$, $q(x)$
(e) $v(x)$
(f) $t(x)$

55 (a)

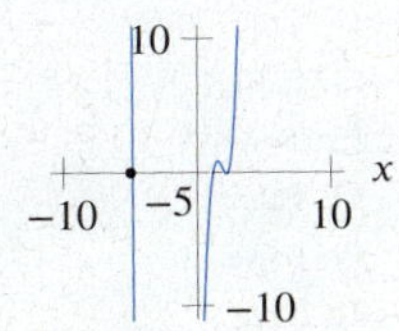


(b) No
(c) 3; $-6 \le x \le 3$, $-3 \le y \le 3$
(d) $f(x) = (x+5)(x-1)(x-2)^2$
(e) 3; No

57 $f(x) = (x+3)/(x+1)$

59 $g(x) = (x+1)^2/(x-1)^2$

61 $p(x) = \frac{7}{1080}(x+3)(x-2)(x-5)(x-6)^2$

63 $w(x) = (-1/4)(x+4)(x-1)(x-3) + 2x + 5$

65 Inversely proportional

67 (a) $P = 4.1 \cdot 10^{-10} d^{3/2}$
(b) 4320 earth days

69 (a)

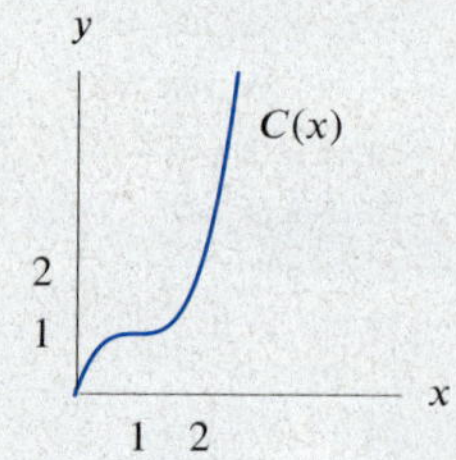


(b) $1000 per unit
(c) Profit: $1000 < x < 2000$;
Break even: $x = 1000$ or $x = 2000$;
Loss: $x > 2000$ or $0 < x < 1000$

71 (a) $T = kR^2D^4$
(b) Increases by factor of 4
(c) Increases by factor of 16
(d) Reduce to 44.4%

73 (a)

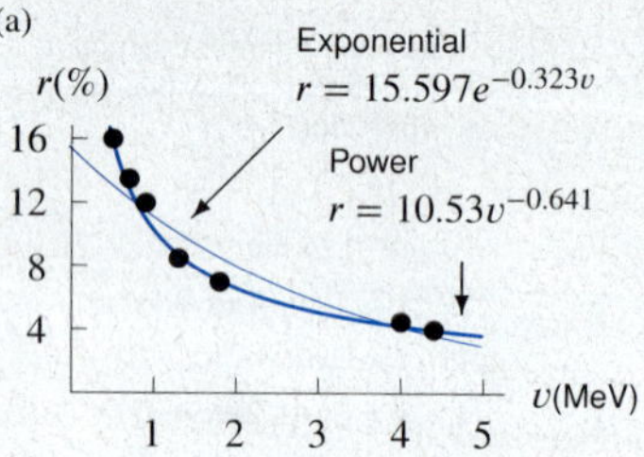


(b) High-energy photons
(c) $r = 15.597e^{-0.323v}$; $r = 10.53v^{-0.641}$ (best fit)
(d) Power function

## Ch. 11 Understanding

1 False
3 True
5 True
7 False
9 True
11 True
13 False
15 True
17 False
19 False
21 True
23 False
25 True
27 True
29 True
31 True
33 True
35 False
37 False
39 True
41 True
43 False
45 False
47 False

## Ch. 11 Skills: Fractions

1 $41/35$
3 $(3-4x)/6x$
5 $-2(1-2y)/yz$
7 $(2-3x)/x^2$
9 $1/18$
11 $x/2$
13 $(4y^3z - 3wx)/(x^2y^4)$
15 $(8(y+4))/(y-4)$
17 $(-27x+44)/((x+1)(3x-4))$
19 $(x+20)/(x^2-16)$
21 $1/2r$
23 $(x-1)/(\sqrt{x})^3 = x\sqrt{x} - \sqrt{x}/x^2$
25 $(4x+1)/(b-a)$
27 $(r_2r_3 + r_1r_3 + r_1r_2)/(r_1r_2r_3)$
29 $(2a+3)/((a+3)(a-3))$
31 $(-2x-h)/\left(x^2(x+h)^2\right)$
33 $-2x-h$
35 $1-(1/a)$
37 $x^2y/(2x+1)$
39 $(2x-4x^4)/(x^3+1)^3$
41 $13/x^2 + 1/(2x^3)$
43 $(2/l^2) + (1/l^3) - 4/(3l^4)$
45 $1/6 - 1/(4x)$
47 $1 - 7/(x+5)$
49 $1 + 1/R$
51 $1 + \sin x/\cos x$
53 False
55 False
57 True

## Section 12.1

1 Scalar
3 Vector
5 Scalar
7 Vector
9 $\vec{p} = 2\vec{w}$
$\vec{q} = -\vec{u}$
$\vec{r} = \vec{u} + \vec{w}$
$\vec{s} = 2\vec{w} - \vec{u}$
$\vec{t} = \vec{u} - \vec{w}$

11

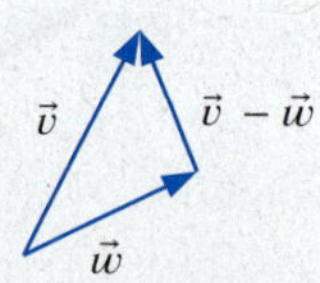


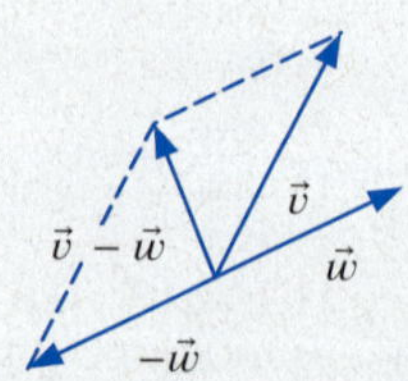


13

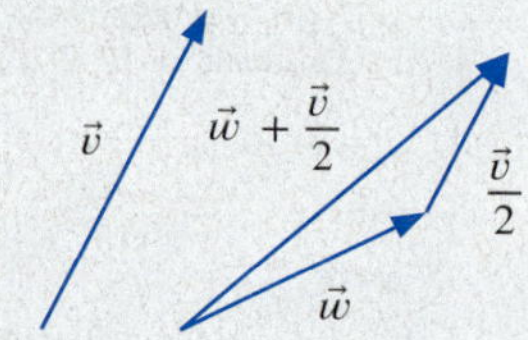


15

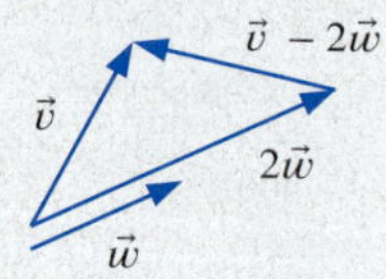


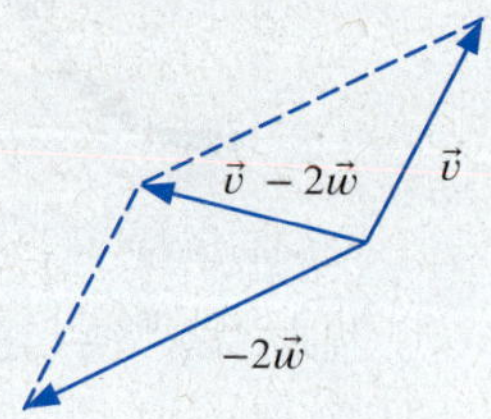


17 (a) 1.710 miles
(b) 5.848 miles

19 5.116 miles; 14.639° east of north

21 14,705 meters;
angle of 17.819° from horizontal

23 (a) 14.3373
(b) Veers right
(c) Not possible

## Section 12.2

1 $-3\vec{i} - 4\vec{j}$
3 $\vec{w} \approx -0.725\vec{i} - 0.95\vec{j}$
5 $\vec{i} + 3\vec{j}$
7 $0.3\vec{i} - 1.8\vec{j} + 0.03\vec{k}$
9 $\sqrt{11} \approx 3.317$
11 7.649
13 $-5\vec{i} + 10\vec{j}$ knots
15 45° or $\pi/4$
17 90° or $\pi/2$
19 $-140.847\vec{i} + 140.847\vec{j} + 18\vec{k}$
21 $21\vec{j} + 35\vec{k}$
23 (a) 50 km/hr
(b) horizontal: 43.301; vertical: 25
25 (a) $3.536(\vec{i} + \vec{j})$.
(b) $3.536\vec{i} + 4.736\vec{j}$
27 (a) (i) $\vec{m} = 3\vec{j}, \vec{h} = 2\vec{j}$

(i)

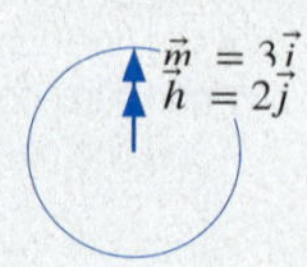

(ii) $\vec{m} = 3\vec{j}, \vec{h} = 2\vec{i}$

(ii)

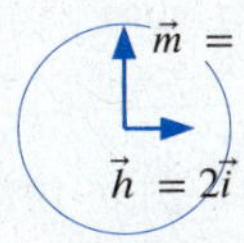


(iii) $\vec{m} = 3\vec{j}, \vec{h} = \vec{i} + \sqrt{3}\vec{j}$

(iii)

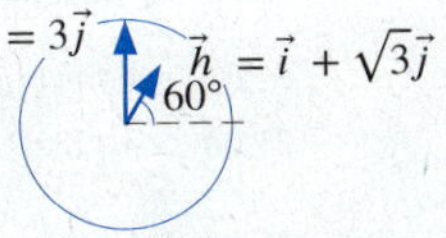


(iv) $\vec{m} = -3\vec{j}, \vec{h} = \sqrt{2}\vec{i} + \sqrt{2}\vec{j}$

(iv)

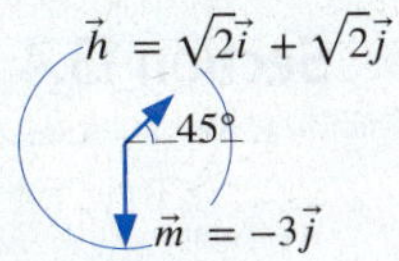


(b) $3\vec{j} - 2\vec{i}$

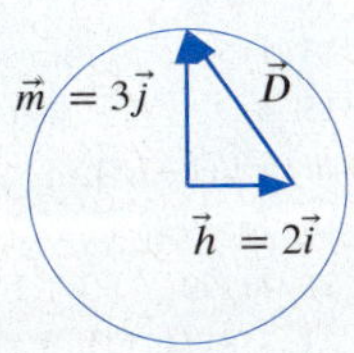


(c) $\sqrt{2}\vec{i} + (\sqrt{2} - 3)\vec{j}$

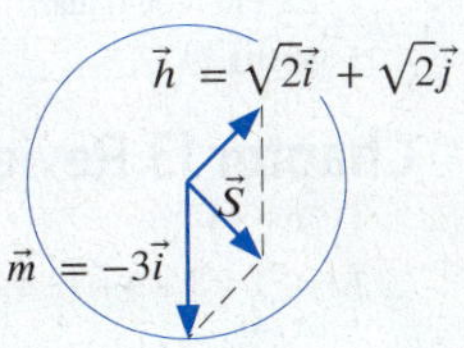


29 $\vec{k}$

31 $\vec{i} + \vec{k}$

## Section 12.3

1 (2, 2, 4, 6, 10, 16)

3 (−4, −5, −5, −5, −4, −2)

5 (5, 6, 7, 8, 9, 10)

7 (13/6, 5/2, 10/3, 25/6, 11/2, 22/3)

9 (3.69, 1.45, 6.67, 1.44, 1.17, 0.75)

11 (3.527, 1.287, 6.507, 1.277, 1.007, 0.587)

13 (79.000, 79.333, 89.000, 68.333, 89.333)

15 3.378° north of east

17 (a) $\vec{v} = 4.330\vec{i} + 2.500\vec{j}$
For the second leg of his journey, $\vec{w} = x\vec{i}$

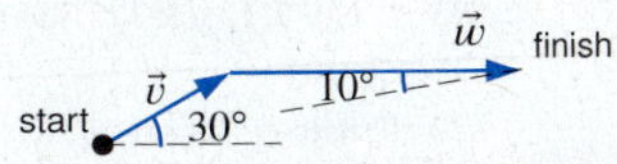


(b) $x = 9.848$
(c) 14.397

19 (a) $\vec{F}_{net} = (8, 7)$
(b) $\vec{F}_4 = (-8, -7)$

21 $\vec{q}_a = 1.065\vec{i} + 1.966\vec{j}$ ; (1.065,1.966)
$\vec{q}_b = 2.703\vec{i} + 3.113\vec{j}$ ; (2.703,3.113)
$\vec{q}_c = 2.129\vec{i} + 3.933\vec{j}$ ; (2.129,3.933)
$\vec{q}_d = 0.491\vec{i} + 2.785\vec{j}$ ; (0.491,2.785)

## Section 12.4

1 −7

3 −38

5 14

7 −2

9 $28\vec{j} + 14\vec{k}$

11 238

13 108.435°

15 2100 ft-lbs

17 1.911 radians (109.471°)

19 For both, max = 11, min = 3

21 No

25 (a) $\vec{a} = (3, 2, 4)$; $\vec{c} = (c_b, c_e, c_m)$
$3c_b + c_e + 4c_m = 40$, or $\vec{a} \cdot \vec{c} = 40$
(c) The "freshness-adjusted" cost is cheaper at Beta

27 43.297°

29 (a) Width
(b) Height
(c) Perimeter

## Section 12.5

1 (a) $\begin{pmatrix} 15 & 35 \\ 10 & -5 \end{pmatrix}$

(b) $\begin{pmatrix} -2 & 10 \\ 0 & -16 \end{pmatrix}$

(c) $\begin{pmatrix} 4 & 2 \\ 2 & 7 \end{pmatrix}$

(d) $\begin{pmatrix} -8 & -26 \\ -6 & 11 \end{pmatrix}$

(e) $\begin{pmatrix} 13 & 45 \\ 10 & -21 \end{pmatrix}$

(f) $\begin{pmatrix} k & -5k \\ 0 & 8k \end{pmatrix}$

3 (a) $\begin{pmatrix} 12 & 8 & 20 & 4 \\ 16 & 24 & 28 & 12 \\ 4 & 36 & 20 & 32 \\ 0 & -8 & 16 & 24 \end{pmatrix}$

(b) $\begin{pmatrix} -2 & -12 & -8 & -4 \\ -6 & -10 & 2 & -14 \\ -18 & -8 & -14 & -6 \\ -4 & -16 & -8 & -10 \end{pmatrix}$

(c) $\begin{pmatrix} 2 & -4 & 1 & -1 \\ 1 & 1 & 8 & -4 \\ -8 & 5 & -2 & 5 \\ -2 & -10 & 0 & 1 \end{pmatrix}$

(d) $\begin{pmatrix} 6 & -12 & 3 & -3 \\ 3 & 3 & 24 & -12 \\ -24 & 15 & -6 & 15 \\ -6 & -30 & 0 & 3 \end{pmatrix}$

(e) $\begin{pmatrix} 4 & 8 & 9 & 3 \\ 7 & 11 & 6 & 10 \\ 10 & 13 & 12 & 11 \\ 2 & 6 & 8 & 11 \end{pmatrix}$

(f) $\begin{pmatrix} 10 & -4 & 12 & 0 \\ 10 & 14 & 30 & -2 \\ -14 & 28 & 6 & 26 \\ -4 & -24 & 8 & 14 \end{pmatrix}$

5 (a) (51, 15, 38)
(b) (−8, −11, 33)
(c) (70, 20, 22)
(d) (11, −6, 17)
(e) 681
(f) $\begin{pmatrix} 24 & 60 & 84 \\ 48 & -72 & 36 \\ 192 & -60 & 0 \end{pmatrix}$

7 (a) Defined
(b) Not defined
(c) Not defined
(d) Not defined
(e) Defined
(f) Not defined

9 (a) $\mathbf{T} = \begin{pmatrix} 0.90 & 0 & 0 \\ 0.10 & 0.50 & 0.02 \\ 0 & 0.50 & 0.98 \end{pmatrix}$
(b) $\vec{p}_1 = (1.8, 0.2, 0)$,
$\vec{p}_2 = (1.62, 0.28, 0.1)$,
$\vec{p}_3 = (1.458, 0.304, 0.238)$

11 (a) $\mathbf{T} = \begin{pmatrix} 0.97 & 0.05 \\ 0.03 & 0.95 \end{pmatrix}$
(b) $\vec{p}_{2011} = (214, 386)$,
$\vec{p}_{2012} = (226.88, 373.12)$.

13 (a) $\vec{v} = \begin{pmatrix} 11 \\ 19 \end{pmatrix}$
(b) $\vec{v} = \begin{pmatrix} 5 \\ 11 \end{pmatrix}$
(c) $\vec{v} = \begin{pmatrix} 2a + b \\ 3a + 2b \end{pmatrix}$

15 (a) $\lambda_2 = -1$
(b) $\lambda_3 = -1$
(c) $\mathbf{A}\vec{v} = \lambda\vec{v}$, and $\mathbf{A}\vec{v}$ is parallel to $\vec{v}$

17 (a) $\begin{pmatrix} 3 & 5 \\ 2 & 4 \end{pmatrix}\begin{pmatrix} a \\ b \end{pmatrix} = a\begin{pmatrix} 3 \\ 2 \end{pmatrix} + b\begin{pmatrix} 5 \\ 4 \end{pmatrix}$
(b) $\vec{v} = \begin{pmatrix} -8.5 \\ 5.5 \end{pmatrix}$
(c) $\vec{v} = -8.5\vec{c}_1 + 5.5\vec{c}_2$

## Chapter 12 Review

1 $(3, 3, 6)$

3 $(-3, -2, 9)$

5 $(7, 8, -21)$

7 $(4, -2, 18)$

9 $\vec{a} = \vec{b} = \vec{c} = 3\vec{k}$
$\vec{d} = 2\vec{i} + 3\vec{k}$
$\vec{e} = \vec{j}$
$\vec{f} = -2\vec{i}$

11 $-4.5\vec{i} + 8\vec{j} + 0.5\vec{k}$

13 13

15 6

17 $6\vec{i} + 6\vec{j} + 6\vec{k}$

19 $\|\vec{u}\| = \sqrt{6}$
$\|\vec{v}\| = \sqrt{5}$

21 (a) Yes
(b) No

23 (a) $\vec{L} = (11, 7, 11, 7, 13)$
(b) $\vec{F} = (32, 36, 21, 8, 4)$,
$\vec{G} = (3, 3, 2, 0, 7)$

25 $F = g \sin\theta$

29 $0.4v\vec{i} + 0.693v\vec{j}$

31 (a) $\overrightarrow{AB} = 2\vec{i} - 2\vec{j} - 7\vec{k}$
$\overrightarrow{AC} = -2\vec{i} + 2\vec{j} - 7\vec{k}$
(b) $\theta = 44.003°$

35 $\overrightarrow{AB} = -\vec{u}$; $\overrightarrow{BC} = 3\vec{v}$; $\overrightarrow{AC} = \overrightarrow{AB} + \overrightarrow{BC} = -\vec{u} + 3\vec{v}$; $\overrightarrow{AD} = 3\vec{v}$

37 $3\vec{n} - 3\vec{m}$;
$3\vec{m} + \vec{n}$;
$4\vec{m} - \vec{n}$;
$\vec{m} - 2\vec{n}$

## Ch. 12 Understanding

1 False

3 False

5 False

7 False

9 True

11 True

13 False

15 True

17 True

19 False

21 True

23 True

## Section 13.1

1 Not arithmetic

3 Arithmetic

5 Arithmetic, $a_n = 3 + 3n$

7 Arithmetic, $a_n = -0.9 - 0.1n$

9 Not geometric

11 Geometric

13 Not geometric

15 Geometric; $4(1/2)^{n-1}$

17 Geometric; $1/(1.2)^{n-1}$

19 1, 5/4, 7/5, 3/2; not geometric

21 −1, 1, −1, 1; geometric

23 $1, 1/\sqrt{2}, 1/\sqrt{3}, 1/2$; not geometric

25 $n \geq 101$

27 10.8, 64.8, $4.8 + 1.2n$

29 7.9, 57.4, $2.4 + 1.1n$

31 1.661, $7(0.75)^{n-1}$

33 486, $2 \cdot 3^{n-1}$

35 (a) 690.1, 706.662, 723.662, 740.989
(b) 590.6, 625.445, 662.347, 701.425
(c) 2015

37 (a) 19.042, 19.151, 19.260
(b) $19.042(1.0057)^n$
(c) 121.951 years

39 Arithmetic, $d > 0$

41 Arithmetic, $d < 0$

43 2, 7, 12, 17; $a_n = -3 + 5n$

45 3, 7, 15, 31;
$a_n = 2^{n-1} \cdot 3 + 2^{n-2} + 2^{n-3} + \cdots + 1$

47 (a) 150, 187.5, 199.219, 199.997, 200, 200; converging

49 (a) \$256
(b) $d_n = 4^n$

## Section 13.2

1 Not arithmetic

3 Not arithmetic

5 $(-1)^2 + 0^2 + 1^2 + 2^2 + 3^2 + 4^2 + 5^2$

7 $1 + 3 + 5 + 7 + 9 + 11$

9 $(-1)^2 + (-1)^3 + (-1)^4 + \cdots + (-1)^{10}$

11 $\sum_{n=1}^{7} 3n$

13 $\sum_{n=1}^{8} (1/2)n$

15 (a) $\sum_{i=1}^{10} 2i$
(b) 110

17 $a_1 = 3, d = 4$

19 $a_1 = 2, d = 9$

21 500,500

23 2625

25 −132

27 −561

29 −111.3

31 150; 2325

33 (a) (i) 248.7, 281.4, 308.7; population at census time
(ii) 22.1, 32.7, 27.3; change in population over the decade.
(iii) 2.73; average yearly population growth over the decade.

35 612

37 9

39 (a) 256 ft, 400 ft, 576 ft
(b) 744 ft, 600 ft, 424 ft

41 7.906 sec

43 Last row: 106
Auditorium: 1360

45 (a) 297

## Section 13.3

1 1,572,768

3 5.997

5 781.248

7 7.199

9 Yes, $a = 1$, ratio $= -1/2$

11 Yes; $a = 5$, ratio $= -2$

13 $\sum_{n=1}^{6} (-1)^{n+1}(3^n)$

15 $\sum_{n=0}^{5} (-1)^n 32(\frac{1}{2})^n$

17 189/32

19 1 if $N$ is even and 0 is $N$ is odd.

21 (a) $\sum_{n=1}^{25} 88(1.007)^{n-1}$
(b) 2395.111 bn barrels

23 (a) \$69,173.67
(b) \$69,274.38

25 (a) Doubles
(b) Less than doubles
(c) More than doubles

27 (a) \$48,804.91
(b) \$5457.18

## Section 13.4

1 Yes, $a = 1$, ratio $= -x$

3 No. Ratio between successive terms is not constant

5 Yes, $a = e^x$, ratio $= e^x$

7 Yes, $a = 1$, ratio $= \sqrt{2}$

9 $y^2/(1 - y), |y| < 1$

11 10

13 1/54

15 4

17 $x^2/(1 - x^2)$

19 235/999

21 11/90

23 3781/4950

25 (a) $P_n = 250(0.04) + 250(0.04)^2 + \cdots + 250(0.04)^{n-1}$
(b) $P_n = 10(1 - (0.04)^{n-1})/(1 - 0.04)$
(c) $P_n = 10.417$

27 22.3 million dollars

29 \$9101.74

## Chapter 13 Review

1 603; 59

3 (a) $1 + 5 + 9 + 13 + 17$
(b) 45

5 No

7 $1/(1 - 2z), |z| < 1/2$

9 315

11 $n(n + 1)/2$

13 24 cans at bottom
3 less per row
8 rows

15 435

17 \$25,503.33

19 (a) 300, 350, 400, 450, 500, 550, 600.
(b) 950 yards
(c) 31st day and after

21 (a) $h_n = 10(3/4)^n$
(b) $D_1 = 10$ feet
$D_2 = h_0 + 2h_1 = 25$ feet

$D_3 = h_0 + 2h_1 + 2h_2 =$ 36.25 feet
$D_4 = h_0 + 2h_1 + 2h_2 + 2h_3 \approx 44.688$ feet

(c) $D_n = 10 + 60\left(1 - (3/4)^{n-1}\right)$

23 (a) \$1250
(b) 12.50

## Ch. 13 Understanding

1 True
3 True
5 True
7 True
9 True
11 False
13 True
15 False
17 False
19 False
21 True
23 False
25 False
27 True
29 False
31 False
33 False

# INDEX

## TABLE OF GREEK LETTERS

| Name | Capital | Lowercase |
|---|---|---|
| Alpha | A | $\alpha$ |
| Beta | B | $\beta$ |
| Gamma | $\Gamma$ | $\gamma$ |
| Delta | $\Delta$ | $\delta$ |
| Epsilon | E | $\epsilon$ |
| Zeta | Z | $\zeta$ |
| Eta | H | $\eta$ |
| Theta | $\Theta$ | $\theta$ |
| Iota | I | $\iota$ |
| Kappa | K | $\kappa$ |
| Lambda | $\Lambda$ | $\lambda$ |
| Mu | M | $\mu$ |
| Nu | N | $\nu$ |
| Xi | $\Xi$ | $\xi$ |
| Omicron | O | o |
| Pi | $\Pi$ | $\pi$ |
| Rho | P | $\rho$ |
| Sigma | $\Sigma$ | $\sigma$ |
| Tau | T | $\tau$ |
| Upsilon | $\Upsilon$ | $\upsilon$ |
| Phi | $\Phi$ | $\phi$ |
| Chi | X | $\chi$ |
| Psi | $\Psi$ | $\psi$ |
| Omega | $\Omega$ | $\omega$ |